Lecture Notes in Computer Science 4207

Commenced Publication in 1973
Founding and Former Series Editors:
Gerhard Goos, Juris Hartmanis, and Jan van Leeuwen

Zoltán Ésik (Ed.)

Computer Science Logic

20th International Workshop, CSL 2006
15th Annual Conference of the EACSL
Szeged, Hungary, September 25-29, 2006
Proceedings

Springer

Volume Editor

Zoltán Ésik
University of Szeged, Department of Computer Science
Arpad ter 2, 6720 Szeged, Hungary
E-mail: ze@inf.u-szeged.hu

Library of Congress Control Number: 2006932724

CR Subject Classification (1998): F.4.1, F.4, I.2.3-4, F.3

LNCS Sublibrary: SL 1 – Theoretical Computer Science and General Issues

ISSN 0302-9743
ISBN-10 3-540-45458-6 Springer Berlin Heidelberg New York
ISBN-13 978-3-540-45458-8 Springer Berlin Heidelberg New York

Springer is a part of Springer Science+Business Media

springer.com

Printed in Germany

Typesetting: Camera-ready by author, data conversion by Scientific Publishing Services, Chennai, India
Printed on acid-free paper SPIN: 11874683 06/3142 5 4 3 2 1 0

Preface

Computer Science Logic (CSL) is the annual conference of the European Association for Computer Science Logic. The conference series started as a programme of International Workshops on Computer Science Logic, and then from its sixth meeting became the Annual Conference of the EACSL. The 15th Annual Conference (and 20th International Workshop), CSL 2006, took place during September 25–29, 2006. It was organized by the Department of Computer Science, University of Szeged. Previous CSL conferences were held in Karlsruhe (1987), Duisburg (1988), Kaiserslautern (1989), Heidelberg (1990), Bern (1991), San Miniato (1992), Swansea (1993), Kazimierz (1994), Padernborn (1995), Utrecht (1996), Aarhus (1997), Brno (1998), Madrid (1999), Fischbachau (2000), Paris (2001), Edinburgh (2002), Vienna (2003), Karpacz (2004) and Oxford (2005).

The suggested topics of the conference included automated deduction and interactive theorem proving, constructive mathematics and type theory, equational logic and term rewriting, automata and formal logics, modal and temporal logic, model checking, logical aspects of computational complexity, finite model theory, computational proof theory, logic programming and constraints, lambda calculus and combinatory logic, categorical logic and topological semantics, domain theory, database theory, specification, extraction and transformation of programs, logical foundations of programming paradigms, verification of security protocols, linear logic, higher-order logic, nonmonotonic reasoning, logics and type systems for biology.

In response to the Call for Papers, a total of 132 abstracts were submitted of which 108 were accompanied by a full paper. The International Programme Committee accepted 37 papers for presentation and inclusion in these proceedings. The Programme Committee invited lectures from Martín Escardó (Birmingham), Paul-André Melliès (Paris), Luke Ong (Oxford), Luc Segoufin (Orsay) and Mirosław Truszczyński (Lexington, KY).

The Ackermann Award is the EACSL Outstanding Dissertation Award for Logic in Computer Science. The 2006 Ackermann Award was presented to Balder ten Cate and Stefan Milius at the conference.

These proceedings contain the texts of 4 invited lectures and the 37 accepted papers and the report of the Ackermann Award Committee whose members were J. Makowsky (President of EACSL), D. Niwiński (Vice-President of EACSL), S. Abramsky, B. Courcelle, E. Grädel, M. Hyland, and A. Razborov.

I would like to thank everybody who submitted a paper to the conference and all members of the Programme Committee and their subreferees for their excellent cooperation in the evaluations of the papers. Finally, I would like to thank my colleagues Zsolt Gazdag, Szabolcs Iván and Zoltán L. Németh for their technical assistance during the preparation of these proceedings.

The conference was sponsored by the Department of Computer Science, University of Szeged, the Hungarian Academy of Science, the Fund for Research and Education in Informatics, the Fund for Szeged and the Nokia Hungary, Ltd.

Szeged, July 2006

Zoltán Ésik

Organization

Programme Committee

Krzysztof Apt (Amsterdam/Singapore)
Matthias Baaz (Vienna)
Michael Benedikt (Chicago)
Pierre-Louis Curien (Paris)
Rocco De Nicola (Florence)
Zoltán Ésik (Szeged, Chair)
Dov Gabbay (London)
Fabio Gadducci (Pisa)
Neil Immerman (Amherst)
Michael Kaminski (Haifa)
Bakhadyr Khoussainov (Auckland)
Ulrich Kohlenbach (Darmstadt)
Marius Minea (Timişoara)
Damian Niwiński (Warsaw)
Ramaswamy Ramanujam (Chennai)
Philip Scott (Ottawa)
Philippe Schnoebelen (Cachan)
Alex Simpson (Edinburgh)

Additional Referees

Luca Aceto
Marco Aiello
Roberto Amadio
Christel Baier
Patrick Baillot
Paolo Baldan
Matteo Baldoni
Bruno Barras
David A. Mix Barrington
Arnold Beckmann
Gianluigi Bellin
Ulrich Berger
Dietmar Berwanger
Lars Birkedal
Roderick Bloem
Stephen L. Bloom
Richard Blute
Mikolaj Bojanczyk
Maria Paola Bonacina
Filippo Bonchi
Guillamue Bonfante
Michele Boreale
Paolo Bottoni
Pierre Boudes
Roberto Bruni
Glenn Bruns
Antonio Bucciarelli
Pedro Cabalar
Andrea Cantini
Venanzio Capretta
Franck Cassez
Maria Luisa Dalla Chiara
Juliusz Chroboczek
Vincenzo Ciancia
Gabriel Ciobanu
Thierry Coquand
Flavio Corradini
Bruno Courcelle
Olivier Danvy
Stéphane Demri
Mariangiola Dezani-Ciancaglini
Roy Dyckhoff
Maribel Fernández
Wan Fokkink
Nissim Francez
David de Frutos-Escrig
Murdoch James Gabbay
Marco Gaboardi
Blaise Genest
Giorgio Ghelli
Silvio Ghilardi
Hugo Gimbert
Jean Goubault-Larrecq
Stefano Guerrini
Esfandiar Haghverdi
Masahiro Hamano
Jerry den Hartog
Masahito Hasegawa
Rolf Hennicker
Claudio Hermida
Daniel Hirschkoff
Pieter Hofstra
Doug Howe
Yuval Ishai
Radha Jagadeesan
Petr Jančar
Alan Jeffrey
Roope Kaivola
Yechiel M. Kimchi
Leszek Aleksander Kołodziejczyk
Beata Konikowska
Eryk Kopczyński
K. Narayan Kumar
Orna Kupferman
Dietrich Kuske
Anna Labella

James Laird
Ivan Lanese
Sławomir Lasota
Stéphane Lengrand
Leonardo Lesmo
Laurenţiu Leuştean
Steven Lindell
Kamal Lodaya
Étienne Lozes
Bas Luttik
Parthasarathy Madhusudan
Johann A. Makowsky
Giulio Manzonetto
Annalisa Marcja
Wim Martens
Fabio Martinelli
Simone Martini
Andrea Masini
Ralph Matthes
Richard Mayr
Hernán Melgratti
Markus Michelbrink
Marino Miculan
Joseph S. Miller
Grigori Mints
Yoram Moses
Madhavan Mukund
Filip Murlak
Peter Møller Neergaard
Linh Anh Nguyen
Hans de Nivelle
Mitsu Okada
Frédéric Olive
C.-H. L. Ong
Jaap van Oosten
Simona Orzan
Eric Pacuit
Vincent Padovani
Valeria de Paiva
Prakash Panangaden
Thomas Perst
Adolfo Piperno
Nir Piterman
Andrew Pitts
Jan von Plato
John Power
Christophe Raffalli
Michael Rathjen
Brian Redmond
Colin Riba
Eike Ritter
Jirí Rosický
Luca Roversi
Antonino Salibra
Ivano Salvo
Luigi Santocanale
Henning Schnoor
Matthias Schröeder
R.A.G. Seely
Jonathan P. Seldin
Géraud Sénizergues
Sharon Shoham
Sunil Easaw Simon
Alan Smaill
Viorica Sofronie-Stokkermans
Jiří Srba
Ian Stark
Frank Stephan
Colin Stirling
Kristian Støevring
Howard Straubing
Ross Street
Thomas Streicher
S. P. Suresh
Ryo Takemura
Tony Tan
Hayo Thielecke
Wolfgang Thomas
Michael Tiomkin
Ralf Treinen
Sergei Tupailo
Christian Urban
Paweł Urzyczyn
Tarmo Uustalu
Betti Venneri
Aymeric Vincent
Heinrich Wansing
Andreas Weiermann
Freek Wiedijk
Thomas Wilke
Bożena Woźna
Shmuel Zaks
Artur Zawłocki
Konrad Zdanowski

Local Organizing Committee

Zoltán Ésik (Szeged, Co-chair)
Zsolt Gazdag (Szeged)
Éva Gombás (Szeged, Co-chair)
Szabolcs Iván (Szeged)
Zsolt Kakuk (Szeged)
Loránd Muzamel (Szeged)
Zoltán L. Németh (Szeged)
Sándor Vágvölgyi (Szeged, Workshop Chair)

Table of Contents

Invited Presentations

Contributions

Appendix

Functorial Boxes in String Diagrams*

Paul-André Melliès

Equipe Preuves, Programmes, Systèmes
CNRS — Université Paris 7 Denis Diderot

Abstract. String diagrams were introduced by Roger Penrose as a handy notation to manipulate morphisms in a monoidal category. In principle, this graphical notation should encompass the various pictorial systems introduced in proof-theory (like Jean-Yves Girard's proof-nets) and in concurrency theory (like Robin Milner's bigraphs). This is not the case however, at least because string diagrams do not accomodate *boxes* — a key ingredient in these pictorial systems. In this short tutorial, based on our accidental rediscovery of an idea by Robin Cockett and Robert Seely, we explain how string diagrams may be extended with a notion of *functorial box* depicting a functor transporting an inside world (its source category) to an outside world (its target category). We expose two elementary applications of the notation: first, we characterize graphically when a faithful balanced monoidal functor $F : \mathbb{C} \longrightarrow \mathbb{D}$ transports a *trace operator* from the category $\mathbb{D}$ to the category $\mathbb{C}$, and exploit this to construct well-behaved *fixpoint operators* in cartesian closed categories generated by models of linear logic; second, we explain how the categorical semantics of linear logic induces that the exponential box of proof-nets decomposes as two enshrined boxes.

1 Introduction

The origins. Although the process was already initiated in the late 1960s and early 1970s, very few people could have foreseen that Logic and Computer Science would converge so harmoniously and so far in the two areas of *proof theory* and *programming language design*. Today, about fourty years later, the two research fields are so closely connected indeed, that any important discovery in one of them will have, sooner or later, an effect on the other one. The very existence of the conference *Computer Science Logic* bears witness of this important and quite extraordinary matter of fact.

The convergence would not have been as successful without the mediation of *category theory* — which made an excellent matchmaker between the two subjects, by exhibiting the algebraic properties underlying the mathematical models (or denotational semantics) of both proof systems and programming languages. At the end of the 1970s, a few people were already aware that:

---

* Research partially supported by the ANR Project INVAL "Invariants algébriques des systèmes informatiques".

Z. Ésik (Ed.): CSL 2006, LNCS 4207, pp. 1–30, 2006.

- intuitionistic logic as articulated in proof theory,
- the λ-calculus as implemented in programming languages,
- cartesian closed categories as investigated in category theory

are essentially the same object in three different guises — see for instance Jim Lambek and Phil Scott's monograph [34]. The idea circulated widely in the community, so that a few years later, in the mid-1980s, the following trilogy of concepts has become prominent:

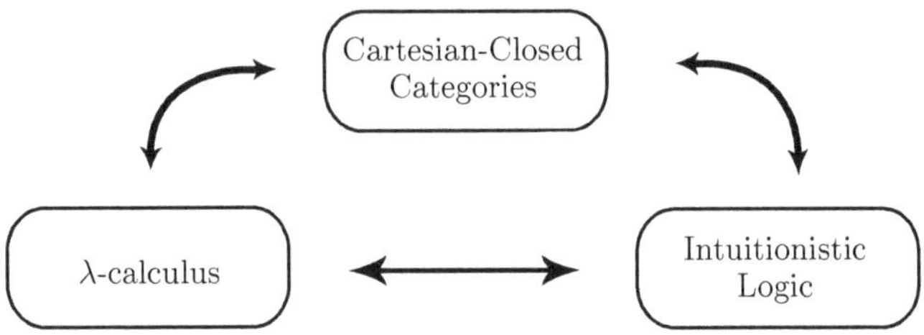

A linear world opens. The year 1985 was then a turning point, with the discovery of linear logic by Jean-Yves Girard. This single discovery had the quite extraordinary effect of refurbishing every part of the subject with new tools, new ideas, and new open problems. In particular, each of the three concepts above was reunderstood in a linear fashion. In effect, Jean-Yves Girard [18,19] introduced simultaneously:

1. a sequent calculus for linear logic, which refines the sequent calculus for Intuitionistic Logic defined by Gerhard Gentzen in the 1930s — in particular, every derivation rule in intuitionistic logic may be translated as a series of more "atomic" derivation rules in linear logic,
2. a graphical syntax of proofs, called *proof-nets*, which refines the term syntax provided by λ-terms — in particular, every simply-typed λ-term may be translated as a proof-net, in such a way that a β-reduction step on the original λ-term is mirrored as a series of more "atomic" cut-elimination steps in the associated proof-net,
3. a denotational semantics of linear logic, based on *coherence spaces* and *cliques*, which refines the model of dI-domains and stable functions defined by Gérard Berry [7] for the purely functional language PCF, a simply-typed λ-calculus extended with a fixpoint operator, a conditional test on booleans, and the main arithmetic operations. People like Robert Seely [45], Yves Lafont [31] and François Lamarche [33] realized very early that the construction amounts to replacing a cartesian closed category (of dI-domains and stable maps) by a *monoidal closed category* (of coherence spaces and cliques) equipped with a particular kind of *comonad* to interpret the *exponential modality* (noted !) of linear logic.

From this followed a new and refined "linear" trilogy, which became prominent in the early 1990s:

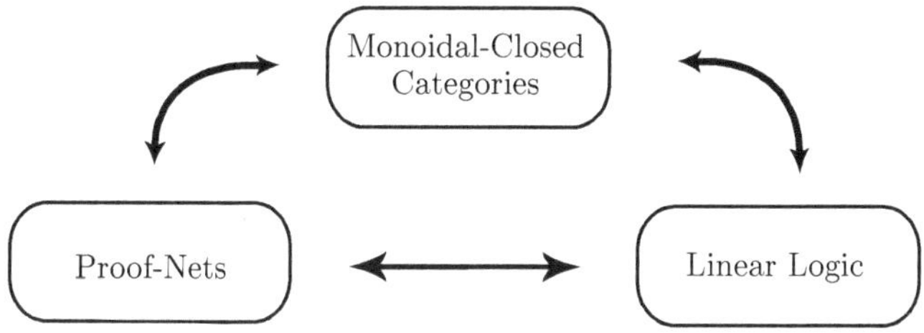

A puzzle in string diagrams. I started my PhD thesis exactly at that time, but in a quite different topic: Rewriting Theory, with Jean-Jacques Lévy at INRIA Rocquencourt. Although I devoted all my energies to exploring the arcanes of my own subject, this culminating in [38,39], I was astonished by the elegance of linear logic, and by the extraordinary perspectives opened by its discovery. Indeed, our emerging field: the semantics of proofs and programs, was suddenly connected to something like mainstream mathematics: linear algebra, representation theory, low-dimensional topology, etc.

My interest was reinforced after a discussion with Yves Lafont, who revealed suddenly to me that *multiplicative* proof-nets, and more generally, his own notion of *interaction nets* [32] are specific instances of a graphical notation invented by Roger Penrose [43,44] to manipulate morphisms in monoidal categories; and that this notation is itself connected to the works by Jean Bénabou on bicategories [4], by Ross Street on computads [47], and by Albert Burroni on polygraphs and higher-dimensional rewriting [13]. Then, André Joyal and Ross Street published at about the same time two remarkable papers [27,28] devoted to *braided monoidal categories* and *string diagrams*. This elegant work finished to convince me... Indeed, I will start this tutorial on string diagrams by giving a very brief and partial account of the two articles [27,28] in Section 2.

Now, it is worth recalling that a proof-net is called *multiplicative* when it describes a proof limited to the *multiplicative* fragment of linear logic. Since multiplicative proof-nets are instances of string diagrams... there remains to understand the "stringy" nature of *general* proof-nets — that is, proof-nets not limited to the multiplicative fragment. A serious difficulty arises at this point: general proof-nets admit *exponential boxes* which depict the action of the exponential modality ! on proofs, by encapsulating them. Recall that the purpose of the modality ! is to transform a "linear" proof which must be used exactly once, into a "multiple" proof which may be repeated or discarded during the reasoning. So, by surrounding a proof, the exponential box indicates that this proof may be duplicated or erased. The trouble is that, quite unfortunately, string diagrams do not admit any comparable notion of "box". Consequently, one would like to extend string diagrams with boxes... But how to proceed?

The lessons of categorical semantics. Interestingly, the solution to this puzzle appears in the categorical semantics of linear logic, in the following way. In the early 1990s, Martin Hyland and Gordon Plotkin initiated together with their students and collaborators Andrew Barber, Nick Benton, Gavin Bierman, Valeria de Paiva, and Andrea Schalk, a meticulous study of the categorical structure defining a model of linear logic [6,8,5,9,3,23]. The research was fruitful in many ways. In particular, it disclosed a common pattern behind the various categorical axiomatizations of linear logic. Indeed, every different axiomatization of linear logic generates what appears to be a *symmetric monoidal* adjunction

$$\mathbb{M} \underset{M}{\overset{L}{\underset{\perp}{\rightleftarrows}}} \mathbb{L} \tag{1}$$

between a symmetric monoidal closed category $\mathbb{L}$ and a cartesian category $\mathbb{M}$. This important notion was introduced and called a Linear-Non-Linear model by Nick Benton [5,37]. Here, it will be simply called a *linear adjunction*. The notations L and M are mnemonics for *Linearize* and *Multiply*. Intuitively, a proof of linear logic is interpreted as a morphism in the category $\mathbb{L}$ or in the category $\mathbb{M}$, depending whether it is "linear" or "multiple". Then,

- the functor M transports a "linear" proof into a "multiple" proof, which may be then replicated or discarded inside the cartesian category $\mathbb{M}$,
- conversely, the functor L transports a "multiple" proof into a "linear" proof, which may be then manipulated inside the symmetric monoidal closed category $\mathbb{L}$.

To summarize: there are two "worlds" or "universes of discourse" noted $\mathbb{L}$ and $\mathbb{M}$, each of them implementing a particular policy, and two functors L and M designed to transport proofs from one world to the other.

An early illustration. Interestingly, this pattern traces back to the very origin of linear logic: coherence spaces. Indeed, Max Kelly notices [25,24] that what one calls "symmetric monoidal adjunction" in (1) is simply an adjunction $L \dashv M$ in the usual sense, in which one requires moreover that the left adjoint functor L transports the cartesian structure of $\mathbb{M}$ to the symmetric monoidal structure of $\mathbb{L}$. The detailed proof of this fact appears in my recent survey on the categorical semantics of linear logic [40]. Such a functor L is called *strong monoidal* in the litterature — the precise definition is recalled in Section 4. Now, the practiced reader will recognize that the linear adjunction (1) describes precisely how the category $\mathbb{M}$ of dI-domains and stable functions is related to the category $\mathbb{L}$ of coherence spaces and cliques. Recall indeed that a coherence space is simply a reflexive graph, and that the functor L transforms every dI-domain D into a coherence space $L(D)$ whose nodes are the elements of D, and in which two nodes $x \in D$ and $y \in D$ are connected by an edge (that is, are coherent) precisely when there exists an element $z \in D$ such that $x \leq z \geq y$. Since the tensor product of coherence spaces is the same thing as the usual product of graphs, the equality follows:

$$L(D \times E) \quad = \quad L(D) \otimes L(E).$$

Although one should check carefully the conditions of Section 4, it is quite immediate that the functor L is strict monoidal — hence strong monoidal. At this point, there only remains to define a right adjoint functor M to the functor L in the way exposed in [18,19,1] in order to find oneself in the situation of a linear adjunction (1).

The exponential modality decomposed. Although the pattern of linear adjunction (1) looks familiar from a semantic point of view, it appears quite unexpected from the point of view of proof-nets — because the exponential modality ! is not a primitive anymore: it is deduced instead as the *comonad*

$$! \quad = \quad L \circ M \tag{2}$$

generated by the linear adjunction (1) in the category $\mathbb{L}$. In other words, the exponential modality ! factors into a pair of more atomic modalities L and M. Nick Benton [5] mirrors this semantic decomposition into a logic and a term language, which he calls Linear-Non-Linear logic. The decomposition may be transposed instead into the pictorial language of proof-nets: it tells then that the exponential box should decompose into a pair of "boxes" interpreting the two modalities L and M. This pictorial decomposition of the box ! should follow the principles of string diagrams, and be nothing more, and nothing less, than a handy graphical notation for the categorical equality (2).

Functorial boxes. Now, the two modalities L and M in the linear adjunction (1) are *monoidal functors* between the monoidal categories $\mathbb{L}$ and $\mathbb{M}$ — where the monoidal structure of $\mathbb{M}$ is provided by its cartesian structure. Hence, monoidal functors are precisely what one wants to depict as "boxes" in string diagrams. The task of Sections 3 and 4 is precisely to explain how monoidal functors are depicted as *functorial boxes* in string diagrams — and what kind of box depicts a lax, a colax or a strong monoidal functor. I rediscover in this way, ten years later, an idea published by Robin Cockett and Richard Seely [15] in their work on linearly distributive categories and functors. See also the related article written in collaboration with Rick Blute [11]. Obviously, all the credit for the idea should go to them. On the other hand, I find appropriate to promote here this graphical notation which remained a bit confidential; and to illustrate how this handy notation for monoidal functors may be applied in other contexts than linear logic or linearly distributive categories.

So, I will discuss briefly in Section 7 how the exponential box ! of linear logic decomposes into a functorial box M enshrined inside a functorial box L. Categorical semantics indicates that the functor L is strong monoidal whereas the functor M is lax monoidal — see Section 4 for a definition. Consequently, the two functorial boxes are of a different nature. One benefit of using string diagrams instead of proof-nets is that the graphical notation mirrors *exactly* the underlying categorical semantics. In particular, I will illustrate how the typical cut-elimination steps in proof-nets are themselves decomposed into sequences of more atomic rewrite steps in string diagrams. Each of these rewrite steps depicts a step in the proof of *soundness* of the categorical semantics of linear logic implemented by Linear-Non-Linear models.

Trace operators in linear logic. In order to interpret recursive calls in a programming language like PCF, one needs a cartesian closed category equipped with a *fixpoint operator*. Recall that a parametric fixpoint operator Fix in a cartesian category $\mathbb{C}$ is a family of functions

$$\mathrm{Fix}_A^U \ : \ \mathbb{C}(A \times U, U) \ \longrightarrow \ \mathbb{C}(A, U)$$

making the diagram below commute

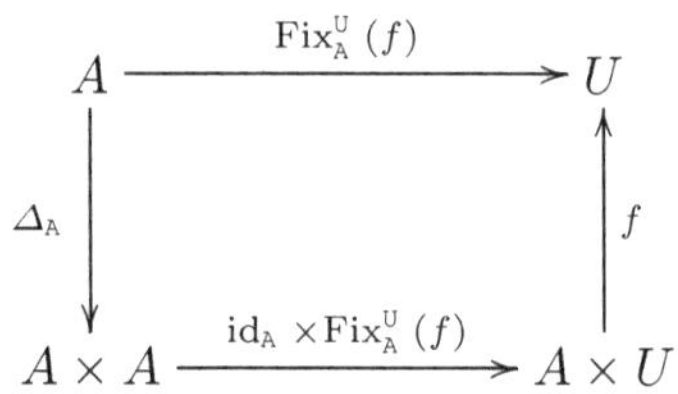

for every morphism $f : A \times U \longrightarrow U$. The diagram expresses that Fix_A^U is a parametric fixpoint of the morphism f. A fixpoint operator should also satisfy a series of naturality properties described in Theorem 3.1 of [20].

A few years ago, Martin Hyland and Masahito Hasegawa [20] have pointed out independently that the notion of fixpoint operator is closely related to the notion of *trace* introduced by André Joyal, Ross Street and Dominic Verity [29] in the context of balanced monoidal categories — a mild refinement of braided monoidal categories, see Section 5 for a definition of trace. More precisely, Martin Hyland and Masahito Hasegawa show that a trace in a cartesian category $\mathbb{C}$ is the same thing as a particularly well-behaved notion of parametric fixpoint, see [20].

Now, it appears that in many existing models of linear logic, formulated here as a linear adjunction (1), the symmetric monoidal closed category $\mathbb{L}$ has a trace. This happens typically when the category $\mathbb{L}$ is *autonomous*, like the category *Rel* of sets and relations (with the usual product of sets as tensor product) or variants recently studied by Nicolas Tabareau [49] of the category of Conway games introduced by André Joyal [26]. An interesting question thus is to understand when a trace in the category $\mathbb{L}$ may be transported to a trace, and thus a fixpoint operator, in the cartesian category $\mathbb{M}$.

A nice example, suggested to me by Masahito Hasegawa, shows that this is not possible in general. Consider the powerset monad T on the usual category *Set* of sets and functions: the monad associates to every set X the set TX of its subsets. The monad T induces an adjunction

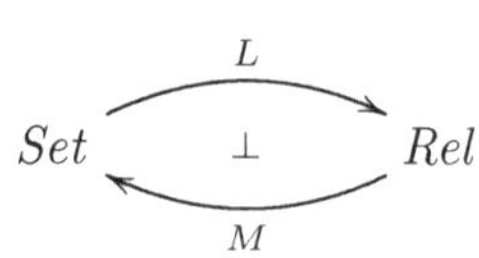

(3)

between the category *Set* and its kleisli category Set_T — which is isomorphic to the category *Rel* of sets and relations. The lifting monad T being commutative, or equivalently, symmetric monoidal (in the lax sense), the adjunction (3) is symmetric monoidal, see [30]. In particular, the kleisli category Set_T inherits its monoidal structure from the cartesian structure of the category *Set*; and the functor L which sends the category *Set* to the subcategory of functions in *Rel*, is strict monoidal. So, the adjunction (3) is linear, and defines a model of linear logic, in which the category $\mathbb{L} = Rel$ is autonomous, and thus has a trace. On the other hand, there is no fixpoint operator, and thus no trace, in the cartesian category $\mathbb{M} = Set$.

At this point, it is worth noticing that the functor L is *faithful* in the typical models of linear logic, because the category $\mathbb{M}$ is either equivalent to a subcategory of commutative comonoids in $\mathbb{L}$, or equivalent to a subcategory of coalgebras of the comonad $! = L \circ M$ — in particular to the category of free coalgebras when $\mathbb{M}$ is the co-kleisli category associated to the comonad. Another equivalent statement is that every component of the unit η of the monad $M \circ L$ is a monomorphism. This observation motivates to characterize in Section 6 when a *faithful* balanced monoidal functor

$$\mathbb{C} \xrightarrow{\quad F \quad} \mathbb{D} \tag{4}$$

between balanced monoidal categories transports a trace in the target category $\mathbb{D}$ to a trace in the source category $\mathbb{C}$. The proof of this result is perfectly elementary, and offers a nice opportunity to demonstrate how string diagrams and functorial boxes may be manipulated in order to produce purely diagrammatic proofs. Of course, the result specializes then to the strong monoidal functor L involved in a typical model of linear logic. This enables to transport a trace in the category $\mathbb{L}$ to a well-behaved parametric fixpoint operator in the category $\mathbb{M}$ in several models of interest — including the relational model of linear logic, and the categories of Conway games mentioned earlier.

String diagrams in computer science and logic: a few perspectives. My ambition in writing this elementary tutorial is to demonstrate in a few pictures that categorical semantics is *also* of a diagrammatic nature. Proof-nets were invented by a genial mind, but they remain an ad'hoc and slightly autarchic artefact of proof-theory. On the other hand, string diagrams flourished in the middle of algebra. Categorical semantics is precisely here to connect the two subjects, with benefits on both sides: logic and computer science on the one hand, categorical algebra on the other hand.

Obviously, much work remains to be done in the area. In many respects, the three concepts appearing in the first trilogy (intuitionistic logic, λ-calculus, cartesian closed categories) were more tightly connected in the mid-1980s than the three concepts appearing in the second trilogy (linear logic, proof-nets, monoidal closed categories) are connected today.

The article published recently by Maria Emilia Maietti, Paola Maneggia, Valeria de Paiva, Eike Ritter [37] is extremely clarifying from that point of view: it is established there that the Linear-Non-Linear term language introduced by Nick Benton [5] is the *internal language* of the category of linear adjunctions (1). Note that the idea of reformulating this result using string diagrams (extended with functorial boxes) instead of a term language motivates implicitly the discussion in Section 7. Another key work in the area was published by Rick Blute, Robin Cockett, Robert Seely and Todd Trimble [12] about coherence in linearly distributive categories. The article describes the free linearly distributive category and the free $*$-autonomous over a given category $\mathbb{C}$, using equations on a variant of Jean-Yves Girard's multiplicative proof-nets.

I am confident that a broader picture will emerge at some point from the current work at the interface of linear logic and categorical algebra. In the near future, we will certainly find natural to extract a language or a logic as the *internal language* of a particular categorical pattern, similar to the linear adjunction (1) and possibly formulated as a 2-dimensional version of Lawvere theory [35,47,13,10,46]. The languages would be expressed alternatively with string diagrams, for handy manipulation, or with terms, for easy implementation. The resulting trilogy of concepts:

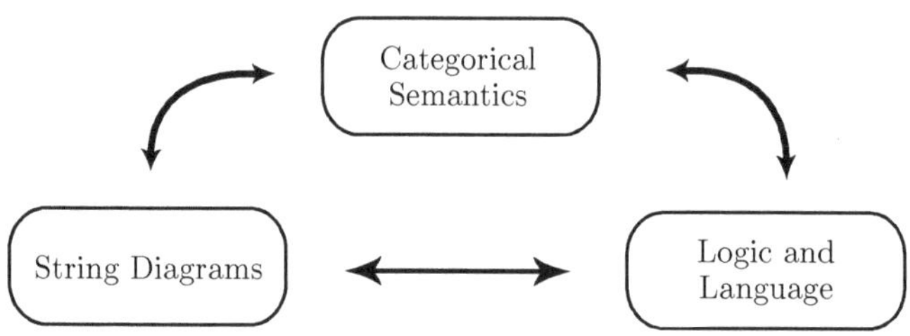

would be broader in scope and more tightly connected than the current one. It would also integrate the algebraic and pictorial systems formulated for concurrency theory, like Robin Milner's bigraphs [41]. The existing categorical semantics of action calculi [22,42,2] indicate a close relationship with the notion of *fibred* functor between *fibred* categories, and with the models of linear logic based on linear adjunctions (1).

I should conclude this introduction by observing that functorial boxes in string diagrams offer a handy 2-dimensional notation for what could be depicted alternatively using Ross Street's 3-dimensional surface diagrams [48]. Surface diagrams are more perspicuous in many ways: for instance, a functor is depicted there as a string, instead of a box. However, the two notations are not extremely different: the practiced reader will easily translate the string diagrams appearing in the tutorial into surface diagrams — in which strings are replaced by ribbons, in order to accomodate the twists. In that respect, this tutorial should be also understood as incentive to carry on in the diagrammatic path, and to depict proofs as surface diagrams. The resulting 3-dimensional notation, added to the observation [16] that a $*$-autonomous category is essentially the same thing as a Frobenius algebra in the autonomous category of small categories and "profunctors" or "distributors" — will certainly offer a revitalizing point of view on linear logic, which remains to be investigated.

2 String Diagrams

In a series of two remarkable papers, André Joyal and Ross Street introduce the notion of *balanced monoidal category* [27] and develop a graphical notation, based on *string diagrams*, to denote morphisms in these categories [28]. Note that from a purely topological point of view, these string diagrams are embedded in the 3-dimensional space. The main task of the second paper [28] is precisely to justify the notation, by showing that any two string diagrams equal modulo continuous deformation denote the same morphism in the balanced monoidal category. The interested reader will find the argument in [28].

Recall that a monoidal category [36] is a category $\mathbb{C}$ equipped with a functor

$$\otimes \;:\; \mathbb{C} \times \mathbb{C} \;\longrightarrow\; \mathbb{C}$$

called the *tensor product*, and an object I called the *unit object*; as well as three natural isomorphisms

$$\alpha_{A,B,C} : (A \otimes B) \otimes C \longrightarrow A \otimes (B \otimes C)$$

$$\lambda_A : I \otimes A \longrightarrow A, \qquad \rho_A : A \otimes I \longrightarrow A$$

called the *associativity*, the *left* and the *right unit constraints* respectively; such that, for all objects A, B, C and D of the category, the following two diagrams called *MacLane's associativity pentagon* and *triangle for unit*, commute:

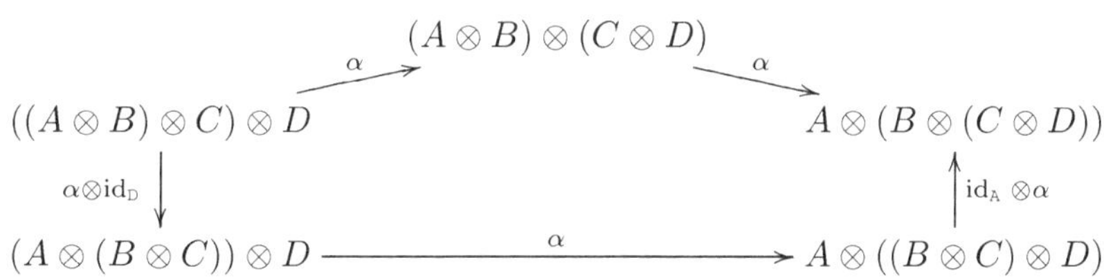

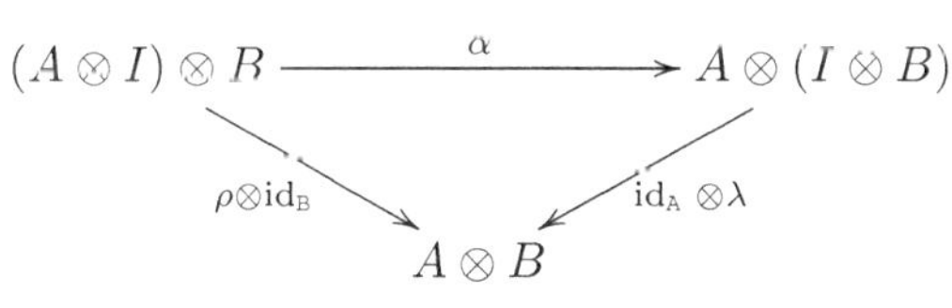

A *braiding* is a natural isomorphism

$$\gamma_{A,B} :\; A \otimes B \longrightarrow B \otimes A$$

such that, for all objects A, B and C of the category, the two hexagonal diagrams below commute:

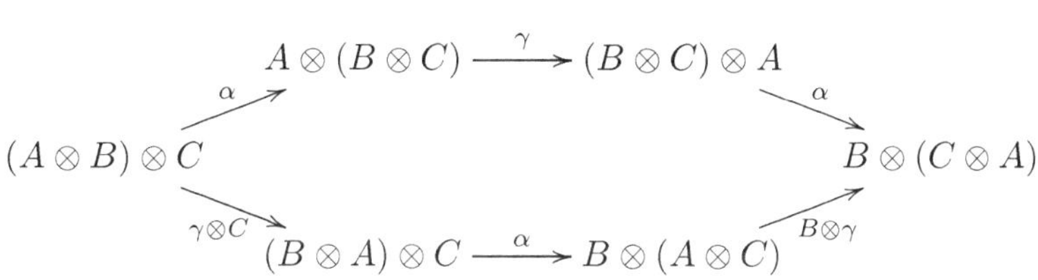

$$\begin{array}{ccccc} & (A\otimes B)\otimes C & \xrightarrow{\gamma} & C\otimes(A\otimes B) & \\ \nearrow^{\alpha^{-1}} & & & & \searrow^{\alpha^{-1}} \\ A\otimes(B\otimes C) & & & & (C\otimes A)\otimes B \\ \searrow_{A\otimes\gamma} & & & & \nearrow_{\gamma\otimes B} \\ & A\otimes(C\otimes B) & \xrightarrow{\alpha^{-1}} & (A\otimes C)\otimes B & \end{array}$$

Finally, a *twist* is a natural isomorphism

$$\theta_A : A \longrightarrow A$$

such that

$$\theta_I = \mathrm{id}_I$$

and, for all objects A and B of the category, the diagram below commutes:

$$\begin{array}{ccc} A\otimes B & \xrightarrow{\gamma_{A,B}} & B\otimes A \\ \downarrow{\scriptstyle \theta_{A\otimes B}} & & \downarrow{\scriptstyle \theta_A\otimes\theta_B} \\ A\otimes B & \xleftarrow[\gamma_{B,A}]{} & B\otimes A \end{array}$$

Definition 1. *A balanced monoidal category is a monoidal category equipped with a braiding and a twist.*

Note that a symmetric monoidal category is a balanced category in which, for all objects A and B of the category, the morphism

$$A\otimes B \xrightarrow{\gamma_{A,B}} B\otimes A \xrightarrow{\gamma_{B,A}} A\otimes B$$

is equal to the identity morphism $\mathrm{id}_{A\otimes B}$; and the twist morphism θ_A coincides with the identity morphism id_A.

From now on, we suppose for legibility that our balanced monoidal category is *strict*: this means that, for all objects A, B and C of the category, the component $\alpha_{A,B,C}$, λ_A and ρ_A of the the associativity and unit isomorphisms, are identity morphisms. We follow the conventions used in [28] and thus depict a morphism $f : A\otimes B\otimes C \longrightarrow D\otimes E$ in string diagrams as:

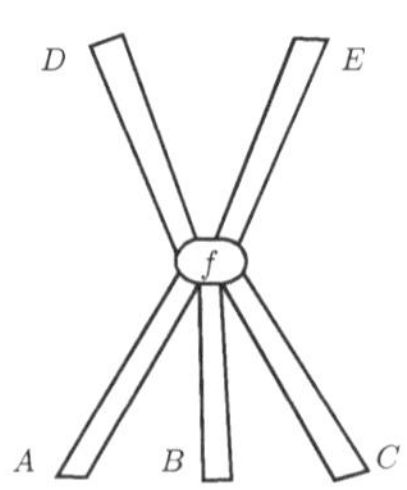

We depict the composite $g\circ f : A \longrightarrow C$ of two morphisms $f : A \longrightarrow B$ and $g : B \longrightarrow C$ as:

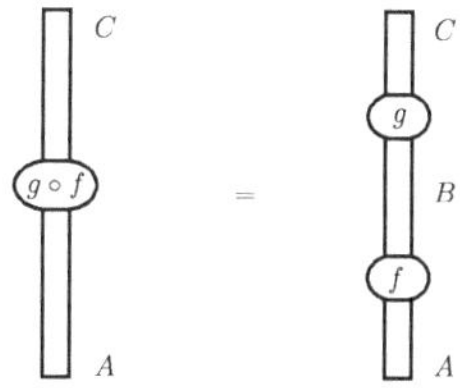

and the tensor product $f \otimes g : A \otimes C \longrightarrow B \otimes D$ of two morphisms $f : A \longrightarrow B$ and $g : C \longrightarrow D$ as:

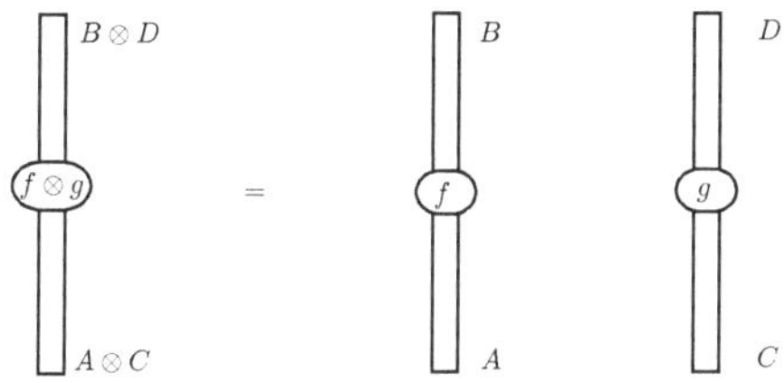

Then, the braiding $\gamma_{A,B}$ and its inverse $\gamma_{A,B}^{-1}$, the twist θ_A and its inverse θ_A^{-1} are depicted respectively as:

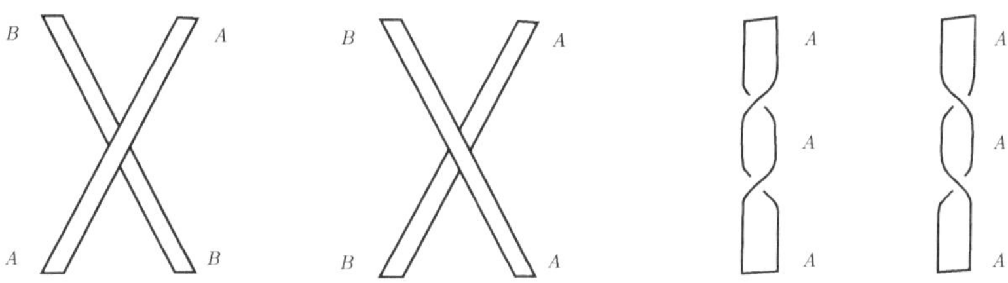

Note that the third dimension of string diagrams enables to depict the braidings, and that drawing ribbons (instead of strings) is convenient to depict the twists.

3 Functors in String Diagrams

Here, we recall the graphical notation introduced by Robin Cockett and Robert Seely [15] in order to depict a usual functor

$$F : \mathbb{C} \longrightarrow \mathbb{D}$$

between balanced monoidal categories. The functor applied to a morphism $f : A \longrightarrow B$ of the category $\mathbb{C}$ is represented as a *box* tagged by the label F, and drawn around the morphism f in the following way:

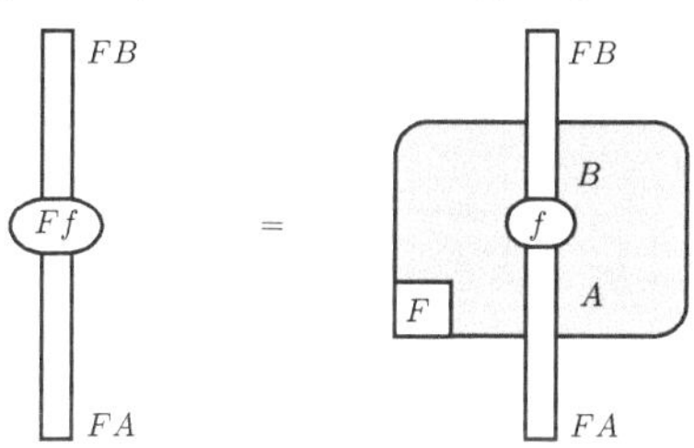

Like any box, the functorial box F is designed to separate an inside world from an outside world: in that case, the inside world is the source category $\mathbb{C}$ and the outside world is the target category $\mathbb{D}$. This explains why a string typed FA outside the box (thus, in the category $\mathbb{D}$) becomes a string typed A (thus, in the category $\mathbb{C}$) when it crosses the frontier and enters the box; and that a string typed B inside the box (in the category $\mathbb{C}$) becomes a string typed FB (in the category $\mathbb{C}$) when it crosses the frontier and leaves the box.

Given a pair of morphisms $f : A \longrightarrow B$ and $g : B \longrightarrow C$, one depicts the two functorial equalities

$$F(g \circ f) \;=\; Fg \circ Ff \qquad\qquad F(\mathrm{id}_A) \;=\; \mathrm{id}_{FA}$$

in the following way:

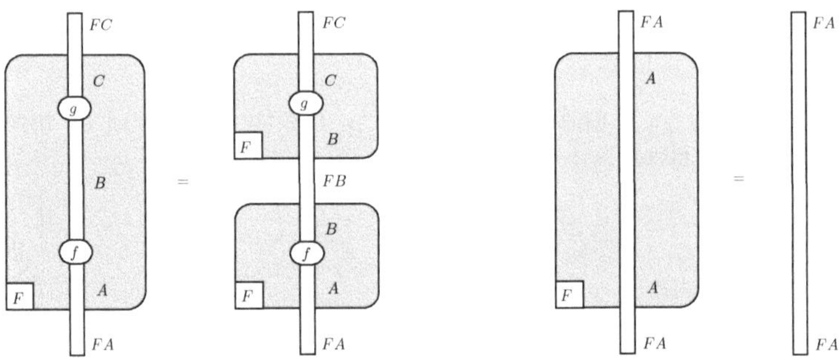

Note that exactly one string enters and exits each functorial box F.

4 Monoidal Functors in String Diagrams

In this section, we recall how the graphical notation for functors introduced in the previous section specializes to monoidal functors, see [15] again. It will appear that a monoidal functor (in the lax sense) implements a particular kind of functorial box in which several strings (possibly none) may enter simultaneously, and from which exactly one string exits. Recall that a lax monoidal functor

$$(F, m) \;:\; \mathbb{C} \;\longrightarrow\; \mathbb{D}$$

between two monoidal categories is a functor F equipped with a morphism

$$m_{[-]} \;:\; I \longrightarrow FI$$

and a natural morphism

$$m_{[A,B]} \;:\; FA \otimes FB \;\longrightarrow\; F(A \otimes B)$$

such that, for all objects A, B and C, the three "coherence" diagrams below commute:

$$\begin{array}{ccc}
(FA \otimes FB) \otimes FC & \xrightarrow{\alpha} & FA \otimes (FB \otimes FC) \\
\downarrow{\scriptstyle m \otimes FC} & & \downarrow{\scriptstyle FA \otimes m} \\
F(A \otimes B) \otimes FC & & FA \otimes F(B \otimes C) \\
\downarrow{\scriptstyle m} & & \downarrow{\scriptstyle m} \\
F((A \otimes B) \otimes C) & \xrightarrow{F\alpha} & F(A \otimes (B \otimes C))
\end{array}$$

$$\begin{array}{ccc}
FA \otimes I & \xrightarrow{\rho} & FA \\
\downarrow{\scriptstyle FA \otimes m} & & \uparrow{\scriptstyle F\rho} \\
FA \otimes FI & \xrightarrow{m} & F(A \otimes I)
\end{array}
\qquad
\begin{array}{ccc}
I \otimes FB & \xrightarrow{\lambda} & FB \\
\downarrow{\scriptstyle m \otimes FB} & & \uparrow{\scriptstyle F\lambda} \\
FI \otimes FB & \xrightarrow{m} & F(I \otimes B)
\end{array}$$

The notion of colax monoidal functor (F, n) is defined in just the same way, except that the coercion morphisms n go in the other direction:

$$n_{[-]} \ : \ FI \ \longrightarrow \ I \qquad\qquad n_{[A,B]} \ : \ F(A \otimes B) \ \longrightarrow \ FA \otimes FB$$

A strong monoidal functor is a lax monoidal functor (F, m) in which the coercion maps m are all isomorphisms; equivalently, it is a colax monoidal functor (F, n) in which the coercion maps n are all isomorphisms.

Let us explain now how we depict monoidal functors in string diagrams. We will suppose for legibility that the two monoidal categories $\mathbb{C}$ and $\mathbb{D}$ are strict. Given k objects in the category $\mathbb{C}$, there may be several ways to construct a morphism

$$m_{[A_1,\cdots,A_k]} \ : \ FA_1 \otimes \cdots \otimes FA_k \ \longrightarrow \ F(A_1 \otimes \cdots \otimes A_k)$$

by applying a series of structural morphisms m. Then, the definition of a lax monoidal functor, and more specifically the coherence diagrams recalled above, are designed to ensure that these various ways define the same morphism $m_{[A_1,\cdots,A_k]}$ in the end. This morphism is depicted in string diagrams as a box F in which k strings labelled $A_1, \cdots, A_k$ enter simultaneously, join together into a unique string labelled $A_1 \otimes \cdots \otimes A_k$ which then exits the box. For instance, the two structural morphisms $m_{[A_1,A_2,A_3]}$ and $m_{[-]}$ are depicted as follows:

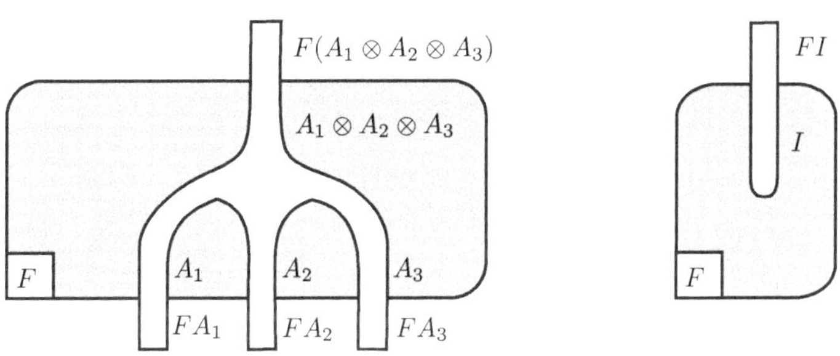

(5)

More generally, given a morphism

$$f : A_1 \otimes \cdots \otimes A_k \longrightarrow B$$

in the source category $\mathbb{C}$, one depicts as the functorial box with k inputs and exactly one output:

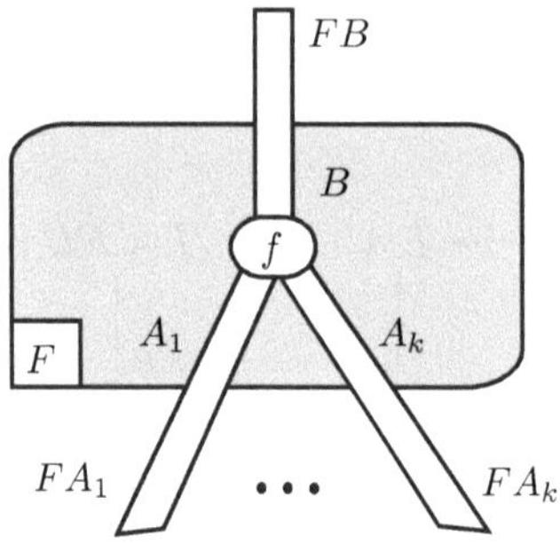

(6)

the morphism

$$F(f) \circ m_{[A_1,\cdots,A_k]} \; : \; FA_1 \otimes \cdots \otimes FA_k \;\longrightarrow\; FB$$

obtained by precomposing the image of f by the functor F in the target category $\mathbb{D}$, with the morphism $m_{[A_1,\cdots,A_k]}$.

Remark. The definition of lax monoidal functor would permit a more general and *delayed* form of fusion between boxes (think of surface diagrams [48] here). Here, we limit ourselves to the specific pattern (5) of k boxes F, each one encapsulating a unique string labelled A_i, for $1 \leq i \leq k$, and joining together *simultaneously* in a box F encapsulating a unique string labelled $A_1 \otimes \cdots \otimes A_k$. This specific pattern generates boxes of the shape (6) which are easy to understand and to manipulate, and sufficient to the purpose of this tutorial.

The coherence properties required by the definition of a monoidal functor ensure that we may safely "merge" two monoidal boxes in a string diagram:

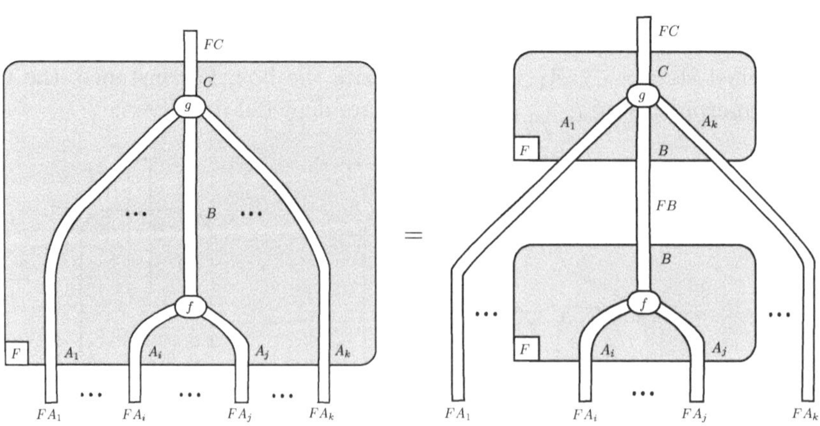

(7)

Note that a colax monoidal functor may be depicated in a similar fashion, as a functorial box in which exactly one string enters, and several strings (possibly none) exit. Now, a strong monoidal functor is at the same time a lax monoidal functor (F, m) and a colax monoidal functor (F, n). It is thus depicted as a functorial box in which several strings may enter, and several strings may exit. Besides, the coercion maps m are inverse to the coercion maps n. Two diagrammatic equalities follow, which enable to split a "strong monoidal" box horizontally:

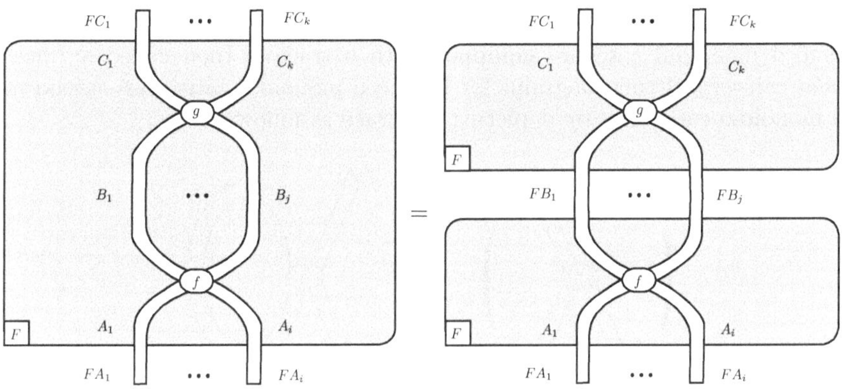

(8)

as well as vertically:

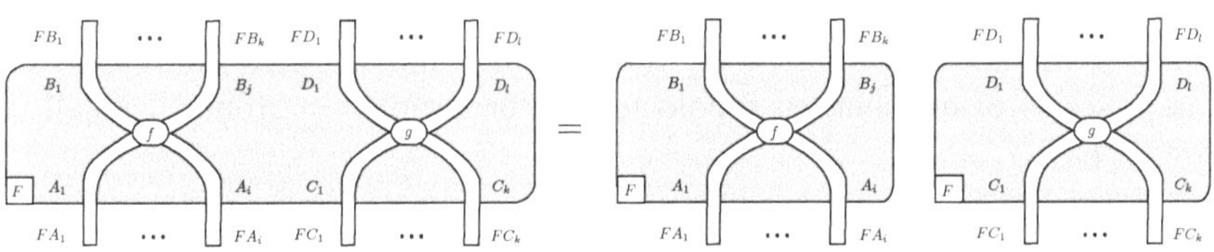

(9)

These equalities will be illustrated in the series of diagrammatic manipulations exposed in Sections 6 and 7.

5 Traced Monoidal Categories

In a remarkable article, André Joyal, Ross Street and Dominic Verity [29] define a *trace* in a balanced monoidal category $\mathbb{C}$ as a natural family of functions

$$\mathrm{Tr}^{U}_{A,B} \;:\; \mathbb{C}(A \otimes U, B \otimes U) \;\longrightarrow\; \mathbb{C}(A, B)$$

satisfying three axioms:

vanishing (monoidality in U)

$$\mathrm{Tr}^{U \otimes V}_{A,B}(g) \;=\; \mathrm{Tr}^{U}_{A,B}(\mathrm{Tr}^{V}_{A \otimes U, B \otimes U}(g)), \qquad \mathrm{Tr}^{I}_{A,B}(f) \;=\; f.$$

superposing

$$\begin{aligned}\mathrm{Tr}^U_{A,B}(f) \otimes g &= \mathrm{Tr}^U_{A\otimes C, B\otimes D}((\mathrm{id}_B \otimes \gamma^{-1}_{D,U}) \circ (f \otimes g) \circ (\mathrm{id}_A \otimes \gamma_{C,U})) \\ &= \mathrm{Tr}^U_{A\otimes C, B\otimes D}((\mathrm{id}_B \otimes \gamma_{D,U}) \circ (f \otimes g) \circ (\mathrm{id}_A \otimes \gamma^{-1}_{C,U}))\end{aligned}$$

yanking

$$\mathrm{Tr}^U_{U,U}(\gamma_{U,U} \circ (\theta^{-1} \otimes \mathrm{id}_U)) \;=\; \mathrm{id}_U \;=\; \mathrm{Tr}^U_{U,U}(\gamma^{-1}_{U,U} \circ (\theta \otimes \mathrm{id}_U)).$$

A balanced monoidal category equipped with a trace is then called a *traced monoidal category.* String diagrams for balanced monoidal categories extend to traced monoidal categories by depicting the trace as follows:

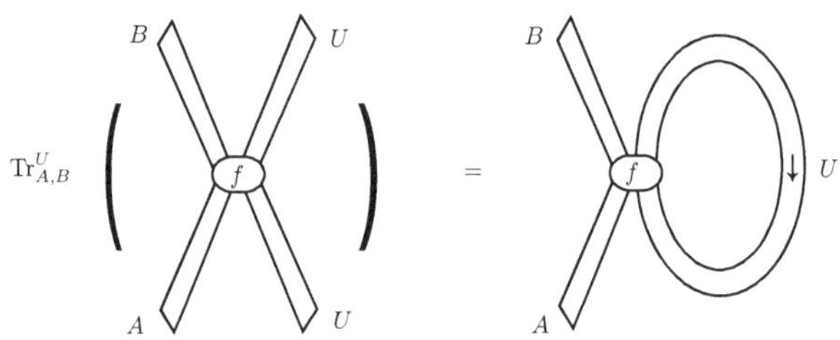

The small arrow embroidered on the ribbon recalls that this part of the string diagram depicts a trace, which expresses intuitively a notion of *feedback.* Thanks to this ingenious notation for traces, the algebraic axioms of a trace are depicted as a series of elementary topological deformations on ribbons, recalled here from [29]:

sliding (naturality in U)

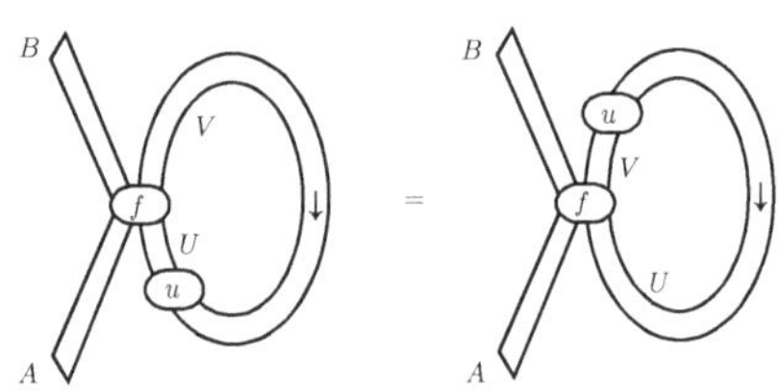

tightening (naturality in A, B)

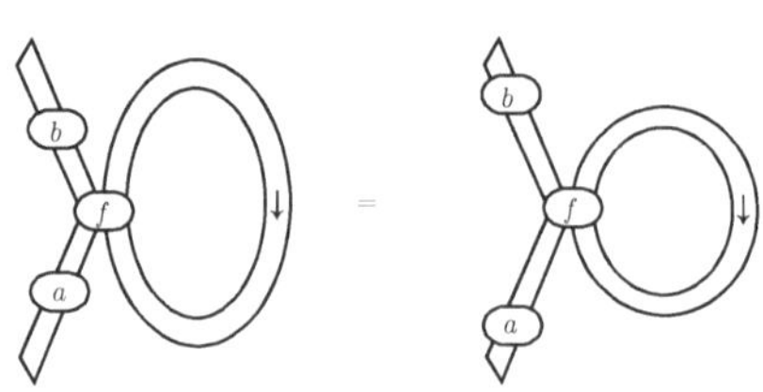

vanishing (monoidality in U)

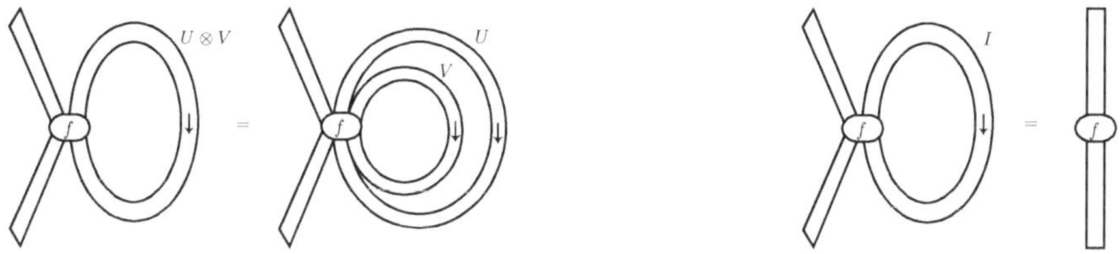

superposing

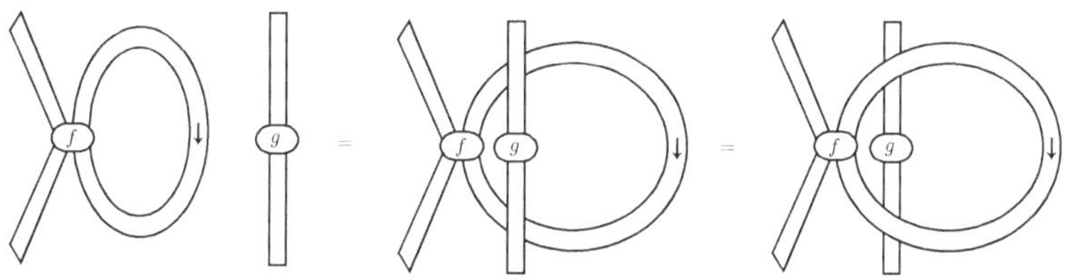

yanking

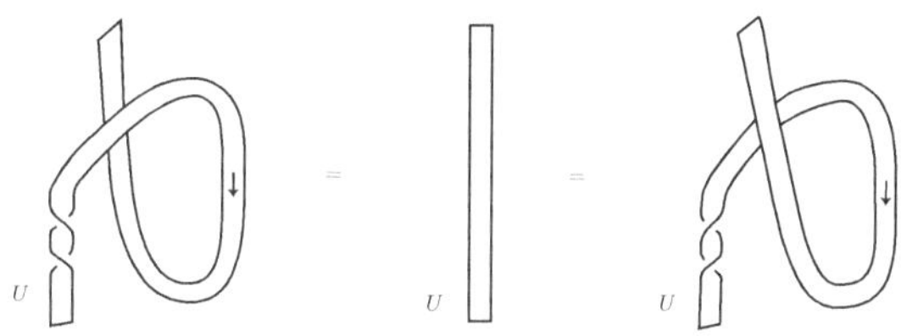

6 Transport of Trace Along a Faithful Functor

Recall [29] that a *balanced* monoidal functor $F : \mathbb{C} \longrightarrow \mathbb{D}$ between balanced monoidal categories is a strong monoidal functor satisfying that, for all objects A and B, the diagram below commutes

$$\begin{array}{ccc} FA \otimes FB & \xrightarrow{\gamma_{A,B}} & FB \otimes FA \\ \downarrow{\scriptstyle m_{A,B}} & & \downarrow{\scriptstyle m_{B,A}} \\ F(A \otimes B) & \xrightarrow{F(\gamma_{A,B})} & F(B \otimes A) \end{array}$$

and the equality $F\theta_A = \theta_{FA}$ holds. This may be depicted as the two equalities:

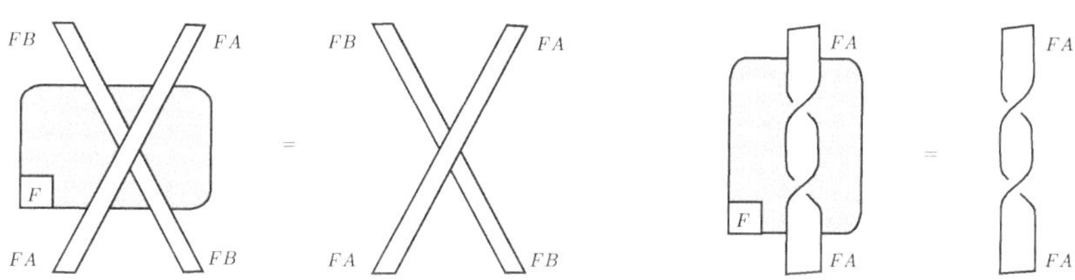

When $\mathbb{C}$ and $\mathbb{D}$ are traced monoidal, one says that $F : \mathbb{C} \longrightarrow \mathbb{D}$ is traced monoidal when F is balanced monoidal, and preserves traces in the expected sense that, for all objects A, B and U and for all morphism $f : A \otimes U \longrightarrow B \otimes U$ of the category $\mathbb{C}$, the following equality holds:

$$F(\mathrm{Tr}^{U}_{A,B}(f)) \;=\; \mathrm{Tr}^{FU}_{FA,FB}(m^{-1}_{[A,B]} \circ Ff \circ m_{[A,B]}).$$

This equality is depicted as follows:

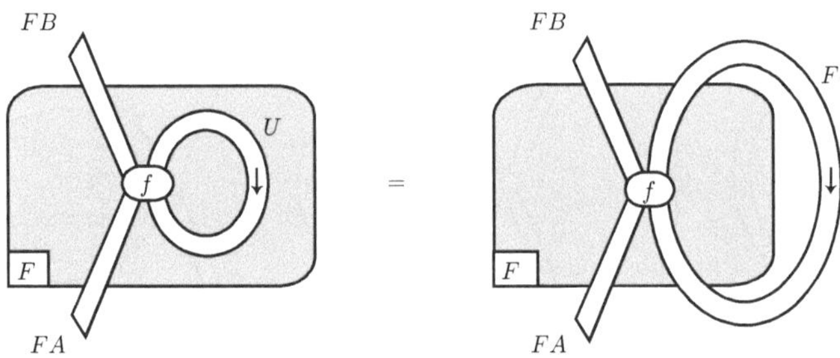

An elementary exercise in string diagrams with functorial boxes follows. It consists in establishing in a purely diagrammatic fashion a mild but useful generalization of a statement (Proposition 2.4) appearing in [29].

Proposition 1 (Characterization of transport along a faithful functor). *Suppose that $F : \mathbb{C} \longrightarrow \mathbb{D}$ is a faithful, balanced monoidal functor with $\mathbb{D}$ traced monoidal. Then, there exists a trace on $\mathbb{C}$ for which F is a traced monoidal functor iff for all objects A, B, U of the category $\mathbb{C}$, and all morphism*

$$f \;:\; A \otimes U \longrightarrow B \otimes U$$

there exists a morphism $g : A \longrightarrow B$ such that

$$F(g) \;=\; \mathrm{Tr}^{FU}_{FA,FB}(m^{-1}_{[A,B]} \circ F(f) \circ m_{[A,B]}) \tag{10}$$

where Tr *denotes the trace in $\mathbb{D}$. The equality is depicted as follows:*

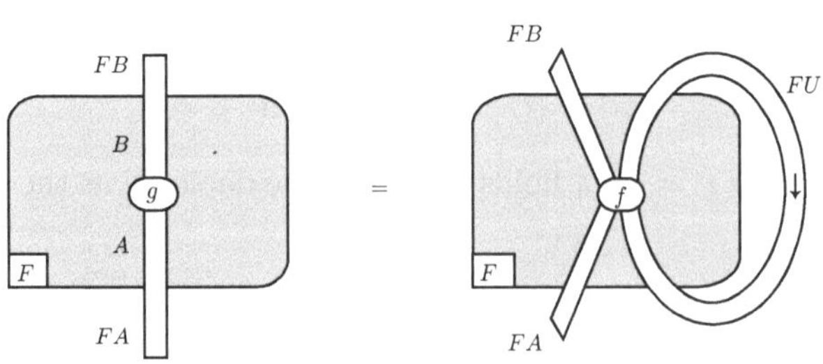

Moreover, if this trace on $\mathbb{C}$ exists, it is unique: it is called the trace on the category $\mathbb{C}$ transported from the category $\mathbb{D}$ along the functor F.

Proof. The direction (⇒) follows immediately from the very definition of a traced monoidal functor. Hence, we only establish here the converse direction (⇐). We suppose from now on that for every morphism $f : A \otimes U \longrightarrow B \otimes U$ there exists a morphism $g : A \longrightarrow B$ satisfying Equation (10). Note that the morphism g is unique because the functor F is faithful. This defines a family of functions noted

$$\mathrm{tr}^U_{A,B} \; : \; \mathbb{C}(A \otimes U, B \otimes U) \; \longrightarrow \; \mathbb{C}(A, B).$$

We establish that tr satisfies the equational axioms of a trace. To that purpose, we introduce a handy notation for the morphism $\mathrm{tr}^U_{A,B}(f)$:

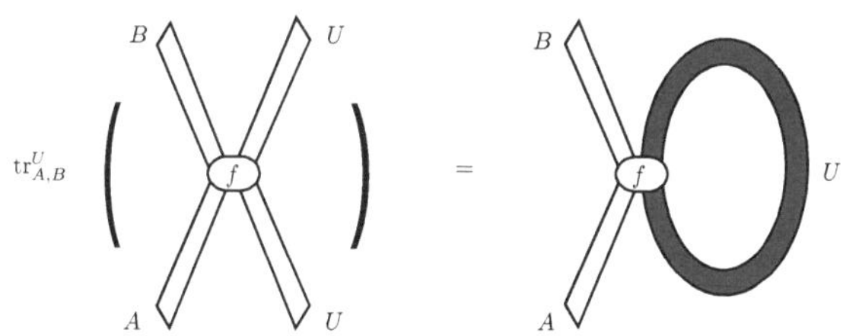

By definition, $\mathrm{tr}^U_{A,B}$ satisfies the diagrammatic equality:

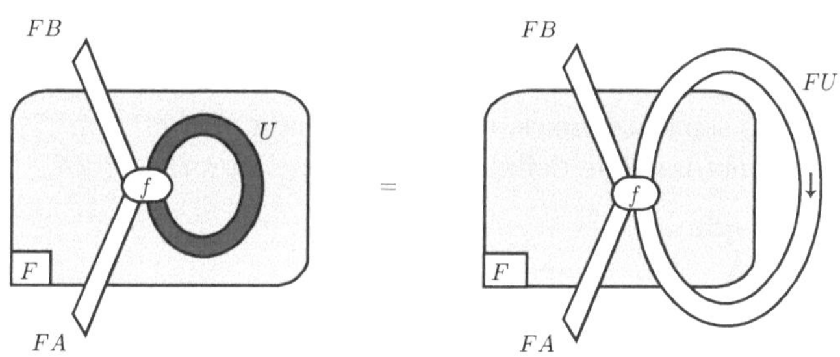

We establish each of the equations by a series of elementary manipulations on string diagrams. Although the proof is diagrammatic, it is absolutely rigorous, and works for weak monoidal categories $\mathbb{C}$ and $\mathbb{D}$ as well as strict ones.

Sliding (naturality in U). We want to show the equality

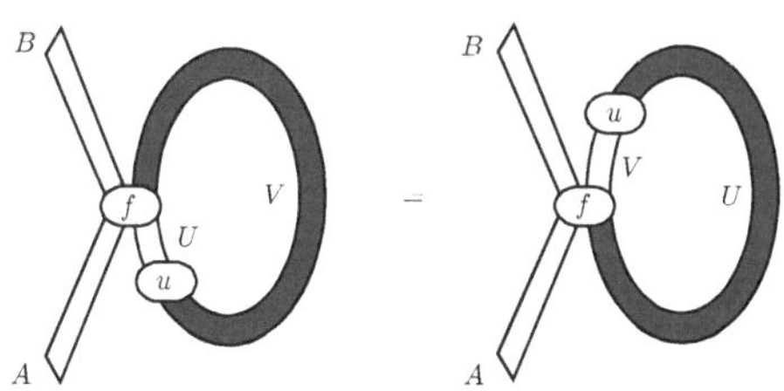

Because the functor F is faithful, it is sufficient to establish that the two morphisms $A \longrightarrow B$ have the same image $FA \longrightarrow FB$ in the target category $\mathbb{D}$:

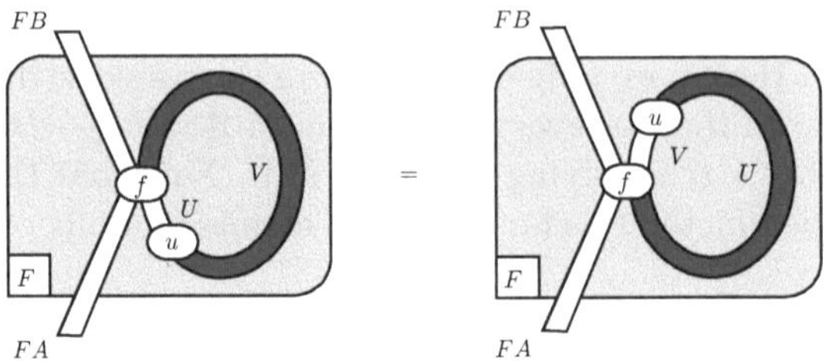

Once the definition of tr applied, we separate the box in two parts, using Equation (7) for the lax monoidal functor F:

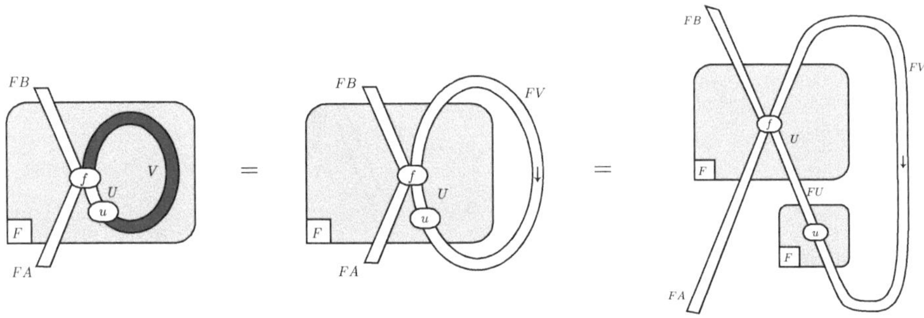

Then, after applying the sliding axiom of the trace Tr in the target category $\mathbb{D}$, we reunify the two separate boxes, using the variant of Equation (7) satisfied by colax monoidal functors. The definition of tr concludes the proof.

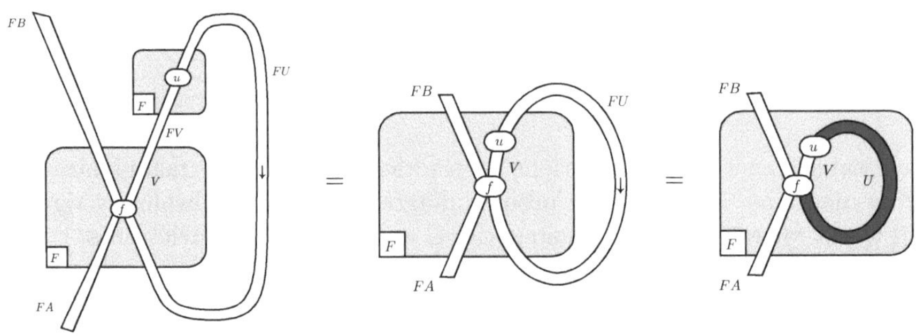

Tightening (Naturality in A and B). The proof is very similar to the proof of the sliding equality. Because the functor F is faithful, we will deduce the equality

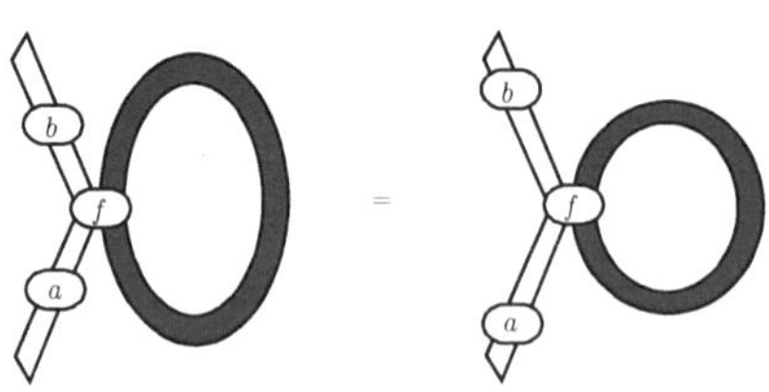

from the equality by F of the image of the two morphisms in the target category:

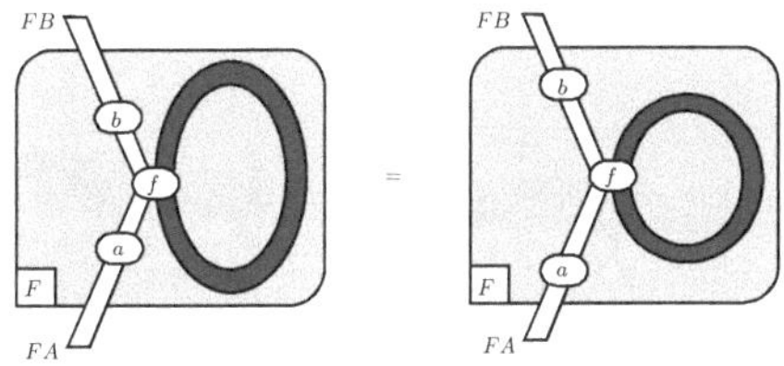

This is established as follows. Once the definition of tr applied, we separate the box in three parts, using Equation (7) for lax monoidal functors, and its colax variant:

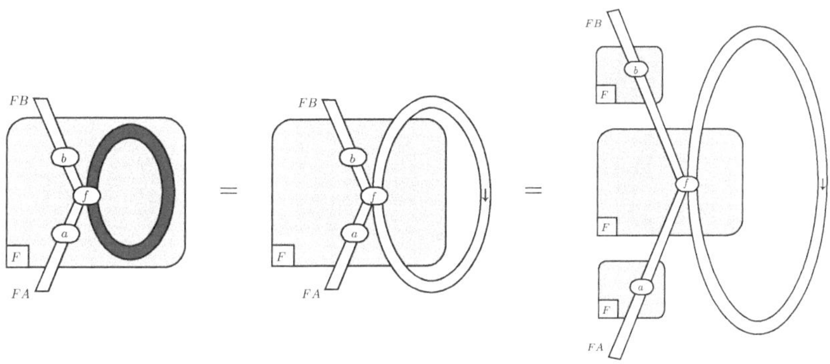

Then, we apply the tightening axiom of the trace Tr in the category $\mathbb{D}$, followed by the definition of tr, and finally reunify the three boxes together, using Equation (7) for lax monoidal functors, and its colax variant.

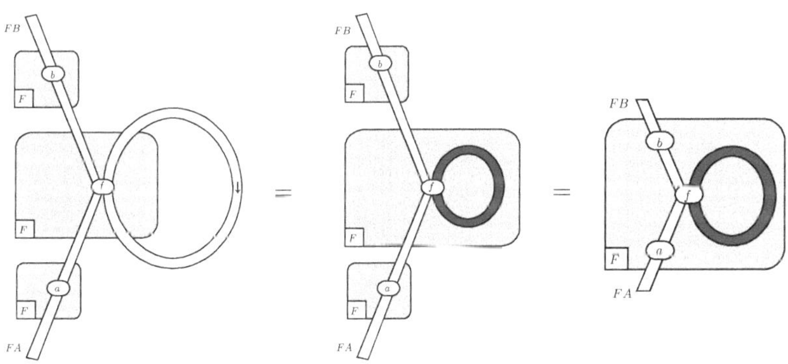

Vanishing (monoidality in U). We will proceed here as in the previous proofs, and deduce the two equalities formulated in the source category $\mathbb{C}$,

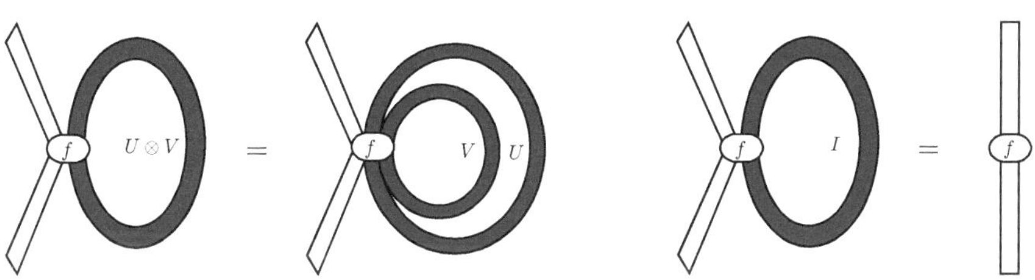

from the two equalities below, formulated in the target category $\mathbb{D}$,

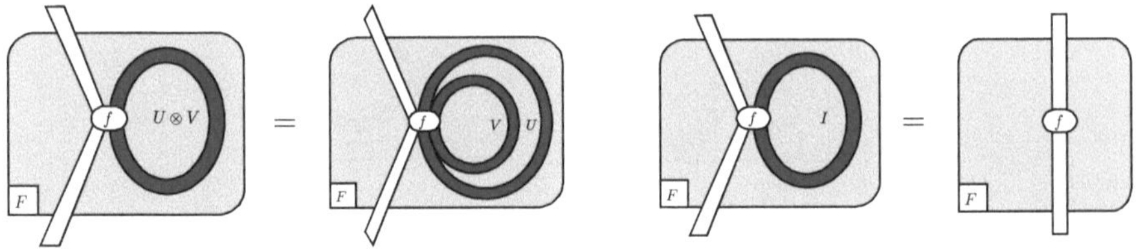

The first equation is established as follows. After applying the definition of tr, we split the string $U \otimes V$ in two strings U and V, then separate the box in two parts, using Equation (8) for the strong monoidal functor F:

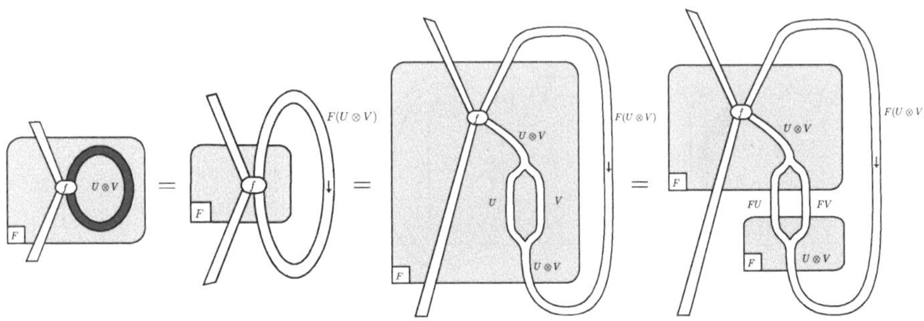

Then, we apply the sliding and vanishing axioms of the trace Tr in the category $\mathbb{D}$, and reunify the two boxes using Equation (8), before concluding by applying the definition of tr twice.

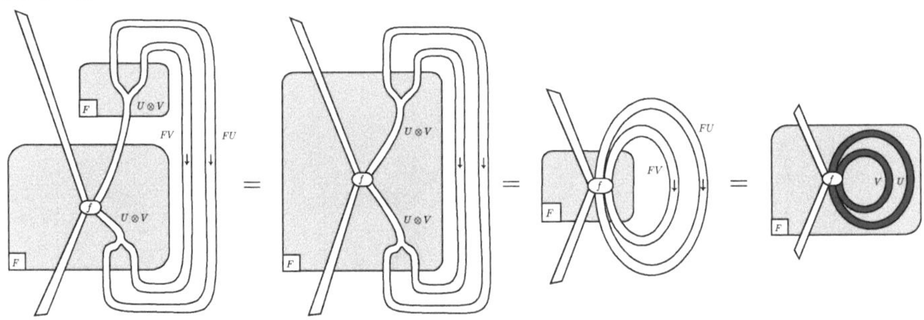

The second equation is established exactly as the previous one, except that we are dealing now with the nullary case instead of the binary one. After applying the definition of tr, we split the string I, and separate the box in two parts, using Equation (8) for the strong monoidal functor F:

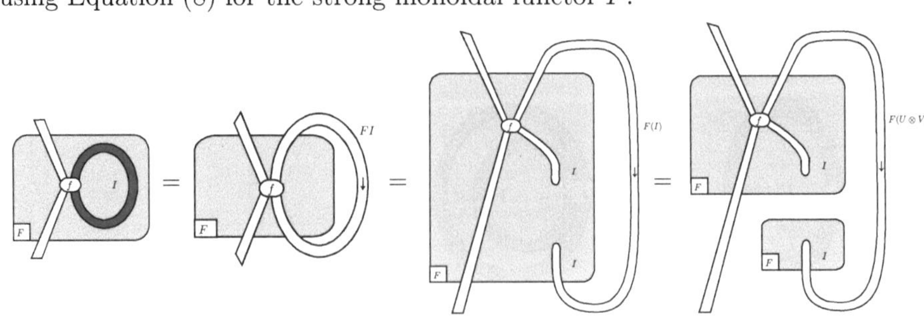

Then, just as for the binary case, we apply the sliding and vanishing axioms of the trace Tr and reunify the two boxes, before concluding.

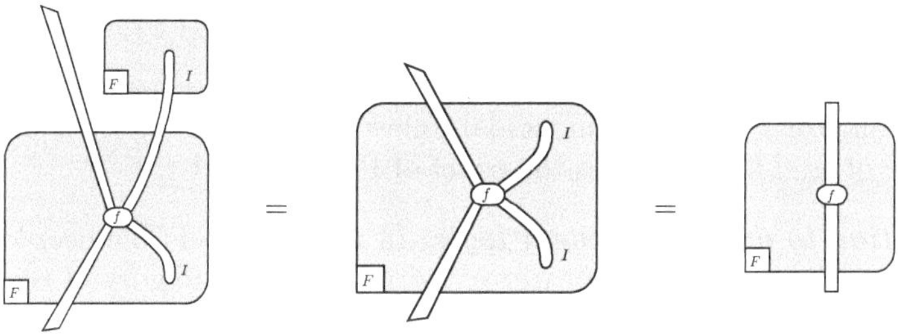

Note that we need the hypothesis that the functor F is *strong* monoidal in order to perform the manipulations for vanishing — while we needed only that it is lax and colax monoidal in the arguments devoted to sliding and tightening.

Superposing. We will establish only the first of the two equalities below — since the second one is proved in exactly the same way.

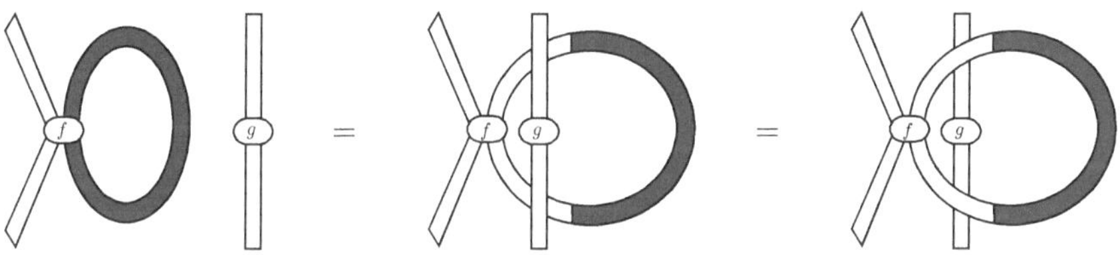

Because the functor F is faithful, this reduces to showing that the two morphisms have the same image in the category $\mathbb{D}$ — which we establish by the series of equalities below. First, we separate the box in two parts, using Equation (9) for the strong monoidal functor F; and apply the definition of tr in one of the two boxes.

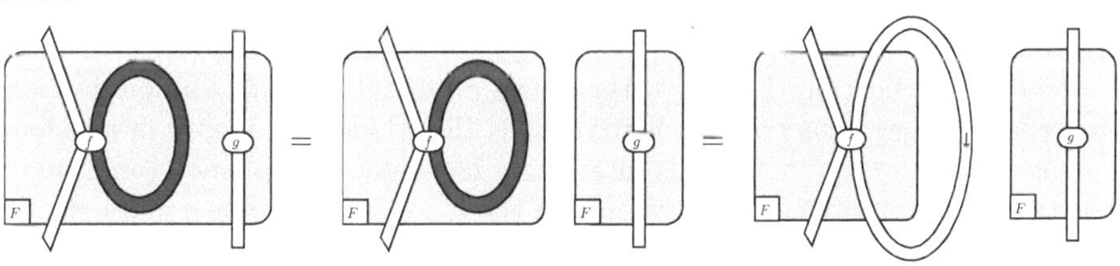

Then, after applying the superposing axiom of the trace Tr in the target category $\mathbb{D}$, we merge the two boxes, using again Equation (9) for the strong monoidal functor F; we insert the two braidings inside the box, using the hypothesis that the functor F is balanced; and finally conclude using the definition of tr.

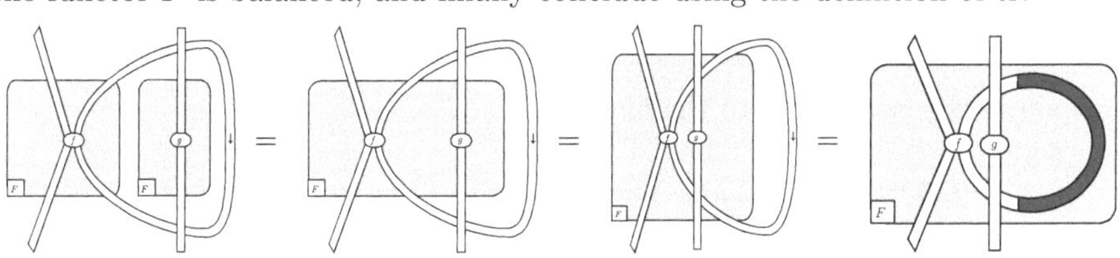

Yanking. The diagrammatic proof follows easily from the hypothesis that the functor F is faithful, and balanced monoidal. The proof is left to the reader as exercise.

From this, we conclude that tr defines a trace in the source category $\mathbb{C}$. The fact that the functor F is traced monoidal follows then immediately from the very definition of tr. This concludes the proof of Proposition 1.

Application to models of linear logic. In a typical model of linear logic based on a linear adjunction (1) the category $\mathbb{M}$ is a full subcategory of the category of Eilenberg-Moore coalgebras of the comonad $! = L \circ M$ in the category $\mathbb{L}$ — and the functor L is the associated forgetful functor. In that case, Proposition 1 ensures that the category $\mathbb{M}$ is traced when the category $\mathbb{L}$ is traced, and when, moreover, the trace

$$\mathrm{Tr}^{LU}_{LA,LB}(f) \quad : \quad LA \longrightarrow LB \tag{11}$$

of every coalgebraic morphism

$$f \quad : \quad LA \otimes LU \longrightarrow LB \otimes LU \tag{12}$$

is coalgebraic. This is precisely what happens in the relational model of linear logic, where:

- $\mathbb{L}$ is the category *Rel* of sets and relations, with tensor product defined as usual set-theoretic product,
- $\mathbb{M}$ is the co-kleisli category of the comonad ! which transports every set A to the free commutative comonoid $!A$ with finite multisets of elements of A as elements, and multiset union as coproduct. Note that the co-kleisli category $\mathbb{M}$ is understood here as the full subcategory of free coalgebras of the exponential comonad.

This establishes that the category $\mathbb{M}$ has a well-behaved fixpoint operator. A similar method applies to construct well-behaved fixpoint operators in categories of games and strategies [49].

Another application: Masahito Hasegawa observed (private communication) that the category $\mathbb{M}$ is traced whenever it is the co-kleisli category of an idempotent comonad $! = L \circ M$. This interesting fact may be explained (and mildly generalized) by applying Proposition 1 in the following way. Let η and ϵ denote the unit of the monad $M \circ L$ and the counit of the comonad $L \circ M$ respectively. For general reasons related to adjunctions, it appears that for every morphism

$$f : A \times U \longrightarrow B \times U \tag{13}$$

in the category $\mathbb{M}$, the morphism

$$h \quad = \quad \mathrm{Tr}^{LU}_{LA,LB}(m^{-1}_{[A,B]} \circ Lf \circ m_{[A,B]}) \quad : \quad LA \longrightarrow LB$$

is equal in the category $\mathbb{L}$ to the morphism

$$LA \xrightarrow{L\eta} LMLA \xrightarrow{LMh} LMLB \xrightarrow{\epsilon_{LB}} LB. \tag{14}$$

The equality is nicely depicted in string diagrams:

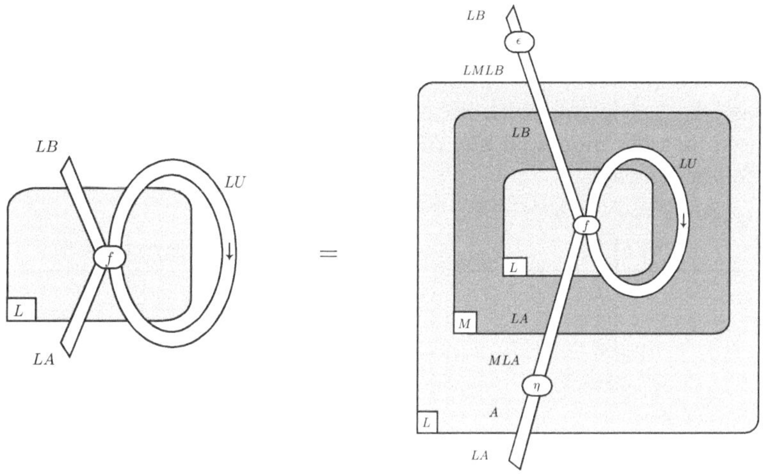

Here, Proposition 1 applies, and the category $\mathbb{M}$ is thus traced, whenever the functor L is faithful, and for every morphism f in (13), the morphism (14) is the image $L(g)$ in the category $\mathbb{L}$ of a morphism $g : A \longrightarrow B$ in the category $\mathbb{M}$.

This is precisely what happens when the category $\mathbb{M}$ is defined as the co-kleisli category associated to an idempotent comonad $! = L \circ M$. In that case, indeed, the morphism $\epsilon_{LB} : LMLB \longrightarrow LB$ is the identity, and the morphism g is defined as:

$$g \quad = \quad A \xrightarrow{\ \eta\ } MLA \xrightarrow{\ Mh\ } MLB.$$

A nice problem remains open. A few years ago, Ryu Hasegawa [21] constructed a trace related to the Lagrange-Good inversion formula, in a category of analytic functors. This category, which is cartesian, is the co-kleisli category associated to a specific model of linear logic. Interestingly, the diagrammatic account exposed in this tutorial does not seem to apply (at least directly) to Ryu Hasegawa's construction. It would be extremely satisfactory to devise alternative algebraic conditions to cover this important example. We leave this open here.

7 Decomposing the Exponential Box of Linear Logic

The decomposition $! = L \circ M$ of the exponential modality of linear logic illustrates the general diagrammatic principle that every functorial box separates an inside world from an outside world, each world implementing his own (eg. cartesian, monoidal closed) policy. We take the freedom of considering here a "balanced" version of linear logic, whose categorical model is defined as a balanced monoidal adjunction

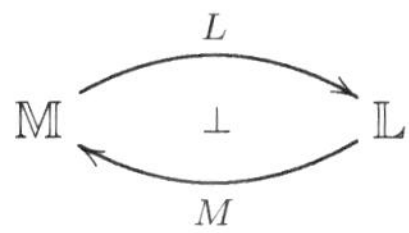

(15)

between a balanced monoidal category $\mathbb{L}$ and a cartesian category $\mathbb{M}$. Note that in such an adjunction, the functor L is balanced monoidal.

In that setting, the exponential box ! with its auxiliary doors labelled by the formulas $!A_1, ..., !A_k$ and with its principal door labelled by the formula $!B$ is translated as a lax monoidal box M enshrined inside a strong monoidal box L, in the following way:

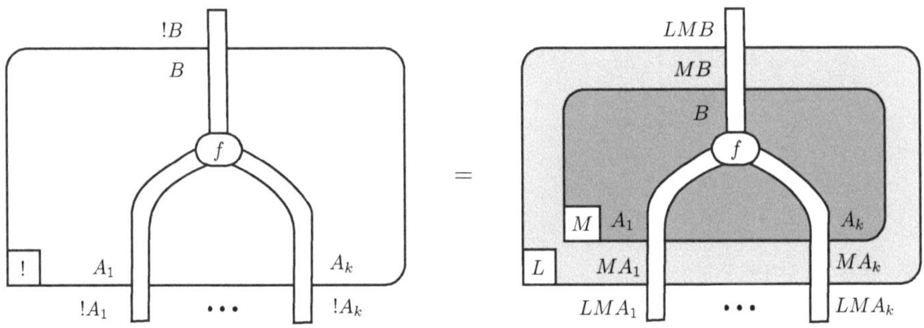

Now, the category $\mathbb{M}$ enshrined "inside" the functorial box L is cartesian, with binary product noted $\times$ here. Hence, every object X of the category $\mathbb{M}$ induces a diagonal morphism

$$\Delta_X : X \longrightarrow X \times X.$$

In particular, every object A of the category $\mathbb{L}$ induces a diagonal morphism

$$\Delta_{MA} : MA \longrightarrow MA \times MA.$$

The *contraction* of linear logic is defined as the morphism $L(\Delta_{MA})$ depicted as the diagonal string Δ_{MA} inside the strong monoidal box L:

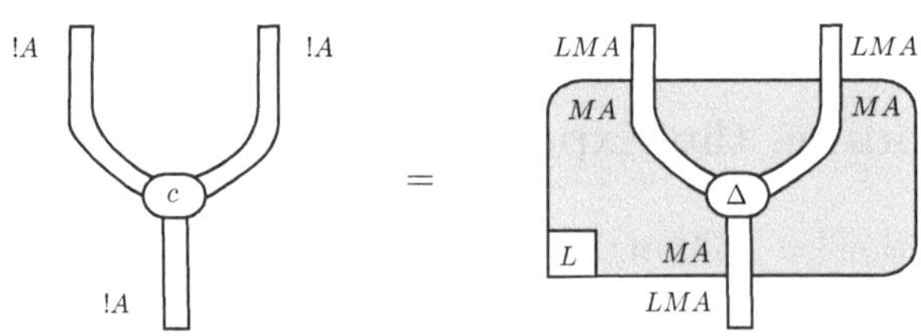

If one translates in string diagrams the usual cut-elimination step of linear logic between a contraction rule and an introduction rule of the exponential box, this decomposes the step in a series of more atomic steps. First, the box L which encapsulates the diagonal Δ_{MA} merges with the box L which encapsulates the content f of the exponential box. This releases the diagonal Δ_{MA} inside the cartesian category $\mathbb{M}$ enshrined in the exponential box.

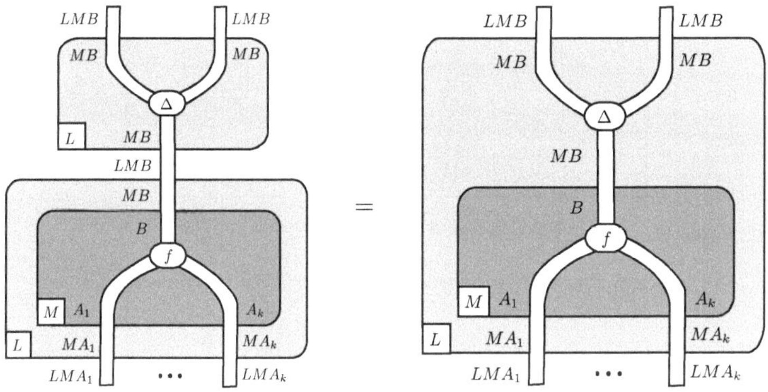

Then, the diagonal Δ_{MA} replicates the content f of the exponential box — or more precisely the morphism f encapsulated by the lax monoidal box M. Note that the duplication step is performed in the cartesian category $\mathbb{M}$ enshrined by the functorial box L.

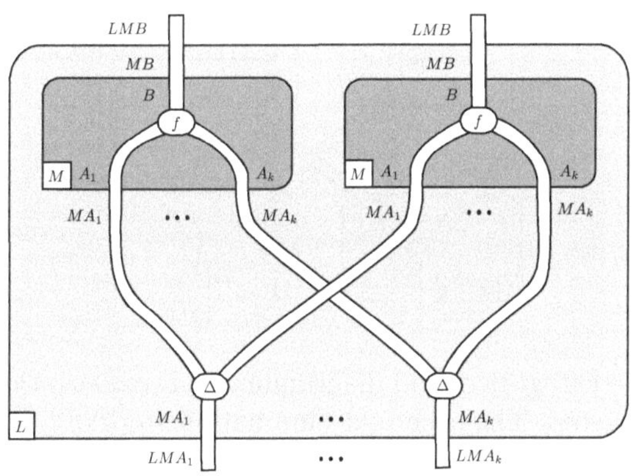

Once the duplication finished, the strong monoidal box is split in three horizontal parts using Equation (8).

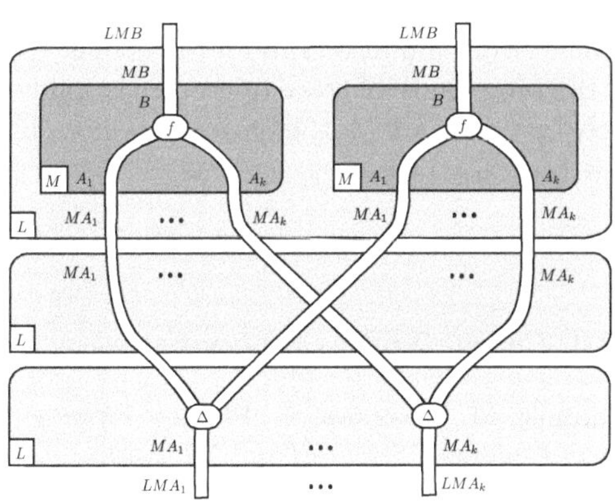

The intermediate box may be removed, because the functor L is balanced.

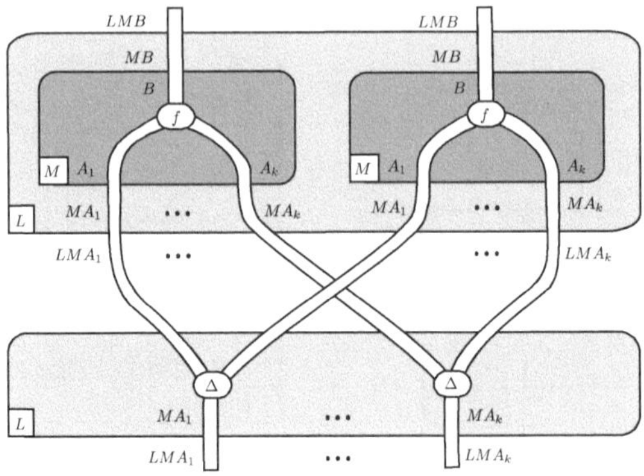

Finally, the remaining monoidal boxes L are split vertically, using Equation (9).

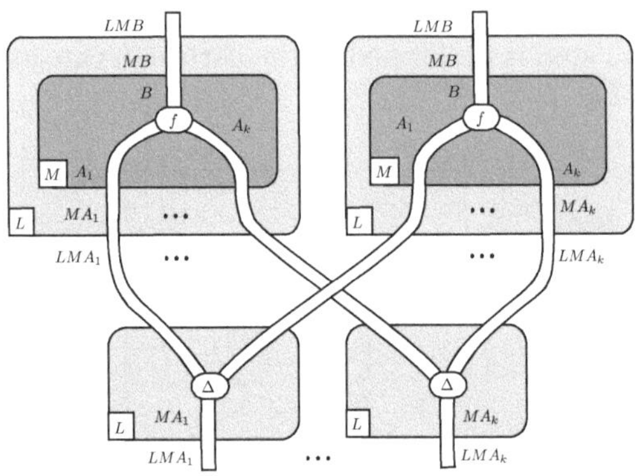

This completes the categorical and diagrammatical transcription of this particular cut-elimination step. The other cut-elimination steps of linear logic involving the exponential box ! are decomposed in a similar fashion.

Acknowledgments. This short tutorial was stimulated and inspired by discussions with Martin Hyland and Masahito Hasegawa on trace operators in models of linear logic, held during the summer 2005 and onwards. I thank them for their precious insights and advice. I am also grateful to Nicolas Tabareau for motivating this work by his ongoing study of trace operators in game semantics. Finally, I thank the Research Institute for Mathematical Science (RIMS) in Kyoto, for hosting me when I wrote this tutorial.

References

1. R. Amadio and P.-L. Curien. *Domains and Lambda-Calculi.* Cambridge University Press, 1998.
2. A. Barber, P. Gardner, M. Hasegawa, G. Plotkin. From action calculi to linear logic. *Proceedings of Computer Science Logic '97*, Aarhus, Denmark. Volume 1414 of Lecture Notes in Computer Science, Springer Verlag. 1997.
3. A. Barber. *Linear Type Theories, Semantics and Action Calculi.* PhD Thesis of the University of Edinburgh. LFCS Technical Report CS-LFCS-97-371. 1997.

4. J. Bénabou. Introduction to bicategories. *Reports of the Midwest Category Seminar.* Volume 47 of Lecture Notes in Mathematics, Springer Verlag. 1967.
5. N. Benton. A Mixed Linear and Non-Linear Logic: Proofs, Terms and Models. *Proceedings of Computer Science Logic '94,* Kazimierz, Poland. Volume 933 of Lecture Notes in Computer Science, Springer Verlag. June 1995.
6. N. Benton, G. Bierman, V. de Paiva, M. Hyland. *Term assignment for intuitionistic linear logic.* Technical Report 262, Computer Laboratory, University of Cambridge, 1992.
7. G. Berry. Stable models of typed lambda-calculi. *Proceedings of the 5th International Colloquium on Automatas, Languages and Programming,* number 62 in Lecture Notes in Computer Science. Springer Verlag 1978.
8. G. Bierman. On intuitionistic linear logic. *PhD Thesis.* University of Cambridge Computer Laboratory, December 1993.
9. G. Bierman. What is a categorical model of intuitionistic linear logic? *Proceedings of the Second International Conference on Typed Lambda Calculus and Applications.* Volume 902 of Lecture Notes in Computer Science, Springer Verlag. Edinburgh, Scotland, April 1995. Pages 73-93.
10. H. Blackwell, M. Kelly, and A. J. Power. Two dimensional monad theory. *Journal of Pure and Applied Algebra,* 59:1–41, 1989.
11. R. Blute, R. Cockett, R. Seely. The logic of linear functors. *Mathematical Structures in Computer Science,*12 (2002)4 pp 513-539.
12. R. Blute, R. Cockett, R. Seely, T. Trimble. Natural Deduction and Coherence for Weakly Distributive Categories. *Journal of Pure and Applied Algebra,* 113(1996)3, pp 229-296.
13. A. Burroni. Higher Dimensional Word Problem, *Category Theory and Computer Science,* Lecture Notes in Computer Science 530, Springer-Verlag, 1991.
14. R. Cockett, R. Seely. Linearly Distributive Categories. *Journal of Pure and Applied Algebra,* 114(1997)2, pp 133-173.
15. R. Cockett, R. Seely. Linear Distributive Functors. In *The Barrfestschrift, Journal of Pure and Applied Algebra,* Volume 143, Issue 1-3, 10 November 1999.
16. B. J. Day and R. Street. Quantum categories, star autonomy, and quantum groupoids. *Galois Theory, Hopf Algebras, and Semiabelian Categories.* Fields Institute Communications 43 (American Math. Soc. 2004) 187-226.
17. G. Gentzen. Investigations into logical deduction (1934). An english translation appears in *The Collected Papers of Gerhard Gentzen.* Edited by M. E. Szabo, North-Holland 1969.
18. J.-Y. Girard. Linear logic. *Theoretical Computer Science,* 50: 1-102, 1987.
19. J.-Y. Girard. Linear logic: its syntax and semantics. In *Advances in Linear Logic,* London Mathematical Society Lecture Note Series 222, pp. 1–42, Cambridge University Press, 1995.
20. M. Hasegawa. Recursion from cyclic sharing: traced monoidal categories and models of cyclic lambda-calculi. *Proceeding of the 3rd International Conference on Typed Lambda-Calculi and Applications,* Springer Verlag, Lecture Notes in Computer Science 1210, (1997).
21. R. Hasegawa. Two applications of analytic functors. *Theoretical Computer Science* 272 (2002) 113-175.
22. C. Hermida and J. Power. Fibrational control structures. *Proceedings of CONCUR 1995.* Springer Lecture Notes in Computer Science 962. pp 117–129, 1995.
23. M. Hyland and A. Schalk. Glueing and orthogonality for models of linear logic. *Theoretical Computer Science* 294(1/2): 183-231, 2003.

24. G. B. Im and M. Kelly. A universal property of the convolution monoidal structure, *J. Pure Appl. Algebra* 43, pp. 75-88, 1986.
25. M. Kelly. Doctrinal adjunction. *Lecture Notes in Math.* 420, pp. 257-280, 1974.
26. A. Joyal. Remarques sur la théorie des jeux à deux personnes. *Gazette des Sciences Mathématiques du Québec*, volume 1, number 4, pp 46–52, 1977.
27. A. Joyal and R. Street. Braided Tensor Categories, *Advances in Mathematics* 102, 20–78, 1993.
28. A. Joyal and R. Street. The geometry of tensor calculus, I. *Advances in Mathematics* 88, 55–112, 1991.
29. A. Joyal, R. Street and D. Verity. Traced monoidal categories. *Math. Proc. Camb. Phil. Soc.* 119, 447–468, 1996.
30. S. Lack. Limits for lax morphisms. *Applied Categorical Structures*. 13(3):189-203, 2005.
31. Y. Lafont. *Logiques, catégories et machines.* PhD thesis, Université Paris 7, 1988.
32. Y. Lafont. From Proof Nets to Interaction Nets, In *Advances in Linear Logic*, London Mathematical Society Lecture Note Series 222, pp. 225–247, Cambridge University Press, 1995.
33. F. Lamarche Sequentiality, games and linear logic, Unpublished manuscript. 1992.
34. J. Lambek and P. Scott. Introduction to Higher Order Categorical Logic. Cambridge Studies in Advanced Mathematics Vol. 7. Cambridge University Press, 1986.
35. F. W. Lawvere. Ordinal sums and equational doctrines. *Springer Lecture Notes in Mathematics No. 80*, Springer, Berlin, 1969, pp. 141-155.
36. S. Mac Lane. Categories for the working mathematician. *Graduate Texts in Mathematics* 5. Springer Verlag 2nd edition, 1998.
37. M. Maietti, P. Maneggia, V. de Paiva, E. Ritter Relating categorical semantics for intuitionistic linear logic. Applied Categorical Structures 13(1): 1-36, 2005.
38. P.-A. Melliès, Typed lambda-calculi with explicit substitutions may not terminate *Proceedings of the Second International Conference on Typed Lambda Calculi and Applications, Edinburgh,* Lecture Notes in Computer Science 902, pp. 328-334, Springer, 1995.
39. P.-A. Melliès, Axiomatic Rewriting 4: a stability theorem in rewriting theory. *Proceedings of Logic in Computer Science 1998*, IEEE Computer Society Press, 1998.
40. P.-A. Melliès, Categorical semantics of linear logic: a survey. To appear in *Panoramas et Synthèses*, Société Mathématique de France, 2007.
41. R. Milner, Pure bigraphs: structure and dynamics. *Information and Computation*, Volume 204, Number 1, January 2006.
42. D. Pavlovic, Categorical logic of names and abstraction in action calculi. *Mathematical Structures in Computer Science*, 7 (6) (1997) 619–637.
43. R. Penrose, Applications of negative dimensional tensors, in *Combinatorial Mathematics and its Applications*, D. J. A. editor, pp. 221-244, Academic Press, New York, 1971.
44. R. Penrose, Spinors and Space-Time, Vol. 1, pp. 68-71, 423-434, Cambridge University Press, Cambridge, U. K., 1984.
45. R. Seely. Linear logic, $*$-autonomous categories and cofree coalgebras. *Applications of categories in logic and computer science*, Contemporary Mathematics, 92, 1989.
46. S. Schanuel and R. Street. The free adjunction. *Cahiers topologie et géométrie différentielle catégoriques* 27 (1986) pp. 81-83.
47. R. Street. Limits indexed by category-valued 2-functors. *J. Pure Appl. Algebra* 8 (1976) 149-181.
48. R. Street. Functorial calculus in monoidal bicategories. *Applied Categorical Structures* 11 (2003) 219-227.
49. N. Tabareau. De l'opérateur de trace dans les jeux de Conway. *Mémoire de Master 2.* Master Parisien de Recherche en Informatique, Université Paris 7, Septembre 2005.

Some Results on a Game-Semantic Approach to Verifying Finitely-Presentable Infinite Structures (Extended Abstract)

C.-H.L. Ong

Oxford University Computing Laboratory
http://users.comlab.ox.ac.uk/luke.ong/

Abstract. We present some results on a game-semantic approach to verifying infinite structures that are generated by higher-order recursion schemes. The key idea is a certain *Transference Principle* from the structure generated by a given recursion scheme to an auxiliary *computation tree*, which is itself generated by a related order-0 recursion scheme. By a structural analysis of the computation tree based on the *innocent game semantics* of the recursion scheme, we can infer certain properties of the generated structure by appropriate algorithmic analysis of the computation tree.

1 Introduction

A basic problem in Verification is to identify classes of finitely-presentable infinite-state systems that have decidable monadic second-order (MSO) theories. This is a question of practical importance because MSO logic is highly expressive: temporal logics that are widely used in computer-aided verification such as LTL, CTL and CTL* are embeddable in the modal mu-calculus, and hence, embeddable in MSO logic. Indeed MSO logic is a kind of gold standard in Verification because it is virtually as strong (a specification language) as can be, in the sense that any obvious extension of the logic would render it undecidable.

Perhaps one of the best known examples of such MSO-decidable structures is the class of *regular trees*, as studied by Rabin [1] in 1969. A notable advance occurred some fifteen years later, when Muller and Shupp [2] proved that the *configuration graphs of pushdown systems* have decidable MSO theories. In the 1990s, as finite-state technologies matured, researchers embraced the challenges of software verification. A highlight in this period was Caucal's result [3] that *prefix-recognisable graphs* have decidable MSO theories. Prefix-recognisable graphs may have unbounded out-degrees; they can be characterized [4] as graphs that are obtained from the configuration graphs of pushdown systems by factoring out the ϵ-transitions.

In 2002 a flurry of discoveries significantly extended and unified earlier developments. In a FOSSACS 2002 paper [5], Knapik, Niwiński and Urzyczyn introduced an infinite hierarchy of (possibly infinite) *Σ-labelled trees* (i.e. ranked and ordered trees whose nodes are labelled by symbols of a ranked alphabet Σ):

Z. Ésik (Ed.): CSL 2006, LNCS 4207, pp. 31–40, 2006.

the nth level of the hierarchy, $\boldsymbol{SafeRecTree}_n\Sigma$, consists of Σ-labelled trees generated by order-n recursion schemes that are *homogeneously typed*[1] and satisfy a syntactic constraint called *safety*[2]. They showed that for every $n \geq 0$, trees in $\boldsymbol{SafeRecTree}_n\Sigma$ have decidable MSO theories; further $\boldsymbol{SafeRecTree}_n\Sigma = \boldsymbol{PushdownTree}_n\Sigma$ i.e. Σ-labelled trees generated by order-n safe recursion schemes are exactly those that are generated by *order-n (deterministic) pushdown automata.* Thus $\boldsymbol{SafeRecTree}_0\Sigma$, the order-0 trees, are the regular trees (i.e. trees generated by finite-state transducers); and $\boldsymbol{SafeRecTree}_1\Sigma$, the order-1 trees, are those generated by deterministic pushdown automata. Later in the year, Caucal [6] introduced an infinite hierarchy of Σ-labelled trees, the nth level of which, $\boldsymbol{CaucalTree}_n\Sigma$, consists of Σ-labelled trees that are obtained from regular Σ-labelled trees by iterating n-times the operation of inverse deterministic rational mapping followed by unravelling. A major result in Caucal's work [6, Theorem 3.5] is that $\boldsymbol{SafeRecTree}_n\Sigma = \boldsymbol{CaucalTree}_n\Sigma$. To summarize:

Theorem 1 (Knapik, Niwinski, Urzyczyn and Caucal 2002). *For any ranked alphabet Σ, and for every $n \geq 0$, we have*

$$\boldsymbol{SafeRecTree}_n\Sigma \ = \ \boldsymbol{PushdownTree}_n\Sigma \ = \ \boldsymbol{CaucalTree}_n\Sigma;$$

further, trees from the class have decidable MSO theories.

Though a rather awkward syntactic constraint, *safety* plays an important algorithmic rôle. Knapik *et al.* have asked [5] if the safety assumption is really necessary for their decidability result. In other words, let $\boldsymbol{RecTree}_n\Sigma$ be the class of Σ-labelled trees generated by order-n recursion schemes (whether safe or not, and whether homogeneously typed or not), the question is:

For which $n \geq 2$ do trees in $\boldsymbol{RecTree}_n\Sigma$ have decidable MSO theories?

A partial answer to the question has recently been obtained by Aehlig, de Miranda and Ong at TLCA 2005 [7]; they showed that all trees in $\boldsymbol{RecTree}_2\Sigma$ have decidable MSO theories. Independently, Knapik, Niwiński, Urzyczyn and Walukiewicz obtained a somewhat sharper result (see their ICALP 2005 paper [8]); they proved that the modal mu-calculus model checking problem for trees generated by order-2 homogeneously-typed recursion schemes (whether safe or not) is 2-EXPTIME complete.

[1] The base type o is *homogeneous*; a function type $A_1 \to (A_2 \to \cdots \to (A_n \to o) \cdots)$ is *homogeneous* just if each A_i is homogeneous, and $ord(A_1) \geq ord(A_2) \geq \cdots \geq ord(A_n)$. A term (or a rewrite rule or a recursion scheme) is *homogeneously typed* just if all types that occur in it are homogeneous.

[2] A homogeneously-typed term of order $k > 0$ is said to be *unsafe* if it contains an occurrence of a parameter of order strictly less than k, otherwise the term is *safe*. An occurrence of an unsafe term t, as a subexpression of a term t', is ***safe*** if it occurs in an operator position (i.e. it is in the context $\cdots (ts) \cdots$), otherwise the occurrence is *unsafe*. A recursion scheme is ***safe*** if no unsafe term has an unsafe occurrence in the righthand side of any rewrite rule.

In this extended abstract, we explain a game-semantic approach to verifying infinite structures that are generated by higher-order recursion schemes. As a major case study, we extend the result of Knapik *et al.* [8] by proving that for all $n \geq 0$, trees in $\boldsymbol{RecTree}_n\Sigma$ have decidable MSO theories; we give an outline of the proof in Section 2. In Section 3 we give an automata-theoretic characterization of trees in $\boldsymbol{RecTree}_n\Sigma$ where $n \geq 0$. We show that for the purpose of generating Σ-labelled tree, recursion schemes are equi-expressive with a new kind of automata called *collapsible pushdown automata*. The same game-semantic approach can be extended to verify certain classes of finitely-presentable infinite graphs. In Section 4, we briefly consider the solution of parity games on configuration graphs of collapsible pushdown automata. Finally we mention a couple of further directions in Section 5.

Technical Preliminaries

Types are generated from the base type o using the arrow constructor $\rightarrow$. Every type A can be written uniquely as $A_1 \rightarrow \cdots \rightarrow A_n \rightarrow o$ (by convention arrows associate to the right), for some $n \geq 0$ which is called its *arity*. We define the *order* of a type by: $ord(o) = o$ and $ord(A \rightarrow B) = \max(ord(A) + 1, ord(B))$. Let Σ be a *ranked alphabet* i.e. each Σ-symbol f has an arity $ar(f) \geq 0$ which determines its type $\underbrace{o \rightarrow \cdots \rightarrow o}_{ar(f)} \rightarrow o$. Further we shall assume that each symbol $f \in \Sigma$ is assigned a finite set $\mathsf{Dir}(f) = \{1, \cdots, ar(f)\}$ of *directions*, and we define $\mathsf{Dir}(\Sigma) = \bigcup_{f \in \Sigma} \mathsf{Dir}(f)$. Let D be a set of directions; a *D-tree* is just a prefix-closed subset of D^*, the free monoid of D. A ***Σ-labelled tree*** is a function $t : \mathsf{Dom}(t) \longrightarrow \Sigma$ such that $\mathsf{Dom}(t)$ is a $\mathsf{Dir}(\Sigma)$-tree, and for every node $\alpha \in \mathsf{Dom}(t)$, the Σ-symbol $t(\alpha)$ has arity k if and only if α has exactly k children and the set of its children is $\{\alpha\,1, \cdots, \alpha\,k\}$ i.e. t is a *ranked* (and ordered) tree.

For each type A, we assume an infinite collection Var^A of variables of type A. A (deterministic) ***recursion scheme*** is a tuple $G = \langle \Sigma, \mathcal{N}, \mathcal{R}, S \rangle$ where Σ is a ranked alphabet of *terminals*; $\mathcal{N}$ is a set of *non-terminals*, each of a fixed type; $S \in \mathcal{N}$ is a distinguished *start symbol* of type o; $\mathcal{R}$ is a finite set of rewrite rules – one for each non-terminal $F : (A_1, \cdots, A_n, o)$ – of the form

$$F\,\xi_1 \cdots \xi_n \;\rightarrow\; e$$

where each ξ_i is in Var^{A_i}, and e is an *applicative term*[3] of type o constructed from elements of $\Sigma \cup \mathcal{N} \cup \{\xi_1, \cdots, \xi_n\}$. The *order* of a recursion scheme is the highest order of its non-terminals.

We use recursion schemes as generators of Σ-labelled trees. The ***value tree*** of (or the tree *generated* by) a recursion scheme G is a possibly infinite applicative term, but viewed as a Σ-labelled tree, *constructed from the terminals in Σ*, that

[3] *Applicative terms* are terms constructed from the generators using the application rule: if $d : A \rightarrow B$ and $e : A$ then $(de) : B$. Standardly we identify finite Σ-labelled trees with applicative terms of type o generated from Σ-symbols endowed with 1st-order types *as given by their arities.*

is obtained by unfolding the rewrite rules of G *ad infinitum*, replacing formal by actual parameters each time, starting from the start symbol S. For $n \geq 0$ we define $\boldsymbol{RecTree}_n \Sigma$ to be the class of Σ-labelled value trees of (arbitrary) order-n recursion schemes. Plainly we have $\boldsymbol{SafeRecTree}_n \Sigma \subseteq \boldsymbol{RecTree}_n \Sigma$ for every $n \geq 0$. It follows from the definition of safety that $\boldsymbol{SafeRecTree}_n \Sigma = \boldsymbol{RecTree}_n \Sigma$ for $n = 0$ and 1, but it is not known whether the inclusion is strict for $n \geq 2$ (see Conjecture 1).

Example 1. Let G be the order-2 (unsafe) recursion scheme with rewrite rules:

$$\begin{aligned} S &\rightarrow H\, a \\ H\, z^{o} &\rightarrow F\,(g\, z) \\ F\, \varphi^{(o,o)} &\rightarrow \varphi\,(\varphi\,(F\, h)) \end{aligned}$$

where the arities of the terminals g, h, a are $2, 1, 0$ respectively. The value tree $[\![\, G \,]\!]$ is the Σ-labelled tree defined by the infinite term $g\, a\,(g\, a\,(h\,(h\,(h\, \cdots))))$:

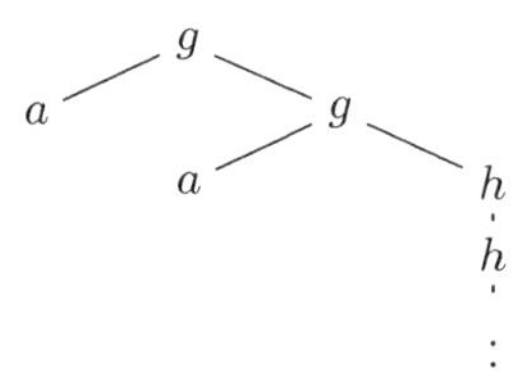

The only infinite *path* in the tree is the node-sequence $\epsilon \cdot 2 \cdot 22 \cdot 221 \cdot 2211 \cdots$ (with the corresponding *trace* $g\, g\, h\, h\, h\, \cdots \in \Sigma^{\omega}$).

2 Trees Generated by Recursion Schemes Have Decidable MSO Theories

We state our first main result as follows:

Theorem 2. *For every $n \geq 0$ the modal mu-calculus model-checking problem for trees in $\boldsymbol{RecTree}_n \Sigma$ is n-EXPTIME complete.*

Since MSO logic and the modal mu-calculus are equi-expressive over trees (see e.g. [9]), it follows that these trees have decidable MSO theories. Our proof of Theorem 2 relies on a certain *Transference Principle* from the *value tree* of a recursion scheme G to an auxiliary *computation tree*, which is itself a tree generated by a related order-0 recursion scheme $\overline{G}$, called the *long transform* of G. By exploiting a structural analysis of the computation tree made possible by the innocent game semantics (in the sense of Hyland and Ong [12]) of the scheme G, we can infer certain properties of the value tree $[\![\, G \,]\!]$ (e.g. acceptance by a given tree automaton) by appropriate algorithmic analysis of the computation tree. A full account of the result can be found in the preprint [10]; see also the summary at LICS 2006 [11].

Removing the safety assumption: a game-semantic approach

Here we briefly explain our approach. First we note that it would be futile to analyse directly the value tree $[\![G]\!]$ since it has no useful structure for our purpose: by definition $[\![G]\!]$ is the *extensional* outcome of a potentially infinite process of rewriting. Rather it is the algorithmics of this computational process that one should seek to understand. Given an order-$(n+1)$ safe recursion scheme G, the approach taken in [5] was to consider an associated tree that is obtained by contracting only (and all) the order-1 (i.e. lowest ordered) β-redexes in the rewrite rules of G. The tree thus generated, written $\beth_G$, coincides with the value tree of a related order-n recursion scheme G^α (i.e. $\beth_G = [\![G^\alpha]\!]$); further the MSO theory of the order-$(n+1)$ tree $[\![G]\!]$ is reducible to that of the order-n tree $[\![G^\alpha]\!]$ in the sense that there is a recursive mapping of MSO sentences $\varphi \mapsto \varphi'$ such that $[\![G]\!] \models \varphi$ iff $[\![G^\alpha]\!] \models \varphi'$ [5, Theorem 3.3]. The safety assumption is crucial to the reduction argument. Roughly speaking, the point is that β-redexes in a safe term can be contracted using *capture-permitting* substitution (i.e. without renaming bound variables). It follows that one can construct the tree $\beth_G$ using only the original variables of the recursion schemes G. Without the safety assumption, the same construction would require an unbounded supply of names!

Our approach to removing the safety assumption stems from an observation due to Klaus Aehlig [7]: by considering the *long transform* of a recursion scheme (which is obtained by expanding the RHS of each rewrite rule to its η-long form, inserting explicit application operators, and then currying the rule), the two constituent actions of the rewriting process, namely, *unfolding* and β-reduction, can be teased out and hence analysed separately. Given an order-n recursion scheme G:

- We first construct the *long transform* $\overline{G}$, which is an order-0 recursion scheme.
- We then build an auxiliary *computation tree* $\lambda(G)$ which is the outcome of performing all of the unfolding[4], but none of the β-reduction, in the $\overline{G}$-rules. As no substitution is performed, no variable-renaming is needed.
- We can now analyse the β-reductions *locally* (i.e. without the *global* operation of substitution) by considering *traversals* over the computation tree, based on innocent game semantics [12].

Note that we do not (need to) assume that the recursion scheme G is safe or type-homogeneous. Formally the *computation tree* $\lambda(G)$ is defined to be the value tree of the long transform $\overline{G}$; the tree $\lambda(G)$ is regular, since $\overline{G}$ is an order-0 recursion scheme.

Correspondence between paths in $[\![G]\!]$ and traversals over $\lambda(G)$

We sketch an outline of the proof of Theorem 2. We are concerned with the decision problem: *Given a modal mu-calculus formula φ and an order-n recursion scheme G, does $[\![G]\!]$ satisfy φ (at the root ϵ)?* The problem is equivalent [14]

[4] I.e. rewriting the LHS of a rule to the RHS with no parameter passing - since the $\overline{G}$ rules, being order-0, have no formal parameters.

to deciding whether a given *alternating parity tree automaton* (APT) $\mathcal{B}$ has an *accepting run-tree* over the Σ-labelled tree $[\![G]\!]$. Recall that an *accepting run-tree* of $\mathcal{B}$ over $[\![G]\!]$ is a certain set of state-annotated paths in $[\![G]\!]$ that respect the transition relation of $\mathcal{B}$, such that every infinite path in the set satisfies the *parity condition*. Instead of analysing paths in $[\![G]\!]$ directly, we use game semantics to establish a strong correspondence between *paths* in the value tree and *traversals* over the computation tree.

Theorem 3 (Correspondence). *Let G be a recursion scheme. There is a one-one correspondence, $p \mapsto t_p$, between maximal paths p in the value tree $[\![G]\!]$ and maximal traversals t_p over the computation tree $\lambda(G)$. Further for every maximal path p in $[\![G]\!]$, we have $t_p \upharpoonright \Sigma^- = p \upharpoonright \Sigma^-$, where $s \upharpoonright \Sigma^-$ denotes the subsequence of s consisting of only Σ^--symbols with $\Sigma^- = \Sigma \setminus \{ \perp \}$.*

The proof of the Theorem is an application of the innocent game semantics of the recursion scheme G (see [13] for a proof of the statement in a more general setting). In the language of game semantics, paths in the value tree correspond exactly to plays in the strategy-denotation of the recursion scheme; a traversal is then (a representation of) the *uncovering* [12] of such a play. The path-traversal correspondence (Theorem 3) allows us to prove the following useful transference result:

Corollary 1. *A given* property[5] *APT has an accepting run-tree over the value tree if and only if it has an accepting* traversal-tree *over the computation tree.*

Relative to a property APT $\mathcal{B}$ over Σ-labelled trees, an (accepting) traversal-tree of $\mathcal{B}$ over $\lambda(G)$ plays the same rôle as an (accepting) run-tree of $\mathcal{B}$ over $[\![G]\!]$. A path in a traversal-tree is a traversal in which each node is annotated by a state of $\mathcal{B}$.

Simulating traversals over $\lambda(G)$ by paths in $\lambda(G)$

Our problem is thus reduced to finding an effective method of recognising certain sets of infinite traversals (over a given computation tree) that satisfy the parity condition. This requires a new idea as a traversal is a sequence of nodes that is most unlike a path; it can jump all over the tree and may even visit certain nodes infinitely often. Our solution again exploits the game-semantic connexion. It is a property of traversals that their *P-views* (in the sense of [12]) are paths (in the computation tree). This allows us to simulate a traversal over a computation tree by (the P-views of its prefixes, which are) annotated paths of a certain kind in the same tree. The simulation is made precise in the notion of *traversal-simulating* APT (associated with a given property APT and a recursion scheme G). We establish the correctness of the simulation by the following result:

Theorem 4 (Simulation). *A given* property *APT has an accepting traversal-tree over the computation tree $\lambda(G)$ if and only if the associated* traversal-simulating *APT has an accepting run-tree over the computation tree.*

[5] *Property* APT because the APT corresponds to the property described by a given modal mu-calculus formula.

Note that decidability of the modal mu-calculus model-checking problem for trees in $\boldsymbol{RecTree}_n \Sigma$ follows at once since computation trees are regular, and the APT acceptance problem for regular trees is decidable [1,14].

To prove n-EXPTIME decidability of the model-checking problem, we first establish a certain *succinctness property* for traversal-simulating APT $\mathcal{C}$: If $\mathcal{C}$ has an accepting run-tree, then it has one with a reduced branching factor. The desired time bound is then obtained by analysing the complexity of solving an associated (finite) acceptance parity game, which is an appropriate product of the traversal-simulating APT and a finite deterministic graph that unravels to the computation tree in question. n-EXPTIME hardness of the model-checking problem follows from a result of Engelfriet [15] (see also [16]).

3 Collapsible Pushdown Automata and Recursion Schemes

In their ICALP 2005 paper [8], Knapik *et al.* showed that order-2 homogeneously-typed recursion schemes are equi-expressive with a variant class of order-2 pushdown automata called *panic automata*. In view of their result, it is natural to ask if there is a corresponding automata-theoretic characterization of arbitrary recursion schemes for all finite orders. In recent joint work with A. S. Murawski [17], we have shown that for the purpose of generating Σ-labelled trees, recursion schemes are equi-expressive with *collapsible pushdown automata*. Precisely:

Theorem 5 (Equi-expressivity). *For each $n \geq 0$, a Σ-labelled tree is generated by an order-n (deterministic) recursion scheme iff it is generated by an order-n (deterministic) collapsible pushdown automaton.*

An order-n *collapsible pushdown automaton* (CPDA) is just an order-n pushdown automaton in which every symbol a in the n-stack S may have a *link* to a necessarily lower-ordered stack situated below a in S, if there is such a stack at that point; if the stack pointed to is of order j, the link is called a $(j+1)$-*link*. For $2 \leq j \leq n$, j-links are introduced by the order-1 push operation $push^a_{1,j}$ where a is a stack symbol: when $push^a_{1,j}$ is applied to an n-stack S, a link is first attached from a copy of a to the $(j-1)$-stack that lies just below the top $(j-1)$-stack of S; the symbol a, together with its link, is then pushed onto the top of the top 1-stack of S. Whenever the top $(j-1)$-stack is duplicated by the order-j operation $push_j$, the link-structure of the top $(j-1)$-stack is preserved. There is a new stack operation called *collapse*, whose effect is to cause the stack S to collapse up to the point indicated by the link emanating from the top_1-element of S i.e. the top $(j-1)$-stack of $collapse(S)$ is the $(j-1)$-stack pointed to by the link emanating from the top_1-element of S.

An outline proof of Theorem 5

Given an order-n recursion scheme G, we construct an order-n collapsible pushdown automaton $\mathsf{CPDA}(G)$ whose stack symbols are given by the (ranked) symbols that label the nodes of the computation tree $\lambda(G)$. We then show that

$\mathsf{CPDA}(G)$ can compute any given traversal over the computation tree $\lambda(G)$, and hence, by the path-traversal correspondence (Theorem 3), it can compute any path in the value tree $[\![G]\!]$ as required.

In the other direction, given an order-n CPDA $\mathcal{A}$ with state-set $\{1, \cdots, m\}$, we construct an order-n recursion scheme $G_{\mathcal{A}}$ whose non-terminals are

$$\mathcal{F}_p^{a,e} \; : \; (n-e)^m \to (n-1)^m \to \cdots \to 0^m \to 0$$

where a ranges over the stack alphabet, $1 \leq p \leq m$, $2 \leq e \leq n$, 0 is the base type (of order 0), and the types $(n+1) = n^m \to n$ (of order $n+1$) are defined by recursion. We use ground-type terms

$$\mathcal{F}_p^{a,e}\,\overline{L}\,\overline{M_{n-1}} \cdots \overline{M_0} \; : \; 0$$

to represent (reachable) configurations (p, S) of $\mathcal{A}$, where S is the n-stack. The idea is that the top_1-element of S – a with an e-link (say) – (more precisely the pair $(p, top_1\, S)$) is coded as $\mathcal{F}_p^{a,e}\,\overline{L} : n$; further for each $1 \leq j \leq n$ and $1 \leq p \leq m$, we have

- $(p, top_j\, S)$ is coded as $\mathcal{F}_p^{a,e}\,\overline{L}\,\overline{M_{n-1}} \cdots \overline{M_{n-j+1}} : n - j + 1$
- $(p, pop_j\, S)$ is coded as $M_{n-j,p}\,\overline{M_{n-j-1}} \cdots \overline{M_0} : 0$
- $(p, collapse\, S)$ is coded as $L_p\,\overline{M_{n-e-1}} \cdots \overline{M_0} : 0$.

Restricted to order-2, CPDA coincide with *second-order pushdown automata with links* (in the sense of Aehlig *et al.* [18]), which are essentially the same as *panic automata* (in the sense of Knapik *et al.* [8]). Our CPDA-to-scheme transformation specialises to exactly the same transformation in [8], when restricted to order-2 CPDA. Our result in this section will be presented elsewhere; a full account can be found in the preprint [17].

4 Parity Games over Configuration Graphs of CPDA

The same game-semantic approach can be carried over to certain classes of finitely-presented graphs such as those given by (non-deterministic) collapsible pushdown automata. Fix an order-n CPDA $\mathcal{A}$ and a parity game over its configuration graph $\mathcal{CG}_{\mathcal{A}}$, and let $G_{\mathcal{A}}$ be the recursion scheme determined by $\mathcal{A}$ (as given by the CPDA-to-scheme transformation in Theorem 5). Paths in the configuration graph correspond exactly to *traversals* over the computation tree $\lambda(G_{\mathcal{A}})$ (or equivalently, traversals over the finite directed graph $\mathrm{Gr}(G_{\mathcal{A}})$ that unravels to $\lambda(G_{\mathcal{A}})$). For any parity game over $\mathcal{CG}_{\mathcal{A}}$, accepting traversal-trees over $\mathrm{Gr}(G_{\mathcal{A}})$ can be recognised by a traversal-simulating APT $\mathcal{C}$ (i.e. a version of Theorem 4); it follows that there is an equivalent *finite* acceptance parity game, which is an appropriate product of $\mathrm{Gr}(G_{\mathcal{A}})$ and $\mathcal{C}$. Hence parity games over the configuration graphs of order-n CPDA are solvable. We intend to extend the approach to the solution of games with ω-regular and non-regular winning conditions. Another interesting problem is the computation of *winning regions* of these games.

5 Further Directions

Does safety constrain expressiveness?

This is the most pressing open problem. In a FOSSACS 2005 paper [18], we have shown that there is no inherently unsafe word language at order 2. More precisely, for every word language that is generated by an order-2 unsafe recursion scheme, there is a safe (but in general non-deterministic) recursion scheme that generates the same language. However it is conjectured that the result does not hold at order 3. Further, for trees, we conjecture that there are already *inherently* unsafe trees at order 2 i.e.

Conjecture 1. $\textbf{\textit{SafeRecTree}}_2\Sigma \subset \textbf{\textit{RecTree}}_2\Sigma$ i.e. there is an unsafe order-2 deterministic recursion scheme whose value tree is not the value tree of any safe deterministic recursion scheme.

The Conjecture is closely related to a word language, which we call *Urzyczyn's language* [18]. The language can be generated by a *deterministic, unsafe* order-2 recursion scheme (and hence, by a *non-deterministic, safe* order-2 recursion scheme). The Conjecture is equivalent to the statement: Urzyczyn's language cannot be generated by any *deterministic, safe* order-2 recursion scheme (or equivalently any order-2 deterministic pushdown automaton).

Semantic vs verification games

We would like to develop further the pleasing mix of Semantics (games) and Verification (games) in the work. A specific project, *pace* [19], is to give a denotational semantics of the lambda calculus "relative to an APT". More generally, construct a cartesian closed category, parameterized by APTs, whose maps are witnessed by the *variable profiles* (see [10]).

References

1. Rabin, M.O.: Decidability of second order theories and automata on infinite trees. Trans. Amer. Maths. Soc **141** (1969) 1–35
2. Muller, D.E., Schupp, P.E.: The theory of ends, pushdown automata, and second-order logic. Theoretical Computer Science **37** (1985) 51–75
3. Caucal, D.: On infinite transition graphs having a decidable monadic theory. In: Proceedings 23rd ICALP, Springer (1996) 194–205 LNCS Vol. 1099.
4. Stirling, C.: Decidability of bisimulation equivalence for pushdown processes. Submitted for publication (2000)
5. Knapik, T., Niwiński, D., Urzyczyn, P.: Higher-order pushdown trees are easy. In: FOSSACS'02, Springer (2002) 205–222 LNCS Vol. 2303.
6. Caucal, D.: On infinite terms having a decidable monadic theory. In: Proc. MFCS'02. Volume 2420 of Lecture Notes in Computer Science. (2002) 165–176
7. Aehlig, K., de Miranda, J.G., Ong, C.H.L.: The monadic second order theory of trees given by arbitrary level two recursion schemes is decidable. In: Proceedings of the 7th International Conference on Typed Lambda Calculi and Applications (TLCA'05). (2005) 39–54 LNCS 3461.

8. Knapik, T., Niwiński, D., Urzyczyn, P., Walukiewicz, I.: Unsafe grammars and panic automata. In: ICALP'05. Volume 3580 of Lecture Notes in Computer Science., Springer-Verlag (2005) 1450–1461
9. Janin, D., Walukiewicz, I.: On the expressive completeness of the propositional mu-calculus with respect to msol. In: Proceedings of CONCUR'96. (1996) 263–277 LNCS 1119.
10. Ong, C.H.L.: On model-checking trees generated by higher-order recursion schemes. Preprint 42 pages (2006)
11. Ong, C.H.L.: On model-checking trees generated by higher-order recursion schemes. In: Proceedings of IEEE Symposium on Logic in Computer Science, Computer Society Press (2006)
12. Hyland, J.M.E., Ong, C.H.L.: On Full Abstraction for PCF: I. Models, observables and the full abstraction problem, II. Dialogue games and innocent strategies, III. A fully abstract and universal game model. Information and Computation **163** (2000) 285–408
13. Ong, C.H.L.: Local computation of beta-reduction by game semantics. Preprint (2006)
14. Emerson, E.A., Jutla, C.S.: Tree automata, mu-calculus and determinacy. In: Proceedings of FOCS'91. (1991) 368–377
15. Engelfriet, J.: Interated stack automata and complexity classes. Information and Computation (1991) 21–75
16. Cachat, T.: Higher order pushdown automata, the Caucal hierarchy of graphs and parity games. In: Proceedings of ICALP 2003. (2003) 556–569 LNCS 2719.
17. Murawski, A.S., Ong, C.H.L.: Collapsible pushdown automata and recursion schemes. Preprint (2006)
18. Aehlig, K., de Miranda, J.G., Ong, C.H.L.: Safety is not a restriction at level 2 for string languages. In: Proceedings of the 8th International Conference on Foundations of Software Science and Computational Structures (FOSSACS'05). (2005) 490–501 LNCS 3411.
19. Aehlig, K.: A finite semantics for simply-typed lambda terms for infinite runs of automata. In: Proceedings of CSL 2006. (2006) LNCS. To appear.

Automata and Logics for Words and Trees over an Infinite Alphabet

Luc Segoufin

INRIA and Université Paris 11
http://www-rocq.inria.fr/~segoufin

Abstract. In a *data word* or a *data tree* each position carries a label from a finite alphabet and a data value from some infinite domain. These models have been considered in the realm of semistructured data, timed automata and extended temporal logics.

This paper survey several know results on automata and logics manipulating data words and data trees, the focus being on their relative expressive power and decidability.

1 Introduction

In many areas there is a need for good abstract models manipulating explicitly data values. We mention two of them here.

In program verification one has to decide statically whether a program satisfies some given specification. Programs may contain several procedures calling each other recursively. Procedures may have parameters and data could be exchanged via the parameters. In program verification, variables over unbounded domains such as integers, arrays, parameters etc. are usually abstracted to finite range domains and configuration graphs of pushdown automata have been used quite successfully in order to model recursive dependencies between procedures. It is then possible to check properties expressed by temporal logics such as LTL or CTL. The modelization using a finite domain has some obvious limitations but it is hard to avoid while remaining decidable. One notable exception is [8] where procedures can have one parameter whose value can range over the integers and the procedures can perform limited arithmetic on that parameter.

In the database context also, most theoretical work on XML and its query languages models XML documents by labeled ordered unranked trees, where the labels are from a finite set. Attribute values are usually ignored. This has basically two reasons, which are not independent. First, the modeling allows to apply automata based techniques, as automata operate on trees of this kind. Second, extending the model by attribute values (data values) quickly leads to languages with undecidable static analysis (see, for instance [1,4,18,27]). Nevertheless, there are examples of decidable static reasoning tasks involving attribute values.

One of them is validation of schema design. A schema contains a structural part and integrity constraints such as *keys* and *inclusion constraints*. In

Z. Ésik (Ed.): CSL 2006, LNCS 4207, pp. 41–57, 2006.

a semistructured model such as XML the structural part is mainly a mechanism for assigning types to nodes of the document tree. It is then natural to ask whether a specification is consistent and whether a set of integrity constraints is minimal or not (*implication problem*). Decidable cases were proposed for XML schemas in [2].

Another one is query optimization and checking whether one query is included into or equivalent to another one. Each of these inclusion tests could be relativized by the presence of a schema. In the context of XML and one of its most popular language XPath several decidable fragments were proposed in [27,4,18,6].

Each of the papers cited above propose a decidable framework where data values are explicitly manipulated. In each of them the decidability was obtained using a non-trivial ad-hoc argument. It is natural to wonder whether there exists a general decidable suitable theoretical framework which could be used to infer all this kind of results. This framework is yet to be discovered and this paper is a modest attempt to gather known results in this direction. We tried to group here interesting models of automata and logical frameworks which could be used in order to code some of the problems mentioned above.

To make this survey finite size we have restricted our attention to an approach that, in our opinion, deserves more attention from the theoretical community. All over this survey we model data values using an infinite domain (like the integers). Moreover our structures have a very simple shape as they are either finite strings or finite trees. More precisely we are given two alphabets over strings and trees, one which is finite and can be used to code names and constants appearing in the programs, in the schemas, or in the queries, and another one which is infinite and which can be used to code data values. We call such structures *data words* or *data trees*.

We present several models of automata and several logics which are evaluated on data words and data trees. We focus on their relative expressive power and on decidability.

When dealing with an infinite domain, like the integers for instance, the allowed arithmetic is a central issue in order to obtain decidability. We avoid this problem here by allowing only the simplest arithmetical operation which is equality test: The only operation that can be performed in our infinite domain is checking whether to values are equal. We will see that this is already enough to make the story interesting.

We have modified slightly several known concepts so that we could present each of them within a uniform framework in order to compare them. We hope that the reader familiar with one of them will not be too much disturbed by this.

This survey contains no proof at all. Those can be found in the references provided all over the paper. We wish we had time to include more references, especially in the verification context.

2 Some Notations

In this paper we consider two kinds of model: *data words* and *data trees*. Let Σ be a finite alphabet of *labels* and D an infinite set of *data values*.

A *data word* $w = w_1 \cdots w_n$ is a finite sequence over $\Sigma \times D$, i.e., each w_i is of the form (a_i, d_i) with $a_i \in \Sigma$ and $d_i \in D$. A *data word language* is a set of data words.

A *data tree* t is a finite unranked, ordered tree where each node is of the form (a_i, d_i) with $a_i \in \Sigma$ and $d_i \in D$. As above a *data tree language* is a set of data trees.

Given a node x of a data tree or a position x of a data word, we denote respectively by $x.d$ and $x.l$ the data value of D and the label of Σ associated to x. The idea is that the alphabet Σ is accessed directly, while data values can only be tested for equality. This amounts to considering words and trees over the finite alphabet Σ endowed with an equivalence relation on the set of positions. With this in mind, given a data word or a data tree w, we will call *class* of w a set of positions/nodes of w having the same data value. Finally, given a data word w (or a data tree t), the *string projection* of w (the *tree projection* of t), denoted by $\text{STR}(w)$ ($\text{TREE}(t)$), is the word (tree) constructed from w (t) by keeping only the label in Σ and projecting out all the data values.

For each integer $n \in \mathbb{N}$ we note $[n]$ the set of all integers from 1 to n. Given a data word w we denote its length by $|w|$.

3 Automata

In this section we present several models of automata over data words and data trees. We present them in details in the word case and only briefly discuss the extension to trees.

3.1 Register Automata

Register automata are finite state machines equiped with a finite number of registers. These registers can be used to store temporarily values from D. When processing a string, an automaton compares the data value of the current position with values in the registers; based on this comparison, the current state and the label of the position it can decide on its action. We stress that the only allowed operation on registers (apart from assignment) is a comparison with the symbol currently being processed. The possible actions are storing the current data value in some register and specifying the new state. This model has been introduced in [20] and was later studied more extensively in [28]. We give here an equivalent formalism that fits with data words.

Definition 3.1. *A* k-register automaton $\mathcal{A}$ *is a tuple* (Q, q_0, F, τ_0, T) *where*

- Q *is a finite set of states;* $q_0 \in Q$ *is the initial state;* $F \subseteq Q$ *is the set of final states;*

– *$\tau_0 : \{1,\ldots,k\} \to D$ is the initial register assignment; and,*
– *T is a finite set of transitions of the forms $(q,a,E) \to q'$ or $(q,a,E) \to (q',i)$. Here, $i \in \{1,\ldots,k\}$, $q,q' \in Q$, $a \in \Sigma$ and $E \subseteq \{1,\ldots,k\}$.*

Given a data word w, a *configuration of $\mathcal{A}$ on w* is a tuple $[j,q,\tau]$ where $0 \leq j \leq |w|$ is the current position in the word, $q \in Q$ is the current state, and $\tau : \{1,\ldots,k\} \to D$ is the current register assignment. The *initial* configuration is $\gamma_0 := [1,q_0,\tau_0]$. A configuration $[j,q,\tau]$ with $q \in F$ is *accepting*. Given $\gamma = [j,q,\tau]$, the transition $(p,a,E) \to \beta$ *applies* to γ iff $p = q$, $w_j.l = a$ and $E = \{l \mid w_j.d = \tau(l)\}$ is the set of registers having the same data value as the current position.

A configuration $\gamma' = [j',q',\tau']$ is a successor of a configuration $\gamma = [j,q,\tau]$ iff there is a transition $(q,a,E) \to q'$ that applies to γ, $\tau' = \tau$, and $j' = j+1$; or there is a transition $(q,a,E) \to (q',i)$ that applies to γ, $j' = j+1$, and τ' is obtained from τ by setting $\tau'(i)$ to $w_j.d$. Based on this, reachable configuration and acceptance of a data word is defined in the standard way. We denote by $L(\mathcal{A})$ the language of data words accepted by the register automata $\mathcal{A}$.

Example 3.2. There exists a 1-register automata which checks whether the input data words contains two positions labeled with a with the same data value (a is not a key): it non-deterministically moves to the first position labeled with a with a duplicated data value, stores the data value in its register and then checks that this value appears again under another position labeled with a.

The complement of this language, all node labeled with a have different data values, which would be useful for key integrity constraints in the database context is not expressible with register automata. To prove this, one can show that for each data word accepted by a given register automata, there exists another accepted data word that have the same string projection but uses a number of data values depending only on the automaton (see [20] for more details).

This shows that register automata are not closed under complementation. They are also not closed under determinization as the first example cannot be achieved with a deterministic register automata.

The main result for register automata is that emptiness is decidable.

Theorem 3.3. *[20] Emptiness of register automata is decidable.*

In term of complexity the problem is PSPACE-complete [14]. Without labels it was shown to be NP-complete in [30]. As illustrated with the examples above, register automata are quite limited in expressive power. This limitation can also be formalized with the following proposition which should be contrasted with the similar one in the case of data automata presented later (Proposition 3.14).

Proposition 3.4. *[20,9] Given a language L of data words accepted by a register automata the language of strings* STR(L) *is regular.*

There exist many obvious extensions of register automata. For instance one could add alternation or 2-wayness. Unfortunately those extensions are undecidable.

Theorem 3.5. *1. Universality of register automata is undecidable [28].*
2. Emptiness of 1-register 2-way automata is undecidable [12].

An immediate corollary of Theorem 3.5 is:

Corollary 3.6. *[28] Inclusion and equivalence of register automata is undecidable.*

There are several variants that weaken or strengthen the model of register automata as presented above without affecting much the results. We only briefly mention two of them. One possible extension is to allow the automata to store a non-deterministically chosen value from D in one of the register. With this extension the language of data words such that the data value of the last position of the word is different from all the other positions can be accepted. Without it the language is not accepted by a register automata. This model was studied in [11,22] and is a little bit more robust than the one presented here. In the model presented above the automata knows the equality type of the current data value with all the registers. One possible weakening of the model is to allow only an equality test of the current data value with a fix register. That is the transitions are of the form $(q, a, i) \longrightarrow q'$ or $(q, a, i) \longrightarrow (q', j)$. This setting was studied in [32]. The language of data words such that the first two data values are distinct can no longer be accepted [21]. On the other hand this model has equivalent regular expressions [21]. Another notable property of this weaker model is that inclusion becomes decidable [32]. To conclude this discussion one should mention that the model of the register automata as presented above have algebraic characterizations, see [9,17] for more details.

Trees. To extend this model to binary trees the technical difficulty is to specify how registers are splited (for the top-down variants) or merged (for the bottom-up variant). In the top-down case the natural solution is to propagate the content of the registers to the children of the current node. In the bottom-up case a function can be included into the specification of the transitions of the automata which specify, for each register, whether the new value should come from some register of the left child or from some register of the right one. It is possible to do this so that the family obtained is decidable and robust in the sense that the bottom-up variant correspond to the top-down variant. This model has all the properties of register automata over data words: emptiness is decidable and the tree projection is regular. We refer to [21] for more details.

3.2 Pebble Automata

Another model of automata uses pebbles instead of registers. The automata can drop and lift pebbles on any position in the string and eventually compare the current value with the ones marked by the pebbles. To ensure a "regular" behavior, the use of pebbles is restricted by a stack discipline. That is, pebbles are numbered from 1 to k and pebble $i + 1$ can only be placed when pebble i is present on the string. A transition depends on the current state, the current label, the pebbles placed on the current position of the head, and the equality type of the current data value with the data values under the placed pebbles. The transition relation specifies change of state, and possibly whether the last pebble is removed or a new pebble is placed. In the following more formal definition we follow [28] and assume that the head of the automata is always the last pebble. This does not make much difference in term of expressive power, at least in the 2-way case which is the most natural for this model, but simplifies a lot the notations:

Definition 3.7. A *k-pebble automaton* $\mathcal{A}$ is a tuple (Q, q_0, F, T) where

- Q is a finite set of *states*; $q_0 \in Q$ is the *initial state*; $F \subseteq Q$ is the set of *final states*; and,
- T is a finite set of *transitions* of the form $\alpha \rightarrow \beta$, where
  - α is of the form (q, a, i, P, V, q), where $i \in \{1, \ldots, k\}$, $a \in \Sigma$, $P, V \subseteq \{1, \ldots, i-1\}$, and
  - β is of the form (q, d) with $q \in Q$ and d is either DROP of LIFT.

Given a data word w, a *configuration of $\mathcal{A}$ on w* is of the form $\gamma = [q, j, \theta]$ where i is the current position in w, $q \in Q$ the current state, $j \in \{1, \ldots, k\}$ the number of pebbles already dropped, and $\theta : \{1, \ldots, j\} \rightarrow [|w|]$ the positions of the pebbles. Note that the current position always correspond to $\theta(j)$. We call θ a pebble assignment. The *initial* configuration is $\gamma_0 := [q_0, 1, \theta_0]$, with $\theta_0(1) = 1$. A configuration $[q, j, \theta]$ with $q \in F$ is *accepting*.

A transition $(p, a, i, P, V, p) \rightarrow \beta$ *applies to a configuration* $\gamma = [q, j, \theta]$, if

1. $i = j$, $p = q$,
2. $V = \{l < j \mid w_{\theta(l)}.d = w_{\theta(j)}.d\}$ is the set of pebbles placed on a position which has the same data value than the current position.
3. $P = \{l < j \mid \theta(l) = \theta(j)\}$ is the set of pebbles placed at the current position, and
4. $w_j = a$.

Intuitively, $(p, a, i, P, V) \rightarrow \beta$ applies to a configuration if pebble i is the current head, p is the current state, V is the set of pebbles that see the same data value as the top pebble, P is the set of pebbles that sit at the same position as the top pebble, and the current label seen by the top pebble is a.

A configuration $[q', j', \theta']$ is a successor of a configuration $\gamma = [q, j, \theta]$ if there is a transition $\alpha \rightarrow (p, d)$ that applies to γ such that $q' = p$ and $\theta'(i) = \theta(i)$, for all $i < j$, and

- if d=DROP, then $j' = j + 1$, $\theta'(j) = \theta(j) = \theta'(j + 1)$,
- if d=LIFT, then $j' = i - 1$.

Note that this implies that when a pebble is lifted, the computation resumes at the position of the previous pebble.

Example 3.8. The languages of data words such that all nodes labeled with a have different data values is accepted by a 2-pebble automata as follows. At each position labeled with a the automata drops pebble 1 and then check that the data value never occurs under a position labeled with a to the right of the pebble. When this is done it comes back to pebble 1 and move to the next position.

Register automata and pebble automata are likely to be incomparable in expressive power. The example above can be expressed by a pebble automata but not by a register automata. On the other hand, pebble automata have a stack discipline constraint on the use of pebbles while register automata can update their register in an arbitrary order. Based on this, examples of languages expressible with register automata but not with pebble automata are easy to construct but we are not aware of any formal proof of this fact.

Strictly speaking the model presented above is not really 1-way as when it lifts a pebble the automata proceed at the position of previous pebble. It is also immediate to extends this model in order to make it 2-way. The advantage of this definition is that the model is robust:

Theorem 3.9. *[28] Pebble automata are determinizable and the 2-way variant has the same expressive power than its 1-way variant.*

But unfortunately the model is too strong in expressive power. In the result below, recall that with our definition the head of the automata counts as 1 pebble. Therefore in a sense the result is optimal.

Theorem 3.10. *[28,12] Emptiness of 2-pebble automata over data words is undecidable.*

Trees. The most natural way to extend this model to trees is to consider tree walking automata with pebbles on trees. This model extend the 2-way variant of pebble automata as presented here in a natural way with moves allowing to "walk" the tree in all possible directions (up, down, left right). Of course, this model remains undecidable.

3.3 Data Automata

This model was introduced in [6,7]. It extends the model of register automata, remains decidable and has better connections with logic. Data automata runs on data words in two passes. During the first one a letter-to-letter transducers is run. This transducers does not have access to the data values and change the label of each position. During the second pass, an automata is run on each sequence of letters having the same data value, in other words, on each class.

A *data automaton* $\mathcal{D} = (\mathcal{A}, \mathcal{B})$ consists of

- a non-deterministic letter-to-letter string transducer $\mathcal{A}$ (the *base automaton*) with input alphabet Σ, for some output alphabet Γ (letter-to-letter means that each transition reads and writes exactly one symbol), and
- a non-deterministic string automaton $\mathcal{B}$ (the *class automaton*) with input alphabet Γ.

A data word $w = w_1 \cdots w_n$ is accepted by $\mathcal{D}$ if there is an *accepting* run of $\mathcal{A}$ on the string projection of w, yielding an output string $b_1 \cdots b_n$, such that, for *each class* $\{x_1, \ldots, x_k\} \subseteq \{1, \ldots, n\}$, $x_1 < \cdots < x_k$, the class automaton accepts $b_{x_1}, \ldots, b_{x_k}$.

Example 3.11. The language of data words such that all positions labeled with a have different data values can be accepted by a data automaton as follows. The base automaton does nothing and only copy its input. The class automata accepts only words containing only one a.

The complement of the above language, the set of all data words containing two positions labeled with a with the same data value, is done as follows. The base automata non-deterministically selects two positions labeled with a and output **1** on each of them and **0** everywhere else. The class automata checks that each class contains either no **1** or exactly 2.

The language of data words such that there exists two positions labeled with a and having the same data value but with no positions labeled with b (no matter what the data value is) in between is accepted by a data automaton. The base automata outputs **0** on all positions excepts two, labeled with a, that it selects non-deterministically and on which it outputs **1**. It also checks that between the two selected positions no b occurs. The class automata checks that each class contains either no **1** or exactly 2.

It is not clear how to do the complement of this language using a data automata. This suggest that data automata are not closed under complementation but we are not aware of any formal proof of this result. It will follow from the results of Section 4 that if they would be closed under complementation then they would not be effectively closed under complementation, by this we mean that there would be no algorithm computing the complement of a data automata.

By just looking at the definitions it is not immediate that data automata extends the model of register automata presented in Section 3.1. But it is indeed the case.

Proposition 3.12. *[5] For each register automata there exists a data automaton accepting the same language.*

The main result on data automata is that emptiness remains decidable. We will use this fact to show decidability of several logics in Section 4.

Theorem 3.13. *[7] Emptiness of data automata is decidable.*

In order to better understand the expressive power of data automata we show that the string projection of languages accepted by data automata correspond to languages accepted to multicounter automata. Recall that for register automata the string projection remains regular (Proposition 3.4).

We first briefly review the definition of multicounter automata. An ϵ-free *multicounter automaton* is a finite automaton extended by a finite set $C = \{1, \ldots, n\}$ of counters. It can be described as a tuple $(Q, \Sigma, C, \delta, q_I, F)$. The set of states Q, finite alphabet Σ, initial state $q_I \in Q$ and final states $F \subseteq Q$ are as in a usual finite automaton. The transition relation δ is a finite subset of $Q \times \Sigma \times (dec^*(i)\, inc^*(i))_{i \in C} \times Q$.

The idea is that in each step, the automaton can change its state and modify the counters, by incrementing or decrementing them, according to the current state and the current letter on the input. In a step, the automaton can apply to each counter $i \in C$ a sequence of decrements, followed by a sequence of increments. Whenever it tries to decrement a counter of value zero the computation stops. Besides this, the transition of a multicounter automaton does not depend on the value of the counters. In particular, it cannot test whether a counter is exactly zero (otherwise the model would be undecidable). Nevertheless, by decrementing a counter k times and incrementing it again afterward it can test whether the value of that counter is at least k.

A *configuration* of such an automaton is a tuple $c = (q, (c_i)_{i \in C}) \in Q \times \mathbb{N}^n$, where q is the current state and c_i is the value of the counter i. A transition

$$(q, a, (dec^{k_i}(i) inc^{l_i}(i))_{i \in C}, q') \in \delta$$

can be applied if the current state is q, the current letter is a and for every counter $i \in C$, the value c_i is at least k_i. The successor configuration is $d = (q', (c(i) - k_i + l_i)_{i \in C})$. A *run* over a word w is a sequence of configurations that is consistent with the transition function δ. The acceptance condition is given by a subset R of the counters C and the final states. A run is *accepting* if it starts in the state q_I with all counters empty and ends in a configuration where all counters in R are empty and the state is final.

Emptiness of multicounter automata is known to be decidable [26,23].

Proposition 3.14. *[7]*

- *If L is a language of data words accepted by a data automata then* STR(L) *a language of strings accepted by a multicounter automata.*
- *If L is a language of strings accepted by a multicounter automata then there exists a language L' of data words accepted by a data automata such that $h(\text{STR}(L')) = L$, where h is an erasing morphism.*

The constructions of Proposition 3.14 are constructive (the time complexity requires a tower of 3 exponentials in the first case and is only polynomial in the second case) and thus Proposition 3.14 implies Theorem 3.13. Therefore the emptiness problem of data automata is elementary equivalent to the emptiness problem of multicounter automata which is not yet known to be elementary (the best known lower-bound being ExpSpace [24] and the best known upper-bound being non-elementary [26,23]). The precise complexity of Theorem 3.13 is therefore still an open issue.

Trees. This model can be extended to unranked ordered trees as follows. Both the base automaton and the class automaton are bottom-up tree automaton. The base automata works as in the word case and reassigns a label to each node. Each class can now be viewed as an unranked ordered tree by contracting the initial tree edges which contains a node not in the class. The class automata is then run on the resulting trees. Unfortunately decidability of data tree automata remains an open issue. Data tree automata still have some connection with multicounter tree automata. It is not clear yet whether the first part of Proposition 3.14 holds or not, but the second one does hold. Therefore showing decidability of data automata would implies showing decidability of multicounter tree automata which has eluded all efforts so far (see [13] and the references therein).

3.4 Other Generalizations

We have consider so far only automata for data words or data trees that are based on the classical notion of finite state automata. It is also possible to consider more powerful devices for manipulating data words such as pushdown automata.

Historically the first models of pushdown automata over infinite alphabet were presented in [3] and [19]. Those models extend the notions of Context-Free grammars and of pushdown automata in the obvious way by allowing infinitely many rules and infinitely many transition functions, one for each letter in D. Hence the models are not even finitely presented and decidability does not make much sense. In term of expressive power, with some technical constraints on the pushdown automata models (see [19]), both the context-free grammars and pushdown automata defines the same class of languages.

The most robust decidable notion so far was given in [11] using ideas similar to register automata. We describe it next. Intuitively the automata has k registers and makes its choices given, the current state, the label of the current node, and the equality type of the current data value with the values present in the registers. The possible actions are: change the current state, update some of the registers with the current data value, push the values stored into some registers into the stack together with some finite content, pop the top of the stack. Moreover, in order to have a robust model with an equivalent context-free grammar, the model allows ϵ transition which can store into one of the registers a new, non-deterministically chosen, data value.

Definition 3.15. *A* k-register pushdown automaton $\mathcal{A}$ *is a tuple* $(Q, q_0, F, \Gamma, \tau_0, T)$ *where*

- Q *is a finite set of states;* $q_0 \in Q$ *is the initial state;* $F \subseteq Q$ *is the set of final states;*
- Γ *is a finite set of labels for stack symbols;*
- $\tau_0 : \{1, \ldots, k\} \rightarrow D$ *is the initial register assignment; and,*
- T *is a finite set of transitions of the form* $(q, a, E, \gamma) \rightarrow (q', i, \overline{s})$ *or* $(q, \epsilon, E, \gamma) \rightarrow (q', i)$.
  Here, $i \in \{1, \ldots, k\}$, $q, q' \in Q$, $a \in \Sigma$, $\gamma \in \Gamma$, $\overline{s} \in (\Gamma \times \{1, \ldots, k\})^*$ *and* $E \subseteq \{\top, 1, \ldots, k\}$.

Given a stack symbol s we denote by $s.l$ and $s.d$ respectively the label in Γ and the data value in D of s. Given a data word w, a *configuration of* $\mathcal{A}$ *on* w is a tuple $[j, q, \tau, \mu]$ where $1 \leq j \leq |w|$ is the current position in the data words,

$q \in Q$ is the current state, $\tau : \{1, \ldots, k\} \to D$ is the current register assignment, and $\mu \in (\gamma \times D)^*$ is the current stack content. The *initial* configuration is $\nu_0 := [1, q_0, \tau_0, \epsilon]$. A configuration $[j, q, \tau, \mu]$ with $q \in F$ is *accepting.* Given $\nu = [j, q, \tau, \mu]$ with top stack symbol s, the transition $(p, a, E, \gamma) \to \beta$ *applies* to ν iff $p = q$, $w_j.l = a$, $s.l = \gamma$, and $E = \{l \mid w_j.d = \tau(l)\} \cup S$ with S is empty if $w_j.d \neq s.d$ and is $\{\top\}$ otherwise.

A configuration $\nu' = [j', q', \tau', \mu']$ is a successor of a configuration $\nu = [j, q, \tau, \mu]$ iff there is a transition $(q, a, E, \gamma) \to (q', i, \overline{s})$ that applies to ν, $\tau'(l) = \tau(l)$ for all $l \neq i$, $\tau'(i) = w_j.d$, $j' = j + 1$ and μ' is μ with the top of the stack removed and replaced by $(a_1, d_{i_1}) \cdots (a_m, d_{i_m})$ where $\overline{s} = (a_1, i_1) \cdots (a_m, i_m)$ and d_l is the current value of register l; or there is a transition $(q, a, E, \gamma) \to (q', i)$ that applies to ν, $j' = j$, $\mu = \mu'$, and τ' is obtained from τ by setting $\tau'(i)$ to some arbitrary value of D. Based on this, reachable configuration and acceptance of a data word is defined in the standard way.

Note that we allow ϵ transitions which can introduce non-deterministically a new symbol in some register. This make the model more robust.

Example 3.16. The language of data words such that the data values form a sequence ww^R where w^R is the reverse sequence of w is accepted by a register data automata in the obvious way by first pushing the data values into the stack and then poping them one by one.

The language of data words such that all positions labeled with a have different data values is not accepted by a register pushdown automata. This uses argument similar than for register automata. See [11] for more details and more example.

This model of register pushdown automata extends in a natural way the classical notion of pushdown automata. It can be shown that the good properties of Context-Free languages can be extended to this model. In particular we have:

Theorem 3.17. *[11] Emptiness of register pushdown automata is decidable.*

This model is also robust and has an equivalent presentation in term of Context-Free grammar that we don't present here. The interested reader will find in [11] more details and more properties of the model. We conclude with a characterization of the projection languages defined by register pushdown automata similar to Proposition 3.4.

Proposition 3.18. *If L is a language accepted by a register pushdown automata then* STR(L) *is Context-Free.*

4 Logics

The previous section was concerned with finding decidable automata for manipulating words and trees containing data values. In this section we are looking for declarative decidable tools such as logics.

Data words and data trees can be seen as models for a logical formula.

In the case of data words, the universe of the model is the set of positions in the word and we have the following built-in predicates: $x \sim y$, $x < y$, $x = y+1$, and a predicate $a(x)$ for every $a \in \Sigma$. The interpretation of $a(x)$ is that the label in position x is a. The order $<$ and successor $+1$ are interpreted in the usual way. Two positions satisfy $x \sim y$ if they have the same data value. Given a formula ϕ over this signature, we write $L(\phi)$ for the set of data words that satisfy the formula ϕ. A formula satisfied by some data word is *satisfiable*.

In the case of data trees, the universe of the structure is the set of nodes of the tree with the following predicates available:

- For each possible label $a \in \Sigma$, there is a unary predicate $a(x)$, which is true for all nodes that have the label a.
- The binary predicate $x \sim y$ holds for two nodes if they have the same data value.
- The binary predicate $E_{\rightarrow}(x,y)$ holds for two nodes if x and y have the same parent node and y is the immediate successor of x in the order of children of that node.
- The binary predicate $E_{\downarrow}(x,y)$ holds if y is a child of x.
- The binary predicates $E_{\Rightarrow}$ and $E_{\Downarrow}$ are the transitive closures of $E_{\rightarrow}$ and $E_{\downarrow}$, respectively.

For both words and trees, we write FO($\sim, <, +1$) for first-order logic over a signature containing all the predicates mentioned above for the corresponding context. We also write FO($\sim, +1$) when the predicates $E_{\Rightarrow}$ and $E_{\Downarrow}$ -in the case of trees- and without $<$ -in the case of words are missing.

A logic L is said to be decidable over a class $\mathcal{M}$ of models if, given a sentence $\varphi \in L$, it is decidable whether there exists a model $M \in \mathcal{M}$ such that $M \models \varphi$.

4.1 First-Order logics

Example 4.1. The language of data words such that all positions labeled with a have different data values can be expressed in FO($\sim, <, +1$) by $\forall x, y \;\; (a(x) \wedge a(y) \wedge x \neq y) \longrightarrow x \not\sim y$.

The complement of the above language, the set of all data words containing two positions labeled with a having the same data value, is thus $\exists x, y \;\; a(x) \wedge a(y) \wedge x \neq y \wedge x \sim y$.

The language of data words where each position labeled with a has a data value which also appears under a position labeled with b (inclusion dependency) is expressed in FO($\sim, <, +1$) by $\forall x \exists y \;\; a(x) \longrightarrow (b(y) \wedge x \sim y)$.

Let L be the language of data words such that (i) any two positions labeled with a have a distinct data value, (ii) any two positions labeled with b have a distinct data value and (iii) the sequence of data values of the positions labeled a is exactly the same as the sequence of data values labeled with b. L can be expressed in FO($\sim, <, +1$) as follows. First we use sentences as given in the first three examples in order to express (i), (ii) and the fact that each data value appears exactly twice under two positions, one labeled a and one labeled b. Then the following sentence shows that the sequences are the same: $\forall x, y, z \;\; (a(x) \wedge a(y) \wedge x < y \wedge b(z) \wedge x \sim z) \longrightarrow (\exists x \;\; b(x) \wedge x \sim y \wedge z < x)$.

The relative expressive power of register and pebble automata with logics has been studied in [28] over data words. In term of expressive power, $\mathrm{FO}(\sim,<,+1)$ is not comparable with register automata and it is strictly included into pebble automata.

As $\mathrm{FO}(\sim,<,+1)$ is a fragment of pebble automata it is natural to wonder whether it forms a decidable fragment. We write FO^k for formulas using at most k variables (possibly reusing them at will). In the examples above note that the first three are definable in $\mathrm{FO}^2(\sim,<,+1)$ while the third one requires three variables. This last example can be pushed a little bit in order to code PCP. Therefore we have:

Theorem 4.2. *[7]* $\mathrm{FO}^3(\sim,<,+1)$ *is not decidable over data words.*

However the two-variable fragment of $\mathrm{FO}(\sim,<,+1)$, which turns out to be often sufficient for our needs, especially in the database context, is decidable.

Theorem 4.3. *[7]* $\mathrm{FO}^2(\sim,<,+1)$ *is decidable over data words.*

Theorem 4.3 follows from the fact that any data word language definable by a formula of $\mathrm{FO}^2(\sim,<,+1)$ is accepted by a data word automata described in Section 3. Actually we can show a stronger result than this. Consider the new binary predicate ± 1 which holds true at position x and y if the positions denoted by x and y have the same data value and y is the successor of x in their class. Note that this predicate is not expressible in $\mathrm{FO}^2(\sim,<,+1)$. Consider now the logic $\mathrm{EMSO}^2(\sim,<,+1,\pm 1)$ which extends $\mathrm{FO}^2(\sim,<,+1,\pm 1)$ by existential monadic second-order predicates quantification in front of $\mathrm{FO}^2(\sim,<,+1,\pm 1)$ formulas.

Theorem 4.4. *[7] For any data word language L the following are equivalent.*

1. *L is definable in* $\mathrm{EMSO}^2(\sim,<,+1,\pm 1)$,
2. *L is accepted by a data word automata.*

Moreover the translations between $\mathrm{EMSO}^2(\sim,<,+1,\pm 1)$ *formulas and data word automata are effective.*

This immediately yields:

Corollary 4.5. $\mathrm{EMSO}^2(\sim,<,+1,\pm 1)$ *is decidable over data words.*

Again the precise complexity is not known as it more or less correspond to deciding emptiness of multicounter automata.

Note that adding a little bit more of arithmetic, like assuming that D is linearly order and that the logic contains an extra binary predicate checking for this order among data values, yields undecidability even for the 2-variables fragment of FO [7].

Trees. The step from data words to data trees seems non trivial and we don't yet know whether Theorem 4.4 holds in the data tree case. In particular we don't know whether languages definable in $\mathrm{FO}^2(\sim,<,+1)$ are always accepted by a data tree automata. On the other hand $\mathrm{FO}^2(\sim,<,+1)$ is expressive enough to simulate multicounter tree automata therefore satisfiability of $\mathrm{FO}^2(\sim,<,+1)$ over data trees is likely to be a difficult problem.

On the other hand we can show:

Theorem 4.6. *[6]* $\mathrm{FO}^2(\sim,+1)$, *and therefore* $\mathrm{EMSO}^2(\sim,+1)$, *is decidable over data trees.*

The proof of Theorem 4.6 is quite involved and does not use automata but rather some kind of *puzzles*. It is shown, using this puzzles, that any satisfiable sentences of $\mathrm{FO}^2(\sim,+1)$ has a tame model and that deciding whether a sentence has a tame model is decidable. The complexity obtained this way is 3NExpTime which is possibly not optimal.

4.2 LTL with Freeze Quantifier

This logic was studied in the context of data words in [14,15]. Roughly speaking it extends the linear temporal logic LTL with a freeze operator which binds the data value of the current position in the data words to a register for later use.

We consider the temporal operators X for Next, X^{-1} for Previous, F for Sometime in the future, F^{-1} for sometime in the past, U for Until, U^{-1} for Previous. As usual we regard G as an abbreviation for $\neg F \neg$. We define $\mathrm{LTL}_n^{\downarrow}$ using the following grammar:

$\mathrm{S} ::= \top \mid a \mid \downarrow_r S \mid S \wedge S \mid \neg S \mid O(S...S) \mid \uparrow_r$

where $r \in [1,n]$, $a \in \Sigma$ and O is a temporal operator in $\{X, X^{-1}, F, F^{-1}, U, U^{-1}\}$.

We only consider well-formed formulas where $\uparrow_r$ is only used in the scope of a $\downarrow_r$. When we want to use only a subset $\mathcal{O}$ of the temporal operators then we write $\mathrm{LTL}_n^{\downarrow}(\mathcal{O})$.

The semantic is classical for the Boolean and temporal operators. The formula a holds at any position whose label is a. The formula $\downarrow_r S$ stores the data value of the current position in the register r and then checks for S starting at the current position. The formula $\uparrow_r$ checks that the data values stored in the register r equals the data value of the current position.

Example 4.7. The language of data words containing two positions labeled with a having the same data value, can be expressed in $\mathrm{LTL}_1^{\downarrow}$ by $F(a \wedge \downarrow_1 XF(a \wedge \uparrow_1))$.

The complement of the above language, data words such that all positions labeled with a have different data values, is thus $\neg F(a \wedge \downarrow_1 XF(a \wedge \uparrow_1))$.

The language of data words where each position labeled with a has a data value which also appears under a position labeled with b (inclusion dependency) is expressed in $\mathrm{LTL}_1^{\downarrow}$ by $G(a \longrightarrow (\downarrow_1 (F(b \wedge \uparrow_1) \vee F^{-1}(b \wedge \uparrow_1))))$.

The language of data words such that any two distinct positions labeled with a and having the same data value have a b in between can be expressed in $\mathrm{LTL}_1^{\downarrow}$ by: $G\ (a \wedge \downarrow_1 F(a \wedge \uparrow_1)) \longrightarrow (\neg(a \wedge \uparrow_1) U b)$

Theorem 4.8. *[14,15]*

1. $\mathrm{LTL}_1^{\downarrow}(X, U)$ *is decidable over data words.*
2. $\mathrm{LTL}_2^{\downarrow}(X, F)$ *is undecidable over data words.*
3. $\mathrm{LTL}_1^{\downarrow}(X, F, F^{-1})$ *is undecidable over data words.*

The logics $\mathrm{LTL}_1^{\downarrow}(X, U)$ and $\mathrm{FO}^2(\sim, <, +1)$ are incomparable. Indeed the third example above (inclusion dependency) is expressible in $\mathrm{FO}^2(\sim, <, +1)$ but not in $\mathrm{LTL}^{\downarrow}$ without past temporal operators. On the other hand the last example above is expressible in $\mathrm{LTL}_1^{\downarrow}(X, U)$ but not in $\mathrm{FO}^2(\sim, <, +1)$ (and most likely this property is also not expressible using a data automaton).

There exists a fragment of $\mathrm{LTL}_1^{\downarrow}$ which correspond to $\mathrm{FO}^2(\sim, <, +1)$. A $\mathrm{LTL}_1^{\downarrow}$ formula is said to be *simple* if any temporal operator is immediately preceded by a $\downarrow_1$ and there are no occurrences of $\downarrow_1$ operators. Then simple-$\mathrm{LTL}_1^{\downarrow}(X, X^{-1}, X^2 F, X^{-2} F^{-1})$ has exactly the same expressive power than $\mathrm{FO}^2(\sim, <, +1)$. This fact was first mentioned in [14]. The translations in both directions are effective and use the same ideas as in [16].

Trees. The extension of this ideas to trees, using languages like CTL, remains to be done.

5 Conclusion

We have presented several models of decidable automata and logics over data words and data trees. The logical approach has the advantage of compositionality and has many other interesting closure properties which makes it easier for coding problems into it. The complexities obtained for logics are usually quite high which makes them quite unsuited for practical applications. However it is possible to obtain lower complexities by imposing some extra constraints, see for instance [6,7].

Several of the results presented in this paper were extended to infinite data words. This is useful in the context of verification in order to code infinite computations. For instance Theorem 4.3 have been extended over infinite data words in [7].

The tree case is usually a lot more difficult than the word case. If several decidability results were obtained over data trees, like for register automata or $\mathrm{FO}^2(\sim, +1)$, many decidability questions remains open, like the decidability of $\mathrm{FO}^2(\sim, <, +1)$.

The decidability result of $\mathrm{EMSO}^2(\sim, +1)$ over data trees presented in Theorem 4.6 was used in a database context in [6] in order to show decidability of the inclusion problem of certain fragments of XPath in the presence of DTDs. It was also used to show decidability of validation of DTDs in the presence of unary key and foreign key constraints [6]. Those two results were obtained via different coding of the problems into data trees and $\mathrm{EMSO}^2(\sim, +1)$. This builds on the fact that any regular tree language (and therefore the structural part of DTDs) can be expressed in $\mathrm{EMSO}^2(\sim, +1)$ (without using the predicate $\sim$) and XPath 1.0 is also intimately related with first-order logics with two variables [25].

Altogether we hope that we have convinced the reader that there is a need for more decidable automata models and more decidable logics over data words and data trees. This is a challenging topic which has a lot of applications, in particular in database and in program verification.

Acknowledgment. We thanks Mikołaj Bojańczik, Anca Muscholl, and Thomas Schwentick for all the interesting discussions we had while writing this paper.

References

1. N. Alon, T. Milo, F. Neven, D. Suciu, and V. Vianu. XML with Data Values: Typechecking Revisited. Journal of Computer and System Sciences, 66(4):688 727, 2003.
2. M. Arenas, W. Fan, and L. Libkin. Consistency of XML specifications. Inconsistency Tolerance, Springer, LNCS vol. 3300, 2005.
3. J.-M. Autebert, J. Beauquier, and L. Boasson. Langages des alphabets infinis. Discrete Applied Mathematics, 2:120, 1980.
4. M. Benedikt, W. Fan, and F. Geerts. XPath satisfiability in the presence of DTDs. In Proc. ACM Symp. on Principles of Database Systems, pages 2536, 2005.
5. H. Björklund and T. Schwentick. Personnal communication.
6. M. Bojańczyk, C. David, A. Muscholl, T. Schwentick, and L. Segoufin. Two- Variable Logic on Data Trees and XML Reasoning. In Proc. ACM Symp. on Principles of Database Systems, 2006.
7. M. Bojańczyk, A. Muscholl, T. Schwentick, L. Segoufin, and C. David. Two- Variable Logic on Words with Data. In Proc. IEEE Conf. on Logic in Computer Science, 2006.
8. A. Bouajjani, P. Habermehl, and R. Mayr. Automatic Verification of Recursive Procedures with one Integer Parameter. Theoretical Computer Science, 295, 2003.
9. P. Bouyer, A. Petit, and D. Thérien. An algebraic approach to data languages and timed languages. Inf. Comput., 182(2):137162, 2003.
10. P. Buneman, S. B. Davidson, W. Fan, C. S. Hara, and W. C. Tan. Reasoning about keys for XML. Inf. Syst., 28(8):10371063, 2003.
11. E. Cheng and M. Kaminski. Context-Free Languages over Infinite Alphabets. Acta Inf., 35(3):245267, 1998.
12. C. David. Mots et données infinis. Master's thesis, Université Paris 7, LIAFA, 2004.
13. P. de Groote, B. Guillaume, and S. Salvati. Vector Addition Tree Automata. In Proc. IEEE Conf. on Logic in Computer Science, pages 6473, 2004.
14. S. Demri and R. Lazić. LTL with the Freeze Quantifer and Register Automata. In Proc. IEEE Conf. on Logic in Computer Science, 2006.
15. S. Demri, R. Lazić, and D. Nowak. On the freeze quantifier in Constraint LTL. In TIMES, 2005.
16. K. Etessami, M. Vardi, and T. Wilke. First-Order Logic with Two Variables and Unary Temporal Logic. Inf. Comput., 179(2):279295, 2002.
17. N. Francez and M. Kaminski. An algebraic characterization of deterministic regular languages over infinite alphabets. Theoretical Computer Science, 306(1-3):155175, 2003.

18. F. Geerts and W. Fan. Satisfiability of XPath Queries with Sibling Axes. In Proc. workshop on Database Programming Language, pages 122137, 2005.
19. J. Idt. Automates a pile sur des alphabets infinis. In STACS, 1984.
20. M. Kaminski and N. Francez. Finite memory automata. Theoretical Computer Science, 134(2):329363, 1994.
21. M. Kaminski and T. Tan. Regular Expressions for Languages over Infinite Alphabets. Fundam. Inform., 69(3):301318, 2006.
22. M. Kaminski and T. Tan. Tree Automata over Infinite Alphabets. Poster at the 11th International Conference on Implementation and Application of Automata, 2006.
23. S. Kosaraju. Decidability of reachability in vector addition systems. In Proc. ACM SIGACT Symp. on the Theory of Computing, pages 267281, 1984.
24. R. Lipton. The reachability problem requires exponential space. Technical report, Dep. of Comp.Sci., Research report 62, Yale University, 1976.
25. M. Marx. First order paths in ordered trees. In Proc. of Intl. Conf. on Database Theory, 2005.
26. E. Mayr. An algorithm for the general Petri net reachability problem. SIAM J. of Comp., 13:441459, 1984.
27. F. Neven and T. Schwentick. XPath Containment in the Presence of Disjunction, DTDs, and Variables. In Proc. of Intl. Conf. on Database Theory, 2003.
28. F. Neven, T. Schwentick, and V. Vianu. Finite state machines for strings over infinite alphabets. ACM Trans. Comput. Logic, 15(3):403435, 2004.
29. F. Otto. Classes of regular and context-free languages over countably infinite alphabet. Discrete and Apllied Mathematics, 12:4156, 1985.
30. H. Sakamoto and D. Ikeda. Intractability of decision problems for finite-memory automata. Theoretical Computer Science, 231:297308, 2000.
31. Y. Shemesh and N. Francez. Finite-State Unification Automata and Relational Languages. Inf. Comput., 114(2):192213, 1994.
32. A. Tal. Decidability of inclusion for unification based automata. Master's thesis, Department of Computer Science, Technion - Israel Institute of Technology, 1999.

Nonmonotonic Logics and Their Algebraic Foundations

Mirosław Truszczyński

Department of Computer Science
University of Kentucky
Lexington, KY 40506-0046, USA

Abstract. The goal of this note is to provide a background and references for the invited lecture presented at Computer Science Logic 2006. We briefly discuss motivations that led to the emergence of nonmonotonic logics and introduce two major nonmonotonic formalisms, default and autoepistemic logics. We then point out to algebraic principles behind the two logics and present an abstract algebraic theory that unifies them and provides an effective framework to study properties of nonmonotonic reasoning. We conclude with comments on other major research directions in nonmonotonic logics.

1 Why Nonmonotonic Logics

In the late 1970s, research on languages for knowledge representation, and considerations of basic patterns of commonsense reasoning brought attention to rules of inference that admit *exceptions* and are used only under the assumption of normality of the world in which one functions or to put it differently, when things are as expected.

For instance, a knowledge base concerning a university should support an inference that, given no information that might indicate otherwise, if Dr. Jones is a professor at that university, then Dr. Jones teaches. Such conclusion might be sanctioned by an inference rule stating that *normally* university professors teach. In commonsense reasoning rules with exceptions are ubiquitous. Planning our day and knowing we are to have lunch with a friend, we might use the following rule: *normally*, lunches end by 1:00pm. If nothing we know indicates that the situation we are in is not normal, we use this rule and conclude that our lunch will be over by 1:00pm.

The problem with such rules is that they do not lend themselves in any direct way to formalizations in terms of first-order logic, unless *all* exceptions are known and explicitly represented — an unrealistic expectation in practice. The reason is that standard logical inference is *monotone*: whenever a sentence α is a consequence of a set T of sentences then α is also a consequence of any set of sentences T' such that $T \subseteq T'$. On the other hand, it is clear that reasoning with normality rules when complete information is unavailable, is not monotone. In our lunch scenario, we may conclude that the lunch will be over by 1:00pm. However, if we learn that our friend will be delayed, the normality assumption is no longer valid our earlier inference is unsupported; we have to withdraw it.

Z. Ésik (Ed.): CSL 2006, LNCS 4207, pp. 58–71, 2006.

Such reasoning, where additional information may invalidate conclusions, is called *nonmonotonic*. As we briefly noted above, it is common. It has been a focus of extensive studies by the knowledge representation community since the early eighties of the last century. This research developed along two major directions.

The first direction is concerned with the design of nonmonotonic logics — formalisms with direct ways to model rules with exceptions and with ways to use them. Arguably, two most studied nonmonotonic formalisms are default logic [1] and autoepistemic logic [2,3]. These two logics are the focus of this note. Our main goal in this paper is to introduce default and autoepistemic logics, identify algebraic principles that underlie them, and show that both logics can be viewed through a single abstract unifying framework of operators on complete lattices.

The second direction focused on studies of nonmonotone inference relations either in terms of classes of models or abstract postulates, the two perspectives being quite closely intertwined. Circumscription [4] and, more generally, preference logics [5] fall in this general research direction, as do studies of abstract properties of nonmonotonic inference relations [6,7,8,9]. Although outside our focus, for the sake of completeness, we will provide a few comments on preference logics and nonmonotonic inference relations in the last section of the paper.

2 Default Logic — An Introduction

In his ground-breaking paper [1] Ray Reiter wrote: *Imagine a first order formalization of what we know about any reasonably complex world. Since we cannot know everything [...] — there will be gaps in our knowledge — this first order theory will be incomplete. [...] The role of a default is to help fill in some of the gaps in the knowledge base [...]. Defaults therefore function somewhat like meta-rules: they are instructions about how to create an* extension *of this incomplete theory. Those formulas sanctioned by the defaults and which extend the theory can be viewed as beliefs about the world. Now in general there are many different ways of extending an incomplete theory, which suggests that the default rules may be nondeterministic. Different applications of the defaults yield different extensions and hence different sets of beliefs about the world.*

According to Reiter defaults are meta-rules of the form "in the absence of any information to the contrary, assume ..." (hence, they admit exceptions), and default reasoning consists of applying them. Reiter's far-reaching contribution is that he provided a formal method to do so.

We will now present basic notions of default logic. We consider the language $\mathcal{L}(At)$ (or simply, $\mathcal{L}$) of propositional logic determined by a set At of propositional variables. A *default* is an expression

$$d = \frac{\alpha\colon \beta_1, \ldots, \beta_k}{\gamma}, \tag{1}$$

where α, β_i, $1 \leq i \leq k$, and γ are formulas from $\mathcal{L}$. We say that α is the *prerequisite*, β_i, $1 \leq i \leq k$, are *justifications*, and γ is the *consequent* of default d. If α is a tautology, we omit it from the notation. For a default d, we write

$p(d)$, $c(d)$ and $j(d)$ for its prerequisite, consequent, and the set of justifications, respectively.

An informal reading of a default (1) is: *conclude* γ *if* α *holds and if all justifications* β_i *are possible.* In other words, to apply a default and assert its consequent, we must derive the prerequisite and establish that all justifications are possible. We will soon formalize this intuition. For now, we note that we can encode the rule arising in the university example by the following default:

$$\frac{prof_J : teaches_J}{teaches_J}$$

saying that if $prof_J$ holds and it is possible that $teaches_J$ holds (no information contradicts $teaches_J$), then $teaches_J$ does hold.

A *default theory* is a pair (D, W), where D is a set of defaults and W is a theory in the language $\mathcal{L}$. The role of W is to represent our knowledge (which is, in general, incomplete) while the role of defaults in D is to serve as "meta-rules" we might use to fill in gaps in what we know.

Let $\Delta = (D, W)$ be a default theory and let S be a propositional theory closed under consequence. If we start with S as our beliefs, Δ could be used to revise them. The revised belief set should contain W. Further, it should be closed under propositional consequence (to be a belief set) and under those defaults whose justifications are not contradicted by the current belief set S (are possible with respect to S). This revision process can be formalized by an operator Γ_Δ such that for a any set S of formulas (not necessarily closed under propositional consequence), $\Gamma_\Delta(S)$ is defined as the inclusion-least set U of propositional formulas satisfying the following conditions:

1. U is closed under propositional provability
2. $W \subseteq U$
3. for every default $d \in D$, if $p(d) \in U$ and for every $\beta \in j(d)$, $S \not\vdash \neg\beta$, then $c(d) \in U$.

Fixpoints of the operator Γ_Δ represent belief sets (by (1) they are indeed closed under propositional consequence) that are in a way *stable* with respect to Δ — they cannot be revised away. Reiter [1] proposed them as belief sets associated with Δ and called them *extensions*.

Definition 1. *Let* Δ *be a default theory. A propositional theory* S *is an* extension *of* Δ *if* $S = \Gamma_\Delta(S)$.

Let us look again at the university scenario, which we expand slightly. We know that Dr. Jones is a professor. We also know that if Dr. Jones is chair of the department then Dr. Jones does not teach. Finally we have the default rule saying that normally Dr. Jones teaches. This knowledge can be captured by a default theory (D, W), where

$$W = \{prof_J, chair_J \supset \neg teaches_J\}$$

and

$$D = \left\{ \frac{prof_J : teaches_J}{teaches_J} \right\}.$$

One can check that this default theory has only one extension and it contains $teaches_J$. However, if we append W by additional information that Dr. Jones is chair of the department ($chair_J$), then the resulting default theory has also one extension but it does not contain $teaches_J$, anymore (it contains $\neg teaches_J$). Thus, default theories with the semantics of extension can model nonmonotonic inferences.

Much of the theory of default logic is concerned with properties of extensions. A detailed studies of extensions can be found in [10,11].

3 Autoepistemic Logic

Autoepistemic logic is a logic in a *modal* propositional language $\mathcal{L}_K(At)$ (or simply, $\mathcal{L}_K$), where At is the a of propositional variables and K stands for the modal operator. It was proposed to formalize how a rational agent with perfect introspection might construct belief sets [2,3].

The first modal nonmonotonic logic was introduced by McDermott and Doyle [12]. They proposed to use modal-free formulas to represent facts about an application domain, and "proper" modal formulas to encode nonmonotonic reasoning patterns. An informal reading of a modal formula $K\alpha$ is "α is believed" or "α is known." It suggests that a formula $\neg K \neg \alpha \supset \beta$ could be read "if $\neg\alpha$ is not believed (or, to put it differently, if α is possible) then β. Given this intuition, McDermott and Doyle [12] proposed to use the formula $\neg K \neg \alpha \supset \beta$ to represent a reasoning pattern "*in the absence of information contradicting α, infer β*" and gave a method to reason with such formulas supporting nonmonotonic inferences.

The logic of McDermott and Doyle was found to have counterintuitive properties [13,2,3]. Moore proposed autoepistemic logic [2,3] as a way to address this problem. As in the case of default logic, the goal was to describe a mechanism to assign to a theory belief sets that can be justified on its basis. Unlike in default logic, a specific objective for autoepistemic logic was to formalize belief sets a rational agent reasoning with perfect introspection might form.

Given a theory $T \subseteq \mathcal{L}_K$, Moore [3] defined an *expansion* of T to be a theory $E \subseteq \mathcal{L}_K$ such that

$$E = Cn(T \cup \{K\alpha \mid \alpha \in E\} \cup \{\neg K\alpha \mid \alpha \notin E\})$$

(Cn stands for the operator of propositional consequence which treats formulas $K\alpha$ as propositional variables). Moore justified this fixpoint equation by arguing that expansions should consist precisely of formulas that can be inferred from T and from formulas obtained by positive and negative introspection on the agent's beliefs.

Moore's expansions of T indeed have properties that make them adequate for modeling belief sets a rational agent reasoning with perfect introspection may built out of a theory T. In particular, expansions satisfy postulates put forth by Stalnaker [14] for belief sets in a modal language:

B1: $Cn(E) \subseteq E$ (rationality postulate)
B2: if $\alpha \in E$, then $K\alpha \in E$ (closure under positive introspection)
B3: if $\alpha \notin E$, then $\neg K\alpha \in E$ (closure under negative introspection).

Although motivated differently, autoepistemic logic can capture similar reasoning patterns as default logic does. For instance, the university example can be described in the modal language by a single theory

$$T = \{prof_J, chair_J \supset \neg teaches_J, Kprof_J \wedge \neg K\neg teaches_J \supset teaches_J\}.$$

This theory has exactly one expansion and it contains $teaches_J$. When extended with $chair_J$, the new theory also has just one expansion but it contains $\neg teaches_J$.

Examples like this one raised the question of the relationship between default and autoepistemic logics. Konolige suggested to encode a default

$$d = \frac{\alpha\colon \beta_1, \ldots, \beta_k}{\gamma}$$

with a modal formula

$$k(d) = K\alpha \wedge \neg K\neg\beta_1 \wedge \ldots \wedge \neg K\neg\beta_k \supset \gamma$$

and to represent a default theory $\Delta = (D, W)$ by a modal theory

$$k(\Delta) = W \cup \{k(d)\colon d \in D\}.$$

The translation seemed intuitive enough. In particular, it worked in the university example in the sense that extension of the default logic representation correspond to expansions of the modal logic representation obtained by translating the default logic one. However, it turned not to align extensions with expansions in general (a default theory $(\{\frac{p:q}{p}\}, \emptyset)$ has one extension but its modal counterpart has two expansions).

4 Default and Autoepistemic Logics — Algebraically

Explaining the relationship between the two logics became a major research challenge. We will present here a recent algebraic account of this relationship [15]. As the first step, we will describe expansions and extensions within the framework of operators on the lattice of possible-world structures.

A *possible-world structure* is a set (possibly empty) of truth assignments to atoms in At. Possible-world structures can be ordered by the *reverse set inclusion*: for $Q, Q' \in \mathcal{W}$, $Q \sqsubseteq Q'$ if $Q' \subseteq Q$. The ordering $\sqsubseteq$ can be thought of as an ordering of increasing knowledge. As we move from one possible-world structure to another, greater with respect to $\sqsubseteq$, some interpretations are excluded and our knowledge of the world improves. We denote the set of all possible-world structures with $\mathcal{W}$. One can check that $\langle \mathcal{W}, \sqsubseteq \rangle$ is a complete lattice.

A possible-world structure Q and an interpretation I, determine the truth function $\mathcal{H}_{Q,I}$ inductively as follows:

1. $\mathcal{H}_{Q,I}(p) = I(p)$, if p is an atom.
2. $\mathcal{H}_{Q,I}(\varphi_1 \wedge \varphi_2) = \mathbf{t}$ if $\mathcal{H}_{Q,I}(\varphi_1) = \mathbf{t}$ and $\mathcal{H}_{Q,I}(\varphi_2) = \mathbf{t}$. Otherwise, $\mathcal{H}_{Q,I}(\varphi_1 \wedge \varphi_2) = \mathbf{f}$.
3. $\mathcal{H}_{Q,I}(\varphi_1 \vee \varphi_2) = \mathbf{t}$ if $\mathcal{H}_{Q,I}(\varphi_1) = \mathbf{t}$ or $\mathcal{H}_{Q,I}(\varphi_2) = \mathbf{t}$. Otherwise, $\mathcal{H}_{Q,I}(\varphi_1 \vee \varphi_2) = \mathbf{f}$.
4. $\mathcal{H}_{Q,I}(\neg\varphi) = \mathbf{t}$ if $\mathcal{H}_{Q,I}(\varphi) = \mathbf{f}$. Otherwise, $\mathcal{H}_{Q,I}(\varphi) = \mathbf{f}$.
5. $\mathcal{H}_{Q,I}(K\varphi) = \mathbf{t}$, if for every interpretation $J \in Q$, $\mathcal{H}_{Q,J}(\varphi) = \mathbf{t}$. Otherwise, $\mathcal{H}_{Q,I}(K\varphi) = \mathbf{f}$.

It is clear that for every formula $\varphi \in \mathcal{L}_K$, the truth value $\mathcal{H}_{Q,I}(K\varphi)$ does not depend on I. Thus, and we will denote it by $\mathcal{H}_Q(K\varphi)$, dropping I from the notation. The *modal theory* of a possible-world structure Q, denoted by $Th_K(Q)$, is the set of all modal formulas that are believed in Q. Formally,

$$Th_K(Q) = \{\varphi \colon \mathcal{H}_Q(K\varphi) = \mathbf{t}\}.$$

The *(modal-free) theory* of Q, denoted $Th(Q)$, is defined by

$$Th(Q) = Th_K(Q) \cap \mathcal{L}.$$

(As an aside, we note here a close relation between possible-world structures and Kripke models with universal accessibility relations.)

Default and autoepistemic logics can both be defined in terms of fixpoints of operators on the lattice $\langle \mathcal{W}, \sqsubseteq \rangle$. A characterization of expansions in terms of fixpoints of an operator on $\mathcal{W}$ has been known since Moore [2]. Given a theory $T \subseteq \mathcal{L}_K$ and a possible-world structure Q, Moore defined a possible-world structure $D_T(Q)$ as follows:

$$D_T(Q) = \{I : \mathcal{H}_{Q,I}(\varphi) = \mathbf{t}, \text{ for every } \varphi \in T\}.$$

The intuition behind this definition is as follows (perhaps not coincidentally, as in the case of default logic, we again refer to belief-set revision intuitions). The possible-world structure $D_T(Q)$ is a revision of a possible-world structure Q. This revision consists of the worlds that are acceptable given the constraints on agent's beliefs captured by T. That is, the revision consists precisely of these worlds that make all formulas in T true (in the context of Q — the current belief state). Fixpoints of the operator D_T represent "stable" belief sets — they cannot be revised any further and so take a special role in the space of belief sets. It turns out [3] that they correspond to expansions!

Theorem 1. *Let $T \subseteq \mathcal{L}_K$. A theory $E \subseteq \mathcal{L}_K$ is an* expansion *of T if and only if $E = Th_K(Q)$, for some possible-world structure Q such that $Q = D_T(Q)$.*

A default theory defines a similar operator. With the Konolige's interpretation of defaults in mind, we first define a truth function on the set of all propositional formulas and defaults. Namely, for a propositional formula φ, we define

$$\mathcal{H}^{dl}_{Q,I}(\varphi) = I(\varphi),$$

and for a default $d = \frac{\alpha:\, \beta_1,\ldots,\beta_k}{\gamma}$, we set

$$\mathcal{H}^{dl}_{Q,I}(d) = \mathbf{t}$$

if at least one of the following conditions holds:

1. there is $J \in Q$ such that $J(\alpha) = \mathbf{f}$.
2. there is i, $1 \leq i \leq k$, such that for every $J \in Q$, $J(\beta_i) = \mathbf{f}$.
3. $I(\gamma) = \mathbf{t}$

(we set $\mathcal{H}^{dl}_{Q,I}(d) = \mathbf{f}$, otherwise).

Given a default theory $\Delta = (D, W)$, for a possible-world structure Q, we define a possible-world structure $D_\Delta(Q)$ as follows:

$$D_\Delta(Q) = \{I : \mathcal{H}_{Q,I}(\varphi) = \mathbf{t},\ \text{for every } \varphi \in W \cup D\}.$$

Do fixpoints of D_Δ correspond to extensions? The answer is no. Fixpoints of D_Δ correspond to *weak extensions* [16], another class of belief sets one can associate with default theories.

To characterize extensions a different operator is needed. The following definition is due (essentially) to Guerreiro and Casanova [17]. Let $\Delta = (D, W)$ be a default theory and let Q be a possible-world structure. We define $\Gamma'_\Delta(Q)$ to be the least possible-world structure Q' (with respect to $\sqsubseteq$) satisfying the conditions:

1. $W \subseteq Th(Q')$
2. for every default $d \in D$, if $p(d) \in Th(Q')$ and for every $\beta \in j(d)$, $\neg\beta \notin Th(Q)$, then $c(d) \in Th(Q')$.

One can show that $\Gamma'_\Delta(Q)$ is well defined. Moreover, for every possible-world structure Q,

$$Th(\Gamma'_\Delta(Q)) = \Gamma_\Delta(Th(Q))$$

Consequently, we have the following result connecting fixpoints of $\Gamma'_\Delta(Q)$ and extensions of Δ [17].

Theorem 2. *Let Δ be a default theory. A theory $S \subseteq \mathcal{L}$ is an extension of Δ if and only if $S = Th(Q)$ for some possible-world structure Q such that $Q = \Gamma'_\Delta(Q)$.*

Several questions arise. Is there a connection between the operators D_Δ and Γ'_Δ? Is there a counterpart to the operator Γ'_Δ in autoepistemic logic? Can these operators, their fixpoints and their interrelations be considered in a more abstract setting? What are abstract algebraic principles behind autoepistemic and default logics? We provide some answers in the next section.

5 Approximation Theory

Possible-world structures form a complete lattice. As we argued, default and autoepistemic theories determine "revision" operators on this lattice. These operators formalize a view of a theory (default or modal) as a device for *revising* possible-world structures. Possible-world structures that are stable under the revision or, more formally, which are fixpoints of the revision operator give a semantics to the theory (of course, with respect to the revision operator used).

Operators on a complete lattice of propositional truth assignments and their fixpoints were used in a similar way to study the semantics of logic programs with negation. Fitting [18,19,20] characterized all major 2-, 3- and 4-valued semantics of logic programs, specifically, supported-model semantics [21], stable-model semantics [22], Kripke-Kleene semantics [18,23] and well-founded semantics [24], in terms of fixpoints of the van Emden-Kowalski operator [25,26] and its generalizations and variants.

These results suggested the existence of more general and abstract principles underlying these characterizations. [27,28] identified them and proposed a comprehensive unifying *abstract* framework of *approximating* operators as an algebraic foundation for nonmonotonic reasoning. We will now outline the theory of approximating operators and use it to relate default and autoepistemic logics. For details, we refer to [27,28].

Let $\langle L, \leq \rangle$ be a poset. An element $x \in L$ is a *pre-fixpoint* of an operator $O\colon L \to L$ if $O(x) \leq x$; x is a *fixpoint* of O if $O(x) = x$. We denote a least fixpoint of O (if it exists) by $lfp(O)$.

An operator $O : L \to L$ is *monotone* if for every $x, y \in L$ such that $x \leq y$, $O(x) \leq O(y)$. Monotone operators play a key role in the algebraic approach to nonmonotonic reasoning. Tarski and Knaster's theorem asserts that monotone operators on complete lattices (from now on L will always stand for a complete lattice) have least fixpoints [29].

Theorem 3. *Let L be a complete lattice and let O be a monotone operator on L. Then O has a least fixpoint and a least pre-fixpoint, and these two elements of L coincide. That is, we have $lfp(O) = \bigwedge\{x \in L\colon O(x) \leq x\}$.*

The *product bilattice* [30] of a complete lattice L is the set $L^2 = L \times L$ with the following two orderings $\leq_p$ and $\leq$:

1. $(x, y) \leq_p (x', y')$ if $x \leq x'$ and $y' \leq y$
2. $(x, y) \leq (x', y')$ if $x \leq x'$ and $y \leq y'$.

Both orderings are complete lattice orderings in L^2. For the theory of approximating operators, the ordering $\leq_p$ is of primary importance.

If $(x, y) \in L^2$ and $x \leq z \leq y$, then $(x, y) \in L^2$ *approximates* z. The "higher" a pair (x, y) in L^2 with respect to $\leq_p$, the more *precise* estimate it provides to elements it approximates. Therefore, we call this ordering the *precision* ordering. Most precise approximations are provided by pairs $(x, y) \in L^2$ for which $x = y$. We call such pairs *exact*.

For a pair $(x, y) \in L^2$, we define its *projections* as:

$$(x, y)_1 = x \qquad \text{and} \qquad (x, y)_2 = y.$$

Similarly, for an operator $A : L^2 \rightarrow L^2$, if $A(x, y) = (x', y')$, we define

$$A(x, y)_1 = x' \qquad \text{and} \qquad A(x, y)_2 = y'.$$

Definition 2. *An operator* $A \colon L^2 \rightarrow L^2$ *is* symmetric *if for every* $(x, y) \in L^2$, $A(x, y)_1 = A(y, x)_2$*;* A *is* approximating *if* A *is symmetric and* $\leq_p$*-monotone.*

Every approximating operator A on L^2 maps exact pairs to exact pairs. Indeed, $A(x, x) = (A(x, x)_1, A(x, x)_2)$ and, by the symmetry of A, $A(x, x)_1 = A(x, x)_2$.

Definition 3. *If* A *is an approximating operator and* O *is an operator on* L *such that for every* $x \in L$ $A(x, x) = (O(x), O(x))$*, then* A *is an* approximating operator for O.

Let $A \colon L^2 \rightarrow L^2$ be an approximating operator. Then for every $y \in L$, the operator $A(\cdot, y)_1$ (on the lattice L) is $\leq$-monotone. Thus, by Theorem 3, it has a least fixpoint. This observation brings us to the following definition.

Definition 4. *Let* $A \colon L^2 \rightarrow L^2$ *be an approximating operator. The* stable operator *for* A*,* $\mathcal{C}_A$*, is defined by*

$$\mathcal{C}_A(x, y) = (C_A(y), C_A(x)),$$

where $C_A(y) = \mathit{lfp}(A(\cdot, y)_1)$ *(or equivalently, as* A *is symmetric,* $C_A(y) = \mathit{lfp}(A(y, \cdot)_2)$*).*

The following result states two key properties of stable operators.

Theorem 4. *Let* $A \colon L^2 \rightarrow L^2$ *be an approximating operator. Then*

1. $\mathcal{C}_A$ *is* $\leq_p$*-monotone, and*
2. *if* $\mathcal{C}_A(x, y) = (x, y)$*, then* $A(x, y) = (x, y)$*.*

Operators A and $\mathcal{C}_A$ are $\leq_p$-monotone. By Theorem 3, they have least fixpoints. We call them the *Kripke-Kleene* and the *well-founded* fixpoints of A, respectively (the latter term is justified by Theorem 4).

Let A be an approximating operator for an operator O. An A*-stable fixpoint* of O is any element x such that (x, x) is a fixpoint of $\mathcal{C}_A$. By Theorem 4, if (x, x) is a fixpoint of $\mathcal{C}_A$ then it is a fixpoint of A and so, x is a fixpoint of O. Thus, our terminology is justified. The following result gathers some basic properties of fixpoints of approximating operators.

Theorem 5. *Let* O *be an operator on a complete lattice* L *and* A *its approximating operator. Then,*

1. *fixpoints of the operator* $\mathcal{C}_A$ *are minimal fixpoints of* A *(with respect to the ordering* $\leq$ *of* L^2*); in particular,* A*-stable fixpoints of* O *are minimal fixpoints of* O

2. *the Kripke-Kleene fixpoint of A approximates all fixpoints of O*
3. *the well-founded fixpoint of A approximates all A-stable fixpoints of O*

How does it all relate to default and autoepistemic logic? In both logics operators D_Δ and D_T have natural generalizations, $\mathcal{D}_\Delta$ and $\mathcal{D}_T$, respectively, defined on the lattice $\mathcal{W}^2$ — the product lattice of the lattice $\mathcal{W}$ of possible-world structures [15]. One can show that $\mathcal{D}_\Delta$ and $\mathcal{D}_T$ are approximating operators for the operators D_Δ and D_T. Fixpoints of operators $\mathcal{D}_\Delta$ and $\mathcal{D}_T$ and their stable counterparts define several classes of belief sets one can associate with default and autoepistemic theories.

Exact fixpoints of the operators $\mathcal{D}_\Delta$ and $\mathcal{D}_T$ (or, more precisely, the corresponding fixpoints of operators D_Δ and D_T) define the semantics of *expansions* (in the case of autoepistemic logic, proposed originally by Moore; in the case of default logic, expansions were known as weak extensions [16]). The stable fixpoints of the operators D_Δ and D_T define the semantics of *extensions* (in the case of default logic, proposed originally by Reiter, in the case of autoepistemic logic the concept was not identified until algebraic considerations in [15] revealed it). Finally, the Kripke-Kleene and the well-founded fixpoints provide three-valued belief sets that approximate expansions and extensions (except for [31], these concepts received essentially no attention in the literature, despite their useful computational properties [15]).

Moreover, these semantics are aligned when we cross from default to autoepistemic logic by means of the Konolige's translation. One can check that the operators $\mathcal{D}_\Delta$ and $\mathcal{D}_{k(\Delta)}$ coincide. The Konolige's translation preserves expansions, extensions, the Kripke-Kleene and the well-founded semantics. However, clearly, it does not align default extensions with autoepistemic expansions. Different principles underlie these two concepts. Expansions are fixpoints of the basic revision operator $\mathcal{D}_\Delta$ or $\mathcal{D}_T$, while extensions are fixpoints of the stable operators for $\mathcal{D}_\Delta$ or $\mathcal{D}_T$, respectively.

Properties of fixpoints of approximating operators we stated in Theorems 4 and 5 specialize to properties of expansions and extensions of default and autoepistemic theories. One can prove several other properties of approximating operators that imply known or new results for default and autoepistemic logics. In particular, one can generalize the notion of stratification of a default (autoepistemic) theory to the case of operators and obtain results on the existence and properties of extensions and expansions of stratified theories as corollaries of more general results on fixpoints of stratified operators [32,33].

Similarly, one can extend to the case of operators concepts of strong and uniform equivalence of nonmonotonic theories and prove characterization results purely in the algebraic setting [34].

6 Additional Comments

In this note, we focused on nonmonotonic logics which use fixpoint conditions to define belief sets and we discussed abstract algebraic principles behind these logics. We will now briefly mention some other research directions in nonmonotonic reasoning.

Default extensions are in some sense minimal (cf. Theorem 5(1)) and minimality was identified early as one of the fundamental principles in nonmonotonic reasoning. McCarthy [4] used it to define *circumscription*, a nonmonotonic logic in the language of first-order logic in which entailment is defined with respect to minimal models only. Circumscription was extensively studied [35,36]. Computational aspects were studied in [37,38,39]; connections to fixpoint-based logics were discussed in [40,41,42].

Preferential models [5,8] generalize circumscription and provide a method to define nonmonotonic inference relations. Inference relations determined by preferential models were shown in [8] to be precisely inference relations satisfying properties of *Left Logical Equivalence*, *Right Weakening*, *Reflexivity*, *And*, *Or* and *Cautious Monotony*. Inference relations determined by *ranked* preferential models were shown in [9] to be precisely those preferential inference relations that satisfy *Rational Monotony*.

Default conditionals that capture statements "if α then normally β" were studied in [43,9,44]. [9,44] introduce the notion of rational closure of sets of conditionals as as a method of inference ([44] uses the term *system Z*).

Default extensions and autoepistemic expansions also define nonmonotonic inference relations. For instance, given a set of defaults D, we might say that a formula β can be inferred from a formula α given D if β is in every extension of the default theory $(D, \{\alpha\})$. A precise relationship (if any) between this and similar inference relations based on the concept of extension or expansion to preferential or rational inference relations is not know at this time. Discovering it is a major research problem.

We conclude this paper by pointing to several research monographs on nonmonotonic reasoning [45,46,10,47,11,48,49,50].

Acknowledgments

We acknowledge the support of NSF grant IIS-0325063 and KSEF grant 1036-RDE-008.

References

1. Reiter, R.: A logic for default reasoning. Artificial Intelligence **13** (1980) 81–132
2. Moore, R.: Possible-world semantics for autoepistemic logic. In: Proceedings of the Workshop on Non-Monotonic Reasoning. (1984) 344–354 Reprinted in: M. Ginsberg, ed., *Readings on Nonmonotonic Reasoning*, pages 137–142, Morgan Kaufmann, 1990.
3. Moore, R.: Semantical considerations on nonmonotonic logic. Artificial Intelligence **25** (1985) 75–94
4. McCarthy, J.: Circumscription — a form of non-monotonic reasoning. Artificial Intelligence **13** (1980) 27–39

5. Shoham, Y.: A semantical approach to nonmonotic logics. In: Proceedings of the Symposium on Logic in Computer Science, LICS-87, IEEE Computer Society (1987) 275–279
6. Gabbay, D.M.: Theoretical foundations for non-monotonic reasoning in expert systems. In: Proceedings of the NATO Advanced Study Institute on Logics and Models of Concurrent Systems, Springer (1989) 439–457
7. Makinson, D.: General theory of cumulative inference. In Reinfrank, M., de Kleer, J., Ginsberg, M., Sandewall, E., eds.: Nonmonotonic reasoning (Grassau, 1988). Volume 346 of Lecture Notes in Computer Science., Berlin-New York, Springer (1989) 1–18
8. Kraus, S., Lehmann, D., Magidor, M.: Nonmonotonic reasoning, preferential models and cumulative logics. Artificial Intelligence Journal **44** (1990) 167–207
9. Lehmann, D., Magidor, M.: What does a conditional knowledge base entail? Artificial Intelligence Journal **55** (1992) 1–60
10. Marek, W., Truszczyński, M.: Nonmonotonic Logic; Context-Dependent Reasoning. Springer, Berlin (1993)
11. Antoniou, G.: Nonmonotonic Reasoning. MIT Press (1997)
12. McDermott, D., Doyle, J.: Nonmonotonic logic I. Artificial Intelligence **13** (1980) 41–72
13. McDermott, D.: Nonmonotonic logic II: nonmonotonic modal theories. Journal of the ACM **29** (1982) 33–57
14. Stalnaker, R.: A note on nonmonotonic modal logic. Unpublished manuscript (1980)
15. Denecker, M., Marek, V., Truszczyński, M.: Uniform semantic treatment of default and autoepistemic logics. Artificial Intelligence Journal **143** (2003) 79–122
16. Marek, W., Truszczyński, M.: Relating autoepistemic and default logics. In: Proceedings of the 1st International Conference on Principles of Knowledge Representation and Reasoning (Toronto, ON, 1989), San Mateo, CA, Morgan Kaufmann (1989) 276–288
17. Guerreiro, R., Casanova, M.: An alternative semantics for default logic. Preprint. The 3rd International Workshop on Nonmonotonic Reasoning, South Lake Tahoe (1990)
18. Fitting, M.C.: A Kripke-Kleene semantics for logic programs. Journal of Logic Programming **2** (1985) 295–312
19. Fitting, M.C.: Bilattices and the semantics of logic programming. Journal of Logic Programming **11** (1991) 91–116
20. Fitting, M.C.: Fixpoint semantics for logic programming – a survey. Theoretical Computer Science **278** (2002) 25–51
21. Clark, K.: Negation as failure. In Gallaire, H., Minker, J., eds.: Logic and data bases. Plenum Press, New York-London (1978) 293–322
22. Gelfond, M., Lifschitz, V.: The stable semantics for logic programs. In: Proceedings of the 5th International Conference on Logic Programming, MIT Press (1988) 1070–1080
23. Kunen, K.: Negation in logic programming. Journal of Logic Programming **4** (1987) 289–308
24. Van Gelder, A., Ross, K., Schlipf, J.: The well-founded semantics for general logic programs. Journal of the ACM **38** (1991) 620–650
25. van Emden, M., Kowalski, R.: The semantics of predicate logic as a programming language. Journal of the ACM **23** (1976) 733–742
26. Apt, K., van Emden, M.: Contributions to the theory of logic programming. Journal of the ACM **29** (1982) 841–862

27. Denecker, M., Marek, V., Truszczyński, M.: Approximations, stable operators, well-founded fixpoints and applications in nonmonotonic reasoning. In Minker, J., ed.: Logic-Based Artificial Intelligence. Kluwer Academic Publishers (2000) 127–144
28. Denecker, M., Marek, V., Truszczyński, M.: Ultimate approximation and its application in nonmonotonic knowledge representation systems. Information and Computation **192** (2004) 84–121
29. Tarski, A.: Lattice-theoretic fixpoint theorem and its applications. Pacific Journal of Mathematics **5** (1955) 285–309
30. Ginsberg, M.: Multivalued logics: a uniform approach to reasoning in artificial intelligence. Computational Intelligence **4** (1988) 265–316
31. Baral, C., Subrahmanian, V.: Dualities between alternative semantics for logic programming and nonmonotonic reasoning (extended abstract). In Nerode, A., Marek, W., Subrahmanian, V., eds.: Logic programming and non-monotonic reasoning (Washington, DC, 1991), Cambridge, MA, MIT Press (1991) 69–86
32. Vennekens, J., Gilis, D., Denecker, M.: Splitting an operator: an algebraic modularity result and its applications to logic programming. In Lifschitz, V., Demoen, B., eds.: Logic programming, Proceedings of the 20th International Conference on Logic Programming, ICLP-04. (2004) 195–209
33. Vennekens, J., Denecker, M.: An algebraic account of modularity in id-logic. In Baral, C., Leone, N., eds.: Proceedings of 8th International Conference on Logic Programming and Nonmonotonic Reasoning, LPNMR-05. LNAI 3662 (2005) 291–303
34. Truszczyński, M.: Strong and uniform equivalence of nonmonotonic theories — an algebraic approach. In Doherty, P., Mylopoulos, J., Welty, C.A., eds.: Proceedings of the 10th International Conference on Principles of Knowledge Representation and Reasoning, KR 2006, AAAI Press (2006) 389–399
35. Lifschitz, V.: Pointwise circumscription. In M.Ginsberg, ed.: Readings in Nonmonotonic Reasoning, Los Altos, CA, Morgan Kaufmann (1987) 179–193
36. Lifschitz, V.: Circumscriptive theories: a logic-based framework for knowledge representation. Journal of Philosophical Logic **17** (1988) 391–441
37. Lifschitz, V.: Computing circumscription. In: Proceedings of IJCAI-85, Los Altos, CA, Morgan Kaufmann (1985) 121–127
38. Ginsberg, M.: A circumscriptive theorem prover. In: Nonmonotonic reasoning (Grassau, 1988). Volume 346 of Lecture Notes in Computer Science., Berlin, Springer (1989) 100–114
39. Przymusiński, T.C.: An algorithm to compute circumscription. Artificial Intelligence **38** (1989) 49–73
40. Gelfond, M., Przymusinska, H.: On the relationship between circumscription and autoepistemic logic. In: Proceedings of the 5th International Symposium on Methodologies for Intelligent Systems. (1986)
41. Bidoit, N., Froidevaux, C.: Minimalism subsumes default logic and circumscription. In: Proceedings of IEEE Symposium on Logic in Computer Science, LICS-87, IEEE Press (1987) 89–97
42. Imieliński, T.: Results on translating defaults to circumscription. Artificial Intelligence **32** (1987) 131–146
43. Lehmann, D.: What does a conditional knowledge base entail? In: Proceedings of the 1st International Conference on Principles of Knowledge Representation and Reasoning, KR-89, Morgan Kaufmann (1989) 212–222

44. Pearl, J.: System Z: A natural ordering of defaults with tractable applications to nonmonotonic reasoning. In: Proceedings of the 3rd Conference on Theoretical Aspects of Reasoning about Knowledge, TARK-90, Morgan Kaufmann (1990) 121–135
45. Besnard, P.: An Introduction to Default Logic. Springer, Berlin (1989)
46. Brewka, G.: Nonmonotonic Reasoning: Logical Foundations of Commonsense. Volume 12 of Cambridge Tracts in Theoretical Computer Science. Cambridge University Press, Cambridge, UK (1991)
47. Brewka, G., Dix, J., Konolige, K.: Nonmonotonic Reasoning, An Overview. CSLI Publications (1997)
48. Bochman, A.: A Logical Theory of Nonmonotonic Inference and Belief Change. Springer, Berlin (2001)
49. Bochman, A.: Explanatory Nonmonotonic Reasoning. Volume 4 of Advances in Logic. World Scientific (2005)
50. Makinson, D.: Bridges from Classical to Nonmonotonic Logic. Volume 5 of Texts in Computing. King's College Publications (2005)

Semi-continuous Sized Types and Termination

Andreas Abel*

Institut für Informatik
Ludwig-Maximilians-Universität München
Oettingenstr. 67, D-80538 München, Germany
abel@tcs.ifi.lmu.de

Abstract. A type-based approach to termination uses sized types: an ordinal bound for the size of a data structure is stored in its type. A recursive function over a sized type is accepted if it is visible in the type system that recursive calls occur just at a smaller size. This approach is only sound if the type of the recursive function is admissible, i.e., depends on the size index in a certain way. To explore the space of admissible functions in the presence of higher-kinded data types and impredicative polymorphism, a semantics is developed where sized types are interpreted as functions from ordinals into sets of strongly normalizing terms. It is shown that upper semi-continuity of such functions is a sufficient semantical criterion for admissibility. To provide a syntactical criterion, a calculus for semi-continuous function is developed.

1 Introduction

Termination of computer programs has received continuous interest in the history of computer science, and classical applications are total correctness and termination of partial evaluation. In languages with a notion of computation on the type-level, such as dependently-typed languages or rich typed intermediate languages in compilers [11], termination of expressions that compute a type is required for type checking and type soundness. Further, theorem provers that are based on the Curry-Howard Isomorphism and offer a functional programming language to write down proofs usually reject non-terminating programs to ensure consistency. Since the pioneering work of Mendler [15], termination analysis has been combined with typing, with much success for strongly-typed languages [14,6,13,19,7,9]. The resulting technique, *type-based termination checking*, has several advantages over a purely syntactical termination analysis: (1) It is *robust* w. r. t. small changes of the analyzed program, since it is working on an abstraction of the program: its type. So if the reformulation of a program (e.g., by introducing a redex) still can be assigned the same sized type, it automatically passes the termination check. (2) In design and justification, type-based termination rests on a technology extensively studied for several decades: types.

* Research supported by the coordination action TYPES (510996) and thematic network Applied Semantics II (IST-2001-38957) of the European Union and the project Cover of the Swedish Foundation of Strategic Research (SSF).

Z. Ésik (Ed.): CSL 2006, LNCS 4207, pp. 72–88, 2006.

(3) Type-based termination is essentially a refinement of the typing rules for recursion and for introduction and elimination of data. This is *orthogonal* to other language constructs, like variants, records, and modules. Thus, a language can be easily enriched without change to the termination module. This is not true if termination checking is a separate static analysis. Orthogonality has an especially pleasing effect: (4) Type-based termination scales to *higher-order functions* and *polymorphism*. (5) Last but not least, it effortlessly creates a termination *certificate*, which is just the typing derivation.

Type-based termination especially plays its strength when combined with higher-order datatypes and higher-rank polymorphism, i. e., occurrence of $\forall$ to the left of an arrow. Let us see an example. We consider the type of generalized rose trees $\mathsf{GRose}\,F A$ parameterized by an element type A and the branching type F. It is given by two constructors:

$$\begin{aligned} &\mathsf{leaf} \;\;: \mathsf{GRose}\,F A \\ &\mathsf{node} : A \to F\,(\mathsf{GRose}\,F A) \to \mathsf{GRose}\,F A \end{aligned}$$

Generalized rose trees are either a leaf or a $\mathsf{node}\,a\,\mathit{fr}$ of a label a of type A and a collection of subtrees fr of type $F\,(\mathsf{GRose}\,F A)$. Instances of generalized rose trees are binary trees ($FA = A \times A$), finitely branching trees ($FA = \mathsf{List}\,A$), or infinitely branching trees ($FA = \mathsf{Nat} \to A$). Programming a generic equality function for generalized rose trees that is polymorphic in F and A, we will end up with the following equations:

$$\mathsf{Eq}\,A = A \to A \to \mathsf{Bool}$$

$$\mathsf{eqGRose} : (\forall A.\, \mathsf{Eq}\,A \to \mathsf{Eq}\,(FA)) \to \forall A.\, \mathsf{Eq}\,A \to \mathsf{Eq}\,(\mathsf{GRose}\,F A)$$

$$\begin{aligned} &\mathsf{eqGRose}\ \mathit{eqF}\ \mathit{eqA}\ \mathsf{leaf}\ \mathsf{leaf} = \mathsf{true} \\ &\mathsf{eqGRose}\ \mathit{eqF}\ \mathit{eqA}\ (\mathsf{node}\ a\ \mathit{fr})\ (\mathsf{node}\ a'\ \mathit{fr}') = (\mathit{eqA}\ a\ a') \wedge \\ &\qquad\qquad\qquad\qquad (\mathit{eqF}\ (\mathsf{eqGRose}\ \mathit{eqF}\ \mathit{eqA})\ \mathit{fr}\ \mathit{fr}') \end{aligned}$$

$$\mathsf{eqGRose}\ \mathit{eqF}\ \mathit{eqA}\ \_\ \_ = \mathsf{false}$$

The generic equality $\mathsf{eqGRose}$ takes two parametric arguments, *eqF* and *eqA*. The second one is a placeholder for an equality test for type A, the first one lifts an equality test for an arbitrary type A to an equality test for the type FA. The equality test for generalized rose trees, $\mathsf{eqGRose}\ \mathit{eqF}\ \mathit{eqA}$, is then defined by recursion on the next two arguments. In the case of two nodes we would expect a recursive call, but instead, the function itself is passed as an argument to *eqF*, one of its own arguments! Nevertheless, $\mathsf{eqGRose}$ is a total function, provided its arguments are total and well-typed. However, with traditional methods, which only take the computational behavior into account, it will be hard to verify termination of $\mathsf{eqGRose}$. This is due to the fact that the polymorphic nature of *eqF* plays a crucial role. It is easy to find an instance of *eqF* of the wrong type which makes the program loop. Take, for instance:

$$\begin{aligned} &\mathit{eqF} : \mathsf{Eq}\,(\mathsf{GRose}\,F\,\mathsf{Nat}) \to \mathsf{Eq}\,(F\,(\mathsf{GRose}\,F\,\mathsf{Nat})) \\ &\mathit{eqF}\ \mathit{eq}\ \mathit{fr}\ \mathit{fr}' = \mathit{eq}\,(\mathsf{node}\,0\,\mathit{fr})\,(\mathsf{node}\,0\,\mathit{fr}') \end{aligned}$$

A type-based termination criterion however passes $\mathsf{eqGRose}$ with ease: Consider the indexed type $\mathsf{GRose}^{\imath}\, F A$ of generalized rose trees whose height is smaller than $\imath$. The types of the constructors are refined as follows:

$$\begin{aligned} \mathsf{leaf} &: \forall F \forall A \forall \imath.\, \mathsf{GRose}^{\imath+1}\, F A \\ \mathsf{node} &: \forall F \forall A \forall \imath.\ A \to \mathsf{GRose}^{\imath}\, F A \to \mathsf{GRose}^{\imath+1}\, F A \end{aligned}$$

When defining $\mathsf{eqGRose}$ for trees of height $< \imath + 1$, we may use $\mathsf{eqGRose}$ on trees of height $< \imath$. Hence, in the clause for two nodes, term $\mathsf{eqGRose}\ \mathit{eqF}\ \mathit{eqA}$ has type $\mathsf{Eq}\,(\mathsf{GRose}^{\imath}\, F A)$, and $\mathit{eqF}\ (\mathit{eqGRose}\ \mathit{eqF}\ \mathit{eqA})$ gets type $\mathsf{Eq}\,(F\,(\mathsf{GRose}^{\imath}\, F A))$, by instantiation of the polymorphic type of eqF. Now it is safe to apply the last expression to fr and fr' which are in $F\,(\mathsf{GRose}^{\imath}\, F A)$, since $\mathsf{node}\, a\, \mathit{fr}$ and $\mathsf{node}\, a'\, \mathit{fr}'$ were assumed to be in $\mathsf{GRose}^{\imath+1}\, F A$.

In essence, type-based termination is a stricter typing of the fixed-point combinator fix which introduces recursion. The unrestricted use, via the typing rule (1), is replaced by a rule with a stronger hypothesis (2):

$$(1)\ \frac{f : A \to A}{\mathsf{fix}\, f : A} \qquad\qquad (2)\ \frac{f : \forall \imath.\, A(\imath) \to A(\imath + 1)}{\mathsf{fix}\, f : \forall n.\, A(n)}$$

Soundness of rule (2) can be shown by induction on n. To get started, we need to show $\mathsf{fix}\, f : A(0)$ which requires $A(\imath)$ to be of a special shape, for instance $A(\imath) = \mathsf{GRose}^{\imath}\, F\, B \to C$ (this corresponds to Hughes, Pareto, and Sabry's *bottom check* [14]). Then $A(0)$ denotes functions which have to behave well for all arguments in $\mathsf{GRose}^{0}\, F\, B$, i. e., for no arguments, since $\mathsf{GRose}^{0}\, F\, B$ is empty. Trivially, any program fulfills this condition. In the step case, we need to show $\mathsf{fix}\, f : A(n+1)$, but this follows from the equation $\mathsf{fix}\, f = f\,(\mathsf{fix}\, f)$ since $f : A(n) \to A(n+1)$, and $\mathsf{fix}\, f : A(n)$ by induction hypothesis.

In general, the index $\imath$ in $A(\imath)$ will be an *ordinal* number. Ordinals are useful when we want to speak of objects of unbounded size, e. g., generalized rose trees of height $< \omega$ that inhabit the type $\mathsf{GRose}^{\omega}\, F A$. Even more, ordinals are required to denote the height of infinitely branching trees: take generalized rose trees with $F A = \mathsf{Nat} \to A$. Other examples of infinite branching, which come from the area of inductive theorem provers, are the W-type, Brouwer ordinals and the accessibility predicate [17].

In the presence of ordinal indices, rule (2) has to be proven sound by transfinite induction. In the case of a limit ordinal λ, we have to infer $\mathsf{fix}\, f : A(\lambda)$ from the induction hypothesis $\mathsf{fix}\, f : \forall \alpha < \lambda.\, A(\alpha)$. This imposes extra conditions on the shape of a so-called *admissible* A, which are the object of this article. Of course, a monotone A is trivially admissible, but many interesting types for recursive functions are not monotone, like $A(\alpha) = \mathsf{Nat}^{\alpha} \to \mathsf{Nat}^{\alpha} \to \mathsf{Nat}^{\alpha}$ (where Nat^{α} contains the natural numbers $< \alpha$). We will show that all types $A(\alpha)$ that are *upper semi-continuous* in α, meaning $\limsup_{\alpha \to \lambda} \mathcal{A}(\alpha) \subseteq \mathcal{A}(\lambda)$ for limit ordinals λ, are admissible. Function types $C(\alpha) = A(\alpha) \to B(\alpha)$ will be admissible if A is *lower semi-continuous* ($\mathcal{A}(\lambda) \subseteq \liminf_{\alpha \to \lambda} \mathcal{A}(\alpha)$) and B is upper semi-continuous. Similar laws will be developed for the other type constructors and put into the form of a kinding system for semi-continuous types.

Before we dive into the mathematics, let us make sure that semi-continuity is really necessary for termination. A type which is not upper semi-continuous is $A(\imath) = (\mathsf{Nat}^\omega \to \mathsf{Nat}^\imath) \to \mathsf{Nat}^\omega$ (see Sect. 4.2). Assuming we can nevertheless use this type for a recursive function, we can construct a loop. First, define successor $\mathsf{succ} : \forall\imath.\, \mathsf{Nat}^\imath \to \mathsf{Nat}^{\imath+1}$ and predecessor $\mathsf{pred} : \forall\imath.\, \mathsf{Nat}^{\imath+1} \to \mathsf{Nat}^\imath$. Note that the size index is an upper bound and ω is the biggest such bound for the case of natural numbers, thus, we have the subtype relations $\mathsf{Nat}^\imath \leq \mathsf{Nat}^{\imath+1} \leq \cdots \leq \mathsf{Nat}^\omega \leq \mathsf{Nat}^{\omega+1} \leq \mathsf{Nat}^\omega$.

We make the following definitions:

$$
\begin{array}{ll}
A(\imath) := (\mathsf{Nat}^\omega \to \mathsf{Nat}^\imath) \to \mathsf{Nat}^\omega & \begin{array}{lcl} f & : & \forall\imath.\, A(\imath) \to A(\imath+1) \\ f & := & \lambda loop \lambda g.\ loop\,(\mathsf{shift}\,g) \end{array} \\
\begin{array}{ll} \mathsf{shift} : & \forall\imath.\,(\mathsf{Nat}^\omega \to \mathsf{Nat}^{\imath+1}) \\ & \quad \to \mathsf{Nat}^\omega \to \mathsf{Nat}^\imath \\ \mathsf{shift} := & \lambda g \lambda n.\, \mathsf{pred}\,(g\,(\mathsf{succ}\,n)) \end{array} & \begin{array}{ll} \mathsf{loop} : & \forall\imath.\, A(\imath) \\ \mathsf{loop} := & \mathsf{fix}\, f \end{array}
\end{array}
$$

Since $\mathsf{Nat}^\omega \to \mathsf{Nat}^0$ is empty, A passes the bottom check. Still, instantiating types to $\mathsf{succ} : \mathsf{Nat}^\omega \to \mathsf{Nat}^\omega$ and $\mathsf{loop} : (\mathsf{Nat}^\omega \to \mathsf{Nat}^\omega) \to \mathsf{Nat}^\omega$ we convince ourselves that the execution of $\mathsf{loop}\ \mathsf{succ}$ indeed runs forever.

1.1 Related Work and Contribution

Ensuring termination through typing is quite an old idea, just think of type systems for the λ-calculus like simple types, System F, System F^ω, or the Calculus of Constructions, which all have the normalization property. These systems have been extended by special recursion operators, like primitive recursion in Gödel's T, or the recursors generated for inductive definitions in Type Theory (e. g., in Coq), that preserve normalization but limit the definition of recursive functions to special patterns, namely instantiations of the recursion scheme dictated by the recursion operator. Taming the general recursion operator fix through typing, however, which allows the definition of recursive functions in the intuitive way known from functional programming, is not yet fully explored. Mendler [15] pioneered this field; he used a certain polymorphic typing of the functional f to obtain primitive (co)recursive functions over arbitrary datatypes. Amadio and Coupet-Grimal [6] and Giménez [13] developed Mendler's approach further, until a presentation using ordinal-indexed (co)inductive types was found and proven sound by Barthe et al. [7]. The system $\lambda\hat{}$ presented in loc. cit. restricts types $A(\imath)$ of recursive functions to the shape $\mu^\imath F \to C(\imath)$ where the domain must be an inductive type $\mu^\imath F$ indexed by $\imath$ and the codomain a type $C(\imath)$ that is monotonic in $\imath$. This criterion, which has also been described by the author [2], allows for a simple soundness proof in the limit case of the transfinite induction, but excludes interesting types like the considered

$$\mathsf{Eq}\,(\mathsf{GRose}^\imath\, F A) = \mathsf{GRose}^\imath\, F A \to \mathsf{GRose}^\imath\, F A \to \mathsf{Bool}$$

which has an antitonic codomain $C(\imath) = \mathsf{GRose}^\imath\, F A \to \mathsf{Bool}$. The author has in previous work widened the criterion, but only for a type system without polymorphism [1]. Other recent works on type-based termination [9,10,8] stick to the restriction of $\lambda\widehat{}$. Xi [19] uses dependent types and lexicographic measures to ensure termination of recursive programs in a call-by-value language, but his indices are natural numbers instead of ordinals which excludes infinite objects we are interested in.

Closest to the present work is the sized type system of Hughes, Pareto, and Sabry [14], *Synchronous Haskell* [16], which admits ordinal indices up to ω. Index quantifiers as in $\forall\imath.\, A(\imath)$ range over natural numbers, but can be instantiated to ω if $A(\imath)$ is *ω-undershooting*. Sound semantical criteria for ω-undershooting types are already present, but in a rather ad-hoc manner. We cast these criteria in the established mathematical framework of semi-continuous functions and provide a syntactical implementation in form of a derivation system. Furthermore, we also allow ordinals up to the ωth uncountable and infinitely branching inductive types that invalidate some criteria for the only finitely branching tree types in *Synchronous Haskell*. Finally, we allow polymorphic recursion, impredicative polymorphism and higher-kinded inductive and coinductive types such as GRose. This article summarizes the main results of the author's dissertation [4].

2 Overview of System $\mathsf{F}_\omega\widehat{}$

In this section we introduce $\mathsf{F}_\omega\widehat{}$, an *a posteriori* strongly normalizing extension of System F^ω with higher-kinded inductive and coinductive types and (co)recursion combinators. Figure 1 summarizes the syntactic entities. Function kinds are equipped with polarities p [18], which are written before the domain or on top of the arrow. Polarity $+$ denotes covariant constructors, $-$ contravariant constructors and $\circ$ mixed-variant constructors [12]. It is well-known that in order to obtain a normalizing language, any constructor underlying an inductive type must be covariant [15], hence, we restrict formation of least fixed-points $\mu^a_\kappa F$ to covariant Fs. (Abel [3] and Matthes [5] provide more explanation on polarities.)

The first argument, a, to μ, which we usually write as superscript, denotes the upper bound for the height of elements in the inductive type. The index a is a constructor of kind ord and denotes an ordinal; the canonical inhabitants of ord are given by the grammar

$$a ::= \imath \mid \mathsf{s}\, a \mid \infty$$

with $\imath$ an ordinal variable. If a actually denotes a finite ordinal (a natural number), then the height is simply the number of data constructors on the longest path in the tree structure of any element of $\mu^a F$. Since a is only an upper bound, $\mu^a F$ is a subtype of $\mu^b F$, written $\mu^a F \leq \mu^b F$ for $a \leq b$, meaning that μ is covariant in the index argument. Finally, $F \leq F'$ implies $\mu^a F \leq \mu^a F'$, so we get the kinding

$$\mu_\kappa : \mathsf{ord} \xrightarrow{+} (\kappa \xrightarrow{+} \kappa) \xrightarrow{+} \kappa$$

for the least fixed-point constructor. The kind κ is required to be *pure*, i.e., a kind not mentioning ord, for cardinality reasons. Only then it is possible to

Polarities, kinds, constructors, kinding contexts.

p	$::= + \mid - \mid \circ$	polarity
κ	$::= * \mid \mathsf{ord} \mid p\kappa \to \kappa'$	kind
κ_*	$::= * \mid p\kappa_* \to \kappa'_*$	pure kind
a, b, A, B, F, G	$::= C \mid X \mid \lambda X\!:\!\kappa.\, F \mid F\, G$	(type) constructor
C	$::= 1 \mid + \mid \times \mid \to \mid \forall_\kappa \mid \mu_{\kappa_*} \mid \nu_{\kappa_*} \mid \mathsf{s} \mid \infty$	constructor constants
Δ	$::= \diamond \mid \Delta, X\!:\!p\kappa$	kinding context

Constructor constants and their kinds ($\kappa \xrightarrow{p} \kappa'$ means $p\kappa \to \kappa'$).

1	$: *$	unit type
$+$	$: * \xrightarrow{+} * \xrightarrow{+} *$	disjoint sum
$\times$	$: * \xrightarrow{+} * \xrightarrow{+} *$	cartesian product
$\to$	$: * \xrightarrow{-} * \xrightarrow{+} *$	function space
$\forall_\kappa$	$: (\kappa \xrightarrow{\circ} *) \xrightarrow{+} *$	quantification
μ_{κ_*}	$: \mathsf{ord} \xrightarrow{+} (\kappa_* \xrightarrow{+} \kappa_*) \xrightarrow{+} \kappa_*$	inductive constructors
ν_{κ_*}	$: \mathsf{ord} \xrightarrow{-} (\kappa_* \xrightarrow{+} \kappa_*) \xrightarrow{+} \kappa_*$	coinductive constructors
s	$: \mathsf{ord} \xrightarrow{+} \mathsf{ord}$	successor of ordinal
∞	$: \mathsf{ord}$	infinity ordinal

Objects (terms), values, evaluation frames, typing contexts.

r, s, t	$::= c \mid x \mid \lambda x t \mid r\, s$	term
c	$::= () \mid \mathsf{pair} \mid \mathsf{fst} \mid \mathsf{snd} \mid \mathsf{inl} \mid \mathsf{inr} \mid \mathsf{case} \mid \mathsf{in} \mid \mathsf{out} \mid \mathsf{fix}^\mu_n \mid \mathsf{fix}^\nu_n$	constant ($n \in \mathbb{N}$)
v	$::= \lambda x t \mid \mathsf{pair}\, t_1\, t_2 \mid \mathsf{inl}\, t \mid \mathsf{inr}\, t \mid \mathsf{in}\, t \mid c \mid \mathsf{pair}\, t \mid \mathsf{fix}^\nabla_n s\, t_{1..m}$	value ($m \leq n$)
$e(\_)$	$::= \_\, s \mid \mathsf{fst}\, \_ \mid \mathsf{snd}\, \_ \mid \mathsf{case}\, \_ \mid \mathsf{out}\, \_ \mid \mathsf{fix}^\mu_n\, s\, t_{1..n}\, \_$	evaluation frame
E	$::= \mathsf{Id} \mid E \circ e$	evaluation context
Γ	$::= \diamond \mid \Gamma, x\!:\!A \mid \Gamma, X\!:\!p\kappa$	typing context

Reduction $t \longrightarrow t'$.

$(\lambda x t)\, s$	$\longrightarrow [s/x]t$	$\mathsf{out}\,(\mathsf{in}\, r)$	$\longrightarrow r$
$\mathsf{fst}\,(r, s)$	$\longrightarrow r$	$\mathsf{fix}^\mu_n\, s\, t_{1..n}\,(\mathsf{in}\, t)$	$\longrightarrow s\,(\mathsf{fix}^\mu_n\, s)\, t_{1..n}\,(\mathsf{in}\, t)$
$\mathsf{snd}\,(r, s)$	$\longrightarrow s$	$\mathsf{out}\,(\mathsf{fix}^\nu_n\, s\, t_{1..n})$	$\longrightarrow \mathsf{out}\,(s\,(\mathsf{fix}^\nu_n\, s)\, t_{1..n})$
$\mathsf{case}\,(\mathsf{inl}\, r)$	$\longrightarrow \lambda x \lambda y.\, x\, r$		
$\mathsf{case}\,(\mathsf{inr}\, r)$	$\longrightarrow \lambda x \lambda y.\, y\, r$	+ closure under all term constructs	

Fig. 1. $\mathsf{F}^{\widehat{}}_\omega$: Syntax and operational semantics

estimate a single *closure ordinal* ∞ at which the fixed-point is reached for *all* inductive types. We have

$$\mu^\infty F = \mu^{\infty+1} F,$$

where $\infty+1$ is a shorthand for $\mathsf{s}\infty$, $\mathsf{s} : \mathsf{ord} \xrightarrow{+} \mathsf{ord}$ being the successor on ordinals. If ord was allowed in the kind of a fixed-point, the closure ordinal of this fixed-point would depend on which ordinals are in the semantics of ord, which in turn would depend on what the closure ordinal for all fixed-points was—a vicious cycle. However, I do not see a practical example where one want to construct the fixed point of a sized-type transformer $F : (\mathsf{ord} \xrightarrow{\circ} \kappa) \xrightarrow{+} (\mathsf{ord} \xrightarrow{\circ} \kappa)$. Note that this does not exclude fixed-points inside fixed-points, such as

$$\mathsf{BTree}^{\imath,\jmath}\, A = \mu^{\imath}\lambda X.\, 1 + X \times (\mu^{\jmath}\lambda Y. 1 + A \times X \times Y),$$

"B-trees" of height $< \imath$ with each node containing $< \jmath$ keys of type A.

Because ∞ is the closure ordinal, the equation $\mathsf{s}\,\infty = \infty$ makes sense. Equality on type constructors is defined as the least congruent equivalence relation closed under this equation and $\beta\eta$.

Example 1 (Some sized types).

$$\begin{array}{ll} \mathsf{Nat} & : \mathsf{ord} \xrightarrow{+} * \\ \mathsf{Nat} & := \lambda\imath.\, \mu^{\imath}\lambda X. 1 + X \end{array} \qquad \begin{array}{ll} \mathsf{GRose} : & \mathsf{ord} \xrightarrow{+} (* \xrightarrow{+} *) \xrightarrow{+} * \xrightarrow{+} * \\ \mathsf{GRose} := & \lambda\imath\lambda F\lambda A.\, \mu^{\imath}\lambda X. 1 + A \times F\, X \end{array}$$

$$\begin{array}{ll} \mathsf{List} & : \mathsf{ord} \xrightarrow{+} * \xrightarrow{+} * \\ \mathsf{List} & := \lambda\imath\lambda A.\, \mu^{\imath}\lambda X. 1 + A \times X \end{array} \qquad \begin{array}{ll} \mathsf{Tree} & : \mathsf{ord} \xrightarrow{+} * \xrightarrow{-} * \xrightarrow{+} * \\ \mathsf{Tree} & := \lambda\imath\lambda B\lambda A.\, \mathsf{GRose}^{\imath}\, (\lambda X. B \to X)\, A \end{array}$$

$$\begin{array}{ll} \mathsf{Stream} : & \mathsf{ord} \xrightarrow{-} * \xrightarrow{+} * \\ \mathsf{Stream} := & \lambda\imath\lambda A.\, \nu^{\imath}\lambda X. A \times X \end{array}$$

The term language of $\mathsf{F}_\omega^{\widehat{}}$ is the λ-calculus plus the standard constants to introduce and eliminate unit (1), sum (+), and product ($\times$) types. Further, there is folding, in, and unfolding, out, of (co)inductive types. Let $\kappa = \boldsymbol{p\kappa} \to *$ a pure kind, $F : +\kappa \to \kappa$, $G_i : \kappa_i$ for $1 \leq i \leq |\boldsymbol{\kappa}|$, $a : \mathsf{ord}$, and $\nabla \in \{\mu, \nu\}$, then we have the following (un)folding rules:

$$\text{TY-FOLD}\ \frac{\Gamma \vdash t : F\, (\nabla^a_\kappa\, F)\, \boldsymbol{G}}{\Gamma \vdash \mathsf{in}\, t : \nabla^{a+1}_\kappa F\, \boldsymbol{G}} \qquad \text{TY-UNFOLD}\ \frac{\Gamma \vdash r : \nabla^{a+1}_\kappa F\, \boldsymbol{G}}{\Gamma \vdash \mathsf{out}\, r : F\, (\nabla^a_\kappa\, F)\, \boldsymbol{G}}$$

Finally, there are fixed-point combinators fix^μ_n and fix^ν_n for each $n \in \mathbb{N}$ on the term level. The term $\mathsf{fix}^\mu_n\, s$ denotes a recursive function with n leading non-recursive arguments; the $n+1$st argument must be of an inductive type. Similarly, $\mathsf{fix}^\nu_n\, s$ is a corecursive function which takes n arguments and produces an inhabitant of a coinductive type.

One-step reduction $t \longrightarrow t'$ is defined by the β-reduction axioms given in Figure 1 plus congruence rules. Interesting are the reduction rules for recursion and corecursion:

$$\begin{array}{rcl} \mathsf{fix}^\mu_n\, s\, t_{1..n}\, (\mathsf{in}\, t) & \longrightarrow & s\, (\mathsf{fix}^\mu_n\, s)\, t_{1..n}\, (\mathsf{in}\, t) \\ \mathsf{out}\, (\mathsf{fix}^\nu_n\, s\, t_{1..n}) & \longrightarrow & \mathsf{out}\, (s\, (\mathsf{fix}^\nu_n\, s)\, t_{1..n}) \end{array}$$

A recursive function is only unfolded if its recursive argument is a value, i. e., of the form $\mathsf{in}\, t$. This condition is required to ensure strong normalization; it is present in the work of Mendler [15], Giménez [13], Barthe et al. [7], and the author [2]. Dually, corecursive functions are only unfolded on demand, i. e., in an evaluation context, the matching one being $\mathsf{out}\ \_$.

As pointed out in the introduction, recursion is introduced by the rule

$$\text{TY-REC}\ \frac{\Gamma \vdash A\ \mathsf{fix}^{\nabla}_{n}\text{-}\mathsf{adm} \qquad \Gamma \vdash a : \mathsf{ord}}{\Gamma \vdash \mathsf{fix}^{\nabla}_{n} : (\forall \imath{:}\mathsf{ord}.\, A\,\imath \to A\,(\imath+1)) \to A\,a}.$$

Herein, ∇ stands for μ or ν, and the judgement $A\ \mathsf{fix}^{\nabla}_{n}\text{-}\mathsf{adm}$ makes sure type A is admissible for (co)recursion, as discussed in the introduction. In the following, we will find out which types are admissible.

3 Semantics

In this section, we provide an interpretation of types as saturated sets of strongly normalizing terms. Let $\mathcal{S}$ denote the set of strongly normalizing terms. We define *safe* (weak head) reduction by these axioms:

$$\begin{array}{llllll}
(\lambda x t)\, s & \rhd [s/x]t & \text{if } s \in \mathcal{S} & \mathsf{case}\,(\mathsf{inl}\, r) & \rhd & \lambda x \lambda y.\, x\, r \\
\mathsf{fst}\,(\mathsf{pair}\, r\, s) & \rhd r & \text{if } s \in \mathcal{S} & \mathsf{case}\,(\mathsf{inr}\, r) & \rhd & \lambda x \lambda y.\, y\, r \\
\mathsf{snd}\,(\mathsf{pair}\, r\, s) & \rhd s & \text{if } r \in \mathcal{S} & \mathsf{fix}^{\mu}_{n} s\, t_{1..n}\,(\mathsf{in}\, r) & \rhd & s\,(\mathsf{fix}^{\mu}_{n} s)\, t_{1..n}\,(\mathsf{in}\, r) \\
\mathsf{out}\,(\mathsf{in}\, r) & \rhd r & & \mathsf{out}\,(\mathsf{fix}^{\nu}_{n} s\, t_{1..n}) & \rhd & \mathsf{out}\,(s\,(\mathsf{fix}^{\nu}_{n} s)\, t_{1..n})
\end{array}$$

Additionally, we close safe reduction under evaluation contexts and transitivity:

$$\begin{array}{lll}
E(t) & \rhd E(t') & \text{if } t \rhd t' \\
t_1 & \rhd t_3 & \text{if } t_1 \rhd t_2 \text{ and } t_2 \rhd t_3
\end{array}$$

The relation is defined such that $\mathcal{S}$ is closed under $\rhd$-expansion, meaning $t \rhd t' \in \mathcal{S}$ implies $t \in \mathcal{S}$. Let ${}^{\rhd}\!\mathcal{A}$ denote the closure of term set $\mathcal{A}$ under $\rhd$-expansion. In general, the *closure* of term set $\mathcal{A}$ is defined as

$$\overline{\mathcal{A}} = {}^{\rhd}(\mathcal{A} \cup \{E(x) \mid x \text{ variable}, E(x) \in \mathcal{S}\}).$$

A term set is *closed* if $\overline{\mathcal{A}} = \mathcal{A}$. The least closed set is the set of neutral terms $\mathcal{N} := \overline{\emptyset} \neq \emptyset$. Intuitively, a neutral term never reduces to a value, it necessarily has a free variable, and it can be substituted into any terms without creating a new redex. A term set $\mathcal{A}$ is *saturated* if $\mathcal{A}$ is closed and $\mathcal{N} \subseteq \mathcal{A} \subseteq \mathcal{S}$.

Interpretation of kinds. The saturated sets form a complete lattice $[\![*]\!]$ with least element $\bot^* := \mathcal{N}$ and greatest element $\top^* := \mathcal{S}$. It is ordered by inclusion $\sqsubseteq^* := \subseteq$ and has set-theoretic infimum $\inf^* := \bigcap$ and supremum $\sup^* := \bigcup$. Let $[\![\mathsf{ord}]\!] := \mathsf{O}$ where $\mathsf{O} = [0; \top^{\mathsf{ord}}]$ is an initial segment of the set-theoretic ordinals. With the usual ordering on ordinals, O constitutes a complete lattice as well. Function kinds $[\![\circ\kappa \to \kappa']\!] := [\![\kappa]\!] \to [\![\kappa']\!]$ are interpreted as set-theoretic function spaces; a covariant function kind denotes just the monotonic functions and a contravariant kind the antitonic ones. For all function kinds, ordering is defined pointwise: $\mathcal{F} \sqsubseteq^{p\kappa\to\kappa'} \mathcal{F}' :\iff \mathcal{F}(\mathcal{G}) \sqsubseteq^{\kappa'} \mathcal{F}'(\mathcal{G})$ for all $\mathcal{G} \in [\![\kappa]\!]$. Similarly, $\bot^{p\kappa\to\kappa'}(\mathcal{G}) := \bot^{\kappa'}$ is defined pointwise, and so are $\top^{p\kappa\to\kappa'}$, $\inf^{p\kappa\to\kappa'}$, and $\sup^{p\kappa\to\kappa'}$.

Limits and iteration. In the following $\lambda \in \mathsf{O}$ will denote a limit ordinal. (We will only consider proper limits, i. e., $\lambda \neq 0$.) For $\mathfrak{L}$ a complete lattice and $f \in \mathsf{O} \to \mathfrak{L}$ we define:

$$\begin{array}{l} \liminf_{\alpha\to\lambda} f(\alpha) := \sup_{\alpha_0<\lambda} \inf_{\alpha_0\le\alpha<\lambda} f(\alpha) \\ \limsup_{\alpha\to\lambda} f(\alpha) := \inf_{\alpha_0<\lambda} \sup_{\alpha_0\le\alpha<\lambda} f(\alpha) \end{array}$$

Using $\inf_\lambda f$ as shorthand for $\inf_{\alpha<\lambda} f(\alpha)$, and analogous shorthands for sup, lim inf, and lim sup, we have $\inf_\lambda f \sqsubseteq \liminf_\lambda f \sqsubseteq \limsup_\lambda f \sqsubseteq \sup_\lambda f$. If f is monotone, then even $\liminf_\lambda f = \sup_\lambda f$, and if f is antitone, then $\inf_\lambda f = \limsup_\lambda f$.

If $f \in \mathfrak{L} \to \mathfrak{L}$ and $g \in \mathfrak{L}$, we define transfinite iteration $f^\alpha(g)$ by recursion on α as follows:

$$\begin{array}{l} f^0 \quad (g) := g \\ f^{\alpha+1}(g) := f(f^\alpha(g)) \\ f^\lambda \quad (g) := \limsup_{\alpha\to\lambda} f^\alpha(g) \end{array}$$

For monotone f, we obtain the usual approximants of least and greatest fixed-points as $\boldsymbol{\mu}^\alpha f = f^\alpha(\bot)$ and $\boldsymbol{\nu}^\alpha f = f^\alpha(\top)$.

Closure ordinal. Let $\beth_n$ be a sequence of cardinals defined by $\beth_0 = |\mathbb{N}|$ and $\beth_{n+1} = |\mathcal{P}(\beth_n)|$. For a pure kind κ, let $|\kappa|$ be the number of $*$s in κ. Since $[\![*]\!]$ consists of countable sets, $|[\![*]\!]| \le |\mathcal{P}(\mathbb{N})| = \beth_1$, and by induction on κ, $|[\![\kappa]\!]| \le \beth_{|\kappa|+1}$. Since an (ascending or descending) chain in $[\![\kappa]\!]$ is shorter than $|[\![\kappa]\!]|$, each fixed point is reached latest at the $|[\![\kappa]\!]|$th iteration. Hence, the closure ordinal for all (co)inductive types can be approximated from above by $\top^{\mathsf{ord}} = \beth_\omega$.

Interpretation of types. For r a term, e an evaluation frame, and $\mathcal{A}$ a term set, let $r \cdot \mathcal{A} = \{r\, s \mid s \in \mathcal{A}\}$ and $e^{-1}\mathcal{A} = \{r \mid e(r) \in \mathcal{A}\}$. For saturated sets $\mathcal{A}, \mathcal{B} \in [\![*]\!]$ we define the following saturated sets:

$$\begin{array}{ll} \mathcal{A} \boxplus \mathcal{B} := \overline{\mathsf{inl} \cdot \mathcal{A}} \cup \overline{\mathsf{inr} \cdot \mathcal{B}} & \boxed{1} := \overline{\{()\}} \\ \mathcal{A} \boxtimes \mathcal{B} := (\mathsf{fst}\ \_)^{-1}\mathcal{A} \cap (\mathsf{snd}\ \_)^{-1}\mathcal{B} & \mathcal{A}^\mu := \overline{\mathsf{in} \cdot \mathcal{A}} \\ \mathcal{A} \boxed{\to} \mathcal{B} := \bigcap_{s\in\mathcal{A}} (\_\ s)^{-1}\mathcal{B} & \mathcal{A}^\nu := (\mathsf{out}\ \_)^{-1}\mathcal{A} \end{array}$$

The last two notations are lifted pointwise to operators $\mathcal{F} \in [\![p\kappa \to \kappa']\!]$ by setting $\mathcal{F}^\nabla(\mathcal{G}) = (\mathcal{F}(\mathcal{G}))^\nabla$, where $\nabla \in \{\mu, \nu\}$.

For a constructor constant $C : \kappa$, the semantics $[\![C]\!] \in [\![\kappa]\!]$ is defined as follows:

$$\begin{array}{llll} [\![+]\!](\mathcal{A}, \mathcal{B} \in [\![*]\!]) & := \mathcal{A} \boxplus \mathcal{B} & [\![1]\!] & := \boxed{1} \\ [\![\times]\!](\mathcal{A}, \mathcal{B} \in [\![*]\!]) & := \mathcal{A} \boxtimes \mathcal{B} & [\![\infty]\!] & := \top^{\mathsf{ord}} \\ [\![\to]\!](\mathcal{A}, \mathcal{B} \in [\![*]\!]) & := \mathcal{A} \boxed{\to} \mathcal{B} & [\![\mathsf{s}]\!](\top^{\mathsf{ord}}) & := \top^{\mathsf{ord}} \\ [\![\mu_\kappa]\!](\alpha)(\mathcal{F} \in [\![\kappa]\!] \xrightarrow{+} [\![\kappa]\!]) & := \boldsymbol{\mu}^\alpha \mathcal{F}^\mu & [\![\mathsf{s}]\!](\alpha < \top^{\mathsf{ord}}) & := \alpha + 1 \\ [\![\nu_\kappa]\!](\alpha)(\mathcal{F} \in [\![\kappa]\!] \xrightarrow{+} [\![\kappa]\!]) & := \boldsymbol{\nu}^\alpha \mathcal{F}^\nu & & \\ [\![\forall_\kappa]\!](\mathcal{F} \in [\![\kappa]\!] \to [\![*]\!]) & := \bigcap_{\mathcal{G}\in[\![\kappa]\!]} \mathcal{F}(\mathcal{G}) & & \end{array}$$

We extend this semantics to constructors F in the usual way, such that if $\Delta \vdash F : \kappa$ and $\theta(X) \in [\![\kappa']\!]$ for all $(X : p\kappa') \in \Delta$, then $[\![F]\!]_\theta \in [\![\kappa]\!]$.

Now we can compute the semantics of types, e. g., $[\![\mathsf{Nat}^\imath]\!]_{(\imath \mapsto \alpha)} = \mathcal{N}at^\alpha = \mu^\alpha(\mathcal{X} \mapsto (\boxed{1} \boxplus \mathcal{X})^\mu)$. Similarly, the semantical versions of List, Stream, etc. are denoted by $\mathcal{L}ist$, $\mathcal{S}tream$, etc.

Semantic admissibility and strong normalization. For the main theorem to follow, we assume semantical soundness of our yet to be defined syntactical criterion of admissibility: If $\Gamma \vdash A\ \mathsf{fix}_n^\nabla\text{-}\mathsf{adm}$ and $\theta(X) \in [\![\kappa]\!]$ for all $(X : \kappa) \in [\![\Gamma]\!]$ then $\mathcal{A} := [\![A]\!]_\theta \in [\![\mathsf{ord}]\!] \to [\![*]\!]$ has the following properties:

1. Shape: $\mathcal{A}(\alpha) = \bigcap_{k \in K} \mathcal{B}_1(k,\alpha) \boxed{\to} \dots \boxed{\to} \mathcal{B}_n(k,\alpha) \boxed{\to} \mathcal{B}(k,\alpha)$ for some K and some $\mathcal{B}_1, \dots, \mathcal{B}_n, \mathcal{B} \in K \times [\![\mathsf{ord}]\!] \to [\![*]\!]$. In case $\nabla = \mu$, $\mathcal{B}(k,\alpha) = \mathcal{I}(k,\alpha)^\mu \boxed{\to} \mathcal{C}(k,\alpha)$ for some $\mathcal{I}, \mathcal{C}$. Otherwise, $\mathcal{B}(k,\alpha) = \mathcal{C}(k,\alpha)^\nu$ for some $\mathcal{C}$.
2. Bottom-check: $\mathcal{I}(k,0)^\mu = \bot^*$ in case $\nabla = \mu$ and $\mathcal{C}(k,0)^\nu = \top^*$ in case $\nabla = \nu$.
3. Semi-continuity: $\limsup_{\alpha \to \lambda} \mathcal{A}(\alpha) \subseteq \mathcal{A}(\lambda)$ for all limit ordinals $\lambda \in [\![\mathsf{ord}]\!] \setminus \{0\}$.

Let $t\theta$ denote the simultaneous substitution of $\theta(x)$ for each $x \in \mathsf{FV}(t)$ in t.

Theorem 1 (Type soundness). *Let $\theta(X) \in [\![\kappa]\!]$ for all $(X : \kappa) \in \Gamma$ and $\theta(x) \in [\![A]\!]\theta$ for all $(x : A) \in \Gamma$. If $\Gamma \vdash t : B$ then $t\theta \in [\![B]\!]\theta$.*

Corollary 1 (Strong normalization). *If $\Gamma \vdash t : B$ then t is strongly normalizing.*

4 Semi-continuity

As motivated in the introduction, only types $\mathcal{C} \in [\![\mathsf{ord}]\!] \to [\![*]\!]$ with $\inf_\lambda \mathcal{C} \sqsubseteq \mathcal{C}(\lambda)$ can be admissible for recursion. Under which conditions on $\mathcal{A}$ and $\mathcal{B}$ can a function type $\mathcal{A}(\alpha) \boxed{\to} \mathcal{B}(\alpha)$ be admissible? It shows that the first choice $\inf_\lambda \mathcal{B} \sqsubseteq \mathcal{B}(\lambda)$ is a requirement too strong: To show $\inf_{\alpha<\lambda}(\mathcal{A}(\alpha) \boxed{\to} \mathcal{B}(\alpha)) \sqsubseteq \mathcal{A}(\lambda) \boxed{\to} \mathcal{B}(\lambda)$ we would need $\mathcal{A}(\lambda) \sqsubseteq \inf_\lambda \mathcal{A}$, which is not even true for $\mathcal{A} = \mathcal{N}at$ at limit ω. However, each type $\mathcal{C}$ with $\limsup_\lambda \mathcal{C} \sqsubseteq \mathcal{C}(\lambda)$ also fulfills $\inf_\lambda \mathcal{C} \sqsubseteq \mathcal{C}(\lambda)$, and the modified condition distributes better over function spaces.

Lemma 1. *If $\mathcal{A}(\lambda) \sqsubseteq \liminf_\lambda \mathcal{A}$ and $\limsup_\lambda \mathcal{B} \sqsubseteq \mathcal{B}(\lambda)$ then $\limsup_\lambda(\mathcal{A}(\alpha) \boxed{\to} \mathcal{B}(\alpha)) \sqsubseteq \mathcal{A}(\lambda) \boxed{\to} \mathcal{B}(\lambda)$.*

The conditions on $\mathcal{A}$ and $\mathcal{B}$ in the lemma are established mathematical terms: They are subconcepts of continuity. In this article, we consider only functions $f \in \mathsf{O} \to \mathfrak{L}$ from ordinals into some lattice $\mathfrak{L}$. For such f, the question whether f is continuous in point α only makes sense if α is a limit ordinal, because only then there are infinite non-stationary sequences which converge to α; and since every strictly decreasing sequence is finite on ordinals (well-foundedness!), it only makes sense to look at *ascending* sequences, i. e., approaching the limit from the left. Hence, function f is *upper semi-continuous* in λ, if $\limsup_\lambda f \sqsubseteq f(\lambda)$, and *lower semi-continuous*, if $f(\lambda) \sqsubseteq \liminf_\lambda f$. If f is both upper and lower semi-continuous in λ, then it is continuous in λ (then upper and lower limit coincide with $f(\lambda)$).

4.1 Positive Results

Basic semi-continuous types. Obviously, any monotone function is upper semi-continuous, and any antitone function is lower semi-continuous. Now consider a monotone f with $f(\lambda) = \sup_\lambda f$, as it is the case for an inductive type $f(\alpha) = \boldsymbol{\mu}^\alpha \mathcal{F}$ (where $\mathcal{F}$ does not depend on α). Since for monotone f, $\sup_\lambda f = \liminf_\lambda f$, f is lower semi-continuous. This criterion can be used to show upper semi-continuity of function types such as $\mathsf{Eq}(\mathsf{GRose}^\imath\, F A)$ (see introduction) and, e. g.,

$$\mathcal{C}(\alpha) = \mathcal{N}at^\alpha \boxed{\rightarrow} \mathcal{L}ist^\alpha(\mathcal{A}) \boxed{\rightarrow} \mathcal{C}'(\alpha)$$

where $\mathcal{C}'(\alpha)$ is any monotonic type-valued function, for instance, $\mathcal{L}ist^\alpha(\mathcal{N}at^\alpha)$, and $\mathcal{A}$ is some constant type: The domain types, $\mathcal{N}at^\alpha$ and $\mathcal{L}ist^\alpha(\mathcal{A})$, are lower semi-continuous according the just established criterion and the monotonic co-domain $\mathcal{C}'(\alpha)$ is upper semi-continuous, hence, Lemma 1 proves upper semi-continuity of $\mathcal{C}$. Note that this criterion fails us if we replace the domain $\mathcal{L}ist^\alpha(\mathcal{A})$ by $\mathcal{L}ist^\alpha(\mathcal{N}at^\alpha)$, or even $\boldsymbol{\mu}^\alpha(\mathcal{F}(\mathcal{N}at^\alpha))$ for some monotone $\mathcal{F}$, since it is not immediately obvious that

$$\boldsymbol{\mu}^\omega(\mathcal{F}(\mathcal{N}at^\omega)) = \sup_{\alpha<\omega} \boldsymbol{\mu}^\alpha(\mathcal{F}(\sup_{\beta<\omega} \mathcal{N}at^\beta)) \stackrel{?}{=} \sup_{\gamma<\omega} \boldsymbol{\mu}^\gamma(\mathcal{F}(\mathcal{N}at^\gamma)).$$

However, domain types where one indexed inductive type is inside another inductive type are useful in practice, see Example 3. Before we consider lower semi-continuity of such types, let us consider the dual case.

For $f(\alpha) = \boldsymbol{\nu}^\alpha \mathcal{F}$, $\mathcal{F}$ not dependent on α, f is antitone and $f(\lambda) = \inf_\lambda f$. An antitone f guarantees $\inf_\lambda f = \limsup_\lambda f$, so f is upper semi-continuous. This establishes upper semi-continuity of a type involved in stream-zipping,

$$\mathcal{S}tream^\alpha(\mathcal{A}) \boxed{\rightarrow} \mathcal{S}tream^\alpha(\mathcal{B}) \boxed{\rightarrow} \mathcal{S}tream^\alpha(\mathcal{C}).$$

The domain types are antitonic, hence lower semi-continuous, and the coinductive codomain is upper semi-continuous. Upper semi-continuity of $\mathcal{S}tream^\alpha(\mathcal{N}at^\alpha)$ and similar types is not yet covered, but now we will develop concepts that allow us to look inside (co)inductive types.

Semi-continuity and (co)induction. Let $f \in \mathfrak{L} \to \mathfrak{L}'$. We say lim sup *pushes through* f, or f is lim sup-*pushable*, if for all $g \in \mathsf{O} \to \mathfrak{L}$, $\limsup_{\alpha\to\lambda} f(g(\alpha)) \sqsubseteq f(\limsup_\lambda g)$. Analogously, f is lim inf-*pullable*, or lim inf *can be pulled out of* f, if for all g, $f(\liminf_\lambda g) \sqsubseteq \liminf_{\alpha\to\lambda} f(g(\alpha))$. These notions extend straightforwardly to fs with several arguments.

Lemma 2 (Facts about limits).

1. $\limsup_{\alpha\to\lambda} f(\alpha,\alpha) \sqsubseteq \limsup_{\beta\to\lambda} \limsup_{\gamma\to\lambda} f(\beta,\gamma)$.
2. $\liminf_{\beta\to\lambda} \liminf_{\gamma\to\lambda} f(\beta,\gamma) \sqsubseteq \liminf_{\alpha\to\lambda} f(\alpha,\alpha)$.
3. $\limsup_{\alpha\to\lambda} \inf_{i\in I} f(\alpha,i) \sqsubseteq \inf_{i\in I} \limsup_{\alpha\to\lambda} f(\alpha,i)$.
4. $\sup_{i\in I} \liminf_{\alpha\to\lambda} f(\alpha,i) \sqsubseteq \liminf_{\alpha\to\lambda} \sup_{i\in I} f(\alpha,i)$.

Strictly positive contexts: $\Pi ::= \diamond \mid \Pi, X : +\kappa_*$.

Semi-continuity $\Delta; \Pi \vdash^{\imath q} F : \kappa$ for $q \in \{\oplus, \ominus\}$.

$$\text{CONT-CO}\ \frac{\Delta, \imath : +\mathsf{ord} \vdash F : \kappa \quad p \in \{+, \circ\}}{\Delta, \imath : p\mathsf{ord}; \Pi \vdash^{\imath\oplus} F : \kappa} \qquad \text{CONT-C'TRA}\ \frac{\Delta, \imath : -\mathsf{ord} \vdash F : \kappa \quad p \in \{-, \circ\}}{\Delta, \imath : p\mathsf{ord}; \Pi \vdash^{\imath\ominus} F : \kappa}$$

$$\text{CONT-IN}\ \frac{\Delta \vdash F : \kappa}{\Delta, \imath : p\mathsf{ord}; \Pi \vdash^{\imath q} F : \kappa} \qquad \text{CONT-VAR}\ \frac{X : p\kappa \in \Delta, \Pi \qquad p \in \{+, \circ\}}{\Delta; \Pi \vdash^{\imath q} X : \kappa}$$

$$\text{CONT-}\forall\ \frac{\Delta; \Pi \vdash^{\imath\oplus} F : \circ\kappa \to *}{\Delta; \Pi \vdash^{\imath\oplus} \forall_\kappa F : *} \qquad \text{CONT-ABS}\ \frac{\Delta, X : p\kappa; \Pi \vdash^{\imath q} F : \kappa'}{\Delta; \Pi \vdash^{\imath q} \lambda X F : p\kappa \to \kappa'}\ X \neq \imath$$

$$\text{CONT-APP}\ \frac{\Delta, \imath : p'\mathsf{ord}; \Pi \vdash^{\imath q} F : p\kappa \to \kappa' \qquad p^{-1}\Delta \vdash G : \kappa}{\Delta, \imath : p'\mathsf{ord}; \Pi \vdash^{\imath q} F\, G : \kappa'}$$

$$\text{CONT-SUM}\ \frac{\Delta; \Pi \vdash^{\imath q} A, B : *}{\Delta; \Pi \vdash^{\imath q} A + B : *} \qquad \text{CONT-PROD}\ \frac{\Delta; \Pi \vdash^{\imath q} A, B : *}{\Delta; \Pi \vdash^{\imath q} A \times B : *}$$

$$\text{CONT-ARR}\ \frac{-\Delta; \diamond \vdash^{\imath\ominus} A : * \qquad \Delta; \Pi \vdash^{\imath\oplus} B : *}{\Delta; \Pi \vdash^{\imath\oplus} A \to B : *}$$

$$\text{CONT-MU}\ \frac{\Delta; \Pi, X : +\kappa_* \vdash^{\imath\ominus} F : \kappa_* \qquad \Delta \vdash^{\imath\ominus} a : \mathsf{ord}}{\Delta; \Pi \vdash^{\imath\ominus} \mu^a_{\kappa_*} \lambda X F : \kappa_*}$$

$$\text{CONT-NU}\ \frac{\Delta; \Pi, X : +\kappa_* \vdash^{\imath\oplus} F : \kappa_* \qquad a \in \{\infty, \mathsf{s}^n \jmath \mid (\jmath : p\mathsf{ord}) \in \Delta \text{ with } p \in \{+, \circ\}\}}{\Delta; \Pi \vdash^{\imath\oplus} \nu^a_{\kappa_*} \lambda X F : \kappa_*}$$

Fig. 2. $\mathsf{F}_{\widehat{\omega}}$: Semi-continuous constructors

Fact 3 states that lim sup pushes through infimum and, thus, justifies rule CONT-∀ in Fig. 2 (see Sect. 5). The dual fact 4 expresses that lim inf can be pulled out of a supremum.

Lemma 3. *Binary sums* $\boxplus$ *and products* $\boxtimes$ *and the operations* $(-)^\mu$ *and* $(-)^\nu$ *are* lim sup*-pushable and* lim inf*-pullable.*

Using monotonicity of the product constructor, the lemma entails that $\mathcal{A}(\alpha) \boxtimes \mathcal{B}(\alpha)$ is upper/lower semi-continuous if $\mathcal{A}(\alpha)$ and $\mathcal{B}(\alpha)$ are. This applies also for $\boxplus$.

A generalization of Lemma 1 is:

Lemma 4 (lim sup **through function space).**
$\limsup_{\alpha \to \lambda} (\mathcal{A}(\alpha) \boxed{\to} \mathcal{B}(\alpha)) \sqsubseteq (\liminf_\lambda \mathcal{A}) \boxed{\to} \limsup_\lambda \mathcal{B}$.

Now, to (co)inductive types. Let $\phi \in \mathsf{O} \to \mathsf{O}$.

Lemma 5. $\boldsymbol{\mu}^{\liminf_\lambda \phi} = \liminf_{\alpha \to \lambda} \boldsymbol{\mu}^{\phi(\alpha)}$ *and* $\limsup_{\alpha \to \lambda} \boldsymbol{\nu}^{\phi(\alpha)} = \boldsymbol{\nu}^{\liminf_\lambda \phi}$.

Lemma 6. *For* $\alpha \in \mathsf{O}$, *let* $\mathcal{F}_\alpha \in \mathfrak{L} \xrightarrow{+} \mathfrak{L}$ *be* lim inf*-pullable and* $\mathcal{G}_\alpha \in \mathfrak{L} \xrightarrow{+} \mathfrak{L}$ *be* lim sup*-pushable. Then for all* $\beta \in \mathsf{O}$, $\boldsymbol{\mu}^\beta(\liminf_\lambda \mathcal{F}) \sqsubseteq \liminf_{\alpha\to\lambda} \boldsymbol{\mu}^\beta \mathcal{F}_\alpha$ *and* $\limsup_{\alpha\to\lambda} \boldsymbol{\nu}^\beta \mathcal{G}_\alpha \sqsubseteq \boldsymbol{\nu}^\beta(\limsup_\lambda \mathcal{G})$.

Proof. By transfinite induction on β.

Corollary 2 (Limits and (co)inductive types).

1. $\boldsymbol{\mu}^{\liminf_\lambda \phi} \liminf_\lambda \mathcal{F} \sqsubseteq \liminf_{\alpha\to\lambda} \boldsymbol{\mu}^{\phi(\alpha)} \mathcal{F}_\alpha$,
2. $\limsup_{\alpha\to\lambda} \boldsymbol{\nu}^{\phi(\alpha)} \mathcal{G}_\alpha \sqsubseteq \boldsymbol{\nu}^{\liminf_\lambda \phi} \limsup_\lambda \mathcal{G}$.

Proof. For instance, the second inclusion can be derived in three steps using Lemma 2.1, Lemma 5, and Lemma 6.

Now, since $\mathcal{G}_\alpha(\mathcal{X}) = (\mathcal{N}at^\alpha \boxtimes \mathcal{X})^\nu$ is lim sup-pushable, we have can infer upper semi-continuity of $\mathcal{S}tream^\alpha(\mathcal{N}at^\alpha) = \boldsymbol{\nu}^\alpha \mathcal{G}_\alpha$. Analogously, we establish lower semi-continuity of $\mathcal{L}ist^\alpha(\mathcal{N}at^\alpha)$.

4.2 Negative Results

Function space and lower semi-continuity. One may wonder whether Lemma 1 can be dualized, i. e., does upper semi-continuity of $\mathcal{A}$ and lower semi-continuity of $\mathcal{B}$ entail lower semi-continuity of $\mathcal{C}(\alpha) = \mathcal{A}(\alpha) \boxed{\rightarrow} \mathcal{B}(\alpha)$? The answer is no, e. g., consider $\mathcal{C}(\alpha) = \mathcal{N}at^\omega \boxed{\rightarrow} \mathcal{N}at^\alpha$. Although $\mathcal{A}(\alpha) = \mathcal{N}at^\omega$ is trivially upper semi-continuous, and $\mathcal{B}(\alpha) = \mathcal{N}at^\alpha$ is lower semi-continuous, $\mathcal{C}$ is not lower semi-continuous: For instance, the identity function is in $\mathcal{C}(\omega)$ but in no $\mathcal{C}(\alpha)$ for $\alpha < \omega$, hence, also not in $\liminf_\omega \mathcal{C}$. And indeed, if this $\mathcal{C}$ was lower semi-continuous, then our criterion would be unsound, because then by Lemma 1 the type $(\mathcal{N}at^\omega \boxed{\rightarrow} \mathcal{N}at^\alpha) \boxed{\rightarrow} \mathcal{N}at^\omega$, which admits a looping function (see introduction), would be upper semi-continuous.

Inductive types and upper semi-continuity. Pareto [16] proves that inductive types are (in our terminology) lim sup-pushable. His inductive types denote only finitely branching trees, but we also consider infinite branching, arising from function space embedded in inductive types. In my thesis [4, Sect. 5.4.3] I show that infinitely branching inductive data types do not inherit upper semi-continuity from their defining body. But remember that inductive types can still be upper semi-continuous if they are covariant in their size index.

5 A Kinding System for Semi-continuity

We turn the results of the last section into a calculus and define a judgement $\Delta; \Pi \vdash^{\imath q} F : \kappa$, where $\imath$ is an ordinal variable $(\imath : \mathsf{pord}) \in \Delta$, the bit $q \in \{\ominus, \oplus\}$ states whether the constructor F under consideration is lower $(\ominus)$ or upper $(\oplus)$ semi-continuous, and Π is a context of *strictly positive* constructor variables

$X : +\kappa'$. The complete listing of rules can be found in Figure 2; in the following, we discuss a few.

$$\text{CONT-CO}\ \frac{\Delta, \imath : +\mathsf{ord} \vdash F : \kappa \qquad p \in \{+, \circ\}}{\Delta, \imath : p\mathsf{ord}; \Pi \vdash^{\imath \oplus} F : \kappa}$$

If $\imath$ appears positively in F, then F is trivially upper semi-continuous. In the conclusion we may choose to set $p = \circ$, meaning that we forget that F is monotone in $\imath$.

$$\text{CONT-ARR}\ \frac{-\Delta; \diamond \vdash^{\imath \ominus} A : * \qquad \Delta; \Pi \vdash^{\imath \oplus} B : *}{\Delta; \Pi \vdash^{\imath \oplus} A \to B : *}$$

This rule incarnates Lemma 1. Note that, because A is to the left of the arrow, the polarity of all ordinary variables in A is reversed, and A may not contain strictly positive variables.

$$\text{CONT-NU}\ \frac{\Delta; \Pi, X : +\kappa_* \vdash^{\imath \oplus} F : \kappa_*}{\Delta; \Pi \vdash^{\imath \oplus} \nu^a \lambda X : \kappa_*. F : \kappa_*}$$

Rule CONT-NU states that strictly positive coinductive types are upper semi-continuous. The ordinal a must be ∞ or $\mathsf{s}^n \jmath$ for some $\jmath : \mathsf{ord} \in \Delta$ (which may also be identical to $\imath$).

Theorem 2 (Soundness of Continuity Derivations). *Let θ a valuation of the variables in Δ and Π, $(X : +\kappa') \in \Pi$, $\mathcal{G} \in [\![\mathsf{ord}]\!] \to [\![\kappa']\!]$, and $\lambda \in [\![\mathsf{ord}]\!]$ a limit ordinal.*

1. *If $\Delta; \Pi \vdash^{\imath \ominus} F : \kappa$ then*
   (a) $[\![F]\!]_{\theta[\imath \mapsto \lambda]} \sqsubseteq \liminf_{\alpha \to \lambda} [\![F]\!]_{\theta[\imath \mapsto \alpha]}$, *and*
   (b) $[\![F]\!]_{\theta[X \mapsto \liminf_\lambda \mathcal{G}]} \sqsubseteq \liminf_{\alpha \to \lambda} [\![F]\!]_{\theta[X \mapsto \mathcal{G}(\alpha)]}$.
2. *If $\Delta; \Pi \vdash^{\imath \oplus} F : \kappa$ then*
   (a) $\limsup_{\alpha \to \lambda} [\![F]\!]_{\theta[\imath \mapsto \alpha]} \sqsubseteq [\![F]\!]_{\theta[\imath \mapsto \lambda]}$, *and*
   (b) $\limsup_{\alpha \to \lambda} [\![F]\!]_{\theta[X \mapsto \mathcal{G}(\alpha)]} \sqsubseteq [\![F]\!]_{\theta[X \mapsto \limsup_\lambda \mathcal{G}]}$

Proof. By induction on the derivation [4, Sect. 5.5]. The soundness of CONT-NU hinges on the fact that strictly positive coinductive types close at ordinal ω.

Now we are able to formulate the syntactical admissibility criterion for types of (co)recursive functions.

$$\begin{array}{l} \Gamma \vdash (\lambda \imath.\ \forall \boldsymbol{X} : \boldsymbol{\kappa}. B_1 \to \cdots \to B_n \to \mu^\imath F \boldsymbol{H} \to C)\ \ \mathsf{fix}^\mu_n\text{-}\mathsf{adm} \\ \text{iff}\quad \Gamma, \imath : \circ\mathsf{ord}, \boldsymbol{X} : \boldsymbol{\kappa}; \diamond \vdash^{\imath \oplus} B_{1..n} \to \mu^\imath F \boldsymbol{H} \to C : * \end{array}$$

$$\begin{array}{l} \Gamma \vdash (\lambda \imath.\ \forall \boldsymbol{X} : \boldsymbol{\kappa}. B_1 \to \cdots \to B_n \to \nu^\imath F \boldsymbol{H})\ \ \mathsf{fix}^\nu_n\text{-}\mathsf{adm} \\ \text{iff}\quad \Gamma, \imath : \circ\mathsf{ord}, \boldsymbol{X} : \boldsymbol{\kappa}; \diamond \vdash^{\imath \oplus} B_{1..n} \to \nu^\imath F \boldsymbol{H} : * \end{array}$$

It is easy to check that admissible types fulfill the semantic criteria given at the end of Section 3.

Example 2 (Inductive type inside coinductive type). Rule CONT-NU allows the type system to accept the following definition, which assigns an informative type to the stream nats of all natural numbers in ascending order:

$$\begin{array}{lll} \mathsf{mapStream} & : & \forall A \forall B.\, (A \to B) \to \forall \imath.\, \mathsf{Stream}^{\imath} A \to \mathsf{Stream}^{\imath} B \\ \mathsf{nats} & : & \forall \imath.\, \mathsf{Stream}^{\imath}\, \mathsf{Nat}^{\imath} \\ \mathsf{nats} & := & \mathsf{fix}^{\nu}_{0} \lambda nats.\, \langle \mathsf{zero},\ \mathsf{mapStream\ succ}\ nats \rangle \end{array}$$

Example 3 (Inductive type inside inductive type). In the following, we describe breadth-first traversal of rose (finitely branching) trees whose termination is recognized by $\mathsf{F}_{\widehat{\omega}}$.

$$\begin{array}{l} \mathsf{Rose} :\ \ \mathsf{ord} \xrightarrow{+} * \xrightarrow{+} * \\ \mathsf{Rose} := \lambda \imath \lambda A.\, \mathsf{GRose}^{\imath}\, \mathsf{List}^{\infty}\, A = \lambda \imath \lambda A.\ \mu^{\imath}_{*} \lambda X.\, A \times \mathsf{List}^{\infty} X \end{array}$$

The step function, defined by induction on $\jmath$, traverses a list of rose trees of height $< \imath + 1$ and produces a list of the roots and a list of the branches (height $< \imath$).

$$\begin{array}{l} \mathsf{step} :\ \ \forall \jmath \forall A \forall \imath.\, \mathsf{List}^{\jmath} (\mathsf{Rose}^{\imath+1}\, A) \to \mathsf{List}^{\jmath}\, A \times \mathsf{List}^{\infty} (\mathsf{Rose}^{\imath}\, A) \\ \mathsf{step} := \mathsf{fix}^{\mu}_{0} \lambda step \lambda l.\, \mathsf{match}\ l\ \mathsf{with} \\ \qquad\qquad \mathsf{nil} \mapsto \langle \mathsf{nil},\ \mathsf{nil} \rangle \\ \qquad\qquad \mathsf{cons}\, \langle a, rs' \rangle\ rs \mapsto \mathsf{match}\ step\ rs\ \mathsf{with} \\ \qquad\qquad\qquad\qquad \langle as,\ rs'' \rangle \mapsto \langle \mathsf{cons}\ a\ as,\ \mathsf{append}\ rs'\ rs'' \rangle \end{array}$$

Now, bf iterates step on a non-empty forest. It is defined by induction on $\imath$.

$$\begin{array}{l} \mathsf{bf} :\ \ \forall \imath \forall A.\, \mathsf{Rose}^{\imath}\, A \to \mathsf{List}^{\infty} (\mathsf{Rose}^{\imath}\, A) \to \mathsf{List}^{\infty} A \\ \mathsf{bf} := \mathsf{fix}^{\mu}_{0} \lambda bf \lambda r \lambda rs.\ \mathsf{match\ step}\, (\mathsf{cons}\ r\ rs)\ \mathsf{with} \\ \qquad\qquad \langle as,\ \mathsf{nil} \rangle \qquad\qquad \mapsto as \\ \qquad\qquad \langle as,\ \mathsf{cons}\ r'\ rs' \rangle \mapsto \mathsf{append}\ as\ (bf\ r'\ rs') \end{array}$$

Function bf terminates because the recursive-call trees in forest $\mathsf{cons}\, r'\, rs$ are smaller than the input trees in forest $\mathsf{cons}\, r\, rs$. This information is available to the type system through the type of step. The type of bf is admissible for recursion since $\mathsf{List}^{\infty} (\mathsf{Rose}^{\imath}\, A)$ is lower semi-continuous in $\imath$—thanks to Cor. 2 and rule CONT-MU.

6 Conclusions

We have motivated the importance of semi-continuity for the soundness of type-based termination checking, explored the realm of semi-continuous functions from ordinals to semantical types, and developed a calculus for semi-continuous types. We have seen a few interesting examples involving semi-continuous types, many more can be found in the author's thesis [4, Ch. 6]. These examples cannot be handled by type-based termination à la Barthe et al. [7,8], but our developments could be directly incorporated into their calculus.

In previous work [1], I have already presented a calculus for admissible recursion types. But the language had neither polymorphism, higher-kinded types, nor semi-continuous types inside each other ($\mathsf{Stream}^{\imath}\,\mathsf{Nat}^{\imath}$). Hughes, Pareto, and Sabry [14] have also given criteria for admissible types similar to ours, but rather ad-hoc ones, not based on the mathematical concept of semi-continuity. Also, a crucial difference is that we also treat *infinitely* branching data structures. To be fair, I should say that their work has been a major source of inspiration for me.

As a further direction of research, I propose to develop a kinding system where semi-continuity is first class, i. e., one can abstract over semi-continuous constructors, and kind arrows can carry the corresponding polarities $\ominus$ or $\oplus$. First attempts suggest that such a calculus is not straightforward, and more fine-grained polarity system will be necessary.

Acknowledgments. I would like to thank my supervisor, Martin Hofmann, for discussions on $\mathsf{F}_{\omega}^{\widehat{}}$. Thanks to John Hughes for lending his ear in difficult phases of this work, for instance, when I was trying to prove upper semi-continuity of inductive types but then found a counterexample. Thanks to the anonymous referees of previous versions of this paper who gave insightful and helpful comments.

References

1. Abel, A.: Termination and guardedness checking with continuous types. In: Hofmann, M., ed., Typed Lambda Calculi and Applications (TLCA 2003), Valencia, Spain, volume 2701 of Lecture Notes in Computer Science. Springer-Verlag (2003), 1–15
2. Abel, A.: Termination checking with types. RAIRO – Theoretical Informatics and Applications **38** (2004) 277–319. Special Issue: Fixed Points in Computer Science (FICS'03)
3. Abel, A.: Polarized subtyping for sized types. In: Grigoriev, D., Harrison, J., Hirsch, E. A., eds., Computer Science Symposium in Russia (CSR 2006), St. Petersburg, June 8-12, 2006, volume 3967 of Lecture Notes in Computer Science. Springer-Verlag (2006), 381–392
4. Abel, A.: A Polymorphic Lambda-Calculus with Sized Higher-Order Types. Ph.D. thesis, Ludwig-Maximilians-Universität München (2006)
5. Abel, A., Matthes, R.: Fixed points of type constructors and primitive recursion. In: Marcinkowski, J., Tarlecki, A., eds., Computer Science Logic, CSL'04, volume 3210 of Lecture Notes in Computer Science. Springer-Verlag (2004), 190–204
6. Amadio, R. M., Coupet-Grimal, S.: Analysis of a guard condition in type theory. In: Nivat, M., ed., Foundations of Software Science and Computation Structures, First International Conference, FoSSaCS'98, volume 1378 of Lecture Notes in Computer Science. Springer-Verlag (1998), 48–62
7. Barthe, G., Frade, M. J., Giménez, E., Pinto, L., Uustalu, T.: Type-based termination of recursive definitions. Mathematical Structures in Computer Science **14** (2004) 1–45

8. Barthe, G., Grégoire, B., Pastawski, F.: Practical inference for type-based termination in a polymorphic setting. In: Urzyczyn, P., ed., Typed Lambda Calculi and Applications (TLCA 2005), Nara, Japan, volume 3461 of Lecture Notes in Computer Science. Springer-Verlag (2005), 71–85
9. Blanqui, F.: A type-based termination criterion for dependently-typed higher-order rewrite systems. In: van Oostrom, V., ed., Rewriting Techniques and Applications, 15th International Conference, RTA 2004, Aachen, Germany, June 3 – 5, 2004, Proceedings, volume 3091 of Lecture Notes in Computer Science. Springer-Verlag (2004), 24–39
10. Blanqui, F.: Decidability of type-checking in the Calculus of Algebraic Constructions with size annotations. In: Ong, C.-H. L., ed., Computer Science Logic, 19th International Workshop, CSL 2005, 14th Annual Conference of the EACSL, Oxford, UK, August 22-25, 2005, Proceedings, volume 3634 of Lecture Notes in Computer Science. Springer-Verlag (2005), 135–150
11. Crary, K., Weirich, S.: Flexible type analysis. In: Proceedings of the fourth ACM SIGPLAN International Conference on Functional Programming (ICFP '99), Paris, France, volume 34 of SIGPLAN Notices. ACM Press (1999), 233–248
12. Duggan, D., Compagnoni, A.: Subtyping for object type constructors (1999). Presented at FOOL 6
13. Giménez, E.: Structural recursive definitions in type theory. In: Larsen, K. G., Skyum, S., Winskel, G., eds., Automata, Languages and Programming, 25th International Colloquium, ICALP'98, Aalborg, Denmark, July 13-17, 1998, Proceedings, volume 1443 of Lecture Notes in Computer Science. Springer-Verlag (1998), 397–408
14. Hughes, J., Pareto, L., Sabry, A.: Proving the correctness of reactive systems using sized types. In: 23rd Symposium on Principles of Programming Languages, POPL'96 (1996), 410–423
15. Mendler, N. P.: Recursive types and type constraints in second-order lambda calculus. In: Proceedings of the Second Annual IEEE Symposium on Logic in Computer Science, Ithaca, N.Y. IEEE Computer Society Press (1987), 30–36
16. Pareto, L.: Types for Crash Prevention. Ph.D. thesis, Chalmers University of Technology (2000)
17. Paulin-Mohring, C.: Inductive definitions in the system Coq—rules and properties. Technical report, Laboratoire de l'Informatique du Parallélisme (1992)
18. Steffen, M.: Polarized Higher-Order Subtyping. Ph.D. thesis, Technische Fakultät, Universität Erlangen (1998)
19. Xi, H.: Dependent types for program termination verification. In: Proceedings of 16th IEEE Symposium on Logic in Computer Science. Boston, USA (2001)

Visibly Pushdown Automata: From Language Equivalence to Simulation and Bisimulation

Jiří Srba*

BRICS**, Department of Computer Science
Aalborg University, Fredrik Bajersvej 7B, 9220 Aalborg, Denmark
srba@cs.aau.dk

Abstract. We investigate the possibility of (bi)simulation-like preorder/equivalence checking on the class of visibly pushdown automata and its natural subclasses visibly BPA (Basic Process Algebra) and visibly one-counter automata. We describe generic methods for proving complexity upper and lower bounds for a number of studied preorders and equivalences like simulation, completed simulation, ready simulation, 2-nested simulation preorders/equivalences and bisimulation equivalence. Our main results are that all the mentioned equivalences and preorders are EXPTIME-complete on visibly pushdown automata, PSPACE-complete on visibly one-counter automata and P-complete on visibly BPA. Our PSPACE lower bound for visibly one-counter automata improves also the previously known DP-hardness results for ordinary one-counter automata and one-counter nets. Finally, we study regularity checking problems for visibly pushdown automata and show that they can be decided in polynomial time.

1 Introduction

Visibly pushdown languages were introduced by Alur and Madhusudan in [4] as a subclass of context-free languages suitable for formal program analysis, yet tractable and with nice closure properties like the class of regular languages. Visibly pushdown languages are accepted by visibly pushdown automata whose stack behaviour is determined by the input symbol. If the symbol belongs to the category of *call actions* then the automaton must push, if it belongs to *return actions* then the automaton must pop, otherwise (for the *internal actions*) it cannot change the stack height. In [4] it is shown that the class of visibly pushdown languages is closed under intersection, union, complementation, renaming, concatenation and Kleene star. A number of decision problems like universality, language equivalence and language inclusion, which are undecidable for context-free languages, become EXPTIME-complete for visibly pushdown languages.

Recently, visibly pushdown languages have been intensively studied and applied to e.g. program analysis [2], XML processing [20] and the language theory

* The author is supported in part by the research center ITI, project No. 1M0545.
** **B**asic **R**esearch in **C**omputer **S**cience, Centre of the Danish National Research Foundation.

Z. Ésik (Ed.): CSL 2006, LNCS 4207, pp. 89–103, 2006.

of this class has been further investigated in [3,6]. Some recent results show for example the application of a variant of vPDA for proving decidability of contextual equivalence (and other problems) for the third-order fragment of Idealized Algol [18].

In this paper we study visibly pushdown automata from a different perspective. Rather than as language acceptors, we consider visibly pushdown automata as devices that generate infinite-state labelled graphs and we study the questions of decidability of behavioral equivalences and preorders on this class. Our results confirm the tractability of a number of verification problems for visibly pushdown automata.

We prove EXPTIME-completeness of equivalence checking on visibly pushdown automata (vPDA) for practically all preorders and equivalences between simulation preorder and bisimulation equivalence that have been studied in the literature (our focus includes simulation, completed simulation, ready simulation, 2-nested simulation and bisimulation). We then study two natural (and incomparable) subclasses of visibly pushdown automata: visibly basic process algebra (vBPA) and visibly one-counter automata (v1CA). In case of v1CA we demonstrate PSPACE-completeness of the preorder/equivalence checking problems and in case of vBPA even P-completeness. For vBPA we provide also a direct reduction of the studied problems to equivalence checking on finite-state systems, hence the fast algorithms already developed for systems with finitely many reachable states can be directly used. All the mentioned upper bounds are matched by the corresponding lower bounds. The PSPACE-hardness proof for v1CA moreover improves the currently known DP lower bounds [13] for equivalence checking problems on ordinary one-counter automata and one-counter nets and some other problems (see Remark 2). Finally, we consider regularity checking for visibly pushdown automata and show P-completeness for vPDA and vBPA, and NL-completeness for v1CA w.r.t. all equivalences between trace equivalence and isomorphism of labelled transition systems.

Related work. The main reason why many problems for visibly pushdown languages become tractable is, as observed in [4], that a pair of visibly pushdown automata can be synchronized in a similar fashion as finite automata. We use this idea to construct, for a given pair of vPDA processes, a single pushdown automaton where we in a particular way encode the behaviour of both input processes so that they can alternate in performing their moves. This is done in such a way that the question of equality of the input processes w.r.t. a given preorder/equivalence can be tested by asking about the validity of particular (and fixed) modal μ-calculus formulae on the single pushdown process. A similar result of reducing weak simulation between a pushdown process and a finite-state process (and vice versa) to the model checking problem appeared in [17]. We generalize these ideas to cover preorders/equivalences between two visibly pushdown processes and provide a generic proof for all the equivalence checking problems. The technical details of our construction are different from [17] and in particular our construction works immediately also for vBPA (as the necessary bookkeeping is stored in the stack alphabet). As a result we thus show how to handle essentially any so far studied equivalence/preorder between simulation and bisimulation in a uniform way for vPDA, vBPA as well as for v1CA.

In [6] the authors study language regularity problems for visibly pushdown automata. Their line of research is orthogonal to ours because they define a visibly pushdown automaton as regular if it is language equivalent to some visibly one-counter automaton. We study the regularity problems in the context of the standard definitions from the concurrency theory, i.e., whether for a given vPDA process there is a behaviorally equivalent finite-state system. Though, as remarked in more detail in the conclusion, questions of finding an equivalent v1CA and in particular vBPA for a given vPDA could be also interesting to investigate.

Note: full version of this paper will appear as BRICS technical report.

2 Definitions

A *labelled transition system* (LTS) is a triple $(S, \mathcal{A}ct, \longrightarrow)$ where S is the set of *states* (or *processes*), $\mathcal{A}ct$ is the set of *labels* (or *actions*), and $\longrightarrow \subseteq S \times \mathcal{A}ct \times S$ is the *transition relation*; for each $a \in \mathcal{A}ct$, we view $\stackrel{a}{\longrightarrow}$ as a binary relation on S where $s \stackrel{a}{\longrightarrow} s'$ iff $(s, a, s') \in \longrightarrow$. The notation can be naturally extended to $s \stackrel{w}{\longrightarrow} s'$ for finite sequences of actions $w \in \mathcal{A}ct^*$. For a process $s \in S$ we define the set of its *initial actions* by $I(s) \stackrel{\text{def}}{=} \{a \in \mathcal{A}ct \mid \exists s' \in S.\ s \stackrel{a}{\longrightarrow} s'\}$.

We shall now define the studied equivalences/preorders which are between simulation and bisimilarity. Given an LTS $(S, \mathcal{A}ct, \longrightarrow)$, a binary relation $R \subseteq S \times S$ is a

- *simulation* iff for each $(s,t) \in R$, $a \in \mathcal{A}ct$, and s' such that $s \stackrel{a}{\longrightarrow} s'$ there is t' such that $t \stackrel{a}{\longrightarrow} t'$ and $(s', t') \in R$,
- *completed simulation* iff R is a simulation and moreover for each $(s,t) \in R$ it holds that $I(s) = \emptyset$ if and only if $I(t) = \emptyset$,
- *ready simulation* iff R is a simulation and moreover for each $(s,t) \in R$ it holds that $I(s) = I(t)$,
- *2-nested simulation* iff R is a simulation and moreover $R^{-1} \subseteq R$, and
- *bisimulation* iff R is a simulation and moreover $R^{-1} = R$.

We write $s \sqsubseteq_s t$ if there is a simulation R such that $(s,t) \in R$, $s \sqsubseteq_{cs} t$ if there is a completed simulation R such that $(s,t) \in R$, $s \sqsubseteq_{rs} t$ if there is a ready simulation R such that $(s,t) \in R$, $s \sqsubseteq_{2s} t$ if there is a 2-nested simulation R such that $(s,t) \in R$, $s \sim t$ if there is a bisimulation R such that $(s,t) \in R$. The relations are called the corresponding *preorders* (except for bisimilarity, which is already an equivalence). For a preorder $\sqsubseteq \in \{\sqsubseteq_s, \sqsubseteq_{cs}, \sqsubseteq_{rs}, \sqsubseteq_{2s}\}$ we define the corresponding equivalence by $s = t$ iff $s \sqsubseteq t$ and $t \sqsubseteq s$. We remind the reader of the fact that $\sim \subseteq \sqsubseteq_{2s} \subseteq \sqsubseteq_{rs} \subseteq \sqsubseteq_{cs} \subseteq \sqsubseteq_s$ and $\sim \subseteq =_{2s} \subseteq =_{rs} \subseteq =_{cs} \subseteq =_s$ and all inclusions are strict.

We shall use a standard game-theoretic characterization of (bi)similarity. A *bisimulation game* on a pair of processes (s_1, t_1) is a two-player game between *Attacker* and *Defender*. The game is played in *rounds* on pairs of states from

$S \times S$. In each round the players change the *current pair of states* (s, t) (initially $s = s_1$ and $t = t_1$) according to the following rule:

1. Attacker chooses either s or t, $a \in \mathcal{A}ct$ and performs a move $s \xrightarrow{a} s'$ or $t \xrightarrow{a} t'$.
2. Defender responds by choosing the opposite process (either t or s) and performs a move $t \xrightarrow{a} t'$ or $s \xrightarrow{a} s'$ under the same action a.
3. The pair (s', t') becomes the (new) current pair of states.

A *play* (of the bisimulation game) is a sequence of pairs of processes formed by the players according to the rules mentioned above. A play is finite iff one of the players gets stuck (cannot make a move); the player who got stuck lost the play and the other player is the winner. If the play is infinite then Defender is the winner.

We use the following standard fact.

Proposition 1. *It holds that $s \sim t$ iff Defender has a winning strategy in the bisimulation game starting with the pair (s, t), and $s \not\sim t$ iff Attacker has a winning strategy in the corresponding game.*

The rules of the bisimulation game can be easily modified in order to capture the other equivalences/preorders.

In the *simulation preorder game*, Attacker is restricted to attack only from the (left-hand side) process s. In the *simulation equivalence game*, Attacker can first choose a side (either s or t) but after that he is not allowed to change the side any more. *Completed/ready simulation game* has the same rules as the simulation game but Defender is moreover losing in any configuration which brakes the extra condition imposed by the definition (i.e. s and t should have the same set of initial actions in case of ready simulation, and their sets of initial actions should be both empty at the same time in case of completed simulation). Finally, in the *2-nested simulation preorder game*, Attacker starts playing from the left-hand side process s and at most once during the play he is allowed to switch sides (the soundness follows from the characterization provided in [1]). In the *2-nested simulation equivalence game*, Attacker can initially choose any side but he is still restricted that he can change sides at most once during the play.

We shall now define the model of pushdown automata. Let $\mathcal{A}ct$ be a finite set of actions, let Γ be a finite set of stack symbols and let Q be a finite set of control states. We assume that the sets $\mathcal{A}ct$, Γ and Q are pairwise disjoint. A *pushdown automaton* (PDA) over the set of actions $\mathcal{A}ct$, stack alphabet Γ and control states Q is a finite set Δ of rules of the form $pX \xrightarrow{a} q\alpha$ where $p, q \in Q$, $a \in \mathcal{A}ct$, $X \in \Gamma$ and $\alpha \in \Gamma^*$.

A PDA Δ determines a labelled transition system $T(\Delta) = (S, \mathcal{A}ct, \longrightarrow)$ where the states are configurations of the form state×stack (i.e. $S = Q \times \Gamma^*$ and configurations like (p, α) are usually written as $p\alpha$ where the top of the stack α is by agreement on the left) and the transition relation is determined by the following prefix rewriting rule.

$$\frac{(pX \xrightarrow{a} q\alpha) \in \Delta, \quad \gamma \in \Gamma^*}{pX\gamma \xrightarrow{a} q\alpha\gamma}$$

A pushdown automaton is called BPA for *Basic Process Algebra* if the set of control states is a singleton set ($|Q| = 1$). In this case we usually omit the control state from the rules and configurations.

A pushdown automaton is called 1CA for *one-counter automaton* if the stack alphabet consists of two symbols only, $\Gamma = \{I, Z\}$, and every rule is of the form $pI \xrightarrow{a} q\alpha$ or $pZ \xrightarrow{a} q\alpha Z$, where $\alpha \in \{I\}^*$. This means that every configuration reachable from pZ is of the form pI^nZ where I^n stands for a sequence of n symbols I and Z corresponds to the bottom of the stack (the value zero). We shall simply denote such a configuration by $p(n)$ and say that it represents the counter value n.

Assume that $\mathcal{A}ct = \mathcal{A}ct_c \cup \mathcal{A}ct_r \cup \mathcal{A}ct_i$ is partitioned into a disjoint union of finite sets of call, return and internal actions, respectively. A *visibly pushdown automaton* (vPDA) is a PDA which, for every rule $pX \xrightarrow{a} q\alpha$, satisfies additional three requirements (where $|\alpha|$ stands for the length of α):

- if $a \in \mathcal{A}ct_c$ then $|\alpha| = 2$ (call),
- if $a \in \mathcal{A}ct_r$ then $|\alpha| = 0$ (return), and
- if $a \in \mathcal{A}ct_i$ then $|\alpha| = 1$ (internal).

Hence in vPDA the type of the input action determines the change in the height of the stack (call by $+1$, return by -1, internal by 0).

The classes of visibly basic process algebra (vBPA) and visibly one-counter automata (v1CA) are defined analogously.

Remark 1. For internal actions we allow to modify also the top of the stack. This model (for vPDA) can be easily seen to be equivalent to the standard one (as introduced in [4]) where the top of the stack does not change under internal actions. However, when we consider the subclass vBPA, the possibility of changing the top of the stack under internal actions significantly increases the descriptive power of the formalism. Unlike in [4], we do not allow to perform return actions on the empty stack.

The question we are interested in is: given a vPDA (or vBPA, or v1CA) and two of its initial configurations pX and qY, can we algorithmically decide whether pX and qY are equal with respect to a given preorder/equivalence and if yes, what is the complexity?

3 Decidability of Preorder/Equivalence Checking

3.1 Visibly Pushdown Automata

We shall now study preorder/equivalence checking problems on the class of visibly pushdown automata. We prove the decidability by reducing the problems to model checking of an ordinary pushdown system against a fixed μ-calculus formula.

Let Δ be a vPDA over the set of actions $\mathcal{A}ct = \mathcal{A}ct_c \cup \mathcal{A}ct_r \cup \mathcal{A}ct_i$, stack alphabet Γ and control states Q. We shall construct a PDA Δ' over the actions $\mathcal{A}ct' \stackrel{\text{def}}{=} \mathcal{A}ct \cup \overline{\mathcal{A}ct} \cup \{\ell, r\}$ where $\overline{\mathcal{A}ct} \stackrel{\text{def}}{=} \{\overline{a} \mid a \in \mathcal{A}ct\}$, stack alphabet $\Gamma' \stackrel{\text{def}}{=} G \times G$ where $G \stackrel{\text{def}}{=} \Gamma \cup (\Gamma \times \Gamma) \cup (\Gamma \times \mathcal{A}ct) \cup \{\epsilon\}$, and control states $Q' \stackrel{\text{def}}{=} Q \times Q$. For notational convenience, elements $(X, a) \in \Gamma \times \mathcal{A}ct$ will be written simply as X_a.

The idea is that for a given pair of vPDA processes we shall construct a single PDA process which simulates the behaviour of both vPDA processes by repeatedly performing a move in one of the processes, immediately followed by a move under the same action in the other process. The actions ℓ and r make it visible, whether the move is performed on the left-hand side or right-hand side. The assumption that the given processes are vPDA ensures that their stacks are kept synchronized.

We shall define a partial mapping $[\,.\,,\,.\,] : \Gamma^* \times \Gamma^* \to (\Gamma \times \Gamma)^*$ inductively as follows ($X, Y \in \Gamma$ and $\alpha, \beta \in \Gamma^*$ such that $|\alpha| = |\beta|$): $[X\alpha, Y\beta] \stackrel{\text{def}}{=} (X, Y)[\alpha, \beta]$ and $[\epsilon, \epsilon] \stackrel{\text{def}}{=} \epsilon$. The mapping provides the possibility to merge stacks.

Assume a given pair of vPDA processes pX and qY. Our aim is to effectively construct a new PDA system Δ' such that for every $\bowtie \in \{\sqsubseteq_s, =_s, \sqsubseteq_{cs}, =_{cs}, \sqsubseteq_{rs}, =_{rs}, \sqsubseteq_{2s}, =_{2s}, \sim\}$ it is the case that $pX \bowtie qY$ in Δ if and only if $(p, q)(X, Y) \models \phi_{\bowtie}$ in Δ' for a fixed μ-calculus formula $\phi_{\bowtie}$. We refer the reader to [16] for the introduction to the modal μ-calculus.

The set of PDA rules Δ' is defined as follows. Whenever $(pX \xrightarrow{a} q\alpha) \in \Delta$ then the following rules belong to Δ':

1. $(p, p')(X, X') \xrightarrow{\ell} (q, p')(\alpha, X'_a)$ for every $p' \in Q$ and $X' \in \Gamma$,
2. $(p', p)(X', X) \xrightarrow{r} (p', q)(X'_a, \alpha)$ for every $p' \in Q$ and $X' \in \Gamma$,
3. $(p', p)(\beta, X_a) \xrightarrow{r} (p', q)[\beta, \alpha]$ for every $p' \in Q$ and $\beta \in \Gamma \cup (\Gamma \times \Gamma) \cup \{\epsilon\}$,
4. $(p, p')(X_a, \beta) \xrightarrow{\ell} (q, p')[\alpha, \beta]$ for every $p' \in Q$ and $\beta \in \Gamma \cup (\Gamma \times \Gamma) \cup \{\epsilon\}$,
5. $(p, p')(X, X') \xrightarrow{a} (p, p')(X, X')$ for every $p' \in Q$ and $X' \in \Gamma$, and
6. $(p', p)(X', X) \xrightarrow{\overline{a}} (p', p)(X', X)$ for every $p' \in Q$ and $X' \in \Gamma$.

From a configuration $(p, q)[\alpha, \beta]$ the rules of the form 1. and 2. select either the left-hand or right-hand side and perform some transition in the selected process. The next possible transition (by rules 3. and 4.) is only from the opposite side of the configuration than in the previous step. Symbols of the form X_a where $X \in \Gamma$ and $a \in \mathcal{A}ct$ are used to make sure that the transitions in these two steps are due to pushdown rules under the same label a. Note that in the rules 3. and 4. it is thus guaranteed that $|\alpha| = |\beta|$. Finally, the rules 5. and 6. introduce a number of self-loops in order to make visible the initial actions of the processes.

Lemma 1. *Let Δ be a vPDA system over the set of actions $\mathcal{A}ct$ and pX, qY two of its processes. Let $(p, q)(X, Y)$ be a process in the system Δ' constructed above. Let*

- $\phi_{\sqsubseteq_s} \equiv \nu Z.[\ell]\langle r\rangle Z$,
- $\phi_{=_s} \equiv \phi_{\sqsubseteq_s} \wedge (\nu Z.[r]\langle \ell\rangle Z)$,

- $\phi_{\sqsubseteq_{cs}} \equiv \nu Z.\big([\ell]\langle r\rangle Z \wedge (\langle \mathcal{A}ct\rangle tt \Leftrightarrow \langle \overline{\mathcal{A}ct}\rangle tt)\big)$,
- $\phi_{=_{cs}} \equiv \phi_{\sqsubseteq_{cs}} \wedge \nu Z.\big([r]\langle \ell\rangle Z \wedge (\langle \mathcal{A}ct\rangle tt \Leftrightarrow \langle \overline{\mathcal{A}ct}\rangle tt)\big)$,
- $\phi_{\sqsubseteq_{rs}} \equiv \nu Z.\big([\ell]\langle r\rangle Z \wedge \bigwedge_{a\in\mathcal{A}ct} (\langle a\rangle tt \Leftrightarrow \langle \overline{a}\rangle tt)\big)$,
- $\phi_{=_{rs}} \equiv \phi_{\sqsubseteq_{rs}} \wedge \nu Z.\big([r]\langle \ell\rangle Z \wedge \bigwedge_{a\in\mathcal{A}ct} (\langle a\rangle tt \Leftrightarrow \langle \overline{a}\rangle tt)\big)$,
- $\phi_{\sqsubseteq_{2s}} \equiv \nu Z.\big([\ell]\langle r\rangle Z \wedge (\nu Z'.[r]\langle \ell\rangle Z')\big)$,
- $\phi_{=_{2s}} \equiv \phi_{\sqsubseteq_{2s}} \wedge \nu Z.\big([r]\langle \ell\rangle Z \wedge (\nu Z'.[\ell]\langle r\rangle Z')\big)$, *and*
- $\phi_{\sim} \equiv \nu Z.[\ell, r]\langle \ell, r\rangle Z$.

For every $\bowtie\, \in \{\sqsubseteq_s, =_s, \sqsubseteq_{cs}, =_{cs}, \sqsubseteq_{rs}, =_{rs}, \sqsubseteq_{2s}, =_{2s}, \sim\}$ *it holds that* $pX \bowtie qY$ *if and only if* $(p,q)(X,Y) \models \phi_{\bowtie}$.

Theorem 1. *Simulation, completed simulation, ready simulation and 2-nested simulation preorders and equivalences, as well as bisimulation equivalence are decidable on vPDA and all these problems are EXPTIME-complete.*

Proof. EXPTIME-hardness (for all relations between simulation preorder and bisimulation equivalence) follows from [17] as the pushdown automaton constructed in the proof is in fact a vPDA.

For the containment in EXPTIME observe that all our equivalence checking problems are reduced in polynomial time to model checking of a pushdown automaton against a fixed size formula of modal μ-calculus. The complexity of the model checking problem for a pushdown automaton with m states and k stack symbols and a formula of the size n_1 and of the alternation depth n_2 is $O((k2^{cmn_1n_2})^{n_2}))$ for some constant c [25]. In our case for a given vPDA system with m states and k stack symbols we construct a PDA system with m^2 states and with $O(k^3 \cdot |\mathcal{A}ct|)$ stack symbols (used in the transition rules). Hence the overall time complexity of checking whether two vPDA processes pX and qY are equivalent is $(k^3 \cdot |\mathcal{A}ct|)2^{O(m^2)}$. □

3.2 Visibly Basic Process Algebra

We shall now focus on the complexity of preorder/equivalence checking for vBPA, a strict subclass of vPDA.

Theorem 2. *Simulation, completed simulation, ready simulation and 2-nested simulation preorders and equivalences, as well as bisimulation equivalence are P-complete on vBPA.*

Proof. Recall that a vBPA process is a vPDA processes with a single control state. By using the arguments from the proof of Theorem 1, the complexity of equivalence checking on vBPA is therefore $O(k^3 \cdot |\mathcal{A}ct|)$ where k is the cardinality of the stack alphabet (and where $m = 1$). P-hardness was proved in [21] even for finite-state systems. □

In fact, for vBPA we can introduce even better complexity upper bounds by reducing it to preorder/equivalence checking on finite-state systems.

$$X \xrightarrow{a} Y$$
$$X \xrightarrow{b} \epsilon$$
$$X \xrightarrow{c} XY$$
$$Y \xrightarrow{b} \epsilon$$

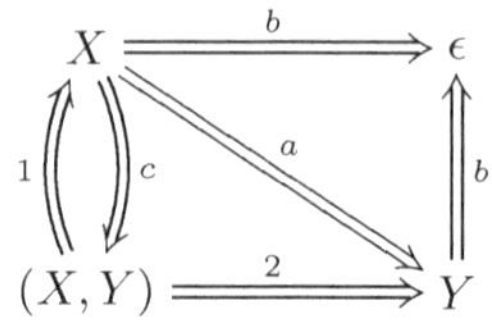

Fig. 1. Transformation of a vBPA into a finite-state system

Theorem 3. *Simulation, completed simulation, ready simulation and 2-nested simulation preorders and equivalences, as well as bisimulation equivalence on vBPA is reducible to checking the same preorder/equivalence on finite-state systems. For any vBPA process Δ (with the natural requirement that every stack symbol appears at least in one rule from Δ), the reduction is computable in time $O(|\Delta|)$ and outputs a finite-state system with $O(|\Delta|)$ states and $O(|\Delta|)$ transitions.*

Proof. Let $\mathcal{A}ct = \mathcal{A}ct_c \cup \mathcal{A}ct_r \cup \mathcal{A}ct_i$ be the set of actions and let Γ be the stack alphabet of a given vBPA system Δ (we shall omit writing the control states as this is a singleton set). Let $S \stackrel{\text{def}}{=} \{(Y, Z) \in \Gamma \times \Gamma \mid \exists(X \stackrel{a}{\longrightarrow} YZ) \in \Delta \text{ for some } X \in \Gamma \text{ and } a \in \mathcal{A}ct_c\}$. We construct a finite-state transition system $T = (\Gamma \cup \{\epsilon\} \cup S, \mathcal{A}ct \cup \{1, 2\}, \Longrightarrow)$ for fresh actions 1 and 2 as follows. For every vBPA rule $(X \stackrel{a}{\longrightarrow} \alpha) \in \Delta$, we add the transitions:

- $X \stackrel{a}{\Longrightarrow} \epsilon$ if $a \in \mathcal{A}ct_r$ (and $\alpha = \epsilon$),
- $X \stackrel{a}{\Longrightarrow} Y$ if $a \in \mathcal{A}ct_i$ and $\alpha = Y$,
- $X \stackrel{a}{\Longrightarrow} (Y, Z)$ if $a \in \mathcal{A}ct_c$ and $\alpha = YZ$,
- $(Y, Z) \stackrel{1}{\Longrightarrow} Y$ if $a \in \mathcal{A}ct_c$ and $\alpha = YZ$, and
- $(Y, Z) \stackrel{2}{\Longrightarrow} Z$ if $a \in \mathcal{A}ct_c$ and $\alpha = YZ$ such that $Y \longrightarrow^* \epsilon$.

Note that the set $\{Y \in \Gamma \mid Y \longrightarrow^* \epsilon\}$ can be (by standard techniques) computed in time $O(|\Delta|)$. Moreover, the finite-state system T has $O(|\Delta|)$ states and $O(|\Delta|)$ transitions. See Figure 1 for an example of the transformation.

Let us now observe that in vBPA systems we have the following decomposition property. It is the case that $X\alpha \sim X'\alpha'$ in Δ (where $X, X' \in \Gamma$ and $\alpha, \alpha' \in \Gamma^*$) if and only if in Δ the following two conditions hold: (i) $X \sim X'$ and (ii) if ($X \longrightarrow^* \epsilon$ or $X' \longrightarrow^* \epsilon$) then $\alpha \sim \alpha'$. Hence for any $X, Y \in \Gamma$ we have that $X \sim Y$ in Δ iff $X \sim Y$ in T. It is easy to check that the fact above holds also for any other preorder/equivalence as stated by the theorem. □

This means that for preorder/equivalence checking on vBPA we can use the efficient algorithms already developed for finite-state systems. For example, for finite-state transition systems with k states and t transitions, bisimilarity can be decided in time $O(t \log k)$ [19]. Hence bisimilarity on a vBPA system Δ is decidable in time $O(|\Delta| \cdot \log |\Delta|)$.

3.3 Visibly One-Counter Automata

We will now continue with studying preorder/equivalence checking problems on v1CA, a strict subclass of vPDA and an incomparable class with vBPA (w.r.t. bisimilarity). We start by showing PSPACE-hardness of the problems. The proof is by reduction from a PSPACE-complete problem of emptiness of one-way alternating finite automata over one-letter alphabet [11].

A *one-way alternating finite automaton over one-letter alphabet* is a 5-tuple $A = (Q_\exists, Q_\forall, q_0, \delta, F)$ where $Q_\exists$ and $Q_\forall$ are finite and disjoint sets of existential, resp. universal control states, $q_0 \in Q_\exists \cup Q_\forall$ is the initial state, $F \subseteq Q_\exists \cup Q_\forall$ is the set of final states and $\delta : Q_\exists \cup Q_\forall \rightarrow 2^{Q_\exists \cup Q_\forall}$ is the transition function.

A *computational tree* for an input word of the form I^n (where n is a natural number and I is the only letter in the input alphabet) is a tree where every branch has exactly $n+1$ nodes labelled by control states from $Q_\exists \cup Q_\forall$ such that the root is labelled with q_0 and every non-leaf node that is already labelled by some $q \in Q_\exists \cup Q_\forall$ such that $\delta(q) = \{q_1, \ldots, q_k\}$ has either

- one child labelled by q_i for some i, $1 \leq i \leq k$, if $q \in Q_\exists$, or
- k children labelled by $q_1, \ldots, q_k$, if $q \in Q_\forall$.

A computational tree is *accepting* if the labels of all its leaves are final (i.e. belong to F). The language of A is defined by $L(A) \stackrel{\text{def}}{=} \{I^n \mid I^n \text{ has some accepting computational tree }\}$.

The emptiness problem for one-way alternating finite automata over one-letter alphabet (denoted as EMPTY) is to decide whether $L(A) = \emptyset$ for a given automaton A. The problem EMPTY is known to be PSPACE-complete due to Holzer [11].

In what follows we shall demonstrate a polynomial time reduction from EMPTY to equivalence/preorder checking on visibly one-counter automata. We shall moreover show the reduction for any (arbitrary) relation between simulation preorder and bisimulation equivalence. This in particular covers all preorders/equivalences introduced in this paper.

Lemma 2. *All relations between simulation preorder and bisimulation equivalence are PSPACE-hard on v1CA.*

Proof. Let $A = (Q_\exists, Q_\forall, q_0, \delta, F)$ be a given instance of EMPTY. We shall construct a visibly one-counter automaton Δ over the set of actions $\mathcal{A}ct_c \stackrel{\text{def}}{=} \{i\}$, $\mathcal{A}ct_r \stackrel{\text{def}}{=} \{d_q \mid q \in Q_\exists \cup Q_\forall\}$, $\mathcal{A}ct_i \stackrel{\text{def}}{=} \{a, e\}$ and with control states $Q \stackrel{\text{def}}{=} \{p, p', t\} \cup \{q, q', t_q \mid q \in Q_\exists \cup Q_\forall\}$ such that

- if $L(A) = \emptyset$ then Defender has a winning strategy from pZ and $p'Z$ in the bisimulation game (i.e. $pZ \sim p'Z$), and
- if $L(A) \neq \emptyset$ then Attacker has a winning strategy from pZ and $p'Z$ in the simulation preorder game (i.e. $pZ \not\sqsubseteq_s p'Z$).

The intuition is that Attacker generates some counter value n in both of the processes pZ and $p'Z$ and then switches into a checking phase by changing the configurations to $q_0(n)$ and $q'_0(n)$. Now the players decrease the counter and change the control states according to the function δ. Attacker selects the successor in any existential configuration, while Defender makes the choice of the successor in every universal configuration. Attacker wins if the players reach a pair of configurations $q(0)$ and $q'(0)$ where $q \in F$.

We shall now define the set of rules Δ. The initial rules allow Attacker (by performing repeatedly the action i) to set the counter into an arbitrary number, i.e., Attacker generates a candidate word from $L(A)$.

$$\begin{array}{ll} pZ \xrightarrow{i} pIZ & p'Z \xrightarrow{i} p'IZ \\ pI \xrightarrow{i} pII & p'I \xrightarrow{i} p'II \\ pZ \xrightarrow{a} q_0Z & p'Z \xrightarrow{a} q'_0Z \\ pI \xrightarrow{a} q_0I & p'I \xrightarrow{a} q'_0I \end{array}$$

Observe that Attacker is at some point forced to perform the action a (an infinite play is winning for Defender) and switch to the checking phase starting from $q_0(n)$ and $q'_0(n)$.

Now for every existential state $q \in Q_\exists$ with $\delta(q) = \{q_1, \ldots, q_k\}$ and for every $i \in \{1, \ldots, k\}$ we add the following rules.

$$qI \xrightarrow{d_{q_i}} q_i \qquad q'I \xrightarrow{d_{q_i}} q'_i$$

This means that Attacker can decide on the successor q_i of q and the players in one round move from the pair $q(n)$ and $q'(n)$ into $q_i(n-1)$ and $q'_i(n-1)$.

Next for every universal state $q \in Q_\forall$ with $\delta(q) = \{q_1, \ldots, q_k\}$ and for every $i \in \{1, \ldots, k\}$ we add the rules

$$\begin{array}{ll} qI \xrightarrow{a} tI & q'I \xrightarrow{a} t_{q_i}I \\ qI \xrightarrow{a} t_{q_i}I & \end{array}$$

and for every $q, r \in Q_\exists \cup Q_\forall$ such that $q \neq r$ we add

$$\begin{array}{ll} tI \xrightarrow{d_q} q & t_qI \xrightarrow{d_q} q' \\ & t_qI \xrightarrow{d_r} r \ . \end{array}$$

These rules are more complex and they correspond to a particular implementation of so called *Defender's Choice Technique* (for further examples see e.g. [15]). We shall explain the idea by using Figure 2. Assume that $q \in Q_\forall$ and $\delta(q) = \{q_1, \ldots, q_k\}$. In the first round of the bisimulation game starting from $q(n)$ and $q'(n)$ where $n > 0$, Attacker is forced to take the move $q(n) \xrightarrow{a} t(n)$. On any other move Defender answers by immediately reaching a pair of syntactically equal processes (and thus wins the game). Defender's answer on Attacker's move $q(n) \xrightarrow{a} t(n)$ is to perform $q'(n) \xrightarrow{a} t_{q_i}(n)$ for some $i \in \{1, \ldots, k\}$. The second round thus starts from the pair $t(n)$ and $t_{q_i}(n)$. Should Attacker choose to play the action d_r for some state r such that $r \neq q_i$ (on either side), Defender can again reach a syntactic equality and wins. Hence Attacker is forced to play the action d_{q_i} on which Defender answers by the same action in the opposite

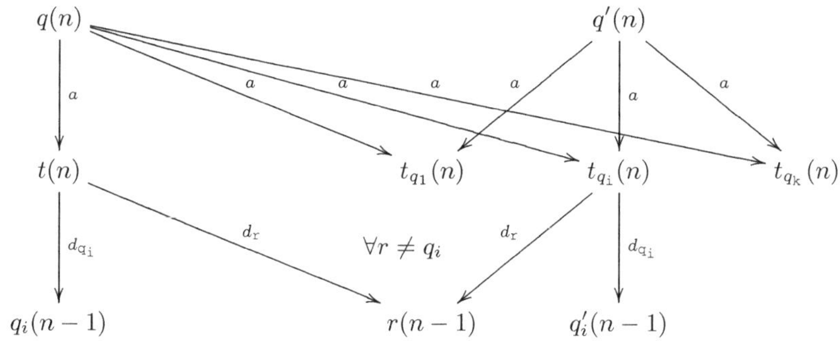

Fig. 2. Defender's Choice: $q \in Q_\forall$ and $\delta(q) = \{q_1, \ldots, q_k\}$

process and the players reach the pair $q_i(n-1)$ and $q'_i(n-1)$. Note that it was Defender who selected the new control state q_i.

Finally, for every $q \in F$ we add the rule

$$qZ \xrightarrow{e} qZ \ .$$

It is easy to see that Δ is a visibly one-counter automaton. Moreover, if $L(A) = \emptyset$ then $pZ \sim p'Z$, and if $L(A) \neq \emptyset$ then $pZ \not\sqsubseteq_s p'Z$. $\square$

Remark 2. The reduction above works also for a strict subclass of one-counter automata called one-counter nets (where it is not allowed to test for zero, see e.g. [13]). It is enough to replace the final rule $qZ \xrightarrow{e} qZ$ with two new rules $q \xrightarrow{e} q$ and $q'I \xrightarrow{e} q'I$ for every $q \in F$. Moreover, a slight modification of the system allows to show PSPACE-hardness of simulation preorder checking between one-counter automata and finite-state systems and vice versa. Hence the previously know DP lower bounds [13] for all relations between simulation preorder and bisimulation equivalence on one-counter nets (and one-counter automata) as well as of simulation preorder/equivalence between one-counter automata and finite-state systems, and between finite-state systems and one-counter automata are raised to PSPACE-hardness.

We are now ready to state the precise complexity of (bi)simulation-like preorders/equivalences on visibly one-counter automata.

Theorem 4. *Simulation, completed simulation, ready simulation and 2-nested simulation preorders and equivalences, as well as bisimulation equivalence are PSPACE-complete on v1CA.*

Proof. PSPACE-hardness follows from Lemma 2. Containment in PSPACE is due to Lemma 1 and due to [23] where it was very recently showed that model checking modal μ-calculus on one-counter automata is decidable in PSPACE. The only slight complication is that the system used in Lemma 1 is not necessarily a one-counter automaton. All stack symbols are of the form (I, I) or

(Z, Z) which is fine, except for the very top of the stack where more stack symbols are used. Nevertheless, by standard techniques, the top of the stack can be remembered in the control states in order to apply the result from [23]. □

4 Decidability of Regularity Checking

In this section we ask the question whether a given vPDA process is equivalent to some finite-state system. Should this be the case, we call the given process *regular* (w.r.t. the considered equivalence). The main result of this section is a semantical characterization of regular vPDA processes via the property of unbounded popping and a polynomial time decision algorithm to test whether a given process satisfies this property.

Let $\mathcal{A}ct = \mathcal{A}ct_c \cup \mathcal{A}ct_r \cup \mathcal{A}ct_i$ be the set of actions of a given vPDA. We define a function $h : \mathcal{A}ct \to \{-1, 0, +1\}$ by $h(a) = +1$ for all $a \in \mathcal{A}ct_c$, $h(a) = -1$ for all $a \in \mathcal{A}ct_r$, and $h(a) = 0$ for all $a \in \mathcal{A}ct_i$. The function h can be naturally extended to sequences of actions by $h(a_1 \ldots a_n) = \sum_{i \in \{1,\ldots,n\}} h(a_i)$. Observe now that for any computation $p\alpha \xrightarrow{w} q\beta$ we have $|\beta| = |\alpha| + h(w)$.

Definition 1. *Let pX be a vPDA configuration. We say that pX provides* unbounded popping *if for every natural number d there is a configuration $q\beta$ and a word $w \in \mathcal{A}ct^*$ such that $h(w) \leq -d$ and $pX \longrightarrow^* q\beta \xrightarrow{w}$.*

Lemma 3. *Let pX be a vPDA configuration which provides unbounded popping. Then pX is not regular w.r.t. trace equivalence.*

Proof (Sketch). By contradiction. Let pX be trace equivalent to some finite-state system A with n states. Let us consider a trace $w_1 w_2$ such that $pX \xrightarrow{w_1} q\beta \xrightarrow{w_2}$ for some $q\beta$ and $h(w_2) \leq -n$. Such a trace must exist because pX provides unbounded popping. The trace $w_1 w_2$ must be executable also in A. However, because A has n states, during the computation on w_2, it must necessarily enter twice the same state such that it forms a loop on some substring w' of w_2. We can moreover assume that $h(w') < 0$. This means that by taking the loop sufficiently many times A can achieve a trace w with $h(w) < 1$. However, this trace is not possible from pX (any word w such that $pX \xrightarrow{w}$ satisfies that $h(w) \geq -1$). This is a contradiction. □

Lemma 4. *Let pX be a vPDA configuration which does not provide unbounded popping. Then pX is regular w.r.t. isomorphism of labelled transition systems.*

Proof. Assume that pX does not provide unbounded popping. In other words, there is a constant d_{max} such that for every process $q\beta$ reachable from pX it is the case that for any computation starting from $q\beta$, the stack height $|\beta|$ cannot be decreased by more than d_{max} symbols. This means that in any reachable configuration it is necessary to remember only d_{max} top-most stack symbols and hence the system can be up to isomorphism described as a finite-state system (in general of exponential size). □

Theorem 5. *Let pX be a vPDA configuration. Then, for any equivalence relation between trace equivalence and isomorphism of labelled transition systems, pX provides unbounded popping if and only if pX is not regular.*

Proof. Directly from Lemma 3 and Lemma 4. □

Theorem 6. *Regularity checking of vPDA w.r.t. any equivalence between trace equivalence and isomorphism of labelled transition systems (in particular also w.r.t. any equivalence considered in this paper) is decidable in deterministic polynomial time. The problems are P-complete for vPDA and vBPA and NL-complete for v1CA.*

Proof (Sketch). We can check for every $q \in Q$ and $Y \in \Gamma$ whether the regular set $post^*(qY) \cap pre^*(\{r\epsilon \mid r \in Q\})$ is infinite. If yes, this means that qY has infinitely many different successors (with higher and higher stacks) such that all of them can be completely emptied. To see whether a given vPDA process pX provides unbounded popping, it is now enough to test whether $pX \in pre^*(qY\Gamma^*)$ for some qY satisfying the condition above. The test can be done in polynomial time because the sets pre^* and $post^*$ are regular and computable in polynomial time as showed e.g. in [7]. The proofs of P-completeness and NL-completeness are in the full version of the paper. □

5 Conclusion

In the following table we provide a comparison of bisimulation, simulation and regularity (w.r.t. bisimilarity) checking on PDA, 1CA, BPA and their subclasses vPDA, v1CA, vBPA. Results achieved in this paper are in bold.

	$\sim$	$\sqsubseteq_s$ and $=_s$	$\sim$-regularity
PDA	decidable [22] EXPTIME-hard [17]	undecidable [10]	? EXPTIME-hard [17,24]
vPDA	**in EXPTIME** EXPTIME-hard [17]	**in EXPTIME** EXPTIME-hard [17]	**P-compl.**
1CA	decidable [12] **PSPACE-hard**	undecidable [14]	decidable [12] P-hard [5,24]
v1CA	**PSPACE-compl.**	**PSPACE-compl.**	**NL-compl.**
BPA	in 2-EXPTIME [8] PSPACE-hard [24]	undecidable [10]	in 2-EXPTIME [9,8] PSPACE-hard [24]
vBPA	**in P** P-hard [5]	**in P** P-hard [21]	**P-compl.**

In fact, our results about EXPTIME-completeness for vPDA, PSPACE-completeness for v1CA and P-completeness for vBPA hold for all preorders and equivalences between simulation preorder and bisimulation equivalence studied in the literature (like completed simulation, ready simulation and 2-nested simulation). The results confirm a general trend seen in the classical language theory of pushdown automata: a relatively minor restriction (from the practical point of view) of being able to distinguish call, return and internal actions often significantly improves the complexity of the studied problems and sometimes even changes undecidable problems into decidable ones, moreover with reasonable complexity upper bounds.

All the upper bounds proved in this paper are matched by the corresponding lower bounds. Here the most interesting result is PSPACE-hardness of preorder/equivalence checking on v1CA for all relations between simulation preorder and bisimulation equivalence. As noted in Remark 2, this proof improves also a number of other complexity lower bounds for problems on standard one-counter nets and one-counter automata, which were previously known to be only DP-hard (DP-hardness is, most likely, a slightly stronger result than NP and co-NP hardness).

Finally, we have proved that for all the studied equivalences, the regularity problem is decidable in polynomial time. Checking whether an infinite-state process is equivalent to some regular one is a relevant question because many problems about such a process can be answered by verifying the equivalent finite-state system and for finite-state systems many efficient algorithms have been developed. A rather interesting observation is that preorder/equivalence checking on vBPA for preorders/equivalences between simulation and bisimilarity can be polynomially translated to verification problems on finite-state systems. On the other hand, the class of vBPA processes is significantly larger than the class of finite-state processes and hence the questions, whether for a given vPDA (or v1CA) process there is some equivalent vBPA process, are of a particular interest. We shall investigate these questions in the future research, as well as a generalization of the unbounded popping property for visibly pushdown automata that enable to perform return actions also on the empty stack.

Acknowledgments. I would like to thank Markus Lohrey for a discussion at ETAPS'06 and for a reference to PSPACE-completeness of the emptiness problem for alternating automata over one-letter alphabet. My thanks go also to the referees of CSL'06 for their useful comments and for suggesting the P-hardness proof of regularity checking for vPDA and vBPA.

References

1. L. Aceto, W. Fokkink, and A. Ingólfsdóttir. 2-nested simulation is not finitely equationally axiomatizable. In *Proc. of STACS'01*, vol. 2010 of *LNCS*, p. 39–50. Springer, 2001.
2. R. Alur, K. Etessami, and P. Madhusudan. A temporal logic of nested calls and returns. In *Proc. of TACAS'04*, vol. 2988 of *LNCS*, p. 467–481. Springer, 2004.
3. R. Alur, V. Kumar, P. Madhusudan, and M. Viswanathan. Congruences for visibly pushdown languages. In *Proc. of ICALP'05*, vol. 3580 of *LNCS*, p. 1102–1114. Springer, 2005.
4. R. Alur and P. Madhusudan. Visibly pushdown languages. In *Proc. of STOC'04*, p. 202–211. ACM Press, 2004.

5. J. Balcazar, J. Gabarro, and M. Santha. Deciding bisimilarity is P-complete. *Formal Aspects of Computing*, 4(6A):638–648, 1992.
6. V. Bárány, Ch. Löding, and O. Serre. Regularity problems for visibly pushdown languages. In *Proc. of STACS'06*, vol. 3884 of *LNCS*, p. 420–431. Springer, 2006.
7. A. Bouajjani, J. Esparza, and O. Maler. Reachability analysis of pushdown automata: Application to model-checking. In *Proc. of CONCUR'97*, vol. 1243 of *LNCS*, p. 135–150. Springer, 1997.
8. O. Burkart, D. Caucal, and B. Steffen. An elementary decision procedure for arbitrary context-free processes. In *Proc. of MFCS'95*, vol. 969 of *LNCS*, p. 423–433. Springer, 1995.
9. O. Burkart, D. Caucal, and B. Steffen. Bisimulation collapse and the process taxonomy. In *Proc. of CONCUR'96*, vol. 1119 of *LNCS*, p. 247–262. Springer, 1996.
10. J.F. Groote and H. Hüttel. Undecidable equivalences for basic process algebra. *Information and Computation*, 115(2):353–371, 1994.
11. M. Holzer. On emptiness and counting for alternating finite automata. In *Proc. of DLT'95*, pages 88–97. World Scientific, 1996.
12. P. Jančar. Decidability of bisimilarity for one-counter processes. *Information and Computation*, 158(1):1–17, 2000.
13. P. Jančar, A. Kučera, F. Moller, and Z. Sawa. DP lower bounds for equivalence-checking and model-checking of one-counter automata. *Information and Computation*, 188(1):1–19, 2004.
14. P. Jančar, F. Moller, and Z. Sawa. Simulation problems for one-counter machines. In *Proc. of SOFSEM'99*, vol. 1725 of *LNCS*, p. 404–413. Springer, 1999.
15. P. Jančar and J. Srba. Highly undecidable questions for process algebras. In *Proc. of TCS'04*, p. 507–520. Kluwer Academic Publishers, 2004.
16. D. Kozen. Results on the propositional μ-calculus. *Theoretical Computer Science*, 27:333–354, 1983.
17. A. Kučera and R. Mayr. On the complexity of semantic equivalences for pushdown automata and BPA. In *Proc. of MFCS'02*, vol. 2420 of *LNCS*, p. 433–445. Springer, 2002.
18. A. Murawski and I. Walukiewicz. Third-order idealized algol with iteration is decidable. In *Proc. of FOSSACS'05*, vol. 3441 of *LNCS*, p. 202–218, 2005.
19. R. Paige and R. E. Tarjan. Three partition refinement algorithms. *SIAM Journal of Computing*, 16(6):973–989, December 1987.
20. C. Pitcher. Visibly pushdown expression effects for XML stream processing. In *Proc. of PLAN-X*, pages 5–19, 2005.
21. Z. Sawa and P. Jančar. P-hardness of equivalence testing on finite-state processes. In *Proc. of SOFSEM'01*, vol. 2234 of *LNCS*, p. 326–345. Springer, 2001.
22. G. Sénizergues. Decidability of bisimulation equivalence for equational graphs of finite out-degree. In *Proc. of FOCS'98*, p. 120–129. IEEE Computer Society, 1998.
23. O. Serre. Parity games played on transition graphs of one-counter processes. In *Proc. of FOSSACS'06*, vol. 3921 of *LNCS*, p. 337–351. Springer, 2006.
24. J. Srba. Strong bisimilarity and regularity of basic process algebra is PSPACE-hard. In *Proc. of ICALP'02*, vol. 2380 of *LNCS*, p. 716–727. Springer, 2002.
25. I. Walukiewicz. Pushdown processes: Games and model-checking. *Information and Computation*, 164(2):234–263, 2001.

A Finite Semantics of Simply-Typed Lambda Terms for Infinite Runs of Automata

Klaus Aehlig

Mathematisches Institut
Ludwig-Maximilians-Universität München
Theresienstr. 39, 80333 München, Germany
aehlig@math.lmu.de

Abstract. Model checking properties are often described by means of finite automata. Any particular such automaton divides the set of infinite trees into finitely many classes, according to which state has an infinite run. Building the full type hierarchy upon this interpretation of the base type gives a finite semantics for simply-typed lambda-trees.

A calculus based on this semantics is proven sound and complete. In particular, for regular infinite lambda-trees it is decidable whether a given automaton has a run or not. As regular lambda-trees are precisely recursion schemes, this decidability result holds for arbitrary recursion schemes of arbitrary level, without any syntactical restriction. This partially solves an open problem of Knapik, Niwinski and Urzyczyn.

1 Introduction and Related Work

The lambda calculus has long been used as a model of computation. Restricting it to simple types allows for a particularly simple set-theoretic semantics. The drawback, however, is that only few functions can be defined in the simply-typed lambda calculus. To overcome this problem one can, for example, add fixed-point combinators Y_σ at every type, or allow infinitary lambda terms. The latter is more flexible, as we can always syntactically unfold fixed points, paying the price to obtain an infinite, but regular, lambda-tree.

Finite automata are a standard tool in the realm of model checking [10]. They provide a concrete machine model for the properties to be verified. In this article we combine automata, and hence properties relevant for model checking, with the infinitary simply-typed lambda calculus, using the fact that the standard set theoretic semantics for the simple types has a free parameter — the interpretation of the base type.

More precisely, we consider the following problem.

> Given a, possibly infinite, simply-typed lambda-tree t of base type, and given a non-deterministic tree automaton $\mathfrak{A}$. Does $\mathfrak{A}$ have a run on the normal form of t?

Z. Ésik (Ed.): CSL 2006, LNCS 4207, pp. 104–118, 2006.

The idea is to provide a "proof" of a run of $\mathfrak{A}$ on the normal form of t by annotating each subterm of t with a semantical value describing how this subterm "looks, as seen by $\mathfrak{A}$". Since, in the end, all the annotations turn out to be out of a fixed finite set, the existence of such a proof is decidable.

So, what does a lambda-tree look like, if seen by an automaton $\mathfrak{A}$? At the base type, a lambda-tree denotes an infinite term. Hence, from $\mathfrak{A}$'s point of view, we have to distinguish for which states there is an infinite run starting in this particular state.

Since we are interested in model checking terms of base type only, we can use any semantics for higher types, as long as it is adequate, that is, sound and complete. So we use the most simple one available, that is, the full set-theoretic semantics with the base type interpreted as just discussed. This yields a finite set as semantical realm for every type.

As an application of the techniques developed in this article, we show that for arbitrary recursion schemes it is decidable whether the defined tree has a property expressible by an automaton with trivial acceptance condition. This gives a partial answer to an open problem by Knapik, Niwinski and Urzyczyn [5].

Infinitary lambda-trees were also considered by Knapik, Niwinski and Urzyczyn [4], who also proved the decidability of the Monadic Second Order (MSO) theory of trees given by recursion schemes enjoying a certain "safety" condition [5]. The fact, that the safety restriction can be dropped at level two has been shown by Aehlig, de Miranda and Ong [2], and, independently, by Knapik, Niwinski, Urzyczyn and Walukiewicz [6]. The work of the former group also uses implicitly the idea of a "proof" that a particular automaton has a run on the normal form of a given infinite lambda-tree.

Recently [9] Luke Ong showed simultaneously and independently that the safety restriction can be dropped for all levels and still decidability for the full MSO theory is obtained. His approach is based on game semantics and is technically quite involved. Therefore, the author believes that his approach, due to its simplicity and straight forwardness, is still of interest, despite showing a weaker result. Moreover, the novel construction of a finite semantics and its adequacy even in a coinductive setting seem to be of independent interest.

2 Preliminaries

Let Σ' be a set of *letters* or *terminals*. We use $\mathfrak{f}$ to denote elements of Σ'. Each terminal $\mathfrak{f}$ is associated an *arity* $\sharp(\mathfrak{f}) \in \mathbb{N}$.

Definition 1. Define $\Sigma = \Sigma' \cup \{\mathcal{R}, \beta\}$ with $\mathcal{R}, \beta$ two new terminals of arity one.

Definition 2. For Δ a set of terminals, a Δ-term is a, not necessarily well-founded, tree labelled with elements of Δ where every node labelled with $\mathfrak{f}$ has $\sharp(\mathfrak{f})$ many children.

Example 3. Let $\Sigma' = \{\mathtt{f}, \mathtt{g}, \mathtt{a}\}$ with $\mathtt{f}$, $\mathtt{g}$ and $\mathtt{a}$ of arities 2, 1, and 0, respectively. Figure 1 shows two Σ'-terms.

Let $\mathfrak{A}$ be a fixed nondeterministic tree automaton with state set Q and transition function $\delta\colon Q \times \Sigma \to \mathfrak{P}((Q \cup \{*\})^N)$ where $N = \max\{\sharp(\mathtt{g}) \mid \mathtt{g} \in \Sigma\}$ is the maximal arity and $\delta(q, \mathtt{g}) \subset Q^{\sharp(\mathtt{g})} \times \{*\}^{N-\sharp(\mathtt{g})}$ whenever $q \in Q$ and $\mathtt{g} \in \Sigma$.

In other words, let $\mathfrak{A}$ be a nondeterministic tree automaton that works on Σ-terms.

Definition 4. We say that $\mathfrak{A}$ *has a run up to level* n, if it has a run that goes at least till all nodes with distance at most n from the root.

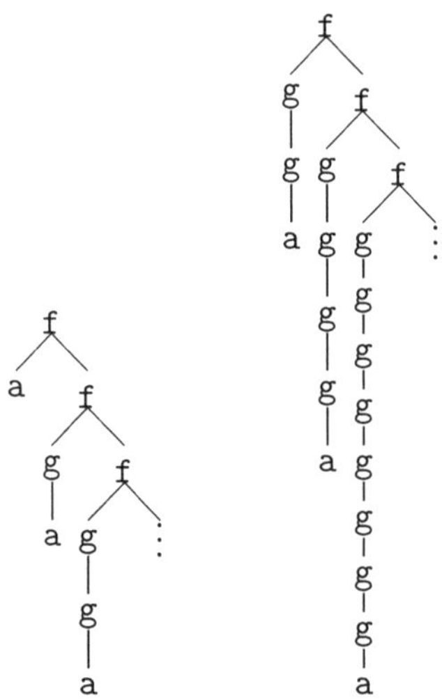

Fig. 1. Two $\{\mathtt{f}, \mathtt{g}, \mathtt{a}\}$-terms

We write $\mathfrak{A}, q \models^n t$ to denote that $\mathfrak{A}$ has a run on t up to level n starting in state q. We write $\mathfrak{A}, q \models^\infty t$ to denote that $\mathfrak{A}$ has an *infinite* run on t starting in state q. Since there are only finitely many ways to extend a run of length n to one of length $n+1$, by König's Lemma we obtain that $\mathfrak{A}, q \models^\infty t$ if and only if $\forall n. \mathfrak{A}, q \models^n t$.

Example 5. Continuing Example 3 consider the property

> "Every maximal chain of letters $\mathtt{g}$ has even length".

It can be expressed by an automaton with two states $Q = \{q_2, q_1\}$ where q_2 means that an even number of $\mathtt{g}$s has been passed on the path so far, where q_1 means that the maximal chain of $\mathtt{g}$s passed has odd length. Then the initial state is q_2 and the transition function is as follows.

$$\begin{array}{ll} \delta(\mathtt{f}, q_2) = \{(q_2, q_2)\} & \delta(\mathtt{f}, q_1) = \emptyset \\ \delta(\mathtt{g}, q_2) = \{(q_1, *)\} & \delta(\mathtt{g}, q_1) = \{(q_2, *)\} \\ \delta(\mathtt{a}, q_2) = \{(*, *)\} & \delta(\mathtt{a}, q_1) = \emptyset \end{array}$$

This automaton can be extended to work on Σ-trees by setting $\delta(q, \mathcal{R}) = \delta(q, \beta) = \{(q, *)\}$. Note that this automaton has an infinite run on the second tree in Figure 1, whereas it has a run only up to level 3 on the first one.

Definition 6. The *simple types*, denoted by ρ, σ, τ, are built from the base type ι by arrows $\rho \to \sigma$. The arrow associates to the right. In particular, $\overrightarrow{\rho} \to \iota$ is short for $\rho_1 \to (\rho_2 \to (\ldots(\rho_n \to \iota)\ldots))$.

Definition 7. *Infinitary simply-typed lambda-trees* over typed terminals Σ' are coinductively given by the grammar

$$r, s \ ::= \ x^\rho \mid (\lambda x^\rho t^\sigma)^{\rho\to\sigma} \mid (t^{\rho\to\sigma} s^\rho)^\sigma \mid \mathfrak{f}^{\iota\to\ldots\to\iota\to\iota}.$$

In other words, they are, not-necessarily well founded, trees built, in a locally type respecting way, from unary λx^ρ-nodes, binary @-nodes representing application, and leaf nodes consisting of typed variables x^ρ of type ρ and typed constants $\mathfrak{f} \in \Sigma'$ of type $\underbrace{\iota \to \ldots \to \iota}_{\sharp(\mathfrak{f})} \to \iota$.

Here λx^ρ binds free occurrences of the variable x^ρ in its body. Trees with all variables bound are called *closed*.

A lambda-tree with only finitely many non-isomorphic subtrees is called *regular*.

We omit type superscripts if they are clear from the context, or irrelevant.

We usually leave out the words "simply typed", tacitly assuming all our lambda-trees to be simply typed and to use terminals from Σ' only. Figure 2 shows two regular lambda-trees. Arrows are used to show where the pattern repeats, or to draw isomorphic subtrees only once. Note that they denote terms (shown in Figure 1) that are not regular. Here, by "denote" we mean the *term reading* of the normal form.

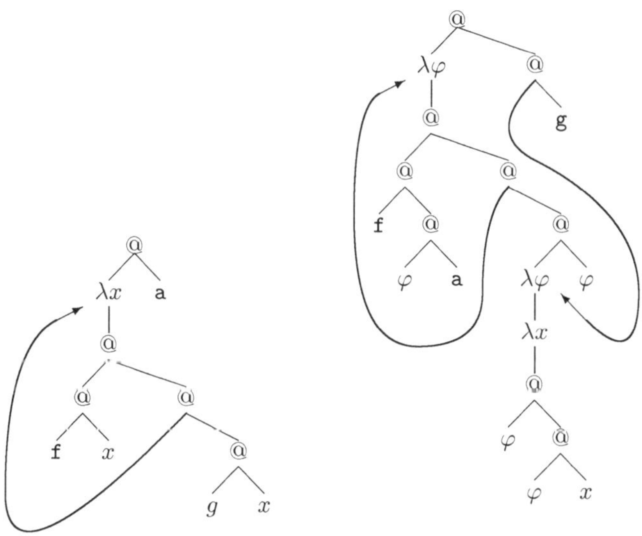

Fig. 2. Two regular lambda-trees with denotation being the $\{\mathtt{f}, \mathtt{g}, \mathtt{a}\}$-terms in Figure 1

Remark 8. It should be noted that in lambda-trees, as opposed to Σ'-terms, all constants and variables, no matter what their type is, occur at leaf positions.

The reason is, that in a lambda-calculus setting the main concept is that of an application. This is different from first order terms, where the constructors are the main concept.

Note that we use lambda-trees to denote Σ'-terms. As these are different concepts, even normal lambda-trees differ from their denotation. For example the lambda-tree 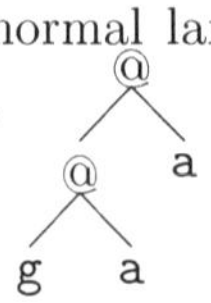

denotes the Σ'-term 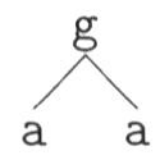

.

3 Recursion Schemes as Means to Define Regular Lambda Trees

The interest in infinitary lambda-trees in the verification community recently arose by the study of recursion schemes. It could be shown [4,5] that under a certain "safety" condition the (infinite) terms generated by recursion schemes have decidable monadic second order theory. For our purpose it is enough to consider recursion schemes as a convenient means to define regular lambda-trees.

Definition 9. *Recursion schemes* are given by a set of first order terminal symbols, simply-typed non-terminal symbols and for every non-terminal F an equation

$$F\overrightarrow{x} = e$$

where e is an expression of ground type built up from terminals, non-terminals and the variables $\overrightarrow{x}$ by type-respecting application. There is a distinguished non-terminal symbol S of ground type, called the *start symbol.*

Definition 10. Each recursion scheme *denotes*, in the obvious way, a partial, in general infinite, term built from the terminals. Starting from the start symbol, recursively replace the outer-most non-terminals by their definitions with the arguments substituted in appropriately.

Definition 11. To every recursion scheme is *associated* a regular lambda-tree in the following way. First replace all equations $F\overrightarrow{x} = e$ by

$$F = \lambda\overrightarrow{x}.e$$

where the right hand side is read as a lambda term.

Then, starting from the start symbol, recursively replace all non-terminals by their definition *without performing any computations.*

Remark 12. Immediately from the definition we note that the β-normal form of the lambda-tree associated with a recursion scheme, when read a term, *is* the term denoted by that recursion scheme.

Example 13. Figure 3 shows two recursion schemes with non-terminals $F\colon \iota \to \iota$, $F'\colon (\iota \to \iota) \to \iota$, $W\colon (\iota \to \iota) \to \iota \to \iota$, and $S, S'\colon \iota$. Their corresponding lambda-trees are the ones shown in Figure 2. The sharing of an isomorphic sub-tree arises as both are translations of the same non-terminal W. As already observed, these recursion schemes denote the terms shown in Figure 1.

$$\begin{aligned} S &= F\mathtt{a} \\ Fx &= \mathtt{f}x(F(\mathtt{g}x)) \end{aligned} \qquad \begin{aligned} S' &= F'(W\mathtt{g}) \\ F'\varphi &= \mathtt{f}(\varphi a)(F'(W\varphi)) \\ W\varphi x &= \varphi(\varphi x) \end{aligned}$$

Fig. 3. Two recursion schemes

Remark 14. The notion of a recursion scheme wouldn't change if we allowed λ-abstractions on the right hand side of the equations; we can always build the closure and "factor it out" as a new non-terminal. For example, the $W\varphi$ in the definition of F' in Figure 3 should be thought of as a factored-out closure of a line that originally looked

$$F'\varphi = \mathtt{f}(\varphi a)(F'(\lambda x.\varphi(\varphi x))) \ .$$

4 Using Continuous Normalisation

As mentioned in the introduction, we are interested in the question, whether $\mathfrak{A}$ has a run on the normal form of some lambda-tree t. Our plan to investigate this question is by analysing the term t.

However, there is no bound on the number of nodes of t that have to be inspected, and no bound on the number of beta-reductions to be carried out, before the first symbol of the normal form is determined — if it ever will be. In fact, it may well be that an infinite simply-typed lambda-tree leaves the normal form undefined at some point.

Whereas the first observation is merely a huge inconvenience, the second observation makes it unclear what it even is supposed to mean that "$\mathfrak{A}$ has a run on the normal form of t" — if there is no such normal form.

Fortunately, it is long known how to overcome this problem. If we don't know any definite answer yet, we just output a "don't know" constructor and carry on. This idea is known as "continuous normalisation" [7,8] and is quite natural [1] in the realm of the lambda calculus.

Definition 15. For t, $\overrightarrow{t}$ closed infinitary simply-typed lambda-trees such that $t\,\overrightarrow{t}$ is of ground type we define a Σ-term $t\underline{@}\overrightarrow{t}$ coinductively as follows.

$$\begin{aligned} (rs)\underline{@}\overrightarrow{t} &= \mathcal{R}(r\underline{@}(s,\overrightarrow{t})) \\ (\lambda x.r)\underline{@}(s,\overrightarrow{t}) &= \beta(r[s/x]\underline{@}\overrightarrow{t}) \\ \mathfrak{f}\underline{@}\overrightarrow{t} &= \mathfrak{f}(t_1^\beta,\ldots,t_n^\beta) \end{aligned}$$

Here we used $r[s/x]$ to denote the substitution of s for x in r. This substitution is necessarily capture free as s is closed. By $\mathfrak{f}(T_1,\ldots,T_n)$ we denote the term with label $\mathfrak{f}$ at the root and $T_1,\ldots,T_n$ as its n children; this includes the case $n = 0$, where $\mathfrak{f}()$ denotes the term consisting of a single node $\mathfrak{f}$. Similar notation is used for $\mathcal{R}(T)$ and $\beta(T)$. Moreover we used r^β as a shorthand for $r\underline{@}()$.

Immediately from the definition we notice that, after removing the $\mathcal{R}$ and β constructors, $r\underline{@}\overrightarrow{s}$ is the term reading of the normal form of $r\,\overrightarrow{s}$, whenever the latter is defined.

The number of β constructors counts the number of reductions necessary to reach a particular letter of the normal form [1]. Therefore, $\mathfrak{A}$ can talk not only about properties of the normal form of t, but also about the computation that led there.

It should be noted that no price has to be paid for this extra expressivity. Given an automaton on Σ' we can extend its transition function δ by setting $\delta(q,\mathcal{R}) = \delta(q,\beta) = \{(q,*,\ldots,*)\}$.

5 Finitary Semantics and Proof System

The main technical idea of this article is to use a finite semantics for the simple types, describing how $\mathfrak{A}$ "sees" an object of that type.

Definition 16. For τ a simple type we define $[\![\tau]\!]$ inductively as follows.

$$\begin{aligned} [\![\iota]\!] &= \mathfrak{P}(Q) \\ [\![\rho \to \sigma]\!] &= [\![\sigma]\!]^{[\![\rho]\!]} \end{aligned}$$

In other words, we start with the powerset of the state set of $\mathfrak{A}$ in the base case, and use the full set theoretic function space for arrow-types.

Remark 17. Obviously all the $[\![\tau]\!]$ are finite sets.

Example 18. Taking the automaton $\mathfrak{A}$ of Example 5, we have $[\![\iota]\!] = \{\emptyset, \{q_2\}, \{q_1\}, Q\}$ and examples of elements of $[\![\iota \to \iota]\!]$ include the identity function id, as well as the "swap function" swap defined by $\mathrm{swap}(\emptyset) = \emptyset$, $\mathrm{swap}(Q) = Q$, $\mathrm{swap}(\{q_2\}) = \{q_1\}$, and $\mathrm{swap}(\{q_1\}) = \{q_2\}$.

Definition 19. $[\![\tau]\!]$ is partially ordered as follows.

- For $R, S \in [\![\iota]\!]$ we set $R \sqsubseteq S$ iff $R \subseteq S$.
- For $f, g \in [\![\rho \to \sigma]\!]$ we set $f \sqsubseteq g$ iff $\forall a \in [\![\rho]\!]. fa \sqsubseteq ga$.

Remark 20. Obviously suprema and infima with respect to $\sqsubseteq$ exist.

We often need the concept "continue with f after reading one $\mathcal{R}$ symbol". We call this $\mathcal{R}$-lifting. Similar for β.

Definition 21. For $f \in [\![\vec{\rho} \to \iota]\!]$ we define the liftings $\mathcal{R}(f), \beta(f) \in [\![\vec{\rho} \to \iota]\!]$ as follows.

$$\begin{aligned} \mathcal{R}(f)(\vec{a}) &= \{q \mid \delta(q,\mathcal{R}) \cap f\vec{a} \times \{*\} \times \ldots \times \{*\} \neq \emptyset\} \\ \beta(f)(\vec{a}) &= \{q \mid \delta(q,\beta) \cap f\vec{a} \times \{*\} \times \ldots \times \{*\} \neq \emptyset\} \end{aligned}$$

Remark 22. If $\mathfrak{A}$ is obtained from an automaton working on Σ'-terms by setting $\delta(q,\mathcal{R}) = \delta(q,\beta) = \{(q,*,\ldots,*)\}$ then $\mathcal{R}(f) = \beta(f) = f$ for all f.

Using this finite semantics we can use it to annotate a lambda-tree by semantical values for its subtrees to show that the denoted term has good properties with respect to $\mathfrak{A}$. We start by an example.

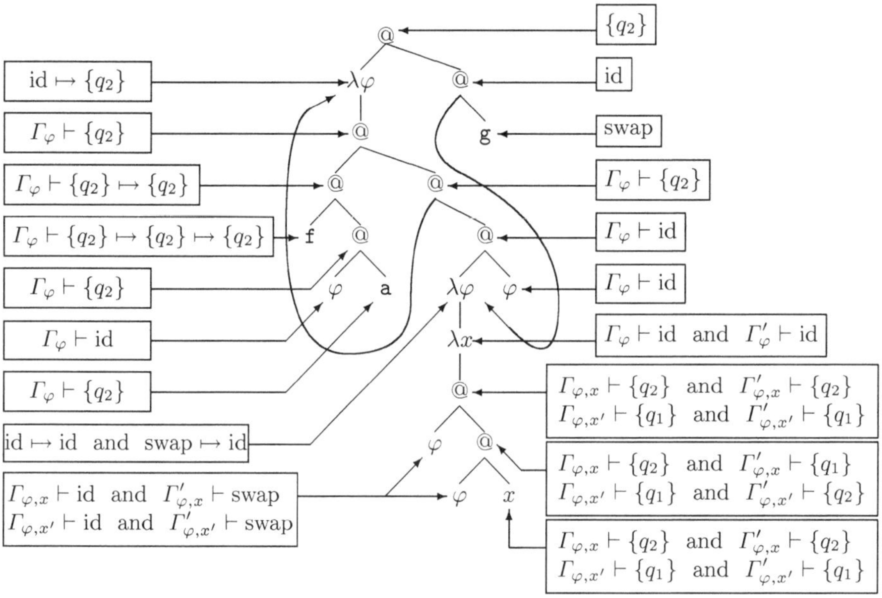

Fig. 4. A proof that $\mathfrak{A}$ has an infinite run starting in q_2 on the denoted term

Example 23. The second recursion scheme in Figure 3 denotes a term where the "side branches" contain $2, 4, 8, \ldots, 2^n, \ldots$ times the letter g. As these are all even numbers, $\mathfrak{A}$ should have a run when starting in q_2.

So we start by assigning the root $\{q_2\} \in [\iota]$. Since the term is an application, we have to guess the semantics of the argument (of type $\iota \to \iota$). Our (correct) guess is, that it keeps the parity of gs unchanged, hence our guess is id; the function side then must be something that maps id to $\{q_2\}$. Let us denote by id $\mapsto \{q_2\}$ the function in ${}^{[\iota\to\iota]}[\iota]$ defined by $(\mathrm{id} \mapsto \{q_2\})(\mathrm{id}) = \{q_2\}$ and $(\mathrm{id} \mapsto \{q_2\})(f) = \emptyset$ if $f \neq \mathrm{id}$.

The next node to the left is an abstraction. So we have to assign the body the value $\{q_2\}$ in a context where φ is mapped to id. Let us denote this context by Γ_φ.

In a similar way we fill out the remaining annotations. Figure 4 shows the whole proof. Here Γ'_φ is the context that maps φ to swap; moreover $\Gamma_{\varphi,x}$, $\Gamma'_{\varphi,x}$, $\Gamma_{\varphi,x'}$, and $\Gamma'_{\varphi,x'}$ are the same as Γ_φ and Γ'_φ but with x mapped to $\{q_2\}$ and $\{q_1\}$, respectively.

It should be noted that a similar attempt to assign semantical values to the other lambda-tree in Figure 2 fails at the down-most x where in the context Γ with $\Gamma(x) = \{q_2\}$ we cannot assign x the value $\{q_1\}$.

To make the intuition of the example precise, we formally define a "proof system" of possible annotations (Γ, a) for a (sub)tree. Since the $[\tau]$ are all finite sets, there are only finitely many possible annotations.

To simplify the later argument on our proof, which otherwise would be coinductive, we add a level n to our notion of proof. This level should be interpreted as "for up to n steps we can pretend to have a proof". This reflects the fact that coinduction is nothing but induction on observations.

Definition 24. For Γ a finite mapping from variables x^σ to their corresponding semantics $[\![\sigma]\!]$, a value $a \in [\![\rho]\!]$, and t an infinitary, maybe open, lambda-tree of type ρ, with free variables among $\mathrm{dom}(\Gamma)$, we define

$$\Gamma \vdash^n_{\mathfrak{A}} a \sqsubseteq t : \rho$$

by induction on the natural number n as follows.

- $\Gamma \vdash^0_{\mathfrak{A}} a \sqsubseteq t : \rho$ always holds.
- $\Gamma \vdash^n_{\mathfrak{A}} a \sqsubseteq x_i : \rho$ holds, provided $a \sqsubseteq \Gamma(x_i)$.
- $\Gamma \vdash^{n+1}_{\mathfrak{A}} a \sqsubseteq st : \sigma$ holds, provided there exists $f \in [\![\rho \to \sigma]\!]$, $u \in [\![\rho]\!]$ such that $a \sqsubseteq \mathcal{R}(fu)$, $\Gamma \vdash^n_{\mathfrak{A}} f \sqsubseteq s : \rho \to \sigma$, and $\Gamma \vdash^n_{\mathfrak{A}} u \sqsubseteq t : \rho$.
- $\Gamma \vdash^{n+1}_{\mathfrak{A}} f \sqsubseteq \lambda x^\rho.s : \rho \to \sigma$ holds, provided for all $a \in [\![\rho]\!]$ there is a $b_a \in [\![\sigma]\!]$ such that $fa \sqsubseteq \beta(b_a)$ and $\Gamma^a_x \vdash^n_{\mathfrak{A}} b_a \sqsubseteq s : \sigma$.
- $\Gamma \vdash^n_{\mathfrak{A}} f \sqsubseteq \mathfrak{f} : \iota \to \ldots \to \iota \to \iota$ holds, provided for all $\vec{a} \in [\![\vec{\iota}]\!]$ we have $f\vec{a} \subset \{q \mid \delta(q, \mathfrak{f}) \cap a_1 \times \ldots \times a_{\sharp(\mathfrak{f})} \times \{*\} \times \ldots \times \{*\} \neq \emptyset\}$.

It should be noted that all the quantifiers in the rules range over finite sets. Hence the correctness of a rule application can be checked effectively (and even by a finite automaton).

We write $\Gamma \vdash^\infty_{\mathfrak{A}} a \sqsubseteq t : \rho$ to denote $\forall n. \Gamma \vdash^n_{\mathfrak{A}} a \sqsubseteq t : \rho$.

Remark 25. Obviously $\Gamma \vdash^{n+1}_{\mathfrak{A}} a \sqsubseteq t : \rho$ implies $\Gamma \vdash^n_{\mathfrak{A}} a \sqsubseteq t : \rho$. Moreover, $a' \sqsubseteq a$ and $\Gamma \vdash^n_{\mathfrak{A}} a \sqsubseteq t : \rho$ imply $\Gamma \vdash^n_{\mathfrak{A}} a' \sqsubseteq t : \rho$. Finally, $\Gamma \vdash^n_{\mathfrak{A}} a \sqsubseteq t : \rho$, if $\Gamma' \vdash^n_{\mathfrak{A}} a \sqsubseteq t : \rho$ for some Γ' which agrees with Γ on the free variables of t.

As already mentioned, for t a term with finitely many free variables, the annotations (Γ, a) come from a fixed finite set, since we can restrict Γ to the set of free variables of t. If, moreover, t has only finitely many different sub-trees, that is to say, if t is regular, then only finitely many terms t have to be considered. So we obtain

Proposition 26. *For t regular, it is decidable whether $\Gamma \vdash^\infty_{\mathfrak{A}} a \sqsubseteq t : \rho$.*

Before we continue and show our calculus in Definition 24 to be sound (Section 6) and complete (Section 7) let us step back and see what we will then have achieved, once our calculus is proven sound and complete.

Proposition 26 gives us decidability for terms denoted by regular lambda-trees, and hence in particular for trees obtained by recursion schemes. Moreover, since the annotations only have to fit locally, individual subtrees of the lambda-tree can be verified separately. This is of interest, as for each non-terminal a separate subtree is generated. In other words, this approach allows for modular verification; think of the different non-terminals as different subroutines. As the semantics is the set-theoretic one, the annotations are clear enough to be meaningful, if we have chosen our automaton in such a way that the individual states can be interpreted extensionally, for example as "even" versus "odd" number of gs.

It should also be noted, that the number of possible annotations only depends on the type of the subtree, and on $\mathfrak{A}$, that is, the property to be investigated. Fixing $\mathfrak{A}$ and the allowed types (which both usually tend to be quite small), the amount of work to be carried out grows only linearly with the representation of t as a regular lambda-tree. For every node we have to make a guess and we have to check whether this guess is consistent with the guesses for the (at most two) child nodes. Given that the number of nodes of the representation of t growth linearly with the size of the recursion scheme, the problem is in fixed-parameter-$\mathcal{NP}$, which doesn't seem too bad for practical applications.

6 Truth Relation and Proof of Soundness

The soundness of a calculus is usually shown by using a logical relation, that is, a relation indexed by a type that interprets the type arrow "$\rightarrow$" as logical arrow "$\Rightarrow$"; in other words, we define partial truth predicates for the individual types [11].

Since we want to do induction on the "observation depth" n of our proof $\cdot \vdash^n_{\mathfrak{A}} \cdot \sqsubseteq \cdot : \tau$ we have to include that depth in the definition of our truth predicates $\cdot \prec\!\!\!\prec^n_{\mathfrak{A}} \cdot : \tau$. For technical reasons we have to build in weakening on this depth as well.

Definition 27. For $f \in [\overrightarrow{\rho} \rightarrow \iota]$, $n \in \mathbb{N}$, t a closed infinitary lambda tree of type $\overrightarrow{\rho} \rightarrow \iota$, the relation $f \prec\!\!\!\prec^n_{\mathfrak{A}} t : \overrightarrow{\rho} \rightarrow \iota$ is defined by induction on the type as follows.

$$\begin{array}{l} f \prec\!\!\!\prec^n_{\mathfrak{A}} t : \overrightarrow{\rho} \rightarrow \iota \quad \text{iff} \\ \forall \ell \leq n \forall \overrightarrow{a} \in [\overrightarrow{\rho}] \forall \overrightarrow{r} : \overrightarrow{\rho} \\ \qquad (\forall i.\ a_i \prec\!\!\!\prec^\ell_{\mathfrak{A}} r_i : \rho_i) \Rightarrow \forall q \in f\overrightarrow{a}\,.\ \mathfrak{A}, q \models^\ell t \underline{@} \overrightarrow{r} \end{array}$$

Remark 28. Immediately from the definition we get the following monotonicity property.

If $f \sqsubseteq f'$ and $f' \prec\!\!\!\prec^n_{\mathfrak{A}} t : \rho$ then $f \prec\!\!\!\prec^n_{\mathfrak{A}} t : \rho$.

Remark 29. In the special case $\overrightarrow{\rho} = \varepsilon$ we get

$$S \prec\!\!\!\prec^n_{\mathfrak{A}} t : \iota \qquad \text{iff} \qquad \forall q \in S. \mathfrak{A}, q \models^n t^\beta$$

Here we used that $\forall \ell \leq n. \mathfrak{A}, q \models^\ell s$ iff $\mathfrak{A}, q \models^n s$.

Immediately from the definition we obtain weakening in the level.

Proposition 30. *If* $f \prec\!\!\!\prec^n_{\mathfrak{A}} t : \rho$ *then* $f \prec\!\!\!\prec^{n-1}_{\mathfrak{A}} t : \rho$.

Theorem 31. *Assume* $\Gamma \vdash^n_{\mathfrak{A}} a \sqsubseteq t : \rho$ *for some* Γ *with domain* $\{x_1, \ldots, x_2\}$. *For all* $\ell \leq n$ *and all closed terms* $\overrightarrow{t} : \overrightarrow{\rho}$, *if* $\forall i.\ \Gamma(x_i) \prec\!\!\!\prec^\ell_{\mathfrak{A}} t_i : \rho_i$ *then* $a \prec\!\!\!\prec^\ell_{\mathfrak{A}} t[\overrightarrow{t}/\overrightarrow{x}] : \rho$.

Proof. Induction on n, cases according to $\Gamma \vdash^n_{\mathfrak{A}} a \sqsubseteq t : \rho$.

We just show the case of the λ-rule. The other cases are similar, and even simpler.

Case $\Gamma \vdash^{n+1}_{\mathfrak{A}} f \sqsubseteq \lambda x^\rho.s : \rho \to \sigma$ thanks to $\forall a \in [\![\rho]\!]\, \exists b_a \in [\![\sigma]\!]$ such that $fa \sqsubseteq \beta(b_a)$ and $\Gamma^a_x \vdash^n_{\mathfrak{A}} b_a \sqsubseteq s : \sigma$.

Let $\ell \leq n+1$ be given and $\overrightarrow{t} : \overrightarrow{\rho}$ with $\Gamma(x_i) \prec\!\!\prec^\ell_{\mathfrak{A}} t_i : \rho_i$.

We have to show $f \prec\!\!\prec^\ell_{\mathfrak{A}} (\lambda x^\rho s^\sigma)\eta : \rho \to \sigma$ where η is short for $[\overrightarrow{t}/\overrightarrow{x}]$.

Let σ have the form $\sigma = \overrightarrow{\sigma} \to \iota$. Let $k \leq \ell$ be given and $r : \rho$, $\overrightarrow{s} : \overrightarrow{\sigma}$, $c \in [\![\rho]\!]$, $c_i \in [\![\sigma_i]\!]$ such that $c \prec\!\!\prec^k_{\mathfrak{A}} r : \rho$, $c_i \prec\!\!\prec^k_{\mathfrak{A}} s_i : \sigma_i$. We have to show for all $q \in fc\overrightarrow{c}$ that $\mathfrak{A}, q \models^k \underbrace{(\lambda xs)\eta@r, \overrightarrow{s}}_{\beta.s\eta^r_x@\overrightarrow{s}}$.

Hence it suffices to show that there is a $\tilde{q} \in \delta(q,\beta)$ such that $\mathfrak{A}, \tilde{q} \models^{k-1} s\eta^r_x@\overrightarrow{s}$.

We know $c \prec\!\!\prec^k_{\mathfrak{A}} r : \rho$; using Proposition 30 we get $c \prec\!\!\prec^{k-1}_{\mathfrak{A}} r : \rho$ and $\forall i.\ \Gamma(x_i) \prec\!\!\prec^{k-1}_{\mathfrak{A}} t_i : \rho_i$. Since $k \leq \ell \leq n+1$ we get $k-1 \leq n$, hence we may apply the induction hypothesis to $\Gamma^a_x \vdash^n_{\mathfrak{A}} b_a \sqsubseteq s : \sigma$ and obtain $b_a \prec\!\!\prec^{k-1}_{\mathfrak{A}} s\eta^r_x : \sigma$.

Since again by Proposition 30 we also know $c_i \prec\!\!\prec^{k-1}_{\mathfrak{A}} s_i : \sigma_i$, we obtain for all $\hat{q} \in b_a\overrightarrow{c}$ that $\mathfrak{A}, \hat{q} \models^{k-1} s\eta^r_x@\overrightarrow{s}$.

Since $fc \sqsubseteq \beta(b_c)$ we get that $\forall q \in fc\overrightarrow{c}\, \exists \tilde{q} \in \delta(q,\beta).\, \tilde{q} \in b_c\overrightarrow{c}$. This, together with the last statement yields the claim.

It should be noted that in the proof of Theorem 31 in the cases of the λ-rule and the application-rule it was possible to use the induction hypothesis due to the fact that we used *continuous* normalisation, as opposed to standard normalisation.

Corollary 32. *For t a closed infinitary lambda term we get immediately from Theorem 31*

$$\emptyset \vdash^n_{\mathfrak{A}} S \sqsubseteq t : \iota \quad \Longrightarrow \quad \forall q \in S.\, \mathfrak{A}, q \models^n t^\beta$$

In particular, if $\emptyset \vdash^\infty_{\mathfrak{A}} S \sqsubseteq t : \iota$ then $\forall q \in S.\, \mathfrak{A}, q \models^\infty t^\beta$.

7 The Canonical Semantics and the Proof of Completeness

If we want to prove that there is an infinite run, then, in the case of an application st, we have to guess a value for the term t "cut out".

We could assume an actual run be given and analyse the "communication", in the sense of game semantics [3], between the function s and its argument t. However, it is simpler to assign each term a "canonical semantics" $\langle\!\langle t \rangle\!\rangle_{\mathfrak{A}\infty}$, roughly the supremum of all values we have canonical proofs for.

The subscript ∞ signifies that we only consider infinite runs. The reason is that the level n in our proofs $\Gamma \vdash^n_{\mathfrak{A}} a \sqsubseteq t : \rho$ is not a tight bound; whenever we have a proofs of level n, then there are runs for at least n steps, but on the other hand, runs might be longer than the maximal level of a proof. This is due to the fact that β-reduction moves subterms "downwards", that is, further away from the root, and in that way may construct longer runs. The estimates in our proof calculus, however, have to consider (in order to be sound) the worst case, that is, that an argument is used immediately.

Since, in general, the term t may also have free variables, we have to consider a canonical semantics $\langle\!\langle t\rangle\!\rangle^{\Gamma}_{\mathfrak{A}\infty}$ with respect to an environment Γ.

Definition 33. By induction on the type we define for t a closed infinite lambda-tree of type $\rho = \overrightarrow{\rho} \to \iota$ its canonical semantics $\langle\!\langle t\rangle\!\rangle_{\mathfrak{A}\infty} \in [\![\rho]\!]$ as follows.

$$\langle\!\langle t\rangle\!\rangle_{\mathfrak{A}\infty}(\overrightarrow{a}) = \{q \mid \exists \overrightarrow{s} : \overrightarrow{\rho}\ .\ \langle\!\langle \overrightarrow{s}\rangle\!\rangle_{\mathfrak{A}\infty} \sqsubseteq \overrightarrow{a}\ \wedge \mathfrak{A}, q \models^{\infty} t\underline{@}\overrightarrow{s}\}$$

Remark 34. For t a closed term of base type we have $\langle\!\langle t\rangle\!\rangle_{\mathfrak{A}\infty} = \{q \mid \mathfrak{A}, q \models^{\infty} t^{\beta}\}$.

Definition 35. For Γ a context, $t\colon \rho$ typed in context Γ of type $\rho = \overrightarrow{\rho} \to \iota$ we define $\langle\!\langle t\rangle\!\rangle^{\Gamma}_{\mathfrak{A}\infty} \in [\![\rho]\!]$ by the following explicit definition.

$$\begin{aligned}\langle\!\langle t\rangle\!\rangle^{\Gamma}_{\mathfrak{A}\infty}(\overrightarrow{a}) = \{q \mid \exists \eta.\ &\mathrm{dom}(\eta) = \mathrm{dom}(\Gamma) \wedge \\ &(\forall x \in \mathrm{dom}(\Gamma).\eta(x) \text{ closed} \wedge \langle\!\langle \eta(x)\rangle\!\rangle_{\mathfrak{A}\infty} \sqsubseteq \Gamma(x)) \wedge \\ &\exists \overrightarrow{s} : \overrightarrow{\rho}.\langle\!\langle \overrightarrow{s}\rangle\!\rangle_{\mathfrak{A}\infty} \sqsubseteq \overrightarrow{a}\ \wedge\ \mathfrak{A}, q \models^{\infty} t\eta\underline{@}\overrightarrow{s}\}\end{aligned}$$

Remark 36. For t a closed term and $\Gamma = \emptyset$ we have $\langle\!\langle t\rangle\!\rangle^{\Gamma}_{\mathfrak{A}\infty} = \langle\!\langle t\rangle\!\rangle_{\mathfrak{A}\infty}$.

Proposition 37. *If s has type $\overrightarrow{\sigma} \to \iota$ in some context compatible with Γ, and η is some substitution with $\mathrm{dom}(\eta) = \mathrm{dom}(\Gamma)$ such that for all $x \in \mathrm{dom}(\Gamma)$ we have $\eta(x)$ closed and $\langle\!\langle \eta(x)\rangle\!\rangle_{\mathfrak{A}\infty} \sqsubseteq \Gamma(x)$, then*

$$\langle\!\langle s\eta\rangle\!\rangle_{\mathfrak{A}\infty} \sqsubseteq \langle\!\langle s\rangle\!\rangle^{\Gamma}_{\mathfrak{A}\infty}$$

Proof. Let $\overrightarrow{a} \in [\![\overrightarrow{\sigma}]\!]$ and $q \in \langle\!\langle s\eta\rangle\!\rangle_{\mathfrak{A}\infty}(\overrightarrow{a})$ be given. Then there are $\overrightarrow{s} : \overrightarrow{\sigma}$ with $\langle\!\langle \overrightarrow{s}\rangle\!\rangle_{\mathfrak{A}\infty} \sqsubseteq \overrightarrow{a}$ such that $\mathfrak{A}, q \models^{\infty} s\eta\underline{@}\overrightarrow{s}$. Together with the assumed properties of η this witnesses $q \in \langle\!\langle s\rangle\!\rangle^{\Gamma}_{\mathfrak{A}\infty}(\overrightarrow{a})$.

Lemma 38. *If r and s are terms of type $\sigma \to \overrightarrow{\rho} \to \iota$ and σ, respectively, in some context compatible with Γ, then we have*

$$\langle\!\langle rs\rangle\!\rangle^{\Gamma}_{\mathfrak{A}\infty} \sqsubseteq \mathcal{R}(\langle\!\langle r\rangle\!\rangle^{\Gamma}_{\mathfrak{A}\infty}\langle\!\langle s\rangle\!\rangle^{\Gamma}_{\mathfrak{A}\infty})$$

Proof. Let $\overrightarrow{a} \in [\![\overrightarrow{\rho}]\!]$ and $q \in \langle\!\langle rs\rangle\!\rangle^{\Gamma}_{\mathfrak{A}\infty}(\overrightarrow{a})$ be given. Then there is η with $\forall x \in \mathrm{dom}(\Gamma).\ \langle\!\langle \eta(x)\rangle\!\rangle_{\mathfrak{A}\infty} \sqsubseteq \Gamma(x)$ and there are $\overrightarrow{s} : \overrightarrow{\rho}$ with $\langle\!\langle \overrightarrow{s}\rangle\!\rangle_{\mathfrak{A}\infty} \sqsubseteq \overrightarrow{a}$ and

$$\mathfrak{A}, q \models^{\infty} \underbrace{(rs)\eta\underline{@}\overrightarrow{s}}_{\mathcal{R}.r\eta\underline{@}s\eta,\overrightarrow{s}}$$

Hence there is a $q' \in \delta(q, \mathcal{R})$ with $\mathfrak{A}, q' \models^{\infty} r\eta\underline{@}s\eta, \overrightarrow{s}$. It suffices to show that for this q' we have $q' \in \langle\!\langle r\rangle\!\rangle^{\Gamma}_{\mathfrak{A}\infty}\langle\!\langle s\rangle\!\rangle^{\Gamma}_{\mathfrak{A}\infty}\overrightarrow{a}$.

By Proposition 37 we have $\langle\!\langle s\eta\rangle\!\rangle_{\mathfrak{A}\infty} \sqsubseteq \langle\!\langle s\rangle\!\rangle^{\Gamma}_{\mathfrak{A}\infty}$ and we already have $\langle\!\langle \overrightarrow{s}\rangle\!\rangle_{\mathfrak{A}\infty} \sqsubseteq \overrightarrow{a}$. So the given η together with $s\eta$ and $\overrightarrow{s}$ witnesses $q' \in \langle\!\langle r\rangle\!\rangle^{\Gamma}_{\mathfrak{A}\infty}\langle\!\langle s\rangle\!\rangle^{\Gamma}_{\mathfrak{A}\infty}\overrightarrow{a}$.

Lemma 39. *Assume that $\lambda x.r$ has type $\sigma \to \vec{\rho} \to \iota$ in some context compatible with Γ. Then*

$$\langle\!\langle \lambda x r \rangle\!\rangle^{\Gamma}_{\mathfrak{A}\infty}(a) \sqsubseteq \beta(\langle\!\langle r \rangle\!\rangle^{\Gamma^a_x}_{\mathfrak{A}\infty})$$

Proof. Let $\vec{a} \in [\![\vec{\rho}]\!]$ and $q \in \langle\!\langle \lambda x r \rangle\!\rangle^{\Gamma}_{\mathfrak{A}\infty}(a, \vec{a})$ be given. Then there is an η with $\forall x \in \mathrm{dom}(\Gamma)$ we have $\eta(x)$ closed and $\langle\!\langle \eta(x) \rangle\!\rangle_{\mathfrak{A}\infty} \sqsubseteq \Gamma(x)$ and there are $s, \vec{s}$ with $\langle\!\langle s \rangle\!\rangle_{\mathfrak{A}\infty} \sqsubseteq a$ and $\langle\!\langle \vec{s} \rangle\!\rangle_{\mathfrak{A}\infty} \sqsubseteq \vec{a}$ such that

$$\mathfrak{A}, q \models^{\infty} \underbrace{(\lambda x r)\eta \underline{@} s, \vec{s}}_{\beta . r_x[s]\eta \underline{@} \vec{s}}$$

So there is a $\tilde{q} \in \delta(q, \beta)$ with $\mathfrak{A}, \tilde{q} \models^{\infty} r_x[s]\eta \underline{@} \vec{s}$. It suffices to show that $\tilde{q} \in \langle\!\langle r \rangle\!\rangle^{\Gamma^a_x}_{\mathfrak{A}\infty}(\vec{a})$.

By the properties of η and since $\langle\!\langle s \rangle\!\rangle_{\mathfrak{A}\infty} \sqsubseteq a$ we know that for all $y \in \mathrm{dom}(\Gamma^a_x)$ we have $\langle\!\langle \eta(y) \rangle\!\rangle_{\mathfrak{A}\infty} \sqsubseteq \Gamma^a_x(y)$. This witnesses $\tilde{q} \in \langle\!\langle r \rangle\!\rangle^{\Gamma^a_x}_{\mathfrak{A}\infty}(\vec{a})$.

Lemma 40. $\langle\!\langle x \rangle\!\rangle^{\Gamma}_{\mathfrak{A}\infty} \sqsubseteq \Gamma(x)$

Theorem 41. $\Gamma \vdash^n_{\mathfrak{A}} \langle\!\langle t \rangle\!\rangle^{\Gamma}_{\mathfrak{A}\infty} \sqsubseteq t : \rho$

Proof. Induction on n, cases on t. Trivial for $n = 0$. So let $n > 0$. We distinguish cases according to t. The cases rs, $\lambda x.r$ and x are immediately from the induction hypotheses and Lemmata 38, 39, and 40, respectively.

So, let $t = \mathfrak{f}$ be a terminal symbol. We have to show $\Gamma \vdash^n_{\mathfrak{A}} \langle\!\langle \mathfrak{f} \rangle\!\rangle^{\Gamma}_{\mathfrak{A}\infty} \sqsubseteq \mathfrak{f} : \iota \to \iota$.

So, let $\vec{S} \in [\![\vec{\iota}]\!]$ and $q \in \langle\!\langle \mathfrak{f} \rangle\!\rangle^{\Gamma}_{\mathfrak{A}\infty}(S)$. Hence there is are $\vec{s}$ of type ι with $\langle\!\langle s_i \rangle\!\rangle_{\mathfrak{A}\infty} \sqsubseteq S_i$ and $\mathfrak{A}, q \models^{\infty} \underbrace{\mathfrak{f} \underline{@} \vec{s}}_{\mathfrak{f}(\vec{s^{\beta}})}$.

So there is $(\tilde{q}_1, \ldots, \tilde{q}_{\sharp(\mathfrak{f})}, *, \ldots, *) \in \delta(q, \mathfrak{f})$ with $\mathfrak{A}, \tilde{q}_i \models^{\infty} s_i^{\beta}$. But then $\tilde{q}_i \in \langle\!\langle s_i \rangle\!\rangle_{\mathfrak{A}\infty} \subset S_i$.

Corollary 42. *If $t\colon \iota$ is closed and of ground type then $\emptyset \vdash^n_{\mathfrak{A}} \{q \mid \mathfrak{A}, q \models^{\infty} t^{\beta}\} \sqsubseteq t : \iota$.*

Finally, let us sum up what we have achieved.

Corollary 43. *For t a closed regular lambda term, and $q_0 \in Q$ it is decidable whether $\mathfrak{A}, q_0 \models^{\infty} t^{\beta}$.*

Proof. By Proposition 26 it suffices to show that $\emptyset \vdash^{\infty}_{\mathfrak{A}} \{q_0\} \sqsubseteq t : \iota$ holds, if and only if $\mathfrak{A}, q_0 \models^{\infty} t^{\beta}$.

The "if"-direction follows from Corollary 42 and the weakening provided by Remark 25. The "only if"-direction is provided by Corollary 32.

8 Model Checking

Formulae of Monadic Second Order Logic can be presented [10] by appropriate tree automata. As mentioned, we consider here only a special case. More precisely, let φ be a property that can be recognised by a non-deterministic tree automaton with trivial acceptance condition, that is, an automaton accepting by having an infinite run. In other words, let φ be such that there is an automaton $\mathfrak{A}_\varphi$ such that $\mathcal{T} \models \varphi \Leftrightarrow \mathfrak{A}_\varphi, q_0 \models^\infty \mathcal{T}$ holds for every Σ'-tree $\mathcal{T}$.

Applying the theory developed above to this setting we obtain the following.

Theorem 44. *Given a tree $\mathcal{T}$ defined by an arbitrary recursion scheme (of arbitrary level) and a property φ that can be recognised by an automaton with trivial acceptance condition it is decidable whether $\mathcal{T} \models \varphi$.*

Proof. Let t be the infinite lambda-tree associated with the recursion scheme. Then t is effectively given as a regular closed lambda term of ground type and $\mathcal{T}$ is the normal form of t.

Let $\mathfrak{A}_\varphi$ be the automaton (with initial state q_0) describing φ. By keeping the state when reading a $\mathcal{R}$ or β it can be effectively extended to an automaton $\mathfrak{A}$ that works on the continuous normal form, rather than on the usual one. So $\mathcal{T} \models \varphi \Leftrightarrow \mathfrak{A}, q_0 \models^\infty t^\beta$. The latter, however, is decidable by Corollary 43.

Remark 45. As discussed after Proposition 26 the complexity is fixed-parameter non-deterministic linear time in the size of the recursion scheme, if we consider φ and the allowed types as a parameter.

References

1. K. Aehlig and F. Joachimski. On continuous normalization. In *Proceedings of the Annual Conference of the European Association for Computer Science Logic (CSL '02)*, volume 2471 of *Lecture Notes in Computer Science*, pages 59–73. Springer Verlag, 2002.
2. K. Aehlig, J. G. d. Miranda, and C. H. L. Ong. The monadic second order theory of trees given by arbitrary level-two recursion schemes is decidable. In P. Urzyczyn, editor, *Proceedings of the 7th International Conference on Typed Lambda Calculi and Applications (TLCA '05)*, volume 3461 of *Lecture Notes in Computer Science*, pages 39–54. Springer-Verlag, Apr. 2005.
3. J. M. E. Hyland and C.-H. L. Ong. On full abstraction for PCF. *Information and Computation*, 163(2):285–408, Dec. 2000.
4. T. Knapik, D. Niwiński, and P. Urzyczyn. Deciding monadic theories of hyperalgebraic trees. In S. Abramsky, editor, *Proceedings of the 5th International Conference on Typed Lambda Caculi and Applications (TLCA '01)*, volume 2044 of *Lecture Notes in Computer Science*, pages 253–267. Springer Verlag, 2001.
5. T. Knapik, D. Niwiński, and P. Urzyczyn. Higher-order pushdown trees are easy. In M. Nielson, editor, *Proceedings of the 5th International Conference Foundations of Software Science and Computation Structures (FOSSACS '02)*, volume 2303 of *Lecture Notes in Computer Science*, pages 205–222, Apr. 2002.
6. T. Knapik, D. Niwiński, P. Urzyczyn, and I. Walukiewicz. Unsafe grammars, panic automata, and decidability. In L. Caires, G. F. Italiano, L. Monteiro, C. Palamidessi, and M. Yung, editors, *32nd International Colloquium on Automata, Languages and Programming (ICALP '05)*, volume 3580 of *Lecture Notes in Computer Science*, pages 1450–1461. Springer Verlag, 2005.

7. G. Kreisel, G. E. Mints, and S. G. Simpson. The use of abstract language in elementary metamathematics: Some pedagogic examples. In R. Parikh, editor, *Logic Colloquium*, volume 453 of *Lecture Notes in Mathematics*, pages 38–131. Springer Verlag, 1975.
8. G. E. Mints. Finite investigations of transfinite derivations. *Journal of Soviet Mathematics*, 10:548–596, 1978. Translated from: Zap. Nauchn. Semin. LOMI 49 (1975). Cited after Grigori Mints. *Selected papers in Proof Theory*. Studies in Proof Theory. Bibliopolis, 1992.
9. C.-H. L. Ong. On model-checking trees generated by higher-order recursion schemes. In *Proceedings of the Twentyfrist Annual IEEE Symposium on Logic in Computer Science (LICS '06)*, 2006. to appear.
10. M. O. Rabin. Decidability of second-order theories and automata on infinite trees. *Transactions of the American Mathematical Society*, 141:1–35, July 1969.
11. W. W. Tait. Intensional interpretations of functionals of finite type. *The Journal of Symbolic Logic*, 32(2):198–212, 1967.

The Power of Linear Functions*

Sandra Alves[1], Maribel Fernández[2], Mário Florido[1], and Ian Mackie[2,3,**]

[1] University of Porto, Department of Computer Science & LIACC, R. do Campo Alegre 823, 4150-180, Porto, Portugal
[2] King's College London, Department of Computer Science, Strand, London WC2R 2LS, U.K.
[3] LIX, École Polytechnique, 91128 Palaiseau Cedex, France

Abstract. The linear lambda calculus is very weak in terms of expressive power: in particular, all functions terminate in linear time. In this paper we consider a simple extension with Booleans, natural numbers and a linear iterator. We show properties of this linear version of Gödel's System $\mathcal{T}$ and study the class of functions that can be represented. Surprisingly, this linear calculus is extremely expressive: it is as powerful as System $\mathcal{T}$.

1 Introduction

One of the many strands of work stemming from Girard's Linear Logic [8] is the area of linear functional programming (see for instance [1,19,14]). These languages are based on a version of the λ-calculus with a type system corresponding to intuitionistic linear logic. One of the features of the calculus (which can be seen as a minimal functional programming language) is that it provides explicit syntactical constructs for copying and erasing terms (corresponding to the exponentials in linear logic).

A question that arises from this work is what exactly is the computational power of a linear calculus *without* the exponentials, i.e., a calculus that is syntactically linear: all variables occur exactly once. This is a severely restricted form of the (simply typed) λ-calculus, and is summarised by just the following three rules:

$$\frac{}{x : A \vdash x : A} \qquad \frac{\Gamma, x : A \vdash t : B}{\Gamma \vdash \lambda x.t : A \multimap B} \qquad \frac{\Gamma \vdash t : A \multimap B \qquad \Delta \vdash u : A}{\Gamma, \Delta \vdash tu : B}$$

Due to the typing constraints—there is no contraction or weakening rule—terms are linear. Reduction is given by the usual β-reduction rule, but since there is no duplication or erasing of terms during reduction, this calculus has limited computational power—all functions terminate in linear time [12].

* Research partially supported by the Treaty of Windsor Grant: "Linearity: Programming Languages and Implementations", and by funds granted to *LIACC* through the *Programa de Financiamento Plurianual, FCT* and *FEDER/POSI*.

** Projet Logical, Pôle Commun de Recherche en Informatique du plateau de Saclay, CNRS, École Polytechnique, INRIA, Université Paris-Sud.

Z. Ésik (Ed.): CSL 2006, LNCS 4207, pp. 119–134, 2006.

Our work builds this language up by introducing: pairs, Booleans with a conditional, and natural numbers with the corresponding iterator, to obtain a *linear* version of Gödel's System $\mathcal{T}$ which we call System $\mathcal{L}$. The study of System $\mathcal{L}$ led us to the discovery of an interesting interplay between linearity and iteration in Gödel's original System $\mathcal{T}$. We will show that there is a great deal of redundancy in Gödel's System $\mathcal{T}$ and the same computational power can be achieved in a much more restricted system. Gödel's System $\mathcal{T}$ is a formalism built from the simply typed λ-calculus, adding numbers and Booleans, and a recursion operator. It is a very simple system, yet has enormous expressive power. We will show that its power comes essentially from primitive recursion combined with *linear* higher-order functions—we can achieve the same power in a calculus which has these two ingredients: System $\mathcal{L}$.

It is interesting to note that, in contrast with previous linear versions of System $\mathcal{T}$ (e.g., [16,13]), System $\mathcal{L}$ accepts iterators on *open* linear functions, since these terms are linear. Reduction is only performed on those terms if the function becomes closed (i.e., reduction does not create non-linear terms). This design choice has an enormous impact in the computation power of the calculus: we show that our calculus is as powerful as System $\mathcal{T}$, whereas previous linear calculi were strictly weaker (see [16]).

From another perspective there have been a number of calculi, again many based on linear logic, for capturing specific complexity classes ([2,7,11,3,15,24,4]). One of the main examples is that of *bounded linear logic* [11], which has as one of its main aims to find a calculus in-between the linear λ-calculus and that with the exponentials (specifically the polynomial time computable functions). There is also previous work that uses linear types to characterise computations with time bounds [13]. Thus our work can be seen as establishing another calculus with good computational properties which does not need the full power of the exponentials, and introduces the non-linear features (copying and erasing) through alternative means.

Summarising, this paper studies the computational power of a linear System $\mathcal{T}$, exposing the structure of the components of Gödel's System $\mathcal{T}$. We show that System $\mathcal{T}$ is intrinsically redundant, in that it has several ways of expressing duplication and erasure. Can one eliminate this redundancy? The answer is yes; in this paper we:

- define a linear λ-calculus with natural numbers and an iterator, and introduce iterative types and the closed reduction strategy for this calculus;
- show that we can define the whole class of primitive recursive functions in this calculus, and more general functions such as Ackermann's;
- demonstrate that this linear System $\mathcal{T}$ has the same computational power as the full System $\mathcal{T}$.

In the next section we recall the background material. In Section 3 we define System $\mathcal{L}$ and in Section 4 we demonstrate that we can encode the primitive recursive functions in this calculus, and even go considerably beyond this class of functions. In Section 5 we show how to encode Gödel's System $\mathcal{T}$. Finally we conclude the paper in Section 6.

2 Background

We assume the reader is familiar with the λ-calculus [5], and with the main notions on primitive recursive functions [23]. In this section we recall some notions on Gödel's System $\mathcal{T}$, for more details we refer to [10].

System $\mathcal{T}$ is the simply typed λ-calculus (with arrow types and products, and the usual β-reduction and projection rules) where two basic types have been added: numbers (built from 0 and S; we write $\bar{n}$ or $\mathsf{S}^n\ 0$ for $\underbrace{\mathsf{S}\ \ldots\ (\mathsf{S}}_{n}\ 0))$ and Booleans with a recursor and a conditional defined by the reduction rules:

$$\begin{array}{lll} \mathsf{R}\,0\;u\;v & \longrightarrow u & \qquad \mathsf{cond\ true}\;u\;v \longrightarrow u \\ \mathsf{R}\,(\mathsf{S}t)\;u\;v & \longrightarrow v\;(\mathsf{R}\;t\;u\;v)\;t & \qquad \mathsf{cond\ false}\;u\;v \longrightarrow v \end{array}$$

System $\mathcal{T}$ is confluent, strongly normalising and reduction preserves types (see [10] for the complete system and results). It is well-known that an iterator has the same computational power as the recursor. We will replace the recursor by a simpler iterator:

$$\mathsf{iter}\;0\;u\;v \longrightarrow u \qquad\qquad \mathsf{iter}\;(\mathsf{S}\;t)\;u\;v \longrightarrow v(\mathsf{iter}\;t\;u\;v)$$

with the following typing rule:

$$\frac{\Gamma \vdash_{\mathcal{T}} t : \mathsf{Nat} \quad \Theta \vdash_{\mathcal{T}} u : A \quad \Delta \vdash_{\mathcal{T}} v : A \to A}{\Gamma, \Theta, \Delta \vdash_{\mathcal{T}} \mathsf{iter}\;t\;u\;v : A}$$

In the rest of the paper, when we refer to System $\mathcal{T}$ it will be the system with iterators rather than recursors (it is also confluent, strongly normalising, and type preserving). We recall the following property, which is used in Section 5:

Lemma 1. – *If $\Gamma \vdash_{\mathcal{T}} \lambda x.u : T$ then $T = A \to B$ and $\Gamma, x : A \vdash_{\mathcal{T}} u : B$ for some A, B.*
– *If $\Gamma \vdash_{\mathcal{T}} \pi_1(s) : T$ then $\Gamma \vdash_{\mathcal{T}} s : T \times B$ for some B.*
– *If $\Gamma \vdash_{\mathcal{T}} \pi_2(s) : T$ then $\Gamma \vdash_{\mathcal{T}} s : A \times T$ for some A.*

We now define a call-by-name evaluation strategy for System $\mathcal{T}$: $t \Downarrow v$ means that the closed term t evaluates to the value v.

$$\frac{v \text{ is a value}}{v \Downarrow v} \qquad \frac{t \Downarrow \lambda x.t' \quad t'[u/x] \Downarrow v}{tu \Downarrow v} \qquad \frac{t \Downarrow \langle s, s' \rangle \quad s \Downarrow v}{\pi_1(t) \Downarrow v} \qquad \frac{t \Downarrow \langle s, s' \rangle \quad s' \Downarrow v}{\pi_2(t) \Downarrow v}$$

$$\frac{t \Downarrow v}{\mathsf{S}\;t \Downarrow \mathsf{S}\;v} \qquad \frac{t \Downarrow \mathsf{S}^n\;0 \quad s^n(u) \Downarrow v}{\mathsf{iter}\;t\;u\;s \Downarrow v} \qquad \frac{b \Downarrow \mathsf{true} \quad t \Downarrow v}{\mathsf{cond}\;b\;t\;u \Downarrow v} \qquad \frac{b \Downarrow \mathsf{false} \quad u \Downarrow v}{\mathsf{cond}\;b\;t\;u \Downarrow v}$$

Values are terms of the form: $\mathsf{S}^n 0$, true, false, $\langle s, s' \rangle$, $\lambda x.s$.

Lemma 2 (Adequacy of $\cdot \Downarrow \cdot$ for System $\mathcal{T}$). *1. If $t \Downarrow v$ then $t \longrightarrow^* v$.*
2. If $\Gamma \vdash t : T$, t closed, then:

$$\begin{array}{ll} T = \mathsf{Nat} \;\Rightarrow t \Downarrow \mathsf{S}(\mathsf{S} \ldots (\mathsf{S}\;0)) & \qquad T = A \times B \;\Rightarrow t \Downarrow \langle u, s \rangle \\ T = \mathsf{Bool} \Rightarrow t \Downarrow \mathsf{true}\ \textit{or}\ t \Downarrow \mathsf{false} & \qquad T = A \to B \Rightarrow\; t \Downarrow \lambda x.s \end{array}$$

A program in System $\mathcal{T}$ is a closed term at base type (Nat or Bool).

3 Linear λ-Calculus with Iterator: System $\mathcal{L}$

In this section we extend the linear λ-calculus [1] with numbers, Booleans, pairs, and an iterator, and we specify a reduction strategy inspired by closed reduction [6,9]. We call this system System $\mathcal{L}$. We begin by defining the set of linear λ-terms, which are terms from the λ-calculus restricted in the following way ($\mathsf{fv}(t)$ denotes the set of free variables of t).

$$\begin{array}{ll} x & \\ \lambda x.t & \text{if } x \in \mathsf{fv}(t) \\ tu & \text{if } \mathsf{fv}(t) \cap \mathsf{fv}(u) = \varnothing \end{array}$$

Note that x is used at least once in the body of the abstraction, and the condition on the application ensures that all variables are used at most once. Thus these conditions ensure syntactic linearity (variables occur exactly once). Next we add to this linear λ-calculus: pairs, Booleans and numbers. Table 1 summarises the syntax of System $\mathcal{L}$.

Pairs:

$$\begin{array}{ll} \langle t, u \rangle & \text{if } \mathsf{fv}(t) \cap \mathsf{fv}(u) = \varnothing \\ \texttt{let } \langle x, y \rangle = t \texttt{ in } u & \text{if } x, y \in \mathsf{fv}(u) \text{ and } \mathsf{fv}(t) \cap \mathsf{fv}(u) = \varnothing \end{array}$$

Note that when projecting from a pair, we use both projections. A simple example of such a term is the function that swaps the components of a pair: $\lambda x.\texttt{let } \langle y, z \rangle = x \texttt{ in } \langle z, y \rangle$.

Booleans: true and false, and a conditional:

$$\texttt{cond}\ t\ u\ v \quad \text{if } \mathsf{fv}(t) \cap \mathsf{fv}(u) = \varnothing \text{ and } \mathsf{fv}(u) = \mathsf{fv}(v)$$

Note that this linear conditional uses the same resources in each branch.

Numbers: 0 and S, and an iterator:

$$\mathsf{iter}\ t\ u\ v \quad \text{if } \mathsf{fv}(t) \cap \mathsf{fv}(u) = \mathsf{fv}(u) \cap \mathsf{fv}(v) = \mathsf{fv}(v) \cap \mathsf{fv}(t) = \varnothing$$

Table 1.

Construction	Variable Constraint	Free Variables (fv)
$0, \mathsf{true}, \mathsf{false}$	—	$\varnothing$
$\mathsf{S}\ t$	—	$\mathsf{fv}(t)$
$\mathsf{iter}\ t\ u\ v$	$\mathsf{fv}(t) \cap \mathsf{fv}(u) = \mathsf{fv}(u) \cap \mathsf{fv}(v) = \mathsf{fv}(t) \cap \mathsf{fv}(v) = \varnothing$	$\mathsf{fv}(t) \cup \mathsf{fv}(u) \cup \mathsf{fv}(v)$
x	—	$\{x\}$
tu	$\mathsf{fv}(t) \cap \mathsf{fv}(u) = \varnothing$	$\mathsf{fv}(t) \cup \mathsf{fv}(u)$
$\lambda x.t$	$x \in \mathsf{fv}(t)$	$\mathsf{fv}(t) \setminus \{x\}$
$\langle t, u \rangle$	$\mathsf{fv}(t) \cap \mathsf{fv}(u) = \varnothing$	$\mathsf{fv}(t) \cup \mathsf{fv}(u)$
$\texttt{let } \langle x, y \rangle = t \texttt{ in } u$	$\mathsf{fv}(t) \cap \mathsf{fv}(u) = \varnothing,\ x, y \in \mathsf{fv}(u)$	$\mathsf{fv}(t) \cup (\mathsf{fv}(u) \setminus \{x, y\})$
$\texttt{cond}\ t\ u\ v$	$\mathsf{fv}(u) = \mathsf{fv}(v), \mathsf{fv}(t) \cap \mathsf{fv}(u) = \varnothing$	$\mathsf{fv}(t) \cup \mathsf{fv}(u)$

Table 2. Closed reduction

Name	Reduction		Condition
Beta	$(\lambda x.t)v$	$\longrightarrow t[v/x]$	$\mathsf{fv}(v) = \varnothing$
Let	$\texttt{let}\ \langle x, y\rangle = \langle t, u\rangle\ \texttt{in}\ v$	$\longrightarrow (v[t/x])[u/y]$	$\mathsf{fv}(t) = \mathsf{fv}(u) = \varnothing$
Cond	$\mathsf{cond\ true}\ u\ v$	$\longrightarrow u$	
Cond	$\mathsf{cond\ false}\ u\ v$	$\longrightarrow v$	
Iter	$\mathsf{iter}\ (\mathsf{S}\ t)\ u\ v$	$\longrightarrow v(\mathsf{iter}\ t\ u\ v)$	$\mathsf{fv}(tv) = \varnothing$
Iter	$\mathsf{iter}\ 0\ u\ v$	$\longrightarrow u$	$\mathsf{fv}(v) = \varnothing$

Definition 1 (Closed Reduction). *Table 2 gives the reduction rules for System $\mathcal{L}$, substitution is a meta-operation defined as usual. Reductions can take place in any context.*

Reduction is weak: for example, $\lambda x.(\lambda y.y)x$ is a normal form. Note that all the substitutions created during reduction (rules *Beta, Let*) are closed; this corresponds to a closed-argument reduction strategy (called ca in [6]). Also note that *Iter* rules are only triggered when the function v is closed.

The following results are proved by induction, by showing that substitution and reduction preserve the variable constraints given in Table 1.

Lemma 3 (Correctness of Substitution). *Let t and u be valid terms, $x \in \mathsf{fv}(t)$, and $\mathsf{fv}(u) = \varnothing$, then $t[u/x]$ is valid.*

Lemma 4 (Correctness of $\longrightarrow$). *Let t be a valid term, and $t \longrightarrow u$, then:*

1. $\mathsf{fv}(t) = \mathsf{fv}(u)$;
2. *u is a valid term.*

Although reduction preserves the free variables of the term, a subterm of the form $\mathsf{iter}\ n\ u\ v$ may become closed after a reduction in a superterm, triggering in this way a reduction with an *Iter* rule.

3.1 Typed Terms

The set of *linear types* is generated by the grammar:

$$A, B ::= \mathsf{Nat} \mid \mathsf{Bool} \mid A \multimap B \mid A \otimes B$$

Definition 2. *Let $A_0, \ldots, A_n$ be a list of linear types (note that $A_0, \ldots, A_n$ cannot be empty). $It(A_0, \ldots, A_n)$ denotes a non-empty set of* iterative types *defined by induction on n:*

$$n = 0 : It(A_0) = \{A_0 \multimap A_0\}$$
$$n = 1 : It(A_0, A_1) = \{A_0 \multimap A_1\}$$
$$n \geq 2 : It(A_0, \ldots, A_n) = It(A_0, \ldots, A_{n-1}) \cup \{A_{n-1} \multimap A_n\}$$

Iterative types will serve to type the functions used in iterators. Note that $It(A_0) = It(A_0, A_0) = It(A_0, \ldots, A_0)$. We associate types to terms in System $\mathcal{L}$

Axiom and **Structural Rule:**

$$\frac{}{x : A \vdash_{\mathcal{L}} x : A}\,(\mathsf{Axiom}) \qquad \frac{\Gamma, x : A, y : B, \Delta \vdash_{\mathcal{L}} t : C}{\Gamma, y : B, x : A, \Delta \vdash_{\mathcal{L}} t : C}\,(\mathsf{Exchange})$$

Logical Rules:

$$\frac{\Gamma, x : A \vdash_{\mathcal{L}} t : B}{\Gamma \vdash_{\mathcal{L}} \lambda x.t : A \multimap B}\,(\multimap\mathsf{Intro}) \qquad \frac{\Gamma \vdash_{\mathcal{L}} t : A \multimap B \quad \Delta \vdash_{\mathcal{L}} u : A}{\Gamma, \Delta \vdash_{\mathcal{L}} tu : B}\,(\multimap\mathsf{Elim})$$

$$\frac{\Gamma \vdash_{\mathcal{L}} t : A \quad \Delta \vdash_{\mathcal{L}} u : B}{\Gamma, \Delta \vdash_{\mathcal{L}} \langle t, u \rangle : A \otimes B}\,(\otimes\mathsf{Intro}) \qquad \frac{\Gamma \vdash_{\mathcal{L}} t : A \otimes B \quad x : A, y : B, \Delta \vdash_{\mathcal{L}} u : C}{\Gamma, \Delta \vdash_{\mathcal{L}} \mathtt{let}\ \langle x, y \rangle = t\ \mathtt{in}\ u : C}\,(\otimes\mathsf{Elim})$$

Numbers

$$\frac{}{\vdash_{\mathcal{L}} 0 : \mathsf{Nat}}\,(\mathsf{Zero}) \qquad \frac{\Gamma \vdash_{\mathcal{L}} n : \mathsf{Nat}}{\Gamma \vdash_{\mathcal{L}} \mathsf{S}\, n : \mathsf{Nat}}\,(\mathsf{Succ})$$

$$\frac{\Gamma \vdash_{\mathcal{L}} t : \mathsf{Nat} \quad \Theta \vdash_{\mathcal{L}} u : A_0 \quad \Delta \vdash_{\mathcal{L}} v : It(A_0, \ldots, A_n) \quad (\star)}{\Gamma, \Theta, \Delta \vdash_{\mathcal{L}} \mathsf{iter}\ t\ u\ v : A_n}\,(\mathsf{Iter})$$

($\star$) where if $t \equiv \mathsf{S}^m\ 0$ then $n = m$ otherwise $n = 0$

Booleans

$$\frac{}{\vdash_{\mathcal{L}} \mathsf{true} : \mathsf{Bool}}\,(\mathsf{True}) \qquad \frac{}{\vdash_{\mathcal{L}} \mathsf{false} : \mathsf{Bool}}\,(\mathsf{False})$$

$$\frac{\Delta \vdash_{\mathcal{L}} t : \mathsf{Bool} \quad \Gamma \vdash_{\mathcal{L}} u : A \quad \Gamma \vdash_{\mathcal{L}} v : A}{\Gamma, \Delta \vdash_{\mathcal{L}} \mathsf{cond}\ t\ u\ v : A}\,(\mathsf{Cond})$$

Fig. 1. Type System for System $\mathcal{L}$

using the typing rules given in Figure 1, where we use the following abbreviations: $\Gamma \vdash_{\mathcal{L}} t : It(A_0, \ldots, A_n)$ iff $\Gamma \vdash_{\mathcal{L}} t : B$ for each $B \in It(A_0, \ldots, A_n)$. We use a Curry-style type system; the typing rules specify how to assign types to untyped terms (there are no type decorations).

Note that the only structural rule in Figure 1 is Exchange, we do not have Weakening and Contraction rules: we are in a linear system. For the same reason, the logical rules split the context between the premises. The rules for numbers are standard. In the case of a term of the form $\mathsf{iter}\ t\ u\ v$, we check that t is a term of type Nat and that v and u are compatible. There are two cases: if t is $\mathsf{S}^n\ 0$ then we require v to be a function that can be iterated n times on u. Otherwise, if t is not (yet) a number, we require v to have a type that allows it to be iterated any number of times (i.e. u has type A and $v : A \multimap A$, for some type A). The typing of iterators is therefore more flexible than in System $\mathcal{T}$, but we will see that this extra flexibility does not compromise the confluence and strong normalisation of the system. Also note that we allow the typing of $\mathsf{iter}\ t\ u\ v$ even if v is open (in contrast with [16,13]), but we do not allow reduction until v is closed. This feature gives our system the full power of System $\mathcal{T}$ (whereas systems that do not allow building iter with v open are strictly weaker [16]).

We denote by $dom(\Gamma)$ the set of variables x_i such that $x_i : A_i \in \Gamma$. Since there are no Weakening and Contraction rules, we have:

Lemma 5. *If $\Gamma \vdash_{\mathcal{L}} t : A$ then $dom(\Gamma) = \mathsf{fv}(t)$.*

Theorem 1 (Subject Reduction). *If $\Gamma \vdash_{\mathcal{L}} M : A$ and $M \longrightarrow N$, then $\Gamma \vdash_{\mathcal{L}} N : A$.*

Proof. By induction on the type derivation $\Gamma \vdash_{\mathcal{L}} M : A$, using a Substitution Lemma: If $\Gamma, x : A \vdash_{\mathcal{L}} t : B$ and $\Delta \vdash_{\mathcal{L}} u : A$ (where $\mathsf{fv}(t) \cap \mathsf{fv}(u) = \varnothing$) then $\Gamma, \Delta \vdash_{\mathcal{L}} t[u/x] : B$. □

3.2 Strong Normalisation

In System $\mathcal{L}$, every sequence of reductions starting from a typable term is finite (i.e. typable terms are strongly normalisable). Note that, although System $\mathcal{L}$ extends the linear λ-calculus (where every term is strongly normalisable), untyped terms of System $\mathcal{L}$ may have infinite reductions. Strong normalisation for System $\mathcal{L}$ is a consequence of strong normalisation for System $\mathcal{T}$. We start by defining a translation from System $\mathcal{L}$ into System $\mathcal{T}$.

Definition 3. *We define the compilation of types and terms in System $\mathcal{L}$ into System $\mathcal{T}$, denoted $[\![\cdot]\!]$, in the following way:*

$$
\begin{array}{llll}
[\![\mathsf{Nat}]\!] = \mathsf{Nat} & & [\![\mathsf{Bool}]\!] = \mathsf{Bool} \\
[\![A \multimap B]\!] = [\![A]\!] \to [\![B]\!] & & [\![A \otimes B]\!] = [\![A]\!] \times [\![B]\!]
\end{array}
$$

$$
\begin{array}{ll}
[\![0]\!] & = 0 \\
[\![\mathsf{true}]\!] & = \mathsf{true} \\
[\![\mathsf{false}]\!] & = \mathsf{false} \\
[\![\mathsf{S}\ t]\!] & = S([\![t]\!]) \\
[\![x]\!] & = x \\
[\![\lambda y.t]\!] & = \lambda y.[\![t]\!] \\
[\![tu]\!] & = [\![t]\!][\![u]\!] \\
[\![\langle t, u\rangle]\!] & = \langle [\![t]\!], [\![u]\!]\rangle \\
[\![\mathtt{let}\ \langle x, y\rangle = t\ \mathtt{in}\ u]\!] & = [\![u]\!][(\pi_1[\![t]\!])/x][(\pi_2[\![t]\!])/y] \\
[\![\mathtt{cond}\ t\ u\ v]\!] & = \mathtt{cond}\ [\![t]\!]\ [\![u]\!]\ [\![v]\!] \\
[\![\mathsf{iter}\ t\ u\ v]\!] & = \begin{cases} [\![v]\!]^m([\![u]\!]) & \textit{if } t = \mathsf{S}^m 0,\ m > 0 \\ \mathsf{iter}\ [\![t]\!]\ [\![u]\!]\ [\![v]\!] & \textit{otherwise} \end{cases}
\end{array}
$$

If $\Gamma = x_1 : A_1, \ldots x_n : A_n$, then $[\![\Gamma]\!] = x_1 : [\![A_1]\!], \ldots x_n : [\![A_n]\!]$. Note that the translation of an iterator where the number of times to iterate is known and positive, develops this iteration. If it is zero or not known we use System $\mathcal{T}$'s iterator.

Theorem 2 (Strong Normalisation). *If $\Gamma \vdash_{\mathcal{L}} t : T$, t is strongly normalisable.*

Proof (Sketch). Strong normalisation for System $\mathcal{L}$ is proved by mapping all the reduction steps in System $\mathcal{L}$ into one or more reduction steps in System $\mathcal{T}$. Notice that reduction steps of the form $\mathsf{iter}\ S^{m+1}0\ u\ v \longrightarrow v(\mathsf{iter}\ S^m 0\ u\ v)$ map into zero reduction steps in System $\mathcal{T}$, but we can prove that any sequence of reduction steps of that form is always terminating. □

3.3 Church-Rosser

System $\mathcal{L}$ is confluent, which implies that normal forms are unique. For typable terms, confluence is a direct consequence of strong normalisation and the fact that the rules are non-overlapping (using Newmann's Lemma [21]). In fact, all System $\mathcal{L}$ terms are confluent even if they are non-terminating: this can be proved using parallel-reductions.

Theorem 3 (Church-Rosser). *If $t \longrightarrow^* t_1$ and $t \longrightarrow^* t_2$, then there is a term t_3 such that $t_1 \longrightarrow^* t_3$ and $t_2 \longrightarrow^* t_3$.*

Theorem 4 (Adequacy). *If t is closed and typable, then one of the following holds:*

- $\vdash_{\mathcal{L}} t : \mathsf{Nat}$ *and* $t \longrightarrow^* \overline{n}$
- $\vdash_{\mathcal{L}} t : \mathsf{Bool}$ *and* $t \longrightarrow^* \mathsf{true}$ *or* $t \longrightarrow^* \mathsf{false}$
- $\vdash_{\mathcal{L}} t : A \multimap B$ *and* $t \longrightarrow^* \lambda x.u$ *for some term* u.
- $\vdash_{\mathcal{L}} t : A \otimes B$ *and* $t \longrightarrow^* \langle u, v\rangle$ *for some terms* u, v.

Proof. By Lemma 5, typing judgements for t have the form $\vdash_{\mathcal{L}} t : T$, and T is either Nat, Bool, $A \multimap B$ or $A \otimes B$. By Subject Reduction, Strong Normalisation, and Lemma 4, we know that t has a closed, typable normal form u. We show the case when $\vdash_{\mathcal{L}} u : \mathsf{Nat}$, the others follow with similar reasoning. Since u is a closed term of type Nat, it cannot be a variable, an abstraction or a pair. Hence u is either an application, a pair projection, a conditional, an iterator or a number.

- Let $u = u_1 u_2 \dots u_n$, $n \geq 2$, such that u_1 is not an application. Then u_1 is closed, and since u is typable, u_1 must have an arrow type. But then by induction u_1 is an abstraction, and then the *Beta* rule would apply, contradicting our assumptions.
- Let $u = \texttt{let}\ \langle x, y\rangle = s\ \texttt{in}\ v$. Then s is closed and $\mathsf{fv}(v) = \{x, y\}$. Since u is typable, s has type $A \otimes B$, and by induction it should be a (closed) pair $\langle s_1, s_2\rangle$. But then the *Let* rule would apply contradicting our assumptions.
- Let $u = \mathsf{cond}\ n\ s\ v$. Then n, t, v are closed. Since u is typable, n must have a Boolean type, and by induction it should be either true or false. But then the *Cond* rule would apply contradicting our assumptions.
- Let $u = \mathsf{iter}\ n\ s\ v$. Since u is closed, so are n, t and v. Since u is typable n must be a term of type Nat, and by induction, n is a number. But then the *Iter* rule would apply contradicting our assumptions.

Thus, if $\vdash_{\mathcal{L}} t : \mathsf{Nat}$ then t reduces to a number. □

4 Primitive Recursive Functions Linearly

In this section we show how to define the primitive recursive functions in System $\mathcal{L}$. We conclude the section indicating that we can encode substantially more than primitive recursive functions.

Erasing linearly. Although System $\mathcal{L}$ is a linear calculus, we can erase numbers. In particular, we can define the projection functions $\mathsf{fst}, \mathsf{snd} : \mathsf{Nat} \otimes \mathsf{Nat} \multimap \mathsf{Nat}$:

$$\begin{aligned} \mathsf{fst} &= \lambda x.\mathtt{let}\ \langle u, v\rangle = x\ \mathtt{in}\ \mathsf{iter}\ v\ u\ (\lambda z.z) \\ \mathsf{snd} &= \lambda x.\mathtt{let}\ \langle u, v\rangle = x\ \mathtt{in}\ \mathsf{iter}\ u\ v\ (\lambda z.z) \end{aligned}$$

Lemma 6. *For any numbers* $\bar{a}$ *and* $\bar{b}$, $\mathsf{fst}\langle \bar{a}, \bar{b}\rangle \longrightarrow^* \bar{a}$ *and* $\mathsf{snd}\langle \bar{a}, \bar{b}\rangle \longrightarrow^* \bar{b}$.

Proof. We show the case for fst. Let $\bar{a} = \mathsf{S}^n\ 0$, $\bar{b} = \mathsf{S}^m\ 0$.

$$\begin{aligned} \mathsf{fst}\langle \bar{a}, \bar{b}\rangle &\longrightarrow (\mathtt{let}\ \langle u, v\rangle = \langle \mathsf{S}^n\ 0, \mathsf{S}^m\ 0\rangle\ \mathtt{in}\ \mathsf{iter}\ v\ u\ \lambda z.z) \\ &\longrightarrow \mathsf{iter}\ (\mathsf{S}^m\ 0)\ (\mathsf{S}^n\ 0)\ \lambda z.z \longrightarrow^* \mathsf{S}^n\ 0 = \bar{a}. \end{aligned}$$ □

Copying linearly. We can also copy natural numbers in this linear calculus. For this, we define a function $C : \mathsf{Nat} \multimap \mathsf{Nat} \otimes \mathsf{Nat}$ that given a number $\bar{n}$ returns a pair $\langle \bar{n}, \bar{n}\rangle$: $C = \lambda x.\mathsf{iter}\ x\ \langle 0, 0\rangle\ (\lambda x.\mathtt{let}\ \langle a, b\rangle = x\ \mathtt{in}\ \langle \mathsf{S}a, \mathsf{S}b\rangle)$.

Lemma 7. *For any number* $\bar{n}$, $C\ \bar{n} \longrightarrow^* \langle \bar{n}, \bar{n}\rangle$.

Proof. By induction on $\bar{n}$.

$$\begin{aligned} C\ 0 &\longrightarrow \mathsf{iter}\ 0\ \langle 0, 0\rangle\ (\lambda x.\mathtt{let}\ \langle a, b\rangle = x\ \mathtt{in}\ \langle \mathsf{S}a, \mathsf{S}b\rangle) \longrightarrow \langle 0, 0\rangle \\ C\ (\mathsf{S}^{t+1}\ 0) &= \mathsf{iter}\ (\mathsf{S}^{t+1}\ 0)\ \langle 0, 0\rangle\ (\lambda x.\mathtt{let}\ \langle a, b\rangle = x\ \mathtt{in}\ \langle \mathsf{S}a, \mathsf{S}b\rangle) \\ &\longrightarrow (\lambda x.\mathtt{let}\ \langle a, b\rangle = x\ \mathtt{in}\ \langle \mathsf{S}a, \mathsf{S}b\rangle) \\ &\qquad (\mathsf{iter}\ (\mathsf{S}^{t}\ 0)\ \langle 0, 0\rangle\ (\lambda x.\mathtt{let}\ \langle a, b\rangle = x\ \mathtt{in}\ \langle \mathsf{S}a, \mathsf{S}b\rangle)) \\ &\longrightarrow^* (\lambda x.\mathtt{let}\ \langle a, b\rangle = x\ \mathtt{in}\ \langle \mathsf{S}a, \mathsf{S}b\rangle)\langle t, t\rangle \\ &\longrightarrow \mathtt{let}\ \langle a, b\rangle = \langle t, t\rangle\ \mathtt{in}\ \langle \mathsf{S}a, \mathsf{S}b\rangle \longrightarrow \langle \mathsf{S}t, \mathsf{S}t\rangle \end{aligned}$$ □

It is easy to apply this technique to other data structures (e.g. linear lists). Note that we do not need iterative types for this (the standard typing of iterators is sufficient). More interestingly, we will show in Section 5 that iterators will indeed allow us to erase any closed term, and moreover copy any closed term.

Primitive Recursive Functions. System $\mathcal{L}$ can express the whole class of primitive recursive functions. We have already shown we can project, and of course we have composition. We now show how to encode a function h defined by primitive recursion from f and g (see Section 2) using iter. First, assume h is defined by the following, simpler scheme (it uses n only once in the second equation):

$$\begin{aligned} h(x, 0) &= f(x) \\ h(x, n+1) &= g(x, h(x, n)) \end{aligned}$$

Given a function $g : \mathsf{Nat} \multimap \mathsf{Nat} \multimap \mathsf{Nat}$, let g' be the term:

$$\lambda y.\lambda z.\mathtt{let}\ \langle z_1, z_2\rangle = C\ z\ \mathtt{in}\ g z_1 (y z_2) : (\mathsf{Nat} \multimap \mathsf{Nat}) \multimap (\mathsf{Nat} \multimap \mathsf{Nat})$$

then $h(x, n)$ is defined by the term $(\mathsf{iter}\ n\ f\ g')x : \mathsf{Nat}$, with $f : \mathsf{Nat} \multimap \mathsf{Nat}$. Indeed, we can show by induction that $(\mathsf{iter}\ n\ f\ g')x$, where x and n are numbers, reduces to the number $h(x, n)$; we use Lemma 7 to copy numbers:

$$\begin{array}{lll}
(\mathsf{iter}\ 0\ f\ g')x & \longrightarrow & (f\ x) = h(x,0) \\
(\mathsf{iter}\ (\mathsf{S}^{n+1}\ 0)\ f\ g')\ x & \longrightarrow^* & (\mathtt{let}\ \langle z_1, z_2\rangle = C\ z\ \mathtt{in}\ gz_1(yz_2)) \\
& & [(\mathsf{iter}\ (\mathsf{S}^n\ 0)\ f\ g')/y, x/z] \\
& \longrightarrow^* & \mathtt{let}\ \langle z_1, z_2\rangle = \langle x, x\rangle\ \mathtt{in}\ gz_1((\mathsf{iter}\ (\mathsf{S}^n\ 0)\ f\ g')z_2) \\
& \longrightarrow & g\ x((\mathsf{iter}\ (\mathsf{S}^n\ 0)\ f\ g')x) = h(x, n+1) \text{ by induction.}
\end{array}$$

Now to encode the standard primitive recursive scheme, which has an extra n in the last equation, all we need to do is copy n: $h(x,n) = \mathtt{let}\ \langle n_1, n_2\rangle = C\ n\ \mathtt{in}\ sx$, where $s = \mathsf{iter}\ n_2\ f\ (\lambda y.\lambda z.\mathtt{let}\ \langle z_1, z_2\rangle = C\ z\ \mathtt{in}\ gz_1(yz_2)n_1)$. Note that the iterator in the encoding of $h(x,n)$ uses an open function, but it will be closed before reduction.

Beyond Primitive Recursion. Ackermann's function is a standard example of a non primitive recursive function:

$$\begin{array}{ll}
ack(0,n) & = \mathsf{S}\ n \\
ack(\mathsf{S}\ n, 0) & = ack(n, \mathsf{S}\ 0) \\
ack(\mathsf{S}\ n, \mathsf{S}\ m) & = ack(n, ack(\mathsf{S}\ n, m))
\end{array}$$

In a higher-order functional language, we have an alternative definition. Let $\mathsf{succ} = \lambda x.\mathsf{S}\ x : \mathsf{Nat} \multimap \mathsf{Nat}$, then $ack(m,n) = a\ m\ n$ where a is the function defined by:

$$\begin{array}{llll}
a\ 0 & = \mathsf{succ} & \qquad A\ g\ 0 & = g(\mathsf{S}\ 0) \\
a\ (\mathsf{S}\ n) & = A\ (a\ n) & \qquad A\ g\ (\mathsf{S}\ n) & = g(A\ g\ n)
\end{array}$$

Lemma 8. *Both definitions are equivalent: For all $x, y : \mathsf{Nat}$, $a\ x\ y = ack(x,y)$.*

Proof. By induction on x, proving first by induction on n that if $g = \lambda y.ack(x,y)$ then $A\ g\ n = ack(\mathsf{S}\ x, n)$. □

We can define a and A in System $\mathcal{L}$ as follows:

$a\ n = \mathsf{iter}\ n\ \mathsf{succ}\ A : \mathsf{Nat} \multimap \mathsf{Nat}$ $\qquad$ $A\ g\ n = \mathsf{iter}\ (\mathsf{S}\ n)\ (\mathsf{S}\ 0)\ g : \mathsf{Nat}$

We show by induction that this encoding is correct:

- $a\ 0 = \mathsf{iter}\ 0\ \mathsf{succ}\ A = \mathsf{succ}$
  $A\ g\ 0 = \mathsf{iter}\ (\mathsf{S}\ 0)\ (\mathsf{S}\ 0)\ g = g(\mathsf{S}\ 0)$
- $a\ (\mathsf{S}\ n) = \mathsf{iter}\ (\mathsf{S}^n\ 0)\ \mathsf{succ}\ A = A(\mathsf{iter}\ n\ \mathsf{succ}\ A) = A(a\ n)$
  $A\ g\ (\mathsf{S}\ n) = \mathsf{iter}\ (\mathsf{S}(\mathsf{S}\ n))\ (\mathsf{S}\ 0)\ g = g(\mathsf{iter}\ (\mathsf{S}\ n)\ (\mathsf{S}\ 0)\ g) = g(A\ g\ n)$.

Then Ackermann's function can be defined in System $\mathcal{L}$ as:

$$ack(m,n) = (\mathsf{iter}\ m\ \mathsf{succ}\ (\lambda gu.\mathsf{iter}\ (\mathsf{S}\ u)\ (\mathsf{S}\ 0)\ g))\ n : \mathsf{Nat}$$

The correctness of this encoding follows directly from the lemma above. Note that $\mathsf{iter}\ (\mathsf{S}\ u)\ (\mathsf{S}\ 0)\ g$ cannot be typed in [16], because g is a free variable. We allow building the term with the free variable g, but we do not allow reduction until it is closed.

5 The Power of System $\mathcal{L}$: System $\mathcal{T}$ Linearly

In this section we show how to compile System $\mathcal{T}$ programs into System $\mathcal{L}$, i.e. we show that System $\mathcal{T}$ and System $\mathcal{L}$ have the same computational power.

Explicit Erasing. In the linear λ-calculus, we are not able to erase arguments. However, terms are consumed by reduction. The idea of erasing by consuming is not new, it is known as Solvability (see [5] for instance). Our goal in this section is to give an analogous result that allows us to obtain a general form of erasing.

Definition 4 (Erasing). *We define the following mutually recursive operations which, respectively, erase and create a System $\mathcal{L}$ term. If $\Gamma \vdash_{\mathcal{L}} t : T$, then $\mathcal{E}(t,T)$ is defined as follows (where $I = \lambda x.x$):*

$$\begin{array}{llll} \mathcal{E}(t,\mathsf{Nat}) &= \mathsf{iter}\; t\; I\; I & \mathcal{E}(t, A \otimes B) &= \mathtt{let}\ \langle x,y\rangle = t\ \mathtt{in}\ \mathcal{E}(x,A)\mathcal{E}(y,B) \\ \mathcal{E}(t,\mathsf{Bool}) &= \mathsf{cond}\; t\; I\; I & \mathcal{E}(t, A \multimap B) &= \mathcal{E}(t\mathcal{M}(A), B) \\ \mathcal{M}(\mathsf{Nat}) &= 0 & \mathcal{M}(A \otimes B) &= \langle \mathcal{M}(A), \mathcal{M}(B)\rangle \\ \mathcal{M}(\mathsf{Bool}) &= \mathsf{true} & \mathcal{M}(A \multimap B) &= \lambda x.\mathcal{E}(x,A)\mathcal{M}(B) \end{array}$$

Lemma 9. *If $\Gamma \vdash_{\mathcal{L}} t : T$ then:*

1. $\mathsf{fv}(\mathcal{E}(t,T)) = \mathsf{fv}(t)$ *and* $\Gamma \vdash_{\mathcal{L}} \mathcal{E}(t,T) : A \multimap A$.
2. $\mathcal{M}(T)$ *is a closed System $\mathcal{L}$ term such that* $\vdash_{\mathcal{L}} \mathcal{M}(T) : T$.

Proof. Simultaneous induction on T. We show two cases:

– $\mathsf{fv}(\mathcal{E}(t, A \otimes B)) = \mathsf{fv}(\mathtt{let}\ \langle x,y\rangle = t\ \mathtt{in}\ \mathcal{E}(x,A)\mathcal{E}(y,B)) = \mathsf{fv}(t)$. By induction: $x : A \vdash_{\mathcal{L}} \mathcal{E}(x,A) : (C \multimap C) \multimap (C \multimap C)$ and $y : B \vdash_{\mathcal{L}} \mathcal{E}(y,B) : C \multimap C$ then $x : A, y : B \vdash_{\mathcal{L}} \mathcal{E}(x,A)\mathcal{E}(y,B) : C \multimap C$. Then $x : A, y : B \vdash_{\mathcal{L}} \mathcal{E}(t, A \otimes B) : C \multimap C$, for any C.
  $\mathsf{fv}(\mathcal{M}(A \otimes B)) = \mathsf{fv}(\langle \mathcal{M}(A), \mathcal{M}(B)\rangle) = \varnothing$ by IH(2), and $\vdash_{\mathcal{L}} \langle \mathcal{M}(A), \mathcal{M}(B)\rangle : A \otimes B$ by IH(2).
– $\mathsf{fv}(\mathcal{E}(t, A \multimap B)) = \mathsf{fv}(\mathcal{E}(t\mathcal{M}(A), B)) = \mathsf{fv}(t\mathcal{M}(A)) = \mathsf{fv}(t)$ by IH (1 and 2). Also, by IH(1) $\Gamma \vdash_{\mathcal{L}} \mathcal{E}(t\mathcal{M}(A), B) : C \multimap C$ for any C, since $\vdash_{\mathcal{L}} \mathcal{M}(A) : A$ by IH(2).
  $\mathsf{fv}(\mathcal{M}(A \multimap B)) = \mathsf{fv}(\lambda x.\mathcal{E}(x,A)\mathcal{M}(B)) = \varnothing$ by IH(1 and 2). Also, $\vdash_{\mathcal{L}} \mathcal{M}(A \multimap B) : A \multimap B$ because by IH(1) $x : A \vdash_{\mathcal{L}} \mathcal{E}(x,A) : B \multimap B$ and by IH(2) $\vdash_{\mathcal{L}} \mathcal{M}(B) : B$. □

Lemma 10. *If $x : A \vdash_{\mathcal{L}} t : T$ and $\vdash_{\mathcal{L}} v : A$ then: $\mathcal{E}(t,T)[v/x] - \mathcal{E}(t[v/x],T)$.*

Proof. By induction on T, using the fact that $\vdash_{\mathcal{L}} t[v/x] : T$. □

Lemma 11 (Erasing Lemma). *If $\vdash_{\mathcal{L}} t : T$ (i.e. t closed) then $\mathcal{E}(t,T) \longrightarrow^* I$.*

Proof. By induction on T, using Theorem 4:

$\mathcal{E}(t,\mathsf{Nat}) = \mathsf{iter}\; t\; I\; I \longrightarrow^* \mathsf{iter}\; (\mathsf{S}^n 0)\; I\; I \longrightarrow^* I$
$\mathcal{E}(t,\mathsf{Bool}) = \mathsf{cond}\; t\; I\; I \longrightarrow^* I$.
If $T = A \otimes B$, then $t \longrightarrow^* \langle a,b\rangle$ and by Theorem 1 and Lemma 4: $\vdash_{\mathcal{L}} a : A$ and $\vdash_{\mathcal{L}} b : B$. By induction, $\mathcal{E}(a,A) \longrightarrow^* I$ and $\mathcal{E}(b,B) \longrightarrow^* I$, therefore $\mathtt{let}\ \langle x,y\rangle = \langle a,b\rangle\ \mathtt{in}\ \mathcal{E}(x,A)\mathcal{E}(y,B) \longrightarrow^* I$.
If $T = A \multimap B$ then $\mathcal{E}(t, A \multimap B) = \mathcal{E}(t\mathcal{M}(A), B)$ and by Lemma 9 $\mathcal{M}(A)$ is a closed System $\mathcal{L}$ term of type A, thus by induction $\mathcal{E}(t\mathcal{M}(A), B) \longrightarrow^* I$. □

Explicit Copying. We have shown how to duplicate numbers in Section 4, but to simulate System $\mathcal{T}$ we need to be able to copy arbitrary terms. The previous technique can be generalised to other data structures, but not to functions. However, the iterator copies (closed) functions. Our aim now is to harness this. We proceed with an example before giving the general principle. Suppose that we want to write $\lambda x.\langle x, x\rangle$. This term can be *linearised*: $\lambda xy.\langle x, y\rangle$. If we now apply this term to two copies of the argument, we are done. Although we don't have the argument yet, we can write a System $\mathcal{L}$ term which will create these copies: $\lambda z.\mathsf{iter}\ (\mathsf{S}^2\ 0)\ (\lambda xy.\langle x, y\rangle)\ (\lambda x.xz)$.

Lemma 12 (Duplication Lemma). *If t is a closed System $\mathcal{L}$ term, then there is a System $\mathcal{L}$ term D such that $Dt = \langle t, t\rangle$.*

Proof. Let $D = \lambda z.\mathsf{iter}\ \mathsf{S}(\mathsf{S}\ 0)\ (\lambda xy.\langle x, y\rangle)\ (\lambda x.xz)$ then

$$\begin{aligned} Dt &\longrightarrow \mathsf{iter}\ \mathsf{S}(\mathsf{S}\ 0)\ (\lambda xy.\langle x, y\rangle)\ (\lambda x.xt) \\ &\longrightarrow^* (\lambda x.xt)((\lambda x.xt)(\lambda xy.\langle x, y\rangle)) \longrightarrow^* \langle t, t\rangle \end{aligned}$$

□

This result also applies to numbers, so we have two different ways of copying numbers in System $\mathcal{L}$.

5.1 Compilation

We now put the previous ideas together to give a formal compilation of System $\mathcal{T}$ into System $\mathcal{L}$.

Definition 5. *System $\mathcal{T}$ types are translated into System $\mathcal{L}$ types using $\langle\cdot\rangle$ defined by:*

$$\begin{array}{llll} \langle \mathsf{Nat}\rangle &= \mathsf{Nat} & \langle \mathsf{Bool}\rangle &= \mathsf{Bool} \\ \langle A \to B\rangle &= \langle A\rangle \multimap \langle B\rangle & \langle A \times B\rangle &= \langle A\rangle \otimes \langle B\rangle \end{array}$$

If $\Gamma = x_1 : T_1, \ldots, x_n : T_n$ then $\langle\Gamma\rangle = x_1 : \langle T_1\rangle, \ldots, x_n : \langle T_n\rangle$.

Definition 6 (Compilation). *Let t be a System $\mathcal{T}$ term such that $\Gamma \vdash_{\mathcal{T}} t : T$. Its compilation into System $\mathcal{L}$ is defined as: $[x_1]\ldots[x_n]\langle t\rangle$ where $\mathsf{fv}(t) = \{x_1, \ldots, x_n\}$, $n \geq 0$, we assume without loss of generality that the variables are processed in lexicographic order, and $\langle\cdot\rangle$, $[\cdot]\cdot$ are defined below. We abbreviate $\mathsf{iter}\ (\mathsf{S}^n\ 0)\ (\lambda x_1 \cdots x_n.t)\ (\lambda z.zx)$ as $C_x^{x_1,\ldots,x_n}\ t$, and $([x]t)[y/x]$ as $A_y^x t$.*

$$\begin{array}{ll} \langle x\rangle &= x \\ \langle tu\rangle &= \langle t\rangle\langle u\rangle \\ \langle \lambda x.t\rangle &= \lambda x.[x]\langle t\rangle, \textit{ if } x \in \mathsf{fv}(t) \\ &= \lambda x.\mathcal{E}(x, \langle A\rangle)\langle t\rangle, \textit{ otherwise, where } \Gamma \vdash_{\mathcal{T}} t : A \to B = T \\ & \quad (\textit{Lemma 1}) \\ \langle 0\rangle &= 0 \\ \langle \mathsf{S}\ t\rangle &= \mathsf{S}\langle t\rangle \\ \langle \mathsf{iter}\ n\ u\ v\rangle &= \mathsf{iter}\ \langle n\rangle\ \langle u\rangle\ \langle v\rangle \\ \langle \mathsf{true}\rangle &= \mathsf{true} \\ \langle \mathsf{false}\rangle &= \mathsf{false} \end{array}$$

$$\begin{array}{ll}\langle \mathsf{cond}\ n\ u\ v\rangle = \mathsf{cond}\ \langle n\rangle\ \langle u\rangle\ \langle v\rangle & \\ \langle\langle t,u\rangle\rangle = \langle\langle t\rangle,\langle u\rangle\rangle & \\ \langle \pi_1 t\rangle = \mathtt{let}\ \langle x,y\rangle = \langle t\rangle\ \mathtt{in}\ \mathcal{E}(y,\langle B\rangle)x, \textit{ where } \Gamma\vdash_{\mathcal{T}} t:A\times B=T\ (\textit{Lemma 1}) & \\ \langle \pi_2 t\rangle = \mathtt{let}\ \langle x,y\rangle = \langle t\rangle\ \mathtt{in}\ \mathcal{E}(x,\langle A\rangle)y, \textit{ where } \Gamma\vdash_{\mathcal{T}} t:A\times B=T\ (\textit{Lemma 1}) &\end{array}$$

and $[\cdot]\cdot$ *is defined as:*

$$\begin{array}{ll}[x](\mathsf{S}\ t) & = \mathsf{S}([x]t)\\ [x]x & = x\\ [x](\lambda y.t) & = \lambda y.[x]t\\ [x](tu) & = \begin{cases} C_x^{x_1,x_2}(A^x_{x_1}t)(A^x_{x_2}u) & x\in\mathsf{fv}(t), x\in\mathsf{fv}(u)\\ ([x]t)u & x\in\mathsf{fv}(t), x\notin\mathsf{fv}(u)\\ t([x]u) & x\in\mathsf{fv}(u), x\notin\mathsf{fv}(t)\end{cases}\\ [x](\mathsf{iter}\ n\ u\ v) & = \begin{cases} \mathsf{iter}\ [x]n\ u\ v & x\in\mathsf{fv}(n), x\notin\mathsf{fv}(uv)\\ \mathsf{iter}\ n\ [x]u\ v & x\notin\mathsf{fv}(nv), x\in\mathsf{fv}(u)\\ \mathsf{iter}\ n\ u\ [x]v & x\notin\mathsf{fv}(nu), x\in\mathsf{fv}(v)\\ C_x^{x,x_2}\mathsf{iter}\ (A^x_{x_1}n)\ (A^x_{x_2}u)\ v & x\in\mathsf{fv}(n)\cap\mathsf{fv}(u), x\notin\mathsf{fv}(v)\\ C_x^{x_1,x_3}\mathsf{iter}\ (A^x_{x_1}n)\ u\ (A^x_{x_3}v) & x\in\mathsf{fv}(n)\cap\mathsf{fv}(v), x\notin\mathsf{fv}(u)\\ C_x^{x_2,x_3}\mathsf{iter}\ n\ (A^x_{x_2}u)\ (A^x_{x_3}v) & x\notin\mathsf{fv}(n), x\in\mathsf{fv}(u)\cap\mathsf{fv}(v)\\ C_x^{x_1,x_2,x_3}\mathsf{iter}\ (A^x_{x_1}n)\ (A^x_{x_2}u)\ (A^x_{x_3}v) & x\in\mathsf{fv}(n)\cap\mathsf{fv}(u)\cap\mathsf{fv}(v)\end{cases}\end{array}$$

$[x](\mathsf{cond}\ n\ u\ v)$ *follows the same structure as* iter *above, replacing* iter *by* cond.

$$[x]\langle t,u\rangle = \begin{cases} C_x^{x_1,x_2}\langle A^x_{x_1}t, A^x_{x_2}u\rangle, x\in\mathsf{fv}(t), x\in\mathsf{fv}(u)\\ \langle [x]t,u\rangle, x\in\mathsf{fv}(t), x\notin\mathsf{fv}(u)\\ \langle t,[x]u\rangle, x\in\mathsf{fv}(u), x\notin\mathsf{fv}(t)\end{cases}$$

$$[x](\mathtt{let}\ \langle y,z\rangle = t\ \mathtt{in}\ u) = \begin{cases} \mathtt{let}\ \langle y,z\rangle = [x]t\ \mathtt{in}\ u & x\in\mathsf{fv}(t), x\notin\mathsf{fv}(u)\\ \mathtt{let}\ \langle y,z\rangle = t\ \mathtt{in}\ [x]u & x\notin\mathsf{fv}(t), x\in\mathsf{fv}(u)\\ C_x^{x_1,x_2}(\mathtt{let}\ \langle y,z\rangle = A^x_{x_1}t\ \mathtt{in}\ A^x_{x_2}u) & \\ & x\in\mathsf{fv}(t), x\in\mathsf{fv}(u)\end{cases}$$

where the variables x_1, x_2 *and* x_3 *above are assumed fresh.*

As an example, we show the compilation of the combinators:

- $\langle\lambda x.x\rangle = \lambda x.x$
- $\langle\lambda xyz.xz(yz)\rangle = \lambda xyz.\mathsf{iter}\ \overline{2}\ (\lambda z_1 z_2.xz_1(yz_2))\ \lambda a.az$
- $\langle\lambda xy.x\rangle = \lambda xy.\mathcal{E}(y,B)x$

Lemma 13. *If* t *is a System* $\mathcal{T}$ *term, then:*

1. $\mathsf{fv}([x_1]\cdots[x_n]\langle t\rangle) = \mathsf{fv}(t)$.
2. $[x_1]\cdots[x_n]\langle t\rangle$ *is a valid System* $\mathcal{L}$ *term (satisfying the constraints in Table 1), if* $\mathsf{fv}(t) = \{x_1,\ldots,x_n\}$.

Proof. The first part is by induction on t using Lemma 9, and the second part by induction on t using the first part. □

We will now prove that the compilation produces a typable term in System $\mathcal{L}$. For this we will need a lemma in which we will use the type system for System $\mathcal{L}$ augmented with weakening and contraction rules for variables in a certain set X. Typing judgements in this system will be denoted $\Gamma \vdash_{L+\mathrm{x}} t : T$. We will denote $\Gamma_{|X}$ the restriction of Γ to the variables in X.

Lemma 14. *If $\Gamma \vdash_{\mathcal{T}} t : T$ and $\{x_1, \ldots, x_n\} \subseteq \mathsf{fv}(t)$ then*

1. $\langle \Gamma_{|\mathsf{fv}(t)} \rangle \vdash_{L+\mathsf{fv}(t)} \langle t \rangle : \langle T \rangle$
2. $\langle \Gamma_{|\mathsf{fv}(t)} \rangle \vdash_{L+\mathrm{x}} [x_1] \ldots [x_n] \langle t \rangle : \langle T \rangle$ *implies* $\langle \Gamma_{|\mathsf{fv}(t)} \rangle \vdash_{L+\mathrm{x}'} [x][x_1] \ldots [x_n] \langle t \rangle : \langle T \rangle$ *where* $X = \mathsf{fv}(t) - \{x_1, \ldots, x_n\}$, $x \in X$, $X' = X - \{x\}$.

Proof. By simultaneous induction on t. □

Corollary 1. *If $\Gamma \vdash_{\mathcal{T}} t : T$ and $\mathsf{fv}(t) = \{x_1, \ldots, x_n\}$ then* $\langle \Gamma_{|\mathsf{fv}(t)} \rangle \vdash_{\mathcal{L}} [x_1] \ldots [x_n] \langle t \rangle : \langle T \rangle$.

We will now prove that we can simulate System $\mathcal{T}$ evaluations. First we prove a substitution lemma.

Lemma 15 (Substitution). *Let t and w be System $\mathcal{L}$ terms such that $\mathsf{fv}(t) = \{x_1, \ldots, x_n\}$, $n \geq 1$, and $\mathsf{fv}(w) = \varnothing$, then* $([x_1] \ldots [x_n] \langle t \rangle)[\langle w \rangle / x_1] \longrightarrow^* [x_2] \ldots [x_n] \langle t[w/x_1] \rangle$.

Proof. By Lemma 13 (Part 1), $\mathsf{fv}(\langle w \rangle) = \varnothing$, and $x_1 \in \mathsf{fv}([x_1] \ldots [x_n] \langle t \rangle)$. We proceed by induction on t. □

Theorem 5 (Simulation). *Let t be a System $\mathcal{T}$ program, then: $t \Downarrow u \Rightarrow \langle t \rangle \longrightarrow^* \langle u \rangle$.*

Proof. By induction on $t \Downarrow u$. We show two cases:
Application. By induction: $\langle tu \rangle = \langle t \rangle \langle u \rangle \longrightarrow^* \langle \lambda x.t' \rangle \langle u \rangle$. There are now two cases:

If $x \in \mathsf{fv}(t')$ then using Lemma 15:

$$\langle \lambda x.t' \rangle \langle u \rangle = (\lambda x.[x]\langle t' \rangle)\langle u \rangle \longrightarrow ([x]\langle t' \rangle)[\langle u \rangle / x] \longrightarrow^* \langle t'[u/x] \rangle \longrightarrow^* \langle v \rangle$$

Otherwise, using Lemmas 10 and 11:

$$\begin{aligned} \langle \lambda x.t' \rangle \langle u \rangle &= (\lambda x.\mathcal{E}(x, A)\langle t' \rangle)\langle u \rangle \longrightarrow^* (\mathcal{E}(\langle u \rangle, \langle A \rangle)\langle t' \rangle) \longrightarrow \langle t' \rangle \\ &= \langle t'[u/x] \rangle \longrightarrow^* \langle v \rangle \end{aligned}$$

Projection. By induction and Lemmas 10 and 11:

$$\begin{aligned} \langle \pi_1 t \rangle &= \texttt{let } \langle x, y \rangle = \langle t \rangle \texttt{ in } \mathcal{E}(y, \langle A \rangle)x \longrightarrow^* \texttt{let } \langle x, y \rangle = \langle \langle u, v \rangle \rangle \texttt{ in } \mathcal{E}(y, \langle A \rangle)x \\ &= \texttt{let } \langle x, y \rangle = \langle \langle u \rangle, \langle v \rangle \rangle \texttt{ in } \mathcal{E}(y, \langle A \rangle)x \longrightarrow \mathcal{E}(\langle v \rangle, \langle A \rangle)\langle u \rangle \longrightarrow^* \langle v \rangle \end{aligned}$$ □

6 Conclusions

We have shown how to build a powerful calculus starting from the (very weak in terms of computational power) linear λ-calculus, by adding Booleans, numbers and linear iterators. We have seen that linear iterators can express much more than primitive recursive functions: the system has the computational power of System $\mathcal{T}$.

We have focused on the computational power of the linear calculus in this paper; there are other interesting aspects that remain to be studied:

- By the Curry-Howard isomorphism, the results can also be expressed as a property of the underlying logic (our translation from System $\mathcal{T}$ to System $\mathcal{L}$ eliminates Weakening and Contraction rules).
- Applications to category theory: It is well-known that a Cartesian closed category (CCC) models the structure of the simply typed λ-calculus (i.e., a CCC is the internal language for the λ-calculus [17,18]). The internal language of a symmetric monoidal closed category (SMCC) is the linear λ-calculus [20]. If we add a natural numbers object (NNO) to this category, then this corresponds to adding natural numbers and an iterator to the calculus. In this setting, a natural question arises : what is the correspondence between CCC+NNO and SMCC+NNO?
- Does the technique extend to other typed λ-calculi, for instance the Calculus of Inductive Constructions [22]?

References

1. S. Abramsky. Computational Interpretations of Linear Logic. *Theoretical Computer Science*, 111:3–57, 1993.
2. A. Asperti. Light affine logic. In *Proc. Logic in Computer Science (LICS'98)*. IEEE Computer Society, 1998.
3. A. Asperti and L. Roversi. Intuitionistic light affine logic. *ACM Transactions on Computational Logic*, 2002.
4. P. Baillot and V. Mogbil. Soft lambda-calculus: a language for polynomial time computation. In *Proc. Foundations of Software Science and Computation Structures (FOSSACS'04)*, LNCS. Springer Verlag, 2004.
5. H. P. Barendregt. *The Lambda Calculus: Its Syntax and Semantics*, volume 103 of *Studies in Logic and the Foundations of Mathematics*. North-Holland Publishing Company, second, revised edition, 1984.
6. M. Fernández, I. Mackie, and F.-R. Sinot. Closed reduction: explicit substitutions without alpha conversion. *Mathematical Structures in Computer Science*, 15(2):343–381, 2005.
7. J. Girard. Light linear logic. *Information and Computation*, 1998.
8. J.-Y. Girard. Linear Logic. *Theoretical Computer Science*, 50(1):1–102, 1987.
9. J.-Y. Girard. Towards a geometry of interaction. In J. W. Gray and A. Scedrov, editors, *Categories in Computer Science and Logic: Proc. of the Joint Summer Research Conference*, pages 69–108. American Mathematical Society, Providence, RI, 1989.

10. J.-Y. Girard, Y. Lafont, and P. Taylor. *Proofs and Types*, volume 7 of *Cambridge Tracts in Theoretical Computer Science*. Cambridge University Press, 1989.
11. J.-Y. Girard, A. Scedrov, and P. J. Scott. Bounded linear logic: A modular approach to polynomial time computability. *Theoretical Computer Science*, 97:1–66, 1992.
12. J. Hindley. BCK-combinators and linear lambda-terms have types. *Theoretical Computer Science (TCS)*, 64(1):97–105, April 1989.
13. M. Hofmann. Linear types and non-size-increasing polynomial time computation. In *Proc. Logic in Computer Science (LICS'99)*. IEEE Computer Society, 1999.
14. S. Holmström. Linear functional programming. In T. Johnsson, S. L. Peyton Jones, and K. Karlsson, editors, *Proceedings of the Workshop on Implementation of Lazy Functional Languages*, pages 13–32, 1988.
15. Y. Lafont. Soft linear logic and polynomial time. *Theoretical Computer Science*, 2004.
16. U. D. Lago. The geometry of linear higher-order recursion. In P. Panangaden, editor, *Proceedings of the Twentieth Annual IEEE Symp. on Logic in Computer Science, LICS 2005*, pages 366–375. IEEE Computer Society Press, June 2005.
17. J. Lambek. From lambda calculus to cartesian closed categories. In J. P. Seldin and J. R. Hindley, editors, *To H. B. Curry: Essays on Combinatory Logic, Lambda Calculus and Formalism*, pages 363–402. Academic Press, London, 1980.
18. J. Lambek and P. J. Scott. *Introduction to Higher Order Categorical Logic*. Cambridge Studies in Advanced Mathematics Vol. 7. Cambridge University Press, 1986.
19. I. Mackie. Lilac: A functional programming language based on linear logic. *Journal of Functional Programming*, 4(4):395–433, 1994.
20. I. Mackie, L. Román, and S. Abramsky. An internal language for autonomous categories. *Journal of Applied Categorical Structures*, 1(3):311–343, 1993.
21. M. Newman. On theories with a combinatorial definition of "equivalence". *Annals of Mathematics*, 43(2):223–243, 1942.
22. C. Paulin-Mohring. Inductive Definitions in the System Coq - Rules and Properties. In M. Bezem and J.-F. Groote, editors, *Proceedings of the conference Typed Lambda Calculi and Applications*, volume 664 of *LNCS*, 1993.
23. I. Phillips. Recursion theory. In S. Abramsky, D. Gabbay, and T. Maibaum, editors, *Handbook of Logic in Computer Science*, volume 1, pages 79–187. Oxford University Press, 1992.
24. K. Terui. Affine lambda-calculus and polytime strong normalization. In *Proc. Logic in Computer Science (LICS'01)*. IEEE Computer Society, 2001.

Logical Omniscience Via Proof Complexity

Sergei Artemov and Roman Kuznets*

CUNY Graduate Center
365 Fifth Ave., New York City, NY 10016, USA
{SArtemov, RKuznets}@gc.cuny.edu

Abstract. The Hintikka-style modal logic approach to knowledge contains a well-known defect of logical omniscience, i.e., the unrealistic feature that an agent knows all logical consequences of her assumptions. In this paper, we suggest the following Logical Omniscience Test (LOT): an epistemic system E is not logically omniscient if for any valid in E knowledge assertion $\mathcal{A}$ of type '*F is known*,' there is a proof of F in E, the complexity of which is bounded by some polynomial in the length of $\mathcal{A}$. We show that the usual epistemic modal logics are logically omniscient (modulo some common complexity assumptions). We also apply LOT to evidence-based knowledge systems, which, along with the usual knowledge operator $\mathrm{K}_i(F)$ ('*agent i knows F*'), contain evidence assertions $t:F$ ('*t is a justification for F*'). In evidence-based systems, the evidence part is an appropriate extension of the Logic of Proofs LP, which guarantees that the collection of evidence terms t is rich enough to match modal logic. We show that evidence-based knowledge systems are logically omniscient w.r.t. the usual knowledge and are not logically omniscient w.r.t. evidence-based knowledge.

1 Introduction

The modal logic approach to knowledge [25] contains a well-known defect of logical omniscience, i.e., the unrealistic feature that an agent knows all logical consequences of her assumptions. In particular, a logically omniscient agent who knows the rules of chess would also know whether White has a non-losing strategy.

The logical omniscience sickness is a major obstacle in the way of applying the logic of knowledge in Computer Science. For example, within the modal logic of knowledge, an agent who knows the product of two primes also knows both of those primes[1], which makes this logic useless in analyzing cryptographic protocols epistemically.

The logical omniscience problem, raised in [14,15,26,38,40], has been studied extensively in logic, epistemology, game theory and economics, distributed systems, artificial intelligence, etc., in a large number of papers, including

* The author is supported in part by a Research Grant for Doctoral Students from the CUNY Graduate Center.

[1] This example is due to Joe Halpern.

Z. Ésik (Ed.): CSL 2006, LNCS 4207, pp. 135–149, 2006.

[1,9,13,14,15,16,22,23,27,33,36,37,41,42,44,45,46,47,48], and many others. Most of them adjust epistemic models to avoid certain features of logical omniscience.

In this paper, we try a general approach based on proof complexity to define and test the logical omniscience property of an epistemic system. This approach was inspired by the Cook-Reckhow theory of proof complexity [11,43].

We see the essence of the logical omniscience problem in a nonconstructive character of modal languages, which are able to symbolically represent knowledge without providing any information about its origin. In a modal language, there are valid knowledge assertions that do not have feasible justifications and hence cannot be regarded valid in any practical sense. We view logical omniscience rather as a syntactic and complexity issue. On the basis of this understanding, we suggest the following test:

> An epistemic system E is *not logically omniscient* if for any valid in E knowledge assertion $\mathcal{A}$ of type F *is known*, there is a proof of F in E, the complexity of which is bounded by some polynomial in the length of $\mathcal{A}$.

We show that the traditional epistemic modal logics do not pass this test and hence are logically omniscient. This complies nicely with the intuition that led to the recognition of the logical omniscience problem in the first place.

The aforementioned test suggests ways of building epistemic systems that are not logically omniscient: one has to alter the syntax of knowledge assertions F *is known* in order to include more information about **why** F *is known*. This added information should be sufficient for recovering a certified justification, e.g. a feasible proof, for F.

We show that recently introduced evidence-based knowledge systems from [3,4,6,8] are not logically omniscient.

In Section 2, we formally introduce the Logical Omniscience Test (LOT). In Section 3, we show that, according to LOT, the traditional epistemic modal logics are logically omniscient. Then, in Section 4, we formulate the system LP, which is a general purpose calculus of evidence terms, and show in Section 5 that LP as an epistemic system is not logically omniscient. Finally, in Section 6, we extend these results to the multi-agent logics with common knowledge and the corresponding evidence-based knowledge systems.

2 Logical Omniscience Test

Let L be a logical theory. According to Cook and Reckhow (cf. [11,43]), a *proof system* for L is a polynomial-time computable function $p\colon \Sigma^* \to L$ from the set of strings in some alphabet, called proofs, onto the set of L-valid formulas. In addition, we consider a measure of size for proofs, which is a function $\ell\colon \Sigma^* \to \mathbb{N}$, and a measure of size for individual formulas $|\cdot|\colon \mathrm{Fm}_L \to \mathbb{N}$.

Logical Omniscience Test (Artemov, 2005). *Let L be a theory capable of expressing knowledge assertions 'formula F is known,' supplied with a proof system p, a measure of size for proofs ℓ, and a measure of size for individual* *formulas* $|\cdot|$. *Theory* L *is* ***not logically omniscient*** *w.r.t. proof system* p *under size measures* ℓ *and* $|\cdot|$ *if there exists a polynomial* P *such that for each valid in* L *knowledge assertion* $\mathcal{A}$ *stating that 'F is known,' formula* F *has a proof* $\mathcal{D} \in \Sigma^*$ *such that*

$$\ell(\mathcal{D}) \leq P(|\mathcal{A}|) \ .$$

Note 1. This test has a proof system and measures of size for proofs and formulas as parameters. With such a freedom, one should be careful when applying this test to real epistemic systems. In particular, in this paper, we consider only complexity measures that are commonly used in proof complexity for various specific types of proofs.

In this paper, we mostly consider Hilbert-style proof systems. The size measures commonly associated with them are

1. the number of formulas in a derivation, i.e., the number of proof steps,
2. the number of logical symbols in a derivation,
3. the bit size of a derivation, i.e., the number of symbols with the size of indices of propositional variables, etc. taken into account. In other words, this is the string length in the alphabet Σ^*.

These are the three measures on which we will concentrate. If the size of a proof $\ell(\mathcal{D})$ is the number of symbols (counting or not counting indices), it seems reasonable to use the same measure for the size of formulas: $|F| = \ell(F)$. But in case 1, i.e., when we only take into account the number of formulas, using the same measure for the size of formulas would yield $|F| = 1$ for any single formula F, which is not a fair measure. So, if the size of a proof is the number of formulas, we will measure the size of individual formulas using number of symbols (again with or without indices). This is the reason why, in general, we need two different measures for proofs and formulas.

3 Modal Epistemic Logics Are Logically Omniscient

It is fairly easy to show that a modal logic, such as S4, is logically omniscient under the bit size measure *w.r.t. any proof system*, modulo a common complexity assumption. Consider S4 with the modality K.

Theorem 1. *Consider any proof system* p *for* S4. *Let the size of a proof (a formula) be the string length of that proof (that formula). Then* S4 *is logically omniscient unless* $PSPACE = NP$.

Proof. Indeed, suppose S4 is not logically omniscient. So for every valid knowledge assertion $\mathrm{K}F$, formula F has a polynomial-size proof in the proof system p, i.e., there exists a polynomial P such that for every knowledge assertion $\mathrm{K}F$ provable in S4 there is a proof $\mathcal{D}_F$ of F with $\ell(\mathcal{D}_F) \leq P(|\mathrm{K}F|)$.

Then we can construct an NP decision procedure for the validity problem in S4. We have S4 $\vdash G$ iff S4 $\vdash \mathrm{K}G$. So to determine whether a formula G is valid, non-deterministically guess its polynomial-size proof in the proof system p. Then, check that it is indeed a proof of G; this can be done in polynomial time of the size of the proof (by definition of a proof system), which, in its turn, is a polynomial in $|\mathrm{K}G| = |G| + 1$.

On the other hand, it is well known that S4 is PSPACE-complete ([30]). Thus, the existence of an NP-algorithm for S4 would ensure that PSPACE $\subseteq$ NP, in which case these two classes coincide. □

If we restrict our attention to the Hilbert-style proofs, there are two more size measures available: the number of proof steps and the number of logical symbols in a derivation. For either of the two, one can show that S4 is logically omniscient (modulo the same common complexity assumption).

Theorem 2. S4 *is logically omniscient w.r.t. the Hilbert proof system with the size of a proof being the number of formulas in it unless PSPACE= NP.*

Proof. Again, we want to construct an NP algorithm for the decision problem in S4. But it is not so easy to NP-guess the whole proof in this case. Although there are only polynomially many formulas, still the proof can *a priori* be exponentially long if the formulas are huge.

We will use unification and modified Robinson's algorithm (see [12]) to do the proof schematically.

Again, for an arbitrary formula G, non-deterministically guess the structure of a Hilbert proof of G, i.e., for each of the polynomially many formulas, guess whether it is an axiom, or a conclusion of a modus ponens rule, or a conclusion of a necessitation rule. For each rule, also guess which of the other formulas was(were) used as its premise(s); for each axiom, guess to which of the finitely many axiom schemes it belongs. This gives us the structure of the derivation tree, in fact, of the derivation dag because in Hilbert proofs, one formula can be used in several rules.

Write each axiom used in the form of the corresponding axiom scheme using variables over formulas (variables in different axioms must be distinct). Then, starting from the axioms, we can restore the proof in a schematic way. Where a necessitation rule needs to be used, just prefix the formula with K. A case of modus ponens is more interesting. Suppose modus ponens is to be used on schemes $X \to Y$ and Z. Then, unify X with Z using modified Robinson's algorithm from [12] and apply the resulting most general unifier (m.g.u.) to Y.

Eventually, at the root of the tree, we will obtain the most general form of formulas that can be proved using derivations with this particular dag structure. Unify this form with the formula G.

All unifications can be done in quadratic time of the size of all the formula dags in the derivation dag; such is the complexity of modified Robinson's algorithm. Each axiom scheme at the beginning has a constant size, and the number of axioms and rules is polynomial in $|\mathrm{K}G|$; hence the whole unification procedure is polynomial.

Again we were able to construct an NP decision algorithm under the assumption that there is a polynomial-step Hilbert derivation. □

So $\mathsf{S4}$ turns out to be logically omniscient w.r.t. an arbitrary proof system under the bit size measure and w.r.t. the Hilbert proofs under any commonly used measure, provided, of course, that PSPACE $\neq$ NP.

It is not hard to generalize this result to the epistemic logic $\mathsf{S4}_n$ of n knowledge agents and the logic of common knowledge $\mathsf{S4}_n^{\mathrm{C}}$. The argument is essentially the same, only for $\mathsf{S4}_n^{\mathrm{C}}$ the effect of it not being logically omniscient would be even more devastating: $\mathsf{S4}_n^{\mathrm{C}}$ is EXPTIME-complete (for $n \geq 2$) (see [23]).

Theorem 3. *1.* $\mathsf{S4}_n$ *is logically omniscient w.r.t. an arbitrary proof system under the bit size measure unless PSPACE = NP.*
2. $\mathsf{S4}_n$ *is logically omniscient w.r.t. the Hilbert proof system with the size of a proof being the number of formulas in it unless PSPACE = NP.*
3. $\mathsf{S4}_n^{\mathrm{C}}$ *is logically omniscient w.r.t. an arbitrary proof system under the bit size measure unless EXPTIME = NP.*
4. $\mathsf{S4}_n^{\mathrm{C}}$ *is logically omniscient w.r.t. the Hilbert proof system with the size of a proof being the number of formulas in it unless EXPTIME = NP.*

Similar results hold for epistemic logics that are co-NP-complete, e.g. $\mathsf{S5}$. Repeating the argument for them would yield NP = co-NP.

4 Logic of Evidence-Based Knowledge LP

The system LP was originally introduced in [2] (cf. [3]) as a logic of formal mathematical proofs. Subsequently, in [4,5,6,7,8,17,19], LP has been used as a general purpose calculus of evidence, which has helped to incorporate justification into formal epistemology, thus meeting a long standing demand in this area.

The issue of having a justification formally presented in the logic of knowledge has been discussed widely in mainstream epistemology, as well as in Computer Science communities [10,20,21,24,31,32,34,39]. This problem can be traced back to Plato who defined knowledge as Justified True Belief (JTB): despite well-known criticism, JTB specification is considered a necessary condition for possessing knowledge. The traditional Hintikka-style modal theory of knowledge does not contain justification and hence has some well-known deficiencies: it does not reflect awareness, agents are logically omniscient, the traditional common knowledge operator effectively ruins logics of knowledge proof-theoretically and substantially increases complexity. Most prominently, however, the traditional modal logic of knowledge lacked expressive tools for discussing evidence and analyzing the reasons for knowledge. According to Hintikka's traditional modal logic of knowledge, an agent i knows F iff F holds in all situations that i considers possible. This approach leaves doors open for a wide range of speculative 'knowledge': occasional, coincidental, not recognizable, etc. The evidence-based approach views knowledge through the prism of justification: the new epistemic atoms here are of the form $t\!:\!F$, "F is known for the reason t." Naturally, this approach required a special theory of justification and the Logic of Proofs revealed the basic structure of evidence. In order to match the expressive power of modal logic, it suffices to have only three manageable operations on evidence: application, union, and evidence checker.

4.1 Axiom System

Evidence terms t are built from evidence constants c_i and evidence variables x_i by means of three operations: unary '!' and binary '+' and '·'

$$t ::= c_i \mid x_i \mid\, !t \mid t \cdot t \mid t + t$$

The axioms of LP_0 are obtained by adding the following schemes to a finite set of axiom schemes of classical propositional logic:

LP1	$s{:}(F \to G) \to (t{:}F \to (s \cdot t){:}G)$	(application)
LP2	$t{:}F \to\, !t{:}t{:}F$	(evidence checker)
LP3	$s{:}F \to (s+t){:}F, \quad t{:}F \to (s+t){:}F$	(union)
LP4	$t{:}F \to F$	(reflexivity)

The only rule of LP_0 is modus ponens. The usual way to define the full LP is to add to LP_0 the rule of axiom necessitation:

If A is a propositional axiom or one of LP1-4 *and c is a constant, infer $c{:}A$.*

The system LP behaves as a normal propositional logic. In particular, LP is closed under substitutions and enjoys the deduction theorem. The standard semantics of proofs for LP considers variables x_i as unspecified proofs, and constants c_i as unanalyzed proofs of "elementary facts," i.e., logical axioms.

A *constant specification* $\mathcal{CS}$ is a set of LP-formulas of form $c : A$, where c is an evidence constant, A is an axiom. Each LP-derivation generates a constant specification that consists of the formulas of form $c{:}A$ introduced by the axiom necessitation rule.

A constant specification is called *injective* if no evidence constant is assigned to two different axioms. In such specifications, each constant carries complete information about the axiom the proof of which it represents.

The *maximal* constant specification is that in which each evidence constant is assigned to every axiom. This corresponds to the situation where there is no restriction on the use of evidence constants in the axiom necessitation rule.

We define $\mathsf{LP}_{\mathcal{CS}}$ as the result of adding constant specification $\mathcal{CS}$ as new axioms to LP_0. LP is $\mathsf{LP}_{\mathcal{CS}}$ for the maximal constant specification $\mathcal{CS}$.

At first glance, LP looks like an explicit version of the modal logic $\mathsf{S4}$ with basic modal axioms replaced by their explicit counterparts. However, some pieces seem to be missing, e.g. the modal necessitation rule $\vdash F \Rightarrow\ \vdash \mathrm{K}F$. The following lemma shows that LP enjoys a clear constructive version of the necessitation rule.

Lifting Lemma 1. ([2,3]) *If $\mathsf{LP} \vdash F$, then there exists a $+$-free ground[2] evidence term t such that $\mathsf{LP} \vdash t{:}F$.*

In fact, the analogy between LP and $\mathsf{S4}$ can be extended to its maximal degree. We define a forgetful mapping as $(t : F)^\circ = \mathrm{K}(F^\circ)$. The following realization theorem shows that $\mathsf{S4}$ is the forgetful projection of LP.

[2] *Ground* here means that no evidence variable occurs within it.

Theorem 4 (Realization Theorem). ([2,3])

1. *If* LP $\vdash G$, *then* S4 $\vdash G^\circ$.
2. *If* S4 $\vdash H$, *then there exists an* LP*-formula* B *(called a realization of* H*) such that* LP $\vdash B$ *and* $B^\circ = H$.

In particular, the Realization Theorem shows that each occurrence of epistemic modality K in a modal epistemic principle H can be replaced by some evidence term, thus extracting the explicit meaning of H. Moreover, it is possible to recover the evidence terms in a Skolem style, namely, by realizing negative occurrences of modality by evidence variables only. Furthermore, any S4-theorem can be realized using injective constant specifications only.

4.2 Epistemic Semantics of Evidence-Based Knowledge

Epistemic semantics for LP was introduced by Fitting in [17,19] based on earlier work by Mkrtychev ([35]). Fitting semantics was extended to evidence-based systems with both knowledge modalities $\mathrm{K}_i F$ and evidence assertions $t : F$ in [4,6,7,8].

A Fitting model for LP is a quadruple $\mathcal{M} = (W, R, \mathcal{E}, V)$, where (W, R, V) is the usual S4 Kripke model and $\mathcal{E}$ is an evidence function defined as follows.

Definition 1. *A* possible evidence function $\mathcal{E}\colon W \times \mathrm{Tm} \to 2^{\mathrm{Fm}}$ *maps worlds and terms to sets of formulas. An* evidence function *is a possible evidence function* $\mathcal{E}\colon W \times \mathrm{Tm} \to 2^{\mathrm{Fm}}$ *that satisfies the following conditions:*

1. Monotonicity: wRu *implies* $\mathcal{E}(w,t) \subseteq \mathcal{E}(u,t)$
2. Closure:
   - Application: $(F \to G) \in \mathcal{E}(w,s)$ *and* $F \in \mathcal{E}(w,t)$ *implies* $G \in \mathcal{E}(w, s\cdot t)$
   - Evidence Checker: $F \in \mathcal{E}(w,t)$ *implies* $t{:}F \in \mathcal{E}(w, !t)$
   - Union: $\mathcal{E}(w,s) \cup \mathcal{E}(w,t) \subseteq \mathcal{E}(w, s+t)$

For a given constant specification $\mathcal{CS}$, *a* $\mathcal{CS}$-evidence function *is an evidence function that respects the constant specification* $\mathcal{CS}$, *i.e.,* $c : A \in \mathcal{CS}$ *implies* $A \in \mathcal{E}(w,c)$. *When speaking about* $\mathcal{CS}$*-evidence functions for the maximal* $\mathcal{CS}$ *(case of* LP*), we will omit prefix* $\mathcal{CS}$ *and simply call them* evidence functions.

Forcing relation $\mathcal{M}, w \Vdash F$ is defined by induction on F.

1. $\mathcal{M}, w \Vdash P$ iff $V(w,P) = t$ for propositional variables P;
2. boolean connectives are classical;
3. $\mathcal{M}, w \Vdash s{:}G$ iff $G \in \mathcal{E}(w,s)$ and $\mathcal{M}, u \Vdash G$ for all wRu.

Again, when speaking about models for LP (case of the maximal $\mathcal{CS}$), we will omit prefix $\mathcal{CS}$ and will simply call them *models* (or *F-models*).

As was shown in [17,19], $\mathsf{LP}_{\mathcal{CS}}$ is sound and complete with respect to $\mathcal{CS}$-models. Mkrtychev models (M-models) are single-world Fitting models. As was shown in [35], $\mathsf{LP}_{\mathcal{CS}}$ is sound and complete with respect to M-models as well.

We are mostly interested in knowledge assertions $t{:}F$. A special calculus for such formulas was suggested in [28].

Definition 2. *The axioms of* logic $\mathsf{rLP}_{\mathcal{CS}}$ *are exactly the set* $\mathcal{CS}$. *The rules are*

$$\frac{t:F}{!t:t:F} \qquad \frac{s:F}{(s+t):F} \qquad \frac{t:F}{(s+t):F} \qquad \frac{s:(F\rightarrow G) \quad t:F}{(s\cdot t):G}$$

Theorem 5. ([28]) $\mathsf{LP}_{\mathcal{CS}} \vdash t:F$ *iff* $\mathsf{rLP}_{\mathcal{CS}} \vdash t:F$.

We will again omit subscript $\mathcal{CS}$ when discussing the maximal constant specification.

5 Evidence-Based Knowledge Is Not Logically Omniscient

Now we are ready to show that evidence-based knowledge avoids logical omniscience. The first question we have to settle is what constitutes a 'knowledge assertion' here. Apparently, the straightforward answer $t : F$, generally speaking, is not satisfactory since both t and F may contain evidence constants, the meaning of which is given only in the corresponding constant specification, thus the latter should be a legitimate part of the input.

Definition 3. *A* comprehensive knowledge assertion *has form*

$$\bigwedge \mathcal{CS} \rightarrow t:F \ ,$$

where $\mathcal{CS}$ *is a finite injective constant specification that specifies all the constants occurring in* t.

Each LP-derivation only uses the axiom necessitation rule finitely many times. Hence, each derivation of F can be turned into an LP_0-derivation of $\bigwedge \mathcal{CS} \rightarrow F$.

Lemma 2. $\mathsf{LP} \vdash t:F$ *iff* $\mathsf{LP}_0 \vdash \bigwedge \mathcal{CS} \rightarrow t:F$ *iff* $\mathsf{rLP}_{\mathcal{CS}} \vdash t:F$ *for some finite constant specification* $\mathcal{CS}$.

In this section we consider all three proof complexity measures: number of formulas, length, and bit size. In all three cases we show that LP is not logically omniscient. In fact, for the number of lines measure we are able to get a stronger result: LP has polynomial-step proofs of F even in the length of $t:F$, i.e., without taking into account constant specifications. For the sake of technical convenience, we begin with this result.

5.1 Number of Formulas in the Proof

Throughout this subsection, the size of a derivation $\ell(\mathcal{D})$ is the number of formulas in the derivation. Moreover, we allow here arbitrary constant specifications, not necessarily injective.

Theorem 6. LP *is not logically omniscient w.r.t. the Hilbert proof system, with the size of a proof being the number of formulas it contains.*

Proof. We show that for each valid knowledge assertion $t{:}F$ there is a Hilbert-style derivation of F that makes a linear number of steps. We will show that actually $3|t|+2$ steps is enough, where $|t|$ is the number of symbols in t.

Indeed, since $\mathsf{LP} \vdash t{:}F$, by Theorem 5 we have $\mathsf{rLP} \vdash t{:}F$. It can be easily seen that a derivation of any formula $t{:}F$ in rLP requires at most $|t|$ steps since each rule increases the size of the outer term by at least 1.

Each axiom of rLP is an instance of an axiom necessitation rule of LP. Each rule of rLP can be emulated in LP by writing the corresponding axiom (LP1 for the $\cdot$-rule, LP2 for the !-rule, or LP3 for the $+$-rule) and by using modus ponens once for each of the second and the third cases or twice for the first case. Thus each step of the rLP-derivation is translated as two or three steps of the corresponding LP-derivation. Finally, to derive F from $t{:}F$ we need to add two formulas: LP4-axiom $t : F \rightarrow F$ and formula F by modus ponens. Hence we need at most $3|t|+2$ steps in this Hilbert-style derivation of F. □

The lower bound on the number of steps in the derivation is also encoded by evidence terms. But here we cannot take an arbitrary term t such that $\mathsf{LP} \vdash t{:}F$. If evidence t corresponds to a very inefficient way of showing validity of F, it would be possible to significantly shorten it. But an efficient evidence term t does give a lower bound on the derivation of F. In what follows, by $\dagger(t)$ we mean the size of the syntactic dag for t, i.e., the number of subterms in t.

Theorem 7. *For a given F, let t be the term smallest in dag-size among all the terms such that $\mathsf{LP} \vdash t{:}F$. Let $\mathcal{D}$ be the shortest Hilbert-style proof of F. Then the number of steps in $\mathcal{D}$ is at least half the number of subterms in t:*

$$\ell(\mathcal{D}) \geq \frac{1}{2}\dagger(t) \ .$$

Proof. Let $\mathcal{D}$ be a derivation of F, minimal in the number of steps $N = \ell(\mathcal{D})$. By Lifting Lemma 1, there exists a $+$-free ground term t' such that $\mathsf{LP} \vdash t'{:}F$. The structure of the derivation tree of $\mathcal{D}$ is almost identical to that of the syntactic tree of t'. The only difference is due to the fact that an axiom necessitation rule $c{:}A$ in a leaf of a derivation tree corresponds to two nodes in the syntactic tree: for c and for $!\,c$. But we are interested in the dag sizes of both. Dag structures may have further differences if one evidence constant was used in $\mathcal{D}$ for several axiom necessitation instances. This would further decrease the size of the dag for t'. Hence, for the dag-smallest term t we have

$$\ell(\mathcal{D}) \geq \frac{1}{2}\dagger(t') \geq \frac{1}{2}\dagger(t) \ . \qquad \square$$

Combining the results of Theorems 6 and 7 we obtain the following

Corollary 1. *Let t be the dag-smallest term such that $\mathsf{LP} \vdash t{:}F$. Let $\mathcal{D}$ be the shortest Hilbert-style proof of F. Then*

$$\frac{1}{2}\dagger(t) \leq \ell(\mathcal{D}) \leq 3|t|+2 \ .$$

Remark 1. Although we were able to obtain both the lower and the upper bound on the size of the derivation, these bounds are not tight as the tree-size (number of symbols) and the dag-size (number of subterms) can differ exponentially. Indeed, consider a sequence $\{t_n\}$ of terms such that $t_1 = c$ and $t_{n+1} = t_n \cdot t_n$. Then $|t_n| = 2^{\dagger(t_n)} - 1$.

5.2 Length and Bit Size of Proofs

Let now $\ell(\mathcal{D})$ stand for either the number of symbols in $\mathcal{D}$ or the number of bits in $\mathcal{D}$. Accordingly, let $|F| = \ell(F)$. We will also assume that constant specifications are injective. This does not limit the scope of LP, since the principal Realization Theorem 4 is established in [2,3] for injective constant specifications as well.

Theorem 8. *Let $\bigwedge\mathcal{CS} \rightarrow t:F$ be a comprehensive knowledge assertion valid in LP_0. Then there exist a polynomial P and a Hilbert-style $\mathsf{LP}_{\mathcal{CS}}$-derivation $\mathcal{D}$ of F such that*

$$\ell(\mathcal{D}) \leq P\left(\left|\bigwedge\mathcal{CS} \rightarrow t:F\right|\right) .$$

Proof. The knowledge assertion $\bigwedge\mathcal{CS} \rightarrow t:F$ is valid, hence $\mathsf{rLP}_{\mathcal{CS}} \vdash t:F$ by Lemma 2. A derivation in $\mathsf{rLP}_{\mathcal{CS}}$ will again consist of at most $|t|$ steps; only here we know exactly which axioms were used in the leaves because of injectivity of $\mathcal{CS}$.

Each formula in this derivation has form $s:G$ where s is a subterm of t; let us call these G's evidenced formulas. We claim that the size of evidenced formulas, $|G|$, is bounded by $\ell(\mathcal{CS}) + |t|^2$. Indeed, the rules for '+' do not change the evidenced formula. The rule for '·' goes from evidenced formulas $A \rightarrow B$ and A to evidenced formula B, which is smaller than $A \rightarrow B$. The only rule that does increase the size of the evidenced formula is the rule for '!': it yields $s:G$ instead of G. Such an increase is by $|s| \leq |t|$ and the number of !-rules is also bounded by $|t|$.

Therefore the $\mathsf{rLP}_{\mathcal{CS}}$-derivation has at most $|t|$ formulas of size at most $\ell(\mathcal{CS})+|t|^2 + |t|$ each. It is clear that the size of the whole derivation is polynomial in $|\bigwedge\mathcal{CS} \rightarrow t{:}F|$.

As before, we convert an $\mathsf{rLP}_{\mathcal{CS}}$-derivation into an $\mathsf{LP}_{\mathcal{CS}}$-derivation as described in the proof of Theorem 6. Evidently, the additional LP-axioms and intermediate results of modus ponens for '·' only yield a polynomial growth of the derivation size.

Finally, we append the $\mathsf{LP}_{\mathcal{CS}}$-derivation with $t:F \rightarrow F$ and F. The resulting derivation of F is polynomial in $|\bigwedge\mathcal{CS} \rightarrow t{:}F|$. □

6 Combining Implicit and Evidence-Based Knowledge

In this section we will extend the Logical Omniscience Test to modal epistemic systems with justifications [4,6,7,8] and show that these systems are logically omniscient w.r.t. the usual (implicit) knowledge, but remain non logically omniscient w.r.t. evidence-based knowledge.

Logic of knowledge with justification S4LP [3] was introduced in [6,7,8]. Along with the usual modality of (implicit) knowledge $\mathrm{K}F$ ('*F is known*'), this system contains evidence-based knowledge assertions $t\!:\!F$ ('*F is known for a reason t*') represented by an LP-style module. S4LP was shown in [6,18] to be sound and complete with respect to F-models, where modality is given the standard Kripke semantics.

In a more general setting, logics $\mathsf{S4}_n\mathsf{LP}$ of evidence-based common knowledge were introduced in [4] to model multiple agents that all agree with the same set of explicit reasons. Its language contains n knowledge modalities K_i along with $t : F$ constructs for the same set of evidence terms as in LP. The axioms and rules of $\mathsf{S4}_n\mathsf{LP}$ are as follows:

1. finitely many propositional axiom schemes and modus ponens rule,
2. standard S4-axioms with necessitation rule for each modality K_i,
3. axioms LP1–LP4 with the axiom necessitation rule,
4. **Connecting principle** $t\!:\!F \rightarrow \mathrm{K}_i F$ for each modality K_i.

The system S4LP is $\mathsf{S4}_n\mathsf{LP}$ for $n = 1$.

Fitting-style models for $\mathsf{S4}_n\mathsf{LP}$ were introduced in [4]. Let W be a non-empty set of worlds. Let $R, R_1, \dots, R_n$ be reflexive and transitive binary relations on W with $R \supseteq R_i$, $i = 1, \dots, n$. Let $\mathcal{E}$ be an evidence function satisfying all the conditions from the definition of F-models, where Monotonicity is formulated with respect to accessibility relation R and constant specification is taken to be the maximal for $\mathsf{S4}_n\mathsf{LP}$. Let V be a valuation in the usual modal sense. An $\mathsf{S4}_n\mathsf{LP}$-model is a tuple $\mathcal{M} = (W, R, R_1, \dots, R_n, \mathcal{E}, V)$ with forcing relation defined as follows:

1. $\mathcal{M}, w \Vdash P$ iff $V(w, P) = t$ for propositional variables P,
2. boolean connectives are classical,
3. $\mathcal{M}, w \Vdash \mathrm{K}_i G$ iff $\mathcal{M}, u \Vdash G$ for all wR_iu.
4. $\mathcal{M}, w \Vdash s\!:\!G$ iff $G \in \mathcal{E}(w, s)$ and $\mathcal{M}, u \Vdash G$ for all wRu.

As was shown in [4], $\mathsf{S4}_n\mathsf{LP}$ is sound and complete with respect to the models described above.

In $\mathsf{S4}_n\mathsf{LP}$, we also have two kinds of knowledge assertions: implicit $\mathrm{K}_i F$ and evidence-based $t\!:\!F$.

Theorem 9. $\mathsf{S4}_n\mathsf{LP}$ *is logically omniscient with respect to usual knowledge assertions (unless PSPACE* $\neq$ *NP) and is not logically omniscient with respect to evidence-based knowledge assertions.*

Proof. Without loss of generality, we will give a proof for $n = 1$, i.e., for S4LP.

1. *Implicit knowledge is logically omniscient* in the same sense as S4 was shown to be in Theorems 1 and 2. The logic S4LP was shown to be PSPACE-complete

[3] It was called LPS4 in [6].

in [29]. It is quite evident that $\mathsf{S4LP} \vdash F$ iff $\mathsf{S4LP} \vdash \mathrm{K}F$. Hence the proof of Theorem 1 remains intact for S4LP and implicit knowledge in S4LP is logically omniscient w.r.t. an arbitrary proof system under the bit size measure.

2. Consider the number of formulas in a Hilbert-style proof as the measure of its size. We show how to adapt the proof of Theorem 2 to S4LP. In addition to axioms, modus ponens and necessitation rules, S4LP-derivations also may have axiom necessitation rules $c{:}A$. For these, we need to guess which of the evidence constants c occurring in $\mathrm{K}F$ are introduced and to which of the axiom schemes those A's belong. Also, for axioms we may need to use variables over evidence terms and unify over them. These are all the changes needed for the proof, and thus implicit knowledge in S4LP is logically omniscient w.r.t. the number of formulas in Hilbert proofs.

3. *Evidence-based knowledge is not logically omniscient.* The primary tool we used in Theorem 6 was N. Krupski's calculus rLP. We need to develop a similar tool for S4LP. It turns out that the calculus in the language of S4LP with the same rules as rLP suffices.

Definition 4. *Let* rS4LP *be the logic in the language of* S4LP *with the same set of rules as* rLP *and with the same maximal constant specification as the set of axioms.*

Lemma 3. $\mathsf{S4LP} \vdash t{:}F$ *iff* $\mathsf{rS4LP} \vdash t{:}F$

Proof. The original proof from [28] remains almost intact. The 'if' part is trivial. For the 'only if' part, it is sufficient to use the minimal evidence function in a single-world F-model instead of one in an M-model as in [28] (see also [29]). □

Now we can take the proof of Theorem 6 word for word, replacing all instances of LP by S4LP and rLP by rS4LP. Thus explicit knowledge in S4LP is not logically omniscient w.r.t. the number of formulas in Hilbert proofs.

4. Similarly, we can define comprehensive knowledge assertions and prove that S4LP is not logically omniscient w.r.t. comprehensive knowledge assertions and Hilbert proofs measured by the number of symbols or number of bits in the proof along the lines of Theorem 8. □

7 Conclusions

We introduced the Logical Omniscience Test for epistemic systems on the basis of proof complexity considerations that were inspired by Cook and Reckhow theory (cf. [11,43]). This test distinguishes the traditional Hintikka-style epistemic modal systems from evidence-based knowledge systems. We show that epistemic systems are logically omniscient with respect to the usual (implicit) knowledge represented by modal statements K_iF (*i-th agent knows* F) whereas none is logically omniscient with respect to evidence-based knowledge assertions $t{:}F$ (F *is known for a reason* t).

One has to be careful when applying the Logical Omniscience Test. One could engineer artificial systems to pass the test by throwing out knowledge assertions from a natural epistemic logic. However, comparing modal epistemic logics with evidence-based systems is fair since, by the Realization Theorem, every knowledge assertion in the former has a representative in the latter. Hence logics of evidence-based knowledge have rich and representative systems of knowledge assertions, both implicit and explicit.

One could try another approach to defining and testing logical omniscience in the spirit of general algorithmic complexity. Consider the following *Strong Logical Omniscience Test* (SLOT): an epistemic system E is not logically omniscient if there is a decision procedure for knowledge assertions $\mathcal{A}$ in E, the time complexity of which is bounded by a polynomial in the length of $\mathcal{A}$. It is obvious that evidence-based knowledge systems are SLOT-logically omniscient w.r.t. the usual implicit knowledge (modulo common complexity assumptions). Furthermore, these systems are not SLOT-logically omniscient w.r.t. the evidence-based knowledge given by +-free terms, i.e., on comprehensive knowledge assertions $\bigwedge \mathcal{CS} \rightarrow t{:}F$, where t is +-free. Note that by the Lifting Lemma 1, for any valid formula F, there is a +-free term t such that $t{:}F$ holds. Unfortunately, the page limit of this paper does not allow us to provide any more details here.

Acknowledgements

We thank Steven Cook, Vladimir Krupski and the anonymous CSL'06 referees of our paper for valuable comments and suggestions. Special thanks to Karen Kletter for proofreading and editing this paper.

References

1. N. Alechina and B. Logan. Ascribing beliefs to resource bounded agents. In *Proceedings of the 1st International Joint Conference on Autonomous Agents & Multiagent Systems (AAMAS-2002), Bologna, Italy, July 15–19, 2002*, volume II, pages 881–888. ACM Press, 2002.
2. S. Artemov. Operational modal logic. Technical Report MSI 95-29, Cornell University, 1995.
3. S. Artemov. Explicit provability and constructive semantics. *Bulletin of Symbolic Logic*, 7(1):1–36, 2001.
4. S. Artemov. Justified Common Knowledge. *Theoretical Computer Science*, 357(1-3):4–22, 2006.
5. S. Artemov, E. Kazakov, and D. Shapiro. Epistemic logic with justifications. Technical Report CFIS 99-12, Cornell University, 1999.
6. S. Artemov and E. Nogina. Logic of knowledge with justifications from the provability perspective. Technical Report TR-2004011, CUNY Ph.D. Program in Computer Science, 2004.
7. S. Artemov and E. Nogina. Introducing justification into epistemic logic. *Journal of Logic and Computation*, 15(6):1059–1073, 2005.
8. S. Artemov and E. Nogina. On epistemic logic with justification. In Ron van der Meyden, editor, *Proceedings of the 10th Conference on Theoretical Aspects of Rationality and Knowledge (TARK-2005), Singapore, June 10–12, 2005*, pages 279–294. National University of Singapore, 2005.

9. R. Aumann. Reasoning about knowledge in economics. In J. Halpern, editor, *Proceedings of the 1st Conference on Theoretical Aspects of Reasoning about Knowledge (TARK-1986), Monterey, CA, USA, March 1986*, page 251. Morgan Kaufmann, 1986.
10. L. Bonjour. The coherence theory of empirical knowledge. *Philosophical Studies*, 30:281–312, 1976. Reprinted in *Contemporary Readings in Epistemology*, M.F. Goodman and R.A. Snyder (eds). Prentice Hall, pp. 70–89, 1993.
11. S. Cook and R. Reckhow. On the lengths of proofs in the propositional calculus (preliminary version). In *Conference Record of 6th Annual ACM Symposium on Theory of Computing (STOC-1974), Seattle, WA, USA, April 30–May 2, 1974*, pages 135–148. ACM Press, 1974.
12. J. Corbin and M. Bidoit. A rehabilitation of Robinson's unification algorithm. In R. E. A. Mason, editor, *Proceedings of the IFIP 9th World Computer Congress (IFIP Congress-1983), Paris, France, September 19–23, 1983*, pages 909–914. North-Holland, 1983.
13. J. Elgot-Drapkin, M. Miller, and D. Perlis. Memory, Reason, and Time: The Step-logic approach. In R. Cummins and J. Pollock, editors, *Philosophy and AI: Essays at the Interface*, pages 79–103. MIT Press, 1991.
14. R. Fagin and J. Halpern. Belief, awareness, and limited reasoning: Preliminary report. In *Proceedings of the Ninth International Joint Conference on Artificial Intelligence (IJCAI-85)*, pages 491–501, 1985.
15. R. Fagin and J. Halpern. Belief, awareness, and limited reasoning. *Artificial Intelligence*, 34(1):39–76, 1988.
16. R. Fagin, J. Halpern, and M. Vardi. A nonstandard approach to the logical omniscience problem. *Artificial Intelligence*, 79(2):203–240, 1995.
17. M. Fitting. A semantics for the logic of proofs. Technical Report TR-2003012, CUNY Ph.D. Program in Computer Science, 2003.
18. M. Fitting. Semantics and tableaus for LPS4. Technical Report TR-2004016, CUNY Ph.D. Program in Computer Science, 2004.
19. M. Fitting. The logic of proofs, semantically. *Annals of Pure and Applied Logic*, 132(1):1–25, 2005.
20. E. Gettier. Is Justified True Belief Knowledge? *Analysis*, 23:121–123, 1963.
21. A. Goldman. A causal theory of knowing. *The Journal of Philosophy*, 64:335–372, 1967.
22. J. Halpern and Y. Moses. A guide to modal logics of knowledge and belief. In *Proceedings of the Ninth International Joint Conference on Artificial Intelligence (IJCAI-85)*, pages 480–490, 1985.
23. J. Halpern and Y. Moses. A guide to completeness and complexity for modal logics of knowledge and beliefs. *Journal of Artificial Intelligence*, 54:319–379, 1992.
24. V. Hendricks. Active agents. *Journal of Logic, Language and Information*, 12(4):469–495, 2003.
25. J. Hintikka. *Knowledge and Belief*. Cornell University Press, 1962.
26. J. Hintikka. Impossible possible worlds vindicated. *Journal of Philosophical Logic*, 4:475–484, 1975.
27. K. Konolige. *A Deductive Model of Belief*. Research Notes in Artificial Intelligence. Morgan Kaufmann, 1986.
28. N. Krupski. On the complexity of the reflected logic of proofs. *Theoretical Computer Science*, 357:136–142, 2006.
29. R. Kuznets. Complexity of evidence-based knowledge. Accepted for publication in proceeding of Rationality and Knowledge workshop of ESSLLI-2006, 2006.

30. R. Ladner. The computational complexity of provability in systems of modal propositional logic. *SIAM Journal on Computing*, 6(3):467–480, 1977.
31. K. Lehrer and T. Paxson. Knowledge: undefeated justified true belief. *The Journal of Philosophy*, 66:1–22, 1969.
32. W. Lenzen. Knowledge, belief and subjective probability. In V. Hendricks, K. Jörgensen, and S. Pedersen, editors, *Knowledge Contributors*. Kluwer, 2003.
33. H. Levesque. A logic of implicit and explicit belief. In R. Brachman, editor, *Proceedings of the National Conference on Artificial Intelligence (AAAI-1984), Austin, TX, USA, August 6–10, 1984*, pages 198–202. AAAI Press, 1984.
34. D. Lewis. Elusive knowledge. *Australian Journal of Philosophy*, 7:549–567, 1996.
35. A. Mkrtychev. Models for the logic of proofs. In S. Adian and A. Nerode, editors, *Logical Foundations of Computer Science '97, Yaroslavl'*, volume 1234 of *Lecture Notes in Computer Science*, pages 266–275. Springer, 1997.
36. R. Montague. Universal Grammar. *Theoria*, 36:373–398, 1970.
37. R. Moore. Reasoning about knowledge in artificial intelligence. In J.Y. Halpern, editor, *Proceedings of the 1st Conference on Theoretical Aspects of Reasoning about Knowledge (TARK-1986), Monterey, CA, USA, March 1986*, page 81. Morgan Kaufmann, 1986.
38. Y. Moses. Resource-bounded knowledge. In M. Vardi, editor, *Proceedings of the 2nd Conference on Theoretical Aspects of Reasoning about Knowledge (TARK-1988), Pacific Grove, CA, USA, March 1988*, pages 261–275. Morgan Kaufmann, 1988.
39. R. Nozick. *Philosophical Explanations*. Harvard University Press, 1981.
40. R. Parikh. Knowledge and the problem of logical omniscience. In Z. Ras and M. Zemankova, editors, *Proceedings of the 2nd International Symposium on Methodologies for Intelligent Systems (ISMIS-1987), Charlotte, NC, USA, October 14–17, 1987*, pages 432–439. North-Holland, 1987.
41. R. Parikh. Logical omniscience. In D. Leivant, editor, *Logical and Computational Complexity. Selected Papers. Logic and Computational Complexity, International Workshop (LCC-1994), Indianapolis, IN, USA, October 13–16, 1994*, volume 960 of *Lecture Notes in Computer Science*, pages 22–29. Springer, 1995.
42. R. Parikh. Logical omniscience and common knowledge: WHAT do we know and what do WE know? In R. van der Meyden, editor, *Proceedings of the 10th Conference on Theoretical Aspects of Rationality and Knowledge (TARK-2005), Singapore, June 10–12, 2005*, pages 62–77. National University of Singapore, 2005.
43. P. Pudlak. The Lengths of Proofs. In S. Buss, editor, *Handbook of Proof Theory*, pages 547–637. Elsevier, 1998.
44. V. Rantala. Impossible worlds semantics and logical omniscience. *Acta Philosophica Fennica*, 35:18–24, 1982.
45. D. Scott. Advice in modal logic. In K. Lambert, editor, *Philosophical Problems in Logic*, pages 143–173. Reidel, 1970.
46. H. Shin and T. Williamson. Representing the knowledge of turing machine. *Theory and Decision*, 37(1):125–146, 1994.
47. M. Vardi. On epistemic logic and logical omniscience. In J. Halpern, editor, *Proceedings of the 1st Conference on Theoretical Aspects of Reasoning about Knowledge (TARK-1986), Monterey, CA, USA, March 1986*, pages 293–305. Morgan Kaufmann, 1986.
48. H. Wansing. A general possible worlds framework for reasoning about knowledge. *Studia Logica*, 49(4):523–539, 1990.

Verification of Ptime Reducibility for System F Terms Via Dual Light Affine Logic

Vincent Atassi[1,⋆], Patrick Baillot[2,⋆], and Kazushige Terui[3,⋆⋆]

[1] LIPN, Univ. Paris 13 / CNRS , France
atassi@lipn.univ-paris13.fr
[2] LIPN, Univ. Paris 13 / CNRS, France
pb@lipn.univ-paris13.fr
[3] National Institute of Informatics, Japan
terui@nii.ac.jp

Abstract. In a previous work we introduced Dual Light Affine Logic (*DLAL*) ([BT04]) as a variant of Light Linear Logic suitable for guaranteeing complexity properties on lambda-calculus terms: all typable terms can be evaluated in polynomial time and all Ptime functions can be represented. In the present work we address the problem of typing lambda-terms in second-order *DLAL*. For that we give a procedure which, starting with a term typed in system F, finds all possible ways to decorate it into a *DLAL* typed term. We show that our procedure can be run in time polynomial in the size of the original Church typed system F term.

1 Introduction

Several works have studied programming languages with intrinsic computational complexity properties. This line of research, Implicit computational complexity (ICC), is motivated both by the perspective of automated complexity analysis, and by foundational goals, in particular to give natural characterizations of complexity classes, like Ptime or Pspace. Different calculi have been used for this purpose coming from primitive recursion, lambda-calculus, rewriting systems (*e.g.* [BC92, MM00, LM93])... A convenient way to see these systems is in general to describe them as a subset of programs of a larger language satisfying certain criteria: for instance primitive recursive programs satisfying safe/ramified recursion conditions, rewriting systems admitting a termination ordering and quasi interpretation, etc...

Inference. To use such ICC systems for programming purpose it is natural to wish to automatize the verification of the criterion. This way the user could stick to a simple programming language and the compiler would check whether the program satisfies the criterion, in which case a complexity property would be guaranteed.

⋆ Partially supported by projects CRISS (ACI), GEOCAL (ACI), NO-CoST (ANR).
⋆⋆ Partially supported by Grant-in-Aid for Scientific Research, MEXT, Japan.

Z. Ésik (Ed.): CSL 2006, LNCS 4207, pp. 150–166, 2006.

In general this decision procedure involves finding a certain *witness*, like a type, a proof or a termination ordering. Depending on the system this witness might be useful to provide more precise information, like an actual bound on the running time, or a suitable strategy to evaluate the program. It might be used as a certificate guaranteeing a particular quantitative property of the program.

Light linear logic. In the present work we consider the approach of Light linear logic (LLL) ([Gir98]), a variant of Linear logic which characterizes polynomial time computation, within the proofs-as-programs correspondence. It includes higher-order and polymorphism, and can be extended to a naive set theory ([Ter04a]), in which the provably total functions correspond to the class of polynomial time functions.

The original formulation of LLL by Girard was quite complicated, but a first simplification was given by Asperti with Light Affine Logic (LAL) ([AR02]). Both systems have two modalities (one more than Linear logic) to control duplication. There is a forgetful map to system F terms (polymorphic types) obtained by erasing some information (modalities) in types; if an LAL typed term t is mapped to an F-typed term M we also say that t is a *decoration* of M.

So an LAL program can be understood as a system F program, together with a typing guarantee that it can be evaluated in polynomial time. As system F is a reference system for the study of polymorphically typed functional languages and has been extensively studied, this seems to offer a solid basis to LAL.

However LAL itself is still difficult to handle and following the previous idea for the application of ICC methods, we would prefer to use plain lambda-calculus as a front-end language, without having to worry about the handling of modalities, and instead to delegate the LAL typing part to a type inference engine. The study of this approach was started in [Bai02]. For it to be fully manageable however several conditions should be fulfilled:

1. a suitable way to execute the lambda-terms with the expected complexity bound,
2. an efficient type inference,
3. a typed language which is expressive enough so that a reasonable range of programs is accepted.

The language LAL presents some drawback for the first point, because the LAL typed terms need to be evaluated with a specific graph syntax, *proof-nets*, in order to satisfy the polynomial bound, and plain beta reduction can lead to exponential blow-up. In a previous work ([BT04]) we addressed this issue by defining a subsystem of LAL, called Dual Light Affine Logic ($DLAL$). It is defined with both linear and non-linear function types. It is complete for Ptime just as LAL and its main advantage is that it is also Ptime sound w.r.t. beta reduction: a $DLAL$ term admits a bound on the length of all its beta reduction sequences. Hence $DLAL$ stands as a reasonable substitute for plain LAL for typing issues.

Concerning point 2, as type inference for system F is undecidable ([Wel99]) we don't try to give a full-fledged type inference algorithm from untyped terms. Instead, to separate the polymorphic part issue from the proper *DLAL* part one, we assume the initial program is already typed in F. Either the system F typing work is left to the user, or one could use a partial algorithm for system F typing for this preliminary phase.

So the contribution of the present work is to define an efficient algorithm to decide if a system F term can be decorated in a *DLAL* typed term. This was actually one of the original motivations for defining *DLAL*. We show here that decoration can be performed in polynomial time. This is obtained by taking advantage of intuitions coming from proof-nets, but it is presented in a standard form with a first phase consisting in generating constraints expressing typability and a second phase for constraints solving. One difficulty is that the initial presentation of the constraints involves disjunctions of linear constraints, for which there is no obvious Ptime bound. Hence we provide a specific resolution strategy.

The complete algorithm is already implemented in ML, in a way that follows closely the specification given in the article. It is modular and usable with any linear constraints solver. The code is commented, and available for public download (Section 6). With this program one might thus write terms in system F and verify if they are Ptime and obtain a time upper bound. It should in particular be useful to study further properties of *DLAL* and to experiment with reasonable size programs.

The point 3 stressed previously about expressivity of the system remains an issue which should be explored further. Indeed the *DLAL* typing discipline will in particular rule out some nested iterations which might in fact be harmless for Ptime complexity. This is related to the line of work on the study of intensional aspects of Implicit computational complexity ([MM00, Hof03]).

However it might be possible to consider some combination of *DLAL* with other systems which could allow for more flexibility, and we think a better understanding of *DLAL* and in particular of its type inference, is a necessary step in that direction.

Related work. Inference problems have been studied for several ICC systems (*e.g.* [Ama05], [HJ03]). Elementary linear logic (*EAL*, [Gir98, DJ03]) in particular is another variant of Linear logic which characterizes Kalmar elementary time and has applications to optimal reduction. Type inference for propositional EAL (without second-order) has been studied in [CM01],[CRdR03],[CDLRdR05] and [BT05] which gives a polynomial time procedure. Type inference for *LAL* was also investigated, in [Bai02, Bai04]. To our knowledge the present algorithm is however the first one for dealing with polymorphic types in a EAL-related system, and also the first one to infer light types in polynomial time.

Due to space constraints some proofs are omitted in this paper, but can be found in [ABT06].

2 From System F to *DLAL*

The language $\mathcal{L}_F$ of system F types is given by:

$$T, U ::= \alpha \mid T \to U \mid \forall \alpha.T \ .$$

We assume that a countable set of term variables $x^T, y^T, z^T, \ldots$ is given for each type T. The terms of system F are built as follows (here we write M^T to indicate that the term M has type T):

$$x^T \quad (\lambda x^T.M^U)^{T \to U} \quad ((M^{T\to U})N^T)^U \quad (\Lambda\alpha.M^U)^{\forall\alpha.U} \quad ((M^{\forall\alpha.U})T)^{U[T/\alpha]}$$

with the proviso that when building a term $\Lambda\alpha.M^U$, α may not occur free in the types of free term variables of M (the *eigenvariable condition*). The set of free variables of M is denoted $FV(M)$.

It is well known that there is no sensible resource bound (i.e. time/space) on the execution of system F terms in general. On the other hand, we are practically interested in those terms which can be executed in polynomial time. Since the class $\mathcal{P}$ of such terms is not recursively enumerable (as can be easily shown by reduction of the complement of Hilbert's 10th problem), we are naturally led to the study of sufficiently large subclasses of $\mathcal{P}$. The system $DLAL$ gives such a class in a purely type-theoretic way.

The language $\mathcal{L}_{DLAL}$ of $DLAL$ types is given by:

$$A, B ::= \alpha \mid A \multimap B \mid A \Rightarrow B \mid \S A \mid \forall\alpha.A \ .$$

We note $\S^0 A = A$ and $\S^{k+1}A = \S\S^k A$. The erasure map $(.)^-$ from $\mathcal{L}_{DLAL}$ to $\mathcal{L}_F$ is defined by: $(\S A)^- = A^-$, $(A \multimap B)^- = (A \Rightarrow B)^- = A^- \to B^-$, and $(.)^-$ commutes with the other connectives. We say $A \in \mathcal{L}_{DLAL}$ is a *decoration* of $T \in \mathcal{L}_F$ if $A^- = T$.

A *declaration* is a pair of the form $x^T : B$ with $B^- = T$. It is often written as $x : B$ for simplicity. A *judgement* is of the form $\Gamma; \Delta \vdash M : A$, where M is a system F term, $A \in \mathcal{L}_{DLAL}$ and Γ and Δ are disjoint sets of declarations. When Δ consists of $x_1 : A_1, \ldots, x_n : A_n$, $\S\Delta$ denotes $x_1 : \S A_1, \ldots, x_n : \S A_n$. The type assignment rules are given on Figure 1. Here, we assume that the substitution $M[N/x]$ used in (§ e) is *capture-free*. Namely, no free type variable α occurring in N is bound in $M[N/x]$. We write $\Gamma; \Delta \vdash_{DLAL} M : A$ if the judgement $\Gamma; \Delta \vdash M : A$ is derivable.

An example of concrete program typable in $DLAL$ is given in Section 6.

Recall that binary words, in $\{0,1\}^*$, can be given in system F the type:

$$W_F = \forall\alpha.(\alpha \to \alpha) \to (\alpha \to \alpha) \to (\alpha \to \alpha) \ .$$

A corresponding type in $DLAL$, containing the same terms, is given by:

$$W_{DLAL} = \forall\alpha.(\alpha \multimap \alpha) \Rightarrow (\alpha \multimap \alpha) \Rightarrow \S(\alpha \multimap \alpha) \ .$$

The *depth* $d(A)$ of a $DLAL$ type A is defined by:

$$d(\alpha) = 0, \qquad d(A \multimap B) = max(d(A), d(B)), \qquad d(\forall\alpha.B) = d(B),$$
$$d(\S A) = d(A) + 1, \ d(A \Rightarrow B) = max(d(A) + 1, d(B)).$$

$$\frac{}{;x^{A^-} : A \vdash x^{A^-} : A} \ (\mathrm{Id})$$

$$\frac{\Gamma; x^{A^-} : A, \Delta \vdash M : B}{\Gamma; \Delta \vdash \lambda x^{A^-}.M : A \multimap B} \ (\multimap \mathrm{i}) \qquad \frac{\Gamma_1; \Delta_1 \vdash M : A \multimap B \quad \Gamma_2; \Delta_2 \vdash N : A}{\Gamma_1, \Gamma_2; \Delta_1, \Delta_2 \vdash (M)N : B} \ (\multimap \mathrm{e})$$

$$\frac{x^{A^-} : A, \Gamma; \Delta \vdash M : B}{\Gamma; \Delta \vdash \lambda x^{A^-}.M : A \Rightarrow B} \ (\Rightarrow \mathrm{i}) \qquad \frac{\Gamma; \Delta \vdash M : A \Rightarrow B \quad ; z : C \vdash N : A}{\Gamma, z : C; \Delta \vdash (M)N : B} \ (\Rightarrow \mathrm{e})\ (*)$$

$$\frac{\Gamma_1; \Delta_1 \vdash M : A}{\Gamma_1, \Gamma_2; \Delta_1, \Delta_2 \vdash M : A} \ (\mathrm{Weak}) \qquad \frac{x_1 : A, x_2 : A, \Gamma; \Delta \vdash M : B}{x : A, \Gamma; \Delta \vdash M[x/x_1, x/x_2] : B} \ (\mathrm{Cntr})$$

$$\frac{; \Gamma, \Delta \vdash M : A}{\Gamma; \S\Delta \vdash M : \S A} \ (\S\ \mathrm{i}) \qquad \frac{\Gamma_1; \Delta_1 \vdash N : \S A \quad \Gamma_2; x : \S A, \Delta_2 \vdash M : B}{\Gamma_1, \Gamma_2; \Delta_1, \Delta_2 \vdash M[N/x] : B} \ (\S\ \mathrm{e})$$

$$\frac{\Gamma; \Delta \vdash M : A}{\Gamma; \Delta \vdash \Lambda\alpha.M : \forall\alpha.A} \ (\forall\ \mathrm{i})\ (**) \qquad \frac{\Gamma; \Delta \vdash M : \forall\alpha.A}{\Gamma; \Delta \vdash (M)B^- : A[B/\alpha]} \ (\forall\ \mathrm{e})$$

(*) $z : C$ can be absent.
(**) α does not occur free in Γ, Δ.

Fig. 1. Typing system F terms in *DLAL*

A type A is said to be Π_1 if it does not contain a negative occurrence of $\forall$; like for instance W_{DLAL}.

The fundamental properties of *DLAL* are the following [BT04]:

Theorem 1

1. *For every function* $f : \{0,1\}^* \longrightarrow \{0,1\}^*$ *in* DTIME$[n^k]$, *there exists a closed term* M *of type* $W_{DLAL} \multimap \S^d W_{DLAL}$ *with* $d = O(\log k)$ *representing* f.
2. *Let* M *be a closed term of system F that has a* Π_1 *type* A *in DLAL. Then* M *can be normalized in* $O(|M|^{2^d})$ *steps by* β*-reduction, where* $d = d(A)$ *and* $|M|$ *is the structural size of* M*. Moreover, the size of any intermediary term occuring in normalization is also bounded by* $O(|M|^{2^d})$.

Although *DLAL* does not capture all Ptime *algorithms* $\mathcal{P}$, the result 1 guarantees that *DLAL* is at least expressive enough to represent all Ptime *functions*. In fact, *DLAL* is as expressive as *LAL* even at the level of algorithms, because there exists a generic translation from *LAL* to *DLAL* given by:

$$(!A)^o = \forall\alpha.((A^o \Rightarrow \alpha) \multimap \alpha), \qquad (.)^o \text{ commutes with other connectives than } !.$$

See [Ter04b] for details.

The result 2 on the other hand implies that if we ignore the embedded types occurring in M, the normal form of M can be computed in polynomial time (*by ordinary* β*-reduction*; that is the difference from *LAL*).

Now, let $M^{W_F \to W_F}$ be a system F typed term and suppose that we know that it has a *DLAL* type $W_{DLAL} \multimap \S^d W_{DLAL}$ for some $d \geq 0$. Then, by the consequence of the above theorem, we know that the term M is Ptime. In fact, given a binary word $w \in \{0,1\}^*$, consider its Church coding $\underline{w}$ of type W_{DLAL}. Then we have that $(M)\underline{w}$ has type $\S^d W_{DLAL}$, and can thus be evaluated in $O(|\underline{w}|^{2^{d+1}})$ steps. Thus by assigning a $DLAL$ type to a given system F term, one can *statically verify* a polynomial time bound for its execution.

In order to use $DLAL$ for resource verification of system F terms, we address the following problem:

Problem 2 (DLAL typing). Given a closed term M^T of system F, determine if there is a decoration A of T such that $\vdash_{DLAL} M : A$.

(Here the closedness assumption is only for readability.)

In the sequel, we show that there is a polynomial time algorithm for solving the $DLAL$ typing problem.

3 Characterizing *DLAL* Typability

3.1 Pseudo-terms

To address the $DLAL$ typing problem, it is convenient to introduce an intermediary syntax which is more informative than system F terms (but not more informative than $DLAL$ derivations themselves).

First we decompose $A \Rightarrow B$ into $!A \multimap B$. The language $\mathcal{L}_{DLAL\star}$ of $DLAL\star$ types is given by:

$$A ::= \alpha \mid D \multimap A \mid \forall\alpha.A \mid \S A\ , \quad D ::= A \mid !A\ .$$

There is a natural map $(.)^\star$ from $\mathcal{L}_{DLAL}$ to $\mathcal{L}_{DLAL\star}$ such that $(A \Rightarrow B)^\star = !A^\star \multimap B^\star$ and commutes with the other operations. The erasure map $(.)^-$ from $\mathcal{L}_{DLAL\star}$ to $\mathcal{L}_F$ can be defined as before. A $DLAL\star$ type is called a *bang type* if it is of the form $!A$, and otherwise called a *linear type*. In the sequel, A, B, C stand for linear types, and D for either bang or linear types.

We assume there is a countable set of term variables $x^D, y^D, z^D, \ldots$ for each $D \in \mathcal{L}_{DLAL\star}$. The *pseudo-terms* are defined by the following grammar:

$$t, u ::= x^D \mid \lambda x^D.t \mid (t)u \mid \Lambda\alpha.t \mid (t)A \mid \S t \mid \bar{\S} t,$$

where A is a linear type and D is an arbitrary one. The idea is that $\S$ corresponds to the main door of a $\S$-box (or a !-box) in *proof-nets* ([Gir87, AR02]) while $\bar{\S}$ corresponds to auxiliary doors. But note that there is no information in the pseudo-terms to link occurrences of $\S$ and $\bar{\S}$ corresponding to the same box, nor distinction between $\S$-boxes and !-boxes.

There is a natural erasure map from pseudo-terms to system F terms, which we will also denote by $(.)^-$, consisting in removing all occurrences of $\S, \bar{\S}$, replacing x^D with x^{D^-} and $(t)A$ with $(t)A^-$. When $t^- = M$, t is called a *decoration* of M.

For our purpose, it is sufficient to consider the class of *regular* pseudo-terms, given by:

$$t ::= \S^m u, \quad u ::= x^D \mid \lambda x^D.t \mid (t)t \mid \Lambda\alpha.t \mid (t)A\ ,$$

where m is an arbitrary value in $\mathbb{Z}$ and $\S^m u$ is $\S\cdots\S u$ (resp. $\bar{\S}\cdots\bar{\S}u$) with m (resp. $-m$) occurrences of $\S$ (resp. $\bar{\S}$) if $m \geq 0$ (resp. $m < 0$). So a pseudo-term is regular if and only if it does not contain any subterm of the form $\S\bar{\S}u$ or $\bar{\S}\S u$.

3.2 Local Typing Condition

We now try to assign types to pseudo-terms in a locally compatible way. A delicate point in $DLAL$ is that it is sometimes natural to associate *two* types to one variable x. For instance, we have $x : A; \vdash_{DLAL} x : \S A$ in $DLAL$, and this can be read as $x : !A \vdash x : \S A$ in terms of $DLAL\star$ types. We thus distinguish between the *input types*, which are inherent to variables, and the *output types*, which are inductively assigned to all pseudo-terms. The condition (i) below is concerned with the output types. In the sequel, D° denotes $\S A$ if D is of the form $!A$, and otherwise denotes D itself.

A pseudo-term t satisfies the *local typing condition* if the following holds:

(i) one can inductively assign a *linear* type to each subterm of t in the following way (here the notation t_A indicates that t has the output type A):

$$(x^D)_{D^\circ} \qquad (\S t_A)_{\S A} \qquad (\bar{\S} t_{\S A})_A \qquad (\lambda x^D.t_B)_{D\multimap B}$$

$$((t_{D\multimap B})u_{D^\circ})_B \qquad (\Lambda\alpha.t_A)_{\forall\alpha.A} \qquad ((t_{\forall\alpha.A})B)_{A[B/\alpha]} \ ,$$

(ii) when a variable x occurs more than once in t, it is typed as $x^{!A}$,

(iii) t satisfies the eigenvariable condition.

We also say that t is *locally typed*.

Notice that when D is a bang type, there is a type mismatch between D and D° in the case of application. For instance, $(t_{!A\multimap B})u_{\S A}$ satisfies (i) whenever t and u do. This mismatch will be settled by the *bang condition* below. Observe also that the local typing rules are syntax-directed.

3.3 Boxing Conditions

We now recall definitions and results from [BT05] giving some necessary conditions for a pseudo-term to be typable (in [BT05] these conditions are used for Elementary Affine Logic typing). We consider words over the language $\mathcal{L} = \{\S, \bar{\S}\}^\star$ and $\leq$ the prefix ordering. If t is a pseudo-term and u is an occurrence of subterm in t, let $\mathsf{doors}(t,u)$ be the word inductively defined as follows. If $t = u$, define $\mathsf{doors}(t,u) = \epsilon$. Otherwise:

$$\begin{aligned}
\mathsf{doors}(\S t, u) &= \S :: (\mathsf{doors}(t,u)),\\
\mathsf{doors}(\bar{\S} t, u) &= \bar{\S} :: (\mathsf{doors}(t,u)),\\
\mathsf{doors}(\lambda y^D.t_1, u) &= \mathsf{doors}(\Lambda\alpha.t_1, u) = \mathsf{doors}((t_1)A, u) = \mathsf{doors}(t_1, u),\\
\mathsf{doors}((t_1)t_2, u) &= \mathsf{doors}(t_i, u) \text{ where } t_i \text{ is the subterm containing } u.
\end{aligned}$$

That is to say, $\mathsf{doors}(t,u)$ collects the modal symbols $\S$, $\bar{\S}$ occurring on the path from the root to the node u in the term tree of t. We define a map $s : \mathcal{L} \to \mathbb{Z}$ by:

$$s(\epsilon) = 0, \qquad s(\S :: l) = 1 + s(l), \qquad s(\bar{\S} :: l) = -1 + s(l).$$

A word $l \in \mathcal{L}$ is *weakly well-bracketed* if $\forall l' \leq l, s(l') \geq 0$, and is *well-bracketed* if this condition holds and moreover $s(l) = 0$: think of § and $\bar{§}$ resp. as opening and closing brackets.

Bracketing condition. Let t be a pseudo-term. We say that t satisfies the *bracketing condition* if:

(i) for any occurrence of free variable x in t, $\mathsf{doors}(t, x)$ is well-bracketed;
moreover for any occurrence of an abstraction subterm $\lambda x.v$ of t,
(ii) $\mathsf{doors}(t, \lambda x.v)$ is weakly well-bracketed, and
(iii) for any occurrence of x in v, $\mathsf{doors}(v, x)$ is well-bracketed.

This condition is sufficient to rule out the canonical morphisms for dereliction and digging, which are not valid in $DLAL$ (nor in EAL):

$$(\lambda x^{§A}.\bar{§}x)_{§A \multimap A}, \qquad (\lambda x^{§A}.§x)_{§A \multimap §§A} .$$

Since $\mathsf{doors}(\bar{§}x, x) = \bar{§}$ and $\mathsf{doors}(§x, x) = §$, they do not satisfy the bracketing condition (iii).

Bang condition. A subterm u is called a *bang subterm* of t if it occurs as $(t'_{!A \multimap B})u_{§A}$ in t. We say that a locally typed pseudo-term t satisfies the *bang condition* if for any bang subterm u of t,

(i) u contains at most one free variable $x^{!C}$, having a bang type $!C$.
(ii) for any subterm v of u such that $v \neq u$ and $v \neq x$, $s(\mathsf{doors}(u, v)) \geq 1$.

This condition is sufficient to rule out the canonical morphisms for monoidalness $!A \otimes !B \multimap !(A \otimes B)$ and $§A \multimap !A$ which are not valid in LAL (the following terms and types are slightly more complicated since $\mathcal{L}_{DLAL\star}$ does not explicitly contain a type of the form $A \multimap !B$):

$$\lambda x^{!(A \multimap B)}.\lambda y^{!B \multimap C}.\lambda z^{!A}.(y)§((\bar{§}x)\bar{§}z) , \qquad \lambda x^{§A}.\lambda y^{!A \multimap B}.(y)§(\bar{§}x) .$$

In the first pseudo-term, the bang subterm $§((\bar{§}x)\bar{§}z)$ contains more than one free variable. In the second pseudo-term, the bang subterm $§(\bar{§}x)$ contains a free variable typed by a linear type. Hence they both violate the bang condition (i).

Λ-Scope condition. The previous conditions, bracketing and bang, would be enough to deal with boxes in the propositional fragment of $DLAL$. For handling second-order quantification though, we need a further condition to take into account the sequentiality enforced by the quantifiers. For instance consider the following two formulas (the second one is known as *Barcan's formula*):

$$(1)\ §\forall\alpha.A \multimap \forall\alpha.§A , \qquad (2)\ \forall\alpha.§A \multimap §\forall\alpha.A .$$

Assuming α occurs free in A, formula (1) is provable while (2) is not. Observe that we can build the following pseudo-terms which are locally typed and have respectively type (1) and (2):

$$t_1 = \lambda x^{§\forall\alpha.A}.\Lambda\alpha.§((\bar{§}x)\alpha) , \qquad t_2 = \lambda x^{\forall\alpha.§A}.§\Lambda\alpha.\bar{§}((x)\alpha) .$$

Both pseudo-terms satisfy the previous conditions, but t_2 does not correspond to a $DLAL$ derivation.

Let u be a locally typed pseudo-term. We say that u *depends on* α if the type of u contains a free variable α. We say that a locally typed pseudo-term t satisfies the *Λ-scope condition* if: for any subterm $\Lambda\alpha.u$ of t and for any subterm v of u that depends on α, $\mathsf{doors}(u, v)$ is weakly well-bracketed.

Coming back to our example: t_1 satisfies the Λ-scope condition, but t_2 does not, because $(x)\alpha$ depends on α and nevertheless $\mathsf{doors}(\bar{\S}((x)\alpha), (x)\alpha) = \bar{\S}$ is not weakly well-bracketed.

So far we have introduced four conditions on pseudo-terms: local typing, bracketing, bang and Λ-scope. Let us call a regular pseudo-term satisfying these conditions *well-structured*. It turns out that the well-structured pseudo-terms exactly correspond to the *DLAL* typing derivations.

Theorem 3. *Let M be a system F term. Then $x_1 : A_1, \ldots, x_m : A_m;\ y_1 : B_1, \ldots, y_n : B_n \vdash M : C$ is derivable in DLAL if and only if there is a decoration t of M with type $C^\star$ and with free variables $x_1^{!A_1^\star}, \ldots, x_m^{!A_m^\star}$, $y_1^{B_1^\star}, \ldots, y_n^{B_n^\star}$ which is well-structured.*

The 'only-if' direction can be shown by induction on the length of the derivation. To show the converse, we observe that whenever pseudo-terms $\lambda x^D.t$, $(t)u$, $\Lambda\alpha.t$, $(t)A$ are well-structured, so are the immediate subterms t and u. The case of $\S t$ is handled by the following key lemma (already used for $EAL^\star$ in [BT05]):

Lemma 4 (Boxing). *If $\S(t_A)$ is a well-structured pseudo-term, then there exist pseudo-terms v_A, $(u_1)_{\S B_1}$, $\ldots$, $(u_n)_{\S B_n}$, unique (up to renaming of v's free variables) such that:*

1. *$FV(v) = \{x_1^{B_1}, \ldots, x_n^{B_n}\}$ and each x_i occurs exactly once in v,*
2. *$\S t = \S v[\bar{\S} u_1/x_1, \ldots, \bar{\S} u_n/x_n]$ (substitution is assumed to be capture-free),*
3. *$v, u_1, \ldots, u_n$ are well-structured.*

As a consequence of Theorem 3, our *DLAL* typing problem boils down to:

Problem 5 (decoration). Given a system F term M, determine if there exists a decoration t of M which is well-structured.

4 Parameterization and Constraints

4.1 Parameterized Terms and Instantiations

To solve the decoration problem (Problem 5), one needs to explore the infinite set of decorations. This can be effectively done by introducing an abstract kind of types and terms with symbolic parameters, and expressing the conditions for such abstract terms to be materialized by boolean and integer constraints over those parameters (like in the related type inference algorithms for *EAL* or *LAL* mentioned in the introduction).

We use two sorts of parameters: *integer parameters* $\mathbf{n}, \mathbf{m}, \ldots$ meant to range over $\mathbb{Z}$, and *boolean parameters* $\mathbf{b_1}, \mathbf{b_2}, \ldots$ meant to range over $\{0, 1\}$. We also use *linear combinations of integer parameters* $\mathbf{c} = \mathbf{n_1} + \cdots + \mathbf{n_k}$, where $k \geq 0$ and each $\mathbf{n_i}$ is an integer parameter. In case $k = 0$, it is written as $\mathbf{0}$.

The set of *parameterized types* (*p-types* for short) is defined by:

$$F ::= \alpha \mid D \multimap A \mid \forall\alpha.A, \qquad A ::= \S^{\mathbf{c}}F, \qquad D ::= \S^{\mathbf{b},\mathbf{c}}F\ ,$$

where $\mathbf{b}$ is a boolean parameter and $\mathbf{c}$ is a linear combination of integer parameters. Informally speaking, in $\S^{\mathbf{b},\mathbf{c}}F$ the $\mathbf{c}$ stands for the number of modalities ahead of the type, while the boolean $\mathbf{b}$ serves to determine whether the first modality, if any, is § or !. In the sequel, A, B, C stand for *linear p-types* of the form $\S^{\mathbf{c}}F$, and D for *bang p-types* of the form $\S^{\mathbf{b},\mathbf{c}}F$, and E for arbitrary p-types.

When A is a linear p-type $\S^{\mathbf{c}}F$, $B[A/\alpha]$ denotes a p-type obtained by replacing each $\S^{\mathbf{c}'}\alpha$ in B with $\S^{\mathbf{c}'+\mathbf{c}}F$ and each $\S^{\mathbf{b},\mathbf{c}'}\alpha$ with $\S^{\mathbf{b},\mathbf{c}'+\mathbf{c}}F$. When $D = \S^{\mathbf{b},\mathbf{c}}F$, D° denotes the linear p-type $\S^{\mathbf{c}}F$.

We assume that there is a countable set of variables $x^D, y^D, \dots$ for each bang p-type D. The *parameterized pseudo-terms* (*p-terms* for short) $t, u \dots$ are defined by the following grammars:

$$t ::= \S^{\mathbf{m}}u, \quad u ::= x^D \mid \lambda x^D.t \mid (t)t \mid \Lambda\alpha.t \mid (t)A\ .$$

We denote by $par^{bool}(t)$ the set of boolean parameters of t, and by $par^{int}(t)$ the set of integer parameters of t. An *instantiation* $\phi = (\phi^b, \phi^i)$ for a p-term t is given by two maps $\phi^b : par^{bool}(t) \to \{0,1\}$ and $\phi^i : par^{int}(t) \to \mathbb{Z}$. The map ϕ^i can be naturally extended to linear combinations $\mathbf{c} = \mathbf{n_1} + \cdots + \mathbf{n_k}$ by $\phi^i(\mathbf{c}) = \phi^i(\mathbf{n_1}) + \cdots + \phi^i(\mathbf{n_k})$. An instantiation ϕ is said to be *admissible* for a p-type E if for any linear combination $\mathbf{c}$ occurring in E, we have $\phi^i(\mathbf{c}) \geq 0$, and moreover whenever $\S^{\mathbf{b},\mathbf{c}}F$ occurs in E, $\phi^b(\mathbf{b}) = 1$ implies $\phi^i(\mathbf{c}) \geq 1$. When ϕ is admissible for E, a type $\phi(E)$ of $DLAL\star$ is obtained as follows:

$$\phi(\S^{\mathbf{c}}F) = \S^{\phi^i(\mathbf{c})}\phi(F), \qquad \begin{aligned}\phi(\S^{\mathbf{b},\mathbf{c}}F) &= \S^{\phi^i(\mathbf{c})}\phi(F) && \text{if } \phi^b(\mathbf{b}) = 0,\\ &= !\S^{\phi^i(\mathbf{c})-1}\phi(F) && \text{otherwise,}\end{aligned}$$

and ϕ commutes with the other connectives. An instantiation ϕ for a p-term t is said to be *admissible* for t if it is admissible for all p-types occurring in t. When ϕ is admissible for t, a regular pseudo-term $\phi(t)$ can be obtained by replacing each $\S^{\mathbf{m}}u$ with $\S^{\phi^i(\mathbf{m})}u$, each x^D with $x^{\phi(D)}$, and each $(t)A$ with $(t)\phi(A)$.

As for pseudo-terms there is an erasure map $(.)^-$ from p-terms to system F terms consisting in forgetting modalities and parameters.

A *free linear decoration* (*free bang decoration*, resp.) of a system F type T is a linear p-type (bang p-type, resp.) E such that (i) $E^- = T$, (ii) each linear combination $\mathbf{c}$ occurring in E consists of a single integer parameter $\mathbf{m}$, and (iii) the parameters occurring in E are mutually distinct. Two free decorations $\overline{T}_1$ and $\overline{T}_2$ are said to be *distinct* if the set of parameters occurring in $\overline{T}_1$ is disjoint from the set of parameters in $\overline{T}_2$.

The *free decoration* $\overline{M}$ of a system F term M (which is unique up to renaming of parameters) is obtained as follows: first, to each type T of a variable x^T used in M, we associate a free bang decoration $\overline{T}$, and to each type U occurring as $(N)U$ in M, we associate a free linear decoration $\overline{U}$ with the following proviso:

(i) one and the same $\overline{T}$ is associated to all occurrences of the same variable x^T;
(ii) otherwise mutually distinct free decorations $\overline{T}_1, \ldots, \overline{T}_n$ are associated to different occurrences of T.

$\overline{M}$ is now defined by induction on the construction of M:

$$\overline{x^T} = \S^{\mathbf{m}} x^{\overline{T}}, \quad \overline{\lambda x^T.M} = \S^{\mathbf{m}} \lambda x^{\overline{T}}.\overline{M}, \quad \overline{(M)N} = \S^{\mathbf{m}}((\overline{M})\overline{N}),$$
$$\overline{\Lambda\alpha.M} = \S^{\mathbf{m}} \Lambda\alpha.\overline{M}, \quad \overline{(M)T} = \S^{\mathbf{m}}((\overline{M})\overline{T}),$$

where all newly introduced parameters $\mathbf{m}$ are chosen to be fresh. The key property of free decorations is the following:

Lemma 6. *Let M be a system F term and t be a regular pseudo-term. Then t is a decoration of M if and only if there is an admissible instantiation ϕ for $\overline{M}$ such that $\phi(\overline{M}) = t$.*

Hence our decoration problem boils down to:

Problem 7 (instantiation). Given a system F term M, determine if there exists an admissible instantiation ϕ for $\overline{M}$ such that $\phi(\overline{M})$ is well-structured.

For that we will need to be able to state the four conditions (local typing, bracketing, bang, and Λ-scope) on p-terms; they will yield some constraints on parameters. We will speak of *linear inequations*, meaning in fact both linear equations and linear inequations.

4.2 Local Typing Constraints

First of all, we need to express the unifiability of two p-types E_1 and E_2. We define a set $\mathcal{U}(E_1, E_2)$ of constraints by

$$\mathcal{U}(\alpha,\alpha) = \emptyset, \qquad \mathcal{U}(D_1 \multimap A_1, D_2 \multimap A_2) = \mathcal{U}(D_1, D_2) \cup \mathcal{U}(A_1, A_2),$$
$$\mathcal{U}(\forall\alpha.A_1, \forall\alpha.A_2) = \mathcal{U}(A_1, A_2), \qquad \mathcal{U}(\S^{\mathbf{c_1}} F_1, \S^{\mathbf{c_2}} F_2) = \{\mathbf{c_1} = \mathbf{c_2}\} \cup \mathcal{U}(F_1, F_2),$$
$$\mathcal{U}(\S^{\mathbf{b_1},\mathbf{c_1}} F_1, \S^{\mathbf{b_2},\mathbf{c_2}} F_2) = \{\mathbf{b_1} = \mathbf{b_2}, \mathbf{c_1} = \mathbf{c_2}\} \cup \mathcal{U}(F_1, F_2).$$

and undefined otherwise. It is straightforward to observe:

Lemma 8. *Let E_1, E_2 be two p-types such that $\mathcal{U}(E_1, E_2)$ is defined, and ϕ be an admissible instantiation for E_1 and E_2. Then $\phi(E_1) = \phi(E_2)$ if and only if ϕ is a solution of $\mathcal{U}(E_1, E_2)$.*

For any p-type E, $\mathcal{M}(E)$ denotes the set $\{\mathbf{c} \geq \mathbf{0} : \mathbf{c} \text{ occurs in } E\} \cup \{\mathbf{b} = \mathbf{1} \Rightarrow \mathbf{c} \geq \mathbf{1} : \S^{\mathbf{b},\mathbf{c}} F \text{ occurs in } E\}$. Then ϕ is admissible for E if and only if ϕ is a solution of $\mathcal{M}(E)$.

Now consider the free decoration $\overline{M}$ of a system F typed term M. We assign to each subterm t of $\overline{M}$ a *linear* p-type B (indicated as t_B) and a set $\mathcal{M}(t)$ of constraints as on Figure 2. Notice that any linear p-type is of the form $\S^{\mathbf{c}} F$. Moreover, since t comes from a system F typed term, we know that F is an implication when t occurs as $(t_{\S^{\mathbf{c}} F})u$, and F is a quantification when t occurs as $(t_{\S^{\mathbf{c}} F})A$. The set $\mathcal{U}(D^\circ, A)$ used in $\mathcal{M}((t)u)$ is always defined, and finally, $\overline{M}$ satisfies the eigenvariable condition.

Let $\mathsf{Ltype}(\overline{M})$ be $\mathcal{M}(\overline{M}) \cup \{\mathbf{b} = \mathbf{1} : x^{\S^{\mathbf{b},\mathbf{c}} F} \text{ occurs more than once in } \overline{M}\}$.

$$
\begin{array}{ll}
(x^D)_{D^\circ} & \mathcal{M}(x) = \mathcal{M}(D) \\
(\S^{\mathbf{m}} t_{\S^{\mathbf{c}} F})_{\S^{\mathbf{m}+\mathbf{c}} F} & \mathcal{M}(\S^{\mathbf{m}} t) = \{\mathbf{m}+\mathbf{c} \geq 0\} \cup \mathcal{M}(t) \\
(\lambda x^D . t_A)_{\S^0 (D \multimap A)} & \mathcal{M}(\lambda x^D . t) = \mathcal{M}(D) \cup \mathcal{M}(t) \\
((t_{\S^{\mathbf{c}} (D \multimap B)}) u_A)_B & \mathcal{M}((t)u) = \{\mathbf{c} = \mathbf{0}\} \cup \mathcal{U}(D^\circ, A) \cup \mathcal{M}(t) \cup \mathcal{M}(u) \\
(\Lambda \alpha . t_A)_{\S^0 \forall \alpha . A} & \mathcal{M}(\Lambda \alpha . t) = \mathcal{M}(t) \\
((t_{\S^{\mathbf{c}} \forall \alpha . B}) A)_{B[A/\alpha]} & \mathcal{M}((t)A) = \{\mathbf{c} = \mathbf{0}\} \cup \mathcal{M}(A) \cup \mathcal{M}(t)
\end{array}
$$

Fig. 2. $\mathcal{M}(t)$ constraints

4.3 Boxing Constraints

In this section we need to recall some definitions from [BT05]. We consider the words over integer parameters $\mathbf{m}, \mathbf{n} \ldots$, whose set we denote by $\mathcal{L}_p$.

Let t be a p-term and u an occurrence of subterm of t. We define, as for pseudo-terms, the word $\mathsf{doors}(t,u)$ in $\mathcal{L}_p$ as follows. If $t = u$, define $\mathsf{doors}(t,u) = \epsilon$. Otherwise:

$$
\begin{array}{ll}
\mathsf{doors}(\S^{\mathbf{m}} t, u) & = \mathbf{m} :: (\mathsf{doors}(t,u)), \\
\mathsf{doors}(\lambda y^D . t_1, u) & = \mathsf{doors}(\Lambda \alpha . t_1, u) = \mathsf{doors}((t_1)A, u) = \mathsf{doors}(t_1, u), \\
\mathsf{doors}((t_1)t_2, u) & = \mathsf{doors}(t_i, u) \text{ when } t_i \text{ is the subterm containing } u.
\end{array}
$$

The sum $s(l)$ of an element l of $\mathcal{L}_p$ is a linear combination of integer parameters defined by: $s(\epsilon) = \mathbf{0}$, $s(\mathbf{m} :: l) = \mathbf{m} + s(l)$. For each list $l \in \mathcal{L}_p$, define $\mathsf{wbracket}(l) = \{s(l') \geq \mathbf{0} \mid l' \leq l\}$ and $\mathsf{bracket}(l) = \mathsf{wbracket}(l) \cup \{s(l) = \mathbf{0}\}$.

Given a system F term M, we define the following sets of constraints:

Bracketing constraints. $\mathsf{Bracket}(\overline{M})$ is the union of the following sets:

(i) $\mathsf{bracket}(\mathsf{doors}(\overline{M}, x))$ for each free variable x in $\overline{M}$,
and for each occurrence of an abstraction subterm $\lambda x.v$ of $\overline{M}$,
(ii) $\mathsf{wbracket}(\mathsf{doors}(\overline{M}, \lambda x.v))$,
(iii) $\mathsf{bracket}(\mathsf{doors}(v, x))$ for each occurrence of x in v.

Bang constraints. A subterm u_A that occurs as $(t_{\S^{\mathbf{c}'}(\S^{\mathbf{b},\mathbf{c}} F \multimap B)}) u_A$ in $\overline{M}$ is called a *bang subterm* of $\overline{M}$ with the *critical parameter* $\mathbf{b}$. Now $\mathsf{Bang}(\overline{M})$ is the union of the following sets: for each bang subterm u of $\overline{M}$ with a critical parameter $\mathbf{b}$,

(i) $\{\mathbf{b} = \mathbf{0}\}$ if u has strictly more than one occurrence of free variable, and
$\{\mathbf{b} = \mathbf{1} \Rightarrow \mathbf{b}' = \mathbf{1}\}$ if u has exactly one occurrence of free variable $x^{\S^{\mathbf{b}',\mathbf{c}'} F'}$.
(ii) $\{\mathbf{b} = \mathbf{1} \Rightarrow s(\mathsf{doors}(u,v)) \geq \mathbf{1} \ : \ v \text{ subterm of } u \text{ such that } v \neq u \text{ and } v \neq x\}$.

Λ-Scope constraints. $\mathsf{Scope}(\overline{M})$ is the union of the following sets:
(i) $\mathsf{wbracket}(\mathsf{doors}(u,v))$ for each subterm $\Lambda \alpha . u$ of $\overline{M}$ and for each subterm v of u that depends on α.

We denote $\mathsf{Const}(\overline{M}) = \mathsf{Ltype}(\overline{M}) \cup \mathsf{Bracket}(\overline{M}) \cup \mathsf{Bang}(\overline{M}) \cup \mathsf{Scope}(\overline{M})$. Then:

Theorem 9. *Let M be a system F term and ϕ be an instantiation for $\overline{M}$. Then: ϕ is admissible for M and $\phi(\overline{M})$ is well-structured if and only if ϕ is a solution of $\mathsf{Const}(\overline{M})$. Moreover, the number of (in)equations in $\mathsf{Const}(\overline{M})$ is quadratic in the size of M.*

5 Solving the Constraints

From a proof-net point of view, naively one might expect that finding a *DLAL* decoration could be decomposed into first finding a suitable *EAL* decoration (that is to say a box structure) and then determining which boxes should be ! ones. This however cannot be turned into a valid algorithm because there can be an infinite number of *EAL* decorations in the first place.

Our method will thus proceed in the opposite way: first solve the boolean constraints, which corresponds to determine which !-boxes are necessary, and then complete the decoration by finding a suitable box structure.

5.1 Solving Boolean Constraints

We split $\mathsf{Const}(\overline{M})$ into three disjoint sets $\mathsf{Const}^b(\overline{M})$, $\mathsf{Const}^i(\overline{M})$, $\mathsf{Const}^m(\overline{M})$:

- A *boolean constraint* $\mathbf{s} \in \mathsf{Const}^b(\overline{M})$ consists of only boolean parameters. $\mathbf{s}$ is of one of the following forms:
  $\mathbf{b_1} = \mathbf{b_2}$ (in $\mathsf{Ltype}(\overline{M})$), $\mathbf{b} = \mathbf{1}$ (in $\mathsf{Ltype}(\overline{M})$),
  $\mathbf{b} = \mathbf{0}$ (in $\mathsf{Bang}(\overline{M})$), $\mathbf{b} = \mathbf{1} \Rightarrow \mathbf{b}' = \mathbf{1}$ (in $\mathsf{Bang}(\overline{M})$).
- A *linear constraint* $\mathbf{s} \in \mathsf{Const}^i(\overline{M})$ deals with integer parameters only. A linear constraint $\mathbf{s}$ is of one of the following forms:
  $\mathbf{c_1} = \mathbf{c_2}$ (in $\mathsf{Ltype}(\overline{M})$), $\mathbf{c} = \mathbf{0}$ (in $\mathsf{Ltype}(\overline{M})$ and $\mathsf{Bracket}(\overline{M})$),
  $\mathbf{c} \geq \mathbf{0}$ (in $\mathsf{Ltype}(\overline{M})$, $\mathsf{Bracket}(\overline{M})$, $\mathsf{Scope}(\overline{M})$).
- A *mixed constraint* $\mathbf{s} \in \mathsf{Const}^m(\overline{M})$ contains a boolean parameter and a linear combination and is of the following form:
  $\mathbf{b} = \mathbf{1} \Rightarrow \mathbf{c} \geq \mathbf{1}$ (in $\mathsf{Ltype}(\overline{M})$ and $\mathsf{Bang}(\overline{M})$).

We consider the set of instantiations on boolean parameters and the extensional order $\leq$ on these maps: $\psi^b \leq \phi^b$ if for any $\mathbf{b}$, $\psi^b(\mathbf{b}) \leq \phi^b(\mathbf{b})$.

Lemma 10. $\mathsf{Const}^b(\overline{M})$ *has a solution if and only if it has a minimal solution* ψ^b*. Moreover one can decide in time polynomial in the cardinality of* $\mathsf{Const}^b(\overline{M})$ *if there exists a solution, and in that case provide a minimal one.*

5.2 Solving Integer Constraints

When ϕ^b is a boolean instantiation, $\phi^b\mathsf{Const}^m(\overline{M})$ denotes the set of linear constraints defined as follows: for any constraint of the form $\mathbf{b} = \mathbf{1} \Rightarrow \mathbf{c} \geq \mathbf{1}$ in $\mathsf{Const}^m(\overline{M})$, $\mathbf{c} \geq \mathbf{1}$ belongs to $\phi^b\mathsf{Const}^m(\overline{M})$ if and only if $\phi^b(\mathbf{b}) = 1$. It is then clear that (*) (ϕ^b, ϕ^i) is a solution of $\mathsf{Const}(\overline{M})$ if and only if ϕ^b is a solution of $\mathsf{Const}^b(\overline{M})$ and ϕ^i is a solution of $\phi^b\mathsf{Const}^m(\overline{M}) \cup \mathsf{Const}^i(\overline{M})$.

Proposition 11. $\mathsf{Const}(\overline{M})$ *admits a solution if and only if it has a solution* $\psi = (\psi^b, \psi^i)$ *such that* ψ^b *is the minimal solution of* $\mathsf{Const}^b(\overline{M})$.

Proof. Suppose that $\mathsf{Const}(\overline{M})$ admits a solution (ϕ^b, ϕ^i). Then by the previous lemma, there is a minimal solution ψ^b of $\mathsf{Const}^b(\overline{M})$. Since $\psi^b \leq \phi^b$, we have $\psi^b\mathsf{Const}^m(\overline{M}) \subseteq \phi^b\mathsf{Const}^m(\overline{M})$. Since ϕ^i is a solution of $\phi^b\mathsf{Const}^m(\overline{M}) \cup \mathsf{Const}^i(\overline{M})$ by (*) above, it is also a solution of $\psi^b\mathsf{Const}^m(\overline{M}) \cup \mathsf{Const}^i(\overline{M})$. This means that (ψ^b, ϕ^i) is a solution of $\mathsf{Const}(\overline{M})$. ∎

Coming back to the proof-net intuition, Proposition 11 means that given a syntactic tree of term there is a most general (minimal) way to place ! boxes (and accordingly ! subtypes in types), that is to say: if there is a *DLAL* decoration for this tree then there is one with precisely this minimal distribution of ! boxes.

Now notice that $\psi^b\mathsf{Const}^m(\overline{M}) \cup \mathsf{Const}^i(\overline{M})$ is a linear inequation system, for which a polynomial time procedure for searching a rational solution is known.

Lemma 12. *$\psi^b\mathsf{Const}^m(\overline{M}) \cup \mathsf{Const}^i(\overline{M})$ has a solution in $\mathbb{Q}$ if and only if it has a solution in $\mathbb{Z}$.*

Theorem 13. *Let M be a System F term. Then one can decide in time polynomial in the cardinality of $\mathsf{Const}(\overline{M})$ whether $\mathsf{Const}(\overline{M})$ admits a solution.*

Proof. First decide if there is a solution of $\mathsf{Const}^b(\overline{M})$, and if it exists, let ψ^b be the minimal one (Lemma 10). Then apply the polynomial time procedure to decide if $\psi^b\mathsf{Const}^m(\overline{M}) \cup \mathsf{Const}^i(\overline{M})$ admits a solution in $\mathbb{Q}$. If it does, then we also have an integer solution (Lemma 12). Otherwise, $\mathsf{Const}(\overline{M})$ is not solvable. ∎

By combining Theorem 3, Lemma 6, Theorems 9 and 13, we finally get:

Theorem 14. *Given a system F term M^T, it is decidable in time polynomial in the size of M whether there is a decoration A of T such that $\vdash_{DLAL} M : A$.*

6 Implementation

Overview. We designed an implementation of the type inference algorithm. The program is written in functional Caml and is quite concise (less than 1500 lines). A running program not only shows the actual feasibility of our method, but also is a great facility for building examples, and thus might allow for a finer study of the algorithm.

Data types as well as functions closely follow the previous description of the algorithm: writing the program in such a way tends to minimise the number of bugs, and speaks up for the robustness of the whole proof development.

The program consists of several successive parts:

1. Parsing phase: turns the input text into a concrete syntax tree. The input is an F typing judgement, in a syntax *à la* Church with type annotations at the binders. It is changed into the de Bruijn notation, and parameterized with fresh parameters. Finally, the abstract tree is decorated with parameterized types at each node.
2. Constraints generation: performs explorations on the tree and generates the boolean, linear and mixed constraints.

3. Boolean constraints resolution: gives the minimal solution of the boolean constraints, or answers negatively if the set admits no solution.
4. Constraints printing: builds the final set of linear constraints.

We use the simplex algorithm to solve the linear constraints. It runs in $O(2^n)$, which comes in contrast with the previous result of polynomial time solving, but has proven to be the best in practice (with a careful choice of the objective function).

Example of execution. Let us consider the reversing function on binary words. It can be defined by a single higher-order iteration, and thus represented by the following system F term, denoted **rev**:

$$\begin{array}{l}\lambda l^W.\Lambda\beta.\lambda so^{\beta\to\beta}.\lambda si^{\beta\to\beta}.(l\ (\beta\to\beta))\\ \quad\lambda a^{\beta\to\beta}.\lambda x^{\beta}.(a)(so)x\\ \quad\lambda a^{\beta\to\beta}.\lambda x^{\beta}.(a)(si)x\ (\Lambda\alpha.\lambda z^{\alpha}.z)\beta\end{array}$$

We apply it to : $\Lambda\alpha.\lambda so^{\alpha\to\alpha}.\lambda si^{\alpha\to\alpha}.\lambda x^{\alpha}.(si)(so)(si)(so)x$, representing the word **1010**, in order to force a meaningful typing. Since **rev** involves higher-order functionals and polymorphism, it is not so straightforward to tell, just by looking at the term structure, whether it works in polynomial time or not.

Given **rev(1010)** as input (coded by ASCII characters), our program produces 177 (in)equations on 79 variables. After constraint solving, we obtain the result:

$$\begin{array}{l}(\lambda l^W.\Lambda\beta.\lambda so^{!(\beta\multimap\beta)}.\lambda si^{!(\beta\multimap\beta)}.\\ \quad\S(\bar{\S}((l\ (\beta\multimap\beta))\\ \quad\S\lambda a^{\beta\multimap\beta}.\lambda x^{\beta}.(a)(\bar{\S}so)x\\ \quad\S\lambda a^{\beta\multimap\beta}.\lambda x^{\beta}.(a)(\bar{\S}si)x)\\ \quad(\Lambda\alpha.\lambda z^{\alpha}.z)\beta)\\ \Lambda\alpha.\lambda so^{!\alpha\to\alpha}.\lambda si^{\alpha\to\alpha}.\S\lambda x^{\alpha}.(\bar{\S}si)(\bar{\S}so)(\bar{\S}si)(\bar{\S}so)x\end{array}$$

It corresponds to the natural depth-1 typing of the term **rev**, with conclusion type $W_{DLAL}\multimap W_{DLAL}$. The solution ensures polynomial time termination, and in fact its depth guarantees normalization in a quadratic number of β-reduction steps. Further examples and the program are available at:
`http://www-lipn.univ-paris13.fr/~atassi/`

7 Conclusion

We showed that typing of system F terms in $DLAL$ can be performed in a feasible way, by reducing typability to a constraints solving problem and designing a resolution algorithm. This demonstrates a practical advantage of $DLAL$ over LAL, while keeping the other important properties. Other typing features could still be automatically infered, like coercions (see [Ata05] for the case of EAL).

This work illustrates how Linear logic proof-net notions like boxes can give rise to techniques effectively usable in type inference, even with the strong boxing discipline of $DLAL$, which extends previous work on EAL. We expect that some of these techniques could be adapted to other variants of Linear logic, existing or to be defined in the future.

Ackowledgements. The authors wish to thank Laurent Régnier for useful discussions related to the topic of this paper.

References

[ABT06] V. Atassi, P. Baillot, and K. Terui. Verification of Ptime reducibility for system F terms via Dual Light Affine Logic. Technical Report HAL ccsd-00021834, july 2006.

[Ama05] R. Amadio. Synthesis of max-plus quasi-interpretations. *Fundamenta Informaticae*, 65:29–60, 2005.

[AR02] A. Asperti and L. Roversi. Intuitionistic light affine logic. *ACM Transactions on Computational Logic*, 3(1):1–39, 2002.

[Ata05] V. Atassi. Inférence de type en logique linéaire élémentaire. Master's thesis, Université Paris 13, 2005.

[Bai02] P. Baillot. Checking polynomial time complexity with types. In *Proceedings of IFIP TCS'02*, Montreal, 2002. Kluwer Academic Press.

[Bai04] P. Baillot. Type inference for light affine logic via constraints on words. *Theoretical Computer Science*, 328(3):289–323, 2004.

[BC92] S. Bellantoni and S. Cook. New recursion-theoretic characterization of the polytime functions. *Computational Complexity*, 2:97–110, 1992.

[BT04] P. Baillot and K. Terui. Light types for polynomial time computation in lambda-calculus. In *Proceedings LICS'04*. IEEE Computer Press, 2004.

[BT05] P. Baillot and K. Terui. A feasible algorithm for typing in elementary affine logic. In *Proceedings of TLCA'05*, volume 3461 of *LNCS*, pages 55–70. Springer, 2005.

[CDLRdR05] P. Coppola, U. Dal Lago, and S. Ronchi Della Rocca. Elementary affine logic and the call-by-value lambda calculus. In *Proceedings of TLCA'05*, volume 3461 of *LNCS*, pages 131–145. Springer, 2005.

[CM01] P. Coppola and S. Martini. Typing lambda-terms in elementary logic with linear constraints. In *Proceedings TLCA'01*, volume 2044 of *LNCS*, 2001.

[CRdR03] P. Coppola and S. Ronchi Della Rocca. Principal typing in Elementary Affine Logic. In *Proceedings TLCA'03*, LNCS, 2003.

[DJ03] V. Danos and J.-B. Joinet. Linear logic and elementary time. *Information and Computation*, 183(1):123–137, 2003.

[Gir87] J.-Y. Girard. Linear logic. *Theoretical Computer Science*, 50:1–102, 1987.

[Gir98] J.-Y. Girard. Light linear logic. *Information and Computation*, 143:175–204, 1998.

[HJ03] M. Hofmann and S. Jost. Static prediction of heap space usage for first-order functional programs. In *Proc. ACM POPL'03*, 2003.

[Hof03] M. Hofmann. Linear types and non-size-increasing polynomial time computation. *Information and Computation*, 183(1):57–85, 2003.

[LM93] D. Leivant and J.-Y. Marion. Lambda-calculus characterisations of polytime. *Fundamenta Informaticae*, 19:167–184, 1993.

[MM00] J.-Y. Marion and J.-Y. Moyen. Efficient first order functional program interpreter with time bound certifications. In *Proceedings of LPAR 2000*, volume 1955 of *LNCS*, pages 25–42. Springer, 2000.

[Ter01] K. Terui. Light Affine Lambda-calculus and polytime strong normalization. In *Proceedings LICS'01*. IEEE Computer Society, 2001. Full version available at http://research.nii.ac.jp/∼ terui.

[Ter04a] K. Terui. Light affine set theory: a naive set theory of polynomial time. *Studia Logica*, 77:9–40, 2004.

[Ter04b] K. Terui. A translation of LAL into DLAL. Preprint, http://research.nii.ac.jp/∼ terui, 2004.

[Wel99] J. B. Wells. Typability and type checking in system F are equivalent and undecidable. *Ann. Pure Appl. Logic*, 98(1-3), 1999.

MSO Queries on Tree Decomposable Structures Are Computable with Linear Delay

Guillaume Bagan

Université de Caen, GREYC, Campus 2, F-14032 Caen cedex - France

Abstract. LINEAR-DELAY$_{lin}$ is the class of enumeration problems computable in two steps: the first step is a precomputation in linear time in the size of the input and the second step computes successively all the solutions with a delay between two consecutive solutions y_1 and y_2 that is linear in $|y_2|$. We prove that evaluating a fixed monadic second order (MSO) query $\varphi(\bar{X})$ (i.e. computing all the tuples that satisfy the MSO formula) in a binary tree is a LINEAR-DELAY$_{lin}$ problem. More precisely, we show that given a binary tree T and a tree automaton Γ representing an MSO query $\varphi(\bar{X})$, we can evaluate Γ on T with a preprocessing in time and space complexity $O(|\Gamma|^3|T|)$ and an enumeration phase with a delay $O(|S|)$ and space $O(max|S|)$ where $|S|$ is the size of the next solution and $max|S|$ is the size of the largest solution. We introduce a new kind of algorithm with nice complexity properties for some algebraic operations on enumeration problems. In addition, we extend the precomputation (with the same complexity) such that the ith (with respect to a certain order) solution S is produced directly in time $O(|S|\log(|T|))$. Finally, we generalize these results to bounded treewidth structures.

1 Introduction

Determining how difficult it is to compute a query written in a given language is an important task in theorical computer sciences. A language which has deserved much attention is the monadic second order logic (MSO). It is well known that a lot of NP-complete problems can be expressed as the model checking of a MSO-sentence. Nevertheless, for particular kinds of structures, the complexity can be improved. Classes of bounded treewidth structures are of particular interest. Courcelle [6] proved that deciding if a MSO-sentence holds in a bounded treewidth structure can be done in linear time in the size of the structure (with a constant factor that highly depends of the size of the formula). Arnborg, Lagergren and Seese[2] proved that given a fixed MSO-formula $\varphi(\bar{X})$ and a bounded treewidth structure $\mathcal{S}$, counting the number of solutions (i.e assignments of variables which satisfy the formula) can be done with the same complexity. Courcelle and Mosbah [8] proved that enumerating all solutions which satisfy a MSO query on a bounded treewidth structure can be done in time polynomial in the size of the structure and the size of the output. Frick, Flum and Grohe [11] improved this by proving that this can be done in linear time in the size of the structure plus the size of the output.

Z. Ésik (Ed.): CSL 2006, LNCS 4207, pp. 167–181, 2006.

Recently, Durand and Grandjean [10] were interested in logical queries viewed as enumeration problems and studied the delay between two solutions instead of the global time of the algorithm.

They introduced the class CONSTANT-DELAY$_{lin}$ which consists of enumeration problems computable in two steps: the first step is a precomputation which is done in time and space complexity linear in the size of the structure, the second step is the enumeration phase which outputs all the solutions with a constant delay between two consecutive ones. They showed that evaluating FO-queries on bounded degree structures is a CONSTANT-DELAY$_{lin}$ problem.

The main goal of this paper is to revisit the complexity of the evaluation of MSO queries on bounded treewidth structures. Meanwhile, the class CONSTANT-DELAY$_{lin}$ contains only problems where the size of each solution is bounded by a constant. We introduce the class LINEAR-DELAY$_{lin}$ which is a generalization of CONSTANT-DELAY$_{lin}$ where the delay between two consecutive solutions is linear in the size of the second one. The main result of this paper is that evaluating MSO-queries on binary trees is a LINEAR-DELAY$_{lin}$ problem. More precisely, by using the well known translation of a MSO formula on binary trees into tree automaton [15], we show, that given a tree T and a tree automaton Γ which represent the formula $\varphi(\bar{X})$, we can do a precomputation in time $O(|\Gamma|^3|T|)$ and then output all the solutions (with respect to a certain order) with a delay $O(|S|)$ where $|S|$ is the size of the next solution S to be computed (that depends neither on the size of the structure nor on the size of the automaton) and space $O(\max\{|S| : S \in \varphi(T)\})$. As a consequence, it can be computed with a total time $O(|\Gamma|^3|T| + \|\varphi(T)\|)$ (where $\|\varphi(T)\|$ is the total size of the outputs that is $\sum_{S\in\varphi(T)} |S|$) and space $O(|\Gamma|^3|T|)$. To obtain this result, we use several tools. First, we introduce some algebraic operations on enumeration problems. Then, we express the problem of evaluation of a MSO-query as a combination of simple enumeration problems by these operations. In addition, we introduce a new kind of precise enumeration algorithm, called tiptop algorithm, which computes a enumeration problem in a lazy way.

Another problem considered in this paper consists, given a MSO formula $\varphi(\bar{X})$ and a binary tree T of producing directly the ith solution S (with respect to the same order than the evaluation problem). We show that this can be done in two part: first we do a precomputation (that does not depend on i) in linear time in the size of the structure and then for each input i, we can produce the ith solution (using the precomputation) with time $O(|S|log(|T|))$. An immediate consequence of this result is an algorithm for the uniform random generation of solutions of a MSO-query with the below complexity.

Finally, we show that these two results can be generalized to structures of bounded treewidth.

Notice that our results improve the similar results of a very recent paper by Courcelle (submitted in Mars 2006 to a special issue of Discrete Mathematics [7]) and have been proved independently and in the same period. By using a notion of DAG structure (AND-OR structure), Courcelle obtains in particular an enumeration algorithm for evaluation of MSO-queries in binary trees with a linear time delay but with a precomputation in $O(|T|log(|T|))$. Our results are simpler and optimal.

The paper is organized as follows. First, basic definitions are given in section 1. In particular, we define the class LINEAR-DELAY$_{lin}$. In section 2, we consider the evaluation problem of MSO formulas over binary trees. In subsection 2.3, we give an efficient implementation of a linear delay algorithm for this problem. In particular, tiptop algorithms and algebraic operations on enumeration problems are introduced. In section 3, we consider the problem of computing directly ith element. Finally, in section 4, we show that these results can be generalized to MSO-queries on bounded treewidth structures.

2 Preliminaries

The reader is expected to be familiar with first-order and monadic second order logic (see e.g. [14]). We use in this paper the Random Access Machine (RAM) model with uniform cost measure (see [1,13,12]).

2.1 Enumeration

Given a binary relation A, the enumeration problem ENUM $- A$ associated with A is the problem described as follows: For each input x, ENUM $- A(x) = \{y|(x, y) \in A\}$. We call ENUM $- A(x)$ the set of solutions of ENUM $- A$ on x. To measure without ambiguity the complexity, we need to be more precise on the structure of the output. We say that an algorithm $\mathcal{A}$ computes ENUM $- A$ if

- for any input x, $\mathcal{A}$ writes sequentially the output $\#y_1\#y_2\#.....\#y_n\#$ where $(y_1, ..., y_n)$ is an enumeration (without repetition) of the set ENUM $- A(x)$,
- it writes the first $\#$ when its starts,
- it stops after writing the last $\#$.

Let $\mathcal{A}$ be an enumeration algorithm and x be an input of $\mathcal{A}$. Let $time_i(x)$ denote the time when the algorithm writes the ith $\#$ if it exists. We define $delay_i(x) = time_{i+1}(x) - time_i(x)$.

Definition 1. *An enumeration algorithm $\mathcal{A}$ is constant delay if there is a constant c such that for any input x and for any i, $delay_i(x) \leq c$ and $\mathcal{A}$ uses space $O(|x|)$.*

An enumeration algorithm $\mathcal{A}$ is linear delay if there is a constant c such that for any input x and for any i, $delay_i(x) \leq c|y_i|$ and $\mathcal{A}$ uses space $O(|x|)$.

An enumeration problem ENUM $- A$ *is said to be computable with linear delay, which is denoted by* ENUM $- A \in$ LINEAR-DELAY *if there is a linear delay algorithm $\mathcal{A}$ which computes* ENUM $- A$.

An enumeration problem ENUM $- A$ *is* LINEAR-DELAY$_{lin}$ *if it is reducible in linear time to a problem in* LINEAR-DELAY.

2.2 Trees and Tree Automata

A Σ-tree T is a pair (D, label) where the tree domain D is a subset of $\{1,2\}^*$ that is prefix-closed and such that if $s \in D$ then either $s.1$ and $s.2$ or none of them belong to D (i. e. T is a locally complete tree) and label is a function from D to Σ (the labelling function). $s.1$ and $s.2$ are the first child and the second child of s. If x is a prefix of y then x is an ancestor of y and y is a descendant of x. ϵ is the root of T. A leaf is a node without children and an internal node is a node with two children. We denote by Leaves(T) the set of leaves of T and by root(T) the root of T. T_x denotes the subtree of T rooted by x. Associate to each alphabet Σ, the first-order signature the signature $\tau_\Sigma = (\text{succ}_1, \text{succ}_2, (P_a)_{a\in\Sigma})$. We do a confusion between a Σ-tree and its encoding as a τ_Σ-structure $T = (D, \text{succ}_1, \text{succ}_2, (P_a)_{a\in\Sigma})$ where P_a is interpreted as $\{s \in D, \text{label}(s) = a\}$ and succ_i is interpreted as $\{(s, s.i) | s, s.i \in D\}$. We denote by Σ – Trees the set of all Σ-trees.

We give some definitions on tree automata, for more details see [5]. A complete deterministic bottom-up tree automaton (tree automaton for short) is a tuple $(\Sigma, Q, q_0, \delta, Q_f)$ where Q is a finite set of states, $q_0 \in Q$ is the initial state, $Q_f \subseteq Q$ is the set of final states, and $\delta : Q \times Q \times \Sigma \to Q$ is the transition function. We abbreviate $\delta(q_0, q_0, x)$ to $\delta_{init}(x)$ Let Γ be a tree automaton and T be a Σ-tree, the run of Γ on T, denoted by run(Γ, T), is the mapping $r : T \to Q$ defined by the following conditions.

- If x is a leaf then $r(x) = \delta_{init}(\text{label}(x))$
- If x is an internal node whose first and second children are y_1 and y_2 then $r(x) = \delta(r(y_1), r(y_2), \text{label}(x))$.

We say that the run r is accepting if $r(\text{root}(T)) \in Q_f$. Γ accepts T if its run is accepting. The set of trees accepted by Γ is called the language recognized by Γ and is denoted by $L(\Gamma)$.

A language L is regular if there is a tree automaton Γ which recognizes L.

Theorem 1. *[15] A language L is regular if and only if it is definable by an MSO-formula. There is an algorithm which, given an MSO[τ_Σ] formula which defines the language L, computes a tree automaton that recognizes the same language L.*

2.3 MSO Queries

For a signature τ and a class of τ-structures $\mathcal{C}$, we consider the following problem Query(MSO, τ, $\mathcal{C}$):
Input: a τ-structure $\mathcal{S} \in \mathcal{C}$ and an MSO[τ] formula $\varphi(\bar{X}, \bar{y})$ where $\bar{X} = (X_1, \ldots, X_k)$ and $\bar{y} = (y_1, \ldots y_l)$.
Parameter: φ
Output: $\varphi(\mathcal{S}) = \{(\bar{A}, \bar{b}) \in \mathcal{P}(D)^k \times D^l | (\mathcal{S}, \bar{A}, \bar{b}) \models \varphi(\bar{X}, \bar{y})\}$ [1].

[1] Here and below, a $\mathcal{P}(D)$ denotes the power set of the set D.

We are interested in the parameterized complexity of the problem (see [9]) with the formula φ as parameter.

From now, we are interested in the specific problem QUERY(MSO, τ_Σ, Σ-TREES).

Without loss of generality, we can restrict the problem to queries without first order variable. Indeed, it suffices to consider a first order variable as a singleton second order variable.

2.4 Specific Notations

For a vector $\bar{v} = (v_1, \ldots, v_k)$ let $\bar{v} \upharpoonright i$ denote its ith component v_i. Let $\bar{S_1}$ and $\bar{S_2}$ be two k-tuples of sets, the combination of $\bar{S_1}$ and $\bar{S_2}$, denoted by $\bar{S_1} \oplus \bar{S_2}$, is defined such that $(\bar{S_1} \oplus \bar{S_2}) \upharpoonright i = (\bar{S_1} \upharpoonright i) \cup (\bar{S_2} \upharpoonright i)$. Let A and B be two sets of k-tuples of sets. The product of A and B, denoted by $A \otimes B$ is the set $\{\bar{S_1} \oplus \bar{S_2} | \bar{S_1} \in A, \bar{S_2} \in B\}$. For an element x of $\mathcal{S}$ and a vector $v \in \{0,1\}^k$, we denote by $\bar{x_v}$ the k-tuple such that

$$\bar{x_v} \upharpoonright i = \begin{cases} \{x\} & \text{if } \bar{v} \upharpoonright i = 1 \\ \emptyset & \text{otherwise} \end{cases}$$

3 Evaluation of MSO-Queries

Theorem 1 applies only to sentences. To adapt it to any MSO formula $\varphi(X_1, \ldots X_k)$, we consider the enriched alphabet $\Sigma' = \Sigma \times \{0,1\}^k$. For any Σ-tree $T = (D, \text{label}_T)$ and any k-tuple of sets $\bar{S}$ [2], we can associate the tree $T_{\bar{S}} = (D, \text{label}_{T'})$ with same domain as T and such that $\text{label}_{T'}(x) = (\text{label}_T(x), \bar{v})$ where $\bar{v}$ is the characteristic vector of $\bar{S}$ i.e. verifying $\bar{v} \upharpoonright i = 1$ if and only $x \in \bar{S} \upharpoonright i$. Therefore, we can consider the language L_φ such that $T_{\bar{S}} \in L_\varphi$ if and only if $(T, \bar{S}) \models \varphi(\bar{X})$. By abuse of notation, we do confusion between the pair $(T, \bar{S})$ and the tree $T_{\bar{S}}$.

For simplicity, we will restrict to Σ-trees whose internal vertices are labeled by a dummy color @ and to formulas where interpretations of second order variables are restricted to subsets of the leaves. The next lemma explains how to reduce the general case to these conditions.

Lemma 1. *Let Σ be a finite alphabet and set $\Sigma' = \Sigma \cup \{@\}$. There is an algorithm which given a $MSO[\tau_\Sigma]$-formula φ computes a $MSO[\tau_{\Sigma'}]$-formula φ' and there is a linear time algorithm which given a Σ-tree T computes a Σ'-tree T' such that*

- *$Dom(T) = Leaves(T')$*
- *the internal nodes of T' are labeled by @*
- *$\varphi(T) = \varphi'(T')$*

---

[2] Here and below, a k-tuple of sets, or for short, a k-tuple $\bar{S}$ is a k-tuple of subsets of the domain D of T.

As internal nodes are labeled by @, we use the notation $q_1, q_2 \to q$ which means $\delta(q_1, q_2, @) = q$ for the transition function δ of an automaton Γ.

We consider now an MSO-query $\varphi(\bar{X})$ that defines a language $L_\varphi = \{T_{\bar{S}} : (T, \bar{S}) \models \varphi(\bar{X})\}$, Let $\Gamma = (\Sigma', Q, q_0, \delta, Q_f)$ be a tree automaton on the alphabet $\Sigma' = \Sigma \times \{0,1\}^k$ which recognizes L_φ. Let $T = (D, \text{label})$ be a Σ-tree. A configuration of (Γ, T) is an element of $D \times Q$. Let a k-tuple $\bar{S} \in \mathcal{P}(\text{Leaves}(T))^k$, an $\bar{S}$-configuration is a configuration (x, q) such that $r(x) = q$ where $r = \text{run}(\Gamma, T, \bar{S})$: it is the unique $\bar{S}$-configuration of x.

We want to enumerate the tuples $\bar{S} \in \mathcal{P}(\text{Leaves}(T))^k$ such that $r(\text{root}(T)) \in Q_f$ for $r = \text{run}(\Gamma, T, \bar{S})$. Clearly, this can be done in a recursive way. For any tuple $\bar{S}$ and any node x of T, let $\bar{S}_x$ denote the restriction of $\bar{S}$ to the subtree T_x. It is clear that the state $q = r(x)$ for $r = run(\Gamma, T, \bar{S})$ only depends on $\bar{S}_x$. For any configuration (x, q) let $\text{stuples}(x, q)$ denote the set of tuples $\bar{S} \in \mathcal{P}(\text{Leaves}(T_x))^k$ such that $r(x) = q$ for $r = run(\Gamma, T, \bar{S})$.

Let us notice two easy but useful facts:

Let x be any internal node of T with children y_1 and y_2

1. It holds $\bar{S_x} = \bar{S_{y_1}} \oplus \bar{S_{y_2}}$ and $(\bar{S_{y_1}} \upharpoonright i) \cap (\bar{S_{y_2}} \upharpoonright i) = \emptyset$ for any $i = 1, \ldots, k$.
2. For each $q \in Q$, it holds

$$\text{stuples}(x, q) = \bigcup_{\substack{(q_1,q_2) \in Q^2 \\ q_1, q_2 \to q}} \text{stuples}(y_1, q_1) \otimes \text{stuples}(y_2, q_2)$$

and this is a disjoint union.

Remark: The disjointness property in (2) is a straightforward consequence of the determinism of the automaton Γ.

At the first glance, properties (1) and (2) above seem to give the very principle of an efficient algorithm that enumerates the elements (i.e., k-tuples) of the set $\text{stuples}(x, q)$ for any given configuration (x, q). Unfortunately, this does not yield the linear delay we hope between consecutive solutions. Indeed such a trivial algorithm may lose much time in performing unions of empty sets, namely in computing $A \otimes B$ when either of the operand sets A and B contains the k-tuples of empty sets $\bar{\emptyset} = (\emptyset, \ldots, \emptyset)$. To overcome this difficulty, we keep apart the "empty" k-tuple $\bar{\emptyset}$ and classify the nodes of the tree T according to each tuple $\bar{S} \in \mathcal{P}(\text{Leaves}(T))^k$, for $\bar{S} \neq \bar{\emptyset}$.

Let $\bar{S} \in \mathcal{P}(\text{Leaves}(T))^k$ and x be a node of T. We say that x is an $\bar{S}$-active node if $(\bar{S} \upharpoonright i) \cap \text{Leaves}(T_x) \neq \emptyset$ for some i. We say otherwise that x is $\bar{S}$-passive. In case x is an internal node with children y_1 and y_2. Then x is $\bar{S}$-transitive if it is $\bar{S}$-active and only one of its children is $\bar{S}$-active. x is $\bar{S}$-fork if its children y_1 and y_2 are both $\bar{S}$-active. A node is $\bar{S}$-useful if it is an $\bar{S}$-fork node or an $\bar{S}$-active leaf. We give similar definitions for configurations. For example, a configuration (x, q) is $\bar{S}$-transitive if it is the $\bar{S}$-configuration of an $\bar{S}$-transitive node.

The reduced tree associated to $(T, \bar{S})$ is the graph denoted by $T_{\bar{S}}^{useful}$ and built as follows:

1. The nodes of $T_{\bar{S}}^{useful}$ are the $\bar{S}$-useful nodes of T
2. We put an edge from x to y if y is a descendant of x in T and there is no $\bar{S}$-useful nodes between them in T.

It is easily seen that $T_{\bar{S}}^{useful}$ is a binary tree.

We notice that, given any vertex x, the $\bar{S}$-configuration (x, q) of x is the same for any tuple $\bar{S}$ such that x is $\bar{S}$-passive. We call q the passive state of x and (x, q) the passive configuration of x.

Example 1. We assume that $S = \{8, 13, 14, 15\}$ for k $= 1$ (one predicate only).The filled circles represent S-useful vertices, bold circles represent S-transitive vertices and normal circles represent S-passive vertices.

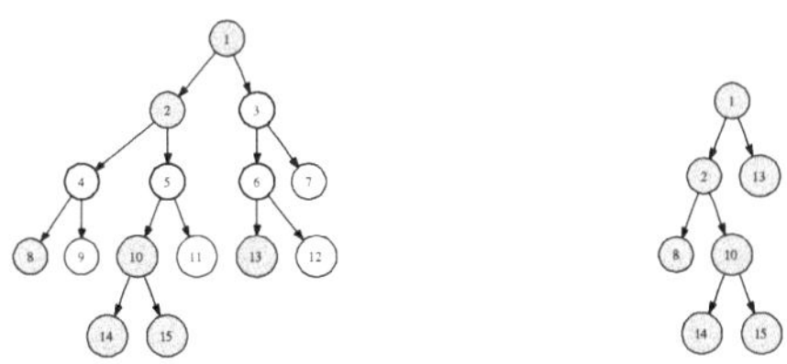

Fig. 1. A tree $T_{\bar{S}}$ and its reduced tree $T_{\bar{S}}^{useful}$

Let (x, q), (y_1, q_1) and (y_2, q_2) be three configurations such that y_1 and y_2 are the first and second children of x, and such that $q_1, q_2 \rightarrow q$. We say that (x, q) is empty-accessible from (y_1, q_1) if (y_2, q_2) is a passive configuration. Similarly (x, q) is empty-accessible from (y_2, q_2) if (y_1, q_1) is a passive configuration We use the notation $(x', q') \overset{\emptyset}{\longmapsto} (x, q)$ to mean that (x, q) is empty-accessible from (x', q').

Let $\overset{*}{\longmapsto}$ denote the reflexive and transitive closure of the relation $\overset{\emptyset}{\longmapsto}$. If we have $(x', q') \overset{*}{\longmapsto} (x, q)$ then we say that (x, q) is transitively accessible from (x', q').

Lemma 2. *1) Let $\bar{S} \in \mathcal{P}(Leaves(T))^k$ and let (x, q) and (x', q') be two $\bar{S}$-active configurations such that x is an ancestor of x'. If there is no $\bar{S}$-fork configuration (except eventually (x', q')) in the unique path between (x, q) and (x', q') then $(x', q') \overset{*}{\longmapsto} (x, q)$.*
2) Let (x, q) and (x', q') be two configurations such that $(x', q') \overset{*}{\longmapsto} (x, q)$. Let $\bar{S} \in \mathcal{P}(Leaves(T_{x'}))^k$. If (x', q') is the $\bar{S}$-active configuration of x' then (x, q) is the $\bar{S}$-configuration of x.

We denote by suseful(x, q) (resp sactive(x, q)) the set of tuples $\bar{S} \in \mathcal{P}$(Leaves $(T_x))^k$ such that (x, q) is a $\bar{S}$-useful ($\bar{S}$-active) configuration. We denote by senum the set of tuples $\bar{S} \neq \bar{\emptyset}$ such that Γ accepts $(T, \bar{S})$.

Lemma 3. *Let (x, q) be a configuration.*

1. *If x is a leaf then*

$$\mathit{suseful}(x,q) = \bigcup_{\substack{\bar{v}\in\{0,1\}^k\setminus\{0\}^k \\ \delta_{init}(\mathit{label}(x),\bar{v})=q}} \{\bar{x_{\bar{v}}}\}$$

and this is a disjoint union.

2. *If x is an internal node of first and second children y_1 and y_2.*

$$\mathit{suseful}(x,q) = \bigcup_{\substack{(q_1,q_2)\in Q^2 \\ q_1,q_2\to q}} \mathit{sactive}(y_1,q_1) \otimes \mathit{sactive}(y_2,q_2) \quad (2)$$

and this is a disjoint union.

3. *It holds*

$$\mathit{sactive}(x,q) = \bigcup_{\substack{(y,q')\in D\times Q \\ (y,q')\overset{*}{\longmapsto}(x,q)}} \mathit{suseful}(y,q') \quad (3)$$

and this is a disjoint union.

4. *It holds*

$$\mathit{senum} = \bigcup_{q\in Q_f} \mathit{sactive}(\mathit{root}(T),q) \quad (4)$$

and this is a disjoint union

Equations (1-4) are not sufficient for our complexity purpose. We need to ensure that we do only non empty unions. We say that a configuration (x,q) is potentially active, (resp potentially useful) if there is a tuple $\bar{S}\in\mathcal{P}(\text{Leaves}(T))^k$ such that (x,q) is $\bar{S}$-active (resp $\bar{S}$-useful). We denote by PUSEFUL the set of potentially useful configurations. Let (x,q) be a configuration such that x is a leaf, and define $\text{PINIT}(x,q) = \{\bar{v}\in\{0,1\}^k - \{0\}^k \delta_{init}(\text{label}(x),\bar{v}) = q\}$. Let (x,q) be a potentially fork configuration and let y_1, y_2 be the children of x. Let $\text{PPAIR}(x,q)$ denote the set of pairs of states (q_1,q_2) such that $q_1,q_2\to q$ and (y_1,q_1) and (y_2,q_2) are potentially active. We denote by PFINAL the set of final states q such that $(\mathit{root}(T),q)$ is potentially active. We can deduce by definition, refinements of the above equations (1-4).

1. If x is a leaf then

$$\text{suseful}(x,q) = \bigcup_{\bar{v}\in\text{PINIT}(x,q)} \{\bar{x_{\bar{v}}}\} \quad (1')$$

2. If x is a internal node with children y_1 and y_2 then

$$\text{suseful}(x,q) = \bigcup_{(q_1,q_2)\in\text{PPAIR}(x,q)} \text{sactive}(y_1,q_1)\otimes\text{sactive}(y_2,q_2) \quad (2')$$

3. If (x, q) is a potentially active configuration then

$$\text{sactive}(x,q) = \bigcup_{\substack{(y,q')\in\text{PUSEFUL}\\ (y,q')\overset{*}{\longmapsto}(x,q)}} \text{suseful}(y,q') \tag{3'}$$

4. It holds

$$\text{senum} = \bigcup_{q\in\text{PFINAL}} \text{sactive}(\text{root}(T), q) \tag{4'}$$

3.1 Transition Tree

The transition tree denoted by $T^\emptyset$ is a graph built as follows

- The domain of $T^\emptyset$ is the set of configurations of (Γ, T).
- We put an edge from c_1 to c_2 if $c_2 \overset{\emptyset}{\longmapsto} c_1$.

As Γ is a deterministic automaton, it is easily seen that $T^\emptyset$ is a forest whose set of roots is $\{root(T)\} \times Q$.

Notice that $T^\emptyset$ is a (non necessarily binary) forest. As it is more convenient to consider $T^\emptyset$ as a tree, we can add to the graph a new dummy configuration and connect it to the root configurations.

By definition of $T^\emptyset$, note that the set of potentially useful configurations (y, q') such that $(y, q') \overset{*}{\longmapsto} (x, q)$ is exactly $\text{PUSEFUL} \cap T^\emptyset_{(x,q)}$ where $T^\emptyset_{(x,q)}$ is the subtree of $T^\emptyset$ of root (x, q). Consider a postfix order σ on the vertices of the tree $T^\emptyset$. By definition of a postfix order, it is easily seen that for every configuration (x, q), the set of nodes of $T^\emptyset_{(x,q)}$ is a segment of σ. Let $\text{PUSEFUL}[i]$ denote the ith element of PUSEFUL with respect to the order σ. For each potentially active configuration (x, q), we call $first(x, q)$ (resp $last(x, q)$) the least (resp the greatest) index i such that $\text{PUSEFUL}[i]$ belongs to $T^\emptyset_{(x,q)}$. As (x, q) is potentially active, $first(x, q)$ and $last(x, q)$ are always defined. Clearly, the potentially useful configurations (x', q') such that $(x', q') \overset{*}{\longmapsto} (x, q)$ are exactly the configurations $\text{PUSEFUL}[i]$ such that $first(x, q) \leq i \leq last(x, q)$. We obtain the following equation which is equivalent to 3'.

For each potentially active configuration (x, q),

$$\text{sactive}(x,q) = \bigcup_{i=first(x,q)}^{last(x,q)} \text{suseful}(\text{PUSEFUL}[i]) \tag{3''}$$

3.2 Efficient Implementation of Our Enumeration Algorithm

Our final enumeration algorithm will implement efficiently the above equations 1',2',3",4' by additional technical tools:

1. two algebraic binary operations to compose enumeration problems and their enumeration algorithms in a modular and uniform manner

2. an efficient data structure to represent outputs : sets $S \subseteq D$ or k-tuples of sets $\bar{S} \in \mathcal{P}(D)^k$.
3. A new kind of precise enumeration algorithm called tiptop algorithm.

Let A and B be two problems of enumeration of k-tuples of sets verifying the following conditions: for any input x and for any $\bar{S_1} \in A(x)$ and $\bar{S_2} \in B(x)$, it holds $(\bar{S_1} \restriction i) \cap (\bar{S_2} \restriction i) = \emptyset$. Then the product of A and B denoted by $A \otimes B$ is defined as follows: $(A \otimes B)(x) = A(x) \otimes B(x)$.

Let A and B be two enumeration problems such that

- A has two inputs x and y and B has one input x.
- for any x, y, we have $A(x, y) \neq \emptyset$
- for any x, y, z such that $y \neq z$, we have $A(x, y) \cap A(x, z) = \emptyset$

then the composition of A and B denoted by $(A \circ B)$ is defined as follows: $(A \circ B)(x) = \bigcup_{y \in B(x)} A(x, y)$.

We rephrase the equations (1'-4') by using our operations $\otimes$ and $\circ$ and by introducing intermediate enumeration problems. We assume implicitly that all the problems have an additionnal input that is the tree T and the automaton Γ.

$$
\begin{aligned}
\text{pfinal} &= \{q \in Q_f : (root(T), q) \text{ is potentially active}\} \\
\text{accessible}(x, q) &= \{\text{PUSEFUL}(i) | first(x, q) \leq i \leq last(x)\} \\
\text{pair}(x, q) &= \{(y_1, q_1, y_2, q_2) | (q_1, q_2) \in \text{PPAIR}(x, q)\} \\
& \quad \text{where } y_1 \text{ and } y_2 \text{ are the first and second child of } x \\
\text{product}(x_1, q_1, x_2, q_2) &= \text{sactive}(x_1, q_1) \otimes \text{sactive}(x_2, q_2) \\
\text{suseful}(x, q) &= \text{sfork}(x, q) \text{ if } x \text{ is an internal node} \\
&= \text{sleaf}(x, q) \text{ otherwise} \\
\text{sleaf}(x, q) &= \{\bar{x_{\bar{v}}} | \bar{v} \in \text{PINIT}(x, q)\} \\
\text{sfork}(x, q) &= \text{product} \circ \text{pair}(x, q) \\
\text{sactive}(x, q) &= \text{suseful} \circ \text{accessible}(x, q) \\
\text{senum} &= \text{sactive} \circ \text{pfinal}
\end{aligned}
$$

It is easily seen that senum computes all tuples $\bar{S} \neq \bar{\emptyset}$ such that Γ accepts $(T, \bar{S})$.

Definition 2. *A tiptop algorithm $\mathcal{A}$ for an enumeration problem is an enumeration (RAM) algorithm using two special instructions denoted by tip and top so that for any input x with exactly m distinct outputs y_i, $i = 1, \ldots, m$*

- *$\mathcal{A}$ runs into exactly m phases, phase(y_i) corresponding to the ith output y_i*
- *phase(y_i) includes exactly one tip instruction called tip(y_i) and one top instruction that is the last one of the phase*
- *tip(y_i) means that the ith output y_i is available in the RAM memory (typically as a data structure using pointers)*
- *at the end of the run that is the mth top, all the RAM memory, except the input memory, is empty*

Moreover A is a tiptop algorithm of delay $D(n)$ and space $S(n)$ if for each y_i, $i = 1, \dots, n$, $time(phase(y_i)) \leq D(|y_i|)$ and $space(phase(y_i)) \leq S(|y_i|)$.

The following lemma concerns the complexity of the enumeration problem $A \otimes B$. It uses in an essential manner an efficient representation of the outputs. A set S will be represented by a binary tree whose leaves are labeled by the elements of S. The size of this tree is linear in the size of S. A k-tuple of sets $\bar{S}$ is represented by an array of binary trees. Using these representations, a disjoint union of two sets A and B can be done in constant time (by creating a new node which is connected to the roots of trees) without side effect and hence the combination $\bar{S} \oplus \bar{S}'$ of two k-tuples of sets $\bar{S}$ and $\bar{S}'$ can also be computed in constant time.

Lemma 4. *Let A and B be two problems of enumeration of k-tuples of sets for which $A \otimes B$ can be defined. Assume that A (resp B) can be enumerated by a tiptop algorithm φ_A (resp φ_B) with delay $f(\bar{S})$ and in space $g(\bar{S})$ then $A \otimes B$ can be computed by a tiptop algorithm such that each tuple $\bar{S} = \bar{S_1} \oplus \bar{S_2}$ is produced with delay $f(\bar{S_1}) + f(\bar{S_2}) + O(1)$ and in space $g(\bar{S_1}) + g(\bar{S_2}) + O(1)$.*

Lemma 5. *Let A and B be two enumeration problems such that $B \circ A$ is defined. Assume that A can be enumerated by a tiptop algorithm φ_A with constant delay and in constant space and that B can be enumerated by a tiptop algorithm φ_B such that an element y is produced with delay $f(y)$ and in space $g(y)$. Then $B \circ A$ is enumerable by a tiptop algorithm $\varphi_{B \circ A}$ such that an element y is produced with delay $f(y) + O(1)$ and in space $g(y) + O(1)$.*

Lemma 6. *There is an algorithm which given a Σ-tree T and a query $\varphi(\bar{X})$ represented by an automaton $\Gamma = (\Sigma \times \{0,1\}^k, Q, q_0, \delta, Q_f)$, computes $\varphi(|T|)$ with a precomputation in time $O(|Q|^3|T|)$, with delay $O(|S|)$ and in space $O(|S|)$.*

Proof. We give first the preprocessing phase called precomp:

1. compute the set PUSEFUL $\subseteq D \times Q$
2. compute the function PPAIR : $D \times Q \rightarrow \mathcal{P}(Q^2)$
3. compute PINIT : Leaves$(T) \times Q \rightarrow \mathcal{P}(\{0,1\}^k \setminus \{0\}^k)$
4. compute the transition-tree $T^\emptyset$
5. compute a postfix order σ of $T^\emptyset$
6. order PUSEFUL with respect to σ
7. compute $first(x, q)$ and $last(x, q)$ for each configuration (x, q)

the set PUSEFUL can be computed in time $O(|Q||T|)$ using a bottom-up approach. Each set PPAIR(x, q) contains at most $|Q|^2$ elements per configuration and can be computed in time $O(|Q|^2)$. Therefore, the total time for computing the function PPAIR is $O(|Q|^3|T|)$. The function PINIT can be computed in $O(2^k|Q||T|)$. Lines 4,7 can easily be done in $O(|Q||T|)$ by bottom up approach. A postfix order can be computed in linear time $O(|T^\emptyset|) = O(|Q||T|)$ by depth first search. We conclude that the time and space of the precomputation is $O(|Q|^3|T|)$.

The enumeration phase is described in the procedure eval. The correctness is immediate. Clearly, with the below precomputation, the problems pfinal, accessible, pair and sleaf are computable with constant delay. We will prove by

Algorithm 1. *eval*(T, Σ, I)

Require: a Σ-tree T, an automaton Γ, data I computed by precomp
Ensure: the set of tuples $\bar{S}$ such that Γ accepts $(T, \bar{S})$

```
output #
start the procedure senum
while senum has not finished do
    continue φ until the next tip
    x̄ := output of senum
    output the tuple of predicates pointed by x̄
    continue senum until the next top
    output #
end while
if Γ accepts (T, ∅̄) then
    output ∅̄
    output #
end if
```

a simultaneous induction that each element of sactive(x, q) is produced with a delay $a(4|S| - 2)$ and each element of suseful(x, q) is produced with a delay $a(4|S| - 3)$.

Let (x, q) be a potentially useful configuration. It is easily seen that the property holds if x is a leaf. Assume that x is an internal node with first and second children y_1 and y_2. Consider a tuple $\bar{S}$ produced by suseful(x, q) with a delay d. Assume that $\bar{S} = \bar{S_1} \oplus \bar{S_2}$ and $\bar{S_1}$ (resp $\bar{S_2}$) is produced by sactive(y_1, q_1) (resp sactive(y_2, q_2)) with delay d_1 (resp d_2).

$$\begin{aligned} d &\leq d_1 + d_2 + O(1) && \text{by Lemmas 4 and 5} \\ &\leq a(4|\bar{S_1}| - 2) + a(4|\bar{S_2}| - 2) + O(1) && \text{by the induction hypothesis} \\ &\leq a(4|\bar{S}| - 3) - a + O(1) && \text{for large enough } a \\ &\leq a(4|\bar{S}| - 3) \end{aligned}$$

Let (x, q) be a potentially active configuration. Let $\bar{S}$ be a tuple produced by sactive(x, q) with a delay d. Assume that $\bar{S}$ is produced by suseful(x', q') with a delay d'.

$$\begin{aligned} d &\leq d' + O(1) \\ &\leq a(4|S| - 3) + O(1) && \text{by the induction hypothesis} \\ &\leq a(4|S| - 2) && \text{for large enough } a \end{aligned}$$

A similar proof can be done for space complexity.

Remark: Although the complexity of the precomputation highly depends on the size of the automaton, the enumeration phase has a delay which depend neither on the size of T nor on the size of the automaton.

Theorem 2. *The problem Query*$(MSO, \tau_\Sigma, \Sigma\text{-TREES})$ *is* LINEAR-DELAY$_{lin}$

Proof. We give the complete algorithm:

Data: An MSO$[\tau_\Sigma]$-formula $\varphi(\bar{X},\bar{y})$ and a τ-tree T

1. Compute a formula $\varphi'(\bar{X},\bar{Y})$ without quantifier-free first-order variables that is "equivalent" to φ
2. Compute a $(\Sigma \cup \{@\})$-tree T' and a formula $\varphi''(\bar{X})$ (as described in Lemma 1)
3. Compute a Σ'-tree automaton Γ associated to φ'' i.e such that ($T_{\bar{S}} \in L(\Gamma)$ if and only if $(T,\bar{S}) \models \varphi''(\bar{X})$ ($\Sigma' = (\Sigma \cup \{@\}) \times \{0,1\}^k$)
4. Perform precomputation phase precomp
5. Call $eval(\Gamma, T)$

Corollary 1. *Let Σ be an alphabet and φ be an MSO-$[\tau_\Sigma]$ formula without second-order free variable, then the evaluation φ on Σ-trees can be computed with linear precomputation and constant delay.*

4 Direct Generation of the ith Solution

We are interested, in this section, in producing directly the ith solution of a query.

Equations 1',2',3",4' give us a characterization of senum which is the set of all solution tuples (except the empty tuple). We want to view all sets (senum, sactive(x, q), ...) as ordered lists. This can be done as follows:

- We give an arbitrary ordering for all finite sets $\mathrm{PPAIR}(x,q)$, PFINAL, $\mathrm{PINIT}(x,q)$
- We consider all unions as ordered unions
- For any sets of k-tuples of sets A and B, $C = A \otimes B$ is viewed as a lexicographic product (i.e. if $C[x] = A[y] \oplus B[z]$ and $C[x'] = A[y'] \oplus B[z']$ then $x \leq x'$ if and only if $(x,y) \leq_{lex} (y',z')$ [3])

We denote

$$\mathrm{sumuseful}(i) = \sum_{k=0}^{i-1} |\mathrm{suseful}(\mathrm{PUSEFUL}[k])|$$

$$\mathrm{sumpair}(x,q,i) = \sum_{\substack{k=0\ldots i-1 \\ (q_1,q_2)=\mathrm{PPAIR}(x,q)[k]}} |\mathrm{sactive}(y_1,q_1)| \times |\mathrm{sactive}(y_2,q_2)|$$

$$\mathrm{sumfinal}(x,q,i) = \sum_{k=0}^{i-1} |\mathrm{sactive}(\mathrm{root}(T),\mathrm{PFINAL}[k])|$$

We are interested in finding, given an index i, the tuple $\bar{S} = \mathrm{senum}[i]$. This can be done by using the following lemma.

[3] Here and below, $A[i]$ denotes the ith element of the ordered list A ($A[0]$ is the first element).

Lemma 7. *1. Let A and B be two ordered sets of k-tuples of sets and $C = A \otimes B$ then $C[i] = A[a] \oplus B[b]$ where a and b are the quotient and the remainder of the integer division of i by $|B|$.*
2. Let $i < |senum|$. Then

$$senum[i] = sactive(root(T), \text{PFINAL}[j])[i - sumuseful(j)]$$

where j is the greatest k such that $sumfinal(k) \leq i$.
3. Let (x, q) be a potentially active configuration and $i < |sactive(x, q)|$. Then

$$sactive(x, q)[i] = suseful(\text{PUSEFUL}[j])[i + sumuseful(first(x, q)) - sumuseful(j)]$$

where j be the greatest k such that $sumuseful(k) \leq sumuseful(first(x, q)) + i$.
4. Let x be an internal node with children y_1 and y_2. Let $i < |suseful(x, q)|$. Then

$$suseful(x, q)[i] = sactive(y_1, q_1)[a] \oplus sactive(y_2, q_2)[b]$$

where j is the greatest k such that $sumpair(x, q, k) \leq i$, $(q_1, q_2) = \text{PPAIR}(x, q)[j]$ and a, b are the quotient and the remainder of the integer division of $i - sumpair(x, q, j)$ by $|sactive(y_2, q_2)|$.

Theorem 3. *Given a fixed MSO[τ_Σ]-formula $\varphi(\bar{X})$ and a Σ-tree T, we can do a precomputation in time $O(|T|)$ and then produce the ith solution $\bar{S}$ (with respect to the below order) in time $O(|\bar{S}| \log(|T|))$.*

Proof. (sketch) The precomputation is an extension of the precomputation for the evaluation problem. In addition, we need to compute sumuseful, sumpair, sumfinal and the cardinality of suseful(x, q) and sactive(x, q) for each configuration (x, q). This can be done in linear time by bottom-up approach. An algorithm to find the ith element is a straightforward application of Lemma 7. We proceed in a recursive way. We need to find given an integer x the greatest element lesser than x in a non decreasing list, this can be done in time $O(\log(|T|))$ by binary search. As the number of recursive calls is linear in the size of the output, we stay within the desired time bound.

5 Structures of Bounded Treewidth

The treewidth of a structure $\mathcal{S}$ is the least k such that $\mathcal{S}$ admits a tree decomposition of width k (for more details, see [3]).

Lemma 8. *[2] For any fixed k, there is an algorithm which given a MSO formula φ computes a formula φ' and an algorithm which given a structure $\mathcal{S}$ and a tree decomposition of $\mathcal{S}$ of width k, computes in linear time a Σ-tree T, such that $\varphi(\mathcal{S}) = \varphi'(T)$*

Bodlaender [4] gives an algorithm which given a structure $\mathcal{S}$ and a fixed k computes a tree-decomposition of $\mathcal{S}$ of width k if it exists in linear time in the size of the $\mathcal{S}$.

Corollary 2. *Let $\mathcal{C}$ be a class of τ-structures of bounded treewidth, then* QUERY *(MSO, τ, $\mathcal{C}$) is* LINEAR-DELAY$_{lin}$.

References

1. Alfred V. Aho, John E. Hopcroft, and Jeffrey D. Ullman. *Data Structures and Algorithms.* Addison-Wesley Series in Computer Science and Information Processing. Addison-Wesley, 1982.
2. Stefan Arnborg, Jens Lagergren, and Detlef Seese. Easy problems for tree-decomposable graphs. *J. Algorithms*, 12(2):308–340, 1991.
3. Hans L. Bodlaender. A tourist guide through treewidth. *Acta Cybernetica*, 11:1–21, 1993.
4. Hans L. Bodlaender. A linear-time algorithm for finding tree-decompositions of small treewidth. *SIAM J. Comput.*, 25(6):1305–1317, 1996.
5. H. Comon, M. Dauchet, R. Gilleron, F. Jacquemard, D. Lugiez, S. Tison, and M. Tommasi. Tree automata techniques and applications. Available on: http://www.grappa.univ-lille3.fr/tata, 1997. release October, 1rst 2002.
6. Bruno Courcelle. Graph rewriting: An algebraic and logic approach. In *Handbook of Theoretical Computer Science, Volume B: Formal Models and Sematics (B)*, pages 193–242. 1990.
7. Bruno Courcelle. Linear delay enumeration and monadic second-order logic, 2006. Submitted.
8. Bruno Courcelle and Mohamed Mosbah. Monadic second-order evaluations on tree-decomposable graphs. *Theor. Comput. Sci.*, 109(1&2):49–82, 1993.
9. R. G. Downey and M. R. Fellows. *Parameterized Complexity.* Springer, 1999.
10. Arnaud Durand and Etienne Grandjean. First-order queries on structures of bounded degree are computable with constant delay. *Transactions on Computational Logic*, To appear.
11. Jörg Flum, Markus Frick, and Martin Grohe. Query evaluation via tree-decompositions. *J. ACM*, 49(6):716–752, 2002.
12. Etienne Grandjean and Frédéric Olive. Graph properties checkable in linear time in the number of vertices. *J. Comput. Syst. Sci.*, 68(3):546–597, 2004.
13. Etienne Grandjean and Thomas Schwentick. Machine-independent characterizations and complete problems for deterministic linear time. *SIAM J. Comput.*, 32(1):196–230, 2002.
14. Leonid Libkin. *Elements of finite model theory.* Springer, 2004. LIB 1 04:1 1.Ex.
15. James W. Thatcher and Jesse B. Wright. Generalized finite automata theory with an application to a decision problem of second-order logic. *Mathematical Systems Theory*, 2(1):57–81, 1968.

Abstracting Allocation
The New `new` Thing

Nick Benton

Microsoft Research
`nick@microsoft.com`

Abstract. We introduce a Floyd-Hoare-style framework for specification and verification of machine code programs, based on relational parametricity (rather than unary predicates) and using both step-indexing and a novel form of separation structure. This yields compositional, descriptive and extensional reasoning principles for many features of low-level sequential computation: independence, ownership transfer, unstructured control flow, first-class code pointers and address arithmetic. We demonstrate how to specify and verify the implementation of a simple memory manager and, independently, its clients in this style. The work has been fully machine-checked within the Coq proof assistant.

1 Introduction

Most logics and semantics for languages with dynamic allocation treat the allocator, and a notion of what has been allocated at a particular time, as part of their basic structure. For example, marked-store models, and those based on functor categories or FM-cpos, have special treatment of locations baked in, as do operational semantics using partial stores, where programs 'go wrong' when accessing unallocated locations. Even type systems and logics for low-level programs, such as TAL [14], hardwire allocation as a primitive.

For high-level languages such as ML in which allocation is observable but largely abstract (no address arithmetic, order comparison or explicit deallocation), building 'well-behaved' allocation into a model seems reasonable. But even then, we typically obtain base models that are far from fully abstract and have to use a second level of non-trivial relational reasoning to validate even the simplest facts about encapsulation.

For low-level languages, hardwiring allocation is less attractive. Firstly, and most importantly, we want to reason about the low-level code that actually implements the storage manager. Secondly, in languages with address arithmetic, such as the while-language with pointers used in separation logic, one is led to treat allocation as a non-deterministic primitive, which is semantically problematic, especially if one tries to reason about refinement, equivalence or imprecise predicates [23,13]. Finally, it just doesn't correspond to the fact that 'machine code programs don't go wrong'. The fault-avoiding semantics of separation logic, for example, is prescriptive, rather than descriptive: one can only prove anything

Z. Ésik (Ed.): CSL 2006, LNCS 4207, pp. 182–196, 2006.

about programs that never step outside their designated footprint, even if they do so in a non-observable way.[1]

We instead start with a completely straightforward operational semantics for an idealized assembly language. There is a single datatype, the natural numbers, though different instructions treat elements of that type as code pointers, heap addresses, integers, etc. The heap is simply a total function from naturals to naturals and the code heap is a total function from naturals to instructions. Computed branches and address arithmetic are perfectly allowable. There is no built-in notion of allocation and no notion of stuckness or 'going wrong': the only observable behaviours are termination and divergence.

Over this simple and permissive model, we aim to develop semantic (defined in terms of observable behaviour) safety properties, and ultimately a program logic, that are rich enough to capture the equational semantics of high-level types as properties of compiled code and also to express and verify the behavioural contracts of the runtime systems, including memory managers, upon which compiled code depends.

Our approach is based on four technical ideas. Firstly, following the interpretation of types as PERs, we work with quantified binary relations rather than the unary predicates more usual in program logics. Program properties are expressed in terms of contextual equivalence, rather than avoidance of some artificial stuck states. Secondly, we use a perping operation, taking relations on states to orthogonal relations on code addresses, to reason about first-class code pointers. Thirdly, we reason modularly about the heap in a style similar to separation logic, but using an explicit notion of the portion of the heap on which a relation depends. Finally, we reason modularly about mutually-recursive program fragments in an assume/guarantee style, using a step-indexing technique similar to that of Appel et al [5,6,3] to establish soundness.

In this paper, we concentrate on the specification and verification of an extremely basic memory allocation module, and an example client. Although the code itself may be simple, the specifications and proofs are rather less so, and provide a non-trivial test case for our general framework, as well as constituting a fresh approach to freshness.

Managing the mind-numbing complexity and detail of specifications and proofs for machine code programs, not to mention keeping oneself honest in the face of changing definitions, seems to call for automated assistance. All the definitions and results presented here have been formalized and checked using the Coq proof assistant.

[1] For example, `skip` and `[10] := [10]` are, under mild assumptions, observationally equivalent, yet do not satisfy exactly the same set of triples. One might reasonably claim that machine code programs *do* go wrong – by segfaulting – and that this justifies faulting semantics and the use of partial stores. But stuck states in most operational semantics, even for low-level code, do not correspond exactly to the places in which segfaults might really occur, and we'd rather not assume or model anything about an operating system for the moment anyway.

2 The Machine

Our idealized sequential machine model looks like:

$$\begin{array}{rcll} s \in & \mathbb{S} & \stackrel{def}{=} \mathbb{N} \to \mathbb{N} & \text{states} \\ l, m, n, b \in & \mathbb{N} & & \text{naturals in different roles} \\ p \in & \mathit{Programs} & \stackrel{def}{=} \mathbb{N} \to \mathit{Instr} & \text{programs} \\ \langle p|s|l\rangle \in & \mathit{Configs} & \stackrel{def}{=} \mathit{Programs} \times \mathbb{S} \times \mathbb{N} & \end{array}$$

The instruction set, *Instr*, includes `halt`, direct and indirect stores and loads, some (total) arithmetic and logical operations, and conditional and unconditional branches.[2] The semantics is given by an obvious deterministic transition relation $\langle p|s|l\rangle \to \langle p|s'|l'\rangle$ between configurations. We write $\langle p|s|l\rangle \Downarrow$ if there exists n,l',s' such that $\langle p|s|l\rangle \to^n \langle p|s'|l'\rangle$ with $p(l') = \texttt{halt}$, and $\langle p|s|l\rangle \Uparrow$ if $\langle p|s|l\rangle \to^\omega$.

The major idealizations compared with a real machine are that we have arbitrary-sized natural numbers as a primitive type, rather than fixed-length words, and that we have separated code and data memory (ruling out self-modifying code and dynamic linking for the moment). Note also that we do not even have any registers.

3 Relations, Supports and Indexing

We work with binary relations on the naturals, $\mathbb{N}$, and on the set of states, $\mathbb{S}$, but need some extra structure. Firstly, the reason for using relations is to express specifications in terms of behavioural equivalences between configurations:

$$\langle p|s|l\rangle \Downarrow \iff \langle p'|s'|l'\rangle \Downarrow$$

and the relations on states and naturals we use to establish such equivalences will generally be functions of the programs p and p' (because they will refer to the sets of code pointers that, in p and p', have particular behaviours). Secondly, to reason modularly about mutually recursive program fragments, we need to restrict attention to relations satisfying an admissibility property, which we capture by step-indexing: relations are parameterized by, and antimonotonic in, the number of computation steps available for distinguishing values (showing they're not in the relation). Formally, an *indexed nat relation*, is a function

$$r : \mathit{Programs} \times \mathit{Programs} \to \mathbb{N} \to \mathbb{P}(\mathbb{N} \times \mathbb{N})$$

such that $(r\,(p, p')\,k) \subseteq (r\,(p, p')\,j)$ whenever $j < k$.

For state relations, we also care about what *parts* of the state our relations depend upon. Separation logic does this implicitly, and sometimes indeterminately, via the existential quantification over splittings of the heap in the definition of

---

[2] The Coq formalization currently uses a shallow embedding of the machine semantics, so the precise instruction set is somewhat fluid.

separating conjunction. Instead, we work with an explicit notion, introduced in [9], of the support of a relation. One might expect this to be a set of locations, but the support is often itself a function of the state (think, for example, of the equivalence of linked lists). However, not all functions $\mathbb{S} \to \mathbb{P}(\mathbb{N})$ make sense as supports: the function itself should not depend on the contents of locations which are not in its result.[3] Formally, we define an *accessibility map* to be a function $A : \mathbb{S} \to \mathbb{P}(\mathbb{N})$ such that

$$\forall s, s'.\ s \sim_{A(s)} s' \implies A(s') = A(s)$$

where, for $L \subseteq \mathbb{N}$ and $s, s' \in \mathbb{S}$, we write $s \sim_L s'$ to mean $\forall l \in L.s(l) = s'(l)$.

Accessibility maps are ordered by $A \subseteq A' \iff \forall s.A(s) \subseteq A'(s)$. Constant functions are accessibility maps, as is the union $A \cup A'$ of two accessibility maps, where $(A \cup A')(s) = A(s) \cup A'(s)$. Despite the name, accessibility maps are about relevance and *not* reachability. Indeed, reachability makes little sense in a model without an inherent notion of pointer.

A *supported indexed state relation* R is a triple $(|R|, A_R, A'_R)$ where

$$|R| : \mathit{Programs} \times \mathit{Programs} \to \mathbb{N} \to \mathbb{P}(\mathbb{S} \times \mathbb{S})$$

satisfies $(|R|\,(p, p')\,k) \subseteq (|R|\,(p, p')\,j)$ for all $(j < k)$, A_R and A'_R are accessibility maps and for all $s_1 \sim_{A_R\,(s_1)} s_2$ and $s'_1 \sim_{A'_R\,(s'_1)} s'_2$,

$$(s_1, s'_1) \in |R|\,(p, p')\,k \implies (s_2, s'_2) \in |R|\,(p, p')\,k.$$

We often elide the $|\cdot|$. The constantly total and empty state relations are each supported by any accessibility maps. The separating product of supported indexed relations is given by

$$\begin{aligned} R_1 \otimes R_2 &= (|R_1 \otimes R_2|,\ A_{R_1} \cup A_{R_2},\ A'_{R_1} \cup A'_{R_2}) \quad \text{where} \\ |R_1 \otimes R_2|\,(p, p')\,k &= (|R_1|\,(p, p')\,k) \ \cap\ (|R_2|\,(p, p')\,k) \ \cap \\ &\quad \{(s, s') \mid A_{R_1}(s) \cap A_{R_2}(s) = \emptyset \wedge A'_{R_1}(s') \cap A'_{R_2}(s') = \emptyset\} \end{aligned}$$

This is associative and commutative with the constantly total relation with empty support, $T_\emptyset$, as unit. The partial order $R_1 \preceq R_2$ on state relations is defined as

$$\forall (s, s') \in |R_1|.\ ((s, s') \in |R_2|) \wedge (A_{R_2}(s) \subseteq A_{R_1}(s)) \wedge (A'_{R_2}(s') \subseteq A'_{R_1}(s'))$$

which has the property that if $R_1 \preceq R_2$ then for any R_I, $|R_1 \otimes R_I| \subseteq |R_2 \otimes R_I|$.

If R is a (supported) indexed state relation, its *perp*, $R^\top$, is an indexed nat relation defined by:

$$\begin{aligned} R^\top\,(p, p')\,k = \{(l, l') \mid \forall j < k. \forall (s, s') \in (R\,(p, p')\,j).& \\ ((\langle p, s, l\rangle \Downarrow_j \implies \langle p', s', l'\rangle \Downarrow) \ \wedge& \\ ((\langle p', s', l'\rangle \Downarrow_j \implies \langle p, s, l\rangle \Downarrow)\}& \end{aligned}$$

[3] In other words, the function should support itself.

where $\Downarrow_j$ means 'converges in fewer than j steps'. Roughly speaking, $R^\top$ relates two labels if jumping to those labels gives equivalent termination behaviour whenever the two initial states are related by R; the indexing lets us deal with cotermination as the limit (intersection) of a sequence of k-step approximants.

If $q \subseteq \mathbb{N} \times \mathbb{N}$, write $\overline{q}$ for the indexed nat relation $\lambda(p, p').\lambda k.q$, and similarly for indexed state relations. If $L \subseteq \mathbb{N}$, A_L is the accessibility map $\lambda s.L$. We write T_L for the supported indexed state relation $(\overline{\mathbb{S} \times \mathbb{S}}, A_L, A_L)$ and write sets of integers $\{m, m+1, \ldots, n\}$ just as mn. If r is an indexed nat relation and $n, n' \in \mathbb{N}$, write $(n, n' \Mapsto r)$ for the supported indexed state relation

$$\lambda(p, p').\lambda k.\,(\{(s, s') \mid (s(n), s'(n')) \in r\,(p, p')\,k\},\ \lambda s.\{n\},\ \lambda s.\{n'\})$$

relating pairs of states that have values related by r stored in locations n and n'. We write the common diagonal case $(n, n \Mapsto r)$ as $(n \mapsto r)$. For $\mathtt{M}$ a program fragment (partial function from naturals to instructions) define

$$\models \mathtt{M} \rhd l : R^\top \stackrel{def}{=} \forall p, p' \supseteq \mathtt{M}.\forall k.\,(l, l) \in (R^\top\,(p, p')\,k)$$

where the quantification is over all (total) programs extending $\mathtt{M}$. We are only considering a single $\mathtt{M}$, so our basic judgement is that a label is related to itself. More generally, define $\boldsymbol{l_i} : \boldsymbol{R_i^\top} \models \mathtt{M} \rhd l : R^\top$ to mean

$$\forall p, p' \supseteq \mathtt{M}.\forall k.\,(\forall i.(l_i, l_i) \in (R_i^\top\,(p, p')\,k)) \implies ((l, l) \in (R^\top\,(p, p')\,k + 1))$$

i.e. for any programs extending $\mathtt{M}$ and for any k, if the hypotheses on the labels $\boldsymbol{l_i}$ are satisfied to index k, then the conclusion about l holds to index $k + 1$.

4 Specification of Allocation

The machine model is very concrete and low-level, so we have to be explicit about details of calling conventions in our specifications. We arbitrarily designate locations 0 - 9 as register-like and, for calling the allocator, will use 0 - 4 for passing arguments, returning results and as workspace. An allocator module is just a code fragment, $\mathtt{M}_a$, which we will specify and verify in just the same way as its clients. There are entry points for initialization, allocation and deallocation.

The code at label `init` sets up the internal data structures of the allocator. It takes a return address in location 0, to which it will jump once initialization is complete. The code at `alloc` expects a return address in location 0 and the size of the requested block in location 1. The address of the new block will be returned in location 0. The code at `dealloc` takes a return address in 0, the size of the block to be freed in 1 and the address of the block to be freed in 2.

After initialization, the allocator owns some storage in which it maintains its internal state, and from which it hands out (transfers ownership of) chunks to clients. The allocator depends upon clients not interfering with, and behaving independently of, both the location and contents of its private state. In particular, clients should be insensitive to the addresses and the initial contents of chunks returned by calls to `alloc`. In return, the allocator promises not to change or depend upon the contents of store owned by the client. All of these independence, non-interference and ownership conditions can be expressed using supported relations. Furthermore, this can be done extensionally, rather than in terms of which locations are read, written or reachable.

There will be some supported indexed state relation R_a for the private invariant of the allocator. The supports of R_a express what store is owned by the allocator; this has to be a function of the state (rather than just a set of locations), because what is owned varies as blocks are handed out and returned. One should think of $|R_a|$ as expressing what configurations of the store owned by the allocator are valid, and which of those configurations are equivalent.

When `init` is called, the allocator takes ownership of some (infinite) part of the store, which we only specify to be disjoint from locations 0-9. On return, locations 0-4 may have been changed, 5-9 will be preserved, and none of 1-9 will observably have been read. So two calls to `init` yield equivalent behaviour when the return addresses passed in location 0 yield equivalent behaviour whenever the states they're started in are as related as `init` guarantees to make them. How related is that? Well, there are no guarantees on 0-4, we'll preserve any relation involving 5-9 *and* we'll establish R_a on a disjoint portion of the heap. Thus, the specification for initialization is that for *any* nat relations $r_5, r_6, \ldots, r_9$,

$$\models \mathtt{M}_a \rhd \mathtt{init} : \left(\left(0 \mapsto (R_a \otimes T_{04} \otimes \bigotimes_{i=5}^{9} (i \mapsto r_i))^{\top}\right) \otimes \bigotimes_{i=5}^{9} (i \mapsto r_i)\right)^{\top} \tag{1}$$

When `alloc` is called, the client (i.e. the rest of the program) will already have ownership of some disjoint part of the heap and its own invariant thereon, R_c. Calls to `alloc` behave equivalently provided they are passed return continuations that behave the same whenever their start states are related by R_c, R_a *and* in each state location 0 points to a block of memory of the appropriate size and disjoint from R_c and R_a. More formally, the specification for allocation is that for *any* n and for *any* R_c

$$\models \mathtt{M}_a \rhd \mathtt{alloc} : (R_{aparms}(n, R_a, R_a) \otimes T_{24} \otimes R_c \otimes R_a)^{\top} \tag{2}$$

where

$$R_{aparms}(n, R_a, R_c) = \left((0 \mapsto (R_{aret}(n) \otimes T_{14} \otimes R_c \otimes R_a)^{\top}) \otimes \left(1 \mapsto \overline{\{(n,n)\}}\right)\right)$$

$$\text{and} \quad R_{aret}(n) = \left(\overline{\{(s,s') \mid s(0) > 9 \wedge s'(0) > 9\}},\ A_{aret}(n),\ A_{aret}(n)\right)$$
$$A_{aret}(n) = \lambda s.\{0\} \cup \{s(0), \ldots, s(0)+n-1\}$$

R_{aret} guarantees that the allocated block will be disjoint from the pseudo-registers, but nothing more; this captures the requirement for clients to behave equivalently whatever block they're returned and whatever its initial contents. A_{aret} includes both location 0, in which the start of the allocated block is returned, and the block itself; the fact that this is tensored with R_a and R_c in the precondition for the return address allows the client to assume that the block is disjoint from both the (updated) internal datastructures of the allocator and the store previously owned by the client. We can avoid saying anything explicit about preservation of locations 5 to 9 here because they can be incorporated into R_c.

When `dealloc` is called, R_a will hold and the client will have an invariant R_c that it expects to be preserved *and* a disjoint block of store to be returned. The client expresses that it no longer needs the returned block by promising that the return address will behave equivalently provided that just R_c and R_a hold. Formally, for *any* R_c and n,

$$\models \mathtt{M}_a \rhd \mathtt{dealloc} : \left(\left(0 \mapsto (T_{04} \otimes R_c \otimes R_a)^{\top}\right) \otimes \left(1 \mapsto \overline{\{(n,n)\}}\right) \otimes T_{34} \otimes R_{fb}(n)\right)^{\top} \tag{3}$$

where

$$R_{fb}(n) = \left(\overline{\{(s,s') \mid s(2) > 9 \wedge s'(2) > 9\}},\, A_{fb}(n),\, A_{fb}(n)\right)$$
$$A_{fb}(n) = \lambda s.\, \{2\} \cup \{s(2), \ldots, s(2)+n-1\}$$

Writing the relations on the RHSs of (1), (2) and (3) as $r_{in}(R_a, r_5, \ldots, r_9)$, $r_{al}(R_a, n, R_c)$ and $r_{de}(R_a, n, R_c)$, respectively, the whole specification of an allocator module is therefore

$$\begin{aligned} \exists R_a.\, \models \mathtt{M}_a \rhd (&\mathtt{init} : \forall r_5, \ldots, r_9.\, r_{in}(R_a, r_5, \ldots, r_9)) \\ &\wedge(\mathtt{alloc} : \forall n.\forall R_c.\, r_{al}(R_a, n, R_c)) \\ &\wedge(\mathtt{dealloc} : \forall n.\forall R_c.\, r_{de}(R_a, n, R_c)) \end{aligned} \tag{4}$$

Note that the existentially-quantified R_a is scoped across the whole module interface: the *same* invariant has to be maintained by the cooperating implementations of all three operations, even though it is abstract from the point of view of clients.

Checking that all the things we have assumed to be accessibility maps and supported relations really *are* is straightforward from the definitions.

5 Verification of Allocation

We now consider verifying the simplest useful allocation module, $\mathtt{M}_a$, shown in Figure 1. Location 10 points to the base of an infinite contiguous chunk of free memory. The allocator owns location 10 and all the locations whose addresses are greater than or equal to the current contents of location 10. Initialization sets the contents of 10 to 11, claiming everything above 10 to be unallocated, and returns. Allocation saves the return address in location 2, copies a pointer to the next currently free location (the start of the chunk to be returned) into 0, bumps location 10 up by the number of locations to be allocated and returns to the saved address. Deallocation is simply a no-op: in this trivial implementation, freed store is actually never reused, though the specification requires that well-behaved clients never rely on that fact.

```
init : [10] ← 11                    // set up free ptr
init + 1 : jmp [0]                  // return

alloc : [2] ← [0]                   // save return address
alloc + 1 : [0] ← [10]              // return value = old free ptr
alloc + 2 : [10] ← [10] + [1]       // bump free ptr by n
alloc + 3 : jmp [2]                 // return to saved address

dealloc : jmp [0]                   // return (!)
```

Fig. 1. The Simplest Allocator Module, $\mathtt{M}_a$

Theorem 1. *The allocator code in Figure 1 satisfies the specification, (4), of the previous section.* □

For this implementation, the relation, R_a, witnessing the existential in the specification is just

$$R_a \stackrel{def}{=} \left(\overline{\{(s,s') \mid (s(10) > 10) \wedge (s'(10) > 10)\}},\, A_a,\, A_a\right)$$

where A_a is $\lambda s.\{10\} \cup \{m|m \geq s(10)\}$. The only invariant this allocator needs is that the next free location pointer is strictly greater than 10, so memory handed out never overlaps either the pseudo registers 0-9 or the allocator's sole bit of interesting private state, location 10 itself. A_a says what storage is owned by the allocator.

The proof of Theorem 1 is essentially forward relational Hoare-style reasoning, using assumed separation conditions to justify the framing of invariants. In particular, the prerelation for `alloc` lets us assume that the support A_c of R_c is disjoint from both $\{0, \ldots, 4\}$ and A_a in each of the related states (s, s') in which we make the initial calls. Since the code only writes to locations coming from those latter two accessibility maps, we know that they are still related by R_c, even though we do not know anything more about what R_c is. More generally, we have the following reasoning principle:

Lemma 1 (Independent Updates). *For any* p, p', k, n, n', v, v', r_{old}, r_{new}, R_{inv}, s, s',

$$(v, v') \in (r_{new}\,(p, p')\,k) \quad \textit{and} \quad (s, s') \in ((n, n' \mapsto r_{old}) \otimes R_{inv})\,(p, p')\,k$$

implies $(s[n \mapsto v],\, s'[n' \mapsto v']) \in ((n, n' \mapsto r_{new}) \otimes R_{inv})\,(p, p')\,k$.

Proof. By assumption, the prestates s and s' are related by $|R_{inv}|$, and the supports $A_{inv}(s)$ and $A'_{inv}(s')$ do not include n and n', respectively. Hence, by the self-supporting property of accessibility maps, $A_{inv}(s[n \mapsto v]) = A_{inv}(s)$, and similarly for A'_{inv}. Thus $s[n \mapsto v] \sim_{A_{inv}(s)} s$ and $s'[n' \mapsto v'] \sim_{A'_{inv}(s')} s'$, so the updated states are still related by $|R_{inv}|$ by the saturation property of supported relations, and the supports of the tensored relations in the conclusion are still disjoint. □

We also use the following for reasoning about individual transitions:

Lemma 2 (Single Steps). *For all p, p', k, R_{pre}, l_{pre}, l'_{pre}, if for all $j < k$ and for all $(s_{pre}, s'_{pre}) \in (R_{pre}\,(p,p')\,j)$*

$$\langle p|s_{pre}|l_{pre}\rangle \to \langle p|s_{post}|l_{post}\rangle \quad \textit{and} \quad \langle p'|s'_{pre}|l'_{pre}\rangle \to \langle p'|s'_{post}|l'_{post}\rangle$$

implies there exists an R_{post} such that

$$(s_{post}, s'_{post}) \in (R_{post}\,(p,p')\,j) \quad \textit{and} \quad (l_{post}, l'_{post}) \in (R_{post}\,(p,p')\,j)^{\top}$$

then $(l_{pre}, l'_{pre}) \in (R_{pre}\,(p,p')\,k)^{\top}$. □

For straight-line code that just manipulates individual values in fixed locations, the lemmas above, together with simple rules of consequence involving $\preceq$, are basically all one needs. The pattern is that one applies the single step lemma to a goal of the form $(l_{pre}, l'_{pre}) \in (R_{pre}\,(p,p')\,k)^{\top}$, generating a subgoal of the form 'transition implies exists R_{post} such that post states are related and (l_{post}, l'_{post}) are in $R_{post}^{\top}$'. One then examines the instructions at l_{pre} and l'_{pre}, which defines the possible post states and values of l_{post} and l'_{post}. One then instantiates R_{post}, yielding one subgoal that the post states (now expressed as functions of the prestates) are related and one about (l_{post}, l'_{post}). In the case that the instruction was an update, one then uses the independent update lemma to discharge the first subgoal, leaving the goal of proving a perp about (l_{post}, l'_{post}), for which the pattern repeats. Along the way, one uses consequence to put relations into the right form for applying lemmas and assumptions.

In interesting cases of ownership transfer, the consequence judgements one has to prove require splitting and recombining relations that have non-trivial supports. This typically involves introducing new existentially quantified logical variables. For example, after the instruction at `alloc+1` we split the state-dependency of the support of R_a by deducing that there exist $b, b' \in \mathbb{N}$, both greater than 10, such that the two intermediate states are related by

$$\left(0 \mapsto \overline{\{b, b'\}}\right) \otimes \left(10 \mapsto \overline{\{b, b'\}}\right) \otimes (\overline{\mathbb{S} \times \mathbb{S}}, A_{old}, A'_{old}) \otimes (\overline{\mathbb{S} \times \mathbb{S}}, A_{new}, A'_{new}) \otimes \cdots$$

where $A_{old}(s) = \{m \mid m \geq b + n\}$, $A'_{old}(s') = \{m \mid m \geq b' + n\}$, $A_{new}(s) = \{m \mid b \leq m < b + n\}$ and $A'_{new}(s') = \{m \mid b' \leq m < b' + n\}$. The first and fourth of these then combine to imply $R_{aret}(n)$, so after the update at `alloc+2` the states are related by

$$R_{aret}(n) \otimes \left(10 \mapsto \overline{\{b+n, b'+n\}}\right) \otimes (\overline{\mathbb{S} \times \mathbb{S}}, A_{old}, A'_{old}) \otimes \cdots$$

the second and third of which then recombine to imply R_a again, eliminating b and b' and establishing the precondition for the return jump at `alloc+3`.

6 Specification and Verification of a Client

We now specify and verify a client of the allocator, using the specification of Section 4. Amongst other things, this shows how we deal modularly with linking, recursion and adaptation. The client specification is intended as a very simple example of how one might express the semantics of types in a high-level language as relations in our low-level logic, expressing the behavioural contracts of code compiled from phrases of those types. In this case, the high-level language is an (imaginary) pure first-order one that, for the purposes of the example, we compile using heap-allocated activation records.

We concentrate on the meaning of the type `nat` → `nat`. From the point of view of the high-level language, the semantics of that type is something like the predomain $\mathbb{N} \to \mathbb{N}_\perp$, or relationally, a PER on some universal domain relating functions that take equal natural number arguments to equal results of type 'natural-or-divergence'. There are many mappings from such a high-level semantics to the low-level, reflecting many different correct compilation schemes. We'll assume values of type `nat` are compiled as the obviously corresponding machine values, so the interpretation $[\![\texttt{nat}]\!]$ is the constantly diagonal relation $\{(n,n) \mid n \in \mathbb{N}\}$.

For functions we choose to pass arguments and return results in location 5, to pass return addresses in 6, to use 7 to point to the activation record, and 0-4 as workspace.[4] Since functions call the allocator, they will also explicitly assume and preserve R_a, as well as some unknown frame R_c for the invariants of the rest of the program. The allocator's invariant is abstract from the point of view of its clients, but they all have to be using the same one, so we parameterize client specs by the allocator's invariant. This leads us to define $[\![\texttt{nat} \to \texttt{nat}]\!]\,(R_a)$ as the following indexed nat relation:

$$\forall R_c.\forall r_7.\begin{pmatrix} T_{04} \otimes (5 \mapsto [\![\texttt{nat}]\!]) \otimes (7 \mapsto r_7) \otimes R_c \otimes R_a \otimes \\ \left(6 \mapsto (T_{04} \otimes (5 \mapsto [\![\texttt{nat}]\!]) \otimes R_c \otimes R_a \otimes T_6 \otimes (7 \mapsto r_7))^\top\right) \end{pmatrix}^\top$$

which one can see as the usual 'equal arguments to equal results' logical relation, augmented with extra invariants that ensure that the code respects the calling convention, uses the allocator properly and doesn't observably read or write any storage that it shouldn't. Although the high-level type is simple, the corresponding low-level specification is certainly non-trivial.

As a concrete example of something that should meet this spec, we (predictably) take an implementation, $\texttt{M}_f$, of the factorial function, shown in Figure 2. The factorial code is mildly optimized: it calls the allocator to allocate its activation record, but avoids the allocation if no recursive call is needed. After a recursive call, the activation record is deallocated using a tail call: `dealloc` returns directly to the caller of `fact`. The ability to reason about optimized code is a benefit of our extensional approach compared with more type-like methods which assume code of a certain shape.

The result we want about the factorial is that it satisfies the specification corresponding to its type whenever it is linked with code satisfying the specification of an allocator. Opening the existential package, this means that for *any*

[4] This differs from the allocator's calling convention because we need to call the allocator to get some space *before* we can save the parameters to a function call.

```
fact    :  brz [5] (fact+17)         // jump to fact+17 if [5]=0
fact+ 1:   [1] <- 3                  // size of activation record
fact+ 2:   [0] <- (fact+4)           // return address for alloc
fact+ 3:   jmp alloc                 // allocate activation record
fact+ 4:   [[0]] <- [5]              // copy arg to frame[0]
fact+ 5:   [[0]+1] <- [6]            // copy ret addr to frame[1]
fact+ 6:   [[0]+2] <- [7]            // copy old frame ptr to frame[2]
fact+ 7:   [7] <- [0]                // new frame ptr in 7
fact+ 8:   [5] <- ([5]-1)            // decrement arg
fact+ 9:   [6] <- (fact+11)          // ret addr for recursive call
fact+10:   jmp fact                  // make recursive call
fact+11:   [5] <- ([5]*[[7]])        // return value = (fact (n-1))*n
fact+12:   [0] <- [[7]+1]            // ret addr for dealloc tail call
fact+13:   [2] <- [7]                // arg for call to dealloc
fact+14:   [7] <- [[7]+2]            // restore old frame ptr
fact+15:   [1] <- 3                  // size of block for dealloc
fact+16:   jmp dealloc               // dealloc frame and tail return
fact+17:   [5] <- 1                  // return value = 1
fact+18:   jmp [6]                   // return
```

Fig. 2. Code for the Factorial Function, $\mathtt{M}_f$

$\mathtt{M}_a$ satisfying (4), there's an R_a such that

$$\models (\mathtt{M}_a \cup \mathtt{M}_f) \rhd (\mathtt{fact} : [\![\mathtt{nat} \to \mathtt{nat}]\!](R_a)) \wedge (\mathtt{alloc} : \forall n.\forall R_c.\, r_{al}(R_a, n, R_c)) \wedge \ldots$$

which is a consequence of the following, quite independent of any particular $\mathtt{M}_a$:

Theorem 2. *For any* R_a*,*

$$\begin{array}{ll} \mathtt{init} : \forall r_5, \ldots, r_9.\, r_{in}(R_a, r_5, \ldots, r_9), & \\ \mathtt{alloc} : \forall n.\forall R_c.\, r_{al}(R_a, n, R_c), & \models \mathtt{M}_f \rhd \mathtt{fact} : [\![\mathtt{nat} \to \mathtt{nat}]\!](R_a) \\ \mathtt{dealloc} : \forall n.\forall R_c.\, r_{de}(R_a, n, R_c) & \end{array}$$ □

This is another Hoare-style derivation, mostly similar to that of Theorem 1. Proving the calls, including the recursive one, requires the universal quantifications over R_c, n and r_7, occurring in the specifications of `alloc`, `dealloc` and `fact`, to be appropriately instantiated ('adapted'). For example, the instantiation of R_c for the recursive call at label `fact+10` is

$$\begin{array}{l} R_c' \otimes (b, b' \Rrightarrow [\![\mathtt{nat}]\!]) \\ \otimes \left(b+1, b'+1 \Rrightarrow ((5 \mapsto [\![\mathtt{nat}]\!]) \otimes T_{04} \otimes R_c' \otimes R_a \otimes T_6 \otimes (7 \mapsto r_7'))^\top\right) \\ \otimes (b+2, b'+2 \Rrightarrow r_7') \end{array}$$

where R_c' and r_7' were the instantiations of the outer call, and b and b' are logical variables standing for the addresses returned by the previous related calls to the allocator at `fact+3`. This clearly expresses how the recursive call has to preserve whatever the outer one had to, plus the frame of the outer call, storing the outer call's argument and return address and the outer call's *caller's* frame pointer from location 7.

Recursion is dealt with in the proof of Theorem 2 as one would expect, by adding $\texttt{fact} : [\![\texttt{nat} \to \texttt{nat}]\!](R_a)$ to the context. This is sound thanks to our use of indexing and interpretation of judgements:

Lemma 3 (Recursion). *For any Γ, l, R and $\texttt{M}$, if $\Gamma, l : R^\top \models \texttt{M} \rhd l : R^\top$ then $\Gamma \models \texttt{M} \rhd l : R^\top$.* □

Lemma 3, proved by a simple induction, suffices for first-order examples but only involves statically known labels.[5] We will discuss 'recursion through the store' in detail in future work, but here give a trivial example to indicate that we already have enough structure to deal with it. Consider *independently* verifying the following code fragments, assuming that $\texttt{wantzero} : (1 \mapsto \overline{\{(0,0)\}})^\top$

```
silly   : brz [1] wantzero
silly+1 : [1] <- [1]-1
silly+2 : jmp [0]
```

```
knot   : [0] <- silly
knot+1 : jmp silly
```

To show $\texttt{knot} : (1 \mapsto [\![\texttt{nat}]\!])^\top$, there are various choices for the specification assumed for `silly` (and proved of its implementation). An obvious one is that `silly` *expects* to be passed itself in 0, but this may be an overspecification. Alternatively, we can use the *recursive* specification $\mu r.\, ((0 \mapsto \overline{r}) \otimes (1 \mapsto [\![\texttt{nat}]\!]))^\top$, the semantics of which is given by well-founded induction: observe that the meaning of $R^\top$ at index k only depends on R at strictly smaller j. In general, we have a fixpoint equation

$$\mu r.\, (R[r])^\top = \left(\lambda(p,p').\lambda k.\, R\left[\mu r.\, (R[r])^\top \; (p,p')\, k\right] \; (p,p')\, k\right)^\top$$

letting us prove the following two judgements, which combine to give the result we wanted about `knot`:

Theorem 3

1. $\texttt{wantzero} : \left(1 \mapsto \overline{\{(0,0)\}}\right)^\top \models \texttt{M}_{silly} \rhd \texttt{silly} : \mu r.\, ((0 \mapsto \overline{r}) \otimes (1 \mapsto [\![\texttt{nat}]\!]))^\top$
2. $\texttt{silly} : \mu r.\, ((0 \mapsto \overline{r}) \otimes (1 \mapsto [\![\texttt{nat}]\!]))^\top \models \texttt{M}_{knot} \rhd \texttt{knot} : (1 \mapsto [\![\texttt{nat}]\!])^\top$ □

7 Discussion

As we said in the introduction, this work is part of a larger project on relational parametricity for low-level code, which one might characterize as *realistic realizability*.[6] It should be apparent that we are drawing on a great deal of earlier work on separation logic [21], relational program logics [19,1,7,23], models and

[5] This is equivalent to the more symmetric linking rule of our previous work [8].

[6] Modulo the use of unbounded natural numbers, etc. Our computational model is clearly only 'morally' realistic, but it's too nice a slogan not to use...

reasoning principles for dynamic allocation [18,20,9], typed assembly language [14], proof-carrying code [15], PER models of types [2], and so on.

Two projects with similar broad goals to ours are the FLINT project at Yale [11] and the Foundational Proof-Carrying Code project at Princeton [4]. The Yale group started with a purely syntactic approach to types for low-level code, and are now combining first-order Hoare-style reasoning using a semantic consequence relation within a more syntactic framework. This is argued to be simpler than techniques based on sophisticated constructions such as indexing, but the treatment of code pointers in [16] seems no less complex, and possibly less useful, than that of the present work. As the syntactic approach never says what types (or higher-order assertions) are supposed to ensure (what they actually *mean*), it seems more difficult to use it to combine proofs generated from different type systems or compilers, link in hand-written and hand-proved fragments or prove optimizations. The Princeton project takee a semantic approach, which is much closer to ours (as we've said, the step-indexing idea that we use comes from work on FPCC), but is still a fixed type system restricted to talking about a single form of memory safety rather than a general logic. FPCC uses a hardwired and rather limited form of allocation and has no deallocation at all [10].

There are other mechanized proofs of storage managers, including one by Yu et al. [24], and one using separation logic by Marti et al. [12]. These both treat more realistic implementations than we do here, but establish intensional 'internal' correctness properties of the implementations, rather than the more extensional and abstract specification used here. In particular, note that our specification uses no 'model variables' for recording the history of allocations.

Note that we make explicit use of second-order quantification over invariants, such as R_c, in our specifications and proofs (this is rather like row-polymorphism in record calculi). In separation logic, by contrast, the tight interpretation of preconditions means that $\{P\}\, C\, \{Q\}$ is semantically equivalent to $\forall I.\{P * I\}\, C\, \{Q * I\}$ so universal quantification over predicates on store outside the footprint of a command can be left implicit, but is still exploitable via the frame rule. Our use of explicit polymorphism is arguably more primitive (especially since procedures and modules require second order quantification anyway), doesn't rule out any programs and is closed under observations. On the other hand, the more modal-style approach of separation logic is simpler for simple programs and its stronger intensional interpretation of separation, whilst being more restrictive, has the significant advantage over ours that it extends smoothly to a concurrent setting.

The proof scripts for the general framework plus the verification of the allocator code and the factorial client currently total about 8,500 lines, which is excessively large. However, this really reflects my own incompetence with Coq, rather than any inherent impracticality of machine-checked proofs in this style. There are dozens of unused lemmas, variations on definitions, cut-and-pasted proofs and downright stupidities that, having learnt more about both Coq and the problem domain, I could now remove. The proofs of actual programs could easily be made an order of magnitude shorter. We have an eye to using this kind of logic in PCC-style scenarios, for which mechanical checkability is certainly necessary. But working in Coq also caught errors in definitions and proofs. For example, we originally took $|R_{aret}(n)|$ to be simply $\overline{\mathbb{S} \times \mathbb{S}}$. The allocator does satisfy that specification, but a failed proof of a simple client revealed that it has a subtle flaw: if the block size is 1, the allocator can return location 0 itself (in location 0) as the free block.

Theorem 2 is only a semantic type soundness result – it does not say that the code *actually* computes factorials. In fact, only a couple of lines need tweaking to add the functional part of the specification too. We presented a type soundness result because that, rather than more general verification, is the direction of our immediate future plans. Once we have refactored our Coq definitions somewhat, we intend to investigate certified compilation of a small functional language in this style. We will also prove a slightly more interesting allocator which actually has a free list.

Although we have so far only focussed on proving a single program, a significant feature of the relational approach is that it can talk about *equivalence* of low-level code modulo a particular contextual contract. For example, one might hope to prove that all (terminating) allocators meeting our specification are observationally equivalent, or to verify the preservation of equational laws from a high-level language. Previous work on modularity, simulation and refinement in separation logic has run into some technical difficulties associated with the non-deterministic treatment of allocation [23,13] which we believe are avoided in our approach. We also need to look more seriously at the adjoint perping operation, taking nat relations to nat×state relations [17,22]. Making all relations be $(\cdot)^{\top\top}$-closed validates more logical principles and may be an alternative to step-indexing.

Thanks to Noah Torp-Smith for helping with an early version of this work, Josh Berdine for many useful discussions about separation, and Georges Gonthier for his invaluable advice and instruction regarding the use of Coq.

References

1. M. Abadi, L. Cardelli, and P.-L. Curien. Formal parametric polymorphism. *Theoretical Computer Science*, 121, 1993.
2. M. Abadi and G. D. Plotkin. A PER model of polymorphism and recursive types. In *Proc. 5th IEEE Symposium on Logic in Computer Science (LICS)*, pages 355–365. IEEE Computer Society Press, June 1990.
3. A. Ahmed. Step-indexed syntactic logical relations for recursive and quantified types. In *Proc. 15th European Symposium on Programming (ESOP)*, 2006.
4. A. Appel. Foundational proof-carrying code. In *Proc. 16th IEEE Symposium on Logic in Computer Science (LICS)*, 2001.
5. A. Appel and A. Felty. A semantic model of types and machine instructions for proof-carrying code. In *Proc. 27th ACM Symposium on Principles of Programming Languages (POPL)*, 2000.
6. A. Appel and D. McAllester. An indexed model of recursive types for foundational proof-carrying code. *ACM Transactions on Programming Languages and Systems (TOPLAS)*, 23(5), 2001.

7. N. Benton. Simple relational correctness proofs for static analyses and program transformations. In *Proc. 31st ACM Symposium on Principles of Programming Languages (POPL)*, January 2004. Revised version available from http://research.microsoft.com/~nick/publications.htm.
8. N. Benton. A typed, compositional logic for a stack-based abstract machine. In *Proc. 3rd Asian Symposium on Programming Languages and Systems (APLAS)*, volume 3780 of *Lecture Notes in Computer Science*, November 2005.
9. N. Benton and B. Leperchey. Relational reasoning in a nominal semantics for storage. In *Proc. 7th International Conference on Typed Lambda Calculi and Applications (TLCA)*, volume 3461 of *Lecture Notes in Computer Science*, 2005.
10. J. Chen, D. Wu, A. W. Appel, and H. Fang. A provably sound TAL for back-end optimization. In *Proc. ACM SIGPLAN Conference on Programming Language Design and Implementation (PLDI)*, 2003.
11. N. A. Hamid, Z. Shao, V. Trifonov, S. Monnier, and Z. Ni. A syntactic approach to foundational proof-carrying code. *Journal of Automated Reasoning*, 31(3-4), 2003.
12. N. Marti, R. Affeldt, and A. Yonezawa. Verification of the heap manager of an operating system using separation logic. In *Proc. 3rd Workshop on Semantics, Program Analysis and Computing Environments for Memory Management (SPACE)*, 2006.
13. I. Mijajlovic, N. Torp-Smith, and P. O'Hearn. Refinement and separation contexts. In *Proc. Annual Conference on Foundations of Software Technology and Theoretical Computer Science (FSTTCS)*, December 2004.
14. G. Morrisett, D. Walker, K. Crary, and N. Glew. From System F to typed assembly language. *ACM Transactions on Programming Languages and Systems (TOPLAS)*, 21(3), 1999.
15. G. Necula. Proof-carrying code. In *Proc. 24th ACM Symposium on Principles of Programming Languages (POPL)*, 1997.
16. Z. Ni and Z. Shao. Certified assembly programming with embedded code pointers. In *Proc. 33rd ACM Symposium on Principles of Programming Languages (POPL)*, 2006.
17. A. M. Pitts. Parametric polymorphism and operational equivalence. *Mathematical Structures in Computer Science*, 10, 2000.
18. A. M. Pitts and I. D. B. Stark. Operational reasoning for functions with local state. In *Higher Order Operational Techniques in Semantics*. CUP, 1998.
19. G. D. Plotkin and M. Abadi. A logic for parametric polymorphism. In *Proc. International Conference on Typed Lambda Calculi and Applications (TLCA)*, volume 664 of *Lecture Notes in Computer Science*, 1993.
20. U. S. Reddy and H. Yang. Correctness of data representations involving heap data structures. *Science of Computer Programming*, 50(1–3):129–160, March 2004.
21. J. C. Reynolds. Separation logic: A logic for shared mutable data structures. In *Proc. 17th IEEE Symposium on Logic in Computer Science (LICS)*, 2002.
22. J. Vouillon and P.-A. Mellies. Semantic types: A fresh look at the ideal model for types. In *Proc. 31st ACM Symposium on Principles of Programming Languages (POPL)*, 2004.
23. H. Yang. Relational separation logic. *Theoretical Computer Science*, 2004. Submitted.
24. D. Yu, N. A. Hamid, and Z. Shao. Building certified libraries for PCC: Dynamic storage allocation. *Science of Computer Programming*, 50, 2004.

Collapsibility in Infinite-Domain Quantified Constraint Satisfaction

Manuel Bodirsky[1] and Hubie Chen[2]

[1] Institut für Informatik
Humboldt-Universität zu Berlin
Berlin, Germany
bodirsky@informatik.hu-berlin.de
[2] Departament de Tecnologia
Universitat Pompeu Fabra
Barcelona, Spain
hubie.chen@upf.edu

Abstract. In this article, we study the quantified constraint satisfaction problem (QCSP) over infinite domains. We develop a technique called collapsibility that allows one to give strong complexity upper bounds on the QCSP. This technique makes use of both logical and universal-algebraic ideas. We give applications illustrating the use of our technique.

1 Introduction

The *constraint satisfaction problem* (CSP) is the problem of deciding the truth of a primitive positive sentence

$$\exists v_1 \ldots \exists v_n (R(v_{i_1}, \ldots, v_{i_k}) \wedge \ldots)$$

over a relational signature, relative to a given relational structure over the same signature. Informally, the goal in an instance of the CSP is to decide if there exists an assignment to a set of variables simultaneously satisfying a collection of constraints. Many search problems in computer science can be naturally formulated as CSPs, such as boolean satisfiability problems, graph homomorphism problems, and the problem of solving a system of equations (over some algebraic structure). The CSP can be equivalently formulated as the relational homomorphism problem [14], or the conjunctive-query containment problem [18].

The ubiquity of the CSP in conjunction with its general intractability has given rise to an impressive research program seeking to identify restricted cases of the CSP that are polynomial-time tractable. In particular, much attention has been focused on identifying those relational structures Γ such that $\mathsf{CSP}(\Gamma)$–the CSP where the relational structure is fixed to be Γ–is polynomial-time tractable. In a problem $\mathsf{CSP}(\Gamma)$, we call Γ the *constraint language*, and use the term *domain* to refer to the universe of Γ. Many recent results have studied the problems $\mathsf{CSP}(\Gamma)$ for finite-domain constraint languages Γ, see for example [8,9,7,6,13] and the references therein. However, it has been recognized that many natural combinatorial problems from areas such as graph theory and temporal reasoning

Z. Ésik (Ed.): CSL 2006, LNCS 4207, pp. 197–211, 2006.

can be expressed as problems of the form $\mathsf{CSP}(\Gamma)$ only if infinite-domain Γ are permitted [1]. This has motivated the study of constraint satisfaction problems $\mathsf{CSP}(\Gamma)$ on infinite domains [1,2,4].

A recent subject of inquiry that builds upon CSP research is the *quantified constraint satisfaction problem (QCSP)*, which is the generalization of the CSP where both existential and universal quantification is allowed, as opposed to just existential quantification. As is well-known, the extra expressiveness of the QCSP comes with an increase in complexity: the finite-domain QCSP is PSPACE-complete, in contrast to the finite-domain CSP, which is NP-complete. Recent work on the QCSP includes that of Börner, Bulatov, Krokhin, and Jeavons [5], Chen [11,10,12], Gottlob, Greco, and Scarcello [15], and Pan and Vardi [20].

In this paper, we consider infinite-domain quantified constraint satisfaction. Our contribution is to introduce, in the infinite-domain setting, a technique called *collapsibility* that allows us to give complexity upper bounds on problems of the form $\mathsf{QCSP}(\Gamma)$, such as NP upper bounds, that are *dramatically* lower than the "obvious" upper bound of PSPACE that typically applies. On a high level, collapsibility allows one to show that, for certain contraint languages Γ, an arbitrary instance of $\mathsf{QCSP}(\Gamma)$ can be reduced to the conjunction of instances of $\mathsf{QCSP}(\Gamma)$ that are simpler in that they have only a constant number of (or no) universally quantified variables; typically, such a conjunction can be cast as an instance of $\mathsf{CSP}(\Gamma')$ for some constraint langauge Γ' with $\mathsf{CSP}(\Gamma')$ in NP, and hence the reduction yields a proof that $\mathsf{QCSP}(\Gamma)$ is in NP.

To develop our collapsibility technique, we make use of a universal-algebraic approach to studying the complexity of constraint languages; this approach associates a set of operations called *polymorphisms* to each constraint language, and uses this set of operations to derive information about complexity. While the present work takes inspiration from technology that was developed in the finite-domain setting [11,10] for similar purposes, there are a number of differences between the infinite and finite settings that necessitate the use of more involved and intricate argumentation in the infinite setting. One is that, while there is a canonical choice for the aforementioned simpler instances in the finite setting, in the infinite setting there is no such canonical choice and indeed often an expansion of the constraint language is required to achieve a reduction from the QCSP to the CSP. Another is that, in the infinite setting, any assignment or partial assignment f to variables induces, via the automorphism group of Γ, an *orbit* of assignments $\{\sigma(f) : \sigma \text{ is an automorphism of } \Gamma\}$. The property of an assignment satisfying constraints over Γ is orbit-invariant, but in the presence of universal quantification, one needs to make inferences about the orbit of an assignment in a careful way (see Lemma 3 and its applications).

2 Preliminaries

When A and B are sets, we use $[A \to B]$ to denote the set of functions mapping from A to B. When $f : A \to B$ is a function and A' is a subset of A, we use $f|_{A'}$ to denote the restriction of f to A'. We extend this notation to sets of functions:

when $F \subseteq [A \to B]$ and A' is a subset of A, we use $F|_{A'}$ to denote the set $\{f|_{A'} : f \in F\}$. When $f : A \to B$ is a function, we use the notation $f[a' \to b']$ to denote the extension of f mapping a' to b'. We will use $[k]$ to denote the first k positive integers, $\{1, \ldots, k\}$.

Relational structures. A *relational language* τ is a (in this paper always finite) set of *relation symbols* R_i, each of which has an associated finite *arity* k_i. A *(relational) structure* Γ over the *(relational) language* τ (also called τ*-structure*) is a set D_Γ (the *domain* or *universe*) together with a relation $R_i \subseteq D_\Gamma^{k_i}$ for each relation symbol R_i from τ. For simplicity, we use the same symbol for a relation symbol and the corresponding relation. If necessary, we write R^Γ to indicate that we are talking about the relation R belonging to the structure Γ. For a τ-structure Γ and $R \in \tau$ it will also be convenient to say that $R(u_1, \ldots, u_k)$ *holds in* Γ iff $(u_1, \ldots, u_k) \in R$. If we add relations to a given structure Γ we call the resulting structure Γ' an *expansion* of Γ, and Γ is called a *reduct* of Γ'.

Homomorphisms. Let Γ and Γ' be τ-structures. A *homomorphism* from Γ to Γ' is a function f from D_Γ to $D_{\Gamma'}$ such that for each n-ary relation symbol R in τ and each n-tuple $(a_1, \ldots, a_n)$, if $(a_1, \ldots, a_n) \in R^\Gamma$, then $(f(a_1), \ldots, f(a_n)) \in R^{\Gamma'}$. In this case we say that the map f *preserves* the relation R. Isomorphisms from Γ to Γ are called *automorphisms*, and homomorphisms from Γ to Γ are called *endomorphisms*. The set of all automorphisms of a structure Γ is a group, and the set of all endomorphisms of a structure Γ is a monoid with respect to composition. When referring to an automorphism of Γ, we sometimes use the term Γ*-automorphism* to make clear the relational structure. An *orbit of k-tuples in* Γ is a set of k-tuples of the form $\{(a(s_1), \ldots, a(s_k)) :$ a is an automorphism of $\Gamma\}$ for some tuple $(s_1, \ldots, s_k)$.

Polymorphisms. Let D be a countable set, and O be the set of *finitary operations* on D, i.e., functions from D^k to D for finite k. We say that a k-ary operation $f \in O$ *preserves* an m-ary relation $R \subseteq D^m$ if whenever $R(x_1^i, \ldots, x_m^i)$ holds for all $1 \leq i \leq k$ in Γ, then $R\big(f(x_1^1, \ldots, x_1^k), \ldots, f(x_m^1, \ldots, x_m^k)\big)$ holds in Γ. If f preserves all relations of a relational τ-structure Γ, we say that f is a polymorphism of Γ. In other words, f is a homomorphism from $\Gamma^k = \Gamma \times \ldots \times \Gamma$ to Γ, where $\Gamma_1 \times \Gamma_2$ is the *(categorical- or cross-) product* of the two relational τ-structures Γ_1 and Γ_2. Hence, the unary polymorphisms of Γ are the endomorphisms of Γ.

Quantified constraint satisfaction. We define a τ-formula to be a *quantified constraint formula* if it has the form $Q_1 v_1 \ldots Q_n v_n (\psi_1 \wedge \ldots \wedge \psi_m)$, where each Q_i is a quantifier from $\{\forall, \exists\}$, and each ψ_i is an atomic τ-formula that can contain variables from $\{v_1, \ldots, v_n\}$.

The quantified constraint satisfaction problem over a τ-structure Γ, denoted by $\mathsf{QCSP}(\Gamma)$, is the problem of deciding, given a quantified constraint formula over τ, whether or not the formula is true under Γ. Note that both the universal and existential quantification is understood to take place over the entire universe of Γ. We use D throughout the paper to denote the universe of a constraint language Γ under discussion. The constraint satisfaction problem over a τ-structure

Γ, denoted by $\mathsf{CSP}(\Gamma)$, is the restriction of $\mathsf{QCSP}(\Gamma)$ to instances only including existential quantifiers.

A *constraint language* is simply a relational structure; we typically refer to a relational structure Γ as a constraint language when we are interested in the computational problem $\mathsf{QCSP}(\Gamma)$ or $\mathsf{CSP}(\Gamma)$. We also refer to Γ as a *template*.

We will illustrate the use of our technique on examples drawn from the following two classes of constraint languages.

Equality constraint languages. An *equality-definable relation* is a relation (on an infinite domain) that can be defined by a boolean combination of atoms of the form $x = y$. An *equality constraint language* is a relational structure having an countably infinite universe D and such that all of its relations are equality-definable relations over D.

When Γ is an equality constraint language with domain D, any permutation of D is an automorphism of Γ, that is, the automorphism group of Γ is the full symmetric group on D. Observe that, if a tuple $t = (t_1, \ldots, t_k)$ is an element of an equality-definable relation $R \subseteq D^k$, then all tuples of the form $(\pi(t_1), \ldots, \pi(t_k))$, where π is a permutation on D, are also contained in R. In studying equality constraint languages, it is therefore natural for us to associate to each tuple $(t_1, \ldots, t_k)$ the equivalence relation ρ on $\{1, \ldots, k\}$ where $i = j$ if and only if $t_i = t_j$. This is because, by our previous observation, a tuple $t = (t_1, \ldots, t_k)$ is in an equality-definable relation R if and only if all k-arity tuples inducing the same equivalence relation as t are in R. We may therefore view an equality-definable relation of arity k as the union of equivalence relations on $\{1, \ldots, k\}$.

It is known that for an equality constraint language Γ, $\mathsf{CSP}(\Gamma)$ is polynomial-time tractable if Γ has a constant unary polymorphism or an injective binary polymorphism, and is NP-complete otherwise [4]. It is also known (and not difficult to verify) that for every equality constraint language Γ, the problem $\mathsf{QCSP}(\Gamma)$ is in PSPACE [3]. In general, the quantified constraint satisfaction problem for equality constraint languages is PSPACE-complete [3]; this is closely related to a result of [21].

Temporal constraint languages. A *temporal relation* is a relation on the domain $\mathbb{Q}$ (the rational numbers) that can be defined by a boolean combination of expressions of the form $x < y$. A *temporal constraint language* is a relational structure having $\mathbb{Q}$ as universe and such that all of its relations are temporal relations. As with equality constraint languages, it is known and not difficult to verify that for every temporal constraint language Γ, the problem $\mathsf{CSP}(\Gamma)$ is in NP, the problem $\mathsf{QCSP}(\Gamma)$ is in PSPACE, and there are temporal constraint languages Γ such that $\mathsf{QCSP}(\Gamma)$ is PSPACE-complete. Temporal constraint languages are well-studied structures in model theory (e.g., they are all ω-categorical; see [16]).

3 Collapsibility

In this section, we present our collapsibility technology. We begin by introducing some notation and terminology.

When Φ is a quantified constraint formula, let V^Φ denote the variables of Φ, let E^Φ denote the existentially quantified variables of Φ, and let U^Φ denote the universally quantified variables of Φ. When $u \in V^\Phi$ is a variable of Φ, we use $V^\Phi_{<u}$ to denote the variables coming strictly before u in the quantifier prefix of Φ, and we use $V^\Phi_{\leq u}$ to denote the variables coming before u (including u) in the quantifier prefix of Φ. When S is a subset of V^Φ, we say that S is an *initial segment* of Φ if $S = \emptyset$ or $S = V^\Phi_{\leq u}$ for a variable $u \in V^\Phi$.

Let us intuitively think of an instance of the QCSP as a game between two players: a *universal player* that sets the universally quantified variables, and an *existential player* that sets the existentially quantified variables. The existential player wants to satisfy all of the constraints. We may formalize the notion of a strategy for the existential player in the following way.

A *strategy* for a quantified constraint formula Φ is a sequence of partial functions $\sigma = \{\sigma_x : [V^\Phi_{<x} \to D] \to D\}_{x \in E^\Phi}$. The intuition behind this definition is that the function σ_x of a strategy describes how to set the variable x given a setting to all of the previous variables. We say that an assignment f to an initial segment of Φ is *consistent* with σ if for every existentially quantified variable x in the domain of f, it holds that $\sigma_x(f|_{V^\Phi_{<x}})$ is defined and is equal to $f(x)$. Intuitively, f is consistent with σ if it could have been reached in a play of the game under σ.

A *playspace* for a quantified constraint formula Φ is a set of mappings $\mathcal{A} \subseteq [V^\Phi \to D]$. We will often be interested in restrictions of a playspace $\mathcal{A}$ of the form $\mathcal{A}|_{V^\Phi_{<u}}$ or $\mathcal{A}|_{V^\Phi_{\leq u}}$; we will use the notation $\mathcal{A}\langle <u \rangle$ and $\mathcal{A}\langle \leq u \rangle$ for these restrictions, respectively. The quantified constraint formula Φ will always be clear from the context. Likewise, for a function f defined on a subset of V^Φ, we will use the notation $f\langle <u \rangle$ and $f\langle \leq u \rangle$ for the restrictions $f|_{V^\Phi_{<u}}$ and $f|_{V^\Phi_{\leq u}}$, respectively.

Intuitively, a playspace will be used to describe a restriction on the actions of the universal player: an existential strategy will be a winning strategy for a playspace as long as it can properly respond to all settings of variables that fall into the playspace. We formalize this in the following way.

Let $\mathcal{A}$ be a playspace for a quantified constraint formula Φ, and let σ be a strategy for the same formula Φ. We say that σ is a *winning* strategy for $\mathcal{A}$ if the following two conditions hold:

- for every variable $x \in E^\Phi$ and every assignment $f \in \mathcal{A}\langle <x \rangle$, if f is consistent with σ, then $\sigma_x(f)$ is defined and $f[x \to \sigma_x(f)] \in \mathcal{A}\langle \leq x \rangle$, and
- every assignment $f \in \mathcal{A}$ consistent with σ satisfies the constraints of Φ.

We call a playspace *winnable* if there exists a winning strategy for it.

Let us say that a playspace $\mathcal{A}$ (for a quantified constraint formula Φ) is $\forall$*-free* ($\exists$*-free*) if for every universally (existentially) quantified variable $u \in V^\Phi$, every domain element $d \in D$, and every function $f \in \mathcal{A}\langle <u \rangle$, the function $f[u \to d]$ is contained in $\mathcal{A}\langle \leq u \rangle$. As a simple example illustrating these notions, observe that for any quantified constraint formula Φ, the playspace $[V^\Phi \to D]$ is both $\forall$-free and $\exists$-free. The notion for $\forall$-freeness yields a characterization of truth for quantified constraint formulas.

Proposition 1. *Let Φ be a quantified constraint formula Φ. The following are equivalent:*

1. *Φ is true.*
2. *The $\forall$-free playspace $[V^{\Phi} \to D]$ has a winning strategy.*
3. *There exists a $\forall$-free playspace for Φ having a winning strategy.*

Having given the basic terminology for collapsibility, we now proceed to develop the technique itself. The following is an outline of the technique. What we aim to show is that for certain templates Γ, an arbitrary instance Φ of $\mathsf{QCSP}(\Gamma)$ is truth-equivalent to (the conjunction of) a collection of "simpler" QCSP instances. These simpler instances will always have the property that the truth of the original instance Φ readily implies the truth of the simpler instances; what is non-trivial is to show that the truth of all of the simpler instances implies the truth of the original instance. We will be able to establish this implication in the following way. First, we will translate the truth of the simpler instances into winnability results on playspaces (for the original instance Φ). Then, we will make use of two tools (to be developed here) that allow us to infer the winnability of larger playspaces based on the winnability of smaller playspaces and the polymorphisms of Φ. These tools will let us demonstrate the winnability of a $\forall$-free playspace, which then implies the truth of Φ by Proposition 1.

We now turn to give the two key tools which allow us to "enlargen" playspaces while still preserving winnability. To illustrate the use of these tools, we will use a running example which will fully develop a collapsibility proof.

Example 2. As a running example for this section, we consider *positive equality constraint languages.* Positive equality constraint languages are equality constraint languages where every relation is definable by a positive combination of atoms of the form $x = y$, that is, definable using such atoms and the boolean connectives $\{\vee, \wedge\}$. A simple example of a positive equality constraint language is $\Gamma = (\mathbb{N}, S)$, where S is the relation

$$S = \{(w, x, y, z) \in \mathbb{N}^4 : (w = x) \vee (y = z)\}.$$

Any equality-definable relation R, viewed as the union of equivalence relations, can be verified to have the following closure property: every equivalence relation ρ' obtainable from an equivalence relation ρ from R by combining two equivalence classes into one is also contained in R. In fact, from this observation, it is not difficult to see that a positive equality constraint language has all unary functions as polymorphisms. (Indeed, the property of having all unary functions as polymorphisms is also sufficient for an equality constraint language to be a positive equality constraint language, and hence yields an algebraic characterization of positive equality constraint languages.)

We will show that, for any positive equality constraint language Γ, the problem $\mathsf{QCSP}(\Gamma)$ reduces to $\mathsf{CSP}(\Gamma \cup \{\neq\})$. In particular, for an instance

$$\Phi = Q_1 v_1 \dots Q_n v_n C$$

of $\mathsf{QCSP}(\Gamma)$, we define the *collapsing of* Φ to be the $\mathsf{CSP}(\Gamma \cup \{\neq\})$ instance

$$\Phi' = \exists v_1 \ldots \exists v_n (\mathcal{C} \wedge \bigwedge \{v_i \neq v_j : i < j, Q_j = \forall\}).$$

That is, the collapsing of Φ is obtained from Φ by adding constraints asserting that each universal variable y is different from all variables coming before y, and then changing all quantifiers to existential. We will show that an instance Φ of $\mathsf{QCSP}(\Gamma)$ is true *if and only if* its collapsing is true. This gives a reduction from a problem whose most obvious complexity upper bound is PSPACE, to a problem in NP. The inclusion of this problem in NP has been previously shown by Kozen [19]; we have elected it as our running example as we believe it allows us to nicely illustrate our technique. Note that our reduction is tight in that there are known NP-hard positive equality constraint languages [3]. (The existence of such NP-hard constraint languages also implies that one cannot hope for a reduction from $\mathsf{QCSP}(\Gamma)$ to $\mathsf{CSP}(\Gamma)$ which does not "augment the template", since for positive equality constraint languages Γ, the problem $\mathsf{CSP}(\Gamma)$ is known to be polynomial-time tractable [4].)

It is readily seen that if an instance Φ is true, then its collapsing Φ' is true. The difficulty in justifying this reduction, then, is in showing that if a collapsing Φ' is true, then the original instance Φ is true. Our first step in showing this is to simply view the truth of Φ' as a winnability result on a playspace. Let $a : \{v_1, \ldots, v_n\} \to D$ be an assignment satisfying the constraints of Φ'. Clearly, the playspace $\{a\}$ is winnable, via the strategy $\sigma = \{\sigma_x\}_{x \in E^\Phi}$ defined by $\sigma_x(a|_{V^\Phi < x}) = a(x)$. We will use the winnability of this playspace to derive the winnability of larger and larger playspaces, ultimately showing the winnability of the largest playspace $[V^\Phi \to D]$, and hence the truth of the formula (by Proposition 1). □

The following lemma allows one to add, to a winnable playspace, tuples from the orbits induced by the tuples already in the playspace, while maintaining the property of winnability.

Lemma 3. *(Orbit Lemma) Let $\mathcal{A}$ be a winnable playspace for a quantified constraint formula Φ over template Γ. Let $y \in U^\Phi$ be a universally quantified variable. There exists a winnable playspace $\mathcal{A}'$ such that the following hold:*

- *for each $t \in \mathcal{A}\langle \leq y \rangle$ and Γ-automorphism σ that fixes every point $\{t(u) : u \in \mathcal{A}\langle < y \rangle\}$, $\sigma(t)$ is in $\mathcal{A}'\langle \leq y \rangle$. Note that here, $\sigma(t)$ is equal to t at all points except (possibly) y.*
- *$\mathcal{A} \subseteq \mathcal{A}' \subseteq \{\tau(t) : t \in \mathcal{A}, \tau$ is a Γ-automorphism$\}$.*
- *$\mathcal{A}\langle < y \rangle = \mathcal{A}'\langle < y \rangle$.*

Proof (idea). Let F be the set of all functions of the form $\sigma(t)$ satisfying the conditions of the first property, that is, t is in $\mathcal{A}\langle \leq y \rangle$ and σ is a Γ-automorphism that fixes every point $\{t(u) : u \in \mathcal{A}\langle < y \rangle\}$. For each element $f \in F \setminus \mathcal{A}\langle \leq y \rangle$, define σ_f and t_f to be such mappings so that $f = \sigma_f(t_f)$. We define $\mathcal{A}'$ to be

$$\mathcal{A} \cup \{\sigma_f(e) : f \in F \setminus \mathcal{A}\langle \leq y \rangle, e \in \mathcal{A}, e\langle \leq y \rangle = t_f\}.$$

Let $\{\rho_x\}$ be a winning strategy for $\mathcal{A}$. We assume without loss of generality that the partial functions ρ_x are only defined on functions $f \in \mathcal{A}\langle <x \rangle$ that are consistent with ρ. We need to extend the ρ_x so that they handle extensions of the functions $f \in F \setminus \mathcal{A}\langle \leq y \rangle$. When g is an extension of such a f, we define $\rho'_x(g)$ as $\sigma_f(\rho_x(\sigma_f^{-1}(g)))$. That is, we translate g back by σ_f and look at the response by ρ_x, and apply σ_f to that response to obtain our response. It is straightforward to verify that the $\{\rho'_x\}$ are a winning strategy for $\mathcal{A}'$. □

Example 4. We continue the discussion of positive equality constraint languages, our running example. We have established the winnability of a size-one playspace $\{a\}$, where for all universally quantified variables y, the value $a(y)$ is different from $a(v)$ for all variables v coming before y in the quantifier prefix. Our goal is to infer the winnability of the largest playspace $[V^\Phi \to D]$, using the winnability of this playspace.

Let us say that a playspace $\mathcal{A}$ is *$\neq$-free* if for every universally quantified variable $y \in V^\Phi$, every function $f \in \mathcal{A}\langle <y \rangle$, and every value $d \in D$ distinct from all values in $\{f(u) : u \in V^\Phi_{<y}\}$, the function $f[y \to d]$ is contained in $\mathcal{A}\langle \leq y \rangle$. Assuming that our original instance Φ contained at least one universally quantified variable y, our playspace $\{a\}$ is *not* $\neq$-free: there is only one extension of $a\langle <y \rangle$ in $\{a\}\langle \leq y \rangle$, namely, $a\langle \leq y \rangle$. However, using the Orbit Lemma, we can expand $\{a\}$ into a $\neq$-free playspace, as follows.

Let y_1 be the first universally quantified variable of Φ. Applying the Orbit Lemma to the playspace $\mathcal{A} = \{a\}$ and variable $y = y_1$, we obtain a playspace $\mathcal{A}_1$ that satisfies the $\neq$-freeness condition at y_1. We demonstrate this as follows. If f is a function in $\mathcal{A}_1\langle <y_1 \rangle$, we have $f \in \mathcal{A}\langle <y_1 \rangle$, since the Orbit Lemma provides $\mathcal{A}\langle <y \rangle = \mathcal{A}'\langle <y \rangle$. Let $h = f[y_1 \to d]$ be any extension of f where d is distinct from all values in the image of f. We want to show that h is contained in $\mathcal{A}_1\langle \leq y_1 \rangle$. We know that there exists an extension $f' = f[y_1 \to d']$ of f such that d' is different from all values in the image of f. (This is because $f \in \mathcal{A}\langle <y_1 \rangle = \{a\}\langle <y_1 \rangle$, and the function a assigns y_1 to a value different from all values assigned to preceding variables.) Let σ be a permutation on D (that is, a Γ-automorphism) that fixes all points in the image of f, but maps d' to d. The Orbit Lemma provides that $\sigma(f') = h$ is in $\mathcal{A}_1\langle \leq y \rangle$. Repeatedly applying the Orbit Lemma to the universally quantified variables $y_1, y_2, \ldots$ of Φ, we obtain an increasing sequence of winnable playspaces $\mathcal{A}_1, \mathcal{A}_2, \ldots$ whose last member is $\neq$-free.

Note that the Orbit Lemma provides, for each i,

$$\mathcal{A}_{i+1} \subseteq \{\tau(t) : t \in \mathcal{A}_i, \tau \text{ is a } \Gamma\text{-automorphism}\}$$

and hence, for each i,

$$\mathcal{A}_i \subseteq \{\tau(t) : t \in \mathcal{A}, \tau \text{ is a } \Gamma\text{-automorphism}\}.$$

From this, we can see that each $\mathcal{A}_i$ has the property that for any universally quantified variable y_j and for any function $f \in \mathcal{A}_i\langle <y_j \rangle$, any extension $f[y_j \to d]$ of f in $\mathcal{A}_i\langle \leq y_j \rangle$ has d distinct from all values in the image of f; this is because

$\mathcal{A}$ has this property, and this property is preserved by adding, to a playspace, permutations of functions already in the playspace.

Summarizing, we have shown that the winnability of the size-one playspace from Example 2 implies the winnability of a $\neq$-free playspace. □

The next theorem allows us to, roughly speaking, use a polymorphism $g : D^k \to D$ of Φ to compose together k winnable playspaces to derive another winnable playspace.

Let $g : D^k \to D$ be an operation. Let $\mathcal{A}, \mathcal{B}_1, \ldots, \mathcal{B}_k$ be playspaces for a quantified constraint formula Φ. We say that $\mathcal{A}$ is *g-composable* from $(\mathcal{B}_1, \ldots, \mathcal{B}_k)$ if for all universally quantified variables $y \in U^\Phi$, the following holds: if $t \in \mathcal{A}\langle<y\rangle$ and $t_1 \in \mathcal{B}_1\langle<y\rangle$, ..., $t_k \in \mathcal{B}_k\langle<y\rangle$ are such that $t = g(t_1, \ldots, t_k)$ pointwise, and $d \in D$ is a value such that $t[y \to d] \in \mathcal{A}\langle\leq y\rangle$, then there exist $d_1, \ldots, d_k \in D$ such that $d = g(d_1, \ldots, d_k)$ and $t_1[y \to d_1] \in \mathcal{B}_1\langle\leq y\rangle$, ..., $t_k[y \to d_k] \in \mathcal{B}_k\langle\leq y\rangle$.

Theorem 5. *Let Φ be a quantified constraint formula, and assume that $g : D^k \to D$ is a polymorphism of all relations in Φ. Assume that $\mathcal{A}, \mathcal{B}_1, \ldots, \mathcal{B}_k$ are playspaces such that $\mathcal{A}$ is $\exists$-free and g-composable from $(\mathcal{B}_1, \ldots, \mathcal{B}_k)$. If each of the playspaces $\mathcal{B}_1, \ldots, \mathcal{B}_k$ is winnable, then $\mathcal{A}$ is winnable.*

Theorem 5 was inspired by machinery developed for finite-domain QCSPs presented in [10, Chapter 4]. Before giving the proof, we give an example application that allows us to conclude our running example.

Example 6. For a QCSP instance Φ over a positive equality constraint language Γ, we have shown, in Examples 2 and 4, the winnability of a $\neq$-free playspace $\mathcal{A}_{\neq}$ based on the truth of the collapsing Φ' of Φ; the collapsing Φ' is a CSP instance (over an equality constraint language). We now complete the justification of our reduction by showing that the winnability of this $\neq$-free playspace implies the winnability of the "full" playspace $[V^\Phi \to D]$.

Let $g : D \to D$ be a surjective unary function such that $g^{-1}(d)$ is of infinite size for every $d \in D$, that is, every point $d \in D$ in the image of g is hit by infinitely many domain points. As noted in Example 2, the function g is a polymorphism of Γ. To show the winnability of the playspace $[V^\Phi \to D]$, we show that it is g-composable from $\mathcal{A}_{\neq}$, from which its winnability follows by appeal to Theorem 5.

Why is the playspace $[V^\Phi \to D]$ g-composable from $\mathcal{A}_{\neq}$? Let $y \in U^\Phi$ be a universally quantified variable, let $t \in \mathcal{A}\langle<y\rangle$, let $t' \in \mathcal{A}_{\neq}\langle<y\rangle$ and suppose that $t = g(t')$ pointwise. It suffices to show that for any value $d \in D$, there exists $d' \in D$ such that $d = g(d')$ and $t'[y \to d'] \in \mathcal{A}_{\neq}$. This holds: one can pick d' to be any point in $g^{-1}(d) \setminus \mathsf{image}(t')$. This set is non-empty as it is the subtraction of a finite set from an infinite set, and for any such d' we have $t'[y \to d'] \in \mathcal{A}_{\neq}\langle\leq y\rangle$ by the $\neq$-freeness of $\mathcal{A}_{\neq}$. □

Proof (Theorem 5). For each $i \in [k]$, let σ^i be a winning strategy for the playspace $\mathcal{B}_i$. We define a sequence of mappings $\sigma = \{\sigma_x\}_{x \in E^\Phi}$ that constitutes a winning strategy for $\mathcal{A}$. We consider each initial segment one by one, in order of increasing size. After the initial segment S has been considered, we will have defined mappings $\{\sigma_x\}_{x \in E^\Phi \cap S}$ having the following properties:

(a) if $S = V^{\Phi}|_{\leq x}$ for an existentially quantified variable x, then for any $f \in \mathcal{A}\langle <x \rangle$ consistent with σ, $\sigma_x(f)$ is defined and $f[x \to \sigma_x(f)] \in \mathcal{A}|_S$.
(b) if $f \in \mathcal{A}|_S$ is consistent with σ, then there exist $f_1 \in \mathcal{B}_1|_S$, ..., $f_k \in \mathcal{B}_k|_S$ such that $f = g(f_1, \ldots, f_k)$ pointwise and f_i is consistent with σ^i for all $i \in [k]$.

This suffices, since after the initial segment $S = V^{\Phi}$ has been considered, the sequence of mappings $\{\sigma_x\}$ constitute a winning strategy. The first requirement in the definition of a winning strategy holds because property (a) holds for all possible initial segments S. The second requirement in the definition of a winning strategy holds: by property (b), any assignment $f \in \mathcal{A}|_{V^{\Phi}}$ consistent with σ is equal to g applied point-wise to assignments $f_1 \in \mathcal{B}_1|_{V^{\Phi}}, \ldots, f_k \in \mathcal{B}_k|_{V^{\Phi}}$ that are consistent with $\sigma^1, \ldots, \sigma^k$, respectively; since the σ^i are winning strategies, each f_i satisfies the constraints of Φ, and since g is a polymorphism of the relations of Φ, f satisfies the constraints of Φ.

We now give the construction.

Let $S' = V^{\Phi}_{\leq u}$ be an initial segment of size $|S'| \geq 1$, and let $S = V^{\Phi}_{<u}$ be the initial segment of size $|S'|-1$. We may assume by induction that the construction has been performed for S. To perform the construction for S', we consider two cases depending on the quantifier of the variable u.

Case 1: u is an $\exists$-quantified variable. We consider each mapping $f \in \mathcal{A}|_S$. If f is not consistent with σ, then we leave $\sigma_u(f)$ undefined. If f is consistent with σ, then in order to satisfy property (a), we need to define $\sigma_u(f)$. Since property (b) holds on S, there exist the described mappings $f_1 \in \mathcal{B}_1|_S$, ..., $f_k \in \mathcal{B}_k|_S$ with $f = g(f_1, \ldots, f_k)$ pointwise and with f_i consistent with σ^i for all $i \in [k]$. Since, for each $i \in [k]$, the σ^i are winning strategies, there is an extension $f'_i \in \mathcal{B}_i|_{S'}$ of f consistent with σ^i. We define $\sigma_u(f)$ as $g(f'_1(u), \ldots, f'_k(u))$. The mapping $f' = f[u \to \sigma_u(f)]$ is in $\mathcal{A}|_{S'}$ by the $\exists$-freeness of $\mathcal{A}$. Now, the mapping f' is consistent with σ, so we need to verify that property (b) holds on f'. It is straightforward to verify that the mappings $f'_1, \ldots, f'_k$ serve at witnesses.

Case 2: u is a $\forall$-quantified variable. Clearly, property (a) is trivially satisfied for S', so we need only consider property (b). Suppose that $f' \in \mathcal{A}|_{S'}$ is consistent with σ. We want to show the existence of the described mappings $f'_1, \ldots, f'_k$. Let $f = f'|_S$. Since property (b) holds for the initial segment S, we know that there exist $f_1 \in \mathcal{B}_1\langle <u \rangle, \ldots, f_k \in \mathcal{B}_k\langle <u \rangle$ such that $f = g(f_1, \ldots, f_k)$ pointwise and f_i is consistent with σ^i for all $i \in [k]$. By the definition of g-composable, there exist extensions $f'_1, \ldots, f'_k$ of $f_1, \ldots, f_k$, respectively, to S', satisfying the conditions of property (b). □

4 Applications

In the previous section, we developed some tools for giving collapsibility proofs, and illustrated their use on positive equality constraint languages. We showed that for any positive equality constraint language Γ, the problem $\mathsf{QCSP}(\Gamma)$ is in NP. In this section, we give further applications of our technique.

4.1 Max-Closed Constraints

We consider temporal constraint languages that are closed under the binary operation $\max : \mathbb{Q} \times \mathbb{Q} \to \mathbb{Q}$ that returns the maximum of its two arguments. We will demonstrate the following theorem.

Theorem 7. *Let Γ be a temporal constraint language having the* max *operation as polymorphism. The problem* QCSP(Γ) *is in* NP.

Example 8. Consider the temporal constraint language $(\mathbb{Q}, R)$ where R is the relation $\{(x, y, z) \in \mathbb{Q}^3 : x < y \text{ or } x < z\}$. This constraint language has the max operation as polymorphism: suppose $(a, b, c), (a', b', c') \in R$. We want to show that $(\max(a, a'), \max(b, b'), \max(c, c')) \in R$. Let us assume without loss of generality that $a > a'$. We know that either $a < b$ or $a < c$. If $a < b$, then $a < \max(b, b')$ and we have $\max(a, a') < \max(b, b')$. If $a < c$, then $a < \max(c, c')$ and we have $\max(a, a') < \max(c, c')$. □

We now prove this theorem. Let Φ be an instance of QCSP(Γ) for a max-closed template Γ. As in the collapsibility proof for positive equality constraint languages, we will show a reduction to a CSP. Whereas in the case of positive languages we gave a direct reduction to a CSP, here, we give a reduction to a conjunction of QCSP instances, each of which has one universally quantified variable; we argue that this ensemble can be formulated as a CSP.

Denote Φ as $Q_1 v_1 \ldots Q_n v_n \mathcal{C}$. (We assume that Φ has at least one universally quantified variable, otherwise, it is an instance of CSP(Γ).) For a universally quantified variable $v_i \in U^\Phi$, we define the v_k*-collapsing* of Φ to be the QCSP instance

$$\Phi' = \exists v_1 \ldots \exists v_{k-1} \forall v_k \exists v_{k+1} \ldots \exists v_n (\mathcal{C} \wedge \bigwedge \{v_i > v_j : i < j, j \neq k, Q_j = \forall\}).$$

That is, the v_k-collapsing of Φ is obtained from Φ by adding constraints asserting that each universal variable y (other than v_k) is less than all variables coming before it, and changing all universal quantifiers to existential except for that of v_k. It is readily verifiable that if the original QCSP(Γ) instance Φ was true, then all of its y-collapsings (with $y \in U^\Phi$) are also true. We show that the converse holds. This suffices to place QCSP(Γ) in NP, by the following lemma.

Lemma 9. *Let $B \subseteq \mathbb{Q}^3$ be the "different-implies-between" relation defined by $B = \{(x, y, z) \in \mathbb{Q}^3 : (x \neq z) \to ((x < y < z) \vee (x > y > z))\}$. Let Γ' be the expansion of a temporal constraint language Γ with B and $<$. Given an instance Φ of* QCSP(Γ)*, there exists an instance Φ' of* CSP(Γ') *that is true if and only if Φ is true. For every constant k, the mapping $\Phi \to \Phi'$ can be computed in polynomial time on those formulas Φ with $|U^\Phi| \leq k$.*

Lemma 9 can be viewed as a strong version of the well-known quantifier elimination property for temporal constraint languages.

We want to show that if all y-collapsings of an instance Φ are true, then Φ itself is true. How will we do this? We will first translate the truth of each y-collapsing into a winnability result on a playspace $\mathcal{A}_y$ for Φ. We will then show that each of these playspaces $\mathcal{A}_y$ can be expanded into a playspace $\mathcal{A}'_y$ that obeys a "freeness" condition but is still winnable. We then compose together the playspaces $\mathcal{A}'_y$ using Theorem 5 to derive the winnability of the full playspace $[V^\Phi \to \mathbb{Q}]$.

We translate the truth of the y-collapsings of Φ into winnability results on playspaces, as follows. Let us say that a playspace $\mathcal{A}$ (for Φ) is *$\forall$-free at $z \in U^\Phi$* if for every assignment $f \in \mathcal{A}\langle <z \rangle$, and every $d \in D$, the function $f[z \to d]$ is contained in $\mathcal{A}\langle \leq z \rangle$. For each $y \in U^\Phi$, it is readily verified that the truth of the y-collapsing of Φ implies the winnability of a playspace $\mathcal{A}_y$ that is $\forall$-free at y, and where for all $t \in \mathcal{A}_y$ it holds that $t(a) > t(b)$ if $b \in U^\Phi \setminus \{y\}$, $a \in V^\Phi$, and a comes before b in the quantifier prefix.

Let $S \subseteq U^\Phi$ be a set of universally quantified variables. We define a playspace $\mathcal{A}$ for Φ to be *$(S, <)$-free* if:

- for every variable $y \in S$, $\mathcal{A}$ is $\forall$-free at y, and
- for every variable $y \in U^\Phi \setminus S$, and every assignment $f \in \mathcal{A}\langle <y \rangle$, there exists an interval $(-\infty, d_y]$ such that for every $d \in (-\infty, d_y]$, the function $f[y \to d]$ is contained in $\mathcal{A}\langle \leq y \rangle$.

Our playspaces $\mathcal{A}_y$ are not $(\{y\}, <)$-free, but via repeated application of the Orbit Lemma, from each playspace $\mathcal{A}_y$ we may obtain a winnable playspace $\mathcal{A}'_y$ that is $(\{y\}, <)$-free.

We prove by induction, on the size of S, that there is a winnable playspace $\mathcal{A}'_S$ that is $(S, <)$-free (for all $S \subseteq U^\Phi$). This suffices to show the winnability of a $(U^\Phi, <)$-free playspace, which is $\forall$-free, implying the truth of Φ by Proposition 1. Suppose $k \geq 1$. By induction, we assume that we have constructed our $\mathcal{A}'_S$ for $|S| \leq k$. Let $S' \subseteq U^\Phi$ be of size $|S'| = k + 1$. We want to show the winnability of a $(S', <)$-free playspace. Pick any element $s_0 \in S'$ and set $S = S' \setminus \{s_0\}$. Suppose that $\mathcal{A}'_{s_0}$ is $(\{s_0\}, <)$-free with respect to $\{d_y\}_{y \in U^\Phi \setminus \{s_0\}}$, and that $\mathcal{A}'_S$ is $(S, <)$-free with respect to $\{e_y\}_{y \in U^\Phi \setminus S}$. We show the winnability of the playspace $\mathcal{A}'_{S'}$ that is $(S', <)$-free with respect to $\{\min(d_y, e_y)\}_{y \in U^\Phi \setminus S'}$, and also $\exists$-free. In particular, we prove that $\mathcal{A}_{S'}$ is max-composable from $(\mathcal{A}'_{s_0}, \mathcal{A}'_S)$. The winnability of $\mathcal{A}_{S'}$ then follows from Theorem 5. Let $y \in U^\Phi$ and consider $t \in \mathcal{A}'_{S'}\langle <y \rangle$, $t_{s_0} \in \mathcal{A}_{s_0}\langle <y \rangle$, $t_S \in \mathcal{A}_S\langle <y \rangle$ such that $t = \max(t_{s_0}, t_S)$ pointwise. Let $d \in \mathbb{Q}$ be a value such that $t[y \to d] \in \mathcal{A}'_{S'}$. We want to find values d_1, d_2 such that $d = \max(d_1, d_2)$ and $t_{s_0}[y \to d_1] \in \mathcal{A}_{s_0}\langle \leq y \rangle$, and $t_S[y \to d_2] \in \mathcal{A}_S\langle \leq y \rangle$. We split into cases.

- $y = s_0$: we select $d_1 = d$ and d_2 to be a value such that $d_2 \leq d$ and $d_2 \leq d_y$. The first inequality guarantees that $d = \max(d_1, d_2)$ and the second guarantees that $t_{s_0}[y \to d_1] \in \mathcal{A}_{s_0}\langle \leq y \rangle$.
- $y \in S$: we select $d_2 = d$ and d_1 to be a value such that $d_1 \leq d$ and $d_1 \leq e_y$. This case is similar to the previous one, except we use the $\forall$-freeness of $\mathcal{A}_S$ at y, whereas in the previous case, we used the $\forall$-freeness of $\mathcal{A}_{s_0}$.
- $y \in U^\Phi \setminus S'$: we select $d_1 = d$ and $d_2 = d$.

4.2 Near-Unanimity Operations

A *near-unanimity operation* is an operation $f : D^k \to D$ of arity $k \geq 3$ satisfying the identities $f(y, x, \ldots, x) = f(x, y, x, \ldots, x) = \cdots = f(x, \ldots, x, y)$ for all $x, y \in D$. Near-unanimity operations have been studied in the finite case in [17,10]. We show that, for a constraint language Γ having a near-unanimity operation as polymorphism, the problem $\mathsf{QCSP}(\Gamma)$ essentially reduces to the problem $\mathsf{CSP}(\Gamma)$.

Theorem 10. *Suppose that Γ is a constraint language having a near-unanimity operation $g : D^k \to D$ as polymorphism. There exists a polynomial-time computable mapping that, given an instance Φ of $\mathsf{QCSP}(\Gamma)$, outputs a set S of instances of $\mathsf{QCSP}(\Gamma)$ such that:*

- *each instance in S has at most $k - 1$ universally quantified variables, and*
- *all instances in S are true if and only if the original instance Φ is true.*

Proof. Let Φ be an instance of $\mathsf{QCSP}(\Gamma)$. In this proof, we define a j-collapsing of Φ to be an instance of $\mathsf{QCSP}(\Gamma)$ obtained from Φ by selecting a subset $S \subseteq U^\Phi$ of universally quantified variables of size $|S| = j$, and changing the quantifiers of the variables $U^\Phi \setminus S$ to existential, and adding constraints $\{y = y' : y, y' \in U^\Phi \setminus S\}$ equating all of the variables in $U^\Phi \setminus S$. (Note that these equalities can subsequently be eliminated by renaming and removing variables.)

Clearly, if Φ is true, all of its j-collapsings are true. We show that if the j-collapsings of Φ are true for all $j \leq k - 1$, then Φ is true. This is obvious if Φ has $k - 1$ or fewer universally quantified variables, so we assume that it has k or more universally quantified variables. It is straightforward to verify that the truth of the j-collapsing of Φ arising from the subset $S \subseteq U^\Phi$ (with $|S| = j$) implies the winnability of a playspace $\mathcal{A}_S$ (for Φ) that is $\forall$-free at all $y \in S$ and such that $f(y') = a$ for all $y \in U^\Phi \setminus S$ for a fixed constant a. (See the proof of Theorem 7 for the definition of $\forall$-*free at* y.)

We prove that for all subsets $S \subseteq U^\Phi$, there is a winnable playspace $\mathcal{A}_S$ (for Φ) that is $\forall$-free at all $y \in S$. This suffices, since then $\mathcal{A}_{U^\Phi}$ is a winnable playspace that is $\forall$-free. We prove this by induction on $|S|$. We have pointed out that this is true when $|S| \leq k - 1$, so assume that $|S| \geq k$. Select k distinct elements $s_1, \ldots, s_k \in S$. For each $i \in [k]$, define $S_i = S \setminus \{s_i\}$. We claim that the playspace $\mathcal{A}_S$ that consists of all functions $f : V^\Phi \to D$ such that $f(y) = a$ for all $y \in U^\Phi \setminus S$, is g-composable from $(\mathcal{A}_{S_1}, \ldots, \mathcal{A}_{S_k})$. The result then follows from Theorem 5. We verify this as follows. Let $y \in U^\Phi$ and suppose that $t \in \mathcal{A}_S\langle <y \rangle$ and $t_i \in \mathcal{A}_{S_i}\langle <y \rangle$ for all $i \in [k]$ are such that $t = g(t_1, \ldots, t_k)$ pointwise, and $d \in D$ is a value such that $t[y \to d] \in \mathcal{A}_S\langle \leq y \rangle$. We want to give values $d_1, \ldots, d_k \in D$ such that $d = g(d_1, \ldots, d_k)$ and $t_i[y \to d_i] \in \mathcal{A}_{S_i}\langle \leq y \rangle$. We split into cases.

- If $y = s_i$ for some $i \in [k]$, we set $d_i = a$ and $d_j = d$ for all other j, that is, $j \in [k] \setminus \{i\}$.
- If $y \in S \setminus \{s_1, \ldots, s_k\}$, we set $d_i = d$ for all $i \in [k]$.
- If $y \in U^\Phi \setminus S$, we set $d_i = a$ for all $i \in [k]$.

□

The following is an example application of Theorem 10. Define the operation $\mathsf{median} : \mathbb{Q}^3 \rightarrow \mathbb{Q}$ to be the operation that returns the median of its three arguments. The operation median is a near-unanimity operation of arity 3.

Theorem 11. *Suppose that Γ is a temporal constraint language having the* median *operation as polymorphism. The problem* $\mathsf{QCSP}(\Gamma)$ *is in* NP.

Proof. We use the reduction of Theorem 10 along with Lemma 9 to obtain a reduction to $\mathsf{CSP}(\Gamma')$ for a temporal constraint language Γ'. □

References

1. Manuel Bodirsky. Constraint satisfaction with infinite domains. Ph.D. thesis, Humboldt-Universität zu Berlin, 2004.
2. Manuel Bodirsky. The core of a countably categorical structure. In Volker Diekert and Bruno Durand, editors, *Proceedings of the 22nd Annual Symposium on Theoretical Aspects of Computer Science (STACS'05), Stuttgart (Germany)*, LNCS 3404, pages 100–110, Springer-Verlag Berlin Heidelberg, 2005.
3. Manuel Bodirsky and Hubie Chen. Quantified equality constraints. Manuscript, 2006.
4. Manuel Bodirsky and Jan Kára. The complexity of equality constraint languages. In *Proceedings of the International Computer Science Symposium in Russia (CSR'06)*, LNCS 3967, pages 114–126, 2006.
5. F. Boerner, A. Bulatov, A. Krokhin, and P. Jeavons. Quantified constraints: Algorithms and complexity. In *Proceedings of CSL'03*, LNCS 2803, pages 58–70, 2003.
6. A. Bulatov and V. Dalmau. A simple algorithm for Mal'tsev constraints. *SIAM J. Comp. (to appear)*.
7. A. Bulatov, A. Krokhin, and P. G. Jeavons. Classifying the complexity of constraints using finite algebras. *SIAM Journal on Computing*, 34:720–742, 2005.
8. Andrei Bulatov, Andrei Krokhin, and Peter Jeavons. The complexity of maximal constraint languages. In *Proceedings of STOC'01*, pages 667–674, 2001.
9. Andrei A. Bulatov. A dichotomy theorem for constraints on a three-element set. In *FOCS'02*, pages 649–658, 2002.
10. Hubie Chen. The computational complexity of quantified constraint satisfaction. Ph.D. thesis, Cornell University, August 2004.
11. Hubie Chen. Collapsibility and consistency in quantified constraint satisfaction. In *AAAI*, pages 155–160, 2004.
12. Hubie Chen. Quantified constraint satisfaction, maximal constraint languages, and symmetric polymorphisms. In *STACS*, pages 315–326, 2005.
13. Victor Dalmau. Generalized majority-minority operations are tractable. In *LICS*, pages 438–447, 2005.
14. T. Feder and M. Vardi. The computational structure of monotone monadic SNP and constraint satisfaction: A study through Datalog and group theory. *SIAM Journal on Computing*, 28:57–104, 1999.
15. Georg Gottlob, Gianluigi Greco, and Francesco Scarcello. The complexity of quantified constraint satisfaction problems under structural restrictions. In *IJCAI 2005*, 2005.
16. Wilfrid Hodges. *A shorter model theory.* Cambridge University Press, 1997.

17. Peter Jeavons, David Cohen, and Martin Cooper. Constraints, consistency and closure. *AI*, 101(1-2):251–265, 1998.
18. Ph. G. Kolaitis and M. Y. Vardi. Conjunctive-query containment and constraint satisfaction. In *Proceedings of PODS'98*, pages 205–213, 1998.
19. Dexter Kozen. Positive first-order logic is NP-complete. *IBM Journal of Research and Development*, 25(4):327–332, 1981.
20. Guoqiang Pan and Moshe Vardi. Fixed-parameter hierarchies inside PSPACE. In *LICS 2006*. To appear.
21. Larry J. Stockmeyer and Albert R. Meyer. Word problems requiring exponential time: Preliminary report. In *STOC*, pages 1–9, 1973.

Towards an Implicit Characterization of NC^k *

G. Bonfante[1], R. Kahle[2], J.-Y. Marion[1], and I. Oitavem[3,**]

[1] Loria - INPL, 615, rue du Jardin Botanique, BP-101, 54602 Villers-lès-Nancy, France
{Jean-Yves.Marion|Guillaume.Bonfante}@loria.fr
[2] Dept. Matemática, Universidade de Coimbra, Apartado 3008, 3001-454 Coimbra, Portugal, and CENTRIA, UNL, 2829-516 Caparica, Portugal
kahle@mat.uc.pt
[3] CMAF, Universidade de Lisboa, Av. Prof. Gama Pinto, 2, 1649-003 Lisboa, Portugal, and DM, UNL, 2829-516 Caparica, Portugal
isarocha@ptmat.fc.ul.pt

Abstract. We define a hierarchy of term systems T^k by means of restrictions of the recursion schema. We essentially use a pointer technique together with tiering. We prove $T^k \subseteq NC^k \subseteq T^{k+1}$, for $k \geq 2$. Special attention is put on the description of T^2 and T^3 and on the proof of $T^2 \subseteq NC^2 \subseteq T^3$. Such a hierarchy yields a characterization of NC.

1 Introduction

The present work enters the field of Implicit Complexity. By Implicit, we mean that focus is done on algorithms rather than functions. In particular, the question of how to compute a function with respect to of one of its algorithms arise.

Apart from its theoretical interest, the present study has some practical consequences. Since the tiering condition we consider on programs is decidable, the term systems can be used for static analysis of programs, that is for certification of bounds on the time/space usage of programs. In an other context, such an approach has been established on the theoretical study of Hofmann [Hof99, Hof02] which found applications in the Embounded Project[1].

The present approach of Implicit Complexity is in the vein of Bellantoni-Cook/Leivant, that is, we use some tiering discipline. Since the seminal papers of Simmons [Sim88], Bellantoni-Cook [BC92], and Leivant-Marion [LM93], the approach has shown to be fruitful. For instance, see Mairson and Neergaard [Nee04] who propose a nice characterization of LOGSPACE by means of a tiering discipline. Another branch of Implicit Complexity use logic, see for instance [Hof99, BM04] for recent work on the subject. We also mention the work of Niggl which covers the crucial issue of imperative programming [Nig05].

* Research supported by the joint french-portuguese project *Theorie et langages de programmation pour les calculs à ressources bornées/Teorias e linguagens de programação para computações com recursos limitados* from *EGIDE – GRICES*.

** The last author was also partially supported by the project POCI/MAT/61720/2004 of FCT.

[1] http://www.embounded.org/

Z. Ésik (Ed.): CSL 2006, LNCS 4207, pp. 212–224, 2006.

In this paper, we try to shape the form of recursion that corresponds to NC^k. In other words, we are working towards an *implicit characterization* of the complexity classes NC^k, $k \in \mathbb{N}$. We discuss the term systems T^k such that $T^k \subseteq NC^k \subseteq T^{k+1}$, for $k \geq 2$. NC^k is the class of languages accepted by uniform boolean circuit families of depth $O(\log^k n)$ and polynomial size with bounded gates; and $NC = \bigcup_k NC^k$.

To motivate the definition of our system we look to NC^k from the point of view of *Alternating Turing Machines* (ATMs). The relation is established by the following theorem.

Theorem 1 (Ruzzo, [Ruz81]). *Let $k \geq 1$. For any language $L \subseteq \{0,1\}^*$, L is recognized by an ATM in $O(\log^k n)$ time and $O(\log n)$ space iff it is in (uniform) NC^k.*

The underlying intuition for our implicit approach is to use ramified recursion to capture the *time* aspect and recursion with *pointers* to capture the *space* aspect. We use linear recursion in the sense of [LM00] in order to stratify the degree of the polylogarithmic time, and recursion with pointers as in [BO04]. We define term systems T^k allowing k ramified recursion of which the lowest one is equipped with pointers.

We work in a sorted context, in the vein of Leivant [Lei95]. For T^k we use $k+1$ tiers:

- tier 0 with no recursion;
- tier 1 for recursion with pointers to capture the *space* aspect;
- tiers 2 to k for ramified recursions which deal with the *time* aspect.

There are two implicit characterization of NC^1, one by Leivant-Marion using *linear ramified recurrence with substitution* [LM00], and one by Bloch [Blo94] in a Bellantoni and Cook [BC92] recursion setting. Clote [Clo90], using bounded recursion schemes, gives a machine-independent characterization of NC^k. Leivant's approach to NC^k [Lei98] is machine and resource independent, however, it is not sharp. It consists of term systems RSR^k for *ramified schematic recurrence*. RSR^k characterizes NC^k only within *three* levels:

$$RSR^k \subseteq NC^k \subseteq RSR^{k+2}, \quad k \geq 2.$$

Our term systems reduce the unsharpness of the characterization of NC^k to *two* levels:

$$T^k \subseteq NC^k \subseteq T^{k+1}, \quad k \geq 2.$$

As related work we mention here [LM95] where alternating computations was captured by mean of ramified recursion with parameter substitution. For NC there exists also an implicit characterization by use of higher type functions in [AJST01].

The structure of the paper is briefly as follows. In Section 2, we present the term systems and state some preliminary results. In Section 3, we describe the upper bounds, that is the way of compiling the term systems in terms of circuits. Section 4 is devoted to the lower bound.

2 The Term Systems T^k

The term systems T^k are formulated in a $k+1$-sorted context, over the tree algebra $\mathbb{T}$. The algebra $\mathbb{T}$ is generated by 0, 1 and $*$ (of arities 0, 0 and 2, respectively), and we use infix notation for $*$. As usual, we introduce three additional constants: L for the left destructor, R for the right destructor and C for the conditional. They are defined as follows: $\mathrm{L}(0)=0$, $\mathrm{L}(1)=1$, $\mathrm{L}(u*v)=u$, $\mathrm{R}(0)=0$, $\mathrm{R}(1)=1$, $\mathrm{R}(u*v)=v$, and $\mathrm{C}(0,x,y,z)=x$, $\mathrm{C}(1,x,y,z)=y$, $\mathrm{C}(u*v,x,y,z)=z$.

Following notation introduced by Leivant in [Lei95], we consider $k+1$ copies of $\mathbb{T}$. Therefore, we formally have $k+1$ copies of the constructors $\mathbb{T}$, and $k+1$ sorts of variables ranging over the different tiers. As usual, we separate different tiers by semicolons.

As initial functions of T^k one considers the constructors, destructors, conditional and projection functions over the $k+1$ tiers. T^k is closed under sorted composition over $k+1$-tiers — $f(\boldsymbol{x}_k;\ldots;\boldsymbol{x}_0)=h(\boldsymbol{g}_k(\boldsymbol{x}_k;\ldots;);\ldots;\boldsymbol{g}_0(\boldsymbol{x}_k;\ldots;\boldsymbol{x}_0))$ — and k schemes of sorted recursion as described below.

We start by define the recursion schemes of T^2 and T^3 and describe only then the general case of T^k. In what follows one should notice that, whenever f is defined by recursion with step function h, h itself cannot be defined by recursion over the same tier as f is defined. Therefore, in T^k we allow, at most, k *(step-)nested* recursions. However, in the base cases the function g can be defined by a further recursion over the same tier as f is defined, i.e., no restriction is imposed on the number of recursions constructed on top of each other (*base-nested* recursions).

2.1 The Term System T^2

According to the underlying idea, the characterization of NC^2 requires three tiers:

- Tier 0: no recursion.
- Tier 1: recursion with pointers (space tier).
- Tier 2: recursion without pointers, modifying tier 1 (time tier).

Tier 1 recursion (with pointers)

$$\begin{aligned}
f(;p,0,\boldsymbol{x};\boldsymbol{w}) &= g(;p,0,\boldsymbol{x};\boldsymbol{w}),\\
f(;p,1,\boldsymbol{x};\boldsymbol{w}) &= g(;p,1,\boldsymbol{x};\boldsymbol{w}),\\
f(;p,u*v,\boldsymbol{x};\boldsymbol{w}) &= h(;;\boldsymbol{w},f(;p*0,u,\boldsymbol{x};\boldsymbol{w}),f(;p*1,v,\boldsymbol{x};\boldsymbol{w})).
\end{aligned}$$

Tier 2 recursion

$$\begin{aligned}
f(0,\boldsymbol{y};\boldsymbol{x};\boldsymbol{w}) &= g(\boldsymbol{y};\boldsymbol{x};\boldsymbol{w}),\\
f(1,\boldsymbol{y};\boldsymbol{x};\boldsymbol{w}) &= g(\boldsymbol{y};\boldsymbol{x};\boldsymbol{w}),\\
f(u*v,\boldsymbol{y};\boldsymbol{x};\boldsymbol{w}) &= h(;\boldsymbol{x};\boldsymbol{w},f(u,\boldsymbol{y};\boldsymbol{x};\boldsymbol{w})).
\end{aligned}$$

Since h can use the variables of lower tier, $\boldsymbol{x}$, to recurse on, we can nest recursions.

In the tier 2 recursion, the recursion input is only used as a counter. In particular, the recursion only takes the height of the tree $u * v$ into account. Therefore, it might be more natural to rewrite this scheme in form of a successor recursion: If one uses in the following scheme the expression $u+1$ as some kind of schematic variable for $u * v$, where v is arbitrary, and 0 as a schematic expression for 1 or the actual 0 (note that $f(1, \boldsymbol{y}; \boldsymbol{x}; \boldsymbol{w})$ is defined as the same as $f(0, \boldsymbol{y}; \boldsymbol{x}; \boldsymbol{w})$), the scheme can be written as follows:

Tier 2 recursion (successor notation)

$$\begin{aligned} f(0, \boldsymbol{y}; \boldsymbol{x}; \boldsymbol{w}) &= g(\boldsymbol{y}; \boldsymbol{x}; \boldsymbol{w}), \\ f(u+1, \boldsymbol{y}; \boldsymbol{x}; \boldsymbol{w}) &= h(; \boldsymbol{x}; \boldsymbol{w}, f(u, \boldsymbol{y}; \boldsymbol{x}; \boldsymbol{w})). \end{aligned}$$

Later on, in the course of a definition of a function using successor notation, $x+1$ can be read as an abbreviation of $x * 1$, and the schematic notation is in accordance with the actual definition in terms of trees.

2.2 The Term System T^3

According to the underlying idea, T^3 has one more tier than T^2 for recursion (without pointers). Note that the recursion for the tiers 1 and 2 differ from those for T^2 only by the extra semicolon needed for the additional tier separation.

Tier 1 recursion (with pointers)

$$\begin{aligned} f(;;p, 0, \boldsymbol{x}; \boldsymbol{w}) &= g(;;p, 0, \boldsymbol{x}; \boldsymbol{w}), \\ f(;;p, 1, \boldsymbol{x}; \boldsymbol{w}) &= g(;;p, 1, \boldsymbol{x}; \boldsymbol{w}), \\ f(;;p, u * v, \boldsymbol{x}; \boldsymbol{w}) &= h(;;; \boldsymbol{w}, f(;;p * 0, u, \boldsymbol{x}; \boldsymbol{w}), f(;;p * 1, v, \boldsymbol{x}; \boldsymbol{w})). \end{aligned}$$

Tier 2 recursion

$$\begin{aligned} f(; 0, \boldsymbol{y}; \boldsymbol{x}; \boldsymbol{w}) &= g(; \boldsymbol{y}; \boldsymbol{x}; \boldsymbol{w}), \\ f(; 1, \boldsymbol{y}; \boldsymbol{x}; \boldsymbol{w}) &= g(; \boldsymbol{y}; \boldsymbol{x}; \boldsymbol{w}), \\ f(; u * v, \boldsymbol{y}; \boldsymbol{x}; \boldsymbol{w}) &= h(;; \boldsymbol{x}; \boldsymbol{w}, f(; u, \boldsymbol{y}; \boldsymbol{x}; \boldsymbol{w})). \end{aligned}$$

Tier 3 recursion

$$\begin{aligned} f(0, \boldsymbol{z}; \boldsymbol{y}; \boldsymbol{x}; \boldsymbol{w}) &= g(\boldsymbol{z}; \boldsymbol{y}; \boldsymbol{x}; \boldsymbol{w}), \\ f(1, \boldsymbol{z}; \boldsymbol{y}; \boldsymbol{x}; \boldsymbol{w}) &= g(\boldsymbol{z}; \boldsymbol{y}; \boldsymbol{x}; \boldsymbol{w}), \\ f(u * v, \boldsymbol{z}; \boldsymbol{y}; \boldsymbol{x}; \boldsymbol{w}) &= h(; \boldsymbol{y}; \boldsymbol{x}; \boldsymbol{w}, f(u, \boldsymbol{z}; \boldsymbol{y}; \boldsymbol{x}; \boldsymbol{w})). \end{aligned}$$

Again the recursion schemes for time can be rewritten in successor notation.

Tier 2 recursion (successor notation)

$$f(;0,\boldsymbol{y};\boldsymbol{x};\boldsymbol{w}) = g(;\boldsymbol{y};\boldsymbol{x};\boldsymbol{w}),$$
$$f(;u+1,\boldsymbol{y};\boldsymbol{x};\boldsymbol{w}) = h(;;\boldsymbol{x};\boldsymbol{w},f(;u,\boldsymbol{y};\boldsymbol{x};\boldsymbol{w})).$$

Tier 3 recursion (successor notation)

$$f(0,\boldsymbol{z};\boldsymbol{y};\boldsymbol{x};\boldsymbol{w}) = g(\boldsymbol{z};\boldsymbol{y};\boldsymbol{x};\boldsymbol{w}),$$
$$f(u+1,\boldsymbol{z};\boldsymbol{y};\boldsymbol{x};\boldsymbol{w}) = h(;\boldsymbol{y};\boldsymbol{x};\boldsymbol{w},f(u,\boldsymbol{z};\boldsymbol{y};\boldsymbol{x};\boldsymbol{w})).$$

2.3 The Term Systems T^k

The extension of the definition of T^2 and T^3 to arbitrary k, $k \geq 4$ is straightforward. In each step from $k-1$ to k we have to add another *time tier*, adapting the notation of the existing recursion schemes to the new number of tiers and adding one more nested recursion for the new tier k of the form:

Tier *k* recursion

$$f(0,\boldsymbol{x_k};\boldsymbol{x_{k-1}};\ldots;\boldsymbol{x_1};\boldsymbol{w}) = g(\boldsymbol{x_k};\boldsymbol{x_{k-1}};\ldots;\boldsymbol{x_1};\boldsymbol{w})$$
$$f(1,\boldsymbol{x_k};\boldsymbol{x_{k-1}};\ldots;\boldsymbol{x_1};\boldsymbol{w}) = g(\boldsymbol{x_k};\boldsymbol{x_{k-1}};\ldots;\boldsymbol{x_1};\boldsymbol{w})$$
$$f(u*v,\boldsymbol{x_k};\boldsymbol{x_{k-1}};\ldots;\boldsymbol{x_1};\boldsymbol{w}) = h(;\boldsymbol{x_{k-1}};\ldots;\boldsymbol{x_1};\boldsymbol{w},f(u,\boldsymbol{x_k};\boldsymbol{x_{k-1}};\ldots;\boldsymbol{x_1};\boldsymbol{w}))$$

In successor notation this scheme reads as follows:

Tier *k* recursion (successor notation)

$$f(0,\boldsymbol{x_k};\boldsymbol{x_{k-1}};\ldots;\boldsymbol{x_1};\boldsymbol{w}) = g(\boldsymbol{x_k};\boldsymbol{x_{k-1}};\ldots;\boldsymbol{x_1};\boldsymbol{w})$$
$$f(u+1,\boldsymbol{x_k};\boldsymbol{x_{k-1}};\ldots;\boldsymbol{x_1};\boldsymbol{w}) = h(;\boldsymbol{x_{k-1}};\ldots;\boldsymbol{x_1};\boldsymbol{w},f(u,\boldsymbol{x_k};\boldsymbol{x_{k-1}};\ldots;\boldsymbol{x_1};\boldsymbol{w}))$$

2.4 The Term System T^1

T^1 is just the restriction of T^2 to two tiers only, and the single recursion scheme:

Recursion of T^1

$$f(p,0,\boldsymbol{x};\boldsymbol{w}) = g(p,0,\boldsymbol{x};\boldsymbol{w})$$
$$f(p,1,\boldsymbol{x};\boldsymbol{w}) = g(p,1,\boldsymbol{x};\boldsymbol{w})$$
$$f(p,u*v,\boldsymbol{x};\boldsymbol{w}) = h(;\boldsymbol{w},f(p*0,u,\boldsymbol{x};\boldsymbol{w}),f(p*1,v,\boldsymbol{x};\boldsymbol{w})).$$

Lemma 2. $T^1 \subseteq NC^1$.

Proof. Let us give the key argument of the proof. Writing $\mathsf{H}(t)$, the height of a term t, it is the case that arguments in tier 1 encountered along the computation have all linear height in the height of the inputs. As a consequence, for a given function, any branch of recursion *for this symbol* is done in logarithmic time in the size of the inputs. We conclude by induction on the definition of functions: a branch of the computation will involve only finitely many such function symbols, and so, can be simulated in NC^1.

By a straightforward induction on the definition of functions, one proves the key fact, that is arguments have linear height in the height of the input. In other words, when computing $f(x_1, \ldots, x_k; \boldsymbol{w})$, for all subcalls of the form $h(x'_1, \ldots, x'_n; \boldsymbol{w'})$, then $\forall x'_i : \mathsf{H}(x') \leq O(\sum_{i \leq k} \mathsf{H}(x_i))$.

Since a function of T^k which has no arguments in tiers greater than 1 can be defined also in T^1 we have the immediate corollary:

Lemma 3. *A function in T^k using only arguments in tier 1 and tier 0 is definable in NC^1.*

As an *ad hoc* designation, in the following, we call NC^1 *function* to a function using only argument in tier 1 and tier 0.

For NC^1 functions we have simultaneous recursion:

Lemma 4. *If $f_1, \ldots, f_n$ are defined by simultaneous recursion over tier 1, then they are definable in T^1.*

The proof is a straightforward adaptation of the corresponding proposition for the system $\mathbf{ST}_{\mathbb{T}}$, characterizing NC, in [Oit04, Proposition 5].

3 The Upper Bound of T^k

For the upper bound we model the computations of T^k by circuits. We start with the exemplary case of T^2.

Theorem 5

$$T^2 \subseteq NC^2.$$

Proof Let $f(\boldsymbol{y}; \boldsymbol{x}; \boldsymbol{w})$ be a function of T^2. We will show that there is a circuit in NC^2 which computes $f(\boldsymbol{y}; \boldsymbol{x}; \boldsymbol{w})$.

The proof is done by induction on the definition of the function. The base cases and composition are straightforward. Recursion for tier 1 only leads to NC^1 functions (Lemma 3). Therefore, we have to consider only the case that a function defined by recursion for tier 2.

We consider first the case where we have only one argument in tier 2. It will serve as a paradigm for the more general case.

BASE CASE: f has only one argument of tier 2, i.e. it is defined by the following scheme:

$$f(0; \boldsymbol{x}; \boldsymbol{w}) = g(; \boldsymbol{x}; \boldsymbol{w}),$$
$$f(u+1; \boldsymbol{x}; \boldsymbol{w}) = h(; \boldsymbol{x}; \boldsymbol{w}, f(u; \boldsymbol{x}; \boldsymbol{w})).$$

where g are h already defined. By Lemma 3 they are both in NC^1. That means there are (uniform) circuits G and H both of polynomial size and $O(\log(n))$ height that compute g and h. The circuit which computes f on $(y; \boldsymbol{x}; \boldsymbol{w})$ is given in figure 1 (with $n = |y|$) .

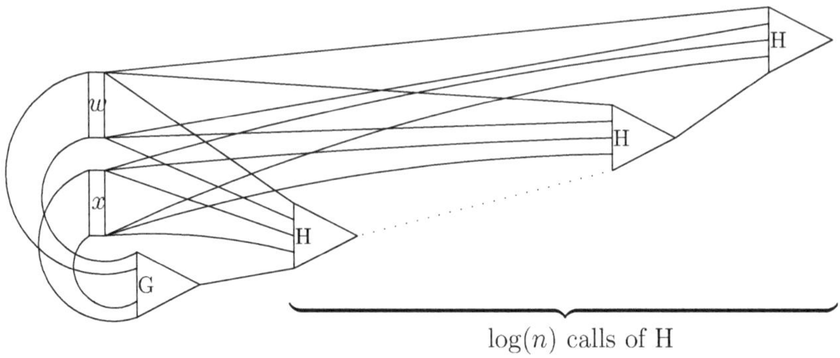

Fig. 1. The circuit F computing f

First of all, observe that the circuit is uniform. Now, the size of this circuit is $|G| + \sum_{i<|n|=\log(n)} |H|$. As the height of the tree is bounded by $O(\log(n))$, the number of H circuits is logarithmic. Since H and G circuits are of polynomial size, it is also the case for F. We end by noting that the height of the circuit is $O(\log(n)) \times O(\log(n)) = O(\log^2(n))$ since it is tree of height $O(\log(n))$ of circuit of height $O(\log(n))$. As a consequence, the circuit is in NC^2.

HIGHER ARITIES: f has $\ell + 1$ argument in tier 2, i.e. it is defined by the following scheme:

$$f(0, y_1, \ldots, y_\ell; \boldsymbol{x}; \boldsymbol{w}) = f'(y_1, \ldots, y_\ell; \boldsymbol{x}; \boldsymbol{w}),$$
$$f(u+1, y_1, \ldots, y_\ell; \boldsymbol{x}; \boldsymbol{w}) = h(; \boldsymbol{x}; \boldsymbol{w}, f(u, y_1, \ldots, y_\ell; \boldsymbol{x}; \boldsymbol{w})).$$

where f' and h are already defined. We already know that h is definable in NC^1.

Suppose that the rule for f' is

$$f'(u+1, y_2, \ldots, y_\ell; \boldsymbol{x}; \boldsymbol{w}) = h'(; \boldsymbol{x}; \boldsymbol{w}, f(u, y_2, \ldots, y_\ell; \boldsymbol{x}; \boldsymbol{w}))$$

There is a circuit H' that computes h' which is in NC^1. As a base case, we have:

$$f'(0, y_2, \ldots, y_\ell; \boldsymbol{x}; \boldsymbol{w}) = f''(y_2, \ldots, y_\ell; \boldsymbol{x}; \boldsymbol{w})$$

and f'' will itself call h'' and f''', etc. After ℓ steps, we get f^ℓ an NC^1 function as in the base case. Let us call it g as above.

The circuitry that computes f is analogous to that given in figure 1. But, in that case, the circuit for F is made of a first layer of height $O(\log(n))$ of H circuit with one leaf (the base case) which is formed by a second layer of a tree of height $O(\log(n))$ of H' circuit, and so on. The circuit remains uniform and its size is of the form $\sum_{i<|y_0|} |H| + \sum_{i<|y_1|} |H'| + \cdots + \sum_{i<|y_\square|} |H^{(\ell)}| + |G|$.

What is the height of the circuit? The first layer has height $O(\log(n)) \times O(\log(n))$ as in the base case. The second layer has also height $O(\log(n)) \times O(\log(n))$ for the same reason. More generally, the height of the tree is $\ell \times O(\log(n)^2) = O(\log(n)^2)$.

Concerning the size of the circuit. The first layer is formed of $\log(n)$ circuits of size $2^{O(\log(n))}$, that is bounded by a polynomial. Actually, all layers have polynomial size. Since there is only a finite number of such layers, there is a polynomial number of circuit of polynomial size.

Following the scheme of this proof, the result can be extended to arbitrary $k \geq 2$ and together with Lemma 2 we have:

Theorem 6. *For every $k \geq 1$:*

$$T^k \subseteq NC^k.$$

4 The (Unsharp) Lower Bound

We adopt the description of NC^k in terms of ATMs, cf. Theorem 1. Here ATMs are assumed to have only one tape. Each machine has a finite number of internal states and each state is classified as either *conjunctive*, *disjunctive*, *oracle*, *accepting* or *rejecting*. Oracle, accepting or rejecting states are *halting states*. Conjunctive or disjunctive states are *action states*. Outputs are single bits — no output device is required. A *configuration* is composed by the tape contents together with the internal state of the machine.

As Leivant in [Lei98] we describe the operational semantics of an ATM M as a two stage process: Firstly, generating an input-independent computation tree; secondly, evaluating that computation tree for a given input. A binary tree T of configurations is a *computation tree* (of M) if each non-leaf of T spawns its children configurations. A computation tree of M is generated as follows: when in a configuration with an action state, depending on the state and bit read, it spawns a pair of successor configurations. These are obtained from the parent by changing the read bit, or/and changing the internal state of the machine. We will be interested in configuration trees which have the initial configuration of M as a root. Each computation tree T maps binary representation of integers (inputs) to a value in $\{0, 1, \perp\}$, where $\perp$ denotes "undefined" — in our term systems $\perp$ will be represented by $0 * 0$. This map is defined accordingly points 1 and 2, below.

1. If T is a single configuration with state q then:
   (a) if q is an accepting [rejecting] state, the returned value is 1 [respectively, 0];
   (b) if q is an action state, the returned value is $\perp$;

(c) if q is an oracle state (i, j), where i is a symbol of the machine's alphabet — 0 or 1 — and j ranges over the number of oracles, the returned value is 1 or 0 depending on whether the nth bit of the jth oracle is i or not, where n is the integer binary represented by the portion of the tape to the right of the current head position.

2. If T is not a single configuration, then the root configuration has a conjunctive or a disjunctive state. We define the value returned by T to be the conjunction, respectively the disjunction, of the values returned by the immediate subtrees.

Conjunctive and disjunctive states may diverge, indicated by the "undetermined value" $\perp$; one understands $0 \vee \perp = \perp$, $1 \vee \perp = 1$, $0 \wedge \perp = 0$, $1 \wedge \perp = \perp$.

Theorem 7

$$NC^2 \subseteq T^3.$$

Proof The proof runs along the lines of the proof of [BO04, Lemma 5.1].

Let M be a ATM working in $O(\log^2 n)$ time and $O(\log n)$ space. Let us say that, for any input $\boldsymbol{X}$, M runs in time $T_M = t_0 \ulcorner \boldsymbol{x} \urcorner + t_1$ and space $S_M = s_0 \ulcorner \boldsymbol{x} \urcorner + s_1$, where $\boldsymbol{x}$ is a minimal balanced tree corresponding to $\boldsymbol{X}$, as in [BO04].

The proof is now based on the idea of *configuration trees.* A configuration tree for M contains as paths all possible configuration codes of M.[2] A configuration tree *for time* t will code on the leafs the values at level t in the bottom-up labeling of the computation tree.

Now, the proof can be split in several steps.

Coding Configurations. A configuration of M is given by a sequence of triples which encode the content of the tape together with information about the position of the head, and, in addition, a encoding of the current state. Padding the tapes with blanks we can assume that we have fixed tape length l. To code the three symbols 0, 1 and blank we will use two bits, $(0, 1)$, $(1, 1)$, and $(0, 0)$ respectively. Now a triple $x_i = (a_i, b_i, c_i)$ in the sequence $x_0, \ldots, x_l$ codes the symbol of cell l by a_l and b_l, and c_l is 1 only for the position of the head at the current state and 0 for all other cells. Finally we add a code w for the current state at the end of this sequence, such that a configuration is uniquely determine by the bit string $x_0, \ldots, x_l, w$ which has a fixed length for all configurations. In the sequent by configuration we mean the path containing the configuration code as described above.

The label^0 function. Given a configuration p and an input x, the function $\text{LABEL}^0(;;p, x;)$ returns 1, 0 or $0{*}0$ depending on the configuration and the input x: 1 if the configuration leads to the acceptance of x, 0 if it leads to rejection, and $0 * 0$ if p is a non-halting configuration. $\text{LABEL}^0(;;p, x;)$ can be defined by composition and simultaneous recursion over tier 1. Since simultaneous tier 1 recursion can be simulated in T^1 (lemma 4), LABEL^0 is a NC^1 function.

[2] In fact, such a tree will have a lot of branches which do not represent configurations; but these branches will not disturb.

Configuration trees. The configuration tree of time 0 is a perfect balanced tree of hight $3s_0\ulcorner \boldsymbol{x} \urcorner + 3s_1 + m$, labeled by 0, 1, or $0 * 0$, according to LABEL^0, where m is the length needed to represent w (the code of the state). Its branches "contain" all possible configurations. A branch p is labeled by 1 if p accepts x, by 0 if p rejects x, and by $0 * 0$ otherwise (i.e. if p has an action state or if it is not a configuration).

We define $\text{CT}^0_{3s_0,3s_1+m}$ by meta-induction on the second index with a side-induction on the first index in the base case. Note that this definition requires space recursion, i.e., recursion in tier 1. It calls in the base case LABEL^0, which also needs a space recursion. Since this is in the base case, we do not need (step-)nested recursion here.

$$\text{CT}^0_{0,0}(;;p,u,x;) = \text{LABEL}^0(;;p,x;), \tag{1}$$

$$\text{CT}^0_{a+1,0}(;;p,0,x;) = \text{CT}^0_{a,0}(;;p,x,x;), \tag{2a}$$

$$\text{CT}^0_{a+1,0}(;;p,1,x;) = \text{CT}^0_{a,0}(;;p,x,x;), \tag{2b}$$

$$\text{CT}^0_{a+1,0}(;;p,u*v,x;) = \text{CT}^0_{a+1,0}(;;p*0,u,x;) * \text{CT}^0_{a+1,0}(;;p*1,v,x;), \tag{2c}$$

$$\text{CT}^0_{a,b+1}(;;p,u,x;) = \text{CT}^0_{a,b}(;;p*0,u,x;) * \text{CT}^0_{a,b}(;;p*1,u,x;). \tag{3}$$

Case (1) and the cases of (2) define $\text{CT}^0_{a,0}$ by meta-induction on a. Within this definition, the cases (2a)–(2c) use tier 1 recursion. Finally, case (3) is the induction step for the definition of $\text{CT}^0_{a,b}$ by meta-induction on b.

Now, we define the initial configuration tree CT^0 as follows.

$$\text{CT}^0(;;x;) = \text{CT}^0_{3s_0,3s_1+m}(;;0,x,x;).$$

Notice, that CT^0 is a NC^1 function.

The idea is now to update this configuration tree along the time the machine is running.

The label^{+1} function. One can define a function LABEL^{+1} which for a configuration p and a configuration tree z, returns 0, 1 or $0*0$ according as configuration p is rejecting, accepting or undetermined, using the labels of the successor configurations of p in z.

LABEL^{+1} uses simultaneous tier 1 recursion.

Note, that LABEL^{+1} is a NC^1 function.

The update function. The aim of the function CT^{+1} is to update a configuration tree for time t to the configuration tree at time $t+1$. Here, we need the space recursion with step function $*$ and base function LABEL^{+1}, in order to build a copy of the given configuration tree where the leaves are updated according to LABEL^{+1}.

We define $\text{CT}^{+1}_{a,b}$ in analogy to $\text{CT}^0_{a,b}$ by meta-induction on a and b.

$$\text{CT}^{+1}_{0,0}(;;p,u,x;z) = \text{LABEL}^{+1}(;;p,x;z)$$

$$\text{CT}^{+1}_{a+1,0}(;;p,0,x;z) = \text{CT}^{+1}_{a,0}(;;p,x,x;z),$$

$$\begin{aligned}
\text{CT}^{+1}_{a+1,0}(;;p,1,x;z) &= \text{CT}^{+1}_{a,0}(;;p,x,x;z),\\
\text{CT}^{+1}_{a+1,0}(;;p,u*v,x;z) &= \text{CT}^{+1}_{a+1,0}(;;p*0,u,x;z)*\text{CT}^{+1}_{a+1,0}(;;p*1,v,x;z),\\
\text{CT}^{+1}_{a,b+1}(;;p,u,x;z) &= \text{CT}^{+1}_{a,b}(;;p*0,u,x;z)*\text{CT}^{+1}_{a,b}(;;p*1,u,x;z).
\end{aligned}$$

The update of a configuration tree for time t is the configuration tree for time $t+1$. For a given configuration tree z, such an update can be performed by the function CT^{+1}:

$$\text{CT}^{+1}(;;x;z) = \text{CT}^{+1}_{3s_0,3s_1+m}(;;0,x,x;z).$$

Note, that CT^{+1} is a NC^1 function.

The iteration. The iteration function iterates the update function $t_0\ulcorner \boldsymbol{x} \urcorner^2 + t_1$ times.

We define it by use of two auxiliary functions IT^1 and IT^2. IT^1 iterates the update function $a\ulcorner \boldsymbol{x} \urcorner$ times; IT^2 iterate then IT^1 $\ulcorner \boldsymbol{x} \urcorner$ times and add b more iterations of the update function.

For a given natural number n, let $\text{CT}^{+n}(;;x;z)$ be the CT^{+1} function composed with itself n times. Thus, in an inductive definition of CT^{+n} we have that $\text{CT}^{+(n+1)}(;;x;z)$ is defined as $\text{CT}^{+1}(;;x;\text{CT}^{+n}(;;x;z))$.

- IT^1 is defined by recursion in tier 2.

$$\begin{aligned}
\text{IT}^1(;0;x;z) &= z,\\
\text{IT}^1(;y+1;x;z) &= \text{CT}^{+t_0}(;;x;\text{IT}^1(;y;x;z)).
\end{aligned}$$

- IT^2 is defined by recursion in tier 3:

$$\begin{aligned}
\text{IT}^2(0;y;x;z) &= \text{CT}^{+t_1}(;;x;z),\\
\text{IT}^2(u+1;y;x;z) &= \text{IT}^1(;y;x;\text{IT}^2(u;y;x;z)).
\end{aligned}$$

Note, that IT^1 and IT^2 are the only non NC^1 functions needed in this proof.

Now, we just have to iterate the CT^{+1} function T_M times, on the initial configuration tree CT^0, in order to obtain the configuration tree CT^{T_M} :

$$\text{CT}^{T_M}(x) = \text{IT}^2(x;x;x;\text{CT}^0(;;x;))$$

Finally, recursing on x we can follow in CT^{T_M} the path corresponding to the initial configuration and we read its label: 0 or 1.

5 Conclusion

Putting together the theorems 5 and 7 one gets:

Theorem 8

$$T^2 \subseteq NC^2 \subseteq T^3.$$

As stated in the proof of theorem 7, only the functions which are performing the iteration are not NC^1 functions. The higher tiers are used only for the iteration. It is straightforward that one additional tier on top enables us to define an iteration of the update function $\log(n)$ times the length of iteration definable by use of the lower tiers. Together with the extension of the upper bound to T^k, stated in theorem 6, one gets:

Theorem 9. *For $k \geq 2$ we have:*

$$T^k \subseteq NC^k \subseteq T^{k+1}.$$

As a corollary we get another characterization of NC as the union of the the term systems T^k, $k \in \mathbb{N}$:

Corollary 10

$$NC = \bigcup_{k \in \mathbb{N}} T^k.$$

References

[AJST01] Klaus Aehlig, Jan Johannsen, Helmut Schwichtenberg, and Sebastiaan Terwijn. Linear ramified higher type recursion and parallel complexity. In R. Kahle, P. Schroeder-Heister, and R. Stärk, editors, *Proof Theory in Computer Science*, volume 2183 of *Lecture Notes in Computer Science*, pages 1–21. Springer, 2001.

[BC92] S. Bellantoni and S. Cook. A new recursion-theoretic characterization of the poly-time functions. *Computational Complexity*, 2:97–110, 1992.

[Blo94] S. Bloch. Function-algebraic characterizations of log and polylog parallel time. *Computational Complexity*, 4(2):175–205, 1994.

[BM04] Patrick Baillot and Virgile Mogbil. Soft lambda-calculus: A language for polynomial time computation. In Igor Walukiewicz, editor, *Foundations of Software Science and Computation Structures, 7th International Conference, FOSSACS 2004*, volume 2987 of *Lecture Notes in Computer Science*, pages 27–41. Springer, 2004.

[BO04] S. Bellantoni and I. Oitavem. Separating NC along the δ axis. *Theoretical Computer Science*, 318:57–78, 2004.

[Clo90] P. Clote. Sequential, machine independent characterizations of the parallel complexity classes *ALogTIME*, AC^k, NC^k and NC. In S. Buss and P. Scott, editors, *Feasible Mathematics*, pages 49–69. Birkhäuser, 1990.

[Hof99] Martin Hofmann. Linear types and non-size-increasing polynomial time computation. In *Symposium on Logic in Computer Science (LICS '99)*, pages 464–473. IEEE, 1999.

[Hof02] Martin Hofmann. The strength of non-size increasing computation. In *POPL '02: Proceedings of the 29th ACM SIGPLAN-SIGACT symposium on Principles of programming languages*, pages 260–269. ACM Press, New York, NY, USA, 2002.

[Lei95] Daniel Leivant. Ramifed recurrence and computational complexity I: Word recurrence and poly-time. In P. Clote and J. B. Remmel, editors, *Feasible Mathematics II*, pages 320–343. Birkhäuser, 1995.

[Lei98] Daniel Leivant. A characterization of NC by tree recursion. In *FOCS 1998*, pages 716–724. IEEE Computer Society, 1998.

[LM93] Daniel Leivant and Jean-Yves Marion. Lambda calculus characterizations of poly-time. *Fundamenta Informaticae*, 19(1/2):167–184, 1993.

[LM95] Daniel Leivant and Jean-Yves Marion. Ramified recurrence and computational complexity II: substitution and poly-space. In L. Pacholski and J. Tiuryn, editors, *Proceedings of CSL 94*, pages 486–500. LNCS 933, Springer Verlag, 1995.

[LM00] Daniel Leivant and Jean-Yves Marion. A characterization of alternating log time by ramified recurrence. *Theoretical Computer Science*, 236(1–2):192–208, 2000.

[Nee04] Peter Møller Neergaard. A functional language for logarithmic space. In *Prog. Lang. and Systems: 2nd Asian Symp. (APLAS 2004)*, volume 3302 of *LNCS*, pages 311–326. Springer-Verlag, 2004.

[Nig05] Karl-Heinz Niggl. Control structures in programs and computational complexity. *Annals of Pure and Applied Logic*, 133(1-3):247–273, 2005. Festschrift on the occasion of Helmut Schwichtenberg's 60th birthday.

[Oit04] Isabel Oitavem. Characterizing *NC* with tier 0 pointers. *Mathematical Logic Quarterly*, 50:9–17, 2004.

[Ruz81] W. L. Ruzzo. On uniform circuit complexity. *J. Comp. System Sci.*, 22:365–383, 1981.

[Sim88] H. Simmons. The realm of primitive recursion. *Archive for Mathematical Logic*, 27:177–188, 1988.

On Rational Trees

Arnaud Carayol and Christophe Morvan

IRISA, Campus de Beaulieu, 35042 Rennes, France
arnaud.carayol@irisa.fr, christophe.morvan@irisa.fr

Abstract. Rational graphs are a family of graphs defined using labelled rational transducers. Unlike automatic graphs (defined using synchronized transducers) the first order theory of these graphs is undecidable, there is even a rational graph with an undecidable first order theory. In this paper we consider the family of rational trees, that is rational graphs which are trees. We prove that first order theory is decidable for this family. We also present counter examples showing that this result cannot be significantly extended both in terms of logic and of structure.

1 Introduction

The algorithmic study of infinite object has achieved many success through the use of finite automata. This concise and efficient model was first introduced to characterize word languages in the late fifties, since then it has been extended and generalized in order to define infinite words, relations, relational structures, group structures, or graphs.

In 1960 Büchi, [Büc60], used finite automata to characterize infinite words, and so proving the decidability of monadic second order logic of the integers with the successor relation. Almost ten years later, this result was extended to the complete binary tree by Rabin [Rab69]. For many years adhoc extensions were proposed. Later on, around the year 1990 Muller and Schupp, then Courcelle and finally Caucal proposed generalizations of Rabin's result based on transformation of the complete binary tree [MS85, Cou90, Cau96].

Another way of using finite automata in the theory of finitely presented infinite objects was introduced by Hodgson [Hod83], simply using finite automata to define relational structures, obtaining the decidability of first order logic. Later on, nurturing from group theory [ECH$^+$92], Khoussainov and Nerode formalized and generalized the notion of automatic structure (and graph) [KN94]. Independently Sénizergue, and later on Pelecq considered a slightly different notion of automatic structure, involving an automatic quotient [Sén92, Pél97]. Several investigations, as well as an extension of first-order logic were conducted by Blumensath and Grädel on automatic structures [BG00]. In 2000 the notion of rational graphs was investigated [Mor00], this general family had already been defined as asynchronous automatic by Khoussainov and Nerode, but it was not very satisfactory from the logical point of view.

Z. Ésik (Ed.): CSL 2006, LNCS 4207, pp. 225–239, 2006.

In most of these cases the decidability of the logic comes from the underlying automaton, or more generally from closure properties. An interesting question is to know whether some structural restriction of these families would yield better decidability results. For automatic structures recently Khoussainov *et alii* considered automatic trees [KRS05], and have been able to disclose properties of these trees, like their Cantor-Bendixson rank, or the existence of a rational infinite path.

In this paper we consider rational trees, that is rational graphs that are trees. We first define carefully this family, state a few basic results, and give simple examples. We then use Gaifman's theorem and compositional methods [She75, Zei94] to prove that their first order logic is decidable. As it is not the case for general rational graphs, it heavily relies on the tree structure, and need a deep investigation. Finally we explore the boundaries of this result by exhibiting a rational directed acyclic graph with an undecidable first-order theory, and also a rational tree with an undecidable first-order theory enriched with rational accessibility.

2 Preliminaries

In this section we will recall the definition of the family of rational graphs. More details can be found in [Mor00, MS01]. We also state some properties of automatic graphs [KN94].

For any set E, its powerset is denoted by 2^E; if it is finite, its size is denoted by $|E|$. Let the set of nonnegative integers be denoted by $\mathbb{N}$, and $\{1, 2, 3, \ldots, n\}$ be denoted by $[n]$. A monoid M is a set equipped with an associative operation (denoted $\cdot$) and a (unique) neutral element (denoted ε). A monoid M is *free* if there exist a finite subset A of M such that $M = A^* := \bigcup_{n\in\mathbb{N}} A^n$ and for each $u \in M$ there exists a unique finite sequence of elements of A, $(u(i))_{i\in[n]}$, such that $u = u(1)u(2)\cdots u(n)$. Elements of a free monoid will be called words. Let u be a word in M, $|u|$ denotes the length of u and $u(i)$ denotes its ith letter.

2.1 Rational Graphs

The family of rational subsets of a monoid $(M, \cdot)$ is the least family containing the finite subsets of M and closed under union, concatenation and iteration.

A transducer is a finite automaton labelled by pairs of words over a finite alphabet X, see for example [AB88] [Ber79]. A transducer accepts a relation in $X^*\times X^*$; these relations are called rational relations as they are rational subsets of the product monoid $(X^*\times X^*, \cdot)$.

Now, let Γ and X be two finite alphabets. A *graph* G is a subset of $X^*\times\Gamma\times X^*$. An *arc* is a triple: $(u, a, v) \in X^* \times \Gamma \times X^*$ (denoted by $u \xrightarrow[G]{a} v$ or simply $u \xrightarrow{a} v$ if G is understood).

Rational graphs, denoted by $Rat(X^* \times \Gamma \times X^*)$, are extensions of rational relations, characterized by *labelled transducers*.

Definition 2.1. A *labelled transducer* $T = (Q, I, F, E, L)$ over X, is composed of a finite set of states Q, a set of initial states $I \subseteq Q$, a set of final states $F \subseteq Q$, a finite set of transitions (or edges) $E \subseteq Q \times X^* \times X^* \times Q$ and a mapping L from F into 2^Γ.

An arc $u \xrightarrow{a} v$ is *accepted* by a labelled transducer T if there is a path from a state in I to a state f in F labelled by (u, v) and such that $a \in L(f)$.

Definition 2.2. A graph in $2^{X^* \times \Gamma \times X^*}$ is *rational* if it is accepted by a labelled rational transducer.

Let G be a rational graph, for each a in Γ we denote by G_a the restriction of G to arcs labelled by a (it defines a rational relation between vertices); let u be a vertex in X^*, we denote by $G_a(u)$ the set of all vertices v such that $u \xrightarrow{a} v$ is an arc of G.

Example 2.3. The graph on the right is generated by the labelled transducer on the left.

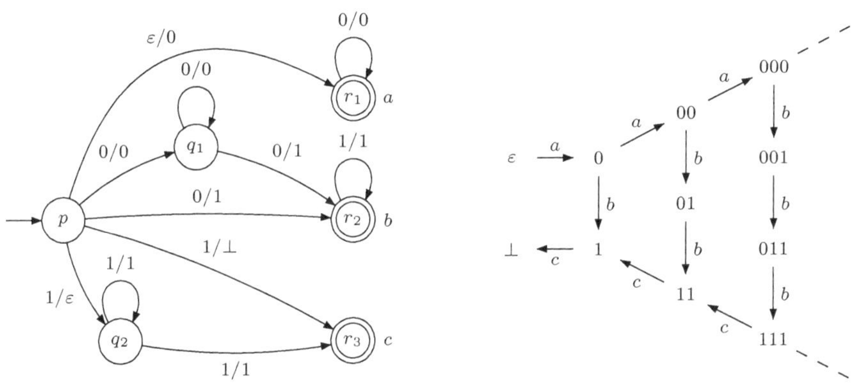

The path $p \xrightarrow{0/0} q_1 \xrightarrow{0/1} r_2 \xrightarrow{1/1} r_2$ accepts the couple $(001, 011)$, the final state r_2 is labelled by b thus there is a arc $001 \xrightarrow{b} 011$ in the graph.

The *trace* of a graph G from an initial vertex i to a final vertex f is the set of path labels labelling a path from i to f. For example the trace of the graph from Example 2.3 between ε and $\perp$ is the set $\{a^n b^n c^n \mid n > 0\}$.

Theorem 2.4 (Morvan, Stirling 01). *The traces of rational graphs from an initial to a final vertex is precisely the context-sensitive languages.*

2.2 Automatic Graphs

A classical subfamily of rational graphs is formed by the set of automatic graphs [KN94, Pél97, BG00].

These graphs are accepted by letter-to-letter transducers with rational terminal functions completing one side of the accepted pairs and assigning a label to the arc.

As the terminal function is rational, it can be introduced in the transducer adding states and transitions. A *left-synchronized transducer* is a transducer such that each path leading from an initial state to a final one can be divided into two parts: the first one contains arcs of the form $p \xrightarrow{A/B} q$ with $A, B \in X$ while the second part contains either arcs of the form $p \xrightarrow{A/\varepsilon} q$ with $A \in X$ or of the form $p \xrightarrow{\varepsilon/B} q$ with $B \in X$ (not both). Right-synchronized transducers are defined conversely.

Definition 2.5. A graph over $X^* \times \Gamma \times X^*$ is *automatic* if it is accepted by a left-synchronized or right-synchronized labelled transducer T.

Example 2.6. The graph defined by Example 2.3 is automatic. The relation G_b is synchronized. And the relations G_a and G_c are right-automatic.

The next result follows from the fact that automatic relations form a boolean algebra.

Proposition 2.7. *The first-order theory of automatic graphs is decidable.*

The Theorem 2.4 was extended to automatic graphs by Rispal in [Ris02].

Theorem 2.8 (Rispal 02). *The traces of rational graphs from an initial to a final vertex is precisely the context-sensitive languages.*

3 Rational Trees, Examples and Boundaries

Trees are natural structures in computer science. A lot of families of trees occurred outside of the study of infinite graphs. For example, regular trees that have only a finite number of sub-trees up to isomorphism, algebraic trees which are the unfolding of regular graphs [Cau02], or also trees that are solutions of higher order recursive program schemes [Dam77].

Definition 3.1. A *rational tree* is a rational graph satisfying these properties:

(i) it is connected;
(ii) every vertex is the target of at most one arc;
(iii) there is a single vertex with in-degree 1, called the *root*.

Each vertex of a rational tree is called a *node*. The *leaves* are vertices that are not source of any arc.

3.1 Elementary Results

The properties *(ii)* and *(iii)* from Definition 3.1 are easy to verify: *(ii)* consists in checking that the relation $\bigcup_{a\in\Gamma}(\xrightarrow{a})^{-1}$ is functional. This is solved using Shutzenberger's theorem, see among others [Ber79]. The condition *(iii)* consist in checking that the rational set $Dom(T) \setminus Im(T)$ has only one element.

In order to prove that it is undecidable to check whether a rational graph is a tree, we use a variation of the classical uniform halting problem for Turing machines.

Proposition 3.2. *Given any deterministic Turing machine M, a deterministic Turing machine M' may be constructed such that: M halts on ε if and only if M' halts from any configuration.*

Proposition 3.3. *Given any deterministic Turing machine a rational (unlabelled) graph $G(M)$ may be constructed in such a way that: M halts from any configuration if and only if $G(M)$ is a tree.*

Proof. Let us consider the deterministic Turing machine $M = (Q, T, \delta, q_0)$, Q is the set of states (with $q_0 \in Q$ the initial state), T the set of tape symbols (including two special symbols $\$$ and $\#$ denoting the extremities of the tape) and $\delta : Q \times T \to Q \times T \times \{l, r, p\}$ the transition function.

We define the configuration of such a machine in the usual way: uqv, with $q \in Q, u \in \$(T + \square)^*, v \in (T + \square)^*\#$, and $\square$ denoting the empty space.

We define $G(M)$ in this way: the *vertices* are precisely the configuration of the machine plus a special vertex $\$\#$.

The *arcs* consist of the transitions of the machine going backwards, and of the set $\{\$\#\} \times \{\$uqAv\# \mid (q, A) \notin Dom(\delta) \wedge u, v \in (T + \square)^*\}$.

The vertex $\$\#$ is the only vertex which is not the target of any arc (condition *(iii)*), and as the machine is deterministic and the arcs go backward, this graph satisfies also the condition *(ii)*. Furthermore this graph is connected if and only if the machine M reaches, from any configuration, a configuration in which there is no possible transition. □

From these two results considering a deterministic Turing machine we construct a second one that halts on every input if the first one stops from the empty word. Now using Proposition 3.3, we construct a rational graphs which is a tree if and only if the second machine halts one every input. This proves the following proposition.

Proposition 3.4. *It is undecidable to know whether a rational graph is a tree.*

We conclude this subsection by a simple result, which is a direct consequence of the rationality of the inverse image of a rational relation, and the fact that all vertices are accessible from the root.

Proposition 3.5. *Given any rational tree, accessibility and rational accessibility are decidable for any given pair of vertices.*

3.2 The 2^n-Tree

We give here a first example of rational tree. Indeed this tree is automatic. It is defined by a line of a's, and the nth vertex of this line is connected to a segment of 2^n b's.

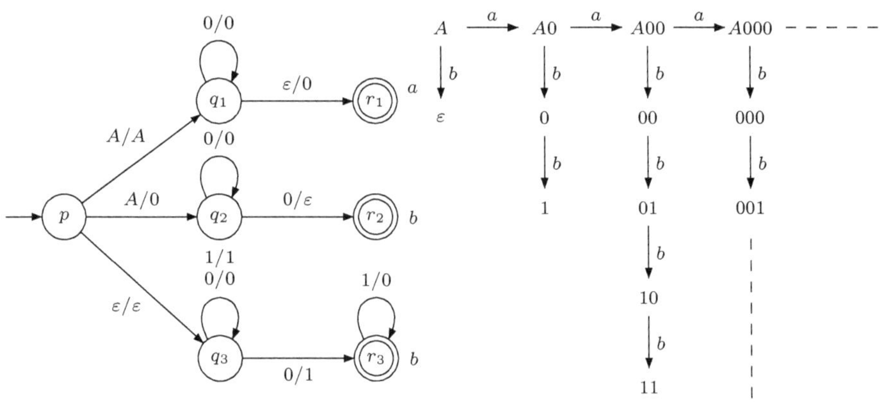

The encoding of the vertices of this tree relies on the fact that there are 2^n n-tuples over $\{0, 1\}$. The transducer performs the binary addition.

3.3 A Non-automatic Rational Tree

We now construct a rational tree of finite, yet unbounded, degree which is not automatic.

This tree is obtained from a rational forest by the adjunction of a line connecting the roots of each connected component. As these roots form a rational set of words, the following lemma allows construct such a line while still obtaining a rational tree.

Lemma 3.6. *Given a rational language L, the graph whose vertices are the words of L connected into a half line in length-lexicographic order is an automatic graph.*

This result is obtained by remarking that the length-lexicographic order (as a relation on words) is an automatic relation, and using closure properties of these relations.

Our example relies on the limit of the growth rate of automatic graphs of finite degree. For such an automatic tree, an obvious counting argument ensures that there exists p, q and s such that there are at most p^{qn+s} vertices at distance n of the root.

Therefore the tree (we call it *simplexp*) such that each vertex of depth n has 2^n sons, has precisely $2^{n(n-1)/2}$ vertices of depth n, and is therefore not automatic.

Still it is the connected component of a rational forest F and the tree constructed from F using the Lemma 3.6 has the same growth and therefore is not automatic, up to isomorphism. For simplicity we only present the forest F and a transducers generating it:

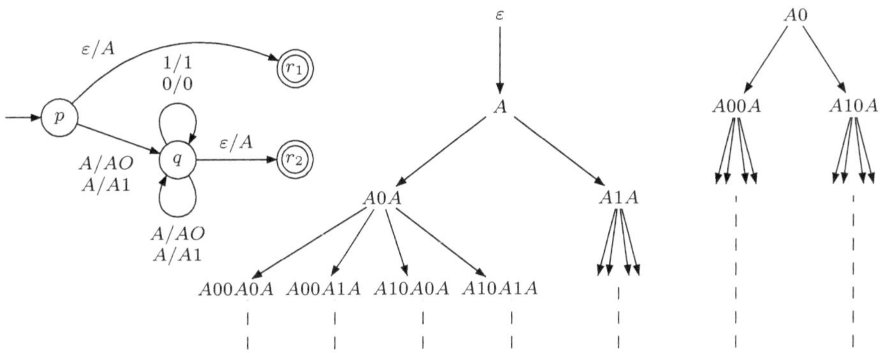

In this forest the connected component of ε is simplexp. Each vertex of depth n has precisely n occurrence of A and thus 2^n sons. Furthermore this transducer is co-functional, and strictly increasing, therefore each connected component is a tree with root. We have, thus, constructed a rational tree of finite degree, which is, up to isomorphism, not automatic.

4 First-Order Theory of Rational Trees Is Decidable

In this section we use Gaifman's theorem (see, e.g., [EF95]) to prove that the first-order theories of rational trees are decidable. This result, which is not true for rational graphs in general, was conjectured in [Mor01]. We will see, in Section 5, that there are no obvious extensions of this result.

4.1 Logical Preliminaries

We introduce basic notations on first-order logic over relational structures.

A *relational signature* Σ is a ranked alphabet. For every symbol $R \in \Sigma$, we write $|R| \geq 1$ the arity of R. A relational structure $\mathcal{M}$ over Σ is given by a tuple $(M, (R^{\mathcal{M}})_{R\in\Sigma})$ where M is the *universe* of $\mathcal{M}$ and where for all $R \in \Sigma$, $R^{\mathcal{M}} \subseteq M^{|R|}$.

Let $\mathcal{V}$ be a countable set of first-order variables. We use $x, y, z \ldots$ to range over first-order variables in $\mathcal{V}$ and $\bar{x}, \bar{y}, \bar{z}, \ldots$ to designate tuples of first-order variables. An atomic formula over Σ is either $R(x_1, \ldots, x_{|R|})$ for $R \in \Sigma$ and $x_1, \ldots, x_{|R|} \in \mathcal{V}$ or $x = y$ for $x, y \in \mathcal{V}$. Formulas over Σ (Σ-formulas) are obtained by closure under conjunction $\wedge$, negation $\neg$ and existential quantification $\exists$ starting from the atomic Σ-formulas. The bounded and free variables of a formula are defined as usual. A formula without free variables is also called a *sentence*. We write $\varphi(\bar{x})$ to indicate that the free variables of φ belong to $\bar{x}$.

For every relational structure $\mathcal{M}$, any formula $\varphi(x_1, \ldots, x_n)$ and $a_1, \ldots, a_n$ in M, we write $\mathcal{M} \models \varphi[a_1, \ldots, a_n]$ if $\mathcal{M}$ satisfies the formula φ when x_i is interpretated as a_i. If φ is a sentence, we simply write $\mathcal{M} \models \varphi$. Two sentences φ and ψ are logically equivalent if for all structure $\mathcal{M}$, $\mathcal{M} \models \varphi$ iff $\mathcal{M} \models \psi$.

The *quantifier rank* $\mathrm{qr}(\varphi)$ of a formula φ is defined by induction on the structure of φ by taking $\mathrm{qr}(\varphi) = 0$ for φ atomic, $\mathrm{qr}(\varphi \wedge \psi) = \max\{\mathrm{qr}(\varphi), \mathrm{qr}(\psi)\}$,

$\mathrm{qr}(\neg\varphi) = \mathrm{qr}(\varphi)$ and $\mathrm{qr}(\exists x\,\varphi) = \mathrm{qr}(\varphi) + 1$. For a fixed signature Σ, there are countably many Σ-sentences of a given quantifier rank. Up to logical equivalence there are only finitely many such sentences, but this equivalence is undecidable. A classical way to overcome this problem is to define a (decidable) syntactical equivalence on formulas such that, up to this equivalence, there are only finitely many formulas of a given quantifier rank (see e.g. [EF95]).

We define for all rank $k \geq 0$ a finite set Norm_k^Σ of normalized Σ-sentences such that for every Σ-sentence φ we can effectively compute a logically equivalent sentence $\mathrm{Norm}(\varphi)$ in Norm_k^Σ. Note that this set is finite and computable.

$$\begin{aligned}
\mathrm{ANA}^\Sigma(\bar{x}) &= \{\, \varphi, \neg\varphi \mid \varphi \text{ atomic over } \Sigma \text{ with free variables in } \bar{x} \,\} \\
\mathrm{Norm}_0^\Sigma(\bar{x}) &= \left\{ \bigvee_{R\in\mathcal{R}} \bigwedge_{\varphi\in R} \varphi \mid \mathcal{R} \subseteq 2^{\mathrm{ANA}^\Sigma(\bar{x})} \right\} \\
\mathrm{Norm}_{k+1}^\Sigma(\bar{x}) &= \left\{ \bigvee_{R\in\mathcal{R}} \bigwedge_{\varphi\in R} \varphi \mid \mathcal{R} \subseteq 2^{\{\, \exists y\varphi, \forall y\varphi \,\mid\, \varphi\in \mathrm{Norm}_k^\Sigma(\bar{x},y) \,\}} \right\}
\end{aligned}$$

where $y \notin \bar{x}$.

The k-theory of a structure $\mathcal{M}$ over Σ is the finite set

$$\mathrm{Thm}_k(\mathcal{M}) := \left\{\, \varphi \mid \varphi \in \mathrm{Norm}_k^\Sigma \text{ and } \mathcal{M} \models \varphi \,\right\}.$$

We write $\mathrm{Thm}_k^\Sigma = 2^{\mathrm{Norm}_k^\Sigma}$ the set of all possible k-theories[1].

4.2 Gaifman's Theorem for Graph Structures

We now focus our attention on graph structures and particularly on trees. A *graph structure* is a relational structure over a signature with symbols of arity 2. To every graph Σ-structure is associated a graph labelled by the symbols of Σ. We say that a graph structure is a *tree structure* if the associated graph is a tree. For all tree structure $\mathcal{T}$, we write $r(\mathcal{T}) \in T$ the root of $\mathcal{T}$. For all $u \in T$, we write $\mathcal{T}_{/u}$ the subtree of $\mathcal{T}$ rooted at u and for all $n \geq 0$, $\mathcal{T}_{/u}^n$ the tree $\mathcal{T}_{/u}$ restricted to the elements of depth at most n.

We recall Gaifman's Theorem, which states that every first-order formula is logically equivalent to a *local formula*.

In order to define local formulas, it is first necessary to define a notion of *distance*. In the following, we write $d(x,y) \leqslant n$ (resp. $d(x,y) < n$) the first-order formula expressing that the distance, without taking the orientation of the arcs into account, between x and y is less or equal to n (resp. less than n).

We denote by $S(r,x)$ the ball of radius r centered at x: $\{y \mid d(x,y) \leqslant r\}$

We now need to restrict a formula $\varphi(x)$ to a ball of radius r centered at x: we denote by $\varphi^{S(r,x)}$ the restriction of formula $\varphi(x)$ to the ball of center x and radius r. This notation is defined by renaming each bound occurrence of x in φ by a new variable, and localizing each quantification:

$$[\exists z\varphi]^{S(r,x)} := \exists z(d(x,z) \leqslant r \wedge \varphi^{S(r,x)})$$

[1] Remark that Thm_k contains elements that are not the k-theory of any structure. For instance, an element of Thm_k may contain both φ and $\neg\varphi$.

A basic local formula is of the the form:

$$\exists x_1 \ldots \exists x_n \bigwedge_{1 \leqslant i < j \leqslant n} (d(x_i, x_j) > 2r \wedge \psi^{S(r,x_i)}(x_i))$$

A local sentence is a boolean combination of basic local sentences.

Theorem 4.1 (Gaifman). *Every first-order sentence is logically equivalent to a local sentence.*

Note that the equivalence stated in this theorem is effective.

4.3 Compositional Results for Trees

We present basic compositional results for trees that will allow us to characterize the center of the balls involved in the definition of basic local formulas. The compositional method is a powerfull way to obtained decidability results mainly developed in [She75] (see [Zei94, Rab06] for a survey). The results presented here are not new and could, for example, be derived from the general templates presented in [Zei94, Rab06].

For every tree structure $\mathcal{T}$ over the signature $\Sigma = \{E_1, \ldots, E_\ell\}$ and for every $k \geq 1$, we define *the reduced tree of* $\mathcal{T}$, the structure $<\mathcal{T}>_k$ over the monadic signature $<\Sigma>_k := \{S_1, \ldots, S_\ell\} \cup \left\{ P_M \mid M \in \mathrm{Thm}_k^\Sigma \right\}$. The universe of $<\mathcal{T}>_k$ is the set of successors of the root of $\mathcal{T}$. The predicats in $<\Sigma>_k$ are interpreted as follows: for all $i \in [\ell]$, $u \in S_i^{<\Sigma>_k}$ iff $(r(\mathcal{T}), u) \in E_i^{\mathcal{T}}$ and for all $M \in \mathrm{Thm}_k^\Sigma$, $u \in P_M$ iff $\mathrm{Thm}(\mathcal{T}_{/u}) = M$.

Example 4.2. In the following picture we illustrate a reduced tree.

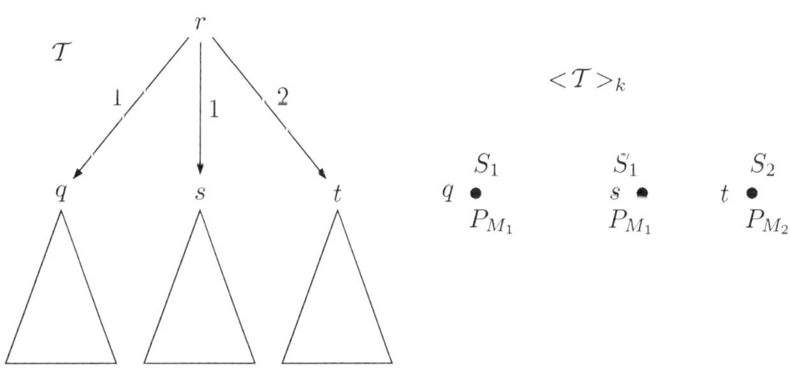

The tree depicted on the left is defined over $\Sigma = \{E_1, E_2\}$, the reduced tree $< \mathcal{T} >_k$ is defined over $< \Sigma >_k := \{S_1, S_2\} \cup \left\{ P_M \mid M \in \mathrm{Thm}_k^\Sigma \right\}$; and $\mathrm{Thm}_k(\mathcal{T}_{/q}) = \mathrm{Thm}_k(\mathcal{T}_{/s}) = M_1, \mathrm{Thm}_k(\mathcal{T}_{/t}) = M_2$

Lemma 4.3. *For all tree structure* $\mathcal{T}$ *over* $\Sigma = \{E_1, \ldots, E_\ell\}$ *and all* $k \geq 1$, $\mathrm{Thm}_k(\mathcal{T})$ *can be effectively computed from* $\mathrm{Thm}_k(<\mathcal{T}>_{k+1})$.

Remark 4.4. As the signature of $<\mathcal{T}>_k$ is monadic, every formula is equivalent to a boolean combinaison of formulas of the form

$$\exists x_1, \ldots, x_\ell \bigwedge_{i \neq j} x_i \neq x_j \ \wedge \bigwedge_{P \in R, i \in [\ell]} P(x_i) \ \wedge \bigwedge_{P \notin R, i \in [\ell]} \neg P(x_i).$$

where $P \subseteq <\Sigma>_k$. See, for example, the Exercise 2.3.12 of [EF95].

The following lemma allows to compute the theory of a ball in a tree from the theories of some subtrees contained in that ball.

Lemma 4.5. *For all tree $\mathcal{T}$ over a signature $\Sigma = \{E_1, \ldots, E_\ell\}$ and any vertex $u \in T$ with a path $u_0 a_1 u_1 \ldots u_m$ (with $u_m = u$), from the root of T and any rank $k \geq 1$ and any depth $n \geq 0$, there exists a constant p effectively computable from m, n and k such that for any formula $\varphi(x)$ with $\mathrm{qr}(\varphi) = k$, we can decide whether $T \models \varphi^{S(n,x)}[u]$ from the sequence of labels $a_1 \ldots a_m$ and from $(\mathrm{Thm}_p(<\mathcal{T}^n_{/u_i}>_p))_{i \in [0,m]}$.*

4.4 First-Order Theory of Rational Trees

We now tackle the proof of the decidabilty of the first-order theories of rational trees using Gaifman's Theorem.

The first step is to use the results from Subsection 4.3 to prove that for all $r \geq 1$ and for all formula $\varphi(x)$, the set of centers of a ball of radius r satisfying $\varphi(x)$ (where x is interpreted as the center of the ball) form a rational set of words.

We start by showing that the set of roots of a subtree of a certain depth having a given k-theory form a rational set of words. In order to apply Lemma 4.3, we need the following key lemma concerning rational trees.

Lemma 4.6. *For all rational tree T labelled by Γ and over X^*, all $i \in \Gamma$ and $L \in \mathrm{Rat}(X^*)$, the set of $u \in \mathrm{Dom}(T)$ having a least ℓ successors by i in L is rational and can be effectively constructed.*

Proof (Sketch). The proof relies on the fact that the in-degree of a tree is of at most one. We use the uniformazition of rational relations [Eil74, Ber79] which states that for every transducer[2] H there exists a functional transducer $\overrightarrow{H}$ such that $\overrightarrow{H} \subseteq H$ and $\mathrm{Dom}(H) = \mathrm{Dom}(\overrightarrow{H})$. As the in-degree of T is at most one, if we restrict H_i (the transducer accepting the i-labelled arcs of T restricted in image to L) to the rational set $X^* \setminus \mathrm{Im}(\overrightarrow{H_i})$ to obtain a transducer H'_i, we have decreased the out-degree of H_i by exactly 1. Hence the set of vertices having at least 2 successors by i is $\mathrm{Dom}(H'_i)$. The proof then follows by a straightforward induction. □

Remark 4.7. Note that this result does not hold when the in-degree is greater than 1. Consider for example, the transducer H depicted bellow. The set of

[2] We do not distinguish between the transducer and the relation it accepts.

words having exactly 1 image by H is the context-free language containing the words having the same number of a's and b's.

Lemma 4.8. *For all rational tree T labelled by $\Gamma = [\ell]$, all $k \geqslant 1$, $n \geqslant 1$, and all sentence φ over $\Sigma = \{E_1, \ldots, E_\ell\}$ or over $<\Sigma>_k$, the sets:*

- $L^{n,k}_\varphi := \{\, u \in \mathrm{Dom}(T) \mid\; <T^n_{/u}>_k \models \varphi \,\}$
- $L^{n}_\varphi := \{\, u \in \mathrm{Dom}(T) \mid T^n_{/u} \models \varphi \,\}$

are rational and effectively computable.

Proof (Sketch). We prove both properties simultaneously by induction on the depth n.

For the basis case $n = 0$, remark that for all rational tree T, T^0 is reduced to a single vertex and for all $k \geqslant 1$, $<T^0>_k$ is empty. As these structures are finite, we can decide for all formula φ if it is satisfied by the structure. Accordingly, L^0_φ and $L^{0,k}_\varphi$ are either the $\emptyset$ or $\mathrm{Dom}(T)$.

For the induction step $n+1$. Let $k \geqslant 1$ be a rank and φ be $<\Sigma>_k$-sentence. By Remark 4.4, we can restrict our attention to formulas stating there exists at least m elements belonging to $S_i^{<T>_k}$ and $P_M^{<T>_k}$ for some $i \in [\ell]$ and $M \in \mathrm{Thm}^\Sigma_k$.

Let $m \geqslant 0$, $i \in [\ell]$, $M \in \mathrm{Thm}^\Sigma_k$ and ψ the corresponding formula. By induction hypothesis, the set of vertices $X := \left\{\, u \in \mathrm{Dom}(T) \mid \mathrm{Thm}_k(T^n_{/u}) = M \,\right\}$ is rational and computable. It is easy to check that for all $u \in \mathrm{Dom}(T)$, $<T^{n+1}_{/u}>_k$ satifies ψ if and only if u has m successors by i belonging to X. By Lemma 4.6, the set $L^{n+1,k}_\psi$ is rational.

The second property follows then by Lemma 4.3. □

It then follows by Lemma 4.5 and 4.8 that:

Lemma 4.9. *For all rational tree T labelled by $\Gamma = [\ell]$, all formula $\varphi(x)$ over $\Sigma = \{E_1, \ldots, E_\ell\}$ and $n \geq 1$, the set $\{\, u \in \mathrm{Dom}(T) \mid T \models \varphi^{S(n,x)}[u] \,\}$ is rational and can be effectively computed.*

Before applying Gaifman's theorem we need a last property of rational trees.

Lemma 4.10. *For all rational tree T with vertices in X^*, $L \subseteq \mathrm{Dom}(T) \in \mathrm{Rat}(X^*)$ and for all $r \geq 1$, we can decide if there exists $u_1, \ldots, u_m \in L$ such that for all $i \neq j \in [m]$ $d(u_i, u_j) > r$.*

We can now use Gaifman's theorem to obtain the decidability of the first-order theory of rational trees.

Proposition 4.11. *Every rational tree has a decidable first-order theory.*

Proof. By Gaifman's theorem 4.1, it is enough to decide basic local sentences. Let T be a rational tree and $\varphi = \exists x_1 \dots \exists x_n \bigwedge_{1 \leq i < j \leq n} (d(x_i, x_j) > 2r \wedge \psi^{S(r,x_i)}(x_i))$ be a basic local sentence.

By Lemma 4.9, the set $L = \{ u \in \mathrm{Dom}(T) \mid \mathcal{T} \models \psi^{S(r,x)}[u] \}$ is rational.

To conclude, by Lemma 4.10, we can decide if there exists $u_1, \dots, u_n \in L$ such that for all $i \neq j \in [n]$ $d(u_i, u_j) > 2r$.

Combinning these two results , we can decide wether T satisfy φ. □

Due to the use of Gaifman's Theorem, the complexity of this decision procedure is non-elementary. However if we only consider rational trees of bounded out-degree, we can obtain an elementary decision procedure using the same technic as for the automatic graphs of bounded degree [Loh03].

5 Discussion on Extension of This Result

In this section, we illustrate that the result we have proved in previous section is in some sense maximal. We will first show that first-order theory together with rational accessibility is undecidable for rational trees. Then we will construct a rational directed acyclic graph with an undecidable first-order theory.

5.1 Finding a Wider Decidable Logic

An obvious extension of first-order logic is first-order logic with accessibility, which is simply first-order theory in the transitive closure of the original structure. A broader extension is first-order logic with rational accessibility. For every rational language $L \in \mathit{Rat}(\Gamma^*)$ we add, to the first-order logic, a binary predicate $\mathbf{reach}_L$ meaning that the first vertex is connected to the second by a path in L.

We now prove that, even though Proposition 3.5 states that accessibility and rational accessibility are decidable for rational trees, first-order logic with rational accessibility is undecidable.

We use the grid (a quarter plane), with backward arcs. It is a rational graph:

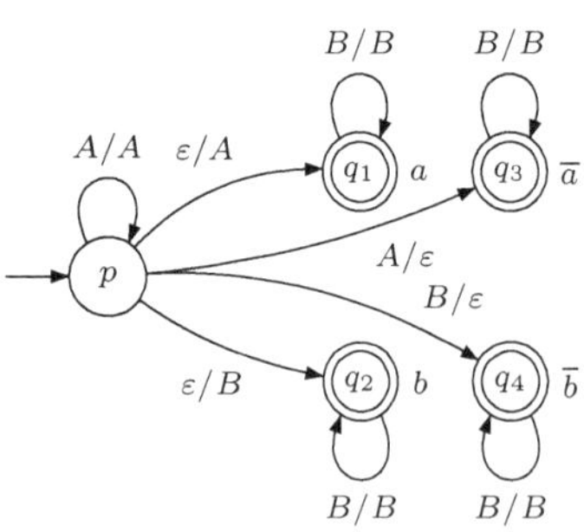

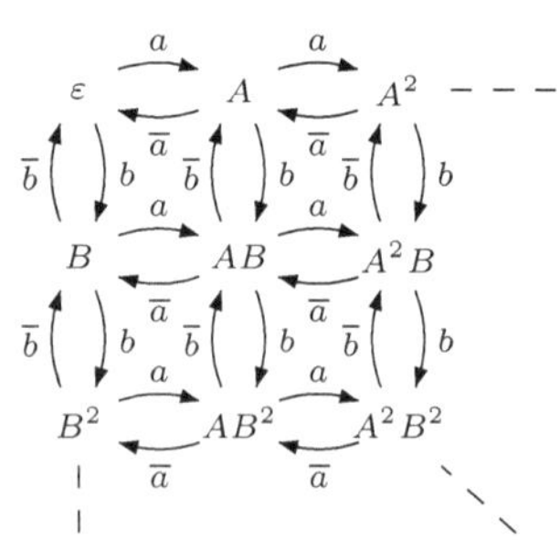

We simulate two counters machines on the unfolding of this graph. As these machines may test for zero, we add a loop on each vertex expressing that either counter, both or none is empty (denoted respectively by $\#_a, \#_b, \#_{ab}, \#$).

In order to unfold the resulting graph we transform the transducer to add the path leading to the vertex. Because the graph is both deterministic and co-deterministic, this yields a deterministic rational forest. This forest is composed of rooted connected components. The connected component with root ε is isomorphic to the unfolding of the grid with backward arcs (like each connected component with root in $\{a, b, \overline{a}, \overline{b}\}^*$). The transducer for arcs labelled a is the following:

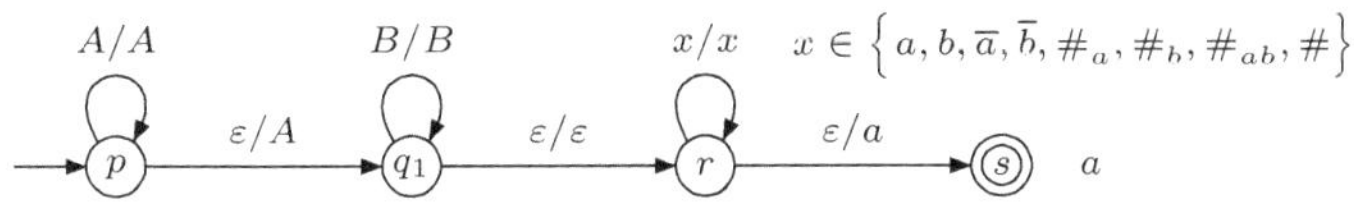

The transducers for $b, \overline{a}$ and $\overline{b}$ are similar. The transducers for $\#_a, \#_b, \#_{ab}, \#$ are the identity for the first part, and correspond to empty A, B, both or none.

Now we have a rational forest. We simply have to transform it into a rational tree. Again we use Lemma 3.6. Finally for each Minsky machine M we define a rational language L_M of its behaviour, and use this first-order formula to check whether it reaches empty counters (which is undecidable):

$$\exists u \exists v (\mathbf{reach}_{L_M}(u, v) \wedge \mathbf{root}(u) \wedge \neg(\exists w (v \xrightarrow{\overline{a}} w \vee v \xrightarrow{\overline{b}} w))).$$

We have, thus, found a rational tree with undecidable first-order theory with rational accessibility.

Remark 5.1. Indeed it is possible to improve this result in creating an ad hoc graph for encoding each machine. In this case it is first-order with accessibility which is undecidable for the whole family (and not just a single graph). Also it is possible to transform the tree in order to have an automatic tree.

5.2 Broaden the Graph Family

The first-order theory of rational graphs is undecidable. Indeed there are rational graphs with an undecidable first-order theory. Now we construct such a graph that is a directed acyclic graph (dag for short). This emphases the fact that the decidability of first-order theory of rational trees is deeply connected to the tree structure of these graphs.

Proposition 5.2. *There exists a rational directed acyclic graph with an undecidable first-order theory.*

Proof (Sketch). The construction of this dag (denoted G_{pcp}) relies on an encoding of *every* instance of the Post correspondence problem (pcp for short).

The precise construction of G_{pcp} is intricate. Thomas gives a similar construction in [Tho02], he construct a rational graph with undecidable first-order theory. It relies on the encoding of a universal Turing machine, and a simple formula detecting a loop depending on the instance of pcp, this example does not translate obviously for dag.

An instance of pcp is a sequence $((u_i, v_i))_{i\in[n]}$, and the problem is to determine whether there is a word w such that $w = u_{i_1}u_{i_2}\ldots u_{i_k} = v_{i_1}v_{i_2}\ldots v_{i_k}$, for some integer k, and a sequence $(i_\ell)_{\ell\in[k]}$ of elements of $[n]$.

The graph G_{pcp} is oriented so that no cycle can occur. There are three components in this graph. The first one is the initialisation that produce all possible sequence of indices. The second part, on one side substitutes k by u_k simultaneously everywhere it occurs, on the other side substitutes k by v_k. These two paths are done separately. The third and final part of the graph joins the u branches to the v branches.

Now for any instance of pcp we construct a first-order sentence whose satisfaction in G_{pcp} implies the existence of a solution of pcp for the corresponding instance. Indeed the formula ensures that the initialisation process is done, that the correct u_i's and v_i's are followed, and that both path meet. □

5.3 Conclusion

In this paper we have investigated some properties of rational trees. The main result is that these graphs have a decidable first-order theory. This result is interesting because it mostly relies on structural properties of this family.

It is well known that the first-order theory of automatic graphs is also decidable. It should be interesting to determine if there are larger families of rational graphs with decidable first-order theory. It would also be interesting to be able to isolate a family having first-order theory with accessibility decidable. It is neither the case for automatic graphs, rational trees, and even automatic trees (see Remark 5.1).

An unexplored aspect of this study is to consider the traces of these graphs. The traces of automatic and rational graphs are context sensitive languages [MS01, Ris02]. Our conjecture is that there are even context-free languages that can not be obtained by rational trees, for instance the languages of words having the same number of a and b.

References

[AB88] J.-M. Autebert and L. Boasson, *Transductions rationelles*, MASSON, 1988.

[Ber79] J. Berstel, *Transductions and context-free languages*, Teubner, 1979.

[BG00] A. Blumensath and E. Grädel, *Automatic Structures*, Proceedings of 15th IEEE Symposium on Logic in Computer Science LICS 2000, 2000, pp. 51–62.

[Büc60] J. R. Büchi, *On a decision method in restricted second order arithmetic*, ICLMPS, Stanford University press, 1960, pp. 1–11.

[Cau96] D. Caucal, *On transition graphs having a decidable monadic theory*, Icalp 96, LNCS, vol. 1099, 1996, pp. 194–205.

[Cau02] ———, *On infinite terms having a decidable monadic theory*, MFCS 02, LNCS, vol. 2420, 2002, pp. 165–176.

[Cou90] B. Courcelle, *Handbook of theoretical computer science*, ch. Graph rewriting: an algebraic and logic approach, Elsevier, 1990.

[Dam77] W. Damm, *Languages defined by higher type program schemes*, ICALP 77 (Arto Salomaa and Magnus Steinby, eds.), LNCS, vol. 52, 1977, pp. 164–179.

[ECH+92] D. Epstein, J.W. Cannon, D.F. Holt, S.V.F. Levy, M.S. Paterson, and Thurston, *Word processing in groups*, Jones and Barlett publishers, 1992.

[EF95] H. D. Ebbinghaus and J. Flum, *Finite model theory*, Springer-Verlag, 1995.

[Eil74] S. Eilenberg, *Automata, languages and machines*, vol. A, Academic Press, 1974.

[Hod83] B. R. Hodgson, *Décidabilité par automate fini*, Ann. Sci. Math. Québec **7** (1983), 39–57.

[KN94] B. Khoussainov and A. Nerode, *Automatic presentations of structures*, LCC (D. Leivant, ed.), LNCS, vol. 960, 1994, pp. 367–392.

[KRS05] B. Khoussainov, S. Rubin, and F. Stephan, *Automatic linear orders and trees*, ACM Trans. Comput. Logic **6** (2005), no. 4, 675–700.

[Loh03] M. Lohrey, *Automatic structures of bounded degree*, Proceedings of LPAR 03, LNAI, vol. 2850, 2003, pp. 344–358.

[Mor00] C. Morvan, *On rational graphs*, Fossacs 00 (J. Tiuryn, ed.), LNCS, vol. 1784, 2000, ETAPS 2000 best theoretical paper Award, pp. 252–266.

[Mor01] ______, *Les graphes rationnels*, Thèse de doctorat, Université de Rennes 1, 2001.

[MS85] D. Muller and P. Schupp, *The theory of ends, pushdown automata, and second-order logic*, Theoretical Computer Science **37** (1985), 51–75.

[MS01] C. Morvan and C. Stirling, *Rational graphs trace context-sensitive languages*, MFCS 01 (A. Pultr and J. Sgall, eds.), LNCS, vol. 2136, 2001, pp. 548–559.

[Pél97] L. Pélecq, *Isomorphismes et automorphismes des graphes context-free, équationnels et automatiques*, Ph.D. thesis, Université de Bordeau I, 1997.

[Rab69] M.O. Rabin, *Decidability of second-order theories and automata on infinite trees*, Trans. Amer. Math. soc. **141** (1969), 1–35.

[Rab06] A. Rabinovich, *Composition theorem for generalized sum*, Personal communication, 2006.

[Ris02] C. Rispal, *Synchronized graphs trace the context-sensitive languages*, Infinity 02 (R. Mayr A. Kucera, ed.), vol. 68, ENTCS, no. 6, 2002.

[She75] S. Shelah, *The monadic theory of order*, Ann. Math. **102** (1975), 379–419.

[Sén92] G. Sénizergues, *Definability in weak monadic second-order logic of some infinite graphs*, Dagstuhl seminar on Automata theory: Infinite computations, Warden, Germany, vol. 28, 1992, p. 16.

[Tho02] W. Thomas, *A short introduction to infinite automata*, DLT 01 (W. Kuich, G. Rozenberg, and A. Salomaa, eds.), LNCS, vol. 2295, 2002, pp. 130–144.

[Zei94] R. S. Zeitman, *The composition method*, Phd thesis, Wayne State University, Michigan, 1994.

Reasoning About States of Probabilistic Sequential Programs*

R. Chadha, P. Mateus, and A. Sernadas

SQIG – IT and IST, Portugal
{rch, pmat, acs}@math.ist.utl.pt

Abstract. A complete and decidable propositional logic for reasoning about states of probabilistic sequential programs is presented. The state logic is then used to obtain a sound Hoare-style calculus for basic probabilistic sequential programs. The Hoare calculus presented herein is the first probabilistic Hoare calculus with a complete and decidable state logic that has truth-functional propositional (not arithmetical) connectives. The models of the state logic are obtained exogenously by attaching sub-probability measures to valuations over memory cells. In order to achieve complete and recursive axiomatization of the state logic, the probabilities are taken in arbitrary real closed fields.

1 Introduction

Reasoning about probabilistic systems is very important due to applications of probability in distributed systems, security, reliability, and randomized and quantum algorithms. Logics supporting such reasoning have branched in two main directions. Firstly, Hoare-style [27,21,6] and dynamic logics [9,17] have been developed building upon denotational semantics of probabilistic programs [16]. The second approach enriches temporal modalities with probabilistic bounds [10, 13, 23].

Our work is in the area of Hoare-style reasoning about probabilistic sequential programs. A Hoare assertion [11] is a triple of the form $\{\xi_1\}\, s\, \{\xi_2\}$ meaning that if program s starts in state satisfying the state assertion formula ξ_1 and s halts then s ends in a state satisfying the state transition formula ξ_2. The formula ξ_1 is known as the pre-condition and the formula ξ_2 is known as the post-condition. For probabilistic programs the development of Hoare logic has taken primarily two different paths. The common denominator of the two approaches is forward denotational semantics of sequential probabilistic programs [16]: program states are (sub)-probability measures over valuations of memory cells and denotations of programs are (sub)-probability transformations.

The first sound Hoare logic for probabilistic programs was given in [27]. The state assertion language is *truth-functional*, *i.e.*, the formulas of the logic are interpreted as either true and false and the truth value of a formulas is determined

* Supported by FCT and FEDER through POCI via CLC QuantLog POCI/MAT/ 55796/2004 Project. Additional support for Rohit Chadha came from FCT and FEDER grant SFRH/BPD/26137/2005.

Z. Ésik (Ed.): CSL 2006, LNCS 4207, pp. 240–255, 2006.

by the truth values of the sub-formulas. The state assertion language in [27] consists of two levels: one classical state formulas γ interpreted over the valuations of memory cells and the second probabilistic state formulas ξ which interpreted over (sub)-probability measures of the valuations. The state assertion language contain terms $(\int \gamma)$ representing probability of γ being true. The language at the probabilistic level is extremely restrictive and is built from term equality using conjunction. Furthermore, the Hoare rule for the alternative if-then-else is incomplete and even simple valid assertions may not be provable.

The reason for incompleteness of the Hoare rule for the alternative composition in [27] as observed in [27,17] is that the Hoare rule tries to combine absolute information of the two alternates truth-functionally to get absolute information of the alternative composition. This fails because the effects of the two alternatives are not independent. In order to avoid this problem, a probabilistic dynamic logic is given in [17] with an *arithmetical* state assertion logic: the state formulas are interpreted as measurable functions and the connectives are arithmetical operations such as addition and subtraction.

Inspired by the dynamic logic in [17], there are several important works in the probabilistic Hoare logic, *e.g.* [14,21], in which the state formulas are either measurable functions or arithmetical formulas interpreted as measurable functions. Intuitively, the Hoare triple $\{f\}\, s\, \{g\}$ means that the expected value of the function g after the execution of s is at least as much as the expected value of the function f before the execution. Although research in probabilistic Hoare logic with arithmetical state logics has yielded several interesting results, the Hoare triples themselves do not seem very intuitive. A high degree of sophistication is required to write down the Hoare assertions needed to verify relatively simple programs. For this reason, it is worthwhile to investigate Hoare logics with truth-functional state logics.

A sound Hoare logic with a truth-functional state logic was presented in [6] and completeness for a fragment of the Hoare-logic is shown for iteration-free programs. In order to deal with alternative composition, a probabilistic sum construct $(\xi_1+\xi_2)$ is introduced in [6]. Intuitively, the formula $(\xi_1+\xi_2)$ is satisfied by a (sub)-probability measure μ if μ can be be written as the sum of two measures μ_1 and μ_2 which satisfy ξ_1 and ξ_2 respectively. The drawback of [6] is that no axiomatization is given for the state assertion logic. The essential obstacle in achieving a complete axiomatization for the state language in [6] is the probabilistic sum construct.

This paper addresses the gap between [27] and [6] and provides a sound Hoare logic for iteration-free probabilistic programs with a truth-functional state assertion logic. Our main contribution is that the Hoare logic herein is the first sound probabilistic Hoare logic with a truth-functional state assertion logic that enjoys a complete and decidable axiomatization.

We tackle the Hoare rule for the alternative composition in two steps. The first step is that our alternative choice construct is a slight modification of the usual if-then-else construct: we mark a boolean memory variable bm with the choice taken at the end of the execution of the conditional branch. Please note that this does not pose any restriction over the expressiveness of the programming language. This modification gives us a handle on the Hoare rule for the alternative construct as all the choices are marked by the appropriate memory variable and thus become independent. Please note that a fixed dedicated boolean register could have been used to mark the choices. However, we decided to use a boolean variable in the syntax because the Hoare rule for the alternative composition refers to the marker.

The second step is that in our state assertion language, we have a *conditional construct* (ξ/γ). Intuitively, the formula (ξ/γ) is satisfied by a (sub)-probability measure μ if ξ is true of the (sub)-probability measure obtained by eliminating the measure of all valuations where γ is false. The conditional formulas (ξ/bm) and $(\xi/(\neg\,\mathsf{bm}))$ in the state logic can then be used to combine information of the alternate choices.

The state assertion logic, henceforth referred to as Exogenous Probabilistic Propositional Logic (EPPL), is designed by taking the exogenous semantics approach to enriching a given logic–the models of the enriched logic are sets of models of the given logic with additional structure. A semantic model of EPPL is a set of possible valuations over memory cells which may result from execution of a probabilistic program along with a discrete (sub)-probability space which gives the probability of each possible valuation.

Unlike most works on probabilistic reasoning about programs, we do not confuse possibility with probability: possible valuations may occur with zero probability. This is not a restriction and we can confuse the two, if desired, by adding an axiom to the proof system. On the other hand, this separation yields more expressivity. The exogenous approach to probabilistic logics first appeared in [24,25] and later in [7,1,20]. EPPL is an enrichment of the probabilistic logic proposed in [20]: the conditional construct (ξ/γ) is not present in [20].

For the sake of convenience, we work with finitely additive, discrete and bounded measures and not just (sub)-probability measures. In order to achieve recursive axiomatization for EPPL, we also assume that the measures take values from an arbitrary *real closed field* instead of the set of real numbers. The first order theory of such fields is decidable [12, 3], and this technique of achieving decidability was inspired by other work in probabilistic reasoning [7, 1].

The programming language is a basic imperative language with assignment to memory variables, sequential composition, probabilistic assignment $(\mathsf{toss}(\mathsf{bm}, r))$ and the marked alternative choice. The statement $\mathsf{toss}(\mathsf{bm}, r)$ assigns bm to true with probability r. The term r is a constant and does not depend on the state of the program. This is not a serious restriction. For instance r is taken to be $\frac{1}{2}$ in probabilistic Turing machines.

One of the novelties of our Hoare logic is the rule for $\mathsf{toss}(\mathsf{bm}, r)$ which gives the weakest pre-condition and is not present in other probabilistic Hoare logics with truth-functional state logics. The corresponding rule in the arithmetical setting is discussed in Section 6. We envisage achieving a complete Hoare logic but this is out of the scope of this paper.

The rest of the paper is organized as follows. The syntax, semantics and the complete recursive axiomatization of EPPL is presented in Section 2. The programming language is introduced in Section 3 and the sound Hoare logic is given in Section 4. We illustrate the Hoare calculus with an example in Section 5. Related work is discussed in detail in Section 6. We summarize the results and future work in Section 7. For lack of space reasons, the proofs in the paper are omitted and are available at `http://wslc.math.ist.utl.pt/ftp/pub/SernadasA/06-CMS-quantlog08s.pdf`. *Acknowledgements.* We would like to thank L. Cruz-Filipe and P. Selinger for useful and interesting discussions. We will also like to thank the anonymous referees whose useful comments have greatly benefited the presentation.

2 Logic of Probabilistic States – EPPL

We assume that in our programming language, there are a finite number of memory cells of two kinds: registers containing real values (with a finite range D fixed once and for all) and registers containing boolean values. In addition to reflecting the usual implementation of real numbers as floating-point numbers, the restriction that real registers take values from a finite range D is also needed for completeness results. Please note that instead of reals, we could have also used any type with finite range.

Any run of a program thus probabilistically assigns values to these registers and such an assignment is henceforth called a *valuation*. If we denote the set of valuations by $\mathcal{V}$ then intuitively a semantic structure of EPPL consists of $V \subseteq \mathcal{V}$, a set of *possible* valuations, along with a finitely additive, discrete and bounded measure μ on $\wp\mathcal{V}$, the power-set of $\mathcal{V}$. A finitely additive, discrete and bounded measure μ on $\wp\mathcal{V}$ is a map from $\wp\mathcal{V}$ to $\mathbb{R}^+$ (the set of non-negative real numbers) such that: $\mu(\emptyset) = 0$; and $\mu(U_1 \cup U_2) = \mu(U_1) + \mu(U_2)$ if $U_1 \cap U_2 = \emptyset$. Loosely speaking, $\mu(U)$ denotes the probability of a possible valuation being in the set U. A measure μ is said to be a probability measure if $\mu(\mathcal{V}) = 1$. We work with general measures instead of just probability measures as it is convenient to do so. We will assume that impossible valuations are improbable, *i.e.*, we require $\mu(U) = 0$ for any $U \subset (\mathcal{V} \setminus V)$. Please note that $\mu(U)$ may be 0 for $U \subset \mathcal{V}$.

Furthermore, in order to obtain decidability, we shall assume that the measures take values from an arbitrary *real closed field* instead of the set of real numbers. An ordered field $\mathcal{K} = (K, +, ., 1, 0, \leq)$ is said to be a real closed field if the following hold: every non-negative element of the K has a square root in K; any polynomial of odd degree with coefficients in K has at least one solution.

Examples of real closed fields include the set of real numbers with the usual multiplication, addition and order relation. The set of computable real numbers with the same operations is another example. A measure that takes values from a real closed field $\mathcal{K}$ will henceforth be called a $\mathcal{K}$-measure.

Any real closed field has a copy of integers and rationals. We can also take square roots and n-th roots for odd n in a real closed field. As a consequence, we shall assume that there is a fixed set $\mathcal{R}$ of "real constants" for our purposes.

A semantic structure of EPPL thus consists of a set of possible valuations, a real closed field $\mathcal{K}$ and a $\mathcal{K}$-measure on $\wp\mathcal{V}$. We will call these semantic structures *generalized probabilistic structures*. We start by describing the syntax of the logic.

2.1 Language

The language consists of formulas at two levels. The formulas at first level, *classical state formulas* reason about individual valuations over the memory cells. The formulas at second level, *probabilistic state formulas*, reason about generalized probabilistic structures. There are two kinds of terms in the language: *real terms* used in classical state formulas to denote elements from the set D, and *probability terms* used in probabilistic state formulas to denote elements in an arbitrary real closed field. The syntax of the language is given in Table 1 using the BNF notation and discussed below.

Table 1. Language of EPPL

Real terms (with the proviso $c \in D$)
$t := \mathsf{xm} \;[\!]\; x \;[\!]\; c \;[\!]\; (t+t) \;[\!]\; (t\,t)$
Classical state formulas
$\gamma := \mathsf{bm} \;[\!]\; b \;[\!]\; (t \leq t) \;[\!]\; \mathsf{ff} \;[\!]\; (\gamma \Rightarrow \gamma)$

Probability terms (with the proviso $r \in \mathcal{R}$)
$p := r \;[\!]\; y \;[\!]\; (\textstyle\int \gamma) \;[\!]\; (p+p) \;[\!]\; (p\,p) \;[\!]\; \tilde{r}$
Probabilistic state formulae:
$\xi := (\Box\gamma) \;[\!]\; (p \leq p) \;[\!]\; (\xi/\gamma) \;[\!]\; \mathsf{fff} \;[\!]\; (\xi \supset \xi)$

Given fixed $\mathbf{m} = \{0, \ldots, m-1\}$, there are two finite disjoint sets of memory variables: $\mathsf{xM} = \{\mathsf{xm}_k : k \in \mathbf{m}\}$ – representing the contents of real registers, and $\mathsf{bM} = \{\mathsf{bm}_k : k \in \mathbf{m}\}$ – representing the contents of boolean registers. We also have two disjoint sets of rigid variables which are useful in reasoning about programs: $\mathsf{B} = \{b_k : k \in \mathbb{N}\}$ – ranging over the truth values $2 = \{\mathsf{ff}, \mathsf{tt}\}$, and $\mathsf{X} = \{x_k : k \in \mathbb{N}\}$ – ranging over elements of D.

The real terms, ranged over by $t, t_1, \ldots$, are built from the sets D, xM and X using the usual addition and multiplication[1]. The classical state formulas, ranged over by $\gamma, \gamma_1, \ldots$, are built from bM, B and comparison formulas $(p_1 \leq p_2)$ using the classical disjunctive connectives ff and $\Rightarrow$. As usual, other classical connectives $(\neg, \vee, \wedge, \Leftrightarrow)$ are introduced as abbreviations. For instance, $(\neg\,\gamma)$ stands for $(\gamma \Rightarrow \mathsf{ff})$.

The probability terms, ranged over by $p, p_1, \ldots$, denote elements of the real closed field in a semantic structure. We also assume a set of variables, $\mathsf{Y} = \{y_k :$

[1] The arithmetical operations addition and multiplication are assumed to be defined so as to restrict them to the range D.

$k \in \mathbb{N}\}$, ranging over elements of the real closed field. The term $(\int \gamma)$ denotes the measure of the set of valuations that satisfy γ. The denotation of the term $\widetilde{r}$ is r if $0 \leq r \leq 1$, 0 if $r \leq 0$ and 1 otherwise.

The probabilistic state formulas, ranged over by $\xi, \xi_1, \ldots$, are built from the *necessity formulas* $(\Box\gamma)$, the comparison formulas $(p_1 \leq p_2)$, and *conditional formulas* (ξ/γ) using the connectives fff and $\supset$. The formula $(\Box\gamma)$ is true when γ is true of every possible valuation in the semantic structure. Intuitively, the conditional (ξ/γ) is true in a generalized probabilistic structure if it is true in the structure obtained by restricting the possible states to the set where γ is true and eliminating the measure of valuations which satisfy $(\neg\gamma)$. Other probabilistic connectives $(\ominus, \cup, \cap, \approx)$ are introduced as abbreviations. For instance, $(\ominus\,\xi)$ stands for $(\xi \supset \mathsf{fff})$. We shall also use $(\Diamond\gamma)$ as an abbreviation for $(\ominus(\Box(\neg\gamma)))$. Please note that the $\Box$ and $\Diamond$ are not modalities[2].

The notion of occurrence of a term p and a probabilistic state formula ξ_1 in the probabilistic state formula ξ can be easily defined. The notion of replacing zero or more occurrences of probability terms and probabilistic formulas can also be suitably defined. For the sake of clarity, we shall often drop parenthesis in formulas and terms if it does not lead to ambiguity.

2.2 Semantics

Formally, by a *valuation* we mean a map that provides values to the memory variables and rigid variables– $v : (\mathsf{xM} \to D, \mathsf{bM} \to 2, \mathsf{X} \to D, \mathsf{B} \to 2)$. The set of all possible valuations is denoted by $\mathcal{V}$. Given a valuation v, the denotation of real terms $[\![t]\!]_v$ and satisfaction of classical state formulas $v \Vdash_{\mathsf{c}} \gamma$ are defined inductively as expected. Given $V \subseteq \mathcal{V}$, the *extent* of γ in V is defined as $|\gamma|_V = \{v \in V : v \Vdash_{\mathsf{c}} \gamma\}$.

A *generalized probabilistic state* is a triple $(V, \mathcal{K}, \mu)$ where V is a (possibly empty) subset of $\mathcal{V}$, $\mathcal{K}$ a real closed field and μ is a finitely additive, discrete and finite $\mathcal{K}$-measure over $\wp\mathcal{V}$ such that $\mu(U) = 0$ for every $U \subseteq (\mathcal{V} \setminus V)$. We denote the set of all generalized states by $\mathcal{G}$.

Given a classical formula γ we also need the following sub-measure of μ: $\mu_\gamma = \lambda U.\ \mu(|\gamma|_U)$. That is, μ_γ is null outside of the extent of γ and coincides with μ inside it.

For interpreting the probabilistic variables, we need the concept of an assignment. Given a real closed field $\mathcal{K}$, a $\mathcal{K}$-*assignment* ρ is a map from Y to $\mathcal{K}$.

Given a generalized state $(V, \mathcal{K}, \mu)$ and a $\mathcal{K}$-assignment ρ, the denotation of probabilistic terms and satisfaction of probabilistic state formulas are defined inductively in Table 2. Please note that the semantics ensures that if V is empty, then $(V, \mathcal{K}, \mu)\rho \Vdash \gamma$ for any γ. The formula $(\Box\gamma)$ is satisfied only if all $v \in V$ satisfy γ. For non-empty V, the formula $(p_1 \leq p_2)$ is satisfied if the term denoted by p_1 is less than p_2. The formula (ξ/γ) is satisfied by $(V, \mathcal{K}, \mu)$ and ρ if $(|\gamma|_V, \mathcal{K}, \mu_\gamma)$ and ρ satisfy ξ. The formula $(\xi_1 \supset \xi_2)$ is satisfied by a semantic model if either ξ_1 is not satisfied by the model or ξ_2 is satisfied by the model.

[2] We do not have formulas such as $\Box(\Box\gamma)$.

Table 2. Semantics of EPPL

Denotation of probability terms

$$
\begin{array}{ll}
[\![r]\!]^{\rho}_{(V,\mathcal{K},\mu)} & = r \\
[\![y]\!]^{\rho}_{(V,\mathcal{K},\mu)} & = \rho(y) \\
[\![(\textstyle\int\gamma)]\!]^{\rho}_{(V,\mathcal{K},\mu)} & = \mu(|\gamma|_V) \\
[\![p_1 + p_2]\!]^{\rho}_{(V,\mathcal{K},\mu)} & = [\![p_1]\!]^{\rho}_{(V,\mathcal{K},\mu)} + [\![p_2]\!]^{\rho}_{(V,\mathcal{K},\mu)} \\
[\![p_1 p_2]\!]^{\rho}_{(V,\mathcal{K},\mu)} & = [\![p_1]\!]^{\rho}_{(V,\mathcal{K},\mu)} \times [\![p_2]\!]^{\rho}_{(V,\mathcal{K},\mu)} \\
[\![\widetilde{r}]\!]^{\rho}_{(V,\mathcal{K},\mu)} & = \max(0, \min(r, 1))
\end{array}
$$

Satisfaction of probabilistic formulas

$$
\begin{array}{ll}
(V,\mathcal{K},\mu)\rho \Vdash (\Box\gamma) & \text{iff } v \Vdash_{\mathsf{c}} \gamma \text{ for every } v \in V \\
(V,\mathcal{K},\mu)\rho \Vdash (p_1 \leq p_2) & \text{iff } V \neq \emptyset \text{ implies } ([\![p_1]\!]^{\rho}_{(V,\mathcal{K},\mu)} \leq [\![p_2]\!]^{\rho}_{(V,\mathcal{K},\mu)}) \\
(V,\mathcal{K},\mu)\rho \Vdash (\xi/\gamma) & \text{iff } (|\gamma|_V, \mathcal{K}, \mu_\gamma)\rho \Vdash \xi \\
(V,\mathcal{K},\mu)\rho \Vdash \mathsf{fff} & \text{iff } V = \emptyset \\
(V,\mathcal{K},\mu)\rho \Vdash (\xi_1 \supset \xi_2) & \text{iff } (V,\mathcal{K},\mu)\rho \Vdash \xi_2 \text{ or } (V,\mathcal{K},\mu)\rho \nVdash \xi_1
\end{array}
$$

Entailment is defined as usual: Ξ entails ξ (written $\Xi \vDash \xi$) if $(V,\mathcal{K},\mu)\rho \Vdash \xi$ whenever $(V,\mathcal{K},\mu)\rho \Vdash \xi_0$ for each $\xi_0 \in \Xi$.

Please note that the $\mathcal{K}$-assignment ρ is sufficient to interpret a useful sub-language of probabilistic state formulas:

$$
\begin{array}{l}
\kappa := (a \leq a) \mathrel{[\!]} \mathsf{fff} \mathrel{[\!]} (\kappa \supset \kappa) \\
a := x \mathrel{[\!]} r \mathrel{[\!]} (a + a) \mathrel{[\!]} (aa) \mathrel{[\!]} \widetilde{r}.
\end{array}
$$

Henceforth, the terms of this sub-language will be called *analytical terms* and the formulas will be called *analytical formulas.*

2.3 The Axiomatization

We need three new concepts for the axiomatization, one of valid state formula, a second one of probabilistic tautology and the third of valid analytical formulas.

A classical state formula γ is said to be valid if it is true of all valuations $v \in \mathcal{V}$. As a consequence of the finiteness of D, the set of valid classical state formulas is recursive.

Consider propositional formulas built from a countable set of propositional symbols Q using the classical connectives $\perp$ and $\rightarrow$. A probabilistic formula ξ is said to be a *probabilistic tautology* if there is a propositional tautology β over Q and a map σ from Q to the set of probabilistic state formulas such that ξ coincides with $\beta_p\sigma$ where $\beta_p\sigma$ is the probabilistic formula obtained from β by replacing all occurrences of $\perp$ by fff, $\rightarrow$ by $\supset$ and $q \in \mathrm{Q}$ by $\sigma(q)$. For instance, the probabilistic formula $((y_1 \leq y_2) \supset (y_1 \leq y_2))$ is tautological (obtained, for example, from the propositional tautology $q \rightarrow q$).

As noted in Section 2.2, if $\mathcal{K}_0$ is the real closed field in a generalized probabilistic structure, then a $\mathcal{K}_0$-assignment is enough to interpret all analytical formulas. We say that κ is a *valid analytical formula* if for any real closed field $\mathcal{K}$ and any $\mathcal{K}$-assignment ρ, κ is true for ρ. Clearly, a valid analytical formula holds for all semantic structures of EPPL. It is a well-known fact from the theory of quantifier elimination [12, 3] that the set of valid analytical formulas so defined is decidable. We shall not go into details of this result as we want to focus on reasoning about probabilistic aspects only.

The axioms and inference rules of EPPL are listed in Table 3 and better understood in the following groups.

Table 3. Axioms for EPPL

Axioms

[**CTaut**] $\vdash (\Box\gamma)$ for each valid state formula γ

[**PTaut**] $\vdash \xi$ for each probabilistic tautology ξ

[**Lift⇒**] $\vdash ((\Box(\gamma_1 \Rightarrow \gamma_2)) \supset (\Box\gamma_1 \supset \Box\gamma_2))$

[**Eqvff**] $\vdash ((\Box\mathsf{ff}) \approx \mathsf{fff})$

[**Ref∧**] $\vdash (((\Box\gamma_1) \cap (\Box\gamma_2)) \supset (\Box(\gamma_1 \wedge \gamma_2)))$

[**RCF**] $\vdash \kappa\{\!|\boldsymbol{y}/\boldsymbol{p}|\!\}$ where κ is a valid analytical formula, $\boldsymbol{y}$ and $\boldsymbol{p}$ are sequences of probability variables and probability terms respectively

[**Meas∅**] $\vdash ((\int \mathsf{ff}) = 0)$

[**FAdd**] $\vdash (((\int(\gamma_1 \wedge \gamma_2)) = 0) \supset ((\int(\gamma_1 \vee \gamma_2)) = (\int\gamma_1) + (\int\gamma_2)))$

[**Mon**] $\vdash ((\Box(\gamma_1 \Rightarrow \gamma_2)) \supset ((\int\gamma_1) \leq (\int\gamma_2)))$

[**Dist⊃**] $\vdash (((\xi_1 \supset \xi_2)/\gamma) \approx ((\xi_1/\gamma) \supset (\xi_2/\gamma)))$

[**Elim1**] $\vdash (((\Box\gamma_1)/\gamma_2) \approx (\Box(\gamma_2 \Rightarrow \gamma_1)))$

[**Elim2**] $\vdash (((p_1 \leq p_2)/\gamma) \approx ((\Diamond\gamma) \supset ((p_1 \leq p_2)|^{(\int\gamma_1)}_{(\int(\gamma_1 \wedge \gamma))})))$

Inference rules

[**PMP**] $\xi_1, (\xi_1 \supset \xi_2) \vdash \xi_2$

[**Cond**] $\vdash (\xi/\gamma)$ whenever $\vdash \xi$

The axiom **CTaut** says that if γ is a valid classical state formula then $(\Box\gamma)$ is an axiom. The axiom **PTaut** says that a probabilistic tautology is an axiom. Since the set of valid classical state formulas and the set of probabilistic tautologies are both recursive, there is no need to spell out the details of tautological reasoning.

The axioms **Lift⇒**, **Eqvff** and **Ref∧** are sufficient to relate (local) classical state reasoning and (global) probabilistic tautological reasoning.

The term $\kappa\{\!|\boldsymbol{y}/\boldsymbol{p}|\!\}$ in the axiom **RCF** is the term obtained by substituting <u>all</u> occurrences of y_i in κ by p_i. The axiom **RCF** says that if κ is a valid analytical formula, then any formula obtained by replacing variables with probability terms is a tautology. We refrain from spelling out the details as the set of valid analytical formulas is recursive.

The axiom **Meas∅** says that the measure of empty set is 0. The axiom **FAdd** is the finite additivity of the measures. The axiom **Mon** relates the classical connectives with probability measures and is a consequence of monotonicity of measures.

The axiom **Dist**$\supset$ says that the connective $\supset$ distributes over the conditional construct. The axioms **Elim1** and **Elim2** eliminate the conditional construct. The probabilistic term $(p_1 \leq p_2)|^{(\int \gamma_1)}_{(\int(\gamma_1 \wedge \gamma))}$ in **Elim2** is the term obtained by replacing all occurrences of $(\int \gamma_1)$ by $(\int(\gamma_1 \wedge \gamma))$ for each classical state formula γ_1.

The inference rule **PMP** is the *modus ponens* for classical and probabilistic implication. The inference rule **Cond** says that if ξ is an theorem. then so is (ξ/γ). The inference rule **Cond** is similar to the generalization rule in modal logics.

As usual we say that a set of formulas Γ *derives* ξ, written $\Gamma \vdash \xi$, if we can build a derivation of ξ from axioms and the inference rules using formulas in Γ as hypothesis. Please note that while applying the rule **Cond**, we are allowed to use only theorems of the logic (and not any hypothesis or any intermediate step in the derivation).

Every probabilistic formula ξ is equivalent to a probabilistic formula η in which there is no occurrence of a conditional construct:

Lemma 1. Let ξ be an EPPL formula. Then, there is a conditional-free formula η such that $\vdash \xi \approx \eta$. Moreover, there is an algorithm to compute η.

Furthermore, the above set of axioms and rules form a recursive axiomatization:

Theorem 1. EPPL is sound and weakly complete. Moreover, the set of theorems is recursive.

3 Basic Probabilistic Sequential Programs

We shall now describe briefly the syntax and semantics of our basic programs.

3.1 Syntax

Assuming the syntax of EPPL, the syntax of the programming language in the BNF notation is as follows (with the proviso $r \in \mathcal{R}$):

- $s := \mathsf{skip} \;[]\; \mathsf{xm} \leftarrow t \;[]\; \mathsf{bm} \leftarrow \gamma \;[]\; \mathsf{toss}(\mathsf{bm}, r) \;[]\; s; s \;[]\; \mathsf{bm\text{–}If}\ \gamma\ \mathsf{then}\ s\ \mathsf{else}\ s.$

The statements $\mathsf{xm} \leftarrow t$ and $\mathsf{bm} \leftarrow \gamma$ are assignments to memory cells xm and bm respectively. For the rest of the paper, by an *expression* we shall mean either the terms t or the classical state formulas γ. Please note that t and γ may contain rigid variables (which may be thought of as input to a program).

The statement $\mathsf{toss}(\mathsf{bm}, r)$ sets bm true with probability $\widetilde{r}$. The command $s; s$ is sequential composition. The statement $\mathsf{bm\text{–}If}\ \gamma\ \mathsf{then}\ s_1\ \mathsf{else}\ s_2$ is the bm–marked alternative choice: if γ is true then s_1 is executed and bm is set to true after the execution of s_1 else s_2 is executed and bm is set to false.

3.2 Semantics

The semantics of the programming language is basically the forward semantics in [17] adapted to our programming language. Given $\mathcal{G}$, the set of generalized probabilistic states, the denotation of a program s is a map $[\![s]\!] : \mathcal{G} \to \mathcal{G}$ defined inductively in Table 4. The definition uses the following notations:

- The denotation of a real term t given a valuation v can be extended to classical state formulas as: $[\![\gamma]\!]_v = \mathsf{tt}$ if $v \Vdash_{\mathsf{C}} \gamma$ otherwise $[\![\gamma]\!]_v = \mathsf{ff}$.
- If m is a memory cell (xm or bm) and e is an expression of the same type (t or γ, respectively) then the map $\delta^{\mathsf{m}}_e : \mathcal{V} \to \mathcal{V}$ is defined as $\delta^{\mathsf{m}}_e(v) = v^{\mathsf{m}}_{[\![e]\!]_v}$, where $v^{\mathsf{m}}_{[\![e]\!]_v}$ assigns the value $[\![e]\!]_v$ to the cell m and coincides with v elsewhere. As usual, $(\delta^{\mathsf{m}}_e)^{-1} : \wp\mathcal{V} \to \wp\mathcal{V}$ is defined by taking each set $U \subset \mathcal{V}$ to the set of its pre-images.
- $(V_1, \mathcal{K}, \mu_1) + (V_2, \mathcal{K}, \mu_2) = (V_1 \cup V_2, \mathcal{K}, \mu_1 + \mu_2)$.
- $r(V, \mathcal{K}, \mu) = (V, \mathcal{K}, r\mu)$.

The denotation of classical assignments, sequential composition and marked alternative are as expected. The probabilistic toss $\mathsf{toss}(\mathsf{bm}, r,)$ assigns bm the value tt with probability $\widetilde{r}$ and the value ff with probability $1 - \widetilde{r}$. Therefore, the denotation of the probabilistic toss is the "weighted" sum of the two assignments $\mathsf{bm} \leftarrow \mathsf{tt}$ and $\mathsf{bm} \leftarrow \mathsf{ff}$.

Table 4. Denotation of programs

$[\![\mathsf{skip}]\!]$	$= \lambda(V, \mathcal{K}, \mu).\,(V, \mathcal{K}, \mu)$
$[\![\mathsf{xm} \leftarrow t]\!]$	$= \lambda(V, \mathcal{K}, \mu).\,(\delta^{\mathsf{xm}}_t(V), \mathcal{K}, \mu \circ (\delta^{\mathsf{xm}}_t)^{-1})$
$[\![\mathsf{bm} \leftarrow \gamma]\!]$	$= \lambda(V, \mathcal{K}, \mu).\,(\delta^{\mathsf{bm}}_\gamma(V), \mathcal{K}, \mu \circ (\delta^{\mathsf{bm}}_\gamma)^{-1})$
$[\![\mathsf{toss}(\mathsf{bm}, r)]\!]$	$= \lambda(V, \mathcal{K}, \mu).\,((1 - \widetilde{r})\,([\![\mathsf{bm} \leftarrow \mathsf{ff}]\!](V, \mathcal{K}, \mu)) + \widetilde{r}\,([\![\mathsf{bm} \leftarrow \mathsf{tt}]\!](V, \mathcal{K}, \mu)))$
$[\![s_1; s_2]\!]$	$= \lambda(V, \mathcal{K}, \mu).\,[\![s_2]\!]\,([\![s_1]\!](V, \mathcal{K}, \mu))$
$[\![\mathsf{bm{-}If}\ \gamma\ \mathsf{then}\ s_1\ \mathsf{else}\ s_2]\!]$	$= \lambda(V, \mathcal{K}, \mu).\,([\![s_1; \mathsf{bm} \leftarrow \mathsf{tt}]\!](\lvert\gamma\rvert_V, \mathcal{K}, \mu_\gamma) + [\![s_2; \mathsf{bm} \leftarrow \mathsf{ff}]\!](\lvert(\neg\gamma)\rvert_V, \mathcal{K}, \mu_{(\neg\gamma)}))$

4 Probabilistic Hoare Logic

We are ready to define the Hoare logic. As expected, the Hoare assertions are :

- $\delta := \xi \;[\!]\; \{\xi\}\, s\, \{\xi\}$.

Satisfaction of Hoare assertions is defined as

- $(V, \mathcal{K}, \mu)\rho \Vdash_h \xi$ if $(V, \mathcal{K}, \mu)\rho \Vdash \xi$,
- $(V, \mathcal{K}, \mu)\rho \Vdash_h \{\xi_1\}\, s\, \{\xi_2\}$ if $(V, \mathcal{K}, \mu)\rho \Vdash \xi_1$ implies $[\![s]\!](V, \mathcal{K}, \mu)\rho \Vdash \xi_2$.

Semantic entailment is defined as expected: we say that Δ entails δ (written $\Delta \vDash \delta$) if $(V, \mathcal{K}, \mu)\rho \Vdash \delta$ whenever $(V, \mathcal{K}, \mu)\rho \Vdash \delta_0$ for each $\delta_0 \in \Delta$.

4.1 Calculus

A sound Hoare calculus for our probabilistic sequential programs is given in Table 5. In the axioms **ASGR** and **ASGB**, the notation ξ_e^m means the formula obtained from ξ by replacing all occurrences (including those in conditionals and probability terms) of the memory variable m by the expression e. The axioms **TAUT**, **SKIP**, **ASGR** and **ASGB** are similar to the ones in the case of sequential programs.

Table 5. Hoare calculus

Axioms

[**TAUT**] $\vdash \xi$ if ξ is an EPPL theorem

[**SKIP**] $\vdash \{\xi\}\,\mathsf{skip}\,\{\xi\}$

[**ASGR**] $\vdash \{\xi_t^{\mathsf{xm}}\}\,\mathsf{xm} \leftarrow t\,\{\xi\}$

[**ASGB**] $\vdash \{\xi_\gamma^{\mathsf{bm}}\}\,\mathsf{bm} \leftarrow \gamma\,\{\xi\}$

[**TOSS**] $\vdash \{\eta\,|^{\Box\gamma}_{\Box(\gamma_{\mathsf{ff}}^{\mathsf{bm}} \wedge \gamma_{\mathsf{tt}}^{\mathsf{bm}})}\,|^{(\int\gamma)}_{(1-\widetilde{r})(\int\gamma_{\mathsf{ff}}^{\mathsf{bm}})+\widetilde{r}(\int\gamma_{\mathsf{tt}}^{\mathsf{bm}})}\}\,\mathsf{toss}(\mathsf{bm},r)\,\{\eta\}$

Inference rules

[**SEQ**] $\{\xi_0\}\,s_1\,\{\xi_1\}, \{\xi_1\}\,s_2\,\{\xi_2\} \vdash \{\xi_0\}\,s_1;s_2\,\{\xi_2\}$

[**IF**] $\{\xi_0\}\,s_1; \mathsf{bm} \leftarrow \mathsf{tt}\,\{\xi_2\}$
$\{\xi_1\}\,s_2; \mathsf{bm} \leftarrow \mathsf{ff}\,\{\xi_3\} \vdash \{\xi_0 \curlyvee_\gamma \xi_1\}\,\mathsf{bm\text{–}If}\ \gamma\ \mathsf{then}\ s_1\ \mathsf{else}\ s_2\,\{\xi_2 \curlyvee_{\mathsf{bm}} \xi_3\}$

[**CONS**] $\xi_0 \supset \xi_1, \{\xi_1\}\,s\,\{\xi_2\}, \xi_2 \supset \xi_3 \vdash \{\xi_1\}\,s\,\{\xi_3\}$

[**OR**] $\{\xi_0\}\,s\,\{\xi_2\}, \{\xi_1\}\,s\,\{\xi_2\} \vdash \{\xi_0 \cup \xi_1\}\,s\,\{\xi_2\}$

[**AND**] $\{\xi_0\}\,s\,\{\xi_1\}, \{\xi_0\}\,s\,\{\xi_2\} \vdash \{\xi_0\}\,s\,\{\xi_1 \cap \xi_2\}$

For the axiom **TOSS**, we do not consider arbitrary probabilistic formulas. Instead, we just have probabilistic formulas η which do not have any conditional sub-terms. This is not a serious limitation as every EPPL formula is equivalent to another EPPL formula without conditionals (see Lemma 1). Furthermore, the formula $\eta\,|^{\Box\gamma}_{\Box(\gamma_{\mathsf{ff}}^{\mathsf{bm}} \wedge \gamma_{\mathsf{tt}}^{\mathsf{bm}})}\,|^{(\int\gamma)}_{(1-\widetilde{r})(\int\gamma_{\mathsf{ff}}^{\mathsf{bm}})+\widetilde{r}(\int\gamma_{\mathsf{tt}}^{\mathsf{bm}})}$ is the formula obtained from η by replacing every occurrence of a necessity formula $(\Box\gamma)$ by $(\Box(\gamma_{\mathsf{ff}}^{\mathsf{bm}} \wedge \gamma_{\mathsf{tt}}^{\mathsf{bm}}))$ and every occurrence of a probability term $(\int\gamma)$ by $(1-\widetilde{r})(\int\gamma_{\mathsf{ff}}^{\mathsf{bm}})+\widetilde{r}(\int\gamma_{\mathsf{tt}}^{\mathsf{bm}})$. Here, the formula γ_e^{bm} is obtained by replacing all occurrences of bm by e. The soundness of this Hoare rule is a consequence of the following lemma:

Lemma 2 (Substitution lemma for probabilistic tosses). *For any formula* η, $([\![\mathsf{toss}(\mathsf{bm},r)]\!]\,(V,\mathcal{K},\mu))\rho \Vdash \eta$ *iff* $(V,\mathcal{K},\mu)\rho \Vdash \eta\,|^{\Box\gamma}_{\Box(\gamma_{\mathsf{ff}}^{\mathsf{bm}} \wedge \gamma_{\mathsf{tt}}^{\mathsf{bm}})}\,|^{(\int\gamma)}_{(1-\widetilde{r})(\int\gamma_{\mathsf{ff}}^{\mathsf{bm}})+\widetilde{r}(\int\gamma_{\mathsf{tt}}^{\mathsf{bm}})}$.

The inference rules **SEQ**, **CONS**, **OR** and **AND** are similar to the ones in sequential programs. For the inference rule **IF**, $\xi \curlyvee_\gamma \xi'$ is an abbreviation for the formula $((\xi/\gamma) \cap (\xi'/\neg\gamma))$. It follows from the definition of semantics of EPPL that $(V,\mathcal{K},\mu)\rho \Vdash \xi \curlyvee_\gamma \xi'$ if and only if $(|\gamma|_V,\mathcal{K},\mu_\gamma)\rho \Vdash \xi$ and $(|\neg\gamma|_V,\mathcal{K},\mu_{\neg\gamma})\rho \Vdash \xi'$. We have:

Theorem 2. The Hoare calculus is sound.

The completeness for the Hoare logic was being worked upon in collaboration with Luís Cruz-Filipe at the time of submission of this paper. The Hoare rule for probabilistic tosses is its weakest pre-condition form as a consequence of the substitution lemma for probabilistic tosses (see Lemma 2 above). For the alternative, we have shown that if ξ_0 and ξ_1 are weakest pre-conditions corresponding to the two marked choices then so is $\xi_0 \curlyvee_\gamma \xi_1$. Furthermore, any EPPL formula ξ is essentially equivalent to a disjunct of formulas of the kind $\xi' \curlyvee_\gamma \xi''$. Roughly, two EPPL formulas are essentially equivalent if the semantic structures satisfying them differ only in the $\mathcal{K}$-assignment part.

5 Example

We illustrate our Hoare calculus on a variation of the quantum one-time pad. A qubit is the basic memory unit in quantum computation (just as a bit is the basic memory unit in classical computation). The state of a qubit is a pair (α, β) of complex numbers such that $|\alpha|^2 + |\beta|^2 = 1$. A quantum one-time pad [2] encrypts a qubit using two key (classical) bits in a secure way: observing the encrypted qubit yields two results, both with equal probability. In the special case that α and β are real numbers one bit key bm_k suffices. We restrict our attention to this special case. If the key $\mathsf{bm}_k = 1$ then the qubit is (unitarily) encrypted as the pair $(\beta, -\alpha)$ otherwise it remains the same. The following program S_{qenc} simulates this process by first generating a random key and then encrypting the qubit $(\mathsf{xm}_1, \mathsf{xm}_2)$: $\mathsf{toss}(\mathsf{bm}_k, \frac{1}{2}); \mathsf{bm{-}If}\ \mathsf{bm}_k\ \mathsf{then}\ \mathsf{Pauli}_{XZ}\ \mathsf{else}\ \mathsf{skip}$, where Pauli_{XZ} is $\mathsf{xm}_3 \leftarrow \mathsf{xm}_1; \mathsf{xm}_1 \leftarrow \mathsf{xm}_2; \mathsf{xm}_2 \leftarrow -\mathsf{xm}_3$ [3].

Assume that the initial values of xm_1 and xm_2 are c_1 and c_2 respectively (with $c_1 \neq c_2$). It follows from quantum information theory that in order to prove the security of the quantum one-time pad, it suffices to show that the probability after the encryption of xm_1 being c_1 is $\frac{1}{2}$ (and hence of xm_1 being c_2 is also $\frac{1}{2}$). We can use our logic to show the above for S_{qenc}. In particular, assuming η_I is $\Box((\mathsf{xm}_1 = c_1) \wedge (\mathsf{xm}_2 = c_2) \wedge (c_1 < c_2))$, we derive the following in our Hoare calculus: $\vdash \{((\int \mathsf{tt}) = 1) \cap \eta_I\}\, S_{\text{qenc}}\, \{(\int(\mathsf{xm}_1 = c_1)) = \frac{1}{2}\}$. Abbreviating the statement $\mathsf{bm{-}If}\ \mathsf{bm}_k\ \mathsf{then}\ \mathsf{Pauli}_{XZ}\ \mathsf{else}\ \mathsf{skip}$ as IF, the derivation is:

1	$\{(\int(\mathsf{xm}_1 = c_1)) = \frac{1}{2}\}\, \mathsf{skip}\, \{(\int(\mathsf{xm}_1 = c_1)) = \frac{1}{2}\}$	SKIP
2	$\{(\int(c_2 = c_1)) = 0\}\, \mathsf{Pauli}_{XZ}\, \{(\int(\mathsf{xm}_1 = c_1)) = 0\}$	ASGR, SEQ
3	$(((\int \mathsf{tt}) = \frac{1}{2}) \cap \eta_I) \supset (\int(\mathsf{xm}_1 = c_1)) = \frac{1}{2}$	TAUT
4	$(((\int \mathsf{tt}) = \frac{1}{2}) \cap \eta_I) \supset (\int(c_2 = c_1)) = 0$	TAUT
5	$\{(((\int \mathsf{tt}) = \frac{1}{2}) \cap \eta_I) \curlyvee_{\mathsf{bm}_k} (((\int \mathsf{tt}) = \frac{1}{2}) \cap \eta_I)\}\, \mathsf{IF}$ $\{(\int(\mathsf{xm}_1 = c_1)) = \frac{1}{2} \curlyvee_{\mathsf{bm}} (\int(\mathsf{xm}_1 = c_1)) = 0\}$	IF, CONS 1-4
6	$((\int(\mathsf{xm}_1 = c_1)) = \frac{1}{2} \curlyvee_{\mathsf{bm}} (\int(\mathsf{xm}_1 = c_1)) = 0) \supset (\int(\mathsf{xm}_1 = c_1)) = \frac{1}{2}$	TAUT

[3] The name Pauli_{XZ} has its roots in quantum mechanics.

7 $(\eta_I \cap ((\int \mathsf{bm}) = \frac{1}{2}) \cap ((\int \neg\, \mathsf{bm}) = \frac{1}{2}))\supset$
$(((\int \mathsf{tt}) = \frac{1}{2}) \cap \eta_I) \curlyvee_{\mathsf{bm}_k} (((\int \mathsf{tt}) = \frac{1}{2}) \cap \eta_I)$ TAUT

8 $\{(\eta_I \cap ((\int \mathsf{bm}) = \frac{1}{2}) \cap ((\int \neg\, \mathsf{bm}) = \frac{1}{2}))\}\, \mathsf{IF}\, \{(\int (\mathsf{xm}_1 = c_1)) = \frac{1}{2}\}$ CONS 5,6,7

9 $\{((\int \mathsf{tt}) = 1) \cap \eta_I\}\{\mathsf{toss}(\mathsf{bm}, \frac{1}{2})\}$
$\{(\eta_I \cap ((\int \mathsf{bm}) = \frac{1}{2}) \cap ((\int \neg\, \mathsf{bm}) = \frac{1}{2}))\}$ TOSS, TAUT

10 $\{((\int \mathsf{tt}) = 1) \cap \eta_I\}\, S\mathrm{qenc}\, \{(\int (\mathsf{xm}_1 = c_1)) = \frac{1}{2}\}$ SEQ 8,9.

6 Related Work

The area of formal methods in probabilistic programs has attracted a lot of work ranging from semantics [16,15,29,22] to logic-based reasoning [9,17,27,10,13,21,23,6].

Our work is in the field of probabilistic dynamic logics. Dynamic logic is a modal logic in which the modalities are of the form $\langle s \rangle \varphi$ where s is a program and φ is a state assertion formula. For probabilistic programs, there are two distinct approaches to dynamic logic. The main difference in the two approaches is that one uses truth-functional state logic while the other one uses state logic with arithmetical connectives.

The first truth-functional probabilistic state logic based works appear in the context of dynamic logic [28,18,26,9,8]. In the context of probabilistic truth-functional dynamic logics, the state language has terms representing probabilities (*e.g.*, $(\int \gamma)$ represents the probability of γ being true). An infinitary complete axiom system for probabilistic dynamic logic is given in [18]. Later, a complete finitary axiomatization of probabilistic dynamic logic was given in [9]. However, the state logic is second-order (to deal with iteration) and no axiomatization of the state logic is achieved. In [8], decidability of a less expressive dynamic logic is achieved.

Hoare logic can be viewed as a fragment of dynamic logic and the first probabilistic Hoare logic with truth-functional propositional state logic appears in [27]. However, as discussed in Section 1, even simple assertions in this logic may not be provable. For instance, the valid Hoare assertion (adapting somewhat the syntax) $\{(\int \mathsf{tt}) = 1\}\, \mathsf{If}\ x = 0\ \mathsf{then\ skip\ else\ skip}\, \{(\int \mathsf{tt}) = 1\}$ is not provable in the logic. As noted in [27,17], the reason for incompleteness is the Hoare rule for the alternative if-then-else which tries to combine absolute information of the two alternatives truth-functionally. The Hoare logic in [6] circumvents the problem of the alternative by defining the probabilistic sum connective as already discussed in Section 1. Although this logic is more expressive than the one in [27] and completeness is achieved for a fragment of the Hoare logic, it is not clear how to axiomatize the probabilistic sum connective [6].

The other approach to dynamic logic uses arithmetical state logic instead of truth-functional state logic [17,15,14,21]. For example, instead of the if-then-else construct the programming language in [17] has the construct $\gamma?s_1 + (\neg\, \gamma)?s_2$ which is closely bound to the forward denotational semantics proposed in [16]. This leads to a probabilistic dynamic logic in which measurable functions are used as state formulas and the connectives are interpreted as arithmetical operations.

In the context of Hoare logics, the approach of arithmetical connectives is the one that has attracted more research. The Hoare triple in this context naturally leads to the definition of *weakest pre-condition* for a measurable function g and a program s: the weakest pre-condition $\mathsf{wp}(g, s)$ is the function that has the greatest expected value amongst all functions f such that $\{f\}\, s\, \{g\}$ is a valid Hoare triple. The weakest pre-condition can thus be thought of as a backward semantics which transforms a post-state g in the context of a program s to a pre-state $\mathsf{wp}(g, s)$. The important result in this area is the duality between the forward semantics and the backwards semantics [14].

Later, [21] extended this framework to address non-determinism and proved the duality between forward semantics and backward semantics. Instead of just using functions f and g as pre-conditions and post-conditions, [21] also allows a rudimentary state language with basic classical state formulas α, negation, disjunction and conjunction. The classical state formula α is interpreted as the function that takes the value 1 in the memory valuations where α is true and 0 otherwise. Conjunction and disjunction are interpreted as minimum and maximum respectively, and negation as subtraction from the constant function 1. For example, the following Hoare assertion is valid in this logic: $\{r\}\, \mathsf{toss}(\mathsf{bm}, r)\, \{\mathsf{bm}\}$. Here r in the pre-condition is the constant function r and bm is the function that take value 1 when bm is true and 0 otherwise. The validity of the above Hoare assertion says that the probability of bm being true after the probabilistic toss is at least r.

We tackle the problem of alternative if-then-else by marking the choices at the end of the execution and by introducing the conditional construct (ξ/γ) in the state logic. The state logic itself is the probabilistic logic in [20] extended with the conditional construct. The logic is designed by the exogenous semantics approach to probabilistic logics [24, 25, 7, 1, 20]. The main difference from the logic in [20] is that the state logic herein has the conditional construct which is not present in [20]. The axioms **Dist**$\supset$, **Elim1** and **Elim2** are used to deal with this conditional construct. Using these, we can demonstrate that every formula is equivalent to another formula without conditionals and the proof of completeness then follows the lines of the proof in [20]. The other difference is that the probabilities in [20] are taken in the set of real numbers and terms contain real computable numbers. The proof of completeness is obtained relative to an (undecidable) oracle for reasoning about reals.

Finally, one main contribution of our paper is the Hoare rule in the weakest pre-condition form for probabilistic toss in the context of truth-functional state logic. The Hoare rule for probabilistic tosses does appear in the context of arithmetical Hoare logics and takes the form

$$\mathsf{wp}(\mathsf{toss}(\mathsf{bm}, r), \alpha) = r \times \mathsf{wp}(\mathsf{bm} \leftarrow \mathsf{tt}, \alpha) + (1 - r) \times \mathsf{wp}(\mathsf{bm} \leftarrow \mathsf{ff}, \alpha).$$

7 Conclusions and Future Work

Our main contribution is a sound probabilistic Hoare calculus with a truth-functional state assertion logic that enjoys recursive axiomatization. The Hoare rule for the if-then-else statement avoids the probabilistic sum construct in [6] by marking the choices taken and by taking advantage of a conditional construct in the state assertion language. Another important contribution is the axiom for probabilistic toss which gives the weakest pre-condition in truth-functional setting and is the counterpart of the weakest pre-condition for probabilistic toss in Hoare logics with arithmetical state logics.

As discussed in Section 4, we are currently working towards complete axiomatization for the Hoare-calculus for the iteration free language. We plan to include the iteration construct and demonic non-determinsim in future work. For iteration, we will investigate completeness using an oracle for arithmetical reasoning.

Our long-term interests are in reasoning about quantum programs and protocols. Probabilities are inevitable in quantum programs because measurements of quantum states yield probabilistic mixtures of quantum states. We aim to investigate Hoare-style reasoning and dynamic logics for quantum programming. Towards this end, we have already designed logics for reasoning about individual quantum states [19,5] and a sound Hoare logic for basic quantum imperative programs [4].

Acknowledgements. We would like to thank L. Cruz-Filipe and P. Selinger for useful and interesting discussions. We will also like to thank the anonymous referees whose useful comments have greatly benefited the presentation.

References

1. M. Abadi and J. Y. Halpern. Decidability and expressiveness for first-order logics of probability. *Information and Computation*, 112(1):1–36, 1994.
2. A. Ambainis, M. Mosca, A. Tapp, and R. de Wolf. Private quantum channels. In *FOCS '00: Proceedings of the 41st Annual Symposium on Foundations of Computer Science*, page 547. IEEE Computer Society, 2000.
3. S. Basu, R. Pollack, and R. Marie-Françoise. *Algorithms in Real Algebraic Geometry*. Springer, 2003.
4. R. Chadha, P. Mateus, and A. Sernadas. Reasoning about quantum imperative programs. *Electronic Notes in Theoretical Computer Science*, 158:19–40, 2006. Invited talk at the Twenty-second Conference on the Mathematical Foundations of Programming Semantics.
5. R. Chadha, P. Mateus, A. Sernadas, and C. Sernadas. Extending classical logic for reasoning about quantum systems. Preprint, CLC, Department of Mathematics, Instituto Superior Técnico, 2005. Invited submission to the Handbook of Quantum Logic.
6. J.I. den Hartog and E.P. de Vink. Verifying probabilistic programs using a hoare like logic. *International Journal of Foundations of Computer Science*, 13(3):315–340, 2002.
7. R. Fagin, J. Y. Halpern, and N. Megiddo. A logic for reasoning about probabilities. *Information and Computation*, 87(1-2):78–128, 1990.
8. Y. A. Feldman. A decidable propositional dynamic logic with explicit probabilities. *Information and Control*, 63((1/2)):11–38, 1984.

9. Y. A. Feldman and David Harel. A probabilistic dynamic logic. *Journal of Computer and System Sciences*, 28:193–215, 1984.
10. H. Hansson and B. Jonsson. A logic for reasoning about time and reliability. *Formal Aspects of Computing*, 6:512–535, 1995.
11. C. Hoare. An axiomatic basis for computer programming. *Communications of the ACM*, 12:576–583, 1969.
12. W. Hodges. *Model Theory.* Cambridge University Press, 1993.
13. M. Huth and M. Kwiatkowska. Quantitative analysis and model checking. In *12th Annual IEEE Symposium on Logic in Computer Science (LICS'97)*, pages 111–122, 1997.
14. C. Jones. *Probabilistic Non-determinism.* PhD thesis, U. Edinburgh, 1990.
15. C. Jones and G. D. Plotkin. A probabilistic powerdomain of evaluations. In *Proceedings of the Fourth Annual Symposium on Logic in Computer Science*, pages 186–195. IEEE Computer Society, 1989.
16. D. Kozen. Semantics of probabilistic programs. *Journal of Computer System Science*, 22:328–350, 1981.
17. D. Kozen. A probabilistic PDL. *Journal of Computer System Science*, 30:162–178, 1985.
18. J.A. Makowsky and M.L.Tiomkin. Probabilistic propositional dynamic logic, 1980. manuscript.
19. P. Mateus and A. Sernadas. Weakly complete axiomatization of exogenous quantum propositional logic. *Information and Computation*, to appear.
20. P. Mateus, A. Sernadas, and C. Sernadas. Exogenous semantics approach to enriching logics. In *Essays on the Foundations of Mathematics and Logic*, volume 1 of *Advanced Studies in Mathematics and Logic*, pages 165–194. Polimetrica, 2005.
21. C. Morgan, A. McIver, and K. Seidel. Probabilistic predicate transformers. *ACM Transactions on Programming Languages and Systems*, 18(3):325–353, 1996.
22. M. A. Moshier and A. Jung. A logic for probabilities in semantics. In *Computer Science Logic*, volume 2471 of *Lecture Notes in Computer Science*, pages 216–231. Springer Verlag, 2002.
23. M. Narasimha, R. Cleaveland, and P. Iyer. Probabilistic temporal logics via the modal mu-calculus. In *Foundations of Software Science and Computation Structures (FOSSACS 99)*, volume 1578 of *Lecture Notes in Computer Science*, pages 288–305. 1999.
24. N. J. Nilsson. Probabilistic logic. *Artificial Intelligence*, 28(1):71–87, 1986.
25. N. J. Nilsson. Probabilistic logic revisited. *Artificial Intelligence*, 59(1-2):39–42, 1993.
26. R. Parikh and A. Mahoney. A theory of probabilistic programs. In *Proceedings of the Carnegie Mellon Workshop on Logic of Programs*, volume 64 of *Lecture Notes in Computer Science*, pages 396–402. Springer-Verlag, 1983.
27. L. H. Ramshaw. *Formalizing the analysis of algorithms.* PhD thesis, Stanford University, 1979.
28. J. H. Reif. Logics for probabilistic programming (extended abstract). In *STOC '80: Proceedings of the twelfth annual ACM symposium on Theory of computing*, pages 8–13, 1980.
29. R. Tix, K. Keimel, and G.D. Plotkin. Semantic domains for combining probability and non-determinism. *Electronic Notes in Theoretical Computer Science*, 129:1–104, 2005.

Concurrent Games with Tail Objectives

Krishnendu Chatterjee

EECS, University of California, Berkeley, USA
c_krish@cs.berkeley.edu

Abstract. We study infinite stochastic games played by two-players over a finite state space, with objectives specified by sets of infinite traces. The games are *concurrent* (players make moves simultaneously and independently), *stochastic* (the next state is determined by a probability distribution that depends on the current state and chosen moves of the players) and *infinite* (proceeds for infinite number of rounds). The analysis of concurrent stochastic games can be classified into: *quantitative analysis*, analyzing the optimum value of the game; and *qualitative analysis*, analyzing the set of states with optimum value 1. We consider concurrent games with tail objectives, i.e., objectives that are independent of the finite-prefix of traces, and show that the class of tail objectives are strictly richer than the ω-regular objectives. We develop new proof techniques to extend several properties of concurrent games with ω-regular objectives to concurrent games with tail objectives. We prove the *positive limit-one* property for tail objectives, that states for all concurrent games if the optimum value for a player is positive for a tail objective Φ at some state, then there is a state where the optimum value is 1 for Φ, for the player. We also show that the optimum values of *zero-sum* (strictly conflicting objectives) games with tail objectives can be related to equilibrium values of *nonzero-sum* (not strictly conflicting objectives) games with simpler reachability objectives. A consequence of our analysis presents a polynomial time reduction of the quantitative analysis of tail objectives to the qualitative analysis for the sub-class of one-player stochastic games (Markov decision processes).

1 Introduction

Stochastic games. Non-cooperative games provide a natural framework to model interactions between agents [13,14]. A wide class of games progress over time and in stateful manner, and the current game depends on the history of interactions. Infinite *stochastic games* [15,9] are a natural model for such dynamic games. A stochastic game is played over a finite *state space* and is played in rounds. In *concurrent games*, in each round, each player chooses an action from a finite set of available actions, simultaneously and independently of the other player. The game proceeds to a new state according to a probabilistic transition relation (stochastic transition matrix) based on the current state and the joint actions of the players. Concurrent games (also known as Blackwell games) subsume the simpler class of *turn-based games*, where at every state at most one player can choose between multiple actions; and Markov decision processes

Z. Ésik (Ed.): CSL 2006, LNCS 4207, pp. 256–270, 2006.

(MDPs), where only one player can choose between multiple actions at every state. Concurrent games also provide the framework to model synchronous reactive systems [6]. In verification and control of finite state reactive systems such games proceed for infinite rounds, generating an infinite sequence of states, called the *outcome* of the game. The players receive a payoff based on a payoff function that maps every outcome to a real number.

Objectives. Payoffs are generally Borel measurable functions [12]. For example, the payoff set for each player is a Borel set B_i in the Cantor topology on S^ω (where S is the set of states), and player i gets payoff 1 if the outcome of the game is a member of B_i, and 0 otherwise. In verification, payoff functions are usually index sets of *ω-regular languages.* The ω-regular languages generalize the classical regular languages to infinite strings, they occur in low levels of the Borel hierarchy (they are in $\mathbf{\Sigma}_3^0 \cap \mathbf{\Pi}_3^0$), and they form a robust and expressive language for determining payoffs for commonly used specifications. The simplest ω-regular objectives correspond to safety ("closed sets") and reachability ("open sets") objectives.

Zero-sum games, determinacy and nonzero-sum games. Games may be *zero-sum*, where two players have directly conflicting objectives and the payoff of one player is one minus the payoff of the other, or *nonzero-sum*, where each player has a prescribed payoff function based on the outcome of the game. The fundamental question for games is the existence of equilibrium values. For zero-sum games, this involves showing a *determinacy* theorem that states that the expected optimum value obtained by player 1 is exactly one minus the expected optimum value obtained by player 2. For one-step zero-sum games, this is von Neumann's minmax theorem [18]. For infinite games, the existence of such equilibria is not obvious, in fact, by using the axiom of choice, one can construct games for which determinacy does not hold. However, a remarkable result by Martin [12] shows that all stochastic zero-sum games with Borel payoffs are determined. For nonzero-sum games, the fundamental equilibrium concept is a *Nash equilibrium* [10], that is, a strategy profile such that no player can gain by deviating from the profile, assuming the other player continue playing the strategy in the profile.

Qualitative and quantitative analysis. The analysis of zero-sum concurrent games can be broadly classified into: (a) *quantitative analysis* that involves analysis of the optimum values of the games; and (b) *qualitative analysis* that involves simpler analysis of the set of states where the optimum value is 1.

Properties of concurrent games. The result of Martin [12] established the determinacy of zero-sum concurrent games for all Borel objectives. The determinacy result sets forth the problem of study and closer understanding of properties and behaviors of concurrent games with different class of objectives. Several interesting questions related to concurrent games are: (1) characterizing certain zero-one laws for concurrent games; (2) relationship of qualitative and quantitative analysis; (3) relationship of zero-sum and nonzero-sum games. The results of [6,7,1] exhibited several interesting properties for concurrent games with

ω-regular objectives specified as parity objectives. The result of [6] showed the positive limit-one property, that states if there is a state with positive optimum value, then there is a state with optimum value 1, for concurrent games with parity objectives. The positive limit-one property has been a key property to develop algorithms and improved complexity bound for quantitative analysis of concurrent games with parity objectives [1]. The above properties can possibly be the basic ingredients for the computational complexity analysis of quantitative analysis of concurrent games.

Outline of results. In this work, we consider *tail objectives*, the objectives that do not depend on any finite-prefix of the traces. Tail objectives subsume canonical ω-regular objectives such as parity objectives and Müller objectives, and we show that there exist tail objectives that cannot be expressed as ω-regular objectives. Hence tail objectives are a strictly richer class of objectives than ω-regular objectives. Our results characterize several properties of concurrent games with tail objectives. The results are as follows.

1. We show the positive limit-one property for concurrent games with tail objectives. Our result thus extends the result of [6] from parity objectives to a richer class of objective that lie in the higher levels of Borel hierarchy. The result of [6] follows from a complementation argument of quantitative μ-calculus formula. Our proof technique is completely different: it uses certain strategy construction procedures and a convergence result from measure theory (Lévy's zero-one law). It may be noted that the positive limit-one property for concurrent games with Müller objectives follows from the positive limit-one property for parity objectives and the reduction of Müller objectives to parity objectives [17]. Since Müller objectives are tail objectives, our result presents a direct proof for the positive limit-one property for concurrent games with Müller objectives.
2. We relate the optimum values of zero-sum games with tail objectives with Nash-equilibrium values of non-zero sum games with reachability objectives. This establishes a relationship between the values of concurrent games with complex tail objectives and Nash equilibrium of nonzero-sum games with simpler objectives. From the above analysis we obtain a polynomial time reduction of quantitative analysis of tail objectives to qualitative analysis for the special case of MDPs. The above result was previously known for the sub-class of ω-regular objectives specified as Müller objectives [4,5,2]. The proof techniques of [4,5,2] use different analysis of the structure of MDPs and is completely different from our proof techniques.

2 Definitions

Notation. For a countable set A, a *probability distribution* on A is a function $\delta : A \to [0,1]$ such that $\sum_{a \in A} \delta(a) = 1$. We denote the set of probability distributions on A by $\mathcal{D}(A)$. Given a distribution $\delta \in \mathcal{D}(A)$, we denote by $\mathrm{Supp}(\delta) = \{x \in A \mid \delta(x) > 0\}$ the *support* of δ.

Definition 1 (Concurrent Games). *A (two-player)* concurrent game structure $G = \langle S, Moves, Mv_1, Mv_2, \delta \rangle$ *consists of the following components:*

- *A finite state space S and a finite set Moves of moves.*
- *Two move assignments $Mv_1, Mv_2 \colon S \to 2^{Moves} \setminus \emptyset$. For $i \in \{1, 2\}$, assignment Mv_i associates with each state $s \in S$ the non-empty set $Mv_i(s) \subseteq Moves$ of moves available to player i at s.*
- *A probabilistic transition function $\delta : S \times Moves \times Moves \to \mathcal{D}(S)$, that gives the probability $\delta(s, a_1, a_2)(t)$ of a transition from s to t when player 1 plays move a_1 and player 2 plays move a_2, for all $s, t \in S$ and $a_1 \in Mv_1(s)$, $a_2 \in Mv_2(s)$.* ∎

An important special class of concurrent games are Markov decision processes (MDPs), where at every state s we have $|Mv_2(s)| = 1$, i.e., the set of available moves for player 2 is singleton at every state.

At every state $s \in S$, player 1 chooses a move $a_1 \in Mv_1(s)$, and simultaneously and independently player 2 chooses a move $a_2 \in Mv_2(s)$. The game then proceeds to the successor state t with probability $\delta(s, a_1, a_2)(t)$, for all $t \in S$. A state s is called an *absorbing state* if for all $a_1 \in Mv_1(s)$ and $a_2 \in Mv_2(s)$ we have $\delta(s, a_1, a_2)(s) = 1$. In other words, at s for all choices of moves of the players the next state is always s. We assume that the players act *non-cooperatively*, i.e., each player chooses her strategy independently and secretly from the other player, and is only interested in maximizing her own reward. For all states $s \in S$ and moves $a_1 \in Mv_1(s)$ and $a_2 \in Mv_2(s)$, we indicate by $\mathrm{Dest}(s, a_1, a_2) = \mathrm{Supp}(\delta(s, a_1, a_2))$ the set of possible successors of s when moves a_1, a_2 are selected.

A *path* or a *play* ω of G is an infinite sequence $\omega = \langle s_0, s_1, s_2, \ldots \rangle$ of states in S such that for all $k \geq 0$, there are moves $a_1^k \in Mv_1(s_k)$ and $a_2^k \in Mv_2(s_k)$ with $\delta(s_k, a_1^k, a_2^k)(s_{k+1}) > 0$. We denote by Ω the set of all paths and by Ω_s the set of all paths $\omega = \langle s_0, s_1, s_2, \ldots \rangle$ such that $s_0 = s$, i.e., the set of plays starting from state s.

Strategies. A *selector* ξ for player $i \in \{1, 2\}$ is a function $\xi : S \to \mathcal{D}(Moves)$ such that for all $s \in S$ and $a \in Moves$, if $\xi(s)(a) > 0$, then $a \in Mv_i(s)$. We denote by Λ_i the set of all selectors for player $i \in \{1, 2\}$. A *strategy* for player 1 is a function $\tau : S^+ \to \Lambda_1$ that associates with every finite non-empty sequence of states, representing the history of the play so far, a selector. Similarly we define strategies π for player 2. We denote by Γ and Π the set of all strategies for player 1 and player 2, respectively.

Once the starting state s and the strategies τ and π for the two players have been chosen, the game is reduced to an ordinary stochastic process. Hence the probabilities of events are uniquely defined, where an *event* $\mathcal{A} \subseteq \Omega_s$ is a measurable set of paths. For an event $\mathcal{A} \subseteq \Omega_s$ we denote by $\Pr_s^{\tau,\pi}(\mathcal{A})$ the probability that a path belongs to $\mathcal{A}$ when the game starts from s and the players follow the strategies τ and π. For $i \geq 0$, we also denote by $\Theta_i : \Omega \to S$ the random variable denoting the i-th state along a path.

Objectives. We specify objectives for the players by providing the set of *winning plays* $\Phi \subseteq \Omega$ for each player. Given an objective Φ we denote by $\overline{\Phi} = \Omega \setminus \Phi$, the complementary objective of Φ. A concurrent game with objective Φ_1 for player 1 and Φ_2 for player 2 is *zero-sum* if $\Phi_2 = \overline{\Phi}_1$. A general class of objectives are the Borel objectives [11]. A *Borel objective* $\Phi \subseteq S^\omega$ is a Borel set in the Cantor topology on S^ω. In this paper we consider *ω-regular objectives* [17], which lie in the first $2^1/_2$ levels of the Borel hierarchy (i.e., in the intersection of $\mathbf{\Sigma}_3^0$ and $\mathbf{\Pi}_3^0$) and *tail objectives* which is a strict superset of ω-regular objectives. The ω-regular objectives, and subclasses thereof, and tail objectives are defined below. For a play $\omega = \langle s_0, s_1, s_2, \ldots \rangle \in \Omega$, we define $\mathrm{Inf}(\omega) = \{ s \in S \mid s_k = s \text{ for infinitely many } k \geq 0 \}$ to be the set of states that occur infinitely often in ω.

- *Reachability and safety objectives.* Given a set $T \subseteq S$ of "target" states, the reachability objective requires that some state of T be visited. The set of winning plays is thus $\mathrm{Reach}(T) = \{ \omega = \langle s_0, s_1, s_2, \ldots \rangle \in \Omega \mid s_k \in T \text{ for some } k \geq 0 \}$. Given a set $F \subseteq S$, the safety objective requires that only states of F be visited. Thus, the set of winning plays is $\mathrm{Safe}(F) = \{\omega = \langle s_0, s_1, s_2, \ldots \rangle \in \Omega \mid s_k \in F \text{ for all } k \geq 0 \}$.
- *Büchi and coBüchi objectives.* Given a set $B \subseteq S$ of "Büchi" states, the Büchi objective requires that B is visited infinitely often. Formally, the set of winning plays is $\text{Büchi}(B) = \{ \omega \in \Omega \mid \mathrm{Inf}(\omega) \cap B \neq \emptyset \}$. Given $C \subseteq S$, the coBüchi objective requires that all states visited infinitely often are in C. Formally, the set of winning plays is $\text{coBüchi}(C) = \{ \omega \in \Omega \mid \mathrm{Inf}(\omega) \subseteq C \}$.
- *Parity objectives.* For $c, d \in \mathbb{N}$, we let $[c..d] = \{ c, c+1, \ldots, d \}$. Let $p : S \to [0..d]$ be a function that assigns a *priority* $p(s)$ to every state $s \in S$, where $d \in \mathbb{N}$. The *Even parity objective* is defined as $\mathrm{Parity}(p) = \{ \omega \in \Omega \mid \min\big(p(\mathrm{Inf}(\omega))\big) \text{ is even } \}$, and the *Odd parity objective* as $\mathrm{coParity}(p) = \{ \omega \in \Omega \mid \min\big(p(\mathrm{Inf}(\omega))\big) \text{ is odd } \}$.
- *Muller objectives.* Given a set $\mathcal{M} \subseteq 2^S$ of subset of states, the *Müller objective* is defined as $\text{Müller}(\mathcal{M}) = \{ \omega \in \Omega \mid \mathrm{Inf}(\omega) \in \mathcal{M} \}$.
- *Tail objectives.* Informally the class of tail objectives is the sub-class of Borel objectives that are independent of all finite prefixes. An objective Φ is a tail objective, if the following condition hold: a path $\omega \in \Phi$ if and only if for all $i \geq 0$, $\omega_i \in \Phi$, where ω_i denotes the path ω with the prefix of length i deleted. Formally, let $\mathcal{G}_i = \sigma(\Theta_i, \Theta_{i+1}, \ldots)$ be the σ-field generated by the random variables $\Theta_i, \Theta_{i+1}, \ldots$. The tail σ-field $\mathcal{T}$ is defined as $\mathcal{T} = \bigcap_{i \geq 0} \mathcal{G}_i$. An objective Φ is a tail objective if and only if Φ belongs to the tail σ-field $\mathcal{T}$, i.e., the tail objectives are indicator functions of events $\mathcal{A} \in \mathcal{T}$.

The Müller and parity objectives are canonical forms to represent ω-regular objectives [16]. Observe that Müller and parity objectives are tail objectives. Note that for a priority function $p : S \to \{ 0, 1 \}$, an even parity objective $\mathrm{Parity}(p)$ is equivalent to the Büchi objective $\text{Büchi}(p^{-1}(0))$, i.e., the Büchi set consists of the states with priority 0. Büchi and coBüchi objectives are special cases of parity objectives and hence tail objectives. Reachability objectives are not necessarily tail objectives, but for a set $T \subseteq S$ of states, if every state $s \in T$ is an absorbing state, then the objective $\mathrm{Reach}(T)$ is equivalent to $\text{Büchi}(T)$ and hence is a tail objective. It may be noted that since σ-fields are closed under complementation, the class of tail objectives are closed under complementation. We give an example to show that the class of tail objectives are richer than ω-regular objectives.[1]

Example 1. Let r be a reward function that maps every state s to a real-valued reward $r(s)$, i.e., $r : S \to \mathbb{R}$. For a constant $c \in \mathbb{R}$ consider the objective Φ_c defined as follows: $\Phi_c = \{\omega \in \Omega \mid \omega = \langle s_1, s_2, s_3, \ldots \rangle, \liminf_{n\to\infty} \frac{1}{n}\sum_{i=1}^{n} r(s_i) \geq c\}$. Intuitively, Φ_c accepts the set of paths such that the "long-run" average of the rewards in the path is at least the constant c. The "long-run" average condition lie in the third-level of the Borel-hierarchy (i.e., in $\mathbf{\Pi}_3^0$ and $\mathbf{\Pi}_3^0$-complete) and cannot be expressed as an ω-regular objective. It may be noted that the "long-run" average of a path is independent of all finite-prefixes of the path. Formally, the class Φ_c of objectives are tail objectives. Since Φ_c are $\mathbf{\Pi}_3^0$-complete objectives, it follows that tail objectives lie in higher levels of Borel hierarchy than ω-regular objectives. ∎

Values. The probability that a path satisfies an objective Φ starting from state $s \in S$, given strategies τ, π for the players is $\Pr_s^{\tau,\pi}(\Phi)$. Given a state $s \in S$ and an objective Φ, we are interested in the maximal probability with which player 1 can ensure that Φ and player 2 can ensure that $\overline{\Phi}$ holds from s. We call such probability the *value of the game* G at s for player $i \in \{1, 2\}$. The value for player 1 and player 2 are given by the functions $\langle\langle 1\rangle\rangle_{val}(\Phi) : S \to [0,1]$ and $\langle\langle 2\rangle\rangle_{val}(\overline{\Phi}) : S \to [0,1]$, defined for all $s \in S$ by $\langle\langle 1\rangle\rangle_{val}(\Phi)(s) = \sup_{\tau\in\Gamma} \inf_{\pi\in\Pi} \Pr_s^{\tau,\pi}(\Phi)$ and $\langle\langle 2\rangle\rangle_{val}(\overline{\Phi})(s) = \sup_{\pi\in\Pi} \inf_{\tau\in\Gamma} \Pr_s^{\tau,\pi}(\overline{\Phi})$. Note that the objectives of the player are complementary and hence we have a zero-sum game. Concurrent games satisfy a *quantitative* version of determinacy [12], stating that for all Borel objectives Φ and all $s \in S$, we have $\langle\langle 1\rangle\rangle_{val}(\Phi)(s) + \langle\langle 2\rangle\rangle_{val}(\overline{\Phi})(s) = 1$. A strategy τ for player 1 is *optimal* for objective Φ if for all $s \in S$ we have $\inf_{\pi\in\Pi} \Pr_s^{\tau,\pi}(\Phi) = \langle\langle 1\rangle\rangle_{val}(\Phi)(s)$. For $\varepsilon > 0$, a strategy τ for player 1 is *ε-optimal* for objective Φ if for all $s \in S$ we have $\inf_{\pi\in\Pi} \Pr_s^{\tau,\pi}(\Phi) \geq \langle\langle 1\rangle\rangle_{val}(\Phi)(s) - \varepsilon$. We define optimal and ε-optimal strategies for player 2 symmetrically. For $\varepsilon > 0$, an objective Φ for player 1 and $\overline{\Phi}$ for player 2, we denote by $\Gamma_\varepsilon(\Phi)$ and $\Pi_\varepsilon(\overline{\Phi})$ the set of ε-optimal strategies for player 1 and player 2, respectively. Even in concurrent games with reachability objectives optimal strategies need not exist [6], and ε-optimal strategies, for all $\varepsilon > 0$, is the best one can achieve. Note that the quantitative determinacy of concurrent games is equivalent to the existence of ε-optimal strategies for objective Φ for player 1 and $\overline{\Phi}$ for player 2, for all $\varepsilon > 0$, at all states $s \in S$, i.e., for all $\varepsilon > 0$, $\Gamma_\varepsilon(\Phi) \neq \emptyset$ and $\Pi_\varepsilon(\overline{\Phi}) \neq \emptyset$.

We refer to the analysis of computing the *limit-sure winning* states (the set of states s such that $\langle\langle 1\rangle\rangle_{val}(\Phi)(s) = 1$) as the *qualitative* analysis of objective Φ. We refer to the analysis of computing the values as the *quantitative* analysis of objective Φ.

[1] Our example shows that there are $\mathbf{\Pi}_3^0$-complete objectives that are tail objectives. It is possible that the tail objectives can express objectives in even higher levels of Borel hierarchy than $\mathbf{\Pi}_3^0$, which will make our results stronger.

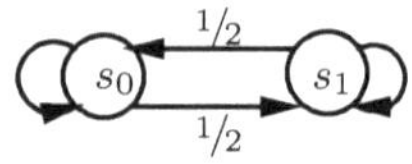

Fig. 1. A simple Markov chain

3 Positive Limit-One Property

The *positive limit-one* property for concurrent games, for a class $\mathcal{C}$ of objectives, states that for all objectives $\Phi \in \mathcal{C}$, for all concurrent games G, if there is a state s such that the value for player 1 is positive at s for objective Φ, then there is a state s' where the value for player 1 is 1 for objective Φ. The property means if a player can win with positive value from some state, then from some state she can win with value 1. The positive limit-one property was proved for parity objectives in [6] and has been one of the key properties used in the algorithmic analysis of concurrent games with parity objectives [1]. In this section we prove the *positive limit-one* property for concurrent games with tail objectives, and thereby extend the positive limit-one property from parity objectives to a richer class of objectives that subsume several canonical ω-regular objectives. Our proof uses a result from measure theory and certain strategy constructions, whereas the proof for the sub-class of parity objectives [6] followed from complementation arguments of quantitative μ-calculus formula. We first show an example that the positive limit-one property is not true for all objectives, even for simpler class of games.

Example 2. Consider the game shown in Fig 1, where at every state s, we have $Mv_1(s) = Mv_2(s) = \{1\}$ (i.e., the set of moves is singleton at all states). From all states the next state is s_0 and s_1 with equal probability. Consider the objective $\bigcirc(s_1)$ which specifies the next state is s_1; i.e., a play ω starting from state s is winning if the first state of the play is s and the second state (or the next state from s) in the play is s_1. Given the objective $\Phi = \bigcirc(s_1)$ for player 1, we have $\langle\langle 1\rangle\rangle_{val}(\Phi)(s_0) = \langle\langle 1\rangle\rangle_{val}(\Phi)(s_1) = 1/2$. Hence though the value is positive at s_0, there is no state with value 1 for player 1. ∎

Notation. In the setting of concurrent games the natural filtration sequence $(\mathcal{F}_n)$ for the stochastic process under any pair of strategies is defined as

$$\mathcal{F}_n = \sigma(\Theta_1, \Theta_2, \ldots, \Theta_n)$$

i.e., the σ-field generated by the random-variables $\Theta_1, \Theta_2, \ldots, \Theta_n$.

Lemma 1 (Lévy's 0-1 law). *Suppose $\mathcal{H}_n \uparrow \mathcal{H}_\infty$, i.e.,$\mathcal{H}_n$ is a sequence of increasing σ-fields and $\mathcal{H}_\infty = \sigma(\cup_n \mathcal{H}_n)$. For all events $\mathcal{A} \in \mathcal{H}_\infty$ we have*

$$\mathrm{E}(\mathbf{1}_\mathcal{A} \mid \mathcal{H}_n) = \Pr(\mathcal{A} \mid \mathcal{H}_n) \to \mathbf{1}_\mathcal{A} \text{ almost-surely, (i.e., with probability 1),}$$

where $\mathbf{1}_\mathcal{A}$ is the indicator function of event $\mathcal{A}$.

The proof of the lemma is available in Durrett (page 262—263) [8]. An immediate consequence of Lemma 1 in the setting of concurrent games is the following lemma.

Lemma 2 (0-1 law in concurrent games). *For all concurrent game structures G, for all events $\mathcal{A} \in \mathcal{F}_\infty = \sigma(\cup_n \mathcal{F}_n)$, for all strategies $(\tau, \pi) \in \Gamma \times \Pi$, for all states $s \in S$, we have*

$$\Pr_s^{\tau,\pi}(\mathcal{A} \mid \mathcal{F}_n) \to \mathbf{1}_{\mathcal{A}} \textit{ almost-surely.}$$

Intuitively, the lemma means that the probability $\Pr_s^{\tau,\pi}(\mathcal{A} \mid \mathcal{F}_n)$ converges almost-surely (i.e., with probability 1) to 0 or 1 (since indicator functions take values in the range $\{0, 1\}$). Note that the tail σ-field $\mathcal{T}$ is a subset of $\mathcal{F}_\infty$, i.e., $\mathcal{T} \subseteq \mathcal{F}_\infty$, and hence the result of Lemma 2 holds for all $\mathcal{A} \in \mathcal{T}$.

Notation. Given strategies τ and π for player 1 and player 2, a tail objective $\overline{\Phi}$, and a state s, for $\beta > 0$, let

$$H_n^{1,\beta}(\tau, \pi, \overline{\Phi}) = \{ \langle s_1, s_2, \ldots, s_n, s_{n+1}, \ldots \rangle \mid \Pr_s^{\tau,\pi}(\overline{\Phi} \mid \langle s_1, s_2, \ldots, s_n \rangle) \geq 1 - \beta \},$$

denote the set of paths ω such that the probability of satisfying $\overline{\Phi}$ given the strategies τ and π, and the prefix of length n of ω is at least $1 - \beta$; and

$$H_n^{0,\beta}(\tau, \pi, \overline{\Phi}) = \{ \langle s_1, s_2, \ldots, s_n, s_{n+1}, \ldots \rangle \mid \Pr_s^{\tau,\pi}(\overline{\Phi} \mid \langle s_1, s_2, \ldots, s_n \rangle) \leq \beta \}.$$

denote the set of paths ω such that the probability of satisfying $\overline{\Phi}$ given the strategies τ and π, and the prefix of length n of ω is at most β.

Proposition 1. *For all concurrent game structures G, for all strategies τ and π for player 1 and player 2, respectively, for all tail objectives $\overline{\Phi}$, for all states $s \in S$, for all $\beta > 0$ and $\varepsilon > 0$, there exists n, such that $\Pr_s^{\tau,\pi}(H_n^{1,\beta}(\tau, \pi, \overline{\Phi}) \cup H_n^{0,\beta}(\tau, \pi, \overline{\Phi})) \geq 1 - \varepsilon$.*

Proof. Let us denote $f_n = \Pr_s^{\tau,\pi}(\overline{\Phi} \mid \mathcal{F}_n)$. It follows from Lemma 2 that $f_n \to \overline{\Phi}$ almost-surely as $n \to \infty$. Since almost-sure convergence implies convergence in probability [8], $f_n \to \overline{\Phi}$ in probability. Formally, we have

$$\forall \beta > 0.\ \Pr_s^{\tau,\pi}(|f_n - \overline{\Phi}| > \beta) \to 0 \qquad \text{as } n \to \infty.$$

Equivalently we have

$$\forall \beta > 0.\ \forall \varepsilon > 0.\ \exists n_0.\ \forall n \geq n_0.\ \Pr_s^{\tau,\pi}(|f_n - \overline{\Phi}| \leq \beta) \geq 1 - \varepsilon.$$

Thus we obtain that $\lim_{n\to\infty} \Pr_s^{\tau,\pi}(H_n^{1,\beta}(\tau, \pi, \overline{\Phi}) \cup H_n^{0,\beta}(\tau, \pi, \overline{\Phi})) = 1$; and hence the result follows. ∎

Theorem 1 (Positive limit-one property). *For all concurrent game structures G, for all tail objectives Φ, if there exists a state $s \in S$ such that $\langle\langle 1 \rangle\rangle_{val}(\Phi)(s) > 0$, then there exists a state $s' \in S$ such that $\langle\langle 1 \rangle\rangle_{val}(\Phi)(s') = 1$.*

Proof. Assume towards contradiction that there exists a state s such that $\langle\langle 1\rangle\rangle_{val}(\Phi)(s) > 0$, but for all states s' we have $\langle\langle 1\rangle\rangle_{val}(\Phi)(s') < 1$. Let $\alpha = 1 - \langle\langle 1\rangle\rangle_{val}(\Phi)(s) = \langle\langle 2\rangle\rangle_{val}(\overline{\Phi})(s)$. Since $0 < \langle\langle 1\rangle\rangle_{val}(\Phi)(s) < 1$, we have $0 < \alpha < 1$. Since $\langle\langle 2\rangle\rangle_{val}(\overline{\Phi})(s') = 1 - \langle\langle 1\rangle\rangle_{val}(\Phi)(s')$ and for all states s' we have $\langle\langle 1\rangle\rangle_{val}(\Phi)(s) < 1$, it follows that $\langle\langle 2\rangle\rangle_{val}(\overline{\Phi})(s') > 0$, for all states s'. Fix η such that $0 < \eta = \min_{s' \in S} \langle\langle 2\rangle\rangle_{val}(\overline{\Phi})(s')$. Also observe that since $\langle\langle 2\rangle\rangle_{val}(\overline{\Phi})(s) = \alpha < 1$, we have $\eta < 1$. Let c be a constant such that $c > 0$, and $\alpha \cdot (1+c) = \gamma < 1$ (such a constant exists as $\alpha < 1$). Also let $c_1 > 1$ be a constant such that $c_1 \cdot \gamma < 1$ (such a constant exists since $\gamma < 1$); hence we have $1 - c_1 \cdot \gamma > 0$ and $1 - \frac{1}{c_1} > 0$. Fix $\varepsilon > 0$ and $\beta > 0$ such that

$$0 < 2\varepsilon < \min\{\, \frac{\eta}{4}, 2c \cdot \alpha, \frac{\eta}{4} \cdot (1 - c_1 \cdot \gamma) \,\}; \qquad \beta < \min\{\, \varepsilon, \frac{1}{2}, 1 - \frac{1}{c_1} \,\}. \quad (1)$$

Fix ε-optimal strategies τ_ε for player 1 and π_ε for player 2. Let $H_n^{1,\beta} = H_n^{1,\beta}(\tau_\varepsilon, \pi_\varepsilon, \overline{\Phi})$ and $H_n^{0,\beta} = H_n^{0,\beta}(\tau_\varepsilon, \pi_\varepsilon, \overline{\Phi})$. Consider n such that $\Pr_s^{\tau_\varepsilon,\pi_\varepsilon}(H_n^{1,\beta} \cup H_n^{0,\beta}) \geq 1 - \frac{\varepsilon}{4}$ (such n exists by Proposition 1). Also observe that since $\beta < \frac{1}{2}$ we have $H_n^{1,\beta} \cap H_n^{0,\beta} = \emptyset$. Let

$$val = \Pr_s^{\tau_\varepsilon,\pi_\varepsilon}(\overline{\Phi} \mid H_n^{1,\beta}) \cdot \Pr_s^{\tau_\varepsilon,\pi_\varepsilon}(H_n^{1,\beta}) + \Pr_s^{\tau_\varepsilon,\pi_\varepsilon}(\overline{\Phi} \mid H_n^{0,\beta}) \cdot \Pr_s^{\tau_\varepsilon,\pi_\varepsilon}(H_n^{0,\beta}).$$

We have

$$val \leq \Pr_s^{\tau_\varepsilon,\pi_\varepsilon}(\overline{\Phi}) \leq val + \frac{\varepsilon}{4}. \quad (2)$$

The first inequality follows since $H_n^{1,\beta} \cap H_n^{0,\beta} = \emptyset$ and the second inequality follows since $\Pr_s^{\tau_\varepsilon,\pi_\varepsilon}(H_n^{1,\beta} \cup H_n^{0,\beta}) \geq 1 - \frac{\varepsilon}{4}$. Since τ_ε and π_ε are ε-optimal strategies we have $\alpha - \varepsilon \leq \Pr_s^{\tau_\varepsilon,\pi_\varepsilon}(\overline{\Phi}) \leq \alpha + \varepsilon$. This along with (2) yield that

$$\alpha - \varepsilon - \frac{\varepsilon}{4} \leq val \leq \alpha + \varepsilon. \quad (3)$$

Observe that $\Pr_s^{\tau_\varepsilon,\pi_\varepsilon}(\overline{\Phi} \mid H_n^{1,\beta}) \geq 1 - \beta$ and $\Pr_s^{\tau_\varepsilon,\pi_\varepsilon}(\overline{\Phi} \mid H_n^{0,\beta}) \leq \beta$. Let $q = \Pr_s^{\tau_\varepsilon,\pi_\varepsilon}(H_n^{1,\beta})$. Since $\Pr_s^{\tau_\varepsilon,\pi_\varepsilon}(\overline{\Phi} \mid H_n^{1,\beta}) \geq 1 - \beta$; ignoring the term $\Pr_s^{\tau_\varepsilon,\pi_\varepsilon}(\overline{\Phi} \mid H_n^{0,\beta}) \cdot \Pr_s^{\tau_\varepsilon,\pi_\varepsilon}(H_n^{0,\beta})$ in val and from the second inequality of (3) we obtain that $(1 - \beta) \cdot q \leq \alpha + \varepsilon$. Since $\varepsilon < c \cdot \alpha, \beta < 1 - \frac{1}{c_1}$, and $\gamma = \alpha \cdot (1 + c)$ we have

$$q \leq \frac{\alpha + \varepsilon}{1 - \beta} < \frac{\alpha \cdot (1 + c)}{1 - (1 - \frac{1}{c_1})} = c_1 \cdot \gamma \quad (4)$$

We construct a strategy $\widehat{\pi}_\varepsilon$ as follows: the strategy $\widehat{\pi}_\varepsilon$ follows the strategy π_ε for the first $n-1$-stages; if a history in $H_n^{1,\beta}$ is generated it follows π_ε, and otherwise it ignores the history and switches to an ε-optimal strategy. Formally, for a history $\langle s_1, s_2, \ldots, s_k\rangle$ we have

$$\widehat{\pi}_\varepsilon(\langle s_1, \ldots, s_k\rangle) = \begin{cases} \pi_\varepsilon(\langle s_1, \ldots, s_k\rangle) & \text{if } k < n; \\ & \text{or } \Pr_s^{\tau_\varepsilon,\pi_\varepsilon}(\overline{\Phi} \mid \langle s_1, s_2, \ldots, s_n\rangle) \geq 1 - \beta; \\ \widetilde{\pi}_\varepsilon(\langle s_n, \ldots, s_k\rangle) & \Pr_s^{\tau_\varepsilon,\pi_\varepsilon}(\overline{\Phi} \mid \langle s_1, s_2, \ldots, s_n\rangle) < 1 - \beta, \text{ and} \\ & k \geq n, \text{ where } \widetilde{\pi}_\varepsilon \text{ is an } \varepsilon\text{-optimal strategy} \end{cases}$$

Since $\widehat{\pi}_\varepsilon$ and π_ε coincides for $n-1$-stages we have $\Pr_s^{\tau_\varepsilon,\widehat{\pi}_\varepsilon}(H_n^{1,\beta}) = \Pr_s^{\tau_\varepsilon,\pi_\varepsilon}(H_n^{1,\beta})$ and $\Pr_s^{\tau_\varepsilon,\widehat{\pi}_\varepsilon}(H_n^{0,\beta}) = \Pr_s^{\tau_\varepsilon,\pi_\varepsilon}(H_n^{0,\beta})$. Moreover, since $\overline{\Phi}$ is a tail objective that is independent of the prefix of length n; $\eta \le \min_{s' \in S} \langle\langle 2 \rangle\rangle_{val}(\overline{\Phi})(s')$ and $\widetilde{\pi}_\varepsilon$ is an ε-optimal strategy, we have $\Pr_s^{\tau_\varepsilon,\widehat{\pi}_\varepsilon}(\overline{\Phi} \mid H_n^{0,\beta}) \ge \eta - \varepsilon$. Also observe that

$$\begin{aligned}\Pr_s^{\tau_\varepsilon,\widehat{\pi}_\varepsilon}(\overline{\Phi} \mid H_n^{0,\beta}) \ge (\eta-\varepsilon) &= \Pr_s^{\tau_\varepsilon,\pi_\varepsilon}(\overline{\Phi} \mid H_n^{0,\beta}) + (\eta - \varepsilon - \Pr_s^{\tau_\varepsilon,\pi_\varepsilon}(\overline{\Phi} \mid H_n^{0,\beta}))\\ &\ge \Pr_s^{\tau_\varepsilon,\pi_\varepsilon}(\overline{\Phi} \mid H_n^{0,\beta}) + (\eta-\varepsilon-\beta),\end{aligned}$$

since $\Pr_s^{\tau_\varepsilon,\pi_\varepsilon}(\overline{\Phi} \mid H_n^{0,\beta}) \le \beta$. Hence we have the following inequality

$$\begin{aligned}
\Pr_s^{\tau_\varepsilon,\widehat{\pi}_\varepsilon}(\overline{\Phi}) &\ge \Pr_s^{\tau_\varepsilon,\widehat{\pi}_\varepsilon}(\overline{\Phi} \mid H_n^{1,\beta}) \cdot \Pr_s^{\tau_\varepsilon,\widehat{\pi}_\varepsilon}(H_n^{1,\beta}) + \Pr_s^{\tau_\varepsilon,\widehat{\pi}_\varepsilon}(\overline{\Phi} \mid H_n^{0,\beta}) \cdot \Pr_s^{\tau_\varepsilon,\widehat{\pi}_\varepsilon}(H_n^{0,\beta})\\
&= \Pr_s^{\tau_\varepsilon,\pi_\varepsilon}(\overline{\Phi} \mid H_n^{1,\beta}) \cdot \Pr_s^{\tau_\varepsilon,\pi_\varepsilon}(H_n^{1,\beta}) + \Pr_s^{\tau_\varepsilon,\widehat{\pi}_\varepsilon}(\overline{\Phi} \mid H_n^{0,\beta}) \cdot \Pr_s^{\tau_\varepsilon,\widehat{\pi}_\varepsilon}(H_n^{0,\beta})\\
&\ge \Pr_s^{\tau_\varepsilon,\pi_\varepsilon}(\overline{\Phi} \mid H_n^{1,\beta}) \cdot \Pr_s^{\tau_\varepsilon,\pi_\varepsilon}(H_n^{1,\beta}) + \Pr_s^{\tau_\varepsilon,\pi_\varepsilon}(\overline{\Phi} \mid H_n^{0,\beta}) \cdot \Pr_s^{\tau_\varepsilon,\pi_\varepsilon}(H_n^{0,\beta})\\
&\quad + (\eta-\varepsilon-\beta)\cdot\left(1-q-\frac{\varepsilon}{4}\right) \qquad \left(\text{as } \Pr_s^{\tau_\varepsilon,\pi_\varepsilon}(H_n^{0,\beta}) \ge 1-q-\frac{\varepsilon}{4}\right)\\
&= val + (\eta-\varepsilon-\beta)\cdot(1-q-\frac{\varepsilon}{4})\\
&\ge \alpha - \frac{5\varepsilon}{4} + (\eta-\varepsilon-\beta)\cdot(1-q-\frac{\varepsilon}{4}) \quad (\text{recall first inequality of (3)})\\
&> \alpha - \frac{5\varepsilon}{4} + (\eta-2\varepsilon)\cdot(1-q-\frac{\varepsilon}{4}) \qquad (\text{as } \beta < \varepsilon \text{ by (1)})\\
&> \alpha - \frac{5\varepsilon}{4} + \frac{\eta}{2}\cdot(1-q-\frac{\varepsilon}{4}) \qquad (\text{as } 2\varepsilon < \tfrac{\eta}{2} \text{ by (1)})\\
&> \alpha - \frac{5\varepsilon}{4} + \frac{\eta}{2}\cdot(1-c_1\cdot\gamma) - \frac{\eta}{2}\cdot\frac{\varepsilon}{4} \qquad (\text{as } q < c_1\cdot\gamma \text{ by (4)})\\
&> \alpha - \varepsilon - \frac{\varepsilon}{4} + 4\varepsilon - \frac{\varepsilon}{8} \qquad (\text{as } 2\varepsilon < \tfrac{\eta}{4}\cdot(1-c_1\cdot\gamma) \text{ by (1),}\\
&\qquad\qquad \text{and } \eta \le 1)\\
&> \alpha + \varepsilon.
\end{aligned}$$

The first equality follows since for histories in $H_n^{1,\beta}$, the strategies π_ε and $\widehat{\pi}_\varepsilon$ coincide. Hence we have $\Pr_s^{\tau_\varepsilon,\widehat{\pi}_\varepsilon}(\overline{\Phi}) > \alpha + \varepsilon$ and $\Pr_s^{\tau_\varepsilon,\widehat{\pi}_\varepsilon}(\Phi) < 1 - \alpha - \varepsilon$. This is a contradiction to the fact that $\langle\langle 1 \rangle\rangle_{val}(\Phi)(s) = 1 - \alpha$ and τ_ε is an ε-optimal strategy. The desired result follows. ∎

Notation. We use the following notation for the rest of the paper:

$$W_1^1 = \{\, s \mid \langle\langle 1 \rangle\rangle_{val}(\Phi)(s) = 1 \,\}; \qquad W_2^1 = \{\, s \mid \langle\langle 2 \rangle\rangle_{val}(\overline{\Phi})(s) = 1 \,\}.$$

$$W_1^{>0} = \{\, s \mid \langle\langle 1 \rangle\rangle_{val}(\Phi)(s) > 0 \,\}; \qquad W_2^{>0} = \{\, s \mid \langle\langle 2 \rangle\rangle_{val}(\overline{\Phi})(s) > 0 \,\}.$$

By determinacy of concurrent games with tail objectives, we have $W_1^1 = S \setminus W_2^{>0}$ and $W_2^1 = S \setminus W_1^{>0}$. We have the following finer characterization of the sets.

Corollary 1. *For all concurrent game structures G, with tail objectives Φ for player 1, the following assertions hold:*

1. *(a) if* $W_1^{>0} \neq \emptyset$*, then* $W_1^1 \neq \emptyset$*; and (b) if* $W_2^{>0} \neq \emptyset$*, then* $W_2^1 \neq \emptyset$*.*
2. *(a) if* $W_1^{>0} = S$*, then* $W_1^1 = S$*; and (b) if* $W_2^{>0} = S$*, then* $W_2^1 = S$*.*

Proof. The first result is a direct consequence of Theorem 1. The second result is derived as follows: if $W_1^{>0} = S$, then by determinacy we have $W_2^1 = \emptyset$. If $W_2^1 = \emptyset$, it follows from part 1 that $W_2^{>0} = \emptyset$, and hence $W_1^1 = S$. The result of part 2 shows that if a player has positive optimum value at every state, then the optimum value is 1 at all states. ∎

4 Zero-Sum Tail Games to Nonzero-Sum Reachability Games

In this section we relate the values of zero-sum games with tail objectives with the Nash equilibrium values of nonzero-sum games with reachability objectives. The result shows that the values of a zero-sum game with complex objectives can be related to equilibrium values of a nonzero-sum game with simpler objectives. We also show that for MDPs the value function for a tail objective Φ can be computed by computing the maximal probability of reaching the set of states with value 1. As an immediate consequence of the above analysis, we obtain a polynomial time reduction of the quantitative analysis of MDPs with tail objectives to the qualitative analysis. We first prove a *limit-reachability* property of ε-optimal strategies: the property states that for tail objectives, if the players play ε-optimal strategies, for small $\varepsilon > 0$, then the game reaches $W_1^1 \cup W_2^1$ with high probability.

Theorem 2 (Limit-reachability). *For all concurrent game structures* G*, for all tail objectives* Φ *for player 1, for all* $\varepsilon' > 0$*, there exists* $\varepsilon > 0$*, such that for all states* $s \in S$*, for all* ε*-optimal strategies* τ_ε *and* π_ε*, we have*

$$\Pr_s^{\tau_\varepsilon,\pi_\varepsilon}(\mathit{Reach}(W_1^1 \cup W_2^1)) \geq 1 - \varepsilon'.$$

Proof. By determinacy it follows that $W_1^1 \cup W_2^1 = S \setminus (W_1^{>0} \cup W_2^{>0})$. For a state $s \in W_1^1 \cup W_2^1$ the result holds trivially. Consider a state $s \in W_1^{>0} \cup W_2^{>0}$ and let $\alpha = \langle\langle 2 \rangle\rangle_{val}(\overline{\Phi})(s)$. Observe that $0 < \alpha < 1$. Let $\eta_1 = \min_{s \in W_2^{>0}} \langle\langle 1 \rangle\rangle_{val}(\Phi)(s)$ and $\eta_2 = \max_{s \in W_2^{>0}} \langle\langle 2 \rangle\rangle_{val}(\overline{\Phi})(s)$, and let $\eta = \min\{\, \eta_1, 1 - \eta_2 \,\}$, and note that $0 < \eta < 1$. Given $\varepsilon' > 0$, fix ε such that $0 < 2\varepsilon < \min\{\, \frac{\eta}{2}, \frac{\eta \cdot \varepsilon'}{12} \,\}$. Fix any ε-optimal strategies τ_ε and π_ε for player 1 and player 2, respectively. Fix β such that $0 < \beta < \varepsilon$ and $\beta < \frac{1}{2}$. Let $H_n^{1,\beta} = H_n^{1,\beta}(\tau_\varepsilon, \pi_\varepsilon, \overline{\Phi})$ and $H_n^{0,\beta} = H_n^{0,\beta}(\tau_\varepsilon, \pi_\varepsilon, \overline{\Phi})$. Consider n such that $\Pr_s^{\tau_\varepsilon,\pi_\varepsilon}(H_n^{1,\beta} \cup H_n^{0,\beta}) = 1 - \frac{\varepsilon}{4}$ (such n exists by Proposition 1), and also as $\beta < \frac{1}{2}$, we have $H_n^{1,\beta} \cap H_n^{0,\beta} = \emptyset$. Let us denote by

$$\mathit{val} = \Pr_s^{\tau_\varepsilon,\pi_\varepsilon}(\overline{\Phi} \mid H_n^{1,\beta}) \cdot \Pr_s^{\tau_\varepsilon,\pi_\varepsilon}(H_n^{1,\beta}) + \Pr_s^{\tau_\varepsilon,\pi_\varepsilon}(\overline{\Phi} \mid H_n^{0,\beta}) \cdot \Pr_s^{\tau_\varepsilon,\pi_\varepsilon}(H_n^{0,\beta}).$$

Similar to inequality (2) of Theorem 1 we obtain that

$$\mathit{val} \leq \Pr_s^{\tau_\varepsilon,\pi_\varepsilon}(\overline{\Phi}) \leq \mathit{val} + \frac{\varepsilon}{4}$$

Since τ_ε and π_ε are ε-optimal strategies, similar to inequality (3) of Theorem 1 we obtain that $\alpha - \varepsilon - \frac{\varepsilon}{4} \leq val \leq \alpha + \varepsilon$.

For $W \subseteq S$, let $\mathrm{Reach}^n(W) = \{ \langle s_1, s_2, s_3 \ldots \rangle \mid \exists k \leq n.\ s_k \in W \}$ denote the set of paths that reaches W in n-steps. We use the following notations: $\overline{\mathrm{Reach}(W_1^1)} = \Omega \setminus \mathrm{Reach}^n(W_1^1)$, and $\overline{\mathrm{Reach}(W_2^1)} = \Omega \setminus \mathrm{Reach}^n(W_2^1)$. Consider a strategy $\widehat{\tau}_\varepsilon$ defined as follows: for histories in $H_n^{1,\beta} \cap \overline{\mathrm{Reach}(W_2^1)}$, $\widehat{\tau}_\varepsilon$ ignores the history after stage n and follows an ε-optimal strategy $\widetilde{\tau}_\varepsilon$; and for all other histories it follows τ_ε. Let $z_1 = \mathrm{Pr}_s^{\tau_\varepsilon,\pi_\varepsilon}(H_n^{1,\beta} \cap \overline{\mathrm{Reach}(W_2^1)})$. Since $\eta_2 = \max_{s \in W_2^{>0}} \langle\langle 2 \rangle\rangle_{val}(\overline{\Phi})(s)$, and player 1 switches to an ε-optimal strategy for histories of length n in $H_n^{1,\beta} \cap \overline{\mathrm{Reach}(W_2^1)}$ and $\overline{\Phi}$ is a tail objective, it follows that for all $\omega = \langle s_1, s_2, \ldots, s_n, s_{n+1}, \ldots \rangle \in H_n^{1,\beta} \cap \overline{\mathrm{Reach}(W_2^1)}$, we have $\mathrm{Pr}_s^{\widehat{\tau}_\varepsilon,\pi_\varepsilon}(\overline{\Phi} \mid \langle s_1, s_2 \ldots, s_n \rangle) \leq \eta_2 + \varepsilon$; where as $\mathrm{Pr}_s^{\tau_\varepsilon,\pi_\varepsilon}(\overline{\Phi} \mid \langle s_1, s_2 \ldots, s_n \rangle) \geq 1 - \beta$. Hence we have

$$val_2 = \mathrm{Pr}_s^{\widehat{\tau}_\varepsilon,\pi_\varepsilon}(\overline{\Phi}) \leq \mathrm{Pr}_s^{\tau_\varepsilon,\pi_\varepsilon}(\overline{\Phi}) - z_1 \cdot (1 - \beta - \eta_2 - \varepsilon) \leq val + \frac{\varepsilon}{4} - z_1 \cdot (1 - \beta - \eta_2 - \varepsilon),$$

since with probability z_1 the decrease is at least by $1 - \beta - \eta_2 - \varepsilon$. Since π_ε is an ε-optimal strategy we have $val_2 \geq \alpha - \varepsilon$; and since $val \leq \alpha + \varepsilon$, we have the following inequality

$$\begin{aligned} z_1 \cdot (1 - \eta_2 - \beta - \varepsilon) &\leq 2\varepsilon + \frac{\varepsilon}{4} < 3\varepsilon \\ \Rightarrow z_1 &< \frac{3\varepsilon}{\eta - \beta - \varepsilon} \qquad (\text{since } \eta \leq 1 - \eta_2) \\ \Rightarrow z_1 &< \frac{3\varepsilon}{\eta - 2\varepsilon} < \frac{6\varepsilon}{\eta} < \frac{\varepsilon'}{4} \qquad (\text{since } \beta < \varepsilon; \varepsilon < \frac{\eta}{4}; \varepsilon < \frac{\eta \cdot \varepsilon'}{24}) \end{aligned}$$

Consider a strategy $\widehat{\pi}_\varepsilon$ defined as follows: for histories in $H_n^{0,\beta} \cap \overline{\mathrm{Reach}(W_1^1)}$, $\widehat{\pi}_\varepsilon$ ignores the history after stage n and follows an ε-optimal strategy $\widetilde{\pi}_\varepsilon$; and for all other histories it follows π_ε. Let $z_2 = \mathrm{Pr}_s^{\tau_\varepsilon,\pi_\varepsilon}(H_n^{0,\beta} \cap \overline{\mathrm{Reach}(W_1^1)})$. Since $\eta_1 = \min_{s \in W_2^{>0}} \langle\langle 2 \rangle\rangle_{val}(\overline{\Phi})(s)$, and player 2 switches to an ε-optimal strategy for histories of length n in $H_n^{0,\beta} \cap \overline{\mathrm{Reach}(W_1^1)}$ and $\overline{\Phi}$ is a tail objective, it follows that for all $\omega = \langle s_1, s_2, \ldots, s_n, s_{n+1}, \ldots \rangle \in H_n^{1,\beta} \cap \overline{\mathrm{Reach}(W_1^1)}$, we have $\mathrm{Pr}_s^{\tau_\varepsilon,\widehat{\pi}_\varepsilon}(\overline{\Phi} \mid \langle s_1, s_2 \ldots, s_n \rangle) \geq \eta_1 - \varepsilon$; where as $\mathrm{Pr}_s^{\tau_\varepsilon,\pi_\varepsilon}(\overline{\Phi} \mid \langle s_1, s_2 \ldots, s_n \rangle) \leq \beta$. Hence we have

$$val_1 = \mathrm{Pr}_s^{\tau_\varepsilon,\widehat{\pi}_\varepsilon}(\overline{\Phi}) \geq \mathrm{Pr}_s^{\tau_\varepsilon,\pi_\varepsilon}(\overline{\Phi}) + z_2 \cdot (\eta_1 - \varepsilon - \beta) \geq val + z_2 \cdot (\eta_1 - \varepsilon - \beta),$$

since with probability z_2 the increase is at least by $\eta_1 - \varepsilon - \beta$. Since τ_ε is an ε-optimal strategy we have $val_1 \leq \alpha + \varepsilon$; and since $val \geq \alpha - \varepsilon + \frac{\varepsilon}{4}$, we have the following inequality

$$\begin{aligned} z_2 \cdot (\eta_1 - \beta - \varepsilon) &\leq 2\varepsilon + \frac{\varepsilon}{4} < 3\varepsilon \\ \Rightarrow z_2 &< \frac{3\varepsilon}{\eta - \beta - \varepsilon} \qquad (\text{since } \eta \leq \eta_1) \\ \Rightarrow z_2 &< \frac{\varepsilon'}{4} \qquad (\text{similar to the inequality for } z_1 < \frac{\varepsilon'}{4}) \end{aligned}$$

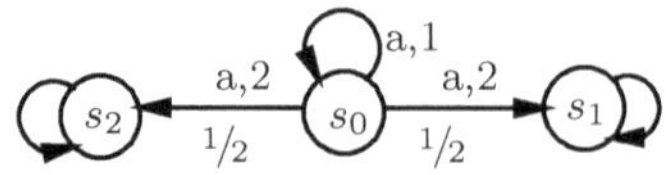

Fig. 2. A game with Büchi objective

Hence $z_1 + z_2 \leq \frac{\varepsilon'}{2}$; and then we have

$$\begin{aligned}
\Pr_s^{\tau_\varepsilon,\pi_\varepsilon}(\text{Reach}(W_1^1 \cup W_2^1)) &\geq \Pr_s^{\tau_\varepsilon,\pi_\varepsilon}(\text{Reach}^n(W_1^1 \cup W_2^1) \cap (H_n^{1,\beta} \cup H_n^{0,\beta})) \\
&= \Pr_s^{\tau_\varepsilon,\pi_\varepsilon}(\text{Reach}^n(W_1^1 \cup W_2^1) \cap H_n^{1,\beta}) \\
&\quad + \Pr_s^{\tau_\varepsilon,\pi_\varepsilon}(\text{Reach}^n(W_1^1 \cup W_2^1) \cap H_n^{0,\beta}) \\
&\geq \Pr_s^{\tau_\varepsilon,\pi_\varepsilon}(\text{Reach}^n(W_1^1) \cap H_n^{1,\beta}) \\
&\quad + \Pr_s^{\tau_\varepsilon,\pi_\varepsilon}(\text{Reach}^n(W_2^1) \cap H_n^{0,\beta}) \\
&\geq \Pr_s^{\tau_\varepsilon,\pi_\varepsilon}(H_n^{1,\beta}) + \Pr_s^{\tau_\varepsilon,\pi_\varepsilon}(H_n^{0,\beta}) - (z_1 + z_2) \\
&\geq 1 - \frac{\varepsilon}{4} + \frac{\varepsilon'}{2} \geq 1 - \varepsilon' \qquad (\text{since } \varepsilon \leq \varepsilon').
\end{aligned}$$

The result follows. ∎

Theorem 2 proves the limit-reachability property for tail objectives, under ε-optimal strategies, for small ε. We present an example to show that Theorem 2 is not true for all objectives, or for tail objectives with arbitrary strategies.

Example 3. Observe that in the game shown in Example 2, the objective was not a tail objective and we had $W_1^1 \cup W_2^1 = \emptyset$. Hence Theorem 2 need not necessarily hold for all objectives. Also consider the game shown in Fig 2. In the game shown s_1 and s_2 are absorbing state. At s_0 the available moves for the players are as follows: $Mv_1(s_0) = \{ a \}$ and $Mv_2(s_0) = \{ 1, 2 \}$. The transition function is as follows: if player 2 plays move 2, then the next state is s_1 and s_2 with equal probability, and if player 2 plays move 1, then the next state is s_0. The objective of player 1 is $\Phi = \text{Büchi}(\{ s_0, s_1 \})$, i.e., to visit s_0 or s_1 infinitely often. We have $W_1^1 = \{ s_1 \}$ and $W_2^1 = \{ s_2 \}$. Given a strategy π that chooses move 1 always, the set $W_1^1 \cup W_2^1$ of states is reached with probability 0; however π is not an optimal or ε-optimal strategy for player 2 (for $\varepsilon < \frac{1}{2}$). This shows that Theorem 2 need not hold if ε-optimal strategies are not considered. In the game shown, for an optimal strategy for player 2 (e.g., a strategy to choose move 2) the play reaches $W_1^1 \cup W_2^1$ with probability 1. ∎

Lemma 3 is immediate from Theorem 2.

Lemma 3. *For all concurrent game structures G, for all tail objectives Φ for player 1 and $\overline{\Phi}$ for player 2, for all states $s \in S$, we have*

$$\lim_{\varepsilon \to 0} \sup_{\tau \in \Gamma_\varepsilon(\Phi), \pi \in \Pi_\varepsilon(\overline{\Phi})} \Pr_s^{\tau,\pi}(\mathit{Reach}(W_1^1 \cup W_2^1)) = 1;$$

$$\lim_{\varepsilon \to 0} \sup_{\tau \in \Gamma_\varepsilon(\Phi), \pi \in \Pi_\varepsilon(\overline{\Phi})} \Pr_s^{\tau,\pi}(\mathit{Reach}(W_1^1)) = \langle\langle 1 \rangle\rangle_{val}(\Phi)(s);$$

$$\lim_{\varepsilon \to 0} \sup_{\tau \in \Gamma_\varepsilon(\Phi), \pi \in \Pi_\varepsilon(\overline{\Phi})} \Pr_s^{\tau,\pi}(\mathit{Reach}(W_2^1)) = \langle\langle 2 \rangle\rangle_{val}(\overline{\Phi})(s).$$

Consider a non-zero sum reachability game G_R such that the states in $W_1^1 \cup W_2^1$ are transformed to absorbing states and the objectives of both players are reachability objectives: the objective for player 1 is Reach(W_1^1) and the objective for player 2 is Reach(W_2^1). Note that the game G_R is not zero-sum in the following sense: there are infinite paths ω such that $\omega \notin$ Reach(W_1^1) and $\omega \notin$ Reach(W_2^1) and each player gets a payoff 0 for the path ω. We define ε-Nash equilibrium of the game G_R and relate some special ε-Nash equilibrium of G_R with the values of G.

Definition 2 (ε-Nash equilibrium in G_R). *A strategy profile $(\tau^*, \pi^*) \in \Gamma \times \Pi$ is an ε-Nash equilibrium at state s if the following two conditions hold:*

$$\Pr_s^{\tau^*,\pi^*}(\mathit{Reach}(W_1^1)) \geq \sup_{\tau \in \Gamma} \Pr_s^{\tau,\pi^*}(\mathit{Reach}(W_1^1)) - \varepsilon$$

$$\Pr_s^{\tau^*,\pi^*}(\mathit{Reach}(W_2^1)) \geq \sup_{\pi \in \Pi} \Pr_s^{\tau^*,\pi}(\mathit{Reach}(W_2^1)) - \varepsilon$$ ∎

Theorem 3 (Nash equilibrium of reachability game G_R). *The following assertion holds for the game G_R.*

1. *For all $\varepsilon > 0$, there is an ε-Nash equilibrium $(\tau_\varepsilon^*, \pi_\varepsilon^*) \in \Gamma_\varepsilon(\Phi) \times \Pi_\varepsilon(\overline{\Phi})$ such that for all states s we have*

$$\lim_{\varepsilon \to 0} \Pr_s^{\tau_\varepsilon^*,\pi_\varepsilon^*}(\mathit{Reach}(W_1^1)) = \langle\langle 1 \rangle\rangle_{val}(\Phi)(s)$$

$$\lim_{\varepsilon \to 0} \Pr_s^{\tau_\varepsilon^*,\pi_\varepsilon^*}(\mathit{Reach}(W_2^1)) = \langle\langle 2 \rangle\rangle_{val}(\overline{\Phi})(s).$$

Proof. It follows from Lemma 3. ∎

Note that in case of MDPs the strategy for player 2 is trivial, i.e., player 2 has only one strategy. Hence in context of MDPs we drop the strategy π of player 2. A specialization of Theorem 3 in case of MDPs yields Theorem 4.

Theorem 4. *For all MDPs G_M, for all tail objectives Φ, we have*

$$\langle\langle 1 \rangle\rangle_{val}(\Phi)(s) = \sup_{\tau \in \Gamma} \Pr_s^{\tau}(\mathit{Reach}(W_1^1)) = \langle\langle 1 \rangle\rangle_{val}(\mathit{Reach}(W_1^1))(s)$$

Since the values in MDPs with reachability objectives can be computed in polynomial time (by linear-programming) [3,9], our result presents a polynomial time reduction of quantitative analysis of tail objectives in MDPs to qualitative analysis. Our results (mainly, Theorem 1 and Theorem 2) can also be used to present simple construction of ε-optimal strategies for ω-regular objectives in concurrent games. These results will be presented in a fuller version of the paper.

Acknowledgments. This research was supported in part by the AFOSR MURI grant F49620-00-1-0327, the NSF ITR grant CCR-0225610 and the SNSF under the Indo-Swiss Joint Research Programme. I thank Tom Henzinger for many insightful discussions. I am grateful to Luca de Alfaro for several key insights on concurrent games that I received from him. I thank David Aldous for a useful discussion. I thank Hugo Gimbert for useful comments on earlier versions of manuscripts.

References

1. K. Chatterjee, L. de Alfaro, and T.A. Henzinger. The complexity of quantitative concurrent parity games. In *SODA 06*, pages 678–687. ACM-SIAM, 2006.
2. K. Chatterjee, L. de Alfaro, and T.A. Henzinger. Trading memory for randomness. In *QEST 04*. IEEE Computer Society Press, 2004.
3. A. Condon. The complexity of stochastic games. *Information and Computation*, 96(2):203–224, 1992.
4. C. Courcoubetis and M. Yannakakis. The complexity of probabilistic verification. *Journal of the ACM*, 42(4):857–907, 1995.
5. L. de Alfaro. *Formal Verification of Probabilistic Systems.* PhD thesis, Stanford University, 1997.
6. L. de Alfaro and T.A. Henzinger. Concurrent omega-regular games. In *LICS 00*, pages 141–154. IEEE Computer Society Press, 2000.
7. L. de Alfaro and R. Majumdar. Quantitative solution of omega-regular games. In *STOC 01*, pages 675–683. ACM Press, 2001.
8. Richard Durrett. *Probability: Theory and Examples.* Duxbury Press, 1995.
9. J. Filar and K. Vrieze. *Competitive Markov Decision Processes.* Springer-Verlag, 1997.
10. J.F. Nash Jr. Equilibrium points in n-person games. *Proceedings of the National Academny of Sciences USA*, 36:48–49, 1950.
11. A. Kechris. *Classical Descriptive Set Theory.* Springer, 1995.
12. D.A. Martin. The determinacy of Blackwell games. *The Journal of Symbolic Logic*, 63(4):1565–1581, 1998.
13. G. Owen. *Game Theory.* Academic Press, 1995.
14. C.H. Papadimitriou. Algorithms, games, and the internet. In *STOC 01*, pages 749–753. ACM Press, 2001.
15. L.S. Shapley. Stochastic games. *Proc. Nat. Acad. Sci. USA*, 39:1095–1100, 1953.
16. W. Thomas. On the synthesis of strategies in infinite games. In *STACS 95*, volume 900 of *LNCS*, pages 1–13. Springer-Verlag, 1995.
17. W. Thomas. Languages, automata, and logic. In *Handbook of Formal Languages*, volume 3, Beyond Words, chapter 7, pages 389–455. Springer, 1997.
18. J. von Neumann and O. Morgenstern. *Theory of games and economic behavior.* Princeton University Press, 1947.

Nash Equilibrium for Upward-Closed Objectives*

Krishnendu Chatterjee

EECS, University of California, Berkeley, USA
c_krish@cs.berkeley.edu

Abstract. We study infinite stochastic games played by n-players on a finite graph with goals specified by sets of infinite traces. The games are *concurrent* (each player simultaneously and independently chooses an action at each round), *stochastic* (the next state is determined by a probability distribution depending on the current state and the chosen actions), *infinite* (the game continues for an infinite number of rounds), *nonzero-sum* (the players' goals are not necessarily conflicting), and undiscounted. We show that if each player has an upward-closed objective, then there exists an ε-Nash equilibrium in memoryless strategies, for every $\varepsilon > 0$; and exact Nash equilibria need not exist. Upward-closure of an objective means that if a set Z of infinitely repeating states is winning, then all supersets of Z of infinitely repeating states are also winning. Memoryless strategies are strategies that are independent of history of plays and depend only on the current state. We also study the complexity of finding values (payoff profile) of an ε-Nash equilibrium. We show that the values of an ε-Nash equilibrium in nonzero-sum concurrent games with upward-closed objectives for all players can be computed by computing ε-Nash equilibrium values of nonzero-sum concurrent games with reachability objectives for all players and a polynomial procedure. As a consequence we establish that values of an ε-Nash equilibrium can be computed in TFNP (total functional NP), and hence in EXPTIME.

1 Introduction

Stochastic games. Non-cooperative games provide a natural framework to model interactions between agents [10]. The simplest class of non-cooperative games consists of the "one-step" games — games with single interaction between the agents after which the game ends and the payoffs are decided (e.g., matrix games). However, a wide class of games progress over time and in stateful manner, and the current game depends on the history of interactions. Infinite *stochastic games* [13,6] are a natural model for such games. A stochastic game is played over a finite *state space* and is played in rounds. In concurrent games, in each round, each player chooses an action from a finite set of available actions, simultaneously and independently of other players. The game proceeds to a new

* This research was supported in part by the NSF grants CCR-0225610 and CCR-0234690, and by the SNSF under the Indo-Swiss Joint Research Programme.

Z. Ésik (Ed.): CSL 2006, LNCS 4207, pp. 271–286, 2006.

state according to a probabilistic transition relation (stochastic transition matrix) based on the current state and the joint actions of the players. Concurrent games subsume the simpler class of *turn-based games*, where at every state at most one player can choose between multiple actions. In verification and control of finite state reactive systems such games proceed for infinite rounds, generating an infinite sequence of states, called the *outcome* of the game. The players receive a payoff based on a payoff function that maps every outcome to a real number.

Objectives. Payoffs are generally Borel measurable functions [9]. The payoff set for each player is a Borel set B_i in the Cantor topology on S^ω (where S is the set of states), and player i gets payoff 1 if the outcome of the game is in B_i, and 0 otherwise. In verification, payoff functions are usually index sets of *ω-regular languages.* The ω-regular languages generalize the classical regular languages to infinite strings, they occur in low levels of the Borel hierarchy (in $\Sigma_3 \cap \Pi_3$), and form a robust and expressive language for determining payoffs for commonly used specifications. The simplest ω-regular objectives correspond to safety ("closed sets") and reachability ("open sets") objectives.

Zero-sum games. Games may be *zero-sum*, where two players have directly conflicting objectives and the payoff of one player is one minus the payoff of the other, or *nonzero-sum*, where each player has a prescribed payoff function based on the outcome of the game. The fundamental question for games is the existence of equilibrium values. For zero-sum games, this involves showing a *determinacy* theorem that states that the expected optimum value obtained by player 1 is exactly one minus the expected optimum value obtained by player 2. For one-step zero-sum games, this is von Neumann's minmax theorem [17]. For infinite games, the existence of such equilibria is not obvious, in fact, by using the axiom of choice, one can construct games for which determinacy does not hold. However, a remarkable result by Martin [9] shows that all stochastic zero-sum games with Borel payoffs are determined.

Nonzero-sum games. For nonzero-sum games, the fundamental equilibrium concept is a *Nash equilibrium* [8], i.e., a strategy profile such that no player can gain by deviating from the profile, assuming the other player continues playing the strategy in the profile. Again, for one-step games, the existence of such equilibria is guaranteed by Nash's theorem [8]. However, the existence of Nash equilibria in infinite games is not immediate: Nash's theorem holds for finite bimatrix games, but in case of stochastic games, the strategy space is not compact. The existence of Nash equilibria is known only in very special cases of stochastic games. In fact, Nash equilibria may not exist, and the best one can hope for is an ε-Nash equilibrium for all $\varepsilon > 0$, where an ε-Nash equilibrium is a strategy profile where unilateral deviation can only increase the payoff of a player by at most ε. Exact Nash equilibria do exist in discounted stochastic games [7]. For concurrent nonzero-sum games with payoffs defined by Borel sets, surprisingly little is known. Secchi and Sudderth [12] showed that exact Nash equilibria do exist when all players have payoffs defined by closed sets ("safety objectives" or Π_1 objectives). In the case of open sets ("reachability objectives" or Σ_1 objectives), the existence of ε-Nash equilibrium for every $\varepsilon > 0$, has been established in [2]. For the special case of two-player games, existence of ε-Nash equilibrium, for every $\varepsilon > 0$, is known for ω-regular objectives [1] and limit-average objectives [15,16]. The existence of ε-Nash equilibrium in n-player concurrent games with objectives in higher levels of Borel hierarchy than Σ_1 and Π_1 has been an intriguing open problem; existence of ε-Nash equilibrium is not even known even when each player has a Büchi objective.

Result and proof techniques. In this paper we show that ε-Nash equilibrium exists, for every $\varepsilon > 0$, for n-player concurrent games with *upward-closed* objectives. However, exact Nash equilibria need not exist. Informally, an objective Ψ is an upward-closed objective, if a play ω that visits a set Z of states infinitely often is in Ψ, then a play ω' that visits $Z' \supseteq Z$ of states infinitely often is also in Ψ. The class of upward-closed objectives subsumes Büchi and generalized Büchi objectives as special cases. For n-player concurrent games our result extends the existence of ε-Nash equilibrium from the lowest level of Borel hierarchy (open and closed sets) to a class of objectives that lie in the higher levels of Borel hierarchy (upward-closed objectives can express objectives in Π_2) and subsumes several interesting class of objectives. Along with the existence of ε-Nash equilibrium, our result presents a finer characterization of ε-Nash equilibrium showing existence of ε-Nash equilibrium in *memoryless strategies* (strategies that are independent of the history of the play and depend only on the current state). Our result is organized as follows.

1. In Section 3 we develop some results on one player version of concurrent games and n-player concurrent games with reachability objectives.
2. In Section 4 we use induction on the number of players, results of Section 3 and analysis of Markov chains to establish the desired result.

Complexity of ε-Nash equilibrium. Computing the values of a Nash equilibria, when it exists, is another challenging problem [11]. For one-step zero-sum games, equilibrium values and strategies can be computed in polynomial time (by reduction to linear programming) [10]. For one-step nonzero-sum games, no polynomial time algorithm is known to compute an exact Nash equilibrium, even in two-player games [11]. From the computational aspects, a desirable property of an existence proof of Nash equilibrium is its ease of algorithmic analysis. We show that our proof for existence of ε-Nash equilibrium is completely algorithmic. Our proof shows that the computation of an ε-Nash equilibrium in n-player concurrent games with upward-closed objectives can be achieved by computing ε-Nash equilibrium of games with reachability objectives and a polynomial time procedure. Our result thus shows that computing ε-Nash equilibrium for upward-closed objectives is no harder than solving ε-Nash equilibrium of n-player games with reachability objectives by a polynomial factor. We then prove that an ε-Nash equilibrium can be computed in TFNP (total functional NP) and hence in EXPTIME.

2 Definitions

Notation. We denote the set of probability distributions on a set A by $\mathcal{D}(A)$. Given a distribution $\delta \in \mathcal{D}(A)$, we denote by $\mathrm{Supp}(\delta) = \{x \in A \mid \delta(x) > 0\}$ the *support* of δ.

Definition 1 (Concurrent game structures). *An n-player* concurrent game structure $\mathcal{G} = \langle S, A, \Gamma_1, \Gamma_2, \ldots, \Gamma_n, \delta \rangle$ *consists of the following components:*

- *A finite state space S and a finite set A of moves.*
- *Move assignments $\Gamma_1, \Gamma_2, \ldots, \Gamma_n : S \to 2^A \setminus \emptyset$. For $i \in \{1, 2, \ldots, n\}$, move assignment Γ_i associates with each state $s \in S$ the non-empty set $\Gamma_i(s) \subseteq A$ of moves available to player i at state s.*
- *A probabilistic transition function $\delta : S \times A \times A \ldots \times A \to \mathcal{D}(S)$, that gives the probability $\delta(s, a_1, a_2, \ldots, a_n)(t)$ of a transition from s to t when player i plays move a_i, for all $s, t \in S$ and $a_i \in \Gamma_i(s)$, for $i \in \{1, 2, \ldots, n\}$.* ∎

We define the size of the game structure $\mathcal{G}$ to be equal to the size of the transition function δ, specifically,

$$|\mathcal{G}| = \sum_{s \in S} \sum_{(a_1, \ldots, a_n) \in \Gamma_1(s) \times \ldots \times \Gamma_n(s)} \sum_{t \in S} |\delta(s, a_1, \ldots, a_n)(t)|,$$

where $|\delta(s, a_1, \ldots, a_n)(t)|$ denotes the space to specify the probability distribution. At every state $s \in S$, each player i chooses a move $a_i \in \Gamma_i(s)$, and simultaneously and independently, and the game then proceeds to the successor state t with probability $\delta(s, a_1, a_2, \ldots, a_n)(t)$, for all $t \in S$. A state s is called an *absorbing state* if for all $a_i \in \Gamma_i(s)$ we have $\delta(s, a_1, a_2, \ldots, a_n)(s) = 1$. In other words, at s for all choices of moves of the players the next state is always s. For all states $s \in S$ and moves $a_i \in \Gamma_i(s)$ we indicate by $\mathrm{Dest}(s, a_1, a_2, \ldots, a_n) = \mathrm{Supp}(\delta(s, a_1, a_2, \ldots, a_n))$ the set of possible successors of s when moves $a_1, a_2, \ldots, a_n$ are selected.

A *path* or a *play* ω of $\mathcal{G}$ is an infinite sequence $\omega = \langle s_0, s_1, s_2, \ldots \rangle$ of states in S such that for all $k \geq 0$, there are moves $a_i^k \in \Gamma_i(s_k)$ and with $\delta(s_k, a_1^k, a_2^k, \ldots, a_n^k)(s_{k+1}) > 0$. We denote by Ω the set of all paths and by Ω_s the set of all paths $\omega = \langle s_0, s_1, s_2, \ldots \rangle$ such that $s_0 = s$, i.e., the set of plays starting from state s.

Randomized strategies. A *selector* ξ_i for player $i \in \{1, 2, \ldots, n\}$ is a function $\xi_i : S \to \mathcal{D}(A)$ such that for all $s \in S$ and $a \in A$, if $\xi_i(s)(a) > 0$ then $a \in \Gamma_i(s)$. We denote by Λ_i the set of all selectors for player $i \in \{1, 2, \ldots, n\}$. A *strategy* σ_i for player i is a function $\sigma_i : S^+ \to \Lambda_i$ that associates with every finite non-empty sequence of states, representing the history of the play so far, a selector. A memoryless strategy is independent of the history of the play and depends only on the current state. Memoryless strategies coincide with selectors, and we often write σ_i for the selector corresponding to a memoryless strategy σ_i. A memoryless strategy σ_i for player i is *uniform memoryless* if the selector of the memoryless strategy is an uniform distribution over its support, i.e., for all states s we have $\sigma_i(s)(a_i) = 0$ if $a_i \notin \mathrm{Supp}(\sigma_i(s))$ and $\sigma_i(s)(a_i) =$

$\frac{1}{|\mathrm{Supp}(\sigma_i(s))|}$ if $a_i \in \mathrm{Supp}(\sigma_i(s))$. We denote by Σ_i, Σ_i^M and Σ_i^{UM} the set of all strategies, set of all memoryless strategies and the set of all uniform memoryless strategies for player i, respectively. Given strategies σ_i for player i, we denote by $\overline{\sigma}$ the strategy profile $(\sigma_1, \sigma_2, \ldots, \sigma_n)$. A strategy profile $\overline{\sigma}$ is memoryless (resp. uniform memoryless) if all the component strategies are memoryless (resp. uniform memoryless).

Given a strategy profile $\overline{\sigma} = (\sigma_1, \sigma_2, \ldots, \sigma_n)$ and a state s, we denote by $\mathrm{Outcome}(s, \overline{\sigma}) = \{\, \omega = \langle s_0, s_1, s_2 \ldots \rangle \mid s_0 = s, \;\text{ for } k \geq 0, \text{ for } i = 1, 2, \ldots, n.\; \exists a_i^k.\; \sigma_i(\langle s_0, s_1, \ldots, s_k \rangle)(a_i^k) > 0. \text{ and } \delta(s_k, a_1^k, a_2^k, \ldots, a_n^k)(s_{k+1}) > 0 \,\}$ the set of all possible plays from s, given $\overline{\sigma}$. Once the starting state s and the strategies σ_i for the players have been chosen, the game is reduced to an ordinary stochastic process. Hence, the probabilities of events are uniquely defined, where an *event* $\mathcal{A} \subseteq \Omega_s$ is a measurable set of paths. For an event $\mathcal{A} \subseteq \Omega_s$, we denote by $\mathrm{Pr}_s^{\overline{\sigma}}(\mathcal{A})$ the probability that a path belongs to $\mathcal{A}$ when the game starts from s and the players follow the strategies σ_i, and $\overline{\sigma} = (\sigma_1, \sigma_2, \ldots, \sigma_n)$.

Objectives. Objectives for the players in nonterminating games are specified by providing the set of *winning plays* $\Psi \subseteq \Omega$ for each player. A general class of objectives are the Borel objectives [9]. A *Borel objective* $\Phi \subseteq S^\omega$ is a Borel set in the Cantor topology on S^ω. The class of ω*-regular objectives* [14], lie in the first $2^1/_2$ levels of the Borel hierarchy (i.e., in the intersection of Σ_3 and Π_3). The ω-regular objectives, and subclasses thereof, can be specified in the following forms. For a play $\omega = \langle s_0, s_1, s_2, \ldots \rangle \in \Omega$, we define $\mathrm{Inf}(\omega) = \{\, s \in S \mid s_k = s \text{ for infinitely many } k \geq 0 \,\}$ to be the set of states that occur infinitely often in ω.

1. *Reachability and safety objectives.* Given a game graph $\mathcal{G}$, and a set $T \subseteq S$ of *target* states, the reachability specification $\mathrm{Reach}(T)$ requires that some state in T be visited. The reachability specification $\mathrm{Reach}(T)$ defines the objective $[\![\mathrm{Reach}(T)]\!] = \{\, \langle s_0, s_1, s_2, \ldots \rangle \in \Omega \mid \exists k \geq 0.\ s_k \in T \,\}$ of winning plays. Given a set $F \subseteq S$ of *safe* states, the safety specification $\mathrm{Safe}(F)$ requires that only states in F be visited. The safety specification $\mathrm{Safe}(F)$ defines the objective $[\![\mathrm{Safe}(F)]\!] = \{\, \langle s_0, s_1, \ldots \rangle \in \Omega \mid \forall k \geq 0.\ s_k \in F \,\}$ of winning of plays.
2. *Büchi and generalized Büchi objectives.* Given a game graph $\mathcal{G}$, and a set $B \subseteq S$ of *Büchi* states, the Büchi specification $\mathrm{Büchi}(B)$ requires that states in B be visited infinitely often. The Büchi specification $\mathrm{Büchi}(B)$ defines the objective $[\![\mathrm{Büchi}(B)]\!] = \{\, \omega \in \Omega \mid \mathrm{Inf}(\omega) \cap B \neq \emptyset \,\}$ of winning plays. Let $B_1, B_2, \ldots, B_n$ be subset of states, i.e., each $B_i \subseteq S$. The generalized Büchi specification is the requires that every Büchi specification $\mathrm{Büchi}(B_i)$ be satisfied. Formally, the generalized Büchi objective is $\bigcap_{i \in \{1,2,\ldots,n\}} [\![\mathrm{Büchi}(B_i)]\!]$.
3. *Müller and upward-closed objectives.* Given a set $M \subseteq 2^S$ of *Müller* set of states, the Müller specification $\mathrm{Müller}(M)$ requires that the set of states visited infinitely often in a play is exactly one of the sets in M. The Müller specification $\mathrm{Müller}(M)$ defines the objective $[\![\mathrm{Müller}(M)]\!] = \{\, \omega \in \Omega \mid \mathrm{Inf}(\omega) \in M \}$ of winning plays. The *upward-closed* objectives form a sub-class

of Müller objectives, with the restriction that the set M is upward-closed. Formally a set $UC \subseteq 2^S$ is upward-closed if the following condition hold: if $U \in UC$ and $U \subseteq Z$, then $Z \in UC$. Given a upward-closed set $UC \subseteq 2^S$, the upward-closed objective is defined as the set $[\![\mathrm{UpClo}(UC)]\!] = \{ \omega \in \Omega \mid \mathrm{Inf}(\omega) \in UC \}$ of winning plays.

The upward-closed objectives subsumes Büchi and generalized Büchi objectives. The upward-closed objectives also subsumes disjunction of Büchi objectives. Since the Büchi objectives lie in the second level of the Borel hierarchy (in Π_2), it follows that upward-closed objectives can express objectives that lie in Π_2. Müller objectives are canonical forms to express ω-regular objectives, and the class of upward-closed objectives form a strict subset of Müller objectives and cannot express all ω-regular properties.

We write Ψ for an arbitrary objective. We write the objective of player i as Ψ_i. The probability that a path satisfies a Müller objective Ψ starting from state $s \in S$ under a strategy profile $\overline{\sigma}$ is denoted as $\Pr_s^{\overline{\sigma}}(\Psi)$.

Notations. Given a strategy profile $\overline{\sigma} = (\sigma_1, \sigma_2, \ldots, \sigma_n)$, we denote by $\overline{\sigma}_{-i} = (\sigma_1, \sigma_2, \ldots, \sigma_{i-1}, \sigma_{i+1}, \ldots, \sigma_n)$ the strategy profile with the strategy for player i removed. Given a strategy $\sigma_i' \in \Sigma_i$, and a strategy profile $\overline{\sigma}_{-i}$, we denote by $\overline{\sigma}_{-i} \cup \sigma_i'$ the strategy profile $(\sigma_1, \sigma_2, \ldots, \sigma_{i-1}, \sigma_i', \sigma_{i+1}, \ldots, \sigma_n)$. We also use the following notations: $\overline{\Sigma} = \Sigma_1 \times \Sigma_2 \times \ldots \times \Sigma_n$; $\overline{\Sigma}^M = \Sigma_1^M \times \Sigma_2^M \times \ldots \times \Sigma_n^M$; $\overline{\Sigma}^{UM} = \Sigma_1^{UM} \times \Sigma_2^{UM} \times \ldots \times \Sigma_n^{UM}$; and $\overline{\Sigma}_{-i} = \Sigma_1 \times \Sigma_2 \times \ldots \Sigma_{i-1} \times \Sigma_{i+1} \times \ldots \Sigma_n$. The notations for $\overline{\Sigma}_{-i}^M$ and $\overline{\Sigma}_{-i}^{UM}$ are similar. For $n \in \mathbb{N}$, we denote by $[n]$ the set $\{ 1, 2, \ldots, n \}$.

Concurrent nonzero-sum games. A concurrent nonzero-sum game consists of a concurrent game structure $\mathcal{G}$ with objective Ψ_i for player i. The zero-sum values for the players in concurrent games with objective Ψ_i for player i are defined as follows.

Definition 2 (Zero-sum values). *Let $\mathcal{G}$ be a concurrent game structure with objective Ψ_i for player i. Given a state $s \in S$ we call the maximal probability with which player i can ensure that Ψ_i holds from s against all strategies of the other players is the* zero-sum value *of player i at s. Formally, the zero-sum value for player i is given by the function $val_i^{\mathcal{G}}(\Psi_i) : S \to [0,1]$ defined for all $s \in S$ by $val_i^{\mathcal{G}}(\Psi_i)(s) = \sup_{\sigma_i' \in \Sigma_i} \inf_{\overline{\sigma}_{-i} \in \overline{\Sigma}_{-i}} \Pr_s^{\overline{\sigma}_{-i} \cup \sigma_i'}(\Psi_i)$.* ■

A two-player concurrent game structure $\mathcal{G}$ with objectives Ψ_1 and Ψ_2 for player 1 and player 2, respectively, is *zero-sum* if the objectives of the players are complementary, i.e., $\Psi_1 = \Omega \setminus \Psi_2$. Concurrent zero-sum games satisfy a *quantitative* version of determinacy [9], stating that for all two-player concurrent games with Müller objectives Ψ_1 and Ψ_2, such that $\Psi_1 = \Omega \setminus \Psi_2$, and all $s \in S$, we have $val_1^{\mathcal{G}}(\Psi_1)(s) + val_2^{\mathcal{G}}(\Psi_2)(s) = 1$. The determinacy also establishes existence of ε-Nash equilibrium, for all $\varepsilon > 0$, in concurrent zero-sum games.

Definition 3 (ε-Nash equilibrium). *Let $\mathcal{G}$ be a concurrent game structure with objective Ψ_i for player i. For $\varepsilon \geq 0$, a strategy profile $\overline{\sigma}^* = (\sigma_1^*, \ldots, \sigma_n^*) \in$*

$\overline{\Sigma}$ is an ε-Nash equilibrium for a state $s \in S$ iff for all $i \in [n]$ we have $\sup_{\sigma_i \in \Sigma_i} \Pr_s^{\overline{\sigma}^*_{-i} \cup \sigma_i}(\Psi_i) \leq \Pr_s^{\overline{\sigma}^*}(\Psi_i) + \varepsilon$. A Nash equilibrium is an ε-Nash equilibrium with $\varepsilon = 0$. ∎

3 Markov Decision Processes and Nash Equilibrium for Reachability Objectives

The section is divided in two parts: first we state some results about one player concurrent game structures and then we state some results about n-player concurrent game structures with reachability objectives. The facts stated in this section will play a key role in the analysis of the later sections.

Markov decision processes. We develop some facts about one player versions of concurrent game structures, known as *Markov decision processes* (MDPs). For $i \in [n]$, a player i-MDP is a concurrent game structure where for all $s \in S$, for all $j \in [n] \setminus \{ i \}$ we have $|\Gamma_j(s)| = 1$, i.e., at every state only player i can choose between multiple moves and the choice for the other players are singleton. If for all states $s \in S$, for all $i \in [n]$, $|\Gamma_i(s)| = 1$, then we have a *Markov chain*. Given a concurrent game structure $\mathcal{G}$, if we fix a memoryless strategy profile $\overline{\sigma}_{-i} = (\sigma_1, \ldots, \sigma_{i-1}, \sigma_{i+1}, \ldots, \sigma_n)$ for players in $[n] \setminus \{ i \}$, then the game structure is equivalent to a player i-MDP $\mathcal{G}_{\overline{\sigma}_{-i}}$ with transition function: $\delta_{\overline{\sigma}_{-i}}(s, a_i)(t) = \sum_{(a_1, a_2, \ldots, a_{i-1}, a_{i+1}, \ldots, a_n)} \delta(s, a_1, a_2, \ldots, a_n)(t) \times \prod_{j \in ([n] \setminus \{ i \})} \sigma_j(s)(a_j)$, for all $s, t \in S$ and $a_i \in \Gamma_i(s)$. Similarly, if we fix a memoryless strategy profile $\overline{\sigma} \in \overline{\Sigma}^M$ for a concurrent game structure $\mathcal{G}$, we obtain a Markov chain, which we denote by $\mathcal{G}_{\overline{\sigma}}$. In an MDP, the sets of states that play an equivalent role to the closed recurrent set of states in Markov chains are called *end components* [3,4]. Without loss of generality, we consider player 1-MDPs and since the set $\overline{\Sigma}_{-1}$ is singleton for player 1-MDPs we only consider strategies for player 1.

Definition 4 (End components and maximal end components). *Given a player 1-MDP $\mathcal{G}$, an* end component *(EC) in $\mathcal{G}$ is a subset $C \subseteq S$ such that there is a memoryless strategy $\sigma_1 \in \Sigma_1^M$ for player 1 under which C forms a closed recurrent set in the resulting Markov chain, i.e., in the Markov chain $\mathcal{G}_{\sigma_1}$. Given a player 1-MDP $\mathcal{G}$, an end component C is a maximal end component, if the following condition hold: if $C \subseteq Z$ and Z is an end component, then $C = Z$, i.e., there is no end component that encloses C.* ∎

Graph of a MDP. Given a player 1-MDP $\mathcal{G}$, the graph of $\mathcal{G}$ is a directed graph (S, E) with the set E of edges defined as follows: $E = \{ (s,t) \mid s, t \in S.\ \exists a_1 \in \Gamma_1(s).\ t \in \mathrm{Dest}(s, a_1) \}$, i.e., $E(s) = \{ t \mid (s,t) \in E \}$ denotes the set of possible successors of the state s in the MDP $\mathcal{G}$.

The following lemma states that in a player 1-MDP, for all strategies of player 1, the set of states visited infinitely often is an end component with probability 1. Lemma 2 follows easily from Lemma 1. Lemma 3 can be proved using the properties of end components.

Lemma 1 **([4,3]).** *Let $\mathcal{C}$ be the set of end components of a player 1-MDP $\mathcal{G}$. For all strategies $\sigma_1 \in \Sigma_1$ and all states $s \in S$, we have* $\Pr_s^{\sigma_1}([\![Müller(\mathcal{C})]\!]) = 1$.

Lemma 2. *Let $\mathcal{C}$ be the set of end components and $\mathcal{Z}$ be the set of maximal end components of a player 1-MDP $\mathcal{G}$. Then for all strategies $\sigma_1 \in \Sigma_1$ and all states $s \in S$, we have* $\Pr_s^{\sigma_1}([\![Reach(L)]\!]) = 1$*, where* $L = \bigcup_{C \in \mathcal{C}} C = \bigcup_{Z \in \mathcal{Z}} Z$*; and*

Lemma 3. *Given a player 1-MDP $\mathcal{G}$ and an end component C, there is a uniform memoryless strategy $\sigma_1 \in \Sigma_1^{UM}$, such that for all states $s \in C$, we have* $\Pr_s^{\sigma_1}(\{\, \omega \mid \mathrm{Inf}(\omega) = C \,\}) = 1$.

Nash equilibrium for reachability objectives. The existence of ε-Nash equilibrium in memoryless strategies in n-player games with reachability objective $[\![\mathrm{Reach}(R_i)]\!]$ for player i, for $R_i \subseteq S$, was shown in [2]. The result can be extended to show the following theorem; we omit the technical details due to lack of space.

Theorem 1 **(ε-Nash equilibrium of full support).** *For every n-player game structure $\mathcal{G}$, with reachability objective $[\![Reach(R_i)]\!]$ for player i, for every $\varepsilon > 0$, there exists a memoryless ε-Nash equilibrium $\overline{\sigma}^* = (\sigma_1^*, \sigma_2^*, \ldots, \sigma_n^*)$ such that for all $s \in S$, for all $i \in [n]$, we have* $\mathrm{Supp}(\sigma_i^*(s)) = \Gamma_i(s)$.

4 Nash Equilibrium for Upward-Closed Objectives

In this section we prove existence of memoryless ε-Nash equilibrium, for all $\varepsilon > 0$, for all n-player concurrent game structures, with upward-closed objectives for all players. The key arguments use induction on the number of players, the results of Section 3 and analysis of Markov chains and MDPs. We present some definitions required for the analysis of the rest of the section.

MDP and graph of a game structure. Given an n-player concurrent game structure $\mathcal{G}$, we define an associated MDP $\overline{\mathcal{G}}$ of $\mathcal{G}$ and an associated graph of $\mathcal{G}$. The MDP $\overline{\mathcal{G}} = (\overline{S}, \overline{A}, \overline{\Gamma}, \overline{\delta})$, where all players unite as a single player, is defined as follows:

- $\overline{S} = S$; $\overline{A} = A \times A \times \ldots \times A = A^n$; and $\overline{\Gamma}(s) = \{\, (a_1, a_2, \ldots, a_n) \mid a_i \in \Gamma_i(s) \,\}$.
- $\overline{\delta}(s, (a_1, a_2, \ldots, a_n)) = \delta(s, a_1, a_2, \ldots, a_n)$.

The graph of the game structure $\mathcal{G}$ is defined as the graph of the MDP $\overline{\mathcal{G}}$.

Games with absorbing states. Given a game structure $\mathcal{G}$ we partition the state space of $\mathcal{G}$ as follows:

1. The set of absorbing states in S are denoted as T, i.e., $T = \{\, s \in C \mid s \text{ is an absorbing state} \,\}$.
2. The set U of states that consists of states s such that $|\Gamma_i(s)| = 1$ for all $i \in [n]$ and $(U \times S) \cap E \subseteq U \times T$. That is at states in U there is no non-trivial choice of moves for the players; thus for any state s in U the game proceeds to the set T according to the probability distribution of the transition function δ at s.
3. $C = S \setminus (U \cup T)$.

Reachable sets. Given a game structure $\mathcal{G}$ and a state $s \in S$, we define $Reachable(s, \mathcal{G}) = \{ t \in S \mid \text{there is a path from } s \text{ to } t \text{ in the graph of } \mathcal{G} \}$ as the set of states that are reachable from s in the graph of the game structure. For a set $Z \subseteq S$, we denote by $Reachable(Z, \mathcal{G})$ the set of states reachable from a state in Z, i.e., $Reachable(Z, \mathcal{G}) = \bigcup_{s \in Z} Reachable(s, \mathcal{G})$. Given a set Z, let $Z_R = Reachable(Z, \mathcal{G})$. We denote by $\mathcal{G} \upharpoonright Z_R$, the sub-game induced by the set Z_R of states. Similarly, given a set $\mathcal{F} \subseteq 2^S$, we denote by $\mathcal{F} \upharpoonright Z_R$ the set $\{ U \mid \exists F \in \mathcal{F}. U = F \cap Z_R \}$.

Terminal non-absorbing maximal end components (TNEC). Given a game structure $\mathcal{G}$, let $\mathcal{Z}$ be the set of maximal end components of the MDP $\overline{\mathcal{G}}$ of $\mathcal{G}$. Let $\mathcal{L} = \mathcal{Z} \setminus T$ be the set of maximal non-absorbing end components and let $H = \bigcup_{L \in \mathcal{L}} L$. A maximal end component $Z \subseteq C$, is a terminal non-absorbing maximal end component (TNEC), if $Reachable(Z, \mathcal{G}) \cap (H \setminus Z) = \emptyset$, i.e., no other non-absorbing maximal end component is reachable from Z.

We consider game structures $\mathcal{G}$ with upward-closed objective $[\![\mathrm{UpClo}(UC_i)]\!]$ for player i. We also denote by $R_i = \{ s \in T \mid \{ s \} \in UC_i \}$ the set of the absorbing states in T that are in UC_i. We now prove the following key result.

Theorem 2. *For all n-player concurrent game structures $\mathcal{G}$, with upward-closed objective $[\![UpClo(UC_i)]\!]$ for player i, one of the following conditions (C1 or C2) hold:*

1. *(Condition C1) There exists a memoryless strategy profile $\overline{\sigma} \in \overline{\Sigma}^M$ such that in the Markov chain $\mathcal{G}_{\overline{\sigma}}$ there is closed recurrent set $Z \subseteq C$, such that $\overline{\sigma}$ is a Nash equilibrium for all states $s \in Z$.*
2. *(Condition C2) There exists a state $s \in C$, such that for all $\varepsilon > 0$, there exists a memoryless ε-Nash equilibrium $\overline{\sigma} \in \overline{\Sigma}^M$ for state s, such that $\mathrm{Pr}_s^{\overline{\sigma}}([\![Reach(T)]\!]) = 1$, and for all $s_1 \in S$, and for all $i \in [n]$, we have $\mathrm{Supp}(\sigma_i(s_1)) = \Gamma_i(s_1)$.*

The proof of Theorem 2 is by induction on the number of players. We first analyze the base case.

Base Case. (One player game structures or MDPs) We consider player 1-MDPs and analyze the following cases:

- (Case 1.) If there in no TNEC in C, then it follows from Lemma 2 that for all states $s \in C$, for all strategies $\sigma_1 \in \Sigma_1$, we have $\mathrm{Pr}_s^{\sigma_1}([\![\mathrm{Reach}(T)]\!]) = 1$, and $\mathrm{Pr}_s^{\sigma_1}([\![\mathrm{Reach}(R_1)]\!]) = \mathrm{Pr}_s^{\sigma_1}([\![\mathrm{UpClo}(UC_1)]\!])$ (recall $R_1 = \{ s \in T \mid \{ s \} \in UC_1 \}$). The result of Theorem 1 yields an ε-Nash equilibrium σ_1 that satisfies condition C2 of Theorem 2, for all states $s \in C$.
- (Case 2.) Else let $Z \subseteq C$ be a TNEC.
  1. If $Z \in UC_1$, fix a uniform memoryless strategy $\sigma_1 \in \Sigma_1^{UM}$ such that for all $s \in Z$, we have $\mathrm{Pr}_s^{\sigma_1}(\{ \omega \mid \mathrm{Inf}(\omega) = Z \}) = 1$ and hence $\mathrm{Pr}_s^{\sigma_1}([\![\mathrm{UpClo}(UC_1)]\!]) = 1$ (such a strategy exists by Lemma 3, since Z is an end component). In other words, Z is a closed recurrent set in the Markov chain $\mathcal{G}_{\sigma_1}$ and the objective of player 1 is satisfied with probability 1. Hence condition C1 of Theorem 2 is satisfied.

2. If $Z \notin UC_1$, then since UC_1 is upward-closed, for all set $Z_1 \subseteq Z$, $Z_1 \notin UC_1$. Hence for any play ω, such that $\omega \in [\![\text{Safe}(Z)]\!]$, we have $\text{Inf}(\omega) \subseteq Z$, and hence $\omega \notin [\![\text{UpClo}(UC_1)]\!]$. Since Z is a TNEC, for all states $s \in Z$ we have

$$\sup_{\sigma_1 \in \Sigma_1} \Pr_s^{\sigma_1}([\![\text{UpClo}(UC_1)]\!]) = \sup_{\sigma_1 \in \Sigma_1} \Pr_s^{\sigma_1}([\![\text{Reach}(R_1)]\!]).$$

If the set of edges from Z to $U \cup T$ is empty (recall $S \setminus C = U \cup T$), then for all strategies σ_1 we have $\Pr_s^{\sigma_1}([\![\text{UpClo}(UC_1)]\!]) = 0$, and hence any uniform memoryless strategy can be fixed and condition C1 of Theorem 2 can be satisfied. Otherwise, the set of edges from Z to $U \cup T$ is non-empty, and then for $\varepsilon > 0$, consider an ε-Nash equilibrium for reachability objective $[\![\text{Reach}(R_1)]\!]$ satisfying the conditions of Theorem 1. Since Z is an end component, for all states $s \in Z$, $\text{Supp}(\sigma_1(s)) = \Gamma_1(s)$, and the set of edges to Z to $U \cup T$ is non-empty it follows that for all states $s \in Z$, we have $\Pr_s^{\sigma_1}([\![\text{Reach}(T)]\!]) = 1$. Thus condition C2 of Theorem 2 is satisfied.

We prove the following lemma, that will be useful for the analysis of the inductive case.

Lemma 4. *Consider a player i-MDP $\mathcal{G}$ with an upward-closed objective UpClo Obj UC_i for player i. Let $\sigma_i \in \Sigma_i^M$ be a memoryless strategy and $Z \subseteq S$ be such that for all $s \in Z$, we have* $\text{Supp}(\sigma_i(s)) = \Gamma_i(s)$ *and Z is a closed recurrent set in the Markov chain $\mathcal{G}_{\sigma_i}$. Then σ_i is a Nash equilibrium (optimal strategy) for all states $s \in Z$.*

Proof. The proof follows from the analysis of two cases.

1. If $Z \in UC_i$, then since Z is a closed recurrent set in $\mathcal{G}_{\sigma_i}$, for all states $s \in S$ we have $\Pr_s^{\sigma_i}(\{\omega \mid \text{Inf}(\omega) = Z\}) = 1$. Hence we have $\Pr_s^{\sigma_i}([\![\text{UpClo}(UC_i)]\!]) = 1$. The result follows.
2. We now consider the case such that $Z \notin UC_i$. Since for all $s \in Z$, we have $\text{Supp}(\sigma_i(s)) = \Gamma_i(s)$, it follows that for all strategies $\sigma_i' \in \Sigma_i$ and for all $s \in Z$, we have $\text{Outcome}(s, \sigma_i') \subseteq \text{Outcome}(s, \sigma_i) \subseteq [\![\text{Safe}(Z)]\!]$ (since Z is a closed recurrent set in $\mathcal{G}_{\sigma_i}$). It follows that for all strategies σ_i' we have $\Pr_s^{\sigma_i'}([\![\text{Safe}(Z)]\!]) = 1$. Hence for all strategies σ_i', for all states $s \in Z$ we have $\Pr_s^{\sigma_i'}(\{\omega \mid \text{Inf}(\omega) \subseteq Z\}) = 1$. Since $Z \notin UC_i$, and UC_i is upward-closed, it follows that for all strategies σ_i', for all states $s \in Z$ we have $\Pr_s^{\sigma_i'}([\![\text{UpClo}(UC_i)]\!]) = 0$. Hence for all states $s \in Z$, we have $\sup_{\sigma_i' \in \Sigma_i} \Pr_s^{\sigma_i'}([\![\text{UpClo}(UC_i)]\!]) = 0 = \Pr_s^{\sigma_i}([\![\text{UpClo}(UC_i)]\!])$. The result follows. ∎

Inductive case. Given a game structure $\mathcal{G}$, consider the MDP $\overline{\mathcal{G}}$: if there are no TNEC in C, then the result follows from analysis similar to Case 1 of the base case. Otherwise consider a TNEC $Z \subseteq C$ in $\overline{\mathcal{G}}$. If for every player i we have $Z \in UC_i$, then fix a uniform memoryless strategy $\overline{\sigma} \in \overline{\Sigma}^{UM}$ such that for all

$s \in Z$, $\Pr_s^{\overline{\sigma}}(\{\omega \mid \mathrm{Inf}(\omega) = Z\}) = 1$ (such a strategy exists by Lemma 3, since Z is an end component in $\overline{\mathcal{G}}$). Hence, for all $i \in [n]$ we have $\Pr_s^{\overline{\sigma}}([\![\mathrm{UpClo}(UC_i)]\!]) = 1$. That is Z is a closed recurrent set in the Markov chain $\mathcal{G}_{\overline{\sigma}}$ and the objective of each player is satisfied with probability 1 from all states $s \in Z$. Hence condition C1 of Theorem 2 is satisfied. Otherwise, there exists $i \in [n]$, such that $Z \notin UC_i$, and without loss of generality we assume that this holds for player 1, i.e., $Z \notin UC_1$. If $Z \notin UC_1$, then we prove Lemma 5 to prove Theorem 2.

Lemma 5. *Consider an n-player concurrent game structure $\mathcal{G}$, with upward-closed objective $[\![UpClo(UC_i)]\!]$ for player i. Let Z be a* TNEC *in $\overline{\mathcal{G}}$ such that $Z \notin UC_1$ and let $Z_R = Reachable(Z, \mathcal{G})$. The following assertions hold:*

1. *If there exists $\sigma_1 \in \Sigma_1^M$, such that for all $s \in Z$, $\mathrm{Supp}(\sigma_1(s)) = \Gamma_1(s)$, and condition C1 of Theorem 2 holds in $\mathcal{G}_{\sigma_1} \upharpoonright Z_R$, then condition C1 Theorem 2 holds in $\mathcal{G}$.*
2. *Otherwise, condition C2 of Theorem 2 holds in $\mathcal{G}$.*

Proof. Given a memoryless strategy σ_1, fixing the strategy σ_1 for player 1, we get an $n-1$-player game structure and by inductive hypothesis either condition C1 or C2 of Theorem 2 holds.

- Case 1. Suppose there is a memoryless strategy $\sigma_1 \in \Sigma_1^M$, such that for all $s \in Z$, $\mathrm{Supp}(\sigma_1(s)) = \Gamma_1(s)$, and condition C1 of Theorem 2 holds in $\mathcal{G}_{\sigma_1} \upharpoonright Z_R$. Let $\overline{\sigma}_{-1} = (\sigma_2, \sigma_3, \ldots, \sigma_n)$ be the memoryless Nash equilibrium and $Z_1 \subseteq Z$ be the closed recurrent set in $\mathcal{G}_{\overline{\sigma}_{-1} \cup \sigma_1}$ satisfying the condition C1 of Theorem 2 in $\mathcal{G}_{\sigma_1}$. Observe that (Z_1, σ_1) satisfy the conditions of Lemma 4 in the MDP $\mathcal{G}_{\overline{\sigma}_{-1}}$. Thus an application of Lemma 4 yields that σ_1 is a Nash equilibrium for all states $s \in Z_1$, in the MDP $\mathcal{G}_{\overline{\sigma}_{-1}}$. Since $\overline{\sigma}_{-1}$ is a Nash equilibrium for all states in Z_1 in $\mathcal{G}_{\sigma_1}$, it follows that $\overline{\sigma} = \overline{\sigma}_{-1} \cup \sigma_1$ and Z_1 satisfy condition C1 of Theorem 2.
- For $\varepsilon > 0$, consider a memoryless ε-Nash equilibrium $\overline{\sigma} = (\sigma_1, \sigma_2, \ldots, \sigma_n)$ in $\mathcal{G}$ with objective $[\![\mathrm{Reach}(R_i)]\!]$ for player i, such that for all $s \in S$, for all $i \in [n]$, we have $\mathrm{Supp}(\sigma_i(s)) = \Gamma_i(s)$ (such an ε-Nash equilibrium exists from Theorem 1). We now prove the desired result analyzing two sub-cases:
  1. Suppose there exists $j \in [n]$, and $Z_j \subseteq Z$, such that $Z_j \in UC_j$, and Z_j is an end component in $\mathcal{G}_{\overline{\sigma}_{-j}}$, then let σ_j' be a memoryless strategy for player j, such that Z_j is a closed recurrent set of states in the Markov chain $\mathcal{G}_{\overline{\sigma}_{-j} \cup \sigma_j'}$. Let $\overline{\sigma}' = \overline{\sigma}_{-j} \cup \sigma_j'$. Since $Z_j \in UC_j$, it follows that for all states $s \in Z_j$, we have $\Pr_s^{\overline{\sigma}'}([\![\mathrm{UpClo}(UC_j)]\!]) = 1$, and hence player j has no incentive to deviate from $\overline{\sigma}'$. Since for all σ_i, for $i \neq j$, and for all states $s \in S$, we have $\mathrm{Supp}(\sigma_i)(s) = \Gamma_i(s)$, and Z_j is a closed recurrent set in $\mathcal{G}_{\overline{\sigma}'}$, it follows from Lemma 4 that for all $j \neq i$, σ_i is a Nash equilibrium in $\mathcal{G}_{\overline{\sigma}'_{-i}}$. Hence we have $\overline{\sigma}'$ is a Nash equilibrium for all states $s \in Z_j$ in $\mathcal{G}$ and condition C1 of Theorem 2 is satisfied.
  2. Hence it follows that if Case 1 fails, for all $i \in [n]$, all end components $Z_i \subseteq Z$, in $\mathcal{G}_{\overline{\sigma}_{-i}}$, we have $Z_i \notin UC_i$. Hence for all $i \in [n]$, for all $s \in Z$, for all $\sigma_i' \in \Sigma_i$, we have $\Pr_s^{\overline{\sigma}_{-i} \cup \sigma_i'}([\![\mathrm{UpClo}(UC_i)]\!]) = \Pr_s^{\overline{\sigma}_{-i} \cup \sigma_i'}([\![\mathrm{Reach}(R_i)]\!])$.

Since $\overline{\sigma}$ is an ε-Nash equilibrium with objectives $[\![\text{Reach}(R_i)]\!]$ for player i in $\mathcal{G}$, it follows that $\overline{\sigma}$ is an ε-Nash equilibrium in $\mathcal{G}$ with objectives $[\![\text{UpClo}(UC_i)]\!]$ for player i. Moreover, if there is an closed recurrent set $Z' \subseteq Z$ in the Markov chain $\mathcal{G}_{\overline{\sigma}}$, then case 1 would have been true (follows from Lemma 4). Hence if case 1 fails, then it follows that there is no closed recurrent set $Z' \subseteq Z$ in $\mathcal{G}_{\overline{\sigma}}$, and hence for all states $s \in Z$, we have $\text{Pr}_s^{\overline{\sigma}}([\![\text{Reach}(T)]\!]) = 1$. Hence condition C2 of Theorem 2 holds, and the result follows. ∎

Inductive application of Theorem 2. Given a game structure $\mathcal{G}$, with upward-closed objective $[\![\text{UpClo}(UC_i)]\!]$ for player i, to prove existence of ε-Nash equilibrium for all states $s \in S$, for $\varepsilon > 0$, we apply Theorem 2 recursively. We convert the game structure $\mathcal{G}$ to $\mathcal{G}'$ as follows.

Transformation 1. If condition C1 of Theorem 2 holds, then let Z be the closed recurrent set that satisfy the condition C1 of Theorem 2.

- In $\mathcal{G}'$ convert every state $s \in Z$ to an absorbing state;
- if $Z \notin UC_i$, for player i, then the objective for player i in $\mathcal{G}'$ is UC_i;
- if $Z \in UC_i$ for player i, the objective for player i in $\mathcal{G}$ is modified to include every state $s \in Z$, i.e., for all $Q \subseteq S$, if $s \in Q$, for some $s \in Z$, then Q is included in UC_i.

Observe that the states in Z are converted to absorbing states and will be interpreted as states in T in $\mathcal{G}'$.

Transformation 2. If condition C2 of Theorem 2 holds, then let $\overline{\sigma}^*$ be an $\frac{\varepsilon}{|S|}$-Nash equilibrium from state s, such that $\text{Pr}_s^{\overline{\sigma}^*}([\![\text{Reach}(T)]\!]) = 1$. The state is converted as follows: for all $i \in [n]$, the available moves for player i at s is reduced to 1, i.e., for all $i \in [n]$, $\Gamma_i(s) = \{ a_i \}$, and the transition function δ' in $\mathcal{G}'$ at s is defined as:

$$\delta(s, a_1, a_2, \ldots, a_n)(t) = \begin{cases} \text{Pr}_s^{\overline{\sigma}^*}([\![\text{Reach}(t)]\!]) & \text{if } t \in T \\ 0 & \text{otherwise.} \end{cases}$$

Note that the state s can be interpreted as a state in U in $\mathcal{G}'$.

To obtain an ε-Nash equilibrium for all states $s \in S$ in $\mathcal{G}$, it suffices to obtain an ε-Nash equilibrium for all states in $\mathcal{G}'$. Also observe that for all states in $U \cup T$, Nash equilibrium exists by definition. Applying the transformations recursively on $\mathcal{G}'$, we proceed to convert every state to a state in $U \cup T$, and the desired result follows. This yields Theorem 3.

Theorem 3. *For all n-player concurrent game structures $\mathcal{G}$, with upward-closed objective $[\![\text{UpClo}(UC_i)]\!]$ for player i, for all $\varepsilon > 0$, for all states $s \in S$, there exists a memoryless strategy profile $\overline{\sigma}^*$, such that $\overline{\sigma}^*$ is an ε-Nash equilibrium for state s.*

Remark 1. Upward-closed objectives are not closed under complementation. Hence Theorem 3 is not a generalization of determinacy result for concurrent

Algorithm 1. UpClCndC1

Input : An n-player game structure $\mathcal{G}$ and upward-closed objective $[\![\text{UpClo}(UC_i)]\!]$ for player i, for all $i \in [n]$.
Output: Either $(Z, \overline{\sigma})$ satisfying condition C1 of Theorem 2 or else $(\emptyset, \emptyset)$.
1. **if** $n = 0$,
   1.1 **if** there is a non-absorbing closed recurrent set Z in the Markov chain $\mathcal{G}$, then **return** $(Z, \emptyset)$.
   1.2 **else return** $(\emptyset, \emptyset)$.
2. $\mathcal{Z}$ =**ComputeMaximalEC**($\overline{\mathcal{G}}$) (i.e., $\mathcal{Z}$ is the set of maximal end components in the MDP of $\mathcal{G}$).
3. **if** there is no TNEC in $\overline{\mathcal{G}}$, **return** $(\emptyset, \emptyset)$.
4. **if** there exists $Z \in \mathcal{Z}$ such that for all $i \in [n]$, $Z \in UC_i$,
   4.1. **return** $(Z, \overline{\sigma})$ such that $\overline{\sigma} \in \overline{\Sigma}^{UM}$ and Z is closed recurrent set in $\mathcal{G}_{\overline{\sigma}}$.
5. Let Z be a TNEC in $\overline{\mathcal{G}}$, and let $Z_R = Reachable(Z, \mathcal{G})$.
6. **else** without loss of generality let $Z \notin UC_n$.
   6.1. Let $\sigma_n \in \Sigma_n^{UM}$ such that for all states $s \in Z_R$, $\sigma_n(s) = \Gamma_n(s)$.
   6.2. $(Z_1, \overline{\sigma})$ = **UpClCndC1** $(\mathcal{G}_{\sigma_n} \upharpoonright Z_R, n-1, [\![\text{UpClo}(UC_i \upharpoonright Z_R)]\!], i \in [n-1])$
   6.3. **if** $(Z_1 = \emptyset)$ **return** $(\emptyset, \emptyset)$; **else return** $(Z_1, \overline{\sigma}_{-n} \cup \sigma_n)$.

zero-sum games with upward-closed objective for one player. For example in concurrent zero-sum games with Büchi objective for a player, ε-optimal strategies require infinite-memory in general, but the complementary objective of a Büchi objective is not upward-closed. In contrast, we show the existence of memoryless ε-Nash equilibrium for n-player concurrent games where each player has an upward-closed objective.

5 Computational Complexity

In this section we present an algorithm to compute an ε-Nash equilibrium for n-player game structures with upward closed objectives, for $\varepsilon > 0$. A key result for the algorithmic analysis is Lemma 6.

Lemma 6. *Consider an n-player concurrent game structure $\mathcal{G}$, with upward-closed objective $[\![UpClo(UC_i)]\!]$ for player i. Let Z be a* TNEC *in $\overline{\mathcal{G}}$ such that $Z \notin UC_n$ and let $Z_R = Reachable(Z, \mathcal{G})$. The following assertion hold.*

- *Suppose there exists $\sigma_n \in \Sigma_n^M$, such that for all $s \in Z$,* $\text{Supp}(\sigma_n(s)) = \Gamma_n(s)$, *and condition C1 of Theorem 2 holds in $\mathcal{G}_{\sigma_n} \upharpoonright Z_R$. Let $\sigma_n^* \in \Sigma_n^{UM}$ such that for all $s \in Z$ we have* $\text{Supp}(\sigma_n^*(s)) = \Gamma_n(s)$ *(i.e., σ_n^* is an uniform memoryless strategy that plays all available moves at all states in Z). Then condition C1 holds in $\mathcal{G}_{\sigma_n^*} \upharpoonright Z_R$.*

Lemma 6 presents the basic principle to identify if condition C1 of Theorem 2 holds in a game structure $\mathcal{G}$ with upward-closed objective $[\![\text{UpClo}(UC_i)]\!]$ for

Algorithm 2. NashEqmCompute

Input : An n-player game structure $\mathcal{G}$ and upward-closed objective$[\![\text{UpClo}(UC_i)]\!]$ for player i, for all $i \in [n]$.
Output: Either $(Z, \overline{\sigma})$ satisfying condition C1 of Theorem 2 or else $(s, \overline{\sigma})$ satisfying condition C2 of Theorem 2.
1. $\mathcal{Z} =$**ComputeMaximalEC**$(\overline{\mathcal{G}})$
2. **if** there is no TNEC in $\overline{\mathcal{G}}$, then **return** $(s,$**ReachEqmFull**$(\mathcal{G}, n, \varepsilon))$ for some $s \in C$.
3. Let Z be a TNEC in $\overline{\mathcal{G}}$, and let $Z_R = \mathit{Reachable}(Z, \mathcal{G})$.
4. Let $(Z_1, \overline{\sigma}) =$ **UpClCndC1** $(\mathcal{G}_{\sigma_n} \restriction Z_R, n-1, [\![\text{UpClo}(UC_i \restriction Z_R)]\!], i \in [n-1])$
5. **if** $(Z_1 \neq \emptyset)$ **return** $(Z_1, \overline{\sigma})$;
6. Let $\overline{\sigma} =$ **ReachEqmFull**$(\mathcal{G}, n, \varepsilon)$.
7. For $s \in C$, if $\overline{\sigma}$ is an ε-Nash equilibrium for s, with objectives $[\![\text{UpClo}(UC_i)]\!]$ for player i, then **return** $(s, \overline{\sigma})$.

player i. An informal description of the algorithm (Algorithm 1) is as follows: the algorithm takes as input a game structure $\mathcal{G}$ of n-players, objectives $[\![\text{UpClo}(UC_i)]\!]$ for player i, and it either returns $(Z, \overline{\sigma})$ satisfying the condition C1 of Theorem 2 or returns $(\emptyset, \emptyset)$. Let $\overline{\mathcal{G}}$ be the MDP of $\mathcal{G}$, and let $\mathcal{Z}$ be the set of maximal end components in $\overline{\mathcal{G}}$ (computed in Step 2 of Algorithm 1). If there is no TNEC in $\overline{\mathcal{G}}$, then condition C1 of Theorem 2 fails and $(\emptyset, \emptyset)$ is returned (Step 3 of Algorithm 1). If there is a maximal end component $Z \in \mathcal{Z}$ such that for all $i \in [n]$, $Z \in UC_i$, then fix an uniform memoryless strategy $\overline{\sigma} \in \overline{\Sigma}^{UM}$ such that Z is a closed recurrent set in $\mathcal{G}_{\overline{\sigma}}$ and return $(Z, \overline{\sigma})$ (Step 4 of Algorithm 1). Else let Z be a TNEC and without of loss of generality let $Z \notin UC_n$. Let $Z_R = \mathit{Reachable}(Z, \mathcal{G})$, and fix a strategy $\sigma_n \in \Sigma_n^{UM}$, such that for all $s \in Z_R$, $\text{Supp}(\sigma_n(s)) = \Gamma_n(s)$. The $n-1$-player game structure $\mathcal{G}_{\sigma_n} \restriction Z_R$ is solved by an recursive call (Step 6.3) and the result of the recursive call is returned. It follows from Lemma 6 and Theorem 2 that if Algorithm 1 returns $(\emptyset, \emptyset)$, then condition C2 of Theorem 2 holds for some state $s \in C$. Let $T(|\mathcal{G}|, n)$ denote the running time of Algorithm 1 on a game structure $\mathcal{G}$ with n-players. Step 2 of the algorithm can be computed in $O(|\mathcal{G}|^2)$ time (see [5] for a $O(|\mathcal{G}|^2)$ time algorithm to compute maximal end components of a MDP). Step 4 can be achieved in time linear in the size of the game structure. Thus we obtain the recurrence: $T(|\mathcal{G}|, n) = O(|\mathcal{G}|^2) + T(|\mathcal{G}|, n-1)$. Hence we have $T(|\mathcal{G}|, n) = O(n \cdot |\mathcal{G}|^2)$.

Basic principle of Algorithm 2. Consider a game structure $\mathcal{G}$ with objective $[\![\text{UpClo}(UC_i)]\!]$ for player i. Let $\overline{\sigma}$ be a memoryless strategy profile such that for all states $s \in S$, for all $i \in [n]$, we have $\text{Supp}(\sigma_i(s)) = \Gamma_i(s)$, and $(s, \overline{\sigma})$ satisfy condition C2 of Theorem 2 for some state $s \in C$. Let $Z_s = \mathit{Reachable}(s, \mathcal{G})$. It follows from the base case analysis of Theorem 2 and Lemma 5, that for all $i \in [n]$, in the MDP $\mathcal{G}_{\overline{\sigma}_{-i}} \restriction Z_s$, for all end components $Z \subseteq Z_s$, $Z \notin UC_i$, and hence in $\mathcal{G}_{\overline{\sigma}_{-i}} \restriction Z_s$, the objective $[\![\text{UpClo}(UC_i)]\!]$ is equivalent to $[\![\text{Reach}(R_i)]\!]$.

It follows that if condition C2 of Theorem 2 holds at a state s, then for all $\varepsilon > 0$, any memoryless ε-Nash equilibrium $\overline{\sigma}$ in $\mathcal{G}$ with objective $[\![\text{Reach}(R_i)]\!]$ for player i, such that for all $s \in S$, for all $i \in [n]$, $\text{Supp}(\sigma_i(s)) = \Gamma_i(s)$, is also an ε-Nash equilibrium in $\mathcal{G}$ with objective $[\![\text{UpClo}(UC_i)]\!]$ for player i. This observation is formalized in Lemma 7. Lemma 7 and Algorithm 1 is the basic principle to obtain a memoryless ε-Nash equilibrium at a non-empty set of states in C.

Lemma 7. *For a game structure $\mathcal{G}$ with objective $[\![UpClo(UC_i)]\!]$ for player i, let $\overline{\sigma}$ be a memoryless strategy profile such for all states $s \in S$, for all $i \in [n]$, we have $\text{Supp}(\sigma_i(s)) = \Gamma_i(s)$, and $(s, \overline{\sigma})$ satisfy condition C2 of Theorem 2 for some state $s \in C$. For $\varepsilon > 0$, any memoryless ε-Nash equilibrium $\overline{\sigma}'$ in $\mathcal{G}$ for state s with objective $[\![Reach(R_i)]\!]$ for player i, such that for all $s \in S$, for all $i \in [n]$, $\text{Supp}(\sigma'_i(s)) = \Gamma_i(s)$, is also an ε-Nash equilibrium in $\mathcal{G}$ for state s with objective $[\![UpClo(UC_i)]\!]$ for player i.*

Description of Algorithm 2. We now describe Algorithm 2 that compute an ε-Nash equilibrium at some state s of a game structure $\mathcal{G}$, with upward-closed objective $[\![\text{UpClo}(UC_i)]\!]$ for player i, for $\varepsilon > 0$. In the algorithm the procedure **ReachEqmFull** returns a strategy $\overline{\sigma} = (\sigma_1, \sigma_2, \ldots, \sigma_n)$ such that for all s, $\text{Supp}(\sigma_i(s)) = \Gamma_i(s)$, and $\overline{\sigma}$ is an ε-Nash equilibrium in $\mathcal{G}$ with reachability objective $[\![\text{Reach}(R_i)]\!]$ for player i, from all states in S. The algorithm first computes the set of maximal end components in $\mathcal{G}$. If there is no TNEC in $\mathcal{G}$, then it invokes **ReachEqmFull**. Otherwise, for some TNEC Z and $Z_R = \mathit{Reachable}(Z, \mathcal{G})$, it invokes Algorithm 1 on the sub-game $\mathcal{G} \upharpoonright Z_R$. If Algorithm 1 returns a non-empty set (i.e., condition C1 of Theorem 2 holds), then the returned value of Algorithm 1 is returned. Otherwise, the algorithm invokes **ReachEqmFull** and returns $(s, \overline{\sigma})$ satisfying condition C2 of Theorem 2. Observe that the procedure **ReachEqmFull** is invoked when: either there is no TNEC in $\mathcal{G}$, or condition C2 holds in $\mathcal{G} \upharpoonright Z_R$. It suffices to compute a memoryless $\frac{\varepsilon}{2}$-Nash equilibrium $\overline{\sigma}' = (\sigma'_1, \sigma'_2, \ldots, \sigma'_n)$ in $\mathcal{G} \upharpoonright Z_R$ with reachability objective $[\![\text{Reach}(R_i)]\!]$ for player i, and then slightly modify $\overline{\sigma}'$ to a memoryless strategy $\overline{\sigma}$ to obtain (s, σ) as desired. Hence it follows that the complexity of **ReachEqmFull** can be bounded by the complexity of a procedure to compute memoryless ε-Nash equilibrium in game structures with reachability objectives. Thus we obtain that the running time of Algorithm 2 is bounded by $O(n \cdot |\mathcal{G}|^2) + \textbf{ReachEqm}(|\mathcal{G}|, n, \varepsilon)$, where **ReachEqm** is the complexity of a procedure to compute memoryless ε-Nash equilibrium in games with reachability objectives.

The inductive application of Theorem 2 to obtain Theorem 3 using transformation 1 and transformation 2 shows that Algorithm 2 can be applied $|S|$-times to compute a memoryless ε-Nash equilibrium for all states $s \in S$. For all constants $\varepsilon > 0$, existence of polynomial witness and polynomial time verification procedure for $\textbf{ReachEqm}(\mathcal{G}, n, \varepsilon)$ has been proved in [2]. It follows that for all constants $\varepsilon > 0$, $\textbf{ReachEqm}(\mathcal{G}, n, \varepsilon)$ is in the complexity class TFNP. The above analysis yields Theorem 4.

Theorem 4. *Given an n-player game structure $\mathcal{G}$ with upward-closed objective $[\![UpClo(UC_i)]\!]$ for player i, a memoryless ε-Nash equilibrium for all $s \in S$ can be computed (a) in TFNP for all constants $\varepsilon > 0$; and (b) in time $O(|S| \cdot n \cdot |\mathcal{G}|^2) + |S| \cdot \mathbf{ReachEqm}(\mathcal{G}, n, \varepsilon)$.*

6 Conclusion

In this paper we establish existence of memoryless ε-Nash equilibrium, for all $\varepsilon > 0$, for all n-player concurrent game structures, with upward-closed objectives for all players, and also present an algotihm to compute an ε-Nash equilibrium. The existence of ε-Nash equilibrium, for all $\varepsilon > 0$, in n-player concurrent game structures with ω-regular objectives, and other class of objectives in the higher levels of Borel hierarchy are interesting open problems.

References

1. K. Chatterjee. Two-player nonzero-sum ω-regular games. In *CONCUR'05*, pages 413–427. LNCS 3653, Springer, 2005.
2. K. Chatterjee, R. Majumdar, and M. Jurdziński. On Nash equilibria in stochastic games. In *CSL'04*, pages 26–40. LNCS 3210, Springer, 2004.
3. C. Courcoubetis and M. Yannakakis. The complexity of probabilistic verification. *JACM*, 42(4):857–907, 1995.
4. L. de Alfaro. *Formal Verification of Probabilistic Systems.* PhD thesis, Stanford University, 1997.
5. Luca de Alfaro. Computing minimum and maximum reachability times in probabilistic systems. In *CONCUR'99*, pages 66–81, 1999.
6. J. Filar and K. Vrieze. *Competitive Markov Decision Processes.* Springer-Verlag, 1997.
7. A.M. Fink. Equilibrium in a stochastic n-person game. *Journal of Science of Hiroshima University*, 28:89–93, 1964.
8. J.F. Nash Jr. Equilibrium points in n-person games. *Proceedings of the National Academny of Sciences USA*, 36:48–49, 1950.
9. D.A. Martin. The determinacy of Blackwell games. *Journal of Symbolic Logic*, 63(4):1565–1581, 1998.
10. G. Owen. *Game Theory.* Academic Press, 1995.
11. C.H. Papadimitriou. On the complexity of the parity argument and other inefficient proofs of existence. *JCSS*, 48(3):498–532, 1994.
12. P. Secchi and W.D. Sudderth. Stay-in-a-set games. *International Journal of Game Theory*, 30:479–490, 2001.
13. L.S. Shapley. Stochastic games. *Proc. Nat. Acad. Sci. USA*, 39:1095–1100, 1953.
14. W. Thomas. Languages, automata, and logic. In *Handbook of Formal Languages*, volume 3, chapter 7, pages 389–455. Springer, 1997.
15. N. Vieille. Two player stochastic games I: a reduction. *Israel Journal of Mathematics*, 119:55–91, 2000.
16. N. Vieille. Two player stochastic games II: the case of recursive games. *Israel Journal of Mathematics*, 119:93–126, 2000.
17. J. von Neumann and O. Morgenstern. *Theory of games and economic behavior.* Princeton University Press, 1947.

Algorithms for Omega-Regular Games with Imperfect Information*,**

Krishnendu Chatterjee[1], Laurent Doyen[2,***], Thomas A. Henzinger[1,3], and Jean-François Raskin[2]

[1] EECS, University of California at Berkeley, U.S.A.
[2] CS, Université Libre de Bruxelles, Belgium
[3] I&C, École Polytechnique Fédérale de Lausanne, Switzerland

Abstract. We study observation-based strategies for two-player turn-based games on graphs with omega-regular objectives. An observation-based strategy relies on imperfect information about the history of a play, namely, on the past sequence of observations. Such games occur in the synthesis of a controller that does not see the private state of the plant. Our main results are twofold. First, we give a fixed-point algorithm for computing the set of states from which a player can win with a deterministic observation-based strategy for any omega-regular objective. The fixed point is computed in the lattice of antichains of state sets. This algorithm has the advantages of being directed by the objective and of avoiding an explicit subset construction on the game graph. Second, we give an algorithm for computing the set of states from which a player can win with probability 1 with a randomized observation-based strategy for a Büchi objective. This set is of interest because in the absence of perfect information, randomized strategies are more powerful than deterministic ones. We show that our algorithms are optimal by proving matching lower bounds.

1 Introduction

Two-player games on graphs play an important role in computer science. In particular, the *controller synthesis* problem asks, given a model for a plant, to construct a model for a controller such that the behaviors resulting from the parallel composition of the two models respects a given specification (e.g., are included in an ω-regular set). Controllers can be synthesized as winning strategies in a graph game whose vertices represent the plant states, and whose players represent the plant and the controller [18,17]. Other applications of graph games

* This research was supported in part by the NSF grants CCR-0225610 and CCR-0234690, by the SNSF under the Indo-Swiss Joint Research Programme, and by the FRFC project "Centre Fédéré en Vérification" funded by the FNRS under grant 2.4530.02.
** A fuller version with proofs is available as UC Berkeley Tech. Rep. EECS-2006-89.
*** Research fellow supported by the Belgian National Science Foundation (FNRS).

Z. Ésik (Ed.): CSL 2006, LNCS 4207, pp. 287–302, 2006.

include realizability and compatibility checking, where the players represent parallel processes of a system, or its environment [1,11,6].

Most results about two-player games played on graphs make the hypothesis of *perfect information*. In this setting, the controller knows, during its interaction with the plant, the exact state of the plant. In practice, this hypothesis is often not reasonable. For example, in the context of hybrid systems, the controller acquires information about the state of the plant using sensors with finite precision, which return imperfect information about the state. Similarly, if the players represent individual processes, then a process has only access to the public variables of the other processes, not to their private variables [19,2].

Two-player games of *imperfect information* are considerably more complicated than games of perfect information. First, decision problems for imperfect-information games usually lie in higher complexity classes than their perfect-information counter-parts [19,14,2]. The algorithmic difference is often exponential, due to a subset construction that, similar to the determinization of finite automata, turns an imperfect-information game into an equivalent perfect-information game. Second, because of the determinization, no symbolic algorithms are known to solve imperfect-information games. This is in contrast to the perfect-information case, where (often) simple and elegant fixed-point algorithms exist [12,8]. Third, in the context of imperfect information, deterministic strategies are sometimes insufficient. A game is *turn-based* if in every state one of the players chooses a successor state. While deterministic strategies suffice to win turn-based games of perfect information, turn-based games of imperfect information require randomized strategies to win with probability 1 (see Example 1). Fourth, winning strategies for imperfect-information games need memory even for simple objectives such as safety and reachability (for an example see the technical-report version of this paper). This is again in contrast to the perfect-information case, where turn-based safety and reachability games can be won with memoryless strategies.

The contributions of this paper are twofold. First, we provide a symbolic fixed-point algorithm to solve games of imperfect information for arbitrary ω-regular objectives. The novelty is that our algorithm is symbolic; it does not carry out an explicit subset construction. Instead, we compute fixed points on the lattice of antichains of state sets. Antichains of state sets can be seen as a symbolic and compact representation for $\subseteq$-downward-closed sets of sets of states.[1] This solution extends our recent result [10] from safety objectives to all ω-regular objectives. To justify the correctness of the algorithm, we transform games of imperfect information into games of perfect information while preserving the existence of winning strategies for every Borel objective. The reduction is only part of the proof, not part of the algorithm. For the special case of parity objectives, we obtain a symbolic EXPTIME algorithm for solving parity games of imperfect

[1] We recently used this symbolic representation of $\subseteq$-downward-closed sets of state sets to propose a new algorithm for solving the universality problem of nondeterministic finite automata. First experiments show a very promising performance; (see [9]).

information. This is optimal, as the reachability problem for games of imperfect information is known to be EXPTIME-hard [19].

Second, we study randomized strategies and winning with probability 1 for imperfect-information games. To our knowledge, for these games no algorithms (symbolic or not) are present in the literature. Following [7], we refer to winning with probability 1 as *almost-sure* winning (*almost* winning, for short), in contrast to *sure* winning with deterministic strategies. We provide a symbolic EXPTIME algorithm to compute the set of almost-winning states for games of imperfect information with Büchi objectives (reachability objectives can be obtained as a special case, and for safety objectives almost winning and sure winning coincide). Our solution is again justified by a reduction to games of perfect information. However, for randomized strategies the reduction is different, and considerably more complicated. We prove our algorithm to be optimal, showing that computing the almost-winning states for reachability games of imperfect information is EXPTIME-hard. The problem of computing the almost-winning states for coBüchi objectives under imperfect information in EXPTIME remains an open problem.

Related work. In [17], Pnueli and Rosner study the synthesis of reactive modules. In their framework, there is no game graph; instead, the environment and the objective are specified using an LTL formula. In [14], Kupferman and Vardi extend these results in two directions: they consider CTL* objectives and imperfect information. Again, no game graph, but a specification formula is given to the synthesis procedure. We believe that our setting, where a game graph is given explicitly, is more suited to fully and uniformly understand the role of imperfect information. For example, Kupferman and Vardi assert that imperfect information comes at no cost, because if the specification is given as a CTL (or CTL*) formula, then the synthesis problem is complete for EXPTIME (resp., 2EXPTIME), just as in the perfect-information case. These hardness results, however, depend on the fact that the specification is given compactly as a formula. In our setting, with an explicit game graph, reachability games of perfect information are PTIME-complete, whereas reachability games of imperfect information are EXPTIME-complete [19]. None of the above papers provide symbolic solutions, and none of them consider randomized strategies.

It is known that for Partially Observable Markov Decision Processes (POMDPs) with boolean rewards and limit-average objectives the quantitative analysis (whether the value is greater than a specified threshold) is EXPTIME-complete [15]. However, almost winning is a qualitative question, and our hardness result for almost winning of imperfect-information games does not follow from the known results on POMDPs. We propose in Section 5 a new proof of the hardness for sure winning of imperfect-information games with reachability objectives, and we extend the proof to almost winning as well. To the best of our knowledge, this is the first hardness result that applies to the qualitative analysis of almost winning in imperfect-information games. A class of *semiperfect*-information games, where one player has imperfect information and the other player has perfect information, is studied in [4]. That class is simpler than the games studied here; it can be solved in NP $\cap$ coNP for parity objectives.

2 Definitions

A *game structure* (*of imperfect information*) is a tuple $G = \langle L, l_0, \Sigma, \Delta, \mathcal{O}, \gamma \rangle$, where L is a finite set of states, $l_0 \in L$ is the initial state, Σ is a finite alphabet, $\Delta \subseteq L \times \Sigma \times L$ is a set of labeled transitions, $\mathcal{O}$ is a finite set of observations, and $\gamma : \mathcal{O} \to 2^L \backslash \emptyset$ maps each observation to the set of states that it represents. We require the following two properties on G: (i) for all $\ell \in L$ and all $\sigma \in \Sigma$, there exists $\ell' \in L$ such that $(\ell, \sigma, \ell') \in \Delta$; and ($ii$) the set $\{\gamma(o) \mid o \in \mathcal{O}\}$ partitions L. We say that G is a game structure of *perfect information* if $\mathcal{O} = L$ and $\gamma(\ell) = \{\ell\}$ for all $\ell \in L$. We often omit $(\mathcal{O}, \gamma)$ in the description of games of perfect information. For $\sigma \in \Sigma$ and $s \subseteq L$, let $\mathsf{Post}_\sigma^G(s) = \{\ell' \in L \mid \exists \ell \in s : (\ell, \sigma, \ell') \in \Delta\}$.

Plays. In a game structure, in each turn, Player 1 chooses a letter in Σ, and Player 2 resolves nondeterminism by choosing the successor state. A *play* in G is an infinite sequence $\pi = \ell_0 \sigma_0 \ell_1 \dots \sigma_{n-1} \ell_n \sigma_n \dots$ such that (i) $\ell_0 = l_0$, and (ii) for all $i \geq 0$, we have $(\ell_i, \sigma_i, \ell_{i+1}) \in \Delta$. The *prefix up to* ℓ_n of the play π is denoted by $\pi(n)$; its length is $|\pi(n)| = n + 1$; and its last element is $\mathsf{Last}(\pi(n)) = \ell_n$. The *observation sequence* of π is the unique infinite sequence $\gamma^{-1}(\pi) = o_0 \sigma_0 o_1 \dots \sigma_{n-1} o_n \sigma_n \dots$ such that for all $i \geq 0$, we have $\ell_i \in \gamma(o_i)$. Similarly, the *observation sequence* of $\pi(n)$ is the prefix up to o_n of $\gamma^{-1}(\pi)$. The set of infinite plays in G is denoted $\mathsf{Plays}(G)$, and the set of corresponding finite prefixes is denoted $\mathsf{Prefs}(G)$. A state $\ell \in L$ is *reachable* in G if there exists a prefix $\rho \in \mathsf{Prefs}(G)$ such that $\mathsf{Last}(\rho) = \ell$. For a prefix $\rho \in \mathsf{Prefs}(G)$, the *cone* $\mathsf{Cone}(\rho) = \{ \pi \in \mathsf{Plays}(G) \mid \rho \text{ is a prefix of } \pi \}$ is the set of plays that extend ρ. The *knowledge* associated with a finite observation sequence $\tau = o_0 \sigma_0 o_1 \sigma_1 \dots \sigma_{n-1} o_n$ is the set $\mathsf{K}(\tau)$ of states in which a play can be after this sequence of observations, that is, $\mathsf{K}(\tau) = \{\mathsf{Last}(\rho) \mid \rho \in \mathsf{Prefs}(G) \text{ and } \gamma^{-1}(\rho) = \tau\}$. For $\sigma \in \Sigma$, $\ell \in L$, and $\rho, \rho' \in \mathsf{Prefs}(G)$ with $\rho' = \rho \cdot \sigma \cdot \ell$, let $o_\ell \in \mathcal{O}$ be the unique observation such that $\ell \in \gamma(o_\ell)$. Then $\mathsf{K}(\gamma^{-1}(\rho')) = \mathsf{Post}_\sigma^G(\mathsf{K}(\gamma^{-1}(\rho))) \cap \gamma(o_\ell)$.

Strategies. A *deterministic strategy* in G for Player 1 is a function $\alpha : \mathsf{Prefs}(G) \to \Sigma$. For a finite set A, a probability distribution on A is a function $\kappa : A \to [0, 1]$ such that $\sum_{a \in A} \kappa(a) = 1$. We denote the set of probability distributions on A by $\mathcal{D}(A)$. Given a distribution $\kappa \in \mathcal{D}(A)$, let $\mathsf{Supp}(\kappa) = \{a \in A \mid \kappa(a) > 0\}$ be the *support* of κ. A *randomized strategy* in G for Player 1 is a function $\alpha : \mathsf{Prefs}(G) \to \mathcal{D}(\Sigma)$. A (deterministic or randomized) strategy α for Player 1 is *observation-based* if for all prefixes $\rho, \rho' \in \mathsf{Prefs}(G)$, if $\gamma^{-1}(\rho) = \gamma^{-1}(\rho')$, then $\alpha(\rho) = \alpha(\rho')$. In the sequel, we are interested in the existence of observation-based strategies for Player 1. A *deterministic strategy* in G for Player 2 is a function $\beta : \mathsf{Prefs}(G) \times \Sigma \to L$ such that for all $\rho \in \mathsf{Prefs}(G)$ and all $\sigma \in \Sigma$, we have $(\mathsf{Last}(\rho), \sigma, \beta(\rho, \sigma)) \in \Delta$. A *randomized strategy* in G for Player 2 is a function $\beta : \mathsf{Prefs}(G) \times \Sigma \to \mathcal{D}(L)$ such that for all $\rho \in \mathsf{Prefs}(G)$, all $\sigma \in \Sigma$, and all $\ell \in \mathsf{Supp}(\beta(\rho, \sigma))$, we have $(\mathsf{Last}(\rho), \sigma, \ell) \in \Delta$. We denote by $\mathcal{A}_G$, $\mathcal{A}_G^O$, and $\mathcal{B}_G$ the set of all Player-1 strategies, the set of all observation-based Player-1 strategies, and the set of all Player-2 strategies in G, respectively. All results of this paper can be proved also if strategies depend on state sequences only, and not on the past moves of a play.

The *outcome* of two deterministic strategies α (for Player 1) and β (for Player 2) in G is the play $\pi = \ell_0\sigma_0\ell_1 \dots \sigma_{n-1}\ell_n\sigma_n \dots \in \mathsf{Plays}(G)$ such that for all $i \geq 0$, we have $\sigma_i = \alpha(\pi(i))$ and $\ell_{i+1} = \beta(\pi(i), \sigma_i)$. This play is denoted $\mathsf{outcome}(G, \alpha, \beta)$. The *outcome* of two randomized strategies α (for Player 1) and β (for Player 2) in G is the set of plays $\pi = \ell_0\sigma_0\ell_1 \dots \sigma_{n-1}\ell_n\sigma_n \dots \in \mathsf{Plays}(G)$ such that for all $i \geq 0$, we have $\alpha(\pi(i))(\sigma_i) > 0$ and $\beta(\pi(i), \sigma_i)(\ell_{i+1}) > 0$. This set is denoted $\mathsf{outcome}(G, \alpha, \beta)$. The *outcome set* of the deterministic (resp. randomized) strategy α for Player 1 in G is the set $\mathsf{Outcome}_i(G, \alpha)$ of plays π such that there exists a deterministic (resp. randomized) strategy β for Player 2 with $\pi = \mathsf{outcome}(G, \alpha, \beta)$ (resp. $\pi \in \mathsf{outcome}(G, \alpha, \beta)$). The outcome sets for Player 2 are defined symmetrically.

Objectives. An *objective* for G is a set ϕ of infinite sequences of observations and input letters, that is, $\phi \subseteq (\mathcal{O} \times \Sigma)^\omega$. A play $\pi = \ell_0\sigma_0\ell_1 \dots \sigma_{n-1}\ell_n\sigma_n \dots \in \mathsf{Plays}(G)$ *satisfies* the objective ϕ, denoted $\pi \models \phi$, if $\gamma^{-1}(\pi) \in \phi$. Objectives are generally Borel measurable: a Borel objective is a Borel set in the Cantor topology on $(\mathcal{O} \times \Sigma)^\omega$ [13]. We specifically consider reachability, safety, Büchi, coBüchi, and parity objectives, all of them Borel measurable. The parity objectives are a canonical form to express all ω-regular objectives [20]. For a play $\pi = \ell_0\sigma_0\ell_1 \dots$, we write $\mathsf{Inf}(\pi)$ for the set of observations that appear infinitely often in $\gamma^{-1}(\pi)$, that is, $\mathsf{Inf}(\pi) = \{o \in \mathcal{O} \mid \ell_i \in \gamma(o)$ for infinitely many i's$\}$.

Given a set $\mathcal{T} \subseteq \mathcal{O}$ of target observations, the *reachability* objective $\mathsf{Reach}(\mathcal{T})$ requires that an observation in $\mathcal{T}$ be visited at least once, that is, $\mathsf{Reach}(\mathcal{T}) = \{ \ell_0\sigma_0\ell_1\sigma_1 \dots \in \mathsf{Plays}(G) \mid \exists k \geq 0 \cdot \exists o \in \mathcal{T} : \ell_k \in \gamma(o) \}$. Dually, the *safety* objective $\mathsf{Safe}(\mathcal{T})$ requires that only observations in $\mathcal{T}$ be visited. Formally, $\mathsf{Safe}(\mathcal{T}) = \{ \ell_0\sigma_0\ell_1\sigma_1 \dots \in \mathsf{Plays}(G) \mid \forall k \geq 0 \cdot \exists o \in \mathcal{T} : \ell_k \in \gamma(o) \}$. The *Büchi* objective $\mathsf{Buchi}(\mathcal{T})$ requires that an observation in $\mathcal{T}$ be visited infinitely often, that is, $\mathsf{Buchi}(\mathcal{T}) = \{ \pi \mid \mathsf{Inf}(\pi) \cap \mathcal{T} \neq \emptyset \}$. Dually, the *coBüchi* objective $\mathsf{coBuchi}(\mathcal{T})$ requires that only observations in $\mathcal{T}$ be visited infinitely often. Formally, $\mathsf{coBuchi}(\mathcal{T}) = \{ \pi \mid \mathsf{Inf}(\pi) \subseteq \mathcal{T} \}$. For $d \in \mathbb{N}$, let $p : \mathcal{O} \to \{ 0, 1, \dots, d \}$ be a *priority function*, which maps each observation to a nonnegative integer priority. The *parity* objective $\mathsf{Parity}(p)$ requires that the minimum priority that appears infinitely often be even. Formally, $\mathsf{Parity}(p) = \{ \pi \mid \min\{ p(o) \mid o \in \mathsf{Inf}(\pi) \}$ is even $\}$. Observe that by definition, for all objectives ϕ, if $\pi \models \phi$ and $\gamma^{-1}(\pi) = \gamma^{-1}(\pi')$, then $\pi' \models \phi$.

Sure winning and almost winning. A strategy λ_i for Player i in G is *sure winning* for an objective ϕ if for all $\pi \in \mathsf{Outcome}_i(G, \lambda_i)$, we have $\pi \models \phi$. Given a game structure G and a state ℓ of G, we write G_ℓ for the game structure that results from G by changing the initial state to ℓ, that is, if $G = \langle L, l_0, \Sigma, \Delta, \mathcal{O}, \gamma \rangle$, then $G_\ell = \langle L, \ell, \Sigma, \Delta, \mathcal{O}, \gamma \rangle$. An *event* is a measurable set of plays, and given strategies α and β for the two players, the probabilities of events are uniquely defined [21]. For a Borel objective ϕ, we denote by $\Pr_\ell^{\alpha,\beta}(\phi)$ the probability ϕ is satisfied in the game G_ℓ given the strategies α and β. A strategy α for Player 1 in G is *almost winning* for the objective ϕ if for all randomized strategies β for

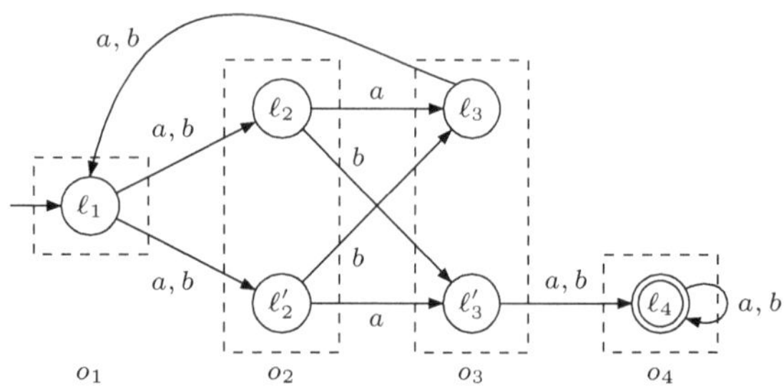

Fig. 1. Game structure G

Player 2, we have $\Pr_{l_0}^{\alpha,\beta}(\phi) = 1$. The set of *sure-winning* (resp. *almost-winning*) *states* of a game structure G for the objective ϕ is the set of states ℓ such that Player 1 has a deterministic sure-winning (resp. randomized almost-winning) observation-based strategy in G_ℓ for the objective ϕ.

Theorem 1 (Determinacy). [16] *For all perfect-information game structures G and all Borel objectives ϕ, either there exists a deterministic sure-winning strategy for Player 1 for the objective ϕ, or there exists a deterministic sure-winning strategy for Player 2 for the complementary objective $\mathsf{Plays}(G) \setminus \phi$.*

Notice that deterministic strategies suffice for sure winning a game: given a randomized strategy α for Player 1, let α^D be the deterministic strategy such that for all $\rho \in \mathsf{Prefs}(G)$, the strategy $\alpha^D(\rho)$ chooses an input letter from $\mathsf{Supp}(\alpha(\rho))$. Then $\mathsf{Outcome}_1(G, \alpha^D) \subseteq \mathsf{Outcome}_1(G, \alpha)$, and thus, if α is sure winning, then so is α^D. The result also holds for observation-based strategies and for perfect-information games. However, for almost winning, randomized strategies are more powerful than deterministic strategies as shown by Example 1.

Example 1. Consider the game structure shown in Fig. 1. The observations o_1, o_2, o_3, o_4 are such that $\gamma(o_1) = \{\ell_1\}$, $\gamma(o_2) = \{\ell_2, \ell_2'\}$, $\gamma(o_3) = \{\ell_3, \ell_3'\}$, and $\gamma(o_4) = \{\ell_4\}$. The transitions are shown as labeled edges in the figure, and the initial state is ℓ_1. The objective of Player 1 is $\mathsf{Reach}(\{o_4\})$, to reach state ℓ_4. We argue that the game is not sure winning for Player 1. Let α be any deterministic strategy for Player 1. Consider the deterministic strategy β for Player 2 as follows: for all $\rho \in \mathsf{Prefs}(G)$ such that $\mathsf{Last}(\rho) \in \gamma(o_2)$, if $\alpha(\rho) = a$, then in the previous round β chooses the state ℓ_2, and if $\alpha(\rho) = b$, then in the previous round β chooses the state ℓ_2'. Given α and β, the play $\mathsf{outcome}(G, \alpha, \beta)$ never reaches ℓ_4. However, the game G is almost winning for Player 1. Consider the randomized strategy that plays a and b uniformly at random at all states. Every time the game visits observation o_2, for any strategy for Player 2, the game visits ℓ_3 and ℓ_3' with probability $\frac{1}{2}$, and hence also reaches ℓ_4 with probability $\frac{1}{2}$. It follows that against all Player 2 strategies the play eventually reaches ℓ_4 with probability 1.

Remarks. First, the hypothesis that the observations form a partition of the state space can be weakened to a covering of the state space, where observations can overlap. In that case, Player 2 chooses both the next state of the game ℓ and the next observation o such that $\ell \in \gamma(o)$. The definitions related to plays, strategies, and objectives are adapted accordingly. Such a game structure G with overlapping observations can be encoded by an equivalent game structure G' of imperfect information, whose state space is the set of pairs (ℓ, o) such that $\ell \in \gamma(o)$. The set of labeled transitions Δ' of G' is defined by $\Delta' = \{((\ell, o), \sigma, (\ell', o')) \mid (\ell, \sigma, \ell') \in \Delta\}$ and $\gamma'^{-1}(\ell, o) = o$. The games G and G' are equivalent in the sense that for every Borel objective ϕ, there exists a sure (resp. almost) winning strategy for Player i in G forϕ if and only if there exists such a winning strategy for Player i in G' for ϕ. Second, it is essential that the objective is expressed in terms of the observations. Indeed, the games of imperfect information with a nonobservable winning condition are more complicated to solve. For instance, the universality problem for Büchi automata can be reduced to such games, but the construction that we propose in Section 3 cannot be used.

3 Sure Winning

We show that a game structure G of imperfect information can be encoded by a game structure G^{K} of perfect information such that for all Borel objectives ϕ, there exists a deterministic observation-based sure-winning strategy for Player 1 in G for ϕ if and only if there exists a deterministic sure-winning strategy for Player 1 in G^{K} for ϕ. We obtain G^{K} by a subset construction. Each state in G^{K} is a set of states of G which represents the knowledge of Player 1. In the worst case, the size of G^{K} is exponentially larger than the size of G. Second, we present a fixed-point algorithm based on antichains of set of states [10], whose correctness relies on the subset construction, but avoids the explicit construction of G^{K}.

3.1 Subset Construction for Sure Winning

Given a game structure of imperfect information $G = \langle L, l_0, \Sigma, \Delta, \mathcal{O}, \gamma\rangle$, we define the *knowledge-based subset construction* of G as the following game structure of perfect information: $G^{\mathsf{K}} = \langle \mathcal{L}, \{l_0\}, \Sigma, \Delta^{\mathsf{K}}\rangle$, where $\mathcal{L} = 2^L \backslash \{\emptyset\}$, and $(s_1, \sigma, s_2) \in \Delta^{\mathsf{K}}$ iff there exists an observation $o \in \mathcal{O}$ such that $s_2 = \mathsf{Post}^G_\sigma(s_1) \cap \gamma(o)$ and $s_2 \neq \emptyset$. Notice that for all $s \in \mathcal{L}$ and all $\sigma \in \Sigma$, there exists a set $s' \in \mathcal{L}$ such that $(s, \sigma, s') \in \Delta^{\mathsf{K}}$.

A (deterministic or randomized) strategy in G^{K} is called a *knowledge-based strategy*. For all sets $s \in \mathcal{L}$ that are reachable in G^{K}, and all observations $o \in \mathcal{O}$, either $s \subseteq \gamma(o)$ or $s \cap \gamma(o) = \emptyset$. By an abuse of notation, we define the *observation sequence* of a play $\pi = s_0\sigma_0 s_1 \dots \sigma_{n-1} s_n \sigma_n \dots \in \mathsf{Plays}(G^{\mathsf{K}})$ as the infinite sequence $\gamma^{-1}(\pi) = o_0\sigma_0 o_1 \dots \sigma_{n-1} o_n \sigma_n \dots$ of observations such that for all $i \geq 0$, we have $s_i \subseteq \gamma(o_i)$; this sequence is unique. The play π *satisfies* an objective $\phi \subseteq (\mathcal{O} \times \Sigma)^\omega$ if $\gamma^{-1}(\pi) \in \phi$. The proof of the following theorem can be found in the technical-report version for this paper.

Theorem 2 (Sure-winning reduction). *Player 1 has a deterministic observation-based sure-winning strategy in a game structure G of imperfect information for a Borel objective ϕ if and only if Player 1 has a deterministic sure-winning strategy in the game structure G^{K} of perfect information for ϕ.*

3.2 Two Interpretations of the μ-Calculus

Form the results of Section 3.1, we can solve a game G of imperfect information with objective ϕ by constructing the knowledge-based subset construction G^{K} and solving the resulting game of perfect information for the objective ϕ using standard methods. For the important class of ω-regular objectives, there exists a fixed-point theory —the μ-calculus— for this purpose [8]. When run on G^{K}, these fixed-point algorithms compute sets of sets of states of the game G. An important property of those sets is that they are *downward closed* with respect to set inclusion: if Player 1 has a deterministic strategy to win the game G when her knowledge is a set s, then she also has a deterministic strategy to win the game when her knowledge is s' with $s' \subseteq s$. And thus, if s is a sure-winning state of G^{K}, then so is s'. Based on this property, we devise a new algorithm for solving games of perfect information.

An *antichain* of nonempty sets of states is a set $q \subseteq 2^L \setminus \emptyset$ such that for all $s, s' \in q$, we have $s \not\subset s'$. Let $\mathcal{C}$ be the set of antichains of nonempty subsets of L, and consider the following partial order on $\mathcal{C}$: for all $q, q' \in \mathcal{C}$, let $q \sqsubseteq q'$ iff $\forall s \in q \cdot \exists s' \in q' : s \subseteq s'$. For $q \subseteq 2^L \setminus \emptyset$, define the set of *maximal* elements of q by $\lceil q \rceil = \{s \in q \mid s \neq \emptyset \text{ and } \forall s' \in q : s \not\subset s'\}$. Clearly, $\lceil q \rceil$ is an antichain. The least upper bound of $q, q' \in \mathcal{C}$ is $q \sqcup q' = \lceil \{s \mid s \in q \text{ or } s \in q'\} \rceil$, and their greatest lower bound is $q \sqcap q' = \lceil \{s \cap s' \mid s \in q \text{ and } s' \in q'\} \rceil$. The definition of these two operators extends naturally to sets of antichains, and the greatest element of $\mathcal{C}$ is $\top = \{L\}$ and the least element is $\bot = \emptyset$. The partially ordered set $\langle \mathcal{C}, \sqsubseteq, \sqcup, \sqcap, \top, \bot \rangle$ forms a complete lattice. We view antichains of state sets as a symbolic representation of $\subseteq$-downward-closed sets of state sets.

A *game lattice* is a complete lattice V together with a *predecessor operator* $\mathsf{CPre} : V \to V$. Given a game structure $G = \langle L, l_0, \Sigma, \Delta, \mathcal{O}, \gamma \rangle$ of imperfect information, and its knowledge-based subset construction $G^{\mathsf{K}} = \langle \mathcal{L}, \{l_0\}, \Sigma, \Delta^{\mathsf{K}} \rangle$, we consider two game lattices: the *lattice of subsets* $\langle \mathcal{S}, \subseteq, \cup, \cap, \mathcal{L}, \emptyset \rangle$, where $\mathcal{S} = 2^{\mathcal{L}}$ and $\mathsf{CPre} : \mathcal{S} \to \mathcal{S}$ is defined by $\mathsf{CPre}(q) = \{s \in \mathcal{L} \mid \exists \sigma \in \Sigma \cdot \forall s' \in \mathcal{L} : \text{if } (s, \sigma, s') \in \Delta^{\mathsf{K}}, \text{ then } s' \in q\}$; and the *lattice of antichains* $\langle \mathcal{C}, \sqsubseteq, \sqcup, \sqcap, \{L\}, \emptyset \rangle$, with the operator $\lceil \mathsf{CPre} \rceil : \mathcal{C} \to \mathcal{C}$ defined by $\lceil \mathsf{CPre} \rceil (q) = \lceil \{s \in \mathcal{L} \mid \exists \sigma \in \Sigma \cdot \forall o \in \mathcal{O} \cdot \exists s' \in q : \mathsf{Post}_\sigma(s) \cap \gamma(o) \subseteq s'\} \rceil$.

The *μ-calculus formulas* are generated by the grammar

$$\varphi ::= o \mid x \mid \varphi \vee \varphi \mid \varphi \wedge \varphi \mid \mathsf{pre}(\varphi) \mid \mu x.\varphi \mid \nu x.\varphi$$

for atomic propositions $o \in \mathcal{O}$ and variables x. We can define $\neg o$ as a shortcut for $\bigvee_{o' \in \mathcal{O} \setminus \{o\}} o'$. A variable is *free* in a formula φ if it is not in the scope of a quantifier μx or νx. A formula φ is *closed* if it contains no free variable. Given a game lattice V, a *valuation* $\mathcal{E}$ for the variables is a function that maps every variable x to an element in V. For $q \in V$, we write $\mathcal{E}[x \mapsto q]$ for the valuation

Lattice of subsets
$[\![o]\!]^{\mathcal{S}}_{\mathcal{E}} = \{s \in \mathcal{L} \mid s \subseteq \gamma(o)\}$
$[\![x]\!]^{\mathcal{S}}_{\mathcal{E}} = \mathcal{E}(x)$
$[\![\varphi_1 \left\{\begin{smallmatrix}\vee \\ \wedge\end{smallmatrix}\right\} \varphi_2]\!]^{\mathcal{S}}_{\mathcal{E}} = [\![\varphi_1]\!]^{\mathcal{S}}_{\mathcal{E}} \left\{\begin{smallmatrix}\cup \\ \cap\end{smallmatrix}\right\} [\![\varphi_2]\!]^{\mathcal{S}}_{\mathcal{E}}$
$[\![\mathsf{pre}(\varphi)]\!]^{\mathcal{S}}_{\mathcal{E}} = \mathsf{CPre}([\![\varphi]\!]^{\mathcal{S}}_{\mathcal{E}})$
$[\![\left\{\begin{smallmatrix}\mu \\ \nu\end{smallmatrix}\right\} x.\varphi]\!]^{\mathcal{S}}_{\mathcal{E}} = \left\{\begin{smallmatrix}\bigcap \\ \bigcup\end{smallmatrix}\right\} \{q \mid q = [\![\varphi]\!]^{\mathcal{S}}_{\mathcal{E}[x \mapsto q]}\}$

Lattice of antichains
$[\![o]\!]^{\mathcal{C}}_{\mathcal{E}} = \{\gamma(o)\}$
$[\![x]\!]^{\mathcal{C}}_{\mathcal{E}} = \mathcal{E}(x)$
$[\![\varphi_1 \left\{\begin{smallmatrix}\vee \\ \wedge\end{smallmatrix}\right\} \varphi_2]\!]^{\mathcal{C}}_{\mathcal{E}} = [\![\varphi_1]\!]^{\mathcal{C}}_{\mathcal{E}} \left\{\begin{smallmatrix}\sqcup \\ \sqcap\end{smallmatrix}\right\} [\![\varphi_2]\!]^{\mathcal{C}}_{\mathcal{E}}$
$[\![\mathsf{pre}(\varphi)]\!]^{\mathcal{C}}_{\mathcal{E}} = \lceil \mathsf{CPre} \rceil([\![\varphi]\!]^{\mathcal{C}}_{\mathcal{E}})$
$[\![\left\{\begin{smallmatrix}\mu \\ \nu\end{smallmatrix}\right\} x.\varphi]\!]^{\mathcal{C}}_{\mathcal{E}} = \left\{\begin{smallmatrix}\bigsqcap \\ \bigsqcup\end{smallmatrix}\right\} \{q \mid q = [\![\varphi]\!]^{\mathcal{C}}_{\mathcal{E}[x \mapsto q]}\}$

that agrees with $\mathcal{E}$ on all variables, except that x is mapped to q. Given a game lattice V and a valuation $\mathcal{E}$, each μ-calculus formula φ specifies an element $[\![\varphi]\!]^V_{\mathcal{E}}$ of V, which is defined inductively by the equations shown in the two tables. If φ is a closed formula, then $[\![\varphi]\!]^V = [\![\varphi]\!]^V_{\mathcal{E}}$ for any valuation $\mathcal{E}$. The following theorem recalls that perfect-information games can be solved by evaluating fixed-point formulas in the lattice of subsets.

Theorem 3 (Symbolic solution of perfect-information games). [8] *For every ω-regular objective ϕ, there exists a closed μ-calculus formula $\mu\mathsf{Form}(\phi)$, called the* characteristic formula *of ϕ, such that for all game structures G of perfect information, the set of sure-winning states of G for ϕ is $[\![\mu\mathsf{Form}(\phi)]\!]^{\mathcal{S}}$.*

Downward closure. Given a set $q \in \mathcal{S}$, the *downward closure* of q is the set $q\!\downarrow = \{s \in \mathcal{L} \mid \exists s' \in q : s \subseteq s'\}$. Observe that in particular, for all $q \in \mathcal{S}$, we have $\emptyset \notin q\!\downarrow$ and $\lceil q \rceil\!\downarrow = q\!\downarrow$. The sets $q\!\downarrow$, for $q \in \mathcal{S}$, are the *downward-closed* sets. A valuation $\mathcal{E}$ for the variables in the lattice $\mathcal{S}$ of subsets is *downward closed* if every variable x is mapped to a downward-closed set, that is, $\mathcal{E}(x) = \mathcal{E}(x)\!\downarrow$.

Lemma 1. *For all downward-closed sets $q, q' \in \mathcal{S}$, we have $\lceil q \cap q' \rceil = \lceil q \rceil \sqcap \lceil q' \rceil$ and $\lceil q \cup q' \rceil = \lceil q \rceil \sqcup \lceil q' \rceil$.*

Lemma 2. *For all μ-calculus formulas φ and all downward-closed valuations $\mathcal{E}$ in the lattice of subsets, the set $[\![\varphi]\!]^{\mathcal{S}}_{\mathcal{E}}$ is downward closed.*

Lemma 3. *For all μ-calculus formulas φ, and all downward-closed valuations $\mathcal{E}$ in the lattice of subsets, we have $\lceil [\![\varphi]\!]^{\mathcal{S}}_{\mathcal{E}} \rceil = [\![\varphi]\!]^{\mathcal{C}}_{\lceil \mathcal{E} \rceil}$, where $\lceil \mathcal{E} \rceil$ is a valuation in the lattice of antichains defined by $\lceil \mathcal{E} \rceil(x) = \lceil \mathcal{E}(x) \rceil$ for all variables x.*

Consider a game structure G of imperfect information and a parity objective ϕ. From Theorems 2 and 3 and Lemma 3, we can decide the existence of a deterministic observation-based sure-winning strategy for Player 1 in G for ϕ without explicitly constructing the knowledge-based subset construction G^{K}, by instead evaluating a fixed-point formula in the lattice of antichains.

Theorem 4 (Symbolic solution of imperfect-information games). *Let G be a game structure of imperfect information with initial state l_0. For every ω-regular objective ϕ, Player 1 has a deterministic observation-based strategy in G for ϕ if and only if $\{l_0\} \sqsubseteq [\![\mu\mathsf{Form}(\phi)]\!]^{\mathcal{C}}$.*

Corollary 1. *Let G be a game structure of imperfect information, let p be a priority function, and let ℓ be a state of G. Whether ℓ is a sure-winning state in G for the parity objective* $\mathsf{Parity}(p)$ *can be decided in* EXPTIME.

Corollary 1 is proved as follows: for a parity objective ϕ, an equivalent μ-calculus formula φ can be obtained, where the size and the fixed-point quantifier alternations of φ is polynomial in ϕ. Thus given G and ϕ, we can evaluate φ in G^{K} in EXPTIME.

4 Almost Winning

Given a game structure G of imperfect information, we first construct a game structure H of perfect information by a subset construction (different from the one used for sure winning), and then establish certain equivalences between randomized strategies in G and H. Finally, we show how the reduction can be used to obtain a symbolic EXPTIME algorithm for computing almost-winning states in G for Büchi objectives. An EXPTIME algorithm for almost winning for coBüchi objectives under imperfect information remains unknown.

4.1 Subset Construction for Almost Winning

Given a game structure of imperfect information $G = \langle L, l_0, \Sigma, \Delta, \mathcal{O}, \gamma \rangle$, we construct game structure of perfect information $H = \mathsf{Pft}(G) = \langle Q, q_0, \Sigma, \Delta_H \rangle$ as follows: $Q = \{ (s, \ell) \mid \exists o \in \mathcal{O} : s \subseteq \gamma(o) \text{ and } \ell \in s \}$; the initial state is $q_0 = (\{l_0\}, l_0)$; the transition relation $\Delta_H \subseteq Q \times \Sigma \times Q$ is defined by $((s, \ell), \sigma, (s', \ell')) \in \Delta_H$ iff there is an observation $o \in \mathcal{O}$ such that $s' = \mathsf{Post}^G_\sigma(s) \cap \gamma(o)$ and $(\ell, \sigma, \ell') \in \Delta$. Intuitively, when H is in state (s, ℓ), it corresponds to G being in state ℓ and the knowledge of Player 1 being s. Two states $q = (s, \ell)$ and $q' = (s', \ell')$ of H are *equivalent*, written $q \approx q'$, if $s = s'$. Two prefixes $\rho = q_0 \sigma_0 q_1 \dots \sigma_{n-1} q_n$ and $\rho' = q'_0 \sigma'_0 q'_1 \dots \sigma'_{n-1} q'_n$ of H are *equivalent*, written $\rho \approx \rho$, if for all $0 \leq i \leq n$, we have $q_i \approx q'_i$, and for all $0 \leq i \leq n-1$, we have $\sigma_i = \sigma'_i$. Two plays $\pi, \pi' \in \mathsf{Plays}(H)$ are *equivalent*, written $\pi_H \approx \pi'_H$, if for all $i \geq 0$, we have $\pi(i) \approx \pi'(i)$. For a state $q \in Q$, we denote by $[q]_\approx = \{ q' \in Q \mid q \approx q' \}$ the $\approx$-equivalence class of q. We define equivalence classes for prefixes and plays similarly.

Equivalence-preserving strategies and objectives. A strategy α for Player 1 in H is *positional* if it is independent of the prefix of plays and depends only on the last state, that is, for all $\rho, \rho' \in \mathsf{Prefs}(H)$ with $\mathsf{Last}(\rho) = \mathsf{Last}(\rho')$, we have $\alpha(\rho) = \alpha(\rho')$. A positional strategy α can be viewed as a function $\alpha : Q \to \mathcal{D}(\Sigma)$. A strategy α for Player 1 in H is *equivalence-preserving* if for all $\rho, \rho' \in \mathsf{Prefs}(H)$ with $\rho \approx \rho'$, we have $\alpha(\rho) = \alpha(\rho')$. We denote by $\mathcal{A}_H$, $\mathcal{A}_H^P$, and $\mathcal{A}_H^{\approx}$ the set of all Player-1 strategies, the set of all positional Player-1 strategies, and the set of all equivalence-preserving Player-1 strategies in H, respectively. We write $\mathcal{A}_H^{\approx(P)} = \mathcal{A}_H^{\approx} \cap \mathcal{A}_H^P$ for the set of equivalence-preserving positional strategies.

An objective ϕ for H is a subset of $(Q \times \Sigma)^\omega$, that is, the objective ϕ is a set of plays. The objective ϕ is *equivalence-preserving* if for all plays $\pi \in \phi$, we have $[\pi]_\approx \subseteq \phi$.

Relating prefixes and plays. We define a mapping $h : \mathsf{Prefs}(G) \to \mathsf{Prefs}(H)$ that maps prefixes in G to prefixes in H as follows: given $\rho = \ell_0\sigma_0\ell_1\sigma_1 \ldots \sigma_{n-1}\ell_n$, let $h(\rho) = q_0\sigma_0 q_1\sigma_1 \ldots \sigma_{n-1}q_n$, where for all $0 \leq i \leq n$, we have $q_i = (s_i, \ell_i)$, and for all $0 \leq i \leq n-1$, we have $s_i = \mathsf{K}(\gamma^{-1}(\rho(i)))$. The following properties hold: (*i*) for all $\rho, \rho' \in \mathsf{Prefs}(G)$, if $\gamma^{-1}(\rho) = \gamma^{-1}(\rho')$, then $h(\rho) \approx h(\rho')$; and (*ii*) for all $\rho, \rho' \in \mathsf{Prefs}(H)$, if $\rho \approx \rho'$, then $\gamma^{-1}(h^{-1}(\rho)) = \gamma^{-1}(h^{-1}(\rho'))$. The mapping $h : \mathsf{Plays}(G) \to \mathsf{Plays}(H)$ for plays is defined similarly, and has similar properties.

Relating strategies for Player 1. We define two strategy mappings $h : \mathcal{A}_H \to \mathcal{A}_G$ and $g : \mathcal{A}_G \to \mathcal{A}_H$. Given a Player-1 strategy α_H in H, we construct a Player-1 strategy $\alpha_G = h(\alpha_H)$ in G as follows: for all $\rho \in \mathsf{Prefs}(G)$, let $\alpha_G(\rho) = \alpha_H(h(\rho))$. Similarly, given a Player-1 strategy α_G in G, we construct a Player-1 strategy $\alpha_H = g(\alpha_G)$ in H as follows: for all $\rho \in \mathsf{Prefs}(H)$, let $\alpha_H(\rho) = \alpha_G(h^{-1}(\rho))$. The following properties hold: (*i*) for all strategies $\alpha_H \in \mathcal{A}_H$, if α_H is equivalence-preserving, then $h(\alpha_H)$ is observation-based; and (*ii*) for all strategies $\alpha_G \in \mathcal{A}_G$, if α_G is observation-based, then $g(\alpha_G)$ is equivalence-preserving.

Relating strategies for Player 2. Observe that for all $q \in Q$, all $\sigma \in \Sigma$, and all $\ell \in L$, we have $|\{ q' = (s', \ell) \mid (q, \sigma, q') \in \Delta_H \}| \leq 1$. Given a Player-2 strategy β_H in H, we construct a Player-2 strategy $\beta_G = h(\beta_H)$ as follows: for all $\rho \in \mathsf{Prefs}(G)$, all $\sigma \in \Sigma$, and all $\ell \in L$, let $\beta_G(\rho, \sigma)(\ell) = \beta_H(h(\rho), \sigma)(s, \ell)$, where $(s, \ell) \in \mathsf{Post}^H_\sigma(\mathsf{Last}(h(\rho)))$. Similarly, given a Player-2 strategy β_G in G, we construct a Player-2 strategy $\beta_H = g(\beta_G)$ in H as follows: for all $\rho \in \mathsf{Prefs}(H)$, all $\sigma \in \Sigma$, and all $q \in Q$ with $q = (s, \ell)$, let $\beta_H(\rho, \sigma)(q) = \beta_G(h^{-1}(\rho), \sigma)(\ell)$.

Lemma 4. *For all $\rho \in \mathsf{Prefs}(H)$, for every equivalence-preserving strategy α of Player 1 in H, and for every strategy β of Player 2 in H, we have* $\Pr_{q_0}^{\alpha,\beta}(\mathsf{Cone}(\rho)) = \Pr_{l_0}^{h(\alpha),h(\beta)}(h^{-1}(\mathsf{Cone}(\rho)))$.

Lemma 5. *For all $\rho \in \mathsf{Prefs}(G)$, for every observational strategy α of Player 1 in G, and for every strategy β of Player 2 in G, we have* $\Pr_{l_0}^{\alpha,\beta}(\mathsf{Cone}(\rho)) = \Pr_{q_0}^{g(\alpha),g(\beta)}(h(\mathsf{Cone}(\rho)))$.

Theorem 5 (Almost-winning reduction). *Let G be a game structure of imperfect information, and let $H = \mathsf{Pft}(G)$ be the game structure of perfect information. For all Borel objectives ϕ for G, all observation-based Player-1 strategies α in G, and all Player-2 strategies β in G, we have $\Pr_{l_0}^{\alpha,\beta}(\phi) = \Pr_{q_0}^{g(\alpha),g(\beta)}(h(\phi))$. Dually, for all equivalence-preserving Borel objectives ϕ for H, all equivalence-preserving Player-1 strategies α in H, and all Player-2 strategies β in H, we have $\Pr_{q_0}^{\alpha,\beta}(\phi) = \Pr_{l_0}^{h(\alpha),h(\beta)}(h^{-1}(\phi))$.*

The proof is as follows: by the Carathéodary unique-extension theorem, a probability measure defined on cones has a unique extension to all Borel objectives. The theorem then follows from Lemma 4.

Corollary 2. *For every Borel objective ϕ for G, we have $\exists \alpha_G \in \mathcal{A}_G^O \cdot \forall \beta_G \in \mathcal{B}_G : \Pr_{\ell_0}^{\alpha_G,\beta_G}(\phi) = 1$ if and only if $\exists \alpha_H \in \mathcal{A}_H^{\approx} \cdot \forall \beta_H \in \mathcal{B}_H : \Pr_{q_0}^{\alpha_H,\beta_H}(h(\phi)) = 1$.*

4.2 Almost Winning for Büchi Objectives

Given a game structure G of imperfect information, let $H = \mathsf{Pft}(G)$ be the game structure of perfect information. Given a set $\mathcal{T} \subseteq \mathcal{O}$ of target observations, let $B_{\mathcal{T}} = \{(s,l) \in Q \mid \exists o \in \mathcal{T} : s \subseteq \gamma(o)\}$. Then $h(\mathsf{Buchi}(\mathcal{T})) = \mathsf{Buchi}(B_{\mathcal{T}}) = \{\pi_H \in \mathsf{Plays}(H) \mid \mathsf{Inf}(\pi_H) \cap B_{\mathcal{T}} \neq \emptyset\}$. We first show that almost winning in H for the Büchi objective $\mathsf{Buchi}(B_{\mathcal{T}})$ with respect to equivalence-preserving strategies is equivalent to almost winning with respect to equivalence-preserving positional strategies. Formally, for $B_{\mathcal{T}} \subseteq Q$, let $Q_{\mathsf{AS}}^{\approx} = \{ q \in Q \mid \exists \alpha \in \mathcal{A}_H^{\approx} \cdot \forall \beta \in \mathcal{B}_H \cdot \forall q' \in [q]_{\approx} : \ \Pr_{q'}^{\alpha,\beta}(\mathsf{Buchi}(B_{\mathcal{T}})) = 1 \}$, and $Q_{\mathsf{AS}}^{\approx(P)} = \{ q \in Q \mid \exists \alpha \in \mathcal{A}_H^{\approx(P)} \cdot \forall \beta \in \mathcal{B}_H \cdot \forall q' \in [q]_{\approx} : \ \Pr_{q'}^{\alpha,\beta}(\mathsf{Buchi}(B_{\mathcal{T}})) = 1 \}$. We will prove that $Q_{\mathsf{AS}}^{\approx} = Q_{\mathsf{AS}}^{\approx(P)}$. Lemma 6 follows from the construction of H from G.

Lemma 6. *Given an equivalence-preserving Player-1 strategy $\alpha \in \mathcal{A}_H$, a prefix $\rho \in \mathsf{Prefs}(H)$, and a state $q \in Q$, if there exists a Player-2 strategy $\beta \in \mathcal{B}_H$ such that $\Pr_q^{\alpha,\beta}(\mathsf{Cone}(\rho)) > 0$, then for every prefix $\rho' \in \mathsf{Prefs}(H)$ with $\rho \approx \rho'$, there exist a Player-2 strategy $\beta' \in \mathcal{B}_H$ and a state $q' \in [q]_{\approx}$ such that $\Pr_{q'}^{\alpha,\beta'}(\mathsf{Cone}(\rho')) > 0$.*

Observe that $Q \setminus Q_{\mathsf{AS}}^{\approx} = \{ q \in Q \mid \forall \alpha \in \mathcal{A}_H^{\approx} \cdot \exists \beta \in \mathcal{B}_H \cdot \exists q' \in [q]_{\approx} : \Pr_{q'}^{\alpha,\beta}(\mathsf{Buchi}(B_{\mathcal{T}})) < 1 \}$. It follows from Lemma 6 that if a play starts in $Q_{\mathsf{AS}}^{\approx}$ and reaches $Q \setminus Q_{\mathsf{AS}}^{\approx}$ with positive probability, then for all equivalence-preserving strategies for Player 1, there is a Player 2 strategy that ensures that the Büchi objective $\mathsf{Buchi}(B_{\mathcal{T}})$ is not satisfied with probability 1.

Notation. For a state $q \in Q$ and $Y \subseteq Q$, let $\mathsf{Allow}(q, Y) = \{ \sigma \in \Sigma \mid \mathsf{Post}_{\sigma}^{H}(q) \subseteq Y \}$. For a state $q \in Q$ and $Y \subseteq Q$, let $\mathsf{Allow}([q]_{\approx}, Y) = \bigcap_{q' \in [q]_{\approx}} \mathsf{Allow}(q', Y)$.

Lemma 7. *For all $q \in Q_{\mathsf{AS}}^{\approx}$, we have $\mathsf{Allow}([q]_{\approx}, Q_{\mathsf{AS}}^{\approx}) \neq \emptyset$.*

Lemma 8. *Given a state $q \in Q_{\mathsf{AS}}^{\approx}$, let $\alpha \in \mathcal{A}_H$ be an equivalence-preserving Player-1 strategy such that for all Player-2 strategies $\beta \in \mathcal{B}_H$ and all states $q' \in [q]_{\approx}$, we have $\Pr_{q'}^{\alpha,\beta}(\mathsf{Buchi}(B_{\mathcal{T}})) = 1$. Let $\rho = q_0\sigma_0 q_1 \ldots \sigma_{n-1} q_n$ be a prefix in $\mathsf{Prefs}(H)$ such that for all $0 \leq i \leq n$, we have $q_i \in Q_{\mathsf{AS}}^{\approx}$. If there is a Player-2 strategy $\beta \in \mathcal{B}_H$ and a state $q' \in [q]_{\approx}$ such that $\Pr_{q'}^{\alpha,\beta}(\mathsf{Cone}(\rho)) > 0$, then $\mathsf{Supp}(\alpha(\rho)) \subseteq \mathsf{Allow}([q]_{\approx}, Q_{\mathsf{AS}}^{\approx})$.*

Notation. We inductively define the *ranks* of states in $Q_{\mathsf{AS}}^{\approx}$ as follows: let $\mathsf{Rank}(0) = B_{\mathcal{T}} \cap Q_{\mathsf{AS}}^{\approx}$, and for all $j \geq 0$, let $\mathsf{Rank}(j+1) = \mathsf{Rank}(j) \cup \{ q \in Q_{\mathsf{AS}}^{\approx} \mid \exists \sigma \in \mathsf{Allow}([q]_{\approx}, Q_{\mathsf{AS}}^{\approx}) : \mathsf{Post}_{\sigma}^{H}(q) \subseteq \mathsf{Rank}(j) \}$. Let $j^* = \min\{ j \geq 0 \mid \mathsf{Rank}(j) = \mathsf{Rank}(j+1) \}$, and let $Q^* = \mathsf{Rank}(j^*)$. We say that the set $\mathsf{Rank}(j+1) \setminus \mathsf{Rank}(j)$ contains the *states of rank* $j+1$, for all $j \geq 0$.

Lemma 9. $Q^* = Q_{\mathsf{AS}}^{\approx}$.

Equivalence-preserving positional strategy. Consider the equivalence-preserving positional strategy α^p for Player 1 in H, which is defined as follows: for a state $q \in Q_{\mathsf{AS}}^{\approx}$, choose all moves in $\mathsf{Allow}([q]_{\equiv}, Q_{\mathsf{AS}}^{\approx})$ uniformly at random.

Lemma 10. *For all states* $q \in Q^{\approx}_{\mathsf{AS}}$ *and all Player-2 strategies* β *in* H, *we have* $\Pr_q^{\alpha^P,\beta}(\mathsf{Safe}(Q^{\approx}_{\mathsf{AS}})) = 1$ *and* $\Pr_q^{\alpha^P,\beta}(\mathsf{Reach}(B_{\mathcal{T}} \cap Q^{\approx}_{\mathsf{AS}})) = 1$.

Proof. By Lemma 9, we have $Q^* = Q^{\approx}_{\mathsf{AS}}$. Let $z = |Q^*|$. For a state $q \in Q^{\approx}_{\mathsf{AS}}$, we have $\mathsf{Post}^H_\sigma(q) \subseteq Q^{\approx}_{\mathsf{AS}}$ for all $\sigma \in \mathsf{Allow}([q]_\approx, Q^{\approx}_{\mathsf{AS}})$. It follows for all states $q \in Q^{\approx}_{\mathsf{AS}}$ and all strategies β for Player 2, we have $\Pr_q^{\alpha^P,\beta}(\mathsf{Safe}(Q^{\approx}_{\mathsf{AS}})) = 1$.

For a state $q \in (\mathsf{Rank}(j+1) \setminus \mathsf{Rank}(j))$, there exists $\sigma \in \mathsf{Allow}([q]_\approx, Q^{\approx}_{\mathsf{AS}})$ such that $\mathsf{Post}^H_\sigma(q) \subseteq \mathsf{Rank}(j)$. For a set $Y \subseteq Q$, let $\Diamond^j(Y)$ denote the set of prefixes that reach Y within j steps. It follows that for all states $q \in \mathsf{Rank}(j+1)$ and all strategies β for Player 2, we have $\Pr_q^{\alpha^P,\beta}(\Diamond^1(\mathsf{Rank}(j))) \geq \frac{1}{|\Sigma|}$. Let $B = B_{\mathcal{T}} \cap Q^{\approx}_{\mathsf{AS}}$. By induction on the ranks it follows that for all states $q \in Q^*$ and all strategies β for Player 2: $\Pr_q^{\alpha^P,\beta}(\Diamond^z(\mathsf{Rank}(0))) = \Pr_q^{\alpha^P,\beta}(\Diamond^z(B)) \geq \left(\frac{1}{|\Sigma|}\right)^z = r > 0$. For $m > 0$, we have $\Pr_q^{\alpha^P,\beta}(\Diamond^{m\cdot z}(B)) \geq 1 - (1-r)^m$. Thus:

$$\Pr_q^{\alpha^P,\beta}(\mathsf{Reach}(B)) = \lim_{m\to\infty} \Pr_q^{\alpha^P,\beta}(\Diamond^{m\cdot z}(B)) \geq \lim_{m\to\infty} 1-(1-r)^m = 1. \qquad \blacksquare$$

Lemma 10 implies that, given the Player-1 strategy α^P, the set $Q^{\approx}_{\mathsf{AS}}$ is never left, and the states in $B_{\mathcal{T}} \cap Q^{\approx}_{\mathsf{AS}}$ are reached with probability 1. Since this happens for every state in $Q^{\approx}_{\mathsf{AS}}$, it follows that the set $B_{\mathcal{T}} \cap Q^{\approx}_{\mathsf{AS}}$ is visited infinitely often with probability 1, that is, the Büchi objective $\mathsf{Buchi}(B_{\mathcal{T}})$ is satisfied with probability 1. This analysis, together with the fact that $[q_0]_\approx$ is a singleton and Corollary 2, proves that $Q^{\approx}_{\mathsf{AS}} = Q^{\approx(P)}_{\mathsf{AS}}$. Theorem 6 follows.

Theorem 6 (Positional almost winning for Büchi objectives under imperfect information). *Let* G *be a game structure of imperfect information, and let* $H = \mathsf{Pft}(G)$ *be the game structure of perfect information. For all sets* $\mathcal{T}$ *of observations, there exists an observation-based almost-winning strategy for Player 1 in* G *for the objective* $\mathsf{Buchi}(\mathcal{T})$ *iff there exists an equivalence-preserving positional almost-winning strategy for Player 1 in* H *for the objective* $\mathsf{Buchi}(B_{\mathcal{T}})$.

Symbolic algorithm. We present a symbolic quadratic-time (in the size of H) algorithm to compute the set $Q^{\approx}_{\mathsf{AS}}$. For $Y \subseteq Q$ and $X \subseteq Y$, let $\mathsf{Apre}(Y, X) = \{q \in Y \mid \exists \sigma \in \mathsf{Allow}([q]_\approx, Y) : \mathsf{Post}^H_\sigma(q) \subseteq X\}$ and $\mathsf{Spre}(Y) = \{q \in Y \mid \mathsf{Allow}([q]_\approx, Y) \neq \emptyset\}$. Note that $\mathsf{Spre}(Y) = \mathsf{Apre}(Y, Y)$. Let $\phi = \nu Y.\mu X.(\mathsf{Apre}(Y, X) \vee (B_{\mathcal{T}} \wedge \mathsf{Spre}(Y))$ and let $Z = [\![\phi]\!]$. It can be shown that $Z = Q^{\approx}_{\mathsf{AS}}$.

Theorem 7 (Complexity of almost winning for Büchi objectives under imperfect information). *Let* G *be a game structure of imperfect information, let* $\mathcal{T}$ *be a set of observations, and let* ℓ *be a state of* G. *Whether* ℓ *is an almost-winning state in* G *for the Büchi objective* $\mathsf{Buchi}(\mathcal{T})$ *can be decided in* EXPTIME.

The facts that $Z = Q^{\approx}_{\mathsf{AS}}$ and that H is exponential in the size of G yield Theorem 7. The arguments for the proofs of Theorem 6 and 7 do not directly extend to coBüchi or parity objectives. In fact, Theorem 6 does not hold for parity objectives in general, for the following reason: in concurrent games with parity objectives with more than two priorities, almost-winning strategies may require infinite memory; for an example, see [5]. Such concurrent games are reducible to semiperfect-information games [4], and semiperfect-information games are reducible to the imperfect-information games we study. Hence a reduction to finite game structures of perfect information in order to obtain randomized positional strategies is not possible with respect to almost winning for general parity objectives. Theorem 6 and Theorem 7 may hold for coBüchi objectives, but there does not seem to be a simple extension of our arguments for Büchi objectives to the coBüchi case. The results that correspond to Theorems 6 and 7 for coBüchi objectives are open.

Direct symbolic algorithm. As in Section 3.2, the subset structure H does not have to be constructed explicitly. Instead, we can evaluate a fixed-point formula on a well-chosen lattice. The fixed-point formula to compute the set $Q^{\approx}_{\mathsf{AS}}$ is evaluated on the lattice $\langle 2^Q, \subseteq, \cup, \cap, Q, \emptyset \rangle$. It is easy to show that the sets computed by the fixed-point algorithm are downward closed for the following order on Q: for $(s, \ell), (s', \ell') \in Q$, let $(s, \ell) \preceq (s', \ell')$ iff $\ell = \ell'$ and $s \subseteq s'$. Then, we can define an antichain over Q as a set of pairwise $\preceq$-incomparable elements of Q, and compute the almost-sure winning states in the lattice of antichains over Q, without explicitly constructing the exponential game structure H.

5 Lower Bounds

We show that deciding the existence of a deterministic (resp. randomized) observation-based sure-winning (resp. almost-winning) strategy for Player 1 in games of imperfect information is EXPTIME-hard already for reachability objectives. The result for sure winning follows from [19], but our new proof extends to almost winning as well.

Sure winning. To show the lower bound, we use a reduction from the membership problem for polynomial-space alternating Turing machines. An *alternating Turing machine* (ATM) is a tuple $M = \langle Q, q_0, g, \Sigma_i, \Sigma_t, \delta, F \rangle$, where Q is a finite set of control states; $q_0 \in Q$ is the initial state; $g : Q \to \{\wedge, \vee\}$; $\Sigma_i = \{0, 1\}$ is the input alphabet; $\Sigma_t = \{0, 1, 2\}$ is the tape alphabet and 2 is the *blank* symbol; $\delta \subseteq Q \times \Sigma_t \times Q \times \Sigma_t \times \{-1, 1\}$ is the transition relation; and $F \subseteq Q$ is the set of accepting states. We say that M is a *polynomial-space* ATM if there exists a polynomial $p(\cdot)$ such that for every word w, the tape space used by M on input w is bounded by $p(|w|)$. Without loss of generality, we assume that the initial control state of the machine is a $\vee$-state, and that transitions connect $\vee$-states to $\wedge$-states, and vice versa. A word w is *accepted* by the ATM M if there exists a run tree of M on w all of whose leaf nodes are accepting configurations (i.e., configurations containing an accepting state); see [3] for details. The *membership problem* is to decide if a given word w is accepted by a given polynomial-space ATM (M, p). This problem is EXPTIME-hard [3].

Sketch of the reduction. Given a polynomial-space ATM M and a word w, we construct a game structure of imperfect information, of size polynomial in the size of (M, w), to simulate the execution of M on w. Player 1 makes choices in $\vee$-states, and Player 2 makes choices in $\wedge$-states. Player 1 is responsible for maintaining the symbol under the tape head. His objective is to reach an accepting configuration of the ATM.

Each turn proceeds as follows. In an $\vee$-state, by choosing a letter (t, a) in the alphabet of the game, Player 1 reveals (i) the transition t of the ATM that he has chosen (this way he also reveals the symbol that is currently under the tape head), and (ii) the symbol a under the next position of the tape head. If Player 1 lies about the current or the next symbol under the tape head, then he should lose the game; otherwise the game proceeds. The machine is now in an $\wedge$-state and Player 1 has no choice: he announces a special symbol ϵ and Player 2, by resolving the nondeterminism on ϵ, chooses a transition of the ATM that is compatible with the current symbol under the tape head revealed by Player 1 at the previous turn. The state of the ATM is updated and the game proceeds. The transition chosen by Player 2 is visible in the next state of the game, and thus Player 1 can update his knowledge about the configuration of the ATM. Player 1 wins whenever an accepting configuration of the ATM is reached.

The difficulty is to ensure that Player 1 loses when he announces a wrong symbol under the tape head. As the number of configurations of the polynomial-space ATM is exponential, we cannot directly encode the full configuration of the ATM in the states of the game. To overcome this difficulty, we use the power of imperfect information as follows. Initially, Player 2 chooses a position k, where $1 \leq k \leq p(|w|)$, on the tape. The chosen number k, as well as the symbol $\sigma \in \{0, 1, 2\}$ that lies in the tape cell with number k, are maintained all along the game in the nonobservable portion of the game states. The pair (σ, k) is thus private to Player 2, and invisible to Player 1. Thus, at any point in the game, Player 2 can check whether Player 1 is lying when announcing the content of cell number k, and go to a sink state if Player 1 cheats (no other states can be reached from there). Since Player 1 does not know which cell is monitored by Player 2 (since k is private), to avoid losing, he must not lie about any of the tape cells, and thus he must faithfully simulate the machine. Then, he wins the game if and only if the ATM accepts the words w.

Almost winning. To establish the lower bound for almost winning, we can use the same reduction. Randomization cannot help Player 1 in this game. Indeed, at any point in the game, if Player 1 takes a chance in either not faithfully simulating the ATM or lying about the symbol under the tape head, then the sink state is reached. In these cases, the probability to reach the sink state is positive, and so the probability to win the game is strictly less than one.

Theorem 8 (Lower bounds). *Let G be a game structure of imperfect information, let $\mathcal{T}$ be a set of observations, and let ℓ be a state of G. Deciding whether ℓ is a sure-winning state in G for the reachability objective* $\mathsf{Reach}(\mathcal{T})$ *is* EXPTIME-*hard. Deciding whether ℓ is an almost-winning state in G for* $\mathsf{Reach}(\mathcal{T})$ *is also* EXPTIME-*hard.*

References

1. M. Abadi, L. Lamport, and P. Wolper. Realizable and unrealizable specifications of reactive systems. In *ICALP*, LNCS 372, pages 1–17. Springer, 1989.
2. R. Alur, T.A. Henzinger, and O. Kupferman. Alternating-time temporal logic. *J. ACM*, 49:672–713, 2002.
3. A.K. Chandra, D. Kozen, and L.J. Stockmeyer. Alternation. *J. ACM*, 28:114–133, 1981.
4. K. Chatterjee and T.A. Henzinger. Semiperfect-information games. In *FSTTCS*, LNCS 3821, pages 1–18. Springer, 2005.
5. L. de Alfaro and T.A. Henzinger. Concurrent ω-regular games. In *Proc. LICS*, pages 141–154. IEEE Computer Society, 2000.
6. L. de Alfaro and T.A. Henzinger. Interface automata. In *Proc. FSE*, pages 109–120. ACM, 2001.
7. L. de Alfaro, T.A. Henzinger, and O. Kupferman. Concurrent reachability games. In *Proc. FOCS*, pages 564–575. IEEE Computer Society, 1998.
8. L. de Alfaro, T.A. Henzinger, and R. Majumdar. From verification to control: Dynamic programs for ω-regular objectives. In *Proc. LICS*, pages 279–290. IEEE Computer Society, 2001.
9. M. De Wulf, L. Doyen, T.A. Henzinger and J.-F. Raskin. Antichains: A new algorithm for checking universality of finite automata. In *CAV*, LNCS (to appear). Springer, 2006.
10. M. De Wulf, L. Doyen, and J.-F. Raskin. A lattice theory for solving games of imperfect information. In *HSCC*, LNCS 3927, pages 153–168. Springer, 2006.
11. D. Dill. *Trace Theory for Automatic Hierarchical Verification of Speed-independent Circuits*. MIT Press, 1989.
12. E.A. Emerson and C.S. Jutla. Tree automata, μ-calculus, and determinacy. In *Proc. FOCS*, pages 368–377. IEEE Computer Society, 1991.
13. A. Kechris. *Classical Descriptive Set Theory*. Springer, 1995.
14. O. Kupferman and M.Y. Vardi. Synthesis with incomplete informatio. In *Advances in Temporal Logic* (H. Barringer et al., eds.), pages 109–127. Kluwer, 1999.
15. M.L. Littman. *Algorithms for Sequential Decision Making*. PhD Thesis, Brown Univ., 1996.
16. D. Martin. Borel determinacy. *Annals of Mathematics*, 102:363–371, 1975.
17. A. Pnueli and R. Rosner. On the synthesis of an asynchronous reactive module. In *ICALP*, LNCS 372, pages 652–671. Springer, 1989.
18. P.J. Ramadge and W.M. Wonham. Supervisory control of a class of discrete-event processes. *SIAM J. Control and Optimization*, 25:206–230, 1987.
19. J.H. Reif. The complexity of two-player games of incomplete information. *J. Computer and System Sciences*, 29:274–301, 1984.
20. W. Thomas. Languages, automata, and logic. In *Handbook of Formal Languages* (G. Rozenberg et al., eds.), volume 3, pages 389–455. Springer, 1997.
21. M.Y. Vardi. Automatic verification of probabilistic concurrent finite-state systems. In *Proc. FOCS*, pages 327–338. IEEE Computer Society, 1985.

Relating Two Standard Notions of Secrecy*

Véronique Cortier, Michaël Rusinowitch, and Eugen Zălinescu

Loria, UMR 7503 & INRIA Lorraine projet Cassis & CNRS, France

Abstract. Two styles of definitions are usually considered to express that a security protocol preserves the confidentiality of a data s. Reachability-based secrecy means that s should never be disclosed while equivalence-based secrecy states that two executions of a protocol with distinct instances for s should be indistinguishable to an attacker. Although the second formulation ensures a higher level of security and is closer to cryptographic notions of secrecy, decidability results and automatic tools have mainly focused on the first definition so far.

This paper initiates a systematic investigation of situations where syntactic secrecy entails strong secrecy. We show that in the passive case, reachability-based secrecy actually implies equivalence-based secrecy for signatures, symmetric and asymmetric encryption provided that the primitives are probabilistic. For active adversaries in the case of symmetric encryption, we provide sufficient (and rather tight) conditions on the protocol for this implication to hold.

1 Introduction

Cryptographic protocols are small programs designed to ensure secure communications. Since they are widely distributed in critical systems, their security is primordial. In particular, verification using formal methods attracted a lot of attention during this last decade. A first difficulty is to formally express the security properties that are expected. Even a basic property such as confidentiality admits two different acceptable definitions namely reachability-based (*syntactic*) secrecy and equivalence-based (*strong*) secrecy. Syntactic secrecy is quite appealing: it says that the secret is never accessible to the adversary. For example, consider the following protocol where the agent A simply sends a secret s to an agent B, encrypted with B's public key.

$$A \to B : \{\mathsf{s}\}_{\mathsf{pub}(B)}$$

An intruder cannot deduce s, thus s is syntactically secret. Although this notion of secrecy may be sufficient in many scenarios, in others, stronger security requirements are desirable. For instance consider a setting where s is a vote and B behaves differently depending on its value. If the actions of B are observable, s remains syntactically secret but an attacker can learn the values of the

* This work has been partially supported by the ACI-SI Satin and the ACI Jeunes Chercheurs JC9005.

Z. Ésik (Ed.): CSL 2006, LNCS 4207, pp. 303–318, 2006.

vote by watching B's actions. The design of equivalence-based secrecy is targeted at such scenarios and intuitively says that an adversary cannot observe the difference when the value of the secret changes. This definition is essential to express properties like confidentiality of a vote, of a password, or the anonymity of participants to a protocol.

Although the second formulation ensures a higher level of security and is closer to cryptographic notions of secrecy, so far decidability results and automatic tools have mainly focused on the first definition. The syntactic secrecy preservation problem is undecidable in general [13], it is co-NP-complete for a bounded number of sessions [17], and several decidable classes have been identified in the case of an unbounded number of sessions [13,10,7,16]. These results often come with automated tools, we mention for example ProVerif [5], CAPSL [12], and Avispa [4]. To the best of our knowledge, the only tool capable of verifying strong secrecy is the resolution-based algorithm of ProVerif [6] and only one decidability result is available: Hüttel [14] proves decidability for a fragment of the spi-calculus without recursion for framed bisimilarity, a related equivalence relation introduced by Abadi and Gordon [2]. Also in [8], Borgström *et al* propose an incomplete decision procedure based on a symbolic bisimulation.

In light of the above discussion, it may seem that the two notions of secrecy are separated by a sizable gap from both a conceptual point of view and a practical point of view. These two notions have counterparts in the cryptographic setting (where messages are bitstrings and the adversary is any polynomial probabilistic Turing machine). Intuitively, the syntactic secrecy notion can be translated into a similar reachability-based secrecy notion and the equivalence-based notion is close to indistinguishability. A quite surprising result [11] states that cryptographic syntactic secrecy actually implies indistinguishability in the cryptographic setting. This result relies in particular on the fact that the encryption schemes are probabilistic thus two encryptions of the same plaintext lead to different ciphertexts.

Motivated by the result of [11] and the large number of available systems for syntactic secrecy verification, we initiate in this paper a systematic investigation of situations where syntactic secrecy entails strong secrecy. Surprisingly, this happens in many interesting cases.

We offer results in both passive and active cases in the setting of the *applied pi calculus* [1]. We first treat in Section 2 the case of passive adversaries. We prove that syntactic secrecy is equivalent to strong secrecy. This holds for signatures, symmetric and asymmetric encryption. It can be easily seen that the two notions of secrecy are not equivalent in the case of deterministic encryption. Indeed, the secret s cannot be deduced from the encrypted message $\{\mathsf{s}\}_{\mathsf{pub}(B)}$ but if the encryption is deterministic, an intruder may try different values for s and check whether the ciphertext he obtained using B's public key is equal to the one he receives. Thus for our result to hold, we require that encryption is probabilistic. This is not a restriction since this is *de facto* the standard in almost all cryptographic applications. Next, we consider the more challenging case of active adversaries. We give sufficient conditions on the protocols for syntactic secrecy to imply strong secrecy (Section 3). Intuitively, we require that the conditional tests are not performed directly on the secret since we have seen above that such tests provide information on the value of this secret. We again exhibit several counter-examples to motivate the introduction of our conditions. An important aspect of our result is that we do not make any assumption on the number of sessions: we put no restriction on the use of replication.

The interest of our contribution is twofold. First, conceptually, it helps to understand when the two definitions of secrecy are actually equivalent. Second, we can transfer many existing results (and the armada of automatic tools) developed for syntactic secrecy. For instance, since the syntactic secrecy problem is decidable for tagged protocols for an unbounded number of sessions [16], by translating the tagging assumption to the applied-pi calculus, we can derive a first decidability result for strong secrecy for an unbounded number of sessions. Other decidable fragments might be derived from [13] for bounded messages (and nonces) and [3] for a bounded number of sessions.

2 Passive Case

Cryptographic primitives are represented by functional symbols. More specifically, we consider the signature $\Sigma = \{\mathsf{enc}, \mathsf{dec}, \mathsf{enca}, \mathsf{deca}, \mathsf{pub}, \mathsf{priv}, \langle\rangle, \pi_1, \pi_2, \mathsf{sign}, \mathsf{check}, \mathsf{retrieve}\}$. $\mathcal{T}(\Sigma, \mathcal{X}, \mathcal{N})$, or simply $\mathcal{T}$, denotes the set of terms built over Σ extended by a set of constants, the infinite set of *names* $\mathcal{N}$ and the infinite set of variables $\mathcal{X}$. A term is *closed* or *ground* if it does not contain any variable. The set of names occurring in a term T is denoted by $\text{fn}(T)$, the set of variables is denoted by $\mathcal{V}(T)$. The *positions* in a term T are defined recursively as usual (*i.e.* as sequences of positive integers), ϵ being the empty sequence. Denote by $\mathbb{N}_+^*$ the set of sequences of positive integers. $\text{Pos}(T)$ denotes the set of positions of T and $\text{Pos}_\text{v}(T)$ the set of positions of variables in T. We denote by $T|_p$ the subterm of T at position p and by $U[V]_p$ the term obtained by replacing in U the subterm at position p by V. We may simply say that a term V *is in* a term U if V is a subterm of U. We denote by $\leq_{st}$ (resp. $<_{st}$) the subterm (resp. strict) order. h_U denotes the function symbol, name or variable at position ϵ in the term U.

We equip the signature with an equational theory E:

$$\begin{cases} \pi_1(\langle z_1, z_2\rangle) = z_1 & \mathsf{deca}(\mathsf{enca}(z_1, \mathsf{pub}(z_2), z_3), \mathsf{priv}(z_2)) = z_1 \\ \pi_2(\langle z_1, z_2\rangle) = z_2 & \mathsf{check}(z_1, \mathsf{sign}(z_1, \mathsf{priv}(z_2)), \mathsf{pub}(z_2)) = \mathsf{ok} \\ \mathsf{dec}(\mathsf{enc}(z_1, z_2, z_3), z_2) = z_1 & \mathsf{retrieve}(\mathsf{sign}(z_1, z_2)) = z_1 \end{cases}$$

The function symbols $\pi_1, \pi_2, \mathsf{dec}, \mathsf{deca}, \mathsf{check}$ and $\mathsf{retrieve}$ are called *destructors*. Let $\mathcal{R}_E$ be the corresponding rewrite system (obtained by orienting the equations from left to right). $\mathcal{R}_E$ is convergent. The normal form of a term T w.r.t. $\mathcal{R}_E$ is denoted by $T\downarrow$. Notice that E is also stable by substitution of names. As usual, we write $U \to V$ if there exists θ, a position p in U and $L \to R \in \mathcal{R}_E$ such that $U|_p = L\theta$ and $V = U[R\theta]_p$.

The symbol $\langle \_, \_ \rangle$ represents the pairing function and π_1 and π_2 are the associated projection functions. The term $\mathsf{enc}(M, K, R)$ represents the message M encrypted with the key K. The third argument R reflects that the encryption is probabilistic: two encryptions of the same messages under the same keys are different. The symbol dec stands for decryption. The symbols enca and deca are very similar but in an asymmetric setting, where $\mathsf{pub}(a)$ and $\mathsf{priv}(a)$ represent respectively the public and private keys of an agent a. The term $\mathsf{sign}(M, K)$ represents the signature of message M with key K. check enables to verify the signature and retrieve enables to retrieve the signed message from the signature.[1]

After the execution of a protocol, an attacker knows the messages sent on the network and also in which order they were sent. Such message sequences are organized as *frames* $\varphi = \nu\widetilde{n}.\sigma$, where $\sigma = \{^{M_1}/_{y_1}, \ldots, ^{M_l}/_{y_l}\}$ is a ground acyclic substitution and $\widetilde{n}$ is a finite set of names. We denote by $\mathrm{dom}(\varphi) = \mathrm{dom}(\sigma) = \{y_1, \ldots, y_l\}$. The variables y_i enable us to refer to each message. The names in $\widetilde{n}$ are said to be *restricted* in φ. Intuitively, these names are *a priori* unknown to the intruder. The names outside $\widetilde{n}$ are said to be *free* in φ. A term M is said *public* w.r.t. a frame $\nu\widetilde{n}.\sigma$ (or w.r.t. a set of names $\widetilde{n}$) if $\mathrm{fn}(M) \cap \widetilde{n} = \emptyset$. The set of restricted names $\widetilde{n}$ might be omitted when it is clear from the context. We usually write $\nu n_1, \ldots, n_k$ instead of $\nu\{n_1, \ldots, n_k\}$.

2.1 Deducibility

Given a frame φ that represents the history of messages sent during the execution of a protocol, we define the *deduction* relation, denoted by $\varphi \vdash M$. Deducible messages are messages that can be obtained from φ by applying functional symbols and the equational theory E.

$$\frac{}{\nu\widetilde{n}.\sigma \vdash x\sigma}\ x \in \mathrm{dom}(\sigma) \qquad \frac{}{\nu\widetilde{n}.\sigma \vdash m}\ m \in \mathcal{N}\backslash\widetilde{n}$$

$$\frac{\nu\widetilde{n}.\sigma \vdash T_1 \quad \cdots \quad \nu\widetilde{n}.\sigma \vdash T_l}{\nu\widetilde{n}.\sigma \vdash f(T_1, \ldots, T_l)} \qquad \frac{\nu\widetilde{n}.\sigma \vdash T \quad T =_E T'}{\nu\widetilde{n}.\sigma \vdash T'}$$

Example 1. k and $\langle k, k' \rangle$ are deducible from the frame $\nu k, k', r.\{^{\mathsf{enc}(k,k',r)}/_x, ^{k'}/_y\}$.

A message is usually said secret if it is not deducible. By opposition to our next notion of secrecy, we say that a term M is *syntactically secret* in φ if $\varphi \not\vdash M$.

2.2 Static Equivalence

Deducibility does not always suffice to express the abilities of an intruder.

Example 2. The set of deducible messages is the same for the frames $\varphi_1 = \nu k, n_1, n_2, r_1.$

[1] Signature schemes may disclose partial information on the signed message. To enforce the intruder capabilities, we assume that messages can always be retrieved out of the signature.

$\{^{\mathsf{enc}(n_1,k,r_1)}/_x, ^{\langle n_1,n_2\rangle}/_y, ^k/_z\}$ and $\varphi_2 = \nu k, n_1, n_2, r_1.\{^{\mathsf{enc}(n_2,k,r_2)}/_x, ^{\langle n_1,n_2\rangle}/_y, ^k/_z\}$, while an attacker is able to detect that the first message corresponds to distinct nonces. In particular, the attacker is able to distinguish the two "worlds" represented by φ_1 and φ_2.

We say that a frame $\varphi = \nu\tilde{n}.\sigma$ *passes the test* (U, V) where U, V are two terms, denoted by $(U = V)\varphi$, if there exists a renaming of the restricted names in φ such that $(\mathrm{fn}(U) \cup \mathrm{fn}(V)) \cap \tilde{n} = \emptyset$ and $U\sigma =_E V\sigma$. Two frames $\varphi = \nu\tilde{n}.\sigma$ and $\varphi' = \nu\tilde{m}.\sigma'$ are *statically equivalent*, written $\varphi \approx \varphi'$, if they pass the same tests, that is $\mathrm{dom}(\varphi) = \mathrm{dom}(\varphi')$ and for all terms U, V such that $(\mathcal{V}(U) \cup \mathcal{V}(V)) \subseteq \mathrm{dom}(\varphi)$ and $(\mathrm{fn}(U) \cup \mathrm{fn}(V)) \cap (\tilde{n} \cup \tilde{m}) = \emptyset$, we have $(U = V)\varphi$ iff $(U = V)\varphi'$.

Example 3. The frames φ_1 and φ_2 defined in Example 2 are not statically equivalent since $(\mathsf{dec}(x, z) = \pi_1(y))\varphi_1$ but $(\mathsf{dec}(x, z) \neq \pi_1(y))\varphi_2$.

Let s be a free name of a frame $\varphi = \nu\tilde{n}.\sigma$. We say that s is *strongly secret* in φ if for every closed public terms M, M' w.r.t. φ, we have $\varphi(^M/_\mathsf{s}) \approx \varphi(^{M'}/_\mathsf{s})$ that is, the intruder cannot distinguish the frames obtained by instantiating the secret s by two terms of its choice. For simplicity we may omit s and write $\varphi(M)$ instead of $\varphi(^M/_\mathsf{s})$.

Of course an intended syntactical secret name s must be restricted, but when talking about instances of s we must consider it (at least) a free name (if not a variable). Hence we compare syntactic secrecy and strong secrecy regarding the same frame modulo the restriction on the secret s. We use the notation $\nu\mathsf{s}.\varphi$ for $\nu(\tilde{n} \cup \{\mathsf{s}\}).\sigma$, where $\varphi = \nu\tilde{n}.\sigma$. Thus s is syntactically secret if $\nu\mathsf{s}.\varphi \nvdash \mathsf{s}$.

2.3 Syntactic Secrecy Implies Strong Secrecy

Syntactic secrecy is usually weaker than strong secrecy! We first exhibit some examples of frames that preserves syntactic secrecy but not strong secrecy. They all rely on different properties.

Probabilistic encryption. The frame $\psi_1 = \nu k, r.\{^{\mathsf{enc}(\mathsf{s},k,r)}/_x, ^{\mathsf{enc}(n,k,r)}/_y\}$ does not preserve the strong secrecy of s. Indeed, $\psi_1(n) \not\approx \psi_1(n')$ since $(x = y)\psi_1(n)$ but $(x \neq y)\psi_1(n')$. This would not happen if each encryption used a distinct randomness, that is, if the encryption was probabilistic.
Key position. The frame $\psi_2 = \nu n.\{^{\mathsf{enc}(\langle n,n'\rangle,\mathsf{s},r)}/_x\}$ does not preserve the strong secrecy of s. Indeed, $\psi_2(k) \not\approx \psi_2(k')$ since $(\pi_2(\mathsf{dec}(x, k)) = n')\,\psi_2(k)$ but $(\pi_2(\mathsf{dec}(x, k)) \neq n')\,\psi_2(k')$. If s occurs in key position in some ciphertext, the intruder may try to decrypt the ciphertext since s is replaced by public terms and check for some redundancy. It may occur that the encrypted message does not contain any verifiable part. In that case, the frame may preserve strong secrecy. It is for example the case for the frame $\nu n.\{^{\mathsf{enc}(n,\mathsf{s},r)}/_x\}$. Such cases are however quite rare in practice.
No destructors. The frame $\psi_3 = \{^{\pi_1(\mathsf{s})}/_x\}$ does not preserve the strong secrecy of s simply because $(x = k)$ is true for $\psi_3(\langle k, k'\rangle)$ while not for $\psi_3(k)$.
Retrieve rule. The $\mathsf{retrieve}(\mathsf{sign}(z_1, z_2)) = z_1$ may seem arbitrary since not all signature schemes enable to get the signed message out of a signature. It is actually crucial for our result. For example, the frame $\psi_4 = \{^{\mathsf{sign}(\mathsf{s},\mathsf{priv}(a))}/_x, ^{\mathsf{pub}(a)}/_y\}$

does not preserve the strong secrecy of s because ($\mathsf{check}(n, x, y) = \mathsf{ok}$) is true for $\psi_4(n)$ but not for $\psi_4(n')$.

In these four cases, the frames preserve the syntactic secrecy of s, that is $\nu\mathsf{s}.\psi_i \not\vdash \mathsf{s}$, for $1 \leq i \leq 4$. This leads us to the following definition.

Definition 1. *A frame $\varphi = \nu\widetilde{n}.\sigma$ is* well-formed *w.r.t. some name* s *if*

1. *Encryption is probabilistic, i.e. for any subterm $\mathsf{enc}(M, K, R)$ of φ, for any term $T \in \varphi$ and position p such that $T|_p = R$ we have $p = q.3$ for some q and $T|_q = \mathsf{enc}(M, K, R)$. In addition, if* s *occurs in M at a position p' such that no encryption appears along the path from the root to p' then R must be restricted, that is $R \in \widetilde{n}$. The same conditions hold for asymmetric encryption. and*
2. s *is not part of a key, i.e. for all $\mathsf{enc}(M, K, R)$, $\mathsf{enca}(M', K', R')$, $\mathsf{sign}(U, V)$, $\mathsf{pub}(W)$, $\mathsf{priv}(W')$ subterms of φ, $\mathsf{s} \notin \mathrm{fn}(K, K', V, W, W', R, R')$.*
3. *φ does not contain destructor symbols.*

Condition 1 requires that each innermost encryption above s contains a restricted randomness. This is not a restriction since s is meant to be a secret value and such encryptions have to be produced by honest agents and thus contain a restricted randomness.

For well-formed frames, syntactic secrecy is actually equivalent to strong secrecy

Theorem 1. *Let φ be a well-formed frame w.r.t.* s*, where* s *is a free name in φ.*

$$\nu\mathsf{s}.\varphi \nvdash \mathsf{s} \quad \textit{if and only if} \quad \varphi({}^{M}/_{\mathsf{s}}) \approx \varphi({}^{M'}/_{\mathsf{s}})$$

for all M, M' closed public terms w.r.t. φ.

Proof. We present the skeleton of the proof; all details can be found in a technical report [18]. Let $\varphi = \nu\widetilde{n}.\sigma$ be a well-formed frame w.r.t. s. If $\nu\mathsf{s}.\varphi \vdash \mathsf{s}$, this trivially implies that s is not strongly secret. Indeed, there exists a public term T w.r.t. φ such that $T\sigma =_E \mathsf{s}$ (this can be easily shown by induction on the deduction system). Let n_1, n_2 be fresh names such that $n_1, n_2 \notin \widetilde{n}$ and $n_1, n_2 \notin \mathrm{fn}(\varphi)$. Since $T\sigma({}^{n_1}/_{\mathsf{s}}) =_E n_1$ the frames $\varphi({}^{n_1}/_{\mathsf{s}})$ and $\varphi({}^{n_2}/_{\mathsf{s}})$ are distinguishable with the test $(T = n_1)$.

We assume now that $\nu\mathsf{s}.\varphi \nvdash \mathsf{s}$. We first show that any syntactic equality satisfied by the frame $\varphi({}^{M}/_{\mathsf{s}})$ is already satisfied by φ.

Lemma 1. *Let $\varphi = \nu\widetilde{n}.\sigma$ be a well-formed frame w.r.t. a free name* s*, U, V terms such that $\mathcal{V}(U), \mathcal{V}(V) \subseteq \mathrm{dom}(\varphi)$ and M a closed term, U, V and M public w.r.t. $\widetilde{n}$. If $\nu\mathsf{s}.\varphi \nvdash \mathsf{s}$ then $U\sigma({}^{M}/_{\mathsf{s}}) = V\sigma({}^{M}/_{\mathsf{s}})$ implies $U\sigma = V\sigma$. Let T be a subterm of a term in σ that does not contain* s*. If $\nu\mathsf{s}.\varphi \nvdash \mathsf{s}$ then $T = V\sigma({}^{M}/_{\mathsf{s}})$ implies $T = V\sigma$.*

The key lemma is that any reduction that applies to a deducible term U where s is replaced by some M, directly applies to U.

Lemma 2. *Let $\varphi = \nu\tilde{n}.\sigma$ be a well-formed frame w.r.t. a free name* s *such that* $\nu\mathsf{s}.\varphi \nvdash \mathsf{s}$. *Let U be a term with $\mathcal{V}(U) \subseteq \mathrm{dom}(\varphi)$ and M be a closed term in normal form, U and M public w.r.t. $\tilde{n}$. If $U\sigma(^M/_{\mathsf{s}}) \rightarrow V$, for some term V, then there exists a well-formed frame $\varphi' = \nu\tilde{n}.\sigma'$ w.r.t.* s

- *extending φ, that is $x\sigma' = x\sigma$ for all $x \in \mathrm{dom}(\sigma)$,*
- *preserving deducible terms: $\nu\mathsf{s}.\varphi \vdash W$ iff $\nu\mathsf{s}.\varphi' \vdash W$,*
- *and such that $V = V'\sigma'(^M/_{\mathsf{s}})$ and $U\sigma \rightarrow V'\sigma'$ for some V' public w.r.t. $\tilde{n}$.*

This lemma allows us to conclude the proof of Theorem 1. Fix arbitrarily two public closed terms M, M'. We can assume w.l.o.g. that M and M' are in normal form. Let $U \neq V$ be two public terms such that $\mathcal{V}(U), \mathcal{V}(V) \subseteq \mathrm{dom}(\varphi)$ and $U\sigma(^M/_{\mathsf{s}}) =_E V\sigma(^M/_{\mathsf{s}})$. Then there are $U_1, \dots, U_k$ and $V_1, \dots, V_l$ such that $U\sigma(^M/_{\mathsf{s}}) \rightarrow U_1 \rightarrow \dots \rightarrow U_k$, $V\sigma(^M/_{\mathsf{s}}) \rightarrow V_1 \rightarrow \dots \rightarrow V_l$, $U_k = U\sigma(^M/_{\mathsf{s}})\downarrow$, $V_l = V\sigma(^M/_{\mathsf{s}})\downarrow$ and $U_k = V_l$.

Applying repeatedly Lemma 2 we obtain that there exist public terms $U'_1, \dots, U'_k$ and $V'_1, \dots, V'_l$ and well-formed frames $\varphi^{u_i} = \nu\tilde{n}.\sigma^{u_i}$, for $i \in \{1, \dots, k\}$ and $\varphi^{v_j} = \nu\tilde{n}.\sigma^{v_j}$, for $j \in \{1, \dots, l\}$ (as in the lemma) such that $U_i = U'_i\sigma^{u_i}(^M/_{\mathsf{s}})$, $U'_i\sigma^{u_i} \rightarrow U'_{i+1}\sigma^{u_{i+1}}$, $V_j = V'_j\sigma^{v_j}(^M/_{\mathsf{s}})$ and $V'_j\sigma^{v_j} \rightarrow V'_{j+1}\sigma^{v_{j+1}}$.

We consider $\varphi' = \nu\tilde{n}.\sigma'$ where $\sigma' = \sigma^{u_k} \cup \sigma^{v_l}$. Since only subterms of φ have been added to φ', it is easy to verify that φ' is still a well-formed frame and for every term W, $\nu\mathsf{s}.\varphi \vdash W$ iff $\nu\mathsf{s}.\varphi' \vdash W$. In particular $\nu\mathsf{s}.\varphi' \nvdash \mathsf{s}$.

By construction we have that $U'_k\sigma^{u_k}(^M/_{\mathsf{s}}) = V'_l\sigma^{v_l}(^M/_{\mathsf{s}})$. Then, by Lemma 1, we deduce that $U'_k\sigma^{u_k} = V'_l\sigma^{v_l}$ that is $U\sigma =_E V\sigma$. By stability of substitution of names, we have $U\sigma(^{M'}/_{\mathsf{s}}) =_E V\sigma(^{M'}/_{\mathsf{s}})$. We deduce that $\varphi(^M/_{\mathsf{s}}) \approx \varphi(^{M'}/_{\mathsf{s}})$.

3 Active Case

To simplify the analysis of the active case, we restrict our attention to pairing and symmetric encryption: the alphabet Σ is now reduced to $\Sigma = \{\mathsf{enc}, \mathsf{dec}, \langle\rangle, \pi_1, \pi_2\}$ and E is limited to the first three equations.

3.1 Modeling Protocols Within the Applied Pi Calculus

The applied pi calculus [1] is a process algebra well suited for modeling cryptographic protocols, generalizing the spi-calculus [2]. We briefly describe its syntax and semantics. This part is mostly borrowed from [1].

Processes, also called plain processes, are defined by the grammar:

P, Q := processes			
$\mathbf{0}$	null process	$\nu n.P$	name restriction
$P \mid Q$	parallel composition	$u(z).P$	message input
$!P$	replication	$\overline{u}\langle M\rangle.P$	message output
if $M = N$ *then* P *else* Q	conditional		

where n is a name, U, V are terms, and u is a name or a variable. The null process $\mathbf{0}$ does nothing. Parallel composition executes the two processes concurrently.

Replication $!P$ creates unboundedly new instances of P. Name restriction $\nu n.P$ builds a new, private name n, binds it in P and then executes P. The conditional *if* $M = N$ *then* P *else* Q behaves like P or Q depending on the result of the test $M = N$. If Q is the null process then we use the notation $[M = N].P$ instead. Finally, the process $u(z).P$ inputs a message and executes P binding the variable z to the received message, while the process $\overline{u}\langle M\rangle.P$ outputs the message M and then behaves like P. We may omit P if it is $\mathbf{0}$. In what follows, we restrict our attention to the case where u is a name since it is usually sufficient to model cryptographic protocols.

Extended processes are defined by the grammar:

$A, B :=$	extended processes		
P	plain process	$\nu n.A$	name restriction
$A \mid B$	parallel composition	$\nu x.A$	variable restriction
$\{^M/_x\}$	active substitution		

Active substitutions generalize *let*, in the sense that $\nu x.(\{^M/_x\}|P)$ corresponds to *let* $x = M$ *in* P, while unrestricted, $\{^M/_x\}$ behaves like a permanent knowledge, permitting to refer globally to M by means of x. We identify variable substitutions $\{^{M_1}/_{x_1}, \ldots, {}^{M_l}/_{x_l}\}$, $l \geq 0$ with extended processes $\{^{M_1}/_{x_1}\}| \ldots |\{^{M_l}/_{x_l}\}$. In particular the empty substitution is identified with the null process.

We denote by $\text{fv}(A)$, $\text{bv}(A)$, $\text{fn}(A)$, and $\text{bn}(A)$ the sets of free and bound variables and free and bound names of A, respectively, defined inductively as usual for the pi calculus' constructs and using $\text{fv}(\{^M/_x\}) = \text{fv}(M) \cup \{x\}$ and $\text{fn}(\{^M/_x\}) = \text{fn}(M)$ for active substitutions. An extended process is *closed* if it has no free variables except those in the domain of active substitutions.

Extended processes built up from the null process (using the given constructions, that is, parallel composition, restriction and active substitutions) are called *frames*[2]. To every extended process A we associate the frame $\varphi(A)$ obtained by replacing all embedded plain processes with $\mathbf{0}$.

An *evaluation context* is an extended process with a hole not under a replication, a conditional, an input or an output.

Structural equivalence ($\equiv$) is the smallest equivalence relation on extended processes that is closed by α-conversion of names and variables, by application of evaluation contexts and such that the standard structural rules for the null process, parallel composition and restriction (such as associativity and commutativity of $|$, commutativity and binding-operator-like behavior of ν) together with the following ones hold.

$$
\begin{array}{rlr}
\nu x.\{^M/_x\} \equiv \mathbf{0} & & \text{ALIAS} \\
\{^M/_x\} \mid A \equiv \{^M/_x\} \mid A\{^M/_x\} & & \text{SUBST} \\
\{^M/_x\} \equiv \{^N/_x\} & \text{if } M =_E N & \text{REWRITE}
\end{array}
$$

If $\widetilde{n}$ represents the (possibly empty) set $\{n_1, \ldots, n_k\}$, we abbreviate by $\nu\widetilde{n}$ the sequence $\nu n_1.\nu n_2 \ldots \nu n_k$. Every closed extended process A can be brought to

[2] We see later in this section why we use the same name as for the notion defined in section 2.

the form $\nu\tilde{n}.\{^{M_1}/_{x_1}\}|\ldots|\{^{M_l}/_{x_l}\}|P$ by using structural equivalence, where P is a plain closed process, $l \geq 0$ and $\{\tilde{n}\} \subseteq \cup_i \mathrm{fn}(M_i)$. Hence the two definitions of frames are equivalent up to structural equivalence on closed extended processes. To see this we apply rule SUBST until all terms are ground (this is assured by the fact that the considered extended processes are closed and the active substitutions are cycle-free). Also, another consequence is that if $A \equiv B$ then $\varphi(A) \equiv \varphi(B)$.

Two semantics can be considered for this calculus, defined by structural equivalence and by *internal reduction* and *labeled reduction*, respectively. These semantics lead to *observational equivalence* (which is standard and not recalled here) and *labeled bisimilarity* relations. The two bisimilarity relations are equal [1]. We use here the latter since it relies on static equivalence and it allows to take implicitly into account the adversary, hence having the advantage of not using quantification over contexts.

Internal reduction is the largest relation on extended processes closed by structural equivalence and application of evaluation contexts such that:

$$\begin{array}{rcll} \overline{c}\langle x\rangle.P \mid c(x).Q & \rightarrow & P \mid Q & \text{COMM} \\ \textit{if } M = M \textit{ then } P \textit{ else } Q & \rightarrow & P & \text{THEN} \\ \textit{if } M = N \textit{ then } P \textit{ else } Q & \rightarrow & Q & \text{ELSE} \end{array}$$

for any ground terms M and N such that $M \neq_E N$

On the other hand, *labeled reduction* is defined by the following rules.

$$c(x).P \xrightarrow{c(M)} P\{^M/_x\} \quad \text{IN} \qquad\qquad \overline{c}\langle u\rangle.P \xrightarrow{\overline{c}\langle u\rangle} P \quad \text{OUT-ATOM}$$

$$\frac{A \xrightarrow{\overline{c}\langle u\rangle} A'}{\nu u.A \xrightarrow{\nu u.\overline{c}\langle u\rangle} A'}\ u \neq c \quad \text{OPEN-ATOM} \qquad\qquad \frac{A \xrightarrow{\alpha} A'}{\nu u.A \xrightarrow{\alpha} \nu u.A'}\ \begin{array}{l}u \text{ does not}\\ \text{occur in } \alpha\end{array} \quad \text{SCOPE}$$

$$\frac{A \xrightarrow{\alpha} A'}{A|B \xrightarrow{\alpha} A'|B}\ (*) \quad \text{PAR} \qquad\qquad \frac{A \equiv B \quad B \xrightarrow{\alpha} B' \quad B' \equiv A'}{A \xrightarrow{\alpha} A'} \quad \text{STRUCT}$$

where c is a name and u is a metavariable that ranges over names and variables, and the condition (*) of the rule PAR is $\mathrm{bv}(\alpha) \cap \mathrm{fv}(B) = \mathrm{bn}(\alpha) \cap \mathrm{fn}(B) = \emptyset$.

Definition 2. Labeled bisimilarity *($\approx_l$) is the largest symmetric relation $\mathcal{R}$ on closed extended processes such that $A\,\mathcal{R}\,B$ implies:*

1. *$\varphi(A) \approx \varphi(B)$;*
2. *if $A \rightarrow A'$ then $B \rightarrow^* B'$ and $A'\,\mathcal{R}\,B'$, for some B';*
3. *if $A \xrightarrow{\alpha} A'$ and $\mathrm{fv}(\alpha) \subseteq \mathrm{dom}(\varphi(A))$ and $\mathrm{bn}(\alpha) \cap \mathrm{fn}(B) = \emptyset$ then $B \rightarrow^* \xrightarrow{\alpha} \rightarrow^* B'$ and $A'\,\mathcal{R}\,B'$, for some B'.*

We denote $A \Rightarrow B$ if $A \rightarrow B$ or $A \xrightarrow{\alpha} B$.

Definition 3. *A frame φ is* valid *w.r.t. a process P if there is A such that $P \Rightarrow^* A$ and $\varphi \equiv \varphi(A)$.*

Definition 4. *Let P be a closed plain process without variables as channels and* s *a free name of P, but not a channel name. We say that* s *is* syntactically secret *in P if, for every valid frame φ w.r.t. P,* s *is not deducible from νs.φ. We say that* s *is* strongly secret *if for any closed terms M, M' such that* $\text{bn}(P) \cap (\text{fn}(M) \cup \text{fn}(M')) = \emptyset$, $P({}^{M}/_{\mathsf{s}}) \approx_l P({}^{M'}/_{\mathsf{s}})$.

Let $\mathcal{M}_o(P)$ be the set of *outputs* of P, that is the set of terms m such that $\overline{c}\langle m \rangle$ is a message output construct for some channel name c in P, and let $\mathcal{M}_t(P)$ be the set of *operands of tests* of P, where a *test* is a couple $M = N$ occurring in a conditional and its *operands* are M and N. Let $\mathcal{M}(P) = \mathcal{M}_o(P) \cup \mathcal{M}_t(P)$ be the set of *messages* of P. Examples are provided at the end of this section.

The following lemma intuitively states that any message contained in a valid frame is an output instantiated by messages deduced from previous sent messages.

Lemma 3. *Let P be a closed plain process, and A be a closed extended process such that $P \Rightarrow^* A$. There are $l \geq 0$, an extended process $B = \nu\widetilde{n}.\sigma_l | P_B$, where P_B is some plain process, and θ a substitution public w.r.t. $\widetilde{n}$ such that: $A \equiv B$, $\{\widetilde{n}\} \subseteq \text{bn}(P)$, for every operand of a test or an output T of P_B there is a message T_0 in P (a operand of a test or an output respectively), such that $T = T_0\theta\sigma_l$, and, $\sigma_i = \sigma_{i-1} \cup \{{}^{M_i\theta_i\sigma_{i-1}}/_{y_i}\}$, for all $1 \leq i \leq l$, where M_i is an output in P, θ_i is a substitution public w.r.t. $\widetilde{n}$ and σ_0 is the empty substitution.*

The proof is done by induction on the number of reductions in $P \Rightarrow^* A$. Intuitively, B is obtained by applying the SUBST rule (from left to right) as much as possible until there are no variables left in the plain process. Note that B is unique up to the structural rules different from ALIAS, SUBST and REWRITE. We say that $\varphi(B)$ is the *standard frame* w.r.t. A.

As a running example we consider the Yahalom protocol:

$$\begin{aligned} A \Rightarrow B &: \ A, N_a \\ B \Rightarrow S &: \ B, \{A, N_a, N_b\}_{K_{bs}} \\ S \Rightarrow A &: \ \{B, K_{ab}, N_a, N_b\}_{K_{as}}, \{A, K_{ab}\}_{K_{bs}} \\ A \Rightarrow B &: \ \{A, K_{ab}\}_{K_{bs}} \end{aligned}$$

In this protocol, two participants A and B wish to establish a shared key K_{ab}. The key is created by a trusted server S which shares the secret keys K_{as} and K_{bs} with A and B respectively. The protocol is modeled by the following process.

$$P_Y(k_{ab}) = \nu k_{as}, k_{bs}.(!P_A)|(!P_B)|(!\nu k.P_S(k))|P_S(k_{ab}) \quad \text{with}$$

$$P_A = \nu n_a.\overline{c}\langle a, n_a\rangle.c(z_a).[b = U_b].[n_a = U_{n_a}].\overline{c}\langle \pi_2(z_a)\rangle$$
$$P_B = c(z_b).\nu n_b, r_b.\overline{c}\langle b, \mathsf{enc}(\langle \pi_1(z_b), \langle \pi_2(z_b), n_b\rangle\rangle, k_{bs}, r_b)\rangle.c(z'_b).[a = \pi_1(\mathsf{dec}(z'_b, k_{bs}))]$$
$$P_S(x) = c(z_s).\nu r_s, r'_s.\overline{c}\langle \mathsf{enc}(\langle \pi_1(z_s), \langle x, V_n\rangle\rangle, k_{as}, r_s), \mathsf{enc}(\langle V_a, x\rangle, k_{bs}, r'_s)\rangle$$

$$\text{and} \quad U_b = \pi_1(\mathsf{dec}(\pi_1(z_a), k_{as})) \quad U_{n_a} = \pi_1(\pi_2(\pi_2(\mathsf{dec}(\pi_1(z_a), k_{as}))))$$
$$V_a = \pi_1(\mathsf{dec}(\pi_2(z_s), k_{bs})) \quad V_n = \pi_2(\mathsf{dec}(\pi_2(z_s), k_{bs})).$$

For this protocol the set of outputs and operands of tests are respectively:

$$\mathcal{M}_o(P_Y) = \{\langle a, n_a\rangle, z_a, \pi_2(z_a), \langle b, \mathsf{enc}(\langle \pi_1(z_b), \langle \pi_2(z_b), n_b\rangle\rangle, k_{bs}, r_b)\rangle, z'_b, \\ \mathsf{enc}(\langle \pi_1(z_s), \langle x, V_n\rangle\rangle, k_{as}, r_s), \mathsf{enc}(\langle V_a, x\rangle, k_{bs}, r'_s)\} \text{ and} \\ \mathcal{M}_t(P_Y) = \{b, U_b, n_a, U_{n_a}, a, \pi_1(\mathsf{dec}(z'_b, k_{bs}))\}.$$

3.2 Our Hypotheses

In what follows, we assume s to be the secret. As in the passive case, destructors above the secret must be forbidden. We also restrict ourself to processes with ground terms in key position. We consider the process $P_1 = \nu k, r, r'.(\bar{c}\langle \mathsf{enc}(\mathsf{s}, k, r)\rangle \mid c(z).\bar{c}\langle \mathsf{enc}(a, \mathsf{dec}(z, k), r')\rangle)$. The name s in P_1 is syntactically secret but not strongly secret. Indeed,

$$\begin{aligned} P_1 &\equiv \nu k, r, r'.(\nu z.(\{^{\mathsf{enc}(\mathsf{s},k,r)}/_z\} \mid \bar{c}\langle z\rangle \mid c(z).\bar{c}\langle \mathsf{enc}(a, \mathsf{dec}(z, k), r')\rangle)) \\ &\rightarrow \nu k, r, r'.(\{^{\mathsf{enc}(\mathsf{s},k,r)}/_z\} \mid \bar{c}\langle \mathsf{enc}(a, \mathsf{s}, r')\rangle) \qquad \text{(COMM rule)} \\ &\equiv \nu k, r, r'.(\nu z'.(\{^{\mathsf{enc}(\mathsf{s},k,r)}/_z, {}^{\mathsf{enc}(a,\mathsf{s},r')}/_{z'}\} \mid \bar{c}\langle z'\rangle)) \\ &\xrightarrow{\nu z'.\bar{c}\langle z'\rangle} \nu k, r, r'.\{^{\mathsf{enc}(\mathsf{s},k,r)}/_z, {}^{\mathsf{enc}(a,\mathsf{s},r')}/_{z'}\} \stackrel{\mathsf{def}}{=} P'_1 \end{aligned}$$

and P'_1 does not preserve the strong secrecy of s (see the frame ψ_2 of Section 2.3).

Without loss of generality with respect to cryptographic protocols, we assume that terms occurring in processes are in normal form and that no destructor appears above constructors. Indeed, terms like $\pi_1(\mathsf{enc}(m, k, r))$ are usually not used to specify protocols. We also assume that tests do not contain constructors. Indeed a test $[\langle M_1, M_2\rangle = N]$ can be rewritten as $[M_1 = N_1].[M_2 = N_2]$ if $N = \langle N_1, N_2\rangle$, and $[M_1 = \pi_1(N)].[M_2 = \pi_2(N)]$ if N does not contain constructors, and will never hold otherwise. Similar rewriting applies for encryption, except for the test $[\mathsf{enc}(M_1, M_2, M_3) = N]$ if N does not contain constructors. It can be rewritten in $[\mathsf{dec}(N, M_2) = M_1]$ but this is not equivalent. However since the randomness of encryption is not known to the agent, explicit tests on the randomness should not occur in general.

This leads us to consider the following class of processes. But first, we say that an occurrence q_{enc} of an encryption in a term T is an *agent encryptions* w.r.t. a set of names $\tilde{n}$ if $t|_{q_{\mathsf{enc}}} = \mathsf{enc}(M, K, R)$ for some M, K, R and $R \in \tilde{n}$.

Definition 5. *A process P is* well-formed *w.r.t. a name s if it is closed and if:*

1. *any occurrence of $\mathsf{enc}(M, K, R)$ in some term $T \in \mathcal{M}(P)$ is an agent encryption w.r.t. $\mathrm{bn}(P)$, and for any term $T' \subset \mathcal{M}(P)$ and position p such that $T'|_p = T$ there is a position q such that $q.3 = p$ and $T'|_q = \mathsf{enc}(M, K, R)$;*
2. *for every term $\mathsf{enc}(M, K, R)$ or $\mathsf{dec}(M, K)$ occurring in P, K is ground;*
3. *any operand of a test $M \in \mathcal{M}_t$ is a name, a constant or has the form $\pi^1(\mathsf{dec}(\ldots \pi^n(\mathsf{dec}(\pi^{n+1}(z), K_l)) \ldots, K_1))$, with $l \geq 0$, where the π^i are words on $\{\pi_1, \pi_2\}$ and z is a variable;*
4. *there are no destructors above constructors, nor above s.*

Conditional tests should not test on s. For example, consider the process $P_3 = \nu k, r.(\bar{c}\langle \mathsf{enc}(\mathsf{s}, k, r)\rangle \mid c(z).[\mathsf{dec}(z, k) = a].\bar{c}\langle \mathsf{ok}\rangle)$ where a is a non restricted name. s in P_3 is syntactically secret but not strongly secret. Indeed, $P_3 \to \nu k, r.(\{^{\mathsf{enc}(\mathsf{s},k,r)}/_z\} \mid [\mathsf{s} = a].\bar{c}\langle \mathsf{ok}\rangle)$. The process $P_3(^a/_{\mathsf{s}})$ reduces further while $P_3(^b/_{\mathsf{s}})$ does not. That is why we have to prevent hidden tests on s. Such tests may occur nested in equality tests. For example, let

$$\begin{aligned} P_4 = \nu k, r, r_1, r_2.(&\bar{c}\langle \mathsf{enc}(\mathsf{s}, k, r)\rangle \mid \bar{c}\langle \mathsf{enc}(\mathsf{enc}(a, k', r_2), k, r_1)\rangle \\ &\mid c(z).[\mathsf{dec}(\mathsf{dec}(z, k), k') = a].\bar{c}\langle \mathsf{ok}\rangle) \\ \to P_4' = \nu k, r, r_1, r_2.(&\{^{\mathsf{enc}(\mathsf{s},k,r)}/_z\} \mid \bar{c}\langle \mathsf{enc}(\mathsf{enc}(a, k', r_2), k, r_1)\rangle \mid [\mathsf{dec}(\mathsf{s}, k') = a].\bar{c}\langle \mathsf{ok}\rangle) \end{aligned}$$

Then $P_4(^{\mathsf{enc}(a,k',r')}/_{\mathsf{s}})$ is not equivalent to $P_4(^n/_{\mathsf{s}})$, since the process $P_4'(^{\mathsf{enc}(a,k',r')}/_{\mathsf{s}})$ emits the message ok while $P_4'(^n/_{\mathsf{s}})$ does not. This relies on the fact that the decryption $\mathsf{dec}(z, k)$ allows access to s in the test.

For the rest of the section we assume that z_0 is a new fixed variable.

To prevent hidden tests on the secret, we compute an over-approximation of the ciphertexts that may contain the secret, by marking with a symbol x all positions under which the secret may appear in clear.

We first introduce a function f_{ep} that extracts the least encryption over s and "clean" the pairing function above s. Formally, we define the partial function

$$f_{ep} \colon \mathcal{T} \times \mathbb{N}_+^* \hookrightarrow \mathcal{T} \times \mathbb{N}_+^*$$

$f_{ep}(U, p) = (V, q)$ where V and q are defined as follows: $q \leq p$ is the position (if it exists) of the lowest encryption on the path p in U. If q does not exist or if p is not a maximal position in U, then $f_{ep}(U, p) = \perp$. Otherwise, V is obtained from $U|_q$ by replacing all arguments of pairs that are not on the path p with new variables. More precisely, let $V' = U|_q$. The subterm V' must be of the form $\mathsf{enc}(M_1, M_2, M_3)$ and $p = q.i.q'$. Then V is defined by $V = \mathsf{enc}(M_1', M_2', M_3')$ with $M_j' = M_j$ for $j \neq i$ and $M_i' = \mathsf{prune}(M_i, q')$ where prune is recursively defined by: $\mathsf{prune}(\langle N_1, N_2\rangle, 1.r) = \langle \mathsf{prune}(N_1, r), x_r\rangle$, $\mathsf{prune}(\langle N_1, N_2\rangle, 2.r) = \langle x_r, \mathsf{prune}(N_2, r)\rangle$ and $\mathsf{prune}(N, \epsilon) = N$.
For example, $f_{ep}(\mathsf{enc}(\mathsf{enc}(\langle\langle a, b\rangle, c\rangle, k_2, r_2), k_1, r_1), 1.1.2) = (\mathsf{enc}(\langle z_\epsilon, c\rangle, k_2, r_2), 1)$.

The function f_e is the composition of the first projection with f_{ep}. With the function f_e, we can extract from the outputs of a protocol P the set of ciphertexts where s appears in clear below the encryption.

$$\mathcal{E}_0(P) = \{f_e(M[\mathrm{x}]_p, p) \mid M \in \mathcal{M}_o(P) \ \wedge \ M|_p = \mathsf{s}\}.$$

For example, $\mathcal{E}_0(P_Y) = \{\mathsf{enc}(\langle z_1, \langle \mathrm{x}, z_{1.2}\rangle\rangle, k_{as}, r_s), \mathsf{enc}(\langle z_1, \mathrm{x}\rangle, k_{bs}, r_s')\}$, where P_Y is the process corresponding to the Yahalom protocol defined in previous section.

However s may appear in other ciphertexts later on during the execution of the protocol after decryptions and encryptions. Thus we also extract from outputs the destructor parts (which may open encryptions). Namely, we define the partial function

$$f_{dp} \colon \mathcal{T} \times \mathbb{N}_+^* \hookrightarrow \mathcal{T} \times \mathbb{N}_+^*$$

$f_{dp}(U,p) = (V,q)$ where V and q are defined as follows: $q \leq p$ is the occurrence of the highest destructor above p (if it exists). Let $r \leq p$ be the occurrence of the lowest decryption above p (if it exists). We have $U|_r = \mathsf{dec}(U_1, U_2)$. Then U_1 is replaced by the variable z_0 that is $V = (U[\mathsf{dec}(\mathrm{z}_0, U_2)]_r)|_q$. If q or r do not exist then $f_{dp}(U,p) = \perp$.

For example, $f_{dp}(\mathsf{enc}(\pi_1(\mathsf{dec}(\pi_2(y), k_1)), k_2, r_2), 1.1.1.1) = (\pi_1(\mathsf{dec}(\mathrm{z}_0, k_1)), 1)$.

The function f_d is the composition of the first projection with f_{dp}. By applying the function f_d to messages of a well-formed process P we always obtain terms D of the form $D = D_1(\dots D_n)$ where $D_i = \pi^i(\mathsf{dec}(\mathrm{z}_0, K_i))$ with $1 \leq i \leq n$, K_i are ground terms and π^i is a (possibly empty) sequence of projections $\pi_{j_1}(\pi_{j_2}(\dots(\pi_{j_l})\dots))$.

With the function f_d, we can extract from the outputs of a protocol P the meaningful destructor part:

$$\mathcal{D}_o(P) = \{f_d(M,p) \mid M \in \mathcal{M}_o(P) \ \wedge \ p \in \mathrm{Pos_v}(M)\}.$$

For example, $\mathcal{D}_o(P_Y) = \{\pi_2(\mathsf{dec}(\mathrm{z}_0, k_{bs})), \pi_1(\mathsf{dec}(\mathrm{z}_0, k_{bs}))\}$.

We are now ready to mark (with $\mathtt{x}$) all the positions where the secret might be transmitted (thus tested). We also define inductively the sets $\mathcal{E}_i(P)$ as follows. For each element E of $\mathcal{E}_i$ we can show that there is an unique term in normal form denoted by $\overline{E}$ such that $\mathcal{V}(\overline{E}) = \{\mathrm{z}_0\}$ and $\overline{E}(E)\downarrow = \mathtt{x}$. For example, let $E_1 = \mathsf{enc}(\langle z_1, \langle \mathtt{x}, z_2\rangle\rangle, k_{as}, r_s)$, then $\overline{E_1} = \pi_1(\pi_2(\mathsf{dec}(\mathrm{z}_0, k_{as})))$. We define

$$\begin{aligned}\overline{\mathcal{E}_i}(P) &= \{U \mid \exists E \in \mathcal{E}_i(P), U \leq_{st} \overline{E} \text{ and } \exists q \in \mathrm{Pos}(U), h_{U|_q} = \mathsf{dec}\},\\ \mathcal{E}_{i+1}(P) &= \{M'[\mathrm{x}]_q \mid \exists M \in \mathcal{M}_o(P), p \in \mathrm{Pos_v}(M) \text{ s.t. } f_{ep}(M,p) = (M',p'),\\ &\qquad f_{dp}(M',p'') = (D,q), p = p'.p'', \text{ and } D_1 \in \overline{\mathcal{E}_i}(P)\}.\end{aligned}$$

For example,

$$\begin{aligned}\overline{\mathcal{E}_0}(P_Y) &= \{\pi_1(\pi_2(\mathsf{dec}(\mathrm{z}_0, k_{as}))), \pi_2(\mathsf{dec}(\mathrm{z}_0, k_{as})), \mathsf{dec}(\mathrm{z}_0, k_{as}),\\ &\qquad \pi_2(\mathsf{dec}(\mathrm{z}_0, k_{bs})), \mathsf{dec}(\mathrm{z}_0, k_{bs})\}\\ \mathcal{E}_1(P_Y) &= \{\mathsf{enc}(\langle z_1, \langle z_{1.2}, \mathtt{x}\rangle\rangle, k_{as})\}\\ \overline{\mathcal{E}_1}(P_Y) &= \{\pi_2(\pi_2(\mathsf{dec}(\mathrm{z}_0, k_{as}))), \pi_2(\mathsf{dec}(\mathrm{z}_0, k_{as})), \mathsf{dec}(\mathrm{z}_0, k_{as})\}\end{aligned}$$

and $\mathcal{E}_i(P_Y) = \emptyset$ for $i \geq 2$.

Note that $\mathcal{E}(P) = \cup_{i \geq 0}\mathcal{E}_i(P)$ is finite up-to renaming of the variables since for every $i \geq 1$, every term $M \in \mathcal{E}_i(P)$, $\mathrm{Pos}(M)$ is included in the (finite) set of positions occurring in terms of $\mathcal{M}_0$.

We can now define an over-approximation of the set of tests that may be applied over the secret.

$$\begin{aligned}\mathcal{M}_t^{\mathsf{s}}(P) = \{M \in \mathcal{M}_t(P) \mid \exists p \in \mathrm{Pos_v}(M) \text{ s.t. } D = D_1(\dots D_n) = f_{dp}(M,p) \neq \perp,\\ \text{and } \exists E \in \mathcal{E}(P), \exists i \text{ s.t. } D_i = \pi^i(\mathsf{dec}(\mathrm{z}_0, K)), E = \mathsf{enc}(U,K,R) \text{ and } \mathtt{x} \in D_i(E)\downarrow\}\end{aligned}$$

For example, $\mathcal{M}_t^{\mathsf{s}}(P_Y) = \{\pi_1(\pi_2(\pi_2(\mathsf{dec}(\pi_1(z_a), k_{as}))))\}$.

Definition 6. *We say that a well-formed process P w.r.t.* s does not test over s *if the following conditions are satisfied:*

1. *for all* $E \in \mathcal{E}(P)$, *for all* $D = D_1(\dots D_n) \in \mathcal{D}_o(P)$, *if* $D_i = \pi^i(\mathsf{dec}(z_0), K)$ *and* $e = \mathsf{enc}(U, K, R)$ *and* $\mathtt{x} \in D_i(E)\downarrow$ *then* $i = 1$ *and* $\overline{E} \not<_{st} D_1$,
2. *if* $M = N$ *or* $N = M$ *is a test of* P *and* $M \in \mathcal{M}_t^{\mathsf{s}}(P)$ *then* N *is a restricted name.*

Note that $\mathcal{E}(P)$ can be computed in polynomial time from P and that whether P *does not test over* s is decidable. We show in the next section that the first condition is sufficient to ensure that frames obtained from P are well-formed. It ensures in particular that there are no destructors right above s. If some D_i cancels some encryption in some E and $\mathtt{x} \in D_i(E)\downarrow$ then all its destructors should reduce in the normal form computation (otherwise some destructors (namely projections from D_i) remain above $\mathtt{x}$). Also we have $i = 1$ since otherwise a D_i may have consumed the lowest encryption above $\mathtt{x}$, thus the other decryption may block, and again there would be destructors left above $\mathtt{x}$.

The second condition requires that whenever a operand of a test $M = N$ is potentially dangerous (that is M or $N \in \mathcal{M}_t^{\mathsf{s}}(P)$) then the other operand should be a restricted name.

3.3 Main Result

We are now ready to prove that syntactic secrecy is actually equivalent to strong secrecy for protocols that are well-formed and do not test over the secret.

Theorem 2. *Let* P *be well-formed process w.r.t. a free name* s, *which is not a channel name, such that* P *does not test over* s. *We have* $\nu\mathsf{s}.\varphi \not\vdash \mathsf{s}$ *for any valid frame* φ *w.r.t.* P *if and only if* $P(^M/_\mathsf{s}) \approx_l P(^{M'}/_\mathsf{s})$, *for all ground terms* M, M' *public w.r.t.* $\mathrm{bn}(P)$.

Proof. Again, we only provide a sketch of the proof. Showing that strong secrecy implies syntactic secrecy is simple so we concentrate here on the converse implication. Let P be well-formed process w.r.t. a free name s with no test over s and assume that P is syntactically secret w.r.t. s.

Let M, M' be to public terms w.r.t. $\mathrm{bn}(P)$. To prove that $P(^M/_\mathsf{s})$ and $P(^{M'}/_\mathsf{s})$ are labeled bisimilar, we need to show that each move of $P(^M/_\mathsf{s})$ can be matched by $P(^{M'}/_\mathsf{s})$ such that the corresponding frames are bisimilar (and conversely). By hypothesis, P is syntactically secret w.r.t. s thus for any valid frame φ w.r.t. P, we have $\nu\mathsf{s}.\varphi \not\vdash \mathsf{s}$. In order to apply our previous result in the passive setting (Theorem 1), we need to show that all the valid frames are well-formed. However, frames may now contain destructors in particular if the adversary sends messages that contain destructors. Thus we first need to extend our definition of well-formedness for frames.

Definition 7. *We say that a frame* $\varphi = \nu\tilde{n}.\sigma$ *is* extended well-formed *w.r.t.* s *if for every occurrence* q_s *of* s *in* $T\downarrow$, *where* $T = x\sigma$ *for some* $x \in \mathrm{dom}(\sigma)$, *there exists an agent encryption w.r.t.* $\tilde{n}$ *above* s. *Let* $q_{\mathsf{enc}} < q_\mathsf{s}$ *the occurrence of the lowest encryption. It must verify that* $h_{T|_q} = \langle\rangle$, *for all positions* q *with* $q_{\mathsf{enc}} < q < q_\mathsf{s}$.

This definition ensures in particular that there is no destructor directly above s.

Theorem 1 can easily be generalized to extended well-formed frames.

Proposition 1. *Let φ be an extended well-formed frame w.r.t.* s*, where* s *is a free name in φ. Then $\nu \mathsf{s}.\varphi \nvdash \mathsf{s}$ iff $\varphi(^{M}/_{\mathsf{s}}) \approx \varphi(^{M'}/_{\mathsf{s}})$ for all M, M' closed public terms w.r.t. φ.*

The first step of the proof of Theorem 2 is to show that any frame produced by the protocol is an extended well-formed frame. We actually prove directly a stronger result, crucial in the proof: the secret s always occurs under an honest encryption and this subterm is an instance of a term in $\mathcal{E}(P)$.

Lemma 4. *Let P be a well-formed process with no test over* s *and $\varphi = \nu\tilde{n}.\sigma$ be a valid frame w.r.t. P such that $\nu \mathsf{s}.\varphi \nvdash \mathsf{s}$. Consider the corresponding standard frame $\nu\tilde{n}.\overline{\sigma} = \nu\tilde{n}.\{^{M_i}/_{y_i} \mid 1 \leq i \leq l\}$. For every i and every occurrence q_{s} of* s *in $M_i\!\downarrow$, we have $f_e(M_i\!\downarrow, q_{\mathsf{s}}) = E[^{W}/_{\mathsf{x}}]$ for some $E \in \mathcal{E}(P)$ and some term W. In addition $\nu\tilde{n}.\sigma_i\!\downarrow$ is an extended well-formed frame w.r.t.* s*.*

The lemma is proved by induction on i and relies deeply on the construction of $\mathcal{E}(P)$.

The second step of the proof consists in showing that any successful test in the process $P(^{M}/_{\mathsf{s}})$ is also successful in P and thus in $P(^{M'}/_{\mathsf{s}})$.

Lemma 5. *Let P be a well-formed process with no test over* s*, $\varphi = \nu\tilde{n}.\sigma$ a valid frame for P such that $\nu \mathsf{s}.\varphi \nvdash \mathsf{s}$ and θ a public substitution. If $T_1 = T_2$ is a test in P, then $T_1\theta\sigma(^{M}/_{\mathsf{s}}) =_E T_2\theta\sigma(^{M}/_{\mathsf{s}})$ implies $T_1\theta\sigma =_E T_2\theta\sigma$.*

This lemma is proved by case analysis, depending on whether $T_1, T_2 \in \mathcal{M}_t^{\mathsf{s}}$ and whether s occurs or not in $\mathrm{fn}(T_1\theta\sigma)$ and $\mathrm{fn}(T_2\theta\sigma)$.

To prove that $P(^{M}/_{\mathsf{s}})$ and $P(^{M'}/_{\mathsf{s}})$ are labeled bisimilar, we introduce the following relation $\mathcal{R}$ between extended processes defined as follows: $A \mathcal{R} B$ if there is an extended process A_0 and terms M, M' such that $P \Rightarrow^* A_0$, $A = A_0(^{M}/_{\mathsf{s}})$ and $B = A_0(^{M'}/_{\mathsf{s}})$. Then we show that $\mathcal{R}$ satisfies the three points of the definition of labeled bisimilarity using in particular Lemma 5. Hence we have also $\mathcal{R} \subseteq \approx_l$. Since we have clearly that $P(^{M}/_{\mathsf{s}}) \mathcal{R} P(^{M'}/_{\mathsf{s}})$, it follows that $P(^{M}/_{\mathsf{s}}) \approx_l P(^{M'}/_{\mathsf{s}})$.

3.4 Examples

We have seen in Section 3.2 that P_Y is a well-formed process w.r.t. k_{ab} and does not test over k_{ab}. Applying Theorem 2, if P_Y preserves the syntactic secrecy of k_{ab}, we can deduce that the Yahalom protocol preserves the strong secrecy of k_{ab} that is $P_Y(^{M}/_{k_{ab}}) \approx_l P_Y(^{M'}/_{k_{ab}})$ for any public terms M, M' w.r.t. $\mathrm{bn}(P_Y)$. We did not formally prove that the Yahalom protocol preserves the syntactic secrecy of k_{ab} but this was done with several tools in slightly different settings (e.g.[9,15]).

We have also verified that the Needham-Schroeder symmetric key protocol and the Wide-Mouthed-Frog protocol are both well-formed process w.r.t. k_{ab}

and do not test over k_{ab}, where k_{ab} is the exchanged key. Again, the syntactic secrecy of k_{ab} has been proved by several tools (e.g. [9]) in slightly different settings for both protocols. Using Theorem 2, we can deduce that they both preserve the strong secrecy of k_{ab}.

References

1. M. Abadi and C. Fournet. Mobile values, new names, and secure communication. In *28th Symp. on Principles of Programming Languages (POPL'01)*. ACM, 2001.
2. M. Abadi and A. D. Gordon. A calculus for cryptographic protocols: The spi calculus. In *4th Conf. on Computer and Communications Security (CCS'97)*, pages 36–47. ACM, 1997.
3. R. Amadio and D. Lugiez. On the reachability problem in cryptographic protocols. In *12th Conf. on Concurrency Theory (CONCUR'00)*, volume 1877 of *LNCS*, 2000.
4. The AVISPA Project. http://www.avispa-project.org/.
5. B. Blanchet. An efficient cryptographic protocol verifier based on prolog rules. In *Computer Security Foundations Workshop (CSFW'01)*, pages 82–96. IEEE Comp. Soc. Press, 2001.
6. B. Blanchet. Automatic Proof of Strong Secrecy for Security Protocols. In *IEEE Symposium on Security and Privacy (SP'04)*, pages 86–100, 2004.
7. B. Blanchet and A. Podelski. Verification of cryptographic protocols: Tagging enforces termination. In *Foundations of Software Science and Computation Structures (FoSSaCS'03)*, volume 2620 of *LNCS*, April 2003.
8. J. Borgström, S. Briais, and U. Nestmann. Symbolic bisimulations in the spi calculus. In *Int. Conf. on Concurrency Theory (CONCUR'04)*, volume 3170 of *LNCS*. Springer, 2004.
9. L. Bozga, Y. Lakhnech, and M. Périn. HERMES: An automatic tool for verification of secrecy in security protocols. In *15th Conference on Computer Aided Verification (CAV'03)*, volume 2725 of *LNCS*, pages 219–222. Springer, 2003.
10. H. Comon-Lundh and V. Cortier. New decidability results for fragments of first-order logic and application to cryptographic protocols. In *Rewriting Techniques and Applications (RTA'2003)*, LNCS 2706, pages 148–164. Springer-Verlag, 2003.
11. V. Cortier and B. Warinschi. Computationally Sound, Automated Proofs for Security Protocols. In *European Symposium on Programming (ESOP'05)*, volume 3444 of *LNCS*, pages 157–171. Springer, April 2005.
12. G. Denker, J. Millen, and H. Rueß. The CAPSL Integrated Protocol Environment. Technical Report SRI-CSL-2000-02, SRI International, Menlo Park, CA, 2000.
13. N. Durgin, P. Lincoln, J. Mitchell, and A. Scedrov. Undecidability of bounded security protocols. In *Workshop on Formal Methods and Security Protocols*, 1999.
14. H. Hüttel. Deciding framed bisimilarity. In *INFINITY'02*, August 2002.
15. L. C. Paulson. Relations between secrets: Two formal analyses of the Yahalom protocol. *Journal of Computer Security*, 9(3):197–216, 2001.
16. R. Ramanujam and S.P.Suresh. Tagging makes secrecy decidable for unbounded nonces as well. In *23rd Conference on Foundations of Software Technology and Theoretical Computer Science (FSTTCS'03)*, Mumbai, 2003.
17. M. Rusinowitch and M. Turuani. Protocol Insecurity with Finite Number of Sessions and Composed Keys is NP-complete. *Theoretical Computer Science*, 299:451–475, 2003.
18. Eugen Zalinescu, Véronique Cortier, and Michaël Rusinowitch. Relating two standard notions of secrecy. Research Report 5908, INRIA, Avril 2006.

Jump from Parallel to Sequential Proofs: Multiplicatives*

Paolo Di Giamberardino[1] and Claudia Faggian[2]

[1] Dip. di Filosofia, Università Roma Tre – Institut de Mathématiques de Luminy
digiambe@uniroma3.it
[2] Dip. di Matematica, Università di Padova – PPS, CNRS et Université Paris 7
claudia@math.unipd.it

Abstract. We introduce a new class of multiplicative proof nets, *J-proof nets*, which are a typed version of Faggian and Maurel's multiplicative L-nets. In J-proof nets, we can characterize nets with different degrees of sequentiality, by gradual insertion of sequentiality constraints. As a byproduct, we obtain a simple proof of the sequentialisation theorem.

1 Introduction

Proof nets have been introduced by Girard [10] as an abstract representation of linear logic proofs; this representation has two main interests: to provide a tool for studying normalization, and to give a canonical representation of proofs.

In proof nets, information about the order in which the rules are performed is reduced to a minimum, only two kinds of information about sequentiality being kept: the one corresponding to the subformula trees and the one providing the axiom links.

To retrieve a sequent calculus derivation from a proof net, we need to recover more information about sequentiality. A sequentialization procedure gives instructions on how to introduce this sequentiality. Such a procedure usually relies on splitting lemmas, which are proved introducing the notion of empire.

In [11], Girard, as part of the correctness criterion for proof nets with quantifiers, introduces a more direct way to represent sequentiality constraints in a proof net, by using *jumps*: a jump is an untyped edge between two nodes (rules) a, b, which expresses a *dependency* relation: a precedes b (bottom-up) in the sequentialisation. Recently, the idea of using jumps as a way to represent sequentiality information has been developed by Faggian and Maurel ([8]) in the abstract context of the L-nets, a parallel variant of Ludics designs.

Here we define a representation of proofs where objects with different degree of parallelism live together, in the spirit of [6,5]; for this purpose we introduce a new class of multiplicative proof nets, *J-proof nets*, that can be considered as a typed, concrete, version of multiplicative L-nets.

* This work has been partially supported by MIUR national project FOLLIA and a grant *Ville de Paris, 2005-2006.*

Z. Ésik (Ed.): CSL 2006, LNCS 4207, pp. 319–333, 2006.

We prove that by gradual insertion of jumps in a J-proof net, one can move in a continuum from J-proof nets of minimal sequentiality to J-proof nets of maximal sequentiality. The former are proof nets in the usual sense, the latter directly correspond to sequent calculus proofs.

In this way, we realize, for the multiplicative fragment of Linear Logic, a proposal put forward by Girard.

Moreover, our technique results into a very simple proof of the sequentialisation theorem. Our main technical result is the *Arborisation Lemma*, which provides the way to add jumps to a J-proof net up to a maximum.

2 Focalization, MLL and HS

In the first part of the paper we consider the multiplicative fragment of the hypersequentialised calculus HS [12,13,9], which is a focussing version of Multiplicative Linear Logic (MLL), as we explain below.

We define proof nets for HS, and then introduce J-proof nets. The strong geometrical properties of HS will allow us to uncover a simple sequentialisation technique for the calculus. We will then be able to apply the same technique to MLL.

2.1 Focalization and MLL

It has been proved by Andreoli [2] that the sequent calculus of Linear Logic enjoys a property called *focalization*: a proof π of a sequent $\vdash \Gamma$ can be transformed into a proof π^{foc} of the same sequent which satisfies a specific discipline, called *focussing discipline,* which we describe below.

Here we stress that π^{foc} is obtained from π solely by permutation of the rules. As a consequence, if we restrict our attention to MLL, there is no difference in the proof net of π and π^{foc}. In fact, we have that:

1. π and π^{foc} are equivalent modulo permutation of the rules;
2. the proof net respectively associated to π and π^{foc} is the same;
3. an MLL proof net has always a focussing proof among its possible sequentialisations.

(2.) is an immediate consequence of (1.), while (3.) is actually the easier way to prove focalization for MLL (as first observed by Andreoli and Maieli in [1]). We revise this below.

Focalization relies on a distinction of Linear Logic connectives into two families, which are as follows.

Positive connectives: $\otimes, \oplus, 1, 0$.

Negative connectives: $⅋, \&, \bot, \top$.

From now on, we only consider the multiplicative fragment of Linear logic; the formulas are hence as follows

$$F ::= A \mid A^{\perp} \mid F \parr F \mid F \otimes F$$

where $A, A^{\perp}$ are atoms.

To understand focalization, it helps to think of MLL proof nets rather than sequent calculus proofs.

Let us partition the nodes which are respectively typed by $\otimes$ and $\parr$ into maximal trees of nodes with the same type (resp. positive and negative trees). We assume that there is at most one negative node which is conclusion of the proof net (otherwise, we put together all negative conclusions by making use of $\parr$).

Consider now sequentialization. That is, we associate a sequent calculus proof to a proof-net; to do this, essentially one has to "find a last rule". The key result in proof-net theory is that this is always possible; a $\otimes$ rule which can be removed from the proof net (and taken as the last rule in the sequent calculus derivation) is called a *splitting* $\otimes$. Let us now choose a specific sequentialization strategy, based on the notion of *hereditarely splitting* $\otimes$, whose existence was proved in ([7]).

- It R has a negative conclusion, we choose that conclusion as last rule of the sequent calculus proof, and remove it from the proof net. We persist until the whole negative tree has been removed.
- If R has only positive conclusions, we choose an *hereditarely splitting* $\otimes$. This means that we can choose a tree of $\otimes$, and persistently remove all the $\otimes$ until the whole positive tree has been removed.

What we obtain is a sequent calculus derivation whose *bottom-up construction* satisfies the focussing discipline below.

Definition 1 (Focussing proofs). *A sequent calculus proof is called* focussing *if its* bottom-up construction *satisfies the following discipline:*

- *First keep decomposing a negative formula (if any) and its subformulas, until one get to atoms or positive subformulas;*
- *choose a positive formula, and keep decomposing it up to atoms or negative subformulas.*

2.2 Synthetic Connectives: HS

Focalization implies that we can consider a maximal tree of connectives of the same polarity (positive or negative) as a single n-ary connective, called a *synthetic connective*, which can be introduced by a specific rule.

In [12] Girard has introduced a new calculus, HS, which uses focalization and synthetic connectives to force a "normal" form for MALL sequent calculus proofs.

HS introduces a polarization on the atoms. This constraint correspond to a hidden decomposition of the atoms, and does not introduce essential differences, while making the geometrical structure strong and clear. For this reason, we will first work with HS; in Section 8 we will then remove the polarization.

3 From Proof Nets to J-Proof Nets

3.1 MHS Sequent Calculus

We indicate by MHS the multiplicative fragment of HS.

Formulas. The formulas of MHS are as follows:

$$\begin{array}{l} N ::= A^{\perp} \mid \parr(P, \ldots, P) \\ P ::= \ A \ \mid \otimes(N, \ldots, N) \end{array}$$

where A denotes a positive atom.

Rules. The rules for proving sequents are the following

$$\frac{}{\vdash A,\ A^{\perp}}(Ax) \qquad \frac{\vdash \Gamma,\ F \quad \vdash \Delta,\ F^{\perp}}{\vdash \Gamma,\ \Delta}(Cut)$$

$$\frac{\vdash \Gamma_1,\ N_1 \ldots \vdash \Gamma_n,\ N_n}{\vdash \Gamma_1, \ldots, \Gamma_n,\ \otimes(N_1, \ldots, N_n)}(+) \quad \frac{\vdash \Gamma,\ P_1, \ldots P_n}{\vdash\ \Gamma,\ \parr(P_1, \ldots, P_n)}(-)$$

$$\frac{\vdash \Gamma \quad \vdash \Delta}{\vdash \Gamma,\ \Delta}(Mix)$$

where all context $\Gamma, \Delta, \ldots$ only contain P formulas.

Remark 1. The calculus admits a unary $\parr$ (resp.$\otimes$) which is a negative (resp. positive) polarity inverter [13,14] This polarity inverter is usually called a *negative (resp. positive) Shift* and denoted by $\uparrow$ (resp. $\downarrow$).

Skeleton of a Sequent Calculus Proof. Let us give a first intuition of our approach. Consider a cut-free proof π, and type each rule application with its active formulas. Observe that if we forget everything but the types, we have a tree, where the nodes are MHS formulas, and the leaves have the form $\{A, A^{\perp}\}$. We can think of this tree as the skeleton of the sequent calculus derivation.

Here we do not push this intuition further, but it is possible to characterizes the trees which correspond to sequent calculus derivations, in the spirit of [13], and extend the approach also to the cut rule.

3.2 MHS Proof-Nets

In this section we define proof-nets for MHS.

Proof-nets provide a graph representation of proofs. Each node represents a rule of the sequent calculus, and it is only concerned with the active formulas.

Definition 2 (Typed structure). *We call typed structure a directed acyclic graph where:*

- *the edges are possibly typed with MHS formulas*
- *the nodes (also called links) are typed either with a MHS formula or with a pair of atoms* $\{A, A^{\perp}\}$.

Given a link, the incoming edges are called the premises *of the link, and the outgoing edges are called* conclusions *of the link. We call positive (resp. negative) a link of type* $\otimes(N_1, \ldots, N_n)$ *(resp.* $\parr(P_1, \ldots, P_n)$*).*

We admit pending edges. *An edge which has no target is called a* conclusion *of the structure, and its source is called a* terminal link.

Definition 3 (Proof structure). *A* proof structure *is a typed structure where the nodes are typed as the conclusions, and the typing satisfies the following constraints:*

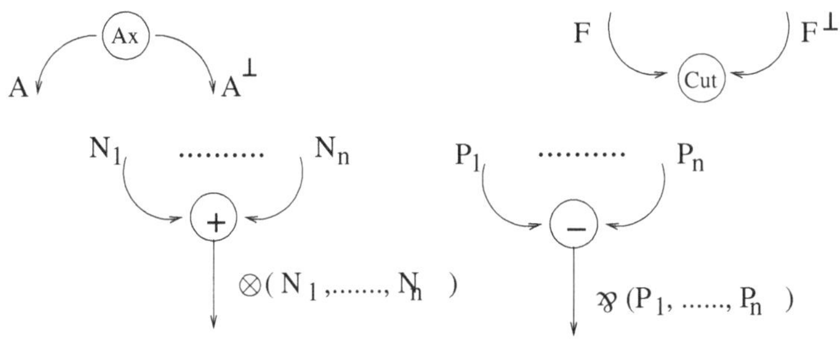

Moreover, we ask that there is at most one negative terminal link.

Definition 4 (Switching path and cycle). *Given a proof-structure, a* switching path *is an unoriented path which never uses two premises of the same negative link.*
A switching cycle *is a switching path which is a cycle.*

Definition 5 (Proof-nets). *A proof structure R is called a proof-net if it has no switching cycles.*

Proposition 1. *Given a sequent calculus proof* π *of MHS, we can associate to it a proof net* π^*.

Proof. We proceed in the standard way.

Definition 6 (Sequentialization). *A proof structure R is* sequentialisable *iff there exists a proof* π *of MHS (that we call a* sequentialisation *of R) s.t.* $\pi^* = R$.

3.3 J-Proof Nets

We enrich proof-nets with jumps, which will allow us to graduate sequentiality.

Definition 7 (J-proof structure)
A J-proof structure *(jumped proof structure) is a proof structure added with untyped edges called* jumps, *which connect a positive to a negative link (the orientation is from positive to negative).*

Definition 8 (Switching path and cycle). *Given a J-proof structure, a* switching path *is an unoriented path which never uses two premises of the same negative link (a jump is also a premise of its target); a* switching cycle *is a switching path which is a cycle.*

Definition 9 (J-Proof nets). *A J-proof structure R is called a* J- proof-net *if it has no switching cycles.*

In Section A we sketch normalization of J-proof nets.

A proof-net is a special case of J-proof net. In the next section we will show that a sequent calculus proof (or rather its skeleton) can also be seen as a special case of J-proof net. This will allow us to define a new technique of sequentialization.

Note. From now on, we only consider J-proof structures without cut links. The cut can be smoothly dealt with essentially by identifying the cut node of premises $F, F^{\perp}$ with the node of which the positive formula is conclusion.

3.4 Partial Order Associated to a J-Proof Net

Since a J-proof net R is a d.a.g., we associate to R in the standard way a strict partial order $\prec_R$ on the typed nodes.

We recall that we can represent a strict partial order as a d.a.g., where we have an edge $a \leftarrow b$ whenever $a <_1 b$ (i.e. $a < b$, and there is no c such that $a < c$ and $c < b$.) Conversely (the transitive closure of) a d.a.g. G induces a strict partial order $\prec_G$ on the nodes of G.

We call **skeleton** of a directed graph G, denoted $Sk(G)$, the minimal graph whose transitive closure is the same as that of G. An edge $a \leftarrow b$ is transitive if there is no node c such that $a \leftarrow c$ and $c \leftarrow b$.

With a slight abuse, we often identify $\prec_G$ and the skeleton of G.

Given a J-proof net R, we call *minimal* (resp. *maximal*) a link c of R which is minimal in $\prec_R$, i.e. there is no node b such that $b \leftarrow c$ (resp. $c \leftarrow b$). Notice that, because of jumps, a node can be terminal, without being minimal.

We call predecessor of a node c, a node which immediately precedes c. Otherwise, we speak of hereditary predecessor. Similarly for the successor.

A strict order r on a set A is *arborescent* when each element has a unique predecessor.

If the order $\prec_R$ associated to a J-proof net R is arborescent, the skeleton of R is a forest.

Finally, we observe that

Remark 2. Let R be a J-proof net.

- $Sk(R)$ is obtained from R by removing the edges which are transitive.
- Only an edge which goes from positive to negative can be transitive.

4 J-Proof Nets and Sequent Calculus

In the next section we will induce a sequentialisation of a proof net by adding jumps. Let us start with an example. Consider the proof-net below. We add a jump from the positive to the negative link, and consider the order induced on the links. We obtain a tree, and such a tree is the skeleton of a sequent calculus proof.

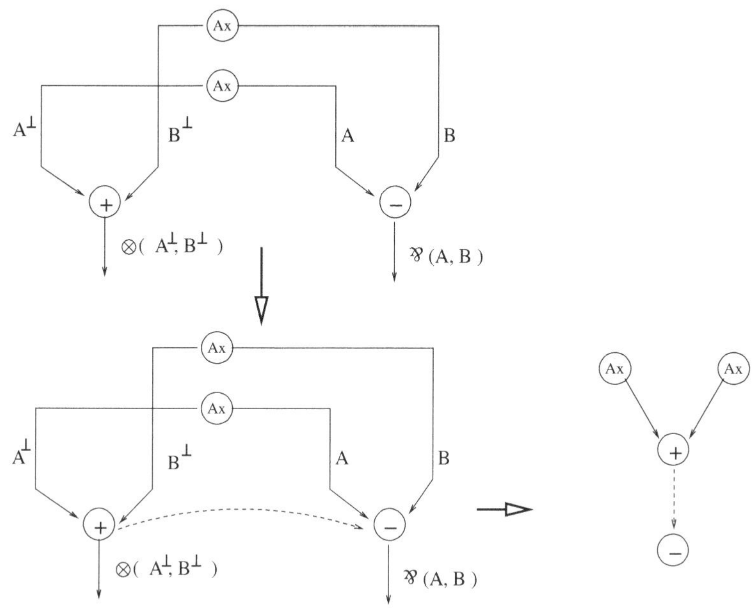

To sequentialize a J-proof net, we will then consider the order associated to a proof net as a directed acyclic graph, and add to it enough jumps, to make the order arborescent, and hence proof-like (Lemma 1).

Let us show that if the order on the nodes of a J-proof net is arborescent, it corresponds to a sequent calculus derivation. (A proof of this is given, in a more general setting and with full details, in [5].)

Proposition 2. *Let R be a J-proof net such that $\prec_R$ is arborescent.*

(i) We can associate to R a proof π^R in the sequent calculus MHS, possibly making use of the Mix Rule.

(ii) Moreover, if the order has a minimum and each negative link has a unique successor (i.e. if the skeleton is a tree which only branches on positive nodes) then π^R does not use the Mix rule.

Proof. The proof is by induction on the number of links. For brevity, we show directly (ii).

1. $n = 1$: The only link of R is an Axiom link of conclusions $A,\ A^{\perp}$, to which we associate $\dfrac{}{\vdash A,\ A^{\perp}}$;

2. $n > 1$: we reason by cases, depending on the type of the minimal link c of R.
   - $⅋(P_1 \dots P_n)$: let $\prec_{R'}$ be the order obtained by erasing c. By induction we associate a proof π' to $\prec_{R'}$. $\pi^{\prec_R}$ is $\dfrac{\pi'}{\vdash \Gamma,\ ⅋_{i\in\{1\dots n\}}(P_i)}$, whose last rule is a $-$ rule on $P_1, \dots, P_n$ ($P_1, \dots, P_n$ are conclusions of π' by construction);
   - $\otimes(N_1, \dots, N_n)$: let $\prec_{R_1}, \dots, \prec_{R_n}$ be the n orders obtained by erasing c. By induction we associate a proof π'_i to each $\prec_{R_i}$; $\pi^{\prec_R}$ is $\dfrac{\pi'_1 \dots \pi'_n}{\vdash \Gamma_1, \dots, \Gamma_n,\ \otimes_{i\in\{1\dots n\}} N_i}$ whose last rule is a $+$ rule on $N_1, \dots, N_n$ (by construction, $N_1, \dots, N_n$ are respectively among the conclusions of $\pi'_1, \dots, \pi'_n$).

5 Sequentialization

Definition 10 (Saturated J-proof net). *A J-proof net R is* saturated *if for every negative link n and for every positive link p, adding a jump between n and p creates a switching cycle or doesn't increase the order $\prec_R$. Given a J-proof net R, a saturation R^J of R is a saturated J-proof net obtained from R by adding jumps.*

Our sequentialisation argument is as follows:

- If the order $\prec_R$ associated to a J-proof net R is arborescent, we can associate to R a proof π^R in the sequent calculus.
- The order associated to a saturated J-proof net is arborescent.
- Any J-proof net can be saturated.

Lemma 1 (Arborisation). *Let R be a J-proof net. If R is saturated then $\prec_R$ is arborescent. Any J-proof net can be saturated.*

Proof. We prove that if $\prec_R$ is not arborescent, then there exists a negative link c and a positive link b s.t. adding a jump between b and c doesn't create switching cycles and makes the order increase.

If $\prec_R$ is not arborescent, then in $\prec_R$ there exists a link a with two immediate predecessors b and c (they are incomparable). Observe that b and c are immediately below a in $Sk(R)$ and also in R.

If a is an Axiom link, then necessarily b and c are respectively a positive link and a negative link; we draw a jump between b and c, this doesn't create a cycle and the order increases.

Otherwise, a is a positive link, and b and c are two negative links; we distinguish two cases:

1. either b or c is terminal in R. Let assume that b is terminal; then c cannot be terminal (by definition of jumped proof structure), and there is a positive link c' which immediately precedes c. If we add a jump between b and c', this doesn't create cycles and the order increases.

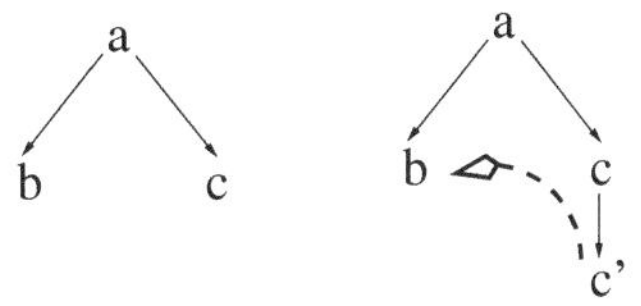

2. Neither b or c are terminal in R. Each of them has an immediate positive predecessor, respectively b' and c'. Suppose that adding a jump from b' to c creates a cycle: we show that adding a jump from c' to b cannot create a cycle. If adding to R the jump $b' \to c$ creates a cycle, that means that there is in R a switching path $r = \langle c, c'....b \rangle$; if adding the jump $c' \to b$ creates a cycle then there is a switching path $r' = \langle b, b'...c \rangle$. Assume that r and r' are disjoint: we exhibit a switching cycle in R $\langle c, c'...b, b'...c \rangle$ by concatenation of r and r'.This contradicts the fact that R is a proof net.

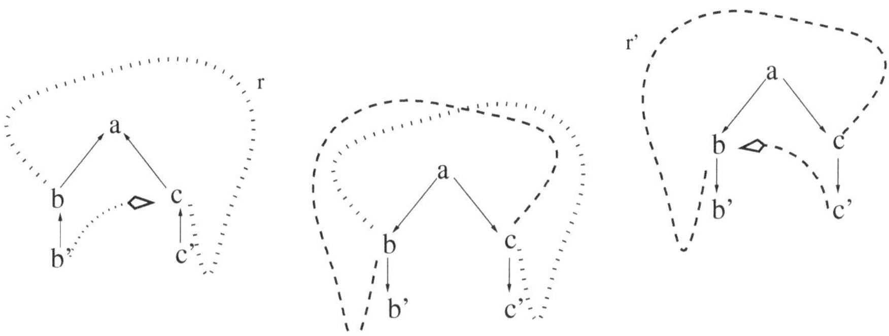

Assume that r and r' are not disjoint. Let x be the first node (starting from b) where r and r' meets. Observe that x must be negative (otherwise there would be a cycle). Each path uses one of the premises, and the conclusion (hence the path meets also in the node below x). From the fact that x is the first point starting from b where r and r' meet it follows that: (i) r' enters in x from one of the premises, and exits from the conclusion; (ii) each of r and r' must use a different premise of x. Then we distinguish two cases:

- r enters x from one of the premises; we build a switching cycle taking the sub path $\langle b, ..., x \rangle$ of r' and the sub path $\langle x, ..., b \rangle$ of r.
- r enters x from the conclusion; then we build a switching cycle composing the sub path of r $\langle c, ..., x \rangle$, the reversed sub path of r' $\langle x, ..., b \rangle$ and the path $\langle b, a, c \rangle$.

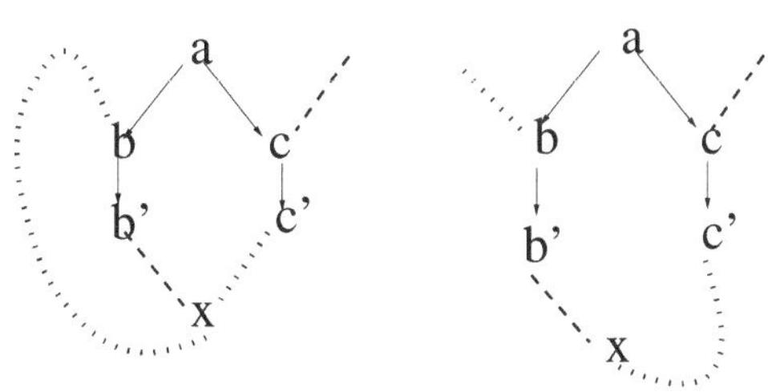

6 Properties

In this section we deal with three standard results one usually has on proof nets. In 6.1 we get rid of the Mix rule, in 6.2 we give an immediate proof of the usual splitting Lemma, in 6.3 we prove that the sequentialization we have defined is correct w.r.t. Definition 6.

The novelty here is the argument. When adding jumps, we gradually transform the skeleton of a graph into a tree. We observe that some properties are invariant under the transformation we consider: adding jumps and removing transitive edges. Our argument is always reduced to simple observations on the final tree (the skeleton of R^J), and on the fact that each elementary graph transformation preserves some properties of the nodes.

6.1 Connectness

Lemma 2. *(i) Two nodes are connected in a d.a.g. G (i.e. there exists a sequence of connected edges between the two nodes) iff they are connected in the skeleton of G.*

(i) If two node are connected in R, then they are connected in R^J.

(iii) If R is connected as a graph so are R^J and Sk(R^J).

Proof. Immediate, because adding edges, or deleting transitive edges, preserves connectness.

We now deal with a more peculiar notion of connectness, to get rid of the mix rule, as is standard in the theory of proof-nets.

Definition 11 (Correction graph). *Given a typed graph R, we call* switching *a function s which associates to every negative node of R one of its premises (again, jumps also are premises of their target); a* correction graph *$s(R)$ is the graph obtained by erasing for every negative node of R the premises not chosen by s.*

Definition 12 (s-connected). *A J-proof net R is s-connected if given a switching of R, its correction graph is connected.*

Remark 3. We only need to check a single switching. The condition that a proof structure has not switching cycles is equivalent to the condition that *all* correction graphs are acyclic.

A simple graph argument shows that assuming that all correction graphs are acyclic, if for a switching s the correction graph $s(R)$ is connected, then for all other switching s' $s'(R)$ is connected.

Proposition 3. *If R is s-connected, then its skeleton is a tree which only branches on positive nodes (i.e., each negative link has a unique successor).*

Proof. First we observe that:

- any switching of R is a switching of R^J, producing the same correction graph. Hence if R is s-connected, R^J is s-connected.

– Given a J-proof net G, any switching of its skeleton is also a switching of G, because the skeleton is obtained by erasing the edges which are transitive. A transitive edge can be premise only of a negative node.

As a consequence, any switching of $Sk(R^J)$ induce a correction graph which is connected. However, $Sk(R^J)$ is a tree, so we cannot erase any edge. Hence each negative link has a unique premise, and the graph has only one switching.

From Proposition 2, it follows that

Proposition 4. *If R is s-connected, and R^J a saturation, we can associate to it a proof π^{R^J} which does not use the Mix rule.*

6.2 Splitting

Observe that a minimal link of S is a root of its skeleton.

Definition 13 (Splitting). *Let R be a typed structure, c a positive link, and $b_1, \ldots, b_n$ the nodes which are immediately above c (the premises of c have the same type as $b_1, \ldots, b_n$). We say that c is* splitting *for R if it is terminal, and removing c there is no more connection (i.e. no sequence of connected edges) between any two of the nodes b_i.*

Remark 4. Assume that R is a connected graph. It is immediate that if R is a J-proof net whose terminal links are all positive, the removal of c splits R into n disjoint connected components $R_1, \ldots, R_n$, and each component is a J-proof net.

Lemma 3 (Splitting lemma). *Let R be a J-proof net whose terminal nodes are all positive, and R^J a saturation; the minimal link c of R^J (i.e. the root of $Sk(R^J)$) is splitting for R.*

Proof. Observe that c is obviously splitting in the skeleton of R^J, because c is the root of a tree. Hence it is splitting in R^J, as a consequence of Lemma 2, (i). Similarly, c must be splitting in R, as a consequence of Lemma 2, (ii).

6.3 Sequentialisation Is Correct

Proposition 5. *Let R be a J-proof-net. For any saturation R^J of R, if $\pi = \pi^{R^J}$ then $(\pi)^* = R$.*

Proof. For brevity, we assume that R is s-connected. Hence, the skeleton of R^J a tree. The proof is by induction on the number of links of R.

1. $n = 1$: then R consists of a single Axiom link, and π is the corresponding Axiom rule.
2. $n > 1$. We consider the minimal link k of R^J.
   Observe that the last rule of π is the rule which correspond to the root k. Let us call $\pi_1, \ldots, \pi_n$ the premises of the rule, and $R_1^J, \ldots, R_n^J$ the subnets obtained from R^J by removing k. By definition, each π_i is the proof associated to an R_i^J.

- Assume k is positive. By the splitting lemma, k is splitting in R. $R_1^J \ldots, R_n^J$ are obviously saturated (we have not erased any jump) so by induction hypothesis on $R_1 \ldots, R_n$ which are the n sub nets obtained by removing k from R, $(\pi^{R_1^J})^* = R_1, \ldots, (\pi^{R_n^J})^* = R_n$; by composing all the $\pi^{R_i^J}$ with the rule corresponding to k , we get a proof which is equal to π^{R^J} and we find that $R = (\pi^{R^J})^*$.
- Assume k is negative. Similarly, we remove k from R and apply induction to obtain the conclusion.

7 Partial Sequentialisation and Desequentialization

The approach we have presented is well suited for *partially* introducing or removing sequentiality, by adding (deleting) a number of jumps.

Actually, it would be straightforward to associate to a sequent calculus proof π a saturated J-proof net. In this way, to π we could associate either a maximal sequential or a maximal parallel J-proof net, on the lines of [6,5].

Given a J-proof net R, let us indicate with $Jump(R)$ ($DeJump(R)$) a J-proof net resulting from (non deterministically) introducing (eliminating) a number of jumps in such a way that every time the order increases (decreases).

The following result apply to a J-proof net of *any* degree of sequentiality.

Theorem 1 (Partial sequentialisation/desequentialization). *Let R, R' be J-proof nets.*

If $R' = Jump(R)$ then there exists $DeJump(R')$ such that $DeJump(R') = R$.

If $R' = Dejump(R)$ then there exists $Jump(R')$ such that $Jump(R') = R$.

Proof. Immediate, since we can reverse any step...

8 MLL

Our sequentialisation proof can now be extended to MLL. It is straightforward to translate an MLL proof net into MHS, however, here we prefer a more direct approach (where the translation is implicit). We proceed in two steps, first by introducing a variant of Andreoli's focussing calculus based on synthetic connectives, and then working directly with MLL.

8.1 *MHS*$^+$

The polarization of HS makes the geometrical structure clean and clear. We now eliminate the polarization constraints, still keeping the calculus focussing.

We call this calculus *MHS*$^+$. The grammar of the formulas is the following:

$$N ::= A \mid A^\perp \mid \parr(P, \ldots, P)$$
$$P ::= A \mid A^\perp \mid \otimes(N, \ldots, N)$$

Remark 5. Observe that now we have *all the formulas* of MLL, modulo clustering/declustering into synthetic connectives. For example, $A \parr A^{\perp}$ is a formula of *MHS*$^+$.

The *sequent calculus rules* are (formally) the same as those of MHS. Observe however that now we consider negative atoms also as P-formulas. This means that the contexts $\Gamma, \Delta, \ldots$ may also contain negative atoms. Moreover, a negative atom can appear in the premises of a negative rule, and a positive atom can appear in the premises of a positive rule.

Proof nets. We modify the Axiom link, by introducing a (formal) decomposition of the atoms. Any atom A can be decomposed into $A^{\surd}$, of opposite polarity (technically, the $A^{\surd}$ has been introduced by Girard, and is used also by Laurent in [14]). Hence we have:

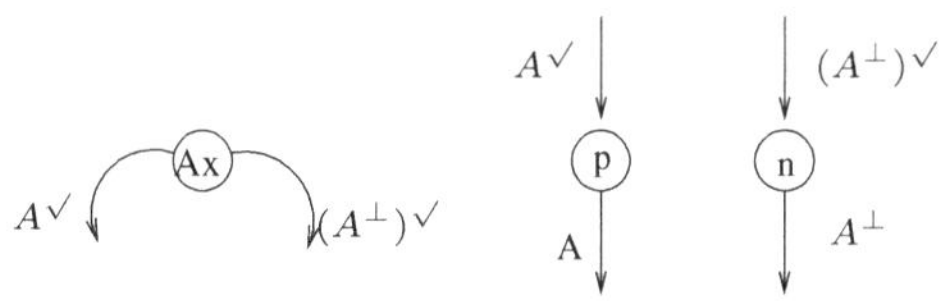

To the identity axiom we associate

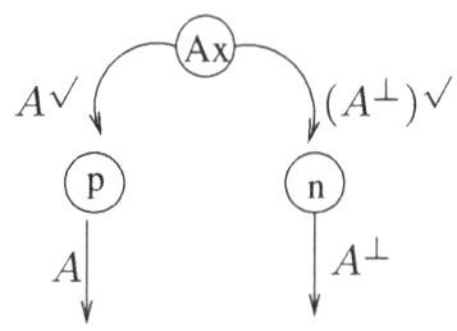

where the n and p links, respectively negative and positive, can be considered as steps of decomposition of the atoms: we call these links *hidden*. They do not appear in the sequent calculus, but provide space for the jumps.

The definitions of the previous sections can be applied to *MHS*$^+$, with this variant: when we associate an order $\prec_R$ to a J-proof net of *MHS*$^+$, we ignore the hidden links. It is straightforward to check that the results of the previous sections (and in particular the Arborisation Lemma) still hold in this case.

Remark 6. *MHS*$^+$ is a variant of MLL^{Foc}: to a proof of *MHS*$^+$ corresponds a proof of MLL^{Foc}, and vice-versa.

Similarly, the proof nets closely correspond to focussing proof nets, as defined by Andreoli [3].

8.2 MLL

It is immediate now that we can

- transform an MLL proof net R into a *MHS*$^+$ proof net R';
- transform *MHS*$^+$ sequent calculus derivation π into an MLL sequent calculus derivation by "declustering" the rules. The sequent calculus derivation which we obtain is focussing.

Observe that this transformations can simply be *"virtual"*. To sequentialize an MLL proof net R, we expand the axiom links, and treat each maximal tree of $\otimes$ as a $+$ link, and each maximal tree of $\parr$ as a $-$ link. Using our procedure, we obtain again an arborescent order on $+$, $-$ and axiom links. Observe that we can substitute each step in Proposition 2 with an expanded version.

9 Conclusions and Future Work

J-proof nets provide a representation of proofs where objects with different degrees of parallelism live together; furthermore, by the use of jumps as sequentiality constraints, we can transform any proof net of MLL in a sequent calculus proof, which seems a very natural way to approach sequentialisation.

Jumps are related with the notion of *empire* of a formula in a proof net; we wish to investigate this relationship, in order to understand the differences between our proof of sequentialisation and the traditional ones.

Also, we would like to understand the relation with work by Banach [4], where the use of an order on the links of the proof net as a tool for sequentialization has a precedent.

As a future research direction, we hope to be able to extend this work to consider a larger fragments of linear logic; recent developements in the theory of L-nets [6,5] seem to make plausible an extension to $MALL$.

Acknowledgements

The authors are indebted with Jean-Yves Girard, for having first suggested the idea of sequentialising with jumps. Many thanks also to Lorenzo Tortora de Falco and Olivier Laurent for useful and fruitful discussions.

References

1. Andreoli, J-M., Maieli, R. : Focusing and Proof nets in Linear and Non-Commutative Logic. In *LPAR99*, 1999.
2. Andreoli, J-M. : Logic Programming with Focusing Proofs in Linear Logic In *Journal of Logic and Computation*, 2(3), 1992.
3. Andreoli, J-M. : Focussing Proof-Net Construction as a Middleware Paradigm. In *Proceedings of Conference on Automated Deduction (CADE)* , 2002
4. Banach, R. : Sequent reconstruction in MLL - A sweepline proof. In *Annals of pure and applied logic*, 73:277-295, 1995
5. Curien, P.-L. and Faggian, C. : An approach to innocent strategies as graphs, submitted to journal.
6. Curien, P.-L. and Faggian, C. : L-nets, Strategies and Proof-Nets. In *CSL 05 (Computer Science Logic)*, LNCS, Springer, 2005.
7. Danos, V. : La Logique Linéaire appliquée l'étude de divers processus de normalisation (principalement du λ-calcul). PhD thesis, Université Paris 7, 1990.
8. Faggian, C. and Maurel, F. : Ludics nets, a game model of concurrent interaction. In *Proc. of LICS'05 (Logic in Computer Science)*, IEEE Computer Society Press, 2005.

9. Girard, J.-Y. : *Le Point Aveugle, Tome I and II.* Cours de théorie de la démonstration, Roma Tre, Octobre-Décembre 2004. Hermann Ed. Available at: http://logica.uniroma3.it/uif/corso,
10. Girard, J.-Y. : Linear Logic. In *Theoretical Computer Science*, 50: 1-102, 1987.
11. Girard, J.-Y. : Quantifiers in Linear Logic II. In *Nuovi problemi della Logica e della Filosofia della scienza*, 1991.
12. Girard, J.-Y. : On the meaning of logical rules II: multiplicative/additive case. In *Foundation of Secure Computation*, NATO series F 175, 183-212. IOS Press, 2000.
13. Girard, J.-Y. : Locus Solum. In *Mathematical Structures in Computer Science*, 11:301-506, 2001.
14. Laurent, O. : Polarized Games. In *Annals of Pure and Applied Logic*, 130(1-3):79-123, 2004.

A Normalization of J-Proof Nets

We can define cut elimination on J-proof nets in the same way as for L-nets:

- *Ax*

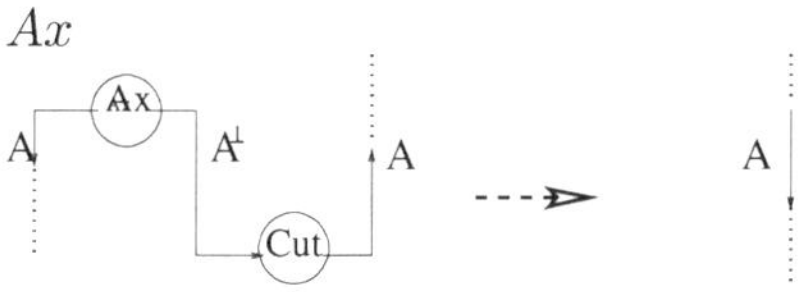

- cut +/−

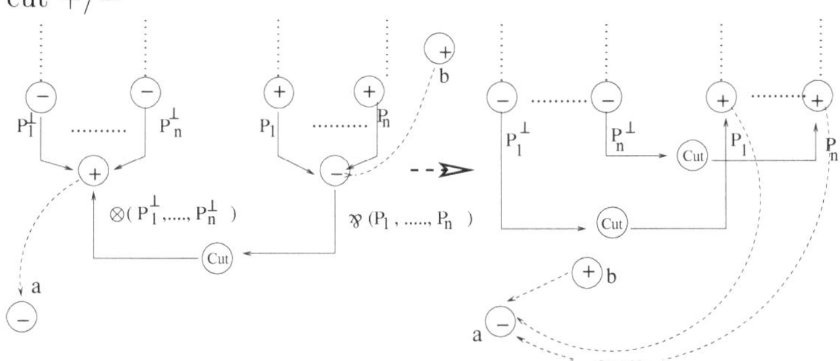

The procedure is confluent and strong normalizing, and preserves correction; furthermore it preserves the order on the links (if a precedes b before the reduction, it still precedes b afterwards). Notice that cuts are oriented from negative to positive: actually, we modify the orientation of the edges of the cut link when we associate an order to a proof net, so to get the above good properties.

First-Order Queries over One Unary Function

A. Durand and F. Olive

[1] Equipe de Logique Mathématique, CNRS UMR 7056 - Université Denis Diderot, 2 place Jussieu, 75251 Paris Cedex 05, France
durand@logique.jussieu.fr
[2] LIF, CNRS UMR 6166 - Université de Provence, CMI, 39 rue Joliot Curie F-13453 Marseille Cedex 13, France
olive@lif.univ-mrs.fr

Abstract. This paper investigates the complexity of query problem for first-order formulas on quasi-unary signatures, that is, on vocabularies made of a single unary function and any number of monadic predicates.

We first prove a form of quantifier elimination result: any query defined by a quasi-unary first-order formula can be equivalently defined, up to a suitable linear-time reduction, by a quantifier-free formula. We then strengthen this result by showing that first-order queries on quasi-unary signatures can be computed with constant delay i.e. by an algorithm that has a precomputation part whose complexity is linear in the size of the structure followed by an enumeration of all solutions (i.e. the tuples that satisfy the formula) with a constant delay (i.e. depending on the formula size only) between each solution. Among other things, this reproves (see [7]) that such queries can be computed in total time $f(|\varphi|).(|S| + |\varphi(S)|)$ where S is the structure, φ is the formula, $\varphi(S)$ is the result of the query and f is some fixed function.

The main method of this paper involves basic combinatorics and can be easily automatized. Also, since a forest of (colored) unranked tree is a quasi-unary structure, all our results apply immediately to queries over that later kind of structures.

Finally, we investigate the special case of conjunctive queries over quasi-unary structures and show that their combined complexity is not prohibitive, even from a dynamical (enumeration) point of view.

1 Introduction

The complexity of logical query languages is a well-studied field of theoretical computer science and database theory. Understanding the complexity of query evaluation for a given language is a good way to measure its expressive power. In this context, first-order logic and its fragments are among the most interesting and studied such languages.

In full generality, the data complexity of first-order queries is in AC^0 hence in polynomial time [15] (see also [11]). However, the size of the formula appears as a major ingredient in the exponent of the polynomial. Hence, taking into account the sizes of both the structure and the formula, the combined complexity becomes highly intractable, even for small formulas. Nevertheless, tractability

Z. Ésik (Ed.): CSL 2006, LNCS 4207, pp. 334–348, 2006.

results have been obtained for natural query problems defined by restricting either the logic or the set of structures. This is the case, for example, for acyclic conjunctive queries [16,13], or for full first-order queries on relational structures of bounded degree [14,5,12] or on tree-decomposable structures [8] (see also [7]).

A quasi-unary signature consists of one unary function and any number of monadic predicates. First-order logic over quasi-unary structures has been often studied and some of its aspects are quite well understood. The satisfiability problem for this kind of first-order formulas is decidable, while by adding just one more unary function symbol in the vocabulary we can interpret graphs, and hence all the first-order logic. In particular, first-order logic over two unary functions structures is undecidable (even for formulas with one variable [9]). In [3], it is proved that the spectrum (i.e. the set of cardinalities of the finite models) of formulas over one unary function are ultimately periodic. This result has been generalized in [10] even to the case of spectra of monadic second-order formulas.

In this paper, we continue the study of first-order logic on quasi-unary vocabularies and show some new structural properties that have interesting consequences on the complexity of query problems for such languages. Our first result shows that it is possible to eliminate variables in first-order formulas on quasi-unary vocabularies at reasonable cost while preserving the answers of the queries. More precisely, given a quasi-unary structure S and a first-order formula φ, one can construct a quantifier-free formula φ' and, in linear time in the size S, a new quasi-unary structure S' such that the results of the queries $\varphi(S)$ and $\varphi'(S')$ are the same. The method used to prove this result is mainly based on combinatorial arguments related to covering problems.

Then, as in [5], we explore the complexity of query evaluation from a dynamical point of view: queries are seen as enumeration problems and the complexity is measured in terms of delay between two successive outputs (i.e. tuples that satisfy the formula). Such an approach provides very precise information on the complexity of query languages: by adding up the delays, it makes it possible to obtain tight complexity bounds on complexity evaluation (in the classical sense) but also to measure how regular this process is. This latter kind of information is useful, for example, for query answering "on demand".

Taking as a starting point the quantifier elimination method, we show that a first-order query φ on a quasi-unary structure S can be computed with constant delay i.e. by an algorithm that has a precomputation part whose complexity is linear in the size of the structure, followed by an enumeration of all the tuples that satisfy the formula with a constant delay (i.e. depending on the formula size only) between two of them. Among other things, this gives an alternative proof that such queries can be computed in total time $f(|\varphi|).(|S| + |\varphi(S)|)$ where f is some fixed function, hence the complexity is linear (see [7] remarking that the graph of one unary function is of tree-width two). One nice feature of quasi-unary structures is their proximity to trees: any forest of (colored) ranked or unranked tree is a quasi-unary structure, hence, all our results immediately apply to queries over trees. Several recent papers have investigated query answering from an enumeration point-of-view for monadic second-order logic on trees (see [1,2]). They mainly prove that the results of MSO queries on binary trees or on tree-like structures can be enumerated with a linear delay (in the size of the next output) between two consecutive solutions. The methods used in these papers rely on tree-automata techniques and some of these results apply to our context. However, our goal in this paper, is to prove strong structural properties of logical formulas (such as quantifier elimination) in this language and show how these properties influence the complexity of query answering.

The paper is organized as follows. In Sect. 2, main definitions about query problems, reductions and enumeration complexity are given. A normal form for formula over quasi-unary vocabularies is also stated. In Sect. 3 the combinatorial material to prove the main results are introduced. Then, in Sect. 4, the variable elimination theorem for first-order formula on quasi-unary vocabulary is proved. Section 5, is devoted to query evaluation and our result about enumeration of query result is proved. Finally, in Sect. 6, the special case of conjunctive queries is investigated.

2 Preliminaries

Definitions. All formulas considered in this paper belong to first-order logic, denoted by FO. The FO-formulas written over a same signature σ are gathered in the class FO^σ. The *arity* of a first-order formula φ, denoted by arity (φ), is the number of free variables occuring in φ. We denote by $\mathrm{FO}(d)$ the class of FO-formulas of arity d. When the prenex form of a FO-formula φ involves at most q quantifiers, we say that it belongs to FO_q. Combining these notations, we get for instance that $\mathrm{FO}_0^E(d)$ is the class of quantifier-free formulas of arity d and of signature $\{E\}$.

Given a signature σ, we denote by $\mathrm{STRUC}(\sigma)$ the class of *finite* σ-structures. The domain of a structure S is denoted by $\mathrm{dom}(S)$. For each $S \in \mathrm{STRUC}(\sigma)$ of domain D and each $\varphi \in \mathrm{FO}^\sigma(d)$, we set:

$$\varphi(S) = \{(a_1, \ldots, a_d) \in D^d : (S, a_1, \ldots, a_d) \models \varphi(x_1, \ldots, x_d)\}.$$

Two σ-formulas φ, ψ of same arity are said *equivalent* if $\varphi(S) = \psi(S)$ for any σ-structure S. We then write $\varphi \sim \psi$.

Let $\mathcal{C} \subseteq \mathrm{STRUC}(\sigma)$ and $\mathcal{L} \subseteq \mathrm{FO}^\sigma$. The *query problem* for $\mathcal{C}$ and $\mathcal{L}$ is the following:

QUERY$(\mathcal{C}, \mathcal{L})$

input: A structure $S \in \mathcal{C}$ and a formula $\varphi(\overline{x}) \in \mathcal{L}$ with free-variables $\overline{x}$;
parameter: the size $|\varphi|$ of $\varphi(\overline{x})$;
output: $\varphi(S)$.

Instances of such a problem are thus pairs $(S, \varphi) \in \mathcal{C} \times \mathcal{L}$. Most of the complexity results of this paper will be expressed in terms of the structure size, considering the size of the formula as a parameter. This explain the definition of a query problem as a parameter problem.

When φ has no free variable then the boolean query problem is also known as the *model-checking problem* for $\mathcal{L}$, often denoted by MC($\mathcal{L}$). Most of the time, $\mathcal{C}$ will simply be the class of all σ-structures STRUC(σ) for a given signature σ and restriction will only concern formulas. In this case, the query problem is denoted QUERY($\mathcal{L}$) and its instances are called *$\mathcal{L}$-queries.*

Reductions, complexity and enumeration problems. The basic model of computation used in this paper is standard Random Access Machine with addition. In the rest of the paper, the *big-O* notation $O_k(m)$ stands for $O(f(k).m)$ for some function f. Such a notation is used to shorten statements of results, especially when k is a parameter whose exact value is difficult to obtain. However, when this value can be made precise, we use the classical *big-O* notation. The definition below specifies the notion of reductions between query problems.

Definition 1. *We say that* QUERY($\mathcal{C}, \mathcal{L}$) linearly reduces *to* QUERY($\mathcal{C}', \mathcal{L}'$) *if there exist two recursive functions $f : \mathcal{C} \times \mathcal{L} \to \mathcal{C}'$ and $g : \mathcal{L} \to \mathcal{L}'$ such that:*

- *if $(S, \varphi) \in \mathcal{C} \times \mathcal{L}$ and $(S', \varphi') = (f(S, \varphi), g(\varphi))$, then dom$(S)$ = dom(S') and arity(φ) = arity(φ');*
- *the results of the queries are the same: $\varphi(S) = \varphi'(S')$;*
- *the function $f : (S, \varphi) \mapsto S'$ is computable in time $O_{|\varphi|}(|S|)$.*

(There is no requirement on g, but its computability.)

This reduction is a kind of fixed-parameter linear reduction with some additional constraints. Notice that this definition differs from the usual notion of interpretation: here, the interpretive structure S' depends both on the structure S *and on the formula* φ to be interpreted (notice also that, in the sequel, S' will always be an extension of S with only new additional monadic predicates).

In Sect. 5, query evaluation is considered as an enumeration problem. A non boolean query problem QUERY($\mathcal{C}, \mathcal{L}$) is *enumerable with constant delay* if one can compute all its solutions in such a way that the delay beween the beginning of the computation and the first solution, between two successive solutions, and between the last solution and the end of the computation are all constant. We denote by CONSTANT-DELAY the class of query problems which are enumerable with constant delay. This class is not robust: a problem which is in CONSTANT-DELAY when represented with a given data-structure may not be in this class anymore for another presentation. Nevertheless, this definition will be relevant to forthcoming technical tasks. The complexity class that is of real interest was first introduced in [5]: the class CONSTANT-DELAY(*lin*) collects query problems that can be enumerated as previously described after a step of preproccessing which costs a time at most linear in the size of the input. This class is robust. A more formal definition of it is given in the following.

Definition 2. *An query problem* QUERY($\mathcal{C}, \mathcal{L}$) *is computable* within constant delay and with linear precomputation *if there exists a RAM algorithm $\mathcal{A}$ which, for any input (S, φ), enumerates the set $\varphi(S)$ in the following way:*

1. $\mathcal{A}$ uses linear space
2. $\mathcal{A}$ can be decomposed into the two following successive steps
   (a) PRECOMP($\mathcal{A}$) *which performs some precomputations in time $O(f(\varphi).|S|)$ for some function f, and*
   (b) ENUM($\mathcal{A}$) *which outputs all elements in $\varphi(S)$ without repetition within a delay bounded by some constant* DELAY($\mathcal{A}$) *which depends only on $|\varphi|$ (and not on S). This delay applies between two consecutive solutions and after the last one.*

The complexity class thus defined is denoted by CONSTANT-DELAY*(lin).*

The following fact is immediate.

Fact 3. *If* QUERY($\mathcal{C},\mathcal{L}$) $\in$ CONSTANT-DELAY*(lin), then each query problem that linearly reduces to* QUERY($\mathcal{C},\mathcal{L}$) *also belongs to* CONSTANT-DELAY*(lin).*

Logical normalization. We now specify the kind of structures and formulas that are considered in this paper. A *unary* signature contains only unary relation symbols. A *quasi-unary* signature is obtained by enriching a unary signature with a single unary function symbol. That is, quasi-unary signatures have the shape $\{f,\overline{U}\}$, where f is a unary function symbol and $\overline{U}$ is a tuple of unary relation symbols.

For any signature σ, a *unary enrichment* of σ is a signature obtained from σ by adding some new unary relation symbols. Therefore, unary enrichments of quasi-unary signatures are also quasi-unary signatures. Structures over quasi-unary (resp. unary) signatures are called *quasi-unary* (resp. *unary*) structures.

Notice that structures over one unary function are disjoint collections of "whirlpools": upside down trees whose roots come together at a cycle (see Fig. 1). Therefore, considering the case where all the cycles consist of a single node, we see that the set of quasi-unary structures includes that of (colored) forests and, *a fortiori*, that of (colored) trees.

Let us consider a quantifier-free disjunctive formula φ over a quasi-unary signature. For any $y \in \text{var}(\varphi)$, we can "break" φ into two pieces ψ and θ, in such a way that y does not occur in ψ while it occurs in θ in a very uniform way. More precisely, each quantifier-free disjunctive formulas over a quasi-unary signature can be written, up to linear reduction, under the form (1) below. This normalization is formally stated in the next proposition. It will provide us with a key tool to eliminate quantified variables in any first-order formulas of quasi-unary signature.

Proposition 4. *Let* $\text{FO}_0^{qu}[\vee]$ *be the class of quantifier-free disjunctive formulas over a quasi-unary signature. Then,* QUERY($\text{FO}_0^{qu}[\vee]$) *linearly reduces to the problem* QUERY($\mathcal{L}_0$)*, where $\mathcal{L}_0$ is the class of* $\text{FO}_0^{qu}[\vee]$*-formulas that fit the shape:*

$$\psi(\overline{x}) \vee [(f^m y = f^n z \wedge U(y)) \rightarrow \bigvee_i f^{m_i}(y) = f^{n_i}(\overline{x})], \tag{1}$$

where ψ is a quantifier-free disjunctive formula, $0 \leq m_1 \leq \cdots \leq m_k \leq m$, $z \in \overline{x}$ and U is a unary relation.

Note that an essential consequence of this normalization is that on the left-hand side of the implication, only one atomic formula involving function f and variable y appears. This will make the elimination of variable y easier.

3 Acyclic Representation

Because of the normal form stated in Proposition 4, a corner-stone of our oncoming results lies on the way we handle formulas of the type:

$$\varphi(\overline{x}) \equiv \forall y : (f^m y = a \wedge U(y)) \rightarrow \bigvee_{i \in [k]} f^{m_i} y = c_i \qquad (2)$$

where $0 \leq m_1 \leq \cdots \leq m_k \leq m$, U is a unary relation and $a, c_1, \ldots, c_k$ are terms in $\overline{x}$. In order to deal efficiently (from a complexity point of view) with such formulas, we introduce in this section a data structure related to the tuple $\overline{f} = (f^{m_1}, \ldots, f^{m_k})$ and the set $X = f^{-m}(a) \cap U$. Then we show that this structure allows for a fast computation of some combinatorial objects – the samples of $\overline{f}$ over X – which are the combinatorial counterparts of the logical assertion (2).

Until the end of this section, f denotes a unary function on a finite domain D, X is a subset of D and $0 \leq m_1 \leq m_2 \leq \cdots \leq m_k$ are nonnegative integers. Furthermore, $\overline{f}$ denotes the tuple $(f^{m_1}, \ldots, f^{m_k})$. We associate a labelled forest to $\overline{f}$ and X in the following way (and we right now refer the reader to Fig. 1 and 2 to follow this definition):

- The set of vertices of the forest is partitioned into $k+1$ sets $L_0, L_1, \ldots, L_k$ corresponding to the sets $X, f^{m_1}(X), \ldots, f^{m_k}(X)$.
- For each i, the label function ℓ is a bijection from L_i to $f^{m_i}(X)$
- There is an edge from y to x if and only if there exists $i \in [0..k]$ such that: $x \in L_i$, $y \in L_{i+1}$ and $f^{m_{i+1}-m_i}\ell(x) = \ell(y)$.

Then we enrich this labelled forest with the data root, next, back, height defined from some depth-first traversal of the forest:

- root is the first root of the forest visited by the depth-first traversal.
- next(s) is the first descendent of s visited after s, if such a node exists. Otherwise, next$(s) = s$.
- back(s) is the first node visited after s which is neither a parent nor a descendant of s, provided that such a node exists. Otherwise, back$(s) = s$.
- height(s) is the height of s in $\mathcal{F}$. That is, height(s) is the distance from s to a leaf.

The resulting structure is called the *acyclic representation of $\overline{f}$ over X*. We denote it by $\mathcal{F}(\overline{f}, X)$. Besides, we denote by height$(\mathcal{F})$ the height of $\mathcal{F}(\overline{f}, X)$, that is the maximum of the values height(s), for s in $\mathcal{F}$, and we call *branch* of the forest any path that connects a root to a leaf (i.e. any maximal path of a tree of $\mathcal{F}$).

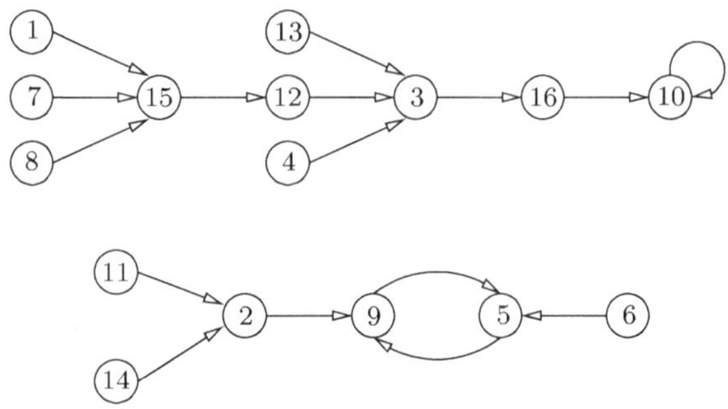

Fig. 1. A function $f : [16] \to [16]$

Example 5. A function $f : D \to D$, where $D = \{1, \ldots, 16\}$, is displayed in Fig. 1. The acyclic representation of $\overline{f} = (f, f, f^2, f^4, f^5, f^7)$ over the set $D \setminus \{3, 8, 10, 14, 16\}$ is given in Fig. 2.

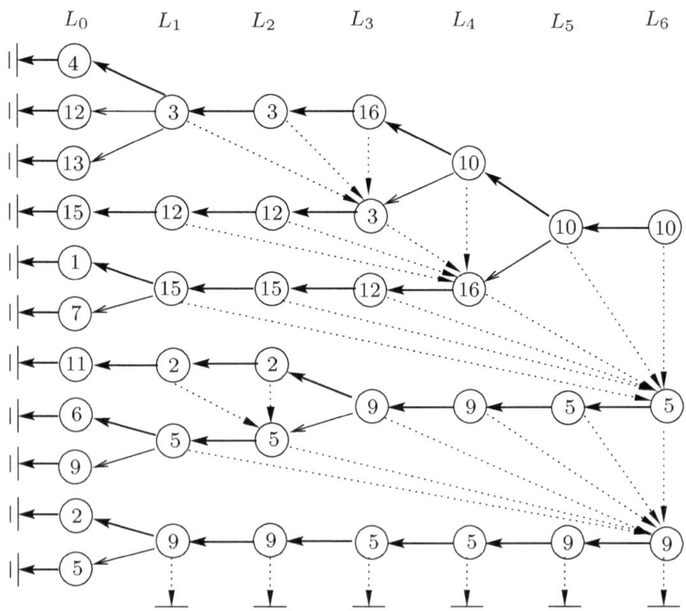

Fig. 2. The acyclic representation of a tuple $\overline{f} = (f^{m_1}, \ldots, f^{m_k})$ over a set X. Here, f is the function drawn in Fig. 1, $\overline{f}$ is the tuple $(f, f, f^2, f^4, f^5, f^7)$ and X is the set $[16] \setminus \{3, 8, 10, 14, 16\}$. Each L_i at the top of the figure corresponds both to the set $f^{m_i}(X)$ and to the set of nodes of height i in the forest.

We let the reader check that the following Lemma is true (the acyclic representation can be easily built by sorting and running through image sets $f^{m_i}(X)$, for all $i \le k$).

Lemma 6. *The acyclic representation of* $\overline{f} = (f^{m_1}, \ldots, f^{m_k})$ *over* X *can be computed in time* $O(k.m_k.|X|)$.

Let us come back to the combinatorial objects (the samples - this terminology comes from the earlier papers [6,4]) mentioned at the beginning of this section. For each $m \geq 0$ and each $c \in D$, we denote by $f^{-m}(c)$ the set of pre-images of c by f^m. That is: $f^{-m}(c) = \{x \in D \mid f^m x = c\}$. We set:

Definition 7. *Let* $P \subseteq [k]$ *and* $(c_i)_{i \in P} \in D^P$. *The tuple* $(c_i)_{i \in P}$ *is a* sample of $\overline{f}$ over X *if*

$$X \subseteq \bigcup_{i \in P} f^{-m_i}(c_i).$$

This sample is minimal *if, moreover, for all* $j \in P$:

$$X \nsubseteq \bigcup_{i \in P \setminus \{j\}} f^{-m_i}(c_i).$$

Notice that each sample contains a minimal sample: if $(c_i)_{i \in P}$ is a sample of $\overline{f}$ over X, then $(c_i)_{i \in P'}$ is a minimal sample of $\overline{f}$ over X, where P' is obtained from P by iteratively eliminating the j's such that $X \subseteq \bigcup_{i \in P \setminus \{j\}} f^{-m_i}(c_i)$.

Samples provide a combinatorial interpretation of some logical assertions. It is easily seen that the assertion $\forall y \in X : \bigvee_{i \in [k]} f^{m_i} y = c_i$ exactly means that $(c_i)_{i \in [k]}$ is a sample of $\overline{f}$ over X. In particular, assertion (2) holds if, and only if, $(c_i)_{i \in [k]}$ is a sample of $(f^{m_i})_{i \in [k]}$ over $f^{-m}(a) \cap U$. This equivalence will yields the variable elimination result of Sect. 4.

Another characterization of samples connects this notion to that of acyclic representation: Let $(x_i)_{i \in P}$ be a sequence of pairwise distinct nodes in $\mathcal{F}(\overline{f}, X)$, where $P \subseteq [k]$. We say that $(x_i)_{i \in P}$ is a *minimal branch marking* of $\mathcal{F}(\overline{f}, X)$ if each branch of the forest contains a unique x_i and if, furthermore, each x_i lies on the level L_i of the forest. Clearly, $(x_i)_{i \in P}$ is a minimal branch marking of $\mathcal{F}(\overline{f}, X)$ iff the tuple $(\ell(x_i))_{i \in P}$ is a minimal sample of $\overline{f}$ over X (recall $\ell(x)$ is the label of x in $\mathcal{F}(\overline{f}, X)$). Roughly speaking: minimal samples of $\overline{f}$ over X are exactly sequences of values that label minimal branch markings of $\mathcal{F}(\overline{f}, X)$.

The next lemma states that both the total number of minimal samples and the time required to compute all of them can be precisely controlled. According to the equivalence above, evaluating all minimal samples of $\overline{f}$ over X amounts to compute all minimal branch markings of $\mathcal{F}(\overline{f}, X)$. This underlies an efficient procedure to carry out such an evaluation. We will not give the proof of the next Lemma: it has already been proved in a more general context in [6,4].

Lemma 8. *There are at most* $k!$ *minimal samples of* $\overline{f}$ *over* X *and their set can be computed in time* $O_k(|X|)$.

We now slightly modify the presentation of minimal samples in order to manipulate them more conveniently inside formulas. The levels (with a few other information) of the acyclic representation are introduced explicitly.

Definition 9. *Let m be an integer, $a \in D$ and $X_a \subseteq f^{-m}(a)$. The sets $L_i^{P,h}$, with $P \subseteq [k]$, $h \leq k!$ and $i \in P$, are defined as the minimal sets such that, for all $a \in f^m(D)$:*

if $(s_i)_{i \in P}$ is the hth minimal sample of $\overline{f}$ over X_a then, for all $i \in P$, $s_i \in L_i^{P,h}$.

Note that if the image set $f^m(D)$ is reduced to a single element i.e. $f^m(D) = \{a\}$, then each $L_i^{P,h}$ contains *at most* one element. We will use this property in the sequel. Note also that each $x \in L_i^{P,h}$ belongs to the level L_i of the acyclic representation of $\overline{f}$: for fixed i, sets $L_i^{P,h}$ are refinements of level L_i.

Not all sets $L_i^{P,h}$ for $P \subseteq [k]$ and $h \leq k!$ are non empty and hence need to be defined. We denote by PH the set of such (P, h) for which the sets $L_i^{P,h}$ are not empty ($i \in P$). The following lemma details how to compute the sets $L_i^{P,h}$.

Lemma 10. *With the notation of Definition 9, the collection of sets $L_i^{P,h}$ for $i \in P$ and $(P, h) \in \mathrm{PH}$ can be computed in time $O_k(|D|)$.*

Proof. We first set $L_i^{P,h} = \emptyset$ for all $P \subseteq [k], h \leq k!$ and $i \in P$. By Lemma 8, for each $a \in f^m(D)$, one can compute the set of the at most $k!$ minimal samples of X_a in time $O_k(|X_a|)$ and assign a different number $h \leq k!$ to each. Now, running through these samples, if $(s_i)_{i \in P}$ is the hth of them, one add each s_i to set $L_i^{P,h}$. This step needs to be repeated for each $a \in f^m(D)$. Since all sets X_a are pairwise disjoints and since their union is included in D, the whole process requires time $O_k(|D|)$ which is the expected time bound. Note that, for indices $(P, h) \notin \mathrm{PH}$, the sets $L_i^{P,h}$ remain empty. □

4 Variable Elimination

We are now in a position to eliminate quantifiers in a first-order formula to be evaluated on a quasi-unary structure. As mentioned in Sect. 3, a first step must be to eliminate y in a formula φ of the following form:

$$\varphi(\overline{x}) \equiv \forall y : (f^m y = a \wedge U(y)) \rightarrow \bigvee_{i \in [k]} f^{m_i} y = c_i$$

where $0 \leq m_1 \leq \cdots \leq m_k \leq m$, U is a unary relation and $a, c_1, \ldots, c_k$ are terms in $\overline{x}$. To deal with this task, we understand $\varphi(\overline{x})$ as meaning:

$$(c_i)_{i \in [k]} \text{ is a sample of } \overline{f} \text{ over } f^{-p}(a) \cap U.$$

That is, since every sample contains a minimal sample:

$$(c_i)_{i \in [k]} \text{ contains a minimal sample of } \overline{f} \text{ over } f^{-m}(a) \cap U$$

or, equivalently:

$$\exists P \subseteq [k] \text{ such that } (c_i)_{i \in P} \text{ is a minimal sample of } \overline{f} \text{ over } f^{-m}(a) \cap U$$

From the previous section, this assertion can now be stated as:

$$\exists h \leq k!\ \exists P \subseteq [k] \text{ such that } \forall i \in P\text{: } L_i^{P,h}(c_i) \text{ and } f^{m-m_i}c_i = a$$

This is more formally written:

$$\bigvee_{(P,h)\in \mathrm{PH}} \bigwedge_{i\in P} (L_i^{P,h}(c_i) \wedge f^{m-m_i}c_i = a).$$

All predicates in this last assertion can be computed in linear time and the disjunction and conjunction deal with a constant number (depending on k) of objects. Variable y is now eliminated. This provides a general framework to iteratively eliminate variables.

Lemma 11. *Every* FO_1*-query of the form* $(\forall y \varphi(\overline{x}, y), S)$*, where* $\varphi(\overline{x}, y)$ *is a quantifier-free disjunctive formula over a quasi-unary signature* σ*, linearly reduces to an* FO_0*-query over a quasi-unary signature.*

Proof. Let S be a quasi-unary structure and $\varphi(\overline{x}, y)$ be a quantifier-free disjunctive formula. Thanks to Proposition 4, $\forall y \varphi(\overline{x}, y)$ may be considered, up to linear reduction, as fitting the form:

$$\psi(\overline{x}) \vee \forall y[\,(f^p y = f^q z \wedge U(y)) \rightarrow \textstyle\bigvee_{i\in[k]} f^{m_i}(y) = f^{n_i}(\overline{x})\,]$$

where $z \in \overline{x} \cup \{y\}$. Assume $z \neq y$ (otherwise the proof gets simpler). Denoting $\overline{f} = (f^{m_1}, \ldots, f^{m_k})$, the second disjunct simply means that $(f^{n_i}(\overline{x}))_{i\in[k]}$ is a sample of $\overline{f}$ over $f^{-p}(f^q(z)) \cap U$. From what has been said before, this implies that there exists $P \subseteq [k]$ such that $(f^{n_i}(\overline{x}))_{i\in P}$ is a minimal sample of $\overline{f}$ over $f^{-p}(f^q(z))$. Recalling Definition 9 and the discussion that followed, one can write:

$$\langle S, (L_i^{P,h})_{\mathrm{PH}}\rangle \models \psi(\overline{x}) \vee \bigvee_{(P,h)\in \mathrm{PH}} \bigwedge_{i\in P} L_i^{P,h}(f^{n_i}(\overline{x})) \wedge f^{p-m_i+n_i}(\overline{x}) = f^q(z)$$

Variable y is well eliminated. From Lemma 10, the collection of sets $L_i^{P,h}$ are linear time computable from structure S. This concludes the proof. □

Theorem 12. *Each non-Boolean (resp. Boolean)* FO*-query over a quasi-unary signature linearly reduces to a* FO_0*-query (resp.* FO_1 *query) over a quasi-unary signature.*

Proof. The proof is by induction on the number k of quantified variables of the query. Most of the difficulty already appears for the case $k = 1$. Let $(\varphi(\overline{z}), S) \in \mathrm{FO}_1^\sigma \times \mathrm{STRUC}(\sigma)$ and $\overline{z}$ be the nonempty tuple of free variables of φ. Since $\exists y\ \psi$ is equivalent to $\neg(\forall x\ \neg\psi)$, one may suppose that φ is of the form $\perp\forall y \varphi(\overline{z}, y)$ (where $\pm$ means that one negation $\neg$ may occur). The conjunctive normal form for the matrix of φ must be computed, and, since universal quantification and conjunction commute, one can put the formula in the form $\varphi(\overline{z}) \equiv \pm \bigwedge_\alpha \forall y \varphi_\alpha(\overline{z}, \overline{x}, y)$, where each φ_α is a quantifier-free disjunctive formula. By Lemma 11, we know that the query $(\bigwedge_\alpha \forall y \varphi_\alpha(\overline{z}, y), S)$ linearly reduces to a quantifier-free query $(\psi(\overline{z}, \overline{x}), S')$ over a quasi-unary signature. This concludes this case.

Assume the result for $k \geq 1$ and prove it for $k+1$. For the same reason as above, $\varphi(\overline{z})$ can be put under the form $\varphi(\overline{z}) \equiv \overline{Q}\overline{x} \pm \bigwedge_\alpha \forall y \varphi_\alpha(\overline{z}, \overline{x}, y)$, where each φ_α is a quantifier-free disjunctive formula and $\overline{x}$ is a tuple of k quantified variables. Again, from Lemma 11 query $(\bigwedge_\alpha \forall y \varphi_\alpha(\overline{z}, \overline{x}, y), S)$ linearly reduces to a quantifier-free query $(\psi(\overline{z}, \overline{x}), S')$ and then, $(\varphi(\overline{z}), S)$ linearly reduces to the k variable query $(\overline{Q}\overline{x}\psi(\overline{z}, \overline{x}), S')$. □

5 Query Enumeration

In this section we prove that each first-order query over a quasi-unary structure $\langle D, f, \overline{U} \rangle$ can be enumerated with constant delay after a linear time preprocessing step. The proof involves the cost of the enumeration of all the elements of D whose images by different powers of f (i.e. by functions of the form f^i) avoid a fixed set of values. It appears that the elements thus defined can be enumerated with constant delay, provided the inputs are presented with the appropriate data structure. Let us formalize the problem, before stating the related enumeration result.

AUTHORIZED VALUES

input: the acyclic representation of a k-tuple $\overline{f} = (f^{m_1}, \ldots, f^{m_k})$ over a set $X \subseteq D$, and a set of *forbidden pairs* $\mathsf{F} \subset [k] \times D$;

parameter: k, $|\mathsf{F}|$;

output: the set $\mathcal{A} = \{y \in X \mid \bigwedge_{(i,c) \in \mathsf{F}} f^{m_i} y \neq c\}$.

Lemma 13. AUTHORIZED VALUES $\in$ CONSTANT-DELAY. *Moreover, the delay between two consecutive solution is* $O(k.|\mathsf{F}|)$.

Proof. Each forbidden pair $(i, c) \in \mathsf{F}$ is either inconsistent (i.e. $f^{-m_i}(c) = \emptyset$) or corresponds to a node s of $\mathcal{F}(\overline{f}, X)$ such that $\mathsf{height}(s) = i$ and $\ell(s) = c$. If we denote by forbid those nodes s of $\mathcal{F}(\overline{f}, X)$ for which $(\mathsf{height}(s), \ell(s)) \in \mathsf{F}$, the set $\mathcal{A}$ to be constructed is exactly the set of values $y \in D$ that label leaves of the forest whose branches avoid all forbidden nodes. Therefore, computing $\mathcal{A}$ amounts to finding all the leaves described above.

This can be done by a depth-first search of the forest, discarding those leaves whose visit led to a forbidden node. Furthermore, the search can be notably sped up by backtracking on each forbidden node: indeed, such a node is the root of a subtree whose all leaves disagree with our criteria. This algorithm clearly runs in linear time. Let us show it enumerates solutions with constant delay:

Consider a sequence of p nodes $s_1, \ldots, s_p$ successively visited by the algorithm. If $p > k$, the sequence must contain a leaf or a forbidden node. Indeed, if it does not contain any node of forbid, the algorithm behaves as a usual DFS between s_1 and s_p. Therefore, one node among $s_2, \ldots, s_{k+1}$ has to be a leaf since $\mathsf{height}(\mathcal{F}(\overline{f}, X)) = k$.

Now, if s, s' are two leaves successively returned by the algorithm and if $f_1, \ldots, f_p$ are the forbidden nodes encountered between these two solutions, then the previous remark ensures that the algorithm did not visit more than k nodes between s and f_1, between f_p and s' or between two successive f_i's. The delay between the outputs s and s' is hence in $O(pk)$ and therefore, in $O(k|\mathsf{forbid}|)$. Furthermore, this reasoning easily extends to the delay between the start of the algorithm and the first solution, or between the last solution and the end of the algorithm. This concludes the proof. □

Now we can prove the main result of this section.

Theorem 14. $\text{Query}(\text{FO}^\sigma) \in \text{Constant-Delay}$*(lin) for any quasi-unary signature* σ.

Proof. Because of Theorem 12 and Fact 3, we just have to prove the result for the quantifier-free restriction of FO^σ. We prove by induction on d that every $\text{FO}_0^\sigma(d)$-query can be enumerated with constant delay and linear time precomputation. (Recall σ is a quasi-unary signature.)

The result is clear for $d = 1$: Given an instance $(S, \varphi(x))$ of $\text{Query}(\text{FO}_0^\sigma(1))$, the set $\varphi(S)$ can be evaluated in time $O(|S|)$ since φ is quantifier-free. Therefore, the following procedure results in a $\text{Constant-Delay}(lin)$-enumeration of $\varphi(S)$:

Algorithm 1. $\text{Enumeration}(1, \varphi(y))$

1: compute $\varphi(S)$
2: enumerate all the values of $\varphi(S)$

Let us now suppose the induction hypothesis is true for $d \geq 1$ and examine the case $d + 1$. This case is divided in several steps. Let $(S, \varphi(\overline{x}, y))$ be a $\text{FO}_0^\sigma(d+1)$-query.

Step **1.** By standard logical techniques, $\varphi(\overline{x}, y)$ can be written in disjunctive normal form as $\bigvee_\alpha \theta_\alpha(\overline{x}, y)$, where the θ_α's are *disjoint* conjunctive (quantifier-free) formulas (i.e., $\theta_\alpha(S) \cap \theta_\beta(S) = \emptyset$ for $\alpha \neq \beta$). And this normalization can be managed in linear time. But it is proved in [5] that the class $\text{Constant-Delay}(lin)$ is stable by disjoint union. Therefore, we do not loose generality when focusing on the $\text{Constant-Delay}(lin)$-enumeration of a query of the form $(S, \theta_\alpha(\overline{x}, y))$.
Step **2.** One can separate parts of formulas that depends exclusively on $\overline{x}$, then a conjunctive $\text{FO}_0(d+1)$-formula $\theta(\overline{x}, y)$ can be written $\theta(\overline{x}, y) \equiv \psi(\overline{x}) \wedge \delta(\overline{x}, y)$. It is essential to consider tuples $\overline{x}$ that satisfy $\psi(\overline{x})$ but that can also be completed by some y such that $\delta(\overline{x}, y)$ holds. Hence, one considers the equivalent formulation :

$$\theta(\overline{x}, y) \equiv \psi(\overline{x}) \wedge \exists y \delta(\overline{x}, y) \wedge \delta(\overline{x}, y)$$

If we set $\psi_1(\overline{x}) \equiv \psi(\overline{x}) \wedge \exists y \delta(\overline{x}, y)$, formula $\theta(\overline{x}, y)$ can now be written, thanks to Proposition 4:

$$\theta(\overline{x}, y) \equiv \psi_1(\overline{x}) \wedge \delta(\overline{x}, y), \tag{3}$$

where $\psi_1(\overline{x})$ is a conjunctive $\mathrm{FO}_1(d)$-formula and $\delta(\overline{x}, y)$ is a conjunctive $\mathrm{FO}_0(d+1)$-formula of the form $f^m y = f^n \overline{x} \wedge Uy \wedge \bigwedge_{i\in[k]} f^{m_i} y \neq f^{n_i} \overline{x}$.

The idea is to enumerate the $(d+1)$-tuples $(\overline{x}, y)$ satisfying θ by enumerating the d-tuples $\overline{x}$ satisfying $\psi_1(\overline{x})$ and, for each such $\overline{x}$, by enumerating the values y fulfilling $\delta(\overline{x}, y)$. Both these enumerations can be done with constant delay: the first, by inductive hypothesis ; the second by Lemma 13. As we made sure that any tuple satisfying $\psi_1(\overline{x})$ can be completed by at least one convenient y, our enumeration procedure will not explore bad directions in depth. Next step makes this precise.

Step **3.** By induction hypothesis, there exists a CONSTANT-DELAY(*lin*)-enumeration algorithm ENUMERATION(d, φ) for formulas φ of arity d. Then, we get the following CONSTANT-DELAY(*lin*)-enumeration algorithm for conjunctive formulas of arity $d+1$, together with its linear time precomputation procedure:

Algorithm 2. PRECOMP$(d+1, \theta(\overline{x}, y)))$

```
1: write θ under the form (3)
2: build the complete acyclic representation of f̄ = (f^{m_1}, ..., f^{m_k}) over X = f^{-m}(f^n x̄) ∩ U
3: PRECOMP(d, ψ_1(x̄))
```

Notice that items 1 and 2 of the above algorithm are carried out in time $O(|X|)$, thanks to Proposition 4 and Lemma 6.

Algorithm 3. ENUMERATION$(d+1, \theta(\overline{x}, y))$

```
PRECOMP(d+1, θ(x̄, y)))
for all x̄ in ENUM(d, ψ_1(x̄)) do
    for all y in AUTHORIZED VALUES (f̄, X, (i, f^{n_i} x̄)_{i∈[k]}) do
        return (x̄, y)
```

Finally, we get a complete CONSTANT-DELAY(*lin*)-enumeration algorithm for $\mathrm{FO}_0(d+1)$-formula by running successively the algorithms for the disjoint conjunctive formulas $\theta_\alpha(\overline{x}, y)$ obtained in the first step. □

6 Conjunctive Queries

So far, we proved that the first-order query problem over quasi-unary signature is "linear-time" computable. However, in full generality the size of the constant (depending on the formula) may be huge: this is essentially due to the variable elimination process that, at each step, may produce an exponentially larger formula. Notice that, once variables are eliminated in formulas, query answering become tractable both in terms of formula and of structure sizes. However, it is not straightforward to know in advance which kind of query admit equivalent formulation in terms of a quantifier-free query of polynomially related size. This amounts to determine the number of expected minimal samples in different situations. This will be the object of further investigation in the extended version of this paper.

In what follows, we examine the very easy particular case of conjunctive queries.

Definition 15. *A first-order formula is existential conjunctive if it uses conjunction and existential quantification only. Conjunctive queries are queries defined by existential conjunctive formulas.*

Conjunctive queries (even union of conjunctive queries) over quasi-unary structures are more tractable from the point of view of answer enumeration. Enumeration of query result is tractable even from the point of view of the constant size. The following result is easy to see by a direct algorithm.

Proposition 16. *The conjunctive query problem over quasi-unary structures belongs to the class* CONSTANT-DELAY *with a delay between two consecutive tuples in* $O(|\varphi|)$.

References

1. G. Bagan. MSO queries on tree decomposable structures are computable with linear delay. Computer Science Logic'06 (this volume), 2006.
2. B. Courcelle. Linear delay enumeration and monadic second-order logic. submitted, 2006.
3. A. Durand, R. Fagin, and B. Loescher. Spectra with only unary function symbols. In M. Nielsen and W. Thomas, editors, *Computer Science Logic, selected paper of CSL'97*, LNCS 1414, pages 111–126. Springer, 1998.
4. A. Durand and E. Grandjean. The complexity of acyclic conjunctive queries revisited. Draft, 2005.
5. A. Durand and E. Grandjean. First-order queries on structures of bounded degree are computable with constant delay. *ACM Transactions on Computational Logic*, 2006. To appear.
6. A. Durand, E. Grandjean, and F. Olive. New results on arity vs. number of variables. Research report 20-2004, LIF, Marseille, France, April 2004.
7. J. Flum, M. Frick, and M. Grohe. Query evaluation via tree decompositions. *Journal of the ACM*, 49(6):716–752, 2002.
8. M. Frick and M. Grohe. Deciding first-order properties of locally tree decomposable structures. *Journal of the ACM*, 48:1184–1206, 2001.
9. Y. Gurevich. The decision problem for standard classes. *J. Symb. Logic*, 41(2):pp.460–464, 1976.
10. Y. Gurevich and S. Shelah. On spectra of one unary function. In *18th IEEE Conference in Logic in Computer Science*, IEEE Computer Society, pages 291–300, 2003.
11. L. Libkin. *Elements of finite model theory*. EATCS Series. Springer, 2004.
12. S. Lindell. A normal form for first-order logic over doubly-linked data structures. *Submitted*, 2006.
13. C. Papadimitriou and M. Yannakakis. On the complexity of database queries. *Journal of Computer and System Sciences*, 58(3):407–427, 1999.

14. D. Seese. Linear time computable problems and first-order descriptions. *Mathematical Structures in Computer Science*, 6(6):505–526, December 1996.
15. M. Y. Vardi. On the complexity of bounded-variable queries. In *ACM Symposium on Principles of Database Systems*, pages 266–276. ACM Press, 1995.
16. M. Yannakakis. Algorithms for acyclic database schemes. In *Proceedings of the 7th International Conference on Very Large Databases*, pages 82–94, 1981.

Infinite State Model-Checking of Propositional Dynamic Logics

Stefan Göller and Markus Lohrey

FMI, Universität Stuttgart, Germany
{goeller, lohrey}@informatik.uni-stuttgart.de

Abstract. Model-checking problems for PDL (propositional dynamic logic) and its extension $\mathrm{PDL}^{\cap}$ (which includes the intersection operator on programs) over various classes of infinite state systems (BPP, BPA, pushdown systems, prefix-recognizable systems) are studied. Precise upper and lower bounds are shown for the data/expression/combined complexity of these model-checking problems.

1 Introduction

Propositional Dynamic Logic (PDL) was introduced by Fischer and Ladner in 1979 as a modal logic for reasoning about programs [10]. In PDL, there are two syntactic entities: formulas and programs. Formulas are interpreted in nodes of a Kripke structure and can be built up from atomic propositions using boolean connectives. Programs are interpreted by binary relations over the node set of a Kripke structure and can be built up from atomic programs using the operations of union, composition, and Kleene hull (reflexive transitive closure). PDL contains two means for connecting formulas and programs: Programs may appear in modalities in front of formulas, i.e., if π is a program and φ is a formula, then $\langle\pi\rangle\varphi$ is true in a node u if there exists a node v, where φ holds and which can be reached from u via the program π. Moreover, PDL allows to construct programs from formulas using the test operator: If φ is a formula, then the program $\varphi?$ is the identity relation on the node set restricted to those nodes where φ holds. Since its invention, many different extensions of PDL were proposed, mainly by allowing further operators on programs, like for instance the converse or intersection operator, see the monograph [13] for a detailed exposition. Recently, PDL, where programs are defined via visibly pushdown automata, was investigated [18]. PDL and its variations found numerous applications, e.g., in program verification, agent-based systems, and XML-querying. In AI, PDL received attention by its close relationship to description logics and epistemic logic, see [16] for references.

In the early days of PDL, researchers mainly concentrated on satisfiability problems and axiomatization of PDL and its variants. With the emergence of automatic verification, also model-checking problems for modal logics became a central research topic, and consequently model-checking problems for PDL attracted attention [16]. In this paper, we start to investigate model-checking problems for PDL over infinite state systems. In recent years, verification of

Z. Ésik (Ed.): CSL 2006, LNCS 4207, pp. 349–364, 2006.

infinite state systems became a major topic in the model-checking community. Usually, infinite state systems, like for instance systems with unbounded communication buffers or unbounded stacks, are modeled by some kind of abstract machine, which defines an infinite transition system (Kripke structure): nodes correspond to system states and state transitions of the system are modeled by labeled edges. Various classes of (finitely presented) infinite transition systems were studied under the model-checking perspective in the past, see e.g. [25] for a survey. In [22] Mayr introduced a uniform classification of infinite state systems in terms of two basic operations: parallel and sequential composition. In this paper, we will mainly follow Mayr's classification.

We believe that model-checking of PDL and its variants over infinite state systems is not only a natural topic, but also a useful and applicable research direction in verification. PDL allows directly to express regular reachability properties, which were studied e.g. in [19,22,30] in the context of infinite state systems. For instance, consider the property that a process can reach a state, where a condition φ holds, via a path on which the action sequence $a_1 a_2 \cdots a_n$ is repeated cyclically. Clearly, this can be expressed in CTL (if φ can be expressed in CTL), but we think that the PDL-formula $\langle (a_1 \circ a_2 \circ \cdots \circ a_n)^* \rangle \varphi$ is a more readable specification. Secondly, and more important, the extension of PDL with the intersection operator on programs [12], PDL$^\cap$ for short, allows to formulate natural system properties that cannot be expressed in the modal μ-calculus (since they do not have the tree model property), like for instance that a system can be reset to the current state (Example 2) or that two forking processes may synchronize in the future (Example 3).

In Section 5 we study model-checking problems for PDL and its variants over infinite state systems. For infinite state systems with parallel composition, PDL immediately becomes undecidable. More precisely, we show that PDL becomes undecidable over BPP (basic parallel processes), which correspond to Petri nets, where every transition needs exactly one token for firing (Proposition 1). This result follows from the undecidability of the model-checking problem for EF (the fragment of CTL, which only contains next-modalities and the "exists finally"-modality) for Petri nets [8]. Due to this undecidability result we mainly concentrate on infinite state systems with only sequential composition. In Mayr's classification these are pushdown systems (PDS) and basic process algebras (BPA), where the latter correspond to stateless pushdown systems. Pushdown systems were used to model the state space of programs with nested procedure calls, see e.g. [9]. Model-checking problems for pushdown systems were studied for various temporal logics (LTL, CTL, modal μ-calculus) [1,9,15,28,29]. We also include prefix-recognizable systems (PRS) into our investigation [3,5], which extend pushdown systems. Model-checking problems for prefix-recognizable systems were studied e.g. in [4,14]. The decidability of PDL and even PDL$^\cap$ for prefix-recognizable systems (and hence also BPA and PDS) follows from the fact that monadic second-order logic (MSO) is decidable for these systems and that PDL$^\cap$ can be easily translated into MSO. But from the viewpoint of complexity, this approach is quite unsatisfactory, since it leads to a nonelementary algorithm. On the other hand, for PDL (without the intersection operator) it turns out that based on the techniques of Walukiewicz for model-checking CTL and EF over pushdown systems [28], we can obtain sharp (elementary) complexity bounds: Whereas test-free PDL behaves w.r.t. to the complexity of the model-checking problem exactly in the same way as EF (PSPACE-complete in most cases), PDL with the test-operator is more difficult (EXP-complete in most cases).

The analysis of PDL$^\cap$ turns out to be more involved. This is not really surprising. PDL$^\cap$ turned out to be notoriously difficult in the past. It does not have the tree model property, and as a consequence the applicability of tree automata theoretic methods is quite limited. Whereas PDL is translatable into the modal μ-calculus, PDL$^\cap$ is orthogonal to the modal μ-calculus with respect to expressiveness. A very difficult result of Danecki states that satisfiability of PDL$^\cap$ is in 2EXP [7]. Only recently, a matching lower bound was obtained by Lange and Lutz [17]. Our main result of this paper states that the expression/combined complexity of PDL$^\cap$ (and also the test-free fragment of PDL$^\cap$) over BPA/PDS/PRS is 2EXP-complete, whereas the data complexity goes down to EXP. For the 2EXP lower bound proof, we use a technique from [28] for describing a traversal of the computation tree of an alternating Turing machine in CTL using a pushdown. The main difficulty that remains is to formalize in PDL$^\cap$ that two configurations of an exponential space alternating Turing machine (these machines characterize 2EXP) are successor configurations. For the upper bound, we transform a PDL$^\cap$ formula φ into a two-way alternating tree automaton A of exponential size, which has to be tested for emptiness. Since emptiness of two-way alternating tree automata can be checked in exponential time [27], we obtain a doubly exponential algorithm. Most of the inductive construction of A from φ uses standard constructions for two-way alternating tree automata. It is no surprise that the intersection operator is the difficult part in the construction of φ. The problem is that two paths from a source node s to a target node t, where the first (resp. second) path is a witness that (s, t) belongs to the interpretation of a program π_1 (resp. π_2) may completely diverge. This makes it hard to check for an automaton whether there is both a π_1-path and a π_2-path from s to t. Our solution is based on a subtle analysis of such diverging paths in pushdown systems.

One might argue that the high complexity (2EXP-completeness) circumvents the application of PDL$^\cap$ model checking for pushdown systems. But note that the data complexity (which is a better approximation to the "real" complexity of model-checking, since formulas are usually small) of PDL$^\cap$ over pushdown systems is only EXP, which is the same as the data complexity of CTL [28]. Moreover, to obtain an exponential time algorithm for PDL$^\cap$ it is not really necessary to fix the formula, but it suffices to bound the nesting depth of intersection operators in programs. One may expect that this nesting depth is small in natural formulas, like in Example 2 or 3 (where it is 1). Table 1 gives an overview on our results. Proofs can be found in the technical report [11].

2 Preliminaries

Let Σ be a finite alphabet and let ε denote the empty word. Let $\Sigma_\varepsilon = \Sigma \cup \{\varepsilon\}$ and let $\overline{\Sigma} = \{\overline{a} \mid a \in \Sigma\}$ be a *disjoint copy* of Σ. For a word $w = a_1 \cdots a_n \in \Sigma^*$ ($a_1, \ldots, a_n \in \Sigma$) let $w^{\text{rev}} = a_n \cdots a_1$. For $L \subseteq \Sigma^*$ let $L^{\text{rev}} = \{w^{\text{rev}} \mid w \in L\}$. Let $R, U \subseteq A \times A$ be binary relations over the set A. Then R^* is the *reflexive and transitive closure of* R. The *composition* of R and U is $R \circ U = \{(a,c) \in A \times A \mid \exists b \in A : (a,b) \in R \wedge (b,c) \in U\}$. Let $f : A \to C$ and $g : B \to C$ be functions, where $A \cap B = \emptyset$. The *disjoint union* $f \uplus g : A \cup B \to C$ of f and g is defined by $(f \uplus g)(a) = f(a)$ for $a \in A$ and $(f \uplus g)(b) = g(b)$ for $b \in B$. Let $A^B = \{f \mid f : B \to A\}$ be the set of all functions from B to A.

We assume that the reader is familiar with standard complexity classes like P (deterministic polynomial time), PSPACE (polynomial space), EXP (deterministic exponential time), and 2EXP (deterministic doubly exponential time), see [24] for more details. Hardness results are always meant w.r.t. logspace reductions. An *alternating Turing machine (ATM)* is a tuple $\mathcal{M} = (Q, \Sigma, \Gamma, q_0, \delta, \Box)$ where (i) $Q = Q_{\text{acc}} \uplus Q_{\text{rej}} \uplus Q_\exists \uplus Q_\forall$ is a finite set of *states* Q which is partitioned into *accepting* (Q_{acc}), *rejecting* (Q_{rej}), *existential* ($Q_\exists$) and *universal* ($Q_\forall$) states, (ii) Γ is a *finite tape alphabet*, (iii) $\Sigma \subseteq \Gamma$ is the *input alphabet*, (iv) $q_0 \in Q$ is the *initial state*, (v) $\Box \in \Gamma \setminus \Sigma$ is the *blank symbol*, and (vi) the map $\delta : (Q_\exists \cup Q_\forall) \times \Gamma \to \text{Moves} \times \text{Moves}$ with $\text{Moves} = Q \times \Gamma \times \{\leftarrow, \rightarrow\}$ assigns to every pair $(q, \gamma) \in (Q_\exists \cup Q_\forall) \times \Gamma$ a pair of *moves*. If $\delta(q,a) = ((q_1, a_1, d_1), (q_2, a_2, d_2))$, then this means that if $\mathcal{M}$ is in state q and reads the symbol a, then the left (right) successor configuration of the current configuration results by writing a_1 (a_2), the read-write head moves in direction d_1 (d_2), and the new state is q_1 (q_2). A configuration of $\mathcal{M}$ is a word from $\Gamma^* Q \Gamma^+$. A configuration c of $\mathcal{M}$, where the current state is q, is *accepting* if (i) $q \in Q_{\text{acc}}$ or (ii) $q \in Q_\exists$ and there exists an accepting successor configuration of c or (iii) $q \in Q_\forall$ and both successor configurations of c are accepting. The machine $\mathcal{M}$ accepts an input w if and only if the initial configuration $q_0 w$ is accepting.

3 Propositional Dynamic Logic and Extensions

Formulas of propositional dynamic logic (PDL) are interpreted over *Kripke structures*: Let $\mathbb{P}$ be a set of *atomic propositions* and let Σ a set of *atomic programs*. A *Kripke structure* over $(\mathbb{P}, \Sigma)$ is a tuple $\mathcal{K} = (S, \{\to_\sigma \mid \sigma \in \Sigma\}, \rho)$ where (i) S is a set of *nodes*, (ii) $\to_\sigma \subseteq S \times S$ is a *transition relation* for all $\sigma \in \Sigma$ and (iii) $\rho : S \to 2^{\mathbb{P}}$ labels every node with a set of atomic propositions. Formulas and programs of the logic $\text{PDL}^\cap$ (PDL with intersection) over $(\mathbb{P}, \Sigma)$ are defined by the following grammar, where $p \in \mathbb{P}$ and $\sigma \in \Sigma$:

$$\varphi ::= p \mid \neg\varphi \mid \varphi_1 \vee \varphi_2 \mid \langle\pi\rangle\,\varphi$$
$$\pi ::= \sigma \mid \pi_1 \cup \pi_2 \mid \pi_1 \cap \pi_2 \mid \pi_1 \circ \pi_2 \mid \pi^* \mid \varphi?$$

We use the abbreviations $(\varphi_1 \wedge \varphi_2) = \neg(\neg\varphi_1 \vee \neg\varphi_2)$ and $[\pi]\varphi = \neg\langle\pi\rangle\neg\varphi$. Moreover, a set $\{a_1, \ldots, a_n\} \subseteq \Sigma$ of atomic programs is identified with the program

$a_1 \cup \cdots \cup a_n$. The *semantic* of PDL$^\cap$ is defined over Kripke structures. Given a Kripke structure $\mathcal{K} = (S, \{\to_\sigma \mid \sigma \in \Sigma\}, \rho)$ over $(\mathbb{P}, \Sigma)$, we define via mutual induction for each PDL$^\cap$ program π a binary relation $[\![\pi]\!]_\mathcal{K} \subseteq S \times S$ and for each PDL$^\cap$ formula φ a subset $[\![\varphi]\!]_\mathcal{K} \subseteq S$ as follows, where $\sigma \in \Sigma$, $p \in \mathbb{P}$, and $\mathrm{op} \in \{\cup, \cap, \circ\}$:

$$\begin{aligned}
[\![\sigma]\!]_\mathcal{K} &= \to_\sigma & [\![p]\!]_\mathcal{K} &= \{s \mid p \in \rho(s)\} \\
[\![\varphi?]\!]_\mathcal{K} &= \{(s,s) \mid s \in [\![\varphi]\!]_\mathcal{K}\} & [\![\neg\varphi]\!]_\mathcal{K} &= S \setminus [\![\varphi]\!]_\mathcal{K} \\
[\![\pi^*]\!]_\mathcal{K} &= [\![\pi]\!]_\mathcal{K}^* & [\![\varphi_1 \vee \varphi_2]\!]_\mathcal{K} &= [\![\varphi_1]\!]_\mathcal{K} \cup [\![\varphi_2]\!]_\mathcal{K} \\
[\![\pi_1 \ \mathrm{op}\ \pi_2]\!]_\mathcal{K} &= [\![\pi_1]\!]_\mathcal{K} \ \mathrm{op}\ [\![\pi_2]\!]_\mathcal{K} & [\![\langle\pi\rangle\varphi]\!]_\mathcal{K} &= \{s \mid \exists t : (s,t) \in [\![\pi]\!]_\mathcal{K} \wedge t \in [\![\varphi]\!]_\mathcal{K}\}
\end{aligned}$$

Note that $[\![\langle\varphi?\rangle\psi]\!]_\mathcal{K} = [\![\varphi \wedge \psi]\!]_\mathcal{K}$. For $s \in S$ we write $(\mathcal{K}, s) \models \varphi$ if and only if $s \in [\![\varphi]\!]_\mathcal{K}$. If the Kripke structure $\mathcal{K}$ is clear from the context we write $[\![\varphi]\!]$ for $[\![\varphi]\!]_\mathcal{K}$. PDL is the fragment of PDL$^\cap$, where the intersection operator $\cap$ on programs is not allowed. *Test-free PDL* (resp. *test-free* PDL$^\cap$) is the fragment of PDL (resp. PDL$^\cap$), where the test-operator "?" is not allowed. The size $|\varphi|$ of a PDL$^\cap$ formula φ and the size $|\pi|$ of a PDL$^\cap$ program π is defined as follows: $|p| = |\sigma| = 1$ for all $p \in \mathbb{P}$ and $\sigma \in \Sigma$, $|\neg\varphi| = |\varphi?| = |\varphi|+1$, $|\varphi \vee \psi| = |\varphi|+|\psi|+1$, $|\langle\pi\rangle\varphi| = |\pi|+|\varphi|$, $|\pi_1 \ \mathrm{op}\ \pi_2| = |\pi_1|+|\pi_2|+1$ for $\mathrm{op} \in \{\cup, \cap, \circ\}$, and $|\pi^*| = |\pi|+1$. The fragment EF of CTL (where only the next and "exists finally" modality is allowed) can be defined as the fragment of test-free PDL, consisting of all formulas φ such that every for every subformula of the form $\langle\pi\rangle\psi$, either $\pi \in \Sigma$ or $\pi = \Sigma^*$.

A PDL program π (where the intersection operator is not allowed) can be viewed as a regular expression and translated into a finite automaton A, where transitions are labeled by symbols from Σ and test formulas $\varphi?$. The semantic $[\![A]\!]$ of this automaton is the union of all relations $[\![c_1]\!] \circ \cdots \circ [\![c_n]\!]$, where $c_1 \cdots c_n$ labels a path from the initial state of A to a final state; note that c_i can be of the form $\varphi?$. This PDL-variant is sometimes called APDL. For PDL$^\cap$ such a translation does not exist. Moreover, PDL$^\cap$ neither possesses the finite model property nor the tree model property in contrast to PDL [13].

Given a class $\mathcal{C}$ of Kripke structures and a logic $\mathcal{L}$ (e.g. PDL or PDL$^\cap$), the *model-checking problem* asks: Given a Kripke structure $\mathcal{K} \in \mathcal{C}$, a node s of $\mathcal{K}$, and a formula $\varphi \in \mathcal{L}$, does $(\mathcal{K}, s) \models \varphi$ hold. Following Vardi [26], we distinguish between three measures of complexity:

- *Data Complexity*: The complexity of verifying for a fixed formula $\varphi \in \mathcal{L}$, whether $(\mathcal{K}, s) \models \varphi$ for a given Kripke structure $\mathcal{K} \in \mathcal{C}$ and a node s of $\mathcal{K}$.
- *Expression Complexity*: The complexity of verifying for a fixed Kripke structure $\mathcal{K} \in \mathcal{C}$ and node s, whether $(\mathcal{K}, s) \models \varphi$ for a given formula $\varphi \in \mathcal{L}$.
- *Combined Complexity*: The complexity of verifying $(\mathcal{K}, s) \models \varphi$ for a given formula $\varphi \in \mathcal{L}$, a given Kripke structure $\mathcal{K} \in \mathcal{C}$, and a node s.

Convention. In the rest of this paper, we will consider PDL$^\cap$ without atomic propositions. A Kripke structure will be just a tuple $\mathcal{K} = (S, \{\to_\sigma \mid \sigma \in \Sigma\})$ where $\to_\sigma \subseteq S \times S$. Formally, we introduce the only atomic proposition true and define

$[\![\text{true}]\!]_{\mathcal{K}} = S$. This is not a restriction, since a Kripke structure $(S, \{\rightarrow_\sigma \mid \sigma \in \Sigma\}, \rho)$ (where $\rho : S \to 2^{\mathbb{P}}$, $\Sigma \cap \mathbb{P} = \emptyset$) can be replaced by the new Kripke structure $(S, \{\rightarrow_\sigma \mid \sigma \in \Sigma \cup \mathbb{P}\})$ where $\rightarrow_p = \{(s,s) \mid p \in \rho(s)\}$ for all $p \in \mathbb{P}$. For the formalisms for specifying infinite Kripke structures that we will introduce in the next section, we will see that (a finite description of) this propositionless Kripke structure can be easily computed from (a finite description of) the original Kripke structure. Moreover, in PDL$^\cap$ formulas, we have to replace every occurrence of an atomic proposition p by the formula $\langle p \rangle$true.

4 Infinite State Systems

In this section, we consider several formalisms for describing infinite Kripke structures. Let Σ be be a set of atomic programs and Γ be a finite alphabet.

A *basic parallel process* (BPP) is a communication free Petri net, i.e., a Petri net, where every transition needs exactly one token for firing. By labeling transitions of a Petri net with labels from Σ, one can associate an infinite Kripke structure $\mathcal{K}(\mathcal{N})$ with a BPP $\mathcal{N}$, see [21] for more details.

A *basic process algebra* (BPA) over Σ is a tuple $\mathcal{X} = (\Gamma, \Delta)$ where $\Delta \subseteq \Gamma_\varepsilon \times \Sigma \times \Gamma^*$ is a finite *transition relation*. The BPA $\mathcal{X}$ describes the Kripke structure $\mathcal{K}(\mathcal{X}) = (\Gamma^*, \{\rightarrow_\sigma \mid \sigma \in \Sigma\})$ over Σ, where $\rightarrow_\sigma = \{(\gamma w, vw) \mid w \in \Gamma^* \text{ and } (\gamma, \sigma, v) \in \Delta\}$ for all $\sigma \in \Sigma$. The *size* $|\mathcal{X}|$ of $\mathcal{X}$ is $|\Gamma| + |\Sigma| + \sum_{(\gamma,\sigma,v)\in\Delta} |v|$. If $(\gamma, \sigma, v) \in \Delta$, we also write $\gamma \xrightarrow{\sigma}_{\mathcal{X}} v$.

Example 1. For a finite alphabet Γ we will use the BPA $\text{Tree}_\Gamma = (\Gamma, \Delta)$ over $\Gamma \cup \overline{\Gamma}$ where $\Delta = \{(\varepsilon, a, a) \mid a \in \Gamma\} \cup \{(a, \overline{a}, \varepsilon) \mid a \in \Gamma\}$. Then $\mathcal{K}(\text{Tree}_\Gamma)$ is the complete tree over Γ with backwards edges.

A *pushdown system* (PDS) over Σ is a tuple $\mathcal{Y} = (\Gamma, P, \Delta)$ where (i) P is a finite set of *control states*, and (ii) $\Delta \subseteq P \times \Gamma_\varepsilon \times \Sigma \times P \times \Gamma^*$ is a finite *transition relation*. The PDS $\mathcal{Y}$ describes the Kripke structure $\mathcal{K}(\mathcal{Y}) = (P\Gamma^*, \{\rightarrow_\sigma \mid \sigma \in \Sigma\})$ over Σ, where $\rightarrow_\sigma = \{(p\gamma w, qvw) \mid w \in \Gamma^* \text{ and } (p, \gamma, \sigma, q, v) \in \Delta\}$ for all $\sigma \in \Sigma$. The *size* $|\mathcal{Y}|$ of $\mathcal{Y}$ is $|\Gamma| + |P| + |\Sigma| + \sum_{(p,\gamma,\sigma,q,v)\in\Delta} |v|$. If $(p, \gamma, \sigma, q, v) \in \Delta$, we also write $p\gamma \xrightarrow{\sigma}_{\mathcal{Y}} qv$. Note that a BPA is just a stateless PDS.

Example 2. Let $\mathcal{K} = (S, \{\rightarrow_\sigma \mid \sigma \in \Sigma\})$ be a deterministic Kripke structure, i.e., for every state $s \in S$ and every $\sigma \in \Sigma$ there is at most one $t \in S$ with $s \rightarrow_\sigma t$. For PDL over BPA and PDS, determinism is no restriction: it can be ensured by choosing a possibly larger set Σ' of atomic programs such that every transition of the BPA (PDS) can be labeled with a unique $\sigma' \in \Sigma'$. Every original atomic program σ can be recovered as a union of some of these new atomic programs (for PRS, this doesn't work). We now want to express that the current state $s \in S$ is a recovery state of the system in the sense that wherever we go from s, we can always move back to s. This property cannot be expressed in the modal μ-calculus unless the state s is somehow uniquely marked, e.g., by a special atomic proposition (but here, we want to define the set of all recovery states). One can show that s is a recovery state if and only if

$$(\mathcal{K}, s) \models [\Sigma^*] \bigwedge_{\sigma \in \Sigma} \Big(\langle \sigma \rangle \text{true} \Rightarrow \langle \text{true?} \cap \sigma \circ \Sigma^* \rangle \text{true} \Big).$$

Note that true? defines the identity relation on S.

Example 3. Let us consider two PDS $\mathcal{Y}_i = (\Gamma, P_i, \Delta_i)$ (with a common pushdown alphabet Γ) over Σ_i ($i \in \{1, 2\}$), where $\Sigma_1 \cap \Sigma_2 = \emptyset$, and such that $\mathcal{K}(\mathcal{Y}_1)$ and $\mathcal{K}(\mathcal{Y}_2)$ are deterministic (which, by the remarks from Example 2, is not a restriction). The systems $\mathcal{Y}_1$ and $\mathcal{Y}_2$ may synchronize over states from the intersection $P_1 \cap P_2$. These two systems can be modeled by the single PDS $\mathcal{Y} = (\Gamma, P_1 \cup P_2, \Delta_1 \cup \Delta_2)$ over $\Sigma_1 \cup \Sigma_2$. In this context, it might be interesting to express that whenever $\mathcal{Y}_1$ and $\mathcal{Y}_2$ can reach a common node s, and from s, $\mathcal{Y}_i$ can reach a node s_i by a local action, then the two systems can reach from s_1 and s_2 again a common node. This property can be expressed by the PDL$^\cap$ formula

$$[\Sigma_1^* \cap \Sigma_2^*] \bigwedge_{a \in \Sigma_1, b \in \Sigma_2} \Big(\langle a \rangle \text{true} \wedge \langle b \rangle \text{true} \Rightarrow \langle a \circ \Sigma_1^* \cap b \circ \Sigma_2^* \rangle \text{true} \Big).$$

Note that $[\![a \circ \Sigma_1^* \cap b \circ \Sigma_2^*]\!]$ is in general not the empty relation, although of course $a \circ \Sigma_1^* \cap b \circ \Sigma_2^* = \emptyset$ when interpreted as a regular expression with intersection.

A relation $U \subseteq \Gamma^* \times \Gamma^*$ is *prefix-recognizable* over Γ, if $U = \bigcup_{i=1}^n R_i$ ($n \geq 1$) and $R_i = \{(uw, vw) \mid u \in U_i, v \in V_i, w \in W_i\}$ for some regular languages $U_i, V_i, W_i \subseteq \Gamma^*$ ($1 \leq i \leq n$). We briefly write $R_i = (U_i \times V_i)W_i$. A *prefix-recognizable system* (PRS) (which should not be confused with Mayr's PRS (process rewrite systems) [21]) over Σ is a pair $\mathcal{Z} = (\Gamma, \alpha)$ where α assigns to every atomic program $\sigma \in \Sigma$ a prefix-recognizable relation $\alpha(\sigma)$ over Γ, which is given by finite automata $\mathcal{A}_1^\sigma, \mathcal{B}_1^\sigma, \mathcal{C}_1^\sigma, \ldots, \mathcal{A}_{n_\sigma}^\sigma, \mathcal{B}_{n_\sigma}^\sigma, \mathcal{C}_{n_\sigma}^\sigma$ such that $\alpha(\sigma) = \bigcup_{i=1}^{n_\sigma} (L(\mathcal{A}_i^\sigma) \times L(\mathcal{B}_i^\sigma)) L(\mathcal{C}_i^\sigma)$. The PRS $\mathcal{Z}$ describes the Kripke structure $\mathcal{K}(\mathcal{Z}) = (\Gamma^*, \{\alpha(\sigma) \mid \sigma \in \Sigma\})$ over Σ. The *size* $|\mathcal{Z}|$ of $\mathcal{Z}$ is $|\Gamma| + |\Sigma| + \sum_{\sigma \in \Sigma} \sum_{i=1}^{n_\sigma} |\mathcal{A}_i^\sigma| + |\mathcal{B}_i^\sigma| + |\mathcal{C}_i^\sigma|$, where $|A|$ is the number of states of a finite automaton A.

Our definition of BPA (resp. PDS) allows transitions of the form $\varepsilon \xrightarrow{\sigma}_{\mathcal{X}} v$ (resp. $p \xrightarrow{\sigma}_{\mathcal{Y}} qv$ for control states p and q). It is easy to see that our definition describes exactly the same class of BPA (resp. PDS) as defined in [20,21] (resp. [2,21,29,28]), and there are logspace translations between the two formalisms.

Usually, in the literature a PDS $\mathcal{Y}$ describes a Kripke structure with atomic propositions from some set $\mathbb{P}$. For this purpose, $\mathcal{Y}$ contains a mapping $\varrho : P \to 2^{\mathbb{P}}$, where P is the set of control states of $\mathcal{Y}$, and one associates with the atomic proposition $\eta \in \mathbb{P}$ the set of all configurations where the current control state p satisfies $\eta \in \varrho(p)$. In our formalism, which does not contain atomic propositions, we can simulate such an atomic propositions η by introducing the new transition rule $p \xrightarrow{\eta} p$ whenever $\eta \in \varrho(p)$, see also the convention from the end of Section 3. Similar remarks apply to BPA and PRS.

Table 1 summarizes our complexity results for PDL and its variants.

Table 1.

		BPA	PDS	PRS
EF PDL\?	data	P-complete		EXP-complete
	expression		PSPACE-complete	
	combined			EXP-complete
PDL	data	P-complete		
	expression		EXP-complete	
	combined			
$\text{PDL}^\cap$ $\text{PDL}^\cap\backslash ?$	data	PSPACE-hard, in EXP	EXP-complete	
	expression		2EXP-complete	
	combined			

5 Model-Checking PDL over Infinite State Systems

It was shown in [8] that the model-checking problem of EF over (the Kripke structures defined by) Petri nets is undecidable. A reduction of this problem to the model-checking problem of test-free PDL over BPP shows:

Proposition 1. *The model-checking problem for test-free PDL over BPP is undecidable.*

Hence, in the following we will concentrate on the (sequential) system classes BPA, PDS, and PRS. Our results for (test-free) PDL without intersection over BPA/PDS/PRS mainly use results or adapt techniques from [2,22,28,29], see Table 1. It turns out that PDL without test behaves in exactly the same way as EF, and that adding the test operator leads in most cases to a complexity jump up to EXP-completeness.

In the rest of the paper, we concentrate on $\text{PDL}^\cap$, for which we prove that the expression and combined complexity over BPA/PDS/PRS is complete for 2EXP. Our lower bound proof uses a technique from [28] for describing a traversal of the computation tree of an alternating Turing machine in CTL using a pushdown. The main difficulty that remains is to formalize in $\text{PDL}^\cap$ that two configurations of an exponential space alternating Turing machine (these machines characterize 2EXP) are successor configurations. For doing this, we adjoin to every tape cell a binary counter, which represents the position of the tape cell. This encoding of configurations is also used in the recent 2EXP lower bound proof of Lange and Lutz for satisfiability of $\text{PDL}^\cap$ [17].

Theorem 1. *There exists a fixed BPA $\mathcal{X}$ such that the following problem is* 2EXP*-hard:*
INPUT: A test-free $\text{PDL}^\cap$-formula φ.
QUESTION: $(\mathcal{K}(\mathcal{X}), \varepsilon) \models \varphi$?

Proof. Since 2EXP equals the class of all languages that can be accepted by an ATM in exponential space [6], we can choose a fixed $2^{p(m)} - 1$ space bounded ATM $\mathcal{M} = (Q, \Sigma_{\mathcal{M}}, \Gamma_{\mathcal{M}}, q_0, \delta, \Box)$ (where $p(m)$ is a polynomial) with a 2EXP-complete acceptance problem. The machine $\mathcal{M}$ satisfies the conventions of Section 2. Let $w \in \Sigma_{\mathcal{M}}^*$ be an input of length n. We construct a fixed BPA $\mathcal{X} = \mathcal{X}(\mathcal{M}) = (\Gamma, \Delta)$ and a test-free PDL$^\cap$-formula $\varphi = \varphi(w, \mathcal{M})$ each over $\Sigma = \Sigma(\mathcal{M})$ such that $w \in L(\mathcal{M})$ if and only if $(\mathcal{K}(\mathcal{X}), \varepsilon) \models \varphi$. Let $N = p(n)$ and $\Omega = Q \cup \Gamma_{\mathcal{M}}$. A configuration c of $\mathcal{M}$ is a word from the language $\bigcup_{0 \leq i \leq 2^N - 2} \Gamma_{\mathcal{M}}^i Q \Gamma_{\mathcal{M}}^{2^N - 1 - i}$. We will represent $c = \gamma_0 \cdots \gamma_{i-1} q \gamma_{i+1} \cdots \gamma_{2^N - 1}$ by the word

$$\gamma_0[0] \cdots \gamma_{i-1}[i-1] q[i] \gamma_{i+1}[i+1] \cdots \gamma_{2^N - 1}[2^N - 1], \tag{1}$$

where $[k]$ denotes the binary representation of k ($0 \leq k \leq 2^N - 1$) with N bits, i.e., $[k] = \beta_0 \cdots \beta_{N-1}$ with $\beta_j \in \{0, 1\}$ and $k = \sum_{j=0}^{N-1} 2^j \cdot \beta_j$. A *cell* is a string $\omega[i]$, where $\omega \in \Omega$ and $0 \leq i \leq 2^N - 1$. Let Moves $= Q \times \Gamma_{\mathcal{M}} \times \{\leftarrow, \rightarrow\}$ be the set of moves of $\mathcal{M}$ and let

$$\begin{aligned} \mathrm{Dir} = \{&L(\mu_1, \mu_2), R(\mu_1, \mu_2) \mid (\mu_1, \mu_2) \in \delta(Q_\forall, \Gamma)\} \cup \\ \{&E(\mu_1), E(\mu_2) \mid (\mu_1, \mu_2) \in \delta(Q_\exists, \Gamma)\} \end{aligned}$$

be the set of *direction markers*. These symbols separate consecutive configurations of the form (1) on the pushdown. As in [28], direction markers are used in order to organize a depth-first left-to-right traversal of the computation tree of the ATM $\mathcal{M}$ on the pushdown. Let $\Gamma = \Omega \cup \{0, 1\} \cup \mathrm{Dir}$ and $\Sigma = \Gamma \cup \overline{\Gamma} \cup \{\lambda\}$, which is a fixed alphabet. We define the fixed BPA $\mathcal{X}$ to be Tree_Γ (see Example 1) together with the rule $(\varepsilon, \lambda, \varepsilon)$, which generates a λ-labeled loop at every node. In order to define the PDL$^\cap$ formula φ, we need several auxiliary programs:

- $\overline{X} = \bigcup_{x \in X} \overline{x}$ for $X \subseteq \Gamma$: Pops a single symbol $x \in X$ from the pushdown.
- $\overline{\mathrm{pop}_i} = \overline{\{0,1\}}^i$ for all $0 \leq i \leq N$: Pops i bits from the pushdown.
- $\overline{\mathrm{cell}} = \overline{\Omega} \circ \mathrm{pop}_N$: Pops a cell $\omega[i]$ from the pushdown.
- $\overline{\mathrm{cell}_0} = \overline{\Omega} \circ \overline{0}^N$: Pops a cell $\omega[0]$ from the pushdown.
- $\overline{\mathrm{cell}_1} = \overline{\Gamma_{\mathcal{M}}} \circ \overline{1}^N$: Pops a cell $\gamma[2^N - 1]$ for $\gamma \in \Gamma_{\mathcal{M}}$ from the pushdown.

Next, we define a program $\overline{\mathrm{inc}}$, which is executable only if on top of the pushdown there is a word of the form $\omega[i]\omega'[i+1]$ for some $\omega, \omega' \in \Omega$ and some $0 \leq i < 2^N - 1$. The program $\overline{\mathrm{inc}}$ pops $\omega[i]$ during it execution. In order to define $\overline{\mathrm{inc}}$ we will use the programs $\chi_{j,\beta}$ ($0 \leq j < N$, $\beta \in \{0, 1\}$) which assure that, after popping j bits of the current cell, a bit β can be popped that matches the bit that can be popped after popping another j bits of the subsequent cell. Afterwards, further bits may be popped:

$$\chi_{j,\beta} = \overline{\text{pop}_j} \circ \overline{\beta} \circ \overline{\{0,1\}}^* \circ \overline{\Omega} \circ \overline{\text{pop}_j} \circ \overline{\beta} \circ \overline{\{0,1\}}^*$$

$$\overline{\text{inc}} = \overline{\text{cell}} \cap \Bigg[(\overline{\text{cell}} \circ \overline{\text{cell}}) \cap \overline{\Omega} \circ \bigcup_{i=0}^{N-1} \Big(\overline{1}^i \circ \overline{0} \circ \overline{\{0,1\}}^* \circ \overline{\text{cell}} \cap$$

$$\overline{\{0,1\}}^* \circ \overline{\Omega} \circ \overline{0}^i \circ \overline{1} \circ \overline{\{0,1\}}^* \cap \bigcap_{j=i+1}^{N-1} (\chi_{j,0} \cup \chi_{j,1})\Big)\Bigg] \circ \Gamma^*$$

The next program $\overline{\text{conf}}$ is only executable if the top of the pushdown is a legal configuration in the sense of (1), i.e.: A word of the form $\omega_0[0]\omega_1[1]\cdots\omega_{2^N-1}[2^N-1]$ is assumed to be on top of the pushdown, for exactly one $0 \leq i \leq 2^N - 2$ we have $\omega_i \in Q$, and for all other i we have $\omega_i \in \Gamma_{\mathcal{M}}$. This top configuration is being popped during execution:

$$\overline{\text{conf}} = (\overline{\text{cell}_0} \circ \overline{\text{cell}}^*) \cap (\overline{\text{inc}}^* \circ \overline{\text{cell}_1}) \cap (\overline{\Gamma_{\mathcal{M}} \cup \{0,1\}}^* \circ \overline{Q} \circ \overline{\Gamma_{\mathcal{M}} \cup \{0,1\}}^*)$$

For all $\omega, \omega' \in \Omega$ the program $\pi_{\omega,\omega'}$ is only executable if the top of the pushdown is a certain suffix of a configuration of $\mathcal{M}$ followed by a direction marker $d \in \text{Dir}$ and a complete configuration of $\mathcal{M}$. More precisely,

$$\omega_k[k]\cdots\omega_{2^N-1}[2^N-1]\, d\, \omega'_0[0]\cdots\omega'_{2^N-1}[2^N-1]$$

with $\omega_k = \omega$ and $\omega'_k = \omega'$ must be on top of the pushdown. During its execution, $\pi_{\omega,\omega'}$ pops $\omega_k[k]$ from the pushdown:

$$\pi_{\omega,\omega'} = \overline{\text{cell}} \cap \Bigg(\bigcap_{i=0}^{N-1} \bigcup_{\beta\in\{0,1\}} \overline{\omega} \circ \overline{\text{pop}_i} \circ \overline{\beta} \circ \overline{\{0,1\}}^* \circ$$

$$\overline{\text{cell}}^* \circ \overline{\text{Dir}} \circ \overline{\text{cell}}^* \circ \overline{\omega'} \circ \overline{\text{pop}_i} \circ \overline{\beta} \circ \overline{\{0,1\}}^*\Bigg) \circ \Gamma^*$$

The program $\pi_= = \bigcup_{\omega\in\Omega} \pi_{\omega,\omega}$ checks whether the content of the top cell $\omega[k]$ equals the content of the k-th cell of the subsequent configuration. Now we define a program check_μ for $\mu \in \text{Moves}$, which is only executable if cdc' is on top of the pushdown, where: (i) c and c' are configurations of $\mathcal{M}$ in the sense of (1), (ii) $d \in \text{Dir}$, and (iii) $\mathcal{M}$ moves from configuration c' to configuration c by the move μ. We restrict ourselves to the case where $\mu = (q, a, \leftarrow)$:

$$\lambda \cap \Big(\overline{\text{conf}} \circ \overline{\text{Dir}} \circ \overline{\text{conf}} \cap \pi_=^* \circ \bigcup_{p\in Q, b,c\in\Gamma_{\mathcal{M}}} (\pi_{q,c} \circ \pi_{c,p} \circ \pi_{a,b}) \circ \pi_=^* \circ \overline{\text{Dir}} \circ \overline{\text{conf}}\Big) \circ \Gamma^*$$

The rest of the proof is analogous to Walukiewicz's proof for the EXP lower bound for CTL over PDS [28]. Using the direction markers, we can define a program traverse, whose execution simulates on the pushdown a single step in a depth-first left-to-right traversal of the computation tree of $\mathcal{M}$. Using a program init, which pushes the initial configuration on the pushdown, we get $(\varepsilon, \varepsilon) \in [\![\text{init} \circ \mathsf{traverse}^*]\!]$ if and only if the initial configuration $q_0 w$ is accepting. □

In the rest of this paper, we sketch a 2EXP upper bound for the combined complexity of $\mathrm{PDL}^\cap$ over PRS. For this, we need the concept of two-way alternating tree automata. Vardi and Kupferman [15] reduced the model-checking problem of the modal μ-calculus over PRS to the emptiness problem for two-way alternating tree automata, and thereby deduced an EXP upper bound for the former problem, see also [4,29]. Our general strategy for model-checking $\mathrm{PDL}^\cap$ over PRS is the same.

Let Γ be a finite alphabet. A *Γ-tree* is a suffix-closed subset $T \subseteq \Gamma^*$, i.e., if $aw \in T$ for $w \in \Gamma^*$ and $a \in \Gamma$, then $w \in T$. Elements of T are called *nodes*. An *infinite path* in the tree T is an infinite sequence $u_1, u_2, \ldots$ of nodes such that $u_1 = \varepsilon$ and for all $i \geq 1$, $u_{i+1} = a_i u_i$ for some $a_i \in \Gamma$. A *Σ-labeled Γ-tree*, where Σ is a finite alphabet, is a pair (T, λ), where T is a Γ-tree and $\lambda : T \to \Sigma$ is a labeling function. The *complete Γ-tree* is the $\Gamma \uplus \{\bot\}$-labeled Γ-tree $(\Gamma^*, \lambda_\Gamma)$ where $\lambda_\Gamma(\varepsilon) = \bot$ and $\lambda_\Gamma(aw) = a$ for $a \in \Gamma$ and $w \in \Gamma^*$. For a finite set X, let $\mathcal{B}^+(X)$ be the set of all *positive boolean formulas over X*; note that true and false are positive boolean formulas. A subset $Y \subseteq X$ *satisfies* $\theta \in \mathcal{B}^+(X)$, if θ becomes true when assigning true to all elements in Y. Let $\mathrm{ext}(\Gamma) = \Gamma \uplus \{\varepsilon, \downarrow\}$ and define for all $u \in \Gamma^*, a \in \Gamma$: $\varepsilon u = u$, $\downarrow au = u$, whereas $\downarrow\varepsilon$ is undefined.

A *two-way alternating tree automaton* (TWATA) over Γ is a triple $\mathcal{T} = (S, \delta, \mathrm{Acc})$, where S is a finite *set of states*, $\delta : S \times (\Gamma \cup \{\bot\}) \to \mathcal{B}^+(S \times \mathrm{ext}(\Gamma))$ is the *transition function*, and $\mathrm{Acc} : S \to \{0, \ldots, m\}$ is the *priority function*, where $m \in \mathbb{N}$. Let $u \in \Gamma^*$ and $s \in S$. An *(s,u)-run* of $\mathcal{T}$ (over the complete Γ-tree $(\Gamma^*, \lambda_\Gamma)$) is a $(S \times \Gamma^*)$-labeled Ω-tree $R = (T_R, \lambda_R)$ for some finite set Ω such that: (i) $\varepsilon \in T_R$, (ii) $\lambda_R(\varepsilon) = (s, u)$, and (iii) if $\alpha \in T_R$ with $\lambda_R(\alpha) = (s', v)$ and $\delta(s', \lambda_\Gamma(v)) = \theta$, then there is a subset $Y \subseteq S \times \mathrm{ext}(\Gamma)$ that satisfies the formula θ and for all $(s'', e) \in Y$ there exists $\omega \in \Omega$ with $\omega\alpha \in T_R$ and $\lambda_R(\omega\alpha) = (s'', ev)$. An (s,u)-run $R = (T_R, \lambda_R)$ of $\mathcal{T}$ is *successful* if for every infinite path $w_1, w_2, \ldots$ of T_R, $\min(\{\mathrm{Acc}(s') \mid \lambda_R(w_i) \in \{s'\} \times \Gamma^*$ for infinitely many $i\})$ is even. Let $[\![\mathcal{T}, s]\!] = \{u \in \Gamma^* \mid$ there is a successful (s,u)-run of $\mathcal{T}\}$. The *size* $|\mathcal{T}|$ of $\mathcal{T}$ is $|\Gamma| + |S| + \sum_{\theta \in \mathrm{ran}(\delta)} |\theta| + |\mathrm{Acc}|$, where $|\mathrm{Acc}| := \max\{\mathrm{Acc}(s) \mid s \in S\}$. For TWATAs $\mathcal{T}_i = (S_i, \delta_i, \mathrm{Acc}_i)$ $(i \in \{1,2\})$ over Γ let $\mathcal{T}_1 \uplus \mathcal{T}_2 = (S_1 \uplus S_2, \delta_1 \uplus \delta_2, \mathrm{Acc}_1 \uplus \mathrm{Acc}_2)$ be their *disjoint union*. Note that in our definition a TWATA over an alphabet Γ only runs on the complete Γ-tree. Hence, our definition is a special case of the definition in [14,15,27], where also runs of TWATAs on arbitrarily labeled trees are considered. Using [27], we obtain:

Theorem 2 ([27]). *For a given TWATA $\mathcal{T} = (S, \delta, \mathrm{Acc})$ and a state $s \in S$, it can be checked in time exponential in $|S| \cdot |\mathrm{Acc}|$ whether $\varepsilon \in [\![\mathcal{T}, s]\!]$.*

It should be noted that the size of a positive boolean formula that appears in the transition function δ of a TWATA $\mathcal{T} = (Q, \delta, \mathrm{Acc})$ can be exponential in $|Q|$, but the size of δ only appears polynomially in the upper bound for emptiness (and not exponentially, which would lead to a 2EXP upper bound for emptiness).

Let $\mathcal{T} = (S, \delta, \mathrm{Acc})$ be a TWATA over Γ. A *nondeterministic finite automaton* (briefly NFA) A over $\mathcal{T}$ is a pair $(Q, \to_A)$ where Q is a finite state set and all transitions are of the form $p \xrightarrow{a}_A q$ for $p, q \in Q, a \in \Gamma \cup \overline{\Gamma}$ or $p \xrightarrow{\mathcal{T}, s}_A q$ for

$p, q \in Q, s \in S$. The latter transitions are called *test-transitions*. Let $A^{\downarrow}$ (resp. $A^{\uparrow}$) be the NFA over $\mathcal{T}$ that results from A by removing all transitions with a label from Γ (resp. $\overline{\Gamma}$), i.e., we only keep test-transitions and transitions with a label from $\overline{\Gamma}$ (resp. Γ). Let $\Rightarrow_A \subseteq (\Gamma^* \times Q) \times (\Gamma^* \times Q)$ be the smallest relation with:

- $(u, p) \Rightarrow_A (au, q)$ whenever $u \in \Gamma^*$ and $p \xrightarrow{a}_A q$
- $(au, p) \Rightarrow_A (u, q)$ whenever $u \in \Gamma^*$ and $p \xrightarrow{\bar{a}}_A q$
- $(u, p) \Rightarrow_A (u, q)$ whenever $u \in [\![\mathcal{T}, s]\!]$ and $p \xrightarrow{\mathcal{T}, s}_A q$

Let $[\![A, p, q]\!] = \{(u, v) \in \Gamma^* \times \Gamma^* \mid (u, p) \Rightarrow_A^* (v, q)\}$ for $p, q \in Q$.

We will inductively transform a given PDL$^{\cap}$-formula (resp. PDL$^{\cap}$-program) into an equivalent TWATA (resp. NFA over a TWATA). In order to handle the intersection operator on programs, we first have to describe a general automata theoretic construction: Let $\mathcal{T} = (S, \delta, \mathrm{Acc})$ be a TWATA over Γ and let $A = (Q, \rightarrow_A)$ be an NFA over $\mathcal{T}$. Let $\mathrm{hop}_A \subseteq \Gamma^* \times Q \times Q$ be the smallest set such that:

- for all $u \in \Gamma^*$ and $q \in Q$ we have $(u, q, q) \in \mathrm{hop}_A$
- if $(au, p', q') \in \mathrm{hop}_A$, $p \xrightarrow{a}_A p'$, and $q' \xrightarrow{\bar{a}}_A q$, then $(u, p, q) \in \mathrm{hop}_A$
- if $(u, p, r), (u, r, q) \in \mathrm{hop}_A$, then $(u, p, q) \in \mathrm{hop}_A$
- if $u \in [\![\mathcal{T}, s]\!]$, $p \xrightarrow{\mathcal{T}, s}_A q$, then $(u, p, q) \in \mathrm{hop}_A$

Intuitively, $(u, p, q) \in \mathrm{hop}_A$ if and only if we can walk from node u of the complete Γ-tree back to u along a path consisting of nodes from $\Gamma^* u$. At the beginning of this walk, the automaton A is initialized in state p, each time we move in the tree from v to av (resp. av to u) we read a (resp. $\bar{a}$) in A, and A ends in state q. Formally, we have:

Lemma 1. *We have $(u, p, q) \in \mathrm{hop}_A$ if and only if there exist $n \geq 1$, $u_1, \ldots, u_n \in \Gamma^* u$, and $q_1, \ldots, q_n \in Q$ such that $u_1 = u_n = u$, $q_1 = p$, $q_n = q$, and $(u_1, q_1) \Rightarrow_A (u_2, q_2) \cdots \Rightarrow_A (u_n, q_n)$.*

The inductive definition of the set hop_A can be translated into a TWATA:

Lemma 2. *There exists a TWATA $\mathcal{U} = (S', \delta', \mathrm{Acc}')$ with state set $S' = S \uplus (Q \times Q)$ such that (i) $[\![\mathcal{U}, s]\!] = [\![\mathcal{T}, s]\!]$ for $s \in S$, (ii) $[\![\mathcal{U}, (p, q)]\!] = \{u \in \Gamma^* \mid (u, p, q) \in \mathrm{hop}_A\}$ for $(p, q) \in Q \times Q$, and (iii) $|\mathrm{Acc}'| = |\mathrm{Acc}|$.*

Define a new NFA $B = (Q, \rightarrow_B)$ over the TWATA $\mathcal{U}$ by adding to A for every pair $(p, q) \in Q \times Q$ the test-transition $p \xrightarrow{\mathcal{U}, (p,q)} q$.

Lemma 3. *Let $u, v \in \Gamma^*$ and $p, q \in Q$. Then the following statements are equivalent:*

- $(u, v) \in [\![A, p, q]\!]$
- $(u, v) \in [\![B, p, q]\!]$

– *there exist a common suffix w of u and v and a state $r \in Q$ with $(u, w) \in [\![B^{\downarrow}, p, r]\!]$ and $(w, v) \in [\![B^{\uparrow}, r, q]\!]$*

Let $\mathrm{drop}_B \subseteq \Gamma^* \times Q \times Q$ be the smallest set such that:

– for all $u \in \Gamma^*$ and $p \in Q$ we have $(u, p, p) \in \mathrm{drop}_B$
– if $(u, p', q') \in \mathrm{drop}_B$, $p \xrightarrow{\bar{a}}_B p'$, and $q' \xrightarrow{a}_B q$, then $(au, p, q) \in \mathrm{drop}_B$
– if $s' \in S'$, $u \in [\![\mathcal{U}, s']\!]$, $p \xrightarrow{\mathcal{U}, s'}_B r$, and $(u, r, q) \in \mathrm{drop}_B$, then $(u, p, q) \in \mathrm{drop}_B$
– if $s' \in S'$, $u \in [\![\mathcal{U}, s']\!]$, $r \xrightarrow{\mathcal{U}, s'}_B q$, and $(u, p, r) \in \mathrm{drop}_B$, then $(u, p, q) \in \mathrm{drop}_B$

Lemma 4. *We have $(u, p, q) \in \mathrm{drop}_B$ if and only if there exist $r \in Q$ and a suffix v of u such that $(u, v) \in [\![B^{\downarrow}, p, r]\!]$ and $(v, u) \in [\![B^{\uparrow}, r, q]\!]$.*

Again, the inductive definition of drop_B can be translated into a TWATA:

Lemma 5. *There exists a TWATA $\mathcal{V} = (S'', \delta'', \mathrm{Acc}'')$ with state set $S'' = S' \uplus (Q \times Q)$ such that: (i) $[\![\mathcal{V}, s']\!] = [\![\mathcal{U}, s']\!]$ for every state $s' \in S'$ of $\mathcal{U}$, (ii) $[\![\mathcal{V}, (p, q)]\!] = \{u \in \Gamma^* \mid (u, p, q) \in \mathrm{drop}_B\}$ for every state $(p, q) \in Q \times Q$, and (iii) $|\mathrm{Acc}''| = |\mathrm{Acc}|$.*

Let $C = (Q, \rightarrow_C)$ be the NFA over the TWATA $\mathcal{V}$ that results from B by adding for every pair $(p, q) \in Q \times Q$ the test-transition $p \xrightarrow{\mathcal{V}, (p,q)} q$. For $u, v \in \Gamma^*$ let $\inf(u, v)$ the longest common suffix of u and v.

Lemma 6. *Let $u, v \in \Gamma^*$ and $p, q \in Q$. Then the following statements are equivalent:*

– *$(u, v) \in [\![A, p, q]\!]$*
– *$(u, v) \in [\![C, p, q]\!]$*
– *there exists $r \in Q$ with $(u, \inf(u, v)) \in [\![C^{\downarrow}, p, r]\!]$ and $(\inf(u, v), v) \in [\![C^{\uparrow}, r, q]\!]$*

Now we are ready to prove the announced 2EXP upper bound for the combined complexity of $\mathrm{PDL}^{\cap}$ over PRS. Let $\mathcal{Z} = (\Gamma, \alpha)$ be a PRS and φ be a $\mathrm{PDL}^{\cap}$ formula each over Σ. We translate $\mathcal{Z}$ and φ into a TWATA $\mathcal{T} = (S, \delta, \mathrm{Acc})$ over Γ together with a state $s \in S$ such that $(\mathcal{K}(\mathcal{Z}), \varepsilon) \models \varphi$ if and only if $\varepsilon \in [\![\mathcal{T}, s]\!]$. The number of states of $\mathcal{T}$ will be exponentially in the size of the formula φ and polynomially in the size of $\mathcal{Z}$ and the size of the priority function Acc will be linear in the size of φ, which proves a 2EXP upper bound by Theorem 2. From now on any occurring TWATA is implicitly over Γ and the size of the priority function is at least 1. The construction of $\mathcal{T}$ is done inductively over the structure of the formula φ. More precisely, (i) for every subformula ψ of φ we construct a TWATA $\mathcal{T}(\psi)$ together with a state s of $\mathcal{T}(\psi)$ such that $[\![\psi]\!] = [\![\mathcal{T}(\psi), s]\!]$ and (ii) for every program π that occurs in φ we construct an NFA $A(\pi)$ over a TWATA $\mathcal{T}(\pi)$ such that $[\![\pi]\!] = [\![A(\pi), p, q]\!]$ for states p and q of $A(\pi)$.

The case $\psi = \mathrm{true}$ is clear, the case $\psi = \psi_1 \wedge \psi_2$ can be skipped since $\psi_1 \wedge \psi_2 \Leftrightarrow \langle \psi_1 ? \rangle \psi_2$. If $\psi = \neg\theta$, then we apply the standard complementation procedure [23], where all positive boolean formulas in the right-hand side of the transition function are dualized and the acceptance condition is complemented by incrementing the priority of every state. If $\psi = \langle\pi\rangle\theta$, then we have already constructed $A(\pi)$, $\mathcal{T}(\pi)$, and $\mathcal{T}(\theta)$ such that $[\![\pi]\!] = [\![A(\pi), p, q]\!]$ for two states p and q of $A(\pi)$ and $[\![\theta]\!] = [\![\mathcal{T}(\theta), s]\!]$ for a state s of $\mathcal{T}(\theta)$. Basically, the TWATA $\mathcal{T}(\psi)$ results from the disjoint union of $A(\pi)$ and $\mathcal{T}(\pi) \uplus \mathcal{T}(\theta)$, additionally $\mathcal{T}(\psi)$ can move from state q to state s. It remains to construct $A(\pi)$ and $\mathcal{T}(\pi)$ for a $\text{PDL}^{\cap}$ subprogram π of φ.

Case $\pi = \psi?$: We can assume that there exists a TWATA $\mathcal{T}(\psi)$ and a state r of $\mathcal{T}(\psi)$ such that $[\![\psi]\!] = [\![\mathcal{T}(\psi), r]\!]$. The TWATA $\mathcal{T}(\pi)$ is $\mathcal{T}(\psi)$. The automaton $A(\pi)$ has two states p and q with the only transition $p \xrightarrow{\mathcal{T},r} q$.

Case $\pi = \sigma \in \Sigma$: Assume that $\alpha(\sigma) = \bigcup_{i=1}^{n}(L(A'_i) \times L(B'_i))L(C'_i)$. Define the homomorphism $h : \Gamma^* \to \overline{\Gamma}^*$ by $h(a) = \overline{a}$ for all $a \in \Gamma$. From the representation of $\alpha(\sigma)$ we can construct finite automata A_i, B_i, C_i such that $L(A_i) = h(L(A'_i))$, $L(B_i) = L(B'_i)^{\text{rev}}$, and $L(C_i) = h(L(C'_i))$. Basically, the automaton $A(\pi)$ first chooses nondeterministically an $i \in \{1, \ldots, n\}$ and then simulates the automaton A_i, until a current tree node $u \in \Gamma^*$ belongs to the language $L(C_i)$. Then it continues by simulating the automaton B_i. Whether the current tree node $u \in \Gamma^*$ belongs to $L(C_i)$ has to be checked by the TWATA $\mathcal{T}(\pi)$, which can be built from the automaton C_i.

Case $\pi = \pi_1 \cup \pi_2$, $\pi = \pi_1 \circ \pi_2$, or $\pi = \chi^*$: We construct $A(\pi)$ by using the standard automata constructions for union, concatenation, and Kleene-star. We set $\mathcal{T}(\pi_1 \cup \pi_2) = \mathcal{T}(\pi_1 \circ \pi_2) = \mathcal{T}(\pi_1) \uplus \mathcal{T}(\pi_2)$ and $\mathcal{T}(\chi^*) = \mathcal{T}(\chi)$.

It remains to construct $A(\pi_1 \cap \pi_2)$ and $\mathcal{T}(\pi_1 \cap \pi_2)$. For this, we use the hop/drop-construction described above: Assume that the NFA $A(\pi_i) = (Q_i, \to_i)$ ($i \in \{1, 2\}$) over the TWATA $\mathcal{T}(\pi_i) = (S_i, \delta_i, \text{Acc}_i)$ is already constructed. Thus, $[\![A(\pi_i), p_i, q_i]\!] = [\![\pi_i]\!]$ for some states $p_i, q_i \in Q_i$. We first construct the NFA $C(\pi_i)$ over the TWATA $\mathcal{V}(\pi_i) = (S''_i, \delta''_i, \text{Acc}''_i)$ as described in Lemma 6. Note that the state set of $C(\pi_i)$ is Q_i (the state set of $A(\pi_i)$) and that $|S''_i| = |S_i| + 2 \cdot |Q_i|^2$. Let $\mathcal{T}(\pi_1 \cap \pi_2) = \mathcal{V}(\pi_1) \uplus \mathcal{V}(\pi_2)$. The NFA $A(\pi_1 \cap \pi_2)$ is the product automaton of $C(\pi_1)$ and $C(\pi_2)$, where test-transitions can be done asynchronously: Let $A(\pi_1 \cap \pi_2) = (Q_1 \times Q_2, \to)$, where for $a \in \Gamma \cup \overline{\Gamma}$ we have $(r_1, r_2) \xrightarrow{a} (r'_1, r'_2)$ if and only if $r_i \xrightarrow{a}_{C(\pi_i)} r'_i$ for $i \in \{1, 2\}$. Finally, for a state s of $\mathcal{V}(\pi_1) \uplus \mathcal{V}(\pi_1)$ we have the test-transition $(r_1, r_2) \xrightarrow{\mathcal{T}(\pi_1 \cap \pi_2), s} (r'_1, r'_2)$ if and only if for some $i \in \{1, 2\}$: s is a state of $\mathcal{V}(\pi_i)$, $r_i \xrightarrow{\mathcal{V}(\pi_i), s}_{C(\pi_i)} r'_i$, and $r_{3-i} = r'_{3-i}$.

Lemma 7. *We have* $[\![A(\pi_1 \cap \pi_2), (p_1, p_2), (q_1, q_2)]\!] = [\![\pi_1 \cap \pi_2]\!]$. *Moreover, if* $A(\pi_i) = (Q_i, \to_i)$, $A(\pi_1 \cap \pi_2) = (Q, \to)$, $\mathcal{T}(\pi_i) = (S_i, \delta_i, \text{Acc}_i)$, *and* $\mathcal{T}(\pi_1 \cap \pi_2) = (S, \delta, \text{Acc})$, *then:* $|Q| = |Q_1| \cdot |Q_2|$, $|S| = |S_1| + |S_2| + 2 \cdot |Q_1|^2 + 2 \cdot |Q_2|^2$, *and* $|\text{Acc}| = \max\{|\text{Acc}_1|, |\text{Acc}_2|\}$.

Recall that we want to check $(\mathcal{K}(\mathcal{Z}), \varepsilon) \models \varphi$ for the PRS $\mathcal{Z}$ and the $\text{PDL}^{\cap}$ formula φ. A careful analysis of the constructions above allows to prove inductively:

Lemma 8. *If* $|\mathcal{Z}|$ *and* $|\varphi|$ *are sufficiently large, then:*

- *For every subformula* ψ *of* φ *with* $\mathcal{T}(\psi) = (S, \delta, \mathrm{Acc})$ *we have* $|S| \leq |\mathcal{Z}|^{2\cdot|\psi|^2}$ *and* $|\mathrm{Acc}| \leq |\psi|$.
- *For every subprogram* π *of* φ *with* $A(\pi) = (Q, \rightarrow)$ *and* $\mathcal{T}(\pi) = (S, \delta, \mathrm{Acc})$ *we have* $|Q| \leq |\mathcal{Z}|^{|\pi|}$, $|S| \leq |\mathcal{Z}|^{2\cdot|\pi|^2}$, *and* $|\mathrm{Acc}| \leq |\pi|$.

From our construction, Lemma 8, and Theorem 2 we get:

Theorem 3. *It can be checked in* 2EXP, *whether* $(\mathcal{K}(\mathcal{Z}), \varepsilon) \models \varphi$ *for a given PRS* $\mathcal{Z}$ *and a given PDL*$^\cap$ *formula* φ. *For a fixed PDL*$^\cap$ *formula, it can be checked in* EXP, *whether* $(\mathcal{K}(\mathcal{Z}), \varepsilon) \models \varphi$ *for a given PRS* $\mathcal{Z}$.

For the data complexity of test-free PDL$^\cap$ over PDA we can prove a matching EXP lower bound, by translating the fixed CTL formula from Walukiewicz's lower bound proof for the data complexity of CTL over PDS [28] into a fixed test-free PDL$^\cap$ formula. For the data complexity of test-free PDL$^\cap$ over BPA we can only prove a lower bound of PSPACE by a reduction from the universality problem from non-deterministic finite automata. Altogether, we obtain the results for PDL$^\cap$ in Table 1.

One might ask, whether an elementary upper bound also holds for the model-checking problem of PDL with the complement operator on programs over pushdown systems. But this model-checking problem allows to express emptiness for a given extended regular expression (i.e., regular expression where the complement operator is allowed), which is a well known nonelementary problem.

6 Open Problems

On the technical side it remains to close the gap between PSPACE and EXP for the data complexity of PDL$^\cap$ over BPA. Another fruitful research direction might be to extend PDL$^\cap$ by a unary fixpoint operator. The resulting logic is strictly more expressive than PDL$^\cap$ and the modal μ-calculus. We are confident that our upper bounds for PDL$^\cap$ from Theorem 3 can be extended to this logic.

References

1. R. Alur, M. Benedikt, K. Etessami, P. Godefroid, T. W. Reps, and M. Yannakakis. Analysis of recursive state machines. *ACM Trans. Program. Lang. Syst*, 27(4):786–818, 2005.
2. A. Bouajjani, J. Esparza, and O. Maler. Reachability analysis of pushdown automata: Application to model-checking. In *Proc. CONCUR'97*, LNCS 1243, pages 135–150. Springer, 1997.
3. A. Blumensath. Prefix-recognizable graphs and monadic second-order logic. Tech. Rep. 2001-06, RWTH Aachen, Germany, 2001.
4. T. Cachat. Uniform solution of parity games on prefix-recognizable graphs. *ENTCS*, 68(6), 2002.
5. D. Caucal. On infinite transition graphs having a decidable monadic theory. *Theor. Comput. Sci.*, 290(1):79–115, 2002.

6. A. K. Chandra, D. C. Kozen, and L. J. Stockmeyer. Alternation. *J. Assoc. Comput. Mach.*, 28(1):114–133, 1981.
7. R. Danecki. Nondeterministic propositional dynamic logic with intersection is decidable. In *Proc. 5th Symp. Computation Theory*, LNCS 208, pages 34–53, 1984.
8. J. Esparza. On the decidabilty of model checking for several mu-calculi and petri nets. In *Proc. CAAP '94*, LNCS 787, pages 115–129. Springer, 1994.
9. J. Esparza, A. Kucera, and S. Schwoon. Model checking LTL with regular valuations for pushdown systems. *Inf. Comput.*, 186(2):355–376, 2003.
10. M. J. Fischer and R. E. Ladner. Propositional dynamic logic of regular programs. *J. Comput. Syst. Sci.*, 18(2):194–211, 1979.
11. S. Göller and M. Lohrey. Infinite State Model-Checking of Propositional Dynamic Logics Technical Report 2006/04, University of Stuttgart, Germany, 2006. ftp://ftp.informatik.uni-stuttgart.de/pub/library/ncstrl.ustuttgart_fi/TR-2006-04/.
12. D. Harel. Recurring dominoes: making the highly undecidable highly understandable. *Ann. Discrete Math.*, 24:51–72, 1985.
13. D. Harel, D. Kozen, and J. Tiuryn. *Dynamic Logic.* Foundations of computing. The MIT Press, 2000.
14. O. Kupferman, N. Piterman, and M. Vardi. Model checking linear properties of prefix-recognizable systems. In *Proc. CAV 2002*, LNCS 2404, pages 371–385, 2002.
15. O. Kupferman and M. Y. Vardi. An automata-theoretic approach to reasoning about infinite-state systems. In *Proc. CAV 2000*, LNCS 1855, pages 36–52. Springer, 2000.
16. M. Lange. Model checking propositional dynamic logic with all extras. *J. Appl. Log.*, 4(1):39–49, 2005.
17. M. Lange and C. Lutz. 2-Exptime lower bounds for propositional dynamic logics with intersection. *J. Symb. Log.*, 70(4):1072–1086, 2005.
18. Ch. Löding and O. Serre. Propositional dynamic logic with recursive programs. In *Proc. FOSSACS 2006*, LNCS 3921 pages 292–306. Springer, 2006.
19. D. Lugiez and P. Schnoebelen. The regular viewpoint on PA-processes. *Theor. Comput. Sci.*, 274(1–2):89–115, 2002.
20. R. Mayr. Strict lower bounds for model checking BPA. *ENTCS*, 18, 1998.
21. R. Mayr. Process rewrite systems. *Inf. Comput.*, 156(1):264–286, 2000.
22. R. Mayr. Decidability of model checking with the temporal logic EF. *Theor. Comput. Sci.*, 256(1-2):31–62, 2001.
23. D. Muller and P. Schupp. Alternating automata on infinite trees. *Theor. Comput. Sci.*, 54(2-3):267–276, 1987.
24. C. H. Papadimitriou. *Computational Complexity.* Addison Wesley, 1994.
25. W. Thomas. Some perspectives of infinite-state verification. In *Proc. ATVA 2005*, LNCS 3707, pages 3–10. Springer, 2005.
26. M. Y. Vardi. The complexity of relational query languages (extended abstract). In *Proc. STOC 1982*, pages 137–146. ACM Press, 1982.
27. M. Y. Vardi. Reasoning about the past with two-way automata. In *Proc. ICALP '98*, LNCS 1443, pages 628–641. Springer, 1998.
28. I. Walukiewicz. Model checking CTL properties of pushdown systems. In *Proc. FSTTCS 2000*, LNCS 1974, pages 127–138. Springer, 2000.
29. I. Walukiewicz. Pushdown processes: Games and model-checking. *Inf. Comput.*, 164(2):234–263, 2001.
30. S. Wöhrle. *Decision problems over infinite graphs: Higher-order pushdown systems and synchronized products.* Dissertation, RWTH Aachen, Germany, 2005.

Weak Bisimulation Approximants

Will Harwood, Faron Moller, and Anton Setzer*

Department of Computer Science, Swansea University
Singleton Park, Sketty, Swansea SA2 8PP, UK
{cswill, f.g.moller, a.g.setzer}@swansea.ac.uk

Abstract. ***Bisimilarity*** $\sim$ and ***weak bisimilarity*** $\approx$ are canonical notions of equivalence between processes, which are defined co-inductively, but may be approached – and even reached – by their (transfinite) inductively-defined ***approximants*** $\sim_\alpha$ and $\approx_\alpha$. For arbitrary processes this approximation may need to climb arbitrarily high through the infinite ordinals before stabilising. In this paper we consider a simple yet well-studied process algebra, the Basic Parallel Processes (BPP), and investigate for this class of processes the minimal ordinal α such that $\approx = \approx_\alpha$.

The main tool in our investigation is a novel proof of ***Dickson's Lemma***. Unlike classical proofs, the proof we provide gives rise to a tight ordinal bound, of ω^n, on the order type of non-increasing sequences of n-tuples of natural numbers. With this we are able to reduce a long-standing bound on the approximation hierarchy for weak bisimilarity $\approx$ over BPP, and show that $\approx = \approx_{\omega^{\square}}$.

1 Introduction

There has been great interest of late in the development of techniques for deciding equivalences between infinite-state processes, particularly for the question of bisimilarity between processes generated by some type of term algebra. Several surveys of this developing area have been published, beginning with [16], and there is now a chapter in the Handbook of Process Algebra dedicated to the topic [2], as well as a website devoted to maintaining an up-to-date comprehensive overview of the state-of-the-art [20].

While questions concerning strong bisimilarity have been successfully addressed, techniques for tackling the question of weak bisimilarity, that is, when silent unobservable transitions are allowed, are still lacking, and many open problems remain. The main difficulty arising when considering weak bisimilarity is that processes immediately become infinite-branching: at any point in a computation, a single action can result in any number of transitions leading to any one of an infinite number of next states. Common finiteness properties fail due to this; in particular, bisimilarity can no longer be characterised by its finite approximations in the way that it can for finite-branching processes. For arbitrary infinite-branching processes, we may need to climb arbitrarily high through

* Supported by EPSRC grant GR/S30450/01.

Z. Ésik (Ed.): CSL 2006, LNCS 4207, pp. 365–379, 2006.

the transfinite approximations to bisimilarity before reaching the bisimulation relation itself.

In this paper we consider the problem of weak bisimilarity for so-called Basic Parallel Processes (BPP), a simple yet well-studied model of concurrent processes. These correspond to commutative context-free processes, or equivalently to communication-free Petri nets. The question as to the decidability of weak bisimilarity between BPP processes remains unsolved (though decidability results for very restricted classes of BPP have been established by Hirshfeld in [10] and Stirling in [21]). It has recently been shown that the problem is at least PSPACE-hard [19], even in the restricted case of so-called normed BPP, but this sheds no light one way or the other as to decidability. Jančar suggests in [14] that the techniques he uses there to establish PSPACE-completeness of strong bisimilarity for BPP might be exploited to give a decision procedure for weak bisimilarity, but three years later this conjecture remains unsubstantiated.

It has long been conjectured that for BPP, weak bisimilarity is characterised by its $(\omega\times 2)$-level approximation. Such a result could provide a way to a decision procedure. However, no nontrivial approximation bound has before now been established; the strength of the $(\omega\times 2)$-conjecture remains rooted only in the fact that no counterexample has been found. In this paper we provide the first non-trivial countable bound on the approximation: for a BPP defined over k variables, weak bisimilarity is reached by the ω^{2k} level; weak bisimilarity is thus reached by the ω^{ω} level for any BPP.

Our argument is based on a new constructive proof of Dickson's Lemma which provides an ordinal bound on the sequences described by the Lemma. This proof is presented in Section 2 of the paper. After this, the definitions necessary for the remainder of the paper are outlined in Section 3 along with a variety of results, and our results on BPP are presented in Section 4. We finish with some concluding observations in Section 5.

2 Ordinal Bounds for Dickson's Lemma

In the sequel we shall use the following notation. We let x, y (with subscripts) range over natural numbers $\mathbb{N} = \{0, 1, 2, \ldots\}$; $\vec{\boldsymbol{x}}$, $\vec{\boldsymbol{y}}$ (with subscripts) range over finite sequences (n-tuples) of natural numbers; and $\vec{\boldsymbol{X}}$, $\vec{\boldsymbol{Y}}$ range over arbitrary (finite or infinite) sequences of such n-tuples. We shall use angle brackets to denote sequences, such as $\vec{\boldsymbol{x}} = \langle x_1, \ldots, x_n\rangle$ and $\vec{\boldsymbol{X}} = \langle \vec{\boldsymbol{x}}_1, \vec{\boldsymbol{x}}_2, \ldots\rangle$, and juxtaposition to represent concatenation; e.g., if $\vec{\boldsymbol{X}}$ is a finite sequence of n-tuples then $\vec{\boldsymbol{X}}\langle\vec{\boldsymbol{x}}\rangle$ is the longer sequence which has the extra n-tuple $\vec{\boldsymbol{x}}$ added to the end. Finally, we shall use the notation $(\cdot)_k$ to select the kth component from a sequence; for example, if $\vec{\boldsymbol{x}}_i = \langle x_1, \ldots, x_n\rangle$ then $(\vec{\boldsymbol{x}}_i)_k = x_k$. (The parentheses are used to avoid confusion with the subscripting allowed in the variable naming the sequence.)

One n-tuple $\vec{\boldsymbol{y}} = \langle y_1, \ldots, y_n\rangle$ of natural numbers ***dominates*** another such n-tuple $\vec{\boldsymbol{x}} = \langle x_1, \ldots, x_n\rangle$ if $\vec{\boldsymbol{x}} \leq \vec{\boldsymbol{y}}$, where $\leq$ is considered pointwise, that is, $x_i \leq y_i$ for each $i \in \{1, \ldots, n\}$. A sequence of n-tuples is a ***non-dominating sequence*** over $\mathbb{N}^n$ if no element of the sequence dominates any of its predecessors in the sequence. A ***tree*** – by which we mean a rooted directed graph with no undirected cycles – with nodes labelled by n-tuples from $\mathbb{N}^n$ is a ***non-dominating tree*** over $\mathbb{N}^n$ if the sequence of labels along any path through the tree is a non-dominating sequence.

Dickson's Lemma [6] asserts that there can be no infinite non-dominating sequences.

Lemma 1 (Dickson's Lemma). *All non-dominating sequences are finite. That is, given an infinite sequence of vectors* $\vec{\boldsymbol{x}}_1, \vec{\boldsymbol{x}}_2, \vec{\boldsymbol{x}}_3, \ldots \in \mathbb{N}^n$*, we can always find indices* i, j *with* $i < j$ *such that* $\vec{\boldsymbol{x}}_i \leq \vec{\boldsymbol{x}}_j$.

The standard proof of this lemma uses a straightforward induction on n: for the base case, any sequence of decreasing natural numbers must be finite; and for the induction step, from an infinite sequence of n-tuples you extract an infinite subsequence in which the last components are nondecreasing (either constant or increasing), and then apply induction on the sequence of $(n-1)$-tuples which arise by ignoring these last components.

The problem with this proof is that it is nonconstructive; in particular, it gives no clue as to the ordinal bound on the lengths of non-dominating sequences. The difficulty with determining an ordinal bound comes from the fact that the domination order is not a total order on n-tuples (as opposed, for example, to lexicographical order). We provide here an alternative constructive proof from which we can extract an ordinal bound on the lengths of such sequences.

Theorem 1 (Constructive Dickson's Lemma). *The order type of the set of non-dominating sequences of* n*-tuples of natural numbers with partial ordering*

$$\vec{\boldsymbol{X}} \prec \vec{\boldsymbol{Y}} \stackrel{def}{\Leftrightarrow} \vec{\boldsymbol{X}} \text{ strictly extends } \vec{\boldsymbol{Y}}$$

is ω^n.

Proof. That the order type is at *least* ω^n is clear: the order type of the set of lexicographically descending sequences with respect to extension is ω^n, and this set is contained in the set of non-dominating sequences.

It remains to show that the order type is at *most* ω^n. This result will follow immediately from the construction of a function

$$f_n : (\mathbb{N}^n)^+ \to \mathbb{N}^n$$

on non-empty finite sequences of n-tuples which satisfies the following property:

> If $\vec{\boldsymbol{X}}\langle\vec{\boldsymbol{x}}\rangle$ is a non-dominating sequence of n-tuples, and $\vec{\boldsymbol{X}}$ is itself non-empty, then $f_n(\vec{\boldsymbol{X}}\langle\vec{\boldsymbol{x}}\rangle) <_{\text{lex}} f_n(\vec{\boldsymbol{X}})$.

We shall inductively define these functions f_n. The base case is straightforward: we can define f_1 by

$$f_1(\langle x_1, \ldots, x_k\rangle) \stackrel{\text{def}}{=} x_k.$$

A non-dominating sequence of natural numbers is simply a decreasing sequence, which has ordinal bound ω.

For illustrative purposes we carry out the construction of the function f_2 for sequences of pairs, and later generalise our construction to sequences of n-tuples.

Given a non-empty finite sequence of pairs $\vec{\boldsymbol{X}} = \langle\langle x_1, y_1\rangle, \ldots, \langle x_k, y_k\rangle\rangle$, define

- $\text{MIN}_x(\vec{\boldsymbol{X}}) \stackrel{\text{def}}{=} \min\{\, x_i : 1 \le i \le k \,\}$,
- $\text{MIN}_y(\vec{\boldsymbol{X}}) \stackrel{\text{def}}{=} \min\{\, y_i : 1 \le i \le k \,\}$, and
- $S_2(\vec{\boldsymbol{X}}) \stackrel{\text{def}}{=} \Big\{ \langle x, y\rangle : \text{MIN}_x(\vec{\boldsymbol{X}}) \le x,\ \text{MIN}_y(\vec{\boldsymbol{X}}) \le y,\ \textit{and}\ \langle x_i, y_i\rangle \not\le \langle x, y\rangle \textit{ for all } i : 1 \le i \le k \Big\}$.

$S_2(\vec{\boldsymbol{X}})$ consists of the pairs with which the sequence $\vec{\boldsymbol{X}}$ can be extended without altering the MIN_x and MIN_y values and yet while maintaining non-domination. Note that $S_2(\vec{\boldsymbol{X}})$ must be finite: if we let i and j be such that $x_i = \text{MIN}_x(\vec{\boldsymbol{X}})$ and $y_j = \text{MIN}_y(\vec{\boldsymbol{X}})$, then in order for $\langle x, y\rangle \not\ge \langle x_i, y_i\rangle$ and $\langle x, y\rangle \not\ge \langle x_j, y_j\rangle$ we must have $x < x_j$ (since $y \ge y_j$) and $y < y_i$ (since $x \ge x_i$).

Suppose that $\vec{\boldsymbol{Y}} = \vec{\boldsymbol{X}}\langle\langle x, y\rangle\rangle$ is a non-dominating sequence, and that $\vec{\boldsymbol{X}}$ is itself non-empty. Then clearly $\text{MIN}_x(\vec{\boldsymbol{Y}}) \le \text{MIN}_x(\vec{\boldsymbol{X}})$ and $\text{MIN}_y(\vec{\boldsymbol{Y}}) \le \text{MIN}_y(\vec{\boldsymbol{X}})$; and if equality holds in both cases then $S_2(\vec{\boldsymbol{Y}}) \subsetneq S_2(\vec{\boldsymbol{X}})$, since $S_2(\vec{\boldsymbol{Y}}) \subseteq S_2(\vec{\boldsymbol{X}})$ yet $\langle x, y\rangle \in S_2(\vec{\boldsymbol{X}}) \setminus S_2(\vec{\boldsymbol{Y}})$. Thus $|S_2(\vec{\boldsymbol{Y}})| < |S_2(\vec{\boldsymbol{X}})|$.

We can then define the function f_2 on non-empty sequences $\vec{\boldsymbol{X}}$ of pairs as follows:

$$f_2(\vec{\boldsymbol{X}}) \stackrel{\text{def}}{=} \langle\, \text{MIN}_x(\vec{\boldsymbol{X}}) + \text{MIN}_y(\vec{\boldsymbol{X}}),\ |S_2(\vec{\boldsymbol{X}})|\,\rangle$$

If $\vec{\boldsymbol{X}}\langle\langle x, y\rangle\rangle$ is a non-dominating sequence and $\vec{\boldsymbol{X}}$ is itself non-empty, then by the above argument we must have that $f_2(\vec{\boldsymbol{X}}\langle\langle x, y\rangle\rangle) <_{\text{lex}} f_2(\vec{\boldsymbol{X}})$.

For the inductive construction of f_n we assume we have constructed the function f_{n-1} as required. For $1 \le i \le n$ we define the function

$$\pi_{\text{-}i}(\langle x_1, \ldots, x_n\rangle) \stackrel{\text{def}}{=} \langle x_1, \ldots, x_{i-1}, x_{i+1}, \ldots, x_n\rangle$$

which simply deletes the ith component from the n-tuple $\langle x_1, \ldots, x_n\rangle$. Next, given a non-empty finite sequence $\vec{\boldsymbol{X}} = \langle \vec{\boldsymbol{x}}_1, \ldots, \vec{\boldsymbol{x}}_k\rangle$ of n-tuples, we define the set

$$\text{ND}_{\text{-}i}(\vec{\boldsymbol{X}}) \stackrel{\text{def}}{=} \Big\{ \langle \pi_{\text{-}i}(\vec{\boldsymbol{x}}_{i_1}), \ldots, \pi_{\text{-}i}(\vec{\boldsymbol{x}}_{i_p})\rangle \ :\ p > 0,\ 0 < i_1 < \cdots < i_p \le k,\ \textit{and}\ \langle \pi_{\text{-}i}(\vec{\boldsymbol{x}}_{i_1}), \ldots, \pi_{\text{-}i}(\vec{\boldsymbol{x}}_{i_p})\rangle \text{ is non-dominating} \Big\}$$

which consists of the non-dominating subsequences of $(n-1)$-tuples of $\vec{\boldsymbol{X}}$ in which the ith components of the n-tuples have been deleted. Finally we make the following definitions:

- $\text{MIN}_{\text{-}i}(\vec{\boldsymbol{X}}) \stackrel{\text{def}}{=} \min_{<_{\text{lex}}}\{\, f_{n-1}(\vec{\boldsymbol{Y}}) \ : \ \vec{\boldsymbol{Y}} \in \text{ND}_{\text{-}i}(\vec{\boldsymbol{X}})\,\}$
- $S_n(\vec{\boldsymbol{X}}) \stackrel{\text{def}}{=} \Big\{ \vec{\boldsymbol{x}} \ : \ \text{MIN}_{\text{-}i}(\vec{\boldsymbol{X}}) = \text{MIN}_{\text{-}i}(\vec{\boldsymbol{X}}\langle\vec{\boldsymbol{x}}\rangle) \textit{ for all } i : 1 \le i \le n, \textit{ and } \vec{\boldsymbol{x}}_i \not\le \vec{\boldsymbol{x}} \textit{ for all } i : 1 \le i \le k \Big\}$

$S_n(\vec{\boldsymbol{X}})$ consists of the n-tuples with which the sequence $\vec{\boldsymbol{X}}$ can be extended without altering the $\text{MIN}_{\text{-}i}$ values and yet while maintaining non-domination. Note that $S_n(\vec{\boldsymbol{X}})$ must be finite. To see this, let $1 \le i \le n$ and $i_1, \ldots, i_p$ be such that

$$\text{MIN}_{\text{-}i}(\vec{\boldsymbol{X}}) = f_{n-1}(\langle \pi_{\text{-}i}(\vec{\boldsymbol{x}}_{i_1}), \ldots, \pi_{\text{-}i}(\vec{\boldsymbol{x}}_{i_p})\rangle),$$

and suppose that $\vec{\boldsymbol{x}} \in S_n(\vec{\boldsymbol{X}})$. If the sequence $\langle \pi_{\text{-}i}(\vec{\boldsymbol{x}}_{i_1}), \ldots, \pi_{\text{-}i}(\vec{\boldsymbol{x}}_{i_p}), \pi_{\text{-}i}(\vec{\boldsymbol{x}})\rangle$ is non-dominating, then by induction we would get that

$$\begin{aligned} \text{MIN}_{\text{-}i}(\vec{\boldsymbol{X}}\langle\vec{\boldsymbol{x}}\rangle) &\le_{\text{lex}} f_{n-1}(\langle \pi_{\text{-}i}(\vec{\boldsymbol{x}}_{i_1}), \ldots, \pi_{\text{-}i}(\vec{\boldsymbol{x}}_{i_p}), \pi_{\text{-}i}(\vec{\boldsymbol{x}})\rangle) \\ &<_{\text{lex}} f_{n-1}(\langle \pi_{\text{-}i}(\vec{\boldsymbol{x}}_{i_1}), \ldots, \pi_{\text{-}i}(\vec{\boldsymbol{x}}_{i_p})\rangle) \\ &= \text{MIN}_{\text{-}i}(\vec{\boldsymbol{X}}) \end{aligned}$$

contradicting $\vec{\boldsymbol{x}} \in S_n(\vec{\boldsymbol{X}})$. Therefore we must have that $\pi_{\text{-}i}(\vec{\boldsymbol{x}}) \ge \pi_{\text{-}i}(\vec{\boldsymbol{x}}_{i_j})$ for some j. But since $\vec{\boldsymbol{x}} \not\ge \vec{\boldsymbol{x}}_{i_j}$ we must then have that $(\vec{\boldsymbol{x}})_i < (\vec{\boldsymbol{x}}_{i_j})_i$.

Suppose that $\vec{\boldsymbol{Y}} = \vec{\boldsymbol{X}}\langle\vec{\boldsymbol{x}}\rangle$ is a non-dominating sequence, and that $\vec{\boldsymbol{X}}$ is itself non-empty. Then $\text{MIN}_{\text{-}i}(\vec{\boldsymbol{Y}}) \le \text{MIN}_{\text{-}i}(\vec{\boldsymbol{X}})$ for all i (since $\text{ND}_{\text{-}i}(\vec{\boldsymbol{X}}) \subseteq \text{ND}_{\text{-}i}(\vec{\boldsymbol{Y}})$); and if equality holds in all cases then $S_n(\vec{\boldsymbol{Y}}) \subsetneq S_n(\vec{\boldsymbol{X}})$ since $S_n(\vec{\boldsymbol{Y}}) \subseteq S_n(\vec{\boldsymbol{X}})$ yet $\vec{\boldsymbol{x}} \in S_n(\vec{\boldsymbol{X}}) \setminus S_n(\vec{\boldsymbol{Y}})$. Thus $|S_n(\vec{\boldsymbol{Y}})| < |S_n(\vec{\boldsymbol{X}})|$.

We can then define the function f_n on non-empty sequences $\vec{\boldsymbol{X}}$ of n-tuples as follows:

$$f_n(\vec{\boldsymbol{X}}) = \Big(\sum_{i=1}^{n} \text{MIN}_{\text{-}i}(\vec{\boldsymbol{X}})\Big)\langle |S_n(\vec{\boldsymbol{X}})|\rangle$$

where the sum is taken component-wise. (This sum is, if we identify $\langle k_1, \ldots, k_n\rangle$ with $\omega^{n-1}\cdot k_1 + \omega^{n-2}\cdot k_2 + \cdots + \omega^0 \cdot k_n$, the natural sum of ordinals.) If $\vec{\boldsymbol{X}}\langle\vec{\boldsymbol{x}}\rangle$ is a non-dominating sequence and $\vec{\boldsymbol{X}}$ is itself non-empty, then by the above argument we must have that $f_n(\vec{\boldsymbol{X}}\langle\vec{\boldsymbol{x}}\rangle) <_{\text{lex}} f_n(\vec{\boldsymbol{X}})$. □

2.1 Ordinal Bounds on Trees

Our constructive version of Dickson's Lemma easily extends to trees, where we take the following definition of the height of a well-founded tree (that is, a tree with no infinite paths).

Definition 1. *The* ***height*** *of a well-founded tree rooted at t is defined by*

$$h(t) \stackrel{def}{=} \sup\{\, h(s) + 1 \ : \ t \longrightarrow s \,\}.$$

(By convention, $\sup \emptyset = 0$.*)*

Theorem 2. *If t is (the root of) a non-dominating tree over $\mathbb{N}^n$, then $h(t) \leq \omega^n$.*

Proof. For each node x of the tree, define $\ell(x) \in \mathbb{N}^n$ by $\ell(x) = f_n(\pi_x)$, where f_n is as defined in the proof of Lemma 1, and π_x is the non-dominating sequence of labels on the path from (the root) t to x. It will suffice then to prove that $h(x) \leq \ell(x)$ (viewing $\ell(x)$ as an ordinal, that is, interpreting the n-tuple $\langle k_1, \ldots, k_n \rangle \in \mathbb{N}^n$ as $\omega^{n-1}k_1 + \omega^{n-2}k_2 + \cdots + \omega^0 k_n$) for all nodes x of the tree. This is accomplished by a straightforward induction on $h(x)$:

$$\begin{aligned} h(x) &= \sup\{\, h(y)+1 : x \rightarrow y \,\} \\ &\leq \sup\{\, \ell(y)+1 : x \rightarrow y \,\} \quad \text{(by induction)} \\ &\leq \ell(x). \end{aligned}$$

□

3 Processes and Bisimilarity

A ***process*** is represented by (a state in) a ***labelled transition system*** defined as follows.

Definition 2. *A* ***labelled transition system (LTS)*** *is a triple $\mathcal{S} = (S, Act, \rightarrow)$ where S is a set of* ***states****, Act is a finite set of* ***actions****, and $\rightarrow \subseteq S \times Act \times S$ is a* ***transition relation****.*

We write $s \xrightarrow{a} t$ instead of $(s, a, t) \in \rightarrow$, thus defining an infix binary relation $\xrightarrow{a} = \{(s,t) : (s,a,t) \in \rightarrow\}$ for each action $a \in Act$.

It is common to admit silent transitions to model the internal unobservable evolution of a system. In standard automata theory these are typically referred to as "epsilon" (or occasionally "lambda") transitions, but in concurrency theory they are commonly represented by a special action $\tau \in Act$. With this, we can then define observable transitions as follows:

$$\begin{aligned} &s \overset{\tau}{\Rightarrow} t \;\text{ iff }\; s \,(\xrightarrow{\tau})^*\, t \;\text{ and} \\ &s \overset{a}{\Rightarrow} t \;\text{ iff }\; s \,(\xrightarrow{\tau})^* \cdot \xrightarrow{a} \cdot (\xrightarrow{\tau})^*\, t \quad \text{for } a \neq \tau. \end{aligned}$$

In general, $\overset{a}{\Rightarrow} \supseteq \xrightarrow{a}$; and over an LTS with no silent transitions, $\overset{a}{\Rightarrow} = \xrightarrow{a}$, and in this case all the relations we define wrt $\Rightarrow$ will be identical to the analogous relations defined wrt $\rightarrow$.

The notion of "behavioural sameness" between two processes (which we view as two states in the same LTS) can be formally captured in many different ways (see, e.g., [8] for an overview). Among those behavioural equivalences, ***bisimilarity*** enjoys special attention. Its formal definition is as follows.

Definition 3. *Let $\mathcal{S} = (S, Act, \rightarrow)$ be an LTS. A binary relation $\mathcal{R} \subseteq S \times S$ is a* ***bisimulation relation*** *iff whenever $(s,t) \in R$, we have that*

- *for each transition $s \xrightarrow{a} s'$ there is a transition $t \xrightarrow{a} t'$ such that $(s', t') \in \mathcal{R}$; and*
- *for each transition $t \xrightarrow{a} t'$ there is a transition $s \xrightarrow{a} s'$ such that $(s', t') \in \mathcal{R}$.*

*Processes s and t are **bisimulation equivalent (bisimilar)**, written $s \sim t$, iff they are related by some bisimulation. Thus $\sim$ is the union, and ergo the largest, of all bisimulation relations.*

*If we replace the transition relation $\rightarrow$ in this definition with the weak transition relation $\Rightarrow$, we arrive at the definition of a **weak bisimulation relation** defining **weak bisimulation equivalence (weak bisimilarity)**, which we denote by $\approx$. In general, $\approx \supseteq \sim$; and over an LTS with no silent transitions, $\approx = \sim$.*

The above definition of (weak) bisimilarity is a co-inductive one, but can be approximated using the following inductively-defined stratification.

Definition 4. *The **bisimulation approximants** $\sim_\alpha$, for all ordinals $\alpha \in \mathcal{O}$, are defined as follows:*

- $s \sim_0 t$ *for all process states s and t.*
- $s \sim_{\alpha+1} t$ *iff*
  - *for each transition $s \xrightarrow{a} s'$ there is a transition $t \xrightarrow{a} t'$ such that $s' \sim_\alpha t'$; and*
  - *for each transition $t \xrightarrow{a} t'$ there is a transition $s \xrightarrow{a} s'$ such that $s' \sim_\alpha t'$.*
- *For all limit ordinals λ, $s \sim_\lambda t$ iff $s \sim_\alpha t$ for all $\alpha < \lambda$.*

*The **weak bisimulation approximants** $\approx_\alpha$ are defined by replacing the transition relation $\rightarrow$ with the weak transition relation $\Rightarrow$ in the above definition.*

The following results are then standard.

Theorem 3

1. *Each $\sim_\alpha$ and $\approx_\alpha$ is an equivalence relation over the states of any LTS.*
2. *Given $\alpha < \beta$, $\sim_\alpha \supseteq \sim_\beta$ and $\approx_\alpha \supseteq \approx_\beta$ over the states of any LTS; and in general these define strictly decreasing hierarchies: given any ordinal α we can provide an LTS with states s and t satisfying $s \sim_\alpha t$ but $s \not\sim_{\alpha+1} t$ (and $s \approx_\alpha t$ but $s \not\approx_{\alpha+1} t$).*
3. *$s \sim t$ iff $s \sim_\alpha t$ for all ordinals $\alpha \in \mathcal{O}$, and $s \approx t$ iff $s \approx_\alpha t$ for all ordinals $\alpha \in \mathcal{O}$. That is, $\sim = \bigcap_{\alpha \in \mathcal{O}} \sim_\alpha$ and $\approx = \bigcap_{\alpha \in \mathcal{O}} \approx_\alpha$.*

Remark 1. *For Part 2 of Theorem 3 we can define an LTS over a singleton alphabet $\{a\}$ ($a \neq \tau$) whose state set is γ for some ordinal γ (that is, each ordinal smaller than γ is a state), and such that $\alpha \xrightarrow{a} \beta$ iff $\beta < \alpha$. Then it is easy to show that for $\alpha < \beta$, $\alpha \sim_\alpha \beta$ but $\alpha \not\sim_{\alpha+1} \beta$. (First we show, by induction on α, that if $\alpha \leq \mu, \nu$ then $\mu \sim_\alpha \nu$; then we show, by induction on α, that if $\alpha < \beta$ then $\alpha \not\sim_{\alpha+1} \beta$.) As this LTS does not have τ actions, and hence $\approx = \sim$, this also gives that $\alpha \approx_\alpha \beta$ but $\alpha \not\approx_{\alpha+1} \beta$.*

If $s \not\sim t$, we must have a least ordinal $\alpha \in \mathcal{O}$ such that $s \not\sim_{\alpha+1} t$, and for this ordinal α we must have $s \sim_\alpha t$. (If $s \not\sim_\lambda t$ for a limit ordinal λ then we must have $s \not\sim_\alpha t$, and hence $s \not\sim_{\alpha+1} t$, for some $\alpha < \lambda$.) We shall identify this value α by writing $s \sim_\alpha^! t$. In the same way we write $s \approx_\alpha^! t$ to identify the least ordinal $\alpha \in \mathcal{O}$ such that $s \not\approx_{\alpha+1} t$.

3.1 Bisimulation Games and Optimal Move Trees

There is a further approach to defining (weak) bisimilarity, one based on games and strategies, whose usefulness is outlined in the tutorial [15]. We describe it here for bisimilarity; its description for weak bisimilarity requires only replacing the transition relation $\rightarrow$ with the weak transition relation $\Rightarrow$, after which all results stated will hold for the weak bisimilarity relations.

A game $\mathcal{G}(s,t)$ corresponding to two states s and t of an LTS is played between two players, **A** and **B**; the first player **A** (the *adversary*) wants to show that the states s and t are different, while the second player **B** (the *bisimulator)* wants to show that they are the same. To this end the game is played by the two players exchanging moves as follows:

- **A** chooses any transition $s \xrightarrow{a} s'$ or $t \xrightarrow{a} t'$ from one of the states s and t;
- **B** responds by choosing a matching transition $t \xrightarrow{a} t'$ or $s \xrightarrow{a} s'$ from the other state;
- the game then continues from the new position $\mathcal{G}(s',t')$.

The second player **B** wins this game if **B** can match every move that the first player **A** makes (that is, if **A** ever cannot make a move or the game continues indefinitely); if, however, **B** at some point cannot match a move made by **A** then player **A** wins. The following is then a straightforward result.

Theorem 4. *$s \sim t$ iff the second player* **B** *has a winning strategy for $\mathcal{G}(s,t)$.*

If $s \not\sim t$, then $s \sim^!_\alpha t$ for some $\alpha \in \mathcal{O}$, and this α in a sense determines how long the game must last, assuming both players are playing optimally, before **B** loses the game $\mathcal{G}(s,t)$:

- Since $s \not\sim_{\alpha+1} t$, **A** can make a move such that, regardless of **B**'s response, the exchange of moves will result in a game $\mathcal{G}(s',t')$ in which $s' \not\sim_\alpha t'$; such a move is an ***optimal move*** for **A**.
- For every $\beta < \alpha$, regardless of the move made by **A**, **B** can respond in such a way that the exchange of moves will result in a game $\mathcal{G}(s',t')$ in which $s' \sim_\beta t'$.

With this in mind, we can make the following definition.

Definition 5. *An* ***optimal move tree*** *is a tree whose nodes are labelled by pairs of non-bisimilar states of an LTS in which an edge $(s,t) \longrightarrow (s',t')$ exists precisely when (s,t) is a node of the tree and the following holds:*

In the game $\mathcal{G}(s,t)$, a single exchange of moves in which **A** *makes an optimal move may result in the game $\mathcal{G}(s',t')$*

The optimal move tree rooted at (s,t) is denoted by ***omt***(s,t).

If $(s,t) \longrightarrow (s',t')$ is an edge in an optimal move tree, then $s \sim^!_\alpha t$ and $s' \sim^!_\beta t'$ for some α and β with $\alpha > \beta$. Hence, every optimal move tree is well-founded. Furthermore, the following result is easily realised.

Lemma 2. *$h(omt(s,t)) = \alpha$ iff $s \sim^!_\alpha t$.*

3.2 Bounded Branching Processes

Over the class of finite-branching labelled transition systems, it is a standard result that $\sim \; = \; \sim_\omega$. We give here a generalisation of this result for infinite-branching processes.

Definition 6. *An infinite cardinal κ is* **regular** *iff it is not the supremum of fewer than κ smaller ordinals.*

Thus for example ω is regular as it is not the supremum of any finite collection of natural numbers.

Definition 7. *A process is* **$<$-κ-branching** *iff all of its states have fewer than κ transitions leading out of them. A tree t is* **$<$-κ-branching** *iff all of its nodes have fewer than κ children.*

Theorem 5. *If κ is a regular cardinal, and t is a well-founded $<$-κ-branching tree, then $h(t) < \kappa$.*

Proof. By (transfinite) induction on $h(t)$. If $t \longrightarrow s$ then $h(s) < h(t)$; and by induction $h(s) < \kappa$ and hence $h(s)+1 < \kappa$. Since $h(t) = \sup\{\, h(s)+1 \;:\; t \longrightarrow s \,\}$, by the regularity of κ we must have that $h(t) < \kappa$. □

The most basic form of this result is König's Lemma: any finite-branching well-founded tree can only have finitely-many nodes (and hence finite height).

The next result follows directly from the fact that $|A\times A| = |A|$ for any infinite set A.

Lemma 3. *If s and t are non-equivalent states of a $<$-κ-branching process, then $omt(s,t)$ is $<$-κ-branching.*

From the above, we arrive at a theorem on approximant collapse, which generalises the standard result that $\sim \; = \; \sim_\omega$ on finite-branching processes as well as a result in [22] concerning countably-branching processes.

Theorem 6. *For regular cardinals κ, $\sim \; = \; \sim_\kappa$ over the class of $<$-κ-branching processes.*

Proof. $\sim \; \subseteq \; \sim_\kappa$ is a given. If on the other hand $s \not\sim t$, then $h(omt(s,t)) = \alpha$ where $s \sim^!_\alpha t$. Thus, by Lemma 3 and Theorem 5, $\alpha < \kappa$, and hence $s \not\sim_\kappa t$. □

4 Basic Parallel Processes

A Basic Process Algebra (BPA) process is defined by a context-free grammar in Greibach normal form. Formally this is given by a triple $G = (V, A, \Gamma)$, where V is a finite set of *variables* (*nonterminal symbols*), A is a finite set of *labels* (*terminal symbols*), and $\Gamma \subseteq V \times A \times V^*$ is a finite set of *rewrite rules* (*productions*); it is assumed that every variable has at least one associated rewrite rule. Such a grammar gives rise to the LTS $\mathcal{S}_G = (V^*, A, \rightarrow)$ in which the states are sequences of variables, the actions are the labels, and the transition relation is given by the rewrite rules extended by the *prefix rewriting rule*: if $(X, a, u) \in \Gamma$ then $Xv \xrightarrow{a} uv$ for all $v \in V^*$. In this way, concatenation of variables naturally represents sequential composition.

A Basic Parallel Processes (BPP) process is defined in exactly the same fashion from such a grammar. However, in this case elements of V^* are read modulo commutativity of concatenation, so that concatenation is interpreted as parallel composition rather than sequential composition. The states of the BPP process associated with a grammar are thus given not by sequences of variables but rather by multisets of variables.

As an example, Figure 1 depicts BPA and BPP processes defined by the same grammar given by the three rules $A \xrightarrow{a} AB$, $A \xrightarrow{c} \varepsilon$ and $B \xrightarrow{b} \varepsilon$.

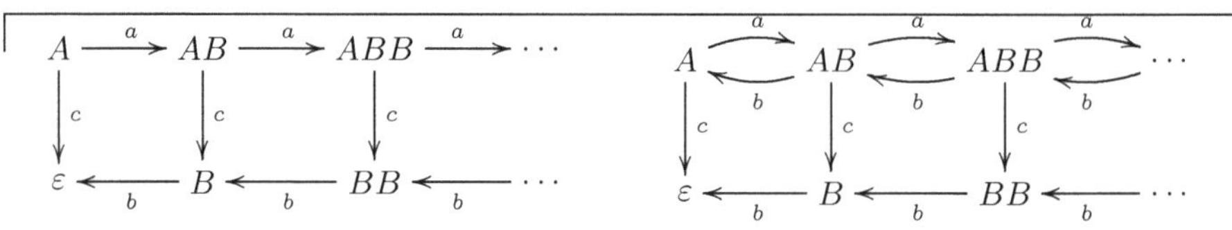

Fig. 1. BPA and BPP processes defined by the grammar $A \xrightarrow{a} AB$, $A \xrightarrow{c} \varepsilon$, $B \xrightarrow{b} \varepsilon$

Decidability results for (strong) bisimilarity checking have been long established for both BPA [5] and BPP [3,4]. For a wide class of interest (normed processes) these problems have even been shown to have polynomial-time solutions [11,12,13]. More recently, the decision problems for full BPA and BPP have been shown to be PSPACE-hard [17,18].

Decidability results for weak bisimilarity are much harder to establish, mainly due to the problems of infinite branching. While over BPA and BPP we have $\sim \, = \bigcap_{n\in\omega} \sim_n$, the infinite-branching nature of the weak transition relations makes this result false. As an example, Figure 2gives a BPP process with states P and Q in which $P \approx_n Q$ for all $n \in \omega$ yet $P \not\approx Q$. In this case we have $P \approx^!_\omega Q$, but from these we can produce BPP process states X_n and Y_n such that $X_n \approx^!_{\omega+n} Y_n$ by adding the following production rules to the defining grammar:

$$X_1 \xrightarrow{a} P \qquad X_{i+1} \xrightarrow{a} X_i \qquad Y_1 \xrightarrow{a} Q \qquad Y_{i+1} \xrightarrow{a} Y_i$$

However, no example BPP states X and Y are known which satisfy $X \approx^!_{\omega\times 2} Y$. This leads to the following long-standing conjecture.

Conjecture (Hirshfeld, Jančar). *Over BPP processes,* $\approx \, = \, \approx_{\omega\times 2}$.

$$A \xrightarrow{a} A \qquad P \xrightarrow{\tau} A \qquad Q \xrightarrow{a} \varepsilon$$
$$P \xrightarrow{\tau} Q \qquad Q \xrightarrow{\tau} QQ$$

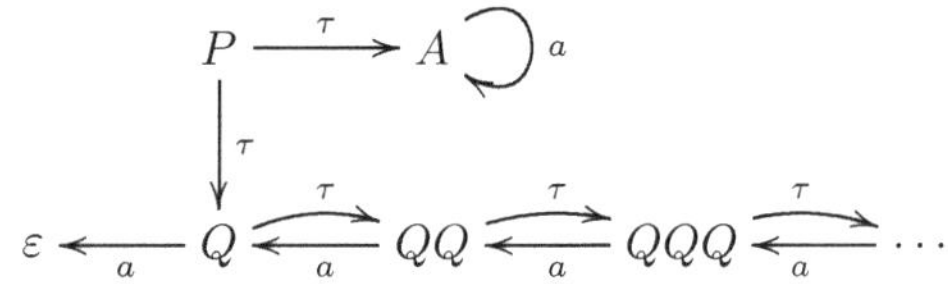

Fig. 2. A BPP process with states P and Q satisfying $P \approx^!_\omega Q$

Remark 2. *The situation is different for BPA, as noted originally in [22]: for any* $\alpha < \omega^\omega$, *we can construct BPA processes* P *and* Q *for which* $P \approx^!_\alpha Q$. *To see this, we consider the grammar* $G = (V, A, \Gamma)$ *in which* $V = \{X_0, X_1, \ldots, X_{n-1}\}$, $A = \{a, \tau\}$, *and* Γ *consists of the following rules:*

$$X_0 \xrightarrow{a} \varepsilon \qquad X_i \xrightarrow{\tau} \varepsilon \qquad X_{i+1} \xrightarrow{\tau} X_{i+1}X_i$$

For each $\alpha < \omega^n$, *with Cantor normal form*

$$\alpha \ = \ \omega^{n-1}a_{n-1} \ + \ \cdots \ + \ \omega^2 a_2 \ + \ \omega a_1 \ + \ a_0,$$

let $P_\alpha = X_0^{a_0} X_1^{a_1} X_2^{a_2} \cdots X_{n-1}^{a_{n-1}}$. *We can show that, for* $\alpha < \beta < \omega^n$, $P_\alpha \approx^!_\alpha P_\beta$. *This will follow from the following sequence of observations which demonstrate a close analogy between the processes* P_α *and the ordinal processes* α *from Remark 1:*

- *If* $P \approx Q$ *then* $RP \approx RQ$ *and* $PR \approx QR$. *The first conclusion is true for every BPA process, while the second conclusion is true for every BPA process in which* $P \overset{\tau}{\Rightarrow} \varepsilon$ *whenever* $P \approx \varepsilon$, *which is certainly the case for the BPA process under consideration since in this case* $P \approx \varepsilon$ *implies that* $P = \varepsilon$. *(Proof:* $\{(RP, RQ) : P \approx Q\}$ *and* $\{(PR, QR) : P \approx Q\}$ *are easily verified to be weak bisimulation relations.)*
- *For* $i > j$: $X_i X_j \approx X_i$. *(Proof: by induction on* $i - j$*.)*
- *Every state* $P \in V^*$ *is weakly bisimilar to some state* P_α. *(Proof: follows directly from the above observations.)*
- $X_k \overset{\tau}{\Rightarrow} P_\alpha$ *for every* $\alpha \leq \omega^k$. *(Proof: easily verified.)*
- $P_\alpha \overset{\tau}{\Rightarrow} P_\beta$ *for every* $\beta \leq \alpha$. *(Proof: generalisation of the above observation.)*
- $P_\alpha \overset{a}{\Rightarrow} P_\beta$ *for every* $\beta < \alpha$. *(Proof: follows from the previous observation and the fact that* $P_{\beta+1} \xrightarrow{a} P_\beta$*.)*
- *If* $P_\alpha \overset{\tau}{\Rightarrow} P$ *then* $P \approx P_\beta$ *for some* $\beta \leq \alpha$. *(Proof: easily verified.)*
- *If* $P_\alpha \overset{a}{\Rightarrow} P$ *then* $P \approx P_\beta$ *for some* $\beta < \alpha$. *(Proof: again easily verified.)*

We thus arrive at the following important observations about the states P_α:

- $P_\alpha \overset{\tau}{\Rightarrow} P_\beta$ *for all* $\beta \leq \alpha$, *and if* $P_\alpha \overset{\tau}{\Rightarrow} P$ *then* $P \approx P_\beta$ *for some* $\beta \leq \alpha$; *and*
- $P_\alpha \overset{a}{\Rightarrow} P_\beta$ *for all* $\beta < \alpha$, *and if* $P_\alpha \overset{a}{\Rightarrow} P$ *then* $P \approx P_\beta$ *for some* $\beta < \alpha$.

This suffices to deduce with little effort, analogously to Remark 1, that if $\alpha<\beta$ *then* $P_\alpha \approx^!_\alpha P_\beta$. *(The conjecture for BPA, though, is that the bound given by this construction is tight:* $\approx\, = \,\approx_{\omega^\omega}$ *.)*

BPP processes with silent moves are countably-branching, and thus by Theorem 6 $\approx\, = \,\approx_{\aleph_1}$. In [22] there is an argument attributed to J. Bradfield which shows that the approximation hierarchy collapses by the level $\approx_{\omega_1^{\mathrm{CK}}}$, the first non-recursive ordinal. (The argument is made there for BPA but clearly holds as well for BPP.) But this is to measure in lightyears what should require centimetres; we proceed here to a more modest bound, based on our ordinal analysis of Dickson's Lemma.

We assume an underlying grammar (V, A, Γ) defining our BPP process, and recall that a state in the associated process is simply a sequence $u \in V^*$ viewed as a multiset. With this, we make the important observation about weak bisimulation approximants over BPP: besides being equivalences, they are in fact congruences.

Lemma 4. *For all* $u, v, w \in V^*$, *if* $u \approx_\alpha v$ *then* $uw \approx_\alpha vw$.

Proof. By a simple induction on α. □

We next observe a result due to Hirshfeld [10].

Lemma 5. *If* $u \approx^!_\alpha v$ *and* $uu' \approx^!_\beta vv'$ *with* $\beta<\alpha$ *then* $uu' \approx^!_\beta uv'$ *and* $vu' \approx^!_\beta vv'$.

Proof. $uu' \approx_\beta uv'$ since $uu' \approx_\beta vv' \approx_\alpha uv'$. On the other hand, if $uu' \approx_{\beta+1} uv'$ then $uu' \approx_{\beta+1} uv' \approx_\alpha vv'$. Thus $uu' \approx^!_\beta uv'$. ($vu' \approx^!_\beta vv'$ can be shown similarly). □

BPP processes, being multisets over the finite variable set V, can be represented as $|V|$-tuples over $\mathbb{N}$. Given non-equivalent BPP states u_0 and v_0, $omt(u_0, v_0)$ can then be viewed as a $\mathbb{N}^{2\cdot|V|}$-labelled tree. In general this tree will not be non-dominating, but the above lemma will enable us to produce a non-dominating $\mathbb{N}^{2\cdot|V|}$-labelled tree from $omt(u_0, v_0)$

Lemma 6. *For BPP processes, if* $u_0 \approx^!_\alpha v_0$ *then there exists a* $\mathbb{N}^{2\cdot|V|}$*-labelled non-dominating tree of height* α.

Proof. We apply the following substitution procedure to each successive level of the weak-transition optimal move tree $omt(u_0, v_0)$ (where the *level* of a node refers to the distance from the root (u_0, v_0) to the node) by induction on the levels:

> For each node x at this level, if x dominates some ancestor node y, that is, if there exists an ancestor node $y = (u, v)$ where $x = (uu', vv')$, then replace the subtree rooted at x with either $x' = omt(uu', uv')$ (if $u <_{\mathrm{lex}} v$) or with $x' = omt(vu', vv')$ (if $v <_{\mathrm{lex}} u$). (If this x' itself then dominates an ancestor node, repeat this action.)

That $<_{\text{lex}}$ is a well-founded relation on $\mathbb{N}^{2\cdot|V|}$ means this repetition must halt; and Lemma 5 implies that this is a height-preserving operation. □

Theorem 7. *Over BPP processes,* $\approx \, = \, \approx_{\omega^\omega}$

Proof. If $u \approx v$ then $u \approx_{\omega^\omega} v$ is a given. If, on the other hand, $u \not\approx v$, then $u \approx^!_\alpha v$ for some α, and by the combination of Lemma 6 and Theorem 2 we must have that $\alpha \leq \omega^{2\cdot|V|}$. Thus, $u \not\approx_{\omega^\omega} v$. □

5 Conclusions

In this paper we provide a bound on the level at which the bisimulation approximation relations collapse over BPP. The bound we give of ω^ω is still a far cry from the widely-accepted conjectured bound of $\omega\times 2$, but it nonetheless represents the first nontrivial countable bound that has been discovered in the decade since this conjecture was first uttered (originally by Hirshfeld and Jančar).

We arrive at our bound through a careful analysis of Dickson's Lemma, and in particular via a novel constructive proof which provides this ordinal bound on non-dominating sequences of n-tuples. (Dickson's Lemma itself merely declares that such sequences are necessarily finite without actually identifying any ordinal bound.) This approach does not immediately seem to be applicable to strengthening the bound, given that this bound on Dickson's Lemma is tight. However, it seems equally likely that by taking into consideration the restricted form of non-dominating sequences produced by BPP transitions we can identify the missing ingredient for the proof of the tighter bound.

There have been other similar constructive proofs of Dickson's Lemma in the area of term rewriting. In particular, Sustik [23] provides a similar proof using an ordinal mapping on sequences in order to mechanically prove Dickson's Lemma using the ACL2 theorem prover. However, the ordinal mapping defined by Sustik gives an inferior bound to the one we provide; in particular, it requires ω^ω already for sequences of pairs.

Blass and Gurevich have very recently (preprint March 2006) written a manuscript [1] in which they define the *stature* of a well partial ordering P to be the order type of nondominating sequences of P, and (amongst other things) derive the same tight bound of ω^n as we have done. Their application of interest lies in program termination, and their proofs, being of more general results, are more complicated than the proof we provide. We therefore feel that our proof, which appeared in an earlier mauscript [9], as well as our application to bisimulation checking is of independent interest.

If the $\omega\times 2$ bound for the weak bisimulation approximation relations over BPP is resolved positively, this can potentially be exploited to resolve the decidability of weak bisimilarity over BPP. Esparza in [7] has shown that weak equivalence is semi-decidable, by demonstrating a semilinear witness of equivalence, so semi-decidability of non-equivalence is all that is required. If $\approx_\omega$ can be shown to be decidable (which is likely a much simpler result to attain than for $\approx$) then it would naturally be expected that the successor relations $\approx_{\omega+1}, \approx_{\omega+1}, \ldots$ would also be decidable, which would give rise to a semi-decision procedure for $\not\approx_{\omega\times 2}$: test each relation $\approx_{\omega+i}$ in turn until one such test fails.

References

1. A. Blass and Y. Gurevich. Program termination and well partial orderings. Microsoft Technical Report MSR-TR-2006-27, 31 pages, March 2006. (Available at ftp://ftp.research.microsoft.com/pub/tr/TR-2006-27.pdf.)
2. O. Burkart, D. Caucal, F. Moller and B. Steffen. Verification over Infinite States. Chapter 9 in the *Handbook of Process Algebra*, pages 545-623, Elsevier Publishers, 2001.
3. S. Christensen, Y. Hirshfeld, and F. Moller. Bisimulation equivalence is decidable for basic parallel processes. In *Proceedings of the 4th International Conference on Concurrency Theory (CONCUR'93)*, LNCS 715, pages 143-157, Springer, 1993.
4. S. Christensen, Y. Hirshfeld, and F. Moller. Decomposability, decidability and axiomatisability for bisimulation equivalence on basic parallel processes. In *Proceedings of the 8th Annual IEEE Symposium on Logic in Computer Science (LICS'93)*, pages 386-396, IEEE Computer Society Press, 1993.
5. S. Christensen, H. Hüttel, and C. Stirling. Bisimulation equivalence is decidable for all context-free processes. In *Proceedings of the 3rd International Conference on Concurrency Theory (CONCUR'92)*, LNCS 630, pages 138-147, Springer, 1992.
6. L.E. Dickson. Finiteness of the odd perfect and primitive abundant numbers with distinct factors. *American Journal of Mathematics* 3:413-422, 1913.
7. J. Esparza. Petri nets, commutative context-free grammars, and basic parallel proceses. *Fundamenta Informaticae* 30:23-41, 1997.
8. R. Glabbeek. The linear time – branching time spectrum I: The semantics of concrete sequential processes. Chapter 1 in the *Handbook of Process Algebra*, pages 3-99, Elsevier Publishers, 2001.
9. W. Harwood and F. Moller. Weak bisimulation approximants. In *Selected Papers from the CALCO Young Researchers Workshop (CALCO-jnr 2005)*, Swansea University Research Report CSR 18-2005, pages 27-40, December 2005. (Available at http://www-compsci.swan.ac.uk/reports/2005.html.)
10. Y. Hirshfeld. Bisimulation trees and the decidability of weak bisimulation. *Electronic Notes in Theoretical Computer Science* 5:2-13, 1997.
11. Y. Hirshfeld, M. Jerrum, and F. Moller. A polynomial algorithm for deciding bisimilarity of normed context-free processes. In *Proceedings of the 35th Annual IEEE Symposium on Foundations of Computer Science (FOCS'94)*, pages 623-631, IEEE Computer Society Press, 1994.
12. Y. Hirshfeld, M. Jerrum, and F. Moller. A polynomial algorithm for deciding bisimilarity of normed context-free processes. *Theoretical Computer Science* 15:143-159, 1996.
13. Y. Hirshfeld, M. Jerrum, and F. Moller. A polynomial algorithm for deciding bisimulation equivalence of normed basic parallel processes. *Mathematical Structures in Computer Science* 6:251-259, 1996.
14. P. Jančar. Strong bisimilarity on Basic Parallel Processes is PSPACE-complete. In *Proceedings of the 18th Annual IEEE Symposium on Logic in Computer Science (LICS'03)*, pages 218-227, IEEE Computer Society Press, 2003.
15. P. Jančar and F. Moller. Techniques for decidability and undecidability of bisimilarity. In *Proceedings of the 10th International Conference on Concurrency Theory (CONCUR'99)*, LNCS 1664, pages 30-45, Springer, 1999.

16. F. Moller. Infinite Results. In *Proceedings of the 7th International Conference on Concurrency Theory (CONCUR'96)*, LNCS 1119, pages 195-216, Springer, 1996.
17. J. Srba. Strong bisimilarity and regularity of Basic Process Algebra is PSPACE-hard. In *Proceedings of the 29th International Conference on Automata, Languages and Programming (ICALP'02)*, LNCS 2380, pages 716-727, Springer, 2002.
18. J. Srba. Strong bisimilarity and regularity of Basic Parallel Processes is PSPACE-hard. In *Proceedings of the 19th International Symposium on Theoretical Aspects of Computer Science (STACS'02)*, LNCS 2285, pages 535-546, Springer, 2002.
19. J. Srba. Complexity of weak bisimilarity and regularity for BPA and BPP. *Mathematical Structures in Computer Science* 13:567-587, 2003.
20. J. Srba. Roadmap of Infinite Results. http://www.brics.dk/~srba/roadmap.
21. C. Stirling. Decidability of weak bisimilarity for a subset of Basic Parallel Processes. In *Proceedings of Foundations of Software Science and Computation Structures (FOSSACS'01)*, LNCS 2030, pages 379-393, 2001.
22. J. Stříbrná. Approximating weak bisimulation on Basic Process Algebras. In *Proceedings of Mathematical Foundations of Computer Science (MFCS'99)*, LNCS 1672, pages 366-375, 1999.
23. M. Sustik. Proof of Dickson's Lemma using the ACL2 theorem prover via an explicit ordinal mapping. Unpublished presentation at the *Fourth International Workshop on the ACL2 Theorem Prover and Its Applications (ACL2-2003)*. (Available from http://www.cs.utexas.edu/users/moore/acl2/workshop-2003.)

Complete Problems for Higher Order Logics

Lauri Hella[1] and José María Turull-Torres[2]

[1] University of Tampere
Department of Mathematics,
Statistics and Philosophy
Kanslerinrinne 1, 33014
Tampere, Finland
Lauri.Hella@uta.fi
[2] Massey University
Information Science Research Centre
Department of Information Systems,
PO Box 756, Wellington, New Zealand
j.m.turull@massey.ac.nz

Abstract. [1] Let $i, j \geq 1$, and let Σ^i_j denote the class of the higher order logic formulas of order $i+1$ with $j-1$ alternations of quantifier blocks of variables of order $i+1$, starting with an existential quantifier block. There is a precise correspondence between the non deterministic exponential time hierarchy and the different fragments of higher order logics Σ^i_j, namely $NEXP^j_i = \Sigma^{i+1}_j$. In this article we present a complete problem for each level of the non deterministic exponential time hierarchy, with a very weak sort of reductions, namely *quantifier-free first order reductions*. Moreover, we don't assume the existence of an order in the input structures in this reduction. From the logical point of view, our main result says that every fragment Σ^i_j of higher order logics can be captured with a first order logic Lindström quantifier. Moreover, as our reductions are quantifier-free first order formulas, we get a normal form stating that each Σ^i_j sentence is equivalent to a single occurrence of the quantifier and a tuple of quantifier-free first order formulas. Our complete problems are a generalization of the well known problem *quantified Boolean formulas* with bounded alternation (QBF_j).

Keywords: Lindström Quantifiers, Complete Problems, Higher Order Logics.

1 Introduction

Descriptive complexity theory is one of the main research areas of finite model theory. It studies the ritchness of logical languages which are needed for expressing computational problems. The starting point was the work by Büchi ([Bu60]), and Fagin gave a decisive impulse to the field when he proved that $\Sigma^1_1 = NP$,

[1] The work presented here was partially supported by grants 04-BRFP-FINL-TURU of the Royal Society of New Zealand and 28139 of The Academy of Finland.

Z. Ésik (Ed.): CSL 2006, LNCS 4207, pp. 380–394, 2006.

that is, a class of finite models is definable by an existential sentence of second order logic (SO) iff it is in NP ([Fag74]). Later, Stockmeyer generalized it to the characterization of full SO ([Sto76]).

Complete problems for given complexity classes play an important role in complexity theory. Complete problems with respect to first order reductions have been previously studied by several authors. Immerman ([Imm83]) proved that, in the presence of a linear order, deterministic transitive closure is complete for LOGSPACE, transitive closure is complete for NLOGSPACE, and alternating transitive closure is complete for PTIME with respect to quantifier free projections, which are weaker than quantifier free first order reductions. Stewart ([Ste91]) studied various NP complete problems, and proved that they remain complete with respect to quantifier free first order reductions, assuming a linear order on the structures.

There are complete problems for NP with respect to first order reductions even without assuming linear order. Indeed, Dahlhaus ([Dah84]) proved that satisfiability is such a problem. However, in the case of PTIME, it is not known whether a complete problem with respect to quantifier free first order reductions exists if order is not assumed. The existence of such a problem would solve the famous open problem in finite model theory, whether PTIME can be captured by an effective logic on the class of all finite models. Dawar ([Daw95]) was able to prove the converse of this observation: if there is any effective logic capturing PTIME, then there is a complete problem with respect to quantifier free first order reductions without assuming order. On the other hand Grohe ([Gro95]) proved that there are complete problems for least fixed point logic and partial fixed point logic with respect to quantifier free first order reductions.

In this paper, we are interested in the descriptive complexity of the classes in the non deterministic exponential time hierarchy. This hierarchy is defined: $NEXP_i^0 = \bigcup_c NTIME(exp(i, O(n^c)))$, and $NEXP_i^j = NEXP_i^{0^{\Sigma^p_{j-1}}}$, where Σ^p_{j-1} is the $(j-1)$-th level of the polynomial hierarchy (PH), and $(exp(i, f(n)))$ is the exponential tower $2^{2^{\cdot^{\cdot^{\cdot^2}}}}$ of height i, and then topped with the function $f(n)$. The non deterministic exponential time hierarchy is captured by higher order logics, in the same way as PH is captured by SO.

The correspondence between the non deterministic exponential time hierarchy and the different fragments of higher order logics Σ^i_j, has been studied by Leivant ([Lei89]) and by Hull and Su ([HS91]), and more recently in ([HT03], [HT06]) where the exact correspondence was proved. That is, for each $i, j \geq 1$, $NEXP_i^j = \Sigma^{i+1}_j$.

From the point of view of complexity theory, our main result is the finding of a complete problem for each level of the non deterministic exponential time hierarchy, with a very weak sort of reductions, namely *quantifier-free first order reductions*. Moreover, we don't assume the existence of an order in the input structures in this reduction.

On the other hand, from the logical point of view, our main result says that every fragment Σ^i_j of higher order logics can be captured with a first order logic Lindström quantifier. Moreover, as our reductions are quantifier-free first order formulas, we get a normal form stating that each Σ^i_j sentence is equivalent to a single occurrence of the quantifier and a tuple of quantifier-free first order formulas.

Our complete problems are a generalization of the well known problem *quantified Boolean formulas* with bounded alternation (QBF_j), which are known to be complete for the corresponding level j of the polynomial hierarchy Σ^p_j. And hence, also complete for the fragment of second order logic Σ^1_j (see [Imm99]).

The paper is organized as follows. In Section 2 we establish the notation and we give the syntax and semantics of higher order logics. In Section 3 we prove that, for every $j \geq 1$, the Σ^i_j theory of the *Boolean model* is complete for Σ^2_j under PTIME reductions. The Boolean model is a two element structure of the *Boolean* signature, i.e., a signature which has no relation (or function) symbols, and which has two constants which are interpreted by the two elements, respectively.

In Section 4, we first define the *extended syntax trees* and the *hyper extended syntax trees* as relational structures which we use to represent second order formulas of the Boolean signature. Then we prove that, for every $j \geq 1$, a certain class of hyper extended syntax trees is complete for the fragment Σ^2_j of third order logic, under quantifier-free first order reductions. Finally, we use those problems as the interpretations of (relativized) Lindström quantifiers, and we give the normal form theorem for the fragments Σ^2_j of third order logic.

In Section 5, we extend our results getting complete problems under quantifier-free first order reductions for the fragments Σ^i_j of higher order logics of all finite orders. Once again, we use those problems as the interpretations of corresponding (relativized) Lindström quantifiers, and we give the normal form theorem for the fragments Σ^i_j of all finite orders.

[HT05] is a Technical Report where the proofs of the results presented in this article can be found.

2 Preliminaries

As usual ([EF99], [AHV95]), we regard a *relational database schema*, as a relational signature, and a *database instance* or simply *database* as a finite structure of the corresponding signature. If $\mathbf{A}$ is a database or structure of some schema σ, we denote its domain as $dom(\mathbf{A})$. If R is a relation symbol in σ of arity r, for some $r \geq 1$, we denote as $R^{\mathbf{A}}$ the (second order) relation of arity r which interprets the relation symbol R in $\mathbf{A}$, with the usual notion of interpretation. We denote as $\mathcal{B}_\sigma$ the class of *finite* σ-structures, or databases of schema σ. In this paper, we consider *total* queries only. Let σ be a schema, let $r \geq 1$, and let R be a relation symbol of arity r. We use the notion of a *logic* in a general sense. A formal definition would only complicate the presentation and is unnecessary for our work. As usual in finite model theory, we regard a logic as a language, that is, as a set of formulas (see [EF99]). We only consider signatures, or vocabularies, which are purely *relational*. We consider *finite* structures only. Consequently, the notion of *satisfaction*, denoted as $\models$, is related to only finite structures. By $\varphi(x_1, \ldots, x_r)$ we denote a formula of some logic whose free variables are *exactly*

$\{x_1, \ldots, x_r\}$. If $\varphi(x_1, \ldots, x_k) \in \mathcal{L}_\sigma$, $\mathbf{A} \in \mathcal{B}_\sigma$, $\bar{a}_k = (a_1, \ldots, a_k)$ is a k-tuple over $\mathbf{A}$, let $\mathbf{A} \models \varphi(x_1, \ldots, x_k)[a_1, \ldots, a_k]$ denote that φ is true, when interpreted by $\mathbf{A}$, under a valuation v where for $1 \leq i \leq k$ $v(x_i) = a_i$. Then we consider the set of all such valuations as follows:

$\varphi^{\mathbf{A}} = \{(a_1, \ldots, a_k) : a_1, \ldots, a_k \in dom(\mathbf{A}) \wedge \mathbf{A} \models \varphi(x_1, \ldots, x_k)[a_1, \ldots, a_k]\}$

That is, $\varphi^{\mathbf{A}}$ is the relation defined by φ in the structure $\mathbf{A}$, and its arity is given by the number of free variables in φ. Formally, we say that a formula $\varphi(x_1, \ldots, x_k)$ of signature σ, *expresses* a query q of schema σ, if for every database $\mathbf{A}$ of schema σ, is $q(\mathbf{A}) = \varphi^{\mathbf{A}}$. Similarly, a sentence φ expresses a Boolean query q if for every database $\mathbf{A}$ of schema σ, is $q(\mathbf{A}) = true$ iff $\mathbf{A} \models \varphi$.

For every $i \geq 2$, in the alphabet of a *Higher Order Logic of order* i, HO^i, besides the usual logical and punctuation symbols, we have a countably infinite set of *individual variables*, and for every arity, and for every order $2 \leq j \leq i$, a countably infinite set of *relation variables*. We use calligraphic letters like $\mathcal{X}$ and $\mathcal{Y}$ for relation variables, and lower case letters like x and y for individual variables. Let σ be a relational vocabulary. We define the set of *atomic formulas* on the vocabulary σ as follows: 1) If R is a relation symbol in σ of arity r, for some $r \geq 1$, and $x_0, \ldots, x_{r-1}$ are individual variables, then $R(x_0, \ldots, x_{r-1})$ is an atomic formula; 2) If x and y are individual variables, then $x = y$ is an atomic formula; 3) If $\mathcal{X}$ is a relation variable of order 2, and of arity r, for some $r \geq 1$, and $x_0, \ldots, x_{r-1}$ are individual variables, then $\mathcal{X}(x_0, \ldots, x_{r-1})$ is an atomic formula; 4) If $\mathcal{X}$ is a relation variable of order j, for some $3 \leq j \leq i$, and of arity r, for some $r \geq 1$, and $\mathcal{Y}_0, \ldots, \mathcal{Y}_{r-1}$ are relation variables of order $j-1$, and of arity r, then $\mathcal{X}(\mathcal{Y}_0, \ldots, \mathcal{Y}_{r-1})$ is an atomic formula; 5) Nothing else is an atomic formula.

We define the set of *well formed formulas* as follows: 1) An atomic formula is a well formed formulas; 2) If φ, ψ are well formed formulas, then the following are also well formed formulas: $(\neg\varphi)$, $(\varphi \wedge \psi)$, $(\varphi \vee \psi)$; 3) If φ is a well formed formulas, and x is an individual variable, then the following are also well formed formulas: $\exists x(\varphi)$, $\forall x(\varphi)$; 4) If φ is a well formed formulas, and $\mathcal{X}$ is a relation variable, then the following are also well formed formulas: $\exists \mathcal{X}(\varphi)$, $\forall \mathcal{X}(\varphi)$; 5) Nothing else is a well formed formula.

Let $r \geq 1$. A *second order relation of arity* r is a relation in the classical sense, i.e., a set of r-tuples of elements of the domain of a given structure. For an arbitrary $i \geq 3$, a *relation of order* i *of arity* r or an i*-th order relation of arity* r is a set of r-tuples of relations of order $i-1$. In general by *higher order relations* we mean relations of order i, for some $i \geq 2$. W.l.o.g., and for the sake of simplicity, we assume that the arity of a higher order relation is propagated downwards, i.e., the relations of order $i-1$ which form the r-tuples for a relation of order i, are themselves of arity r, and so on, all the way down to the second order relations, which are also of arity r. Note that we could also allow relations of order $< i-1$ to form r-tuples for relations of order i. Again, for the sake of simplicity, and w.l.o.g., we choose not to do so.

We can now define the semantics for higher order logics. Let σ be a relational vocabulary. A *valuation* v on a σ-structure $\mathbf{A}$, is a function which assigns to each individual variable x an element in $dom(\mathbf{A})$, and to each relation variable $\mathcal{X}$ of order j, for some $2 \leq j \leq i$, and of arity r, for some $r \geq 1$, a relation of order j and of arity r on $dom(\mathbf{A})$. Let v_0, v_1 be two valuations on a σ-structure $\mathbf{A}$, and let V be a variable of whichever kind, we say that v_0 and v_1 are V-*equivalent* if they coincide in every variable of whichever kind, with the possible exception of variable V. We also use the notion of equivalence w.r.t. sets of variables. Let $\mathbf{A}$ be a σ-structure, and let v be a valuation on $\mathbf{A}$.

Next, we define inductively the notion of *satisfaction*. Besides the usual rules for FO, we have the following: 1) $\mathbf{A}, v \models \mathcal{X}(x_0, \ldots, x_{r-1})$, where $\mathcal{X}$ is a relation variable of order 2 and of arity r, for some $r \geq 1$, and $x_0, \ldots, x_{r-1}$ are individual variables, iff the r-tuple $(v(x_0), \ldots, v(x_{r-1}))$ belongs to the second order relation $v(\mathcal{X})$; 2) $\mathbf{A}, v \models \mathcal{X}(\mathcal{Y}_0, \ldots, \mathcal{Y}_{r-1})$, where $\mathcal{X}$ is a relation variable of order j, for some $3 \leq j \leq i$, and of arity r, for some $r \geq 1$, and $\mathcal{Y}_0, \ldots, \mathcal{Y}_{r-1}$ are relation variables of order $j-1$ and of arity r, iff the r-tuple of relations of order $j-1$, $(v(\mathcal{Y}_0), \ldots, (v(\mathcal{Y}_{r-1}))$ belongs to the relation of order j $v(\mathcal{X})$; 3) $\mathbf{A}, v \models \exists \mathcal{X}(\varphi)$, where $\mathcal{X}$ is a relation variable, and φ is a well formed formula, iff there is a valuation v', which is $\mathcal{X}$-equivalent to v, such that $\mathbf{A}, v' \models \varphi$; 4) $\mathbf{A}, v \models \forall \mathcal{X}(\varphi)$, where $\mathcal{X}$ is a relation variable, and φ is a well formed formula, iff for every valuation v', which is $\mathcal{X}$-equivalent to v, $\mathbf{A}, v' \models \varphi$.

Let $i, j \geq 1$, as it is usual in classical Logic we denote by Σ^i_j the class of formulas $\varphi \in HO^{i+1}$ of the form $\exists \mathcal{X}_{11} \ldots \exists \mathcal{X}_{1s_1} \forall \mathcal{X}_{21} \ldots \forall \mathcal{X}_{2s_2} \exists \mathcal{X}_{31} \ldots \exists \mathcal{X}_{3s_3} \ldots Q\mathcal{X}_{j1} \ldots Q\mathcal{X}_{js_j}(\psi)$, where $\psi \in HO^i$, Q is either $\exists$ or $\forall$, depending on whether j is odd or even, respectively, and for $k \geq 1$ it is $s_k \geq 1$. That is, Σ^i_j is the class of HO^{i+1} formulas with $j-1$ alternations of quantifiers blocks of variables of order $i+1$, starting with an existential quantifier. Similarly, we denote by Π^i_j the class of formulas $\varphi \in HO^{i+1}$ of the form $\forall \mathcal{X}_{11} \ldots \forall \mathcal{X}_{1s_1} \exists \mathcal{X}_{21} \ldots \exists \mathcal{X}_{2s_2} \forall \mathcal{X}_{31} \ldots \forall \mathcal{X}_{3s_3} \ldots Q\mathcal{X}_{j1} \ldots Q\mathcal{X}_{js_j}(\psi)$, where $\psi \in HO^i$, Q is either $\forall$ or $\exists$, depending on whether j is odd or even, respectively, and for $k \geq 1$ it is $s_k \geq 1$. That is, Π^i_j is the class of HO^{i+1} formulas with $j-1$ alternations of quantifiers blocks of variables of order $i+1$, starting with an universal quantifier. We say that the formula φ is in *generalized Skolem normal form*, or *GSNF* if it belongs to either Σ^i_j or Π^i_j, for some $i, j \geq 1$. Note that, unfortunately, in the notations Σ^i_j and Π^i_j the index i denotes the order $i+1$. The following lemma is well known.

Lemma 1. (folklore) *For every $i \geq 2$, and for every formula $\varphi \in HO^i$ there is a formula $\hat{\varphi} \in HO^i$ which is in GSNF and which is equivalent to φ.*

We define next the *non-deterministic exponential hierarchy.*

Definition 1. *For every $i \geq 0$, let $NEXPH^0_i = \bigcup_{c \in N} NTIME(exp(i, O(n^c)))$; for every $j \geq 1$, let $NEXPH^j_i = NEXPH^{0^{\Sigma^p_{j-1}}}_i$, where Σ^p_{j-1} is defined as usual in the Polynomial Hierarchy (i.e., $\Sigma^p_0 = P$, and $\Sigma^p_j = NP^{\Sigma^p_{j-1}}$, for $j \geq 1$, see [BDG95]). That is, $NEXPH^j_i$ is the class of non deterministic Turing machines in the class $NEXPH^0_i$ with an oracle in Σ^p_{j-1}.*

Theorem 1. *([HT03]) Let σ be a relational signature. For every $i, j \geq 1$, the classes of σ-structures which are finitely axiomatizable in Σ^i_j are exactly those classes which belong to the complexity class $NEXPH^{j-1}_{i-1}$.*

3 A Complete Problem for Σ^2_j

3.1 Reduction of Structures to Formulas

Let ϕ be a formula of third order logic. W.l.o.g. and to simplify our exposition, we assume that the arity of all third order and second order variables is r, for some $r \geq 1$ and moreover, the arity of all second order variables used as components in third order atomic formulas is also r. All predicates in the signature of ϕ are also of arity r. We also assume that no variable in ϕ is quantified more than once.

We will define a translation function f_ϕ that maps interpretations of ϕ into formulas of second order logic. Recall that an interpretation for ϕ is a pair $\mathcal{A} = \langle \mathbf{A}, \alpha \rangle$, where $\mathbf{A}$ is a structure of the vocabulary τ_ϕ consisting of the relation symbols and constant symbols occurring in ϕ, and α is a mapping that interprets the free variables of ϕ in the domain A of $\mathbf{A}$. For each model $\mathbf{A}$, we define $f_\phi(\langle \mathbf{A}, \alpha \rangle)$ by induction on ϕ simultaneously for the possible variable assignments α. In addition we will define a function g that maps each interpretation $\mathcal{A}$ of ϕ to an interpretation $g(\mathcal{A})$ of the second order formula $f_\phi(\mathcal{A})$.

Before going into the definitions of $f_\phi(\mathcal{A})$ and $g(\mathcal{A})$, we need to fix a mapping that assigns for each higher order variable in ϕ a corresponding (sequence of) variable(s) in $f_\phi(\mathcal{A})$. We assign a first order variable $x_{V,\boldsymbol{a}}$ for each second order variable V in ϕ and each tuple $\boldsymbol{a} \in A^r$. All these variables are assumed to be distinct: $x_{U,\boldsymbol{a}} = x_{V,\boldsymbol{b}}$ if and only if $U = V$ and $\boldsymbol{a} = \boldsymbol{b}$. For each third order variable $\mathcal{T}$ of ϕ we assign a distinct second order variable $U_\mathcal{T}$ of arity $r \cdot |A|^r$.

We use the notation $\boldsymbol{x}_V$ for the tuple $(x_{V,\boldsymbol{a}_1}, \ldots, x_{V,\boldsymbol{a}_m})$, where $m = |A|^r$ and $\boldsymbol{a}_1, \ldots, \boldsymbol{a}_m$ is a fixed listing of the elements of A^r.

Let $\top$ be a sentence that is always true, and $\bot$ a sentence that is always false (e.g., $\top$ and $\bot$ can be chosen to be $\forall x\,(x = x)$ and $\exists x\,(x \neq x)$, respectively).

We are now ready to give the inductive definition for the function f_ϕ. For atomic formulas ϕ the definition is as follows: 1) if ϕ is $s = t$, where s and t are terms, then $f_\phi(\mathcal{A}) := \top$ if $s^\mathcal{A} = t^\mathcal{A}$, and $\bot$ if $s^\mathcal{A} \neq t^\mathcal{A}$; 2) if ϕ is $R(\boldsymbol{t})$ for a relation symbol R in τ_ϕ and a tuple $\boldsymbol{t}$ of terms, then $f_\phi(\mathcal{A}) := \top$ if $\boldsymbol{t}^\mathcal{A} \in R^\mathbf{A}$, and $\bot$ if $\boldsymbol{t}^\mathcal{A} \notin R^\mathbf{A}$; 3) if ϕ is $V(\boldsymbol{t})$ for a second order variable V and a tuple $\boldsymbol{t}$ of terms, then $f_\phi(\mathcal{A}) := (x_{V,\boldsymbol{a}} = 1)$, where $\boldsymbol{a} = \boldsymbol{t}^\mathcal{A}$; 4) if ϕ is $\mathcal{T}(\boldsymbol{V})$ for a third order variable $\mathcal{T}$ and a tuple of second order variables $\boldsymbol{V} = (V_1, \ldots, V_r)$, then $f_\phi(\mathcal{A}) := U_\mathcal{T}(\boldsymbol{x}_{V_1}, \ldots, \boldsymbol{x}_{V_r})$.
Assume then that f_ψ has been defined for all subformulas ψ of ϕ. Then f_ϕ is defined as follows: 1) if ϕ is $\neg\psi$, then $f_\phi(\mathcal{A}) := \neg f_\psi(\mathcal{A})$; 2) if ϕ is $\psi \vee \theta$, then $f_\phi(\mathcal{A}) := f_\psi(\mathcal{A}) \vee f_\theta(\mathcal{A})$; 3) if ϕ is $\psi \wedge \theta$, then $f_\phi(\mathcal{A}) := f_\psi(\mathcal{A}) \wedge f_\theta(\mathcal{A})$; 4) if ϕ is $\exists x\,\psi$, then $f_\phi(\mathcal{A}) := \bigvee_{a\in A} f_\psi(\mathcal{A}[a/x])$; 5) if ϕ is $\forall x\,\psi$, then $f_\phi(\mathcal{A}) := \bigwedge_{a\in A} f_\psi(\mathcal{A}[a/x])$; 6) if ϕ is $\exists V\,\psi$, then $f_\phi(\mathcal{A}) := \exists \boldsymbol{x}_V\, f_\psi(\mathcal{A})$; 7) if ϕ is $\forall V\,\psi$,

then $f_\phi(\mathcal{A}) := \forall \boldsymbol{x}_V \, f_\psi(\mathcal{A})$; 8) if ϕ is $\exists \mathcal{T} \, \psi$, then $f_\phi(\mathcal{A}) := \exists U_{\mathcal{T}} \, f_\psi(\mathcal{A})$; 9) if ϕ is $\forall \mathcal{T} \, \psi$, then $f_\phi(\mathcal{A}) := \forall U_{\mathcal{T}} \, f_\psi(\mathcal{A})$.

Note that the quantification used in this transformation for first order variables ($\exists \boldsymbol{x}_V$ and $\forall \boldsymbol{x}_V$) is done in such a way that we always use the same order in the occurrences of the individual variables in the block, which in turn is also the order in which variables are used in the atomic formulas $U_{\mathcal{T}}(\boldsymbol{x}_{V_1}, \ldots, \boldsymbol{x}_{V_r})$. We call these kind of formulas *Block Quantified Boolean Formulas* and we denote it by BBF. Later we will make use of this property.

The interpretation for the second order formula $f_\phi(\mathcal{A})$ is $g(\mathcal{A}) = \langle \mathbf{B}, \beta_{\mathcal{A}} \rangle$, where $\mathbf{B}$ is the two-element model $\langle \{0,1\}, 0, 1 \rangle$ (i.e., both elements are interpretations of constant symbols).

Before defining the interpretation $\beta_{\mathcal{A}}$ of free variables we observe that each relation $S \subseteq A^r$ can be represented on the model $\mathbf{B}$ as a binary string $\boldsymbol{b}_S = (b_1, \ldots, b_m) \in \{0,1\}^m$, where for each j, $b_j = 1 \iff \boldsymbol{a}_j \in S$. And vice versa, each binary string $\boldsymbol{b} \in \{0,1\}^m$ can be seen as a binary encoding of a relation $S_{\boldsymbol{b}} \subseteq A^r$.

The mapping $\beta_{\mathcal{A}}$ is defined as follows: 1) for each second order variable V and $\boldsymbol{a} \in A^r$, $\beta_{\mathcal{A}}(x_{V,\boldsymbol{a}}) = 1$ if $\boldsymbol{a} \in V^{\mathcal{A}}$, and 0 if $\boldsymbol{a} \notin V^{\mathcal{A}}$; 2) for each third order variable $\mathcal{T}$, $\beta_{\mathcal{A}}(U_{\mathcal{T}}) = \{(\boldsymbol{b}_{S_1}, \ldots, \boldsymbol{b}_{S_r}) \in \{0,1\}^{r \cdot m} \mid S_1, \ldots, S_r \subseteq A^r \text{ and } (S_1, \ldots, S_r) \in \mathcal{T}^{\mathcal{A}}\}$.

We make the following observations, which can be easily proved by induction given the definitions of $f_\phi(\mathcal{A})$ and $g(\mathcal{A})$: 1) the interpretation $g(\mathcal{A})$ does not depend on the interpretations $\alpha(x)$ of first order variables: $g(\mathcal{A}[a/x]) = g(\mathcal{A})$ for all $a \in A$; 2) the translation $f_\phi(\mathcal{A})$ does not depend on the interpretations $\alpha(U)$ and $\alpha(\mathcal{T})$ of second and third order variables: $f_\phi(\mathcal{A}[S/U]) = f_\phi(\mathcal{A})$ for all $S \subseteq A^r$, and $f_\phi(\mathcal{A}[\mathcal{R}/\mathcal{T}]) = f_\phi(\mathcal{A})$ for all $\mathcal{R} \subseteq (\mathcal{P}(A^r))^r$.

Theorem 2. *If ϕ is a third order formula and $\mathcal{A}$ is an interpretation for ϕ, then $\mathcal{A} \models \phi \iff g(\mathcal{A}) \models f_\phi(\mathcal{A})$.* ◇

Corollary 1. *For every $j \geq 1$ Σ^1_j-Th($\mathbf{B}$) is hard for Σ^2_j under P reductions.* ◇

Theorem 3. *For every $j \geq 1$ Σ^1_j-Th($\mathbf{B}$) is in Σ^2_j.* ◇

4 Completeness with First Order Reductions

4.1 Interpretation of Formulas as Structures

In this subsection we will only consider second order formulas which are BBF (see remark in previous subsection). Clearly each such second order Boolean formula can be represented as a structure by describing its syntax tree with suitable relations. In a straightforward description each variable corresponds to a different unary relation. However, since our aim is to find a class of structures that is complete for Σ^2_j, we need a uniform representation of formulas in a fixed signature. This can be achieved by representing variables by equivalence classes of an equivalence relation. We also need to identify the position in which blocks of variables occur in a given atomic formula. That can be done by using a constant number of special nodes as indices, and by then linking each block of variables with the index which corresponds to the position in which the block occurs in its respective atomic formula.

In addition we will need a binary relation that links each quantifier in a formula to the (occurrences of) variables it binds. For this purpose we have to work with an extended notion of syntax tree, in which each occurrence of a variable is considered as a separate node. More precisely, if an atomic formula $X(\boldsymbol{x_1}, \ldots, \boldsymbol{x_r})$ occurs in a formula ψ, then each block of variables $\boldsymbol{x_i}$, for $1 \leq i \leq r$, is regarded as an immediate subformula of $X(\boldsymbol{x_1}, \ldots, \boldsymbol{x_r})$ in the *extended syntax tree* of ψ. And furthermore each variable x_{ij}, for $1 \leq i \leq r$ and $1 \leq j \leq m$, is regarded as an immediate subformula of the block of variables $\boldsymbol{x_i}$ of length m. On the other hand, we will treat blocks of similar quantifiers as single nodes in the extended syntax tree; for example $\exists V_1 \ldots \exists V_n\, \theta$ corresponds to the node $\exists \boldsymbol{V}\, \theta$.

The formal definition of the translation of formulas into structures goes as follows. Let ψ be a second order Boolean formula which is in negation normal form. Then $\mathbf{C}_\psi$ is the structure $\langle C_\psi, S_\psi, E_\psi, R_\psi, I_\psi, L_{1_\square}, \ldots, L_{15_\square} \rangle$, where:

1) C_ψ is the union of the set of all subformulas in the extended syntax tree of ψ (different occurrences of the same subformula are considered as different elements) and a set of nodes used as indices (see predicate L_{15} below);

2) $S_\psi \subseteq (C_\psi)^2$ is the immediate subformula relation in the extended syntax tree of ψ;

3) $E_\psi \subseteq (C_\psi)^2$ is the set of all pairs (a, b) such that a and b are occurrences in the extended syntax tree of ψ of either the same first order variable, or atomic formulas with the same second order variable;

4) $R_\psi \subseteq (C_\psi)^2$ is the set of all pairs (a, b) such that a is a subformula of ψ of the form $\exists \boldsymbol{V}\, \theta$ or $\forall \boldsymbol{V}\, \theta$, ($\exists \boldsymbol{x}\, \theta$ or $\forall \boldsymbol{x}\, \theta$) and b is an occurrence of an atomic formula $V_i(\boldsymbol{x})$ in a, where V_i is component of $\boldsymbol{V}$ (b is an occurrence of a component of $\boldsymbol{x}$ in a, respectively);

5) $I_\psi \subseteq (C_\psi)^2$ is the set of all pairs (a, b) such that a is a block of variables $\boldsymbol{x}$, b is the index node which we designated as the j-th index, and the block of variables $\boldsymbol{x}$ occurs in position j in the atomic formula of which $\boldsymbol{x}$ is the immediate subformula;

6) $L_1, \ldots, L_{15}$ are unary relations encoding the "labels" of the nodes: a) L_1 is the set of all a that are of the form $\bigvee_{i \in I} \theta_i$; b) L_2 is the set of all a that are of the form $\bigwedge_{i \in I} \theta_i$; c) L_3 is the set of all a that are of the form $\exists \boldsymbol{V}\, \theta$; d) L_4 is the set of all a that are of the form $\forall \boldsymbol{V}\, \theta$; e) L_5 is the set of all a that are of the form $\exists \boldsymbol{x}\, \theta$; f) L_6 is the set of all a that are of the form $\forall \boldsymbol{x}\, \theta$; g) L_7 is the set of all a that are of the form $V(\boldsymbol{x})$; h) L_8 is the set of all a that are of the form $\boldsymbol{x}$; i)L_9 is the set of all a that are of the form x; j) L_{10} is the set of all a that are of the form $\neg\theta$; k) L_{11} is the set of all a that are of the form $\theta_1 \wedge \theta_2$; l) L_{12} is the set of all a that are of the form $\theta_1 \vee \theta_2$; m) L_{13} is the set of all a that are of the form $x_i = x_j$; n) L_{14} is the set of all a that are of the form $x_i = 1$; o) L_{15} is the set of index nodes.

Lemma 2. *Let C be an extended syntax tree such that there are Σ^2_j sentences ϕ_1, ϕ_2, of some vocabulary τ_ϕ, for some $j \geq 1$, and τ_ϕ structures $\mathcal{A}_1$, $\mathcal{A}_2$, such that C is isomorphic to both $\mathbf{C}_{f_{\phi_1}(\mathcal{A}_1)}$ and $\mathbf{C}_{f_{\phi_2}(\mathcal{A}_2)}$. Then $\mathcal{A}_1 \models \phi_1 \iff \mathcal{A}_2 \models \phi_2$.* $\diamond$

4.2 A First Order Reduction

Let ϕ be a Σ^2_j sentence, for some $j \geq 1$, of vocabulary τ_ϕ. Let $\mathcal{A}$ be a τ_ϕ structure. We will now sketch an interpretation of $\mathcal{A}$ which will define the extended syntax tree of the Σ^1_j sentence $f_\phi(\mathcal{A})$, $\mathbf{C}_{f_\phi(\mathcal{A})}$. As the structure $\mathcal{A}$ is not necessarily rigid, we cannot represent the nodes in $\mathbf{C}_{f_\phi(\mathcal{A})}$ as individual elements from $\mathcal{A}$. Instead, we will use *identity types* to distinguish the nodes of the syntax tree of the original sentence ϕ. For a tuple $\boldsymbol{v} = (v_1, \ldots, v_k)$, we denote as $itp(\boldsymbol{v})$ the identity type of $\boldsymbol{v}$, that is, the (unique) quantifier free FO formula with variables $\{x_1, \ldots, x_k\}$, which is a conjunction of equalities and inequalities, and which is true when $x_1, \ldots, x_k$ are substituted by $v_1, \ldots, v_k$, respectively. Note that $\mathbf{C}_{f_\phi(\mathcal{A})}$ has more nodes than the syntax tree of ϕ. To include them in the interpretation, in addition to the identity types, we use three more tuples of elements as coordinates, as follows.

With each node e in $\mathbf{C}_{f_\phi(\mathcal{A})}$ we associate tuples of elements $(\boldsymbol{v}_e, \boldsymbol{y}_e, \boldsymbol{z}_e, \boldsymbol{w}_e)$, where $\boldsymbol{v} = (v_1, \ldots, v_k)$, $\boldsymbol{y} = (y_1, \ldots, y_m)$, $\boldsymbol{z} = (z_1, \ldots, z_r)$, $\boldsymbol{w} = (w_1, \ldots, w_t)$, m is the quantifier rank of ϕ, and r is the arity of all relation variables in ϕ. k is a positive integer, big enough to allow us to assign one identity type for each node in the syntax tree of ϕ. For instance, if $k = 3$ we have the following identity types: $v_1 = v_2 = v_3$, $v_1 = v_2 \neq v_3$, $v_1 \neq v_2 = v_3$, $v_1 = v_3 \neq v_2$, and $v_1 \neq v_2 \neq v_3$. And t is a positive integer, big enough to allow us to have r identity types.

Note that for a given node in the syntax tree of ϕ, we might have more than one node in $\mathbf{C}_{f_\phi(\mathcal{A})}$. This expansion can be originated in five different situations:

1) For each node a of type $\exists x\, \theta$ ($\forall x\, \theta$) in ϕ which is in the path from the root to e in the syntax tree of ϕ, we have a node of type $\bigvee_{i \in I} \theta_i$ ($\bigwedge_{i \in I} \theta_i$) in $\mathbf{C}_{f_\phi(\mathcal{A})}$ which has $|A|$ successors. And all those nodes in $\mathbf{C}_{f_\phi(\mathcal{A})}$ which are originated from the same node in the syntax tree of ϕ, have the same identity type of $\boldsymbol{v}$. The sub-tuple $\boldsymbol{y}_e$ for a given node e, keeps track of the corresponding branch in each node of type $\bigvee_{i \in I} \theta_i$ or $\bigwedge_{i \in I} \theta_i$ in the path from the root to e in $\mathbf{C}_{f_\phi(\mathcal{A})}$. The maximum number of such types of nodes in the path from the root to any given node is given by the quantifier rank of ϕ (m).

2) For each node a of type $\mathcal{T}(\boldsymbol{V})$ in ϕ, for a third order variable $\mathcal{T}$ and a tuple of second order variables $\boldsymbol{V} = (V_1, \ldots, V_r)$, we have a node of type $U_{\mathcal{T}}(\boldsymbol{x}_{V_1}, \ldots, \boldsymbol{x}_{V_r})$ in $\mathbf{C}_{f_\phi(\mathcal{A})}$. Furthermore, that node has r successors in $\mathbf{C}_{f_\phi(\mathcal{A})}$, corresponding to the blocks of variables $\boldsymbol{x}_{V_1}, \ldots, \boldsymbol{x}_{V_r}$, which in turn have each $|A|^r$ successors, of type $x_{V_i, \boldsymbol{a}}$, for each r-tuple $\boldsymbol{a}$ in $\mathcal{A}$. We assign a different identity type of $\boldsymbol{v}$ for each of those three levels. The node of type $U_{\mathcal{T}}(\boldsymbol{x}_{V_1}, \ldots, \boldsymbol{x}_{V_r})$ in $\mathbf{C}_{f_\phi(\mathcal{A})}$ has one identity type of $\boldsymbol{v}$. All the nodes of type $\boldsymbol{x}_{V_i}$ (that is, those in the second level) have the same identity type of $\boldsymbol{v}$, but different from the one assigned to the node of type $U_{\mathcal{T}}(\boldsymbol{x}_{V_1}, \ldots, \boldsymbol{x}_{V_r})$. And all the nodes of type $x_{V_i, \boldsymbol{a}}$ (that is, those in the third level) have the same identity type of $\boldsymbol{v}$, but different from the one assigned to the node of type $U_{\mathcal{T}}(\boldsymbol{x}_{V_1}, \ldots, \boldsymbol{x}_{V_r})$, and the

one assigned to all its successor nodes of type $\boldsymbol{x}_{V_i}$. The sub-tuple $\boldsymbol{w}_e$ for a given node e of type $\boldsymbol{x}_{V_i}$, keeps track of the corresponding branch in the node of type $U_T(\boldsymbol{x}_{V_1},\ldots,\boldsymbol{x}_{V_r})$ of which e is successor in $\mathbf{C}_{f_\phi(\mathcal{A})}$. There are r branches in such node. The sub-tuple $\boldsymbol{z}_e$ for a given node e of type $x_{V_i,\boldsymbol{a}}$, keeps track of the corresponding branch in the node of type $\boldsymbol{x}_{V_i}$ of which e is successor in $\mathbf{C}_{f_\phi(\mathcal{A})}$.

3) For each node a of type $s = t$ in ϕ, if $s^{\mathcal{A}} = t^{\mathcal{A}}$ we have a node of type $\forall x\,\theta$ in $\mathbf{C}_{f_\phi(\mathcal{A})}$. Furthermore, that node has a successor in $\mathbf{C}_{f_\phi(\mathcal{A})}$, of type $x = x$, which in turn has a successor, of type x. We assign a different identity type of $\boldsymbol{v}$ for each of those three nodes.

4) For each node a of type $s = t$ in ϕ, if $s^{\mathcal{A}} \neq t^{\mathcal{A}}$ we have a node of type $\forall x\,\theta$ in $\mathbf{C}_{f_\phi(\mathcal{A})}$. Furthermore, that node has a successor in $\mathbf{C}_{f_\phi(\mathcal{A})}$, of type $\neg\theta$, which has a successor of type $x = x$, which in turn has a successor, of type x. We assign a different identity type of $\boldsymbol{v}$ for each of those four nodes.

5) For each node a of type $V(\boldsymbol{t})$ in ϕ, where $\boldsymbol{a} = \boldsymbol{t}^{\mathcal{A}}$ we have a node of type $x_{V,\boldsymbol{a}} = 1$ in $\mathbf{C}_{f_\phi(\mathcal{A})}$. Furthermore, that node has a successor in $\mathbf{C}_{f_\phi(\mathcal{A})}$, of type x. We assign a different identity type of $\boldsymbol{v}$ for each of those two nodes.

If c is a node in $\mathbf{C}_{f_\phi(\mathcal{A})}$ of type $x_{V,\boldsymbol{a}}$ which is a successor of a node of type $x_{V,\boldsymbol{a}} = 1$, we denote by $y^c_{i_1},\ldots,y^c_{i_r}$ the components in $\boldsymbol{y}_c$ which correspond to the components of the tuple $\boldsymbol{a}$ in $\mathcal{A}$. Note that each of these components in $\boldsymbol{y}_c$ corresponds to a particular branch after a node of type $\bigvee_{i\in I}\theta_i$ or $\bigwedge_{i\in I}\theta_i$ in the circuit, and all together determine the tuple $\boldsymbol{a}$.

As we are using identity types as a part of the identifiers of the nodes for $\mathbf{C}_{f_\phi(\mathcal{A})}$, we are getting multiple copies of each node of $\mathbf{C}_{f_\phi(\mathcal{A})}$ in the structure that we are *actually defining* with our reduction. Then we define the *hyper extended syntax tree* of the formula $f_\phi(\mathcal{A})$, denoted by $\mathbf{H}_{f_\phi(\mathcal{A})}$ as a structure of the same vocabulary as $\mathbf{C}_{f_\phi(\mathcal{A})}$, plus an additional binary relation symbol whose interpretation is a congruence relation, such that the quotient under that congruence relation, is isomorphic to the extended syntax tree of the formula $f_\phi(\mathcal{A})$, $\mathbf{C}_{f_\phi(\mathcal{A})}$.

Let b, c be two nodes in a structure $\mathbf{H}_{f_\phi(\mathcal{A})}$, for a given Σ^1_j sentence $f_\phi(\mathcal{A})$, for some $j \geq 1$, with identifiers $(\boldsymbol{v}_b,\boldsymbol{y}_b,\boldsymbol{z}_b,\boldsymbol{w}_b)$ and $(\boldsymbol{v}_c,\boldsymbol{y}_c,\boldsymbol{z}_c,\boldsymbol{w}_c)$, respectively We denote by $\boldsymbol{y}_b \doteq_{bc} \boldsymbol{y}_c$ the fact that $\boldsymbol{y}_b$ and $\boldsymbol{y}_c$ agree in the first h components, where h is the smaller number of nodes of type $\bigvee_{i\in I}\theta_i$ or $\bigwedge_{i\in I}\theta_i$ which occur in the path from the root to the nodes b, c in $\mathbf{C}_{f_\phi(\mathcal{A})}$, *excluding* the nodes b, c.

Note that if the nodes b, c are of type $\bigvee_{i\in I}\theta_i$ or $\bigwedge_{i\in I}\theta_i$ (which in turn are originated by a node of type $\exists x\theta$ and $\forall x\theta$, respectively, in the original formula ϕ) the value for y_h is ignored in the relation $\boldsymbol{y}_b \doteq_{bc} \boldsymbol{y}_c$ for the nodes b, c, but it is considered in $\boldsymbol{y}_b \doteq_{bc} \boldsymbol{y}_c$ for the successor nodes in $\mathbf{C}_{f_\phi(\mathcal{A})}$ (which correspond to the sub-formula θ).

We now define the equivalence relation $\equiv$ for $\mathbf{H}_{f_\phi(\mathcal{A})}$:

$b \equiv c$ iff the following holds: 1) b, c are nodes of type *index* and $itp(\boldsymbol{v}_b) = itp(\boldsymbol{v}_c)$; 2) b, c are nodes of type $x_{V,\boldsymbol{a}}$, which are successors of corresponding nodes of type $\boldsymbol{x}_V$ in $\mathbf{C}_{f_\phi(\mathcal{A})}$, $itp(\boldsymbol{v}_b) = itp(\boldsymbol{v}_c)$, $\boldsymbol{y}_b \doteq_{bc} \boldsymbol{y}_c$, $\boldsymbol{z}_b = \boldsymbol{z}_c$, and $\boldsymbol{w}_b = \boldsymbol{w}_c$; 3) b, c are nodes of type $x_{V,\boldsymbol{a}}$, which are successors of corresponding nodes of type

$x_{V,\boldsymbol{a}} = 1$ in $\mathbf{C}_{f_\blacksquare(\mathcal{A})}$, $itp(\boldsymbol{v}_b) = itp(\boldsymbol{v}_c)$, and $\boldsymbol{y}_b \doteq_{bc} \boldsymbol{y}_c$; 4) b, c are nodes of type $\boldsymbol{x}_V$, $itp(\boldsymbol{v}_b) = itp(\boldsymbol{v}_c)$, $\boldsymbol{y}_b \doteq_{bc} \boldsymbol{y}_c$, and $\boldsymbol{w}_b = \boldsymbol{w}_c$; 5) b, c are nodes of any other type, $itp(\boldsymbol{v}_b) = itp(\boldsymbol{v}_c)$, and $\boldsymbol{y}_b \doteq_{bc} \boldsymbol{y}_c$.

Note that the equivalence classes of $\equiv$ can be mapped to elements of $\mathbf{C}_{f_\blacksquare(\mathcal{A})}$ in a cannonical way by using the identity types which form the tuples assigned as identifiers to the nodes in $\mathbf{H}_{f_\blacksquare(\mathcal{A})}$. Hence, we can define all the relations in $\mathbf{H}_{f_\blacksquare(\mathcal{A})}$ in a unique way such that the quotient structure $\mathbf{H}_{f_\blacksquare(\mathcal{A})}/ \equiv_{\mathbf{H}_{f_\blacksquare(\mathcal{A})}}$ is isomorphic to $\mathbf{C}_{f_\blacksquare(\mathcal{A})}$.

Definition 2. *For every $j \geq 1$, we define the class $\mathcal{H}^2_j$ as the class of all structures $\mathbf{H}$ in the vocabulary of the hyper extended syntax trees such that the quotient structure $\mathbf{H}/ \equiv_{\mathbf{H}}$ is isomorphic to $\mathbf{C}_{f_\blacksquare(\mathcal{A})}$ for some $\phi \in \Sigma^2_j$, and τ_ϕ interpretation $\mathcal{A}$, such that $\langle\{0,1\},0,1\rangle \models f_\phi(\mathcal{A})$.*

Theorem 4. *For every $j \geq 1$, the class $\mathcal{H}^2_j$ is hard for Σ^2_j under quantifier free first order reductions.* ◇

Theorem 5. *For every $j \geq 1$, the class $\mathcal{H}^2_j$ is in Σ^2_j.* ◇

4.3 Capturing Σ^2_j by a Lindström Quantifier

Observe that the classes $\mathcal{H}^2_j$, for every $j \geq 1$, are by definition closed under isomorphisms. Hence in the model theoretical setting they can be considered as the interpretations of (relativized) Lindström quantifiers (see [EF99], [Lib04]) which we denote by Q_{2j}. From Theorems 4 and 5 it follows that, for each $j \geq 1$, the extension of FO with the vectorized quantifiers Q^k_{2j}, for $k \in \omega$, captures Σ^2_j. Thus it is possible to capture fragments of third order logic with first order generalized quantifiers.

More precisely, as our reduction is quantifier-free, we obtain the following normal form theorem for Σ^2_j.

Corollary 2. *Let $j \geq 1$. For every signature τ, a class of τ structures is definable in Σ^2_j iff it is definable by a sentence of the form*

$$Q^k_{2j}\, \boldsymbol{x}_C, \ldots, \boldsymbol{x}_{L_{15}}(\alpha_C(\boldsymbol{x}_C), \ldots, \alpha_{L_{15}}(\boldsymbol{x}_{L_{15}}))$$

where $k \geq 1$ and $\alpha_C, \ldots, \alpha_{L_{15}}$ are quantifier-free FO formulas. ◇

Note that the formulas $\alpha_C(\boldsymbol{x}_1), \ldots, \alpha_{L_{15}}(\boldsymbol{x}_{L_{15}})$ in the previous corollary are those used for the translation of τ structures to hyper extended syntax trees. In particular, the formula α_C defines the universe to which the quantifier Q^k_{2j} is relativized.

The normal form is very strong, in the sense that there is a single application of the quantifier Q^k_{2j}. If we consider the extension of FO as it is usually defined, $FO(Q^\omega_{2j})$, then we get a logic which is probably stronger than Σ^2_j. In particular, $FO(Q^\omega_{2j})$, is closed under negations.

5 Beyond Third Order Logic

5.1 Translation of the Formula

We will now see that actually all our results hold for *all* orders $i \geq 2$. We will sketch the alterations to the corresponding observations and proofs for the case of fourth order. The generalization for all other orders then follows the same strategy as for fourth order.

We will use $\mathcal{X}$ for a fourth order relation variable (of arity r), and $\Re$ for a fourth order relation (of arity r).

Let ϕ be a Σ^i_j sentence, for arbitrary $i \geq 2$ and $j \geq 1$. We also assume, w.l.o.g., that all variables of all orders are of arity r, which in turn is propagated downward to the corresponding lower order variables which form the atomic formulas in ϕ. As we did before, we will define a translation function f_ϕ that maps interpretations of ϕ of the form $\mathcal{A} = \langle \mathbf{A}, \alpha \rangle$, where $\mathbf{A}$ is a structure of the vocabulary τ_ϕ, into formulas of Σ^{i-1}_j. We will also use a function g that maps each interpretation $\mathcal{A}$ of ϕ to an interpretation $g(\mathcal{A})$ of the Σ^{i-1}_j formula $f_\phi(\mathcal{A})$.

In addition to our definitions in Section 2, we need the following (for fourth order logic): for each fourth order variable $\mathcal{X}$ of ϕ we assign a distinct third order variable $\mathcal{T}_\mathcal{X}$ of arity r.

For atomic formulas ϕ we need the following additional definition (for fourth order logic): if ϕ is $\mathcal{X}(\boldsymbol{\mathcal{T}})$ for a fourth order variable $\mathcal{X}$ and a tuple of third order variables $\boldsymbol{\mathcal{T}} = (\mathcal{T}_1, \ldots, \mathcal{T}_r)$, then $f_\phi(\mathcal{A}) := \mathcal{T}_\mathcal{X}(U_{\mathcal{T}_1}, \ldots, U_{\mathcal{T}_r})$, where the second order variables $U_{\mathcal{T}_1}, \ldots, U_{\mathcal{T}_r}$ are all of arity $r \cdot |A|^r$ (see Section 2).
Assume then that f_ψ has been defined for all subformulas ψ of ϕ. Then we need the following additional definitions for f_ϕ (for fourth order logic): 1) if ϕ is $\exists \mathcal{X}\, \psi$, then $f_\phi(\mathcal{A}) := \exists \mathcal{T}_\mathcal{X}\, f_\psi(\mathcal{A})$; 2) if ϕ is $\forall \mathcal{X}\, \psi$, then $f_\phi(\mathcal{A}) := \forall \mathcal{T}_\mathcal{X}\, f_\psi(\mathcal{A})$.

For the mapping $\beta_\mathcal{A}$ we need the following additional definition (for fourth order logic): for each fourth order variable $\mathcal{X}$,

$\beta_\mathcal{A}(\mathcal{T}_\mathcal{X}) = \{(T_{\mathcal{R}_1}, \ldots, T_{\mathcal{R}_r}) \mid \mathcal{R}_1, \ldots, \mathcal{R}_r \subseteq (\mathcal{P}(A^r))^r \text{ and } (\mathcal{R}_1, \ldots, \mathcal{R}_r) \in \mathcal{X}^\mathcal{A}\}$
where for each $1 \leq i \leq r$,

$T_{\mathcal{R}_i} = \{(\boldsymbol{b}_{S_1}, \ldots, \boldsymbol{b}_{S_r}) \in \{0,1\}^{r \cdot m} \mid S_1, \ldots, S_r \subset A^r \text{ and } (S_1, \ldots, S_r) \in \mathcal{R}_i\}$.
We make the following additional observation (for fourth order logic), which can be easily proved by induction given the definitions of $f_\phi(\mathcal{A})$ and $g(\mathcal{A})$:

the translation $f_\phi(\mathcal{A})$ does not depend on the interpretations $\alpha(\mathcal{X})$ of fourth order variables: $f_\phi(\mathcal{A}[\Re/\mathcal{X}]) = f_\phi(\mathcal{A})$ for all $\Re \subseteq [\mathcal{P}[(\mathcal{P}(A^r))^r]]^r$.

Theorem 6. *If ϕ is a HO^i formula, for some $i \geq 3$, and $\mathcal{A}$ is an interpretation for ϕ, then $\mathcal{A} \models \phi \iff g(\mathcal{A}) \models f_\phi(\mathcal{A})$.* ◇

Corollary 3. *For every $i \geq 1$ and $j \geq 1$ Σ^i_j-Th($\mathbf{B}$) is hard for Σ^{i+1}_j under P reductions.* ◇

Theorem 7. *For every $i \geq 1$ and $j \geq 1$ Σ^i_j-Th($\mathbf{B}$) is in Σ^{i+1}_j.* ◇

5.2 Completeness with First Order Reductions

As to the defintion of the extended syntax tree for a fourth order sentence ψ, we must do the following alterations to the relations in $\mathbf{C}_\psi$, where $\mathcal{T}$ is a third order relation variable and V is an SO relation variable. Note that the extended syntax tree will now also contain nodes for third order atoms and also nodes for SO variables as successors of them.

1) $E_\psi \subseteq (C_\psi)^2$: We must include the pairs (a, b) such that a and b are occurrences in the extended syntax tree of ψ of either the same SO variable, or atomic formulas with the same third order variable, or $a(b)$ is an atomic formula with an SO variable V and $b(a)$ is an occurrence of the variable V.

2) $R_\psi \subseteq (C_\psi)^2$: We must include the pairs (a, b) such that a is a subformula of ψ of the form $\exists \boldsymbol{\mathcal{T}}\, \theta$ or $\forall \boldsymbol{\mathcal{T}}\, \theta$ and b is an occurrence of an atomic formula $\mathcal{T}_i(\boldsymbol{V})$ in a, where $\mathcal{T}_i$ is component of $\boldsymbol{\mathcal{T}}$. We must also include the pairs (a, b) such that a is a subformula of ψ of the form $\exists V\, \theta$ or $\forall V\, \theta$ and b is an occurrence of the variable V.

3) $I_\psi \subseteq (C_\psi)^2$: We must include the pairs (a, b) such that a is an SO variable V, b is the index node which we designated as the j-th index, and the variable V occurs in position j in the atomic formula of which V is the immediate subformula.

4) We must add the following ‘labels" of the nodes: a) L_{16} is the set of all a that are of the form $\exists \boldsymbol{\mathcal{T}}\, \theta$; b) L_{17} is the set of all a that are of the form $\forall \boldsymbol{\mathcal{T}}\, \theta$; c) L_{18} is the set of all a that are of the form V; d) L_{19} is the set of all a that are of the form $\mathcal{T}(\boldsymbol{V})$.

The following result is immediate.

Lemma 3. *Let $i \geq 1$, and $j \geq 1$. Let C be an extended syntax tree such that there are Σ^{i+1}_j sentences ϕ_1, ϕ_2, of some vocabulary τ_ϕ, and τ_ϕ structures $\mathcal{A}_1$, $\mathcal{A}_2$, such that C is isomorphic to both $\mathbf{C}_{f_{\blacksquare 1}(\mathcal{A}_1)}$ and $\mathbf{C}_{f_{\blacksquare 2}(\mathcal{A}_2)}$. Then $\mathcal{A}_1 \models \phi_1 \iff \mathcal{A}_2 \models \phi_2$.* ◇

As to the equivalence relation in the hyper extended syntax tree $\mathbf{H}_{f_\blacksquare(\mathcal{A})}$, we must consider an additional case (for fourth order logic): b, c are nodes of type V, for a SO variable V, $itp(\boldsymbol{v}_b) = itp(\boldsymbol{v}_c)$, $\boldsymbol{y}_b \doteq_{bc} \boldsymbol{y}_c$, and $\boldsymbol{w}_b = \boldsymbol{w}_c$.

Now we must generalize the definition of the classes $\mathcal{H}^2_j$.

Definition 3. *For every $i \geq 1$, and $j \geq 1$, we define the class $\mathcal{H}^i_j$ as the class of all structures $\mathbf{H}$ in the vocabulary of the hyper extended syntax trees such that the quotient structure $\mathbf{H}/\equiv_\mathbf{H}$ is isomorphic to $\mathbf{C}_{f_\blacksquare(\mathcal{A})}$ for some $\phi \in \Sigma^i_j$, and τ_ϕ structure $\mathcal{A}$, such that $\langle \{0, 1\}, 0, 1\rangle \models f_\phi(\mathcal{A})$.*

Theorem 8. *For every $i \geq 1$, and $j \geq 1$, the class $\mathcal{H}^i_j$ is hard for Σ^i_j under quantifier free first order reductions.* ◇

Theorem 9. *For every $i \geq 1$, and $j \geq 1$, the class $\mathcal{H}^i_j$ is in Σ^i_j.* ◇

5.3 Capturing Σ_j^i by a Lindström Quantifier

The observations we made before are also valid for the general case. Then, from Theorems 8 and 9 it follows that, for each $i \geq 1$, and $j \geq 1$, the extension of FO with the vectorized quantifiers which we now denote by Q_{ij}^k, for $k \in \omega$, captures Σ_j^i. Thus, even fragments of i-th order logic, for an arbitrary order i, can be captured with first order generalized quantifiers.

We also obtain in the general case a normal form theorem for Σ_j^i.

Corollary 4. *For every $i \geq 1$ there is a constant n_i such that, for every $j \geq 1$, and every signature τ, a class of τ structures is definable in Σ_j^i iff it is definable by a sentence of the form*

$$Q_{ij}^k\, \boldsymbol{x}_1, \ldots, \boldsymbol{x}_{n_i}(\alpha_1(\boldsymbol{x}_1), \ldots, \alpha_{n_i}(\boldsymbol{x}_{n_i}))$$

where $k \geq 1$ and $\alpha_1, \ldots, \alpha_{n_i}$ are quantifier-free FO formulas. ◇

Note that the formulas $\alpha_1(\boldsymbol{x}_1), \ldots, \alpha_{n_i}(\boldsymbol{x}_{n_i})$ in the previous corollary are those used for the translation of τ structures to hyper extended syntax trees, so that for every $i \geq 1$, n_i is the number of relation symbols in the vocabulary of the hyper extended syntax trees for order i. In particular, the formula α_1 defines the universe to which the quantifier Q_{ij}^k is relativized.

Also in the general case the normal form is very strong, in the sense that there is a single application of the quantifier Q_{ij}^k. If we consider the extension of FO as it is usually defined, $FO(Q_{ij}^\omega)$, then we get a logic which is probably stronger than Σ_j^i. In particular, $FO(Q_{ij}^\omega)$ is closed under negations.

References

[AHV95] Abiteboul, S., Hull, R. and Vianu, V., Foundations of Databases, Addison-Wesley, 1995.

[BDG95] Balcázar, J., Díaz, J., Gabarró, J., Structural Complexity I, Springer, 2nd. ed., 1995.

[Bu60] Büchi, J. R., "Weak Second-Order Arithmetic and Finite Automata", Zeitschrift für Mathematische Logik und Grundlagen der Mathematik 6, pp. 66–92, 1960.

[Dah84] Dahlhaus, E., "Reduction to NP-Complete Problems by Interpretations", in "Logic and Machines: Decision Problems and Complexity", E. Börger, G. Hasenjaeger, D. Rödding (eds), Lecture Notes in Computer Science 171, Springer, pp. 357–365, 1984.

[Daw95] Dawar, A., "Generalized Quantifiers and Logical Reducibilities", Journal of Logic and Computation 5, pp. 213–226, 1995.

[EF99] Ebbinghaus, H., Flum, J., Finite Model Theory, Springer, 2nd. ed., 1999.

[Fag74] Fagin, R., "Generalized First Order Spectra and Polynomial Time Recognizable Sets", in "Complexity of Computation", ed. R. Karp, SIAM AMS Proceedings 7, pp. 43–73, 1974.

[Gro95] Grohe, M., "Complete Problems for Fixed-Point Logics", Journal of Symbolic Logic 60, pp. 517–527, 1995.

[HS91] Hull, R., Su, J., "On the Expressive Power of Database Queries with Intermediate Types", Journal of Computer and System Sciences 43(1), pp. 219–267, 1991.

[HT03] Hella, L., Turull Torres, J. M., "Expressibility of Higher Order Logics", Electronic Notes in Theoretical Computer Science, Vol. 84, 2003.

[HT05] Hella, L., Turull Torres, J. M., "Complete Problems for Higher Order Logics", Technical Report 12/2005, 27 pp., Information Systems Department, Massey University, 2005.

[HT06] Hella, L., Turull Torres, J. M., "Computing Queries with Higher Order Logics", Theoretical Computer Science 355(2), pp. 197-214, 2006.

[Imm83] Immerman, N., "Languages which Capture Complexity Classes", 15th ACM STOC Symposium, pp. 347–354, 1983. Revised version: "Languages that Capture Complexity Classes", SIAM Journal of Computing 16(4), pp. 760–778, 1987.

[Imm99] Immerman, N., Descriptive Complexity, Springer, 1999.

[Lei89] Leivant, D., "Descriptive Characterizations of Computational Complexity", Journal of Computer and System Sciences 39, pp. 51–83, 1989.

[Lib04] Libkin, L, Elements of Finite Model Theory, Springer, 2004.

[Ste91] Stewart, I., "Comparing the Expressibility of Languages Formed Using NP-Complete Operators", Journal of Logic and Computation 1(3), pp. 305–330, 1991.

[Sto76] Stockmeyer, L., "The Polynomial Time Hierarchy", Theoretical Computer Science 3(1), pp. 1–22, 1976.

Solving Games Without Determinization*

Thomas A. Henzinger and Nir Piterman

EPFL, Switzerland

Abstract. The synthesis of reactive systems requires the solution of two-player games on graphs with ω-regular objectives. When the objective is specified by a linear temporal logic formula or nondeterministic Büchi automaton, then previous algorithms for solving the game require the construction of an equivalent *deterministic* automaton. However, determinization for automata on infinite words is extremely complicated, and current implementations fail to produce deterministic automata even for relatively small inputs. We show how to construct, from a given nondeterministic Büchi automaton, an equivalent *nondeterministic* parity automaton $\mathcal{P}$ that is *good for solving games* with objective $\mathcal{P}$. The main insight is that a nondeterministic automaton is good for solving games if it fairly simulates the equivalent deterministic automaton. In this way, we omit the determinization step in game solving and reactive synthesis. The fact that our automata are nondeterministic makes them surprisingly simple, amenable to symbolic implementation, and allows an incremental search for winning strategies.

1 Introduction

One of the most ambitious goals in formal methods is to automatically produce designs from specifications, a process called *synthesis*. We are interested in *reactive systems*, i.e., systems that continuously interact with other programs, users, or their environment (like operating systems or CPUs). The complexity of a reactive system does not arise from computing a complicated function but rather from the fact that it has to be able to react to all possible inputs and maintain its behavior forever. There are two (essentially equivalent) approaches to solving the synthesis problem. The first is by reducing it to the emptiness problem of tree automata [Rab72], and the second, by reducing it to solving infinite-duration two-player games [BL69]. We consider the second view. The two players in the game are the system and its environment. The environment tries to violate the specification and the system tries to satisfy it. The system wins the game if it has a strategy such that all infinite outcomes satisfy the specification. The winning strategy, the way in which the system updates its internal variables, is then translated into an implementation that satisfies the specification when interacting with any possible environment.

More formally, a *game* is a directed graph where the vertices are partitioned between player 0 (system) and player 1 (environment). A *play* proceeds by moving a token along the edges of the graph. If the token is on a vertex of player

* This research was supported in part by the Swiss National Science Foundation.

Z. Ésik (Ed.): CSL 2006, LNCS 4207, pp. 395–410, 2006.

0, she gets to choose to which successor to move the token. If the token is on a vertex of player 1, she chooses the successor. When they continue in this fashion ad infinitum, the token passes an infinite sequence of vertices. We determine who wins the play by looking at this infinite outcome. We define winning plays either by conditions (such as parity or Rabin conditions) on the locations that occur infinitely often along a play, or by recognizers (such as linear temporal logic formulas or Büchi automata) of infinite words over the alphabet of locations. In either case, we are interested in *solving* the game. That is, we wish to determine from which locations of the game, player 0 has a winning strategy, i.e., a way to resolve her decisions so that the resulting plays are winning. For example, when the winning condition is a parity condition [EJ91], the problem of solving the game is in NP∩co-NP [EJS93] and the current best complexity for solving such games is $O(t \cdot g^{\lfloor \frac{k}{2} \rfloor})$, where g, t, and k are the number of locations and transitions in the game and priorities in the parity condition, respectively [Jur00, JV00].

In the context of synthesis, we consider an interaction of the system and the environment as winning for the system if it satisfies the specification. Thus, it makes more sense to consider games where the winning condition is given as a *linear temporal logic* (LTL) formula or *nondeterministic Büchi word automaton* (NBW). The way to solve such games is by reducing the problem to the solution of simpler games such as parity or Rabin. As part of this reduction, before taking the product of the game with the winning condition, we have to construct a deterministic automaton for the winning condition. This is because every sequence of choices made in the game has to satisfy the specification.

The first problem we encounter when we come to determinize automata on infinite words is that the Büchi acceptance condition is not strong enough [Lan69]. We have to use stronger acceptance conditions like parity or Rabin. Indeed, Safra suggested a determinization construction that takes an NBW and constructs a deterministic Rabin automaton [Saf88]. Recently, Piterman suggested a variant of this construction with a smaller number of states that results in a deterministic parity automaton [Pit06]. Specifically, starting from an NBW with n states, he constructs a deterministic parity automaton with n^{2n+2} states and $2n$ priorities. When we combine the game with the deterministic automaton, we get a game with $g \cdot n^{2n+2}$ locations and $t \cdot n^{2n+2}$ transitions, where g and t are the number of locations and transitions in the original game. The overall complexity of solving this game, therefore, is $O(t \cdot n^{2n+2} \cdot (g \cdot n^{2n+2})^n)$. This theory is not applicable in practice, because determinization is extremely complex. Every state of the deterministic automaton is a tree of subsets of states of the original automaton. A transition moves states between different nodes of the tree, adds and removes nodes, and changes the names of the nodes. Only recently, 16 years after the publications of Safra's construction, it was finally implemented [THB95, KB05, ATW05]. These implementations are limited to determinize automata with approximately 10 states[1]. One possible

[1] Piterman's variant maintains the tree structure of Safra that proved hard to implement. It reduces the amount of information associated with every node, however, at the price of giving the nodes dynamic names, which may prove hard to implement.

solution is to consider restricted specifications that can be handled more efficiently (cf. [RW89, RA03, KPP05]). Another possible solution is to use nondeterministic specification automata, which make the approach sound but incomplete [HRS05, JGB05].

Here we do pursue complete solutions for general ω-regular specifications. While we cannot improve the worst-case complexity of synthesis, it is desirable to have an algorithm that performs well in many cases that occur in practice, even if they involve a large number of states. In particular, we wish to use two heuristics that have had great success in formal verification, but cannot be used when applying determinization. The first is to reason *symbolically* about sets of states, rather than explicitly about individual states [McM93]. Using a symbolic state representation in Safra's construction seems difficult. Second, we wish to be able to find a winning strategy in a game that uses a small amount of memory, if such a strategy exists. The memory used by a strategy corresponds to the number of states of a parity or Rabin specification automaton. Thus, small memory is not possible if we construct the deterministic automaton as the first step of the synthesis algorithm. Instead, we want to *incrementally* increase, as much as necessary, the memory provided to strategies.

For this purpose we propose a general solution that does not involve determinization. We define *good for games* automata (GFG, for short), which are the class of nondeterministic automata that can be used in the context of games. The main idea is that if an automaton can resolve its nondeterminism in a stepwise fashion, then it is good enough for reasoning about games. The formal definition of a GFG automaton considers a game played on the structure of the automaton in which the opponent chooses input letters, one at a time, and the automaton resolves its nondeterminism for each input letter. The automaton wins if whenever the infinite word chosen by the opponent is in the language of the automaton, then the run chosen by the automaton is accepting. The automaton is GFG if it has a winning strategy in this game. We show that a nondeterministic specification automaton with this property can indeed be used for solving games without prior determinization. That is, in the product of a game with a GFG automaton, the winning states correspond to the winning states of the original game. In order to check if an automaton is GFG, we give an alternative characterization: an automaton is GFG iff it fairly simulates [HKR97] a deterministic automaton for the same language.

Our main contribution is a construction that takes an NBW and produces a GFG automaton for the same language. Given an NBW with n states, we construct a nondeterministic parity automaton with $2^n \cdot n^{2n}$ states and $2n$ priorities. The resulting overall complexity is $O(t \cdot (2^n \cdot n^{2n})^2 \cdot (g \cdot 2^n \cdot n^{2n})^n)$ for synthesis. We generalize the $n!$ lower bound on determinization [Mic88] to the size of GFG automata, establishing that our construction is essentially optimal.

The most important feature of our nondeterministic GFG automaton is its simplicity. The automaton basically follows n different sets of subsets of the original automaton. This leads to a simple structure and even simpler transitions, which are amenable to symbolic implementations. Another attractive advantage of this approach is that it offers a natural hierarchy of nondeterministic automata of increasing complexity that converge to the full GFG solution. That is, given a game and an NBW specification automaton, we can try first solving the game with a small automaton for the winning condition. If we succeed, we are done, having found a winning strategy with small memory for the particular game we are solving. If we fail, we increase the size of the automaton (and thus the memory size we consider), and try again. In the worst case, we get to the full GFG construction, whose memory suffices to win every game with that winning condition. If the GFG automaton fails, then we know that the original specification is not realizable. In Section 6, we give a family of game graphs and winning conditions for which this incremental approach indeed leads to considerable savings.

In addition, simple modifications of our construction lead to nondeterministic parity automata whose number of states ranges between n^{2n} and n^{3n}. Using the smallest possible automaton, the theoretical upper bound reduces to $O(t{\cdot}(n^{2n})^2{\cdot}(g{\cdot}n^{2n})^n)$, which almost matches the upper bound using the deterministic automaton. Recall that our automata are intended for symbolic implementation. Thus, it makes no sense to count the exact upper bound but rather to check which variant of the construction works best in practice. In addition, our hope for synthesis is that in many practical cases it would perform better than the worst-case theoretical upper bound. Using our construction it is possible to search for smaller strategies. Indeed, even for small values of n the time complexity $O(n^{2n^2})$ is impossible.

Recently, Kupferman and Vardi suggested another construction that avoids determinization in certain situations [KV05]. Their algorithm shows how to solve the emptiness problem of alternating parity tree automata through a reduction to the emptiness problem of nondeterministic Büchi tree automata. In order to use their construction to solve games, one has to be able to express the winning condition of the opponent by an NBW. Thus, their algorithm can be applied to synthesis for LTL specifications, because given an LTL winning condition, we negate the LTL formula to get the winning condition of the opponent. On the other hand, when the winning condition is given as an NBW, there is no easy way to complement it, and their algorithm cannot be applied. Furthermore, the worst-case complexity of their algorithm may be quadratically worse, and the size of the produced strategy may be exponentially larger than our algorithm.

2 Preliminaries

Nondeterministic Automata. A *nondeterministic automaton* is $\mathcal{N} = \langle \Sigma, S, \delta, s_0, \alpha \rangle$, where Σ is a finite alphabet, S is a finite set of states, $\delta : S \times \Sigma \to 2^S$ is a transition function, $s_0 \in S$ is an initial state, and α is an acceptance condition to be defined below. A *run* of $\mathcal{N}$ on a word $w = w_0 w_1 \ldots$ is an infinite sequence of states $t_0 t_1 \ldots \in S^\omega$ such that $t_0 = s_0$ and for all $i \geq 0$, we have $t_{i+1} \in \delta(t_i, w_i)$. For a run $r = t_0 t_1 \ldots$, let $\mathit{inf}(r) = \{s \in S \mid s = t_i \text{ for infinitely many } i\text{'s}\}$ be the set of all states occurring infinitely often in the run. We consider two acceptance conditions. A *parity* condition α is a partition $\{F_0, \ldots, F_k\}$ of S. We call k the *index* of the parity condition. The run r is *accepting* according to the parity condition α if for some even i we have $inf(r) \cap F_i \neq \emptyset$, and for all $j < i$, we have $inf(r) \cap F_j = \emptyset$. That is, the minimal set that is visited infinitely often is even. A *Büchi* condition is a set $F \subseteq S$ of states. The run r is *accepting* according to the Büchi condition F if $inf(r) \cap F \neq \emptyset$. That is, the run visits infinitely often states from F. A word w is *accepted* by $\mathcal{N}$ if there exists some accepting run of $\mathcal{N}$ over w. The *language* of $L(\mathcal{N})$ is the set of words accepted by $\mathcal{N}$. Two automata $\mathcal{N}_1$ and $\mathcal{N}_2$ are *equivalent* if they have the same language, i.e., $L(\mathcal{N}_1) = L(\mathcal{N}_2)$.

Given a set $S' \subseteq S$ of states and a letter $\sigma \in \Sigma$, we denote by $\delta(S', \sigma)$ the set $\bigcup_{s \in S'} \delta(s, \sigma)$. The automaton $\mathcal{N}$ is *deterministic* if for every state $s \in S$ and letter $\sigma \in \Sigma$, we have $|\delta(s, \sigma)| = 1$. In that case we write $\delta : S \times \Sigma \to S$.

We use the acronyms NBW, DPW, and NPW to denote automata. NBW stands for nondeterministic Büchi word automaton, DPW for deterministic parity word automaton, and NPW for nondeterministic parity word automaton.

Games. A *game* is $G = \langle V, V_0, V_1, \rho, W \rangle$, where V is a finite set of locations, V_0 and V_1 are a partition of V into locations of player 0 and player 1, respectively, $\rho \subseteq V \times V$ is a transition relation, and $W \subseteq V^\omega$ is a winning condition.

A *play* in G is a maximal sequence $\pi = v_0 v_1 \ldots$ of locations such that for all $i \geq 0$, we have $(v_i, v_{i+1}) \in \rho$. The play π is *winning* for player 0 if $\pi \in W$, or π is finite and the last location of π is in V_1 (i.e., player 1 cannot move from the last location in π). Otherwise, player 1 wins.

A *strategy* for player 0 is a partial function $f : V^* \times V_0 \to V$ such that if $f(\pi \cdot v)$ is defined, then $(v, f(\pi \cdot v)) \in \rho$. A play $\pi = v_0 v_1 \ldots$ is *f-conform* if whenever $v_i \in V_0$, we have $v_{i+1} = f(v_0 \ldots v_i)$. The strategy f is *winning from* a location $v \in V$ if every f-conform play that starts in v is winning for player 0. We say that player 0 *wins from* v if she has a winning strategy from v. The *winning region* of player 0 is the set of locations from which player 0 wins. We denote the winning region of player 0 by W_0. Strategies, winning strategies, winning, and winning regions are defined dually for player 1. We *solve* a game by computing the winning regions W_0 and W_1. For the kind of games considered in this paper, W_0 and W_1 form a partition of V [Mar75].

We consider parity winning conditions. A parity condition α is a partition $\{F_0, \ldots, F_k\}$ of V. The parity condition α defines the set W of infinite plays in G such that the minimal set that is visited infinitely often is even, i.e., $\pi \in W$ if there exists an even i such that $inf(\pi) \cap F_i \neq \emptyset$, and for all $j < i$, we have $inf(\pi) \cap F_j = \emptyset$. The complexity of solving parity games is as follows:

Theorem 1. [Jur00] *Given a parity game G with g locations, t transitions, and index k, we can solve G in time $O(t \cdot g^{\lfloor \frac{k}{2} \rfloor})$.*

We are also interested in more general winning conditions. We define W using an NBW over the alphabet V (or some function of V). Consider a game $G = \langle V, V_0, V_1, \rho, W \rangle$ and an NBW $\mathcal{N}$ over the alphabet V such that $W = L(\mathcal{N})$. We abuse notation and write $G = \langle V, V_0, V_1, \rho, \mathcal{N} \rangle$, or just $G = \langle V, \rho, \mathcal{N} \rangle$.

The common approach to solving such games is by reducing them to parity games. Consider a game $G = \langle V, V_0, V_1, \rho, W\rangle$ and a deterministic automaton $\mathcal{D} = \langle V, S, \delta, s_0, \alpha\rangle$, whose alphabet is V such that $L(\mathcal{D}) = W$. We define the *product* of G and $\mathcal{D}$ to be the game $G \times \mathcal{D} = \langle V \times S, V_0 \times S, V_1 \times S, \rho', W'\rangle$ where $((v, s), (v', s')) \in \rho'$ iff $(v, v') \in \rho$ and $s' = \delta(s, v)$ and W' contains all plays whose projections onto the second component are accepting according to α. A *monitor for* G is a deterministic automaton $\mathcal{D}$ such that for all locations v of G, player 0 wins from v in G iff player 0 wins from (v, s_0) in $G \times \mathcal{D}$.

The common way to solve a game $G = \langle V, \rho, \mathcal{N}\rangle$ where $\mathcal{N}$ is an NBW, is by constructing an equivalent DPW $\mathcal{D}$ [Pit06] and solving the product game $G\times\mathcal{D}$. Unfortunately, determinization has defied implementation until recently, and it cannot be implemented symbolically [THB95, ATW05, KB05]. This means that theoretically we know very well how to solve such games, however, practically we find it very difficult to do so. Formally, we have the following.

Theorem 2. *Consider a game G whose winning condition $\mathcal{N}$ is an NBW with n states. We can construct a DPW $\mathcal{D}$ equivalent to $\mathcal{N}$ with n^{2n+2} states and index $2n$. The parity game $G \times \mathcal{D}$ can be solved in time $O(t{\cdot}n^{2n+2}{\cdot}(g{\cdot}n^{2n+2})^n)$, where g and t are the number of locations and transitions of G.*

It is common wisdom that nondeterministic automata cannot be used for game monitoring. In this paper we show that this claim is false. We define nondeterministic automata that can be used for game monitoring; we call such automata *good for games* (GFG). Our main result is a construction that takes an NBW and produces a GFG NPW. Though our NPW may be slightly bigger than the equivalent DPW (see Section 5), it is much simpler, amenable to symbolic implementation, and suggests a natural hierarchy of automata of increasing complexity that lead to the full solution.

3 Good for Games Automata

In this section we define when an automaton can be used as a monitor for games. In order to define GFG automata, we consider the following game. Let $\mathcal{N} = \langle \Sigma, S, \delta, s_0, \alpha\rangle$ be an automaton. The *monitor game* for $\mathcal{N}$ is a game played on the set S of states. The game proceeds in rounds in which player 1 chooses a letter, and player 0 answers with a successor state reading that letter. Formally, a *play* is a maximal sequence $\pi = t_0\sigma_0t_1\sigma_1\ldots$ such that $t_0 = s_0$ and for all $i \geq 0$, we have $t_{i+1} \in \delta(t_i, \sigma_i)$. That is, an infinite play produces an infinite word $w(\pi) = \sigma_0\sigma_1\ldots$, and a run $r(\pi) = t_0t_1\ldots$ of $\mathcal{N}$ on $w(\pi)$. A play π is *winning* for player 0 if π is infinite and, in addition, either $w(\pi)$ is not in $L(\mathcal{N})$ or $r(\pi)$ is an accepting run of $\mathcal{N}$ on $w(\pi)$. Otherwise, player 1 wins. That is, player 0 wins if she never gets stuck and, in addition, either the resulting word constructed by player 1 is not in the language or (the word is in the language and) the resulting run of $\mathcal{N}$ is accepting.

A *strategy* for player 0 is a partial function $f : (S{\times}\Sigma)^+{\rightarrow}S$ such that if $f(\pi{\cdot}(s, \sigma))$ is defined, then $f(\pi{\cdot}(s, \sigma)) \in \delta(s, \sigma)$. A play $\pi = t_0\sigma_0t_1\sigma_1\ldots$ is f-*conform* if for all $i \geq 0$, we have $t_{i+1} = f(t_0\ldots\sigma_i)$. The strategy f is *winning from* a state $s \in S$ if every f-conform play that starts in s is winning for player 0. We say that *player 0 wins from* s if she has a winning strategy from s. The automaton $\mathcal{N}$ is good for games (GFG) if player 0 wins from the initial state s_0.

We show how to use GFG automata for game monitoring. Consider a GFG automaton $\mathcal{N} = \langle V, S, \delta, s_0, \alpha \rangle$ and a game $G = \langle V, V_0, V_1, E, \mathcal{N} \rangle$. We construct the following extended product game. Let $G \otimes \mathcal{M} = \langle V', V'_0, V'_1, E', W' \rangle$, where the components of $G \otimes \mathcal{N}$ are as follows:

- $V' = V \times S \times \{0, 1\}$, where $V'_0 = (V \times S \times \{0\}) \cup (V_0 \times S \times \{1\})$ and $V'_1 = V_1 \times S \times \{1\}$.
  Given a location $(v, s, i) \in V'$, let $v \Downarrow_S = s$ be the projection onto the state in S. We extend $\Downarrow_S$ to sequences of locations in the natural way.
- $E' = \{((v, s, 0), (v, s', 1)) \mid s' \in \delta(s, v)\} \cup \{((v, s, 1), (v', s, 0)) \mid (v, v') \in E\}$.
- $W' = \{\pi \in V'^{\omega} \mid \pi \Downarrow_S$ is accepting according to $\alpha\}$.

W.l.o.g., we assume that the acceptance condition of $\mathcal{N}$ is closed under finite stuttering (which is true for Büchi and parity). When $\mathcal{N}$ is GFG, we can use $G \otimes \mathcal{N}$ to solve G.

Theorem 3. *Player 0 wins from location v in the game G iff she wins from location $(v, s_0, 0)$ in the game $G \otimes \mathcal{N}$, where s_0 is the initial state of $\mathcal{N}$.*

A win in $G \otimes \mathcal{N}$ is easily translated into a win in G by forgetting the $\mathcal{N}$ component. In the other direction, a winning strategy in G is combined with a winning strategy in the monitor game for $\mathcal{N}$ to produce a strategy in $G \otimes \mathcal{N}$. As the strategy in G is winning, the projection of a resulting play onto the locations of G is a word accepted by $\mathcal{N}$. As the strategy in the monitor game is winning, the projection of the play onto the states of $\mathcal{N}$ is an accepting run of $\mathcal{N}$.

4 Checking the GFG Property

In this section we suggest one possible way of establishing that an automaton is GFG. We prove that an automaton is GFG by showing that it fairly simulates another GFG automaton for the same language. By definition, every deterministic automaton is GFG. This follows from the fact that player 0 does not have any choices in the monitor component of the game. Hence, if an automaton fairly simulates the deterministic automaton for the same language, it is GFG.

4.1 Fair Simulation

We define *fair simulation* [HKR97]. Consider two automata $\mathcal{N} = \langle \Sigma, S, \delta, s_0, \alpha \rangle$ and $\mathcal{R} = \langle \Sigma, T, \eta, t_0, \beta \rangle$ with the same alphabet. In order to define fair simulation, we define the *fair-simulation game*. Let $G_{\mathcal{N},\mathcal{R}} = \langle V, V_0, V_1, \rho, W \rangle$ be the game with the following components:

- $V = (S \times T) \cup (S \times T \times \Sigma)$, where $V_0 = S \times T \times \Sigma$ and $V_1 = S \times T$.
- $\rho = \{((s, t), (s', t, \sigma)) \mid s' \in \delta(s, \sigma)\} \cup \{((s, t, \sigma), (s, t')) \mid t' \in \eta(t, \sigma)\}$.

Given an infinite play π of $G_{\mathcal{N},\mathcal{R}}$ we define π_1 to be the projection of π onto the states in S and π_2 the projection of π onto the states in T. Player 0 *wins* a play π if π is infinite and whenever π_1 is an accepting run of $\mathcal{N}$, then π_2 is an accepting run of $\mathcal{R}$ (w.l.o.g., the acceptance conditions α and β are closed under finite stuttering). If player 0 wins the fair-simulation game from a location (s,t), then the state t of $\mathcal{R}$ *fairly simulates* the state s of $\mathcal{N}$, denoted by $s \leq_f t$. If $s_0 \leq_f t_0$ for the initial states s_0 and t_0, then $\mathcal{R}$ *fairly simulates* $\mathcal{N}$, denoted $\mathcal{N} \leq_f \mathcal{R}$.

4.2 Proving an Automaton GFG

Here we show that if an automaton $\mathcal{N}$ fairly simulates a GFG automaton $\mathcal{D}$ for the same language, then $\mathcal{N}$ is a GFG automaton as well.

Theorem 4. *Let $\mathcal{N}$ be a nondeterministic automaton and $\mathcal{D}$ a GFG automaton equivalent to $\mathcal{N}$. Then $\mathcal{D} \leq_f \mathcal{N}$ implies $\mathcal{N}$ is GFG.*

Proof. Let $\mathcal{N}=\langle \Sigma, N, \delta, n_0, \alpha\rangle$ and $\mathcal{D}=\langle \Sigma, D, \eta, d_0, \beta\rangle$. Assume that $\mathcal{D} \leq_f \mathcal{N}$. Let $G_{\mathcal{D},\mathcal{N}} = \langle V, V_0, V_1, \rho, W\rangle$ be the fair-simulation game between $\mathcal{D}$ and $\mathcal{N}$. Let $f : V^* \times V_0 \rightarrow V$ be a winning strategy for player 0 in $G_{\mathcal{D},\mathcal{N}}$. We denote the monitor game for $\mathcal{D}$ by G_1, and the monitor game for $\mathcal{N}$ by G_2. Let $h : (D \times \Sigma)^+ \rightarrow D$ be the winning strategy of player 0 in G_1. We compose f and h to resolve the choices of player 0 in G_2: we use the choices of player 1 in G_2 to simulate choices of player 1 in G_1, and then h instructs us how to simulate player 1 in $G_{\mathcal{D},\mathcal{N}}$, and the choice of f in $G_{\mathcal{D},\mathcal{N}}$ translates to the choice of player 0 in G_2. Accordingly, we construct plays in the three games that adhere to the following invariants:

- The plays in $G_{\mathcal{D},\mathcal{N}}$ and G_1 are f-conform and h-conform, respectively.
- The projection of the play in G_2 onto Σ is the projection onto Σ of the plays in G_1 and $G_{\mathcal{D},\mathcal{N}}$.
- The projection of the play in G_1 onto the states of $\mathcal{D}$ is the projection of the play in $G_{\mathcal{D},\mathcal{N}}$ onto the states of $\mathcal{D}$.
- The projection of the play in $G_{\mathcal{D},\mathcal{N}}$ onto the states of $\mathcal{N}$ is the projection of the play in G_2 onto the states of $\mathcal{N}$.

We call such plays *matching*. Consider the following plays of length one. The initial position in G_1 is d_0, The initial position in G_2 is n_0, and the initial position in $G_{\mathcal{D},\mathcal{N}}$ is (d_0, n_0). Obviously, these are matching plays.

Let $\pi_2 = n_0\sigma_0\ n_1\sigma_1\ \ldots\ n_i$ be a play in G_2, let $\pi_1 = d_0\sigma_0\ d_1\sigma_1\ \ldots\ d_i$ be a play in G_1, and let $\pi_s = (d_0,n_0)\ (d_1,n_0,\sigma_0)\ (d_1,n_1)\ \ldots\ (d_i,n_i)$ be a play in $G_{\mathcal{D},\mathcal{N}}$. Assume that π_1, π_2, and π_s are matching. Let σ_i be the choice of player 1 in G_2. We set d_{i+1} to $h(\pi_1\sigma_i)$, and set $\pi_1' = \pi_1\sigma_i d_{i+1}$. Let (d_{i+1}, n_{i+1}) be $f(\pi_s(d_{i+1}, n_i, \sigma_i))$, and set $\pi_s' = \pi_s(d_{i+1}, n_i, \sigma_i)(d_{i+1}, n_{i+1})$. Finally, we play n_{i+1} in G_2. By definition of $G_{\mathcal{D},\mathcal{N}}$, it follows that $n_{i+1} \in \delta(n_i, \sigma_i)$. The plays π_1', π_s', and π_2' are matching. Clearly, we can extend the plays according to this strategy to infinite plays.

Let π_1, π_s, and π_2 be some infinite plays constructed according to the above strategy. Let w be the projection of π_2 onto Σ. If $w \notin L(\mathcal{N})$, then player 0 wins in G_2. Assume that $w \in L(\mathcal{N})$. As h is a winning strategy in G_1, we conclude that the projection of π_1 onto D is an accepting run of $\mathcal{D}$. As f is a winning strategy in $G_{\mathcal{D},\mathcal{N}}$, we conclude that the projection of π_s onto N is also accepting. As the projections of π_2 and π_s onto N are equivalent, we are done.

The above condition is not only sufficient, but also necessary. Given two equivalent GFG automata, we can use the strategies in the respective monitor games to construct a winning strategy in the fair-simulation game. In fact, all GFG automata that recognize the same language fairly simulate each other.

5 Constructing GFG Automata

In this section we describe our main contribution, a construction of a GFG automaton for a given language. We start with an NBW and end up with a GFG NPW. In order to prove that our NPW is indeed a GFG, we prove that it fairly simulates the DPW for the same language.

5.1 From NBW to NPW

The idea behind the construction of the NPW is to mimic the determinization construction [Pit06]. We replace the tree structure by nondeterminism. We simply follow the sets maintained by the Safra trees without maintaining the tree structure. In addition we have to treat acceptance. This is similar to the conversion of alternating Büchi word automata to NBW [MH84]: a subset is marked accepting when *all* the paths it follows visit the acceptance set at least once; when this happens we start again. The result is a simple GFG NPW.

Let $\mathcal{N} = \langle \Sigma, S, \delta, s_0, \alpha \rangle$ be an NBW such that $|S| = n$. We construct a GFG NPW $\mathcal{P} = \langle \Sigma, Q, \eta, q_0, \alpha' \rangle$ with the following components.

- The set Q of states is an n-tuple of annotated subsets of S.
  Every state in a subset is annotated by 0 or 1. The annotation 1 signifies that this state is reachable along a path that visited the acceptance set α of $\mathcal{N}$ recently. When a state s is annotated 1, we say that it is *marked*, and when it is annotated 0, we say that it is *unmarked*. Such an annotated subset can be represented by an element $C \in \{0, 1, 2\}^S$, where $C(s) = 0$ means that s is not in the set, $C(s) = 1$ means that s is in the set and is unmarked, and $C(s) = 2$ means that s is in the set and is marked. For simplicity of notation, we represent such an annotated set C by a pair $(A, B) \in 2^S \times 2^S$ of sets with $B \subseteq A$, such that $s \in B$ means $C(s) = 2$, $s \in A - B$ means $C(s) = 1$, and $s \notin A$ means $C(s) = 0$. We abuse notation and write $(A, B) \in \{0, 1, 2\}^S$. We write $(A, B) \subseteq (C, D)$ to denote $A \subseteq C$ and $B \subseteq D$. We sometimes refer to an annotated set as a set.

  In addition, we demand that a set is contained in the B part of some previous set and disjoint from all sets between the two. If some set is empty, then all sets after it are empty as well. Formally,

$$Q = \left\{ \langle (A_1, B_1), \ldots, (A_n, B_n) \rangle \;\middle|\; \begin{array}{l} \forall i \,.\, (A_i, B_i) \in \{0,1,2\}^S, \\ \forall i \,.\, A_i = \emptyset \text{ implies } A_{i+1} = \emptyset, \\ \forall i < j \,.\, \begin{bmatrix} \text{either } A_i \cap A_j = \emptyset \\ \text{or } A_j \subseteq B_i \end{bmatrix} \end{array} \right\}.$$

– $q_0 = \langle (\{s_0\}, \{s_0\} \cap \alpha), (\emptyset, \emptyset), \ldots, (\emptyset, \emptyset) \rangle$.
That is , the first set is initialized to the set that contains the initial state of $\mathcal{N}$. All other sets are initialized to the empty set.
– In order to define the transition function η we need a few definitions.
For $(A, B) \in \{0,1,2\}^S$, $\sigma \in \Sigma$, and $i \in \{0,1\}$, let $succ((A, B), \sigma, i)$ denote the set defined as follows:

$$succ((A,B),\sigma,i) = \begin{cases} \left\{ (A', B') \;\middle|\; \begin{array}{l} A' \subseteq \delta(A, \sigma) \text{ and} \\ B' \subseteq (\delta(B, \sigma) \cap A') \cup (A' \cap \alpha) \end{array} \right\} & \text{If } B \neq A \text{ and } i = 0 \\ \left\{ (A', B') \;\middle|\; \begin{array}{l} A' \subseteq \delta(A, \sigma) \text{ and} \\ B' \subseteq A' \cap \alpha \end{array} \right\} & \text{If } B = A \text{ and } i = 0 \\ \left\{ (A', B') \;\middle|\; \begin{array}{l} A' \subseteq \delta(A, \sigma) \text{ and} \\ B' \subseteq A' \end{array} \right\} & \text{If } i = 1 \end{cases}$$

That is, given a set $(A, B) \subseteq \{0,1,2\}^S$, the possible successors (A', B') are subsets of the states reachable from (A, B). We add to the marked states all visits to α, and if all states are marked (that is, if $A{=}B$), then we unmark them.[2] In the case that $i = 1$, we are completely free in the choice of B'.
For $(A, B), (C, D) \in \{0,1,2\}^S$ and $\sigma \in \Sigma$, let $trans((A, B), \sigma, (C, D))$ be:

$$trans((A,B),\sigma,(C,D)) = \begin{cases} succ((A,B),\sigma,0) & \text{If } A \neq \emptyset \\ succ((C,D),\sigma,1) & \text{If } A = \emptyset \end{cases}$$

That is, we may choose a successor of either (A, B) or (C, D). We may use (C, D) only if (A, B) is empty. In this case, we may choose to initialize the set of markings as we wish. As $succ((A, B), \sigma, i)$ includes every subset of $\rho(A, \sigma)$, it is always possible to choose the empty set, and in the next step, to choose a subset of (the successors of) (C, D).
The transition function η is defined for every state $q \in Q$ and letter $\sigma \in \Sigma$ as follows. Let $q = \langle (A_1, B_1), \ldots, (A_n, B_n) \rangle$. Then:

$$\eta(q, \sigma) = Q \cap \prod_{i=1}^{n} trans((A_i, B_i), \sigma, (A_1, B_1))$$

Intuitively, (A_1, B_1) holds the set of states that are reachable from the initial state. The other sets correspond to guesses as to which states from (A_1, B_1) to follow in order to ignore the non-accepting runs. Whenever one of the

[2] The decision to allow the set B to decrease nondeterministically may seem counter-intuitive. This is equivalent to 'forgetting' that some of the followed paths visited α. This is more convenient and allows more freedom. It also simplifies proofs.

sets becomes empty, it can be *loaded* by a set of successors of (A_1, B_1). It follows that in order to change a guess, the automaton has to empty the respective set and in the next move load a new set. Notice that emptying a set forces the automaton to empty all the sets after it and load them again from (A_1, B_1).

– Consider a state $q = \langle (A_1, B_1), \ldots, (A_n, B_n) \rangle$. We define $ind_E(q)$ to be the minimal value k in $[2..n]$ such that $A_k = \emptyset$, or $n + 1$ if no such value exists. Formally, $ind_E(q) = min\{k, n+1 \mid 1 < k \leq n \text{ and } A_k = \emptyset\}$. Similarly, $ind_F(q)$ is the minimal value k in $[2..n]$ such that $A_k = B_k$ and $A_k \neq \emptyset$, or $n + 1$ if no such value exists. Formally, $ind_F(q) = min\{k, n+1 \mid 1 < k \leq n \text{ and } A_k = B_k \neq \emptyset\}$.

  The parity condition α' is $\langle F_0, \ldots, F_{2n-1} \rangle$, where

  - $F_0 = \{q \in Q \mid A_1 = B_1 \text{ and } A_1 \neq \emptyset\}$;
  - $F_{2i+1} = \{q \in Q \mid ind_E(q) = i+2 \text{ and } ind_F(q) \geq i+2\}$;
  - $F_{2i+2} = \{q \in Q \mid ind_F(q) = i+2 \text{ and } ind_E(q) > i+2\}$.

  As all sets greater than $ind_E(q)$ are empty, the odd sets require that for all sets $A_i \neq B_i$ or $A_i = \emptyset$. In these cases $ind_F(q) = n + 1$. Notice that we do not consider the case that (A_1, B_1) is empty. This is a rejecting sink state.

This completes the definition of $\mathcal{P}$.

We first show that $\mathcal{N}$ and $\mathcal{P}$ are equivalent. We show that $L(\mathcal{P})$ contains $L(\mathcal{N})$ by tracing the runs of $\mathcal{N}$. For each run r of $\mathcal{N}$, we use the first set in a state of $\mathcal{P}$ to follow singletons from r. The proof that $L(\mathcal{P})$ is contained in $L(\mathcal{N})$ is similar to the proof that the DPW constructed by Piterman is contained in the language of $\mathcal{N}$ [Pit06].

Lemma 1. $L(\mathcal{P}) = L(\mathcal{N})$.

Let $\mathcal{D}$ be the DPW constructed by Piterman [Pit06]. We show that $\mathcal{P}$ fairly simulates $\mathcal{D}$. The proof proceeds by showing how to choose a state of $\mathcal{P}$ that maintains the same sets as labels of the nodes in Safra trees, but without maintaining the parenthood function. In fact, $\mathcal{D}$ also fairly simulates $\mathcal{P}$. This follows immediately from the equivalence of the two and $\mathcal{D}$ being deterministic [HKR97].

Lemma 2. $\mathcal{D} \leq_f \mathcal{P}$.

5.2 Complexity Analysis

We count the number of states of the GFG automaton $\mathcal{P}$ and analyze the complexity of using it for solving games.

Theorem 5. *Given an NBW $\mathcal{N}$ with n states, we can construct an equivalent GFG NPW $\mathcal{P}$ with $2^n \cdot n^{2n}$ states and index $2n$.*

Proof. We represent a state of $\mathcal{P}$ as a tree of subsets (or sometimes a forest). The pair (A_i, B_i) is a son of the pair (A_j, B_j) such that $A_i \subseteq B_j$. This tree structure is represented by the parenthood function $p : [n] \rightarrow [n]$ (here $[n]$ is $\{1, \ldots, n\}$). We map every state of $\mathcal{N}$ to the minimal node in the tree (according to the parenthood function) to which it belongs. Thus, the partition into $A_1, \ldots, A_n$ is represented by a function $l : S \rightarrow [n]$. Every state of $\mathcal{N}$ that appears in a pair (A_j, B_j) and also in some son (A_i, B_i) belongs to B_j. In addition, we have to remember all states of $\mathcal{N}$ that appear in some set A_i, in no descendant of A_i, and also appear in B_i. It suffices to remember the subset of all these states.

To summarize, there are at most n^n parenthood functions, n^n state labelings, and 2^n subsets of S. This gives a total of $2^n \cdot n^{2n}$ states for $\mathcal{P}$.

We note that the GFG NPW is larger than the DPW constructed in [Pit06] by a factor of 2^n. However, the NPW is much simpler than the DPW. Although Piterman's variant is slightly simpler than Safra's construction, it still maintains the tree structure that proved hard to implement. Existing implementations of Safra's construction [ATW05, KB05] enumerate the states. We believe that this would be the case also with Piterman's variant. The structure of the NPW above is much simpler and amenable to symbolic methods. In order to represent a set of NBW states, we associate a Boolean variable with every state of the NBW. A BDD over n variables can represent a set of sets of states. In order to represent tuples of n sets, we need a BDD over n^2 variables.

We note that very simple modifications can be made to the NPW without harming its GFG structure. We could remove the restrictions on the containment order between the labels in the sets, or tighten them to be closer to the restrictions imposed on the trees in the DPW. This would result in increasing or reducing the number of states between n^{2n} and n^{3n}. The best structure may depend not on the final number of states, but rather on which structure is most efficiently represented symbolically. It may be the case that looser structures may have a better symbolic representation and work better in practice.

We compare the usage of our automata in the context of game solving to other methods. Consider a game $G = \langle V, V_0, V_1, \rho, W \rangle$, where W is given by an NBW $\mathcal{N} = \langle V, S, \delta, s_0, \alpha \rangle$. Let $|S| = n$, and let g and t be the number of locations and transitions of G, respectively. Using Piterman's construction, we construct a DPW $\mathcal{P}$ with n^{2n+2} states and index $2n$. According to Theorem 1, we can solve the resulting parity game in time $O(t \cdot n^{2n+2} \cdot (g \cdot n^{2n+2})^n)$. When we combine our GFG NPW with the game G, the resulting structure may have $t \cdot (2^n \cdot n^{2n})^2$ transitions and $g \cdot 2^n \cdot n^{2n}$ states. That is, we can solve the resulting parity game in time $O(t \cdot 2^{2n} \cdot n^{4n} \cdot (g \cdot 2^n \cdot n^{2n})^n)$. Note also that the construction of Kupferman and Vardi cannot be applied directly [KV05]. This is because Kupferman and Vardi's construction requires an NBW for the complement of the winning condition. On the other hand, in the context of LTL games (i.e., games with LTL winning conditions), Kupferman and Vardi's construction can be applied. Their construction leads to a time complexity of $O(t \cdot n^{2n+2} \cdot (g \cdot n^{2n+2})^{2n})$, with $2n$ in the exponent instead of n. The memory used by the extracted strategy is bounded by $2^{O(n^2)}$ while it is bounded by $2^n \cdot n^{2n}$ in our case.

We note that for checking the emptiness of alternating parity tree automata, our GFG construction cannot be applied. The reason is similar to the reason why Kupferman and Vardi's method cannot be used for solving games with NBW winning conditions. In this case, we have to construct a GFG NPW for the complement language, which we do not know how to do.

We can use the lower bound on the memory needed for winning strategies to show that our construction is in some sense optimal. For this purpose, we generalize Michel's lower bound on the size of determinization [Mic88, Löd98]. That is, we construct a game with an NBW acceptance condition whose winning strategies require $n!$ memory. Given that our GFG automaton can be used as the memory of a winning strategy and that the resulting game is a parity game that requires no additional memory, this proves that every GFG automaton for the given language has at least $n!$ states.

6 Incremental Construction

Our automata have a natural incremental structure. We simply choose how many sets of states to follow in a state of the GFG automaton. Consider a game $G = \langle V, \rho, \mathcal{N} \rangle$, where $\mathcal{N}$ is an NBW. Let $\mathcal{N}'$ be a nondeterministic automaton such that $L(\mathcal{N}) = L(\mathcal{N}')$ and let s_0 be the initial state of $\mathcal{N}'$. It is simple to see that if player 0 wins from $(v, s_0, 0)$ in $G \otimes \mathcal{N}'$, then player 0 wins from v in G. Indeed, this is the basis of the incomplete approaches described in [HRS05, JGB05]. Using this fact, we suggest the following incremental approach to solving games with ω-regular winning conditions.

Let n be the number of states of $\mathcal{N}$. We apply the construction from Section 5 on $\mathcal{N}$ but use only 2 sets (i.e., restrict the sets $3, \ldots, n$ to the empty set), let $\mathcal{P}_1$ denote this automaton. We then solve the game $G \otimes \mathcal{P}_1$. It is simple to see that $\mathcal{P}_1$ is equivalent to $\mathcal{N}$. Thus, if $(v, q_0, 0)$ is winning for player 0 in $G \otimes \mathcal{P}_1$ (where q_0 is the initial state of $\mathcal{P}_1$), then v is winning for player 0 in G. It follows, that by solving $G \otimes \mathcal{P}_1$ we recognize a subset $W_0' \subseteq W_0$ of the winning region of player 0 in G. Sometimes, we are not interested in the full partition of G to W_0 and W_1, we may be interested in a winning strategy from some set of initial locations in G. If this set of locations is contained in W_0', then we can stop here. Otherwise, we try a less restricted automaton with 3 sets, then 4 sets, etc. For every number of sets used, the resulting automaton may not be GFG, but it recognizes the language of $\mathcal{N}$. A strategy winning in the combination of G and such an automaton is winning also in the original game G (with winning condition $\mathcal{N}$). If we increase the number of sets to n, and still find that the states that interest us are losing, then we conclude that the game is indeed lost. The result is a series of games of increasing complexity. The first automaton has $2^n \cdot 2^{n+2}$ states and index four, resulting in complexity $O(t \cdot (2^n \cdot n^{n+2})^2 \cdot (g \cdot n^2 \cdot n^{n+2})^2)$, where g and t are the number of locations and transitions in G, respectively. In general, the ith automaton has $2^n \cdot i^{n+i}$ states and index $2i$, resulting in complexity $O(t \cdot (2^n \cdot i^{n+i})^2 \cdot (g \cdot 2^n \cdot i^{n+i})^i)$.

We give a family of games and automata that require almost the full power of our construction. Furthermore, we identify several sets of edges in each game

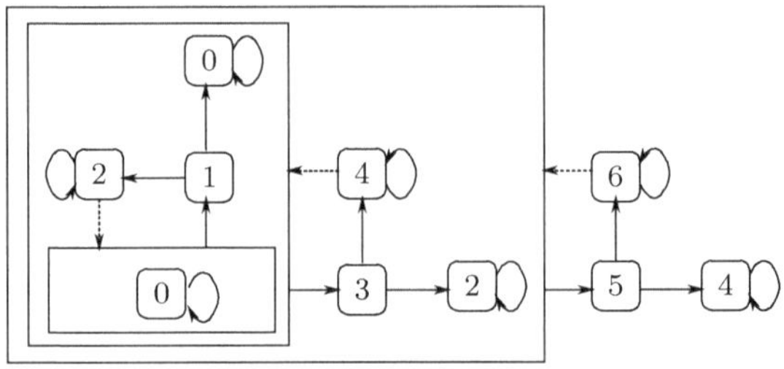

Fig. 1. The game G_3

such that removing one set of edges allows to remove one set from the GFG automaton and identify the winning regions correctly.

We give a recursive definition of the game G_i. Let G_0 be $\langle V^0, \emptyset, V^0, \rho^0, \mathcal{N}^0 \rangle$, where $V^0 = S^0 = \{s_0^0\}$ and $\rho^0 = \{(s_0^0, s_0^0)\}$. The acceptance condition is given with respect to a labeling of the states of the game, to be defined below. The game G_i is $\langle V^i, \emptyset, V^i, \rho^i, \mathcal{N}^i \rangle$, where $V^i = V^{i-1} \cup S^i$, and $S^i = \{s_1^i, s_2^i, s_3^i\}$, and $\rho^i = \rho^{i-1} \cup T^i \cup R^i$, and $T^i = \{(s_1^i, s_2^i), (s_1^i, s_3^i), (s_2^i, s_2^i), (s_3^i, s_3^i)\} \cup ((\bigcup_{j<i} S^j) \times \{s_1^i\})$, and $R^i = \{s_3^i\} \times (\bigcup_{j<i} S^j)$. The labeling on the states of the game is defined as follows. We set $L(s_0^0) = 0$ and for all $i \geq 1$, we set $L(s_1^i) = 2i-1$, $L(s_2^i) = 2i-2$, and $L(s_3^i) = 2i$. The graph depicted in Figure 1 is G_3. An edge from a rectangle to a state s is a shorthand for edges from all states in the rectangle to s, and similarly for edges from states to rectangles. Note that G_{i-1} is contained in G_i.

The winning condition is $\mathcal{N}^i = \langle [2i+2], [2i+2], \eta, 2i+2, [2i+2]^{\text{even}} \rangle$ where $[n]$ is $\{1, \ldots, n\}$, and $[n]^{\text{even}} = \{i \in [n] \mid i \text{ is even}\}$, and η is as follows:

$$\eta(2k, j) = \begin{cases} \emptyset & j{>}2k \\ \{2k\} & j{=}2k \\ \{j, j{+}1, j{+}3, \ldots, 2k{-}1\} & j{<}2k \text{ even} \\ \{j, j{+}2, \ldots, 2k-1\} & j{<}2k \text{ odd} \end{cases}$$

$$\eta(2k+1, j) = \begin{cases} \emptyset & j{>}2k{+}2 \\ \{2k{+}2\} & j{=}2k{+}2 \\ \{j, j{+}1, j{+}3, \ldots, 2k{+}1\} & j{<}2k{+}2 \text{ is even} \\ \{j, j{+}2, \ldots, 2k{+}1\} & j{<}2k{+}2 \text{ is odd} \end{cases}$$

It is also the case that $\mathcal{N}_{i-1}$ is contained in $\mathcal{N}_i$.

We show that for all $i \geq 0$, player 0 wins from every state in G_i. Furthermore, in order to use our GFG construction from Section 5, we have to use $i{+}1$ sets. That is, if we take the product of the graph G_i with the GFG that uses $i{+}1$ sets (denoted $\mathcal{P}_i$), then player 0 wins form every state in the resulting parity game. We further show that this does not hold for the GFG with i sets. That is, player 1 wins from some of the states in the product of G_i and $\mathcal{P}_{i-1}$. Finally, the edges in G_i are $\rho_0 \cup \bigcup_{j \leq i} T^j \cup R^j$. Consider a set of edges R^j for $j < i$. We show that if we remove R^j from G_i, then we can remove one set from the GFG. If we now remove R^k for $k < i$, then we can remove another set from the GFG, and so on. This is summarized in the following lemmata.

Lemma 3. *For all $i \geq 0$, the following are true:*
- *player 0 wins from every location in G_i,*
- *player 0 wins the parity game $G_i \otimes \mathcal{P}_i$, and*
- *if $i \geq 1$, then player 1 wins from $(s_0^0, q_0, 0)$ in $G_i \otimes \mathcal{P}_{i-1}$.*

Consider some set $I \subseteq [i]$ such that $i \in I$. Let G_i^I denote the game with locations S^i and transitions $(\bigcup_{i' \leq i} T^i) \cup (\bigcup_{i' \in I} R^{i'})$. That is, G_i^I includes only the transitions in $R^{i'}$ for $i' \in I$.

Lemma 4. *For all $I \subseteq [i]$ such that $i \in I$ and $|I| = j$ the following are true:*

- *player 0 wins the parity game $G_i^I \otimes \mathcal{P}_j$, and*
- *player 1 wins from $(s_0^0, q_0, 0)$ in $G_i^I \times \mathcal{P}_{j-1}$.*

7 Conclusion and Future Work

We introduced a definition of nondeterministic automata that can be used for game monitoring. Our main contribution is a construction that takes an NBW and constructs a GFG NPW with $2^n \cdot n^{2n}$ states. In comparison, the DPW constructed by Piterman has n^{2n+2} states. However, the structure of the NPW is much simpler, and we suggest that it be implemented symbolically.

We also suggest an incremental approach to solving games. The algorithm of Kupferman and Vardi also shares this property [KV05] (though it cannot be used directly for games with NBW winning conditions). In addition, their algorithm allows to reuse the work done in the earlier stages of the incremental search. We believe that the symmetric structure of our automata will allow similar savings. Another interesting problem is to find a property of game graphs that determines the number of sets required in the GFG construction.

Starting from a Rabin or a parity automaton, it is easy to construct an equivalent Büchi automaton. This suggests that we can apply our construction to Rabin and parity automata as well. Recently, it has been shown that tailored determinization constructions for these types of automata can lead to great savings in the number of states. A similar question is open for GFG automata, as well as for Streett automata.

Finally, we mention that our GFG automaton cannot be used for applications like emptiness of alternating tree automata. The reason is that emptiness of alternating tree automata requires co-determinization, i.e., producing a deterministic automaton for the complement of the original language. We are searching for ways to construct a GFG automaton for the complement language.

Acknowledgment

We thank M.Y. Vardi for pointing out the disadvantages of the construction in [KV05].

References

[ATW05] C.S. Althoff, W. Thomas, and N. Wallmeier. Observations on determinization of büchi automata. In *CIAA*, LNCS. Springer, 2005.

[BL69] J.R. Büchi and L.H.G. Landweber. Solving sequential conditions by finite-state strategies. *Trans. AMS*, 138:295–311, 1969.

[EJ91] E.A. Emerson and C. Jutla. Tree automata, μ-calculus and determinacy. In *FOCS*, pp. 368–377, IEEE, 1991.

[EJS93] E.A. Emerson, C. Jutla, and A.P. Sistla. On model checking for fragments of μ-calculus. In *CAV*, LNCS 697, pp. 385–396, Springer, 1993.

[HKR97] T.A. Henzinger, O. Kupferman, and S. Rajamani. Fair simulation. In *CONCUR*, LNCS 1243, pp. 273–287, Springer, 1997.

[HRS05] A. Harding, M. Ryan, and P.Y. Schobbens. A new algorithm for strategy synthesis in LTL games. In *TACAS*, LNCS 3440, pp. 477–492. 2005.

[JGB05] B. Jobstmann, A. Griesmayer, and R. Bloem. Program repair as a game. In *CAV*, LNCS 3576, pp. 226–238, Springer, 2005.

[Jur00] M. Jurdzinski. Small progress measures for solving parity games. In *STACS*, LNCS 1770, pp. 290–301. Springer, 2000.

[JV00] M. Jurdzinski J. Voge. A discrete strategy improvement algorithm for solving parity games. In *CAV*, LNCS 1855, pp. 202–215, Springer, 2000.

[KB05] J. Klein and C. Baier. Experiments with deterministic ω-automata for formulas of linear temporal logic. In *CIAA*, LNCS. Springer, 2005.

[KPP05] Y. Kesten, N. Piterman, and A. Pnueli. Bridging the gap between fair simulation and trace containment. *I&C*, 200:35–61, 2005.

[KV05] O. Kupferman and M.Y. Vardi. Safraless decision procedures. In *FOCS*, IEEE, 2005.

[Lan69] L.H. Landweber. Decision problems for ω–automata. *MST*, 3:376–384, 1969.

[Löd98] C. Löding. Methods for the transformation of ω-automata: Complexity and connection to second-order logic. Master Thesis, University of Kiel, 1998.

[Mar75] D.A. Martin. Borel determinacy. *Annals of Mathematics*, 65:363–371, 1975.

[McM93] K.L. McMillan. *Symbolic Model Checking*. Kluwer, 1993.

[MH84] S. Miyano and T. Hayashi. Alternating finite automata on ω-words. *Theoretical Computer Science*, 32:321–330, 1984.

[Mic88] M. Michel. Complementation is more difficult with automata on infinite words. CNET, Paris, 1988.

[Pit06] N. Piterman. From nondeterministic Büchi and Streett automata to deterministic parity automata. In *LICS*, IEEE, 2006. To appear.

[RA03] P. Madhusudan R. Alur, and S. La Torre. Playing games with boxes and diamonds. In *CONCUR*, LNCS 2761, pp. 127–141, Springer, 2003.

[Rab72] M.O. Rabin. Automata on infinite objects and Church's problem. *American Mathematical Society*, 1972.

[RW89] P.J.G. Ramadge and W.M. Wonham. The control of discrete event systems. *Trans. Control Theory*, 77:81–98, 1989.

[Saf88] S. Safra. On the complexity of ω-automata. In *FOCS*, IEEE, 1988.

[THB95] S. Tasiran, R. Hojati, and R.K. Brayton. Language containment using non-deterministic ω-automata. In *CHARME*, LNCS 987, 1995.

Game Quantification on Automatic Structures and Hierarchical Model Checking Games*

Łukasz Kaiser

Mathematische Grundlagen der Informatik, RWTH Aachen
Ahornstasse 55, 52074 Aachen, Germany
kaiser@informatik.rwth-aachen.de

Abstract. Game quantification is an expressive concept and has been studied in model theory and descriptive set theory, especially in relation to infinitary logics. Automatic structures on the other hand appear very often in computer science, especially in program verification. We extend first-order logic on structures on words by allowing to use an infinite string of alternating quantifiers on letters of a word, the game quantifier. This extended logic is decidable and preserves regularity on automatic structures, but can be undecidable on other structures even with decidable first-order theory. We show that in the presence of game quantifier any relation that allows to distinguish successors is enough to define all regular relations and therefore the game quantifier is strictly more expressive than first-order logic in such cases. Conversely, if there is an automorphism of atomic relations that swaps some successors then we prove that it can be extended to any relations definable with game quantifier. After investigating it's expressiveness, we use game quantification to introduce a new type of combinatorial games with multiple players and imperfect information exchanged with respect to a hierarchical constraint. It is shown that these games on finite arenas exactly capture the logic with game quantifier when players alternate their moves but are undecidable and not necessarily determined in the other case. In this way we define the first model checking games with finite arenas that can be used for model checking first-order logic on automatic structures.

1 Introduction

Game quantification, the use of infinite strings of quantifiers $Q_1 x_1 Q_2 x_2 \ldots$ with $Q_i = \forall$ or $\exists$, is an intuitive and expressive concept and has interesting connections to model theory, infinitary logics and descriptive set theory [10]. A formula with game quantifiers, e.g.

$$(\exists x_1 \forall y_1 \exists x_2 \forall y_2 \ldots) R(x_1, y_1, x_2, y_2, \ldots),$$

where R is a set of infinite sequences, is normally interpreted using Gale-Stewart games. In the corresponding game $G(\exists\forall, R)$ two players alternatively choose

* This research has been partially supported by the European Community Research Training Network "Games and Automata for Synthesis and Validation" (GAMES).

Z. Ésik (Ed.): CSL 2006, LNCS 4207, pp. 411–425, 2006.

elements of the structure and the first player wins (and the formula is true) if the resulting sequence belongs to R.

Traditionally game quantification was investigated on open or closed sets R, i.e. sets that are defined as infinite disjunctions or conjunctions of finitary relations, $R(\overline{x}) = \bigvee_n R_n(x_1, \ldots, x_n)$. In such cases the formulas with alternating quantifiers can be identified with the monotone open game quantifier $\mathcal{G}^\exists$ or the dual closed game quantifier $\mathcal{G}^\forall$. The duality of these quantifiers ($X \in \mathcal{G}^\exists \iff X \notin \mathcal{G}^\forall$) results from the determinacy of Gale-Stewart games for open and closed sets [7], which was extended by Martin to any Borel set [11].

We are going to introduce game quantification for presentations of automatic structures, i.e. for structures over finite or infinite sequences of letters chosen from a finite alphabet where each relation R is recognised by a finite Muller automaton. Automatic structures, for example Presburger arithmetic, are often used in computer science. They appear in verification problems, have decidable first-order theory [4] and are actively investigated (see [9,1] and references therein). Automatic relations are Borel, so we can use the duality result mentioned before, but we look more closely at the games that appear in this setting. It turns out that we can not only bring the formulas to negation normal form, but we can as well give a computable procedure to construct the automaton recognising the set defined by any formula with game quantifiers and thus show that such formulas are decidable.

The expressive power of game quantification is traditionally compared to infinitary logics over the structure of elements and is most evident in the formula that allows to compare order types of two elements with respect to given orderings. In our case the alphabet is finite and therefore our reference point will be first-order logic over finite or infinite sequences of letters, i.e. over the considered presentation of an automatic structure. It turns out that a formula similar in some way to the one comparing order types allows us to compare the length of common prefixes of words. Using this we are able to show that on some automatic structures game quantification is indeed stronger than first-order logic and we investigate its expressiveness in more detail. On the other hand, it follows from the decidability result that the logic with game quantifier collapses to first-order logic on complete-automatic structures.

To gain deeper insight into definability in the presence of game quantifier on weaker automatic structures we look for automorphisms of structures that are invariant for the logic we study. Similar to the action of permutations of ω on countable models of sentences in infinitary logic studied by invariant descriptive set theory, we define a family of inductive automorphisms where permutation of the alphabet is applied on each position separately and show that these are invariant for the logic with game quantification. This completes the picture of the dependency between expressibility in logic with game quantification and possibility to distinguish different successors.

After analysing the logic with game quantifier we define a family of multiplayer Muller games with imperfect information shared in a hierarchical way. Such games, even when played on a small arenas, can be used to model complex interactions between players and can be used for model checking. Expressing the semantic of a logic by means of games has proved fruitful for developing model checking algorithms [8], especially for μ-calculus which corresponds to parity games [6]. Additionally, the game semantic is quite intuitive and we use multi-player Muller games with imperfect information [2], which is interesting as these types of games have so far not been widely used for model-checking.

We start investigating this class of games by showing that they are not necessarily determined and undecidable if players are not forced to alternate their moves. On the other hand, when players alternate moves we prove the exact correspondence between the games and the logic with game quantification. More precisely, the games can be used as model checking games on automatic structures for first-order logic with game quantifier and at the same time the winning region can be defined in this logic. It follows that deciding the winner is non-elementary in this case. Still, we argue that these games can give rise to efficient algorithms for model checking on complex structures, since recently developed algorithms for games with semiperfect information [5] could be used in practical cases.

2 Preliminaries

In this paper we will be working mainly with structures on words, finite or infinite sequences of letters from a finite alphabet Σ. We denote by Σ^* the set of finite words over Σ and by Σ^ω the set of infinite words, $\Sigma^{\leq\omega} = \Sigma^* \cup \Sigma^\omega$. We normally assume that Σ is fixed and that it contains at least two elements, in our examples usually $\Sigma = \{a, b\}$, and when we need an element not in the alphabet we denote it by $\Box \notin \Sigma$.

Let us fix the notation used for operations on words. For any sequence or word w let us denote by $w|_n$ the finite word composed of the first n letters of w, with $w|_0 = \varepsilon$, the empty word or sequence, and by $w[n]$ the nth letter or element of w for $n = 1, 2, \ldots$. We say that $v \sqsubseteq w$ if v is a prefix of w and in such case we denote by $w - v$ the word u such that $v \cdot u = w$. For an infinite word $w \in \Sigma^\omega$ the set of letters that appear infinitely often in this word is denoted by $\mathrm{Inf}(w)$. We sometimes extend all the notations presented here to vectors of words, so for example if $\overline{x}$ is a tuple of words then $\overline{x}[n]$ is a tuple consisting of the nth letter of each word in $\overline{x}$.

2.1 Automatic Structures

We are going to analyse inductive structures modelled over finite and infinite words, so formally we consider the following structure:

$$(\Sigma^{\leq\omega}, R_1, \ldots, R_K),$$

where each relation R_i has arity $\mathrm{ar}(i)$, so $R_i \subseteq (\Sigma^{\leq\omega})^{\mathrm{ar}(i)}$. Sometimes we want the relations to be recognised by automata and in such cases we will consider them as ω-languages over the tuple-alphabet extended with $\Box$ for finite words, $\otimes R_i \subseteq ((\Sigma \cup \{\Box\})^{\mathrm{ar}(i)})^\omega$.

To define the relations $\otimes R_i$ we have to compose infinite words over the tuple-alphabet $(\Sigma \cup \{\Box\})^{\mathrm{ar}(i)}$ from finite and infinite words over Σ. In such case, if we have a number of words $w^1 = x_1^1 x_2^1 \dots$ and so up to $w^k = x_1^k x_2^k \dots$, then we denote the composed word by $\otimes \overline{w} =$

$$w^1 \otimes \ldots \otimes w^k = \begin{bmatrix} x_1^1 \\ \vdots \\ x_1^k \end{bmatrix} \begin{bmatrix} x_2^1 \\ \vdots \\ x_2^k \end{bmatrix} \ldots \in ((\Sigma \cup \{\Box\})^k)^\omega,$$

whereas if some w^l was finite, $w^l = x_1^l x_2^l \dots x_L^l$, then we prolong it with $\Box^\omega$, i.e. $x_{L+i}^l = \Box$. This allows us to define $\otimes R_i$ with respect to R_i by $R_i(w_1, \dots, w_k) \iff \otimes R_i(w_1 \otimes \ldots \otimes w_k)$.

To speak about presentations of ω-automatic structures we will use Muller automata to recognise ω-regular languages. A (deterministic) *Muller automaton* over $\Gamma = \Sigma \cup \{\Box\}$ is a tuple $\mathcal{A} = (Q, \delta, q_0, \mathcal{F})$ where Q is a finite set of states, δ is a state transition function $\delta : Q \times \Gamma \to Q$, $q_0 \in Q$ is the initial state and $\mathcal{F} \subseteq \mathcal{P}(Q)$ is the acceptance condition. A *run* of $\mathcal{A}$ on $w \in \Gamma^\omega$ is a sequence

$$\rho_{\mathcal{A}}(w) = q_0, q_1, \ldots \in Q^\omega \text{ such that } q_i = \delta(q_{i-1}, w[i]).$$

The word w is *accepted* by $\mathcal{A}$ if the set of states appearing infinitely often during the run is in the acceptance condition, also when $\mathrm{Inf}(\rho_{\mathcal{A}}(w)) \in \mathcal{F}$, and a language $L \subseteq \Gamma^\omega$ is ω-regular if there is a Muller automaton $\mathcal{A}$ that accepts exactly the words $w \in L$. A structure is automatic, or actually, as we consider only structures on words, is a presentation of an automatic structure, if for each relation R_i in this structure the language $\otimes R_i$ is ω-regular over $(\Sigma \cup \{\Box\})^{\mathrm{ar}(i)}$.

You should note that since we allow both finite and infinite words all our words when interpreted over $\Sigma \cup \{\Box\}$ have the property that if a $\Box$ appears then $\Box^\omega$ follows.

2.2 Alternating Automata

We have introduced the standard notion of automata, but we still need to present alternating Muller automata which are an important tool in our proofs. The intuition behind alternating automata is that, unlike in the deterministic case when only one run on a given word is possible, we have more possibilities of transitions from each state for a given letter. Moreover, we do not only want to accept when there *exists* an accepting run among all possible ones (nondeterministic automata), or when *all* possible runs are accepting (universal automata), but we want to be able to alternate the conditions with respect to states of the automaton, also to have both existential and universal branching choices.

An *alternating (Muller) automaton* is formally a tuple $\mathcal{A} = (Q, \delta, q_0, \mathcal{F})$ where as before Q is the set of states, q_0 is the initial state, $\mathcal{F} \subseteq \mathcal{P}(Q)$ is the acceptance condition, but this time δ does not point to a single next state but specifies a whole boolean condition, $\delta : Q \times \Gamma \to \mathcal{B}^+(Q)$, where $\mathcal{B}^+(Q)$ denotes positive boolean formulas over Q. Intuitively a *correct run* of $\mathcal{A}$ on a word w is an infinite tree labelled with Q where the successors of each node form a satisfying set for the boolean condition related to the state in this node and to the corresponding letter in w. We will not present this intuition in a more formal and complete way here, let us only mention here that for every alternating automaton one can find a deterministic Muller automaton accepting the same language [3]. This deterministic automaton does not need to be bigger than double exponential in the size of the alternating one [12].

3 Game Quantifier on Automatic Structures

We want to extend first-order logic to make explicit use of the inductive structure of the words and therefore let us introduce $\Game$, the game quantifier. The meaning of the formula $\Game xy\ \varphi(x, y)$ is that φ can be satisfied when the arguments are constructed stepwise by two players, i.e. first the first letter of x, then the first letter of y given by the second player, another letter of x by the first player and so on. Formally the play will be infinite so to capture finite words we have to define it on $\Gamma = \Sigma \cup \{\Box\}$ by

$$\Game xy\ \varphi(x, y) \iff (\exists \text{ well-formed } f : \Gamma^* \times \Gamma^* \to \Gamma)$$

$$(\forall \text{ well-formed } g : \Gamma^* \times \Gamma^* \to \Gamma)\ \varphi(x_{fg}, y_{fg}),$$

where x_{fg} and y_{fg} are the Σ-words constructed inductively using f and g up to the first appearance of $\Box$,

$$x_{fg}[n+1] = f(x_{fg}|_n, y_{fg}|_n),$$

$$y_{fg}[n+1] = g(x_{fg}|_{n+1}, y_{fg}|_n),$$

and well-formedness means that if any of the functions f resp. g outputs $\Box$ then the word x_{fg} resp. y_{fg} is considered to be finite and the function must then continue to output $\Box$ infinitely, formally h is well-formed when

$$h(w, u) = \Box \implies (\forall w' \sqsupseteq w)\ (\forall u' \sqsupseteq u)\ h(w', u') = \Box.$$

Please note that this direct definition coincides with the traditional one that uses infinite string of quantifiers,

$$\Game xy\ \varphi(x, y) \iff (\exists a_1 \forall b_1 \exists a_2 \forall b_2 \ldots)\ \varphi(a_1 a_2 \ldots, b_1 b_2 \ldots).$$

Moreover, using our notation, $\Game xy\ \varphi(x)$ is equivalent to $\exists x\ \varphi(x)$ as we can always forget opponent moves and play letters from x or conversely use any g to obtain the witness x. Similarly $\Game xy\ \varphi(y)$ is equivalent to $\forall y\ \varphi(y)$. Thus, we do not need to consider the standard quantifiers when the game quantifier is present.

On some structures it is possible to encode a pair of words into a single one, but that is not always the case. Therefore we might sometimes need to use the game quantifier with more variables:

$$\Game x_1 \ldots x_k y_1 \ldots y_m\ \varphi(\overline{x}, \overline{y}) \iff$$

$$(\exists f : (\Gamma^*)^k \times (\Gamma^*)^m \to \Gamma^k)\ (\forall g : (\Gamma^*)^k \times (\Gamma^*)^m \to \Gamma^m)\ \varphi(\overline{x}_{fg}, \overline{y}_{fg}),$$

where again the functions must be well–formed in each column and

$$\overline{x}_{fg}[n+1] = f(\overline{x}_{fg}|_n, \overline{y}_{fg}|_n),\ \overline{y}_{fg}[n+1] = g(\overline{x}_{fg}|_{n+1}, \overline{y}_{fg}|_n).$$

As an example of the use of game quantifier let us consider the following relation R given by the formula:

$$R(u, w, s, t) := \Game xy\ (y = u \to x = s) \wedge (y = w \to x = t).$$

We claim, that R means that the common prefix of s and t is longer than the common prefix of u and w. Denoting by $v \sqcap r$ the common prefix of v and r and by $|v|$ the length of v we can say, that

$$R(u, w, s, t) \equiv |u \sqcap w| < |s \sqcap t|$$

for $u \neq w$ and $s \neq t$. The way we think about evaluating such formula is by means of a game played by two players – the Verifier for x and the Falsifier for y. To see the above equivalence, let us assume that indeed the common prefix of s and t is longer than the common prefix of u and w. In this case the Falsifier will have to choose $y = u$ or $y = w$ before the Verifier chooses if $x = s$ or if $x = t$, and therefore the Verifier is going to win. In the other case, the Falsifier can make the formula false as he knows if $x = s$ or if $x = t$ before choosing whether $y = u$ or $y = w$.

3.1 Basic Properties of FO+$\Game$

The two most important properties of FO+$\Game$ that interest us are the decidability of it on ω-automatic structures and the existence of negation normal form, which semantically corresponds to the determinacy of the underlying games.

To be able to clearly state the existence of negation normal form let us introduce another variation of game quantifier, namely one where it is the Falsifier who makes the moves first. Formally, let

$$\Game^{\forall} xy\ \varphi(x, y) \iff (\exists f : \Gamma^* \times \Gamma^* \to \Gamma)\ (\forall g : \Gamma^* \times \Gamma^* \to \Gamma)\ \varphi(x^{\forall}_{fg}, y^{\forall}_{fg}),$$

where again the functions must be well-formed and this time the words are constructed in reverse order,

$$y^{\forall}_{fg}[n+1] = g(x^{\forall}_{fg}|_n, y^{\forall}_{fg}|_n),\ x^{\forall}_{fg}[n+1] = f(x^{\forall}_{fg}|_n, y^{\forall}_{fg}|_{n+1}).$$

If we denote the game quantifier introduced before by $\Game^{\exists}$ then the intended relation that leads to negation normal form can be stated as follows:

$$\Game^{\exists} xy\ \varphi(x, y) \equiv \neg\Game^{\forall} yx\ \neg\varphi(x, y).$$

Please note that when the relation of prefixing with a letter is present, the quantifier $\Game^{\forall}$ is superfluous and can be eliminated by adding one arbitrary letter,

$$\Game^{\forall} xy\ \varphi(x, y) \iff \Game^{\exists} zy\ \exists x\ z = ax \wedge \varphi(x, y).$$

To verify this equivalence, please note that on the right side the Verifier must start with an a and later play a strategy that satisfies φ, so the same strategy without the first a can be used on the left side. Conversely, if Verifier's strategy on the left side is given then playing an a and later the same strategy is winning for the right side.

To prove decidability and the existence of negation normal form we actually need one crucial lemma, namely that if we begin with ω-regular relations then anything defined in the FO+$\supset$ logic remains ω-regular. The proof relies on the fact that, when used on an automaton, the game quantifier indeed constructs a game and changes the automaton to an alternating one.

Lemma 1. *If the relation $R(\overline{x}, \overline{y}, \overline{z})$ is ω-regular over $\overline{x} \otimes \overline{y} \otimes \overline{z}$ then the relation $S(\overline{z}) \iff \supset \overline{xy}\ R(\overline{x}, \overline{y}, \overline{z})$ is ω-regular over $\otimes \overline{z}$.*

Proof. Let us take the deterministic automaton $\mathcal{A}_R$ for R over $\overline{x} \otimes \overline{y} \otimes \overline{z}$ and construct an alternating automaton $\mathcal{A}_S$ for S over $\otimes \overline{z}$ in the following way. The set of states, acceptance condition and initial state remain the same and the new transition relation is defined by

$$\delta_S(q, \overline{z}) = \bigvee_{\overline{x} \in \Gamma^k} \bigwedge_{\overline{y} \in \Gamma^l} \delta_R(q, \overline{x} \otimes \overline{y} \otimes \overline{z}),$$

where k is the number of elements of $\overline{x}$ and l is the number of elements of $\overline{y}$.

By definition, the semantic of the relation S is

$$S(\overline{z}) \iff (\exists f : (\Gamma^*)^k \times (\Gamma^*)^l \to \Gamma^k)(\forall g : (\Gamma^*)^k \times (\Gamma^*)^l \to \Gamma^l)\ \varphi(\overline{x}_{fg}, \overline{y}_{fg}, \overline{z}).$$

One can see that the function f in this definition corresponds to the choice of the letters for x in the boolean formula when selecting the run of the alternating automaton and that the function g corresponds to the choice of the branch of the run, as all need to be accepted. ∎

This lemma immediately gives us decidability of FO+$\supset$ on automatic structures and also allows us to use determinacy of Muller games for the proof for game quantifier inversion.

Corollary 1. *FO+$\supset$ is decidable on ω-automatic structures, all relations definable in it are ω-automatic and for a fixed number of quantifier alternations it has elementary complexity.*

Corollary 2. *For each FO+$\supset$ formula φ on every automatic structure $\mathfrak{A}$*

$$\mathfrak{A}, \overline{z} \models \supset^{\exists} \overline{xy}\ \varphi(\overline{x}, \overline{y}, \overline{z}) \iff \mathfrak{A}, \overline{z} \models \neg \supset^{\forall} \overline{yx}\ \neg\varphi(\overline{x}, \overline{y}, \overline{z}).$$

The last corollary follows from the determinacy of finitely coloured Muller games. You should note that because of $\overline{z}$ the game arena itself might be infinite, but the number of colours depends only on the size of Muller automaton for φ and is therefore finite. As was already mentioned the determinacy of Muller games can be derived from a more general result by Martin [11] which can be used to generalise the corollary to a wider class of structures, namely all where the relations are Borel sets.

4 Expressive Power of Game Quantification

Some automatic structures are known to be *complete*, meaning that every regular relation over such structure can be defined in first-order logic. For structures over finite and infinite words the canonical example of such structure is the binary tree, $\mathcal{T} = (\{a,b\}^{\leq\omega}, \sigma_a, \sigma_b, \sqsubseteq, \mathrm{el})$, where σ_a and σ_b denote a and b-successors of a finite word (i.e. $\sigma_a(u, ua)$), $\sqsubseteq$ is the prefix relation and $\mathrm{el}(x,y)$ means that x and y have equal length. Each automatic and ω-automatic relation over $\{a,b\}^{\leq\omega}$ can be described by an FO formula over this structure, so since FO+⅁ relations are automatic by Lemma 1, then FO+⅁ is as strong as FO in such case.

This situation changes when $\sqsubseteq$ and el are not given as then FO+⅁ can be used to define them using just σ_a and σ_b and is therefore complete and stronger than FO.

Fact 1. *On the structure* $(\{a,b\}^{\leq\omega}, \sigma_a, \sigma_b)$ *the logic FO+⅁ can define all regular relations and is therefore stronger than FO.*

The proof of this fact does not make any use of the successor relations to define $\sqsubseteq$ and el. Let us now take a weaker structure, namely $(\{a,b\}^{\leq\omega}, S_a)$ where $S_a(x,y)$ is any relation with the property that for each $x \in \{a,b\}^*$ it holds $S_a(x, xa)$ but $S_a(x, xb)$ does *not* hold. We did not specify how the relation S_a behaves on words of bigger difference in length, but this can be compensated for using $\sqsubseteq$ and el. Therefore with game quantifier the relation S_a is enough to express successors in the following way:

$$|x| = |y| + 1 \equiv |x| < |y| \wedge \forall z\ |z| < |y| \rightarrow |z| \leq |x|,$$

$$\sigma_a(x) = (y \equiv S_a(x,y) \wedge |y| = |x| + 1),\ \ \sigma_b(x) = (y \equiv \neg S_a(x,y) \wedge |y| = |x| + 1).$$

When one considers encoding natural numbers as binary words and analysing such structure, it is necessary to have a relation EQ that defines the equality between numbers as opposed to equality over words which might have redundant zeros, $\mathrm{EQ}(x,y) \equiv (x = n0^k$ and $y = n0^l)$. You can see that the relation EQ, definable in the natural presentation of numbers, satisfies the constraints that we put on S_0. Therefore the game quantifier is enough to define all regular relations in the binary presentation of $(\mathbb{N}, =)$. This can as well be used to define $+$ in such presentation so if we add some stronger non-regular relation then model checking becomes undecidable.

Corollary 3. *On the binary presentation of* $(\mathbb{N}, =)$ *the logic FO+⅁ can define all regular relations and therefore the binary presentations of* $(\mathbb{N}, =), (\mathbb{N}, s), (\mathbb{N}, <), (\mathbb{N}, +)$ *are complete-automatic for FO+⅁.*

Corollary 4. *The logic FO+⅁ is undecidable on the binary presentation of Skolem arithmetic* $(\mathbb{N}, \cdot)$.

4.1 Inductive Automorphisms

After analysing what can be expressed in FO+$\Game$ we want to look for methods to establish what relations can *not* be expressed in this logic. For example one could ask if a^ω can be expressed in FO+$\Game$ without any relations other than equality of words. We are going to develop a general method to answer such questions by showing that there is a class of automorphisms of a structure that extend to all relations definable in FO+$\Game$.

First of all please note that not all automorphisms extend to relations definable in FO+$\Game$. For example you can take the bijection of $\Sigma^{\leq\omega}$ that swaps a^ω with b^ω and leaves other elements untouched. The relation $|s \sqcap t| < |u \sqcap w|$ is definable in FO+$\Game$ just with equality, but you can see that this bijection does not extend to an automorphism of the set with this relation as

$$|b^\omega \sqcap ab^\omega| < |a^\omega \sqcap ab^\omega| \text{ but } |a^\omega \sqcap ab^\omega| > |b^\omega \sqcap ab^\omega|.$$

To define the class of *inductive automorphisms* that do extend to relations definable in FO+$\Game$ we are going to restrict the bijections of $\Sigma^{\leq\omega}$ only to a special form.

Definition 1. *The bijection* $\pi : \Sigma^{\leq\omega} \to \Sigma^{\leq\omega}$ *is* inductive *when it does not change the length of the words,* $|\pi(u)| = |u|$ *for every word* u*, and additionally there exists a family of permutations*

$$\{\pi_w\}_{w\in\Sigma^*} \quad \pi_w : \Sigma \to \Sigma,$$

so that for each word u *with at least* n *letters the* n*th letter of* $\pi(u)$ *is given by the appropriate permutation*

$$\pi(u)[n] = \pi_{u|_{n-1}}(u[n]).$$

Please note that the inverse automorphism ϕ^{-1} of any inductive automorphism ϕ is again inductive as inverse permutations $\{\pi_w^{-1}\}$ can be used.

It turns out that if we restrict our attention only to an automorphism ϕ that is an inductive bijection then the structure can be extended with any FO+$\Game$ definable relation and ϕ will still be an automorphism of the extended structure.

Theorem 1. *If* ϕ *is an inductive automorphism of a structure* $(\Sigma^{\leq\omega}, R_1, \ldots, R_k)$ *and* R *is a relation definable by an FO+$\Game$ formula,* $R(\overline{x}) \iff \varphi(\overline{x})$ *for some* $\varphi \in$ FO+$\Game$*, then* ϕ *is an automorphism of the extended structure* $(\Sigma^{\leq\omega}, R_1, \ldots, R_k, R)$.

Proof. Clearly when we proceed by induction on the structure of formulas it is enough to consider the inductive step for game quantifier, i.e. to show that if for a formula φ it holds that $\varphi(\overline{x}, \overline{y}, \overline{z}) \iff \varphi(\overline{\phi}(x), \overline{\phi}(y), \overline{\phi}(y))$ then for $\psi(\overline{z}) = \Game xy\ \varphi(\overline{x}, \overline{y}, \overline{z})$ it holds $\psi(\overline{z}) \iff \psi(\overline{\phi}(z))$ (the converse follows with the inverse automorphism ϕ^{-1}).

To show the above let us first define for any strategies f of the Verifier and g of the Verifier used in $\Game xy\ \varphi(\overline{x}, \overline{y}, \overline{z})$ the transposed strategies f_ϕ, g_ϕ in the following way:

$$f_\phi(x,y) = \pi_{\phi^{-1}(x)} h(\phi^{-1}(x), \phi^{-1}(y)), \; g_\phi(x,y) = \pi_{\phi^{-1}(y)} h(\phi^{-1}(x), \phi^{-1}(y)),$$

where π_w is the permutation for word w associated with ϕ. You should observe that when the players play with strategies f_ϕ, g_ϕ then the resulting words are exactly the images of the words that result from using f and g,

$$x_{f_\phi g_\phi} = \phi(x_{fg}), \; y_{f_\phi g_\phi} = \phi(y_{fg}).$$

In this way we can use the winning strategy f for the first player in $\psi(\overline{z})$ and play with f_ϕ in $\psi(\overline{\phi}(z))$. If the opponent chooses to play g then at the end the formula $\varphi(\overline{x}_{f_\phi g}, \overline{y}_{f_\phi g}, \overline{\phi}(z))$ will be evaluated, but

$$\varphi(\overline{x}_{f_\phi g}, \overline{y}_{f_\phi g}, \overline{\phi}(z)) \equiv \varphi(\overline{\phi}(x_{f g_{\phi^{-1}}}), \overline{\phi}(y_{f g_{\phi^{-1}}}), \overline{\phi}(z)) \equiv \varphi(\overline{x}_{f g_{\phi^{-1}}}, \overline{y}_{f g_{\phi^{-1}}}, \overline{z}),$$

which is true as f is winning against any strategy, in particular against $g_{\phi^{-1}}$. ■

The above theorem gives a general method to investigate definability in FO+∂. For example we can answer the question we stated at the beginning and say that a^ω is not definable in FO+∂ just with equality, because a bijection of $\Sigma^{\leq\omega}$ that swaps the first letter is an inductive bijections and moves a^ω to ba^ω. Together with the fact proved in the previous section that a relation distinguishing successors is enough to define all regular relations in FO+∂ we get a detailed picture of what can and what can not be defined in this logic.

5 Muller Games with Information Levels

To define model checking games that capture first-order and game quantification on automatic structures we need to go beyond two-player perfect information games and use multi-player games with imperfect information. Therefore these games will be played by two coalitions, I and II, each consisting of N players,

$$\Pi = (1,\mathrm{I}), (2,\mathrm{I}), \ldots, (N,\mathrm{I}), (1,\mathrm{II}), (2,\mathrm{II}), \ldots, (N,\mathrm{II}),$$

taking actions described as letters in Σ. The arena of the game is therefore given by the pairwise disjoint sets of positions belonging to each player $V_{1,\mathrm{I}}, \ldots, V_{N,\mathrm{I}}$, $V_{1,\mathrm{II}}, \ldots, V_{N,\mathrm{II}}$ and the function μ defining the moves. Positions of coalition I are denoted by $V_\mathrm{I} = V_{1,\mathrm{I}} \cup \ldots \cup V_{N,\mathrm{I}}$ and of coalition II by $V_\mathrm{II} = V_{1,\mathrm{II}} \cup \ldots \cup V_{N,\mathrm{II}}$ with all positions $V = V_\mathrm{I} \cup V_\mathrm{II}$. In any position v the player can choose an action a from Σ and then move to the position $\mu(v,a)$ as $\mu : V \times \Sigma \to V$. The objective of the game is given by a Muller winning condition $\mathcal{F}$.

The *(general) Muller game with information levels* or *hierarchical Muller game* is therefore given by the tuple

$$(V_{1,\mathrm{I}}, \ldots, V_{N,\mathrm{I}}, V_{1,\mathrm{II}}, \ldots, V_{N,\mathrm{II}}, \; \mu, \; \mathcal{F} \subseteq \mathcal{P}(V)).$$

In such game *play actions* are the sequence of actions taken by the players during a play, so formally it is an infinite word $\alpha \in \Sigma^\omega$. The play corresponding to play actions α and starting in position v_0 is an infinite sequence of positions resulting from taking the moves as described by α,

$$\pi_\alpha(v_0) = v_0 v_1 \ldots \iff v_i = \mu(v_{i-1}, \alpha[i]), \quad i = 1, 2, \ldots.$$

During the play $\pi_\alpha(v_0)$ we encounter a sequence of players that take the moves and let us denote this sequence by $\Pi_\alpha(v_0) = p_0 p_1 \ldots \Leftrightarrow v_i \in V_{p_i}$.

When we want to play the game each of the $2N$ players has to decide on a strategy $s_p : \Sigma^* \to \Sigma$. In a game with perfect information we would say that play actions α are coherent with the strategy s_p in a play starting in v_0 when for each move i taken by player p, also $v_i \in V_p$, the action taken is given by the strategy acting on the history of actions, $\alpha[i+1] = s_p(\alpha|_i)$.

But since the players do not have perfect information, we assume additionally that for each player p there is a function ν_p that extracts from the history of play actions the information visible for this player. More precisely $\nu_p : (\Sigma \times \Pi)^* \to \Sigma^*$ extracts the information visible to player p from the history of play actions together with players that took the moves. Therefore play actions α in a play starting in v_0 are coherent with s_p when for each i such that $v_i \in V_p$ it holds

$$\alpha[i+1] = s_p(\nu_p((a_1, p_0)(a_2, p_1) \ldots (a_i, p_{i-1}))),$$

where $\alpha = a_1, a_2, \ldots$ and $\Pi_\alpha(v_0) = p_0, p_1, \ldots$.

The above definition of views of play history is very general and we will use only a concrete special case of *hierarchical* view functions. The hierarchical information views are defined so that in each coalition the player i is able to see the moves of players $1, \ldots, i$ in both coalitions, but can not see the moves of players with numbers $j > i$. More formally $\nu_{i,c}((a_1, p_0)(a_2, p_1) \ldots) = a_{i_1}, a_{i_2}, \ldots$ when the indices i_k are precisely those for which $p_{i_k - 1} = (j, d)$ with $j \leq i$.

To define when coalition I wins such a hierarchical game we can not require from coalition I to put forth their winning strategies before II does (as usual in such definitions), because as you saw the player with higher numbers have strictly more information and their advantage would be lost if they disclosed their strategies too early. Therefore we use the following definition that requires that strategies are given stepwise, level by level going through the levels of information visibility.

Definition 2. *Coalition* I *wins the hierarchical game*

$$(V_{1,\mathrm{I}}, \ldots, V_{N,\mathrm{I}}, V_{1,\mathrm{II}}, \ldots, V_{N,\mathrm{II}}, \ \mu, \ \mathcal{F})$$

starting from position v_0 when the following condition holds. There exists a strategy $s_{1,\mathrm{I}}$ for player $1,\mathrm{I}$ *such that for each strategy $s_{1,\mathrm{II}}$ of player* $1,\mathrm{II}$ *there exists a strategy $s_{2,\mathrm{I}}$ such that for each strategy $s_{2,\mathrm{II}}$ … there exists a strategy $s_{N,\mathrm{I}}$ such that for each strategy $s_{N,\mathrm{II}}$ the play actions sequence α, starting from v_0 and coherent with all strategies $s_{1,\mathrm{I}}, s_{1,\mathrm{II}}, \ldots, s_{N,\mathrm{I}}, s_{N,\mathrm{II}}$, results in a play $\pi_\alpha(v_0)$ winning for* I, *i.e. such that* $\mathrm{Inf}(\pi_\alpha(v_0)) \in \mathcal{F}$.

As you can expect, the definition for coalition II is dual, i.e. it says that there exists a $s_{1,\mathrm{II}}$ so that for all $s_{1,\mathrm{I}}$, …, the play is not winning, $\mathrm{Inf}(\pi_\alpha(v_0)) \notin \mathcal{F}$.

In general, determining the winner of hierarchical games is undecidable, what can be proved by reducing the Post correspondence problem. Let us state it as a theorem.

Theorem 2. *The question whether coalition* I *wins in a general Muller game with information levels is undecidable.*

We will improve the situation by restricting only to games where players alternate their moves in the next section. We just want to note, that in general hierarchical games are not even determined, but it is not the case when players alternate their moves.

5.1 Model Checking with Hierarchical Games

To connect the logic FO+⅁ to the games with information levels let us restrict our attention only to such games where players alternate their moves in order of information visibility. More precisely let an *alternating game with information levels* be such a game, where for each letter $a \in \Sigma$ and each level $i = 1, \dots, N$ the following alternation conditions hold:

$$v_i \in V_{i,\mathrm{I}} \implies \mu(v_i, a) \in V_{i,\mathrm{II}},\ v_i \in V_{i,\mathrm{II}} \implies \mu(v_i, a) \in V_{(i \bmod N)+1,\mathrm{I}}.$$

In an alternating game every infinite play actions sequence can be divided into sequences of $2N$ actions, each taken by a different player,

$$\alpha = a_1^{1,\mathrm{I}} a_1^{1,\mathrm{II}} a_1^{2,\mathrm{I}} a_1^{2,\mathrm{II}} \dots a_1^{N,\mathrm{I}} a_1^{N,\mathrm{II}} a_2^{1,\mathrm{I}} \dots a_2^{N,\mathrm{II}} a_3^{1,\mathrm{I}} \dots .$$

Let the $2N$-*split* of these play actions be the tuple of words played by each of the players,

$$\mathrm{split}_{2N}(\alpha) = (a_1^{1,\mathrm{I}} a_2^{1,\mathrm{I}} \dots,\ \{a_i^{1,\mathrm{II}}\},\ \dots,\ \{a_i^{N,\mathrm{I}}\},\ \{a_i^{N,\mathrm{II}}\}).$$

You should note that, since the set of plays starting from a fixed v_0 that are winning for I is ω-regular, then also the set of corresponding $2N$-splits of play actions is ω-regular. This can be seen by taking only each $2N$th state of the Muller automaton recognising the plays and making a product with Σ^{2N} to store the states that were omitted from the path in the original automaton. For an alternating hierarchical Muller game G let us denote the $2N$ary relation recognising the $2N$-split of plays winning for I by

$$W_{\mathrm{I}}^{G,v_0}(a_1, \dots, a_{2N}) \Leftrightarrow \forall \alpha\ (\mathrm{split}_{2N}(\alpha) = (a_1, \dots, a_{2N}) \Rightarrow \mathrm{Inf}(\pi_\alpha(v_0)) \in \mathcal{F}^G).$$

The definition for coalition II is analogous, just with $\mathrm{Inf}(\pi_\alpha(v_0)) \notin \mathcal{F}^G$.

You can now note that the condition that coalition I (resp. II) wins in an alternating hierarchical Muller game can be expressed in FO+⅁ using the relation W_{I}^{G,v_0}, which results in the following theorem.

Theorem 3. *For any alternating game with information levels G and the relations W_{I}^{G,v_0} and W_{II}^{G,v_0} defined as above, coalition* I *(resp.* II*) wins the game G starting from v_0 exactly if the following formula holds in $(\Sigma^\omega, W_{\mathrm{I}}^{G,v_0})$:*

$$⅁ x_1 y_1 \dots ⅁ x_N y_N\ W_{\mathrm{I}}^{G,v_0}(x_1, y_1, \dots, x_N, y_N).$$

After we captured winning in alternating games in FO+$\Game$ let us do the converse and construct a model checking game for a given FO+$\Game$ formula on an automatic structure. At first we will restrict ourself to formulas in the simple form

$$\varphi = \Game x_1 y_1 \Game x_2 y_2 \ldots \Game x_N y_N \; R(x_1, y_1, \ldots, x_N, y_N)$$

and just construct a game so that the *split* of the winning plays will allow us to use the previous theorem.

The construction can be understood intuitively as prefixing each variable with all possible letters in the order of information hierarchy and making a step of the automaton when all variables are prefixed. To define these games precisely let us take the automaton for R, namely $\mathcal{A}_R = (Q, q_0, \delta, \mathcal{F}_R)$, and construct the model checking game G_φ for φ in the following way. For each even tuple of letters $c_1, d_1, c_2, d_2, \ldots, c_M, d_M$, with $0 \le M < N$, and for every state $q \in Q$, we will have in our game the position

$$R^q(c_1 x_1, d_1 y_1, \ldots, c_M x_M, d_M y_M, x_{M+1}, \ldots, y_N), \tag{1}$$

and for each uneven tuple $c_1, d_1, c_2, d_2, \ldots, c_M, d_M, c_{M+1}$, $0 \le M < N$, the position

$$R^q(c_1 x_1, \ldots, d_M y_M, c_{M+1} x_{M+1}, y_{M+1}, \ldots, y_N). \tag{2}$$

In each of these positions the next move is made by the player corresponding to the next variable that is not yet prefixed by a letter, e.g. in position 1 it is the player $M+1$ of coalition I who makes the move for x_{M+1} and in position 2 it is the player $M+1$ of coalition II. We can now formally define the set of positions for players on each level i as $V_{i,\mathrm{I}} = Q \times \Sigma^{2(i-1)}$, $V_{i,\mathrm{II}} = Q \times \Sigma^{2i-1}$.

The moves in G_φ are defined in an intuitive way — the player chooses a letter to prefix his variable with, so for $0 \le M < N$

$$\mu(R^q(c_1 x_1, \ldots, d_M y_M, x_{M+1}, \ldots, y_N), c_{M+1}) =$$

$$= R^q(c_1 x_1, \ldots, d_M y_M, c_{M+1} x_{M+1}, y_{M+1}, \ldots, y_N),$$

and for $0 < M < N-1$

$$\mu(R^q(c_1 x_1, \ldots, c_{M+1} x_{M+1}, y_{M+1}, \ldots, y_N), d_{M+1}) -$$

$$= R^q(c_1 x_1, \ldots, c_{M+1} x_{M+1}, d_{M+1} y_{M+1}, x_{M+2}, \ldots, y_N).$$

The only special case is the final position $R^q(c_1 x_1, d_1 y_1, \ldots, c_N x_N, y_N)$. When the player N, II chooses the final letter d_N then it will not be appended, but instead all the prefixing letters will be removed and the state of the automaton will be changed (here $\overline{\alpha} = c_1 d_1 \ldots c_N d_N$):

$$\mu(R^q(c_1 x_1, d_1 y_1, \ldots, c_N x_N, y_N), d_N) = R^{\delta(q, \overline{\alpha})}(x_1, y_1, \ldots, x_N, y_N).$$

The winning condition $\mathcal{F}$ in the game is defined to correspond to the acceptance condition $\mathcal{F}_R$ of the automaton for R in such way, that we look only at the state component of each position.

To see that the game G_φ is indeed the model checking game for φ we can use Theorem 3 again, just observe that the $2N$-split of the winning paths in G_φ is exactly the relation R, $W_{\mathrm{I}}^{G_\varphi, R^{q_0}(x_1,y_1,\ldots,x_N,y_N)} = R$.

In this way we know how to construct the model checking game for formulas in simple form. As we have seen, any formula in FO+$\Game$ can be written in negation normal form and additionally, by renaming variables, it can be reduced to prenex normal form. Let us therefore consider now a general formula in the form $\varphi = \Game x_1 y_1 \Game x_2 y_2 \ldots \Game x_N y_N\ \psi(x_1, y_1, \ldots, x_N, y_N)$, where ψ is in negation normal form and does not contain quantifiers. Let us construct the game G_φ inductively with respect to ψ.

In the case of $\psi(\overline{x}) = R(\overline{x})$ or $\psi(\overline{x}) = \neg R(\overline{x})$ the solution was already presented, when considering $\neg R$ we just have to complement the acceptance condition of the automaton for R. Let us show how to construct the game for boolean connectives, i.e. for $\psi_1 \wedge \psi_2$ and for $\psi_1 \vee \psi_2$. We want to adhere to the usual convention of model checking games and to have only one additional position for any junctor. The game for $\psi_1 \circ \psi_2$, where $\circ = \wedge, \vee$, is therefore constructed as follows: we take the two games for ψ_1 and ψ_2 and we add one more position on higher level of information that has two possible moves — to the starting position of ψ_1 and to the starting position of ψ_2. The new position belongs to coalition I when $\circ = \vee$ and to coalition II when $\circ = \wedge$ and in both cases the other coalition does not play on that information level. With the construction described above we face a problem, as the game is not strictly alternating any more, but it is not a significant obstacle.

To formally prove that the resulting games are indeed model checking games for formulas with boolean connectives you can just replace the connectives with a new variable and the formula with one relation where only the first letter of connective-variables is considered. Then the automata for such relation corresponds to the defined game and Theorem 3 can be used again.

The exact correspondence of alternating hierarchical games and FO+$\Game$ makes it possible to use our knowledge about this logic. In particular we can transfer the results about complexity including the non-elementary lower bound on deciding first-order logic on automatic structures.

Corollary 5. *The question whether coalition* I *(resp.* II*) wins in an alternating Muller game with information levels on a finite arena is decidable, non-elementary when the number of levels is not fixed and it can be done in* 2K-EXPTIME *for K information levels.*

The possibility to get negation normal form for FO+$\Game$ can as well be translated and gives the proof of determinacy of alternating hierarchical games.

Corollary 6. *Alternating Muller games with information levels are determined.*

6 Conclusions and Future Work

We described how game quantification can be used on automatic structures and the resulting logic turned out to be very interesting. It is decidable and the defined relations remain regular, which might be used in the study of presentations of automatic structures. On the other hand the logic is strictly more expressive than first-order on some weaker structures. Most notably on the binary tree and on presentations of natural numbers it is possible to define all regular relations when game quantification is allowed. The methods that we used, for example inductive automorphisms, might be extended to morphisms between presentations of the same automatic structure and used to study intrinsic regularity.

On the other hand, it might be interesting to ask what is the expressive power of FO+$\Game$ on formulas with just one game quantifier, i.e. $\Game \overline{xy}\ \varphi(\overline{x}, \overline{y})$ where φ is quantifier-free. Such formulas may be more expressive than just existential or universal fragment of first-order logic even on complete-automatic structures and can be decided with double exponential complexity.

Game quantification made it possible to define an expressive class of model checking games that we used for checking first-order logic on automatic structures. These games use multiple players and imperfect information in a novel way and might be used to derive more efficient algorithms for verification, especially if the efficient algorithms from [5] can be generalised to hierarchical games.

References

1. V. Barany, *Invariants of Automatic Presentations and Semi-Synchronous transductions*, Proceedings of STACS 06, vol. 3884 of Lecture Notes in Computer Science, pp. 289-300, 2006.
2. J. C. Bradfield, *Parity of Imperfection* or *Fixing Independence*, vol. 2803 of Lecture Notes in Computer Science, pp. 72-85, 2003.
3. J. R. Büchi, *On Decision Method in Restricted Second Order Arithmetic*, Proceedings of the International Congress on Logic, Methodology and Philosophy of Science, pp. 1-11, 1962.
4. A. Blumensath, E. Grädel, *Finite Presentations of Infinite Structures: Automata and Interpretations*, vol. 37 of Theory of Computing Systems, pp. 641-674, 2004.
5. K. Chatterjee, T. A. Henzinger, *Semiperfect-Information Games*, Proceedings of FSTTCS 2005, vol. 3821 of Lecture Notes in Computer Science, pp. 1-18, 2005.
6. A. Emerson, C. Jutla, *Tree automata, mu-calculus and determinacy*, in Proc. 32nd IEEE Symp. on Foundations of Computer Science, pp. 368-377, 1991.
7. D. Gale, F.M. Stewart. *Infinite Games with Perfect Information*, in *H. W. Kuhn, A. W. Tucker* eds. Contributions to the Theory of Games, Volume II, pp. 245-266, Annals of Mathematics Studies, Princeton University Press, 1953.
8. E. Grädel, *Model Checking Games*, Proceedings of WOLLIC 02, vol. 67 of Electronic Notes in Theoretical Computer Science, 2002.
9. B. Khoussainov, A. Nerode, *Automatic Presentations of Structures*, Proceedings of LCC 94, vol. 960 of Lecture Notes in Computer Science, pp. 367-392, 1995.
10. Ph. G. Kolaitis *Game Quantification*, in *J. Barwise, S. Feferman* eds. Model Theoretic Logics, pp. 365-422, 1985
11. D. Martin, *Borel determinacy*, Annals of Mathematics (102), pp. 336-371, 1975.
12. S. Miyano, T. Hayashi, *Alternating Finite Automata on ω-words*, vol. 32 of Theoretical Computer Science, pp. 321-330, 1984.

An Algebraic Point of View on the Crane Beach Property

Clemens Lautemann, Pascal Tesson[1,⋆], and Denis Thérien[2,⋆]

[1] Département d'Informatique et de Génie Logiciel, Université Laval
pascal.tesson@ift.ulaval.ca
[2] School of Computer Science, McGill University
denis@cs.mcgill.ca

Abstract. A letter $e \in \Sigma$ is said to be neutral for a language L if it can be inserted and deleted at will in a word without affecting membership in L. The Crane Beach Conjecture, which was recently disproved, stated that any language containing a neutral letter and definable in first-order with arbitrary numerical predicates ($\mathbf{FO}[Arb]$) is in fact $\mathbf{FO}[<]$ definable and is thus a regular, star-free language. More generally, we say that a logic or a computational model has the Crane Beach property if the only languages with neutral letter that it can define/compute are regular.

We develop an algebraic point of view on the Crane Beach properties using the program over monoid formalism which has proved of importance in circuit complexity. Using recent communication complexity results we establish a number of Crane Beach results for programs over specific classes of monoids. These can be viewed as Crane Beach theorems for classes of bounded-width branching programs. We also apply this to a standard extension of **FO** using modular-counting quantifiers and show that the boolean closure of this logic's $\mathbf{\Sigma}_1$ fragment has the CBP.

We would like to dedicate this paper to the memory of Clemens Lautemann. Clemens passed away in April 2005 after a battle with cancer. He raised many of the questions underlying this paper but unfortunately did not live to see their final resolution. He was also a founding member of the EACSL which organizes the present conference. We will miss Clemens as a collaborator and as a friend.

1 Introduction

A number of results of the last ten years indicate that weak models of computation are considerably handicapped when they are used to recognize languages with a so-called *neutral letter*. A letter $e \in \Sigma$ is said to be *neutral* for the language $L \subseteq \Sigma^*$ if it can be inserted and deleted at will in a word without affecting membership in L, i.e. if for all $u, v \in \Sigma^*$ it holds that $uv \in L \Leftrightarrow uev \in L$. It is natural to think that languages with neutral letters pose particular problems for

⋆ Research supported in part by NSERC and FQRNT.

Z. Ésik (Ed.): CSL 2006, LNCS 4207, pp. 426–440, 2006.

models of computation that rely heavily on the precise location of certain input bits or are too weak to preprocess their input by removing the blank symbols.

To the best of our knowledge, the first result along those lines appears in work of Barrington and Straubing on short bounded-width branching programs: any language recognized by a width k branching program of length $o(n \log \log n)$ is in fact regular and belongs to some well defined class $\mathcal{L}_k$ of regular languages [4].

In light of this result, Lautemann and Thérien conjectured that any language with a neutral letter in the circuit class AC^0 is in fact regular and star-free (i.e. it can be defined by a regular expression built from the letters of the alphabet, the empty set symbol, union and complementation but no Kleene star). This statement, which came to be known as the *Crane Beach conjecture*, has a nice logical formulation. A language is known to belong to AC^0 iff it can be defined in $\mathbf{FO}[Arb]$, i.e. by a first-order sentence using arbitrary numerical predicates [12]. Restrictions on the set of numerical predicates available can be interpreted as *uniformity restrictions* on circuits [3,13,22] (see Section 2). When order is the only available numerical predicate ($\mathbf{FO}[<]$), the class of definable languages corresponds exactly to the star-free regular languages (cf. [22,33]). Thus if $\mathcal{L}_e$ denotes the class of languages with a neutral letter, the Crane Beach conjecture postulated that $\mathbf{FO}[Arb] \cap \mathcal{L}_e = \mathbf{FO}[<] \cap \mathcal{L}_e$. The underlying intuition was the apparent inability to take advantage of complicated numerical predicates in the presence of a neutral letter. But the Crane Beach conjecture was refuted in [2].

On the other hand [2] show that the boolean closure of the $\mathbf{\Sigma}_1$ fragment of $\mathbf{FO}[Arb]$ does have the Crane Beach property in the sense that $\mathbf{B\Sigma}_1[Arb] \cap \mathcal{L}_e = \mathbf{B\Sigma}_1[<] \cap \mathcal{L}_e$. Such Crane Beach results for fragments of $\mathbf{FO}$ are related to so-called *collapse results* in database theory [8].

We also consider the logic $\mathbf{FO}+\mathbf{MOD}$ which is a standard extension of first-order logic on words in which *modular counting* quantifiers are introduced. The expressive power of $\mathbf{FO}+\mathbf{MOD}[<]$ is limited to a specific class of regular languages [25] whereas $\mathbf{FO}+\mathbf{MOD}[Arb]$ captures the circuit class ACC^0.

We say that a logic or a computation model has the *Crane Beach property* if all languages with a neutral letter that it can define or compute are regular. We develop an algebraic point of view on the Crane Beach property by considering finite monoids as language recognizers. Two methods of recognition by finite monoids have been introduced in the literature: recognition via morphisms and recognition via *programs over monoids*. The first is more classical and is a key ingredient in a number of major results concerning regular languages and in particular about the expressive power of fragments of $\mathbf{FO}+\mathbf{MOD}[<]$. Programs are more powerful and yield algebraic characterizations of AC^0, ACC^0 and NC^1 [1,6].

The extra expressive power obtained by moving from recognition via morphisms over monoids to recognition by programs over monoids is in many ways similar to the extra power afforded to $\mathbf{FO}$ or $\mathbf{FO}+\mathbf{MOD}$ when we go from sentences that use only $<$ as a numerical predicate to sentences using arbitrary numerical predicates. We show that there are classes of monoids for which the presence of a neutral letter nullifies this advantage in the sense that any language with a neutral letter recognized by a program over a monoid in the class can in fact be recognized by a morphism over a monoid in the same class.

These results allow us to show that the boolean closure of the $\mathbf{\Sigma}_1$ fragment of $\mathbf{FO+MOD}$, which we formally denote as $\mathbf{B\Sigma}_1^{(s,p)}$ has the Crane Beach property, i.e. $\mathbf{B\Sigma}_1^{(s,p)}[Arb] \cap \mathcal{L}_e = \mathbf{B\Sigma}_1^{(s,p)}[<] \cap \mathcal{L}_e$. They also provide some additional insight into a possible dividing line between classes of monoids or fragments of $\mathbf{FO+MOD}$ which exhibit the Crane Beach property from those which do not.

We begin by reviewing in section 2 the necessary background in circuit complexity, descriptive complexity and the study of finite monoids as language recognizers. In section 3 we obtain our results on the Crane Beach property for programs over monoids and study their application to logic in section 4. We conclude with a discussion on how our results fit with other results concerning languages with a neutral letter.

Due to space limitations, some technical results have been omitted but an extended version of the paper is available from the authors' websites. One of our results in section 4 is credited in [2] to an unpublished manuscript of Clemens Lautemann and his former student Andrea Krol which we were not able to reach during the preparation of the present paper.

2 Logic, Circuits, Programs over Monoids and Automata

2.1 Circuits

An n-input boolean circuit $\mathcal{C}_n$ is a directed acyclic graph with a single node of out-degree 0 called the *output gate*. The *input gates* have in-degree 0 and are labeled either with an input variable x_i, its complement $\overline{x_i}$ or one of the boolean constants $0, 1$. When the inputs are not boolean but take values in some finite alphabet Σ, input nodes are labeled by $x_i = a$ for some $a \in \Sigma$. Finally, any non-input gate g is labeled by some symmetric boolean function f_g taken from some predetermined set. Our focus will be on the case where these functions are either the AND or the OR function, or the function MOD_q which outputs 1 if the sum of its entries is divisible by q. The *depth* and *size* of a circuit $\mathcal{C}_n$ are, respectively, the longest path from an input node to the output node and the number of gates. A circuit naturally computes a function $f_{\mathcal{C}_n} : \Sigma^n \to \{0,1\}$ and we define the *language accepted by* $\mathcal{C}_n$ as $L_{\mathcal{C}_n} = \{x : f_{\mathcal{C}_n} = 1\}$. This language is a subset of Σ^n: in order to recognize subsets of Σ^* we use families of circuits $\mathcal{C} = \{\mathcal{C}_n\}_{n\geq 0}$ where each $\mathcal{C}_n$ is an n-input circuit which is used to process inputs of that particular length. We can then consider the depth and size of $\mathcal{C}$ as a function of n and study its asymptotics.

The class ACC^0 consists of languages which can be accepted by a family of circuits having polynomial-size and bounded-depth and constructed with AND gates, OR gates and MOD_q gates for some $q \geq 2$. We further define AC^0 as the restriction of ACC^0 where only AND and OR gates are allowed and CC^0 as the restriction of ACC^0 where only MOD_q gates (for some $q \geq 2$) are used. All of these classes lie in the class NC^1 of languages recognized by circuits of depth $O(\log n)$ constructed with AND and OR gates of bounded fan-in.

We have not imposed any sort of restriction on the effective constructibility of the circuit families and the circuit classes are correspondingly dubbed 'non-uniform'. It makes sense to require that the nth circuit of a family $\mathcal{C}$ be constructible efficiently and such requirements are called *uniformity conditions*. For any complexity class $\mathcal{D}$, we say that a family of circuits $\mathcal{C} = \{C_n\}$ is $\mathcal{D}$-*uniform* if there is an algorithm in $\mathcal{D}$ which on input n computes a representation of C_n (see e.g. [3] for a formal discussion). DLOGTIME-*uniformity* is widely accepted as the desired 'correct' notion of uniformity for subclasses of NC^1: roughly speaking it requires that there exists an algorithm which on input $\langle t, a, b, n\rangle$ can check in time $O(\log n)$ whether the ath gate of C_n is of type t (i.e. which function it computes) and feeds into the bth gate [3].

2.2 Logic over Words

We are interested in considering first-order logical sentences defining sets of finite words over an alphabet Σ. We only briefly overview this logical apparatus and refer the reader to [22,15] for a more thorough and formal discussion.

Let us start with an example. Over the alphabet $\{a, b\}$ we view the sentence

$$\exists x \exists y \;\; Q_a x \wedge Q_b y \wedge (y = x + x)$$

as defining the set of words in which there exists a position x holding an a such that the position $2x$ holds a b. The variables in the sentence stand for positions in a finite word and the access to the content of these positions is provided by the unary predicates Q_a and Q_b.

More generally, for any alphabet Σ we construct sentences using two types of atomic formulas. First, for each $a \in \Sigma$, we include a *content predicate* $Q_a x$ which is interpreted as true of a finite word w if the position x in w holds the letter a. The second atomic formulas are *numerical predicates* $P(x_1, \ldots x_k)$. The truth of $P(x_1, \ldots, x_k)$ depends only on the values $x_1, \ldots, x_k$ and on the length of w but not on the actual letters in w. For a set $\mathcal{P}$ of numerical predicates, we denote as $\mathbf{FO}[\mathcal{P}]$ the class of sentences which can be constructed from the atomic formulas $Q_a x$ and $P(x_1, \ldots, x_k)$ with $P \in \mathcal{P}$ using existential and universal quantifiers and boolean connectives. For $\phi \in \mathbf{FO}\,[\mathcal{P}]$ we further denote as L_ϕ the language in Σ^* defined by ϕ i.e. the set of finite words such that $w \models \phi$.

We also consider the case where first-order is extended by the introduction of modular-counting quantifiers. The formula $\exists^{i \pmod p} x \; \psi(x)$ is true if the number of positions x such that $\psi(x)$ is equal to i modulo p. We denote as $\mathbf{FO}+\mathbf{MOD}\,[\mathcal{P}]$ the class of sentences constructed from the atomic formulas, boolean connectives and both modular and existential/universal quantifiers.

The case where $\mathcal{P}$ contains only the order relation $<$ has been thoroughly investigated [22,33,32]. A corollary of Büchi's theorem about monadic second-order logic over words establishes that $\mathbf{FO}+\mathbf{MOD}\,[<]$ contains only regular languages. In fact these can be characterized as languages whose syntactic monoid is solvable (see next subsection). The expressive power of various fragments of

FO+MOD $[<]$ can also be characterized using algebraic automata theory and in particular **FO** $[<]$ captures exactly the star-free languages which in turn correspond to languages with aperiodic syntactic monoids.

On the other end of the spectrum, let Arb be the set of all numerical predicates. The classes **FO** $[Arb]$ and **FO+MOD** $[Arb]$ correspond exactly to non-uniform AC^0 and ACC^0 respectively. Restrictions on the set of allowed numerical predicates translate in many natural cases into uniformity restrictions on the circuits [3,7]. Most notably, **FO** $[+,*]$ and **FO+MOD** $[+,*]$ correspond to the DLOGTIME-uniform versions of AC^0 and ACC^0.

The class $\mathcal{R}eg$ of regular numerical predicates has also been the focus of some attention. A numerical predicate is said to be *regular* if it can be defined by an **FO+MOD** $[<]$ formula. By definition **FO+MOD** $[\mathcal{R}eg]$ has the same expressive power as **FO+MOD** $[<]$ and thus contains only regular languages. It is also known that a language L is definable in **FO** $[\mathcal{R}eg]$ iff it is regular and can be recognized by an AC^0 circuit. In other words **FO** $[Arb] \cap \text{REG} =$ **FO** $[\mathcal{R}eg]$, where REG denotes the class of regular languages. For a number of other fragments of **FO+MOD** it has been shown that when defining regular languages arbitrary numerical predicates hold no expressive advantage over regular predicates.

For technical reasons it is convenient to have a quantifier-free description of regular numerical predicates. For integers $t \geq 0$ and $q \geq 2$ and any $n < t+q$ we define a binary relation $\delta_{n,t,q}$ as follows: if $n < t$ then $x \, \delta_{n,t,q} \, y$ if $x - y = n$ and if $n \geq n$ then $x \, \delta_{n,t,q} \, y$ if $x - y \geq t$ and $x - y \equiv n \mod q$. We further define for any $n < t + q$ a unary relation $\kappa_{n,t,q}$ by setting $\kappa_{n,t,q}(x) \Leftrightarrow x \, \delta_{n,t,q} \, 0$. A numerical predicate is regular iff it can be defined as a boolean combination of $\delta_{n,t,q}$, $\kappa_{n,t,q}$ and $<$ (see e.g. [17]).

2.3 Programs over Monoids

We now turn to an algebraic characterization of the circuit classes presented earlier. We refer the reader to [31] for a more thorough discussion of the links between complexity and the algebraic theory of regular languages.

A *monoid* is a set M equipped with a binary associative operation $\cdot_M$ and a distinguished identity element 1_M. A class of finite monoids forms a *variety* (or more precisely a *pseudovariety*) if it is closed under direct product, formation of submonoids and morphic images.

The *free monoid* Σ^* over the alphabet Σ is the set of finite words over Σ with concatenation as the monoid operation. The empty word ϵ acts as the identity element in this case. With the exception Σ^*, all monoids considered in this paper are finite and we view these algebraic objects as language recognizers.

We say that a language $L \subseteq \Sigma^*$ is *recognized via morphism* or simply *recognized* by the finite monoid M if there exists a morphism $\phi : \Sigma^* \to M$ and a set $F \subseteq M$ such that $L = \phi^{-1}(M)$. This definition simply restates algebraically the notion of acceptance by a finite automaton and a simple variant of Kleene's theorem shows that a language is regular if and only if it can be recognized by some finite monoid. For every regular language L, the syntactic monoid $M(L)$ of L is the smallest monoid recognizing L and $M(L)$ is in fact isomorphic to the transition monoid of L's minimal automaton. For a variety $\mathbf{V}$ we denote as $\mathbf{L}(\mathbf{V})$ the class of regular languages with syntactic monoids in $\mathbf{V}$. These classes (which form *language varieties*) are a natural unit of classification for regular languages and are at the heart of the algebraic theory of regular languages [18].

We give a list of varieties that bear importance in this paper but also in other applications of algebraic automata theory [18,31].

- The variety $\mathbf{A}$ consists of *aperiodic* or *group-free* monoids, i.e. monoids having no non-trivial subgroup.
- The variety $\mathbf{G_{nil}}$ consists of *nilpotent groups*, i.e. groups which are direct products of p-groups. An alternate and in our case more useful definition of nilpotency can be given as follows. For a finite group G and any $g, h \in G$, the *commutator* $[g, h]$ of g and h is the element $g^{-1}h^{-1}gh$. For any subgroups $H_1, H_2 \subseteq G$ we denote as $[H_1, H_2]$ the subgroup generated by the commutators $[h_1, h_2]$ with $h_1 \in H_1$ and $h_2 \in H_2$. Now define inductively the chain of subgroups of G by $G_0 = G$ and $G_{i+1} = [G, G_i]$. We say that a group is *nilpotent of class k* if G_k is the trivial group and denote as $\mathbf{G_{nil,k}}$ the variety of such groups. A group is *nilpotent* if it is nilpotent of class k for some k. Note that a group is nilpotent of class 1 iff it is Abelian.
- The variety $\mathbf{G_{sol}}$ of solvable groups and the variety $\mathbf{M_{sol}}$ of *solvable monoids*, i.e. monoids whose subgroups are solvable.
- For any variety of groups $\mathbf{H}$, we denote as $\overline{\mathbf{H}}$ the variety of monoids whose subgroups all belong to $\mathbf{H}$.
- The variety $\mathbf{DO}$ consists of monoids which for some $n \geq 1$ satisfy the identity $(xy)^n(yx)^n(xy)^n = (xy)^n$.
- The variety $\mathbf{DA}$ consists of monoids which satisfy $(xy)^n y (xy)^n = (xy)^n$ for some n. In fact $\mathbf{DA}$ is the intersection of $\mathbf{DO}$ and $\mathbf{A}$.
- The variety $\mathbf{J}$ of $\mathcal{J}$-trivial monoids consists of aperiodic monoids which satisfy $(xy)^n = (yx)^n$ for some n.

For any variety $\mathbf{V}$ in the above list, the corresponding class of regular languages $\mathbf{L}(\mathbf{V})$ admits nice descriptions [18,31] and the varieties $\mathbf{DA}, \mathbf{DO}$ and $\mathbf{G_{nil}}$ are often central in investigations in the complexity of regular languages and their logical descriptions [29,30,32].

The *program over monoid* formalism introduced by Barrington and Thérien provides a slight extension of a finite monoid's computing power. An n-input program ϕ_n over M of length ℓ is a sequence of *instructions* $\phi_n : (i_1, f_1) \ldots (i_\ell, f_\ell)$ with $1 \leq i_j \leq n$ and where each f_i is a function from the input alphabet Σ to M. Given an input $w \in \Sigma^n$ a program produces a string of ℓ monoid elements $\phi_n(w) = f_1(w_{i_1}) \ldots f_\ell(w_{i_\ell})$ which are then multiplied in M. We abuse notation and also denote as $\phi_n(w)$ the product $f_1(w_{i_1}) \cdot_M \cdots \cdot_M f_\ell(w_{i_\ell})$. By specifying a set of accepting elements $F \subseteq M$ we can use such a program to recognize a subset of Σ^n and subsets of Σ^* can be recognized through families of programs. As is the case for circuits, one can consider uniformity restrictions on these families.

A result of Barrington [1] shows that a language L can be recognized by a polynomial-length family of programs over a finite monoid iff L belongs to NC^1. We denote as $\mathbf{P}(\mathbf{V})$ the class of languages which can be recognized by a program of polynomial length over a monoid in $\mathbf{V}$. Further refinements of Barrington's theorem appear in [6]: L belongs to AC^0 iff L lies in $\mathbf{P}(\mathbf{A})$, L belongs to CC^0 iff it lies in $\mathbf{P}(\mathbf{G_{sol}})$ and L belongs to ACC^0 iff it lies in $\mathbf{P}(\mathbf{M_{sol}})$. These results are robust with respect to many standard uniformity restrictions [3].

A variety $\mathbf{V}$ of finite monoids forms a *program-variety* if every regular language with a neutral letter in $\mathbf{P}(\mathbf{V})$ is in $\mathbf{L}(\mathbf{V})$. Alternatively, we can introduce the notion as follows: say that the multiplication of a monoid M can be program-simulated by a monoid N if for every element $m \in M$ the language $L_m \subseteq M^*$ defined as $L_m = \{m_1 m_2 \ldots m_n : m_1 \cdot m_2 \cdot \cdots \cdot m_n = m\}$ can be recognized by a polynomial-length program over N. Now $\mathbf{V}$ forms a program-variety if any M which can be simulated by some $N \in \mathbf{V}$ is in fact in $\mathbf{V}$ itself [16,23].

The lower bounds for AC^0 circuits computing the MOD_p function [21] can be rephrased as showing that the aperiodic monoids form a program-variety. Many of the important questions in circuit complexity can similarly be rephrased in algebraic terms: for instance ACC^0 is strictly contained in NC^1 iff the solvable monoids form a program-variety.

Programs over finite monoids are closely related to bounded-width branching programs (BWBP). An n-input BWBP of width k and length ℓ over the input alphabet Σ is a leveled directed graph with the following structure. Each level $1 \leq i < \ell$ is associated with an input variable x_{j_i} and contains k nodes that each have $|\Sigma|$ outgoing edges (to level $i+1$) labeled by the possible values of the input variable x_{j_i}. Moreover, the first level contains a distinguished start node while the last level contains an accepting and a rejecting node. Any word $w \in \Sigma^n$ naturally traces out a unique path in this graph and the language accepted by the BWBP is the set of w leading to the accepting node.

Note that in a BWBP a letter $a \in \Sigma$ induces a function $f_{i,a}$ from the k nodes of level i to the k nodes of level $(i+1)$. It is not hard to see that the difference between BWBP and programs over monoids is essentially cosmetic since a program over M can immediately be rewritten as a BWBP of width $|M|$ while, conversely, a BWBP of width k can be rewritten as a program over the finite monoid generated by the functions $f_{i,a}$. The algebraic point of view provides a finer analysis of the BWBP model by parameterizing its power in terms of the algebraic structure of the $f_{i,a}$.

3 The Crane Beach Property

We say that a class $\mathcal{L}$ of languages has the *Crane Beach property* (or CBP) if every language with a neutral letter in $\mathcal{L}$ is regular. As we mentioned in the introduction, it was conjectured but later disproved [2] that $\mathbf{FO}[Arb]$ has the CBP and one can infer from [4] that BWBP of length $o(n \log \log n)$ have the CBP.

For a class of languages having the Crane Beach property, it is also interesting to understand exactly what regular languages with a neutral letter belong to the class. In the case of a logical fragment of $\mathbf{FO}+\mathbf{MOD}$ using numerical predicates in some class $\mathcal{P}$ we are often most interested in cases where the presence of the neutral letter reduces the expressive power to that obtained with the *same* fragment but using $<$ as the sole numerical predicate. For instance, $\mathbf{B\Sigma}_1[Arb]$ has the CBP [2] and the regular languages with a neutral letter definable in this fragment are exactly those definable in $\mathbf{B\Sigma}_1[<]$. We will usually refer to such theorems as *strong Crane Beach results.*

3.1 A Communication Complexity Crane Beach Theorem

The "input on the forehead" model of communication complexity, first introduced in [9] has found a wide variety of applications in numerous areas of complexity theory [14]. It involves k parties wishing to compute a function f of k variables $x_1, \ldots, x_k$: the ith player receives access to all the inputs except x_i so that one can conveniently picture this player as having x_i written on his forehead. The players want to minimize the number of bits that need to be exchanged when computing f on the worst-case input. When the function to be computed is not explicitly given as a k variable function, we further assume that input bits are partitioned in a way that is known to the different parties but chosen adversarially. The k-party *communication complexity of a language* L is the function $D_k(L) : \mathbb{N} \to \mathbb{N}$ giving for each n the minimum number of bits that k parties need to exchange to compute membership in L of the worst-case input w of length n under the worst-case partition of the letters in w. The following theorem which combines two results of [10] establishes a Crane Beach property for the k-party model.

Theorem 1
a) If L is a language with a neutral letter such that $D_k(L) = O(1)$ for some fixed $k \geq 2$ then L is regular.
b) If L is a regular language with a neutral letter then $M(L)$ lies in $\mathbf{DO} \cap \overline{\mathbf{G_{nil}}}$ iff there exists some k such that $D_k(L) = O(1)$.

We define the k-party *communication complexity of a finite monoid* M (denoted $D_k(M)$) as the complexity for k parties to evaluate the product in M of n elements $m_1, \ldots, m_n$ distributed on their foreheads. Because the identity element 1_M acts as a neutral letter for this problem, it can be shown that, up to a constant factor, the worst input-partition in this case gives to player i access to all elements except those with an index congruent to i modulo k. Underlying the previous theorem is the result of [26] that a monoid lies in $\mathbf{DO} \cap \overline{\mathbf{G_{nil}}}$ iff there exists a k s.t. $D_k(M) = O(1)$.

Suppose that k players want to test if a word w of length n belongs to a language L which is recognized by a program ϕ of length $\ell(n)$ over a finite monoid M. The output of an instruction querying bit x_i can be computed privately by any of the $k - 1$ players having access to x_i and the output of the program $\phi(w)$ can then be evaluated using a protocol which evaluates the product of the monoid elements resulting from individual instructions. Hence, the k-party communication complexity of L on inputs of length n is at most the communication complexity of M on strings of length $\ell(n)$. These observations lead to the following lemma [19,26]:

Lemma 2. *Let $k \geq 2$ be some integer and let $f : \mathbb{N} \to \mathbb{N}$ be such that $f = O(\log^c n)$ for some $c \geq 0$. The class* $\mathbf{V}$ *of monoids M such that $D_k(M) = O(f)$ is a program-variety.*

We omit the proof. This lemma provides a way to use recent results on the communication complexity of finite monoids [19,29,10,26] to identify program-varieties.

Corollary 3. *The following are program-varieties:* $\mathbf{DA}, \mathbf{G_{nil}}, \mathbf{DO} \cap \overline{\mathbf{G_{nil}}}$ *and* $\mathbf{G_{nil,k}}$ *for each $k \geq 1$.*

Proof (Sketch). It is not hard to see that the intersection of two program-varieties also forms a program-variety. We know that aperiodic monoids form a program-variety. Moreover, by Lemma 2, the class of monoids with 2-party communication complexity $O(\log n)$ also forms a program-variety. By results of [29], an aperiodic monoid has 2-party communication complexity $O(\log n)$ if it belongs to $\mathbf{DA}$ and so $\mathbf{DA}$ is a program-variety.

The statement for $\mathbf{DO} \cap \overline{\mathbf{G_{nil}}}$ follows directly from lemma 2 and theorem 1.

The statement for $\mathbf{G_{nil}}$ is a consequence of the work of [5]. Once we have that $\mathbf{G_{nil}}$ is a program-variety however, we can again use lemma 2 to obtain that each $\mathbf{G_{nil,k}}$ is a program-variety because [19] shows that a group has bounded $k+1$-party communication complexity iff it is nilpotent of class k. □

3.2 The Crane Beach Property for Programs over Monoids

Definition 4. *A variety of monoids* $\mathbf{V}$ *is said to have the* weak Crane Beach property *if* $\mathbf{P}(\mathbf{V}) \cap \mathcal{L}_e \subseteq$ REG, *where REG denotes the class of regular languages. Furthermore,* $\mathbf{V}$ *has the* strong CBP *if* $\mathbf{P}(\mathbf{V}) \cap \mathcal{L}_e \subseteq \mathbf{L}(\mathbf{V})$, *that is if polynomial length programs over* $\mathbf{V}$ *are no more powerful than morphisms over* $\mathbf{V}$ *in the presence of a neutral letter.*

Note that if $\mathbf{V}$ has the weak CBP then any subvariety of $\mathbf{V}$ also has this property. However, the same does not hold in general for the strong CBP but we prove the following simple lemma in the full paper.

Lemma 5. *If* $\mathbf{V}$ *has the weak CBP and* $\mathbf{W} \subseteq \mathbf{V}$ *is a program-variety then* $\mathbf{W}$ *has the strong CBP.*

We can use the communication complexity results cited earlier to obtain:

Theorem 6. *The variety* $\mathbf{DO} \cap \overline{\mathbf{G_{nil}}}$ *has the strong CBP.*

Proof. The result for $\mathbf{DO} \cap \overline{\mathbf{G_{nil}}}$ stems from Theorem 1. Indeed, we need to show that any language L with a neutral letter which is recognized by a program ϕ over some $M \in \mathbf{DO} \cap \overline{\mathbf{G_{nil}}}$ is regular and has its syntactic monoid in $\mathbf{DO} \cap \overline{\mathbf{G_{nil}}}$.

Since L can be recognized by a program over a monoid that has bounded k-party communication complexity for some k, it follows from our comments preceding Lemma 2 that $D_k(L) = O(1)$. Since L has a neutral letter part *a*) of theorem 1 guarantees that L is regular. Now, using part *b*), L is a regular language with a neutral letter and has bounded k-party complexity so we must have $M(L) \in \mathbf{DO} \cap \overline{\mathbf{G_{nil}}}$. □

In fact, this theorem is not the first indication that programs over $\mathbf{DO} \cap \overline{\mathbf{G_{nil}}}$ are weak. It was shown in [26], building on work of [27] that this variety has the *polynomial-length property* in the sense that any program ϕ over $M \in \mathbf{DO} \cap \overline{\mathbf{G_{nil}}}$ is equivalent to a program ψ over M that has polynomial length. In contrast, a monoid M is said to be *universal* if any language can be recognized by some program over M of possibly exponential length. A simple counting argument shows that any monoid having the polynomial length property cannot be universal and there are indications that the two notions are in fact dual [27].

A direct application of Corollary 3 and Lemma 5, the following subvarieties of $\mathbf{DO} \cap \overline{\mathbf{G_{nil}}}$ also have the strong CBP.

Corollary 7. *The following varieties have the strong CBP:* $\mathbf{G_{nil}}, \mathbf{J}, \mathbf{DA}$ *and* $\mathbf{G_{nil,k}}$ *for any* $k \geq 1$.

It is possible to exhibit varieties that do not have even the weak CBP. In particular, a main result of [2] can be rephrased algebraically as stating that the variety of aperiodic monoids does not have the weak CBP. Furthermore, Barrington showed that polynomial length programs over any non-solvable group are as powerful as NC^1 which in particular contains languages with a neutral letter which are not regular. We thus obtain:

Theorem 8. *If* $\mathbf{V}$ *is a variety containing a non-solvable group or containing the variety* $\mathbf{A}$ *of aperiodic finite monoids, then* $\mathbf{V}$ *does not have the weak CBP.*

We conjecture that in fact any variety containing a universal monoid fails to have the CBP. In particular, we believe that there are non-regular languages with a neutral letter definable in $\mathbf{\Sigma}_2\,[Arb]$.

4 An Application to Logic

The study of Crane Beach properties was foremost motivated by logical considerations and we use the results of the preceding section to describe fragments of $\mathbf{FO}+\mathbf{MOD}$ which have the CBP.

For any $s \geq 0$ and $p \geq 2$ we denote as $\mathbf{\Sigma}_1^{(s,p)}$ the fragment of $\mathbf{FO}+\mathbf{MOD}$ which consists of sentences of the form $\exists^{t,i \pmod p}(x_1, \ldots, x_k)\ \phi(x_1, \ldots, x_k)$ where ϕ is quantifier-free and where the quantifier $\exists^{t,i \pmod p}$, which ranges over k-tuples of variables, is true if the number of k-tuples satisfying ϕ is either equal to $t < s$ or congruent to i modulo p. Note that if $s = 0$, this fragment does not have the ability to simulate an existential quantifier. For a sentence $\phi \in \mathbf{B\Sigma}_1^{(s,p)}$, we define the *maximum arity* of ϕ to be the maximum arity of any of the quantifiers in ϕ.

The expressive power of the $\mathbf{\Sigma}_1^{(s,p)}$ fragment was studied in depth in [24]. In particular, it is recalled that a language L is definable in $\mathbf{\Sigma}_1^{(s,p)}[<]$ iff it is regular and the syntactic monoid of L lies in $\mathbf{J} \vee \mathbf{G_{nil}}$, the variety generated by $\mathbf{J}$ and $\mathbf{G_{nil}}$. This is not surprising given the existing combinatorial descriptions of languages in $\mathbf{L}(\mathbf{J} \vee \mathbf{G_{nil}})$ which we describe next.

We say that a word $u = u_1 \dots u_k$ is a *subword* of w if w can be factorized as $w = \Sigma^* u_1 \Sigma^* \dots \Sigma^* u_k \Sigma^*$ and we denote as $\binom{w}{u}$ the number of such factorizations. A language L is *piecewise-testable* if there exists a k such that membership of $w \in L$ depends only on the set of subwords of length at most k that occur in w. It is not hard to see that L is a piecewise testable language iff it is definable in $\mathbf{B\Sigma}_1[<]$. A theorem of Simon moreover shows that L is piecewise testable iff its syntactic monoid lies in $\mathbf{J}$.

Similarly, we say that a language L *counts subwords of length k modulo p* if membership of a word w in L only depends on the values $\binom{w}{u_1}, \dots, \binom{w}{u_n}$ modulo p for some words $u_1, \dots, u_n$ of length at most k. Again, it is not hard to see that L is of that form iff it can be defined in $\mathbf{\Sigma}_1^{(0,p)}[<]$. It can also be shown that this class corresponds to languages with syntactic monoids in $\mathbf{G_{nil,k}}$.

We say that two words v and w have the same number of subwords of length k *up to threshold s and modulo p* and write $v \sim_{k,s,p} w$ if for any u of length at most k we have either $\binom{v}{u} \leq s$ and $\binom{v}{u} = \binom{w}{u}$ or $\binom{v}{u} > t$ and $\binom{v}{u} = \binom{w}{u}$ modulo p. It can be shown that this relation is a congruence on Σ^* and that a regular language L has a syntactic monoid in $\mathbf{J} \vee \mathbf{G_{nil}}$ iff L is a union of $\sim_{k,s,p}$-classes for some k, s, p (see e.g. [24]). Straubing also establishes the following:

Lemma 9 ([24]). *If L is a regular language definable in $\mathbf{B\Sigma}_1^{(s,p)}[Arb]$ then L is in fact definable in $\mathbf{B\Sigma}_1^{(s,p)}[\mathcal{R}eg]$.*

Note that the above statement does not assume that the language has a neutral letter. A stronger result can be proved under that hypothesis and the next lemma resolves an open problem of [24].

Lemma 10. *If L is a regular language with a neutral letter and is definable in $\mathbf{B\Sigma}_1^{(s,p)}[Arb]$ then it is in fact definable in $\mathbf{B\Sigma}_1^{(s,p)}[<]$.*

Proof (Sketch). By lemma 9 we already know that L is definable in $\mathbf{B\Sigma}_1^{(s,p)}[\mathcal{R}eg]$ and we will furthermore show that it lies in $\mathbf{B\Sigma}_1^{(s,p)}[<]$ by proving that its syntactic monoid $M(L)$ belongs to $\mathbf{J} \vee \mathbf{G_{nil}}$. Let ϕ be the $\mathbf{B\Sigma}_1^{(s,p)}[\mathcal{R}eg]$ defining L and let k be the maximum arity of any quantifier in ϕ. Further let r be the maximum of s, p and any t or q occurring in any $\delta_{n,t,q}$ and any $\kappa_{n,t,q}$ (see section 2.2) needed to express the regular numerical predicates occurring in ϕ.

Let $e \in \Sigma$ be a neutral letter for L and let v and w be two words in $(\Sigma - \{e\})^*$ such that $v \sim_{k,s,p} w$. In particular, $|v| = |w|$ up to treshold s and modulo p. We claim that v is in L iff w is in L. This suffices to establish our result since L is then a union of $\sim$ classes and $M(L)$ thus lies in $\mathbf{J} \vee \mathbf{G_{nil}}$.

To establish the claim, we build words $V = e^{r!-1} v_1 e^{r!-1} \dots e^{r!-1} v_{|v|} e^{r!-1}$ and $W = e^{r!-1} w_1 e^{r!-1} \dots e^{r!-1} w_{|w|} e^{r!-1}$. Since e is neutral v (resp. w) is in L iff V (resp. W) is in L. Let $\psi(x_1, \dots, x_k)$ be a quantifier-free formula constructed from the content predicates, the order predicate and $\delta_{n,t,q}$ or $\kappa_{n,t,q}$ predicates with $t, q \leq r$. We claim that the number of tuples $(x_1, \dots, x_k)$ s.t. $\psi(x_1, \dots, x_k)$ is true on V is equal up to threshold s and modulo q to the number of tuples such that ψ is true on W. This is sufficient to show that V and W satisfy the same sentences in $\mathbf{B\Sigma}_1^{(s,p)}$ sentences and are thus either both in L or both not in L.

We can rewrite $\psi(x_1,\ldots,x_k)$ as a boolean combination of formulas

$$\phi(x_1,\ldots,x_k) : \bigwedge_i (Q_{a_i}x_i \wedge \kappa_{n_i,t_i,q_i}x_i) \wedge \bigwedge_{i\neq j}((x_i * x_j) \wedge x_i\delta_{n_{ij},t_{ij},q_{ij}}x_j)$$

where $a_i \in \Sigma$; $t_i, t_{ij}, q_i, q_{ij} \leq r$ and $*$ is one of $\{<,>,=\}$.

Suppose for simplicity that for $i < j$ the formula ϕ requires $x_i < x_j$. To evaluate the number of tuples satisfying ϕ over V, we let $I = \{i : a_i \in \Sigma - \{e\}\}$ and first choose values for the *non-neutral positions* x_i, i.e. the x_i such that $i \in I$ and such that position x_i in V holds the desired non-neutral letter a_i. We denote as u the word of length $|I| \leq k$ formed by these letters. Note that each non-neutral x_i is congruent to 0 modulo $r!$ because non-neutral letters only occur at such positions. Hence, we can discard any $\delta_{n,t,q}$ predicates involving two such x_i and any $\kappa_{n,t,q}$ involving one such x_i.

Any tuple of non-neutral positions $\overline{D}$ in V corresponds naturally to a tuple of positions $\overline{d} = (d_1,\ldots,d_{|I|})$ in the original word v which specifies an occurrence of the subword u in v.

When $|I| = k$, the number of tuples satisfying $\phi(x_1,\ldots,x_k)$ is simply the number of occurrences of the subword $u = a_1\ldots a_k$ in v which is equal up to threshold s and modulo p to $\binom{w}{u}$. When $|I| < k$ then the number of tuples satisfying ϕ over V depends not only on the number of occurrences of u in v but also on the number of ways in which each choice of non-neutral positions can be completed by a choice for the positions lying outside of I. It can be shown that for any $\overline{D}$, the number of such completions depends only on the *mod p signature* of $\overline{d}$, i.e. on the vector $(d_1, d_2,\ldots,d_{|I|}) \mod p$. The sum of the number of $\overline{d}$ of any possible signature is again $\binom{v}{u}$ which is equal up to treshold s and modulo p to $\binom{w}{u}$. This can be used to prove the claim that the total number of tuples (treshold s, modulo p) $(x_1,\ldots,x_k)$ satisfying $\phi(x_1,\ldots,x_k)$ over V is some function of the number of occurrences threshold t modulo p of u in v. Since the same holds for W and $v \sim_{k,s,p} w$ we have that V and W and thus v and w either both lie in L or lie both outside L. □

In the $(k+1)$-party communication game, any k-tuple of input letters is fully accessible to at least one of the k parties and this immediately yields the following lemma whose simple proof is omitted.

Lemma 11. *If L is definable by a boolean combination of sentences of the form $\phi : \exists^{(t,i \pmod p)}(x_1,\ldots,x_k).\ \psi(x_1,\ldots,x_k)$ in which the quantifier has arity at most k then the $k+1$-party communication complexity of L is $O(1)$.*

Combining this result with theorem 1 and lemma 10 we obtain:

Theorem 12. *The boolean closure of $\mathbf{\Sigma}_1^{(s,p)}$ has the strong CBP.*

Proof. Let L be a language with a neutral letter definable in $\mathbf{B\Sigma}_1^{(s,p)}[Arb]$. By lemma 11, L has bounded k-party communication complexity for some k and thus, by theorem 1, L is regular. Finally, by lemma 10 any regular language with a neutral letter definable in $\mathbf{B\Sigma}_1^{(s,p)}[Arb]$ is in fact definable in $\mathbf{B\Sigma}_1^{(s,p)}[<]$. □

As a corollary, we get an alternative proof of the following theorem of [2].

Corollary 13. *The boolean closure of* $\mathbf{\Sigma}_1$ *has the strong CBP.*

Proof. Let L be a language with a neutral letter definable in $\mathbf{B\Sigma}_1[Arb]$. By Theorem 12, L is regular and its syntactic monoid $M(L)$ lies in $\mathbf{J} \vee \mathbf{G_{nil}}$. Moreover, the only regular languages with a neutral letter definable in $\mathbf{FO}[Arb]$ are those whose syntactic monoid is aperiodic and so $M(L) \in \mathbf{A}$. A simple semigroup-theory argument shows that the intersection of $\mathbf{J} \vee \mathbf{G_{nil}}$ and $\mathbf{A}$ is $\mathbf{J}$. □

Readers familiar with the Ehrenfeucht-Fraïssé approach of [2] might find it surprising that our alternative proof seems to avoid the use of Ramsey-theoretic arguments. In fact, the communication complexity result of [10] which is so crucial in our method relies on the Ramsey-like theorem of Hales-Jewett.

Of course our main theorem also specializes to the other extreme case of $\mathbf{\Sigma}^{(s,p)}$ sentences in which existential and universal quantifiers do not appear: $\mathbf{B\Sigma}_1^{(0,p)}$. Moreover, the maximum arity of the quantifiers can be preserved.

Corollary 14. *For each* $k \geq 1$, $\mathbf{B\,\Sigma}_1^{(0,p)}$ *of max-arity* k *has the strong CBP.*

Proof. One can show that programs over $\mathbf{G_{nil,k}}$ have exactly the same expressive power as $\mathbf{B\Sigma}^{(0,p)}[Arb]$ of maximum arity k while morphisms over $\mathbf{G_{nil,k}}$ have exactly the same expressive power as $\mathbf{B\Sigma}^{(0,p)}[<]$ of maximum arity k (see full paper, or [26]). The statement of the corollary then follows simply by the fact that each $\mathbf{G_{nil,k}}$ has the strong CBP. □

5 Conclusion

The algebraic perspective on the Crane Beach property allows for a more detailed study of the fine line separating computational models or logical fragments which possess a Crane Beach-like property. Moreover it enables a systematic use of the powerful communication complexity results of [10]. Theorem 12 about the Crane Beach property of $\mathbf{\Sigma}_1^{(s,p)}$ can be obtained by an ad hoc argument using Ramsey theory and Ehrenfeucht-Fraïssé games reminiscent of the techniques for the $\mathbf{\Sigma}_1$ case [2]. The proof given here is considerably simpler if less transparent.

Our results about varieties of monoids exhibiting the Crane Beach property also provide a nice complement to previously existing work comparing the power of programs and morphisms over finite monoids in the presence of a neutral letter. Programs over monoids are generally much more expressive than morphism. This extra power of programs stems from (or is limited by) three sources: the algebraic structure of their underlying monoid, their length and their degree of non-uniformity. However, in the presence of a neutral letter the advantages of programs over monoids can disappear:

- results of [4] show that length $\Omega(n \log\log n)$ is necessary to recognize non-regular languages with a neutral letter, regardless of the underlying monoid or the degree of non-uniformity;
- results of [2,20] show that in the presence of a neutral letter polynomial length programs over solvable monoids which are too uniform cannot break the regular barrier.
- Our results complete the picture: programs over monoids whose structure is unsophisticated cannot recognize non-regular languages with a neutral letter, regardless of their length or their degree of non-uniformity.

We have used the algebraic point of view on the CBP to extend the result of [2] on the CBP for $\mathbf{\Sigma}_1$ to $\mathbf{\Sigma}_1^{(s,p)}$. Some of the more general results of Section 3, however, do not seem to have such simple logical applications. In particular, we have shown that a number of important subvarieties of $\mathbf{DO} \cap \overline{\mathbf{G_{nil}}}$ have the CBP but these do not capture any significant logical fragment of $\mathbf{FO}+\mathbf{MOD}$. The case of the variety $\mathbf{DA}$ is particularly intriguing, given its importance in applications of semigroup theory to complexity [28]. Programs over $\mathbf{DA}$ have the same expressive power as decision trees of bounded-rank [11] and so this model also has the CBP. The languages recognized via morphism by monoids in $\mathbf{DA}$ are exactly those definable by both a $\mathbf{\Sigma}_2[<]$ and a $\mathbf{\Pi}_2[<]$ sentence and also those definable in the restriction of $\mathbf{FO}[<]$ to sentences using only two variables ($\mathbf{FO}_2[<]$) but programs over $\mathbf{DA}$ are not known to correspond to the intersection of $\mathbf{\Sigma}_2[Arb]$ and $\mathbf{\Pi}_2[Arb]$ or to $\mathbf{FO}_2[Arb]$. The latter two classes are important candidates for logical fragments of $\mathbf{FO}$ that may possess the CBP.

References

1. D. A. Barrington. Bounded-width polynomial-size branching programs recognize exactly those languages in NC^1. *J. Comput. Syst. Sci.*, 38(1):150–164, 1989.
2. D. A. M. Barrington, N. Immerman, C. Lautemann, N. Schweikardt, and D. Thérien. First-order expressibility of languages with neutral letters or: The Crane Beach conjecture. *J. Comput. Syst. Sci.*, 70(2):101–127, 2005.
3. D. A. M. Barrington, N. Immerman, and H. Straubing. On uniformity within NC^1. *J. Comput. Syst. Sci.*, 41(3):274–306, 1990.
4. D. A. M. Barrington and H. Straubing. Superlinear lower bounds for bounded-width branching programs. *J. Comput. Syst. Sci.*, 50(3):374–381, 1995.
5. D. A. M. Barrington, H. Straubing, and D. Thérien. Non-uniform automata over groups. *Information and Computation*, 89(2):109–132, 1990.
6. D. A. M. Barrington and D. Thérien. Finite monoids and the fine structure of NC^1. *Journal of the ACM*, 35(4):941–952, 1988.
7. C. Behle and K.-J. Lange. **FO**-uniformity. In *Proc. 21st Conf. on Computational Complexity (CCC'06)*, 2006.
8. M. Benedikt and L. Libkin. Expressive power: The finite case. In *Constraint Databases*, pages 55–87, 2000.
9. A. K. Chandra, M. L. Furst, and R. J. Lipton. Multi-party protocols. In *Proc. 15th ACM Symp. on Theory of Computing (STOC'83)*, pages 94–99, 1983.
10. A. Chattopadhyay, A. Krebs, M. Koucký, M. Szegedy, P. Tesson, and D. Thérien. Functions with bounded multiparty communication complexity. Submitted, 2006.

11. R. Gavaldà and D. Thérien. Algebraic characterizations of small classes of boolean functions. In *Proc. of Symp. on Theoretical Aspects of Comp. Sci. (STACS'03)*, 2003.
12. Y. Gurevich and H. Lewis. A logic for constant-depth circuits. *Information and Control*, 61(1):65–74, 1984.
13. N. Immerman. Languages that capture complexity classes. *SIAM J. Comput.*, 16(4):760–778, 1987.
14. E. Kushilevitz and N. Nisan. *Communication Complexity*. Cambridge University Press, 1997.
15. L. Libkin. *Elements of Finite Model Theory*. Springer Verlag, 2004.
16. P. McKenzie, P. Péladeau, and D. Thérien. NC^1: The automata theoretic viewpoint. *Computational Complexity*, 1:330–359, 1991.
17. P. Péladeau. Formulas, regular languages and boolean circuits. *Theor. Comput. Sci.*, 101(1):133–141, 1992.
18. J.-E. Pin. Syntactic semigroups. In *Handbook of language theory*, volume 1, chapter 10, pages 679–746. Springer Verlag, 1997.
19. J.-F. Raymond, P. Tesson, and D. Thérien. An algebraic approach to communication complexity. In *Proc. 25th Int. Coll. on Automata Languages and Programming (ICALP'98)*, pages 29–40, 1998.
20. A. Roy and H. Straubing. Definability of languages by generalized first-order formulas over $(\mathbf{N}, +)$. In *23rd Symp. on Theoretical Aspects of Comp. Sci. (STACS'06)*, pages 489–499, 2006.
21. R. Smolensky. Algebraic methods in the theory of lower bounds for boolean circuit complexity. In *Proc. 19th ACM STOC*, pages 77–82, 1986.
22. H. Straubing. *Finite Automata, Formal Logic and Circuit Complexity*. Boston: Birkhauser, 1994.
23. H. Straubing. When can one monoid simulate another? In *Algorithmic Problems in Groups and Semigroups*, pages 267–288. Birkhäuser, 2000.
24. H. Straubing. Languages defined by modular quantifiers. *Information and Computation*, 166:112–132, 2001.
25. H. Straubing, D. Thérien, and W. Thomas. Regular languages defined by generalized quantifiers. *Information and Computation*, 118:289–301, 1995.
26. P. Tesson. *Computational Complexity Questions Related to Finite Monoids and Semigroups*. PhD thesis, McGill University, 2003.
27. P. Tesson and D. Thérien. The computing power of programs over finite monoids. *Journal of Automata, Languages and Combinatorics*, 7(2):247–258, 2002.
28. P. Tesson and D. Thérien. Diamonds are forever: the variety **DA**. In *Semigroups, Algorithms, Automata and Languages*. WSP, 2002.
29. P. Tesson and D. Thérien. Complete classifications for the communication complexity of regular languages. *Theory of Computing Systems*, 38(2):135–159, 2005.
30. P. Tesson and D. Thérien. Restricted two-variable sentences, circuits and communication complexity. In *Proc. 32nd Int. Conf. on Automata, Languages and Programming (ICALP'05)*, pages 526–538, 2005.
31. P. Tesson and D. Thérien. Bridges between algebraic automata theory and complexity theory. *The Complexity Column, Bull. EATCS*, 88:37–64, 2006.
32. P. Tesson and D. Thérien. Logic meets algebra: the case of regular languages. 2006. Submitted.
33. W. Thomas. *Languages, Automata and Logic*, volume III, chapter 7, pages 389–455. Springer, 1997.

A Sequent Calculus for Type Theory

Stéphane Lengrand[1,2], Roy Dyckhoff[1], and James McKinna[1]

[1] School of Computer Science, University of St Andrews, Scotland
[2] PPS, Université Paris 7, France
{sl, rd, james}@dcs.st-and.ac.uk

Abstract. Based on natural deduction, Pure Type Systems (PTS) can express a wide range of type theories. In order to express proof-search in such theories, we introduce the Pure Type Sequent Calculi (PTSC) by enriching a sequent calculus due to Herbelin, adapted to proof-search and strongly related to natural deduction.

PTSC are equipped with a normalisation procedure, adapted from Herbelin's and defined by local rewrite rules as in Cut-elimination, using explicit substitutions. It satisfies Subject Reduction and it is confluent. A PTSC is logically equivalent to its corresponding PTS, and the former is strongly normalising if and only if the latter is.

We show how the conversion rules can be incorporated inside logical rules (as in syntax-directed rules for type checking), so that basic proof-search tactics in type theory are merely the root-first application of our inference rules.

Keywords: Type theory, PTS, sequent calculus, proof-search, strong normalisation.

1 Introduction

In this paper, we apply to the framework of *Pure Type Systems* [Bar92] the insights into the relationship between sequent calculus and natural deduction as developed in previous papers by Herbelin [Her94, Her95], the second author and others [DP99b, PD00, DU03]

In sequent calculus the proof-search space is often the cut-free fragment, since the latter usually satisfies the subformula property. Herbelin's sequent calculus LJT has the extra advantage of being closer to natural deduction, in that it is permutation-free, and it makes proof-search more deterministic than a Gentzen-style sequent calculus. This makes LJT a natural formalism to organise proof-search in intuitionistic logic [DP99a], and, its derivations being close to the notion of uniform proofs, LJT can be used to describe proof-search in pure Prolog and some of its extensions [MNPS91]. The corresponding term assignment system also expresses the intimate details of β-normalisation in λ-calculus in a form closer to abstract (stack-based) machines for reduction (such as Krivine's [Kri]).

The framework of *Pure Type Systems* (PTS) [Bar92] exploits and generalises the Curry-Howard correspondence, and accounts for many systems already existing, starting with *Barendregt's Cube*. Proof assistants based on them, such

Z. Ésik (Ed.): CSL 2006, LNCS 4207, pp. 441–455, 2006.

as the Coq system [Coq] or the Lego system [LP92], feature interactive proof construction methods using proof-search tactics. Primitive tactics display an asymmetry between introduction rules and elimination rules of the underlying natural deduction calculus: the tactic Intro corresponds to the right-introduction rule for the Π-type (whether in natural deduction or in sequent calculus), but the tactics Apply in Coq or Refine in Lego are much closer (in spirit) to the left-introduction of Π-types (as in sequent calculus) than to elimination rules of natural deduction.

Although encodings from natural deduction to sequent calculus and vice-versa have been widely studied [Gen35, Pra65, Zuc74], the representation in sequent calculus of type theories is relatively undeveloped compared to the literature about type theory in natural deduction. An interesting approach to Pure Type Systems using sequent calculus is in [GR03]. Nevertheless, only the typing rules are in a sequent calculus style, whereas the syntax is still in a natural deduction style: in particular, proofs are denoted by λ-terms, the structure of which no longer matches the structure of proofs.

However, proofs in sequent calculus *can* be denoted by terms; for instance, a construction $M \cdot l$, representing a list of terms with head M and tail l, is introduced in [Her94, Her95] to denote the left-introduction of implication (in the sequent calculus LJT):

$$\frac{\Gamma \vdash M:A \quad \Gamma; B \vdash l:C}{\Gamma; A \rightarrow B \vdash M \cdot l:C}$$

This approach is extended to the corner of the Cube with dependent types and type constructors in [PD00], but types are still built with λ-terms, so the system extensively uses conversion functions from sequent calculus to natural deduction and back.

With such term assignment systems, cut-elimination can be done by means of a rewrite system, cut-free proofs being thus denoted by terms in normal form. In type theory, not only is the notion of proof-normalisation/cut-elimination interesting on its own, but it is even necessary to define the notion of typability, as soon as types depend on terms.

In this paper we enrich Herbelin's sequent calculus LJT into a collection of systems called *Pure Type Sequent Calculi* (PTSC), capturing the traditional PTS, with the hope to improve the understanding of implementation of proof systems based on PTS in respect of:

- having a direct analysis of the basic tactics, which could then be moved into the kernel, rather than requiring a separate type-checking layer for correctness,
- opening the way to improve the basic system with an approach closer to abstract machines to express reductions, both in type-checking *and* in execution (of extracted programs),
- studying extensions to systems involving inductive types/families (such as the Calculus of Inductive Constructions).

Inspired by the fact that, in type theory, implication and universal quantification are just a dependent product, we modify the inference rule above to obtain the left-introduction rule for a Π-type in a PTSC:

$$\frac{\Gamma \vdash M{:}A \quad \Gamma; \langle M/x\rangle B \vdash l{:}C}{\Gamma; \Pi x^A.B \vdash M{\cdot}l{:}C}\,\Pi\mathsf{l}$$

We use here explicit substitutions, whose natural typing rule are cuts [BR95]. From our system a version with implicit substitutions can easily be derived, but this does not allow cuts on an arbitrary formula of a typing environment Γ. Also, explicit substitutions allow the definition of a normalisation procedure by local (small-step) rewrite rules in the spirit of Gentzen's cut-elimination.

Derivability of sequents in a PTSC is denoted by $\vdash$, while derivability in a PTS is denoted by $\vdash_{\mathsf{PTS}}$. We establish the logical equivalence between a PTSC and its corresponding PTS by means of type-preserving encodings. We also prove that the former is strongly normalising if and only if the latter is. The proof is based on mutual encodings that allow the normalisation procedure of one formalism to be simulated by that of the other. Part of the proof also uses a technique by Bloo and Geuvers [BG99], introduced to prove strong normalisation properties of an explicit substitution calculus and later used in [DU03].

In order to show the convenience for proof-search of the sequent calculus approach, we then present a system that is syntax-directed for proof-search, by incorporating the conversion rules into the typing rules that correspond to term constructions. This incorporation is similar to the constructive engine of [Hue89], but different in that proof search takes Γ and A as inputs and produces a (normal) term M such that $\Gamma \vdash M : A$, while the constructive engine takes Γ and M as inputs and produces A. Derivability in the proof-search system is denoted by $\vdash_{\mathsf{PS}}$.

Section 2 presents the syntax of a PTSC and gives the rewrite rules for normalisation. Section 3 gives the typing system with the parameters specifying the PTSC, and a few properties are stated such as Subject Reduction. Section 4 establishes the correspondence between a PTSC and its corresponding PTS, from which we derive confluence. Section 5 presents the strong normalisation result. Section 6 discusses proof-search in a PTSC.

2 Syntax and Operational Semantics of a PTSC

The syntax of a PTSC depends on a given set $\mathcal{S}$ of sorts, written $s, s', \ldots$, and a denumerable set $\mathcal{X}$ of variables, written $x, y, z, \ldots$. The set $\mathcal{T}$ of *terms* (denoted $M, N, P, \ldots$) and the set $\mathcal{L}$ of *lists* (denoted $l, l', \ldots$) are inductively defined as

$$\begin{aligned} M, N, A, B &::= \Pi x^A.B \mid \lambda x^A.M \mid s \mid x\,l \mid M\,l \mid \langle M/x\rangle N \\ l, l' &::= [] \mid M{\cdot}l \mid l@l' \mid \langle M/x\rangle l \end{aligned}$$

$\Pi x^A.M$, $\lambda x^A.M$, and $\langle N/x\rangle M$ bind x in M, and $\langle M/x\rangle l$ binds x in l, thus defining the free variables of terms and lists as well as α-conversion. The set of free variables of a term M (resp. a list l) is denoted $\mathsf{FV}(M)$ (resp. $\mathsf{FV}(l)$). We use Barendregt's convention that no variable is free and bound in a term in order to avoid variable capture when reducing it. Let $A \rightarrow B$ denote $\Pi x^A.B$ when $x \notin \mathsf{FV}(B)$.

This syntax is an extension of Herbelin's $\overline{\lambda}$ [Her95] (with type annotations on λ-abstractions). Lists are used to represent series of arguments of a function, the terms $x\ l$ (resp. $M\ l$) representing the application of x (resp. M) to the list of arguments l. Note that a variable alone is not a term, it has to be applied to a list, possibly the empty list, denoted $[]$. The list with head M and tail l is denoted $M{\cdot}l$, with a typing rule corresponding to the left-introduction of Π-types (c.f. Section 3). Successive applications give rise to the concatenation of lists, denoted $l@l'$, and $\langle M/x\rangle N$ and $\langle M/x\rangle l$ are explicit substitution operators on terms and lists, respectively. They will be used in two ways: first, to instantiate a universally quantified variable, and second, to describe explicitly the interaction between the constructors in the normalisation process, which is adapted from [DU03] and shown in Fig. 1. Side-conditions to avoid variable capture can be inferred from the reduction rules and are ensured by Barendregt's convention. Confluence of the system is proved in section 4. More intuition about Herbelin's calculus, its syntax and operational semantics is given in [Her95].

$$
\begin{array}{llll}
& \mathsf{B} & (\lambda x^A.M)\ (N{\cdot}l) \longrightarrow & (\langle N/x\rangle M)\ l \\
\mathsf{x}\left\{\begin{array}{l} \\ \\ \\ \\ \\ \\ \\ \\ \\ \\ \end{array}\right. &
\begin{array}{ll}
\mathsf{B1} & M\ [] \\
\mathsf{B2} & (x\ l)\ l' \\
\mathsf{B3} & (M\ l)\ l' \\
\mathsf{A1} & (M{\cdot}l')@l \\
\mathsf{A2} & []@l \\
\mathsf{A3} & (l@l')@l''
\end{array} &
\begin{array}{l}
\longrightarrow M \\
\longrightarrow x\ (l@l') \\
\longrightarrow M\ (l@l') \\
\longrightarrow M{\cdot}(l'@l) \\
\longrightarrow l \\
\longrightarrow l@(l'@l'')
\end{array} & \\
\mathsf{xsubst:}\left\{\begin{array}{l} \\ \\ \\ \\ \\ \end{array}\right. &
\begin{array}{ll}
\mathsf{C1} & \langle P/y\rangle\lambda x^A.M \\
\mathsf{C2} & \langle P/y\rangle(y\ l) \\
\mathsf{C3} & \langle P/y\rangle(x\ l) \\
\mathsf{C4} & \langle P/y\rangle(M\ l) \\
\mathsf{C5} & \langle P/y\rangle\Pi x^A.B \\
\mathsf{C6} & \langle P/y\rangle s \\
\mathsf{D1} & \langle P/y\rangle[] \\
\mathsf{D2} & \langle P/y\rangle(M{\cdot}l) \\
\mathsf{D3} & \langle P/y\rangle(l@l')
\end{array} &
\begin{array}{l}
\longrightarrow \lambda x^{\langle P/y\rangle A}.\langle P/y\rangle M \\
\longrightarrow P\ \langle P/y\rangle l \\
\longrightarrow x\ \langle P/y\rangle l \quad \text{if } x \neq y \\
\longrightarrow \langle P/y\rangle M\ \langle P/y\rangle l \\
\longrightarrow \Pi x^{\langle P/y\rangle A}.\langle P/y\rangle B \\
\longrightarrow s \\
\longrightarrow [] \\
\longrightarrow (\langle P/y\rangle M){\cdot}(\langle P/y\rangle l) \\
\longrightarrow (\langle P/y\rangle l)@(\langle P/y\rangle l')
\end{array} &
\end{array}
$$

Fig. 1. Reduction Rules

We denote by $\longrightarrow_G$ the contextual closure of the reduction relation defined by any system G of rewrite rules (such as $\mathsf{B}, \mathsf{xsubst}, \mathsf{x}$). The transitive closure of $\longrightarrow_G$ is denoted by $\longrightarrow^+_G$, its reflexive and transitive closure is denoted by $\longrightarrow^*_G$, and its symmetric reflexive and transitive closure is denoted by $\longleftrightarrow^*_G$. The set of strongly normalising elements (those from which no infinite $\longrightarrow_G$ -

reduction sequence starts) is SN^G. When not specified, G is assumed to be the system B,x from Fig. 1.

A simple polynomial interpretation shows that system x is terminating. If we add rule B, then the system fails to be terminating unless we only consider terms that are typed in a particular typing system.

3 Typing System and Properties

Given the set of sorts $\mathcal{S}$, a particular PTSC is specified by a set $\mathcal{A} \subseteq \mathcal{S}^2$ and a set $\mathcal{R} \subseteq \mathcal{S}^3$. We shall see an example in section 5.

Definition 1 (Environments)

- Environments *are lists of pairs from* $\mathcal{X} \times \mathcal{T}$ *denoted* $(x : A)$.
- *We define the* domain *of an environment and the* application of a substitution to an environment *as follows:*

$$\begin{array}{ll} \mathit{Dom}([]) = \emptyset & \mathit{Dom}(\Gamma, (x : A)) = \mathit{Dom}(\Gamma) \cup \{x\} \\ \langle P/y \rangle([]) = [] & \langle P/y \rangle(\Gamma, (x : A)) = \langle P/y \rangle \Gamma, (x : \langle P/y \rangle A) \end{array}$$

- *We define the following* inclusion relation *between environments:* $\Gamma \sqsubseteq \Delta$ *if for all* $(x : A) \in \Gamma$*, there is* $(x : B) \in \Delta$ *with* $A \longleftrightarrow^* B$

The inference rules in Fig. 2 inductively define the derivability of three kinds of judgement: some of the form Γ wf, some of the form $\Gamma \vdash M : A$ and some of the form $\Gamma; B \vdash l : A$. In the latter case, B is said to be in the *stoup* of the sequent. Side-conditions are used, such as $(s_1, s_2, s_3) \in \mathcal{R}$, $x \notin \mathsf{Dom}(\Gamma)$, $A \longleftrightarrow^* B$ or $\Delta \sqsubseteq \Gamma$, and we use the abbreviation $\Delta \sqsubseteq \Gamma$ wf for $\Delta \sqsubseteq \Gamma$ and Γ wf.

Since the substitution of a variable in an environment affects the rest of the environment (which could depend on the variable), the two rules for explicit substitutions Cut_2 and Cut_4 must have a particular shape that is admittedly complex: thinning (Lemma 3) is built-in by allowing a controlled change of environment. This may appear artificial, but simpler versions that we have tried failed the thinning property. More generally, typing rules for explicit substitutions in type theory are known to be a tricky issue (see for instance [Blo01]), often leading to the failure of subject reduction (Theorem 1). The rules here are sound in that respect, but more elegant alternatives are still to be investigated, possibly by enriching the structure of environments as in [Blo01].

The case analysis for C' in the rule Cut_4 is only necessary for Lemma 1.2 to hold in the presence of top sorts (untyped sorts), and is avoided in [Blo01] by not using explicit substitutions for types in sequents. Here we were appealed by the uniformity of using them everywhere, the use of implicit substitutions for C' and the stoup of the third premiss of $\Pi\mathsf{l}$ being only a minor variant.

There are three conversion rules conv_r, conv'_r, and conv_l in order to deal with the two kinds of judgements and, for one of them, convert the type in the stoup. The lemmas of this section are proved by induction on typing derivations:

$$\frac{}{[\,]\ \mathsf{wf}}\ \mathsf{empty} \qquad \frac{\Gamma \vdash A{:}s \quad x \notin \mathsf{Dom}(\Gamma)}{\Gamma, (x{:}A)\ \mathsf{wf}}\ \mathsf{extend}$$

$$\frac{\Gamma \vdash A{:}s}{\Gamma; A \vdash [\,]{:}A}\ \mathsf{axiom} \qquad \frac{\Gamma \vdash \Pi x^A.B{:}s \quad \Gamma \vdash M{:}A \quad \Gamma; \langle M/x\rangle B \vdash l{:}C}{\Gamma; \Pi x^A.B \vdash M{\cdot}l{:}C}\ \Pi\mathsf{l}$$

$$\frac{\Gamma; C \vdash l{:}A \quad \Gamma \vdash B{:}s \quad A \longleftrightarrow^* B}{\Gamma; C \vdash l{:}B}\ \mathsf{conv}'_r \qquad \frac{\Gamma; A \vdash l{:}C \quad \Gamma \vdash B{:}s \quad A \longleftrightarrow^* B}{\Gamma; B \vdash l{:}C}\ \mathsf{conv}_l$$

$$\frac{\Gamma; C \vdash l'{:}A \qquad \Gamma; A \vdash l{:}B}{\Gamma; C \vdash l'@l{:}B}\ \mathsf{Cut}_1$$

$$\frac{\Gamma \vdash P{:}A \qquad \Gamma, (x{:}A), \Delta; B \vdash l{:}C \quad \Gamma, \langle P/x\rangle\Delta \sqsubseteq \Delta'\ \mathsf{wf}}{\Delta'; \langle P/x\rangle B \vdash \langle P/x\rangle l{:}\langle P/x\rangle C}\ \mathsf{Cut}_2$$

$$\frac{\Gamma\ \mathsf{wf} \quad (s, s') \in \mathcal{A}}{\Gamma \vdash s{:}s'}\ \mathsf{sorted} \qquad \frac{\Gamma \vdash A{:}s_1 \quad \Gamma, (x{:}A) \vdash B{:}s_2 \quad (s_1, s_2, s_3) \in \mathcal{R}}{\Gamma \vdash \Pi x^A.B{:}s_3}\ \Pi\mathsf{wf}$$

$$\frac{\Gamma \vdash \Pi x^A.B{:}s \quad \Gamma, (x{:}A) \vdash M{:}B}{\Gamma \vdash \lambda x^A.M{:}\Pi x^A.B}\ \Pi\mathsf{r} \qquad \frac{\Gamma; A \vdash l{:}B \quad (x{:}A) \in \Gamma}{\Gamma \vdash x\,l{:}B}\ \mathsf{Select}_x$$

$$\frac{\Gamma \vdash M{:}A \quad \Gamma \vdash B{:}s \quad A \longleftrightarrow^* B}{\Gamma \vdash M{:}B}\ \mathsf{conv}_r$$

$$\frac{\Gamma \vdash M{:}A \qquad \Gamma; A \vdash l{:}B}{\Gamma \vdash M\,l{:}B}\ \mathsf{Cut}_3$$

$$\frac{\Gamma \vdash P{:}A \qquad \Gamma, (x{:}A), \Delta \vdash M{:}C \quad \Gamma, \langle P/x\rangle\Delta \sqsubseteq \Delta'\ \mathsf{wf}}{\Delta' \vdash \langle P/x\rangle M{:}C'}\ \mathsf{Cut}_4$$

where either ($C' = C \in \mathcal{S}$) or $C \notin \mathcal{S}$ and $C' = \langle P/x\rangle C$

Fig. 2. Typing rules of a PTSC

Lemma 1 (Properties of typing judgements). *If* $\Gamma \vdash M{:}A$ *(resp.* $\Gamma; B \vdash l : C$*) then* $\mathsf{FV}(M) \subseteq \mathsf{Dom}(\Gamma)$ *(resp.* $\mathsf{FV}(l) \subseteq \mathsf{Dom}(\Gamma)$*), and the following judgements can be derived with strictly smaller typing derivations:*

1. Γ *wf*
2. $\Gamma \vdash A\!:\!s$ *for some* $s \in \mathcal{S}$*, or* $A \in \mathcal{S}$
   (resp. $\Gamma \vdash B\!:\!s$ *and* $\Gamma \vdash C\!:\!s'$ *for some* $s, s' \in \mathcal{S}$*)*

Corollary 1 (Properties of well-formed environments)

1. *If* $\Gamma, x : A, \Delta$ *wf then* $\Gamma \vdash A\!:\!s$ *for some* $s \in \mathcal{S}$ *with* $x \notin \mathsf{Dom}(\Gamma) \cup \mathsf{Dom}(\Delta)$ *and* $\mathsf{FV}(A) \subseteq \mathsf{Dom}(\Gamma)$ *(and in particular* $x \notin \mathsf{FV}(A)$*)*
2. *If* Γ, Δ *wf then* Γ *wf.*

Lemma 2 (Weakening). *Suppose* Γ, Γ' *wf and* $\mathsf{Dom}(\Gamma') \cap \mathsf{Dom}(\Delta) = \emptyset$*.*

1. *If* $\Gamma, \Delta \vdash M\!:\!B$ *then* $\Gamma, \Gamma', \Delta \vdash M\!:\!B$*.*
2. *If* $\Gamma, \Delta; C \vdash l\!:\!B$*, then* $\Gamma, \Gamma', \Delta; C \vdash l\!:\!B$*.*
3. *If* Γ, Δ *wf, then* Γ, Γ', Δ *wf.*

We can also strengthen the *weakening property* into the *thinning property* by induction on the typing derivation. This allows to weaken the environment, change its order, and convert the types inside, as long as it remains well-formed:

Lemma 3 (Thinning). *Suppose* $\Gamma \sqsubseteq \Gamma'$ *wf.*

1. *If* $\Gamma \vdash M\!:\!B$ *then* $\Gamma' \vdash M\!:\!B$*.*
2. *If* $\Gamma; C \vdash l\!:\!B$*, then* $\Gamma'; C \vdash l\!:\!B$*.*

Using all of the results above, Subject Reduction can be proved (see [LDM]).

Theorem 1 (Subject Reduction in a PTSC)

1. *If* $\Gamma \vdash M\!:\!A$ *and* $M \longrightarrow M'$*, then* $\Gamma \vdash M'\!:\!A$
2. *If* $\Gamma; A \vdash l\!:\!B$ *and* $l \longrightarrow l'$*, then* $\Gamma; A \vdash l'\!:\!B$

4 Correspondence with Pure Type Systems

There is a logical correspondence between a PTSC given by the sets $\mathcal{S}$, $\mathcal{A}$ and $\mathcal{R}$ and the PTS given by the same sets.

We briefly recall the syntax and semantics of the PTS. The terms have the following syntax:

$$t, u, v, T, U, V, \ldots ::= x \mid s \mid \Pi x^T.t \mid \lambda x^T.t \mid t\ u$$

which is equipped with the β-reduction rule $(\lambda x^v.t)\ u \longrightarrow_\beta t\{x = u\}$, in which the substitution is implicit, i.e. is a meta-operation.

The terms are typed by the typing rules in Fig. 3, which depend on the sets $\mathcal{S}$, $\mathcal{A}$ and $\mathcal{R}$. PTS are confluent and satisfy subject reduction and thinning [Bar92].

In order to encode the syntax of a PTS into that of a PTSC, it is convenient to re-express the syntax of a PTS with a grammar closer to that of a PTSC as follows:

$$\begin{aligned} w &::= s \mid \Pi x^T.U \mid \lambda x^T.t \\ t, u, v, T, U, V, \ldots &::= w \mid x\ \overrightarrow{t} \mid w\ u\ \overrightarrow{t} \end{aligned}$$

$$\frac{}{[]\ \mathsf{wf}} \qquad \frac{\Gamma \vdash_{\mathsf{PTS}} T\!:\!s \quad x \notin \mathsf{Dom}(\Gamma)}{\Gamma,(x:T)\ \mathsf{wf}} \qquad \frac{\Gamma\ \mathsf{wf} \quad (x:T)\in\Gamma}{\Gamma \vdash_{\mathsf{PTS}} x\!:\!T}$$

$$\frac{\Gamma\ \mathsf{wf} \quad (s,s')\in\mathcal{A}}{\Gamma \vdash_{\mathsf{PTS}} s\!:\!s'} \qquad \frac{\Gamma \vdash_{\mathsf{PTS}} U\!:\!s_1 \quad \Gamma,(x:U) \vdash_{\mathsf{PTS}} T\!:\!s_2 \quad (s_1,s_2,s_3)\in\mathcal{R}}{\Gamma \vdash_{\mathsf{PTS}} \Pi x^U.T\!:\!s_3}$$

$$\frac{\Gamma \vdash_{\mathsf{PTS}} \Pi x^U.V\!:\!s \quad \Gamma,(x:U) \vdash_{\mathsf{PTS}} t\!:\!V}{\Gamma \vdash_{\mathsf{PTS}} \lambda x^U.t\!:\!\Pi x^U.V} \qquad \frac{\Gamma \vdash_{\mathsf{PTS}} t\!:\!\Pi x^T.U \quad \Gamma \vdash_{\mathsf{PTS}} u\!:\!T}{\Gamma \vdash_{\mathsf{PTS}} t\ u\!:\!U\{x=u\}}$$

$$\frac{\Gamma \vdash_{\mathsf{PTS}} t\!:\!U \quad \Gamma \vdash_{\mathsf{PTS}} V\!:\!s \quad U \longleftrightarrow^*_\beta V}{\Gamma \vdash_{\mathsf{PTS}} t\!:\!V}$$

Fig. 3. Typing rules of a PTS

$$\begin{array}{|lcl|}
\hline
\mathsf{h_w}(s) & = & s \\
\mathsf{h_w}(\Pi x^v.v') & = & \Pi x^{\mathsf{h}(v)}.\mathsf{h}(v') \\
\mathsf{h_w}(\lambda x^v.t) & = & \lambda x^{\mathsf{h}(v)}.\mathsf{h}(t) \\
\hline
\mathsf{h}(w) & = & \mathsf{h_w}(w) \\
\mathsf{h}(x\ \overrightarrow{t}) & = & x\ \mathsf{h}_l(\overrightarrow{t}) \\
\mathsf{h}(w\ u\ \overrightarrow{t}) & = & \mathsf{h_w}(w)\ (\mathsf{h}(u)\cdot\mathsf{h}_l(\overrightarrow{t})) \\
\hline
\mathsf{h}_l(\overrightarrow{\emptyset}) & = & [] \\
\mathsf{h}_l(\overrightarrow{u_1\ \ldots\ u_i}) & = & \mathsf{h}(u_1)\cdot\mathsf{h}_l(\overrightarrow{u_2\ \ldots\ u_i}) \\
\hline
\end{array}$$

From a PTS to a PTSC

$$\begin{array}{|lcll|}
\hline
\mathsf{k}(\Pi x^A.B) & = & \Pi x^{\mathsf{k}(A)}.\mathsf{k}(B) & \\
\mathsf{k}(\lambda x^A.M) & = & \lambda x^{\mathsf{k}(A)}.\mathsf{k}(M) & \\
\mathsf{k}(s) & = & s & \\
\mathsf{k}(x\ l) & = & \mathsf{k}^z(l)\{z=x\} & z \text{ fresh} \\
\mathsf{k}(M\ l) & = & \mathsf{k}^z(l)\{z=\mathsf{k}(M)\} & z \text{ fresh} \\
\mathsf{k}(\langle P/x\rangle M) & = & \mathsf{k}(M)\{x=\mathsf{k}(P)\} & \\
\hline
\mathsf{k}^y([]) & = & y & \\
\mathsf{k}^y(M\cdot l) & = & \mathsf{k}^z(l)\{z=y\ \mathsf{k}(M)\} & z \text{ fresh} \\
\mathsf{k}^y(l@l') & = & \mathsf{k}^z(l')\{z=\mathsf{k}^y(l)\} & z \text{ fresh} \\
\mathsf{k}^y(\langle P/x\rangle l) & = & \mathsf{k}^y(l)\{x=\mathsf{k}(P)\} & \\
\hline
\end{array}$$

From a PTSC to a PTS

Fig. 4. Mutual encodings between a PTS and a PTSC

where $\overrightarrow{t}$ represents a list of "t-terms" of arbitrary length. The grammar is sound and complete with respect to the usual one presented at the begining of the section, and it has the advantage of isolating redexes in one term construction in a way similar to a PTSC.

Given in the left-hand side of Fig. 4, the encoding into the corresponding PTSC, is threefold: $\mathsf{h_w}$ applies to "w-terms", h applies to "t-terms", and h_l to lists of "t-terms". The right-hand side of Fig. 4 shows the encoding from a PTSC to a PTS. We prove the following theorem by induction on t and M:

Theorem 2 (Properties of the encodings)

1. *$\mathsf{h}(t)$ is always an* x*-normal form, and* $\langle \mathsf{h}(t)/x\rangle \mathsf{h}(u) \longrightarrow^*_{\mathsf{x}} \mathsf{h}(u\{x=t\})$.
2. $\mathsf{k}(\mathsf{h}(t)) = t$

3. *If M is an x-normal form, then $M = \mathsf{h}(\mathsf{k}(M))$*
4. $M \longrightarrow^*_{\mathsf{x}} \mathsf{h}(\mathsf{k}(M))$
5. *If $t \longrightarrow_\beta u$ then $\mathsf{h}(t) \longrightarrow^+ \mathsf{h}(u)$*
6. *If $M \longrightarrow N$ then $\mathsf{k}(M) \longrightarrow^*_\beta \mathsf{k}(N)$*

Now we use Theorem 2 to prove the confluence of PTSC and the equivalence of the equational theories.

Corollary 2 (Confluence). $\longrightarrow_{\mathsf{x}}$ *and* $\longrightarrow$ *are confluent.*

Proof. We use the technique of simulation: consider two reduction sequences starting from a term in a PTSC. They can be simulated through k by β-reductions, and since a PTS is confluent, we can close the diagram. Now the lower part of the diagram can be simulated through h back in the PTSC, which closes the diagram there as well. The diagrams can be found in [LDM]. □

Corollary 3 (Equational theories)
$t \longleftrightarrow^*_\beta u$ *if and only if* $\mathsf{h}(t) \longleftrightarrow^* \mathsf{h}(u)$
$M \longleftrightarrow^* N$ *if and only if* $\mathsf{k}(M) \longleftrightarrow^*_\beta \mathsf{k}(N)$

Regarding typing, we first define the following operations on environments:

$$\begin{array}{ll} \mathsf{h}([]) = [] & \mathsf{k}([]) = [] \\ \mathsf{h}(\Gamma, (x : v)) = \mathsf{h}(\Gamma), (x : \mathsf{h}(v)) & \mathsf{k}(\Gamma, (x : A)) = \mathsf{k}(\Gamma), (x : \mathsf{k}(A)) \end{array}$$

Preservation of typing is proved by induction on the typing derivations:

Theorem 3 (Preservation of typing 1)

1. *If $\Gamma \vdash_{\mathsf{PTS}} t : T$ then $\mathsf{h}(\Gamma) \vdash \mathsf{h}(t) : \mathsf{h}(T)$*
2. *If $(\Gamma \vdash_{\mathsf{PTS}} t_i : T_i\{x_1 = t_1\} \ldots \{x_{i-1} = t_{i-1}\})_{i=1\ldots n}$ and $\mathsf{h}(\Gamma) \vdash \mathsf{h}(\Pi x_1{}^{T_1}.\ldots \Pi x_n{}^{T_n}.T) : s$ then $\mathsf{h}(\Gamma); \mathsf{h}(\Pi x_1{}^{T_1}.\ldots \Pi x_n{}^{T_n}.T) \vdash \mathsf{h}_l(t_1 \ldots t_n) : \mathsf{h}(T\{x_1 = t_1\} \ldots \{x_n = t_n\})$*
3. *If Γ wf then $\mathsf{h}(\Gamma)$ wf*

Theorem 4 (Preservation of typing 2)

1. *If $\Gamma \vdash M : A$ then $\mathsf{k}(\Gamma) \vdash_{\mathsf{PTS}} \mathsf{k}(M) : \mathsf{k}(A)$*
2. *If $\Gamma; B \vdash l : A$ then $\mathsf{k}(\Gamma), y : \mathsf{k}(B) \vdash_{\mathsf{PTS}} \mathsf{k}^y(l) : \mathsf{k}(A)$ for a fresh y*
3. *If Γ wf then $\mathsf{k}(\Gamma)$ wf*

5 Equivalence of Strong Normalisation

Theorem 5. *A PTSC given by the sets $\mathcal{S}$, $\mathcal{A}$, and $\mathcal{R}$ is strongly normalising if and only if the PTS given by the same sets is.*

Proof. Assume that the PTSC is strongly normalising, and let us consider a well-typed t of the corresponding PTS, i.e. $\Gamma \vdash_{\mathsf{PTS}} t : T$ for some Γ, T. By Theorem 3, $\mathsf{h}(\Gamma) \vdash \mathsf{h}(t) : \mathsf{h}(T)$ so $\mathsf{h}(t) \in \mathsf{SN}$. Now by Theorem 2, any reduction sequence starting from t maps to a reduction sequence of at least the same length starting from $\mathsf{h}(t)$, but those are finite.

Now assume that the PTS is strongly normalising and that $\Gamma \vdash M : A$ in the corresponding PTSC. We shall now apply Bloo and Geuvers' technique from [BG99]. By subject reduction, any N such that $M \longrightarrow^* N$ satisfies $\Gamma \vdash N{:}A$ and any sub-term P (resp. sub-list l) of any such N is also typable. By Theorem 4, for any such P (resp. l), $\mathsf{k}(P)$ (resp. $\mathsf{k}^y(l)$) is typable in the PTS, so it is strongly normalising by assumption and we denote by $\sharp\mathsf{k}(P)$ (resp. $\sharp\mathsf{k}^y(l)$) the length of the longest β-reduction sequence reducing it.

We now encode any such P and l into a first-order syntax given by the following ordered infinite signature:

$$\star \prec \mathsf{i}(\_) \prec \mathsf{ii}(\_,\_) \prec \mathsf{cut}^n(\_,\_) \prec \mathsf{sub}^n(\_,\_)$$

for all integers n. Moreover, we set $\mathsf{sub}^n(\_,\_) \prec \mathsf{cut}^m(\_,\_)$ if $n < m$. The order is well-founded, and the *lexicographic path ordering* (lpo) that it induces on the first-order terms is also well-founded (definitions and results can be found in [KL80]). The encoding is given in Fig 5. An induction on terms shows that reduction decreases the lpo. □

$\mathcal{T}(s)$		$= \star$	
$\mathcal{T}(\lambda x^A.M)$	$= \mathcal{T}(\Pi x^A.M)$	$= \mathsf{ii}(\mathcal{T}(A), \mathcal{T}(M))$	
$\mathcal{T}(x\ l)$		$= \mathsf{i}(\mathcal{T}(l))$	
$\mathcal{T}(M\ l)$		$= \mathsf{cut}^{\sharp\mathsf{k}(M\ l)}(\mathcal{T}(M), \mathcal{T}(l))$	
$\mathcal{T}(\langle M/x\rangle N)$		$= \mathsf{sub}^{\sharp\mathsf{k}(\langle M/x\rangle N)}(\mathcal{T}(M), \mathcal{T}(N))$	
$\mathcal{T}([])$		$= \star$	
$\mathcal{T}(M{\cdot}l)$		$= \mathsf{ii}(\mathcal{T}(M), \mathcal{T}(l))$	
$\mathcal{T}(l@l')$		$= \mathsf{ii}(\mathcal{T}(l), \mathcal{T}(l'))$	
$\mathcal{T}(\langle M/x\rangle l)$		$= \mathsf{sub}^{\sharp\mathsf{k}^y(\langle M/x\rangle l)}(\mathcal{T}(M), \mathcal{T}(l))$	where y is fresh

Fig. 5. First-order encoding

Examples of strongly normalising PTS are the systems of Barendregt's Cube, including the *Calculus of Constructions* [CH88] on which the proof-assistant Coq is based [Coq] (but it also uses inductive types and local definitions), for all of which we now have a corresponding PTSC that can be used for proof-search.

6 Proof-Search

In contrast to propositional logic where cut is an admissible rule of sequent calculus, terms in normal form may need a cut-rule in their typing derivation. For instance in the rule Πl, a type which is not normalised ($\langle M/x\rangle B$) must appear in the stoup of the third premiss, so that cuts might be needed to type it inside the derivation. However if we modify Πl by now using an *implicit* substitution $B\{x = M\}$, normal forms can then be typed not using Cut_2 and Cut_4, but still using Cut_1 and Cut_3.

In this section we present a system for proof-search that avoids all cuts, is complete and is sound provided that types are checked independently. In proof-search, the inputs are an environment Γ and a type A, henceforth called *goal*, and the output is a term M such that $\Gamma \vdash M : A$. When we look for a list, the type in the stoup is also an input. The inference rules now need to be directed by the shape of the goal (or of the type in the stoup), and the proof-search system (PS, for short) can be obtained by optimising the use of the conversion rules as shown in Fig. 6. The incorporation of the conversion rules into the other rules is similar to that of the constructive engine in natural deduction [Hue89, JMP94]; however the latter was designed for type synthesis, for which the inputs and outputs are not the same as in proof-search, as mentioned in the introduction.

$$\frac{A \longleftrightarrow^* A'}{\Gamma; A \vdash_{\mathsf{PS}} [\,] : A'}\ \mathsf{axiom}_{\mathsf{PS}} \qquad \frac{D \longrightarrow^* \Pi x^A.B \quad \Gamma \vdash_{\mathsf{PS}} M : A \quad \Gamma; \langle M/x \rangle B \vdash_{\mathsf{PS}} l : C}{\Gamma; D \vdash_{\mathsf{PS}} M \cdot l : C}\ \Pi\mathsf{l}_{\mathsf{PS}}$$

$$\frac{C \longrightarrow^* s_3 \quad (s_1, s_2, s_3) \in R \quad \Gamma \vdash_{\mathsf{PS}} A : s_1 \quad \Gamma, (x : A) \vdash_{\mathsf{PS}} B : s_2}{\Gamma \vdash_{\mathsf{PS}} \Pi x^A.B : C}\ \Pi\mathsf{wf}_{\mathsf{PS}}$$

$$\frac{C \longrightarrow^* s' \quad (s, s') \in \mathcal{A}}{\Gamma \vdash_{\mathsf{PS}} s : C}\ \mathsf{sorted}_{\mathsf{PS}} \qquad \frac{(x : A) \in \Gamma \quad \Gamma; A \vdash_{\mathsf{PS}} l : B}{\Gamma \vdash_{\mathsf{PS}} x\ l : B}\ \mathsf{Select}_x$$

$$\frac{C \longrightarrow^* \Pi x^A.B \quad A \text{ is a normal form} \quad \Gamma, (x : A) \vdash_{\mathsf{PS}} M : B}{\Gamma \vdash_{\mathsf{PS}} \lambda x^A.M : C}\ \Pi\mathsf{r}_{\mathsf{PS}}$$

Fig. 6. Rules for Proof-search

Notice than in PS there are *no* cut-rules. Indeed, even though in the original typing system cuts are required in typing derivations of normal forms, they only occur to check that types are well-typed themsleves. Here we removed those type-checking constraints, relaxing the system, because types are the input of proof-search, and they would be checked before starting the search. PS is sound and complete in the following sense:

Theorem 6

1. *(Soundness) Provided $\Gamma \vdash A : s$, if $\Gamma \vdash_{\mathsf{PS}} M : A$ then $\Gamma \vdash M : A$ and M is a normal form.*
2. *(Completeness) If $\Gamma \vdash M : A$ and M is a normal form, then $\Gamma \vdash_{\mathsf{PS}} M : A$.*

Proof. Both proofs are done by induction on typing derivations, with similar statements for lists. For Soundness, the type-checking proviso is verified every time we need the induction hypothesis. For Completeness, the following lemma is required (and also proved inductively): assuming $A \longleftrightarrow^* A'$ and $B \longleftrightarrow^* B'$, if $\Gamma \vdash_{\mathsf{PS}} M : A$ then $\Gamma \vdash_{\mathsf{PS}} M : A'$, and if $\Gamma; B \vdash_{\mathsf{PS}} l : A$ then $\Gamma; B' \vdash_{\mathsf{PS}} l : A'$. □

Note that neither part of the theorem relies on the unsolved problem of *expansion postponement* [JMP94, Pol98]. Indeed, PS *does not* check types. When recovering a full derivation tree from a PS one by the soundness theorem, expansions and cuts might be introduced at any point, coming from the derivation of the type-checking proviso.

The condition that A is in normal form in rule $\Pi\mathsf{r}_{\mathsf{PS}}$ is not problematic for completeness: whether or not the PTSC is strongly normalising, such a normal form is given as the type annotation of the λ-abstraction, in the term $\bar{M}$ of the hypothesis of completeness. On the other hand, the condition allows the soundness theorem to state that all terms typable in system PS are normal forms. Without it, terms would be in normal forms but for their type annotations in λ-abstractions.

Basic proof-search can be done in the proof-search system simply by reducing the goal, or the type in the stoup, and then, depending on its shape, trying to apply one of the inference rules bottom-up.

There are three points of non-determinism in proof-search:

- The choice of a variable x for applying rule Select_x, knowing only Γ and B (this corresponds in natural deduction to the choice of the head-variable of the proof-term). Not every variable of the environment will work, since the type in the stoup will eventually have to be unified with the goal, so we still need back-tracking.
- When the goal reduces to a Π-type, there is an overlap between rules $\Pi\mathsf{r}_{\mathsf{PS}}$ and Select_x; similarly, when the type in the stoup reduces to a Π-type, there is an overlap between rules $\Pi\mathsf{l}_{\mathsf{PS}}$ and $\mathsf{axiom}_{\mathsf{PS}}$. Both overlaps disappear when Select_x is restricted to the case when the goal does not reduce to a Π-type (and sequents with stoups never have a goal reducing to a Π-type). This corresponds to looking only for η-long normal forms in natural deduction. This restriction also brings the derivations in LJT (and in our PTSC) closer to the notion of uniform proofs. Further work includes the addition of η to the notion of conversion in PTSC.
- When the goal reduces to a sort s, three rules can be applied (in contrast to the first two points, this source of non-determinism does not already appear in the propositional case).

The non-determinism is already present in natural deduction, but the sequent calculus version conveniently identifies where it occurs exactly.

We now give the example of a derivation in PS. We consider the PTSC equivalent to system F, i.e. the one given by the sets:
$\mathcal{S} = \{\mathsf{Type}, \mathsf{Kind}\}$, $\mathcal{A} = \{(\mathsf{Type}, \mathsf{Kind})\}$, and $\mathcal{R} = \{(\mathsf{Type}, \mathsf{Type}), (\mathsf{Kind}, \mathsf{Type})\}$.

For brevity we omit the types on λ-abstractions, we abbreviate $x\ []$ as x for any variable x and simplify $\langle N/x \rangle P$ as P when $x \notin \mathsf{FV}(P)$. We also write $A \wedge B$ for $\Pi Q^{\mathsf{Type}}.(A \to (B \to Q)) \to Q$. Trying to find a term M such that $A : \mathsf{Type}, B : \mathsf{Type} \vdash M : (A \wedge B) \to (B \wedge A)$, we get the PS-derivation below:

$$
\dfrac{\dfrac{\dfrac{\dfrac{\pi_B}{\Gamma \vdash_{\mathsf{PS}} N_B : B} \qquad \dfrac{\dfrac{\pi_A}{\Gamma \vdash_{\mathsf{PS}} N_A : A} \qquad \dfrac{}{\Gamma; Q \vdash_{\mathsf{PS}} [] : Q}\,\mathsf{axiom}_{\mathsf{PS}}}{\Gamma; A \to Q \vdash_{\mathsf{PS}} N_A\cdot[] : Q}\,\Pi\mathsf{l}_{\mathsf{PS}}}{\Gamma; B \to (A \to Q) \vdash_{\mathsf{PS}} N_B\cdot N_A\cdot[] : Q}\,\Pi\mathsf{l}_{\mathsf{PS}}}{\Gamma \vdash_{\mathsf{PS}} y\ N_B\cdot N_A\cdot[] : Q}\,\mathsf{Select}_y}{A : \mathsf{Type}, B : \mathsf{Type} \vdash_{\mathsf{PS}} \lambda x.\lambda Q.\lambda y.y\ N_B\cdot N_A\cdot[] : (A \wedge B) \to (B \wedge A)}\,\Pi\mathsf{r}_{\mathsf{PS}}
$$

where $\Gamma = A : \mathsf{Type}, B : \mathsf{Type}, x : A \wedge B, Q : \mathsf{Type}, y : B \to (A \to Q)$, and π_A is the following derivation ($N_A = x\ A\cdot(\lambda x'.\lambda y'.x')\cdot[]$):

$$
\dfrac{\dfrac{\dfrac{\dfrac{}{\Gamma; \mathsf{Type} \vdash_{\mathsf{PS}} [] : \mathsf{Type}}}{\Gamma \vdash_{\mathsf{PS}} A : \mathsf{Type}} \qquad \dfrac{\dfrac{\dfrac{\dfrac{}{\Gamma, x' : A, y' : B; A \vdash_{\mathsf{PS}} [] : A}}{\Gamma, x' : A, y' : B \vdash_{\mathsf{PS}} x' : A}}{\Gamma \vdash_{\mathsf{PS}} \lambda x'.\lambda y'.x' : A \to (B \to A)} \qquad \dfrac{}{\Gamma; A \vdash_{\mathsf{PS}} [] : A}}{\Gamma; \langle A/Q\rangle(A \to (B \to Q)) \to Q \vdash_{\mathsf{PS}} (\lambda x'.\lambda y'.x')\cdot[] : A}}{\Gamma; A \wedge B \vdash_{\mathsf{PS}} A\cdot(\lambda x'.\lambda y'.x')\cdot[] : A}}{\Gamma \vdash_{\mathsf{PS}} x\ A\cdot(\lambda x'.\lambda y'.x')\cdot[] : A}
$$

and π_B is the derivation similar to π_A ($N_B = x\ B\cdot(\lambda x'.\lambda y'.y')\cdot[]$) with conclusion $\Gamma \vdash_{\mathsf{PS}} x\ B\cdot(\lambda x'.\lambda y'.y')\cdot[] : B$.

This example shows how the non-determism of proof-search is sometimes quite constrained by the need to eventually unify the type in the stoup with the goal. For instance in π_A (resp. π_B), the resolution of $\Gamma \vdash Q : \mathsf{Type}$ by A (resp. B) could be inferred from the unification in the right-hand side branch.

In Coq [Coq], the proof-search tactic `apply x` can be decomposed into the bottom-up application of Select_x followed by a series of bottom-up applications of $\Pi\mathsf{l}_{\mathsf{PS}}$ and finally $\mathsf{axiom}_{\mathsf{PS}}$, but it either delays the resolution of sub-goals or automatically solves them from the unification attempt, often avoiding obvious back-tracking.

In order to mimic even more closely this basic tactic, delaying the resolution of sub-goals can be done by using meta-variables, to be instantiated later with the help of the unification constraint. By extending PTSC with meta-variables, we can go further and express a sound and complete algorithm for type inhabitant enumeration (similar to Dowek's [Dow93] and Muñoz's [Mun01] in natural deduction) simply as the bottom-up construction of derivation trees in sequent calculus.

Proof-search tactics in natural deduction simply depart from the simple bottom-up application of the typing rules, so that their readability and usage would be made more complex. Just as in propositional logic [DP99a], sequent calculi can be a useful theoretical approach to study and design those tactics, in the hope to improve semi-automated reasoning in proof-assistants such as Coq.

7 Conclusion and Further Work

We have defined a parameterised formalism that gives a sequent calculus for each PTS. It comprises a syntax, a rewrite system and typing rules. In constrast to previous work, the syntax of both types and proof-terms of PTSC is in a sequent-calculus style, thus avoiding the use of implicit or explicit conversions to natural deduction [GR03, PD00].

A strong correspondence with natural deduction has been established (regarding both the logic and the strong normalisation), and we derive from it the confluence of each PTSC. We can give as examples the corners of Barendregt's Cube, for which we now have a elegant theoretical framework for proof-search: We have shown how to deal with conversion rules so that basic proof-search tactics are simply the root-first application of the typing rules.

Further work includes studying direct proofs of strong normalisation (such as Kikuchi's for propositional logic [Kik04]), and dealing with inductive types such as those used in Coq. Their specific proof-search tactics should also clearly appear in sequent calculus. Finally, the latter is also more elegant than natural deduction to express classical logic, so it would be interesting to build classical Pure Type Sequent Calculi.

References

[Bar92] H. P. Barendregt. Lambda calculi with types. In S. Abramsky, D. M. Gabby, and T. S. E. Maibaum, editors, Hand. Log. Comput. Sci., volume 2, chapter 2, pages 117–309. Oxford University Press, 1992.

[BG99] R. Bloo and H. Geuvers. Explicit substitution: on the edge of strong normalization. Theoret. Comput. Sci., 211(1-2):375–395, 1999.

[Blo01] R. Bloo. Pure type systems with explicit substitution. Math. Structures in Comput. Sci., 11(1):3–19, 2001.

[BR95] R. Bloo and K. H. Rose. Preservation of strong normalisation in named lambda calculi with explicit substitution and garbage collection. In Computing Science in the Netherlands (CSN '95), pages 62–72, Koninklijke Jaarbeurs, Utrecht, 1995.

[CH88] T. Coquand and G. Huet. The calculus of constructions. Inf. Comput., 76(2–3):95–120, 1988.

[Coq] The Coq Proof Assistant. http://coq.inria.fr/.

[Dow93] G. Dowek. A complete proof synthesis method for type systems of the cube. J. Logic Comput., 1993.

[DP99a] R. Dyckhoff and L. Pinto. Proof search in constructive logics. In Sets and proofs (Leeds, 1997), pages 53–65. Cambridge Univ. Press, Cambridge, 1999.

[DP99b] R. Dyckhoff and L. Pinto. Permutability of proofs in intuitionistic sequent calculi. Theoret. Comput. Sci., 212(1–2):141–155, 1999.

[DU03] R. Dyckhoff and C. Urban. Strong normalization of Herbelin's explicit substitution calculus with substitution propagation. J. Logic Comput., 13(5):689–706, 2003.

[Gen35] G. Gentzen. Investigations into logical deduction. In Gentzen collected works, pages 68–131. Ed M. E. Szabo, North Holland, (1969), 1935.

[GR03] F. Gutiérrez and B. Ruiz. Cut elimination in a class of sequent calculi for pure type systems. In R. de Queiroz, E. Pimentel, and L. Figueiredo, editors, ENTCS, volume 84. Elsevier, 2003.

[Her94] H. Herbelin. A lambda-calculus structure isomorphic to Gentzen-style sequent calculus structure. In L. Pacholski and J. Tiuryn, editors, Computer Science Logic, 8th Int. Work., CSL '94, volume 933 of LNCS, pages 61–75. Springer, 1994.

[Her95] H. Herbelin. Séquents qu'on calcule. PhD thesis, Université Paris 7, 1995.

[Hue89] G. Huet. The constructive engine. World Scientific Publishing, Commemorative Volume for Gift Siromoney, 1989.

[Kik04] K. Kikuchi. A direct proof of strong normalization for an extended Herbelin's calculus. In Y. Kameyama and P. J. Stuckey, editors, Proc. of the 7th Int. Symp. on Functional and Logic Programming (FLOPS'04), volume 2998 of LNCS, pages 244–259. Springer-Verlag, 2004.

[KL80] S. Kamin and J.-J. Lévy. Attempts for generalizing the recursive path orderings. Handwritten paper, University of Illinois, 1980.

[Kri] J.-L. Krivine. Un interpréteur du λ-calcul. available at `http://www.pps.jussieu.fr/~krivine/`.

[LDM] S. Lengrand, R. Dyckhoff, and J. McKinna. A sequent calculus for type theory - longer version. available at `http://www.pps.jussieu.fr/~lengrand/Work/Reports/Proofs.ps`.

[LP92] Z. Luo and R. Pollack. LEGO Proof Development System: User's Manual. Technical Report ECS-LFCS-92-211, School of Informatics, University of Edinburgh, 1992.

[MNPS91] D. Miller, G. Nadathur, F. Pfenning, and A. Scedrov. Uniform proofs as a foundation for logic programming. Ann. Pure Appl Logic, 51:125–157, 1991.

[Mun01] C. Munoz. Proof-term synthesis on dependent-type systems via explicit substitutions. Theor. Comput. Sci., 266(1-2):407–440, 2001.

[PD00] L. Pinto and R. Dyckhoff. Sequent calculi for the normal terms of the $\Lambda\Pi$ and $\Lambda\Pi\Sigma$ calculi. In D. Galmiche, editor, ENTCS, volume 17. Elsevier, 2000.

[Pol98] E. Poll. Expansion Postponement for Normalising Pure Type Systems. J. Funct. Programming, 8(1):89–96, 1998.

[Pra65] D. Prawitz. Natural deduction, a proof-theoretical study. In Acta Universitatis Stockholmiensis, volume 3. Almqvist & Wiksell, 1965.

[JMP94] B. van Benthem Jutting, J. McKinna, and R. Pollack. Checking Algorithms for Pure Type Systems. In H. Barendregt and T. Nipkow, editors, Types for Proofs and Programs, volume 806 of LNCS. Springer-Verlag, 1994.

[Zuc74] J. Zucker. The correspondence between cut-elimination and normalization. Annals of Mathematical Logic, 7:1–156, 1974.

Universality Results for Models in Locally Boolean Domains

Tobias Löw and Thomas Streicher

TU Darmstadt, Schloßgartenstraße 7, D-64289 Darmstadt

Abstract. In [8] J. Laird has shown that an infinitary sequential extension of PCF has a fully abstract model in his category of locally boolean domains (introduced in [10]). In this paper we introduce an extension SPCF_∞ of his language by recursive types and show that it is universal for its model in locally boolean domains.

Finally we consider an infinitary target language CPS_∞ for (the) CPS translation (of [18]) and show that it is universal for a model in locally boolean domains which is constructed like Dana Scott's D_∞ where $D = \{\bot, \top\}$.

1 Introduction

In [5] Cartwright, Curien and Felleisen have shown that for SPCF, an extension of PCF with error elements and a **catch** construct, one can construct extensional fully abstract models whose induced theory in the finitary case (i.e. over base type boolean) is still decidable and thus much simpler than the fully abstract models for PCF (see [1,7,15]) as demonstrated by Loader's result [11]. The model of [5] consists of error-propagating sequential algorithms between concrete data structures (with errors). About a decade later in [10] J. Laird has arrived at a reformulation of this model in terms of a category **LBD** of *locally boolean domains* (lbds) and *sequential maps* between them.

In the current paper we show that in **LBD** one can interpret an infinitary variant SPCF_∞ of the language SPCF of [5]. Roughly speaking, the language SPCF_∞ is an extension of simply typed λ-calculus by countable sums and products, error elements $\top$ for each type, a control construct **catch** and recursive types. For SPCF_∞ without recursive types it has been shown in [8] that the **LBD** model is fully abstract, i.e. that all finite elements arise as denotations of programs. We show that actually *all* elements of (possibly recursive) SPCF_∞ types can be denotes by SPCF_∞ terms, i.e. that SPCF_∞ is universal for its **LBD** model. In the proof we first show that every SPCF_∞ type can be obtained as an SPCF_∞ definable retract of the first order type $\mathbf{U} = \mathbf{N} \to \mathbf{N}$ (adapting an analogous result in [12] for ordinary sequential algorithms without error elements) and then conclude by observing that every element of $\mathbf{U}$ is (trivially) SPCF_∞ definable.

In [18] it has been observed that 0_∞, i.e. Scott's D_∞ with $D = 0 = \{\bot, \top\}$, can be obtained as bifree solution (cf. [16]) of $D = [D^\omega \to 0]$. Since solutions of recursive type equations are available in **LBD** (see section 2) we may consider also

Z. Ésik (Ed.): CSL 2006, LNCS 4207, pp. 456–470, 2006.

the bifree solution of the equation for D in **LBD**. Canonically associated with this type equation is the language CPS_∞ whose terms are given by the grammar

$$M ::= x \mid \lambda\vec{x}.M\langle\vec{M}\rangle \mid \lambda\vec{x}.\top$$

where $\vec{x}$ ranges over infinite lists of pairwise disjoint variables and $\vec{M}$ over infinite lists of terms. Notice that CPS_∞ is more expressive than (untyped) λ-calculus with an error element $\top$ in the respect that one may apply a term to an infinite list of arguments. Consider e.g. the term $\lambda\vec{x}.x_0\langle\vec{\bot}\rangle$ whose interpretation retracts D to $\mathsf{0}$ (i.e. sends $\top$ to $\top$ and everything else to $\bot$) whereas this retraction is not expressible in λ-calculus with a constant $\top$. We show that CPS_∞ is universal for its model in D. For this purpose we proceed as follows.

We first observe that the finite elements of D all arise from simply typed λ-calculus over $\mathsf{0}$. Since the latter is universal for its **LBD** model (as shown in [8]) and all retractions of D to finite types are CPS_∞ definable it follows that all finite elements of D are definable in CPS_∞. Then borrowing an idea from [9] we show that the supremum of stably bounded elements of D is CPS_∞ definable. Using this we show that the supremum of every chain of finite elements increasing w.r.t. $\leq_s$ is CPS_∞ and thus every element of D is CPS_∞ definable as well.

Although interpretation of CPS_∞ in D is surjective it happens that interpretation in D may identify terms with different infinite normal form, i.e. the interpretation is not faithful. Finally, we discuss a way how this shortcoming can be avoided, namely to extend CPS_∞ with a parallel construct $\|$ and refining the observation type $\mathsf{0}$ to $\widetilde{\mathsf{0}} \cong \mathsf{List}(\widetilde{\mathsf{0}})$. This amounts to a "qualitative" reformulation of a "quantitative" method introduced by F. Maurel in his Thesis [14].

2 Locally Boolean Domains

This section contains a short introduction to the theory of lbds and sequential maps (cf. [10]).

Definition 1. *A* locally boolean order (lbo) *is a triple $A = (|A|, \sqsubseteq, \neg)$ where $(A, \sqsubseteq)$ is a partial order and $\neg : |A| \to |A|$ is antitonic and an involution (i.e. $x \sqsubseteq y \Rightarrow \neg y \sqsubseteq \neg x$ and $\neg\neg x = x$ for all $x, y \in |A|$) such that*

(1) *for every $x \in A$ the set $\{x, \neg x\}$ has a least upper bound $x^\top = x \sqcup \neg x$ (and, therefore, also a greatest lower bound $x_\bot = \neg(x^\top) = x \sqcap \neg x$)*
(2) *whenever $x \sqsubseteq y^\top$ and $y \sqsubseteq x^\top$ (notation $x \uparrow y$) then $\{x, y\}$ has a supremum $x \sqcup y$ and an infimum $x \sqcap y$.*

A is complete *if $(|A|, \sqsubseteq)$ is a cpo, i.e. every directed subset X has a supremum $\bigsqcup X$. A is* pointed *if it has a least element $\bot$ (and thus also a greatest element $\top = \neg\bot$).* ◇

We write $x \downarrow y$ as an abbreviation for $\neg x \uparrow \neg y$, and $x \updownarrow y$ for $x \uparrow y$ and $x \downarrow y$. Notice that $x \updownarrow y$ iff $x_\bot = y_\bot$ iff $x^\top = y^\top$. A subset $X \subseteq A$ is called *stably coherent* (notation $\uparrow X$) iff $x \uparrow y$ for all $x, y \in X$. Analogously, X is called *costably coherent* (notation $\downarrow X$) iff $x \downarrow y$ for all $x, y \in X$. Finally, X is called *bistably coherent* (notation $\updownarrow X$) iff $\uparrow X$ and $\downarrow X$.

Definition 2. *For a lbo A and $x, y \in A$ we define*

stable order: $x \leq_s y$ *iff* $x \sqsubseteq y$ *and* $x \uparrow y$
costable order: $x \leq_c y$ *iff* $x \sqsubseteq y$ *and* $x \downarrow y$ (*iff* $\neg y \leq_s \neg x$)
bistable order: $x \leq_b y$ *iff* $x \leq_s y$ *and* $x \leq_c y$ ◇

For the definition of locally boolean domains we introduce the notion of *finite* and *prime* elements.

Definition 3. *Let A be a lbo.*
An element $p \in |A|$ is called prime *iff*

$$\forall x, y \in A. \left((x \uparrow y \vee x \downarrow y) \wedge p \sqsubseteq x \sqcup y \right) \rightarrow (p \sqsubseteq x \vee p \sqsubseteq y)$$

We write $P(A)$ for the set $\{p \in |A| \mid p \text{ is prime}\}$ and $P(x)$ for the set $\{p \in P(A) \mid p \leq_s x\}$.
An element $e \in |A|$ is called finite *iff the set $\{x \in A \mid x \leq_s e\}$ is finite. We put*

$$F(A) := \{e \in |A| \mid e \text{ is finite}\} \quad \text{and} \quad F(x) := \{e \in F(A) \mid e \leq_s x\}.$$

For handling finite primes, *i.e. elements that are finite and prime, we define $FP(A) := P(A) \cap F(A)$ and $FP(x) := P(x) \cap F(A)$.* ◇

Definition 4. *A* locally boolean domain (lbd) *is a pointed, complete lbo A such that for all $x \in A$*

(1) $x = \bigsqcup FP(x)$ *and*
(2) *all finite primes in A are compact w.r.t. $\sqsubseteq$, i.e. for all $p \in FP(A)$ and directed sets X with $p \sqsubseteq \bigsqcup X$ there is an $x \in X$ with $p \sqsubseteq x$.* ◇

One can show that stably coherent subsets X of a lbd A have a supremum $\bigsqcup X$ which is a supremum also w.r.t. $\leq_s$. Moreover, if X is also nonempty then X has an infimum $\bigsqcap X$ which is an infimum also w.r.t. $\leq_s$. For costably coherent subsets the dual claims hold. Further, we have the following property of maximal bistably coherent subsets.

Lemma 5. *Let A be a lbd and $x \in A$. Then $[x]_\updownarrow := \{y \in A \,|\, y \updownarrow x\}$ with $\sqcap$, $\sqcup$ and $\neg$ restricted to $[x]_\updownarrow$ forms a complete atomic boolean algebra.*

The following lemma is needed for showing that our definition of locally boolean domain is equivalent with the original one given by J. Laird in [10].

Lemma 6. *Let x and y be elements of a lbd A then the following are equivalent*

(1) $x \sqsubseteq y$
(2) $\forall p \in FP(x). \exists q \in FP(y). p \sqsubseteq q$
(3) $\forall c \in F(x). \exists d \in F(y). c \sqsubseteq d$

Thus A is a coherently complete dI*-domain (cf. [2]) w.r.t. the stable order $\leq_s$.*

Next we define an appropriate notion of sequential map between lbds.

Definition 7. *Let A and B be lbds. A* sequential *map from A to B is a Scott continuous function $f : (|A|, \sqsubseteq) \to (|B|, \sqsubseteq)$ such that for all $x \updownarrow y$ it holds that $f(x) \updownarrow f(y)$, $f(x \sqcap y) = f(x) \sqcap f(y)$ and $f(x \sqcup y) = f(x) \sqcup f(y)$.* $\diamond$

We denote the ensuing category of lbds and sequential maps by **LBD**. The category **LBD** is cpo-enriched w.r.t. $\sqsubseteq$ and $\leq_s$ and order extensional w.r.t. $\sqsubseteq$, i.e. in particular well-pointed. In [10] J. Laird has shown that the category **LBD** is equivalent to the category **OSA** of *observably sequential algorithms* which has been introduced in [5] where it was shown that it gives rise to a fully abstract model for the language SPCF, an extension of PCF by error elements and a control operator **catch**.

The category **LBD** enjoys all the properties required for interpreting the language SPCF_∞ introduced subsequently in Section 3, namely that **LBD** is cartesian closed, has countable products and bilifted sums and inverse limits of ω-chains of projections. We just give the construction of these lbds, for a detailed verification of their characterising properties see [13].

Cartesian Products. For each family of lbds $(A_i)_{i \in I}$ the cartesian product $\prod_{i \in I} A_i$ is constructed as follows: $(\prod_{i \in I} |A_i|, \sqsubseteq, \neg)$ with $\sqsubseteq$ and $\neg$ defined pointwise.

Exponentials. For lbds A, B the function space $[A \to B]$ is constructed as follows: $|[A \to B]| = \mathbf{LBD}(A, B)$, the extensional order is defined pointwise and negation is given by $(\neg f)(x) := \bigsqcup \{\neg f(\neg c) \mid c \in F(x)\}$.

Terminal Object. The object 1 is given by $(\{*\}, \sqsubseteq, \neg)$.

Bilifted Sum. For each family of lbds $(A_i)_{i \in I}$ the bilifted sum $\sum_{i \in I} A_i$ is constructed as follows: $(\bigcup_{i \in I} \{i\} \times |A_i| \cup \{\bot, \top\}, \sqsubseteq, \neg)$ with

$$x \sqsubseteq y \Leftrightarrow x{=}\bot \vee y{=}\top \vee (\exists i{\in}I. \exists x', y'{\in}A_i.\, x{=}(i, x') \wedge y{=}(i, y') \wedge x' {\sqsubseteq_i} y')$$

and negation given by $\neg\bot = \top$ and $\neg(i, x) = (i, \neg_i x)$.

Natural Numbers. The data type $\mathsf{N} = \sum_{i \in \omega} 1$ will serve as the type of bilifted natural numbers. More explicitly N can be described as the lbd $(\mathbb{N} \cup \{\bot, \top\}, \sqsubseteq, \neg)$ with $x \sqsubseteq y$ iff $x = \bot$ or $y = \top$ or $x = y$, and negation is given by $\neg\bot = \top$ and $\neg n = n$ for all $n \in \mathbb{N}$.

Type of Observations. The type of observations $\mathsf{0} = \sum_{i \in \emptyset}$. More explicitly $\mathsf{0}$ can be described as the lbd $(\{\bot, \top\}, \sqsubseteq, \neg)$ with $\bot \sqsubseteq \top$ and $\neg\bot = \top$. Notice that $[A \to \mathsf{0}]$ separates points in A for any lbd A.

The exponential transpose of functions is defined as usual and since evaluation is sequential it follows that the category **LBD** is cartesian closed.

Notice that for exponentials we cannot simply define negation of a sequential map by $(\neg f)(x) = \neg f(\neg x)$ as the following example shows that sequentiality does not imply cocontinuity w.r.t. $\leq_c$.

Example 8. Let $F : [\mathsf{N} \to \mathsf{0}^\mathsf{0}] \to \mathsf{0}$ be defined recursively as

$$F(f) = f(0)(F(\lambda n. f(n{+}1)))\,.$$

Let $f = \lambda n.\mathrm{id}_O$ and $f_n(k) = \mathrm{id}_O$ for $k < n$ and $f(k) = \lambda n.\top$ for $k \geq n$. Obviously, the set $X := \{f_n \mid n \in \mathbb{N}\}$ is costably coherent, codirected w.r.t. $\leq_c$ and $f = \bigsqcap X$. As f is a minimal solution of its defining equation we have $F(f) = \bot$ and $F(f_n) = \top$ for all n. Thus, we have $f(\bigsqcap X) = \bot$ whereas $\bigsqcap f[X] = \top$, i.e. F fails to be cocontinuous w.r.t. $\leq_c$.

Nevertheless we always have

Lemma 9. *Let $f : A \to B$ be a* **LBD** *morphism and $x \in A$. Then $(\neg f)(x) \sqsubseteq \neg f(\neg x)$*

Proof. For all $c \in F(x)$ we have $\neg x \sqsubseteq \neg c$, thus, $f(\neg x) \sqsubseteq f(\neg c)$ and $\neg f(\neg c) \sqsubseteq \neg f(\neg x)$. Hence, it follows that $(\neg f)(x) = \bigsqcup\{\neg f(\neg c) \mid c \in F(x)\} \sqsubseteq \neg f(\neg x)$.

For the construction of recursive types in **LBD** we have to introduce an appropriate notion of embedding/projection pairs for lbds.

Definition 10. *An* embedding/projection pair (ep-pair) *from X to Y in* **LBD** *(notation $(\iota, \pi) : X \to Y$) is a pair of* **LBD** *morphisms $\iota : X \to Y$ and $\pi : Y \to X$ with $\pi \circ \iota = \mathrm{id}_X$ and $\iota \circ \pi \leq_s \mathrm{id}_Y$.*

If $(\iota, \pi) : X \to Y$ and $(\iota', \pi') : Y \to Z$ then their composition is defined as $(\iota', \pi') \circ (\iota, \pi) = (\iota' \circ \iota, \pi \circ \pi')$. We write $\mathbf{LBD}^E$ *for the ensuing category of embedding/projection pairs in* **LBD**. ◇

Notice that this is the usual definition of ep-pair when viewing **LBD** as order enriched by the stable and not by the extensional order.

Next we describe the construction of inverse limits of ω-chains of ep-pairs in **LBD**. The underlying cpos are constructed as usual. However, it needs some care to define negation appropriately (since in general projections do not preserve negation).

Theorem 11. *Given a functor $A : \omega \to \mathbf{LBD}^E$ its inverse limit of the projections is given by $(A_\infty, \sqsubseteq, \neg)$ where*

$$A_\infty = \{x \in \prod_{n\in\omega} A_n \mid x_n = \pi_{n,n+1}(x_{n+1}) \textit{ for all } n \in \omega\}$$

the extensional order $\sqsubseteq$ is defined pointwise and

$$(\neg x)_n = \bigsqcap\nolimits_{k\geq n} \pi_{n,k}(\neg x_k)$$

for all $n \in \omega$.

Notice that the full subcategory of *countably based* lbds, i.e. lbds A where $FP(A)$ is countable, is closed under the above constructions as long as products and bilifted sums are assumed as countable.

3 The Language SPCF_∞

The language SPCF_∞ is an infinitary version of SPCF as introduced in [5]. More explicitly, it is obtained from simply typed λ-calculus by adding (countably) infinite sums and products, error elements, a control operator **catch** and recursive types. For a detailed presentation of SPCF_∞ see Table 1.

Table 1. The language SPCF_∞

Types: $\sigma ::= \alpha \mid \sigma \to \sigma \mid \mu\alpha.\sigma \mid \Sigma_{i \in I}\sigma \mid \Pi_{i \in I}\sigma$ with I countable,
$\mathbb{1} := \Pi_\emptyset, \quad \mathbf{0} := \Sigma_\emptyset, \quad \mathbf{N} := \Sigma_{i \in \omega}\mathbb{1}, \quad \sigma^\omega := \Pi_{i \in \omega}\sigma$

Environments: $\Gamma \equiv x_1 : \sigma_1, \ldots, x_n : \sigma_n$ for closed types σ_i

Terms: $t ::= x \mid (\lambda x : \sigma.t) \mid (tt) \mid \langle t_i \rangle_{i \in I} \mid \mathbf{pr}_i(t) \mid \mathbf{in}_i(t) \mid \mathbf{case}\, t\, \mathbf{of}\, (\mathbf{in}_i\, x \Rightarrow t_i)_{i \in I} \mid \mathbf{fold}(t) \mid \mathbf{unfold}(t) \mid \top \mid \mathbf{catch}(t)$

Values $v ::= (\lambda x : \sigma.t) \mid \langle t_i \rangle_{i \in I} \mid \mathbf{in}_i(t) \mid \mathbf{fold}(t) \mid \top$

Abbreviations / Combinators: $\underline{n} := \mathbf{in}_n\langle\rangle$ for all $n \in \omega$

$\mathbf{catch}^{\sigma_1 \to \ldots \to \sigma_n \to \mathbf{N}} := \lambda f.\, \mathbf{catch}(\lambda x : \mathbf{0}^\omega.\, \mathbf{case}$
$f(e_1(\mathbf{pr}_1\, x), \ldots, e_n(\mathbf{pr}_n\, x))\, \mathbf{of}\, (\mathbf{in}_i\, y \Rightarrow \mathbf{pr}_{i+n}\, x)_{i \in \omega})$
with $e_i := \lambda x : \mathbf{0}.\, \mathbf{case}^{\mathbf{0}, \sigma_i}\, x\, \mathbf{of}\, ()$ for all $i \in \{1, \ldots, n\}$

$\mathbf{Y}_\sigma := k(\mathbf{fold}^\tau(k))$ with $\tau := \mu\alpha.(\alpha \to (\sigma \to \sigma) \to \sigma)$
and $k := \lambda x : \tau.\lambda f : \sigma \to \sigma.f(\mathbf{unfold}^\tau(x)xf)$

Typing rules:

$$\frac{}{\Gamma, x : \sigma, \Delta \vdash x : \sigma} \qquad \frac{}{\Gamma \vdash \top^{\Sigma_{i \in I}\sigma_i} : \Sigma_{i \in I}\sigma_i} \qquad \frac{\Gamma \vdash t : \mathbf{0}^\omega \to \mathbf{0}}{\Gamma \vdash \mathbf{catch}(t) : \mathbf{N}}$$

$$\frac{\Gamma, x : \sigma \vdash t : \tau}{\Gamma \vdash (\lambda x : \sigma.t) : \sigma \to \tau} \qquad \frac{\Gamma \vdash t : \sigma \to \tau \quad \Gamma \vdash s : \sigma}{\Gamma \vdash (ts) : \tau} \qquad \frac{\Gamma \vdash t_i : \sigma_i \quad \text{for all } i \in I}{\Gamma \vdash \langle t_i \rangle_{i \in I} : \Pi_{i \in I}\sigma_i}$$

$$\frac{\Gamma \vdash t : \Pi_{i \in I}\sigma_i}{\Gamma \vdash \mathbf{pr}_i^{\Pi_{i \in I}\sigma_i}(t) : \sigma_i} \qquad \frac{\Gamma \vdash t : \sigma[\mu\alpha.\sigma/\alpha]}{\Gamma \vdash \mathbf{fold}^{\mu\alpha.\sigma}(t) : \mu\alpha.\sigma} \qquad \frac{\Gamma \vdash t : \mu\alpha.\sigma}{\Gamma \vdash \mathbf{unfold}^{\mu\alpha.\sigma}(t) : \sigma[\mu\alpha.\sigma/\alpha]}$$

$$\frac{\Gamma \vdash t : \Sigma_{i \in I}\sigma_i \quad \Gamma, x : \sigma_i \vdash s_i : \tau \quad \text{for all } i \in I}{\Gamma \vdash \mathbf{case}^{\Sigma_{i \in I}\sigma_i, \tau}\, t\, \mathbf{of}\, (\mathbf{in}_i\, x \Rightarrow s_i)_{i \in I} : \tau} \qquad \frac{\Gamma \vdash t : \sigma_i}{\Gamma \vdash \mathbf{in}_i^{\Sigma_{i \in I}\sigma_i}(t) : \Sigma_{i \in I}\sigma_i}$$

The operational semantics of SPCF_∞ is given in Table 2. Notice that each SPCF_∞ term t which is not already a value has a unique decomposition into an evaluation context E and a redex t' with $E[t'] \equiv t$.

The interpretation of SPCF_∞ in locally boolean domains can be found in Table 3. The interpretation of recursive types is done as usual via inverse limits whose existence is guaranteed by Theorem 11. One can prove adequacy of the model like in [16,17].

Since by definition sequential maps preserve infima and suprema of bistably coherent arguments a sequential map from $\mathsf{0}^\omega$ to $\mathsf{0}$ is either constant (with value

Table 2. Operational semantics of SPCF_∞

Evaluation contexts:

$$E ::= [\,] \mid Et \mid \mathbf{pr}_i(E) \mid \mathbf{unfold}(E) \mid \mathbf{case}\, E\, \mathbf{of}\, (\mathbf{in}_i \Rightarrow t)_{i\in I} \mid \mathbf{catch}(\lambda x : \mathbf{0}^\omega.E)$$

Redex reduction:

$$(\lambda x : \sigma.t)s \to_{\mathsf{red}} t[s/x] \qquad \mathbf{case}\,\mathbf{in}_i\, s\,\mathbf{of}\,(\mathbf{in}_i\, x \Rightarrow t_i) \to_{\mathsf{red}} t_i[s/x]$$

$$\mathbf{pr}_i(\langle t_i\rangle_{i\in I}) \to_{\mathsf{red}} t_i \qquad \mathbf{unfold}(\mathbf{fold}(t)) \to_{\mathsf{red}} t$$

Evaluation context reduction:

$$E[t] \to_{\mathsf{op}} E[t'] \qquad \text{if } t \to_{\mathsf{red}} t'$$

$$E[\top] \to_{\mathsf{op}} \top \qquad \text{if } E \neq [\,]$$

$$E[\mathbf{catch}\, t] \to_{\mathsf{op}} t\langle E[\underline{n}]\rangle_{n\in\omega}$$

$\bot$ or $\top$) or is a projection $\pi_i : 0^\omega \to 0$. For this reason there exists an isomorphism $\mathsf{catch} : [0^\omega \to 0] \xrightarrow{\cong} \mathsf{N}$ with

$$\mathsf{catch}(f) = i \quad \text{iff} \quad f \text{ is } i\text{-strict, i.e. } f(x) = \bot \Leftrightarrow \pi_i(x) = \bot$$

which will serve as interpretation of the control operator **catch** of SPCF_∞.

Table 3. Interpretation of SPCF_∞

$$[\![x_1 : \sigma_1, \ldots, x_n : \sigma_n \vdash x_i : \sigma_i]\!] := \pi_i$$

$$[\![\Gamma \vdash \top : \Sigma_{i\in I}\sigma_i]\!] := x^{[\![\Gamma]\!]} \mapsto \top_{[\![\Sigma_{i\in I}\sigma_i]\!]}$$

$$[\![\Gamma \vdash (\lambda x : \sigma.t) : \sigma\to\tau]\!] := \mathsf{curry}_{[\![\Gamma]\!],[\![\sigma]\!]}([\![\Gamma, x : \sigma \vdash t : \tau]\!])$$

$$[\![\Gamma \vdash ts : \tau]\!] := \mathsf{eval} \circ \langle [\![\Gamma \vdash t : \sigma\to\tau]\!], [\![\Gamma \vdash s : \sigma]\!]\rangle$$

$$[\![\Gamma \vdash \langle t_i\rangle_{i\in I}^{\Pi_{i\in I}\sigma_i} : \Pi_{i\in I}\sigma_i]\!] := \langle [\![\Gamma \vdash t_i : \sigma_i]\!]\rangle_{i\in I}$$

$$[\![\Gamma \vdash \mathbf{pr}_i(t) : \sigma]\!] := \pi_i \circ [\![\Gamma \vdash t : \sigma]\!]$$

$$[\![\Gamma \vdash \mathbf{case}^{\Sigma_{i\in I}\tau_i,\sigma}\, t\, \mathbf{of}\, (\mathbf{in}_i\, x \Rightarrow t_i) : \sigma]\!] := \mathsf{case} \circ \langle [\![\Gamma \vdash t]\!], \langle [\![\Gamma \vdash (\lambda x : \tau_i.t_i) : \tau_i \to\sigma]\!]\rangle_{i\in I}\rangle$$

$$[\![\Gamma \vdash \mathbf{in}_i(t) : \Sigma_{i\in I}\sigma_i]\!] := \iota_i \circ [\![\Gamma \vdash t : \sigma_i]\!]$$

$$[\![\Gamma \vdash \mathbf{catch}(t) : \mathbf{N}]\!] := \mathsf{catch} \circ [\![\Gamma \vdash t : \mathbf{0}^\omega \to \mathbf{0}]\!]$$

$$[\![\Gamma \vdash \mathbf{fold}^{\mu\alpha.\sigma}(t) : \mu\alpha.\sigma]\!] := \mathsf{fold} \circ [\![\Gamma \vdash t : \sigma[\mu\alpha.\sigma/\alpha]]\!]$$

$$[\![\Gamma \vdash \mathbf{unfold}^{\mu\alpha.\sigma}(t) : \sigma[\mu\alpha.\sigma/\alpha]]\!] := \mathsf{unfold} \circ [\![\Gamma \vdash t : \mu\alpha.\sigma]\!]$$

4 Universality for SPCF_∞

In this section we show that the first order type $\mathbf{U} = \mathbf{N}\to\mathbf{N}$ is universal for the language SPCF_∞ by proving that every type is a SPCF_∞ definable retract of $\mathbf{U}$. Since all elements of the lbd $[\![\mathbf{U}]\!]$ can be defined syntactically we get universality of SPCF_∞ for its model in **LBD**.

Definition 12. *A closed* SPCF_∞ *type* σ *is called a* SPCF_∞ definable retract *of a* SPCF_∞ *type* τ *(denoted* $\sigma \lhd \tau$*) iff there exist closed terms* $e : \sigma\to\tau$ *and* $p : \tau\to\sigma$ *with* $[\![p]\!] \circ [\![e]\!] = \mathrm{id}_{[\![\sigma]\!]}$. ◇

Theorem 13. *Every* SPCF_∞ *type appears as* SPCF_∞ *definable retract of the type* $\mathbf{U} := \mathbf{N} \to \mathbf{N}$.

Proof. It suffices to show that for all $I \in \omega + 1$ the types

$$\mathbf{U} \to \mathbf{U} \qquad \Pi_{i \in I}\mathbf{U} \qquad \Sigma_{i \in I}\mathbf{U}$$

are SPCF_∞ definable retracts of $\mathbf{U}$.

The SPCF_∞ programs exhibiting $\Pi_{i \in I}\mathbf{U}$ as definable retract of $\mathbf{U}$ are given in Table 5. Using this we get $\Sigma_{i \in I}\mathbf{U} \lhd \mathbf{U}$ since obviously $\Sigma_{i \in I}\mathbf{U} \lhd \mathbf{U} \times \Sigma_{i \in I}\mathbb{1}$.

By currying we have $\mathbf{U} \to \mathbf{U} \cong (\mathbf{U} \times \mathbf{N}) \to \mathbf{N}$. As $\mathbf{U} \times \mathbf{N} \lhd \mathbf{U} \times \mathbf{U} \lhd \mathbf{U}$ it suffices to construct a retraction $\mathbf{U} \to \mathbf{N} \lhd \mathbf{U}$ for showing that $\mathbf{U} \to \mathbf{U} \lhd \mathbf{U}$ holds. For this purpose we adapt an analogous result given by J. Longley in [12] for ordinary sequential algorithms without error elements. The programs establishing the retraction are given in Table 6. The function p interprets elements of $\mathbf{U}$ as sequential algorithms for functionals of type $\mathbf{U} \to \mathbf{N}$ as described in [12]. For a given $F : \mathsf{U} \to \mathsf{N}$ the element $[\![e]\!](F) : \mathsf{N} \to \mathsf{N}$ is a strategy / sequential algorithm for computing F. This is achieved by computing sequentiality indices iteratively using **catch**. □

Since all sequential function from N to N can be programmed using a (countably infinite) case analysis (available by the **case**-construct for index set ω) it follows that

Theorem 14. *The language* SPCF_∞ *is universal for its model in* **LBD**.

In a sense the elements denotable by finite SPCF terms may be considered as the "computable" ones but it is not clear how this somewhat *ad hoc* notion of computability can be rephrased in terms of recursive enumerability of finite approximations (see [3] for a discussion).

Table 4. The language CPS_∞

Contexts: $\Gamma \equiv [x_i \,|\, i \in I]$ with $I \in \omega + 1$

Terms: $M ::= x \mid \lambda\vec{x}.t \qquad \vec{x} \equiv (x_i)_{i \in \omega}$

$t ::= \top \mid M\langle\vec{M}\rangle \qquad \vec{M} \equiv (M_i)_{i \in \omega}$

Terms-in-context:

$$\frac{}{[x_i \,|\, i \in I] \vdash x_i} \qquad \frac{}{\Gamma \vdash \lambda\vec{x}.\top} \qquad \frac{\Gamma \cup [x_i \,|\, i \in I] \vdash M \quad \Gamma \cup [x_i \,|\, i \in I] \vdash N_i}{\Gamma \vdash \lambda\vec{x}.M\langle\vec{N}\rangle}$$

Operational semantics:

$$\frac{}{\top \Downarrow \top} \qquad \frac{t[M_i/x_i]_{i \in \omega} \Downarrow \top}{(\lambda\vec{x}.t)\langle\vec{M}\rangle \Downarrow \top}$$

Table 5. Retraction $\Pi_{i\in I}\mathbf{U} \lhd \mathbf{U}$

$$e := \lambda f : \Pi_{i\in I}\mathbf{U}.\lambda n : \mathbf{N}.\,\mathbf{case}\,\mathbf{pr}_0(\beta_2^* n)\,\mathbf{of}\begin{pmatrix}\mathbf{in}_0\,x \Rightarrow \mathbf{catch}^{\mathbf{N}\to\mathbf{N}}(\mathbf{pr}_i\,f)\\ \mathbf{in}_1\,x \Rightarrow \mathbf{case}\,\mathbf{pr}_0(\beta_I^*(\mathbf{pr}_1(\beta_2^* n)))\,\mathbf{of}\\ \qquad (\underline{j} \Rightarrow (\mathbf{pr}_j\,f)(\mathbf{pr}_1(\beta_I^*(\mathbf{pr}_1(\beta_2^* n)))))_{j\in I}\end{pmatrix}$$

$$p := \lambda f : \mathbf{U}.\langle \mathbf{case}\,f(\beta_2\langle\underline{0},\underline{i}\rangle)\,\mathbf{of}\begin{pmatrix}\mathbf{in}_0\,x \Rightarrow \lambda n : \mathbf{N}.f(\beta_2\langle\underline{1},\beta_I\langle\underline{i},n\rangle\rangle)\\ \mathbf{in}_{j+1}\,x \Rightarrow \lambda n : \mathbf{N}.\underline{j}\end{pmatrix}_{j\in\omega}\rangle_{i\in I}$$

with $\beta_I : ((\Sigma_{i\in I}\mathbb{1})\times\mathbf{N})\to\mathbf{N}$ and $\beta_I^* : \mathbf{N}\to((\Sigma_{i\in I}\mathbb{1})\times\mathbf{N})$ satisfying $\beta_I^*(\beta_I\langle\underline{i},\underline{n}\rangle) = \langle\underline{i},\underline{n}\rangle$ for all $i \in I$ and $n \in \omega$

Table 6. Retraction $\mathbf{U}\to\mathbf{N} \lhd \mathbf{U}$

$$e := \lambda F : \mathbf{U}\to\mathbf{N}.\lambda n : \mathbf{N}.\,\mathbf{case}\,\alpha^*(n)\,\mathbf{of}\begin{pmatrix}\mathbf{in}_0\,t \Rightarrow \mathbf{case}\,\mathbf{catch}^{\mathbf{U}\to\mathbf{N}}(F)\,\mathbf{of}\,R\\ \mathbf{in}_1\,t \Rightarrow \alpha(\mathbf{in}_1(F(\lambda x : \mathbf{N}.t)))\\ \mathbf{in}_2\,t \Rightarrow \mathbf{case}\,S\,\mathbf{of}\begin{pmatrix}\mathbf{in}_{2i}\,x \Rightarrow \alpha(\mathbf{in}_1\,\underline{i})\\ \mathbf{in}_{2i+1}\,x \Rightarrow \alpha(\mathbf{in}_2\,\underline{i})\end{pmatrix}_{i\in\omega}\end{pmatrix}$$

$$R := \begin{pmatrix}\mathbf{in}_0\,x \Rightarrow \alpha(\mathbf{in}_0\,x)\\ \mathbf{in}_{i+1}\,x \Rightarrow \alpha(\mathbf{in}_1\,\underline{i})\end{pmatrix}_{i\in\omega}$$

$$S := \mathbf{catch}(\lambda x : \mathbf{0}^\omega\to\mathbf{0}.\,\mathbf{case}\,F(\lambda n : \mathbf{N}.\;\mathbf{case}\,\mathsf{find}(t,n)\,\mathbf{of}\begin{pmatrix}\mathbf{in}_0\,s \Rightarrow s\\ \mathbf{in}_1\,s \Rightarrow \mathbf{case}^{\mathbf{0},\mathbf{N}}(\mathbf{pr}_{2i+1}\,x)\,\mathbf{of}\,()\end{pmatrix})\;\mathbf{of}\,(\mathbf{in}_i\,s \Rightarrow \mathbf{pr}_{2i}\,x)_{i\in\omega})$$

$$p := \lambda r : \mathbf{N}\to\mathbf{N}.\lambda f : \mathbf{N}\to\mathbf{N}.\,\mathbf{case}\,r(\alpha(\mathbf{in}_0\langle\rangle))\,\mathbf{of}\begin{pmatrix}\mathbf{in}_0\,t \Rightarrow T\\ \mathbf{in}_{i+1}\,t \Rightarrow \underline{i}\end{pmatrix}_{i\in\omega}$$

$$T := \mathbf{case}\,\mathbf{catch}^{\mathbf{N}\to\mathbf{N}}(f)\,\mathbf{of}\begin{pmatrix}\mathbf{in}_0\,t \Rightarrow U(\mathsf{nil})\\ \mathbf{in}_{i+1}\,t \Rightarrow \mathbf{case}\,\alpha^*(r(\alpha(\mathbf{in}_1\,\underline{i})))\,\mathbf{of}\begin{pmatrix}\mathbf{in}_0\,t \Rightarrow \bot\\ \mathbf{in}_1\,t \Rightarrow t\\ \mathbf{in}_2\,t \Rightarrow \bot\end{pmatrix}\end{pmatrix}_{i\in\omega}$$

$$U := \mathbf{Y}_{\mathbf{N}\to\mathbf{N}}(\lambda h : \mathbf{N}\to\mathbf{N}.\lambda g : \mathbf{N}.\,\mathbf{case}\,\alpha^*(r(\alpha(\mathbf{in}_2\,g)))\;\mathbf{of}\begin{pmatrix}\mathbf{in}_0\,t \Rightarrow \bot\\ \mathbf{in}_1\,t \Rightarrow t\\ \mathbf{in}_2\,t \Rightarrow h(\mathsf{cons}(g,(t,f(t))))\end{pmatrix})$$

with $\alpha : (\mathbb{1}+\mathbf{N}+\mathbf{N})\to\mathbf{N}$ and $\alpha^* : \mathbf{N}\to(\mathbb{1}+\mathbf{N}+\mathbf{N})$ satisfying $\alpha^*(\alpha(\mathbf{in}_0\langle\rangle)) = \mathbf{in}_0\langle\rangle$ and $\alpha^*(\alpha(\mathbf{in}_i\,n)) = \mathbf{in}_i\,n$ for $i = 1, 2$ and $n \in \omega$, and the following auxiliary list-handling functions in Haskell-style where γ encodes lists of natural numbers as natural numbers

$$\mathsf{nil} := \gamma(\texttt{[]})$$

$$\mathsf{cons}(g,(x,y)) := \gamma(\mathsf{cons}((x,y),\gamma^{-1}(g)))$$

$$\mathsf{find}(g,x) := \texttt{case}\;\gamma^{-1}(g)\;\texttt{of}\quad \begin{array}{ll}\texttt{[]} & \texttt{->}\;\mathbf{in}_1\,\underline{0}\\ ((x,y):r) & \texttt{->}\;\mathbf{in}_0\,y\\ (\_:r) & \texttt{->}\;\mathsf{find}(\gamma(r),x)\end{array}$$

5 Universality for an Infinitary Untyped CPS Target Language CPS_∞

The interpretation of the SPCF_∞ type $\delta := \mu\alpha.(\alpha^\omega \to \mathbf{0})$ (where for arbitrary types σ we henceforth write σ^ω as an abbreviation for $\Pi_{i\in\omega}\sigma$) is the minimal solution of the domain equation $D \cong [D^\omega \to \mathsf{0}]$. Obviously, we have $D \cong [D \to D]$. Moreover, it has been shown in [18] that D is isomorphic to $\mathsf{0}_\infty$, i.e. what one obtains by performing D. Scott's D_∞ construction in **LBD** when instantiating D by $\mathsf{0}$.

We now describe an untyped infinitary language CPS_∞ canonically associated with the domain equation $D \cong [D^\omega \to \mathsf{0}]$. The precise syntax of CPS_∞ is given in Table 4. We interpret CPS_∞ terms in context Γ, i.e. a set of variables, as sequential maps from D^Γ to D in the obvious way.

The language CPS_∞ is an extension of pure untyped λ-calculus since applications MN can be expressed by $\lambda\vec{x}.M\langle N, \vec{x}\rangle$ with fresh variables $\vec{x}$ and abstraction $\lambda x.M$ by $\lambda x\vec{y}.M\langle\vec{y}\rangle$ with fresh variables $\vec{y}$. Thus, CPS_∞ allows for recursion and we can define recursion combinators in the usual way.

Notice that CPS_∞ is more expressive than pure untyped λ-calculus since the latter does not contain a term semantically equivalent to

$$\lambda\vec{x}.x_0\langle\vec{\bot}\rangle$$

which sends $\top_D$ to $\top_D$ and all other elements of D to $\bot_D$.[1] Since the retraction of D to $\mathsf{0}$ is CPS_∞ definable all other retractions to the finite approximations of D (which are isomorphic to simple types over $\mathsf{0}$) are definable as well.

Lemma 15. *The lbds* N *and* U *are both* CPS_∞ *definable retract of the lbd* D.

Proof. Since we can retract the lbd D to the lbd $\mathsf{0}$ and $[\mathsf{0}^\omega \to \mathsf{0}] \cong \mathsf{N}$ it follows that N is a CPS_∞ definable retract of D. As $[D \to D]$ is a CPS_∞ definable retract of D it follows that $\mathsf{U} = [\mathsf{N} \to \mathsf{N}]$ is a CPS_∞ definable retract of D. □

Thus, we can do arithmetic within CPS_∞. Natural numbers are encoded by $\underline{n} \equiv \lambda\vec{x}.x_n\langle\vec{\bot}\rangle$ and a function $f{:}\mathsf{N}\to\mathsf{N}$ by its graph, i.e. $\underline{f} \equiv \lambda x\vec{y}.x\langle\lambda\vec{z}.\underline{f(i)}\langle\vec{y}\rangle\rangle_{i\in\omega}$. Notice that CPS_∞ allows for the implementation of an infinite case construct.

Universality for CPS_∞ will be shown in two steps. First we argue why all finite elements of D are CPS_∞ definable. Then adapting a trick from [9] we show that suprema of chains increasing w.r.t. $\leq_s$ are CPS_∞definable, too.

Lemma 16. *All finite elements of the lbd* D *are* CPS_∞ *definable.*

Proof. In [8] Jim Laird has shown that the language $\Lambda^\top_\bot$, i.e. simply typed λ-calculus over the base type $\{\bot, \top\}$ is universal for its model in **LBD**. Thus, since all retractions of D to its finitary approximations are CPS_∞ definable it follows that all finite elements of D are CPS_∞ definable. □

---

[1] Since $[\![\lambda\vec{x}.x_0\langle\vec{\bot}\rangle]\!]$ is certainly "computable" pure λ-calculus with constant $\top$ cannot denote all "computable" elements.

Next we show that for all $f : A \to \mathsf{0}$ in **LBD** the map $\widetilde{f} : A \to [\mathsf{0}{\to}\mathsf{0}]$ with

$$\widetilde{f}(a)(u) := \begin{cases} u & \text{if } f(a^\top) = \bot_\mathsf{0} \text{ and} \\ f(a) & \text{otherwise} \end{cases} \tag{1}$$

is an **LBD** morphism as well.

Lemma 17. *If $f : A \to \mathsf{0}$ is a sequential map between lbds then the function $\widetilde{f} : A \to [\mathsf{0}{\to}\mathsf{0}]$ given by* (1) *is sequential.*

Proof. For showing monotonicity suppose $a_1, a_2 \in A$ with $a_1 \sqsubseteq a_2$ and $u \in \mathsf{0}$. We proceed by case analysis on $f(a_1^\top)$.

Suppose $f(a_1^\top) = \bot_\mathsf{0}$. Thus, $\widetilde{f}(a_1)(u) = u$. If $f(a_2^\top) = \bot_\mathsf{0}$ then $\widetilde{f}(a_2)(u) = u$, and we get $\widetilde{f}(a_1)(u) = u = \widetilde{f}(a_2)(u)$. If $f(a_2^\top) = \top_\mathsf{0}$ then $\widetilde{f}(a_2)(u) = f(a_2)$. As $f(a_1^\top) = \bot_\mathsf{0}$ it follows that $f(\neg a_1) = \bot_\mathsf{0}$ and $f(\neg a_2) = \bot_\mathsf{0}$ (because $\neg a_2 \sqsubseteq \neg a_1$). As $\top_\mathsf{0} = f(a_2^\top) = f(a_2) \sqcup f(\neg a_2)$ it follows that $f(a_2) = \top_\mathsf{0}$ as desired.

If $f(a_1^\top) = \top_\mathsf{0}$ then $\widetilde{f}(a_1)(u) = f(a_1)$. W.l.o.g. assume $f(a_1) = \top_\mathsf{0}$. Then $\top_\mathsf{0} = f(a_1) \sqsubseteq f(a_2) \sqsubseteq f(a_2^\top)$. Hence, $f(a_2) = \top_\mathsf{0} = f(a_2^\top)$ and we get $\widetilde{f}(a_2)(u) = f(a_2) = \top_\mathsf{0}$.

Next we show that $\widetilde{f}$ is bistable. Let $a_1 \updownarrow a_2$, thus (†) $a_1^\top = a_2^\top = (a_1 \sqcap a_2)^\top$.

If $f(a_1^\top) = f(a_2^\top) = \bot_\mathsf{0}$ then $\widetilde{f}(a_1) = \mathrm{id}_\mathsf{0} = \widetilde{f}(a_2)$. If $f(a_1^\top) = f(a_2^\top) = \top_\mathsf{0}$ then $\widetilde{f}(a_i) = \lambda x{:}\mathsf{0}.\, f(a_i)$ for $i \in \{1,2\}$. Since $\lambda x{:}\mathsf{0}.\, \bot_\mathsf{0} \updownarrow \lambda x{:}\mathsf{0}.\, \top_\mathsf{0}$ it follows that $\widetilde{f}$ preserves bistable coherence.

Finally we show that $\widetilde{f}$ preserves bistably coherent suprema. If $f((a_1 \sqcap a_2)^\top) = \bot_\mathsf{0}$ then $\widetilde{f}(a_1 \sqcap a_2)(u) = u = \widetilde{f}(a_1)(u) \sqcap \widetilde{f}(a_2)(u)$ (since $f(a_1^\top) = f(a_2^\top) = \bot_\mathsf{0}$ by (†)). Otherwise, if $f((a_1 \sqcap a_2)^\top) = \top_\mathsf{0}$ then $\widetilde{f}(a_1 \sqcap a_2)(u) = f(a_1 \sqcap a_2) = f(a_1) \sqcap f(a_2) = \widetilde{f}(a_1)(u) \sqcap \widetilde{f}(a_2)(u)$ (since f is bistable and $f(a_1^\top) = f(a_2^\top) = \top_\mathsf{0}$ by (†)).

Analogously, it follows that $\widetilde{f}$ preserves bistably coherent suprema. □

The following observation is useful when computing with functions of the form $\widetilde{f}$.

Lemma 18. *If $f : A \to \mathsf{0}$ is a* **LBD** *morphism then $\widetilde{f}(a)(\bot_\mathsf{0}) = f(a)$.*

Proof. If $f(a) = \bot_\mathsf{0}$ then $\widetilde{f}(a)(\bot_\mathsf{0}) = \bot_\mathsf{0} = f(a)$ since $\bot$ and $f(a)$ are the only possible values of $\widetilde{f}(a)(\bot_\mathsf{0})$. If $f(a) = \top_\mathsf{0}$ then $f(a^\top) = \top_\mathsf{0}$ and thus $\widetilde{f}(a)(\bot_\mathsf{0}) = f(a)$ as desired.

If $f \in D \cong [D^\omega{\to}\mathsf{0}]$ the we write $\widehat{f}$ for that element of D with

$$\widehat{f}(d, \vec{d}) = \begin{cases} \widetilde{f}(\vec{d})(\top_\mathsf{0}) & \text{if } d \neq \bot \\ \widetilde{f}(\vec{d})(\bot_\mathsf{0}) & \text{if } d = \bot \end{cases}$$

Lemma 19. *For every finite f in D the element $\widehat{f}$ is also finite and thus* CPS_∞ *definable.*

Proof. If A is a finite lbd then for every $f : A \to 0$ the **LBD** map $\widetilde{f} : A \to [0\to 0]$ is also finite. This holds in particular for f in the finite type hierarchy over 0.

Since embeddings of lbds preserves finiteness of elements we conclude that for every finite f in D the element $\widehat{f}$ is finite as well. Thus, by Lemma 16 the element $\widehat{f}$ is CPS_∞ definable. □

Lemma 20. *For $f, g : A \to 0$ with $f \leq_s g$ it holds that $\widetilde{g} = \lambda a{:}A.\, \widetilde{f}(a) \circ \widetilde{g}(a)$.*

Proof. Suppose $f \leq_s g$. Let $a \in A$ and $u \in 0$. We have to show that $\widetilde{g}(a)(u) = \widetilde{f}(a)(\widetilde{g}(a)(u))$.

If $g(a^\top) = \bot_0$ then $f(a^\top) = \bot_0$ (since $f \leq_s g$) and thus $\widetilde{g}(a)(u) = \widetilde{f}(a)(\widetilde{g}(a)(u))$.

Thus, w.l.o.g. suppose $g(a^\top) = \top_0$. Then $\widetilde{g}(a)(u) = g(a)$.

If $f(a) = \top_0$ then $f(a^\top) = \top_0 = g(a)$ and, therefore, we have $\widetilde{f}(a)(\widetilde{g}(a)(u)) = f(a) = \top_0 = g(a) = \widetilde{g}(a)(u)$.

Now suppose $f(a) = \bot_0$.

If $g(a) = \bot_0$ then we have $\widetilde{f}(a)(\widetilde{g}(a)(u)) = \widetilde{f}(a)(g(a)) = \widetilde{f}(a)(\bot_0) = \bot_0$ where the last equality holds by Lemma 18.

Now suppose $g(a) = \top_0$. We proceed by case analysis on the value of $f(a^\top)$. If $f(a^\top) = \bot_0$ then $\widetilde{f}(a)(\widetilde{g}(a)(u)) = \widetilde{g}(a)(u)$. We show that $f(a^\top) = \top_0$ cannot happen.

Suppose $f(a^\top) = \top_0$ holds. Then by bistability we have $\top_0 = f(a^\top) = f(a) \sqcup f(\neg a) = \bot_0 \sqcup f(\neg a) = f(\neg a)$ and thus also $\neg f(\neg a) = \bot_0$. Since $f \leq_s g$ we have $g \sqsubseteq f^\top$. Moreover, by Lemma 9 we have $(\neg f)(a) \sqsubseteq \neg f(\neg a)$. Thus, we have $\top_0 = g(a) \sqsubseteq f^\top(a) = f(a) \sqcup (\neg f)(a) = (\neg f)(a) \sqsubseteq \neg f(\neg a) = \bot_0$ which clearly is impossible. □

Now we are ready to prove our universality result for CPS_∞.

Theorem 21. *All elements of the lbd D are CPS_∞ definable.*

Proof. Suppose $f \in D$. Then $f = \bigsqcup f_n$ for some increasing (w.r.t. $\leq_s$) chain $(f_n)_{n\in\omega}$ of finite elements. Since by Lemma 19 all $\widehat{f_n}$ are CPS_∞ definable there exists a CPS_∞ term F with $[\![F\underline{n}]\!] = \widehat{f}_n$ for all $n \in \omega$.

Since recursion is available in CPS_∞ one can exhibit a CPS_∞ term Ψ such that

$$\Psi g = \lambda x.\, g(\underline{0})(\Psi(\lambda n.\, g(n{+}1))x) = \bigsqcup_{n\in\omega} (g(\underline{0}) \circ \cdots \circ g(\underline{n}))(\bot)$$

Thus, the term $M_f \equiv \lambda\vec{x}.\, \Psi(\lambda y.\lambda z.F\langle y, z, \vec{x}\rangle)$ denotes f since

$$\begin{aligned} M_f(\vec{d}) &= \Psi(\lambda y.\lambda z.F(y, z, \vec{d})) \\ &= \bigsqcup_{n\in\omega} ((\lambda z.F\underline{0}(z, \vec{d})) \circ \cdots \circ (\lambda z.F\underline{n}(z, \vec{d})))(\bot) \\ &= \bigsqcup_{n\in\omega} ((\lambda z.\widehat{f_0}(z, \vec{d})) \circ \cdots \circ (\lambda z.\widehat{f_n}(z, \vec{d})))(\bot) \end{aligned}$$

$$= \bigsqcup_{n\in\omega} ((\lambda z.\widetilde{f_0}(\vec{d})(z)) \circ \cdots \circ (\lambda z.\widetilde{f_n}(\vec{d})(z)))(\bot)$$

$$= \bigsqcup_{n\in\omega} (\widetilde{f_0}(\vec{d}) \circ \cdots \circ \widetilde{f_n}(\vec{d}))(\bot)$$

$$= \bigsqcup_{n\in\omega} (\widetilde{f_n}(\vec{d}))(\bot) \qquad \text{(by Lemma 20)}$$

$$= \bigsqcup_{n\in\omega} f_n(\vec{d}) \qquad \text{(by Lemma 18)}$$

$$= f(\vec{d})$$

for all $\vec{d} \in D^\omega$. □

6 Faithfulness of the Interpretation

In the previous section we have shown that the interpretation of closed CPS_∞ terms in the lbd D is surjective. There arises the question whether the interpretation is also faithful. Recall that infinite normal forms for CPS_∞ are given by the grammar

$$N ::= x \mid \lambda\vec{x}.\top \mid \lambda\vec{x}.x\langle\vec{N}\rangle$$

understood in a coinductive sense.

Definition 22. *We call a model* faithful *iff for all normal forms N_1, N_2 if $[\![N_1]\!] = [\![N_2]\!]$ then $N_1 = N_2$.* ◇

We will show that the **LBD** model of CPS_∞ is not faithful. For a closed CPS_∞ term M consider

$$M^* \equiv \lambda\vec{x}.x_0\langle\bot, \lambda\vec{y}.x_0\langle M, \vec{\bot}\rangle, \vec{\bot}\rangle$$

Lemma 23. *For closed CPS_∞ terms M_1, M_2 it follows that $[\![M_1^*]\!] = [\![M_2^*]\!]$.*

Proof. We will show that for all terms M the term M^* is semantically equivalent to $\lambda\vec{x}.x_0\langle\vec{\bot}\rangle$, i.e. for all $\vec{d} \in D^\omega$ we have $[\![M^*]\!](\vec{d}) = \top$ iff $d_0 = \top$. Suppose $d_0 \neq \top$. Then $d_0 = \bot$ or there is an n such that d_0 evaluates the n-th argument first. If $n = 1$ then $d_0\langle M, \vec{\bot}\rangle = \bot$, thus

$$d_0\langle\bot, \lambda\vec{y}.d_0\langle M, \vec{\bot}\rangle, \vec{\bot}\rangle = \bot$$

which is also the case if $n \neq 1$. □

Suppose N_1 and N_2 are different infinite normal forms. Then N_1^* and N_2^* have different infinite normal forms and we get $[\![N_1^*]\!] = [\![N_2^*]\!]$ by the above consideration. Thus, the **LBD** model of CPS_∞ is not faithful.

Lemma 24. *There exist infinite normal forms N_1, N_2 in CPS_∞ that can not be separated.*

Notice that in pure untyped λ-calculus different normal forms can always be separated. (cf. [4])

We think that the lack of faithfulness of CPS_∞ can be overcome by extending the language by a parallel construct and refining the observation type $\mathsf{0}$ to $\mathsf{0}' \cong \mathsf{List}(\mathsf{0}')$. The language $\mathsf{CPS}_\infty^{\|}$ associated with the domain equation $D \simeq D^{\mathsf{N}} \to \mathsf{0}'$ is given by

$$M ::= x \mid \lambda\vec{x}.t$$
$$t ::= \top \mid M\langle\vec{M}\rangle \mid (\!| t \parallel \ldots \parallel t |\!)$$

the syntactic values are given by the grammar $V ::= \top \mid (\!| V \| \ldots \| V |\!)$ operational semantics of $\mathsf{CPS}_\infty^{\|}$ is the operational semantics of CPS_∞ extended by the rule

$$\frac{(\lambda\vec{x}.t_i)\langle\vec{M}\rangle \Downarrow V_i \quad \text{for all } i \in \{1,\ldots,n\}}{(\lambda\vec{x}.(\!| t_1 \| \ldots \| t_n |\!))\langle\vec{M}\rangle \Downarrow (\!| V_1 \| \ldots \| V_n |\!)}$$

and the normal forms of $\mathsf{CPS}_\infty^{\|}$ are given by the grammar

$$N ::= x \mid \lambda\vec{x}.t$$
$$t ::= \top \mid x\langle\vec{N}\rangle \mid (\!| t \| \ldots \| t |\!)$$

understood in a coinductive sense.

Obviously, separation of normal forms can be shown for an affine version of CPS_∞ by substituting the respective projections for head variables. Using the parallel construct $(\!| \ldots \| \ldots |\!)$ of $\mathsf{CPS}_\infty^{\|}$ we can substitute for a head variable quasi simultaneously *both* the respective projection *and* the head variable itself. Since the interpretation of $\mathsf{CPS}_\infty^{\|}$ is faithful w.r.t. the parallel construct $(\!| \ldots \| \ldots |\!)$ we get separation for $\mathsf{CPS}_\infty^{\|}$ normal forms as in the affine case. This kind of argument can be seen as as a "qualitative" reformulation of a related "quantitative" method introduced by F. Maurel in his Thesis [14] albeit in the somewhat more complex context of J.-Y. Girard's *Ludics*.

In a sense this is not surprising since our parallel construct introduced above allows one to make the same observations as with parallel-or. The only difference is that our parallel construct keeps track of all possibilities simultaneously whereas the traditional semantics of parallel-or takes their supremum thus leading out of the realm of sequentiality. This is avoided by our parallel construct at the price of a more complicated domain of observations. For an approach in a similar spirit see [6].

References

1. Samson Abramsky, Radha Jagadeesan, and Pasquale Malacaria. Full abstraction for PCF. *Inf. Comput.*, 163(2):409–470, 2000.
2. Roberto M. Amadio and Pierre-Louis Curien. *Domains and lambda-calculi*. Cambridge University Press, New York, NY, USA, 1998.

3. Andrea Asperti. Stability and computability in coherent domains. *Inf. Comput.*, 86(2):115–139, 1990.
4. H. P. Barendregt. *The Lambda Calculus - its syntax and semantics.* North Holland, 1981, 1984.
5. R. Cartwright, P.-L. Curien, and M. Felleisen. Fully abstract models of observably sequential languages. *Information and Computation*, 111(2):297–401, 1994.
6. Russell Harmer and Guy McCusker. A fully abstract game semantics for finite nondeterminism. In *LICS*, pages 422–430, 1999.
7. J. M. E. Hyland and C.-H. L. Ong. On full abstraction for PCF: I. models, observables and the full abstraction problem, ii. dialogue games and innocent strategies, iii. a fully abstract and universal game model. *Information and Computation*, 163:285–408, 2000.
8. J. Laird. Bistable biorders: a sequential domain theory. Submitted, 2005.
9. J. Laird. *A semantic analysis of control.* PhD thesis, University of Edinburgh, 1998.
10. J. Laird. Locally boolean domains. *Theoretical Computer Science*, 342:132 – 148, 2005.
11. Ralph Loader. Finitary PCF is not decidable. *Theor. Comput. Sci.*, 266(1-2):341–364, 2001.
12. John Longley. The sequentially realizable functionals. *Ann. Pure Appl. Logic*, 117(1-3):1–93, 2002.
13. T. Löw. *Locally Boolean Domains and Curien-Lamarche Games.* PhD thesis, Technical University of Darmstadt, 2006. in prep., preliminary version available from `http://www.mathematik.tu-darmstadt.de/~loew/lbdclg.pdf`.
14. F. Maurel. *Un cadre quantitatif pour la Ludique.* PhD thesis, Université Paris 7, Paris, 2004.
15. Peter W. O'Hearn and Jon G. Riecke. Kripke logical relations and PCF. *Information and Computation*, 120(1):107–116, 1995.
16. Andrew M. Pitts. Relational properties of domains. *Information and Computation*, 127(2):66–90, 1996.
17. G. D. Plotkin. Lectures on predomains and partial functions. Course notes, Center for the Study of Language and Information, Stanford, 1985.
18. B. Reus and T. Streicher. Classical logic, continuation semantics and abstract machines. *J. Funct. Prog.*, 8(6):543–572, 1998.

Universal Structures and the Logic of Forbidden Patterns

Florent Madelaine

Department of Computer Science, University of Durham,
Science Labs, South Road, Durham DH1 3LE, U.K.
f.r.madelaine@durham.ac.uk

Abstract. Forbidden Patterns Problems (FPPs) are a proper generalisation of Constraint Satisfaction Problems (CSPs). However, we show that when the input belongs to a proper minor closed class, a FPP becomes a CSP. This result can also be rephrased in terms of expressiveness of the logic MMSNP, introduced by Feder and Vardi in relation with CSPs. Our proof generalises that of a recent paper by Nešetřil and Ossona de Mendez. Note that our result holds in the general setting of problems over arbitrary relational structures (not just for graphs).

Keywords: Finite Model theory, Monadic Second Order Logic, Constraint Satisfaction, Graph Homomorphism and Duality.

1 Introduction

Graph homomorphisms and related problems have received considerable attention in the recent years not only as a topic in combinatorics and graph theory but also in relation with constraint satisfaction. A lot of possible directions of research in the area have been opened by different motivations, and have been explored (the very recent monograph [1] serves as a good survey of the area).

The present work was motivated mainly by constraint satisfaction problems (CSPs) and their generalisations, forbidden patterns problems (FPPs), introduced in [2]. Our main result falls into the category of so-called (restricted) duality theorems, and has the same flavour as results from combinatorics and graph theory such as [3,4,5,6,7], (some of which will be explained shortly). Our main result can also be presented in terms of the expressiveness of the logic MMSNP, a fragment of monadic second order logic, that is closely related to FPPs (every sentence of MMSNP expresses a finite union of FPPs). For arbitrary structures, MMSNP captures problems that are not CSPs. However, when the input is restricted, it may be that MMSNP collapses to the class of CSPs (with the same input restriction): for example, when the input is connected and of bounded degree [6]. The main result of this paper states that this is also the case when the input is connected and belongs to some *proper minor closed class* (that is, a class that is defined by forbidding a finite set of graphs as minors, *e.g.* planar graphs). In the rest of this introduction, we will need to elaborate on work and ideas from two different areas, descriptive complexity and combinatorics.

Z. Ésik (Ed.): CSL 2006, LNCS 4207, pp. 471–485, 2006.

Let us start with MMSNP, a fragment of monadic second order logic that was introduced in [8], motivated by the search of a logic that exhibits a dichotomy (every problem is either in $\mathcal{P}$ or $\mathcal{NP}$-complete). It is still open whether MMSNP has a dichotomy but Feder and Vardi proved that every problem Ω in MMSNP is equivalent to a CSP Ω' (recall that, for a fixed structure H, the *Constraint Satisfaction Problem* with *template* H is the class of structures G such that there exists a homomorphism from G to H). Feder and Vardi's reduction from Ω' to Ω was randomised but has been recently derandomised by Kun [9]. Hence, MMSNP has a dichotomy if, and only if, the class of CSPs has a dichotomy. The dichotomy conjecture for CSPs is still open but is supported by numerous results (see *e.g.* [10,11,12,13]). So, one could argue that MMSNP and the class of CSPs are essentially the same. However, it is known that the logic MMSNP is too strong and captures problems which are not CSPs [8,14]. Note also that in general the signature of Ω' is exponentially larger than the signature of Ω and that Kun's derandomised reduction involves gadgets that are built from graph-expanders which are themselves obtained via the random method (and as such their size is some unknown constant). Moreover, Bodirsky *et al.* [15,16] showed that problems in MMSNP are in fact examples of CSPs with a countable template. So, in the context of descriptive complexity, we argue that MMSNP and CSPs are rather different. The combinatorial counterpart of MMSNP is directly related to *Forbidden Patterns Problems* (FPPs): every sentence of MMSNP captures a finite union of FPPs. In [17,2], an exact characterisation of FPPs that are CSPs was proved: given a forbidden patterns problem, we can decide whether it is a CSP with a finite or an infinite template and in the former case, we can compute effectively this finite template. Since the transformation of a sentence of MMSNP into a finite union of FPPs is also effective, as a corollary, we can decide whether a given sentence of MMSNP captures a finite union of (finite) CSPs, or captures a finite union of infinite CSPs. The characterisation of FPPs that are finite CSPs subsumes the characterisation of duality pairs obtained by Tardif and Nešetřil [3]. A *duality pair* is a pair (F, H) of structures such that for every structure G, there is no homomorphism from F to G (we say that G is F-mote) if, and only if, there is a homomorphism from G to H (we say that G is homomorphic to H). The word *duality* is used in this context to describe the inversion of the direction of homomorphisms. Forbidden patterns problems generalise F-moteness by involving coloured structures (see Section 2 for definition and examples) and the property of F-moteness corresponds to FPPs with a single colour and a single forbidden pattern (namely F). So a characterisation of duality pairs corresponds to a characterisation of those rather restricted FPPs which are also CSPs. Using a similar language, we could describe the characterisation of those FPPs that are CSPs as a *coloured duality theorem*.

Häggkvist and Hell [5] showed a different kind of duality result for graph homomorphism (or CSPs for graphs). They built a universal graph H for the class of F-mote graphs of bounded degree b. Thus, any degree-d graph G is F-mote if, and only if, there is a homomorphism from G to H. Note that the usual notion of universal graph, as used by Fraïssé [18] is for induced substructures, not homomorphism, and that the universal graph is usually infinite. In our context, the word universal graph refers to a *finite* graph and is universal with respect to the existence of homomorphisms. The result of Häggkvist and Hell can be seen as a restricted form of *duality*, as they quantify only over bounded degree graphs, rather than all graphs. This type of result is called a *restricted duality theorem*. In the last decade, a series of papers built upon Häggkvist and Hell's technique. In particular, in [4], where the authors investigated more generally the existence of universal graphs for the class of graphs that are simultaneously F-mote and H-colourable. In [6], we recently extended this last result by proving that FPPs restricted to connected inputs of bounded-degree become CSPs. This result could be described as a *restricted coloured duality theorem*.

A well-known extension of monadic second order logic consists in allowing monadic predicates to range over tuples of elements, rather than just elements of a structure. This extension is denoted by MSO_2 whereas the more standard logic with monadic predicates ranging over elements only is denoted by MSO_1. In general, MSO_2 is strictly more expressive than MSO_1. However Courcelle [19] proved that MSO_2 collapses to MSO_1, for some restriction of the input: for graphs of bounded degree, for partial k-trees (for fixed k), for planar graphs and, more generally for graphs belonging to a *proper minor closed class* (a class $\mathcal{K}$ such that: a minor of a graph G belongs to $\mathcal{K}$ provided that G belongs to $\mathcal{K}$; and, $\mathcal{K}$ does not contain all graphs).

It is perhaps worth mentioning that in [6] and the present paper we assume a definition of FPPs that is more general than the original one in [2] in that the new definition allows colourings not only of the vertices but also of the edges of the input graph. Essentially, this means that we are now considering problems related to MMSNP_2, the extension of MMSNP with edge quantification (for clarity, from now on, we denote by MMSNP_1 the original logic, i.e. without edge quantification). Thus, in [6] we obtain as a corollary that MMSNP_1 and MMSNP_2 collapse to CSP when restricted to connected inputs of bounded degree.

Nešetřil and Ossona de Mendez [7] introduced the notion of *tree-depth* of a graph and, using a result of De Vos *et al.* [20] on low tree-width decomposition for proper minor closed class, they proved a similar result using tree-depth rather than tree-width. This allowed them to prove a restricted duality theorem for proper minor closed classes of graphs. In this paper, we prove a *restricted coloured duality theorem for proper minor closed classes*: we extend the notion of tree-depth to arbitrary structures and prove this theorem for arbitrary structures, not only for graphs. As a corollary, we get the main result of this paper: we prove that MMSNP_1 and MMSNP_2 collapse to CSP when restricted to connected structures whose Gaifman graphs belong to a proper minor closed class. Note that the proof of our previous result [6] uses a different technique: for bounded degree d input, the construction of the universal graph relies on the fact that every subgraph induced by the vertices at distance at most p from a given vertex in a bounded degree graph is finite and its size depends only of d and p (p is a constant that is fixed according to the size of the forbidden patterns). This is not the case when the input comes from a proper minor closed class.

The paper is organised as follows. In Section 2, we define CSPs, FPPs and give some examples. Then, we define MMSNP_1 and recall some known results and extend them to MMSNP_2. Next, we prove that problems in MMSNP_2 are also infinite CSP in the sense of Bodirsky. We conclude Section 2 by stating our main result, the collapse to CSPs of MMSNP_1 and MMSNP_2 when restricted to connected structures from a proper minor closed class. We illustrate this result by a few examples. We also give a brief overview of the proof structure to guide the reader in the following technical sections. In Section 3, we introduce the notion of tree-depth for structures. Then, we show that coloured structures of bounded tree-depth have bounded cores. Finally, we show that structures that belong to a (fixed) proper minor closed class have a low tree-depth decomposition. In Section 4, we build a finite universal coloured structure for forbidden patterns for problems that are restricted to a class of structure that have a low tree-depth decomposition. We conclude this section by a proof of our main result. Most proofs have been omitted due to space restrictions and are available in the online appendix [21].

2 Preliminaries

2.1 CSP and FPP

Let σ be a signature that consists of finitely many relation symbols. From now on, unless otherwise stated, every structure considered will be a σ-structure. Let S and T be two structures. A *homomorphism* h from S to T is a mapping from $|S|$ (the domain of S) to $|T|$ such that for every r-ary relation symbol R in σ and every elements $x_1, x_2, \ldots, x_r$ of S, if $R(x_1, x_2, \ldots, x_r)$ holds in S then $R(h(x_1), h(x_2), \ldots, h(x_r))$ holds in T.

Constraint Satisfaction Problems. The *constraint satisfaction problem* with *template* T is the decision problem with,

- input: a finite structure S; and,
- question: does there exist a homomorphism from S to T?

We denote by CSP the class of constraint satisfaction problems with a *finite* template.

Example 1. The constraint satisfaction problem with template K_3 (the clique with three elements, *i.e.* a triangle) is nothing else than the 3-colourability problem from graph theory.

Let $\mathcal{V}$ (respectively, $\mathcal{E}$) be a finite set of *vertex colours* (respectively, *edge colours*). A *coloured structure* is a triple $(S, s^{\mathcal{V}}, s^{\mathcal{E}})$, where S is a structure, $s^{\mathcal{V}}$ is a mapping from $|S|$ to $\mathcal{V}$ and $s^{\mathcal{E}}$ is a mapping from $E(S)$ to $\mathcal{E}$ where,

$$E(S) := \bigcup_{R\in\sigma} \{(R, x_1, x_2, \ldots, x_r) \text{ s.t. } R(x_1, x_2, \ldots, x_r) \text{ holds in } S\}.$$

Let $(S, s^{\mathcal{V}}, s^{\mathcal{E}})$ and $(S', s'^{\mathcal{V}}, s'^{\mathcal{E}})$ be two coloured structures. A *colour-preserving homomorphism* h from S to S' is a homomorphism from S to S' such that $s'^{\mathcal{V}} \circ h = s$ and for every tuple $t = (R, x_1, x_2, \ldots, x_r)$ in $E(S)$, $s'^{\mathcal{E}}(t') = s^{\mathcal{E}}(t)$, where $t' := \big(R, h(x_1), h(x_2), \ldots, h(x_r)\big)$. When the colours are clear from the context, we simply write that h *preserves colours*. Note that the composition of two homomorphism that preserve colours is also a homomorphism that preserves colours.

A structure S is *connected* if it can not be partitioned into two disjoint induced substructures. A *pattern* is a finite coloured structure $(F, f^{\mathcal{V}}, f^{\mathcal{E}})$ such that F is connected. In this paper, patterns are used to model constraints in a negative fashion and consequently, we refer to them as *forbidden* patterns. Let $\mathfrak{F}$ be a finite set of forbidden patterns. We say that a coloured structure $(S, s^{\mathcal{V}}, s^{\mathcal{E}})$ is *valid* with respect to $\mathfrak{F}$ if, and only if, for every forbidden pattern $(F, f^{\mathcal{V}}, f^{\mathcal{E}})$ in $\mathfrak{F}$, there does not exist any colour-preserving homomorphism h from F to S.

Forbidden Patterns Problems. The *problem with forbidden patterns* $\mathfrak{F}$ is the decision problem with,

- input: a finite structure S
- question: does there exist $s^{\mathcal{V}} : |S| \to \mathcal{V}$ and $s^{\mathcal{E}} : E(S) \to \mathcal{E}$ such that $(S, s^{\mathcal{V}}, s^{\mathcal{E}})$ is valid with respect to $\mathfrak{F}$?

We denote by FPP_1 the class of forbidden patterns problem with vertex colour only (that is for which $\mathcal{E}$ has size one) and by FPP_2 the class of forbidden patterns problems.

Example 2. Let G be an undirected graph. It is usual to represent G as a relational structure with a single binary relation E that is symmetric. However, the logics considered in this paper are monotone and we can not express that E is symmetric and we use a different representation to encode graphs. We say that a structure S with one binary relation symbol E *encodes* G, if $|S| = V(G)$ and for any x and y in $V(G)$, x and y are adjacent in G if, and only if, $E(x, y)$ or $E(y, x)$ holds in S. Note that this encoding is not bijective. Modulo this encoding, the following graph problems are Forbidden Patterns Problems.

1. Vertex-No-Mono-Tri: consists of the graphs for which there exists a partition of the vertex set in two sets such that no triangle has its three vertices occurring in a single partition. It was proved in [8,14] that this problem is not in CSP and in [22] that it is $\mathcal{NP}$-complete.
2. Tri-Free-Tri: consists of the graphs that are both three colourable (tripartite) and in which there is no triangle. It was proved in [14] that this problem is not in CSP.
3. Edge-No-Mono-Tri: consists of the graphs for which there exists a partition of the edge set in two sets such that no triangle has its three edges occurring in a single partition. It is known to be $\mathcal{NP}$-complete (see [23]).

2.2 $MMSNP_1$ and $MMSNP_2$

The class of forbidden patterns problems with colours over the vertices only, corresponds to the problems that can be expressed by a formula in Feder and Vardi's MMSNP (Monotone Monadic SNP without inequalities, see [8,2]). Note that allowing colours over the edges does not amount to drop the hypothesis of monadicity altogether. Rather, it corresponds to a logic, let's call it $MMSNP_2$, which is similar to MMSNP but allows *first-order* variables over edges (just like Courcelle's MSO (Monadic Second Order logic) and MSO_2, see [19]).

This section is organised as follows: first, we introduce formally $MMSNP_1$ and recall some known results; and, secondly, we introduce $MMSNP_2$ and prove that it captures finite union of problems in FPP_2. Finally, we prove that problems in FPP_2 are infinite CSP in the sense of Bodirsky [15].

Definition 3. Monotone Monadic SNP without inequality, *$MMSNP_1$, is the fragment of ESO consisting of those formulae Φ of the following form:*

$$\exists \mathbf{M} \forall \mathbf{t} \bigwedge_i \neg\big(\alpha_i(\sigma, \mathbf{t}) \wedge \beta_i(\mathbf{M}, \mathbf{t})\big),$$

where $\mathbf{M}$ is a tuple of monadic relation symbols (not in σ), $\mathbf{t}$ is a tuple of (first-order) variables and for every negated conjunct $\neg(\alpha_i \wedge \beta_i)$:

- *α_i consists of a conjunction of positive atoms involving relation symbols from σ and variables from $\mathbf{t}$; and*
- *β_i consists of a conjunction of atoms or negated atoms involving relation symbols from $\mathbf{M}$ and variables from $\mathbf{t}$.*

(Notice that the equality symbol does not occur *in Φ.)*

The negated conjuncts $\neg(\alpha \wedge \beta)$ correspond to (partially coloured) forbidden structures (and this is the reason why we use such a notation in the definition rather than using implications or clausal form). To get forbidden patterns problems, we need to restrict sentences so that negated conjuncts correspond precisely to coloured connected structures. Such a restriction was introduced in [2] as follows.

Definition 4. *Moreover, Φ is* primitive *if, and only if, for every negated conjunct $\neg(\alpha \wedge \beta)$:*

- *for every first-order variable x that occurs in $\neg(\alpha \wedge \beta)$ and for every monadic symbol C in $\mathbf{M}$, exactly one of $C(x)$ and $\neg C(x)$ occurs in β;*
- *unless x is the only first-order variable that occurs in $\neg(\alpha \wedge \beta)$, an atom of the form $R(\mathbf{t})$, where x occurs in $\mathbf{t}$ and R is a relation symbol from σ, must occur in α; and,*
- *the structure induced by α is connected.*

Theorem 5. *[2] The class of problems captured by the primitive fragment of the logic $MMSNP_1$ is exactly the class FPP_1 of forbidden patterns problems with vertex colours only.*

It is only a technical exercise to relate any sentence of MMSNP_1 with its primitive fragment.

Proposition 6. *[2] Every sentence of* $\mathbf{MMSNP}_1$ *is logically equivalent to a finite disjunction of primitive sentences.*

Consequently, we have the following characterisation.

Corollary 7. *Every sentence Φ in $MMSNP_1$ captures a problem Ω that is the union of finitely many problems in FPP_1.*

Remark 8. We have altered slightly the definitions w.r.t. [2]. We now require a pattern to be connected and we have amended the notion of a primitive sentence accordingly.

The logic MMSNP_2 is the extension of the logic MMSNP_1 where in each negated conjunct $\neg(\alpha \wedge \beta)$, we allow a monadic predicate to range over a tuple of elements of the structure, that is we allow new "literals" of the form $M(R(x_1, x_2, \ldots, x_n))$ in β, where R is n-ary relation symbol from the signature σ. We also insists that whenever such a literal occurs in β then $R(x_1, x_2, \ldots, x_n)$ appears in α. The semantic of a monadic predicate M is extended and is defined as both a subset of the domain and a subset of the set of tuples that occur in some input relation: that is, for a structure S, $M^S \subset |S| \cup E(S)$. We say that a sentence of MMSNP_2 is *primitive* if each negated conjunct $\neg(\alpha \wedge \beta)$ satisfies the same conditions as in Definition 4 and a further condition:

- if $R(x_1, x_2, \ldots, x_n)$ occurs in α then for every (existentially quantified) monadic predicate C exactly one of $C(R(x_1, x_2, \ldots, x_n))$ or $\neg C(R(x_1, x_2, \ldots, x_n))$ occurs in β.

It is only a technical exercise to extend all the previous results of this section concerned with MMSNP_1 and FPP_1 to MMSNP_2 and FPP_2. In particular, we have the following result.

Corollary 9. *Every sentence Φ in $MMSNP_2$ captures a problem Ω that is the union of finitely many problems in FPP_2.*

2.3 Infinite CSP and MMSNP_2

Bodirsky *et al.* have investigated constraint satisfaction problems, where the template is infinite and have proposed restrictions that ensure that the problems are decidable (and in $\mathcal{NP}$): when the template is countable and *homogeneous* in [24], and more recently to a more general case when the template is ω-categorical in [15] (a countable structure Γ is *ω-categorical* if all countable models of its first-order theory of Γ are isomorphic to Γ). Denote by $\mathrm{CSP}^\star$ the set of constraint satisfaction problems that have a ω-categorical countable template and belong to $\mathcal{NP}$. Using a recent result due to Cherlin, Shelah and Chi [25], Bodirsky and Dalmau proved the following result.

Theorem 10. *[16] Every non-empty problem in MMSNP$_1$ that is closed under disjoint union belongs to CSP*.*

It follows directly from its definition that every problem in FPP$_1$ is closed under disjoint union. Hence, we get the following result.

Corollary 11. *Every problem in FPP$_1$ is in CSP*. Consequently, every problem in MMSNP$_1$ is the union of finitely many problems in CSP*.*

Since the ω-categoricity is preserved by first-order interpretation, we can prove the following.

Theorem 12. *Every problem in FPP$_2$ is in CSP*. Consequently, every problem in MMSNP$_2$ is the union of finitely many problems in CSP*.*

Remark 13. As pointed out in [15], there are problems that are in CSP* but not in MMSNP$_1$. For example, the problem over directed graphs with template induced by the linear order over $\mathbb{Q}$. Unfortunately, this problem is not expressible in MMSNP$_2$ either. In fact, we do not know whether MMSNP$_1$ is strictly contained in MMSNP$_2$. Indeed, to the best of our knowledge, none of the problems used to separate MSO$_1$ from MSO$_2$ are expressible in MMSNP$_2$. We suspect that the problem EDGE-NO-MONO-TRI is not expressible in MMSNP$_1$ and that MMSNP$_1$ is strictly contained in MMSNP$_2$.

2.4 Main Result: Collapse of MMSNP$_1$ and MMSNP$_2$

Let S be a structure. The *Gaifman graph* of S, which we denote by $\mathcal{G}_S$ is the graph with vertices $|S|$ in which two distincts elements x and y of S are adjacent if, and only if, there exists a tuple in a relation of S in which they occur simultaneously. Given a class $\mathcal{K}$ of structures, we denote by $\mathcal{G}_\mathcal{K}$ the class of their Gaifman graphs. A class of graphs $\mathcal{G}$ is said to be a *proper minor closed class* if the following holds: first, for any graph G and any minor H of G, if G belongs to $\mathcal{G}$ then so does H; and, secondly $\mathcal{G}$ does not contain all graphs. Alternatively, $\mathcal{G}$ is proper minor closed if it excludes at least one fixed graph as a minor, a so-called *forbidden minor*. We say that a class of structures $\mathcal{K}$ is a *proper minor closed class* if, and only if $\mathcal{G}_\mathcal{K}$ is a proper minor closed class.

The restricted duality theorem for proper minor closed classes of graphs proved in [7] can be rephrased as follows.

Theorem 14. *Let $\mathcal{K}$ be a proper minor closed class of graphs. Let $\mathfrak{F}$ be a finite set of connected graphs. There exists a universal graph U such that for any graph G in $\mathcal{K}$, the graph G is F-mote (there is no homomorphism from F to G) for every graph F in $\mathfrak{F}$ if, and only if, there exists a homomorphism from G to U.*

We can use this result and *ad hoc* techniques from [6] to show that two of our examples are CSP when restricted to a proper minor closed class $\mathcal{K}$. Let U be the universal graph for $\mathfrak{F} = \{K_3\}$. Let U' be the product of U with K_3. It is not difficult to check that TRI-FREE-TRI, restricted to $\mathcal{K}$ is the CSP with template U'. Similarly VERTEX-NO-MONO-TRI is the CSP with template U', where U'' is the graph that consists of two copies of U, such that every pair of elements from different copies are adjacent. Note that technically, our problems being defined over structures with a single binary relation E, we have not really expressed them as a CSP. However, according to our encoding, replacing any two adjacent vertices x and y in the above graphs by two arcs $E(x, y)$ and $E(y, x)$ provides us with a suitable template. The last problem EDGE-NO-MONO-TRI is also a CSP when restricted to a proper minor closed class. However, we do not know how to get a template directly from the above result. For this, we need to use the main result of this paper which is a *restricted coloured duality theorem for proper minor closed class* and is formulated as follows.

Theorem 15. *The logic* $MMSNP_1$ *and* $MMSNP_2$ *have the same expressive power when restricted to any proper minor closed class* $\mathcal{K}$*: they both express finite union of (finite) CSPs. Furthermore, the logic* $MMSNP_2$ *(and,* a fortiori $MMSNP_1$*) collapses to CSP, when restricted to connected structures that belong to any proper minor closed class.*

We wish to show that FPPs are CSPs when the input belongs to a proper minor closed class (Theorem 15 follows directly from this fact using results of Section 3). To prove this fact, given a forbidden patterns problem, we need to build a structure U such that, for every input structure S (that comes from a fixed proper minor closed class $\mathcal{K}$), S is a yes-instance of the given forbidden patterns problem if, and only if, S is homomorphic to U. So we wish to build a structure U that is *universal* with respect to our forbidden patterns problem. Recall that we use the word *universal structure* with a different meaning to that of Fraïssé, in particular U has to be finite rather than infinite and is universal w.r.t. the existence of homomorphisms rather than induced substructures (*i.e.* existence of embeddings). Assume for now that the forbidden patterns problem in question has a single colour. The key property to build such a finite U is that of *bounded tree-depth.* It turns out that though the size of a structure of bounded tree-depth may be arbitrarily large, the size of its *core* is bounded. The core of a structure S is the smallest structure that is homomorphically equivalent to S. Let Y_p be the disjoint union of all cores of structures of tree-depth at most p that are valid (w.r.t. our fixed forbidden patterns problem). Note that Y_p is finite and for any structure S of tree-depth at most p, S is homomorphic to Y_p if, and only if, S is valid. The reason why we impose that the input comes from a *proper minor closed class* is that any such structure can be decomposed into a finite number of parts, say q, so that any $p \leq q$ parts induce a structure of tree-depth at most p. So, given some input S together with such a decomposition, if the largest forbidden pattern has size p then it suffices to check that for any choice of p parts of the input, the structure induced by these p-parts is valid, or equivalently, that it is homomorphic to Y_p. Finally, we use the key concept of pth *truncated product.* This concept allows us to translate the existence of homomorphisms to a structure T, for any $p-1$ parts of a p partitioned input S, to the existence of a homomorphism to the pth truncated product of T. Hence, by taking a sequence of suitable truncated products of Y_p, we get the desired finite universal structure U. Note that we assumed that the forbidden patterns problem had a single colour. In order to get our result in general, we adapt the above ideas and concepts to coloured structures.

3 Tree-Depth, Cores and Low Tree-depth Decomposition

3.1 Tree-Depth and Elimination Tree of a Structure

Following the theory of tree-depth of graphs introduced in [7], we develop elements of a theory of tree-depth for structures. Let S be a structure. We denote by $\mathcal{H}_S$ the *hypergraph induced by* S, that has the same domain as S and whose hyperedges are the sets that consists of the elements that occur in the tuples of the relations of S. If $\mathcal{H}$ is an hypergraph and r is an element of the domain of $\mathcal{H}$ then we denote by $\mathcal{H} \setminus \{r\}$ the hypergraph obtained from $\mathcal{H}$ by deleting r from the domain of $\mathcal{H}$ and removing r from every hyperedge in which it occurs (e.g. $\{a, b, r\}$ is replaced by $\{a, b\}$). A connected component $\mathcal{H}_i$ of $\mathcal{H}_S \setminus \{r\}$ induces a substructure S_i of S in a natural way: S_i is the induced substructure of S with the same domain as $\mathcal{H}_i$. If S is connected, then we say that a rooted tree (r, Y) is an *elimination tree* for S if, and only if, either $|S| = \{r\}$ and $|Y| = r$, or for every component S_i of S $(1 \leq i \leq p)$ induced by the connected components $\mathcal{H}_i$ of $\mathcal{H}_S \setminus \{r\}$, Y is the tree with root r adjacent to subtrees (r_i, Y_i), where (r_i, Y_i) is an elimination tree of S_i. Let F be a rooted forest (disjoint union of rooted trees). We define the *closure of* F $\operatorname{clos}(F, \sigma)$ to be the structure with domain $|F|$ and all tuples $R_i(x_1, x_2, \ldots, x_{r_i})$ such that the elements mentioned in this tuple $\{x_i | 1 \leq i \leq r_i\}$ form a chain w.r.t. $\leq_F$, where $\leq_F$ is the partial order induced by F, i.e. $x \leq_F y$ if, and only if, x is an ancestor of y in F. The *tree-depth* of S, denoted by $\operatorname{td}(S)$, is the minimum height of a rooted forest F such that S is a substructure of the closure of F, $\operatorname{clos}(F, \sigma)$.

These notions are closely related.

Lemma 16. *Let S be a connected structure. A rooted tree (r, Y) is an elimination tree for S if, and only if, S is a substructure of* $\operatorname{clos}(Y, \sigma)$. *Consequently, the tree-depth of S is the minimum height of an elimination tree.*

3.2 Tree-Depth and Cores

We show that a coloured structures of bounded tree-depth $\mathcal{K}$ has a core of bounded size. Recall that a *retract* of a structure S is an induced substructure S' of S for which there exists a homomorphism from S to S'. A minimal retract of a structure S is called a *core* of S. Since it is unique up to isomorphism, we may speak of *the* core of a structure [1]. This notion extends naturally to coloured structures.

We say that a (colour-preserving) automorphism μ of a (coloured structure) S has the *fixed-point property* if, for every connected substructure T of S, either $\mu(T) \cap T = \emptyset$ or there exist an element x in T such that $\mu(x) = x$. We say that μ is *involuting* if $\mu \circ \mu$ is the identity.

Theorem 17. *There exists a function $\eta : \mathbb{N} \times \mathbb{N} \to \mathbb{N}$ such that, for any coloured structure $(S, s^{\mathcal{V}}, s^{\mathcal{E}})$ such that $|S| > \eta(N, \operatorname{td}(S))$ and any mapping g :*

$|S| \to \{1, 2, \ldots, N\}$, *there exists a non-trivial involuting g-preserving automorphism* μ *of* $(S, s^{\mathcal{V}}, s^{\mathcal{E}})$ *with the fixed-point property.*

Theorem 18. *Let* $(S, s^{\mathcal{V}}, s^{\mathcal{E}})$ *be a coloured structure. If* $|S| > \eta(1, \mathrm{td}(S))$ *then* $(S, s^{\mathcal{V}}, s^{\mathcal{E}})$ *maps homomorphically into one of its proper induced substructure* $(S_0, s_0^{\mathcal{V}}, s_0^{\mathcal{E}})$. *Consequently,* $(S, s^{\mathcal{V}}, s^{\mathcal{E}})$ *has a core of size at most* $\eta(1, \mathrm{td}(S))$.

Thus, we get the following result.

Corollary 19. *Let* $\mathcal{K}$ *be any set of coloured structures of bounded tree-depth* k. *Then the set* $\mathcal{K}'$ *of cores of structures from* $\mathcal{K}$ *(up to isomorphism) is finite.*

3.3 Decomposition of Low Tree-Depth

In [20] De Vos *et al.* proved that for any proper minor closed class $\mathcal{K}$ and any integer $p \geq 1$, there exists an integer q such that for every graph G in $\mathcal{K}$ there exists a vertex partition of G into q parts such that any subgraph of G induced by at most $j \leq p$ parts has tree-width at most $p-1$. Using this result, Nešetřil and Ossona de Mendez prove a similar result for bounded tree-depth in [7], which can be rephrased as follows.

Theorem 20. *Let* $p \geq 1$. *Let* $\mathcal{K}$ *be a proper minor closed class of graphs. There exists an integer* q *such that any graph in* $\mathcal{K}$ *has a proper* q*-colouring in which any* p *colours induce a graph of tree-depth at most* p.

We extend this result to arbitrary structure using the following lemma.

Lemma 21. *A rooted tree* (r, Y) *is an elimination tree for a structure* S *if, and only if, it is an elimination-tree for its Gaifman graph* $\mathcal{G}_S$.

Corollary 22. *Let* $p \geq 1$. *Let* $\mathcal{K}$ *be a proper minor closed class of structures. There exists an integer* q *such that for any structure* S *in* $\mathcal{K}$, *there exists a partition of* $|S|$ *into* q *sets such that any substructure of* S *induced by at most* p *of these sets has tree-depth at most* p.

4 Restricted Coloured Dualities

4.1 Truncated Product

We extend the definition of truncated product and adapt two lemmas from [7] to coloured structures. Let $(S, s^{\mathcal{V}}, s^{\mathcal{E}})$ be a coloured structure and $p \geq 2$ be an integer. We define the pth *truncated product* of $(S, s^{\mathcal{V}}, s^{\mathcal{E}})$, to be the coloured structure $(S', {s'}^{\mathcal{V}}, {s'}^{\mathcal{E}})$ defined as follows.

- Its domain is a subset of $\bigcup_{i=1}^{p} W^i$ where,

$$W^i := \{(x_1, x_2, \ldots, x_{i-1}, \star, x_{i+1}, \ldots, x_p) \text{ s.t.} x_k \in |S|, 1 \leq k \leq p, k \neq i\}$$

($\star$ denotes a new element, *i.e.* $\star \notin |S|$).

- We restrict further W^i to $\tilde{W}^i$ that consists of elements

$$w^i = (x_1, x_2, \ldots, x_{i-1}, \star, x_{i+1}, \ldots, x_p)$$

of W^i such that there exists $v \in \mathcal{V}$ such that for every $1 \leq k \leq p$ with $k \neq i$ we have $s^{\mathcal{V}}(x_k) = v$ and we set $|S'| := \cup_{i=1}^{p} \tilde{W}^i$ and $s'^{\mathcal{V}}(w^i) := v$.
- For every relation symbol R of arity r, and every tuple $(w^{i_1}, w^{i_2}, \ldots, w^{i_r})$ where for every $1 \leq k \leq p$, w^{i_k} belongs to $\tilde{W}^{i_k}$ and

$$w^{i_k} = (x_1^{i_k}, x_2^{i_k}, \ldots, x_{i_k-1}^{i_k}, \star, x_{i_k+1}^{i_k}, \ldots, x_p^{i_k})$$

satisfies,
  - for every $1 \leq k < k' \leq r$, we have $i_k \neq i_{k'}$;
  - for every $1 \leq i \leq p$, such that $i \notin \{i_1, i_2, \ldots, i_r\}$, $R(t_i)$ holds in S, where $t_i = (x_i^{i_1}, x_i^{i_2}, \ldots, x_i^{i_r})$; and,
  - there exists $e \in \mathcal{E}$ such that for every $1 \leq i \leq p$, $s^{\mathcal{E}}(t_i) = e$,

we set $R(w^{i_1}, w^{i_2}, \ldots, w^{i_r})$ to hold in S' and we set $s'^{\mathcal{E}}(w^{i_1}, w^{i_2}, \ldots, w^{i_r}) := e$.

We denote the pth truncated product by $(S, s^{\mathcal{V}}, s^{\mathcal{E}})^{\Uparrow p}$.

This product has two important properties: it preserves validity w.r.t. a forbidden patterns problem (for a suitable p); and, the existence of all "partial" colour-preserving homomorphisms is equivalent to the existence of a homomorphism to the truncated product (see Lemma 23 and Lemma 24 below).

Lemma 23. *Let $p \geq 2$. Let $\mathfrak{F}$ be a set of forbidden patterns such that for every $(F, f^{\mathcal{V}}, f^{\mathcal{E}})$ in $\mathfrak{F}$, we have $|F| < p$. If a coloured structure $(S, s^{\mathcal{V}}, s^{\mathcal{E}})$ is valid w.r.t $\mathfrak{F}$ then its p-truncated product $(S', s'^{\mathcal{V}}, s'^{\mathcal{E}})$ is also valid w.r.t. $\mathfrak{F}$.*

Lemma 24. *Let $(U, u^{\mathcal{V}}, u^{\mathcal{E}})$ be a coloured structure and let p be an integer greater than the arity of any symbol in σ. Let $(S, s^{\mathcal{V}}, s^{\mathcal{E}})$ be a coloured structure. If there exists a partition $V_1, V_2, \ldots, V_p$ of $|S|$ such that for every substructure $(\tilde{S}_i, \tilde{s}^{\mathcal{V}}, \tilde{s}^{\mathcal{E}})$ of $(S, s^{\mathcal{V}}, s^{\mathcal{E}})$ induced by $|S| \setminus V_i$ there exist a colour-preserving homomorphism $\tilde{s}_i$ from $(\tilde{S}_i, \tilde{s}^{\mathcal{V}}, \tilde{s}^{\mathcal{E}})$ to $(U, u^{\mathcal{V}}, u^{\mathcal{E}})$ then there exists a colour-preserving homomorphism $\tilde{s}$ from $(\tilde{S}_i, \tilde{s}^{\mathcal{V}}, \tilde{s}^{\mathcal{E}})$ to $(U, u^{\mathcal{V}}, u^{\mathcal{E}})^{\Uparrow p}$.*

Using the two previous lemma, we get the following result.

Proposition 25. *Let p be an integer greater than the arity of any symbol in σ. Let $\mathfrak{F}$ be a set of forbidden patterns such that for every $(F, f^{\mathcal{V}}, f^{\mathcal{E}})$ in $\mathfrak{F}$, we have $|F| < p$. Let $q \geq p$. Let $(U, u^{\mathcal{V}}, u^{\mathcal{E}})$ be a coloured structure that is valid w.r.t $\mathfrak{F}$. Let $(S, s^{\mathcal{V}}, s^{\mathcal{E}})$ be a coloured structure.*

Assume that there exists a partition $V_1, V_2, \ldots, V_q$ of $|S|$ such that for every substructure $(\tilde{S}_i, \tilde{s}_i^{\mathcal{V}}, \tilde{s}_i^{\mathcal{E}})$ of $(S, s^{\mathcal{V}}, s^{\mathcal{E}})$ induced by p subsets $V_{i_1}, V_{i_2}, \ldots, V_{i_p}$ there exist a colour-preserving homomorphism $\tilde{s}_i$ from $(\tilde{S}_i, \tilde{s}_i^{\mathcal{V}}, \tilde{s}_i^{\mathcal{E}})$ to $(U, u^{\mathcal{V}}, u^{\mathcal{E}})$.

Then there exists a colour-preserving homomorphism $\tilde{s}$ from $(\tilde{S}_i, \tilde{s}^{\mathcal{V}}, \tilde{s}^{\mathcal{E}})$ to $(U', u'^{\mathcal{V}}, u'^{\mathcal{E}})$ and $(U', u'^{\mathcal{V}}, u'^{\mathcal{E}})$ is valid with respect to $\mathfrak{F}$, where

$$(U', u'^{\mathcal{V}}, u'^{\mathcal{E}}) := (U, u^{\mathcal{V}}, u^{\mathcal{E}})^{\Uparrow(p+1)\Uparrow(p+2)\ldots\Uparrow q}.$$

4.2 Universal Structure

We say that a coloured structure $(U, u^{\mathcal{V}}, u^{\mathcal{E}})$ is *universal* for a forbidden patterns problem restricted to a class $\mathcal{K}$ of structures if for every valid coloured structure there exists a colour-preserving homomorphism to $(U, u^{\mathcal{V}}, u^{\mathcal{E}})$, and *vice versa.*

Theorem 26. *Let p be an integer greater than the arity of any symbol in σ. Let $\mathfrak{F}$ be a set of forbidden patterns such that for every $(F, f^{\mathcal{V}}, f^{\mathcal{E}})$ in $\mathfrak{F}$, we have $|F| < p$. Let $\mathcal{K}$ be a proper minor closed class. There exists a universal coloured structure $(U, u^{\mathcal{V}}, u^{\mathcal{E}})$ for the restriction to $\mathcal{K}$ of the forbidden patterns problems with representation $\mathfrak{F}$.*

Proof. Let $(S, s^{\mathcal{V}}, s^{\mathcal{E}})$ be a coloured structure such that S belongs to $\mathcal{K}$. By Corollary 22, there exists an integer N such that every structure S in $\mathcal{K}$ can be partitioned into q parts, such that every p parts induce a substructure of S of tree-depth at most p. By Theorem 18, the core of the coloured structure induced by p parts is bounded. Let $(U, u^{\mathcal{V}}, u^{\mathcal{E}})$ be the disjoint union of all such cores that are valid w.r.t. $\mathfrak{F}$ (there are only finitely many). Since the forbidden patterns have size at most p, it suffices to check every substructure of S of size at most p and *a fortiori* $(S, s^{\mathcal{V}}, s^{\mathcal{E}})$ is valid w.r.t. $\mathfrak{F}$ if, and only if, every of its coloured substructure induced by p parts is valid, or equivalently if every such coloured substructure maps homomorphically into $(U, u^{\mathcal{V}}, u^{\mathcal{E}})$.

By Proposition 25, if $(S, s^{\mathcal{V}}, s^{\mathcal{E}})$ is valid w.r.t. $\mathfrak{F}$ then it is homomorphic to $(U, u^{\mathcal{V}}, u^{\mathcal{E}})^{\Uparrow(p+1)\Uparrow(p+2)\dots\Uparrow q}$; and, since $(U, u^{\mathcal{V}}, u^{\mathcal{E}})^{\Uparrow(p+1)\Uparrow(p+2)\dots\Uparrow q}$ is valid w.r.t. $\mathfrak{F}$ the converse holds as forbidden patterns problems are closed under inverse homomorphism. □

We have now all the elements to settle the problem of restricted coloured dualities for proper minor closed class.

Corollary 27. *Every forbidden patterns problem restricted to a proper minor closed class is a CSP.*

Proof. Let $(U, u^{\mathcal{V}}, u^{\mathcal{E}})$ be the universal coloured structure. We may take U as a template. A homomorphism from a structure S to U induces mapping $s^{\mathcal{V}}$ and $s^{\mathcal{E}}$ such that $(S, s^{\mathcal{V}}, s^{\mathcal{E}})$ is valid. Conversely, if S is not homomorphic to U then for every colour maps $s^{\mathcal{V}}$ and $s^{\mathcal{E}}$, there is no colour-preserving homomorphism from $(S, s^{\mathcal{V}}, s^{\mathcal{E}})$ to $(U, u^{\mathcal{V}}, u^{\mathcal{E}})$ and $(S, s^{\mathcal{V}}, s^{\mathcal{E}})$ is not valid. □

We are now ready to prove our main result.

Proof. (of Theorem 15). By Corollary 7 (resp. Corollary 9) every problem in MMSNP_1 (resp. MMSNP_2) is equivalent to a finite union of problems from FPP_1 (resp. FPP_2). By the previous theorem, when restricted to a proper minor closed class , every forbidden patterns problem is a restricted CSP. This proves that MMSNP_1 and MMSNP_2 collapse to finite union of CSP.

Moreover, if the problem is restricted to elements of a proper minor closed class that are connected then by taking the disjoint union of the template of each CSP, we get a template for our problem. □

5 Conclusion and Open Problems

We have proved that every forbidden patterns (with colours on both edges and vertices) problem is in fact a constraint satisfaction problem, when restricted to inputs that belong to a proper minor closed class. We derived from this result that the logic MMSNP_2 (and MMSNP_1) coincides with the class of CSPs on connected inputs that belong to a proper minor closed class. We proved, using a different method, a similar result for bounded degree graphs in [6]. So, together these results cover the restrictions considered by Courcelle in [19] under which MSO_1 and MSO_2 have the same expressive power. However, we still have not proved that for arbitrary input, MMSNP_2 is more expressive than MMSNP_1.

Another interesting open question is related to $\text{CSP}^\star$, the class of (well-behaved) infinite CSPs. We know that any problem in MMSNP_2 is a finite union of problems from $\text{CSP}^\star$. However, there are problems in $\text{CSP}^\star$ that are not expressible in MMSNP_2, which yields the following question. *Which extension of $MMSNP_2$ captures precisely $CSP^\star$?*

Courcelle [26] has recently shown that MSO_1 and MSO_2 have the same expressiveness for *uniformly k-sparse graphs*. This notion subsumes the restrictions from [19] mentioned above, which allows for a unified proof. Thus, it is reasonable to ask ourselves the following question: *Do $MMSNP_1$ and $MMSNP_2$ have the same expressiveness over uniformly k-sparse graphs?* And more precisely, *do forbidden patterns problems become constraint satisfaction problems when restricted to uniformly k-sparse graphs?* Nešetřil and Ossona de Mendez [27] have recently unified the proofs of the restricted duality theorems for bounded degree and proper minor closed classes using a more general concept, that of *bounded expansion*. Moreover, it turns out that graphs of bounded expansion are examples of (very well behaved) sparse graphs. So if we cannot settle the previous question in its full generality, we can at least hope to settle it in the case of graphs of bounded expansion.

A final point concerns the notion of a proper minor closed class of *structures*. In the present paper, we use the Gaifman graph to define this concept. However, it would be more natural and perhaps preferable to define a notion of *minor for structures*. The following definition seems reasonable. A minor of a structure S is obtained from S by performing a finite sequence of the following operations: taking a (not necessarily induced) substructure; and, identifying some elements, provided that they all occur in some tuple of some relation. This new definition subsumes the definition used in this paper and provokes our final question: *Do the results of this paper hold under this new definition?*

References

1. Hell, P., Nešetřil, J.: Graphs and homomorphisms. Oxford University Press (2004)
2. Madelaine, F., Stewart, I.A.: Constraint satisfaction, logic and forbidden patterns. submitted. (2005)
3. Nešetřil, J., Tardif, C.: Duality theorems for finite structures (characterising gaps and good characterisations). Journal of Combin. Theory Ser. B **80** (2000) 80–97

4. Paul A. Dreyer Jr, Malon, C., Nešetřil, J.: Universal H-colourable graphs without a given configuration. Discrete Mathematics **250**(1-3) (2002) 245–252
5. Häggkvist, R., Hell, P.: Universality of A-mote graphs. European Journal of Combinatorics **14** (1993) 23–27
6. Dantchev, S., Madelaine, F.: Bounded-degree forbidden-pattern problems are constraint satisfaction problems. In: International Computer Science Symposium in Russia June 8-12, St.Petersburg). (2006) CSR06.
7. Nešetřil, J., patrice Ossona de Mendez: Tree-depth, subgraph coloring and homomorphism bounds. European Journal of Combinatorics (2005)
8. Feder, T., Vardi, M.Y.: The computational structure of monotone monadic SNP and constraint satisfaction: a study through datalog and group theory. SIAM J. Comput. **28** (1999) 57–104
9. Kun, G.: Constraints, MMSNP and expander structures. manuscript (2006)
10. Schaefer, T.J.: The complexity of satisfiability problems. In: STOC. (1978)
11. J. Nešetřil, Hell, P.: On the complexity of H-coloring. J. Combin. Theory Ser. B **48** (1990)
12. Bulatov, A.: A dichotomy theorem for constraints on a three-element set. In: FOCS. (2002)
13. Bulatov, A., Jeavons, P., Krokhin, A.: Classifying the complexity of constraints using finite algebras. SIAM J. Comput. **34**(3) (2005)
14. Madelaine, F., Stewart, I.A.: Some problems not definable using structures homomorphisms. Ars Combinatoria **LXVII** (2003)
15. Bodirsky, M.: Constraint Satisfaction with Infinite Domains. PhD thesis, Humboldt-Universität zu Berlin (2004)
16. Bodirsky, M., Dalmau, V.: Datalog and constraint satisfaction with infinite templates. In: STACS. (2006) 646–659
17. Madelaine, F.: Constraint satisfaction problems and related logic. PhD thesis, University of Leicester, Department of Mathematics and Computer Science (2003)
18. Fraïssé, R.: Sur l'extension aux relations de quelques propriétés des ordres. Ann. Sci. Ecole Norm. Sup. (3) **71** (1954) 363–388
19. Courcelle, B.: The monadic second order logic of graphs VI: On several representations of graphs by relational structures. Discrete Applied Mathematics **54** (1994) 117–149
20. DeVos, M., Ding, G., Oporowski, B., Sanders, D.P., Reed, B., Seymour, P., Vertigan, D.: Excluding any graph as a minor allows a low tree-width 2-coloring. J. Combin. Theory Ser. B **91**(1) (2004) 25–41
21. Madelaine, F.: Online appendix of this paper. electronic form (2006) available from www.dur.ac.uk/f.r.madelaine.
22. Achlioptas, D.: The complexity of G-free colourability. Discrete Mathematics **165-166** (1997) Pages 21–30
23. Garey, M., Johnson, D.: Computers and intractability: a guide to NP-completeness. Freeman, San Francisco, California (1979)
24. Bodirsky, M., jaroslav Nešetřil: Constraint satisfaction with countable homogeneous templates. In: CSL. Volume 2803 of LNCS. (2003) 44–57
25. Cherlin, G., Shelah, S., Shi, N.: Universal graphs with forbidden subgraphs and algebraic closure. Adv. in Appl. Math. **22**(4) (1999) 454–491
26. Courcelle, B.: The monadic second-order logic of graphs. XIV. Uniformly sparse graphs and edge set quantifications. Theoret. Comput. Sci. **299**(1-3) (2003) 1–36
27. Nešetřil, J., patrice Ossona de Mendez: Grad and classes with bounded expansion iii. restricted dualities. Technical Report 2005-741, KAM-DIMATIA (2005)

On the Expressive Power of Graph Logic

Jerzy Marcinkowski

Institute of Computer Science,
University of Wrocław,
Przesmyckiego 20, 51151 Wrocław, Poland
jma@ii.uni.wroc.pl

Abstract. Graph Logic, a query language being a sublogic of Monadic Second Order Logic is studied in [CGG02]. In the paper [DGG04] the expressiveness power of Graph Logic is examined, and it is shown, for many MSO properties, how to express them in Graph Logic. But despite of the positive examples, it is conjectured there that Graph Logic is strictly less expressive than MSO Logic. Here we give a proof of this conjecture.

1 Introduction

1.1 Previous Work and Our Contribution

There are no good tools known for proving, for a given sublogic $\mathcal{L}$ of Monadic Second Order Logic, that $\mathcal{L}$ is strictly less expressive than the whole MSOL.

Take for example the prefix subclasses of MSOL. The case of the so called *monadic NP*, the class of properties expressible by formulas with the quantifier prefix of the form $\boldsymbol{\exists}^*(\forall\exists)^*$, is well understood[1], and several cute techniques are known, for showing that an MSO Logic property is not in this class. But this is no longer the case if more second order quantification is allowed. In [AFS98] a programme was proposed of studying *closed monadic NP*, the class of properties definable (on finite structures) by formulas with the prefix of the form $[\boldsymbol{\exists}^*(\forall\exists)^*]^*$. Some complicated MSO properties not expressible in $(\forall\exists)^*\boldsymbol{\exists}^*(\forall\exists)^*$ were constructed in [AFS98]. In [M99] we answered a question from [AFS98], showing that directed reachability is not expressible in $(\forall\exists)^*\boldsymbol{\exists}^*(\forall\exists)^*$. In [JM01] a set of simple tools was presented for constructing MSO properties not in the last class. Further simplification of the tools is presented in [KL04]. But, despite of many efforts, we are not even able to show that there is any monadic second order property of finite structures not being expressible by a formula with the quantifier prefix $\boldsymbol{\exists}^*(\forall\exists)^*\boldsymbol{\exists}^*(\forall\exists)^*$. The question, whether closed monadic NP is closed under complementation, appears to be well beyond our reach.

[1] We use the symbols $\boldsymbol{\exists}$ and $\boldsymbol{\forall}$ for monadic second order quantifiers and the symbols $\exists$ and $\forall$ for first order quantifiers.

Z. Ésik (Ed.): CSL 2006, LNCS 4207, pp. 486–500, 2006.

Another sublogic of MSOL, called Graph Logic, is studied in [CGG02] and in [DGG04]. The second order quantification is restricted here not in terms of the quantifier prefix, but by the way the formula following a quantifier can be written. Second order quantification in Graph Logic is always of the form: *there exists a set S of edges such that $\phi \wedge \psi$*, where ϕ can only concern edges[2] **in** S while ψ can only concern edges **not in** S. So, in a sense, each second order quantifier splits the (set of edges of the) structure into two parts, which are invisible from each other. This construction may appear strange, but it comes from some sort of application: it is explained in [CGG02] how, and why, they want to use Graph Logic as a query language. But does this restriction really change anything? In [DGG04] the expressive power of this logic is studied. It turns out that surprisingly many MSO properties are expressible in Graph Logic (although the GL formulas usually turn out to be much longer than the respective MSOL formulas), for example 2-colorability and 4-colorability are expressible (but 3-colorability is not known to be). All regular word languages are expressible (but regular tree languages are not known to be). There are also MSO properties on any level of Polynomial Hierarchy[3] which can be expressed in Graph Logic. Another thing which is easy to show, is that all the properties concerning connectivity, usually used to separate classes inside MSOL, are definable in GL. Despite the positive results, the conjecture which was left open in [DGG04], is that Graph Logic is strictly less expressive than Monadic Second Order Logic. In this paper we give a proof of this conjecture.

There is however one more subtle point. For reasons coming from the intended applications, the Graph Logic, as defined in [CGG02] uses an additional construct: quantification over labels of edges. This is not what we usually have in MSOL. In [DGG04] the authors want to treat GL as a sublogic of MSOL, so they enrich Monadic Second Order Logic with the construct. In our paper we call this richer logic MSO+. What is conjectured in [DGG04], and proved in Section 3 of this paper, is exactly that MSO+ is strictly more expressive than GL. Our believe is that already this proof is not quite trivial, but we admit that the really interesting problem, and probably a more difficult one, would be to separate Graph Logic without edge labeling (we call this logic GL–) from the standard MSO Logic. In Section 4 we show why this cannot be done by a natural modification of the technique from Section 3. A careful reader will notice that the methods of Section 4 can also prove that MSO is exponentially more succinct than GL–.

---

[2] The signature here consists of a single graph relation, and the monadic quantification is Courcelle style (see [C90]): we quantify over sets of edges rather than over sets of vertices.

[3] The second order quantification in GL, as we described it, is only existential, but we can go beyond NP using negation on top of it, and nesting the quantifiers: the logic is closed under all boolean connectives, under first order quantification and under the restricted form of second order quantification.

1.2 Informal Introduction to the Technical Part

As we said above, we are not even able to show that there is any Monadic Second Order property of finite structures not expressible by a formula with the quantifier prefix $\exists^*(\forall\exists)^* \exists^*(\forall\exists)^*$.

The reason of this humiliating state of affairs is that (almost) the only technique we know to prove non-expressibility of some property $\mathcal{P}$ is using Ehrenfeucht–Fraïssé games. To use this technique, two structures A, B must be constructed, with A having property $\mathcal{P}$ and B not having this property, and a strategy for a player called Duplicator must be shown, allowing him to pretend that A and B are isomorphic. The game consists of *colouring rounds*, one round for each block of monadic second order quantifiers, and *first order rounds*, one round for each first order quantifier. We understand quite well what first order types are, and in what circumstances Duplicator will have a winning strategy in a game consisting of first order rounds. But the same cannot be said about second order types and colouring rounds. There are two ideas known that can help the Duplicator's case. One is called Ajtai-Fagin game. In such a game Duplicator commits on the structure B only after the Spoiler's colouring of A is known. This works well, but only when the game begins with a colouring round, and there are no more colouring rounds, which means that this trick is exactly tailored for monadic NP. The second idea (from [M99]) is to take A and B to be structures with many symmetries, so that Duplicator does not need to commit, until the colouring round, which objects in B counterpart which objects in A. Some of the symmetries are lost during the initial first order rounds, but still with this trick we can afford first order rounds before a colouring round. But again, the technique works only for games with a single colouring round. After Spoiler colours the structure there is no way to preserve any symmetries.

Imagine, for example, structures which are unions of no more than 2^m disjoint strings of length m, for some natural m. There are not many MSO properties that can be defined on such structures, but Duplicator's life is hard anyway: consider a colouring round here. Before such a round is played, a structure has plenty of automorphisms. But then Spoiler can colour each string in different way, making the structure rigid (notice that if a first order round was played instead, just one of the strings would get fixed, and all the automorphisms which do not move this string would survive).

Actually, our proof in Section 3 is more or less about the above example. We show (in Lemma 2 and Lemma 4) that if the colouring is understood in the Graph Logic sense, as splitting the structure into two substructures which are invisible from each other, then whatever Spoiler will do, each of the resulting two substructures will have many symmetries. This observation is then iterated (Lemma 5), and that gives a winning strategy for Duplicator in a monadic game with arbitrarily many colouring rounds.

2 Preliminaries

By *a structure* we will understand, until the end of Section 3, a triple $\mathcal{G} = \langle V, E, \mathcal{L} \rangle$, where V is a set of vertices, $E \subseteq V \times V$ is an edge relation, and $\mathcal{L}$ is a family of subsets of $V \times V$, called *edge labels*.[4]

Let $\mathcal{G} = \langle V, E, \mathcal{L} \rangle$ be a structure. We will use the notation $\mathcal{G} = \mathcal{G}_1 | \mathcal{G}_2$ to say that:

1. $\mathcal{G}_1 = \langle V, E_1, \mathcal{L} \rangle$, $\mathcal{G}_2 = \langle V, E_2, \mathcal{L} \rangle$,
2. $E_1 \cap E_2 = \emptyset$
3. $E_1 \cup E_2 = E$

The notation $E = E_1 | E_2$ will be used instead of $\mathcal{G} = \mathcal{G}_1 | \mathcal{G}_2$ if V and $\mathcal{L}$ are clear from the context.

2.1 The Logics Under Consideration

Consider a language with 3 sorts of variables: standard first order variables (to denote them we will use letters x, y, z and their variants), first order labeling variables (denoted by $l_1, l_2 \ldots$) and monadic second order variables ($P, Q \ldots$).

All the logics we are going to consider are defined by the following constructs.

1. First order logic constructs. If x, y are first order variables then $E(x, y), x = y$ are atomic formulas, and x, y are free in them. If ϕ, ψ are formulas then $\phi \wedge \psi$, $\phi \vee \psi$, $\neg \psi$, $\exists x\ \phi$ and $\forall x\ \phi$ are formulas, with the usual convention concerning free variables. Semantics of the constructions is standard.
2. Edge labeling. If x, y are first order variables and l is a first order labeling variable then $l(x, y)$ is a formula, with l, x, y being free variables. If l, x, y are interpreted in a structure $\mathcal{G}$ as $\bar{x}, \bar{y} \in V$, $\bar{l} \in \mathcal{L}$ then $\mathcal{G} \models l(x, y)$ if $[\bar{x}, \bar{y}] \in E \cap \bar{l}$, that is if "the edge $E(\bar{x}, \bar{y})$ exists and is labeled by $\bar{l}$".
   If l is free in ϕ then $\exists l\ \phi$ is a formula, which is true if there exists a label in $\mathcal{L}$ making ϕ true.
3. Graph logic quantification. If ϕ, ψ are formulas, then $\phi | \psi$ is a formula, whose free variables are the variables which are free in ϕ or in ψ. $\mathcal{G} \models \phi | \psi$ if there exist $\mathcal{G}_1$ and $\mathcal{G}_2$ such that $\mathcal{G} = \mathcal{G}_1 | \mathcal{G}_2$, that $\mathcal{G}_1 \models \phi$ and that $\mathcal{G}_2 \models \psi$.
4. Monadic second order quantification. If x, y are first order variables and P is a monadic second order variable then $P(x, y)$ is a formula, with P, x, y being

[4] We found it convenient to understand Graph Logic labels as sets of edges. But this implies that our "family of subsets" is itself a multiset: two different labels can be equal as sets. (The names for) labels, as we define them, are not (as they are in [CGG02]) elements of the signature. If they were, then Definition 1 would be even more complicated: since individual names for labels would be available for Spoiler, he would be able to force Duplicator, in some circumstances, to respond in his move in a labeling round, with exactly the same label as Spoiler used for his move. That would however not affect our non-expressibility proofs: our Duplicator always responds with the Spoiler's label in the labeling rounds anyway, keeping all his tricks for the colouring rounds.

free variables. If ϕ is a formula, and P is free in ϕ then $\exists P\ \phi$ and $\forall P\ \phi$ are formulas. The semantics is standard, but remember that what we consider is monadic quantification over sets of edges. This means that we are only allowed to quantify over *existing* edges, so $P(x,y)$ can only be true is $E(x,y)$ is true.

By Graph Logic (GL) we will mean the logic defined by the constructs from items (i)-(iii). By Monadic Second Order Logic (MSO) we will mean the logic defined in items (i) and (iv). Notice that to make MSO and GL comparable, a version of MSO where quantifiers range over sets of edges rather than sets of vertices is used here. By Monadic Second Order Logic with Labeling (MSO+) we will mean the logic defined in items (i),(ii) and (iv). By Weak Graph Logic (GL–) we will mean the logic defined by the constructs from items (i) and (iii).

Example. Let ρ be the first order formula: $\forall s,w,z\ \ \neg(E(s,w) \wedge E(w,z))$. Then the GL– formula: $\exists x \neq y\ \ [(x$ and y are isolated $\wedge \rho)|\rho]$ defines, over the set of all strings, the class of strings with odd number of edges.

The question left open by [DGG04], and answered in this paper, can be now stated as: *Is* MSO+ *strictly more expressive than* GL?

2.2 Ehrenfeucht – Fraïssé Games for Graph Logic

In this subsection we define the Ehrenfeucht–Fraïssé style game for Graph Logic, and state a lemma asserting the equivalence between logic and games. There is nothing counterintuitive in the lemma for anyone having some experience with games for Monadic Second Order Logic (see for example [EF95], and thus we decided to skip its proof.

Game for graph logic is played by two players, Spoiler and Duplicator and takes r rounds.

Each round is played according to a protocol, whose input and output are of the form $\langle \mathcal{G}; c_1, c_2, \ldots c_p; l_1, l_2, \ldots l_q\rangle$, $\langle \mathcal{G}'; c'_1, c'_2, \ldots c'_p; l'_1, l'_2, \ldots l'_q\rangle$, for some $p, q \geq 0$, where $\mathcal{G} = \langle V, E, \mathcal{L}\rangle$, and $\mathcal{G}' = \langle V', E', \mathcal{L}'\rangle$ are two structures, where $c_1, c_2, \ldots c_l \in V$ and $c'_1, c'_2, \ldots c'_l \in V'$ and where $l_1, l_2, \ldots l_q \in \mathcal{L}$ and $l'_1, l'_2, \ldots l'_q \in \mathcal{L}'$.

The following definition gives the protocol of a single round of the game for graph logic GL. Nothing, possibly except of the description of the colouring round, is going to surprise the reader here:

Definition 1. *Each round is of one of four kinds:* negation round *or* first order round, *or* first order labeling round, *or* colouring round. *At the beginning of each round Spoiler decides which kind of round he wishes this one to be.*

If the round is decided to be a negation round, then Spoiler gets a pair $\langle \mathcal{G}; c_1, c_2, \ldots c_p; l_1, l_2, \ldots l_q\rangle$, $\langle \mathcal{G}'; c'_1, c'_2, \ldots c'_p; l'_1, l'_2, \ldots l'_q\rangle$, *and the output is the pair* $\langle \mathcal{G}'; c'_1, c'_2, \ldots c'_p; l'_1, l'_2, \ldots l'_q\rangle$, $\langle \mathcal{G}; c_1, c_2, \ldots c_p; l_1, l_2, \ldots l_q\rangle$. *Duplicator does not move in a negation round.*

If the round is decided to be a first order round, then Spoiler gets a pair $\langle \mathcal{G}; c_1, c_2, \ldots c_p; l_1, l_2, \ldots l_q \rangle$, $\langle \mathcal{G}'; c'_1, c'_2, \ldots c'_p; l'_1, l'_2, \ldots l'_q \rangle$ *as the input, and picks one* $c \in V$. *Duplicator answers Spoiler's move by picking one* $c' \in V'$, *and the output is the pair:* $\langle \mathcal{G}; c, c_1, c_2, \ldots c_p; l_1, l_2, \ldots l_q \rangle$, $\langle \mathcal{G}'; c', c'_1, c'_2, \ldots c'_p; l'_1, l'_2, \ldots l'_q \rangle$.

If the round is a first order labeling round, then Spoiler gets, as the input, a pair $\langle \mathcal{G}; c_1, c_2, \ldots c_p; l_1, l_2, \ldots l_q \rangle$, $\langle \mathcal{G}'; c'_1, c'_2, \ldots c'_p; l'_1, l'_2, \ldots l'_q \rangle$, *and picks one* $l \in \mathcal{L}$. *Duplicator answers this move by picking one* $l' \in \mathcal{L}'$. *The output of the round is then the pair:* $\langle \mathcal{G}; c_1, c_2, \ldots c_p; l, l_1, l_2, \ldots l_q \rangle$, $\langle \mathcal{G}'; c'_1, c'_2, \ldots c'_p; l', l'_1, l'_2, \ldots l'_q \rangle$.

Things are more complicated if the round is a colouring round. Spoiler gets a pair $\langle \mathcal{G}; c_1, c_2, \ldots c_p; l_1, l_2, \ldots l_q \rangle$, $\langle \mathcal{G}'; c'_1, c'_2, \ldots c'_p; l'_1, l'_2, \ldots l'_q \rangle$. *He chooses a pair of structures* $\mathcal{G}_1, \mathcal{G}_2$, *such that* $\mathcal{G} = \mathcal{G}_1 | \mathcal{G}_2$. *Duplicator answers with a pair of structures* $\mathcal{G}'_1, \mathcal{G}'_2$, *such that* $\mathcal{G}' = \mathcal{G}'_1 | \mathcal{G}'_2$. *Then Spoiler decides if he wants the output of the round to be the pair* $\langle \mathcal{G}_1; c_1, c_2, \ldots c_p; l_1, l_2, \ldots l_q \rangle$, $\langle \mathcal{G}'_1; c'_1, c'_2, \ldots c'_p; l'_1, l'_2, \ldots l'_q \rangle$, *or if he prefers it to be* $\langle \mathcal{G}_2; c_1, c_2, \ldots c_p; l_1, l_2, \ldots l_q \rangle$, $\langle \mathcal{G}'_2; c'_1, c'_2, \ldots c'_p; l'_1, l'_2, \ldots l'_q \rangle$.

Definition 2. *A pair* $\langle \mathcal{G}; c_1, c_2, \ldots c_p; l_1, l_2, \ldots l_q \rangle$, $\langle \mathcal{G}'; c'_1, c'_2, \ldots c'_p; l'_1, l'_2, \ldots l'_q \rangle$, *with* $\mathcal{G} = \langle V, E, \mathcal{L} \rangle$, $\mathcal{G}' = \langle V', E', \mathcal{L}' \rangle$, *is called a* winning position for Duplicator *in the game for GL if:*

- *for each* $1 \leq i, j \leq p$ *it holds that* $E(c_i, c_j)$ *if and only if* $E'(c'_i, c'_j)$;
- *for each* $1 \leq i, j \leq p$ *such that* $E(c_i, c_j)$, *and for each* $1 \leq k \leq q$ *it holds that* $l_k(c_i, c_j)$ *if and only if* $l'_k(c'_i, c'_j)$.

Once a protocol for a single round is defined, and the winning condition is stated, the notion of a winning strategy for a player in an r-round game is clear. Game for GL– is analogous to game for GL, with the only difference that Spoiler is not allowed to use first order labeling rounds in the game for GL–.

Lemma 1. *The property* $\mathcal{W}$ *of structures is non-expressible in graph logic GL (in graph logic GL–) if and only if, for each natural number* r, *there exist two structures* $\mathcal{G}_1, \mathcal{G}_2$, *such that* $\mathcal{G}_1 \in \mathcal{W}$, $\mathcal{G}_2 \notin \mathcal{W}$ *and Duplicator has a winning strategy in the* r*-round game for GL (or, respectively, in the* r*-round game for GL–) on* $\langle \mathcal{G}_1; ; \rangle$ *and* $\langle \mathcal{G}_2; ; \rangle$.

We will sometimes write $\mathcal{G}$ instead of $\langle \mathcal{G}; ; \rangle$.

3 MSO+ vs. GL

We will use a sort of algebraic notation to talk about structures. Fix a natural number m. Let $M = \{1, 2, \ldots m\}$ and let $l_1, l_2 \ldots l_m$ be a fixed set of m names for labels.

We will identify a set $U \subseteq M$ with a structure[5] whose vertices are $u_0, u_1, \ldots u_m$, whose edges are $E(u_{i-1}, u_i)$ for each $i \in U$ and whose label l_i consists of the single pair $\langle u_{i-1}, u_i \rangle$.

[5] The structure is defined up to isomorphism. We usually identify isomorphic structures.

Definition 3. *1. Let $U \subseteq M$. Then (the structure identified with) U is an* m-fibre.

2. *Each* m-fibre *is an* m-fibre structure
3. *Suppose S_1, S_2 are m-fibre structures, with $S_1 = \langle V_1, E_1, \{l_1^1, l_2^1, \dots l_m^1\}\rangle$ and $S_2 = \langle V_2, E_2, \{l_1^2, l_2^2, \dots l_m^2\}\rangle$, such that $V_1 \cap V_2 = \emptyset$. Then $S = \langle V_1 \cup V_2, E_1 \cup E_2, \{l_1^1 \cup l_1^2, l_2^1 \cup l_2^2, \dots l_m^1 \cup l_m^2\}\rangle$ is an m-fibre structure denoted by $S_1 +_m S_2$. So $S_1 +_m S_2$ is the structure, whose set of vertices, set of edges, and each label are disjoint unions of the sets of vertices, sets of edges, and the respective labels of S_1 and S_2.*
4. *If $U_1, U_2, \dots U_k$ are isomorphic m-fibre structures, then we will write $k *_m U_1$ instead of $U_1 +_m U_2 +_m \dots +_m U_k$.*
5. *A structure is called a* fibre structure *if it is an m-fibre structure for some m.*

Even if the names of labels $l_1, l_2, \dots l_m$ are not part of the language, having them in the structure we can quantify over them, and express (by the formula $\exists l\ l(a,b) \wedge l(c,d) \wedge a \neq c$) the property that edges $E(a,b)$ and $E(c,d)$ are "in the same place" in two fibres. The proof of Claim (i) of Theorem 1 will rely on that.[6]

The following theorem states the separation result conjectured in [DGG04]:

Theorem 1. *Let $\mathcal{W}$ be the set of fibre structures which are of the form $s *_m M$, for some m and some $s \leq 2^{2m}$. Then:*

1. *$\mathcal{W}$ is definable in MSO+, over the set of all fibre structures which are of the form $s *_m M$, for some m and some s.*
2. *$\mathcal{W}$ is not definable in GL over the set of all fibre structures which are of the form $s *_m M$, for some m and some s.*

The property of being a fibre structure is definable in MSO+. It is also easy to see that the property of being of the form $s *_m M$, for some m and some s, is first order definable over the set of all fibre structures. So, if we wanted, we could state the first claim of the above theorem simply as "$\mathcal{W}$ is definable in MSO+". We however do not need it for the separation result: to prove separation result for two logics it is enough to prove that they differ on some fixed class of structures.

Proof. Claim (i) of the theorem is very easy to show.

There are 2^m ways of colouring M with a single predicate, and 2^{2m} ways if we have two predicates. So in order to say, that there are no more than 2^{2m} copies of M it is enough to write an MSO+ formula, with two existentially quantified monadic second order variables P, P' saying that each copy of M in the structure (a copy is represented by some edge from u to t) is coloured in a different way by P and P' (and the edges labeled by l_d are where the difference can be spotted).

[6] Mathematics is about using definitions. But the way we exploit the definition of GL here is a win on technicalities, so shameless that the author briefly considered applying for a job in one of the Law Departments.

One more predicate Q is needed to express the fact that two vertices are in the same copy of M:

$\mathbf{\exists} P, P' \ \forall u \neq u' \ \forall t, t', l \ \exists x, y, x', y', l_d \ \ [l(u,t) \wedge l(u',t')] \Rightarrow$
$[\ \ l_d(x,y) \wedge l_d(x',y')$
$\wedge \ \neg([P(x,y) \Leftrightarrow P(x',y')] \wedge [P'(x,y) \Leftrightarrow P'(x',y')])$
$\wedge$ $\mathbf{\exists} Q$ Q is a path from u to y or Q is a path from x to t
$\wedge$ $\mathbf{\exists} Q$ Q is a path from u' to y' or Q is a path from x' to t']

Where "Q is a path from u to y" is the formula:

$(u = y \vee \exists x \ Q(x,y))$
$\wedge \ \forall x_1 \ \ \forall x_2 \neq y \ \ \exists x_3 \ \ Q(x_1,x_2) \Rightarrow Q(x_2,x_3)$
$\wedge \ \forall x_1 \neq u \ \ \forall x_2 \ \ \exists x_0 \ \ Q(x_1,x_2) \Rightarrow Q(x_0,x_1)$

The rest of Section 3 will be devoted to the proof of Claim (ii).

3.1 Crosswords Lemma

A (k,n)-*crossword* is a matrix with k columns and n rows, and with each element being either the constant *white* or the constant *black*.

For two (k,n)-crosswords C and C' we say that they are *one step equivalent* if one of the following two conditions holds:

Swapping. C' is a result of swapping two rows in C.

Crossing-over. (see Fig. 1) There exist $1 \leq i < k, 1 \leq j < n$ such that:

1. $C(i,j) = C(i,j+1) \neq C(i+1,j) = C(i+1,j+1)$,
2. if $j \neq p \neq j+1$ then $C(q,p) = C'(q,p)$, for each $1 \leq q \leq k$,
3. if $q \leq i$ then $C(q,p) = C'(q,p)$, for each $1 \leq p \leq n$
4. if $q \geq i+1$ then $C(q,j) = C'(q,j+1)$ and $C(q,j+1) = C'(q,j)$

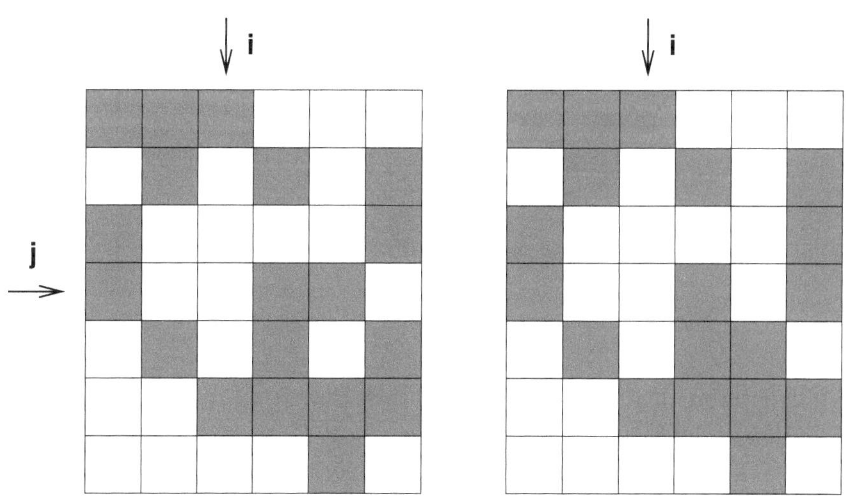

Fig. 1. One step equivalence by crossing-over

Let $\mathcal{R}$ be the smallest equivalence relation containing one step equivalence. The following lemma is going be our main tool:

Lemma 2. *For each (k, n)-crossword C, with $k \leq 2^s$ for some natural number s, there exists a C', such that $\mathcal{R}(C, C')$ and that the first $\frac{n}{2k^s}$ rows of C' are equal to each other.*

Proof. By induction on s. If $s = 0$ then $k = 1$ and there obviously are at least $n/2$ equal rows in C.

Now the induction step. Without loss of generality we can assume that $C(k/2, p)$ is *white* for each $p \leq n/2$. Now, for each $p \leq n/2$ let $left(p)$ be the greatest $q \leq k/2$ such that $C(q, p)$ is *black*, and let $right(p)$ be the smallest $q \geq k/2$ such that $C(q, p)$ is *black*. Let $A(l, r) = \{p \leq n/2 : left(p) = l, right(p) = r\}$. Obviously, since there are only $k/2$ possible choices for l and $k/2$ possible choices for r, there must exist l and r, such that $A(l, r)$ has some $a \geq \frac{2n}{k^2}$ elements. Let C_A be the result of swapping, to the top of C, the rows of C with numbers from $A(l, r)$. Consider the $(l - 1, a)$-crossword C_L defined as $C_L(i, j) = C_A(i, j)$, and the $(k - r, a)$-crossword C_R defined as $C_R(i, j) = C_A(i + r, j)$. One can imagine C_L and C_R to be – respectively – in the top left and top right corner of C_A (see Fig. 2).

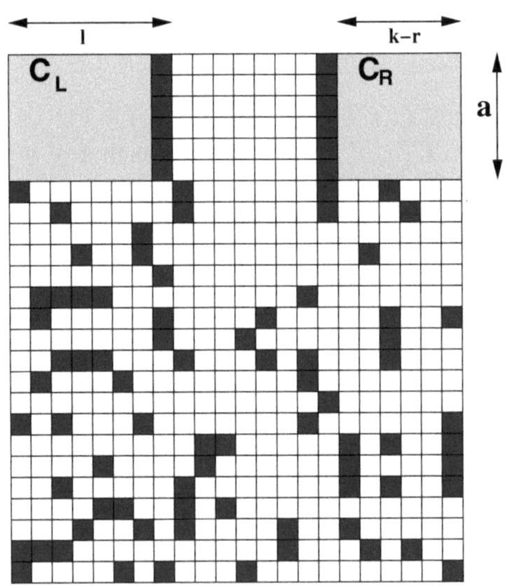

Fig. 2. The crossword C_A from the proof of Lemma 2

By hypothesis there exist C'_L and C'_R such that $\mathcal{R}(C_L, C'_L)$, $\mathcal{R}(C_R, C'_R)$ and that both C'_L and C'_R have all the first $\frac{a}{2(k/2)^{s-1}} \geq \frac{n}{2k^s}$ rows equal. Let $C'(i, j)$ be $C_A(i, j)$ if $j > a$ or if $l \leq i \leq r$. For $j \leq a$ and $i < l$ define $C'(i, j)$ as $C'_L(i, j)$, and finally for $j \leq a$ and $i > r$ define $C'(i, j)$ as $C'_R(i + r, j)$. So C' is C_A with C_L substituted by C'_L and with C_R substituted by C'_R. The top $\frac{a}{2(k/2)^{s-1}}$ rows of C' are equal. To finish the proof notice that l and $r-1$ where both defined in such a way that they can serve as the i from the crossing-over condition. From this we can get that $\mathcal{R}(C_A, C')$ holds, and in consequence also that $\mathcal{R}(C, C')$. ∎

3.2 One Colouring Round

Let r be the number of rounds, as in Section 2.2. Take m such that $\frac{m}{\log^2 m} \geq r$, which implies that $2^{2m} \geq 2(2m^{\log m})^r$. Define $S = 2^{2m} *_m M$ and $S' = (2^{2m} + 1) *_m M$. Let E be the set of edges of S and E' be the set of edges of S'. Of course $S \in \mathcal{W}$ and $S' \notin \mathcal{W}$.

What remains to be proved is that Duplicator has a winning strategy in the r-round game on $\langle S; ; \rangle$ and $\langle S'; ; \rangle$.

To illustrate the idea of the strategy we consider a colouring round, played as the first round of the game, on structures $\langle S; ; \rangle$ and $\langle S'; ; \rangle$. Spoiler commits on some subset $B \subseteq E$. Notice that there is a natural bijection *cross* between subsets of E and $(m, 2^{2m})$-crosswords: for $B \subseteq E$ the crosswords $cross(B)$ is such that $cross(B)(i,j)$ is *black* if and only if the edge from u_{i-1} to u_i, in the j'th copy of M in S, is in B. For $B \subseteq E$ let $\bar{B} \subseteq E$ be such that $E = B|\bar{B}$ (so that $cross(\bar{B})$ is a "negative" of $cross(B)$).

To keep the notation light we will identify each subset of E (or of E') with an m-fibre structure, having the same vertices as S has (or as S' has).

Lemma 3. *If $\mathcal{R}(cross(B), cross(B_0))$ then the m-fibre structures B and B_0 are isomorphic. Also the m-fibre structures $\bar{B}$ and $\bar{B}_0$ are isomorphic then.*

Proof. By induction it is enough to prove that the lemma holds for one step equivalence. Since $+_m$ is commutative, there is nothing to prove if $cross(B)$ and $cross(B_0)$ are one step equivalent by swapping. If they are equivalent by crossing-over, then notice that we can make use of the following observation: if $A_1, A_2 \subseteq \{1, 2, \ldots l-1\}$, $A_3, A_4 \subseteq \{l+1, l+2 \ldots m\}$ for some natural number l, then $(A_1 \cup A_3) +_m (A_2 \cup A_4)$ is isomorphic to $(A_1 \cup A_4) +_m (A_2 \cup A_3)$. ∎

From the last lemma and from Lemma 2 we get:

Lemma 4. *For each $B \subseteq E$ there exists a set $B_0 \subseteq E$ such that: the m-fibre structures B and B_0 are isomorphic, the m-fibre structures $\bar{B}$ and $\bar{B}_0$ are isomorphic, and that $B_0 = (l *_m A) +_m D$ for some m-fibre A, some m-fibre structure D and some $l \geq \frac{2^{2m}}{2m^{\log m}}$*

Notice that, if B_0 is as in the lemma, then $\bar{B_0} = (l *_m (M \setminus A) +_m F$, for some F.

The strategy of Duplicator is now straightforward. Remember that he wants to hide the fact that, compared to S, there is an extra copy of a fibre in S'. Given the Spoiler's choice of $B, \bar{B}$, Duplicator takes B_0 as in last lemma and answers with a pair $B', \bar{B}'$ such that $B'|\bar{B}' = E'$ and $B' = ((l+1) *_m A) +_m D$. Now Spoiler decides whether he wants to continue the game on B, B' or if he likes the pair $\bar{B}, \bar{B}'$ more. But, by last lemma this means playing either on B_0, B'

or $\bar{B}_0, \bar{B}'$ respectively. So whatever choice Spoiler makes, the input to the next round is a pair of the form $\langle (l *_m A) +_m D; \; ; \; \rangle$, $\langle ((l+1) *_m A) +_m D; \; ; \; \rangle$. This is good news for Duplicator: there is still huge crowd of isomorphic fibres in which the additional copy of A can melt.

3.3 Beyond the First Round

Now we extend the idea from Section 3.2 beyond the first round.
Let $\langle \mathcal{G}; c_1, c_2, \dots c_t; l_1, l_2, \dots l_q \rangle$, $\langle \mathcal{G}'; c'_1, c'_2, \dots c'_{t'}; l'_1, l'_2, \dots l'_{q'} \rangle$ be two m-fibre structures with some distinguished vertices and labels (like the considered in Section 2.2). By $\langle \mathcal{G}; c_1, c_2, \dots c_t; l_1, l_2, \dots l_q \rangle +_m \langle \mathcal{G}'; c'_1, c'_2, \dots c'_{t'}; l'_1, l'_2, \dots l'_{q'} \rangle$ we will denote the structure: $\langle \mathcal{G} +_m \mathcal{G}'; c_1, c_2, \dots c_t, c'_1, c'_2, \dots c'_{t'}; l_1, l_2, \dots l_q, l'_1, l'_2, \dots l'_{q'} \rangle$.
This means that the distinguished vertices and labels are inherited from both the structures joined by the operation $+_m$.

Lemma 5. *Duplicator has a strategy in which, after round p, the position of the game is:*
$(k *_m A) +_m \langle D, c_1, c_2, \dots c_t; l_1, l_2, \dots l_q \rangle$, $(k' *_m A) +_m \langle D, c_1, c_2, \dots c_t; l_1, l_2, \dots l_q \rangle$
for some m-fibre A, some m-fibre structure D with distinguished vertices and labels, and some $k \geq \frac{2^{2m}}{2^p m^{p \log m}}$. *Moreover,* $k' = k + 1$ *or* $k' = k - 1$.

Notice that, according to the lemma, the position after round p is winning for Duplicator. So, once the lemma is proved, the proof of Claim (ii) of Theorem 1 will be finished.

Proof. By induction of p. Suppose the claim is true after p rounds, and the position from the lemma is the input to round $p+1$. We need to show how Duplicator can make sure that the output of the round will be of the form $(j *_m A') +_m \langle D'; c_1, c_2, \dots c_{t'}; l_1, l_2, \dots l_{q'} \rangle$, $(j' *_m A') +_m \langle D'; c_1, c_2, \dots c_{t'}; l_1, l_2, \dots l_{q'} \rangle$ for some m-fibre A', some m-fibre structure D', some $j \geq \frac{2^{2m}}{2^{p+1} m^{(p+1) \log m}}$, and $j' = j + 1$ or $j' = j - 1$.

Negation round. Nothing happens here, just k and k' are swapped.

First order labeling round. Again nothing interesting happens: Spoiler picks a label in the first structure and Duplicator answers with the same label in the second structure, so that $q' = q + 1$ and nothing changes except of it.

First order round. If Spoiler picks a vertex from D then Duplicator answers with a corresponding vertex in the isomorphic copy of D in his structure. If Spoiler picks a vertex c in one of the copies of A, then Duplicator picks a corresponding vertex c in one of the copies of A in his structure and the output of round $p+1$ is: $((k-1) *_m A) +_m \langle A; c; \; \rangle +_m \langle D; c_1, c_2, \dots c_t; l_1, l_2, \dots l_q \rangle$, $((k'-1) *_m A) +_m \langle A; c; \; \rangle +_m \langle D; c_1, c_2, \dots c_t; l_1, l_2, \dots l_q \rangle$, which clearly is of the form we wanted.

Colouring round. Let $B, \bar{B}$ be the Spoiler's choice, for some $B, \bar{B}$ such that $B|\bar{B} = (k *_m A) +_m D$. Decompose B as $B_A +_m B_D$, where $B_A \subseteq (k *_m A)$ and

$B_D \subseteq D$. Since D is the same in both structures, Duplicator feels no need of cheating there, and so his answer will be $B'_A +_m B_D$, for some $B'_A \subseteq k' *_m A$.

Let $A_1, A_2 \dots A_d$ be all the connected components of A (which means here that if $i \neq j$ and $x \in A_i$ for some number $x \in M$, then none of $x-1, x, x+1$ can be in A_j, and if $x \leq y \leq z$ and $x, z \in A_i$ for some i then also $y \in A_i$). Let $a_i = |A_i| - 1$ be the number of edges in A_i. We will repeat the construction from Section 3.2 for each A_i.

$B_A = B^1 \cup B^2 \cup \dots \cup B^d$ for some sets such that $B^i \subseteq k *_m A_i$ (for $1 \leq i \leq d$). Like in Section 3.2 we can view each $B^i \subseteq k *_m A_i$ as a (a_i,k)-crossword, and find B^i_0, as in Lemma 4, with $l_i = \frac{k}{2a_i^{\log a_i}} \geq \frac{k}{2m^{\log m}}$ equal fibres. Take $j = \frac{2^{2m}}{2^{p+1} m^{(p+1)\log m}}$. Then, by hypothesis, $l_i \geq j$, for each i. So $B^i_0 = (j *_m F_i) + H_i$ for some fibre F_i and some fibre structure F_i. Since the sets A_i,for $1 \leq i \leq d$ are pairwise not connected, the observation from the proof of Lemma 3 applies, and B_A is isomorphic to $B^0_A = (j *_m F) +_m H$ where $F = F_1 \cup F_2 \cup \dots \cup F_d$ and $H = H_1 \cup H_2 \cup \dots \cup H_d$. Also $(k *_m A) \setminus B_A$ is isomorphic to $(k *_m A) \setminus B^0_A$. Now, $B'_A, \bar{B'_A}$, where $B'_A = (j + k' - k) *_m F +_m H$ and $B'_A | \bar{B'_A} = (k' *_m A) +_m D$ is the answer which Duplicator was looking for. ■

4 MSO vs. GL–. Why a Counting Argument Fails

Our intuition behind the proof in Section 3 was that it exploits Graph Logic's poor ability to transmit information.

But maybe things are simpler than that, and what we really use in the proof is just a counting argument?

Suppose we have n isomorphic copies of some structure of size m, and the structures are such that we can address their edges in some monadic second order way (as we could do, using the labels and the relation S, in the case of the fibre structures of Section 3). For each fixed natural number k we can then express, in MSO, the property that $n = 2^{km}$. On the other hand, as one could see in the proof of claim (ii) of Lemma 1, using GL we were not able to count to any number higher than superpolynomial. Why was that? One could think that there is a simple counting reason behind it: while in the game for MSO, after r colouring rounds (in each round the players commit on one predicate), the structure of size m can be coloured in 2^{rm} ways, in the game for GL (or GL–) after r rounds we get one of just 2^m possible structures, regardless of r. This is since the edges do not have colours in the game for GL. They simply, at each stage of the game, either exist or do not exist.

If the above explanation were true, we could simplify the construction from Section 3 getting rid of labels and using the same edges as the arena of counting and as the mean of addressing. Such a simple construction is studied in this Section.

All the structures considered in this Section will be bipartite graphs. Moreover, they will be all *directed* in the sense that each vertex will be either isolated, or an out-vertex (with in-degree 0 and non-zero out-degree), or will be an in-vertex (its out-degree will be 0, and in-degree will be non-zero). The out-vertices will, in some sense, play the role of fibres from Section 3, and the in-vertices will play the role of labels. A structure will be called *full* if for each its out-vertex x and each in-vertex y there is an edge from x to y. Notice, that a full structure can still have some isolated vertices.

By m, n-structure we will mean a full structure with m out-vertices and n in-vertices

For a function $f : \mathcal{N} \to \mathcal{N}$ define $\mathcal{W}_{f(n)}$ to be the set of all m, n-structures with $m \leq f(n)$.

Of course we can use the method from the proof of Claim (i) of Theorem 1 to show that for each constant c the property $\mathcal{W}_{2^{cn}}$ is definable in MSO.

But it turns out that the analogue of Claim (ii) of Theorem 1 is no longer true in this context, or at least it cannot be proved by an analogous argument. This is since GL– can now count more than 2^{cn} out-vertices for any fixed c. The counting power of GL– equals the counting power of MSO.

Definition 4. *Let* $g_1(n) = 2^n$ *and let* $g_{r+1}(n) = \sum_{k=0}^{n} g_r(k)\binom{n}{k}$.

Lemma 6. *(i)* $r^n \leq g_r(n)$
(ii) $g_r(n) \leq 2^n r^n$

So, in other words, $2^{n \log r} \leq g_r(n) \leq 2^{n(1+\log r)}$.

Proof. Both claims are proved by induction on r. The case $r = 1$ is obvious.

(i) The induction step follows from the fact that $\sum_{k=0}^{n} r^k \binom{n}{k} = (r+1)^n$. To see that the equality holds true, notice that its both sides represent the number of ways in which a set of n elements can be split into $r+1$ subsets (on the left hand side we begin from selecting the first set , with $n-k$ elements, and then split the remaining elements into r sets).
(ii) $g_{r+1}(n) = \sum_{k=0}^{n} g_r(k)\binom{n}{k} \leq \sum_{k=0}^{n} 2^k r^k \binom{n}{k} \leq 2^n \sum_{k=0}^{n} r^k \binom{n}{k} = 2^n (r+1)^n$ ■

Theorem 2. *For each fixed r, the property* $\mathcal{W}_{g_r(n-1)}$ *is expressible in GL–.*

We will write a GL– formula for $\mathcal{W}_{g_r(n-1)}$, but need some taxonomy first:

Definition 5. *Let* $\mathcal{G} = \langle V, E \rangle$ *be a structure, and let c be a constant, interpreted as an in-vertex* $c \in V$.

1. *An out-vertex* $x \in V$ *is called* proper *if* $E(x, c)$ *holds.* $\mathcal{G}$ *is* proper *if all its out-vertices are proper.*
2. *For an out-vertex* $x \in V$ *by* $Im(x)$ *we mean the set* $\{y \neq c : E(x, y)\}$
3. *Let* $x, y \in V$ *be two out-vertices. We will say that x and y are* $\mathcal{G}$*-equivalent if* $Im(x) = Im(y)$*. By* $Eq(x)$ *we denote the equivalence class of x.*
4. *For a function* $f : \mathcal{N} \to \mathcal{N}$ *we say that* $\mathcal{G}$ *is f-sparse if* $|Eq(x)| \leq f(|Im(x)|)$ *for each* $x \in V$.

Lemma 7. *For each natural r there exists formulas* ϕ_r *and* ψ_r*, (over the signature consisting of E and c) such that:*

1. *ϕ_r expresses the property that $\mathcal{G}$ is a proper g_r-sparse structure and all the out-vertices of $\mathcal{G}$ are equivalent.*
2. *ψ_r expresses the property that $\mathcal{G}$ is a proper g_r-sparse structure.*

Proof. First, suppose that ϕ_r is defined. Then ψ_r is $\rho \wedge \neg((\rho \wedge \rho_1 \wedge \neg\phi_r)|\rho)$, where ρ is a first order formula, saying that the structure is proper, and ρ_1 is a first order logic formula, saying that all the out-vertices of the structure are equivalent. In other words, ψ_r says that the structure is proper and it cannot be split into two proper structures, the first of them being full and not g_r-sparse.

We leave it as an exercise for the reader to see why we need the constant c, and the notion of a proper structure here.

Now ϕ_r will be defined by induction on r.

Take ϕ_1 as $\rho \wedge \rho_1 \wedge (\phi_0|\rho_2)$ where ϕ_0 says that there are no two equivalent out-vertices in the structure, and ρ_2 is a first order logic formula, saying that c is isolated.

For the induction step, suppose that we have ψ_r. Then ϕ_{r+1} is $\rho_1 \wedge (\rho_2|\psi_r)$, where ρ_1 and ρ_2 are as above. Notice how the definition of function g is used here: if we have no more than $g_{r+1}(n)$ copies of some set A, with $|A| = n$, then some elements can be removed from some of the copies in such a way that among the resulting sets we do not have more than $g_r(m)$ copies of any subset of A with m elements. ∎

Proof of Theorem 2: It is easy to see that $\exists c\ \phi_r$ is the formula we wanted. ∎

Notice, that we **do not** prove in Theorem 2 that $\mathcal{W}_{2^{cn}}$ can be expressed in the logic GL–. Actually, we do not really know if it can be. What we show is just that the non-expressibility proof cannot depend only on an argument of the sort "the number 2^{cn} is too big for GL–".

On the other hand (and we leave it as an exercise for a careful reader to show it[7]) a strategy for Duplicator can be constructed, proving that a GL– formula of quantifier depth at least r is needed to express $\mathcal{W}_{f(n)}$ in GL–, for each function $f > g_r$. Since $c+2$ quantifiers are enough to express $W_{2^{cn}}$ in MSOL, this implies, by Lemma 6 that GL is exponentially less succinct than MSO (see for example [GS03] for definition of succinctness). Both the logics have the same power to count, but GL needs exponentially more quantifiers to do it. As we said in Section 1.1, many MSO properties are defined in [DGG04] as Graph Logic properties, but the GL formulas are much longer than their MSO counterparts. The last result gives some sort of explanation of this phenomenon.

Acknowledgment

This paper has been prepared when the author visited Isaac Newton Institute for Mathematical Sciences in Cambridge, as a participant of the programme *Logic*

[7] Hint: Take two full structures with n in-vertices each, one of them with $f(n)$ out-vertices and another with $f(n)+1$ vertices. With l rounds remaining Duplicator should make sure that neither of the 2 structures is g_l-sparse.

and Algorithms. The blackboards of Newton Institute are, up to my knowledge, the world's best pitch for playing Ehrenfeucht – Fraïssé games, and Albert Atserias, Anuj Dawar, Lauri Hella and Timos Antonopoulos were the most demanding Spoilers I have ever a chance to play against as Duplicator. It was also Albert, who generously paid for the beer[8] I drank when the proof of the first version of Lemma 2 came to my mind.

References

[AFS98] M.Ajtai, R.Fagin, L.Stockmeyer *The Closure of Monadic $\mathcal{NP}$*, Proc. of 13th STOC, pp 309-318, 1998;

[CGG02] L. Cardelli, P. Gardner and G. Ghelli, *A spatial logic for querying graphs*, ICALP 2002, Springer LNCS, vol. 2380, pp. 597-610.

[C90] B. Courcelle, *Graph rewriting: An algebraic and logic approach.* In J. van Leeuwen, editor, Handbook of Theoretical Computer Science, volume 2, pages 194–242. Elsevier Science Publishers, 1990. 12

[DGG04] A. Dawar, P. Gardner and G. Ghelli, *Expressiveness and Complexity of Graph Logic*, Imperial College, Department of Computing Technical Report 2004/3, to appear in *Information and Computation*;

[EF95] H-D. Ebinghaus, J. Flum, *Finite Model Theory*, Springer 1995;

[GS03] M. Grohe, N. Schweikardt, *Comparing the succinctness of monadic query languages over finite trees* Proceedings of CSL 2003, Springer LNCS, vol 2803, pp.226-240;

[JM01] D. Janin, J. Marcinkowski, *A Toolkit for First Order Extensions of Monadic Games*, Proceedings of STACS 2001, Springer LNCS 2010, pp 353-364;

[KL04] H. Jerome Keisler and Wafik Boulos Lotfallah, *First order quantifiers in monadic second order logic*, J. Symbolic Logic 69, iss. 1 (2004), 118-136;

[M99] J. Marcinkowski, *Directed Reachability: From Ajtai-Fagin to Ehrenfeucht-Fraisse games*, Proceedings of the Annual Conference of the European Association of Computer Science Logic (CSL 99) Springer LNCS 1683, pp 338-349;

[8] A pint of bitter.

Hoare Logic in the Abstract

Ursula Martin, Erik A. Mathiesen, and Paulo Oliva

Department of Computer Science
Queen Mary, University of London
Mile End Road
London E1 4NS, UK
{uhmm, erikarne, pbo}@dcs.qmul.ac.uk

Abstract. We present an abstraction of Hoare logic to traced symmetric monoidal categories, a very general framework for the theory of systems. We first identify a particular class of functors – which we call 'verification functors' – between traced symmetric monoidal categories and subcategories of Preord (the category of preordered sets and monotone mappings). We then give an abstract definition of Hoare triples, parametrised by a verification functor, and prove a single soundness and completeness theorem for such triples. In the particular case of the traced symmetric monoidal category of while programs we get back Hoare's original rules. We discuss how our framework handles extensions of the Hoare logic for while programs, e.g. the extension with pointer manipulations via separation logic. Finally, we give an example of how our theory can be used in the development of new Hoare logics: we present a new sound and complete set of Hoare-logic-like rules for the verification of linear dynamical systems, modelled via stream circuits.

1 Introduction

A few years ago one of the authors spent some time with an aeronautical engineer learning how flight control systems are designed. Standard engineering tools like Simulink represent the differential equations that model the system as a box-and-arrow diagram, incorporating composition and feedback. The engineer's informal explanation of how an input signal is transformed to an output signal as it passes through each component of the diagram was strikingly similar to a Hoare logic presentation of a program proof, and inspired the development of a sound, if ugly, ad-hoc Hoare-like logic for a fragment of linear differential equations [7].

So obvious questions (for us, if not the engineer) are how it is that such disparate things as programs and differential equations can both support Hoare-like logics, and whether there is a principled way of developing such logics in new situations. This paper provides the answer: both are instances of a particular kind of traced symmetric monoidal category, and for any such instance a sound and complete set of Hoare-logic-like rules can be read off from the categorical structure.

Under the general label of *Hoare logic*, the early work of Floyd [10] and Hoare [12] on axiom systems for flowcharts and while programs has been applied to various other domains, such as recursive procedures [2], pointer programs [19], and

Z. Ésik (Ed.): CSL 2006, LNCS 4207, pp. 501–515, 2006.

higher-order languages [4]. Our goal in this paper is to identify a minimal structure supporting soundness and (relative) completeness results, in the manner of Cook's presentation for *while programs* [8]. This is achieved via an abstraction of Hoare logic to the theory of traced symmetric monoidal categories [13], a very general framework for the theory of systems. Traced symmetric monoidal categories precisely capture the intrinsic structure of both flowcharts and dynamical systems, namely: sequential and parallel 'composability', and infinite behaviour. In fact, traced symmetric monoidal categories are closely related to Bainbridge's work [3] on the duality between flowcharts and networks. The scope of traced symmetric monoidal categories, however, is much broader, being actively used, for instance, for modelling computation (e.g. [21]) and in connection with Girard's geometry of interaction (e.g. [11]).

The main feature of Hoare logic is the use of pre- and post-conditions to specify the behaviour of a program, and the order relation between pre- and post-condition given by logical implication. Let Preord be the category of preordered sets and monotone mappings. We first identify a particular class of functors – which we call *verification functors* – between traced symmetric monoidal categories and subcategories of Preord. We then give an abstract definition of Hoare triples, parametrised by a verification functor, and prove a single soundness and completeness (in the sense of Cook) theorem for such triples. In the particular case of the traced symmetric monoidal category of while programs (respectively, pointer programs) this embedding gives us back Hoare's original logic [12] (respectively, O'Hearn and Reynolds logic [19]). In order to illustrate the generality of our framework, we also derive new sound and complete Hoare-logic-like rules for the verification of linear dynamical systems, in our case, modelled via stream circuits.

In general, our abstraction of Hoare logic provides a 'categorical' recipe for the development of new (automatically) sound and complete Hoare-logic-like rules for any class of systems having the underlying structure of a traced symmetric monoidal category. Moreover, Hoare logic notions such as expressiveness conditions, relative completeness [8] and loop invariants, have a clear cut correspondence to some of our abstract notions.

The chief contributions of this paper are as follows: (i) The definition of a verification functor, between traced symmetric monoidal categories and the category Preord (Section 3). (ii) An abstraction of Hoare triples in terms of verification functors (Definition 3). (iii) Sound and complete rules for our abstract notion of Hoare triples, over a fixed verification functor (Theorem 1). (iv) Three concrete instances of our abstraction, namely: while programs, pointer programs, and stream circuits (Section 4). In Section 5, we discuss the link between our work and other abstractions of Hoare logic.

For the rest of this article we will assume some basic knowledge of category theory. For a readable introduction see [15]. Given a category $\mathcal{S}$, we will denote its objects by $\mathcal{S}_o$ and its morphisms by $\mathcal{S}_m$. Composition between two morphisms will be denoted as usual by $(g \circ f) : X \to Z$, if $f : X \to Y$ and $g : Y \to Z$.

Fig. 1. Trace diagrammatically

2 Traced Symmetric Monoidal Categories

A pair of a category $\mathcal{S}$ and a functor $\otimes$ is called a *monoidal category* if for some particular object of $\mathcal{S}$ (the identity element of the monoid) $\otimes$ satisfies the monoidal axioms of associativity and identity (for details, see Chapter 11 of [15]). A monoidal category is called symmetric if there exists a family of natural isomorphisms $c_{X,Y} : X \otimes Y \to Y \otimes X$ satisfying the two braiding axioms plus the symmetry axiom $c_{X,Y} \circ c_{Y,X} = \mathsf{id}_{X \otimes Y}$.

In [13], the notion of *traced symmetric monoidal category* is introduced[1], for short *tmc*. A symmetric monoidal category is traced if for any morphism $f : X \otimes Z \to Y \otimes Z$ there exists a morphism $\mathsf{Tr}^Z_{X,Y}(f) : X \to Y$ satisfying the trace axioms (see [13] for details). We will normally omit the decoration in $\mathsf{Tr}^Z_{X,Y}$ whenever it is clear over which object the trace is being applied. Morphisms of a tmc can be represented diagrammatically as input-output boxes, so that, if $f : X \otimes Z \to Y \otimes Z$ then $\mathsf{Tr}(f) : X \to Y$ corresponds to a *feedback* over the 'wire' Z, as shown in Figure 1.

The two most common examples of traced symmetric monoidal categories are the category of sets and relations where $\otimes$ is either disjoint union, with trace defined as

$$x \,\mathsf{Tr}(f)\, y \;:\equiv\; \exists \boldsymbol{z}(\langle 0, x\rangle f \langle 1, z_0\rangle \wedge \ldots \langle 1, z_i\rangle f \langle 1, z_{i+1}\rangle \ldots \wedge \langle 1, z_n\rangle f \langle 0, y\rangle)$$

or cartesian product, with trace defined as

$$x \,\mathsf{Tr}(f)\, y \;:\equiv\; \exists z(\langle x, z\rangle f \langle y, z\rangle)$$

Note that we abbreviate a tuple of variables $z_0, \ldots, z_n \in Z$ by $\boldsymbol{z}$.

Definition 1 (Basic morphisms). *Let $M \subseteq \mathcal{S}_m$ be a subset of the morphisms of a traced symmetric monoidal category $\mathcal{S}$. We define $\mathsf{cl}(M)$, the* traced symmetric monoidal closure of M, *as the smallest subset of $\mathcal{S}_m$ containing M which is closed under sequential composition, monoidal closure, and trace. If M is such that $\mathsf{cl}(M) = \mathcal{S}_m$ then M is called a set of* basic morphisms *for the category $\mathcal{S}$.*

Intuitively, we view each morphism in a tmc as representing a particular system. The categorical composition models the sequential composition of systems, while

[1] In fact, [13] introduces the theory of traces for a more general class of monoidal categories, so-called balanced monoidal categories, of which symmetric monoidal categories are a special case.

the monoidal operation is related to the 'parallel' composition of two systems. Finally, the trace corresponds to feedback, allowing infinite behaviour. The basic morphisms represent the basic systems, out of which complex systems are built. Note that the whole set of morphisms is trivially a set of basic morphisms. We are interested, however, in tmc in which the set of basic morphisms (systems) forms a strict subset of the set of all morphisms, as in the following three examples.

The categories we describe below are purely *semantical*. Nevertheless, we build the categories in such a way that each *syntactic* while program or stream circuit is denoted by a morphism in the category, and *vice versa* each morphism has a representation as a syntactic while program or stream circuit.

2.1 Example 1: Flowcharts

The first example we consider is that of *flowcharts*. Let Store be the set of stores, i.e. mappings $\rho : \mathsf{Var} \to \mathbb{Z}$ from program variables $\mathsf{Var} = \{x, y, \ldots\}$ to integers. Let $\mathcal{F}_o$ be the set (of sets) containing the empty set $\emptyset$ and Store, and closed under *disjoint union*, i.e. $\mathcal{F}_o \equiv \{\emptyset, \mathsf{Store}, \mathsf{Store} \uplus \mathsf{Store}, \ldots\}$. Consider also the following family of functions $\mathcal{F}_b$ between sets in $\mathcal{F}_o$:

- *Skip*, $\mathsf{id} : \mathsf{Store} \to \mathsf{Store}$
  $\mathsf{id}(\rho) :\equiv \rho$
- *Assignment*, $(x := t) : \mathsf{Store} \to \mathsf{Store}$
  $(x := t)(\rho) :\equiv (\rho)[t_\rho / x]$
- *Joining*, $\Delta : \mathsf{Store} \uplus \mathsf{Store} \to \mathsf{Store}$
  $\Delta(\rho) :\equiv \mathsf{proj}(\rho)$
- *Forking*, $\nabla_b : \mathsf{Store} \to \mathsf{Store} \uplus \mathsf{Store}$
  $\nabla_b(\rho) :\equiv \begin{cases} \mathsf{inj}_0(\rho) & \text{if } \neg b_\rho \\ \mathsf{inj}_1(\rho) & \text{otherwise} \end{cases}$

The conditional forking (∇_b) and the assignment ($x := t$) are parametrised by functions b and t (intuitively, expressions) from Store to the boolean lattice $\mathbb{B}$ and $\mathbb{Z}$, respectively, so that b_ρ and t_ρ denote their value on a given store ρ. We use inj_0 and inj_1 for left and right injections into $\mathsf{Store} \uplus \mathsf{Store}$, while proj is the projection. We then close the set of basic functions $\mathcal{F}_b$ under sequential composition of functions, disjoint union of functions and the standard trace for disjoint union, to form the set of partial functions $\mathcal{F}_m$. It is easy to see that $\mathcal{F} \equiv (\mathcal{F}_o, \mathcal{F}_m, \uplus, \mathsf{Tr})$ forms a tmc, if we add to the set of basic morphisms a family of functions $c_{X,Y} : X \uplus Y \to Y \uplus X$, which simply flip the flag of the disjoint union. Note that $\mathcal{F}_b$ is by construction a set of basic morphisms for $\mathcal{F}$.

2.2 Example 2: Pointer Programs

The second example of tmc we consider is an extension of $\mathcal{F}$ to a category of programs that manipulate both stores and heaps, which we will refer to as the traced symmetric monoidal category of *pointer programs* $\mathcal{P}$. Let $\mathsf{State} :\equiv (\mathsf{Store} \times \mathsf{Heap}) \cup \{\mathsf{abort}\}$, where Store is as in the previous section and Heap is the set of partial functions (from $\mathbb{N}$ to $\mathbb{Z}$) with finite domain. We view the elements of the heap as pairs consisting of a function $h : \mathbb{N} \to \mathbb{Z}$ and a finite set $d \in \mathcal{P}_{\mathsf{fin}}(\mathbb{N})$ describing the valid domain of h. We then form the set (of sets) $\mathcal{P}_o$ as the set containing the empty set $\emptyset$ and State, and closed under *disjoint union*, i.e. $\mathcal{P}_o \equiv \{\emptyset, \mathsf{State}, \mathsf{State} \uplus \mathsf{State}, \ldots\}$. Each of the basic functions of $\mathcal{F}_b$ can be lifted to the extended type structure, by simply ignoring the 'heap' component. The set of functions $\mathcal{P}_b$ is an extension of the set of (the lifting of the) functions $\mathcal{F}_b$ with the following family of functions:

- *Look up*, $(x := [t]) : \mathsf{State} \to \mathsf{State}$
$$(x := [t])(\rho, h, d) :\equiv \begin{cases} (\rho[h(t_\rho)/x], h, d) & t_\rho \in d \\ \mathsf{abort} & \mathsf{otherwise} \end{cases}$$
- *Mutation*, $([t] := s) : \mathsf{State} \to \mathsf{State}$
$$([t] := s)(\rho, h, d) :\equiv \begin{cases} (\rho, h[t_\rho \mapsto s_\rho], d) & t_\rho \in d \\ \mathsf{abort} & \mathsf{otherwise} \end{cases}$$
- *Allocation*, $x := \mathsf{new}(\boldsymbol{t}) : \mathsf{State} \to \mathsf{State}$
$$(x := \mathsf{new}(\boldsymbol{t}))(\rho, h, d) :\equiv (\rho[x \mapsto i], h[i + j \mapsto t_j], d \uplus \{i, \ldots, i + n\})$$
- *Deallocation*, $\mathsf{disp}(t) : \mathsf{State} \to \mathsf{State}$
$$(\mathsf{disp}(t))(\rho, h, d) :\equiv \begin{cases} (\rho, h, d \backslash \{t_\rho\}) & t_\rho \in d \\ \mathsf{abort} & \mathsf{otherwise} \end{cases}$$

As done in the case of flowcharts above, we can then close the set of functions $\mathcal{P}_b$ under sequential composition, disjoint union and trace, to form the set of partial functions $\mathcal{P}_m$. We are assuming that the functions are strict with respect to abort, i.e. on the abort state all programs will return abort. The category $\mathcal{P} \equiv (\mathcal{P}_o, \mathcal{P}_m, \uplus, \mathsf{Tr})$, with the standard trace for disjoint union, forms another example of a tmc with the extra basic morphisms look up, mutation, allocation and deallocation.

2.3 Example 3: Stream Circuits

Finally, we give an example of a tmc (the category of *stream circuits*) based on cartesian product, rather than disjoint union. Let Σ denote the set of streams[2] of real numbers, i.e. all functions $\mathbb{N} \to \mathbb{R}$. Let $\mathcal{C}_o$ be the set containing the singleton set $\{\varepsilon\}$ and Σ, and closed under cartesian product, i.e. $\mathcal{C}_o \equiv \{\{\varepsilon\}, \Sigma, \Sigma \times \Sigma, \ldots\}$. Consider the following set $\mathcal{C}_b$ of functions between the sets in $\mathcal{C}_o$:

- *Wire*, $\mathsf{id} : \Sigma \to \Sigma$
$$\mathsf{id}(\upsilon) :\equiv \upsilon$$
- *Scalars*, $(a\times) : \Sigma \to \Sigma$
$$(a\times)(\sigma) :\equiv [a\sigma_0, a\sigma_1, \ldots]$$
- *Register*, $R : \Sigma \to \Sigma$
$$R(\sigma) :\equiv [0, \sigma_0, \sigma_1, \ldots]$$
- *Copy*, $\mathsf{c} : \Sigma \to \Sigma \times \Sigma$
$$\mathsf{c}(\sigma) :\equiv \langle \sigma, \sigma \rangle$$
- *Sum*, $(+) : \Sigma \times \Sigma \to \Sigma$
$$(+)\langle \sigma, \sigma' \rangle :\equiv [\sigma_0 + \sigma'_0, \sigma_1 + \sigma'_1, \ldots]$$

We then form the set of *relations* $\mathcal{C}_m$ by viewing the basic functions $\mathcal{C}_b$ above as relations (via their graph) and closing this set under relational composition, cartesian product of relations, and the standard trace for the cartesian product (cf. Section 2). The diagrams in Figure 2 illustrate two stream circuits which do

[2] As shown in [9], much of mathematical analysis can be developed co-inductively using streams: the stream corresponding to an analytic function f is just $[f(0), f'(0), f''(0), \ldots]$ which is related to the coefficients of the Taylor series expansion of f.

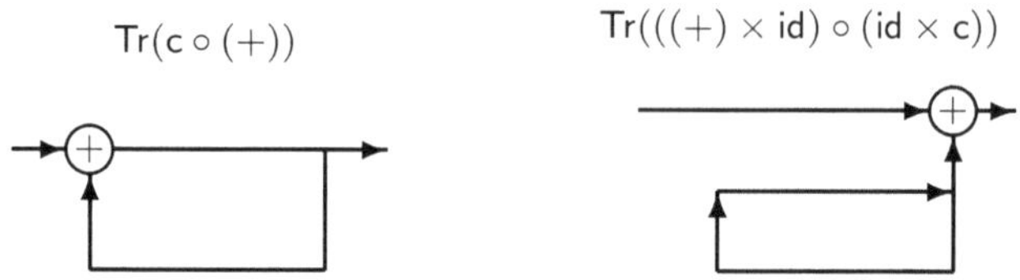

Fig. 2. Stream circuits not defining functions

not define a function (but define a relation). In the circuit on the left there is no fixed point for the input stream $\sigma_a :\equiv [a, a, \ldots]$, for $a \neq 0$. On the other hand, in the circuit on the right any stream τ is a fixed point of the trace, so that any input stream σ is related to an arbitrary (output) stream. In order to avoid such pathological circuits (i.e. circuits not defining a function), one normally puts a further restriction on the formation of stream circuits. A stream circuit $f : X \to Y$ is called *valid* if in every sub-circuit of the form $\mathsf{Tr}^Z_{X',Y'}(f') : X' \to Y'$ the path from X' to the output Z goes through a register. This requirement for forming loops is standard in stream circuits (cf. [20]). Although $\mathcal{C}$ is a category of relations, valid circuits are denoted by *total* functions, since the fixed points of a valid circuit exist and are unique.

It is again easy to see that $\mathcal{C} \equiv (\mathcal{C}_o, \mathcal{C}_m)$ forms a category. The category $\mathcal{C}$ is symmetric monoidal, with the monoidal structure of cartesian product, if we add to the set of basic morphisms the family of relations $c_{X,Y} : X \times Y \to Y \times X$, which simply flip the two components of the input pair, i.e. $\langle x, y \rangle c_{X,Y} \langle y, x \rangle$. Finally, with the standard family of trace relations (defined in Section 2), $\mathcal{C}$ forms a traced symmetric monoidal category.

3 Abstract Hoare Logic

In this section we present an abstraction of Hoare logic, including pre- and post-conditions and Hoare triples as normally used in Hoare logic for while programs. Our goal here is to isolate the minimal structure needed from these pre- and post-conditions, for the development of a sound and complete Hoare logic.

Recall that a mapping $f : X \to Y$ between two preordered sets X, Y is called *monotone* if $P \sqsubseteq_X Q$ implies $f(P) \sqsubseteq_Y f(Q)$. Let Preord denote the category of preordered sets and monotone mappings. It is easy to see that Preord can be considered a symmetric monoidal category, with the monoidal operation as cartesian product, since the cartesian product of two preordered sets X, Y forms again a preordered set with the order on $X \times Y$ defined coordinatewise.

For the rest of this section, let $\mathcal{S}$ be a fixed tmc. A subset $\mathcal{S}'_m$ of the morphisms $\mathcal{S}$ is called *subsystem closed* if $f, g \in \mathcal{S}'_m$, whenever $\mathsf{Tr}(f)$, $f \circ g$ or $f \otimes g$ is in $\mathcal{S}'_m$.

Definition 2 (Verification functor). *Let $\mathcal{S}'_m$ be a subsystem closed subset of $\mathcal{S}_m$. A strict monoidal functor $H : \mathcal{S} \to \mathsf{Preord}$ is called a* verification functor *for $\mathcal{S}'_m$ if*

$$H(\mathsf{Tr}(f))(P) \sqsubseteq R \;\Leftrightarrow\; \exists Q \in H(Z)(H(f)\langle P, Q \rangle \sqsubseteq \langle R, Q \rangle) \tag{1}$$

for all $f : X \otimes Z \to Y \otimes Z$ *such that* $\mathsf{Tr}(f) \in \mathcal{S}'_m$, *and* $P \in H(X)$, $R \in H(Y)$. *If* $\mathcal{S}'_m = \mathcal{S}_m$ *we simply call* H *a verification functor.*

Let a verification functor $H : \mathcal{S} \to \mathsf{Preord}$ for $\mathcal{S}'_m$ be fixed. For simplicity assume $\mathcal{S}'_m = \mathcal{S}_m$ (see Remark 1). Each object $X \in \mathcal{S}_o$ corresponds to a preordered set $H(X)$, and each morphism $f \in \mathcal{S}_m$ corresponds to a monotone mapping $H(f)$.

Definition 3 (Abstract Hoare Triples). *Let* $f : X \to Y$ *be a morphism (system) in* $\mathcal{S}$, $P \in H(X)$ *and* $Q \in H(Y)$. *Define* abstract Hoare triples *as*

$$\{P\}\ f\ \{Q\} \quad :\equiv \quad H(f)(P) \sqsubseteq_{H(Y)} Q \tag{2}$$

Although we use the same notation as the standard Hoare triple, it should be noted that the meaning of our abstract Hoare triple can only be given once the verification functor H is fixed. The usual Hoare logic meaning of *if* P *holds before the execution of* f *then, if* f *terminates,* Q *holds afterwards* will be one of the special cases of our general theory. See Section 4 for other meanings of $\{P\}\ f\ \{Q\}$.

Let a tmc $\mathcal{S}$ and verification functor $H : \mathcal{S} \to \mathsf{Preord}$ be fixed. Moreover, let $\mathcal{S}_b$ be a set of basic morphisms for $\mathcal{S}$. We will denote by $\mathbf{H}(\mathcal{S}, H)$ the following set of rules:

$$\frac{\{P\}\ f\ \{Q\} \quad \{Q\}\ g\ \{R\}}{\{P\}\ g \circ f\ \{R\}}\ (\circ) \qquad \frac{}{\{P\}\ f\ \{H(f)(P)\}}\ (f \in \mathcal{S}_b)$$

$$\frac{\{P\}\ f\ \{Q\} \quad \{R\}\ g\ \{S\}}{\{\langle P, R\rangle\}\ f \otimes g\ \{\langle Q, S\rangle\}}\ (\otimes) \qquad \frac{\{P\}\ f\ \{Q\}}{\{P'\}\ f\ \{Q'\}}\ \left(\begin{matrix} P' \sqsubseteq P \\ Q \sqsubseteq Q' \end{matrix}\right)$$

$$\frac{\{\langle P, Q\rangle\}\ f\ \{\langle R, Q\rangle\}}{\{P\}\ \mathsf{Tr}_{\mathcal{S}}(f)\ \{R\}}\ (\mathsf{Tr})$$

The formal system $\mathbf{H}(\mathcal{S}, H)$ should be viewed as a *syntactic* axiomatisation of the ternary relation $\{P\}\ f\ \{Q\}$. The verification functor H gives the *semantics* of the Hoare triples (and rules). By soundness and completeness of the system $\mathbf{H}(\mathcal{S}, H)$ we mean that syntax corresponds precisely to semantics, i.e. a syntactic Hoare triple $\{P\}\ f\ \{Q\}$ is provable in $\mathbf{H}(\mathcal{S}, H)$ if and only if the inequality $H(f)(P) \sqsubseteq Q$ is true in Preord.

Theorem 1 (Soundness and completeness). *The system* $\mathbf{H}(\mathcal{S}, H)$ *is sound and complete.*

Proof. Soundness is trivially true for the axioms. The consequence rule is sound by the monotonicity of $H(f)$ and transitivity of $\sqsubseteq$. Soundness of the composition rule also uses the monotonicity of $H(g)$, the transitivity of $\sqsubseteq$ and the fact that H respects composition, i.e. $H(g)(H(f)(P)) = H(g \circ f)(P)$. Soundness of the cartesian product rule uses that the functor H is monoidal, i.e.

$H(f \otimes g)\langle P, Q\rangle = (H(f) \times H(g))\langle P, Q\rangle = \langle H(f)(P), H(g)(Q)\rangle$. For the soundness of the trace rule, assume $H(f)\langle P, Q\rangle \sqsubseteq \langle R, Q\rangle$. By the definition of a verification functor we have $H(\mathsf{Tr}(f))(P) \sqsubseteq R$. We argue now about completeness. By the consequence rule and the axioms it is easy to verify that all true statements of the form $\{P\}\ f\ \{Q\}$, for $f \in \mathcal{S}_b$, are provable. It remains to show that if $\{P\}\ f\ \{Q\}$ is true, for an arbitrary morphism f, then there exists a premise of the corresponding rule (depending on the structure of f) which is also true. If $\{P\}\ g \circ f\ \{R\}$ is true then so it is $\{P\}\ f\ \{H(f)(P)\}$, by the reflexivity of the ordering, and $\{H(f)(P)\}\ g\ \{R\}$, by the fact that H respects composition. If $\{\langle P, R\rangle\}\ f \otimes g\ \{\langle Q, S\rangle\}$ is true then so it is $\{P\}\ f\ \{Q\}$ and $\{R\}\ g\ \{S\}$, by the fact that H is monoidal. Finally, if $\{P\}\ \mathsf{Tr}(f)\ \{R\}$ is true, then so is $\{\langle P, Q\rangle\}\ f\ \{\langle R, Q\rangle\}$, for some Q, by the definition of a verification functor. □

The abstract proof of soundness and completeness presented above is both short and simple, using only the assumption of a verification functor and basic properties of the category Preord. As we will see, the laborious work is pushed into showing that H is a verification functor. That will be clear in Section 4.1, where we build a verification functor appropriate for while programs, using Cook's expressiveness condition [8].

Remark 1. It is easy to see that all arguments in the proof of Theorem 1 will go through if we restrict ourselves to a particular subsystem closed set of morphisms $\mathcal{S}'_m$. This follows from the fact that in the completeness proof we simply use that any morphism can be 'generated' from the basic morphisms, which is also true for morphisms in $\mathcal{S}'_m$. If H is a verification functor for $\mathcal{S}'_m \subset \mathcal{S}_m$, the rules of $\mathbf{H}(\mathcal{S}, H)$ should be restricted to morphisms in $\mathcal{S}'_m$. Similarly, the soundness and completeness theorems apply only to systems in $\mathcal{S}'_m$. See Section 4.3 for an instantiation of the general framework where a special class of systems $\mathcal{S}'_m$ is singled out.

4 Instantiations

We will now look at some concrete examples of verification functors for the traced symmetric monoidal categories described in Section 2. For each of the instantiations of tmc $\mathcal{S}$ we have described we present a monoidal functor $H : \mathcal{S} \to \mathsf{Preord}$ and show that it satisfies condition (1).

First we will consider the traced symmetric monoidal categories of flowcharts and pointer programs. These instantiations will produce, respectively, the original Hoare logic [12], and Reynold's [19] axiomatisation based on separation logic. We then present a new variation of Hoare logic for stream circuits, obtained through our abstraction.

4.1 Flowcharts (Forward Reasoning)

In this section we present a verification embedding of the tmc of flowcharts $\mathcal{F}$ (see Section 2.1). This will give us soundness and (relative) completeness of Hoare's original verification logic [12] for partial correctness (using forward reasoning).

Let us now define the monoidal functor $H : \mathcal{F} \to \mathsf{Preord}$. Let a Cook-expressive[3] first-order theory $\mathcal{L}$ be fixed. On the objects $X \in \mathcal{F}_o$ we let $H(X)$ be the preordered set of formulas of $\mathcal{L}$. The ordering $P \sqsubseteq R$ on the elements of $H(X)$ is taken to be $\mathcal{L} \vdash P \to R$. On the morphisms (flowcharts) $f \in \mathcal{F}_m$ we let $H(f)$ be the predicate transformer producing strongest post-conditions for any pre-condition P, i.e.

$$H(f)(P) :\equiv \mathsf{SPC}(f, P)$$

where $\mathsf{SPC}(f, P)$ is a formula expressing the strongest post-condition of f under P. Such formula exists by our assumption that the theory $\mathcal{L}$ is Cook-expressive. It is also easy to see what those are for the basic morphisms of $\mathcal{F}$

$$\begin{aligned}
&\mathsf{SPC}(\mathsf{id}, P) &&:\equiv P\\
&\mathsf{SPC}(x := t, P) &&:\equiv \exists v (P[v/x] \wedge x = t[v/x])\\
&\mathsf{SPC}(\nabla_b, P) &&:\equiv \langle P \wedge \neg b, P \wedge b\rangle\\
&\mathsf{SPC}(\Delta, \langle P, R\rangle) &&:\equiv P \vee R
\end{aligned}$$

The functor H is monoidal because a formula P describing a subset of $X_0 \uplus X_1$ can be seen as a pair of formulas $\langle P_0, P_1\rangle$ such that each P_i describes a subset of X_i, i.e. $H(X \uplus Y)$ is isomorphic to $H(X) \times H(Y)$. Similarly, there is a one-to-one correspondence between strongest post-condition transformer for a parallel composition of flowcharts $f \uplus g$ and pairs of predicate transformers $H(f) \times H(g)$. We argue now in two steps that H is also a verification functor.

Lemma 1 ([8]). *Let $f : X \uplus Z \to Y \uplus Z$ in $\mathcal{F}$ and formulas $P \in H(X)$ and $R \in H(Y)$ be fixed. If $\mathsf{SPC}(\mathsf{Tr}(f), P) \to R$ then $\mathsf{SPC}(f, \langle P, Q\rangle) \to \langle R, Q\rangle$, for some formula Q.*

Proof. We construct the formula Q such that it is a fixed point for f on P, i.e. $\mathsf{SPC}(f, \langle P, Q\rangle) \leftrightarrow \langle R', Q\rangle$, for some formula R'. We also argue that $\mathsf{SPC}(\mathsf{Tr}(f), P) \leftrightarrow R'$. By our hypothesis $\mathsf{SPC}(\mathsf{Tr}(f), P) \to R$ it follows that R' implies R, as desired. The fixed point Q is essentially the strongest loop invariant, and is constructed as follows. Given the partial function f we build a new partial function $f' : X \uplus Z \to Y \uplus Z \uplus Z$ where the internal states of Z can be observed, even after the trace is applied. Let $f' :\equiv (\mathsf{id} \uplus \nabla_{\boldsymbol{z}=\boldsymbol{y}}) \circ f$ (see Figure 3) where $\boldsymbol{z}$ is the finite sequence of variables mentioned in f (by construction of the category $\mathcal{F}$, each morphism only changes finitely many variables), and $\boldsymbol{y}$ is a fresh tuple of variables of same length. Notice that $\mathsf{Tr}^Z_{X,Y \uplus Z}(f')$ behaves almost as $\mathsf{Tr}^Z_{X,Y}(f)$ except that $\mathsf{Tr}(f')$ might 'terminate' earlier (meaning that the fixed point sequence is shorter) if the state $\boldsymbol{z}$ matches $\boldsymbol{y}$. Let $\mathsf{SPC}(\mathsf{Tr}(f'), P) = \langle Q_0, Q_1(\boldsymbol{y})\rangle$. It is easy to see that the formula $Q \equiv \exists \boldsymbol{y}\, Q_1(\boldsymbol{y})$ characterises the possible internal states $\boldsymbol{z}$ in the trace of f on an input satisfying P, i.e. Q is a fixed point for

[3] Recall that a logic is Cook-expressive if for any program f and pre-condition P the strongest post-condition of f under P is expressible by a formula in $\mathcal{L}$ (cf. [8]).

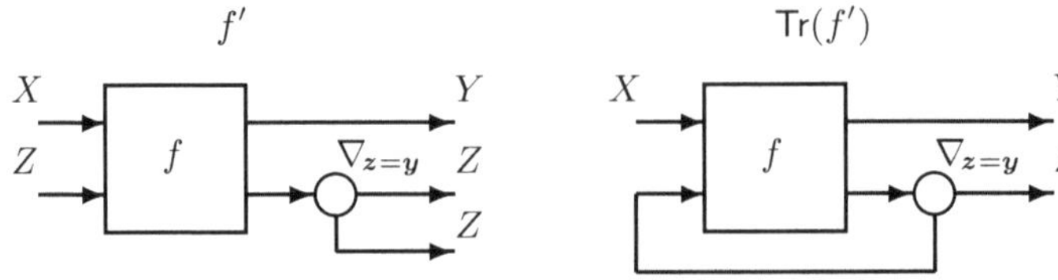

Fig. 3. Cook's construction

$H(f)$ on P, $\mathsf{SPC}(f, \langle P, Q \rangle) \leftrightarrow \langle R', Q \rangle$, for some formula R'. It also follows that $\mathsf{SPC}(\mathsf{Tr}(f), P) \leftrightarrow R'$, since Q characterises precisely the internal states of the trace $\mathsf{Tr}(f)$ on inputs satisfying P. □

Theorem 2. *$H : \mathcal{F} \to \mathsf{Preord}$, as defined above, is a verification functor.*

Proof. By Lemma 1, it remains to be shown that whenever

(*i*) $\mathsf{SPC}(f, \langle P, Q' \rangle) \to \langle R, Q' \rangle$,

for some formula Q', then $\mathsf{SPC}(\mathsf{Tr}(f), P) \to R$. Assume (*i*) and $\mathsf{SPC}(\mathsf{Tr}(f), P)(\rho)$, for some store value ρ. We must show $R(\rho)$. By the definition of the strongest post-condition there exists a sequence of stores $\rho', \rho_0, \ldots, \rho_n$ such that $P(\rho')$ and

$$f\langle 0, \rho' \rangle = \langle 1, \rho_0 \rangle, \ldots, f\langle 1, \rho_k \rangle = \langle 1, \rho_{k+1} \rangle, \ldots, f\langle 1, \rho_n \rangle = \langle 0, \rho \rangle.$$

By a simple induction, using the assumption (*i*), we get that all ρ_k satisfy Q' and that ρ satisfies R, as desired. □

The system $\mathbf{H}(\mathcal{F}, H)$ obtained via our embedding H is a refinement of the system given by Hoare in [12], i.e. Hoare's rules are derivable from ours. See, for instance, the case of the while loop rule in Figure 4, given that a while loop $\mathsf{while}_b(f)$ can be represented in $\mathcal{F}$ as $\mathsf{Tr}((\mathsf{id} \uplus f) \circ \nabla_b \circ \Delta)$. Moreover, the soundness and (relative) completeness of the Hoare logic rules for while programs follow easily from Theorems 1 and 2.

4.2 Pointer Programs (Backward Reasoning)

In the previous section we showed how one can obtain the standard Hoare logic for while programs as an instantiation of our general framework. We have presented a verification embedding that gives rise to the usual Hoare triple semantics for *forward reasoning*. In this section we show how the verification functor can be changed so that one automatically gets Hoare logic rules for *backward reasoning*, from the same categorical rules presented in Section 3. We will illustrate backward reasoning using an extension of the category of flowcharts, namely the category of pointer programs $\mathcal{P}$ (see Section 2.2).

Consider the following functor $H : \mathcal{P} \to \mathsf{Preord}$. On the objects $X \in \mathcal{P}_o$ the embedding H returns the preordered set of formulas (see below for the description of the logical language used) describing subsets of $X \backslash \{\mathsf{abort}\}$. The

$$\dfrac{\dfrac{(\Delta \in \mathcal{F}_b)}{\{\langle P, P\rangle\}\ \Delta\ \{P\}} \qquad \dfrac{\dfrac{(\nabla_b \in \mathcal{F}_b)}{\{P\}\ \nabla_b\ \{\langle P \wedge \neg b, P \wedge b\rangle\}} \quad \dfrac{\dfrac{(\mathsf{id} \in \mathcal{F}_b)}{\{P \wedge \neg b\}\ \mathsf{id}\ \{P \wedge \neg b\}} \quad \{\mathbf{P} \wedge \mathbf{b}\}\ \mathbf{f}\ \{\mathbf{P}\}}{\{\langle P \wedge \neg b, P \wedge b\rangle\}\ \mathsf{id} \uplus f\ \{\langle P \wedge \neg b, P\rangle\}}}{\{P\}\ (\mathsf{id} \uplus f) \circ \nabla_b\ \{\langle P \wedge \neg b, P\rangle\}}}{\dfrac{\dfrac{\{\langle P, P\rangle\}\ (\mathsf{id} \uplus f) \circ \nabla_b \circ \Delta\ \{\langle P \wedge \neg b, P\rangle\}}{\{P\}\ \mathsf{Tr}((\mathsf{id} \uplus f) \circ \nabla_b \circ \Delta)\ \{P \wedge \neg b\}}\ (\mathsf{Tr})}{\{\mathbf{P}\}\ \mathsf{while}_{\mathbf{b}}(\mathbf{f})\ \{\mathbf{P} \wedge \neg \mathbf{b}\}}\ (\mathrm{def})}$$

Fig. 4. Derivation of Hoare's while loop rule in $\mathbf{H}(\mathcal{F}, H)$

preorder on $H(X)$ is: $R \sqsubseteq P$ iff $P \to R$, i.e. the logical equivalent of *reverse set inclusion*. On the morphisms of $\mathcal{P}$ we define

$$H(f)(P) :\equiv \mathsf{WPC}(f, P)$$

where $\mathsf{WPC}(f, P)$ is the weakest liberal pre-condition of the partial function f on post-condition P. We are assuming the dual of Cook's expressiveness condition, namely, that weakest liberal pre-conditions are expressible in the language.

According to our definition, abort is not an element of $H(X)$, which reflects the fact that Hoare triples for pointer programs ensure that programs do not crash (see [19]).

It has been shown in [17,19], that the weakest liberal pre-conditions for the new basic statements can be concisely expressed in *separation logic* as (see [19] for notation)

$$\begin{array}{ll}
\mathsf{WPC}(x := [t], P) & :\equiv \exists v'((t \mapsto v') * ((t \mapsto v') -\!\!* P[v'/x])) \\
\mathsf{WPC}([t] := s, P) & :\equiv (t \mapsto -) * ((t \mapsto s) -\!\!* P) \\
\mathsf{WPC}(x := \mathsf{new}(\boldsymbol{t}), P) & :\equiv \forall i((i \mapsto \boldsymbol{t}) -\!\!* P[i/x]) \\
\mathsf{WPC}(\mathsf{disp}(t), P) & :\equiv (t \mapsto -) * P
\end{array}$$

Similarly to Lemma 1 and Theorem 2, one can show that the H defined above is a verification functor. The system $\mathbf{H}(\mathcal{P}, H)$, which we then obtain from our abstract approach is basically the one presented in Reynolds [19], for *global backward reasoning*, using separation logic as an 'oracle' for the consequence rule.

4.3 Stream Circuits

In [7], a Hoare logic was suggested for the frequency analysis of linear control systems with feedback, modelled as linear differential equations. Here we propose a different approach, based on the modelling of linear differential equations as stream circuits (also called signal flow graphs, see [20]).

We now present an embedding of the tmc $\mathcal{C}$ of stream circuits, described in Section 2.3, into Preord. Let $\mathcal{C}'_m$ be the set of (functions denoting) valid stream

$$
\cfrac{\cfrac{\cfrac{\cfrac{\cfrac{\cfrac{(\mathsf{id}\in\mathcal{F}_b)}{\{s\}\ \mathsf{id}\ \{s\}}\quad \{\mathbf{s}+\mathbf{t}\}\ \mathbf{f}\ \{\mathbf{t}\}}{\{\langle s,s+t\rangle\}\ \mathsf{id}\times f\ \{\langle s,t\rangle\}}(\times)\quad \cfrac{((+)\in\mathcal{F}_b)}{\{\langle s,t\rangle\}\ (+)\ \{s+t\}}}{\{\langle s,s+t\rangle\}\ (+)\circ(\mathsf{id}\times f)\ \{s+t\}}(\circ)\quad \cfrac{((\mathsf{c})\in\mathcal{F}_b)}{\{s+t\}\ (\mathsf{c})\ \{\langle s+t,s+t\rangle\}}}{\{\langle s,s+t\rangle\}\ (\mathsf{c})\circ(+)\circ(\mathsf{id}\times f)\ \{\langle s+t,s+t\rangle\}}(\circ)}{\{s\}\ \mathsf{Tr}((\mathsf{c})\circ(+)\circ(\mathsf{id}\times f)\ \{s+t\}}(\mathsf{Tr})}{\{\mathbf{s}\}\ \mathsf{fdback}(\mathbf{f})\ \{\mathbf{s}+\mathbf{t}\}}(\mathrm{def})
$$

Fig. 5. Derivation of feedback rule in $\mathbf{H}(\mathcal{C}, H)$

circuits. Notice that the class of valid circuits is closed under sub-circuits. We will show that our embedding is a verification functor for $\mathcal{C}'_m$, so that for valid circuits a set of sound and complete Hoare-logic-like rules is derived.

The monoidal functor $H : \mathcal{C} \to \mathsf{Preord}$ is defined as follows. For each of the objects $X \in \mathcal{C}_o$ (X is of the form $\Sigma \times \ldots \times \Sigma$) we define $H(X)$ as the preordered set of all finite approximations (prefixes) of elements in X. We will use the variables q, r, s, t to range over elements of the objects of Preord. The preorder $t \sqsubseteq s$ on element of $H(X)$ is defined as $s \preceq t$ on $H(X)$, i.e. s is a prefix of t. The top of the preorder is the (tuple of) empty stream(s) ε. Intuitively, each element t of the preordered set corresponds to a subset of X having a common prefix t, with ε corresponding to whole set X. On the morphisms (stream circuits) $f : X \to Y$ in $\mathcal{C}_m$ and $t \in H(X)$, we define $H(f)(t)$ as

$$H(f)(t) :\equiv \mathsf{LCP}\{\sigma \ : \ (\tau f \sigma) \wedge (t \preceq \tau)\}$$

where, for a set of streams S, $\mathsf{LCP}(S)$ denotes the longest common prefix of all streams in that set. $H(f)$ is clearly a monotone function. For the basic morphisms f of $\mathcal{C}$, $H(f)$ can be easily described as:

$$
\begin{aligned}
H(a\times)[t_0, \ldots, t_n] \quad &:\equiv [at_0, \ldots, at_n] \\
H(R)[t_0, \ldots, t_n] \quad &:\equiv [0, t_0, \ldots, t_n] \\
H(\mathsf{c})[t_0, \ldots, t_n] \quad &:\equiv \langle [t_0, \ldots, t_n], [t_0, \ldots, t_n] \rangle \\
H(+)\langle [t_0, \ldots, t_n], [r_0, \ldots, r_m] \rangle &:\equiv [t_0 + r_0, \ldots, t_{\min\{n,m\}} + r_{\min\{n,m\}}]
\end{aligned}
$$

It is easy to check that H respects the monoidal structure of $\mathcal{C}$, since the prefixes of $X \times Y$, i.e. $H(X \times Y)$, can be seen as pairs of prefixes, i.e. elements of $H(X) \times H(Y)$. We now argue that H is also a verification functor.

Lemma 2. *Let $f : X \times Z \to Y \times Z$ be such that $\mathsf{Tr}(f) \in \mathcal{C}'_m$. Moreover, let streams $t \in H(X)$ and $r \in H(Y)$ be fixed. If $r \preceq H(\mathsf{Tr}(f))(t)$ then, for some $q \in H(Z)$, $\langle r, q \rangle \preceq H(f)\langle t, q \rangle$.*

Proof. Notice that the mappings $H(f)$ are continuous, i.e. a finite prefix of the output only requires a finite prefix of the input. Moreover, by the condition that

$\mathsf{Tr}(f)$ is a valid circuit, a finite prefix of the output of a trace only requires a finite prefix of the fixed points in the following sense: for a fixed $t \in H(X)$ the increasing chain

- $q_0 :\equiv \varepsilon$
- $q_{k+1} :\equiv \pi_2(H(f)\langle t, q_k \rangle)$

where π_2 denotes the second projection, is such that for some k we must have $\langle r', q_k \rangle \preceq H(f)\langle t, q_k \rangle$, where $r' = H(\mathsf{Tr}(f))(t)$. By our assumption that $r \preceq H(\mathsf{Tr}(f))(t)$ this implies $\langle r, q_k \rangle \preceq H(f)\langle t, q_k \rangle$, as desired. □

Theorem 3. *$H : \mathcal{C} \to \mathsf{Preord}$, as defined above, is a verification functor for valid circuits.*

Proof. By Lemma 2, it remains to be shown that if (i) $\langle r, q' \rangle \preceq H(f)\langle t, q' \rangle$, for some q', then $r \preceq H(\mathsf{Tr}(f))(t)$. Let τ be such that $t \prec \tau$. By the definition of $\mathsf{Tr}(f)$ (and the fact that f is a valid circuit) there exists a unique fixed point σ such that $\langle \tau, \sigma \rangle f \langle \nu, \sigma \rangle$. By the uniqueness of the fixed point we have that $q' \preceq \sigma$ (otherwise q' could be extended into a different fixed point). Finally, by our assumption (i) it follows that $r \preceq \nu \, (= \mathsf{Tr}(f)(t))$. □

This gives rise to a sound and complete Hoare-logic system for reasoning about *valid* stream circuits. Notice that the rules would not be sound had we not restricted ourselves to valid circuits. For instance, assuming that $a \neq 0$ we have that

$$\{\langle [a], \varepsilon \rangle\} \; \mathsf{c} \circ (+) \; \{\langle \varepsilon, \varepsilon \rangle\}$$

holds but it is not true that $\{[a]\} \; \mathsf{Tr}(\mathsf{c} \circ (+)) \; \{\varepsilon\}$, as argued in Section 2.3.

In this instantiation, the Hoare triples $\{t\} \; f \; \{s\}$ stand for $s \preceq H(f)(t)$. Given that streams represent the Taylor expansion of analytic functions, in this particular example, the pre- and post-conditions in the Hoare logic range over partial sums for these Taylor expansions. We have then categorically obtained a sound and complete formal system $\mathbf{H}(\mathcal{C}, H)$, for reasoning about valid stream circuits and their input-output behaviour over classes of functions with a common partial Taylor sum. Given that a feedback circuit $\mathsf{fdback}(f)$ can be represented in $\mathcal{S}$ as $\mathsf{Tr}((\mathsf{c}) \circ (+) \circ (\mathsf{id} \times f)$, a rule for stream circuit feedbacks can then be derived (see Figure 5) in a similar fashion to the derivation of the rule for while loops (cf. Figure 4).

5 Conclusion and Related Work

In this final section, we discuss other abstractions of Hoare logic in the light of our work.

Kozen's [14] *Kleene Algebra with Test*, KAT, consists essentially of a Kleene algebra with a Boolean subalgebra. Hoare triples $\{P\} \; f \; \{Q\}$ are modelled as equations $Pf = PfQ$, using the multiplication of KAT. The rules of Hoare logic are then obtained as consequences of the equational theory of KAT. Although our work is based on similar ideas (reducing Hoare triples to preorder statements) there does not seem to be a clear cut connection between the two approaches. Whereas Kozen relies on the rich theory of KAT to derive the usual rules of Hoare logic, in our development we use a minimal theory of preordered sets for obtaining soundness and completeness.

Kozen's work is closely related to works on iteration theory [6,16] and dynamic logic [18]. All these, however, focus on the semantics of Hoare logic over flowcharts and while programs, where the intrinsic monoidal structure is *disjoint union*. As we have shown in Section 4.3, our approach is more general including systems with an underlying *cartesian structure* as well.

Abramsky et al. [1] have also studied the categorical structure of Hoare logic, using the notion of *specification structures*. It is easy to see that a tmc $\mathcal{S}$ together with a verification functor $H : \mathcal{S} \to \mathsf{Preord}$ gives rise to a specification structure: H maps objects $X \in \mathcal{S}_o$ to sets $H(X)$, and $H(f)(P) \sqsubseteq Q$ defines a ternary relation $H(X) \times \mathcal{S}(X, Y) \times H(Y)$. The extra structure of preorder and trace, however, allows us to prove an abstract completeness theorem, which does not seem to be the focus of [1].

Blass and Gurevich [5] considered the *underlying logic of Hoare logic*. Since Cook's celebrated completeness result [8], Cook-expressive first-order logics have been used in proofs of relative completeness for Hoare logic. Blass and Gurevich have shown that existential fixed-point logic **EFL** is sufficient for proving Cook's completeness result, without the need for Cook's expressiveness condition. **EFL** contains the necessary constructions to ensure that the functor $H(f)(P)$ of Section 4.1 (producing strongest post-conditions of f on P) can be inductively built, rather than assumed to exist. The fixed-point construction is used in order to produce the fixed point Q of Lemma 1.

Much remains to be done: our main interest is in alternative verification embeddings H of the tmc $\mathcal{C}$ of stream circuits, which will allow other (hopefully more practical) Hoare logics for dynamical systems. One instance of traced symmetric monoidal categories that we have not explored here is that of finite dimensional Hilbert spaces and bounded linear mappings, where trace is defined as a generalisation of matrix trace. This seems significant given its connection with quantum computing and Girard's geometry of interaction [11].

Acknowledgements. The authors gratefully acknowledge support of the UK EPSRC grant GR/S31242/01. We also wish to thank Peter O'Hearn, Nick Benton, Rob Arthan and the anonymous referees for many relevant comments on an earlier version of this paper.

References

1. S. Abramsky, S. Gay, and R. Nagarajan. Specification structures and propositions-as-types for concurrency. In G. Birtwistle and F. Moller, editors, *Logics for Concurrency: Structure vs. Automata*, pages 5–40. Springer-Verlag, 1996.

2. K. R. Apt. Ten years of Hoare's logic: A survey – Part 1. *ACM Transactions on Programming Languages and Systems*, 3(4):431–483, October 1981.
3. E. S. Bainbridge. Feedback and generized logic. *Information and Control*, 31:75–96, 1976.
4. M. Berger, K. Honda, and N. Yoshida. A logical analysis of aliasing in imperative higher-order functions. In *ICFP 2005*, pages 280–293, 2005.
5. A. Blass and Y. Gurevich. The underlying logic of Hoare logic. *Bull. of the Euro. Assoc. for Theoretical Computer Science*, 70:82–110, 2000.
6. S. L. Bloom and Z. Ésik. Floyd-Hoare logic in iteration theories. *J. ACM*, 38(4):887–934, October 1991.
7. R. J. Boulton, R. Hardy, and U. Martin. A Hoare logic for single-input single-output continuous time control systems. *In Proceedings of the 6th International Workshop on Hybrid Systems, Computation and Control, Springer LNCS*, 2623:113–125, 2003.
8. S. A. Cook. Soundness and completeness of an axiom system for program verification. *SIAM J. Comput.*, 7(1):70–90, February 1978.
9. M. H. Escardó and D. Pavlovic. Calculus in coinductive form. In *LICS'1998*, Indiana, USA, June 1998.
10. R. W. Floyd. Assigning meanings to programs. *Proc. Amer. Math. Soc. Symposia in Applied Mathematics*, 19:19–31, 1967.
11. E. Haghverdi and P. Scott. Towards a typed geometry of interaction. In L. Ong, editor, *CSL'2005, LNCS*, volume 3634, pages 216–231, 2005.
12. C. A. R. Hoare. An axiomatic basis for computer programming. *Communications of the ACM*, 12(10):576–585, October 1969.
13. A. Joyal, R. Street, and D. Verity. Traced monoidal categories. *Mathematical Proceedings of the Cambridge Philosophical Society*, 119:447–468, 1996.
14. D. Kozen. On Hoare logic and Kleene algebra with tests. *ACM Transactions on Computational Logic (TOCL)*, 1(1):60–76, 2000.
15. S. Mac Lane. *Categories for the Working Mathematician*, volume 5 of *Graduate texts in mathematics*. Springer, 2nd ed., 1998.
16. E. G. Manes and M. A. Arbib. *Algebraic Approaches to Program Semantics*. AKM series in theoretical computer science. Springer-Verlag, New York, NY, 1986.
17. P. O'Hearn, J. Reynolds, and H. Yang. Local reasoning about programs that alter data structures. In L. Fribourg, editor, *CSL'2001, LNCS*, volume 2142, pages 1–19. Springer-Verlag, 2001.
18. V. R. Pratt. Semantical considerations on Floyd-Hoare logic. In *FoCS'1976*, pages 109–121, 1976.
19. J. C. Reynolds. Separation logic: A logic for shared mutable data structures. In *LICS'2002*, pages 55–74, 2002.
20. J. J. M. M. Rutten. An application of stream calculus to signal flow graphs. *Lecture Notes in Computer Science*, 3188:276–291, 2004.
21. A. K. Simpson and G. D. Plotkin. Complete axioms for categorical fixed-point operators. In *LICS'2000*, pages 30–41, 2000.

Normalization of IZF with Replacement

Wojciech Moczydłowski*

Department of Computer Science, Cornell University, Ithaca, NY, 14853, USA
wojtek@cs.cornell.edu

Abstract. IZF is a well investigated impredicative constructive version of Zermelo-Fraenkel set theory. Using set terms, we axiomatize IZF with Replacement, which we call IZF_R, along with its intensional counterpart IZF_R^-. We define a typed lambda calculus λZ corresponding to proofs in IZF_R^- according to the Curry-Howard isomorphism principle. Using realizability for IZF_R^-, we show weak normalization of λZ by employing a reduction-preserving erasure map from lambda terms to realizers. We use normalization to prove disjunction, numerical existence, set existence and term existence properties. An inner extensional model is used to show the properties for full, extensional IZF_R.

1 Introduction

Four salient properties of constructive set theories are:

- Numerical Existence Property (NEP): From a proof of a statement "there exists a natural number x such that ..." a witness $n \in \mathbb{N}$ can be extracted.
- Disjunction Property (DP): If a disjunction is provable, then one of the disjuncts is provable.
- Set Existence Property (SEP): If $\exists x.\ \phi(x)$ is provable, then there is a formula $\psi(x)$ such that $\exists! x.\ \phi(x) \land \psi(x)$ is provable, where both ϕ and ψ are term-free.
- Term Existence Property (TEP): If $\exists x.\ \phi(x)$ is provable, then $\phi(t)$ is provable for some term t.

How to prove these properties for a given theory? There are a variety of methods applicable to constructive theories. Cut-elimination, proof normalization, realizability, Kripke models.... Normalization proofs, based on Curry-Howard isomorphism, have the advantage of providing an explicit method of witness and program extraction from the proofs. They also provide information about the behaviour of the proof system.

We are interested in intuitionistic set theory IZF. It is essentially what remains of ZF set theory after excluded middle is carefully taken away. An important decision to make on the way is whether to use Replacement or Collection axiom schema. We will call the version with Collection IZF_C and the version with Replacement IZF_R. In the literature, IZF usually denotes IZF_C. Both theories

* Partly supported by NSF grants DUE-0333526 and 0430161.

Z. Ésik (Ed.): CSL 2006, LNCS 4207, pp. 516–530, 2006.

extended with excluded middle are equivalent to ZF. They are not equivalent ([1]). While the proof-theoretic power of IZF_C is equivalent to ZF, the exact power of IZF_R is unknown. Arguably IZF_C is less constructive, as Collection, similarly to Choice, asserts the existence of a set without defining it.

Both versions have been investigated thoroughly. Results up to 1985 are presented in [2] and in [3], later research was concentrated on weaker subsystems, in particular on predicative constructive set theory CZF. [4] describes the set-theoretic apparatus available in CZF and provides further references.

We axiomatize IZF_R, along with its intensional version IZF_R^-, using set terms. We define typed lambda calculus λZ corresponding to proofs in IZF_R^-. We also define realizability for IZF_R^-, in the spirit of [5]. We show weak normalization of λZ by employing a reduction-preserving erasure map from lambda terms to realizers. Strong normalization of λZ does not hold; moreover, we show that in non-well-founded IZF even weak normalization fails.

With normalization in hand, the properties NEP, DP, SEP and TEP follow easily. To show these properties for full, extensional IZF_R, we define an inner model T of IZF_R, consisting of what we call transitively L-stable sets. We show that a formula is true in IZF_R iff its relativization to T is true in IZF_R^-. Therefore IZF_R is interpretable in IZF_R^-. This allows us to use properties proven for IZF_R^-. More detailed proofs of our results can be found in [6].

The importance of these properties in the context of computer science stems from the fact that they make it possible to extract programs from constructive proofs. For example, suppose $IZF_R \vdash \forall n \in \mathbb{N} \exists m \in \mathbb{N}.\ \phi(n, m)$. From this proof a program can be extracted — take a natural number n, construct a proof $IZF_R \vdash \overline{n} \in \mathbb{N}$. Combine the proofs to get $IZF_R \vdash \exists m \in \mathbb{N}.\ \phi(\overline{n}, m)$ and apply NEP to get a number m such that $IZF_R \vdash \phi(\overline{n}, \overline{m})$. We present in details program extraction from IZF_R proofs in [7].

There are many provers with the program extraction capability. However, they are usually based on a variant of type theory, which is a foundational basis very different from set theory. This makes the process of formalizing program specification more difficult, as an unfamiliar new language and logic have to be learned from scratch. [8] strongly argues *against* using type theory for the specification purposes, instead promoting standard set theory.

IZF_R provides therefore the best of both worlds. It is a set theory, with familiar language and axioms. At the same time, programs can be extracted from proofs. Our λZ calculus and the normalization theorem make the task of constructing the prover based on IZF_R not very difficult.

This paper is organized as follows. In section 2 we define IZF_R along with its intensional version IZF_R^-. In section 3 we define a lambda calculus λZ corresponding to IZF_R^- proofs. Realizability for IZF_R^- is defined in section 4 and used to prove normalization of λZ in section 5. We prove the properties in section 6, and show how to derive them for IZF_R in section 7. Comparison with other results can be found in section 8.

2 IZF$_R$

Intuitionistic set theory IZF$_R$ is a first-order theory. We postpone the detailed definition of the logic to Section 3.2, stating at this moment only that equality is not a primitive in the logic. IZF$_R$ is equivalent to ZF, if extended with excluded middle. It's a definitional extension of term-free versions presented in [9], [2] and [1] among others. The signature consists of one binary relational symbol $\in$ and function symbols used in the axioms below. We will generally use letters a, b, c, d, e, f to denote logic variables and t, u, s to denote logic terms. The relational symbol $t = u$ is an abbreviation for $\forall z.\ z \in t \leftrightarrow z \in u$ and $\phi \leftrightarrow \psi$ abbreviates $(\phi \to \psi) \land (\psi \to \phi)$. Function symbols 0 and $S(t)$ are abbreviations for $\{x \in \omega \mid \bot\}$ and $\bigcup\{t, \{t, t\}\}$. Bounded quantifiers and the quantifier $\exists! a$ (there exists exactly one a) are also abbreviations defined in the standard way. The axioms are as follows:

- (PAIR) $\forall a, b \forall c.\ c \in \{a, b\} \leftrightarrow c = a \lor c = b$
- (INF) $\forall c.\ c \in \omega \leftrightarrow c = 0 \lor \exists b \in \omega.\ c = S(b)$
- (SEP$_{\phi(a,\overline{f})}$) $\forall \overline{f} \forall a \forall c.\ c \in S_{\phi(a,\overline{f})}(a, \overline{f}) \leftrightarrow c \in a \land \phi(c, \overline{f})$
- (UNION) $\forall a \forall c.\ c \in \bigcup a \leftrightarrow \exists b \in a.\ c \in b$
- (POWER) $\forall a \forall c.\ c \in P(a) \leftrightarrow \forall b.\ b \in c \to b \in a$
- (REPL$_{\phi(a,b,\overline{f})}$) $\forall \overline{f} \forall a \forall c. c \in R_{\phi(a,b,\overline{f})}(a, \overline{f}) \leftrightarrow (\forall x \in a \exists! y. \phi(x, y, \overline{f}))) \land ((\exists x \in a.\ \phi(x, c, \overline{f}))$
- (IND$_{\phi(a,\overline{f})}$) $\forall \overline{f}.\ (\forall a. (\forall b \in a. \phi(b, \overline{f})) \to \phi(a, \overline{f})) \to \forall a. \phi(a, \overline{f})$
- (L$_{\phi(a,\overline{f})}$) $\forall \overline{f}, \forall a, b.\ a = b \to \phi(a, \overline{f}) \to \phi(b, \overline{f})$

Axioms SEP$_\phi$, REPL$_\phi$, IND$_\phi$ and L$_\phi$ are axiom schemas, and so are the corresponding function symbols — there is one for each formula ϕ. Formally, we define formulas and terms by mutual induction:

$$\phi ::= t \in t \mid t = t \mid \ldots \qquad t ::= a \mid \{t, t\} \mid S_{\phi(a,\overline{f})}(t, \overline{t}) \mid R_{\phi(a,b,\overline{f})}(t, \overline{t}) \mid \ldots$$

IZF$_R^-$ will denote IZF$_R$ without the Leibniz axiom schema L$_\phi$. IZF$_R^-$ is an intensional version of IZF$_R$ — even though extensional equality is used in the axioms, it does not behave as the "real" equality.

Axioms (PAIR), (INF), (SEP$_\phi$), (UNION), (POWER) and (REPL$_\phi$) all assert the existence of certain classes and have the same form: $\forall \overline{a}. \forall c.\ c \in t_A(\overline{a}) \leftrightarrow \phi_A(\overline{a}, c)$, where t_A is a function symbol and ϕ_A a corresponding formula for the axiom A. For example, for (POWER), t_{POWER} is P and ϕ_{POWER} is $\forall b.\ b \in c \to b \in a$. We reserve the notation t_A and ϕ_A to denote the term and the corresponding formula for the axiom A.

3 The λZ Calculus

We present a lambda calculus λZ for IZF$_R^-$, based on the Curry-Howard isomorphism principle. The purely logical part is essentially $\lambda P1$ from [10].

The lambda terms in λZ will be denoted by letters M, N, O, P. Letters x, y, z will be used for lambda variables. There are two kinds of lambda abstractions, one used for proofs of implications, the other for proofs of universal quantification. Since variables in the latter abstractions correspond very closely to the logic variables, we also use letters a, b, c for them. Letters t, s, u are reserved for IZF_R terms. The types in the system are IZF_R formulas.

$$M ::= x \mid M\ N \mid \lambda a.\ M! \mid \lambda x : \phi.\ M \mid \mathrm{inl}(M) \mid \mathrm{inr}(M) \mid \mathrm{fst}(M) \mid \mathrm{snd}(M) \mid [t, M] \mid M t$$

$$\langle M, N\rangle \mid \mathrm{case}(M, x.N, x.O) \mid \mathrm{magic}(M) \mid \mathrm{let}\ [a, x : \phi] = M\ \mathrm{in}\ N \mid \mathrm{ind}_{\phi(a,\overline{b})}(M, \overline{t})$$

$$\mathrm{pairProp}(t, u_1, u_2, M) \mid \mathrm{pairRep}(t, u_1, u_2, M) \mid \mathrm{unionProp}(t, u, M)$$

$$\mathrm{unionRep}(t, u, M) \mid \mathrm{sep}_{\phi(a,\overline{f})}\mathrm{Prop}(t, u, \overline{u}, M) \mid \mathrm{sep}_{\phi(a,\overline{f})}\mathrm{Rep}(t, u, \overline{u}, M)$$

$$\mathrm{powerProp}(t, u, M) \mid \mathrm{powerRep}(t, u, M) \mid \mathrm{infProp}(t, M) \mid \mathrm{infRep}(t, M)$$

$$\mathrm{repl}_{\phi(a,b,\overline{f})}\mathrm{Prop}(t, u, \overline{u}, M) \mid \mathrm{repl}_{\phi(a,b,\overline{f})}\mathrm{Rep}(t, u, \overline{u}, M)$$

The ind term corresponds to the (IND) axiom, and Prop and Rep terms correspond to the respective axioms. To avoid listing all of them every time, we adopt a convention of using axRep and axProp terms to tacitly mean all Rep and Prop terms, for ax being one of pair, union, sep, power, inf and repl. With this convention in mind, we can summarize the definition of the Prop and Rep terms as:

$$\mathrm{axProp}(t, \overline{u}, M) \mid \mathrm{axRep}(t, \overline{u}, M),$$

where the number of terms in the sequence $\overline{u}$ depends on the particular axiom.

The free variables of a lambda term are defined as usual, taking into account that variables in λ, case and let terms bind respective terms. The relation of α-equivalence is defined taking this information into account. We consider α-equivalent terms equal. We denote all free variables of a term M by $FV(M)$ and the free logical variables of a term by $FV_L(M)$. Free (logical) variables of a context Γ are denoted by $FV(\Gamma)$ ($FV_L(\Gamma)$) and defined in a natural way.

3.1 Reduction Rules

The deterministic reduction relation $\to$ arises from the following reduction rules and evaluation contexts:

$$(\lambda x : \phi.\ M)N \to M[x := N] \qquad (\lambda a.\ M)t \to M[a := t] \qquad \mathrm{fst}(\langle M, N\rangle) \to M$$

$$\mathrm{case}(\mathrm{inl}(M), x.N, x.O) \to N[x := M] \qquad \mathrm{case}(\mathrm{inr}(M), x.N, x.O) \to O[x := M]$$

$$\mathrm{snd}(\langle M, N\rangle) \to N \qquad \mathrm{let}\ [a, x : \phi] = [t, M]\ \mathrm{in}\ N \to N[a := t][x := M]$$

$$\mathrm{axProp}(t, \overline{u}, \mathrm{axRep}(t, \overline{u}, M)) \to M$$

$$\mathrm{ind}_{\phi(a,\overline{b})}(M, \overline{t}) \to \lambda c.\ M\ c\ (\lambda b.\lambda x : b \in c.\ \mathrm{ind}_{\phi(a,\overline{b})}(M, \overline{t})\ b)$$

$$[\circ] ::= \mathrm{fst}([\circ]) \mid \mathrm{snd}([\circ]) \mid \mathrm{case}([\circ], x.M, x.N) \mid \mathrm{axProp}(t, \overline{u}, [\circ])$$

$$\mathrm{let}\ [a, y : \phi] = [\circ]\ \mathrm{in}\ N \mid [\circ]\ M \mid \mathrm{magic}([\circ])$$

In the reduction rules for ind terms, the variable x is new. In other words, the reduction relation arises by lazily evaluating the rules above.

Definition 1. *We write $M \downarrow$ if the reduction sequence starting from M terminates. We write $M \downarrow v$ if we want to state that v is the term at which this reduction sequence terminates. We write $M \to^* M'$ if M reduces to M' in some number of steps.*

We distinguish certain λZ terms as values. The values are generated by the following abstract grammar, where M is an arbitrary term. Clearly, there are no reductions possible from values.

$$V ::= \lambda a.\ M \mid \lambda x : \phi.\ M \mid \mathrm{inr}(M) \mid \mathrm{inl}(M) \mid [t, M] \mid \langle M, N\rangle \mid \mathrm{axRep}(t, \overline{u}, M)$$

3.2 Types

The type system for λZ is constructed according to the principle of the Curry-Howard isomorphism for IZF_R^-. Types are IZF_R formulas, and terms are λZ terms. Contexts Γ are finite sets of pairs (x_1, ϕ_i). The *range* of a context Γ is the corresponding first-order logic context that contains only formulas and is denoted by $rg(\Gamma)$. The proof rules follow:

$$\frac{}{\Gamma, x : \phi \vdash x : \phi} \qquad \frac{\Gamma \vdash M : \phi \to \psi \quad \Gamma \vdash N : \phi}{\Gamma \vdash M\ N : \psi} \qquad \frac{\Gamma, x : \phi \vdash M : \psi}{\Gamma \vdash \lambda x : \phi . M : \phi \to \psi}$$

$$\frac{\Gamma \vdash M : \phi \quad \Gamma \vdash N : \psi}{\Gamma \vdash \langle M, N\rangle : \phi \wedge \psi} \quad \frac{\Gamma \vdash M : \phi \wedge \psi}{\Gamma \vdash \mathrm{fst}(M) : \phi} \quad \frac{\Gamma \vdash M : \phi \wedge \psi}{\Gamma \vdash \mathrm{snd}(M) : \psi} \quad \frac{\Gamma \vdash M : \bot}{\Gamma \vdash \mathrm{magic}(M) : \phi}$$

$$\frac{\Gamma \vdash M : \phi}{\Gamma \vdash \mathrm{inl}(M) : \phi \vee \psi} \qquad \frac{\Gamma \vdash M : \psi}{\Gamma \vdash \mathrm{inr}(M) : \phi \vee \psi} \qquad \frac{\Gamma \vdash M : \phi}{\Gamma \vdash \lambda a.\ M : \forall a.\ \phi}\ a \notin FV_L(\Gamma)$$

$$\frac{\Gamma \vdash M : \phi \vee \psi \quad \Gamma, x : \phi \vdash N : \vartheta \quad \Gamma, x : \psi \vdash O : \vartheta}{\Gamma \vdash \mathrm{case}(M, x.N, x.O) : \vartheta} \qquad \frac{\Gamma \vdash M : \forall a.\ \phi}{\Gamma \vdash M\ t : \phi[a := t]}$$

$$\frac{\Gamma \vdash M : \exists a.\ \phi \quad \Gamma, x : \phi \vdash N : \psi}{\Gamma \vdash \mathrm{let}\ [a, x : \phi] := M\ in\ N : \psi}\ a \notin FV_L(\Gamma, \psi) \qquad \frac{\Gamma \vdash M : \phi[a := t]}{\Gamma \vdash [t, M] : \exists a.\ \phi}$$

The rules above correspond to the first-order logic. Formally, we *define* the first-order logic we use by erasing lambda-terms from the typing judgments above and replacing every context by its range. The rest of the rules corresponds to IZF_R^- axioms:

$$\frac{\Gamma \vdash M : \phi_A(t, \overline{u})}{\Gamma \vdash \mathrm{axRep}(t, \overline{u}, M) : t \in t_A(\overline{u})} \qquad \frac{\Gamma \vdash M : t \in t_A(\overline{u})}{\Gamma \vdash \mathrm{axProp}(t, \overline{u}, M) : \phi_A(t, \overline{u})}$$

$$\frac{\Gamma \vdash M : \forall c.\ (\forall b.\ b \in c \to \phi(b, \overline{t})) \to \phi(c, \overline{t})}{\Gamma \vdash \mathrm{ind}_{\phi(b,\overline{c})}(M, \overline{t}) : \forall a.\ \phi(a, \overline{t})}$$

Lemma 1 (Curry-Howard isomorphism). *If $\Gamma \vdash O : \phi$ then $IZF_R^- + rg(\Gamma) \vdash \phi$. If $IZF_R^- + \Gamma \vdash \phi$, then there exists a term M such that $\{(x_\phi, \phi) \mid \phi \in \Gamma\} \vdash M : \phi$.*

Proof. Straightforward. Use

$$\lambda\overline{a}\lambda c.\langle \lambda x : c \in t_A(\overline{a}).\ \mathrm{axProp}(c, \overline{a}, x), \lambda x : \phi_A(c, \overline{a}).\ \mathrm{axRep}(c, \overline{a}, x)\rangle$$

and $\lambda\overline{f}\lambda x : (\forall a.(\forall b.\ b \in a \to \phi(b, \overline{f})) \to \phi(a, \overline{f})).\ \mathrm{ind}_{\phi(b,\overline{c})}(x, \overline{f})$ to witness IZF_R^- axioms.

Lemma 2 (Canonical forms). *Suppose M is a value and $\vdash M : \vartheta$. Then:*

- *If $\vartheta = \phi \vee \psi$, then ($M = \mathrm{inl}(N)$ and $\vdash N : \phi$) or ($M = \mathrm{inr}(N)$ and $\vdash N : \psi$).*
- *If $\vartheta = \exists a.\ \phi$ then $M = [t, N]$ and $\vdash N : \phi[a := t]$.*
- *If $\vartheta = t \in t_A(\overline{u})$ then $M = \mathrm{axRep}(t, \overline{u}, N)$ and $\vdash N : \phi_A(t, \overline{u})$.*

Lemma 3 (Progress). *If $\vdash M : \phi$, then either M is a value or there is a N such that $M \to N$.*

Proof. By induction on $\vdash M : \phi$.

Lemma 4 (Subject reduction). *If $\Gamma \vdash M : \phi$ and $M \to N$, then $\Gamma \vdash N : \phi$.*

Proof. By induction on the definition of $M \to N$, using appropriate substitution lemmas on the way.

Corollary 1. *If $\vdash M : \phi$ and $M \downarrow v$, then $\vdash v : \phi$ and v is a value.*

4 Realizability for IZF_R^-

In this section we work in ZF.

We use terms of type-free version of lambda calculus for realizers. We call this calculus $\lambda\overline{Z}$. The terms of $\lambda\overline{Z}$ are generated by the following grammar and are denoted by $\Lambda_{\overline{Z}}$. The set of $\lambda\overline{Z}$ values is denoted by $\lambda\overline{Z}_v$.

$$M ::= x \mid M\ N \mid \lambda x.\ M \mid \mathrm{inl}(M) \mid \mathrm{inr}(M) \mid \mathrm{magic}(M) \mid \mathrm{fst}(M) \mid \mathrm{snd}(M) \mid \langle M, N\rangle$$

$$\mathrm{case}(M, x.N, x.O) \mid \mathrm{axRep}(M) \mid \mathrm{axProp}(M) \mid \mathrm{ind}(M) \mid \mathrm{app}(M, N)$$

The term $\mathrm{app}(M, N)$ denotes call-by-value application with the evaluation context $\mathrm{app}(M, [\circ])$ and the reduction rule $\mathrm{app}(M, v) \to M\ v$. Essentially, $\lambda\overline{Z}$ results from λZ by erasing of all first-order information. This can be made precise by the definition of the erasure map $\overline{M}$ from terms of λZ to $\lambda\overline{Z}$:

$$\overline{x} = x \qquad \overline{M\ N} = \overline{M}\ \overline{N} \qquad \overline{\lambda a.M} = \overline{M} \qquad \overline{\lambda x : \tau.M} = \lambda x.\ \overline{M} \qquad \overline{\mathrm{inl}(M)} = \mathrm{inl}(\overline{M})$$

$$\overline{[t, M]} = \overline{M} \qquad \overline{\langle M, N\rangle} = \langle \overline{M}, \overline{N}\rangle \qquad \overline{\mathrm{inr}(M)} = \mathrm{inr}(\overline{M}) \qquad \overline{\mathrm{fst}(M)} = \mathrm{fst}(\overline{M})$$

$$\overline{\mathrm{snd}(M)} = \mathrm{snd}(\overline{M}) \qquad \overline{\mathrm{magic}(M)} = \mathrm{magic}(\overline{M}) \quad \overline{\mathrm{let}\,[a, y] = M \text{ in } N} = \mathrm{app}(\lambda y.\ \overline{N}, \overline{M})$$

$$\overline{\mathrm{axRep}(t, \overline{u}, M)} = \mathrm{axRep}(\overline{M}) \quad \overline{\mathrm{axProp}(t, \overline{u}, M)} = \mathrm{axProp}(\overline{M})$$

$$\overline{\mathrm{ind}_\phi(M, \overline{t}, u)} = \mathrm{ind}(\overline{M})$$

We call a λZ reduction *atomic* if it is of the form $(\lambda a.\ M)\ t \to M[a := t]$. The reduction rules and values in $\lambda\overline{Z}$ are induced in an obvious way from λZ, so that if $M \to M'$ is a nonatomic reduction in λZ, then $\overline{M} \to \overline{M'}$, if $M \to M'$ is an atomic reduction in λZ, then $\overline{M} = \overline{M'}$ and if M is a value in λZ not of the form $\lambda a.\ N$, then $\overline{M}$ is a value in $\lambda\overline{Z}$. In particular $\mathrm{ind}(M) \to M\ (\lambda x.\ \mathrm{ind}(M))$.

Lemma 5. *If $\overline{M}$ normalizes, so does M.*

Proof. Any infinite chain of reductions starting from M must contain an infinite number of nonatomic reductions, which map to reductions in $\overline{M}$ in a natural way.

We now move to the definition of the language for the realizability relation.

Definition 2. *A set A is a λ-name iff A is a set of pairs (v, B) such that $v \in \lambda\overline{Z}_v$ and B is a λ-name.*

In other words, λ-names are sets hereditarily labelled by $\lambda\overline{Z}$ values.

Definition 3. *The class of λ-names is denoted by V^λ.*

Formally, V^λ is generated by the transfinite inductive definition on ordinals:

$$V_\alpha^\lambda = \bigcup_{\beta<\alpha} P(\lambda\overline{Z}_v \times V_\beta^\lambda) \qquad V^\lambda = \bigcup_{\alpha \in ORD} V_\alpha^\lambda$$

The λ-*rank* of a λ-name A is the smallest α such that $A \in V_\alpha^\lambda$.

Definition 4. *For any $A \in V^\lambda$, A^+ denotes $\{(M, B) \mid M \downarrow v \wedge (v, B) \in A\}$.*

Definition 5. *A (class-sized) first-order language L arises by enriching the IZF_R signature with constants for all λ-names.*

From now on until the end of this section, symbols M, N, O, P range exclusively over $\lambda\overline{Z}$-terms, letters a, b, c vary over logical variables in the language, letters A, B, C vary over λ-names and letter ρ varies over finite partial functions from logic variables in L to V^λ. We call such functions *environments*.

Definition 6. *For any formula ϕ of L, any term t of L and ρ defined on all free variables of ϕ and t, we define by metalevel mutual induction a realizability relation $M \Vdash_\rho \phi$ in an environment ρ and a meaning of a term $[\![t]\!]_\rho$ in an environment ρ:*

1. $[\![a]\!]_\rho \equiv \rho(a)$
2. $[\![A]\!]_\rho \equiv A$
3. $[\![\omega]\!]_\rho \equiv \omega'$, *where ω' is defined by the means of inductive definition: ω' is the smallest set such that:*
   - $(\mathrm{infRep}(N), A) \in \omega'$ *if* $N \downarrow \mathrm{inl}(O)$, $O \Vdash A = 0$ *and* $A \in V_\omega^\lambda$.
   - *If* $(M, B) \in \omega'^+$, *then* $(\mathrm{infRep}(N), A) \in \omega'$ *if* $N \downarrow \mathrm{inr}(O)$, $O \downarrow \langle M, P\rangle$, $P \Vdash A = S(B)$ *and* $A \in V_\omega^\lambda$.

   It is easy to see that any element of ω' is in V_α^λ for some finite α and so that $\omega' \in V_{\omega+1}^\lambda$.
4. $[\![t_A(\overline{u})]\!]_\rho \equiv \{(\mathrm{axRep}(N), B) \in \lambda\overline{Z}_v \times V_\gamma^\lambda \mid N \Vdash_\rho \phi_A(B, [\![\overline{u}]\!]_\rho)\}$
5. $M \Vdash_\rho \bot \equiv \bot$
6. $M \Vdash_\rho t \in s \equiv M \downarrow v \wedge (v, [\![t]\!]_\rho) \in [\![s]\!]_\rho$

7. $M \Vdash_\rho \phi \wedge \psi \equiv M \downarrow \langle M_1, M_2\rangle \wedge M_1 \Vdash_\rho \phi \wedge M_2 \Vdash_\rho \psi$
8. $M \Vdash_\rho \phi \vee \psi \equiv (M \downarrow \mathrm{inl}(M_1) \wedge M_1 \Vdash_\rho \phi) \vee (M \downarrow \mathrm{inr}(M_1) \wedge M_1 \Vdash_\rho \psi)$
9. $M \Vdash_\rho \phi \to \psi \equiv (M \downarrow \lambda x.\ M_1) \wedge \forall N.\ (N \Vdash_\rho \phi) \to (M_1[x := N] \Vdash_\rho \psi)$
10. $M \Vdash_\rho \forall a.\ \phi \equiv \forall A \in V^\lambda.\ M \Vdash_\rho \phi[a := A]$
11. $M \Vdash_\rho \exists a.\ \phi \equiv \exists A \in V^\lambda.\ M \Vdash_\rho \phi[a := A]$

Note that $M \Vdash_\rho A \in B$ iff $(M, A) \in B^+$.

The definition of the ordinal γ in item 4 depends on $t_A(\overline{u})$. This ordinal is close to the rank of the set denoted by $t_A(\overline{u})$ and is chosen so that Lemma 8 can be proven. Let $\overline{\alpha} = \overline{rank(\llbracket u\rrbracket_\rho)}$. Case $t_A(\overline{u})$ of:

- $\{u_1, u_2\}$ — $\gamma = max(\alpha_1, \alpha_2)$
- $P(u)$ — $\gamma = \alpha + 1$.
- $\bigcup u$ — $\gamma = \alpha$
- $S_{\phi(a,\overline{f})}(u, \overline{u})$ — $\gamma = \alpha_1$.
- $R_{\phi(a,b,\overline{f})}(u, \overline{u})$. Tedious. Use Collection in the metatheory to get the appropriate ordinal. Details can be found in [6].

Lemma 6. *The definition of realizability is well-founded.*

Proof. Use the measure function m which takes a clause in the definition and returns a triple of integers, with lexicographical order in $\mathbb{N}^3$.

- $m(M \Vdash_\rho \phi)$ = ("number of constants ω in ϕ", "number of function symbols in ϕ", "structural complexity of ϕ")
- $m(\llbracket t\rrbracket_\rho)$ = ("number of constants ω in t", "number of function symbols in t", 0)

Since the definition is well-founded, (metalevel) inductive proofs on the definition of realizability are justified.

Lemma 7. *If $A \in V^\lambda_\alpha$, then there is $\beta < \alpha$ such that for all B, if $M \Vdash_\rho B \in A$, then $B \in V^\lambda_\beta$. Also, if $M \Vdash_\rho B = A$, then $B \in V^\lambda_\alpha$.*

The following lemma states the crucial property of the realizability relation.

Lemma 8. $(M, A) \in \llbracket t_A(\overline{u})\rrbracket_\rho$ *iff* $M = \mathrm{axRep}(N)$ *and* $N \Vdash_\rho \phi_A(A, \llbracket\overline{u}\rrbracket_\rho)$.

Proof. For all terms apart from ω, the left-to-right direction is immediate. For the right-to-left direction, suppose $N \Vdash_\rho \phi_A(A, \llbracket\overline{u}\rrbracket_\rho)$ and $M = \mathrm{axRep}(N)$. To show that $(M, A) \in \llbracket t_A(\overline{u})\rrbracket_\rho$, we need to show that $A \in V^\lambda_\gamma$. The proof proceeds by case analysis on $t_A(\overline{u})$. Let $\overline{\alpha} = \overline{rank(\llbracket u\rrbracket_\rho)}$. Case $t_A(\overline{u})$ of:

- $\{u_1, u_2\}$. Suppose that $N \Vdash_\rho A = \llbracket u_1\rrbracket_\rho \vee A = \llbracket u_2\rrbracket_\rho$. Then either $N \downarrow \mathrm{inl}(N_1) \wedge N_1 \Vdash_\rho A = \llbracket u_1\rrbracket_\rho$ or $N \downarrow \mathrm{inr}(N_1) \wedge N_1 \Vdash_\rho A = \llbracket u_2\rrbracket_\rho$. By Lemma 7, in the former case $A \in V^\lambda_{\alpha_1}$, in the latter $A \in V^\lambda_{\alpha_2}$, so $A \in V^\lambda_{max(\alpha_1,\alpha_2)}$.
- $P(u)$. Suppose that $N \Vdash_\rho \forall c.\ c \in A \to c \in \llbracket u\rrbracket_\rho$. Then $\forall C.\ N \Vdash_\rho C \in A \to C \in \llbracket u\rrbracket_\rho$, so $\forall C.\ N \downarrow \lambda x.\ N_1$ and $\forall O.\ (O \Vdash C \in A) \Rightarrow N_1[x := O] \Vdash_\rho C \in \llbracket u\rrbracket_\rho$. Take any $(v, B) \in A$. Then $v \Vdash_\rho B \in A$. So $N_1[x := v] \Vdash_\rho B \in \llbracket u\rrbracket_\rho$. Thus any such B is in V^λ_α, so $A \in V^\lambda_{\alpha+1}$.

- $\bigcup u$. Suppose $N \Vdash_\rho \exists c.\ c \in [\![u]\!]_\rho \wedge A \in c$. It is easy to see that $A \in V^\lambda_\alpha$.
- $S_{\phi(a,\overline{f})}(u, \overline{u})$. Suppose $N \Vdash_\rho A \in [\![u]\!]_\rho \wedge \ldots$. It follows that $A \in V^\lambda_{\alpha_1}$.
- $R_{\phi(a,\overline{f})}(u, \overline{u})$. Tedious. For details, see [6].

For ω, for the left-to-right direction proceed by induction on the definition of ω'. The right-to-left direction is easy, using Lemma 7.

The following sequence of lemmas lays ground for the normalization theorem. They are proven either by induction on the definition of terms and formulas or by induction on the definition of realizability.

Lemma 9. $[\![t[a := s]]\!]_\rho = [\![t]\!]_{\rho[a:=[\![s]\!]_\rho]}$ *and* $M \Vdash_\rho \phi[a := s]$ *iff* $M \Vdash_{\rho[a:=[\![s]\!]_\rho]} \phi$.

Lemma 10. $[\![t[a := s]]\!]_\rho = [\![t[a := [\![s]\!]_\rho]]\!]_\rho$ *and* $M \Vdash_\rho \phi[a := s]$ *iff* $M \Vdash_\rho \phi[a := [\![s]\!]_\rho]$.

Lemma 11. *If* $(M \Vdash_\rho \phi)$ *then* $M \downarrow$.

Lemma 12. *If* $M \to^* M'$ *then* $M' \Vdash_\rho \phi$ *iff* $M \Vdash_\rho \phi$.

Lemma 13. *If* $M \Vdash_\rho \phi \to \psi$ *and* $N \Vdash_\rho \phi$, *then* $M\ N \Vdash_\rho \psi$.

5 Normalization

In this section, environments ρ map lambda variables to $\lambda\overline{Z}$ terms and logic variables to sets in V^λ. Any such environment can be used as a realizability environment by ignoring the mapping of lambda variables.

Definition 7. *For a sequent* $\Gamma \vdash \phi$, $\rho \models \Gamma \vdash \phi$ *means that* $\rho : FV(\Gamma, \phi) \to (V^\lambda \cup \Lambda_{\overline{Z}})$, *for all* $a \in FV_L(\Gamma, \phi)$, $\rho(a) \in V^\lambda$ *and for all* $(x_i, \phi_i) \in \Gamma$, $\rho(x_i) \Vdash_\rho \phi_i$.

Note that if $\rho \models \Gamma \vdash \phi$, then for any term t in Γ, ϕ, $[\![t]\!]_\rho$ is defined and so is the realizability relation $M \Vdash_\rho \phi$.

Definition 8. *For a sequent* $\Gamma \vdash \phi$, *if* $\rho \models \Gamma \vdash \phi$ *and* $M \in \Lambda_{\overline{Z}}$, *then* $M[\rho]$ *is* $M[x_1 := \rho(x_1), \ldots, x_n := \rho(x_n)]$.

Theorem 1. *If* $\Gamma \vdash M : \vartheta$ *then for all* $\rho \models \Gamma \vdash \vartheta$, $\overline{M}[\rho] \Vdash_\rho \vartheta$.

Proof. For any $\lambda\overline{Z}$ term M, M' in the proof denotes $\overline{M}[\rho]$ and IH abbreviates inductive hypothesis. We proceed by metalevel induction on $\Gamma \vdash M : \vartheta$. We show the interesting cases. Case $\Gamma \vdash M : \vartheta$ of:

-
$$\frac{\Gamma \vdash M : \phi_A(t, \overline{u})}{\Gamma \vdash \mathrm{axRep}(t, \overline{u}, M) : t \in t_A(\overline{u})}$$
By IH, $M' \Vdash_\rho \phi_A(t, \overline{u})$. By Lemma 10 this is equivalent to $M' \Vdash_\rho \phi_A([\![t]\!]_\rho, \overline{[\![u]\!]_\rho})$. By Lemma 8 $(\mathrm{axRep}(M'), [\![t]\!]_\rho) \in [\![t_A(\overline{u})]\!]_\rho$, so $\mathrm{axRep}(M') \Vdash_\rho t \in t_A(\overline{u})$, so also $\overline{\mathrm{axRep}(t, \overline{u}, M)}[\rho] \Vdash_\rho t \in t_A(\overline{u})$.

–
$$\frac{\Gamma \vdash M : t \in t_A(\overline{u})}{\Gamma \vdash \text{axProp}(t, \overline{u}, M) : \phi_A(t, \overline{u})}$$

By IH, $M' \Vdash_\rho t \in t_A(\overline{u})$. This means that $M' \downarrow v$ and $(v, [\![t]\!]_\rho) \in [\![t_A(\overline{u})]\!]$. By Lemma 8, $v = \text{axRep}(N)$ and $N \Vdash_\rho \phi_A([\![t]\!]_\rho, \overline{[\![u]\!]_\rho})$. By Lemma 10, $N \Vdash_\rho \phi_A(t, \overline{u})$. Moreover, $\overline{\text{axProp}(t, \overline{u}, M)} = \text{axProp}(M') \to^* \text{axProp}(\text{axRep}(N)) \to N$. Lemma 12 gives us the claim.

–
$$\frac{\Gamma \vdash M : \phi}{\Gamma \vdash \lambda a.\ M : \forall a.\ \phi}$$

By IH, for all $\rho \models \Gamma \vdash M : \phi$, $\overline{M}[\rho] \Vdash \phi$. We need to show that for all $\rho \models \Gamma \vdash \lambda a.\ M : \forall a.\ \phi$, $\overline{\lambda a.\ M} = \overline{M}[\rho] \Vdash_\rho \forall a.\ \phi(a)$. Take any such ρ. We need to show that $\forall A.\ \overline{M}[\rho] \Vdash_\rho \phi[a := A]$. Take any A. By Lemma 9, it suffices to show that $\overline{M}[\rho] \Vdash_{\rho[a:=A]} \phi$. However, $\rho[a := A] \models \Gamma \vdash M : \phi$, so we get the claim by IH.

–
$$\frac{\Gamma \vdash M : \forall a.\ \phi}{\Gamma \vdash M\ t : \phi[a := t]}$$

By IH, $M' \Vdash_\rho \forall a.\ \phi$, so $\forall A. M' \Vdash_\rho \phi[a := A]$. in particular $M' \Vdash_\rho \phi[a := [\![t]\!]_\rho]$, so by Lemma 10 $M' = \overline{(M\ t)}[\rho] \Vdash_\rho \phi[a := t]$.

–
$$\frac{\Gamma \vdash M : \forall c.\ (\forall b.\ b \in c \to \phi(b, \overline{t})) \to \phi(c, \overline{t})}{\Gamma \vdash \text{ind}_{\phi(b,\overline{c})}(M, \overline{t}) : \forall a.\ \phi(a, \overline{t})}$$

We need to show that $\text{ind}(M') \Vdash_\rho \forall a.\ \phi(a, \overline{t})$, that is, that for all A, $\text{ind}(M') \Vdash_\rho \phi(A, \overline{t})$. We proceed by induction on λ-rank of A. Since $\text{ind}(M') \to M'\ (\lambda x.\ \text{ind}(M'))$, by Lemma 12 it suffices to show that $M'\ (\lambda x.\ \text{ind}(M')) \Vdash_\rho \phi(A, \overline{t})$. By IH, we have $M' \Vdash_\rho \forall c.\ (\forall b.\ b \in c \to \phi(b, \overline{t})) \to \phi(c, \overline{t})$, so for all C, $M' \Vdash_\rho (\forall b.\ b \in C \to \phi(b, \overline{t})) \to \phi(C, \overline{t})$. If we take $C = A$, then by Lemma 13 it suffices to show that $\lambda x.\ \text{ind}(M') \Vdash_\rho \forall b.\ b \in A \to \phi(b, \overline{t})$. Take any B. It suffices to show that $\lambda x.\ \text{ind}(M') \Vdash_\rho B \in A \to \phi(B, \overline{t})$. Take any $N \Vdash_\rho B \in A$. By Lemma 7, the λ-rank of B is smaller than the λ-rank of A and so by inner inductive hypothesis $\text{ind}(M') \Vdash_\rho \phi(B, t)$. Since x is new in the reduction rule, $\text{ind}(M') = \text{ind}(M')[x := N]$ and we get the claim.

Corollary 2 (Normalization). *If* $\vdash M : \phi$, *then* $M \downarrow$.

Proof. By Theorem 1, for any $\rho \models (\vdash M : \phi)$, we have $\overline{M}[\rho] \Vdash_\rho \phi$. Take any such ρ, for example mapping all free logic variables of M and ϕ to $\emptyset$. By Lemma 11, $\overline{M}[\rho] \downarrow$, and since $\overline{M} = \overline{M}[\rho]$, $\overline{M} \downarrow$. Lemma 5 gives us the claim.

As the reduction system is deterministic, the distinction between strong and weak normalization does not exist. If the reduction system is extended to allow reductions anywhere inside of the term, the Corollary 2 shows only weak normalization. Strong normalization then, surprisingly, does not hold. One reason, trivial, are ind terms. However, even without them, the system would not strongly normalize, as the following counterexample, invented by Crabbé and adapted to our framework shows:

Theorem 2 (Crabbé's counterexample). *There is a formula ϕ and term M such that $\vdash M : \phi$ and M does not strongly normalize.*

Proof. Let $t = \{x \in 0 \mid x \in x \to \bot\}$. Consider the terms:

$$N \equiv \lambda y : t \in t.\ \mathrm{snd}(\mathrm{sepProp}(t, 0, y))\ y \quad M \equiv \lambda x : t \in 0.\ N\ (\mathrm{sepRep}(t, 0, \langle x, N\rangle))$$

Then it is easy to see that $\vdash N : t \in t \to \bot$, $\vdash M : t \in 0 \to \bot$ and that M does not strongly normalize.

Moreover, a slight (from a semantic point of view) modification to IZF_R^-, namely making it non-well-founded, results in a system which is not even weakly normalizing. A very small fragment is sufficient for this effect to arise. Let T be an intuitionistic set theory consisting of 2 axioms:

- (C) $\forall a.\ a \in c \leftrightarrow a = c$
- (D) $\forall a.\ a \in d \leftrightarrow a \in c \wedge a \in a \to a \in a$.

The constant c denotes a non-well-founded set. The existence of d can be derived from separation axiom: $d = \{a \in c \mid a \in a \to a \in a\}$. The lambda calculus corresponding to T is defined just as for IZF_R^-.

Theorem 3. *There is a formula ϕ and term M such that $\vdash_T M : \phi$ and M does not weakly normalize.*

Proof. It is relatively easy to find a term N such that $\vdash_T N : d \in c$. Take $\phi = d \in d \to d \in d$. The term M below proves the claim.

$$O \equiv \lambda x : d \in d.\ \mathrm{snd}(\mathrm{dRep}(d, c, x))\ x \qquad M \equiv O\ (\mathrm{dProp}(d, c, \langle N, O\rangle)).$$

We believe all these results could be formalized in IZF_C (Collection seems to be necessary for the definition of the realizability set corresponding to the Replacement term in Section 4). Powell has shown in [11] that the notion of rank can be defined meaningfully in intuitionistic set theories, so it should be possible to carry out the developments in Section 4 with the notion of λ-rank which makes sense in IZF_C. We haven't carried out the detailed check, though.

6 Applications

The normalization theorem provides immediately several results.

Corollary 3 (Disjunction Property). *If $IZF_R^- \vdash \phi \vee \psi$, then $IZF_R^- \vdash \phi$ or $IZF_R^- \vdash \psi$.*

Proof. Suppose $\mathrm{IZF}_R^- \vdash \phi \vee \psi$. By Curry-Howard isomorphism, there is a λZ term M such that $\vdash M : \phi \vee \psi$. By Corollary 1, $M \downarrow v$ and $\vdash v : \phi \vee \psi$. By Canonical

Forms, either $v = \mathrm{inl}(N)$ and $\vdash N : \phi$ or $v = \mathrm{inr}(N)$ and $\vdash N : \psi$. By applying the other direction of Curry-Howard isomorphism we get the claim.

Corollary 4 (Term Existence Property). *If $IZF_R^- \vdash \exists x.\ \phi(x)$, then there is a term t such that $IZF_R^- \vdash \phi(t)$.*

Proof. By Curry-Howard isomorphism, there is a λZ-term M such that $\vdash M : \exists x.\ \phi$. By normalizing M and applying Canonical Forms, we get $[t, N]$ such that $\vdash N : \phi(t)$. and thus by Curry-Howard isomorphism $\mathrm{IZF}_R^- \vdash \phi(t)$.

Corollary 5 (Set Existence Property). *If $IZF_R^- \vdash \exists x.\ \phi(x)$ and $\phi(x)$ is term-free, then there is a term-free formula $\psi(x)$ such that $IZF_R^- \vdash \exists! x.\ \phi(x) \land \psi(x)$.*

Proof. Take t from Term Existence Property. It is not difficult to see that there is a term-free formula $\psi(x)$, defining t, so that $\mathrm{IZF}_R^- \vdash (\exists! x.\ \psi(x)) \land \psi(t)$. Then $\mathrm{IZF}_R^- \vdash \exists! x.\ \phi(x) \land \psi(x)$ can be easily derived.

To show NEP, we first define an extraction function F which takes a proof $\vdash M : t \in \omega$ and returns a natural number n. F works as follows:

It normalizes M to natRep(N). By Canonical Forms, $\vdash N : t = 0 \lor \exists y \in \omega.\ t = S(y)$. F then normalizes N to either $\mathrm{inl}(O)$ or $\mathrm{inr}(O)$. In the former case, F returns 0. In the latter, $\vdash O : \exists y. y \in \omega \land t = S(y)$. Normalizing O it gets $[t_1, P]$, where $\vdash P : t_1 \in \omega \land t = S(t_1)$. Normalizing P it gets Q such that $\vdash Q : t_1 \in \omega$. Then F returns $F(\vdash Q : t_1 \in \omega) + 1$.

To show that F terminates for all its arguments, consider the sequence of terms $t, t_1, t_2, \ldots$ obtained throughout the life of F. We have $\mathrm{IZF}_R^- \vdash t = S(t_1)$, $\mathrm{IZF}_R^- \vdash t_1 = S(t_2)$ and so on. Thus, the length of the sequence is at most the rank of the set denoted by t, so F must terminate after at most $rank([\![t]\!])$ steps.

Corollary 6 (Numerical existence property). *If $IZF_R^- \vdash \exists x \in \omega.\ \phi(x)$, then there is a natural number n and term t such that $IZF_R^- \vdash \phi(t) \land t = \overline{n}$.*

Proof. As before, use Curry-Howard isomorphism to get a value $[t, M]$ such that $\vdash [t, M] : \exists x.\ x \in \omega \land \phi(x)$. Thus $M \vdash t \in \omega \land \phi(t)$, so $M \downarrow \langle M_1, M_2 \rangle$ and $\vdash M_1 : t \in \omega$. Take $n = F(\vdash M_1 : t \in \omega)$. It's easy to see that patching together in an appropriate way proofs obtained throughout the execution of F, a proof of $t = \overline{n}$ for some natural number n can be produced.

This version of NEP differs from the one usually found in the literature, where in the end $\phi(\overline{n})$ is derived. However, IZF_R^- does not have the Leibniz axiom for the final step. We conjecture that it is the only version which holds in non-extensional set theories.

7 Extensional IZF_R

We will show that we can extend our results to full IZF_R. We work in IZF_R^-.

Lemma 14. *Equality is an equivalence relation.*

Definition 9. *A set C is* L-stable*, if $A \in C$ and $A = B$ implies $B \in C$.*

Definition 10. *A set C is* transitively L-stable *($TLS(C)$ holds) if it is L-stable and every element of C is transitively L-stable.*

This definition is formalized in a standard way, using transitive closure, available in IZF_R^-, as shown e.g. in [4]. We denote the class of transitively L-stable sets by T. The statement $V = T$ means that $\forall A.\ TLS(A)$. Class T in IZF_R^- plays a similar role to the class of well-founded sets in ZF without Foundation. By $\in$-induction we can prove:

Lemma 15. *$IZF_R \vdash V = T$.*

The restriction of a formula ϕ to T, denoted by ϕ^T, is defined as usual, taking into account the following translation of terms:

$$a^T \equiv a \quad \{t,u\}^T \equiv \{t^T, u^T\} \qquad \omega^T \equiv \omega \quad (\bigcup t)^T \equiv \bigcup t^T \qquad (P(t))^T \equiv P(t^T) \cap T$$

$$(S_{\phi(a,\overline{f})}(u,\overline{u}))^T \equiv S_{\phi^T(a,\overline{f})}(u^T, \overline{u^T}) \quad (R_{\phi(a,b,\overline{f})}(t,\overline{u}))^T \equiv R_{b \in T \wedge \phi^T(a,b,\overline{f})}(t^T, \overline{u^T})$$

The notation $T \models \phi$ means that ϕ^T holds. It is not very difficult to show:

Theorem 4. *$T \models IZF_R$.*

Lemma 16. *$IZF_R \vdash \phi$ iff $IZF_R^- \vdash \phi^T$.*

Corollary 7. *IZF_R satisfies DP and NEP.*

Proof. For DP, suppose $\mathrm{IZF}_R \vdash \phi \vee \psi$. By Lemma 16, $\mathrm{IZF}_R^- \vdash \phi^T \vee \psi^T$. By DP for IZF_R^-, either $\mathrm{IZF}_R^- \vdash \phi^T$ or $\mathrm{IZF}_R^- \vdash \psi^T$. Using Lemma 16 again we get either $\mathrm{IZF}_R \vdash \phi$ or $\mathrm{IZF}_R \vdash \psi$.

For NEP, suppose $\mathrm{IZF}_R \vdash \exists x.\ x \in \omega \wedge \phi(x)$. By Lemma 16, $\mathrm{IZF}_R^- \vdash \exists x.\ x \in T \wedge x \in \omega^T.\ \phi^T(x)$, so $\mathrm{IZF}_R^- \vdash \exists x \in \omega^T.\ x \in T \wedge \phi^T(x)$. Since $\omega^T = \omega$, using NEP for IZF_R^- we get a natural number n such that $\mathrm{IZF}_R^- \vdash \exists x.\ \phi^T(x) \wedge x = \overline{n}$. By Lemma 16 and $\overline{n} = \overline{n}^T$, we get $\mathrm{IZF}_R \vdash \exists x.\ \phi(x) \wedge x = \overline{n}$. By the Leibniz axiom, $\mathrm{IZF}_R \vdash \phi(\overline{n})$.

We cannot establish TEP and SEP as easily, since it is not the case that $t^T = t$ for all terms t. However, a simple modification to the axiomatization of IZF_R yields these results too. It suffices to guarantee that whenever a set is defined, it must be in T. To do this, we modify three axioms and add one new, axiomatizing transitive closure. Let $PTC(a,c)$ be a formula that says: $a \subseteq c$ and c is transitive. The axioms are:

- (SEP'$_{\phi(a,\overline{f})}$) $\forall \overline{f} \forall a \forall c.\ c \in S_{\phi(a,\overline{f})}(a, \overline{f}) \leftrightarrow c \in a \wedge \phi^T(c, \overline{f})$
- (POWER') $\forall a \forall c. c \in P(a) \leftrightarrow c \in T \wedge \forall b.\ b \in c \to b \in a$
- (REPL'$_{\phi(a,b,\overline{f})}$) $\forall \overline{f} \forall a \forall c. c \in R_{\phi(a,b,\overline{f})}(a, \overline{f}) \leftrightarrow (\forall x \in a \exists! y \in T. \phi^T(x, y, \overline{f})) \wedge (\exists x \in a.\ \phi^T(x, c, \overline{f}))$
- (TC) $\forall a, c.\ c \in TC(a) \leftrightarrow (c \in a \vee \exists d \in TC(a).\ c \in d) \wedge \forall d.\ PTC(a,d) \to c \in d.$

In the modified axioms, the definition of T is written using TC and relativization of formulas to T this time leaves terms intact, we set $t^T \equiv t$ for all terms t.

It is not difficult to see that this axiomatization is equivalent to the old one and is still a definitional extension of term-free versions of [9], [2] and [1].We can therefore adopt it as the official axiomatization of IZF_R. All the developments in sections 4-8 can be done for the new axiomatization in the similar way. In the end we get:

Corollary 8. *IZF_R satisfies DP, NEP, TEP and SEP.*

A different technique to tackle the problem of the Leibniz axiom bas been used by Friedman in [12]. He defines new membership ($\in^*$) and equality ($\sim$) relations in an intensional universe from scratch, so that $(V, \in^*, \sim)$ interprets his intuitionistic set theory along with the Leibniz axiom. Our T, on the other hand, utilizes the existing $\in, =$ relations. We plan to present an alternative normalization proof, where the method to tackle the Leibniz axiom is closer to Friedman's ideas, in the forthcoming [13].

8 Related Work

In [9], DP, NEP, SEP are proven for IZF_R without terms. TEP is proven for comprehension terms, the full list of which is not recursive. It is easy to see that IZF_R is a definitional extension of Myhill's version. Our results therefore improve on [9], by providing an explicit recursive list of terms corresponding to IZF_R axioms to witness TEP.

In [14] strong normalization of a constructive set theory without induction and replacement axioms is shown using Girard's method. As both normalization and theory are defined in a nonstandard way, it is not clear if the results could entail any of DP, NEP, SEP and TEP for the theory.

[15] defines realizability using lambda calculus for classical set theory conservative over ZF. The types for the calculus are defined. However, it seems that the types correspond more to the truth in the realizability model than to provable statements in the theory. Moreover, the calculus doesn't even weakly normalize.

In [16], a set theory without the induction and replacement axioms is interpreted in the strongly normalizing lambda calculus with types based on $F\omega.2$. This has been extended with conservativeness result in [17].

In [18], DP and NEP along with other properties are derived for CZF using a combination of realizability and truth. The technique likely extends to IZF_C, but it does not seem to be strong enough to prove SEP and TEP for IZF_R.

Acknowledgements

I would like to thank my advisor, Bob Constable, for giving me the idea for this research and support, Richard Shore for helpful discussions, Daria Walukiewicz-Chrząszcz for the higher-order rewriting counterexample, thanks to which I could prove Theorem 3 and anonymous referees for helpful comments.

References

1. Friedman, H., Ŝĉedrov, A.: The lack of definable witnesses and provably recursive functions in intuitionistic set theories. Advances in Mathematics **57** (1985) 1–13
2. Beeson, M.: Foundations of Constructive Mathematics. Springer-Verlag (1985)
3. Ŝĉedrov, A.: Intuitionistic set theory. In: Harvey Friedman's Research on the Foundations of Mathematics, Elsevier (1985) 257–284
4. Aczel, P., Rathjen, M.: Notes on constructive set theory. Technical Report 40, Institut Mittag-Leffler (The Royal Swedish Academy of Sciences) (2000/2001)
5. McCarty, D.: Realizability and Recursive Mathematics. D.Phil. Thesis, University of Oxford (1984)
6. Moczydłowski, W.: Normalization of IZF with Replacement. Technical Report 2006-2024, Computer Science Department, Cornell University (2006)
7. Constable, R., Moczydłowski, W.: Extracting Programs from Constructive HOL Proofs via IZF Set-Theoretic Semantics. (2006) Submitted to IJCAR 2006.
8. Lamport, L., Paulson, L.C.: Should your specification language be typed? ACM-TOPLAS: ACM Transactions on Programming Languages and Systems **21** (1999)
9. Myhill, J.: Some properties of intuitionistic Zermelo-Fraenkel set theory. In: Cambridge Summer School in Mathematical Logic. Volume 29., Springer (1973) 206–231
10. Sørensen, M., Urzyczyn, P.: Lectures on the Curry-Howard isomorphism. DIKU rapport 98/14, DIKU (1998)
11. Powell, W.: Extending Gödel's negative interpretation to ZF. Journal of Symbolic Logic **40** (1975) 221–229
12. Friedman, H.: The consistency of classical set theory relative to a set theory with intuitionistic logic. Journal of Symbolic Logic **38** (1973) 315–319
13. Moczydłowski, W.: Normalization of IZF with Replacement and Inaccessible Sets. Submitted for publication (2006)
14. Bailin, S.C.: A normalization theorem for set theory. J. Symb. Log. **53**(3) (1988) 673–695
15. Louis Krivine, J.: Typed lambda-calculus in classical Zermelo-Fraeænkel set theory. Archive for Mathematical Logic **40**(3) (2001) 189–205
16. Miquel, A.: A Strongly Normalising Curry-Howard Correspondence for IZF Set Theory. In Baaz, M., Makowsky, J.A., eds.: CSL. Volume 2803 of Lecture Notes in Computer Science., Springer (2003) 441–454
17. Dowek, G., Miquel, A.: Cut elimination for Zermelo's set theory. (2006) Manuscript, available from the web pages of the authors.
18. Rathjen, M.: The disjunction and related properties for constructive Zermelo-Fraenkel set theory. Journal of Symbolic Logic **70** (2005) 1233–1254

Acyclicity and Coherence in Multiplicative Exponential Linear Logic

Michele Pagani*

Dipartimento di Filosofia – Università Roma Tre
Via Ostiense 231 – 00124 – Roma
pagani@uniroma3.it

Abstract. We give a geometric condition that characterizes $MELL$ proof structures whose interpretation is a clique in non-uniform coherent spaces: visible acyclicity.

We define the *visible paths* and we prove that the proof structures which have no visible cycles are exactly those whose interpretation is a clique. It turns out that visible acyclicity has also nice computational properties, especially it is stable under cut reduction.

1 Introduction

Proof nets are a graph-theoretical presentation of linear logic proofs, that gives a more geometric account of logic and computation. Indeed, proof nets are in a wider set of graphs, that of *proof structures*. Specifically, proof nets are those proof structures which correspond to logically correct proofs, i.e. sequent calculus proofs.

A striking feature of the theory of proof nets is the characterization by geometric conditions of such a logical correctness. In multiplicative linear logic (MLL) Danos and Regnier give (see [1]) a very simple correctness condition, which consists in associating with any proof structure a set of subgraphs, called *switchings*, and then in characterizing the proof nets as those structures whose switchings are connected and acyclic.

Later Fleury and Retoré relax in [2] Danos-Regnier's condition, proving that acyclicity alone characterizes those structures which correspond to the proofs of MLL enlarged with the mix rule (Figure 4). More precisely, the authors define the *feasible paths* as those paths which are "feasible" in the switchings, then they prove that a proof structure is associated with a proof of MLL plus mix if and only if all its feasible paths are acyclic.

Proof structures are worthy since cut reduction is defined straight on them, not only on proof nets. We can thus consider a concrete denotational semantics for proof structures, as for example the relational semantics. Here the key notion is that of *experiment*, introduced by Girard in [3]. Experiments allow to associate with any proof structure (not only proof net) π a set of points $[\![\pi]\!]$ invariant under the reduction of the cuts in π.

* This work has been partially supported by MIUR national project FOLLIA.

Z. Ésik (Ed.): CSL 2006, LNCS 4207, pp. 531–545, 2006.

In [4] Bucciarelli and Ehrhard provide a notion of coherence between the points of the relational semantics, so introducing *non-uniform coherent spaces* and *non-uniform cliques*. Such spaces are called non-uniform since the web of their exponentials does not depend on the coherence relation, as it does instead in case of usual (uniform) coherent spaces (see [3]).[1]

Hence we have a geometrical notion – feasible acyclicity – dealing with logical correctness, and a semantical one – clique – defined in the framework of coherent spaces. Such notions are deeply related in MLL: from [3] it is known that for any proof structure π, if π is feasible acyclic then its interpretation $[\![\pi]\!]$ is a clique; conversely Retoré proves in [5] that for any cut-free π, if $[\![\pi]\!]$ is a clique then π is feasible acyclic.

Such results tighten the link between coherent spaces and multiplicative proof nets: as a corollary we derive in [6] the full-completeness of coherent spaces for MLL with mix.

What happens to this correspondence in presence of the exponentials, i.e. in multiplicative exponential linear logic ($MELL$)?

The notion of feasible path can be easily extended to $MELL$ (Definition 5). In this framework feasible acyclicity characterizes the proof structures which correspond to the proofs of $MELL$ sequent calculus (Figure 3) enlarged with the rules of mix and daimon (Figure 4). However the link between feasible acyclicity and coherent spaces fails: there are proof structures which are associated with cliques even if they have feasible cycles (for example Figure 5).

The main novelty of our paper is to find out a geometrical condition on $MELL$ proof structures, which recovers the missed link with coherent spaces. In Definition 6 we introduce the *visible paths*, which are an improvement of the feasible paths in presence of exponentials. Then we prove in Theorems 2 and 3:

- for any $MELL$ proof structure π, if π is visible acyclic, then $[\![\pi]\!]$ is a non-uniform clique;
- for any $MELL$ cut-free proof structure π, if $[\![\pi]\!]$ is a non-uniform clique then π is visible acyclic.

Finally, it turns out that visible acyclicity has also nice computational properties, especially it is stable under cut reduction (Theorem 4), moreover it assures confluence and strong normalization.

2 Proof Structures

The formulas of $MELL$ are defined by the following grammar:

$$F ::= X \mid X^\perp \mid F \parr F \mid F \otimes F \mid ?F \mid !F$$

As usual we set $(A \parr B)^\perp = A^\perp \otimes B^\perp$, $(A \otimes B)^\perp = A^\perp \parr B^\perp$, $(?F)^\perp = !F^\perp$ and $(!F)^\perp = ?F^\perp$. We denote by capital Greek letters $\Sigma, \Pi, \ldots$ the multisets of

[1] Actually the difference between uniform and non-uniform spaces has a relevance only in presence of exponentials: in MLL we can speak simply of coherent spaces and cliques.

formulas. We write $A_1 \odot \ldots \odot A_{n-1} \odot A_n$ for $A_1 \odot (\ldots \odot (A_{n-1} \odot A_n) \ldots)$, where $\odot$ is $\parr$ or $\otimes$.

Proof structures are directed graphs with boxes – highlighted subgraphs – and pending edges – edges without target. Edges are labeled by $MELL$ formulas and nodes (called *links*) are labeled by $MELL$ rules. Links are defined together with both an arity (the number of incident edges, called the *premises of the link*) and a coarity (the number of emergent edges, called the *conclusions of the link*). The links of $MELL$ are ax, cut, $\otimes$, $\parr$, $!$, $?d$, $?c$, $?w$, as defined in Figure 1. The orientation of the edges will be always from top to bottom, so that we may omit the arrows of the edges.

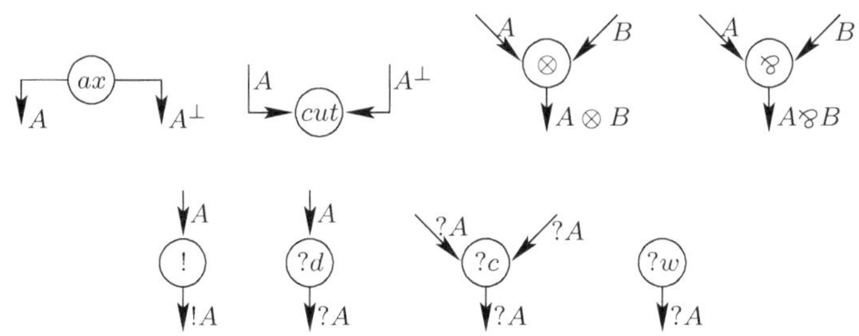

Fig. 1. $MELL$ links

Definition 1 (Proof structure, [3]). *A* proof structure π *is a directed graph whose nodes are MELL links and such that:*

1. *every edge is conclusion of exactly one link and premise of at most one link. The edges which are not premise of any link are the* conclusions *of the proof structure;*
2. *with every link ! o is associated a unique subgraph of π, denoted by π^o, satisfying condition 1 and such that one conclusion of π^o is the premise of o and all further conclusions of π^o are labeled by ?-formulas. π^o is called the* exponential box of o *(or simply the box of o) and it is represented by a dash frame. The conclusion of o is called* the principal door of π^o, *the conclusions of π^o labeled by ?-formulas are called* the auxiliary doors of π^o*;*
3. *two exponential boxes are either disjoint or included one in the other.*

The depth of a link *in π is the number of boxes in which it is contained. The* depth of an edge *a is 0 in case a is a conclusion of π, otherwise it is the depth of the link of which a is premise.*

The depth of π *is the maximum depth of its links, the* size of π *is the number of its links, the* cosize of π *is the number of its links ?c.*

A link l is terminal *when either l is a ! and all the doors of π^l are conclusions of π, or l is not a ! and all the conclusions of l are conclusions of π.*

Proof structures are denoted by Greek letters: $\pi, \sigma, \ldots$, edges by initial Latin letters: $a, b, c \ldots$ and links by middle-position Latin letters: $l, m, n, o \ldots$. We write $a : A$ if a is an edge labeled by the formula A.

A proof structure without cuts is called *cut-free*. The cut reduction rules are graph rewriting rules which modify a proof structure π, obtaining a proof structure π' with the same conclusions as π. We do not give here the cut reduction rules, which are completely standard (see [3]). However we remark that at the level of proof structures there exist cuts, called *deadlocks*, which are irreducible. These are the cuts between the two premises of an axiom, and the cuts between two doors of an exponential box (see Figure 2).

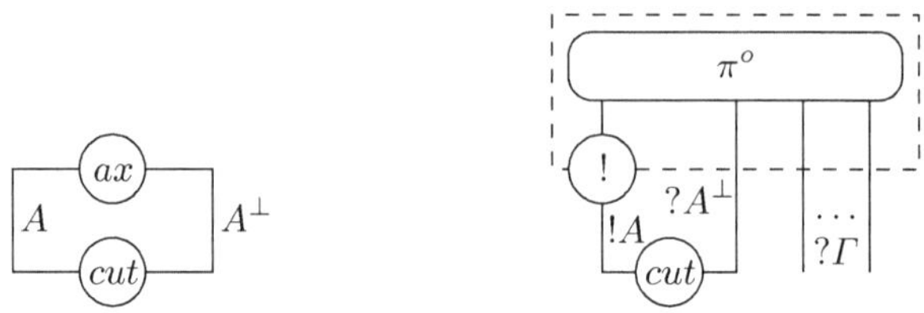

Fig. 2. Examples of deadlocks

We denote by $\pi \rightsquigarrow_\beta \pi'$ whenever π' is the result of the reduction of a cut in π. As always, $\rightarrow_\beta$ is the reflexive and transitive closure of $\rightsquigarrow_\beta$ and $=_\beta$ is the symmetrical closure of $\rightarrow_\beta$.

3 Non-uniform Coherent Spaces

We recall that a *multiset* v is a set of elements in which repetitions can occur. We denote multisets by square brackets, for example $[a, a, b]$ is the multiset containing twice a and once b. The plus symbol $+$ denotes the disjoint union of multisets, whose neutral element is the empty multiset $\emptyset$. If n is a number and v a multiset, we denote by nv the multiset $\underbrace{v + \ldots + v}_{n \text{ times}}$. The support of v, denoted by $Supp(v)$, is the set of elements of v, for example $Supp([a, a, b]) = \{a, b\}$.

If $\mathcal{C}$ is a set, by $M(\mathcal{C})$ (resp. $M_{fin}(\mathcal{C})$) we mean the set of all multisets (resp. finite multisets) of $\mathcal{C}$.

Definition 2 (Non-uniform coherent space, [4]). *A* non-uniform coherent space *$\mathcal{X}$ is a triple $(|\mathcal{X}|, \overset{\frown}{\smile}, \equiv)$, where $|\mathcal{X}|$ is a set, called the* web *of $\mathcal{X}$, while $\overset{\frown}{\smile}$ and $\equiv$ are two binary symmetric relations on $|\mathcal{X}|$, such that for every $x, y \in \mathcal{X}$, $x \equiv y$ implies $x \overset{\frown}{\smile} y$. $\overset{\frown}{\smile}$ (resp. $\equiv$) is called the* coherence *(resp. the* neutrality*) of $\mathcal{X}$.*

A clique *of $\mathcal{X}$ is a subset $\mathcal{C}$ of $|\mathcal{X}|$ such that for every $x, y \in \mathcal{C}$, $x \overset{\frown}{\smile} y$.*

Remark the difference from Girard's (uniform) coherent spaces: we do not require the relation $\overset{\frown}{\smile}$ to be also reflexive.

We will write $x \overset{\frown}{\smile} y\,[\mathcal{X}]$ and $x \equiv y\,[\mathcal{X}]$ if we want to state explicitly which coherent space $\overset{\frown}{\smile}$ and $\equiv$ refer to. We introduce the following notation, well-known in the framework of coherent spaces:

strict coherence: $x \frown y\,[\mathcal{X}]$, if $x \overset{\frown}{\smile} y\,[\mathcal{X}]$ and $x \not\equiv y\,[\mathcal{X}]$;
incoherence: $x \overset{\smile}{\frown} y\,[\mathcal{X}]$, if not $x \frown y\,[\mathcal{X}]$;
strict incoherence: $x \smile y\,[\mathcal{X}]$, if $x \overset{\smile}{\frown} y\,[\mathcal{X}]$ and $x \not\equiv y\,[\mathcal{X}]$.

Notice that we may define a non-uniform coherent space by specifying its web and two well chosen relations among $\equiv$, $\overset{\frown}{\smile}$, $\frown$, $\overset{\smile}{\frown}$, $\smile$.

Let $\mathcal{X}$ be a non-uniform coherent space, the *non-uniform coherent model on* $\mathcal{X}$ ($\mathfrak{nuCoh}^{\mathcal{X}}$) associates with formulas non-uniform coherent spaces, by induction on the formulas, as follows:

- with X it associates $\mathcal{X}$;
- with $A^{\perp}$ it associates the following $\mathcal{A}^{\perp}$: $|\mathcal{A}^{\perp}| = |\mathcal{A}|$, the neutrality and coherence of $\mathcal{A}^{\perp}$ are as follows:
  - $a \equiv a'\,[\mathcal{A}^{\perp}]$ iff $a \equiv a'\,[\mathcal{A}]$,
  - $x \overset{\frown}{\smile} y\,[\mathcal{A}^{\perp}]$ iff $x \overset{\smile}{\frown} y\,[\mathcal{A}]$;
- with $A \otimes B$ it associates the following $\mathcal{A} \otimes \mathcal{B}$: $|\mathcal{A} \otimes \mathcal{B}| = |\mathcal{A}| \times |\mathcal{B}|$, the neutrality and coherence of $\mathcal{A} \otimes \mathcal{B}$ are as follows:
  - $< a, b > \equiv < a', b' > [\mathcal{A} \otimes \mathcal{B}]$ iff $a \equiv a'\,[\mathcal{A}]$ and $b \equiv b'\,[\mathcal{B}]$,
  - $< a, b > \overset{\frown}{\smile} < a', b' > [\mathcal{A} \otimes \mathcal{B}]$ iff $a \overset{\frown}{\smile} a'\,[\mathcal{A}]$ and $b \overset{\frown}{\smile} b'\,[\mathcal{B}]$;
- with $!A$ it associates the following $!\mathcal{A}$: $|!\mathcal{A}| = M_{fin}(|\mathcal{A}|)$, the strict incoherence and neutrality of $!\mathcal{A}$ are as follows:
  - $v \smile u\,[!\mathcal{A}]$ iff $\exists a \in v$ and $\exists a' \in u$, s.t. $a \smile a'\,[\mathcal{A}]$,
  - $v \equiv u\,[!\mathcal{A}]$ iff not $v \smile u\,[!\mathcal{A}]$ and there is an enumeration of v (resp. of u) $v = [a_1, \dots, a_n]$ (resp. $u = [a'_1, \dots, a'_n]$), s.t. for each $i \leq n$, $a_i \equiv a'_i\,[\mathcal{A}]$.

Of course the space $\mathcal{A} \invamp \mathcal{B}$ is defined by $(\mathcal{A}^{\perp} \otimes \mathcal{B}^{\perp})^{\perp}$ as well as $?\mathcal{A}$ is defined by $(!\mathcal{A}^{\perp})^{\perp}$.

Remark that $!\mathcal{A}$ may have elements strictly incoherent with themselves, i.e. $\overset{\frown}{\smile}$ is not reflexive. For example, suppose a, b are two elements in $\mathcal{A}$ such that $a \smile b\,[\mathcal{A}]$, then the multiset $[a, b]$ is an element of $!\mathcal{A}$ such that $[a, b] \smile [a, b]\,[!\mathcal{A}]$.

For each proof structure π, we define the *interpretation of* π *in* $\mathfrak{nuCoh}^{\mathcal{X}}$, denoted by $[\![\pi]\!]_{\mathcal{X}}$, where the index $_{\mathcal{X}}$ is omitted in case it is clear which coherent space is associated with X. $[\![\pi]\!]$ is defined by using the notion of experiment, introduced by Girard in [3].

We define an experiment e on π by induction on the exponential depth of π.[2]

Definition 3 (Experiment). *A* $\mathfrak{nuCoh}^{\mathcal{X}}$ *experiment* e *on a proof structure* π, *denoted by* $e : \pi$, *is a function which associates with every link* $!\,o$ *at depth* 0 *a multiset* $[e_1^o, \dots, e_k^o]$ *(for* $k \geq 0$*) of experiments on* π^o, *and with every edge* $a : A$ *at depth* 0 *an element of* $\mathcal{A}$, *s.t.:*

[2] Definition 3 is slightly different from the usual one (see for example [7]), namely e is defined only on the edges at depth 0 of π. Such a difference however does not play any crucial role in the proof of our result.

- *if a, b are the conclusions (resp. the premises) of a link ax (resp. cut) at depth* 0*, then $e(a) = e(b)$;*
- *if c is the conclusion of a link $\parr$ or $\otimes$ at depth* 0 *with premises a, b, then $e(c) =< e(a), e(b) >$;*
- *if c is the conclusion of a link $?d$ with premise a, then $e(c) = [e(a)]$; if c is the conclusion of a link $?c$ with premises a, b, then $e(c) = e(a) + e(b)$; if c is the conclusion of a link $?w$, then $e(c) = \emptyset$;*
- *if c is a door of a box associated with a link $!\ o$ at depth* 0*, let a be the premise of o and $e(o) = [e_1^o, \ldots, e_k^o]$. If c is the principal door then $e(c) = [e_1^o(a), \ldots, e_k^o(a)]$, if c is an auxiliary door then $e(c) = e_1^o(c) + \ldots + e_k^o(c)$;*

If $c_1, \ldots, c_n$ are the conclusions of π, then the result of e*, denoted by $|e|$, is the element $< e(c_1), \ldots, e(c_n) >$. The* interpretation of π in $\mathfrak{nuCoh}^{\mathcal{X}}$ *is the set of the results of its experiments:*

$$[\![\pi]\!]_{\mathcal{X}} = \left\{ |e| \ \ s.t.\ e \text{ is a } \mathfrak{nuCoh}^{\mathcal{X}} \text{ experiment on } \pi \right\}$$

The interpretation of a proof structure is invariant under cut reduction:

Theorem 1 (Soundness of $\mathfrak{nuCoh}^{\mathcal{X}}$). *For every proof structures π, π', $\pi =_\beta \pi'$ implies $[\![\pi]\!]_{\mathcal{X}} = [\![\pi']\!]_{\mathcal{X}}$.*

Proof (Sketch). It is a straightforward variant of the original proof given by Girard in [3] for proof nets. □

For concluding we explain why we choose the non-uniform variant of coherent spaces for proving our result.

The main difference between uniform and non-uniform coherent spaces is in the definition of the web of $!\mathcal{A}$. The non-uniform web of $!\mathcal{A}$ contains all finite multisets of elements in $\mathcal{A}$, while the uniform web of $!\mathcal{A}$ contains only those finite multisets whose support is a clique in $\mathcal{A}$ (see [3]).

Uniform webs thus have less elements than the non-uniform ones. Less elements means less experiments for proof structures. Indeed uniform experiments must satisfy besides the conditions of Definition 3 also a *uniformity condition*, namely the elements associated with edges of type $!A$ (or $?A^\perp$) have to be in the uniform web of $!\mathcal{A}$ (see [7]).

Concerning our result, we believe that the experiment e_ϕ, defined in the proof of Lemma 2, satisfies also the uniformity condition, but we haven't proved yet. Thus we conjecture that Theorem 3 holds also for uniform coherent spaces, but the proof should be quite harder.[3]

[3] Uniform coherent spaces are special cases of non-uniform spaces, where $\equiv$ coincides with the identity. Denote by $[\![\pi]\!]_{\mathfrak{Coh}^{\mathcal{X}}}$ (resp. $[\![\pi]\!]_{\mathfrak{nuCoh}^{\mathcal{X}}}$) the uniform (resp. non-uniform) interpretation of π based on a space $\mathcal{X}$. One can prove $[\![\pi]\!]_{\mathfrak{Coh}^{\mathcal{X}}} \subseteq [\![\pi]\!]_{\mathfrak{nuCoh}^{\mathcal{X}}}$. This means that if $[\![\pi]\!]_{\mathfrak{nuCoh}^{\mathcal{X}}}$ is a clique so is $[\![\pi]\!]_{\mathfrak{Coh}^{\mathcal{X}}}$, while the vice-versa does not hold in general. Hence remark that Theorem 3 (resp. Theorem 2) is stronger (resp. weaker) when it refers to $\mathfrak{Coh}^{\mathcal{X}}$ instead of $\mathfrak{nuCoh}^{\mathcal{X}}$.

4 Paths and Acyclicity

In Figure 3 we present the sequent calculus for $MELL$. As we noticed in the Introduction, sequent proofs can be translated into proof structures (see [3]). Such a translation is the gate to a geometry of logic and computation, since it makes possible to describe several properties of proofs by means of paths and graph-theoretical conditions such as connectivity or acyclicity.

In this paper, in particular, we consider paths with the following features:

orientation: a path is oriented, i.e. it crosses an edge a either upward, from the link a is conclusion to the link a is premise, or downward, from the link a is premise to the link a is conclusion;

black-box principle: a path never crosses the frame of an exponential box, i.e. for a path a box is a node, whose emergent edges are the doors of the box.

$$\frac{}{\vdash X, X^{\perp}}\, ax \qquad \frac{\vdash \Gamma, A \qquad \vdash \Delta, A^{\perp}}{\vdash \Gamma, \Delta}\, cut \qquad \frac{\vdash ?\Gamma, A}{\vdash ?\Gamma, !A}\, ! \qquad \frac{\vdash \Gamma, A}{\vdash \Gamma, ?A}\, d$$

$$\frac{\vdash \Gamma, A \qquad \vdash \Delta, B}{\vdash \Gamma, \Delta, A \otimes B}\, \otimes \qquad \frac{\vdash \Gamma, A, B}{\vdash \Gamma, A ⅋ B}\, ⅋ \qquad \frac{\vdash \Gamma}{\vdash \Gamma, ?A}\, w \qquad \frac{\vdash \Gamma, ?A, ?A}{\vdash \Gamma, ?A}\, c$$

Fig. 3. $MELL$ sequent calculus

Definition 4 (Path). *An* oriented edge *is an edge a together with a direction* upward, *denoted by $\uparrow a$, or* downward, *denoted by $\downarrow a$. We write $\updownarrow a$ in case we do not want to specify if we mean $\uparrow a$ or $\downarrow a$. An* oriented path *(or simply path) from $\updownarrow a_0$ to $\updownarrow a_n$ is a sequence of oriented edges $<\updownarrow a_0, \ldots, \updownarrow a_n>$ such that for any $i < n$, $\updownarrow a_i, \updownarrow a_{i+1}$ have the same depth and:*

- *if $\updownarrow a_i = \uparrow a_i$ and $\updownarrow a_{i+1} = \uparrow a_{i+1}$, then a_i is conclusion of a link l and a_{i+1} is premise of l;*
- *if $\updownarrow a_i = \uparrow a_i$ and $\updownarrow a_{i+1} = \downarrow a_{i+1}$, then either a_i and a_{i+1} are different conclusions of the same link ax l, or they are different doors of the same exponential box associated with a link $!$ l;*
- *if $\updownarrow a_i = \downarrow a_i$ and $\updownarrow a_{i+1} = \downarrow a_{i+1}$, then a_i is the premise of a link l and a_{i+1} is conclusion of l;*
- *if $\updownarrow a_i = \downarrow a_i$ and $\updownarrow a_{i+1} = \uparrow a_{i+1}$, then a_i and a_{i+1} are different premises of the same link l.*

In any case we say that the path $<\updownarrow a_0, \ldots, \updownarrow a_n>$ crosses *the edges a_i, a_{i+1} and the link l. We denote by $*$ the concatenation of sequences of oriented edges.*

A cycle *is a path in which occurs twice $\updownarrow a$ for an edge a.*

Remark that a_i and a_{i+1} must be different edges, i.e. we do not consider bouncing paths as $<\uparrow a, \downarrow a>$.

We denote paths by Greek letters $\phi, \tau, \psi, \ldots$. We write $\updownarrow a \in \phi$ to mean that $\updownarrow a$ occurs in ϕ.

4.1 Feasible Paths

Feasible paths have been introduced by Fleury and Retoré in [2] as the paths "feasible" in a switching of a proof structure.

Definition 5 (Feasible path, [2]). *A path is* feasible *whenever it does not contain two premises of the same link* $\parr$ *or* $?c$.

A proof structure is feasible acyclic *whenever it does not contain any feasible cycle.*

Feasible acyclicity characterizes the proof structures which corresponds to the proofs of $MELL$ sequent calculus enlarged by the rules of mix and daimon of Figure 4. The proofs of standard $MELL$ instead are not easily characterizable by a geometrical condition, because of the weakening link. We do not enter in the details of the problem, for which we refer to [7].

$$\frac{\vdash \Gamma \qquad \vdash \Delta}{\vdash \Gamma, \Delta}\,mix \qquad\qquad \frac{}{\vdash ?A}\,dai$$

Fig. 4. Mix and daimon rules

Let us compare feasible acyclicity and coherent spaces. As written before, they are tightly related in MLL by the following results:

Girard's Theorem, [3]: let π be an MLL proof structure, $\mathcal{X}$ be a (uniform) coherent space. If π is feasible acyclic, then $[\![\pi]\!]_{\mathcal{X}}$ is a clique.

Retoré's Theorem, [5]: let π be a cut-free MLL proof structure, $\mathcal{X}$ be a (uniform) coherent space with $x, y, z \in |\mathcal{X}|$ s.t. $x \frown y\,[\mathcal{X}]$ and $x \smile z\,[\mathcal{X}]$. If $[\![\pi]\!]_{\mathcal{X}}$ is a clique, then π is feasible acyclic.

However the situation changes in $MELL$: it is still true that any feasible acyclic proof structure is interpreted with a clique, but there are proof structures associated with cliques even if they have feasible cycles. For example take the proof structure π of Figure 5.

π has the feasible cycle $<\uparrow a, \downarrow b, \uparrow a>$, nevertheless $[\![\pi]\!]_{\mathcal{X}}$ is a clique for any coherent space $\mathcal{X}$. In fact, let e_1, e_2 be two experiments on π, we show that $|e_1| \frown\!\!\!\!\smile |e_2|\,[(?\mathcal{I} \otimes ?\mathcal{I}) \parr !?\mathcal{X}]$, where $I = X \parr X^{\perp}$. Suppose $e_1(o) = \left[e_1^1, \ldots, e_1^n\right]$ and $e_2(o) = \left[e_2^1, \ldots, e_2^m\right]$. Remark that for any experiments e_i^l, e_j^h, $e_i^l(a) \frown\!\!\!\!\smile e_j^h(a)\,[?\mathcal{I}]$ as well as $e_i^l(b) \frown\!\!\!\!\smile e_j^h(b)\,[?\mathcal{I}]$. We split in two cases. In case $n = m$, then we deduce $e_1(a) \frown\!\!\!\!\smile e_2(a)$ and $e_1(b) \frown\!\!\!\!\smile e_2(b)$, hence $e_1(c) \frown\!\!\!\!\smile e_2(c)$. Of course $e_1(d) \equiv e_2(d)$,

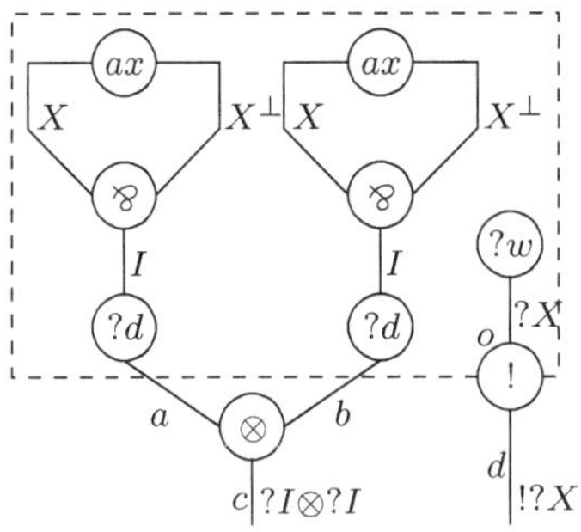

Fig. 5. Example of feasible cycle invisible by coherent spaces

thus $|e_1| \stackrel{\frown}{\smile} |e_2| \, [(?\mathcal{I}\otimes ?\mathcal{I})⅋!?\mathcal{X}]$. In case $n \neq m$, then $e_1(d) \frown e_2(d)$, thus $|e_1| \stackrel{\frown}{\smile} |e_2| \, [(?\mathcal{I}\otimes ?\mathcal{I})⅋!?\mathcal{X}]$. We conclude that $[\![\pi]\!]$ is a clique.[4]

The failure of the correspondence between feasible acyclicity and coherent spaces shows that these latter read the exponential boxes in a different way than feasible paths do. Indeed the cycle $<\uparrow a, \downarrow b, \uparrow a>$ is due to the box associated with o: if we erase o and the frame of its box, we would get a feasible acyclic proof structure. Coherent spaces do not read the boxes as feasible paths do, but it is not true that they do not read the boxes at all. For example, consider the proof structure π' in figure 6.

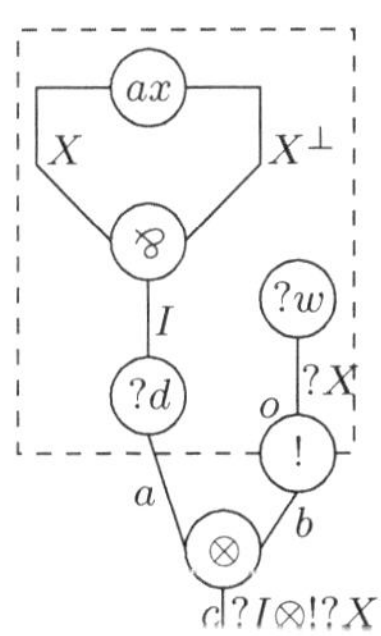

Fig. 6. Example of feasible cycle visible by coherent spaces

π' has the feasible cycle $<\uparrow a, \downarrow b, \uparrow a>$, which is due to the box of o, as in the example before. However in this case the cycle is visible by coherent spaces, i.e. $[\![\pi']\!]$ is not a clique. In fact, let e_1, e_2 be two experiments on π', s.t. $e_1(o) = \emptyset$ and $e_2(o) = [e']$, for an experiment e' on the box of o. Clearly $e_1(a) \smile e_2(a) \, [?\mathcal{I}]$, which implies $e_1(c) \smile e_2(c) \, [?\mathcal{I}\otimes !?\mathcal{X}]$.

4.2 Visible Paths

Here is our main definition, that of *visible paths* (Definition 6).

[4] Remark that the same argument applies for usual Girard's coherent spaces.

Let ϕ be a path, π^o be a box associated with a link $!\,o$. A *passage of ϕ through π^o* is any sequence $<\uparrow a, \downarrow b>$ in ϕ, for a, b doors of π^o.

Notice that a feasible path can pass through an exponential box by means of any pair of its doors; the following definition forbids instead some of such passages:

Definition 6 (Visible path). *Let π be a proof structure. By induction on the depth of π, we define its* visible paths*:*

- *if π has depth 0, then the visible paths of π are exactly the feasible paths;*
- *if π has depth $n+1$, let π^o be a box associated with a link $!\,o$, a, b be doors of π^o, we say that:*
  - a is in the orbit of o *if either a is the principal door or there is a visible path in π^o from the premise of o to a;*
  - a leads to b *if either b is in the orbit of o or there is a visible path in π^o from a to b;*

  then a visible path in π is a feasible path s.t. for any passage $<\uparrow a, \downarrow b>$ through an exponential box, a leads to b.

A proof structure is visible acyclic *whenever it does not contain any visible cycle.*

Visible paths introduce two noteworthy novelties with respect to the feasible paths:

1. they partially break the black box principle: the admissible passages through an exponential box depend on what is inside the box, i.e. changing the contents of a box may alter the visible paths outside it;
2. they are sensitive to the direction: if ϕ is visible from a to b, the same path done in the opposite direction from b to a may be no longer visible. For example recall the proof structure of Figure 6: the path $<\uparrow a, \downarrow b, \uparrow a>$ is visible, but $<\downarrow a, \uparrow b, \downarrow a>$ isn't, since b does not lead to a.

Of course if π is feasible acyclic then it is visible acyclic, but the converse does not hold. For example recall the proof structure of Figure 5, which is visible acyclic although it contains a feasible cycle. However it is remarkable that the two notions match in the polarized fragment of $MELL$, as we show in the following subsection.

4.3 Feasible and Visible Paths in Polarized Linear Logic

The formulas of the polarized fragment of $MELL$ are of two kinds, *negatives* (N) and *positives* (P), and are defined by the following grammar:

$$N ::= X \mid N \parr N \mid ?P$$
$$P ::= X^\perp \mid P \otimes P \mid !N$$

A proof structure π is *polarized* whenever all its edges are labeled by polarized formulas. An edge of π is called *positive* (resp. *negative*) if it is labeled by a positive (resp. negative) formula.

The notion of polarization has a key role in proof theory since it is at the core of the translations of intuitionistic and classical logic into linear logic. We do not enter in the details, for which we refer to [8]. We limit ourself to notice the following proposition:

Proposition 1. *Let π be a polarized proof structure. π is visible acyclic iff π is feasible acyclic.*

5 Visible Acyclicity and Coherent Spaces

In this section we present the main theorems of the paper, stating the link between visible acyclicity and coherent spaces:

Theorem 2. *Let π be a MELL proof structure, $\mathcal{X}$ be a non-uniform coherent space.*

If π is visible acyclic, then $[\![\pi]\!]_{\mathcal{X}}$ is a clique.

Theorem 3. *Let π be a cut-free MELL proof structure, $\mathcal{X}$ be a non-uniform coherent space with $x, y, z \in |\mathcal{X}|$ s.t. $x \frown y\,[\mathcal{X}]$, $x \smile z\,[\mathcal{X}]$ and $x \equiv x\,[\mathcal{X}]$.*

If $[\![\pi]\!]_{\mathcal{X}}$ is a clique, then π is visible acyclic.

Subsection 5.1 (resp. 5.2) gives a sketch of the proof of Theorem 2 (resp. Theorem 3).

5.1 Proof of Theorem 2

Theorem 2 is an immediate consequence of the following lemma:

Lemma 1. *Let π be a visible acyclic proof structure. If $d : D$ is a conclusion of π and e_1, e_2 are two experiments s.t. $e_1(d) \smile e_2(d)\,[\mathcal{D}]$, then there is a visible path ϕ from $\uparrow d$ to a conclusion of π $\downarrow d' : D'$ s.t. $e_1(d') \frown e_2(d')\,[\mathcal{D}']$.*

Proof (Sketch). Let $e_1(d) \smile e_2(d)\,[\mathcal{D}]$. The lemma is proved by induction on the depth of π. The proof provides a procedure which defines a sequence of visible paths $\phi_1 \subset \phi_2 \subset \phi_3 \subset \ldots$, such that ϕ_1 is exactly $<\uparrow d>$, and for each ϕ_j:

1. ϕ_j is a visible path at depth 0;
2. for every edge $c : C$, if $\uparrow c \in \phi_j$ then $e_1(c) \smile e_2(c)\,[\mathcal{C}]$, if $\downarrow c \in \phi_j$ then $e_1(c) \frown e_2(c)\,[\mathcal{C}]$.

Since π is visible acyclic, no ϕ_j is a cycle. Hence the sequence $\phi_1, \phi_2, \phi_3, \ldots$ will meet eventually a conclusion d' of π, so terminating in a path ϕ_k satisfying the lemma. □

A straight consequence of Lemma 1 is that whenever π is visible acyclic, the results of the experiments on π are pairwise coherent, i.e. $[\![\pi]\!]$ is a clique.

We underline that the proof of Lemma 1 is a generalization in the framework of visible paths of the proof of the Compatibility Lemma (Lemma 3.18.1 in [3]).

5.2 Proof of Theorem 3

The proof of Theorem 3 is based on the key Lemma 2. In some sense Lemma 2 is the converse of Lemma 1: Lemma 1 associates with two experiments e_1, e_2 a visible path proving $|e_1| \mathrel{\overset{\frown}{\smile}} |e_2|$; Lemma 2 instead is used in the proof of Lemma 3 to associate with a visible cycle (morally) two experiments s.t. $|e_1| \smile |e_2|$.

However Lemma 2 has to take care of a typical difficulty of the links $?c$. In order to prove the lemma we need to manage the coherence/incoherence relationship between the values of e_1 and e_2. Unfortunately the links $?c$ soon make such a relationship unmanageable. In fact, if l is a link $?c$ with conclusion c and premises a, b, the incoherence $e_1(c) \mathrel{\overset{\smile}{\frown}} e_2(c)$ holds if and only if $e_1(a) \mathrel{\overset{\smile}{\frown}} e_2(a)$, $e_1(a) \mathrel{\overset{\smile}{\frown}} e_2(b)$, $e_1(b) \mathrel{\overset{\smile}{\frown}} e_2(a)$ and $e_1(b) \mathrel{\overset{\smile}{\frown}} e_2(b)$ hold: such an exponential explosion of the number of incoherences soon becomes unmanageable.

Luckily there is a solution that avoids this problem. Namely we noticed that one of the two experiments e_1, e_2 can be chosen to be very simple, i.e. e_1 can be a (x, n)-*simple experiment* (see Definition 9). The key property of a (x, n)-simple experiment is that all of its possible values on an arbitrary edge of type A are semantically characterized, precisely they are (x, n)-*simple elements of* $\mathcal{A}$ *with degree less or equal to* wn^d, where d is the depth of π and w is the cosize of π (Definition 8 and Proposition 2). In this way we may define the second experiment e_2 not by looking at the particular value that e_1 takes on an edge of type A, but by referring in general to the (x, n)-simple elements of $\mathcal{A}$ with degree less or equal to wn^d. So whenever we consider the premises $a :?A, b :?B$ of a link $?c$, instead of taking care of the four incoherences $e_1(a) \mathrel{\overset{\smile}{\frown}} e_2(a)$, $e_1(a) \mathrel{\overset{\smile}{\frown}} e_2(b)$, $e_1(b) \mathrel{\overset{\smile}{\frown}} e_2(a)$ and $e_1(b) \mathrel{\overset{\smile}{\frown}} e_2(b)$, we will check only that for each (x, n)-simple element $v \in ?\mathcal{A}$ with degree less or equal to wn^d, $v \mathrel{\overset{\smile}{\frown}} e_2(a)$ and $v \mathrel{\overset{\smile}{\frown}} e_2(b)$.

Definition 7 ([7]). *Let $\mathcal{C}$ be the $\mathfrak{nuCoh}^{\mathcal{X}}$ interpretation of a formula C and $v \in \mathcal{C}$. For every occurrence of a subformula A of C we define the* projection of v on A *(denoted as $|v|_A$) as the multiset defined by induction as follows:*

- *if $C = A$, then $|v|_A = [v]$;*
- *if $C = D \parr E$ or $C = D \otimes E$, $v = < v', v'' >$ and A is a subformula of D (resp. of E), then $|v|_A = |v'|_A$ (resp. $|v|_A = |v''|_A$);*
- *if $C = ?D$ or $C = !D$, $v = [v_1, \ldots, v_n]$ and A is a subformula of D, then $|v|_A = |v_1|_A + \ldots + |v_n|_A$.*

Definition 8. *Let $n \in \mathbb{N}$, x be an element of a non-uniform coherent space $\mathcal{X}$ and $\mathcal{C}$ be the $\mathfrak{nuCoh}^{\mathcal{X}}$ interpretation of a formula C. An element $v \in \mathcal{C}$ is a* (x, n)-simple element *if:*

1. *for any atomic subformula X, $X^{\perp}$ of C, $Supp(|v|_X) = Supp(|v|_{X^{\perp}}) = \{x\}$;*
2. *for any !-subformula $!A$ of C, if $u \in |v|_{!A}$ then $u = n\,[u']$;*

Moreover v is stable *if also:*

3. *for any ?-subformula $?A$ of C, if $u \in |v|_{?A}$ then $u = n\,[u']$.*

The degree of a (x, n)-simple element v, *denoted by* $d(v)$, *is the number:*

$$d(v) = max\,\{m \mid \exists ?A \text{ subform. of } C, \exists u \in |v|_{?A} \text{ s.t. } u = [u_1, \ldots, u_m]\}$$

Remark that an element is (x, n)-simple in both $\mathcal{C}$ and $\mathcal{C}^\perp$ only if it is also stable. Moreover, notice that for any formula C, there is only one stable (x, n)-simple element in $\mathcal{C}$.

Definition 9 ([7]). *Let π be a proof structure, $n \in \mathbb{N}$, x be an element of a non-uniform coherent space $\mathcal{X}$.* The (x, n)-simple experiment on π, *denoted by* $e^\pi_{(x,n)}$, *is defined as follows (by induction on the depth of π):*

- *for each conclusion $a : A$ of an axiom at depth 0, $e^\pi_{(x,n)}(a) = s$, where s is the stable (x, n)-simple element of $\mathcal{A}$;*
- *for each link ! o at depth 0, let π^o be the box of o, $e^\pi_{(x,n)}(o) = n\left[e^{\pi^o}_{(x,n)}\right]$.*

Proposition 2. *Let π be a proof structure of depth d and cosize w. Let $e^\pi_{(x,n)}$ be the (x, n)-simple experiment on π. For any edge $c : C$ at depth 0, $e^\pi_{(x,n)}(c)$ is a (x, n)-simple element of $\mathcal{C}$ with degree at most wn^d.*

Lemma 2. *Let $\mathfrak{nuCoh}^{\mathcal{X}}$ be defined from a coherent space $\mathcal{X}$ s.t. $\exists x, y, z \in \mathcal{X}$, $x \equiv x\,[\mathcal{X}]$, $x \frown y\,[\mathcal{X}]$ and $x \smile z\,[\mathcal{X}]$.*

Let π be a cut-free proof structure, k be the maximal number of doors of a box of π. Let ϕ be a visible path at depth 0 from a conclusion $\uparrow a$ to a conclusion $\downarrow b$, s.t. ϕ is not a cycle.

For any $n, m \in \mathbb{N}$, $m \geq n \geq k$, there is an experiment $e_\phi : \pi$, s.t. for any edge $c : \mathcal{C}$ at depth 0 and any (x, n)-simple element v in $\mathcal{C}$ with degree less or equal m:

1. *if there is c' equal or above c s.t. $\updownarrow c' \in \phi$, then $e_\phi(c) \not\equiv v\,[\mathcal{C}]$;*
2. *if $\downarrow c \notin \phi$, then $e_\phi(c) \smallsmile\!\!\!\!\smallfrown v\,[\mathcal{C}]$.*

Proof (Sketch). The lemma is proved by induction on the depth of π. The proof is divided in two steps: firstly, e_ϕ is defined by giving its values on the links ax and ! at depth 0; secondly, e_ϕ is proved to satisfy conditions 1, 2 for any edge c, by induction on the number of edges at depth 0 above c. □

Lemma 3. *Let $\mathfrak{nuCoh}^{\mathcal{X}}$ be defined from a coherent space $\mathcal{X}$ s.t. $\exists x, y, z \in \mathcal{X}$, $x \equiv x\,[\mathcal{X}]$, $x \frown y\,[\mathcal{X}]$ and $x \smile z\,[\mathcal{X}]$.*

Let π be a cut-free proof structure with conclusions Π, k be the maximal number of doors of an exponential box in π. If π has a visible cycle, then for any $n, m \in \mathbb{N}$, $m \geq n \geq k$, there is an experiment $e : \pi$, such that for any (x, n)-simple element v in $\bindnasrepma\Pi$ with degree less or equal to m, $|e| \smile v\,[\bindnasrepma\Pi]$.

Proof (Sketch). The proof is by induction on the size of π. The induction step splits in seven cases, one for each type of link (except *cut*). The crucial case deals with the link $\otimes$, since removing it may erase visible cycles. In this case Lemma 2 will be used. □

The proof of Theorem 3 is straight once we have Lemma 3. Let π be a cut-free proof structure with conclusions Π, depth d, cosize w and let k be the maximal number of doors of a box of π. If π has a visible cycle then by Lemma 3 there is an experiment $e : \pi$ such that for any (x, n)-simple element v in $\parr\Pi$ with degree less or equal to m, $|e| \smile v\, [\parr\Pi]$, where $n = k$, $m = wn^d$.

Let $e^{\pi}_{(x,n)}$ be the (x, n)-simple experiment on π. By Proposition 2, $|e^{\pi}_{(x,n)}|$ is a (x, n)-simple element in $\parr\Pi$ with degree less or equal to m. So $|e| \smile |e^{\pi}_{(x,n)}|\, [\parr\Pi]$, i.e. $[\![\pi]\!]_{\mathcal{X}}$ is not a clique in $\parr\Pi$.

6 Visible Acyclicity and Cut Reduction

It turns out that visible acyclicity has also nice computational properties, especially it is stable under cut reduction:

Theorem 4 (Stability). *Let π, π' be MELL proof structures. If $\pi \rightarrow_\beta \pi'$ and π is visible acyclic then π' is visible acyclic.*

Proof (Sketch). Indeed if $\pi \rightsquigarrow_\beta \pi'$, then every visible cycle in π' is the "residue" of a visible cycle in π. □

Remark that a proof structure with deadlocks is not visible acyclic, hence Theorem 4 assures that the cut reduction of visible acyclic proof structures never produces deadlocks.

Moreover, visible acyclicity guarantees also confluence and strong normalization of $\rightarrow_\beta$. Here we give only the statements of the theorems, which will be treated in details in a forthcoming paper:

confluence: let π be a visible acyclic *MELL* proof structure, if $\pi \rightarrow_\beta \pi_1$ and $\pi \rightarrow_\beta \pi_2$ then there is π_3 s.t. $\pi_1 \rightarrow_\beta \pi_3$ and $\pi_2 \rightarrow_\beta \pi_3$;

strong normalization: let π be a visible acyclic *MELL* proof structure, there is no infinite sequence of proof structures $\pi_0, \pi_1, \pi_2, \ldots$ s.t. $\pi_0 = \pi$ and $\pi_i \rightsquigarrow_\beta \pi_{i+1}$.

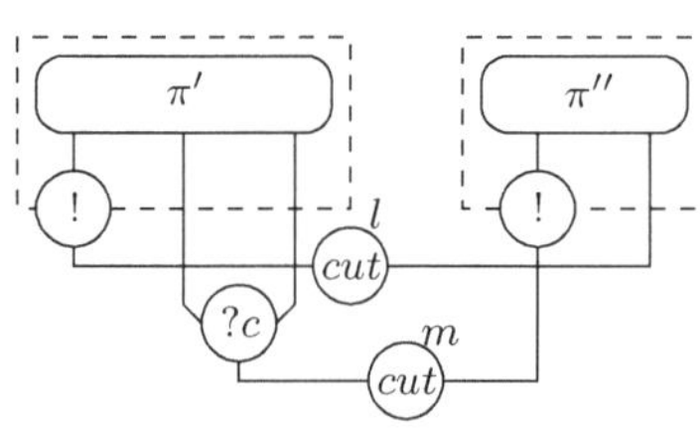

Fig. 7. Counter-example of confluence and strong normalization of cut reduction on proof structures

Remark that both confluence and strong normalization are false on proof structures with visible cycles. For example consider the proof structure π in Figure 7. If you reduce the *cut* link l, then m becomes a deadlock, while if you reduce the *cut* link m and its residues m_1, m_2, l becomes a deadlock: i.e. π is a counter-example to the confluence. Concerning strong normalization, notice that reducing m, then l, then a residue of m, then a residue of l and so on ... we get an infinite sequence of cut reductions.

Acknowledgments. I would like to thank Lorenzo Tortora de Falco for his invaluable support and Paolo Di Giamberardino, whose Mémoire de D.E.A. has been the starting point of this work.

References

1. Danos, V., Regnier, L.: The structure of multiplicatives. Archive for Mathematical Logic **28** (1989) 181-203
2. Fleury, A., Retoré, C.: The mix rule. Mathematical Structures in Computer Science **4(2)** (1994) 173-185
3. Girard, J.-Y.: Linear Logic. Theoretical Computer Science **50** (1987) 1-102
4. Bucciarelli, A., Ehrhard, T.: On phase semantics and denotational semantics: the exponentials. Annals of Pure and Applied Logic **109** (2001) 205-241
5. Retoré, C.: A semantic characterization of the correctness of a proof net. Mathematical Structures in Computer Science **7(5)** (1997) 445-452
6. Pagani, M.: Proofs, Denotational Semantics and Observational Equivalences in Multiplicative Linear Logic. To appear in: Mathematical Structures in Computer Science
7. Tortora de Falco, L.: Réseaux, cohérence et expériences obsessionnelles. Thèse de doctorat, Université Paris VII (2000)
8. Laurent, O.: Étude de la polarisation en logique. Thèse de doctorat, Université Aix-Marseille II (2002)

Church Synthesis Problem with Parameters

Alexander Rabinovich

Dept. of CS, Tel Aviv Univ.
rabinoa@post.tau.ac.il

Abstract. The following problem is known as the Church Synthesis problem:
Input: an *MLO* formula $\psi(X, Y)$.
Task: Check whether there is an operator $Y = F(X)$ such that

$$Nat \models \forall X \psi(X, F(X)) \tag{1}$$

and if so, construct this operator.
Büchi and Landweber proved that the Church synthesis problem is decidable; moreover, they proved that if there is an operator F which satisfies (1), then (1) can be satisfied by the operator defined by a finite state automaton. We investigate a parameterized version of the Church synthesis problem. In this version ψ might contain as a parameter a unary predicate P. We show that the Church synthesis problem for P is computable if and only if the monadic theory of $\langle Nat, <, P\rangle$ is decidable. We also show that the Büchi-Landweber theorem can be extended only to ultimately periodic parameters.

1 Introduction

Two fundamental results of classical automata theory are decidability of the monadic second-order logic of order (MLO) over $\omega = (Nat, <)$ and computability of the Church synthesis problem. These results have provided the underlying mathematical framework for the development of formalisms for the description of interactive systems and their desired properties, the algorithmic verification and the automatic synthesis of correct implementations from logical specifications, and advanced algorithmic techniques that are now embodied in industrial tools for verification and validation.

Büchi [Bu60] proved that the monadic theory of $\omega = \langle Nat, <\rangle$ is decidable. Even before the decidability of the monadic theory of ω has been proved, it was shown that the expansions of ω by "interesting" functions have undecidable monadic theory. In particular, the monadic theory of $\langle Nat, <, +\rangle$ and the monadic theory of $\langle Nat, <, \lambda x.2 \times x\rangle$ are undecidable [Rob58, Trak61]. Therefore, most efforts to find decidable expansions of ω deal with expansions of ω by monadic predicates.

Elgot and Rabin [ER66] found many interesting predicates $\mathbf{P}$ for which *MLO* over $\langle Nat, <, \mathbf{P}\rangle$ is decidable. Among these predicates are the set of factorial

Z. Ésik (Ed.): CSL 2006, LNCS 4207, pp. 546–561, 2006.

numbers $\{n! : n \in Nat\}$, the sets of k-th powers $\{n^k : n \in Nat\}$ and the sets $\{k^n : n \in Nat\}$ (for $k \in Nat$).

The Elgot and Rabin method has been generalized and sharpened over the years and their results were extended to a variety of unary predicates (see e.g., [Ch69, Th75, Sem84, CT02]). In [Rab05] we provided necessary and sufficient conditions for the decidability of monadic (second-order) theory of expansions of the linear order of the naturals ω by unary predicates.

Let Spec be a specification language and Pr be an implementation language. The synthesis problem for these languages is stated as follows: find whether for a given specification $S(I, O) \in$SPEC there is a program $\mathcal{P}$ which implements it, i.e., $\forall I(S(I, \mathcal{P}(I))$.

The specification language for the Church Synthesis problem is the Monadic second-order Logic of Order. An *MLO* formula $\varphi(X, Y)$ specifies a binary relation on subsets of *Nat*. Note that every subset P of *Nat* is associated with its characteristic ω-string u_P (where $u_P(i) = 1$ if $i \in P$ and otherwise $u_P(i) = 0$). Hence, $\varphi(X, Y)$ can be considered as a specification of a binary relation on ω-strings.

As implementations, Church considers functions from the set $\{0,1\}^\omega$ of ω-strings over $\{0,1\}$ to $\{0,1\}^\omega$. Such functions are called operators. A machine that computes an operator at every moment $t \in Nat$ reads an input symbol $X(t) \in \{0,1\}$ and produces an output symbol $Y(t) \in \{0,1\}$. Hence, the output $Y(t)$ produced at t depends only on inputs symbols $X(0), X(1), \ldots, X(t)$. Such operators are called *causal* operators (C-operators); if the output $Y(t)$ produced at t depends only on inputs symbols $X(0), X(1), \ldots, X(t-1)$, the corresponding operator is called *strongly causal* (SC-operator). The sets of recursive causal and strongly causal operators are defined naturally; a C-or a SC-operator is a *finite state* operator if it is computable by a finite state automaton (for precise definitions, see Subsection 2.3).

The following problem is known as the Church Synthesis problem.

Church Synthesis problem
Input: an *MLO* formula $\psi(X, Y)$.
Task: Check whether there is a C-operator F such that
$Nat \models \forall X \psi(X, F(X))$ and if so, construct this operator.

The Church Synthesis problem is much more difficult than the decidability problem for *MLO* over ω. Büchi and Landweber [BL69] proved that the Church synthesis problem is computable. Their main theorem is stated as follows:

Theorem 1. *For every MLO formula $\psi(X,Y)$ either there is a finite state C-operator F such that $Nat \models \forall X \psi(X, F(X))$ or there is a finite state SC-operator G such that $Nat \models \forall Y \neg\psi(G(Y), Y)$. Moreover, it is decidable which of these cases holds and a corresponding operator is computable from ψ.*

In this paper we consider natural generalizations of the Church Synthesis Problem over expansions of ω by monadic predicates, i.e., over the structures $\langle Nat, <, \mathbf{P}\rangle$.

For example, let **Fac** $= \{n! : n \in Nat\}$ be the set of factorial numbers, and let $\varphi(X, Y, \mathbf{Fac})$ be a formula which specifies that $t \in Y$ iff $t \in \mathbf{Fac}$ and $(t' \in X) \leftrightarrow (t' \in \mathbf{Fac})$ for al all $t' \leq t$. It is easy to observe that there is no finite state C-operator F such that $\forall X \varphi(X, F(X), \mathbf{Fac})$. However, there is a recursive C-operator H such that $\forall X \varphi(X, H(X), \mathbf{Fac})$. It is also easy to construct a finite state C-operator $G(X, Z)$ such that $\forall X \varphi(X, G(X, \mathbf{Fac}), \mathbf{Fac})$. It was surprising for us to discover that it is decidable whether for a formula $\psi(X, Y, \mathbf{Fac})$ there is a C-operator F such that $\forall X \varphi(X, F(X), \mathbf{Fac})$ and if such an operator exists, then it is recursive and computable from ψ.

Here is the summary of our results. We investigate a parameterized version of the Church synthesis problem. In this version ψ might contain as a parameter a unary predicate P. Below five synthesis problems with a parameter $\mathbf{P} \subseteq Nat$ are stated.

Synthesis Problems for $\mathbf{P} \subseteq Nat$

Input: an *MLO* formula $\psi(X, Y, P)$.

Problem 1: Check whether there is a C-operator $Y = F(X, P)$ such that $Nat \models \forall X \psi(X, F(X, \mathbf{P}), \mathbf{P})$ and if there is such a recursive operator - construct it.

Problem 2: Check whether there is a recursive C-operator $Y = F(X, P)$ such that $Nat \models \forall X \psi(X, F(X, \mathbf{P}), \mathbf{P})$ and if so - construct this operator.

Problem 3: Check whether there is a recursive C-operator $Y = F(X)$ such that $Nat \models \forall X \psi(X, F(X), \mathbf{P})$ and if so - construct this operator.

Problem 4: Replace "recursive" by "finite state" in *Problem 2.*

Problem 5: Replace "recursive" by "finite state" in *Problem 3.*

We show

Theorem 2. *Let* $\mathbf{P}$ *be a subset of* Nat. *The following conditions are equivalent:*

1. *Problem 1 for* $\mathbf{P}$ *is computable.*
2. *Problem 2 for* $\mathbf{P}$ *is computable.*
3. *Problem 3 for* $\mathbf{P}$ *is computable.*
4. *The monadic theory of* $\langle Nat, <, \mathbf{P}\rangle$ *is decidable.*
5. *For every MLO formula* $\psi(X, Y, P)$ *either there is a recursive C-operator* F *such that* $Nat \models \forall X \psi(X, F(X), \mathbf{P})$ *or there is a recursive SC-operator* G *such that* $Nat \models \forall Y \neg\psi(G(Y), Y, \mathbf{P})$. *Moreover, it is decidable which of these cases holds and the (description of the) corresponding operator is computable from* ψ.

The difficult part of this theorem is the implication (4)$\Rightarrow$(5). In [BL69] a reduction of the Church synthesis problem to infinite two-player games on finite graphs was provided. Our proof is based on a reduction of the Church synthesis problem with parameters to infinite two-player games on infinite graphs. The main part of the proof shows that the reduction is "uniform" in the parameters and the corresponding infinite graph can be interpreted in $(Nat, <, \mathbf{P})$. Lemma 13 from [Rab05] also plays a key role in this proof.

The trivial examples of predicates with decidable monadic theory are ultimately periodic predicates. Recall that a predicate $\mathbf{P}$ is ultimately periodic if there is $p, d \in Nat$ such that $(n \in \mathbf{P} \leftrightarrow n + p \in \mathbf{P})$ for all $n > d$. Ultimately periodic predicates are *MLO* definable. Hence, for these predicates computability of Problems 1-5 can be derived from Theorem 1.

We prove that the Büchi-Landweber theorem can be extended only to ultimately periodic parameters.

Theorem 3. *Let* $\mathbf{P}$ *be a subset of Nat. The following conditions are equivalent and imply computability of Problem 4:*

1. $\mathbf{P}$ *is ultimately periodic.*
2. *For every MLO formula* $\psi(X, Y, P)$ *either there is a finite state C-operator* F *such that* $Nat \models \forall X \psi(X, F(X, \mathbf{P}), \mathbf{P})$ *or there is a finite state SC-operator* G *such that* $Nat \models \forall Y \neg\psi(G(Y, \mathbf{P}), Y, \mathbf{P})$.

The paper is organized as follows. In Section 2 notations are fixed and standard definitions and facts about automata and logic are recalled. In Section 3 parity games and their connection to the synthesis problems are discussed. Büchi and Landweber [BL69] used the determinacy of such games as one of the main tools to prove computability of the Church synthesis problem and derive Theorem 1. We prove here that the question about the existence of a C-operator F such that $Nat \models \forall X \psi(X, F(X, \mathbf{P}), \mathbf{P})$ can be effectively reduced to the decidability of monadic theory of $\langle Nat, <, \mathbf{P}\rangle$.

In Section 4 a proof of Theorem 2 is provided. In Section 5 finite state synthesis problems with parameters are considered and Theorem 3 is proved. Finally, in Section 6 some open problems are stated and related works are discussed.

2 Preliminaries

2.1 Notations and Terminology

We use k, l, m, n, i for natural numbers; *Nat* for the set of natural numbers and capital bold letters $\mathbf{P}$, $\mathbf{S}$, $\mathbf{R}$ for subsets of *Nat*. We identify subsets of a set A and the corresponding unary (monadic) predicates on A.

The set of all (respectively, non-empty) finite strings over an alphabet Σ is denoted by Σ^* (respectively, by Σ^+). The set of ω-strings over Σ is denoted by Σ^ω.

Let $a_0 \dots a_k \dots$ and $b_0 \dots b_k \dots$ be ω-strings. We say that these ω-strings coincide on an interval $[i, j]$ if $a_k = b_k$ for $i \leq k \leq j$. A function F from Σ_1^ω to Σ_2^ω will be called an operator of type $\Sigma_1 \to \Sigma_2$. An operator F is called *causal* (respectively, *strongly causal*) operator, if $F(X)$ and $F(Y)$ coincide on an interval $[0, t]$, whenever X and Y coincide on $[0, t]$ (respectively, on $[0, t)$). We will refer to causal (respectively, strongly causal) operators as C-operators (respectively, SC-operators).

Every SC-operator F of type $\Sigma \to \Sigma$ has a unique fixed point, i.e., there is a unique $X \in \Sigma^\omega$ such that $X = F(X)$.

Let $G : \Sigma^\omega \rightarrow \Delta^\omega$ be an operator. In the case Σ is the Cartesian product $\Sigma_1 \times \Sigma_2$ we will identify F with the corresponding operator $F : \Sigma_1^\omega \times \Sigma_2^\omega \rightarrow \Delta^\omega$. An operator $F : \Sigma_1^\omega \times \Sigma_2^\omega \rightarrow \Delta^\omega$ is said to be SC-operator (C-operator) if G is SC-operator (respectively, C-operator).

There exists a one-one correspondence between the set of all ω-strings over the alphabet $\{0,1\}^n$ and the set of all n-tuples $\langle \mathbf{P}_1, \ldots, \mathbf{P}_n \rangle$ of unary predicates over the set of natural numbers. With an n-tuple $\langle \mathbf{P}_1, \ldots, \mathbf{P}_n \rangle$ of unary predicates over Nat, we associate the ω-string $a_0 a_1 \ldots a_k \ldots$ over alphabet $\{0,1\}^n$ defined by $a_k =_{def} \langle b_1^k, \ldots b_n^k \rangle$ where b_i^k is 1 if $\mathbf{P}_i(k)$ holds and b_i^k is 0 otherwise. Let $Q = \{q_1, \ldots, q_m\}$ be a finite set of state. There is a natural one-one correspondence between subset of $Q \times Nat$ and the set of m-tuples of unary predicates over Nat: with $U \subseteq Q \times Nat$ we associate the m-tuple $\langle \mathbf{P}_1, \ldots, \mathbf{P}_m \rangle$ defined as $i \in \mathbf{P}_j$ iff $U(q_j, i)$ (for $i \in Nat$ and $j \leq m$).

Similarly, there is a one-one correspondence between the set of all strings of length m over the alphabet $\{0,1\}^n$ and the set of all n-tuples $\langle \mathbf{P}_1, \ldots, \mathbf{P}_n \rangle$ of unary predicates over the set $\{0, \ldots, m-1\}$.

A linearly ordered set will be called a chain. A chain with n monadic predicates over its domain will be called an n-labelled chain; whenever n is clear from the context, n-labelled chains will be called labelled chains.

We will sometimes identify an n-labelled chain $M = \langle Nat, <, \mathbf{P}_1, \ldots, \mathbf{P}_n \rangle$ with the ω-string over the alphabet $\{0,1\}^n$ which corresponds to the n-tuple $\langle \mathbf{P}_1, \ldots, \mathbf{P}_n \rangle$; this ω-string will be called the *characteristic* ω-string (or ω-word) of M. Similarly, we will identify finite n-labelled chains with corresponding strings over $\{0,1\}^n$.

2.2 Monadic Second-Order Logic and Monadic Logic of Order

Let σ be a relational signature. Atomic formulas of the monadic second-order logic over σ are $R(t_1, ..., t_n)$, $t_1 = t_2$, and $t_1 \in X$ where $t_1, \ldots, t_n$ are individual variables, $R \in \sigma$ is an n-are relational symbol, and X is a set variable. Formulas are obtained from atomic formulas by conjunction, negation, and quantification $\exists t$ and $\exists X$ for t an individual and X a set variable. The satisfaction relation $M, \tau_1, \ldots \tau_k; \mathbf{S}_1, \ldots, \mathbf{S}_m \models \varphi(t_1, \ldots, t_k; X_1, \ldots, X_m)$ is defined as usual with the understanding that set variables range over subsets of M.

We use standard abbreviations, e.g., we write $X \subseteq X'$ for $\forall t.\ X(t) \rightarrow X'(t)$; we write $X = X'$ for $\forall t.\ X(t) \leftrightarrow X'(t)$; symbols "$\exists^{\leq 1}$" and "$\exists!$" stands for "there is at most one" and "there is a unique".

If a signature σ contains one binary predicate $<$ which is interpreted as a linear order, and all other predicates are unary, the monadic second-order logic for this signature is called Monadic Logic of Order (MLO). The formulas of MLO are interpreted over labelled chains.

The *monadic theory* of a labelled chain M is the set of all MLO sentences which hold in M.

We will deal with the expansions of ω by monadic predicates, i.e., with the structures of the form $M = \langle Nat, <, \mathbf{P}_1, \ldots, \mathbf{P}_n \rangle$. We say that a chain $M = \langle Nat, <, \mathbf{P}_1, \ldots, \mathbf{P}_n \rangle$ is *recursive* if all $\mathbf{P}_i$ are recursive subsets of Nat.

An ω-language L is said to be *defined* by an *MLO* formula $\psi(X_1, \ldots, X_n)$ if the following condition holds: an ω string is in L iff the corresponding n-tuple of unary predicates satisfies ψ.

2.3 Automata

A *deterministic transition system* D is a tuple $\langle \mathcal{Q}, \Sigma, \delta, q_{init} \rangle$, consisting of a set $\mathcal{Q}$ of *states*, an *alphabet* Σ, a *transition function* $\delta : \mathcal{Q} \times \Sigma \rightarrow \mathcal{Q}$ and *initial state* $q_{init} \in \mathcal{Q}$. The transition function is extended as usual to a function from $\mathcal{Q} \times \Sigma^*$ to $\mathcal{Q}$ which will be also denoted by δ. The function $\delta_{init} : \Sigma^* \rightarrow \mathcal{Q}$ is defined as $\delta_{init}(\pi) = \delta(q_{init}, \pi)$. A transition systems is finite if $\mathcal{Q}$ and Σ are finite. D is recursive if (1) the sets of states and the alphabet are at most countable and (2) there are enumerations of the sets of states $\mathcal{Q} = \{q_i : i \in Nat\}$ and the alphabet $\Sigma = \{a_i : i \in Nat\}$ such that the function $\lambda i \lambda j.\delta(q_i, a_j)$ is recursive.

A *finite deterministic automaton* $\mathcal{A}$ is a tuple $\langle \mathcal{Q}, \Sigma, \delta, q_{init}, F \rangle$, where $\langle \mathcal{Q}, \Sigma, \delta, q_{init} \rangle$ is a finite deterministic transition system and F is a subset of $\mathcal{Q}$. A string $\pi \in \Sigma^*$ is accepted by $\mathcal{A}$ if $\delta_{init}(\pi) \in F$. The language accepted (or defined) by $\mathcal{A}$ is the set of string accepted by $\mathcal{A}$.

A *Mealey automaton* is a tuple $\langle \mathcal{Q}, \Sigma, \delta, q_{init}, \Delta, out \rangle$, where $\langle \mathcal{Q}, \Sigma, \delta, q_{init} \rangle$ is a deterministic transition system, Δ is an alphabet and $out : \mathcal{Q} \rightarrow \Delta$ is an output function. With a Mealey automaton $\mathcal{A} = \langle \mathcal{Q}, \Sigma, \delta, q_{init}, \Delta, out \rangle$ we associate a function $h_{\mathcal{A}} : \Sigma^* \rightarrow \Delta$ and an operator $F_{\mathcal{A}} : \Sigma^\omega \rightarrow \Delta^\omega$ defined as follows:

$$h_{\mathcal{A}}(a_0 \ldots a_{i-1}) = out(\delta_{init}(a_0 \ldots a_{i-1}))$$

$$F_{\mathcal{A}}(a_0 \ldots a_i \ldots) = b_0 \ldots b_i \ldots \text{ iff } b_i = h_{\mathcal{A}}(a_0 \ldots a_{i-1}))$$

It is easy to see that an operator is strongly causal (SC-operator) iff it is definable by a Mealey automaton. We say that a SC-operator $F : \Sigma^\omega \rightarrow \Delta^\omega$ is finite state (respectively, recursive) iff it is definable by a finite state (respectively, by a recursive) Mealey automaton.

A (deterministic) *parity automaton* $\mathcal{A}$ is a finite Mealey automaton $\langle \mathcal{Q}, \Sigma, \rightarrow, \delta, q_{init}, \Delta, col \rangle$, where the output alphabet Δ is a (finite) subset of Nat; the output function *col* is usually named coloring function.

With an ω-string $a_0 a_1 \ldots a_i \cdots \in \Sigma^\omega$ we associate the ω sequence $\delta_{init}(a_0)\delta_{init}(a_1) \ldots \delta_{init}(q_i) \ldots$ of states and the set $\mathtt{Inf}$ of all $q \in \mathcal{Q}$ that appear infinitely many times in this sequence. An ω-string is accepted by $\mathcal{A}$ if the minimal element of the set $\{col(q) : q \in \mathtt{Inf}\}$ is even. The ω-language *accepted* (or defined) by $\mathcal{A}$ is the set of all ω-strings accepted by $\mathcal{A}$.

Sometimes the alphabet Σ of $\mathcal{A}$ will be the Cartesian product $\Sigma_1 \times \Sigma_2 \times \Sigma_3$ of other alphabets. In this case we say that $\mathcal{A}$ defines a relation $R_{\mathcal{A}} \subseteq \Sigma_1^\omega \times \Sigma_2^\omega \times \Sigma_3^\omega$; a triplet $\langle a, b, c \rangle$ of ω-strings is in $R_{\mathcal{A}}$ iff the ω string $(a_0, b_0, c_0)(a_1, b_1, c_1) \ldots (a_i, b_i, c_i) \ldots$ is accepted by $\mathcal{A}$.

Here is the classical theorem due to Büchi, Elgot and Trakhtenbrot.

Theorem 4. *1. A language is accepted by a finite deterministic automaton iff it is definable by an MLO formula.*

2. An ω-language is accepted by a parity automaton iff it is definable by an MLO formula.

3. *Moreover, there is an algorithm which for every formula $\varphi(X_1, \dots, X_m)$ computes an equivalent deterministic automaton $\mathcal{A}$ i.e., the language definable by φ is accepted by $\mathcal{A}$. There is an algorithm which for every deterministic automaton $\mathcal{A}$ computes an equivalent MLO formula. Similarly, there are translation algorithms between formulas and deterministic parity automata.*

A *Moore automaton* is a tuple $\langle \mathcal{Q}, \Sigma, \delta, q_{init}, \Delta, out \rangle$, where $\langle \mathcal{Q}, \Sigma, \delta, q_{init} \rangle$ is a deterministic transition system, Δ is an alphabet and $out : \mathcal{Q} \times \Sigma \rightarrow \Delta$ is an output function.

With a Moore automaton $\mathcal{A} = \langle \mathcal{Q}, \Sigma, \delta, q_{init}, \Delta, out \rangle$ we associate a function $h_{\mathcal{A}} : \Sigma^+ \rightarrow \Delta$ and an operator $F_{\mathcal{A}} : \Sigma^\omega \rightarrow \Delta^\omega$ defined as follows:

$$h_{\mathcal{A}}(a_0 \dots, a_i) = out(\delta_{init}(a_0 \dots a_{i-1}), a_i)$$

$$F_{\mathcal{A}}(a_0 \dots a_i \dots) = b_0 \dots b_i \dots \text{ iff } b_i = h_{\mathcal{A}}(a_0 \dots, a_i)$$

It is easy to see that an operator is causal (C-operator) iff it is definable by a Moore automaton.

We say that a C-operator $F : \Sigma^\omega \rightarrow \Delta^\omega$ is finite state (respectively, recursive) iff it is definable by a finite state (respectively, by a recursive) Moore automaton.

3 Parity Games and the Synthesis Problem

In the next subsection we provide standard definitions and facts about infinite two-player perfect information games. In [BL69] a reduction of the Church synthesis problem to infinite two-player games on finite graphs was provided. In subsection 3.2 we provide a reduction of the Church synthesis problem with parameters to infinite two-player games on infinite graphs; this reduction is "uniform" in the parameters. The main technical results needed for the proof of Theorem 2 are Lemmas 8 and 9 which roughly speaking state that and the corresponding infinite graph can be interpreted in $(Nat, <, \mathbf{P})$.

3.1 Parity Games

We consider here two-player perfect information games played on graphs in which each player chooses in turn a vertex adjacent to a current vertex. The presentation is based on [PP04].

A (directed) bipartite graph $G = (V_1, V_2, E)$ is called a *game arena* if the outdegree of every vertex is at least one. If G is an arena, a game on G is defined by an initial node $v_{init} \in V_1$ and a set of winning ω-paths $\mathcal{F}$ from this node.

Player I plays on vertices in V_1 and Player II on vertices in V_2. A play from a node v is an infinite path in G which starts at v. Player I wins if the play belongs to $\mathcal{F}$.

A strategy f for Player I (Player II) is a function which assigns to every path of even (odd) length a node adjacent to the last node of the path. A play $v_{init}v_2v_3 \dots$ is played according to a strategy f_1 of Player I (strategy f_2 of Player II) if for every prefix $\pi = v_{init}v_2 \dots v_n$ of even (odd) length $v_{n+1} = f_1(\pi)$ (respectively, $v_{n+1} = f_2(\pi)$). A strategy is winning for Player I (respectively, for

Player II) if all the plays played according to this strategy are in $\mathcal{F}$ (respectively, in the complement of $\mathcal{F}$). A strategy is *memoryless* if it depends only on the last nodes in the path.

Parity games are games on graphs in which the set of winning paths are defined by parity conditions. More precisely, let $G = (V_1, V_2, E)$ be a game graph and let $c : V_1 \cup V_2 \rightarrow \{0, 1, \dots m\}$ be a coloring.

Let $\rho = v_1 v_2 \dots$ be a play. With such a play ρ, we associate the set of colors C_ρ that appear infinitely many times in the ω-sequence $col(v_1)col(v_2)\dots$; a play ρ is winning if the minimal element of C_ρ is odd. The following theorem [EJ91, GTW02, PP04] is fundamental:

Theorem 5. *In a parity game, one of the players has a memoryless winning strategy.*

3.2 Games and the Church Synthesis Problem

Let $\mathcal{A} = \langle \mathcal{Q}, \Sigma, \delta_{\mathcal{A}}, q_{init}, col\rangle$ be a deterministic parity automaton over the alphabet $\Sigma = \{0,1\} \times \{0,1\} \times \{0,1\}$, let $R_{\mathcal{A}} \subseteq \{0,1\}^\omega \times \{0,1\}^\omega \times \{0,1\}^\omega$ be the relation definable by $\mathcal{A}$ and let $\mathbf{P}$ be a subset of Nat. We will define a parity game $G_{\mathcal{A},\mathbf{P}}$ such that

1. Player I has a winning strategy in $G_{\mathcal{A},\mathbf{P}}$ iff there is a SC-operator $G : \{0,1\}^\omega \rightarrow \{0,1\}^\omega$ such that $R_{\mathcal{A}}(G(Y), Y, \mathbf{P})$ holds for every Y.
2. Player II has a winning strategy in $G_{\mathcal{A},\mathbf{P}}$ iff there is a C-operator $F : \{0,1\}^\omega \rightarrow \{0,1\}^\omega$ such that $\neg R_{\mathcal{A}}(X, F(X), \mathbf{P})$ holds for every X.

The arena $G(V_1, V_2, E)$ of $G_{\mathcal{A},\mathbf{P}}$ is defined as follows:

1. $V_1 = \mathcal{Q} \times Nat$ and $V_2 = \mathcal{Q} \times \{0,1\} \times Nat$.
2. From $\langle q, n\rangle \in V_1$ there exit two edges; one to $\langle q, 0, n\rangle \in V_2$, and the second to $\langle q, 1, n\rangle \in V_2$. We will assign labels to these edges. The first one will be labeled by 0 and the second one will be labeled by 1. These edge labels play no role in the game on our graph; however, it will be convenient to refer to them later.
3. From $\langle q, a, n\rangle \in V_2$ there exit two edges defined as follows: Let c be 1 if $n \in \mathbf{P}$ and 0 if $n \not\in \mathbf{P}$; and for $b \subset \{0,1\}$ Let q_b be $\delta_{\mathcal{A}}(q, \langle a, b, c\rangle)$. One edge from $\langle q, a, n\rangle$ is connected to $\langle q_0, n+1\rangle$ and the second one to $\langle q_1, n+1\rangle$. We label the first edge by 0 and the second one by 1.

The color of a node of the arena is defined by the color of its automaton's component, i.e., $c(\langle q, n\rangle) = c(\langle q, a, n\rangle = col(q)$.

Every node of the game graph for $G_{\mathcal{A},\mathbf{P}}$ has two successors. The subsets of V_1 (respectively, of V_2) can be identified with the memoryless strategies of Player I (respectively, of Player II). For a subset $U_1 \subseteq V_1$, the corresponding memoryless strategy f_{U_1} is defined as

$$f_{U_1}(\langle q, n\rangle) = \begin{cases} \langle q, 1, n\rangle & \text{if } \langle q, n\rangle \in U_1 \\ \langle q, 0, n\rangle & \text{otherwise} \end{cases}$$

In other words, for $v \in V_1$ the strategy f_{U_1} chooses the nodes reachable from v by the edge with label $U_1(v)$.

To a subset $U_1 \subseteq V_1$, correspond a function $h_{U_1} : \{0,1\}^* \to V_1$ and a SC-operator $F_{U_1} : \{0,1\}^\omega \to \{0,1\}^\omega$. First, we provide the definition for h_{U_1}, and later for F_{U_1}.

h_{U_1} is defined as follows:

$$h_{U_1}(\epsilon) = \langle q_{init}, 0\rangle$$

$$\text{For } \pi \in \{0,1\}^* \text{ and } b \in \{0,1\} \ : \ h_{U_1}(\pi b) = \langle \delta_{\mathcal{A}}(q,a,b,c), n+1\rangle, \text{ where}$$
$$\langle q,n\rangle = h_{U_1}(\pi), \ a = 1 \text{ iff } \langle q,n\rangle \in U_1 \text{ and } \ c = 1 \text{ iff } n \in \mathbf{P}.$$

Some explanation might be helpful at this point. Let G_{U_1} be the subgraph of $G_{\mathcal{A},\mathbf{P}}$ obtained by removing from every node $v \in V_1$ the edge labeled by $\neg U_1(v)$ and removing the label from the other edge exiting v. In this graph every V_1 node has outdegree one and every V_2 node has two exiting edges; one is labeled by 0 and the other is labeled by 1. For every π in $\{0,1\}^*$ there is a unique path from $\langle q_{init}, 0\rangle$ to a state $v_1 \in V_1$ such that π is the sequence of labels on the edges of this path; this node v_1 is h_{U_1} image of π.

Now a SC-operator $F_{U_1} : \{0,1\}^\omega \to \{0,1\}^\omega$ which corresponds to U_1 is defined as follows. Let $\pi = b_0 b_1 \dots$ be an ω-string. There is a unique ω-path ρ from $\langle q_{init}, 0\rangle$ in G_{U_1} such that π is the sequence labels on the edges of this path. Let $v_1 v_2 \dots$ be the sequence of V_1 nodes on ρ and let $a_i = 1$ if $v_i \in U_1$ and 0 otherwise. The ω sequence $a_0 a_1 \dots$ is defined as the F_{U_1} image of π.

Similarly, $U_2 \subseteq V_2$ corresponds to a memoryless strategy f_{U_2} for Player II and C-operator F_{U_2}. The following lemmas are easy

Lemma 6. *1. Let U_1 be a subset of V_1. The memoryless strategy defined by U_1 is winning for Player I iff $R_{\mathcal{A}}(F_{U_1}(Y), Y, \mathbf{P})$ holds for every Y.*
2. Let U_2 be a subset of V_2. The memoryless strategy defined by U_2 is winning for Player II iff $\neg R_{\mathcal{A}}(X, F_{U_2}(X), \mathbf{P})$ holds for every X.

Lemma 7. *If $\mathbf{P}$ is recursive, then the game $G_{\mathcal{A},\mathbf{P}}$ is recursive. If in addition $U_1 \subseteq Q \times Nat$ (respectively, $U_2 \subseteq Q \times \{0,1\} \times Nat$) is recursive, then the corresponding operator F_{U_1} (respectively, F_{U_2}) is recursive.*

Recall that every subset U of $Q \times Nat$ corresponds to a tuple $W_1, \dots, W_{|Q|}$ of subsets of Nat such that $\langle q,m\rangle \in U$ iff $m \in W_q$.

The next Lemma shows that the set of winning memoryless strategies for each of the players in $G_{\mathcal{A},\mathbf{P}}$ is definable in the structure $\langle Nat, <, \mathbf{P}\rangle$.

Lemma 8. *Let $\mathcal{A} = \langle Q, \Sigma, \delta_{\mathcal{A}}, q_{init}, col\rangle$, be a deterministic parity automaton over the alphabet $\Sigma = \{0,1\} \times \{0,1\} \times \{0,1\}$.*

1. *There is an MLO formula $Win_{\mathcal{A}}(Z_1, \dots, Z_{|Q|}, Z)$ such that for every $\mathbf{P} \subseteq Nat$ and $W_1, \dots, W_{|Q|} \subseteq Nat$:*
   $Nat \models Win_{\mathcal{A}}(W_1, \dots, W_{|Q|}, \mathbf{P})$ iff the corresponding subset $U \subseteq Q \times Nat$ defines a winning memoryless strategy for Player I in $G_{\mathcal{A},\mathbf{P}}$.

2. *There is an MLO formula* $Lose_{\mathcal{A}}(Z_1, \ldots, Z_{|\mathcal{Q}|}, Z'_1, \ldots, Z'_{|\mathcal{Q}|}, Z)$ *such that for every* $\mathbf{P} \subseteq Nat$ *and* $W_1, \ldots, W_{|\mathcal{Q}|}, W'_1, \ldots, W'_{|\mathcal{Q}|} \subseteq Nat$*:*

   $Nat \models Lose_{\mathcal{A}}(W_1, \ldots, W_{|\mathcal{Q}|}, W'_1, \ldots, W'_{|\mathcal{Q}|}, \mathbf{P})$ *iff the corresponding subset* $U \subseteq \mathcal{Q} \times \{0,1\} \times Nat$ *defines a winning memoryless strategy for Player II in* $G_{\mathcal{A},\mathbf{P}}$*.*
3. *Moreover, there is an algorithm that computes formulas* $Win_{\mathcal{A}}$ *and* $Lose_{\mathcal{A}}$ *from* $\mathcal{A}$*.*

Proof. (Sketch) The game arena $G_{\mathcal{A},\mathbf{P}}$ can be considered as a logical structure M for the signature $\tau_{\mathcal{A}} = \{R_i \ : \ i \in \mathcal{Q} \cup \mathcal{Q} \times \{0,1\}\} \cup \{P, E_0, E_1\}$, where R_i and P are unary predicates and E_0, E_1 are binary predicates with the interpretation

$$\mathbf{R}_i^M = \begin{cases} \{\langle q, j\rangle \ : \ j \in Nat\} & \text{for } i = q \in \mathcal{Q} \\ \{\langle q, a, j\rangle \ : \ j \in Nat\} & \text{for } i = \langle q, a\rangle \in \mathcal{Q} \times \{0,1\} \end{cases}$$

$$\mathbf{P}^M = \{\langle q, m\rangle \ : \ m \in \mathbf{P}\} \cup \{\langle q, a, m\rangle \ : \ a \in \{0,1\} \text{ and } m \in \mathbf{P}\}$$

$$\mathbf{E}_0^M(v_1, v_2) \text{ (respectively, } \mathbf{E}_1^M(v_1, v_2)) \text{ holds}$$

iff there is an edge labeled by 0 (respectively, by 1) from v_1 to v_2.

Every set S of nodes in M corresponds to the tuple $\langle \ldots, W_i, \ldots\rangle$ where $i \in \mathcal{Q} \cup \mathcal{Q} \times \{0,1\}$) of subsets of Nat such that $\langle q, m\rangle \in S$ iff $m \in W_q$ and $\langle q, a, m\rangle \in S$ iff $m \in W_{\langle q,a\rangle}$.

It can be shown that there is an algorithm which for every formula $\psi(X)$ in the second order monadic logic over the signature $\tau_{\mathcal{A}}$ with a free monadic variable X constructs a formula $\varphi(Y, \ldots, Z_i, \ldots)$ with free monadic variables $\{Y\} \cup \{Z_i \ : \ i \in \mathcal{Q} \cup \mathcal{Q} \times \{0,1\}\}$, such that for every $\mathbf{P} \subseteq Nat$ and a tuple $\langle \ldots, W_i, \ldots\rangle$ where $i \in \mathcal{Q} \cup \mathcal{Q} \times \{0,1\}$) of subsets of Nat the following equivalence holds:

$$Nat \models \varphi(\mathbf{P}, \ldots, W_i, \ldots), \text{ iff } M \models \psi(S),$$

where S is the subset of nodes in $G_{\mathcal{A},\mathbf{P}}$, which corresponds to $\langle \ldots, W_i, \ldots\rangle$.

Lemma 8 follows from the existence of the above algorithm and from the observation that the set of subsets of $G_{\mathcal{A},\mathbf{P}}$ which correspond to memoryless winning strategies for Player I (respectively, Player II) in $G_{\mathcal{A},\mathbf{P}}$ can be defined by a second order monadic formula $\psi(X)$. □

Finally, note that the operator $\lambda\langle U_1, X\rangle.F_{U_1}(X)$ is causal both in U_1 and in X and its construction is uniform in $\mathbf{P}$. More precisely,

Lemma 9. *1. Let* $\mathcal{A} = \langle \mathcal{Q}_{\mathcal{A}}, \Sigma, \delta_{\mathcal{A}}, q_{init}, col\rangle$ *be a finite parity automaton over the alphabet* $\Sigma = \{0,1\} \times \{0,1\} \times \{0,1\}$*. There is a finite state Mealey automaton* $\mathcal{B} = \langle \mathcal{Q}_{\mathcal{B}}, \Sigma_{\mathcal{B}}, \delta_{\mathcal{B}}, q'_{init}, \Delta, out\rangle$*, where* $\Sigma_{\mathcal{B}} = \{0,1\}^{|\mathcal{Q}_{\mathcal{A}}|} \times \{0,1\} \times \{0,1\}$ *and* $\Delta = \{0,1\}$*, such that for every* $\mathbf{P} \subseteq Nat$ *and every subset* $U_1 \subseteq \mathcal{Q}_{\mathcal{A}} \times Nat$ *of the first Player's positions in the game* $G_{\mathcal{A},\mathbf{P}}$

the SC-operator $\lambda Y.F_{U_1}(Y)$ defined by U_1 in $G_{\mathcal{A},\mathbf{P}}$ is the same as the operator $\lambda Y.F_{\mathcal{B}}(W_1, W_2, \ldots, W_{|Q_{\mathcal{A}}|}, Y, \mathbf{P})$, where the tuple $W_1, \ldots, W_{|Q_{\mathcal{A}}|} \subseteq Nat$ corresponds to U_1.

Moreover, $\mathcal{B}$ is computable from $\mathcal{A}$.

2. *The assertion similar to (1) for the C-operators defined by the subsets of the second Player's positions.*

4 Proof of Theorem 2

Our proof of Theorem 2 is organized as follows:

1. The implications (1)⇒(4), (2)⇒(4) and (3)⇒(4) follow from Lemma 11.
2. The implication (4)⇒(5) follows from Theorem 14.
3. The implications (5)⇒(1), (5)⇒(2) and (5)⇒(3) are proved in Lemma 15.

First note

Theorem 10. *For every $\mathbf{P} \subseteq Nat$ and every MLO formula $\psi(X, Y, Z)$ either there is a C-operator F such that $Nat \models \forall X \psi(X, F(X), \mathbf{P})$ or there is a SC-operator G such that $Nat \models \forall Y \neg\psi(G(Y), Y, \mathbf{P})$.*

Proof. Let $\mathcal{A}$ be a parity automaton equivalent to $\neg\psi(Y, X, Z)$. By Theorem 5, one of the players has a memoryless winning strategy in the game $G_{\mathcal{A},\mathbf{P}}$. A memoryless winning strategy U of Player I (respectively, of Player II) defines SC-operator (respectively, C-operator) F_U. Hence, by Lemma 6, either there is a C-operator F such that $Nat \models \forall X \psi(X, F(X), \mathbf{P})$ or there is a SC-operator G such that $Nat \models \forall Y \neg\psi(G(Y), Y, \mathbf{P})$.

Lemma 11. *If one of the Problems 1-5 is computable for $\mathbf{P}$, then the monadic theory of $\langle Nat, <, \mathbf{P}\rangle$ is decidable.*

Proof. Let $\beta(P)$ be a sentence in *MLO* and let $\psi_\beta(X, Y, P)$ be defined as $\big(\beta \rightarrow (Y = \{0\})\big) \wedge \big(\neg\beta \rightarrow (X = \emptyset)\big)$.

Observe that $Nat \models \beta(\mathbf{P})$ iff there is a C-operator F such that $Nat \models \forall X \psi_\beta(X, F(X, \mathbf{P}), \mathbf{P})$ iff $Nat \models \forall X \psi_\beta(X, H(X, \mathbf{P}), \mathbf{P})$ where H is a constant C-operator defined as $H = \lambda\langle X, P\rangle.10^\omega$

Hence, if one of the Problems 1-5 is computable for $\mathbf{P}$, then we can decide whether $Nat \models \beta(\mathbf{P})$. □

The proof of Lemma 11 also implies that if the following Problem 1′ is decidable for $\mathbf{P}$, then the monadic theory of $\langle Nat, <, \mathbf{P}\rangle$ is decidable.

> *Decision Problem 1′ for* $\mathbf{P} \subseteq Nat$
> *Input:* an *MLO* formulas $\psi(X, Y, P)$.
> Question: Check whether there is a C-operator $Y = F(X, P)$ such that $Nat \models \forall X \psi(X, F(X, \mathbf{P}), \mathbf{P})$.

Problem 1′ is actually Problem 1 without construction part.

The implication (4)⇒(5) of Theorem 2 is its difficult part. Its proof relies on the following Lemmas:

Lemma 12. *If the monadic theory of* $M = \langle Nat, <, \mathbf{P}_1, \ldots, \mathbf{P}_n \rangle$ *is decidable, then* $\mathbf{P}_1, \ldots, \mathbf{P}_n$ *are recursive.*

Lemma 13. *If* $M = \langle Nat, <, \mathbf{P}_1, \ldots, \mathbf{P}_n \rangle$ *is recursive and* $M \models \exists X_1 \ldots \exists X_m \alpha(X_1, \ldots, X_m)$*, then there are recursive sets* $\mathbf{S}_1, \ldots, \mathbf{S}_m$ *such that* $M \models \alpha(\mathbf{S}_1, \ldots, \mathbf{S}_m)$*. Moreover, if the monadic theory of* M *is decidable, then there is an algorithm for computing (programs for)* $\mathbf{S}_1, \ldots, \mathbf{S}_m$ *from* α*.*

Lemma 12 is trivial. Lemma 13 is stated by Siefkes (cf. Lemma 3 [Sie75]) without "Moreover clause". It was shown in [Sie75] that there is no algorithm that computes programs for $\mathbf{S}_1, \ldots, \mathbf{S}_m$ from programs for $\mathbf{P}_1, \ldots, \mathbf{P}_n$ and α. The "Moreover clause" was proved in [Rab05]. The algorithm in [Rab05] can even cover arbitrary unary predicates $\mathbf{P}_1, \ldots, \mathbf{P}_n$ and not only the recursive ones. It is effective when given an oracle which supplies the truth values of any *MLO* sentence on M.

Now the implication (4)⇒(5) is a consequence of Lemma 12 and the following

Theorem 14. *Let* $\mathbf{P}$ *be a unary recursive predicate over* Nat*. For every MLO formula* $\psi(X, Y, \mathbf{P})$ *either there is a recursive C-operator* F *such that* $Nat \models \forall X \psi(X, F(X), \mathbf{P})$ *or there is a recursive SC-operator* G *such that* $Nat \models \forall Y \neg\psi(G(Y), Y, \mathbf{P})$*. Moreover, if the monadic theory of* $\langle Nat, <, \mathbf{P} \rangle$ *is decidable, then it is decidable which of these cases holds and the (description of the) corresponding operator is computable from* ψ*.*

Proof. Let $\varphi(X, Y, Z)$ be the formula obtained from $\psi(X, Y, P)$ when the predicate name P is replaced by variable Z. Let $\mathcal{A} = \langle \mathcal{Q}, \Sigma, \delta_{\mathcal{A}}, q_{init}, col \rangle$, be a deterministic parity automaton equivalent to φ. By Theorem 5, one of the players has a memoryless winning strategy in the parity game $G_{\mathcal{A},\mathbf{P}}$.

Assume that Player I has a memoryless winning strategy. Then by Lemma 8(2), there are $W_1, \ldots, W_{|\mathcal{Q}|} \subseteq Nat$ such that $Nat \models Win_{\mathcal{A}}(W_1, \ldots, W_{|\mathcal{Q}|}, \mathbf{P})$. By Lemma 13, there are recursive $W_1, \ldots, W_{|\mathcal{Q}|} \subseteq Nat$ such that $Nat \models Win_{\mathcal{A}}(W_1, \ldots, W_{|\mathcal{Q}|}, \mathbf{P})$. Moreover, programs for $W_1, \ldots, W_{|\mathcal{Q}|}$ are computable from ψ. Hence, the corresponding winning strategy U_1 for Player I in $G_{\mathcal{A},\mathbf{P}}$ is recursive. Therefore, by Lemma 7, F_{U_1} is recursive and $Nat \models \forall Y \neg\psi(G(Y), Y, \mathbf{P})$ holds by Lemma 6 and the definition of $\mathcal{A}$.

Similar arguments show that in the case when Player II has a memoryless winning strategy, there is a recursive C-operator F_{U_2} such that $Nat \models \forall X \psi(X, F(X), \mathbf{P})$. □

Finally, we have

Lemma 15. *The implications (5)⇒(1), (5)⇒(2) and (5)⇒(3) hold.*

Proof. Let $\psi(X, Y, P)$ be a formula. By (5) either there is a recursive C-operator F such that $Nat \models \forall X \psi(X, F(X), \mathbf{P})$ or there is a recursive SC-operator G such that $Nat \models \forall Y \neg\psi(G(Y), Y, \mathbf{P})$. Moreover, it is decidable which of these cases holds and the corresponding operator is computable from ψ.

In the first case, the answer to Problems 1-3 is positive and F is a corresponding operator.

In the second case, the answer to Problems 1-3 is negative.

Indeed, for the sake of contradiction, assume that there is a C-operator (even non-recursive) F such that $Nat \models \forall X \psi(X, F(X, \mathbf{P}), \mathbf{P})$. Observe that F is a C-operator and G is a SC-operator. Hence, $H = \lambda X.G(F(X, \mathbf{P})$ is a SC-operator. Every SC-operator has a fixed point. Let $\mathbf{X}_0$ be a fixed point of H and let $\mathbf{Y}_0 = F(\mathbf{X}_0, \mathbf{P})$. Then we have: $\mathbf{X}_0 = G(\mathbf{Y}_0)$. Therefore, we obtain

$$Nat \models \psi(\mathbf{X}_0, \mathbf{Y}_0, \mathbf{P})$$

because $Nat \models \forall X \psi(X, F(X, \mathbf{P}), \mathbf{P})$, and $Nat \models \neg\psi(\mathbf{X}_0, \mathbf{Y}_0, \mathbf{P})$ because $Nat \models \forall Y \neg\psi(G(Y), Y, \mathbf{P})$. Contradiction. □

5 Finite State Synthesis Problems with Parameters

Recall that a predicate $\mathbf{P} \subseteq Nat$ is ultimately periodic if there is $p, d \in Nat$ such that $(n \in \mathbf{P} \leftrightarrow n+p \in \mathbf{P})$ for all $n > d$. Ultimately periodic predicates are *MLO* definable. Hence, for every ultimately periodic predicate $\mathbf{P}$ the monadic theory of $\langle Nat, <, \mathbf{P}\rangle$ is decidable.

The next theorem implies Theorem 3 and shows that Theorem 1 can be extended only to ultimately periodic predicates.

Theorem 16. *Let* $\mathbf{P}$ *be a subset of* Nat. *The following conditions are equivalent and imply computability of Problem 4:*

1. $\mathbf{P}$ *is ultimately periodic.*
2. *For every MLO formula* $\psi(X, Y, P)$ *either there is a finite state C-operator* F *such that* $Nat \models \forall X \psi(X, F(X, \mathbf{P}), \mathbf{P})$ *or there is a finite state C-operator* G *such that* $Nat \models \forall Y \neg\psi(G(Y, \mathbf{P}), Y, \mathbf{P})$.
3. $\mathbf{P}$ *satisfies the following selection condition:*
   For every formula $\alpha(X, P)$ *such that* $Nat \models \exists X \alpha(X, \mathbf{P})$ *there is a finite state C-operator* $H : \{0,1\}^\omega \rightarrow \{0,1\}^\omega$ *such that* $Nat \models \alpha(H(\mathbf{P}), \mathbf{P})$.

Proof. The implication (1)⇒ (2) follows from Theorem 1 and the fact that every ultimately periodic predicate is definable by an *MLO* formula. The implication (2)⇒(3) is trivial.

The implication (3)⇒(1) is derived as follows. Let $\alpha(X, P)$ be $\forall t\big(X(t) \leftrightarrow P(t+1)\big)$. Note $Nat \models \exists X \alpha(X, \mathbf{P})$ for every $\mathbf{P} \subseteq Nat$. Therefore, if $\mathbf{P}$ satisfies selection condition, then there is C-operator $H : \{0,1\}^\omega \rightarrow \{0,1\}^\omega$ such that $Nat \models \alpha(H(\mathbf{P}), \mathbf{P})$.

Assume that a finite state Moore automaton $\mathcal{A}$ computes H and has n states. We are going to show that $\mathbf{P}$ is ultimately periodic with period at most $2n+1$. For $i \in Nat$ let a_i be one if $i \in \mathbf{P}$ and a_i be zero otherwise. Let $q_0 q_1 \dots q_{2n+1} \dots$ be the sequence states passed by $\mathcal{A}$ on the input $a_0 a_1 \dots a_{2n+1} \dots$. There are $i < j < 2n$ such that $a_i = a_j$ and $q_i = q_j$. Observe that $q_{i+1} = \delta_{\mathcal{A}}(q_i, a_i) = \delta_{\mathcal{A}}(q_j, a_j) = q_{j+1}$ and $a_{i+1} = out_{\mathcal{A}}(q_i, a_i) = out_{\mathcal{A}}(q_j, a_j) = a_{j+1}$. And by induction we get that $q_{i+m} = q_{j+m}$ and $a_{i+m} = a_{j+m}$ for all $m \in Nat$. Therefore, $\mathbf{P}$ is an ultimately periodic with a period $j - i < 2n$.

Note that this theorem does not imply that Problem 4 is computable only for ultimately periodic predicates. The next theorem can be established by the same arguments.

Theorem 17. *The following conditions are equivalent and imply computability of Problem 5:*

1. $\mathbf{P}$ *is ultimately periodic.*
2. *For every MLO formula* $\psi(X, Y, P)$ *either there is a finite state C-operator* F *such that* $Nat \models \forall X \psi(X, F(X), \mathbf{P})$ *or there is a finite state SC-operator* G *such that* $Nat \models \forall Y \neg\psi(G(Y), Y, \mathbf{P})$. *Moreover, it is decidable which of these cases holds and the corresponding operator is computable from* ψ.

6 Conclusion and Related Work

We investigated the Church synthesis problem with parameters. We provided the necessary and sufficient conditions for computability of Synthesis problems 1-3.

Rabin [Rab72] provided an alternative proof for computability of the Church synthesis problem. This proof used an automata on infinite trees as a natural tool for treating the synthesis problem. A C-operator $F : \{0,1\}^\omega \rightarrow \{0,1\}^\omega$ can be represented by a labelled infinite full binary tree $\langle T_2, <, \mathbf{S}\rangle$, where $\mathbf{S}$ is a subset of the tree nodes. Namely, the branches of the tree represent $X \in \{0,1\}^\omega$ and the sequence of values assigned by $\mathbf{S}$ to the nodes along the branch X represents $F(X) = Y \in \{0,1\}^\omega$. Also, the fact that $\mathbf{S}$ represents a C-operator F which uniformizes $\varphi(X, Y)$ can be expressed by an *MLO* formula $\psi(Z)$ (computable from $\varphi(X, Y)$): $T_2 \models \psi(\mathbf{S})$ iff $Nat \models \forall\varphi(X, F_{\mathbf{S}}(X))$, where $F_{\mathbf{S}}$ is the C-operator that corresponds to $\mathbf{S}$. Hence, the question whether there exists a C-operator which uniformizes φ is reduced to the problem whether $T_2 \models \exists Z \psi(Z)$. Now, the Rabin basis theorem states that if $T_2 \models \exists Z \psi(Z)$ then there is a regular subset $\mathbf{S} \subseteq T_2$ such that $T_2 \models \psi(\mathbf{S})$. The C-operator which corresponds to a regular set $\mathbf{S}$ is computable by a finite state automaton. Hence, the Büchi and Landweber theorem is obtained as a consequence of the decidability of the monadic logic of order of the full binary tree and the basis theorem.

One could try to apply the Rabin method to the Church synthesis problem with parameters. The reduction which is similar to Rabin's reduction shows that the decidability of Problem 1′ for $\mathbf{P} \subseteq Nat$ is reduced to the decidability of the monadic theory of the labelled full binary tree $\langle T_2, <, \mathbf{Q}\rangle$ where a node is in $\mathbf{Q}$, if its distance from the root is in $\mathbf{P}$. The decidability of the latter problem can be reduced by Shelah-Stupp Muchnick Theorem [Shel75, Wal02, Th03] to the decidability of $\langle Nat, <, \mathbf{P}\rangle$. Now in order to establish computability of problems 1-3, one can try to prove the basis theorem for $\langle T_2, < .\mathbf{Q}\rangle$. Unfortunately, arguments similar to the proof of Theorem 16 show that for $\mathbf{P}, \mathbf{Q}$, as above, the following version of the basis theorem

> for every $\psi(Z, Q)$ such that $T_2 \models \exists Z \psi(Z, \mathbf{Q})$ there is a finite state operator $F(Y, U)$ such that the set which corresponds to the C-operator $\lambda X F(X, \mathbf{P})$ satisfies $\psi(Z, \mathbf{Q})$

holds only for ultimately periodic **P**. Our proof of Theorem 2 implies that if the monadic theory of $\langle Nat, <, \mathbf{P}\rangle$ is decidable, then "finite state" can be replaced by "recursive" in the above version of the basis theorem.

Open Question: Are the following assertions equivalent?

1. The monadic theory of $\langle T_2, < \mathbf{S}\rangle$ is decidable.
2. For every $\psi(Z, U)$ such that $T_2 \models \exists Z\psi(Z, \mathbf{S})$ there is a recursive set $\mathbf{Q} \subset T_2$ such that $T_2 \models \psi(\mathbf{Q}, \mathbf{S})$.

Siefkes [Sie75] proved that there is a recursive set **S** for which the second assertion fails.

The conditions of Theorem 16 and Theorem 17 are sufficient, but are not necessary for computability of Synthesis problems 4-5. For example, let $\mathbf{Fac} = \{n! \ : \ n \in Nat\}$ be the set of factorial numbers. We can show that Problems 4-5 are computable for this predicate **Fac** [Rab06]. These computability results can be extended to the class of predicates for which decidability of the monadic theory of $\langle Nat, <, \mathbf{P}\rangle$ was shown by Elgot and Rabin [ER66]. Among these predicates are the sets of k-th powers $\{n^k \ : \ n \in Nat\}$ and the sets $\{k^n \ : \ n \in Nat\}$ (for $k \in Nat$). We can also show that Problems 4-5 are computable for each unary predicate in the class $\mathcal{K}$ considered by Carton and Thomas [CT02].

It is an open question whether decidability of $\langle Nat, <, \mathbf{P}\rangle$ is a sufficient condition for computability of Synthesis problems 4-5.

Games over pushdown graphs and operators computable by pushdown automata were recently studied in the literature [Th95, Wal01, CDT02, GTW02, Se04]. It is a natural question to consider the synthesis Problems 1-3 where "recursive" is replaced by "computable by pushdown automata".

Kupferman and Vardi [KV97] considered the synthesis problem with incomplete information for the specifications described by temporal logics LTL and CTL*. Their main results deal with the complexity of this synthesis problem. The decidability of the synthesis problem with incomplete information for LTL (respectively, for CTL*) can be easily derived from the Büchi-Landweber (respectively, Rabin) theorem. It seems that there are no interesting connections between the synthesis problems with incomplete information and the synthesis problems with parameters considered here.

In [RT98] a program for the relativization of finite automata theory was proposed. Our results can be seen as the first step in this direction. This step corresponds to the case where oracles are C-operators without inputs.

Acknowledgments. I would like to thank Wolfgang Thomas for bringing to my attention the paper by D. Siefkes [Sie75]. I am grateful to the anonymous referees for their suggestions.

References

[Bu60] J. R. Büchi. On a decision method in restricted second order arithmetic In *Proc. International Congress on Logic, Methodology and Philosophy of Science*, E. Nagel at al. eds, Stanford University Press, pp 1-11, 1960.

[BL69] J. R. Büchi and L. H. Landweber. Solving sequential conditions by finitestate strategies. Transactions of the AMS, 138(27):295–311, 1969.

[CDT02] T. Cachat, J. Duparc, and W. Thomas. Solving pushdown games with a Σ_3 winning condition. In CSL02, LNCS vol. 2471, pp. 322-336, 2002.

[CT02] O. Carton and W.Thomas. The Monadic Theory of Morphic Infinite Words and Generalizations. Inf. Comput. 176(1), pp. 51-65, 2002.

[Ch69] Y. Choueka. Finite Automata on Infinite Structure. Ph.D Thesis, Hebrew University, 1970.

[ER66] C. Elgot and M. O. Rabin. Decidability and Undecidability of Extensions of Second (First) Order Theory of (Generalized) Successor. J. Symb. Log., 31(2), pp. 169-181, 1966.

[EJ91] E. A. Emerson, C. S. Jutla: Tree Automata, Mu-Calculus and Determinacy (Extended Abstract) FOCS91: 368-377, 1991.

[GTW02] E. Grädel, W. Thomas and T. Wilke. Automata, Logics, and Infinite Games, LNCS 2500, 2002.

[KV97] O. Kupferman and M.Y. Vardi, Synthesis with incomplete information, In 2nd International Conference on Temporal Logic, pp 91–106, 1997.

[PP04] D. Perrin and J. E. Pin. Infinite Words Automata, Semigroups, Logic and Games. Pure and Applied Mathematics Vol 141 Elsevier, 2004.

[Rab72] M. O. Rabin. Automata on Infinite Objects and Church's Problem Amer. Math. Soc. Providence, RI, 1972.

[RT98] A. Rabinovich and B.A. Trakhtenbrot From Finite Automata toward Hybrid Systems Proceddings of Fundamentals of Computation Theory. Lecture Notes in Computer Science 1450, pp. 411-422, Springer, 1998.

[Rab05] A. Rabinovich. On decidability of Monadic logic of order over the naturals extended by monadic predicates. Submitted, 2005.

[Rab06] A. Rabinovich. The Church problem over ω expanded by factorial numbers. In preparation, 2006.

[Sem84] A. Semenov. Logical theories of one-place functions on the set of natural numbers. Mathematics of the USSR - Izvestia, vol. 22, pp 587-618, 1984.

[Rob58] R. M. Robinson. Restricted Set-Theoretical Definitions in Arithmetic. In Proceedings of the AMS Vol. 9, No. 2. pp. 238-242, 1958.

[Se04] O. Serre. Games With Winning Conditions of High Borel Complexity. In ICALP 2004, LNCS volume 3142, pp. 1150-1162, 2004.

[Shel75] S. Shelah. The monadic theory of order. *Ann. of Math.* **102**:379–419, 1975.

[Sie75] D. Siefkes. The recursive sets in certain monadic second order fragments of arithmetic. Arch. Math. Logik, pp71-80, 17(1975).

[Th75] W. Thomas. Das Entscheidungsproblem für einige Erweiterungen der Nachfalger-Arithmetic. Ph. D. Thesis Albert-Ludwigs Universität, 1975.

[Th95] W. Thomas. On the synthesis of strategies in infinite games. In STACS '95, LNCS vo. 900, pp. 1-13. 1995.

[Th03] W. Thomas. Constructing infinite graphs with a decidable MSO-theory. In MFCS03, LNCS 2747, 2003.

[Trak61] B. A. Trakhtenbrot. Finite automata and the logic of one-place predicates. (Russian version 1961). In AMS Transl. 59, 1966, pp. 23-55.

[Wal01] I. Walukiewicz. Pushdown processes: games and model checking. Information and Computation 164 pp. 234-263, 2001.

[Wal02] I. Walukiewicz. Monadic second order logic on tree-like structures.TCS 1:275, pp 311-346, 2002.

Decidable Theories of the Ordering of Natural Numbers with Unary Predicates*

Dedicated to Boris A. Trakhtenbrot on the occasion of his 85th birthday

Alexander Rabinovich[1] and Wolfgang Thomas[2]

[1] Tel Aviv University, Department of Computer Science
rabino@math.tau.ac.il
[2] RWTH Aachen, Lehrstuhl Informatik 7, 52056 Aachen, Germany
thomas@informatik.rwth-aachen.de

Abstract. Expansions of the natural number ordering by unary predicates are studied, using logics which in expressive power are located between first-order and monadic second-order logic. Building on the model-theoretic composition method of Shelah, we give two characterizations of the decidable theories of this form, in terms of effectiveness conditions on two types of "homogeneous sets". We discuss the significance of these characterizations, show that the first-order theory of successor with extra predicates is not covered by this approach, and indicate how analogous results are obtained in the semigroup theoretic and the automata theoretic framework.

1 Introduction

In [1], Büchi showed that the monadic theory of the ordering $(\mathbb{N}, <)$ of the natural numbers is decidable. Many authors studied the question for which expansions of $(\mathbb{N}, <)$ this decidability result can be preserved. For most examples of natural functions or binary relations it turned out that the corresponding monadic theory is undecidable, usually shown via an interpretion of first-order arithmetic. This applies, for instance, to the double function $\lambda x.2x$ ([9,20]).

For the expansion of $(\mathbb{N}, <)$ by unary predicates, the situation is different: Many examples P of such predicates are known such that the monadic theory of $(\mathbb{N}, <, P)$ is decidable, among them – as shown by Elgot and Rabin [5] – the set of factorial numbers, the set of powers of k and the set of k-th powers (for fixed k). A larger class of such predicates was presented in [3,4]; another comprehensive study is [11]. Contrary to the case of functions, no "natural" recursive predicate P is known such that the monadic theory of $(\mathbb{N}, <, P)$ is undecidable. Moreover, it is known that in the cases where undecidability holds, the undecidability proof cannot be done via an interpretation of first-order arithmetic (see [2,16]).

* This paper was written during a visit of the first author in Aachen in March 2006, funded by the European Science Foundation ESF in the Research Networking Programme AutoMathA (Automata: From Mathematics to Applications).

Z. Ésik (Ed.): CSL 2006, LNCS 4207, pp. 562–574, 2006.

The approach introduced by Elgot and Rabin [5] for showing decidability of the monadic theory of a structure $(\mathbb{N}, <, P)$ is built on a method to decompose this structure in a "periodic" way, together with the translation of monadic formulas to Büchi automata. By this translation, the monadic theory of $(\mathbb{N}, <, P)$ is decidable iff the following decision problem Acc_{u_P} is decidable for the characteristic ω-word u_P associated with P (where $u_P(i) = 1$ if $i \in P$ and otherwise $u_P(i) = 0$):

(Acc_{u_P}): Given a Büchi automaton $\mathcal{A}$, does $\mathcal{A}$ accept u_P?

Considering the predicate F of factorial numbers as an example, Elgot and Rabin defined for a given Büchi automaton $\mathcal{A}$ a "contraction" $c(u_F)$ of u_F which is accepted by $\mathcal{A}$ iff u_F is accepted by $\mathcal{A}$. The contraction $c_{\mathcal{A}}(u_F)$ is obtained from u_P by applying a pumping argument to the 0-segments between successive letters 1. The word $c_{\mathcal{A}}(u_F)$ has 0-segments of bounded length and turns out to be ultimately periodic; so one can decide whether $\mathcal{A}$ accepts $c_{\mathcal{A}}(u_F)$ and hence u_F. Also the method of [3,4] follows this pattern: Decidability of Acc_{u_P} is reduced to the question whether, given a finite semigroup (replacing the Büchi automaton as used by Elgot and Rabin), one can compute a representation of an ultimately periodic word which can replace u_P for answering the question about u_P. An abstract version of this "effective reduction to ultimately periodic predicates" is given in our main theorem below. As a key tool we use Ramsey's Theorem on the existence of homogeneous sets over colored orderings of order type ω (as already Büchi did in [1]).

In [8] this "non-uniform" procedure of reducing u_P to ultimately periodic sequences, depending on the monadic formula, the Büchi automaton, or the semigroup under consideration, was replaced by a "uniform" periodicity condition on P, thus settling a conjecture raised in [4]. The main result of [8] states that the monadic theory of $(\mathbb{N}, <, P)$ is decidable iff a recursive predicate P' exists which is "homogeneous for P". This predicate captures, in some sense, all the ultimately periodic structures that arise from the non-uniform approach mentioned above.

The purpose of the present paper is to give a streamlined proof of the result of [8], clarifying the connection to the "non-uniform" method, and at the same time generalizing it from monadic logic to a class of logics between first-order logic and monadic logic. We also discuss the case where the successor relation S is considered instead of the ordering $<$ (a modification which is irrelevant when monadic logic is considered). As in [8], we present the proofs in a logical framework, avoiding the use of automata or semigroups, and building on composition theorems in the spirit of Shelah [12] (see also [6,18]). As explained in Section 4.3, however, the arguments do not depend on this framework and can be transferred easily to the domains of automata, respectively semigroups.

The present work represents a merge of ideas of the two independently written and so far unpublished papers [8,15] (of 2005 and 1975, respectively).

In the subsequent section we introduce the background theory and state the main result. Section 3 is devoted to the proof. In Section 4, we discuss several aspects of the main theorem: its significance, its failure for the first-order theory of successor, and the alternative frameworks of semigroups and automata in place of logic. A summary and outlook conclude the paper.

2 Logical Background and Main Result

The structures considered in this paper are of the form $M = (\mathbb{N}, <, P_1, \ldots, P_m)$ with $P_i \subseteq \mathbb{N}$. We call them m-labelled ω-chains. These structures are in one-to-one correspondence with ω-words over the alphabet $\{0,1\}^m$. The ω-word $u_{\overline{P}} = u_{\overline{P}}(0)u_{\overline{P}}(1)\ldots$ corresponding to $\overline{P} = (P_1, \ldots, P_m)$ has value 1 in the j-th component of $u_P(i)$ iff $i \in P_j$. By an m-labelled chain we mean a linear ordering $(A, <, P_1, \ldots, P_m)$ with finite A or $A = \mathbb{N}$.

Let us recall some standard logical systems; here we assume that the signature is chosen according to the type of structure above (and in our notation we do not distinguish, for example, between the relation $<$ and the relation symbol denoting it). The system of first-order logic FO[<] has, besides equality, the mentioned relation symbols $<, P_1, \ldots, P_m$. The atomic formulas are of the form $x = y, x < y, P_i(x)$ with first order variables x, y. Formulas are built up using boolean connectives and the first-order quantifiers $\exists, \forall$. In the first-order logic FO[S], the successor relation S is used in place of $<$.

It is known that over labelled chains one can increase the expressive power of first-order logic by adjoining "modular counting quantifiers" $\exists^{r,q}$ (with $0 \leq r < q$), where $\exists^{r,q}x\varphi(x)$ means that the number of elements x satisfying φ is finite and equal to r modulo q. We denote this logic by FO[<]+MOD. A detailed introduction is given in [13].

Still more expressive are the logical systems MSO of monadic second-order logic and WMSO of weak monadic second-order logic. They arise from FO[<] by adding unary second-order variables $X, Y, \ldots$ and corresponding atomic formulas (written $X(y)$ etc.). In MSO, quantification over set variables ranges over the subsets of $\mathbb{N}$, in WMSO only over the finite subsets of $\mathbb{N}$. Over labelled ω-orderings, WMSO and MSO have the same expressive power, which however exceeds that of FO[<]+MOD (cf. [17,13]).

In the sequel the letter L stands for any of the logics introduced above. The L-theory of $(\mathbb{N}, <, \overline{P})$ is the set of L-sentences which are true in $(\mathbb{N}, <, \overline{P})$.

For the analysis of the L-theory of $(\mathbb{N}, <, \overline{P})$ we use the composition method which was developed by Shelah [12] for monadic second-order logic. We recall the facts underlying the composition method.

Two m-labelled chains M, M' are called k-equivalent for L (written: $M \equiv_k^L M'$) if $M \models \varphi \Leftrightarrow M' \models \varphi$ for every L-sentence φ of quantifier depth k. This is an equivalence relation between labelled chains; its equivalence classes are called k-types for L (and for the given signature with $<$ and m unary predicate symbols). Let us list some fundamental and well-known properties of k-types for any of the logics L above; here we suppress the reference to L for simplicity of notation.

Proposition 1. *1. For every m and k there are only finitely k-types of m-labelled chains. (In the case of FO[<]+MOD, we assume that also a maximal divisor q is fixed in advance.)*

2. *For each k-type t there is a sentence (called "characteristic sentence") which defines t (i.e., is satisfied by a labelled m-chain iff it belongs to t). For given k and m, a finite list of characteristic sentences for all the possible k-types can be computed. (We take the characteristic sentences as the canonical representations of k-types. Thus, for example, transforming a type into another type means to transform sentences.)*
3. *Each sentence φ is equivalent to a (finite) disjunction of characteristic sentences; moreover, this disjunction can be computed from φ.*

The proofs of these facts can be found in several sources, we mention [12,18,19] for MSO and FO, and [13] for FO[<]+MOD.

As a simple consequence we note that the L-theory of an m-labelled chain M is decidable iff the function which associates to each k the k-type of M for L is computable.

Given m-labelled chains M_0, M_1 we write $M_0 + M_1$ for their concatenation (ordered sum). In our context, M_0 will always be finite and M_1 finite or of order type ω. Similarly, if for $i \geq 0$ the chain M_i is finite, the model $\Sigma_{i \in \mathbb{N}} M_i$ is obtained by the concatenation of all M_i in the order given by the index structure $(\mathbb{N}, <)$.

We need the following composition theorem on ordered sums:

Theorem 2 (Composition Theorem). *Let L be any of the logics considered above.*

(a) *The k-types of m-labelled chains M_0, M_1 for L determine the k-type of the ordered sum $M_0 + M_1$ for L, which moreover can be computed from the k-types of M_0, M_1.*

(b) *If the m-labelled chains $M_0, M_1, \ldots$ all have the same k-type for L, then this k-type determines the k-type of $\Sigma_{i \in \mathbb{N}} M_i$, which moreover can be computed from the k-type of M_0.*

Part (a) of the theorem justifies the notation $s + t$ for the k-type of an m-chain which is the sum of two m-chains of k-types s and t, respectively. Similarly, we write $t * \omega$ for the k-type of a sum $\Sigma_{i \in \mathbb{N}} M_i$ where all M_i are of k-type t.

Let us call a logic L *compositional* if the Composition Theorem above with parts (a) and (b) holds. All logics L listed above are compositional. For FO[<] and WMSO this goes back to Läuchli, for MSO to Shelah [12], and for FO[S] and FO[<]+MOD one may consult [13].

The fundamental fact which enters all decidability proofs below (and which underlies also Büchi's work [1]) is the following: The two parts (a) and (b) of the Composition Theorem suffice to generate the k-types of arbitrary (even non-periodic) m-labelled chains $M = (\mathbb{N}, <, P_1, \ldots, P_m)$. This is verified by decomposing M into segments such that all of them except possibly the first one have the same k-type. The elements (numbers) that separate the segments of such a decomposition form a "homogeneous set". Given $M = (\mathbb{N}, <, \overline{P})$, let us write $M[i, j)$ for the m-labelled chain with domain $[i, j) = \{i, \ldots, j-1\}$ and the predicates $<$ and $\overline{P}$ restricted to $[i, j)$.

Definition 3 (k-homogeneous set). *A set $H = \{h_0 < h_1 < \ldots\}$ is called* k-homogeneous for $M = (\mathbb{N}, <, \overline{P})$ *with respect to L, if all segment models $M[h_i, h_j)$ for $i < j$ (and hence all segment models $M[h_i, h_{i+1})$ for $i \geq 0$) have the same k-type for L.*

In the main theorem below, a stronger notion of homogeneity [8] enters:

Definition 4 (uniformly homogeneous set). *A set $H = \{h_0 < h_1 < \ldots\}$ is called* uniformly homogeneous for $M = (\mathbb{N}, <, \overline{P})$ *with respect to L if for each k the set $H_k = \{h_k < h_{k+1} < \ldots\}$ is k-homogeneous with respect to L.*

The existence of uniformly homogeneous sets will be shown in the next section, while the existence of k-homogeneous sets is well-known (see e.g. [17]):

Proposition 5 (Ramsey). *Let f be a function from $\mathbb{N}^2$ into a finite set C. Then there is $c \in C$ and an infinite set H such that $f(i, j) = c$ for all $i < j \in H$.*

In particular, if L is a logic satisfying item 1 of Proposition 1, and M an m-labelled ω-chain, there is a k-homogeneous set for M with respect to L.

Given a k-homogeneous set $H = \{h_0 < h_1 < \ldots\}$ for $M = (\mathbb{N}, <, \overline{P})$ with respect to L, the Composition Theorem implies that the k-type for L of $M = (\mathbb{N}, <, \overline{P})$ can be computed from the k-types for L of $M[0, h_0)$ and of $M[h_0, h_1)$; note that all the segment models $M[h_i, h_{i+1})$ have the same k-type for L.

For two k-types s, t (for L) of m-labelled chains consider the following condition:

$\mathrm{Hom}^L_{s,t}$:

There is a k-homogeneous set $H = \{h_0 < h_1 < \ldots\}$ with respect to L such that $M[0, h_0)$ has k-type s and $M[h_0, h_1)$ has k-type t for L.

If $\mathrm{Hom}^L_{s,t}$ is true in $M = (\mathbb{N}, <, \overline{P})$, the k-type of $M = (\mathbb{N}, <, \overline{P})$ for L is computable as the type $s + t * \omega$. Thus, Ramsey's Theorem reduces the decision problem for the L-theory of $M = (\mathbb{N}, <, \overline{P})$ to the problem of deciding, for each k and k-types s, t, whether the statement $\mathrm{Hom}^L_{s,t}$ holds in M. Ramsey's Theorem guarantees that for given M and k such a pair (s, t) of k-types exist.

For an m-labelled ω-chain M, let $\mathrm{RecRamsey}^L(M)$ be the following condition:

$\mathrm{RecRamsey}^L(M)$:

There is a recursive function assigning to each k a pair of k-types s and t for L such that $\mathrm{Hom}^L_{s,t}$ holds in M.

We call the logic L *expressive for the existence of homogeneous sets* if for any k-types s, t for L, there is an L-sentence which expresses $\mathrm{Hom}^L_{s,t}$.

We can now state our main result.

Theorem 6. *Let L be a logic which is both compositional and expressive for the existence of homogeneous sets. Then the following are equivalent for any given m-labelled ω-chain $M = (\mathbb{N}, <, \overline{P})$ with recursive sets $\overline{P}$:*

1. *The L-theory of M is decidable.*
2. *RecRamsey^L(M).*
3. *There is a recursive uniformly homogeneous set for M with respect to L.*

Let us verify that the theorem covers all the logics L mentioned above, excepting FO[S]. For this it remains to show that FO[<], FO[<]+MOD, WMSO, MSO are expressive for the existence of homogeneous sets.

This is obvious for MSO; in a straightforward formalization of $\mathrm{Hom}^L_{s,t}$ one uses an existential set quantifier $\exists X$ and relativizes the characteristic sentences for s and t to the segments from 0 to the first element of X, respectively to the segments enclosed by successive X-elements. For the remaining logics it suffices to show the following (see e.g. [17]).

Proposition 7. *FO[<] is expressive for the existence of homogeneous sets.*

Proof. We write $T_k[x, y) = t$ for a formula expressing that the k-type of the segment $[x, y)$ (for FO[<]) is t. The proof covers all logics L considered here which extend FO[<]; in our notation we suppress the reference to FO[<] or to such L. Note that $\mathrm{Hom}_{s,t}$ can only hold for a k-type t with $t = t + t$. We show that $\mathrm{Hom}_{s,t}$ holds iff

$$\exists x(T_k[0, x) = s \ \wedge \ \forall y \ \exists z z' \ (\ y < z < z' \ \wedge \ T_k[x, z) = t \ \wedge T_k[z, z') = t \ )$$

The direction from left to right is trivial; take, e.g., for x the minimal element of the homogeneous set given by the assumption.

For the direction from right to left choose a number x as given by the formula, and apply its latter clause by choosing a sequence of numbers $z_1 < z'_1 < z_2 < z'_2 < \ldots$ such that $T_k[x, z_i) = T_k[z_i, z'_i) = t$ and hence (note that $t + t = t$) $T_k[x, z'_i) = t$. We shall find a subsequence $z_1 < z_{i_1} < z_{i_2} \ldots$of $z_1 < z_2 < \ldots$ such that $T_k[x, z_{i_m}) = T_k[z_{i_m}, z_{i_n}) = t$ for all $m < n \in \mathbb{N}$.

Define a coloring col from $\mathbb{N}^2$ to the set T_k of all k-types as follows: $\mathrm{col}(i, j) = T_k[z'_i, z_j)$. By Ramsey's theorem there is $t_1 \in T_k$ and an infinite set $i_1 < i_2 \ldots$ such that $\mathrm{col}(i_m, i_n) = t_1$ for all $m < n$. Note that for $m < n \in \mathbb{N}$:

$$t = T_k[x, z_{i_n}) = T_k[x, z'_{i_m}) + T_k[z'_{i_m}, z_{i_n}) = t + t_1$$

Hence,

$$T_k[z_{i_m}, z_{i_n}) = T_k[z_{i_m}, z'_{i_m}) + T_k[z'_{i_m}, z_{i_n}) = t + \mathrm{col}(i_m, i_n) = t + t_1 = t.$$

□

Let us address the relation of Theorem 6 to the main result of [8]. It is shown there that the following conditions are equivalent:

1. The monadic (second-order) theory of $M = (\mathbb{N}, <, \overline{P})$ is decidable.
2. There is a recursive uniformly homogeneous set for M with respect to the monadic (second-order) logic.

The proof of the implication (1)⇒(2) in [8] relies on the expressive power of MSO-logic and proceeds as follows. For $k = 1, 2, \ldots$ an MSO-formula $H_k(X, Y, \overline{P})$ is constructed (effectively from k and the number of predicates in $\overline{P}$) which defines for any infinite subset Y of $M = (\mathbb{N}, <, \overline{P})$ an infinite set $X \subseteq Y$ which is k-homogeneous for M. (The uniqueness proof for X requires a nontrivial uniformization result, using [7].) Hence, the set Q_1 such that $M \models H_1(Q_1, \mathbb{N}, \overline{P})$ is 1-homogeneous for M, and more generally the set Q_{k+1} such that $M \models H_{k+1}(Q_{k+1}, Q_k, \overline{P})$ is $(k+1)$-homogeneous for M. The sets $Q_1 \supseteq Q_2 \supseteq \ldots$ are definable by formulas and therefore are recursive (in the monadic theory of M). Hence, the set $H = \{a_k \ : \ a_k$ is k-th element of $Q_k\}$ is recursive and uniformly homogeneous for M.

In the present paper, the proof of Theorem 6 relates the conditions of non-uniform and uniform homogeneity in a direct way, covers more logics (between FO[<] and MSO) than MSO, and is somewhat simpler since it does not involve the uniformization result of [7].

3 Proof of Theorem 6

For the conditions

(1) The L-theory of M is decidable
(2) RecRamsey$^L(M)$
(3) There is a recursive uniformly homogeneous set for M with respect to L

we show the implication chain (3) ⇒ (2) ⇒ (1) ⇒ (3).

(3)⇒(2). Assume that $H = \{h_0 < h_1 < \ldots\}$ is recursive and uniformly homogeneous for M with respect to L.

Let k be a natural number. If s is the k-type of $M[0, h_k)$ and t is the k-type of $M[h_k, h_{k+1})$ then $M \models Hom^L_{s,t}$. Let t_i be the k-type of one element chain $M[i, i]$. Note that t_i is computable because M is recursive. The k-type of $M[0, h_k)$ is $s = \sum_{i=0}^{i=h_k-1} t_i$ and the k type of $t = M[h_k, h_{k+1})$ is $\sum_{i=h_k}^{i=h_{k+1}-1} t_i$. These sums are computable from the Composition Theorem.

(2)⇒(1). Let ψ be a sentence. In order to check whether ψ holds in M we proceed as follows:

1. Let k be the quantifier depth of ψ.
2. By RecRamsey$^L(M)$ we can compute k-types s and t for L such that $M \models Hom^L_{s,t}$.
3. Hence, the k-type t_1 of M can be computed as $t_1 = s + t * \omega$.
4. In order to check whether $t_1 \rightarrow \psi$ is valid, we can compute a finite disjunction of k-characteristic sentences which is equivalent to ψ, and note that $t_1 \rightarrow \psi$ holds iff t_1 is one of these disjuncts.
5. Finally, $t_1 \rightarrow \psi$ iff ψ holds in M.

(1)⇒(3) Assume that the L-theory of M is decidable. We present an algorithm which enumerates in increasing order the numbers of a recursive homogeneous set $H = \{n_1 < n_2 < n_3 < \ldots\}$ for M. We use T_k^L for the (finite) set of k-types of the language L.

Algorithm

Basis

1. Find $t_1, s_1 \in T_1^L$ such that $t_1 = t_1 + t_1$ and $M \models Hom^L_{s_1,t_1}$. Note that such s_1 and t_1 exist by the Ramsey Theorem. Moreover, there is an algorithm to find s_1 and t_1, because of finiteness of T_1^L, the assumption that $Hom^L_{s_1,t_1}$ is expressible in L, and decidability of the L-theory of M.
2. Let n_1 be the minimal n such that

$$s_1 \text{ is the 1-type of } M[0,n) \text{ and} \tag{1}$$

$$M[n,\infty) \models Hom^L_{t_1,t_1} \tag{2}$$

This number n_1 can be computed as follows. Let $\alpha_s(v)$ be a formula which expresses $T_k[0,v) = s$, and let $\beta_t(v)$ be a formula obtained from the sentence $Hom_{t,t}$ by relativizing all quantifiers to the interval $[v,\infty)$. It is clear that n_1 defined above is the unique natural number that satisfies

$$\gamma(v) =_{def} \alpha_{s_1}(v) \wedge \beta_{t_1}(v) \wedge \forall u((0 < u < v) \rightarrow \neg(\alpha_{s_1}(u) \wedge \beta_{t_1}(u))).$$

From the fact that the L theory of M is decidable and that every natural number n is defined by an L formula $\psi_n(v)$ (computable from n) we can compute this n_1 by finding the minimal number n such that $M \models \exists v(\psi_n(v) \wedge \gamma(v))$.

Inductive step $k \mapsto k+1$

1. Find $t_{k+1}, s_{k+1} \in T^L_{k+1}$ such that $t_{k+1} \rightarrow t_k$ and $s_{k+1} \rightarrow t_k$ are valid and $t_{k+1} = t_{k+1} + t_{k+1}$ and $M[n_k,\infty) \models Hom^L_{s_{k+1},t_{k+1}}$. The arguments similar to the arguments in the step 1 of the basis show that t_{k+1}, s_{k+1} are computable.
2. Let n_{k+1} be the minimal $n > n_k$ such that

$$s_{k+1} \text{ is the k+1 type of } M[n_k,n) \text{ and} \tag{3}$$

$$M[n,\infty) \models Hom^L_{t_{k+1},t_{k+1}} \tag{4}$$

The arguments similar to the arguments in the step 2 of the basis show that t_{k+1}, s_{k+1} and n_{k+1} are computable.

It is clear that the set $H = \{n_1 < n_2 < \ldots\}$ generated by our algorithm is recursive. We show that it is uniformly homogeneous:

By our construction for every k the k-type of $M[n_k, n_{k+1})$ is t_k.

Since $s_i \rightarrow t_k$ and $t_i \rightarrow t_k$ for $i > k$ we obtain that the k-type of $M[n_i, n_{i+1})$ is also t_k for all $i > k$. Since $t_k + t_k = t_k$, we obtain that the k type of $M[n_i, n_j)$ is also t_k for all $j > i > k$. This proves the uniform homogeneity of H.

4 Discussion

4.1 Comments on Uniform Homogeneity

The main theorem provides two reductions of the decision problem for the L-theory of a structure $M = (\mathbb{N}, <, \overline{P})$: With the first reduction one can transform the question "Is the sentence φ true in M?" to a problem to determine a decomposition of M into a sequence of segments, which depends only on the complexity k of φ. This decomposition gives two L-types s_k and t_k from which one can infer by a an algorithmic procedure whether φ is implied. The decision problem for the L-theory of M is thus reduced to the question whether the function $k \mapsto s_k, t_k$ is recursive.

The second reduction captures this recursiveness by the recursiveness of a *single* decomposition of M into segments. This single decomposition results from an infinite refinement process of the types s_k, t_k mentioned above, and correspondingly it leads to a sequence of decomposition segments which satisfy k-types for larger and larger k.

In a more general formulation on the existence of uniformly homogeneous sets we can cover arbitrary unary predicates $\overline{P}$ rather than just recursive ones. Consider an m-chain $M = (\mathbb{N}, <, \overline{P})$. Note that the Algorithm of Section 3 is effective when given an oracle which can supply the truth value of any condition $\mathrm{Hom}^L_{s,t}$. So we obtain the following result from the proof of the Theorem 6:

Theorem 8. *Let L be compositional and expressive.*

1. *For each structure $M = (\mathbb{N}, <, \overline{P})$ there is a uniformly homogeneous set H which is recursive in the L-theory of M.*
2. *For each structure $M = (\mathbb{N}, <, \overline{P})$ and each uniformly homogeneous set H for M, the L-theory of M is recursive in the recursion theoretic join of $(\overline{P}, H)$.*

We can refine this result by a bound on the recursion theoretic complexity of H relative to $\overline{P}$. By Proposition 7, $\mathrm{Hom}^L_{s,t}$ is a Σ^0_3 statement over the recursion-theoretic join of the predicates in $\overline{P}$, which implies that $\mathrm{Hom}^L_{s,t}$ is recursive in $\overline{P}'''$, the third jump of the recursion-theoretic join of the predicates in $\overline{P}$. (For recursion theoretic terminology see [10].) Thus in the first part of Theorem 8, H can be chosen to be recursive in $\overline{P}'''$. As shown in [2,17], the quantifier structure of the formula that expresses $\mathrm{Hom}^L_{s,t}$ can be simplified even to a boolean combination of Σ^0_2-formulas. So the recursion theoretic bound on H can be sharpened to "truth-table reducible to $\overline{P}''$". By [16] this is optimal in the sense that bounded truth-table reducibility does not suffice.

While our main theorem provides two characterizations of the decidable L-theories of structures $(\mathbb{N}, <, \overline{P})$, it is not easily applicable in order to find interesting predicates P such that, say, the first-order or the monadic second-order theory of $(\mathbb{N}, <, P)$ is decidable. Let us first compare the condition $\mathrm{RecRamsey}^L(M)$ with the classical method of Elgot and Rabin [5], taking the factorial predicate F as an example. Elgot and Rabin proposed a deterministic procedure to transform F into an ultimately periodic predicate, depending on the given formula (or automaton). The condition $\text{RecRamsey}^L(M)$ only involves the existence of a procedure and does not provide one in concrete examples. However, the decomposition ensured by $\text{RecRamsey}^L(M)$ is "stronger" in the sense that it provides an ultimately constant (and not just periodic) sequence of types.

The uniformly homogeneous sets given by the third clause of the theorem also do not settle (immediately) the decision problem for concrete theories of structures $(\mathbb{N}, <, P)$. A prominent example is the predicate $\mathbb{P}$ of the prime numbers. The open twin prime hypothesis is easily expressible already in FO[<] (we use here for simplicity also the successor relation S, which is definable in FO[<]):

$$\varphi_0 := \forall x \exists y_0 \exists y_1 \exists y_2 (x < y_0 \wedge \mathbb{P}(y_0) \wedge S(y_0, y_1) \wedge S(y_1, y_2) \wedge \mathbb{P}(y_2))$$

Now $k = 5$ is the quantifier depth of an FO[<]-sentence which avoids this abbreviation with S. Taking the uniformly homogeneous set H for $\mathbb{P}$ with respect to FO[<], one could decide φ_0 by inspecting the segment from the 5-th to the 6-th element of H: There are infinitely many twin primes iff a pair of primes of distance 2 occurs in this segment; otherwise all twin primes would be included in the initial segment before. It is clear that H encodes this information not only about the twin primes but all other conceivable configurations of primes that are MSO-definable, for instance patterns within segments of some bounded length. Thus H encodes a lot of known and unknown number theory.

4.2 The Successor Theory

In the main result Theorem 6 we excluded the logic FO[S]. For example, the proof of Proposition 7, which shows that FO[<] is expressive for the existence of homogeneous sets, uses the $<$ relation in an essential way. Indeed, we can show that the main theorem fails for FO[S].

It turns out that a recursive uniformly homogeneous set H encodes more information than needed for deciding FO[S]-sentences. While H supplies information about the infinite occurrence of certain segment types, FO[S]-sentences can only express such occurrences in numbers up to a certain finite bound. Indeed, it is well known that the FO-theory of $M = (\mathbb{N}, S, P)$ is decidable iff for each isomorphism type τ of finite segments and each m, one can decide whether τ occurs $\geq m$ times in M (see, e.g., [16,19]).

Theorem 9. *There is a recursive predicate P such that the FO-theory of* $(\mathbb{N}, S, P)$ *is decidable but there is no recursive uniformly homogeneous set for* $(\mathbb{N}, S, P)$.

Proof. We use a predicate P, presented in [16], for which the FO-theory of $(\mathbb{N}, S, P)$ is decidable whereas the FO-theory of $(\mathbb{N}, <, P)$ is undecidable.

Suppose $\mathcal{P}$ is a procedure which runs through all pairs (i, j) of natural numbers in some order; we write (i_n, j_n) for the n-th pair in this order. $\mathcal{P}$ generates a bit sequence as follows: When treating (i_n, j_n), it checks whether the i_n-th Turing machine runs for at least j_n steps when started on the empty tape. $\mathcal{P}$ outputs $10^n10^{i_n}$ in this case, and otherwise generates just 10^n. The resulting bit sequence u is obtained as the concatenation of the $\mathcal{P}$-outputs. Clearly u is recursive, and it has the property that for each given $w \in \{0,1\}^*$ and threshold number m one can decide whether w occurs m times in u. (To verify this, note that by construction of u, the only segment types that occur infinitely often are in the languages 0^*, 0^*10^*, and, for certain values of i, $0^*10^i10^*$. To test whether a segment of the latter type occurs m times, check the output of procedure $\mathcal{P}$ up to the m-th treatment of Turing machine M_i.) From this fact one infers that the first-order theory of $(\mathbb{N}, S, P)$ is decidable (cf. [16]).

By construction of u, the i-th Turing machine does not halt on the empty tape iff the segment 10^i1 occurs infinitely often in u. We show that the latter can be decided if there is a recursive uniformly homogeneous set H for $(\mathbb{N}, S, P)$.

Let $H = \{h_0 < h_1 < \ldots\}$ be a recursive uniformly homogeneous set for $(\mathbb{N}, S, P)$. Given i choose k large enough such that from a k-type of a 1-labelled chain one can infer whether the following holds:

> $(*)$ there is a sequence of $i+2$ successive elements such that its first and last element are in P but the others are not.

Consider the segment $M[h_k, h_{k+1})$, which can be obtained effectively by recursiveness of H. $M[h_k, h_{k+1})$ satisfies $(*)$ iff 10^i1 occurs infinitely often in u. □

4.3 Algebraic and Automata Theoretic Types

In this section we discuss alternative ways of introducing "k-types", using semigroups or automata rather than formulas to describe properties of words. When referring to a logic L, we assume that it is compositional and expressive for the existence of homogeneous sets.

Recall that for such a logic L, for each k the set T_k^L of k-types of L with the $+$ operation is a finite semigroup. Let $\mathcal{S}_L$ be the family of finite semigroups defined as follows:

$S \in \mathcal{S}_L$ iff there is $k \in \mathbb{N}$ and a semigroup homomorphism from T_L^k onto S

Note that $\mathcal{S}_{WMSO} = \mathcal{S}_{MSO}$ is the family of all finite semigroups and that $\mathcal{S}_{FO}$ is the family of finite aperiodic semigroups.

Let $\mathcal{S}$ be a family of finite semigroups. Define an equivalence relation $\sim_k^{\mathcal{S}}$ on Σ^+ as follows:

> $w_1 \sim_k^{\mathcal{S}} w_2$ iff $h(w_1) = h(w_2)$ for every $S \in \mathcal{S}$ of size at most k and for every morphism $h : \Sigma^+ \to S$.

The following lemma is technical but straightfoward.

Lemma 10. 1. *For every $k \in \mathbb{N}$ there is $m \in \mathbb{N}$ computable from k such that if $w_1 \sim_m^{\mathcal{S}_L} w_2$ then $w_1 \equiv_k^L w_2$.*
2. *For every $m \in \mathbb{N}$ there is $k \in \mathbb{N}$ computable from m such that if $w_1 \equiv_k^L w_2$ then $w_1 \sim_m^{\mathcal{S}_L} w_2$.*

As representations of such semigroups one may take the transformation semigroups of finite automata extended by information about visited states. Then the parameter k can be set to be the cardinality of automata rather than of semigroups. Formally, one refers to a class $\mathcal{A}$ of finite automata and uses the congruences $\sim_k^{\mathcal{A}}$ over Σ^+ defined as follows: For $w_1, w_2 \in \Sigma^+$, define $w_1 \sim_k^{\mathcal{A}} w_2$ iff for any automaton $A \in \mathcal{A}$ with k states, any states p, q of A and any set $\mathcal{P}$ of states of A, there is a run of A on w_1 from p to q with states forming the set $\mathcal{P}$ iff this holds for w_2.

Definition 11. *An ω-word u is* effectively homogeneous *for a family $\mathcal{S}$ of finite semigroups if there is a recursive ω-sequence $w_1, w_2, \ldots$ of finite words such that $u = w_1 w_2 \ldots$ and for every $k \in \mathbb{N}$ and semigroup $S \in \mathcal{S}$ of size at most k and morphism $h: \Sigma^+ \to S$ there is $s \in S$ such that $h(w_i) = s$ for all $i > k$.*

The following theorem is an immediate corollary of Theorem 6 and Lemma 10.

Theorem 12. *Let L be a logic which is both compositional and expressive for the existence of homogeneous sets. The L-theory of an ω-word u is decidable iff u is effectively homogeneous for $\mathcal{S}_L$.*

Hence we have

Corollary 13. *The FO-theory of an ω-word u is decidable iff u is effectively homogeneous for the family of finite aperiodic semigroups. The WMSO-theory and the MSO-theory of an ω-word u is decidable iff u is effectively homogeneous for the family of finite semigroups.*

5 Conclusion

We analyzed, for some natural logics L including first-order and monadic second-order logic, the decision problem for the L-theories of structures $M = (\mathbb{N}, <, \overline{P})$ where $\overline{P}$ is a tuple of unary predicates. Our main result gave two characterizations of the decidable theories of this form, using recursiveness conditions on two different versions of "homogeneous sets".

As already mentioned, it seems hard to apply the main theorem of this paper as a tool to find new predicates P where, say, the monadic-second theory of $(\mathbb{N}, <, P)$ is decidable, or to establish even an interesting predicate where this theory is undecidable.

Another kind of application, left open in this paper, is the generation of concrete classes of predicates P (by certain closure properties) such that say the MSO theory of $(\mathbb{N}, <, P)$ is decidable. This kind of application would yield decidability results via the transformation of uniformly homogeneous sets.

References

1. J. R. Büchi. On a decision method in restricted second order arithmetic In *Proc. International Congress on Logic, Methodology and Philosophy of Science*, E. Nagel at al. eds, Stanford University Press, pp 1-11, 1960.
2. J.R. Büchi, L.H. Landweber, Definability in the monadic second-order theory of successor. *J. Symb. Logic* 34, 166-170, 1969.
3. O. Carton and W.Thomas. The Monadic Theory of Morphic Infinite Words and Generalizations. In *Proc. MFCS 2000*, Springer LNCS 1893 (2000), 275-284.

4. O. Carton and W.Thomas. The Monadic Theory of Morphic Infinite Words and Generalizations. *Inf. Comput. 176*(1), pp. 51-65, 2002.
5. C. Elgot and M. O. Rabin. Decidability and Undecidability of Extensions of Second (First) Order Theory of (Generalized) Successor. *J. Symb. Logic*, 31(2), pp. 169-181, 1966.
6. Y. Gurevich. Monadic second order theories. In J. Barwise and S. Feferman eds.,em Model Theoretic Logics pp. 479-506, Springer Verlag, 1986.
7. S. Lifsches and S. Shelah. Uniformization and Skolem Functions in the Class of Trees. *J. Symb. Logic*, 63, 103–127, 1998.
8. A. Rabinovich. On decidability of monadic logic of order over the naturals extended by monadic predicates, manuscript, 2005, submitted.
9. R. M. Robinson. Restricted Set-Theoretical Definitions in Arithmetic. *Proceedings of the American Mathematical Society*, Vol. 9, No. 2. pp. 238-242, 1958.
10. H.R. Rogers, *Theory of Recursive Functions and Effective Computability*, McGraw-Hill, New York 1967.
11. A. Semenov. Logical theories of one-place functions on the set of natural numbers. *Mathematics of the USSR - Izvestia*, vol. 22, pp 587-618, 1984.
12. S. Shelah. The monadic theory of order. *Ann. of Math.* 102, 349-419, 1975.
13. H. Straubing. *Finite Automata, Formal Logic, and Circuit Complexity*, Birkhäuser, Boston, 1994.
14. D. Siefkes. Decidable extensions of monadic second-order successor arithmetic. In *Automatentheorie und Formale Sprachen*, J. Dörr and G. Hotz, Eds., pp 441-472, BI-Wissenschaftsverlag, Mannheim 1970.
15. W. Thomas. *Das Entscheidungsproblem für einige Erweiterungen der Nachfolger-Arithmetik.* Ph. D. Thesis Albert-Ludwigs Universität, Freiburg 1975.
16. W. Thomas. The theory of successor with an extra predicate. *Math. Ann.* 237, 121-132, 1978.
17. W. Thomas. Automata on infinite objects. In: *Handbook of Theoretical Computer Science* (J. v. Leeuwen, ed.), Vol. B, Elsevier, Amsterdam 1990, pp. 135-191.
18. W. Thomas. Ehrenfeucht Games, the composition method, and the monadic theory of ordinal words. In *Structures in Logic and Computer Science: A Selection of Essays in Honor of A. Ehrenfeucht, Lecture Notes in Computer Science* 1261:118-143, 1997, Springer-Verlag.
19. W. Thomas. Languages, automata, and logic. In: Handbook of Formal Languages (G. Rozenberg, A. Salomaa, eds.), Vol 3. Springer 1997, pp. 389-455.
20. B. A. Trakhtenbrot. Finite automata and the logic of one-place predicates. (Russian version 1961). In AMS Transl. 59, 1966, pp. 23-55.

Separation Logic for Higher-Order Store

Bernhard Reus[1] and Jan Schwinghammer[2]

[1] Department of Informatics, University of Sussex, Brighton, UK
[2] Programming Systems Lab, Saarland University, Saarbrücken, Germany

Abstract. Separation Logic is a sub-structural logic that supports local reasoning for imperative programs. It is designed to elegantly describe sharing and aliasing properties of heap structures, thus facilitating the verification of programs with pointers. In past work, separation logic has been developed for heaps containing records of basic data types. Languages like C or ML, however, also permit the use of code pointers. The corresponding heap model is commonly referred to as "higher-order store" since heaps may contain commands which in turn are interpreted as partial functions between heaps.

In this paper we make Separation Logic and the benefits of local reasoning available to languages with higher-order store. In particular, we introduce an extension of the logic and prove it sound, including the Frame Rule that enables specifications of code to be extended by invariants on parts of the heap that are not accessed.

1 Introduction and Motivation

Since the beginning of program verification for high-level languages [7], pointers (and the aliasing they cause) have presented a major stumbling block for formal correctness proofs. Some of the pain of verifying pointer programs has been removed in recent years with the introduction of *Separation Logic*, developed by Reynolds, O'Hearn and others [25,9,14]. This is a variant of Hoare logic where assertions may contain the *separation conjunction*: The assertion $P * Q$ states that P and Q hold for disjoint parts of the heap store – in particular, there is no sharing between these regions. The separation connective allows for the elegant formulation of a *frame rule* which is key to local reasoning: In a triple $\{P\}\, c\, \{Q\}$, the assertions P and Q need to specify the code c only in terms of the heap cells that are actually used (the "footprint"). Clients can add invariants R for disjoint heap areas in a modular fashion, to obtain $\{P * R\}\, c\, \{Q * R\}$ without reproving c.

Some impressive results have been obtained within this formalism, including the verification of several algorithms operating on heap-based graph structures such as the Schorr-Waite graph marking algorithm [29,4]. Separation logic has been extended in several directions, covering shared-variable concurrent programs [13], modules [16] and higher-order procedures [5]. However, in all cases only values of *basic data types* can be stored. On the other hand, languages like C, ML, Scheme, and (implicitly) Java provide *code pointers*. In object-oriented

Z. Ésik (Ed.): CSL 2006, LNCS 4207, pp. 575–590, 2006.

programs, stored procedures are commonly used as callbacks. Moreover, code pointers "also appear in low-level programs that use techniques of higher-order or object-oriented programming" [25].

In this paper we address the problem of extending Separation Logic to languages that allow the storage of procedures. Reynolds emphasized the importance of code pointers in [25], speculating that the marriage of separation logic with continuation semantics could provide a way to reason about them. A step in this direction has been taken in [28] (although mutual dependencies of stored procedures were initially excluded) and [12]. Building on our results in [23,22,21] we suggest a much more direct extension of Separation Logic, by using a denotational semantics instead of an operational one. This allows us to model code pointers by means of a higher-order store, i.e., as a (mixed-variant) recursively defined domain where stores map locations to basic values *or to state transformers* (denoting partial maps from store to store).

The starting point for our work is [23] where a Hoare-style logic for a language with higher-order store is presented. This language assumes a global store and does not provide explicit means to allocate or dispose memory. The logic in [23] extends traditional Hoare logic by rules to reason about the *(mutual) recursion through the store* that becomes possible with command storage [10].

We extend the language of [23] with memory allocation constructs, and the logic with the rules of Separation Logic. The semantics of dynamically allocated memory raises a subtle point in connection with Separation Logic: soundness of the frame rule relies on the fact that the choice of a fresh location made by the allocation mechanism is irrelevant, as far as the logic is concerned. To the best of our knowledge, in all previous approaches this requirement has been enforced by making allocation *non-deterministic* so that valid predicates cannot possibly depend on assumptions about particular locations. However, in the presence of higher-order store where we have to solve recursive domain equations we found the use of (countable) non-determinism quite challenging (for instance, programs would no longer denote ω-continuous functions, see also [6,2]). Standard techniques [18] for proving the existence of recursively defined predicates over recursively defined domains are not immediately applicable.

Instead, our technical development takes place in a functor category so that the semantic domains are indexed by sets w of locations. Intuitively, w contains all the locations that are in use, and we can define a deterministic memory allocator. Non-determinism is not needed, due to the following observations:

- A renaming $f : w \to w'$ between location sets gives rise to a corresponding transformation in the semantics of programs.
- We can identify a class of predicates (over stores) that are *invariant* under location renamings. This property captures the irrelevance of location names.

In contrast to previous uses of possible worlds models [24,17,15,11] our semantics is not "tight" in the sense that stores may have allocated only a subset of the locations in w. Thus runtime errors are still possible by dereferencing dangling pointers. Memory faults are unavoidable because the language includes a *free* operation that may create dangling pointers. Moreover, once stores are "taken

Table 1. Syntax of expressions and commands

$x, y \in \text{VAR}$		variables
$b, e \in \text{EXP} ::=$	$\mathsf{true} \mid \mathsf{false} \mid e_1 \leq e_2 \mid \neg b \mid b_1 \wedge b_2 \mid \ldots \mid$	boolean expressions
	$0 \mid -1 \mid 1 \mid \ldots \mid x \mid e_1 + e_2 \mid \ldots \mid$	integer expressions
	'c'	quote (command as expression)
$c \in \text{COM} ::=$	$\mathsf{skip} \mid c_1;c_2 \mid \mathsf{if}\ b\ \mathsf{then}\ c_1\ \mathsf{else}\ c_2\ \mathsf{fi} \mid$	no op, composition, conditional
	$\mathsf{let}\ x = \mathsf{new}\ e\ \mathsf{in}\ c \mid \mathsf{free}\ x \mid$	memory allocation, disposal
	$[x]{:=}e \mid \mathsf{let}\ y{=}[x]\ \mathsf{in}\ c \mid$	assignment, lookup
	$\mathsf{eval}\ e$	unquote

apart" according to the separation conjunction, the concept of incomplete stores is convenient, even in the context of a statically typed language [20]. Nevertheless, with respect to the logic, *proved* programs do not yield memory faults.

In summary, we extend Separation Logic to higher-order store, thereby facilitating reasoning about code pointers. Technically, this is achieved by developing a functor category semantics that provides explicit location renamings, instead of using a non-determistic computation model. We believe this latter aspect is also of interest independently of the presence of higher-order store.

Structure of the paper. In Section 2 we present the syntax of programming language and logic, along with the proof rules. Section 3 develops the necessary background to interpret the language and logic, the semantics itself is given in Section 4. Section 5 concludes with an outlook on related and future work.

2 Programming Language and Logic

We present a variant of the language considered by Reus and Streicher [23], but extended with constructs for the dynamic allocation and disposal of memory cells. Two assumptions on the language simplify our presentation: Firstly, we follow [5] in the slightly non-standard adoption of (ML-like) immutable identifiers. That is, all mutation takes place in the heap, whereas the stack variables are immutable. Secondly, expressions only depend on the stack but not on the heap. As a consequence there is no need for *modifies clauses* in the proof rules.

2.1 Programming Language

The syntax of the language is given in Table 1. The set EXP of expressions includes boolean and integer expressions. Additionally, a command c can be turned into an expression (delaying its execution), via the *quote* operation 'c'.

The set COM of commands consists of the usual no op, sequential composition, and conditional constructs. Because stack variables are not mutable, new memory is allocated by $\mathsf{let}\ x = \mathsf{new}\ e\ \mathsf{in}\ c$ that introduces an identifier with local scope c that is bound to (the location of) the new memory cell. We stress that the initial contents e may be a (quoted) command. This is also the case for an update, $[x]{:=}e$. The command $\mathsf{free}\ x$ disposes the memory cell that x denotes, and $\mathsf{let}\ y{=}[x]\ \mathsf{in}\ c$ introduces a new stack variable y bound to the cell contents.

Table 2. Syntax of assertions

$A, B \in \text{pAssn} ::= \mathbf{true} \mid e_1 \leq e_2 \mid$	pure basic predicates
$\neg A \mid A \wedge B \mid \forall x.\, A$	predicate logic connectives
$P, Q \in \text{Assn} ::= x \mapsto e \mid \mathbf{emp} \mid P * Q \mid$	separation logic connectives
$A \mid P \wedge Q \mid P \vee Q \mid \forall x.\, P \mid \exists x.\, P$	predicate logic connectives

Table 3. Specific proof rules

$$\text{Frame}\ \frac{\{P\}\, c\, \{Q\}}{\{P * R\}\, c\, \{Q * R\}} \qquad \text{Free}\ \frac{}{\{x \mapsto \_\}\ \mathsf{free}\ x\ \{\mathbf{emp}\}} \qquad \text{New}\ \frac{\{P * x \mapsto e\}\, c\, \{Q\}}{\{P\}\ \mathsf{let}\ x = \mathsf{new}\ e\ \mathsf{in}\ c\ \{Q\}}\, x \notin fv(e, P, Q)$$

$$\text{Eval}\ \frac{\{P\}\, c\, \{Q\}}{\{P\}\ \mathsf{eval}\ `c'\ \{Q\}} \qquad \text{Assign}\ \frac{}{\{x \mapsto \_\}\ [x] := e\ \{x \mapsto e\}} \qquad \text{Deref}\ \frac{\{P * x \mapsto e\}\ c[e/y]\ \{Q\}}{\{P * x \mapsto e\}\ \mathsf{let}\ y = [x]\ \mathsf{in}\ c\ \{Q\}}$$

$$\text{Rec}\ \frac{\bigwedge_{1 \leq i \leq n} \{P_1\}\ \mathsf{eval}\ x_1\ \{Q_1\} \ldots \{P_n\}\ \mathsf{eval}\ x_n\ \{Q_n\} \vdash \{P_i\}\ c_i\ \{Q_i\}}{\{P_j[`\mathbf{c}'/\boldsymbol{x}]\}\ \mathsf{eval}\ `c_j'\ \{Q_j[`\mathbf{c}'/\boldsymbol{x}]\}}\, 1 \leq j \leq n$$

Finally, $\mathsf{eval}\ e$ is the "unquote" command, i.e., if e denotes a quoted command c then c is executed. We give a formal semantics of this language in Section 4, after developing the necessary machinery in Section 3.

Note that Table 1 does not include any looping constructs – recursion can be expressed "through the store" [10]. Here is a simple example of a non-terminating command: $[x]{:=}`\mathsf{let}\ y{=}[x]\ \mathsf{in}\ \mathsf{eval}\ y'$; $\mathsf{let}\ y{=}[x]\ \mathsf{in}\ \mathsf{eval}\ y$.

2.2 Assertions and Proof Rules

The assertions used in Hoare triples are built from the formulae of predicate logic and the additional separation logic assertions that describe the heap ($\_ \mapsto \_$, $\mathbf{emp}$ describing the empty heap, and $P * Q$; cf. [25]). Note that in our language variables can also be bound to quoted code. The syntax of the assertions is given in Table 2. It is important to note that we distinguish between "pure" assertions pAssn, i.e., those that do not depend on the heap, and normal assertions Assn which do depend on the heap. Only the former allow negation. The reason for this will become clear when we give the semantics in Section 4.1. As usual, assertion $e_1 \leq e_2 \wedge e_2 \leq e_1$ is abbreviated $e_1 = e_2$ and assertion $\exists z.\, x \mapsto z$ is abbreviated to $x \mapsto \_$. For pure assertions pAssn the predicate **false** and connectives $\vee$, $\Rightarrow$, and $\exists x.A$ can be derived as usual using negation.

The inference rules of our program logic contain the standard Hoare rules (for skip, conditional, sequential composition, weakening and substitution), a standard axiomatization of predicate logic as well as an axiomatization of the Separation Logic connectives stating associativity and symmetry of $*$, neutrality of $\mathbf{emp}$ with respect to $*$, and some distributive laws (see e.g. [25]). The rules specific to our programming language are given in Table 3.

The frame rule extends triples by invariants for inaccessible parts of the heap. Rules (FREE) and (ASSIGN) specify the corresponding heap operations "tightly". The inferences (NEW) and (DEREF) combine heap allocation and dereferencing, resp., with local variable definitions (and hence are not tight). Unlike [5] our (NEW) permits (non-recursive) initializations (and self-reference can be introduced by assignment as in Section 2.3). Substitution on c is used in the premiss of (DEREF) to avoid equations of the form $y = e$ that would be problematic when e is a stored procedure. Rule (EVAL) is reminiscent of standard non-recursive procedure call rules; it expresses that evaluating a quoted command has the same effect as the command itself. Indeed, (EVAL) is a degenerated case of rule (REC) that deals with recursion through the store. It is similar to the standard rule for Hoare-calculus with (mutually) recursive procedures, but since procedures are stored on the heap, they have to be accounted for in the assertions which leads to the substitution in the conclusion.

2.3 Example

Let Σn be the sum $\sum_{0\leq i\leq n} i$, let c_P be the command

$$\mathsf{let}\ !y{=}\,[y]\ \mathsf{in}\ \mathsf{let}\ !x{=}\,[x]\ \mathsf{in}\ \mathsf{if}\ !x{\leq}0\ \mathsf{then\ skip\ else}\ [y]{:=}!y{+}!x;\ [x]{:=}!x{-}1;\ \mathsf{let}\ c{=}\,[f]\ \mathsf{in\ eval}\ c\ \mathsf{fi}$$

and observe that $!x$ and $!y$ are stack variable names representing the values in the cells denoted by (pointers) x and y, respectively. If c_P is stored in f, the program is defined by recursion through the store since c_P calls the procedure stored in heap cell f. This is also referred to as a "knot in the store." We prove below that c_P adds to $!y$ the sum $\Sigma!x$ of natural numbers up to and including $!x$. In the presentation we omit various applications of the weakening rule (which are easy to insert).

$$
\dfrac{
\textsc{Seq}\ \dfrac{
\textsc{Frame}\ \dfrac{\textsc{Assign}\ \{f\mapsto\text{`skip'}\}\,[f]{:=}\text{`}c_P\text{'}\,\{f\mapsto\text{`}c_P\text{'}\}}{\{x\mapsto n * y\mapsto 0 * f\mapsto\text{`skip'}\}\,[f]{:=}\text{`}c_P\text{'}\,\{x\mapsto\_ * y\mapsto 0 * f\mapsto\text{`}c_P\text{'}\}}
\qquad
\textsc{Deref}\ \dfrac{\textsc{Subst}\ \dfrac{\textsc{Rec}\ \dfrac{\alpha}{\{x\mapsto n * y\mapsto m * f\mapsto\text{`}c_P\text{'}\}\ \mathsf{eval}\ \text{`}c_P\text{'}\ \{x\mapsto\_ * y\mapsto\Sigma n{+}m * f\mapsto\text{`}c_P\text{'}\}}}{\{x\mapsto n * y\mapsto 0 * f\mapsto\text{`}c_P\text{'}\}\ \mathsf{eval}\ \text{`}c_P\text{'}\ \{x\mapsto\_ * y\mapsto\Sigma n * f\mapsto\text{`}c_P\text{'}\}}}{\{x\mapsto n * y\mapsto 0 * f\mapsto\text{`}c_P\text{'}\}\ \mathsf{let}\ c{=}\,[f]\ \mathsf{in\ eval}\ c\,\{x\mapsto\_ * y\mapsto\Sigma n * f\mapsto\text{`}c_P\text{'}\}}
}{\{x\mapsto n * y\mapsto 0 * f\mapsto\text{`skip'}\}\,[f]{:=}\text{`}c_P\text{'};\ \mathsf{let}\ c{=}\,[f]\ \mathsf{in\ eval}\ c\,\{x\mapsto\_ * y\mapsto\Sigma n * f\mapsto\text{`}c_P\text{'}\}}
}{\{x\mapsto n * y\mapsto 0\}\ \mathsf{let}\ f{=}\mathsf{new}\ \text{`skip'}\ \mathsf{in}\ [f]{:=}\text{`}c_P\text{'};\ \mathsf{let}\ c{=}\,[f]\ \mathsf{in\ eval}\,c\,\{x\mapsto\_ * y\mapsto\Sigma n * f\mapsto\text{`}c_P\text{'}\}}
$$

Here, the last inference is by (NEW). For the derivation tree α we let x_P denote 'c_P' and assume

$$\{x\mapsto n * y\mapsto m * f\mapsto x_P\}\ \mathsf{eval}\ \ x_P\,\{x\mapsto n * y\mapsto\Sigma n{+}m * f\mapsto x_P\} \qquad (\dagger)$$

and prove $\{x\mapsto n * y\mapsto m * f\mapsto x_P\}\,c_P\,\{x\mapsto 0 * y\mapsto\Sigma n{+}m * f\mapsto x_P\}$.

$$
\dfrac{\textsc{If}\ \dfrac{\beta_t \qquad \beta_f}{\{x\mapsto n * y\mapsto m * f\mapsto x_P\}\ \mathsf{if}\ n{\leq}0\ \mathsf{then\ skip\ else}\ \ldots\ \mathsf{fi}\,\{x\mapsto\_ * y\mapsto\Sigma n{+}m * f\mapsto x_P\}}}{\{x\mapsto n * y\mapsto m * f\mapsto x_P\}\ \underbrace{\mathsf{let}\ !y{=}\,[y]\ \mathsf{in}\ \ldots [y]{:=}!y{+}!x\ldots}_{c_P}\,\{x\mapsto\_ * y\mapsto\Sigma n{+}m * f\mapsto x_P\}}
$$

The final inference is by two applications of (DEREF), and β_t and β_f, resp., are:

$$\textsc{Weak}\ \frac{\textsc{Skip}\ \{x{\mapsto}n * y{\mapsto}m * f{\mapsto}x_P \wedge n \leq 0\}\, \mathsf{skip}\, \{x{\mapsto}n * y{\mapsto}m * f{\mapsto}x_P \wedge n \leq 0\}}{\{x{\mapsto}n * y{\mapsto}m * f{\mapsto}x_P \wedge n \leq 0\}\, \mathsf{skip}\, \{x{\mapsto}\_ * y{\mapsto}\Sigma n{+}m * f{\mapsto}x_P\}}$$

$$\textsc{Seq}^2\ \frac{\gamma \qquad \delta \qquad \epsilon}{\{x{\mapsto}n{*}y{\mapsto}m{*}f{\mapsto}x_P \wedge n{>}0\}\, [y]{:=}\ldots; [x]{:=}\ldots; \mathsf{let}\ldots \{x{\mapsto}\_{*}y{\mapsto}\Sigma n{+}m{*}f{\mapsto}x_P\}}$$

The derivations γ, δ and ϵ are as follows:

$$\textsc{Frame}\ \frac{\textsc{Assign}\ \{y{\mapsto}m\}\, [y]{:=}m{+}n\, \{y{\mapsto}m{+}n\}}{\{x{\mapsto}n * y{\mapsto}m * f{\mapsto}x_P\}\, [y]{:=}m{+}n\, \{x{\mapsto}n * y{\mapsto}m{+}n * f{\mapsto}x_P\}}$$

$$\textsc{Frame}\ \frac{\textsc{Assign}\ \{x{\mapsto}n\}\, [x]{:=}n{-}1\, \{x{\mapsto}n{-}1\}}{\{x{\mapsto}n * y{\mapsto}m{+}n * f{\mapsto}x_P\}\, [x]{:=}n{-}1\, \{x{\mapsto}n{-}1 * y{\mapsto}m{+}n * f{\mapsto}x_P\}}$$

$$\textsc{Deref}\ \frac{\textsc{Subst}\ \dfrac{(\dagger) \equiv \{x{\mapsto}n * y{\mapsto}m * f{\mapsto}x_P\}\, \mathsf{eval}\ x_P\, \{x{\mapsto}\_ * y{\mapsto}\Sigma n{+}m * f{\mapsto}x_P\}}{\{x{\mapsto}n{-}1 * y{\mapsto}m{+}n * f{\mapsto}x_P\}\, \mathsf{eval}\ x_P\, \{x{\mapsto}\_ * y{\mapsto}\Sigma n{+}m * f{\mapsto}x_P\}}}{\{x{\mapsto}n{-}1 * y{\mapsto}m{+}n * f{\mapsto}x_P\}\, \mathsf{let}\, c{=}\,[f]\ \mathsf{in\ eval}\ c\, \{x{\mapsto}\_ * y{\mapsto}\Sigma n{+}m * f{\mapsto}x_P\}}$$

Note how the Frame Rule is used to peel off those predicates of the assignment rule that do not relate to the memory cell affected.

3 A Model of Dynamic Higher-Order Store

This section defines the semantic domains in which the language of Section 2 finds its interpretation. The semantic properties (*safety monotonicity* and *frame property*) that programs must satisfy to admit local reasoning [25] are rephrased, using the renamings made available by the functor category machinery. Due to the higher-order character of stores, these predicates are recursive and their existence must be established. The framework of Pitts is used [18,11].

3.1 Worlds

Fix a well-ordered, countably infinite set $\mathbb{L}$ of *locations* (e.g., the natural numbers). Let $\mathbb{W}$ be the category consisting of finite subsets $w \subseteq \mathbb{L}$ as objects and injections $f : w_1 \to w_2$ as morphisms. We call the objects w of $\mathbb{W}$ *worlds*. The intuition is that $w \in \mathbb{W}$ describes (a superset of) the locations currently in use; in particular, every location *not* in w will be fresh. The inclusion $w \subseteq w'$ is written $\iota_w^{w'}$, and the notation $f : w_1 \xrightarrow{\sim} w_2$ is used to indicate that f is a bijection.

The injections formalise a possible *renaming* of locations, as well as an *extension* of the set of available locations because of allocation.

3.2 Semantic Domains: Stores, Values and Commands

Let **pCpo** be the category of cpos (partial orders closed under taking least upper bounds of countable chains, but not necessarily containing a least element) and partial continuous functions. For a partial continuous function g we write $g(a)\!\downarrow$ if the application is defined, and $g(a)\!\uparrow$ otherwise. By $g; h$ we denote composition in diagrammatic order. Let **Cpo** be the subcategory of **pCpo** where the morphisms are *total* continuous functions. For cpos A and B we write $A \rightharpoonup B$ and $A \to B$ for the cpos of partial and total continuous functions from A to B, respectively, each with the pointwise ordering. For a family (A_i) of cpos, $\sum_i A_i$ denotes their disjoint union; we write its elements as $\langle i, a\rangle$ where $a \in A_i$.

For every $w \in \mathbb{W}$ we define a cpo of *w-stores* as records of values whose domain is a subset of w (viewed as discrete cpo). The fields of such a store contain values that may refer to locations in w:

$$St(w) = \mathsf{Rec}_w(Val(w)) = \sum_{w' \subseteq w} (w' \to Val(w)) \tag{1}$$

We abuse notation to write s for $\langle w', s\rangle \in St(w)$; we set $\mathsf{dom}(s) = w'$ and may use record notation $\{\!|l = v_l|\!\}_{l \in \mathsf{dom}(s)}$ where $s(l) = v_l$. The order on (1) is defined in the evident way, by $r \sqsubseteq s$ iff $\mathsf{dom}(r) = \mathsf{dom}(s)$ and $r(l) \sqsubseteq s(l)$ for all $l \in \mathsf{dom}(r)$.

A value (over locations $w \in \mathbb{W}$) is either a *basic value* in $BVal$, a location $l \in w$, or a *command*, i.e.,

$$Val(w) = BVal + w + Com(w) \tag{2}$$

We assume $BVal$ is a discretely ordered cpo that contains integers and booleans.

Commands $c \in Com(w)$ operate on the store; given an initial store the command may either diverge, terminate abnormally or terminate with a result store. Abnormal termination is caused by dereferencing dangling pointers which may refer to memory locations that either have not yet been allocated, or have already been disposed of. Thus, in contrast to [23] where store could not vary dynamically, we need to have a defined result error to flag undefined memory access. The possibility of *dynamic memory allocation* prompts a further refinement compared to [23]: a command should work for extended stores, too, and may also extend the store itself.

Formally, the collection of commands is given as a functor $Com : \mathbb{W} \longrightarrow \mathbf{Cpo}$, defined on objects by

$$Com(w) = \prod_{i:w \to w'} (St(w') \rightharpoonup (\mathsf{error} + \sum_{j:w' \to w''} St(w''))) \tag{3}$$

and on morphisms by the obvious restriction of the product,

$$Com(f : w_1 \to w_2)(c)_{i:w_2 \to w_3} = c_{(f;i)}$$

Viewing commands this way is directly inspired by Levy's model of an ML-like higher-order language with general references [11].

By considering $BVal$ as constant functor, and locations as the functor $\mathbb{W} \longrightarrow \mathbf{Cpo}$ that acts on $f : w_1 \to w_2$ by sending $l \in w_1$ to $f(l) \in w_2$, Val can also be seen as a functor $\mathbb{W} \longrightarrow \mathbf{Cpo}$. Note that, by expanding the requirements (1), (2) and (3), Val is expressed in terms of a mixed-variant recursion. In Section 3.3 we address the issue of well-definedness of Val.

One might want to exclude run-time memory errors statically (which is possible assuming memory is never disposed, so that there is no way of introducing dangling pointers from within the language). An obvious solution to model this is by defining w-stores as $\prod_w \mathit{Val}(w)$, i.e., all locations occurring in values are guaranteed to exist in the current store. However, this approach means that there is no canonical way to extend stores, nor can values be restricted to smaller location sets. Consequently St is neither co- nor contravariantly functorial[1]. In contrast, our more permissive definition of stores (that may lead to access errors) *does* allow a functorial interpretation of St, as follows. For an injection $f : w_1 \to w_2$ we write $f^{-1} : \mathsf{im} f \to w_1$ for the right-inverse to f, and let

$$\mathit{St}(f) : \mathsf{Rec}_{w_1}(\mathit{Val}(w_1)) \to \mathsf{Rec}_{w_2}(\mathit{Val}(w_2))$$
$$\mathit{St}(f) = \lambda \langle w \subseteq w_1, s\rangle.\, \langle f w, f^{-1}; s; \mathit{Val}(f)\rangle$$

The case where f is a bijection then corresponds to a consistent renaming of the store and its contents. We will make some use of the functoriality of St in the following, to lift recursively defined predicates from values to stores.

For $s_1, s_2 \in \mathit{St}(w)$ we write $s_1 \perp s_2$ if their respective domains $w_1, w_2 \subseteq w$ are disjoint. In this case, their composition $s_1 * s_2 \in \mathit{St}(w)$ is defined by conjoining them in the obvious way, it is undefined otherwise. Observe that for $f : w \to w'$ we have $\mathit{St}(f)(s_1 * s_2) = \mathit{St}(f)(s_1) * \mathit{St}(f)(s_2)$; the right-hand side is defined because f is injective.

3.3 Domain Equations and Relational Structures on Bilimit-Compact Categories

This section briefly summarises the key results from [11,27] about the solution of recursive domain equations in bilimit-compact categories. We will make use of the generalisation of Pitts' techniques [18] for establishing the well-definedness of (recursive) predicates, as outlined in [11].

Definition 1 (Bilimit-Compact Category [11]). *A category* $\mathbb{C}$ *is* bilimit-compact *if*

- $\mathbb{C}$ *is* **Cpo**-*enriched and each hom-cpo* $\mathbb{C}(A, B)$ *has a least element* $\perp_{A,B}$ *such that* $\perp \circ f = \perp = g \circ \perp$*;*
- $\mathbb{C}$ *has an initial object; and*
- *in the category* $\mathbb{C}^E$ *of embedding-projection pairs of* $\mathbb{C}$*, every* ω*-chain* $\Delta = D_0 \to D_1 \to \ldots$ *has an O-colimit [27]. More precisely, there exists a cocone* $(e_n, p_n)_{n<\omega} : \Delta \to D$ *in* $\mathbb{C}^E$ *such that* $\bigsqcup_{n<\omega}(p_n; e_n) = id_D$ *in* $\mathbb{C}(D, D)$*.*

It follows that every locally continuous functor $F : \mathbb{C}^{op} \times \mathbb{C} \longrightarrow \mathbb{C}$ has a minimal invariant, i.e., an object D and isomorphism i in $\mathbb{C}$ such that $i : F(D, D) \cong D$ (unique up to unique isomorphism) and id_D is the least fixed point of the

[1] Levy [11] makes this observation for a similar, typed store model.

Table 4. Solving the domain equation: $F_{Val}, F_{Com} : \mathbb{C}^{op} \times \mathbb{C} \longrightarrow \mathbb{C}$ and $F_{St} : \mathbb{C} \longrightarrow \mathbb{C}$

On $\mathbb{C}$-objects A^-, A^+, B, worlds $w, w' \in \mathbb{W}$ and $f : w \to w'$,

$$F_{Val}(A^-,A^+)(w) = BVal + w + F_{Com}(A^-,A^+)(w)$$

$$F_{Val}(A^-,A^+)(f) = \lambda v. \begin{cases} v & \text{if } v \in BVal \\ f(v) & \text{if } v \in w \\ F_{Com}(A^-,A^+)(f)(v) & \text{if } v \in F_{Com}(A^-,A^+)(w) \end{cases}$$

$$F_{Com}(A^-,A^+)(w) = \prod_{i:w\to w'} (F_{St}(A^-)(w') \rightharpoonup (\mathsf{error} + \textstyle\sum_{j:w'\to w''} F_{St}(A^+)(w'')))$$

$$F_{Com}(A^-,A^+)(f) = \lambda c \lambda i.\, c_{(f;i)}$$

$$F_{St}(B)(w) = \textstyle\sum_{w_1 \subseteq w} (w_1 \to B(w))$$

$$F_{St}(B)(f) = \lambda \langle w_1, s\rangle . \langle f w_1, f^{-1}; s; B(f)\rangle$$

On $\mathbb{C}$-morphisms $h = (h_w) : B^- \dot\rightharpoonup A^-$ and $k = (k_w) : A^+ \dot\rightharpoonup B^+$,

$$F_{Val}(h,k)_w = \lambda v. \begin{cases} v & \text{if } v \in BVal \text{ or } v \in w \\ F_{Com}(h,k)_w(v) & \text{if } v \in F_{Com}(A^-,A^+) \end{cases}$$

$$F_{Com}(h,k)_w = \lambda c \lambda i{:}w \to w' \lambda s.$$

$$\begin{cases} \text{undefined} & \text{if } F_{St}(h)_{w'}(s)\uparrow \\ & \text{or } F_{St}(h)_{w'}(s)\downarrow \wedge c_i(F_{St}(h)_{w'}(s))\uparrow \\ & \text{or } c_i(F_{St}(h)_{w'}(s)) = \langle j : w' \to w'', s'\rangle \wedge F_{St}(k)_{w''}(s')\uparrow \\ \mathsf{error} & \text{if } F_{St}(h)_{w'}(s)\downarrow \wedge c_i(F_{St}(h)_{w'}(s)) = \mathsf{error} \\ \langle j, F_{St}(k)_{w''}(s')\rangle & \text{if } c_i(F_{St}(h)_{w'}(s)) = \langle j : w' \to w'', s'\rangle \wedge F_{St}(k)_{w''}(s')\downarrow \end{cases}$$

$$F_{St}(k)_w = \lambda\langle w_1, s\rangle . \begin{cases} \langle w_1, s; k_w\rangle & \text{if } \forall l \in w_1.\, k_w(s(l))\downarrow \\ \text{undefined} & \text{otherwise} \end{cases}$$

continuous endofunction $\delta : \mathbb{C}(D,D) \to \mathbb{C}(D,D)$ defined by $\delta(e) = i^{-1}; F(e,e); i$ [18]. To ease readability the isomorphism i is usually omitted below.

To solve the domain equation of the preceding subsection we shall be interested in the case where $\mathbb{C}$ denotes the category $[\mathbb{W}, \mathbf{Cpo}]$ of functors $\mathbb{W} \longrightarrow \mathbf{Cpo}$ and *partial* natural transformations, i.e., a morphism $e : A \dot\rightharpoonup B$ in $\mathbb{C}$ is a family $e = (e_w)$ of partial continuous functions $e_w : A(w) \rightharpoonup B(w)$ such that $A(f); e_{w'} = e_w; B(f)$ for all $f : w \to w'$.

Lemma 1 (Bilimit-Compactness [11]). $\mathbb{C} = [\mathbb{W}, \mathbf{Cpo}]$ *is bilimit-compact.*

Thus, for well-definedness of Val it suffices to show that requirements (1), (2) and (3) induce a locally continuous functor $\mathbb{C}^{op} \times \mathbb{C} \longrightarrow \mathbb{C}$ for which Val is the minimal invariant. Table 4 defines such a functor F_{Val} in the standard way [18], by separating positive and negative occurrences of St in (3).

Lemma 2 (Minimal Invariant). $F_{Val} : \mathbb{C}^{op} \times \mathbb{C} \longrightarrow \mathbb{C}$ *is locally continuous. In particular, the minimal invariant* $Val = F_{Val}(Val, Val)$ *exists.*

From this we can then define $Com = F_{Com}(Val, Val)$ and $St = F_{St}(Val)$ which satisfy (3) and (1). The minimal invariant in fact lives in the category of functors $\mathbb{W} \longrightarrow \mathbf{Cpo}$ and natural transformations that are *total*, i.e., those $e = (e_w)$: $A \dot\to B$ where each e_w is a total continuous function $A(w) \to B(w)$. This is because F_{Val} restricts to this category, which is sub-bilimit-compact within $\mathbb{C}$ in the sense of [11]. A (normal) relational structure $\mathcal{R}$ on $\mathbb{C}$ in the sense of Pitts [18] is given as follows.

Definition 2 (Kripke Relational Structure). *For each $A : \mathbb{W} \longrightarrow$ **Cpo** let $\mathcal{R}(A)$ consist of the $\mathbb{W}$-indexed families $R = (R_w)$ of admissible predicates $R_w \subseteq A(w)$ such that for all $f : w \to w'$ and $a \in A(w)$,*

$$a \in R_w \quad \Longrightarrow \quad A(f)(a) \in R_{w'} \qquad \text{(KripkeMon)}$$

For each natural transformation $e = (e_w) : A \stackrel{\cdot}{\rightharpoonup} B$ and all $R \in \mathcal{R}(A)$, $S \in \mathcal{R}(B)$,

$$e : R \subset S \quad :\Longleftrightarrow \quad \forall w \in \mathbb{W}\, \forall a \in A(w).\, a \in R_w \ \wedge\ e_w(a)\downarrow \Longrightarrow e_w(a) \in S_w$$

Note that (KripkeMon) in particular covers the case where $f : w \to w$ is a bijection, i.e., the Kripke relations are invariant under permutation of locations.

For an object $A : \mathbb{W} \longrightarrow$ **Cpo** and $\mathcal{R}$-relation $R = (R_w) \in \mathcal{R}(A)$ we let $St(R) \in \mathcal{R}(F_{St}(A))$ be the relation where $s \in St(R)_w$ if and only if $s(l) \in R_w$ for all $l \in \mathsf{dom}(s)$. It is easy to check admissibility and (KripkeMon). Two elementary properties are stated in the following lemma.

Lemma 3 (Relations over St). *Let $A, B : \mathbb{W} \longrightarrow$ **Cpo**, $R \in \mathcal{R}(A)$, $S \in \mathcal{R}(B)$. Let $e : A \stackrel{\cdot}{\rightharpoonup} B$ and $w \in \mathbb{W}$.*

1. *If $e : R \subset S$ then $F_{St}(e) : St(R) \subset St(S)$.*
2. *If $s_1, s_2 \in F_{St}(A)(w)$ and $s_1 \perp s_2$ then $s_1 \in St(R)_w$ and $s_2 \in St(R)_w$ if and only if $s_1 * s_2 \in St(R)_w$.*

Theorem 1 (Invariant Relation [18]). *Let F_{Val} be the locally continuous functor for which Val is the minimal invariant. Suppose Φ maps $\mathcal{R}$-relations to $\mathcal{R}$-relations such that for all $R, R', S, S' \in \mathcal{R}(Val)$ and $e \sqsubseteq id_{Val}$,*

$$e : R' \subset R \ \wedge\ e : S \subset S' \quad \Longrightarrow \quad F(e, e) : \Phi(R, S) \subset \Phi(R', S')$$

Then there exists a unique $\Delta \in \mathcal{R}(Val)$ such that $\Phi(\Delta, \Delta) = \Delta$.

Proof. By [11], the proof of Pitts' existence theorem [18, Thm. 4.16] generalises from **Cppo** (pointed cpos and strict continuous maps) to arbitrary bilimit-compact categories. Since the $\mathcal{R}$-relations of Definition 2 are admissible in the sense of [18] and $\mathcal{R}$ has inverse images and intersections, the theorem follows. □

3.4 Safety Monotonicity and Frame Property

Safety monotonicity is the observation that if executing a command in heap h does not result in a memory fault, then this is also true when running the command in a heap that extends h. The second key semantic principle underlying separation logic is the observation that if execution of a command does not result in a memory fault (i.e., no dangling pointers are dereferenced), then running the command in an extended heap does not influence its observable behaviour — in particular, the additional heap region remains unaffected. The *frame property* [25] formalises this idea. Since the actual results of these executions may differ in the action of the memory allocator, the choice of locations is taken into account.

As the store may contain commands itself (which may be executed), both safety monotonicity and frame property must already be *required* to hold of the data in the initial store. In order to give a sufficiently strong induction hypothesis later, we additionally require that the properties are *preserved* by the execution of commands. Unfortunately, we cannot adopt separate definitions for safety monotonicity and frame property (like [25]) but have to *combine* them. The reason is that safety monotonicity is not preserved by composition of commands, unless commands additionally satisfy the frame property.[2] Because of the higher-order store, both properties are expressed by mixed-variant recursive definitions, and existence of a predicate satisfying these definitions requires a proof. It is in this proof that both properties are needed simultaneously.

For this reason the following property LC (for "local commands") is proposed, subsuming both safety and frame property: For $R, S \in \mathcal{R}(Val)$ let $\Phi(R, S)$ be the $\mathbb{W}$-indexed family of relations where

$$
\begin{array}{l}
c \in \Phi(R,S)_w \quad :\Longleftrightarrow \quad c \in Com(w) \implies \\
\quad \forall f{:}w \to w_2\, \forall i{:}w_2 \tilde{\to} w_2'\, \forall g{:}w_2 \to w_3\, \forall s_1, s_2 \in St(R)_{w_2}\, \forall s' \in St(w_3).\ s_1 \perp s_2 \implies \\
\qquad \left.\begin{array}{l} c_f(s_1)\uparrow \implies c_{f;i}(St(i)(s_1 * s_2))\uparrow \\ \wedge\ c_f(s_1 * s_2) = \mathsf{error} \implies c_{f;i}(St(i)(s_1)) = \mathsf{error} \end{array}\right\} \textit{safety mon.} \\
\qquad \left.\begin{array}{l} \wedge\ c_f(s_1) \neq \mathsf{error} \wedge c_f(s_1 * s_2) = \langle g, s' \rangle \implies \\ \quad \exists g'{:}w_2' \to w_3'\, \exists j{:}w_3' \tilde{\to} w_3\, \exists s_1' \in St(w_3')\, \exists s_2' \in St(w_2). \\ \qquad c_{f;i}(St(i)(s_1)) = \langle g', s_1' \rangle \ \wedge\ s_2' \sqsubseteq s_2 \ \wedge\ i; g'; j = g \\ \qquad \wedge\ s' = St(j)(s_1' * St(i; g')(s_2')) \ \wedge\ s' \in St(S)_{w_3} \end{array}\right\} \textit{frame property}
\end{array}
$$

and define the predicate $LC \in \mathcal{R}(Val)$ on values as the fixpoint $LC = \Phi(LC, LC)$ of this functional.

This definition is complex so some remarks are in order. Besides combining safety and frame property, Φ strengthens the obvious requirements by allowing the use of a renaming i on the initial store as well. This provides a strong invariant that we need for the proof of Theorem 2 below in the case of sequential composition. To obtain the fixed point of Φ, Lemma 4 appeals to Theorem 1 which forced us to weaken the frame property to an inequality ($s_2' \sqsubseteq s_2$). This extends conservatively the usual notion of [25] to the case of higher-order stores.

Lemma 4 (Existence). *LC is well-defined, i.e., there exists a unique $LC \in \mathcal{R}(Val)$ such that $LC = \Phi(LC, LC)$.*

Proof. One checks that Φ maps Kripke relations to Kripke relations, i.e. for all $R, S \in \mathcal{R}(Val)$, $\Phi(R, S) \in \mathcal{R}(Val)$. By Theorem 1 it remains to show for all $e \sqsubseteq id_{Val}$, if $e : R' \subset R$ and $e : S \subset S'$ then $F_{Val}(e, e) : \Phi(R, S) \subset \Phi(R', S')$. □

2 As pointed out to us by Hongseok Yang, this is neither a consequence of using a denotational semantics, nor of our particular formulation employing renamings rather than non-determinism; counter-examples can easily be constructed in a relational interpretation of commands.

4 Semantics of Programs and Logic

Table 5 contains the interpretation of the language. Commands and expressions depend on environments because of free (stack) variables, so that $\mathcal{E}[\![e]\!] : Env \dot{\rightarrow} Val$ and $\mathcal{C}[\![c]\!] : Env \dot{\rightarrow} Com$ where the functor Env is Val^{VAR}. The semantics of boolean and integer expressions is standard and omitted from Table 5; because of type mismatches (negation of integers, addition of booleans,...) expressions may denote error. The semantics of quote refers to the interpretation of commands and uses the injection of Com into Val. Sequential composition is interpreted by composition in the functor category but also propagates errors and non-termination. Conditional and skip are standard. The semantics of the memory commands is given in terms of auxiliary operations *extend* and *update*.

The following theorem shows the main result about the model: commands of the above language satisfy (and preserve) the locality predicate LC.

Theorem 2 (Locality). *Let $w \in \mathbb{W}$ and $\rho \in Env(w)$ such that $\rho(x) \in LC_w$ for all $x \in \text{VAR}$. Let $c \in \text{COM}$. Then $[\![c]\!]_w\, \rho \in LC_w$.*

Proof. By induction on c. The case of sequential composition relies on LC taking safety monotonicity and frame property into account simultaneously. □

4.1 Interpretation of the Logic

The assertions of the logic are interpreted as predicates over St that are compatible with the possible-world structure. In contrast to the $\mathcal{R}$-relations of Section 3.3 they depend on environments, and downward-closure is required to prove the frame rule sound. This is made precise by the following relational structure $\mathcal{S}$.

Definition 3 (dclKripke Relational Structure). *Let $\mathcal{S}$ consist of the $\mathbb{W}$-indexed families $p = (p_w)$ of predicates $p_w \subseteq Env(w) \times St(w)$ such that for all $f : w \to w'$, $\rho \in Env(w)$ and $s \in St(w)$,*

Kripke Monotonicity *if $(\rho, s) \in p_w$ then $(Env(f)(\rho), St(f)(s)) \in p_{w'}$;*
Downward Closure *$\{s \in St(w) \mid (\rho, s) \in p_w\}$ is downward-closed in $St(w)$.*

For each natural transformation $e = (e_w) : Val \dot{\rightharpoonup} Val$ and $p, q \in \mathcal{S}$ we write $e : p \subset q$ if for all $w \in \mathbb{W}$, $\rho \in Env(w)$ and $s \in St(w)$,

$$(\rho, s) \in p_w \wedge (F_{Env}(e)_w(\rho)\downarrow \vee F_{St}(e)_w(s)\downarrow) \implies (F_{Env}(e)_w(\rho), F_{St}(e)_w(s)) \in q_w$$

where $F_{Env}(e) = F_{Val}{}^{\text{VAR}}(e, e)$.

Assertions $P \in \text{ASSN}$ are interpreted by $\mathcal{S}$-relations $\mathcal{A}[\![P]\!]$. Some cases of the definition are given in Table 6. All assertions are indeed downward-closed in the store component, and pure assertions denote either true or false since they do not depend on the heap. The interpretation shows that $\leq$ is not supposed to compare code (but yield false instead). Correspondingly, we assume the non-standard axiom $\neg(\text{`}c_1\text{'} \leq e_2) \wedge \neg(e_1 \leq \text{`}c_2\text{'})$ for the comparison operator.

We can now give the semantics of Hoare triples. Correctness is only ensured if the command in question is run on stores that contain local procedures only.

Table 5. Semantics of expressions and commands

$$
\begin{array}{ll}
\mathcal{E}[\![e]\!] : \mathit{Env} \dot{\rightarrow} \mathit{Val} + \mathsf{error} & \text{where } f : w \rightarrow w' \\
\mathcal{E}[\![\text{`}c\text{'}]\!]_w\, \rho & = \mathcal{C}[\![c]\!]_w\, \rho \\
\mathcal{C}[\![c]\!] : \mathit{Env} \dot{\rightarrow} \mathit{Com} & \text{where } f : w \rightarrow w' \text{ and } s \in \mathit{St}(w') \\
(\mathcal{C}[\![\mathsf{skip}]\!]_w\, \rho)_f(s) & = \langle \mathit{id}, s \rangle \\
(\mathcal{C}[\![c_1;c_2]\!]_w\, \rho)_f(s) & = \begin{cases} \text{undefined} & \text{if } (\mathcal{C}[\![c_1]\!]_w\, \rho)_f s \uparrow \\ \mathsf{error} & \text{if } (\mathcal{C}[\![c_1]\!]_w\, \rho)_f s = \mathsf{error} \\ (\mathcal{C}[\![c_2]\!]_w\, \rho)_{(f;g)} s' & \text{if } (\mathcal{C}[\![c_1]\!]_w\, \rho)_f s = \langle g, s' \rangle \end{cases} \\
(\mathcal{C}[\![\mathsf{if}\, b\, \mathsf{then}\, c_1\, \mathsf{else}\, c_2]\!]_w\, \rho)_f(s) & = \begin{cases} (\mathcal{C}[\![c_1]\!]_w\, \rho)_f s & \text{if } \mathcal{E}[\![b]\!]_w\, \rho = \mathit{true} \\ (\mathcal{C}[\![c_2]\!]_w\, \rho)_f s & \text{if } \mathcal{E}[\![b]\!]_w\, \rho = \mathit{false} \\ \mathsf{error} & \text{otherwise} \end{cases} \\
(\mathcal{C}[\![\mathsf{let}\ x{=}\mathsf{new}\ e\ \mathsf{in}\ c]\!]_w\, \rho)_f(s) & = (\mathcal{C}[\![c]\!]_{w''}\, (\mathit{Env}(f; \iota_{w'}^{w''})(\rho))[x{:=}l])_{id}\, s'; \mathit{shift}_{(f;\iota_{w'}^{w''})} \text{ where} \\
 & \quad l{=} \min \mathbb{L} \backslash w', w''{=}w' \cup \{l\}, s'{=}t_{w',l}(\mathit{Val}(f)(\mathcal{E}[\![e]\!]_w\, \rho), s) \\
(\mathcal{C}[\![\mathsf{free}\ x]\!]_w\, \rho)_f(s) & = \begin{cases} \langle \mathit{id}, \{\![l' = s(l')]\!\}_{l' \in \mathsf{dom}(s),\, l' \neq l} \rangle \\ \qquad \text{if } \exists l \in w.\, \mathcal{E}[\![x]\!]_w\, \rho = l \text{ and } f(l) \in \mathsf{dom}(s) \\ \mathsf{error} \quad \text{otherwise} \end{cases} \\
(\mathcal{C}[\![[x]{:=}e]\!]_w\, \rho)_f(s) & = \begin{cases} \langle \mathit{id}, \mathit{update}_{w'}(f(l),\ s,\ \mathit{Val}(f)(\mathcal{E}[\![e]\!]_w\, \rho)) \rangle \\ \qquad \text{if } \exists l \in w.\, \mathcal{E}[\![x]\!]_\rho = l \text{ and } f(l) \in \mathsf{dom}(s) \\ \qquad \text{and } \mathcal{E}[\![e]\!]_w\, \rho \in \mathit{Val}(w) \\ \mathsf{error} \quad \text{otherwise} \end{cases} \\
(\mathcal{C}[\![\mathsf{let}\ y = [x]\ \mathsf{in}\ c]\!]_w\, \rho)_f(s) & = \begin{cases} (\mathcal{C}[\![c]\!]_{w'}\, (\mathit{Env}(f)(\rho))[y := s(f(l))])_{id}(s); \mathit{shift}_f \\ \qquad \text{if } \exists l \in w.\, \mathcal{E}[\![x]\!]_w\, \rho = l \text{ and } f(l) \in \mathsf{dom}(s) \\ \mathsf{error} \quad \text{otherwise} \end{cases} \\
(\mathcal{C}[\![\mathsf{eval}\ e]\!]_w\, \rho)_f(s) & = \begin{cases} (\mathcal{E}[\![e]\!]_w\, \rho)_f s & \text{if } \mathcal{E}[\![e]\!]_w\, \rho \in \mathit{Com}(w) \\ \mathsf{error} & \text{otherwise} \end{cases}
\end{array}
$$

$$
\begin{array}{ll}
\multicolumn{2}{l}{\mathit{shift}_f : (\mathsf{error} + \sum_{g' : w' \rightarrow w''} \mathit{St}(w'')) \rightarrow (\mathsf{error} + \sum_{g : w \rightarrow w''} \mathit{St}(w''))} \\
\mathit{shift}_f(v) & = \begin{cases} \mathsf{error} & \text{if } v = \mathsf{error} \\ \langle f; g', s' \rangle & \text{if } v = \langle g', s' \rangle \end{cases} \\
\multicolumn{2}{l}{t_{w,l} : \mathit{Val}(w) \times \mathit{St}(w) \rightarrow \mathit{St}(w \uplus \{l\})} \\
t_{w,l}(v, s) & = \mathit{St}(\iota_w^{w \cup \{l\}})(s) * \{\![l = \mathit{Val}(\iota_w^{w \cup \{l\}})(v)]\!\} \\
\multicolumn{2}{l}{\mathit{update}_w : w \times \mathit{St}(w) \times \mathit{Val}(w) \rightarrow \mathit{St}(w)} \\
\multicolumn{2}{l}{\mathit{update}_w(l, s, v) = \{\![l{=}v]\!\} * \{\![l'{=}s(l')]\!\}_{l' \in \mathsf{dom}(s), l' \neq l}}
\end{array}
$$

Definition 4 (Validity). *Let $w \in \mathbb{W}$, $\rho \in \mathit{Env}(w)$, $s \in \mathit{St}(LC)_w$, $c \in \mathit{Com}(w) \cap LC_w$ and $p, q \in \mathcal{S}$. An auxiliary meaning of "semantical triples" with respect to a fixed world, written $(\rho, s) \models_w \{p\}\, c\, \{q\}$, holds if and only if for all $f : w \rightarrow w_1$,*

$$
\begin{aligned}
\forall g : w_1 \rightarrow w_2\, \forall s' \in \mathit{St}(w_2).\ & (\mathit{Env}(f)(\rho), \mathit{St}(f)(s)) \in p_{w_1} \wedge \\
& c_f(\mathit{St}(f)(s)) = \langle g, s' \rangle \quad \Longrightarrow \quad (\mathit{Env}(f; g)(\rho), s') \in q_{w_2}
\end{aligned}
$$

Observe that $\{p\}\, c\, \{\mathbf{true}\}$ means that, assuming p for the initial state, the command does not lead to a memory fault. Validity *of syntactic triples in context*

Table 6. Interpretation of assertions

$\mathcal{A}[\![P]\!] : \mathcal{S}$	
$(\rho, s) \in \mathcal{A}[\![\mathbf{true}]\!]_w$	$:\Longleftrightarrow$ true
$(\rho, s) \in \mathcal{A}[\![e_1 \leq e_2]\!]_w$	$:\Longleftrightarrow \mathcal{E}[\![e_i]\!]_w\,\rho \notin Com(w) \ \wedge\ \mathcal{E}[\![e_1]\!]_w\,\rho \ \leq\ \mathcal{E}[\![e_2]\!]_w\,\rho$
$(\rho, s) \in \mathcal{A}[\![\neg A]\!]_w$	$:\Longleftrightarrow (\rho, s) \notin \mathcal{A}[\![A]\!]_w$
$(\rho, s) \in \mathcal{A}[\![P \wedge Q]\!]_w$	$:\Longleftrightarrow (\rho, s) \in \mathcal{A}[\![P]\!]_w \ \wedge\ (\rho, s) \in \mathcal{A}[\![Q]\!]_w$
$(\rho, s) \in \mathcal{A}[\![\forall x.\, P]\!]_w$	$:\Longleftrightarrow \forall v\in Val(w).\ (\rho[x \mapsto v], s) \in \mathcal{A}[\![P]\!]_w$
$(\rho, s) \in \mathcal{A}[\![\mathbf{emp}]\!]_w$	$:\Longleftrightarrow \mathsf{dom}(s) = \emptyset$
$(\rho, s) \in \mathcal{A}[\![P_1 * P_2]\!]_w$	$:\Longleftrightarrow \exists s_1, s_2 \in St(w).\, s = s_1 * s_2 \ \wedge\ (\rho, s_i) \in \mathcal{A}[\![P_i]\!]_w$
$(\rho, s) \in \mathcal{A}[\![x \mapsto e]\!]_w$	$:\Longleftrightarrow \mathsf{dom}(s) = \{\rho(x)\} \ \wedge\ s(\rho(x)) \sqsubseteq \mathcal{E}[\![e]\!]_w\,\rho$

of assumptions is written $\models \{P_1\}\,c_1\,\{Q_1\}, \ldots, \{P_n\}\,c_n\,\{Q_n\} \vdash \{P\}\,c\,\{Q\}$ *and holds if and only if for all* $w \in \mathbb{W}$,

$$\forall\rho\in Env(w)\,\forall s\in St(LC)_w.\ \bigwedge_{1\leq i\leq n} (\rho, s) \models_w \{\mathcal{A}[\![P_i]\!]\}\,\mathcal{C}[\![c_i]\!]_w\,\rho\,\{\mathcal{A}[\![Q_i]\!]\}$$
$$\Longrightarrow\ (\rho, s) \models_w \{\mathcal{A}[\![P]\!]\}\,\mathcal{C}[\![c]\!]_w\,\rho\,\{\mathcal{A}[\![Q]\!]\}$$

For an empty context we simply write $\models \{P\}\,c\,\{Q\}$ *instead of* $\models\ \vdash \{P\}\,c\,\{Q\}$.

Theorem 3 (Soundness). *The logic presented in Section 2.2 is sound with respect to our semantics.*

Proof. Lack of space permits only a sketch for the two most interesting rules.

Soundness of the frame rule (Frame). Except for exploiting the renaming of locations the proof uses the standard argument: Suppose $\{P\}\,c\,\{Q\}$ is valid, let $w_1 \in \mathbb{W}$, $\rho \in Env(w_1)$ and $s \in St(LC)_{w_1}$ such that $(Env(f)(\rho), St(f)(s)) \in \mathcal{A}[\![P * R]\!]_{w_2}$ and $c_f(St(f)(s))\!\downarrow$, where $f : w_1 \to w_2$. Thus, $St(f)(s) = s_1 * s_2$ for some s_1, s_2 with $(Env(f)(\rho), s_1) \in \mathcal{A}[\![P]\!]_{w_2}$ and $(Env(f)(\rho), s_2) \in \mathcal{A}[\![R]\!]_{w_2}$.

Now if $c_f(St(f)(s)) = \mathsf{error}$ then, by assumption $c \in LC_w$, also $c_f(s_1) = \mathsf{error}$ which contradicts validity of $\{P\}\,c\,\{Q\}$. Thus, $c_f(St(f)(s)) = \langle g, s'\rangle$ for some $g : w_2 \to w_3$ and $s' \in St(w_3)$. By $c \in LC_w$ there exist $g' : w_2 \to w_3'$, $j : w_3' \xrightarrow{\sim} w_3$, $s_1' \in St(w_3')$ and $s_2' \in St(w_2)$ such that $s_2' \sqsubseteq s_2$, $c_f(s_1) = \langle g', s_1'\rangle$ and

$$s' = St(j)(s_1') * St(g';j)(s_2') \tag{4}$$

Downward-closure of $\mathcal{A}[\![R]\!]$ entails $(Env(f)(\rho), s_2') \in \mathcal{A}[\![R]\!]_{w_2}$, and therefore $(Env(f;g';j)(\rho), St(g';j)(s_2')) \in \mathcal{A}[\![R]\!]_{w_3}$ by Kripke monotonicity. By validity of $\{P\}\,c\,\{Q\}$ we have $(Env(f;g')(\rho), s_1') \in \mathcal{A}[\![Q]\!]_{w_3'}$. Kripke monotonicity of $\mathcal{A}[\![Q]\!]$ and (4) entail $(Env(f;g)(\rho), s') \in \mathcal{A}[\![Q * R]\!]_{w_3}$, proving $\models \{P * R\}\,c\,\{Q * R\}$.

Soundness of the recursion rule (Rec). This is proved along the lines of [23]: Pitts' technique (cf. Theorem 1) is used to establish existence of a suitable recursive $\mathcal{S}$-relation containing the commands defined by mutual recursion. As in [23] one shows for all assertions P that $\mathcal{A}[\![P]\!] \in \mathcal{S}$ satisfies the following properties: for all $w \in \mathbb{W}$, the set $\{s \mid (\rho, s) \in \mathcal{A}[\![P]\!]_w\}$ is downward closed, the set $\{\rho \mid (\rho, s) \in \mathcal{A}[\![P]\!]_w\}$ is upward closed, and $e : \mathcal{A}[\![P]\!] \subset \mathcal{A}[\![P]\!]$ for all $e \sqsubseteq id_{Val}$.

The first property is built into the definition of $\mathcal{S}$-relations, the latter two can be established by induction on assertions. Note that the way $x \mapsto e$ and $e_1 \leq e_2$ are defined in Table 6 is essential for this result. In particular, $e_1 \leq e_2$ had to be defined differently in [23] where the extra level of locations was absent. □

5 Conclusions and Further Work

We have presented a logic for higher-order store that admits a local reasoning principle in form of the (first-order) frame rule. Soundness relies on a denotational semantics employing powerful constructions known from domain theory.

Our reasoning principle for recursion through the store (Rec) is based on explicitly keeping track of the code in pre- and postconditions. Instead of code, Honda et al. [8] use abstract specifications of code, in terms of nested triples in assertions. Their logic is for programs of an ML-like imperative higher-order language, with dynamic memory allocation and function storage. In contrast to our work, it builds on operational techniques and does not address local reasoning. Consequently, an improvement of our logic would be the integration of nested triples in assertions while admitting a frame rule that is proved sound employing the semantical approach presented here.

Stored procedures are particularly important for object-oriented programming, and we are currently investigating how a separation logic for higher-order store can be extended to simple object-based languages like the object calculus to obtain a logic that combines the power of local reasoning with the principle ideas of Abadi and Leino's logic [1,21]. To achieve that, our results need to be generalised from Hoare triples to more general *transition relations*. Separation conjunction in such a framework has been considered in [19].

There are several possibilities for further improvements. It would be interesting to see if the FM models of [26,3], rather than a presheaf semantics, can simplify the semantics. It also needs to be investigated whether a higher-order frame rule can be proven sound in our setting analogous to [16,5].

References

1. M. Abadi and K. R. M. Leino. A logic of object-oriented programs. In *Verification: Theory and Practice. Essays Dedicated to Zohar Manna on the Occasion of his 64th Birthday*, pages 11–41. Springer, 2004.
2. K. R. Apt and G. D. Plotkin. Countable nondeterminism and random assignment. *Journal of the ACM*, 33(4):724–767, 1986.
3. N. Benton and B. Leperchey. Relational reasoning in a nominal semantics for storage. In Proc. TLCA'05, volume 3461 of *LNCS*, pages 86–101. Springer, 2005.
4. L. Birkedal, N. Torp-Smith, and J. C. Reynolds. Local reasoning about a copying garbage collector. In *Proc. 31st POPL*, pages 220–231. ACM Press, 2004.
5. L. Birkedal, N. Torp-Smith, and H. Yang. Semantics of separation-logic typing and higher-order frame rules. In *Proc. 20th LICS*. IEEE Press, 2005.
6. P. di Gianantonio, F. Honsell, and G. D. Plotkin. Uncountable limits and the lambda calculus. *Nordic Journal of Computing*, 2(2):126–145, 1995.

7. C. A. R. Hoare. An Axiomatic Basis of Computer Programming. *Communications of the ACM*, 12:576–580, 1969.
8. K. Honda, N. Yoshida, and M. Berger. An observationally complete program logic for imperative higher-order functions. In *Proc. 20th LICS*, pages 270–279. IEEE Computer Society Press, 2005.
9. S. S. Ishtiaq and P. W. O'Hearn. BI as an assertion language for mutable data structures. *ACM SIGPLAN Notices*, 36(3):14–26, 2001.
10. P. J. Landin. The mechanical evaluation of expressions. *Computer Journal*, 6(4):308–320, 1964.
11. P. B. Levy. *Call-By-Push-Value. A Functional/Imperative Synthesis*, volume 2 of *Semantic Structures in Computation*. Kluwer, 2004.
12. Z. Ni and Z. Shao. Certified assembly programming with embedded code pointers. In *Proc. 33rd POPL*, pages 320–333. ACM Press, 2006.
13. P. W. O'Hearn. Resources, concurrency and local reasoning. In *Proc. CONCUR'04*, volume 3170 of *LNCS*, pages 49–67. Springer, 2004.
14. P. W. O'Hearn, J. C. Reynolds, and H. Yang. Local reasoning about programs that alter data structures. In *Proc. CSL'01*, volume 2142 of *LNCS*, pages 1–18. Springer, 2001.
15. P. W. O'Hearn and R. D. Tennent. Semantics of local variables. In *Applications of Categories in Computer Science*, volume 177 of *London Mathematical Society Lecture Note Series*, pages 217–238. Cambridge University Press, 1992.
16. P. W. O'Hearn, H. Yang, and J. C. Reynolds. Separation and information hiding. In *Proc. 31st POPL*, pages 268–280. ACM Press, 2004.
17. F. J. Oles. *A Category-theoretic approach to the semantics of programming languages*. PhD thesis, Syracuse University, 1982.
18. A. M. Pitts. Relational properties of domains. *Information and Computation*, 127:66–90, 1996.
19. A. Podelski and I. Schaefer. Local reasoning for termination. In *Informal Workshop Proc. Verification of Concurrent Systems with Dynamically Allocated Heaps*, 2005.
20. U. S. Reddy and H. Yang. Correctness of data representations involving heap data structures. *Science of Computer Programming*, 50(1–3):129–160, 2004.
21. B. Reus and J. Schwinghammer. Denotational semantics for a program logic of objects. *Mathematical Structures in Computer Science*, 16(2):313–358, 2006.
22. B. Reus and T. Streicher. Semantics and logic of object calculi. *Theoretical Computer Science*, 316:191–213, 2004.
23. B. Reus and T. Streicher. About Hoare logics for higher-order store. In *Proc. ICALP'05*, volume 3580 of *LNCS*, pages 1337–1348. Springer, 2005.
24. J. C. Reynolds. The essence of Algol. In J. W. deBakker and J. C. van Vliet, editors, *Algorithmic Languages*, pages 345–372. North-Holland, 1981.
25. J. C. Reynolds. Separation logic: A logic for shared mutable data structures. In *Proc. 17th LICS*, pages 55–74. IEEE Computer Society, 2002.
26. M. R. Shinwell and A. M. Pitts. On a monadic semantics for freshness. *Theoretical Computer Science*, 342:28–55, 2005.
27. M. B. Smyth and G. D. Plotkin. The category-theoretic solution of recursive domain equations. *SIAM Journal on Computing*, 11(4):761–783, Nov. 1982.
28. H. Thielecke. Frame rules from answer types for code pointers. In *Proc. 33rd POPL*, pages 309–319. ACM Press, 2006.
29. H. Yang. An example of local reasoning in BI pointer logic: the Schorr-Waite graph marking algorithm. In *Proc. 2nd SPACE workshop*, 2001.

Satisfiability and Finite Model Property for the Alternating-Time μ-Calculus*

Sven Schewe and Bernd Finkbeiner

Universität des Saarlandes, 66123 Saarbrücken, Germany
{schewe, finkbeiner}@cs.uni-sb.de

Abstract. This paper presents a decision procedure for the alternating-time μ-calculus. The algorithm is based on a representation of alternating-time formulas as automata over concurrent game structures. We show that language emptiness of these automata can be checked in exponential time. The complexity of our construction meets the known lower bounds for deciding the satisfiability of the classic μ-calculus. It follows that the satisfiability problem is EXPTIME-complete for the alternating-time μ-calculus.

1 Introduction

In the design of distributed protocols, we are often interested in the strategic abilities of certain agents. For example, in a contract-signing protocol, it is important to ensure that while Alice and Bob can cooperate to sign a contract, Bob never has a strategy to obtain Alice's signature unless, at the same time, Alice has a strategy to obtain Bob's signature as well (cf. [10]). Such properties can be expressed in the alternating-time μ-calculus (AMC) [1], which extends the classic μ-calculus with modalities that quantify over the strategic choices of a group of agents. The models of AMC are a special type of labeled transition systems, called *concurrent game structures*, where each transition results from a set of decisions, one for each agent.

In this paper, we present the first decision procedure for the satisfiability of AMC formulas. The *satisfiability* problem asks for a given AMC formula φ whether there exists a concurrent game structure that satisfies φ. Previous research has focused on the model checking problem [1,2], which asks whether a *given* concurrent game structure satisfies its specification. By contrast, our procedure checks whether a specification can be implemented at all. For example, we can automatically prove the classic result that it is *impossible* to implement fair contract-signing without a trusted third party [6].

We introduce an automata-theoretic framework for alternating-time logics. *Automata over concurrent game structures* (ACGs) are a variant of alternating

* This work was partly supported by the German Research Foundation (DFG) as part of the Transregional Collaborative Research Center "Automatic Verification and Analysis of Complex Systems" (SFB/TR 14 AVACS).

Z. Ésik (Ed.): CSL 2006, LNCS 4207, pp. 591–605, 2006.

tree automata, where the atoms in the transition function do not refer to individual successors in the input structure, but instead quantify universally or existentially over all successors that result from the agents' decisions. Specifically, a *universal* atom $(\Box, A')$ refers to *all* successor states for *some* decision of the agents in a set A', and an *existential* atom $(\Diamond, A')$ refers to *some* successor state for *each* decision of the agents *not* in A'. In this way, the automaton can run on game structures with arbitrary, even infinite, branching degree. Every AMC formula can be translated into an automaton that accepts exactly the models of the formula. Satisfiability of AMC formulas thus corresponds to language nonemptiness of ACGs.

The core result of the paper is the finite model property for ACGs. We first prove that, given any game structure accepted by an ACG $\mathcal{G}$, we can find a *bounded* game structure that is also accepted by $\mathcal{G}$. In the bounded game structure, the number of possible decisions of each agent is limited by some constant m, determined by the size of $\mathcal{G}$.

The emptiness problem of ACGs thus reduces to the emptiness problem of alternating tree automata and, since non-empty automata over finitely-branching structures always accept some finite structure [15,14], there must exist a finite game structure in the language of $\mathcal{G}$. The drawback of this reduction is that the trees accepted by the alternating tree automaton branch over the decisions of *all* agents: the number of directions is therefore exponential in the number of agents. Since the emptiness problem of alternating tree automata is exponential in the number of directions, this results in a double-exponential decision procedure.

We show that it is possible to decide emptiness in single-exponential time. Instead of constructing an *alternating* automaton that accepts exactly the m-bounded game structures in the language of the ACG, we construct a *universal* automaton that only preserves emptiness. Unlike alternating automata, universal automata can be reduced to deterministic automata with just a single exponential increase in the number of states. For deterministic automata, the complexity of the emptiness problem is only linear in the number of directions.

Our approach is constructive and yields a tight complexity bound: the satisfiability problem for AMC is EXPTIME-complete. If the AMC formula is satisfiable, we can synthesize a finite model within the same complexity bound. Since AMC subsumes the alternating-time temporal logic ATL* [4,1], we obtain a decision procedure for this logic as well.

Related work. The automata-theoretic approach to the satisfiability problem was initiated in the classic work by Büchi, McNaughton, and Rabin on monadic second-order logic [3,13,15]. For linear-time temporal logic, satisfiability can be decided by a translation to automata over infinite words [18]; for branching-time logics, such as CTL* and the modal μ-calculus, by a translation to automata over infinite trees that branch according to inputs and nondeterministic choices [12,5,11,20]. For alternating-time temporal logics, previous decidability results have been restricted to ATL [17,19], a sublogic of ATL*.

Automata over concurrent game structures, introduced in this paper, provide an automata-theoretic framework for alternating-time logics. Automata over concurrent game structures extend symmetric alternating automata [20], which have been proposed as the automata-theoretic framework for the classic μ-calculus. Symmetric automata branch universally into all successors or existentially into some successor.

2 Preliminaries

2.1 Concurrent Game Structures

Concurrent game structures [1] generalize labeled transition systems to a setting with multiple agents. A *concurrent game structure* (CGS) is a tuple $\mathcal{C} = (P, A, S, s_0, l, \Delta, \tau)$, where

- P is a finite set of atomic propositions,
- A is a finite set of agents,
- S is a set of states, with a designated initial state $s_0 \in S$,
- $l : S \to 2^P$ is a labeling function that decorates each state with a subset of the atomic propositions,
- Δ is a set of possible decisions for every agent, and
- $\tau : S \times \Delta^A \to S$ is a transition function that maps a state and the decisions of the agents to a new state.

A concurrent game structure is called *bounded* if the set Δ of decisions is finite, *m-bounded* if $\Delta = \mathbb{N}_m = \{1, \ldots, m\}$, and *finite* if S and Δ are finite.

Example. As a running example, we introduce a simple CGS $\mathcal{C}_0$ with an infinite number of states and an infinite number of possible decisions. In every step, two agents each pick a real number and move to the state ${d_2}^2 - {d_1}^2$, where d_1 is the decision of agent a_1 and d_2 is the decision of agent a_2. We use two propositions, p_1 and p_2, where p_1 identifies the non-negative numbers and p_2 the rational numbers. Let $\mathcal{C}_0 = (P, A, S, s_0, l, \Delta, \tau)$, with $P = \{p_1, p_2\}$, $A = \{a_1, a_2\}$, $S = \mathbb{R}$, $s_0 = 0$, $p_1 \in l(s)$ iff $s \geq 0$, $p_2 \in l(s)$ iff $s \in \mathbb{Q}$, $\Delta = \mathbb{R}$, and $\tau : (s, (d_1, d_2)) \mapsto {d_2}^2 - {d_1}^2$. It is easy to see that in all states of this CGS, agent a_1 can enforce that p_1 eventually always holds true. Additionally, if agent a_1 decides before agent a_2, agent a_2 can always respond with a decision such that p_2 holds in the following state.

2.2 Alternating-Time μ-Calculus

The *alternating-time μ-calculus* (AMC) extends the classical μ-calculus with modal operators which express that an agent or a coalition of agents has a strategy to accomplish a goal. AMC formulas are interpreted over concurrent game structures.

AMC Syntax. AMC contains the modality $\Box_{A'}\varphi$, expressing that a set $A' \subseteq A$ of agents can enforce that a property φ holds in the successor state, and the modality $\Diamond_{A'}\varphi$, expressing that it cannot be enforced against the agents A' that φ is violated in the successor state. Let P and B denote disjoint finite sets of atomic propositions and bound variables, respectively. Then

- *true* and *false* are AMC formulas.
- p, $\neg p$ and x are AMC formulas for all $p \in P$ and $x \in B$.
- If φ and ψ are AMC formulas then $\varphi \wedge \psi$ and $\varphi \vee \psi$ are AMC formulas.
- If φ is an AMC formula and $A' \subseteq A$ then $\Box_{A'}\varphi$ and $\Diamond_{A'}\varphi$ are AMC formulas.
- If $x \in B$ and φ is an AMC formula where x occurs only free, then $\mu x.\varphi$ and $\nu x.\varphi$ are AMC formulas.

The set of subformulas of a formula φ is denoted by $sub(\varphi)$ and its alternation depth by $alt(\varphi)$ (for simplicity we use the syntactic alternation of least and greatest fixed-point operators).

AMC Semantics. An AMC formula φ with atomic propositions P is interpreted over a CGS $\mathcal{C} = (P, A, S, s_0, l, \Delta, \tau)$. $\|\varphi\|_{\mathcal{C}} \subseteq S$ denotes the set of states where φ holds. A CGS $\mathcal{C} = (P, A, S, s_0, l, \Delta, \tau)$ is a *model* of a specification φ with atomic propositions P iff $s_0 \in \|\varphi\|_{\mathcal{C}}$.

- Atomic propositions are interpreted as follows: $\|false\|_{\mathcal{C}} = \emptyset$ and $\|true\|_{\mathcal{C}} = S$, $\|p\|_{\mathcal{C}} = \{s \in S \mid p \in l(s)\}$ and $\|\neg p\|_{\mathcal{C}} = \{s \in S \mid p \notin l(s)\}$.
- Conjunction and disjunction are interpreted as intersection and union, respectively: $\|\varphi \wedge \psi\|_{\mathcal{C}} = \|\varphi\|_{\mathcal{C}} \cap \|\psi\|_{\mathcal{C}}$ and $\|\varphi \vee \psi\|_{\mathcal{C}} = \|\varphi\|_{\mathcal{C}} \cup \|\psi\|_{\mathcal{C}}$.
- A state $s \in S$ is in $\|\Box_{A'}\varphi\|_{\mathcal{C}}$ iff the agents A' can make a decision $\upsilon \in \Delta^{A'}$ such that, for all decisions $\upsilon' \in \Delta^{A \smallsetminus A'}$, φ holds in the successor state: $\|\Box_{A'}\varphi\|_{\mathcal{C}} = \{s \in S \mid \exists \upsilon \in \Delta^{A'}.\ \forall \upsilon' \in \Delta^{A \smallsetminus A'}.\, \tau(s, (\upsilon, \upsilon')) \in \|\varphi\|_{\mathcal{C}}\}$.
- A state $s \in S$ is in $\|\Diamond_{A'}\varphi\|_{\mathcal{C}}$ iff for all decisions $\upsilon \in \Delta^{A \smallsetminus A'}$ of the agents not in A', the agents in A' have a counter decision $\upsilon' \in \Delta^{A'}$ which ensures that φ holds in the successor state: $\|\Diamond_{A'}\varphi\|_{\mathcal{C}} = \{s \in S \mid \forall \upsilon' \in \Delta^{A \smallsetminus A'}.\ \exists \upsilon \in \Delta^{A'}.\, \tau(s, (\upsilon, \upsilon')) \in \|\varphi\|_{\mathcal{C}}\}$.
- The least and greatest fixed points are interpreted as follows: $\|\mu x.\varphi\|_{\mathcal{C}} = \bigcap\{S_x \subseteq S \mid \|\varphi\|_{\mathcal{C}_x^{S_x}} \subseteq S_x\}$, $\|\nu x.\varphi\|_{\mathcal{C}} = \bigcup\{S_x \subseteq S \mid \|\varphi\|_{\mathcal{C}_x^{S_x}} \supseteq S_x\}$, where $\mathcal{C}_x^{S_x} = (P \cup \{x\}, A, S, s_0, l_x^{S_x}, \Delta, \tau)$ denotes the modified CGS with the labeling function $l_x^{S_x} : S \to 2^{P \cup \{x\}}$ with $l_x^{S_x}(s) \cap P = l(s)$ and $x \in l_x^{S_x}(s) \Leftrightarrow s \in S_x \subseteq S$. Since the bound variable x occurs only positive in φ, $\|\varphi\|_{\mathcal{C}_x^{S_x}}$ is monotone in S_x and the fixed points are well-defined.

AMC contains the classic μ-calculus with the modal operators $\Box$ and $\Diamond$, which abbreviate $\Box_\emptyset$ and $\Diamond_A$, respectively. AMC also subsumes the temporal logic ATL* [1], which is the alternating-time extension of the branching-time temporal logic CTL*. ATL* contains the path quantifier $\langle\!\langle A' \rangle\!\rangle$, which ranges over all paths the players in A' can enforce. There is a canonical translation from ATL* to AMC [4].

Example. As discussed in Section 2.1, the example CGS $\mathcal{C}_0$ has the property that in all states, agent a_1 can enforce that p_1 eventually always holds true, and agent a_2 can respond to any decision of agent a_1 with a counter decision such that p_2 holds in the following state. This property is expressed by the AMC formula $\psi = \nu x.(\mu y.\nu z.\Box_{\{a_1\}}(p_1 \wedge z \vee y)) \wedge \Diamond_{\{a_2\}}p_2 \wedge \Diamond_\emptyset x$.

2.3 Automata over Finitely Branching Structures

An *alternating parity automaton* with a finite set Υ of directions is a tuple $\mathcal{A} = (\Sigma, Q, q_0, \delta, \alpha)$, where Σ is a finite alphabet, Q is a finite set of states, $q_0 \in Q$ is a designated initial state, δ is a transition function, and $\alpha : Q \to C \subset \mathbb{N}$ is a coloring function. The transition function $\delta : Q \times \Sigma \to \mathbb{B}^+(Q \times \Upsilon)$ maps a state and an input letter to a positive boolean combination of states and directions.

In the context of this paper, we consider alternating parity automata that run on bounded CGSs with a fixed set P of atomic propositions ($\Sigma{=}2^P$), a fixed set A of agents and a fixed finite set Δ of decisions ($\Upsilon{=}\Delta^A$). The acceptance mechanism is defined in terms of run trees. As usual, an Υ-tree is a prefix-closed subset $Y \subseteq \Upsilon^*$ of the finite words over the set Υ of directions. For given sets Σ and Υ, a *Σ-labeled Υ-tree* is a pair $\langle Y, l \rangle$, consisting of a tree $Y \subseteq \Upsilon^*$ and a labeling function $l : Y \to \Sigma$ that maps every node of Y to a letter of Σ. If Υ and Σ are not important or clear from the context, $\langle Y, l \rangle$ is called a tree.

A *run tree* $\langle R, r \rangle$ on a given CGS $\mathcal{C} = (P, A, S, s_0, l, \Delta, \tau)$ is a $Q{\times}S$-labeled tree whose root is decorated with $r(\varepsilon) = (q_0, s_0)$, and for each node $n \in R$ decorated with a label $r(n) = (q, s)$, there is a set $\mathfrak{A}_n \subseteq Q \times \Upsilon$ that satisfies $\delta(q, l(s))$, such that (q', υ) is in $\mathfrak{A}_n$ iff some child of n is decorated with a label $(q', \tau(s, \upsilon))$.

A run tree is *accepting* iff all infinite paths fulfill the *parity condition*. An infinite path fulfills the parity condition iff the highest color of the states appearing infinitely often on the path is even. A CGS is *accepted* by the automaton iff it has an accepting run tree. The set of CGSs accepted by an automaton $\mathcal{A}$ is called its *language* $\mathcal{L}(\mathcal{A})$. An automaton is empty iff its language is empty.

The acceptance of a given CGS $\mathcal{C}$ can also be viewed as the outcome of a *game* played over $Q{\times}S$, starting in (q_0, s_0). When the game reaches a position (q, s), player *accept* first chooses a set $\mathfrak{A} \subseteq Q{\times}\Upsilon$ of atoms that satisfies $\delta(q, l(s))$. Player *reject* then chooses one atom (q', υ) from $\mathfrak{A}$ and the game continues in $(q', \tau(s, \upsilon))$. An infinite sequence $(q_0, s_0)(q_1, s_1)(q_2, s_2)\ldots$ of game positions is called a *play*. A play is *winning* for player *accept* iff it satisfies the parity condition. A *strategy* for player *accept* (*reject*) maps each history of decisions of both players to a decision of player *accept* (*reject*). A pair of strategies determines a play. A strategy for player *accept* is *winning* iff, for all strategies of player *reject*, the play determined by the strategies is winning for player *accept*. The CGS $\mathcal{C}$ is accepted iff player *accept* has a winning strategy.

An automaton is *universal* iff the image of δ consists only of conjunctions, *nondeterministic* iff the image of δ consists only of formulas that, when rewritten into disjunctive normal form, contain in each disjunct exactly one element of $Q \times \{\upsilon\}$ for each $\upsilon \in \Upsilon$, and *deterministic* iff it is universal and nondeterministic.

For nondeterministic automata, emptiness can be checked with an *emptiness game* over Q where, instead of considering the letter $l(s)$ on some state s of a given CGS, the letter is chosen by player *accept*. The nondeterministic automaton is non-empty iff player *accept* has a winning strategy in the emptiness game.

3 Automata over Concurrent Game Structures

In this section, we introduce *automata over concurrent game structures* (ACGs) as an automata-theoretic framework for the alternating-time μ-calculus. The automata over finitely branching structures described in Section 2.3 do not suffice for this purpose, because they are limited to bounded CGSs. Generalizing symmetric automata [20], ACGs contain *universal atoms* $(\square, A')$, which refer to *all* successor states for *some* decision of the agents in A', and *existential atoms* $(\diamondsuit, A')$, which refer to *some* successor state for *each* decision of the agents *not* in A'. In this way, ACGs can run on CGSs with an arbitrary, even infinite, number of decisions.

An ACG is a tuple $\mathcal{G} = (\Sigma, Q, q_0, \delta, \alpha)$, where Σ, Q, q_0, and α are defined as for alternating parity automata in the previous section. The transition function $\delta : Q \times \Sigma \rightarrow \mathbb{B}^+(Q \times ((\{\square, \diamondsuit\} \times 2^A) \cup \{\varepsilon\}))$ now maps a state and an input letter to a positive boolean combination of three types of atoms: $(\square, A')$ is a universal atom, $(\diamondsuit, A')$ is an existential atom, and ε is an ε-transition, where only the state of the automaton is changed and the state of the CGS remains unchanged. If an ACG has no ε-transitions, it is called *ε-free*.

A run tree $\langle R, r\rangle$ on a given CGS $\mathcal{C} = (P, A, S, s_0, l, \Delta, \tau)$ is a $Q\times S$-labeled tree where the root is labeled with (q_0, s_0) and where, for a node n with a label (q, s) and a set $L = \{r(n \cdot \rho) \mid n \cdot \rho \in R\}$ of labels of its successors, the following property holds: there is a set $\mathfrak{A} \subseteq Q \times (\{\square, \diamondsuit\} \times 2^A \cup \{\varepsilon\})$ of atoms satisfying $\delta(q, l(s))$ such that

- for all universal atoms $(q', \square, A')$ in $\mathfrak{A}$, there exists a decision $\upsilon \in \Delta^{A'}$ of the agents in A' such that, for all counter decisions $\upsilon' \in \Delta^{A \smallsetminus A'}$, $(q', \tau(s, (\upsilon, \upsilon'))) \in L$,
- for all existential atoms $(q', \diamondsuit, A')$ in $\mathfrak{A}$ and all decisions $\upsilon' \in \Delta^{A \smallsetminus A'}$ of the agents not in A', there exists a counter decision $\upsilon \in \Delta^{A'}$ such that $(q', \tau(s, (\upsilon, \upsilon'))) \in L$, and
- for all ε-transitions (q', ε) in $\mathfrak{A}$, $(q', s) \in L$.

As before, a run tree is accepting iff all paths satisfy the parity condition, and a CGS is accepted iff there exists an accepting run tree.

The acceptance of a CGS can again equivalently be defined as the outcome of a game over $Q \times S$, starting in (q_0, s_0). Each round of the game now consists of two stages. In the first stage, player *accept* chooses a set $\mathfrak{A}$ of atoms satisfying $\delta(q, l(s))$, and player *reject* picks one atom from $\mathfrak{A}$. If the result of the first stage is an ε-transition (q', ε), then the round is finished and the game continues in (q', s) with the new state of the automaton. If the result of the first stage is a universal atom $(q', (\square, A'))$, the second stage begins by player *accept* making the decisions $\upsilon \in \Delta^{A'}$ for the agents in A', followed by player *reject* making the decisions $\upsilon' \in \Delta^{A \smallsetminus A'}$ for the remaining agents. Finally, if the result of the first stage is an existential atom $(q, (\diamondsuit, A'))$, the order of the two choices is reversed: first, player *reject* makes the decisions $\upsilon' \in \Delta^{A \smallsetminus A'}$ for the agents in $A \smallsetminus A'$; then, player *accept* makes the decisions $\upsilon \in \Delta^{A'}$ for the players in A'. After the decisions are made, the game continues in $(q', \tau(s, (\upsilon, \upsilon')))$.

A winning strategy for player *accept* uniquely defines an accepting run tree, and the existence of an accepting run tree implies the existence of a winning strategy. The game-theoretic characterization of acceptance is often more convenient than the characterization through run trees, because parity games are memoryless determined [5]. A CGS is therefore accepted by an ACG iff player *accept* has a memoryless winning strategy in the acceptance game, i.e., iff she has a strategy where her choices only depend on the state of the game and the previous decisions in the current round.

As an additional complexity measure for an ACG $\mathcal{G}$, we use the set $atom(\mathcal{G}) \subseteq Q \times \{\Box, \Diamond, \varepsilon\} \times 2^A$ of atoms that actually occur in some boolean function $\delta(q, \sigma)$. The elements of $atom(\mathcal{G})$ are called the *atoms of* $\mathcal{G}$.

Example. The CGSs that satisfy the AMC formula $\psi = \nu x.(\mu y.\nu z.\Box_{\{a_1\}}(p_1 \wedge z \vee y)) \wedge \Diamond_{\{a_2\}} p_2 \wedge \Diamond_{\emptyset} x$ from Section 2.2 are recognized by the ACG $\mathcal{G}_\psi = (\Sigma, Q, q_0, \delta, \alpha)$, where $\Sigma = 2^{\{p_1, p_2\}}$ and $Q = \{q_0, q_\mu, q_\nu, q_{p_2}\}$. The transition function δ maps

- (q_{p_2}, σ) to *true* if $p_2 \in \sigma$, and to *false* otherwise,
- (q_μ, σ) and (q_ν, σ) to $(q_\nu, \Box, \{a_1\})$ if $p_1 \in \sigma$, and to $(q_\mu, \Box, \{a_1\})$ otherwise, and
- (q_0, σ) to $\delta(q_\mu, \sigma) \wedge (q_{p_2}, \Diamond, \{a_2\}) \wedge (q_0, \Diamond, \emptyset)$.

The coloring function α maps q_μ to 1 and the remaining states to 0.

Consider again the example CGS $\mathcal{C}_0$ from Section 2.1, which satisfies ψ. In the acceptance game of $\mathcal{G}_\psi$ for $\mathcal{C}_0$, player *accept* has no choice during the first stage of each move, and can win the game by making the following decisions during the second stage:

- If one of the atoms $(q_\mu, \Box, \{a_1\})$ or $(q_\nu, \Box, \{a_1\})$ is the outcome of the first stage, agent a_1 makes the decision 0.
- If the atom $(q_{p_2}, \Diamond, \{a_2\})$ is the outcome of the first stage and agent a_1 has made the decision d_1, agent a_2 chooses $d_2 = d_1$.
- For all other atoms $(q, \circ, A')$, the decision for all agents in A' is 0.

3.1 From AMC Formulas to Automata over Concurrent Game Structures

The following theorem provides a translation of AMC formulas to equivalent ACGs. It generalizes the construction for the modal μ-calculus suggested in [20] and can be proved analogously.

Theorem 1. *Given an AMC formula φ, we can construct an ACG $\mathcal{G}_\varphi^\varepsilon = (2^V, sub(\varphi), \varphi, \delta, \alpha)$ with $|sub(\varphi)|$ states and atoms and $O(|alt(\varphi)|)$ colors that accepts exactly the models of φ.*

Construction: W.l.o.g., we assume that the bound variables have been consistently renamed to ensure that for each pair of different subformulas $\lambda x.\psi$ and $\lambda' x'.\psi'$ $(\lambda, \lambda' \in \{\mu, \nu\})$ of φ, the bound variables are different $(x \neq x')$.

- The transition function δ is defined, for all free variables p and all bound variables x, by
  - $\delta(p,\sigma) = \mathit{true}$, $\delta(\neg p,\sigma) = \mathit{false}$ $\forall p \in \sigma$;
  - $\delta(\neg p,\sigma) = \mathit{true}$, $\delta(p,\sigma) = \mathit{false}$ $\forall p \in P \smallsetminus \sigma$;
  - $\delta(\varphi \wedge \psi,\sigma) = (\varphi,\varepsilon) \wedge (\psi,\varepsilon)$ and $\delta(\varphi \vee \psi,\sigma) = (\varphi,\varepsilon) \vee (\psi,\varepsilon)$;
  - $\delta(\Box_{A'}\varphi,\sigma) = (\varphi,(\Box,A'))$ and $\delta(\Diamond_{A'}\varphi,\sigma) = (\varphi,(\Diamond,A'))$;
  - $\delta(x,\sigma) = (\lambda x.\varphi,\varepsilon)$ and $\delta(\lambda x.\varphi,\sigma) = (\varphi,\varepsilon)$ $\lambda \in \{\mu,\nu\}$.
- The coloring function α maps every subformula that is not a fixed point formula to 0. The colors of the fixed point formulas are defined inductively:
  - Every least fixed point formula $\mu p.\psi$ is colored by the smallest odd color that is greater or equal to the highest color of each subformula of ψ.
  - Every greatest fixed point formula $\nu p.\psi$ is colored by the smallest even color that is greater or equal to the highest color of each subformula of ψ. □

3.2 Eliminating ε-Transitions

Given an ACG $\mathcal{G}^\varepsilon = (\Sigma, Q, q_0, \delta, \alpha)$ with ε-transitions, we can find an ε-free ACG that accepts the same language. The idea of our construction is to consider the sequences of transitions from some position of the acceptance game that *exclusively* consist of ε-transitions: if the sequence is infinite, we can declare the winner of the game without considering the rest of the game; if the sequence is finite, we skip forward to the next non-ε-atom.

The construction is related to the elimination of ε-transitions in ordinary alternating automata [20] and will be included in the full version.

Lemma 1. *Given an ACG $\mathcal{G}^\varepsilon$ with n states, c colors and a atoms, we can construct an equivalent ε-free ACG with at most $c \cdot n$ states, c colors and $c \cdot a$ atoms.* □

4 Bounded Models

We now show that for every ACG $\mathcal{G}$ there exists a bound m such that $\mathcal{G}$ is empty if and only if $\mathcal{G}$ does not accept any m-bounded CGSs. Consider an ε-free ACG $\mathcal{G}$ and a CGS $\mathcal{C} = (P, A, S, s_0, l, \Delta, \tau)$ accepted by $\mathcal{G}$. In the following, we define a finite set Γ of decisions and a transition function $\tau' : S \times \Gamma^A \to S$, such that the resulting bounded CGS $\mathcal{C}' = (P, A, S, s_0, l, \Gamma, \tau')$ is also accepted by $\mathcal{G}$. Before we formally define the construction in the proof of Theorem 2 below, we first give an informal outline.

Let us begin with the special case where all atoms of $\mathcal{G}$ are of the form $(q, \Box, \{a\})$, i.e., a *universal* atom with a *single* agent. We use the set of atoms as the new set of decisions of each agent. The new transition function is obtained by first mapping the decision of each agent in $\mathcal{C}'$ to a decision in $\mathcal{C}$, and then applying the old transition function.

To map the decisions, we fix a memoryless winning strategy for player *accept* in the acceptance game for $\mathcal{C}$. After an atom $(q, \Box, \{a\})$ has been chosen in the first stage of the acceptance game, player *accept* begins the second stage by selecting a decision d_a for agent a. We map each decision $(q, \Box, \{a\})$ in $\mathcal{C}'$ to this decision d_a in $\mathcal{C}$.

Player *accept* wins the acceptance game for $\mathcal{C}'$ with the following strategy: In the first stage of each move, we apply the winning strategy of player *accept* in the acceptance game for $\mathcal{C}$. In the second stage, we simply select the atom $(q', \Box, \{a'\})$ that was chosen in the first stage as the decision for agent a'. Since the strategy for $\mathcal{C}$ wins for all possible decisions of the agents in $A \smallsetminus \{a'\}$, it wins in particular for the decisions selected in the transition function.

Suppose next that we still have only *universal* atoms $(q, \Box, A')$, but that the set A' of agents is not required to be singleton. There is no guarantee that the decisions of the agents in A' are consistent: an agent a may choose an atom $(q, \Box, A')$ where A' does not contain a or contains some other agent a' who made a different decision. For the purpose of computing the transition function, we therefore *harmonize* the decisions by replacing, in such cases, the decision of agent a with a fixed decision $(q_0, \Box, \{a\})$.

To win the acceptance game for $\mathcal{C}'$, player *accept* selects, after an atom $(q, \Box, A')$ has been chosen in the first stage, this atom $(q, \Box, A')$ for all agents in A'. The selection is therefore consistent for all agents in A'. Since the strategy wins for all decisions of the agents in $A \smallsetminus A'$, it does not matter if some of their decisions have been replaced. Note that, this way, only decisions of player *reject* are changed in the harmonization.

Finally, suppose that $\mathcal{G}$ contains *existential* atoms. If an existential atom $(q, \Diamond, A')$ is the outcome of the first stage of the acceptance game, player *accept* only decides *after* the decisions of the agents in $A \smallsetminus A'$ have been made by player *reject*. To implement this order of the choices in the computation of the transition function, we allow the player who chooses the last existential atom to override all decisions for existential atoms of his opponent. We add the natural numbers $\leq |A|$ as an additional component to the decisions of the agents. For a given combined decision of the agents, the sum over the numbers in the decisions of the agents, modulo $|A|$, then identifies one *favored* agent $a_0 \in A$. In this way, whichever player chooses *last* can determine the favored agent. Given the decision of agent a_0 for some atom $(q'', \Diamond, A'')$ or $(q'', \Box, A'')$, we replace each decision for an existential atom by an agent in $A \smallsetminus A''$ by the fixed decision $(q_0, \Box, \{a\})$.

To win the acceptance game for $\mathcal{C}'$, the strategy for player *accept* makes the following choice after an atom $(q', \Diamond, A')$ has been chosen in the first stage and player *reject* has made the decisions for all agents in $A \smallsetminus A'$: for all agents in A', she selects the atom $(q', \Diamond, A')$, combined with some number that ensures that the favored agent a_0 is in A'.

Example. Consider again the CGS $\mathcal{C}_0$, which is accepted by the ACG $\mathcal{G}_\psi$ with the winning strategy for player *accept* described in Section 3.1. The new transition function consists of two steps:

In the first step, we harmonize the given combined decision of the agents by replacing the inconsistent decisions. In the acceptance game, this may change the decisions of the agents controlled by player *reject*. If, for example, the atom $(q_{p_2}, \Diamond, \{a_2\})$ is the outcome of the first stage of the acceptance game and player *reject* makes the decision $(q_0, \Diamond, \emptyset, 1)$ for agent a_1, player *accept* responds by making the decision $(q_{p_2}, \Diamond, \{a_2\}, 1)$ for agent a_2. The sum of the natural numbers $(1+1)$ identifies agent a_2, and all existential choices for groups of agents *not* containing a_2 are overridden. The resulting choices are $(q_0, \Box, \{a_1\})$ for agent a_1 and $(q_{p_2}, \Diamond, \{a_2\})$ for agent a_2.

In the second step, the decisions of the agents are mapped to decisions in the CGS $\mathcal{C}_0$. First, the universal choices are evaluated: The winning strategy maps $(q_0, \Box, \{a_1\})$ to the decision $d_1 = 0$ for agent a_1. Then, the existential choice is evaluated: The winning strategy maps $(q_{p_2}, \Diamond, \{a_2\})$ and the decision $d_1 = 0$ for agent a_1 to the decision $d_2 = d_1 = 0$ for agent a_2.

The resulting bounded CGS is very simple: the new transition function maps all decisions to state 0.

Theorem 2. *An ε-free ACG $\mathcal{G} = (\Sigma, Q, q_0, \delta, \alpha)$ is non-empty iff it accepts a $(|atom(\mathcal{G})| \cdot |A|)$-bounded CGS.*

Proof. If $\mathcal{C} = (P, A, S, s_0, l, \Delta, \tau)$ is accepted by the ε-free ACG $\mathcal{G} = (2^P, Q, q_0, \delta, \alpha)$, then player *accept* has a memoryless winning strategy in the acceptance game for $\mathcal{C}$. We fix such a memoryless strategy and use it to construct the bounded CGS $\mathcal{C}' = (P, A, S, s_0, l, \Gamma, \tau')$.

Decisions. For convenience, we assume that the set A of agents is an initial sequence of the natural numbers. The new set of decisions $\Gamma = atom(\mathcal{G}) \times A$ consists of pairs of atoms and numbers. If the first component is an existential atom $(q, \Diamond, A')$, then the sum of the second components of the decisions of all agents is used to validate the choice.

We say that two decisions $d_1, d_2 \in \Gamma$ are *equivalent* if they agree on their first component: $(\mathbf{a}_1, a_1') \sim (\mathbf{a}_2, a_2') :\Leftrightarrow \mathbf{a}_1 = \mathbf{a}_2$.

We say that a combined decision $\upsilon \in \Gamma^A$ *favors* an agent $a \in A$, $\upsilon \rightarrowtail a$, if the sum of the second arguments, modulo $|A|$, of this combined decision is equal to a.

We say that the decision $d_a \in \Gamma$ of agent a *prevails* in the combined decision $\upsilon \in \Gamma^A$ if the following conditions hold for $d_a = ((q, \circ, A'), a'')$, $\circ \in \{\Box, \Diamond\}$:

- $a \in A'$,
- all agents $a' \in A'$ have made a decision $d_{a'} \sim d_a$ equivalent to the decision of a, and
- if $\circ = \Diamond$, then a cooperates with the agent favored by the combined decision υ ($\upsilon \rightarrowtail a' \in A'$).

Harmonization. Let $\mathbf{A} = Q \times \{\Box, \Diamond\} \times 2^A$. The harmonization $h : \Gamma^A \to \mathbf{A}^A$ maps the decision of the agents to a harmonic decision. Harmonic decisions are elements of $\mathbf{A}^A$ such that

- each agent $a \in A$ chooses an atom $(q, \circ, A')$ with $q \in Q$, $\circ \in \{\Box, \Diamond\}$, and $a \in A' \subseteq A$,

- if an agent $a \in A$ chooses an atom $(q, \circ, A') \in \mathbf{A}$, then all agents $a' \in A'$ choose the same atom, and
- if an agent $a \in A$ chooses an existential atom $(q, \Diamond, A') \in \mathbf{A}$, then all agents $a' \notin A'$ choose universal atoms.

For prevailing decisions, the harmonization h only deletes the second component. Non-prevailing decisions of an agent a are replaced by the fixed decision $(q_0, \Box, \{a\})$ (which is not necessarily in $atom(\mathcal{G})$).

Direction. We define the function $f_s : \mathbf{A}^A \to \Delta^A$ that maps a harmonic decision to a direction $\upsilon \in \Delta^A$ in $\mathcal{C}$. f_s depends on the state $s \in S$ of $\mathcal{C}$ and is determined by the second stage of the fixed memoryless strategy.

First, the universal decisions are evaluated: if an agent makes the harmonic decision $(q, \Box, A')$, then $\upsilon' \in \Delta^{A'}$ is determined by the choice of player *accept* in the second stage of the winning strategy in state s, when confronted with the atom $(q, \Box, A')$.

Then, the existential decisions are evaluated: If an agent makes the harmonic decision $(q, \Diamond, A')$ then $\upsilon' \in \Delta^{A'}$ is determined by the choice of player *accept* in the second stage of the winning strategy in state s, when confronted with the atom $(q, \Diamond, A')$ and the decision $\upsilon'' \in \Delta^{A \smallsetminus A'}$ fixed by the evaluation of the universal harmonic decisions.

The new transition function $\tau' : S \times \Gamma^A \to S$ is defined as $\tau' : (s, \upsilon) \mapsto \tau(s, f_s(h(\upsilon)))$.

Acceptance. In the acceptance game for $\mathcal{C}'$, player *accept* has the following strategy: in the first stage of each round, she applies the winning strategy of the acceptance game for $\mathcal{C}$. The strategy for the second stage depends on the outcome of the first stage:

- If an atom $(q, \Box, A')$ is chosen in the first stage, player *accept* fixes the prevailing decision $((q, \Box, A'), 1)$ for all agents $a \in A'$.
- If an atom $(q, \Diamond, A')$ with $A' \neq \emptyset$ is chosen in the first stage and player *reject* has made the decisions d_a for all agents $a \notin A'$, player *accept* fixes the prevailing decisions $((q, \Diamond, A'), n_a)$ for the agents $a \in A'$ such that an agent $a' \in A'$ is favored.
- If an atom $(q, \Diamond, \emptyset)$ is chosen in the first stage, then player *accept* does not participate in the second stage.

We now show that the run tree $\langle R', r' \rangle$ defined by this strategy is accepting. Let $\langle R, r \rangle$ be the run tree defined by the winning strategy in the acceptance game for $\mathcal{C}$. In the following, we argue that for each branch labeled $(q_0, s_0)\,(q_1, s_1)\,(q_2, s_2)\,\ldots$ in $\langle R', r' \rangle$, there is an identically labeled branch in $\langle R, r \rangle$. Since all branches of $\langle R, r \rangle$ satisfy the parity condition, $\langle R', r' \rangle$ must be accepting as well.

The root of both run trees is labeled by (q_0, s_0). If a node labeled (q_i, s_i) in $\langle R', r' \rangle$ has a child labeled (q_{i+1}, s_{i+1}), then there must be an atom $(q_{i+1}, \circ, A') \in Q \times \{\Box, \Diamond\} \times 2^A$ in the set of atoms chosen by player *accept*, such that following holds: for the decision $\upsilon' \in \Gamma^{A'}$ defined by the strategy of player *accept*, there is a decision $\upsilon'' \in \Gamma^{A \smallsetminus A'}$ such that $s_{i+1} = \tau'(s_i, (\upsilon', \upsilon'')) = \tau(s_i, f_{s_i}(h(\upsilon', \upsilon'')))$.

Now consider a node labeled (q_i, s_i) in $\langle R, r\rangle$. Since the strategy of player *accept* in the first stage of each round is identical for the two acceptance games, the atom $(q_{i+1}, \circ, A')$ is also included in the set of atoms chosen by player *accept* in the acceptance game for $\mathcal{C}$. Player *reject* can enforce the decision $\upsilon = f_{s_i}(h(\upsilon', \upsilon''))$ as follows:

- If $\circ = \Box$, player *accept* chooses the $\Delta^{A'}$ part of υ under the fixed memoryless strategy for the acceptance game of $\mathcal{C}$, and player *reject* can respond by choosing the $\Delta^{A \smallsetminus A'}$ part of υ.
- If $\circ = \Diamond$, player *reject* can choose the $\Delta^{A \smallsetminus A'}$ part of υ, and player *accept* will react by choosing the $\Delta^{A'}$ part of υ under the fixed memoryless strategy for the acceptance game of $\mathcal{C}$, guaranteeing that $\upsilon = f_{s_i}(h(\upsilon', \upsilon''))$.

In both cases, the new state $s_{i+1} = \tau(s_i, \upsilon)$ is chosen. The node labeled (q_i, s_i) in $\langle R, r\rangle$ must therefore have a child labeled (q_{i+1}, s_{i+1}). □

5 Satisfiability and Complexity

An AMC formula is satisfiable if and only if the language of its ACG is nonempty. A simple procedure for deciding emptiness of ACGs is immediately suggested by Theorem 2: since we can restrict our attention to m-bounded CGSs with fixed $m = |atom(\mathcal{G})| \cdot |A|$, we can replace $(q, (\Box, A'))$ and $(q, (\Diamond, A'))$ by the corresponding positive boolean combinations: the resulting automaton accepts exactly the m-bounded concurrent game structures in the language of $\mathcal{G}$. To decide emptiness, we nondeterminize the automaton [7,14] and then solve the emptiness game. The complexity of this construction is double-exponential: solving the emptiness game of the nondeterministic automaton is exponential in the number of directions, which is already exponential in the number of agents ($m^{|A|}$).

We now describe an alternative algorithm with only single-exponential complexity. Instead of going through an alternating automaton to a nondeterministic automaton, we go through a universal automaton to a *deterministic* automaton. The advantage of solving the emptiness game for a deterministic automaton instead of for a nondeterministic automaton is that the set of atoms chosen by player *accept* is uniquely determined by the input letter; this reduces the number of choices from exponential in the number of directions to linear in the size of the input alphabet.

The construction of the universal automaton is based on the observation that the winning strategy of player *accept* that we defined in the previous section can be represented by assigning a function $f_s : Q \to 2^{atom(\mathcal{G})}$ to each state s of $\mathcal{C}$. The set $f_s(q)$ contains the atoms that player *accept* chooses in the first stage of the game at position (q, s). Since the strategy for the second stage depends only on the chosen atom and not on the states of $\mathcal{C}$ and $\mathcal{G}$, f_s determines the entire strategy of player *accept*.

The universal automaton runs on bounded CGSs that are annotated by this function f_s; i.e., we extend the alphabet from Σ to $\Sigma \times (Q \to 2^{atom(\mathcal{G})})$ and a CGS is accepted iff f_s identifies a winning strategy for player *accept* in the acceptance game of the ACG. The construction does not change the set of states and increases the input alphabet by an exponential factor in the number of states and atoms of the ACG.

Lemma 2. *Given an ε-free ACG $\mathcal{G} = (\Sigma, Q, q_0, \delta, \alpha)$ and a set A of agents, we can construct a universal parity automaton $\mathcal{U} = (\Sigma \times (Q \to 2^{atom(\mathcal{G})}), Q, q_0, \delta', \alpha)$ on $\Sigma \times (Q \to 2^{atom(\mathcal{G})})$-labeled CGSs with the set $atom(\mathcal{G}) \times A$ of decisions such that $\mathcal{U}$ has the following properties:*

- *If $\mathcal{U}$ accepts a CGS $\mathcal{C} = (P, A, S, s_0, l \times strat_1, atom(\mathcal{G}) \times A, \tau)$ then $\mathcal{G}$ accepts its Σ projection $\mathcal{C}' = (P, A, S, s_0, l, atom(\mathcal{G}) \times A, \tau)$.*
- *If $\mathcal{U}$ is empty, then $\mathcal{G}$ is empty.*

Proof. We denote with $strat_2$ the function that maps each atom $\mathbf{a}$ of $\mathcal{G}$ to the set $D \subseteq (atom(\mathcal{G}) \times A)^A$ of decisions that are the outcome of the second stage of the acceptance game for some strategy of player reject, when the outcome of the first stage is $\mathbf{a}$ and player *accept* follows the simple strategy for the second stage described in the proof of Theorem 2. Generalizing $strat_2$ to sets of atoms, we define the transition function δ' of $\mathcal{U}$ by setting $\delta'(q; \sigma, s)$ to *false* if $s(q)$ does not satisfy $\delta(q, \sigma)$, and to a conjunction over $strat_2(s(q))$ otherwise.

If $\mathcal{U}$ accepts a CGS $\mathcal{C} = (P, A, S, s_0, l \times strat_1, atom(\mathcal{G}) \times A, \tau)$, then player *accept* has a winning strategy for $\mathcal{C}' = (P, A, S, s_0, l, atom(\mathcal{G}) \times A, \tau)$ in the acceptance game of $\mathcal{G}$, where the strategy in the first stage is defined by $strat_1$ and the strategy in the second stage is as defined in the proof of Theorem 2.

If $\mathcal{G}$ accepts a CGS $\mathcal{C}$, then there exists, as described in the proof of Theorem 2, a CGS $\mathcal{C}' = (P, A, S, s_0, l, atom(\mathcal{G}) \times A, \tau)$, such that player *accept* wins the acceptance game using some memoryless strategy $strat_1$ in the first stage and the canonical strategy in the second stage. The CGS $\mathcal{C}'' = (P, A, S, s_0, l \times strat_1, atom(\mathcal{G}) \times A, \tau)$ is accepted by $\mathcal{U}$. □

We transform the universal parity automaton $\mathcal{U}$ into a deterministic parity automaton by first transforming $\mathcal{U}$ into a universal co-Büchi automaton with $O(c \cdot n)$ states and then using Safra's construction [16,7].

Lemma 3. *Given a universal automaton $\mathcal{U}$ with n states and c colors, we can construct an equivalent deterministic parity automaton $\mathcal{D}$ with $n^{O(c \cdot n)}$ states and $O(c \cdot n)$ colors.* □

Our transformation of the ACG to the deterministic automaton $\mathcal{D}$ thus increases both the number of states and the size of the input alphabet to at most exponential in the number of states of the ACG. The emptiness game of $\mathcal{D}$ is solved in polynomial time both in the number of states and in the size of the input alphabet, providing an exponential-time procedure for deciding emptiness of an ACG.

Lemma 4. *Given a deterministic parity automaton $\mathcal{D} = (\Sigma, Q, q_0, \delta, \alpha)$ with n states and c colors, we can, in time $(n \cdot |\Sigma|)^{O(c)}$, decide emptiness and, if $\mathcal{L}(\mathcal{D}) \neq \emptyset$, construct a finite CGS $\mathcal{C} \in \mathcal{L}(\mathcal{D})$.*

Proof. The emptiness problem can be reduced to a bipartite parity game with $n\cdot(1+|\Sigma|)$ positions and c colors: Player *accept* owns the positions Q and chooses a label $\sigma \in \Sigma$. Player *reject* owns the resulting pairs $Q \times \Sigma$ and can move from a position (q,σ) with $\delta(q,\sigma) = \bigwedge_{\upsilon\in\Upsilon}(q_\upsilon,\upsilon)$ to a position q_υ (intuitively by choosing a direction $\upsilon \in \Upsilon$). The colors of the positions owned by player *accept* are defined by the coloring function α, while all states owned by player *reject* are colored by the minimum color in the mapping of α. This parity game can be solved in time $(n \cdot |\Sigma|)^{O(c)}$ [8].

$\mathcal{D}$ is empty iff player *reject* has a winning strategy, and the Σ-projection of a memoryless winning strategy for player *accept* defines a CGS in the language of $\mathcal{D}$. □

Combining Lemma 1, Theorem 2, and Lemmata 2, 3 and 4, we obtain the finite model property of automata over concurrent game structures.

Theorem 3. *Every non-empty ACG with n states, c colors, a atoms and a' agents accepts some finite CGS with $a \cdot a'$ directions and at most $n^{O(c^3\cdot n^2\cdot a^2\cdot a')}$ states, which can be constructed in time $n^{O(c^3\cdot n^2\cdot a^2\cdot a')}$.* □

Combining Theorem 3 with Theorem 1, we furthermore obtain the finite model property for the alternating-time μ-calculus:

Theorem 4. *Given an AMC formula φ with alternation depth d, n subformulas, and a agents, we can decide satisfiability of φ and, if φ is satisfiable, construct a model of φ in time $n^{O(d^3\cdot n^4\cdot a)}$.* □

Matching lower bounds for the AMC satisfiability and synthesis problems are given by the lower bounds for the classic μ-calculus [9,11].

Corollary 1. *The satisfiability and synthesis problems for the alternating-time μ-calculus are EXPTIME-complete.* □

References

1. R. Alur, T. A. Henzinger, and O. Kupferman. Alternating-time temporal logic. *Journal of the ACM*, 49(5):672–713, 2002.
2. R. Alur, T. A. Henzinger, F. Y. C. Mang, S. Qadeer, S. K. Rajamani, and S. Tasiran. Mocha: Modularity in model checking. In *Proc. CAV*, pages 521–525. Springer-Verlag, June 1998.
3. J. R. Büchi. On a decision method in restricted second order arithmetic. *Logic, Methodology and Philosophy of Science*, pages 1–11, 1962.
4. L. de Alfaro, T. A. Henzinger, and R. Majumdar. From verification to control: Dynamic programs for omega-regular objectives. In *Proc. LICS*, pages 279–290. IEEE Computer Society Press, June 2001.
5. E. A. Emerson and C. S. Jutla. Tree automata, μ-calculus and determinacy. In *Proc. FOCS*, pages 368–377. IEEE Computer Society Press, October 1991.
6. S. Even and Y. Yacobi. Relations among public key signature systems. Technical Report 175, Technion, Haifa, Israel, March 1980.

7. B. Finkbeiner and S. Schewe. Uniform distributed synthesis. In *Proc. LICS*, pages 321–330. IEEE Computer Society Press, June 2005.
8. M. Jurdziński. Small progress measures for solving parity games. In *Proc. STACS*, pages 290–301. Springer-Verlag, 2000.
9. D. Kozen and R. J. Parikh. A decision procedure for the propositional μ-calculus. In *Proc. Logic of Programs*, pages 313–325. Springer-Verlag, 1983.
10. S. Kremer and J.-F. Raskin. A game-based verification of non-repudiation and fair exchange protocols. *Journal of Computer Security*, 11(3):399–430, 2003.
11. O. Kupferman and M. Y. Vardi. μ-calculus synthesis. In *Proc. MFCS*, pages 497–507. Springer-Verlag, 2000.
12. O. Kupferman, M. Y. Vardi, and P. Wolper. An automata-theoretic approach to branching-time model checking. *Journal of the ACM*, 47(2):312–360, March 2000.
13. R. McNaughton. Testing and generating infinite sequences by a finite automaton. *Information and Control*, 9(5):521–530, October 1966.
14. D. E. Muller and P. E. Schupp. Simulating alternating tree automata by nondeterministic automata: new results and new proofs of the theorems of Rabin, McNaughton and Safra. *Theor. Comput. Sci.*, 141(1-2):69–107, 1995.
15. M. O. Rabin. *Automata on Infinite Objects and Church's Problem*, volume 13 of *Regional Conference Series in Mathematics*. Amer. Math. Soc., 1972.
16. S. Safra. On the complexity of the ω-automata. In *Proc. FoCS*, pages 319–327. IEEE Computer Society Press, 1988.
17. G. van Drimmelen. Satisfiability in alternating-time temporal logic. In *Proc. LICS*, pages 208–217. IEEE Computer Society Press, June 2003.
18. M. Y. Vardi and P. Wolper. Reasoning about infinite computations. *Journal of Information and Computation*, 115(1):1–37, May 1994.
19. D. Walther, C. Lutz, F. Wolter, and M. Wooldridge. Atl satisfiability is indeed exptime-complete. *Journal of Logic and Computation*, 2006. To appear.
20. T. Wilke. Alternating tree automata, parity games, and modal μ-calculus. *Bull. Soc. Math. Belg.*, 8(2), May 2001.

Space-Efficient Computation by Interaction
A Type System for Logarithmic Space

Ulrich Schöpp

Ludwig-Maximilians-Universität München
Oettingenstraße 67, D-80538 München, Germany
schoepp@tcs.ifi.lmu.de

Abstract. We introduce a typed functional programming language for logarithmic space. Its type system is an annotated subsystem of Hofmann's polytime LFPL. To guide the design of the programming language and to enable the proof of LOGSPACE-soundness, we introduce a realisability model over a variant of the Geometry of Interaction. This realisability model, which takes inspiration from Møller-Neergaard and Mairson's work on BC_{ε}^{-}, provides a general framework for modelling space-restricted computation.

1 Introduction

Many important complexity classes can be captured by programming languages and logics [11]. Such implicit characterisations of complexity classes are desirable for many reasons. For instance, one may want to avoid technical details in the construction and manipulation of Turing Machines, or one may want to get insight into which kinds of high-level programming principles are available in certain complexity classes. It has also become apparent that implicit characterisation of complexity classes can help in devising methods for analysing the resource-consumption of programs and in finding compilation strategies that guarantee certain resource bounds [19].

In this paper we address the question of how to design a functional programming language that captures LOGSPACE. Existing functional characterisations of LOGSPACE, such as [18,17,4,13,14], are all minimal by design. Although minimal languages are well-suited for theoretical study, they tend not to be very convenient for writing programs in. Here we take the first steps towards extending the ideas in loc. cit. to a less minimal programming language for LOGSPACE. One of the main issues in carrying out such an extension lies in the manageability of the computation model. While for the small languages in loc. cit. it is possible to show LOGSPACE-soundness by considering the whole language, this becomes increasingly difficult the more constructs are added to the programming language. Hence, an important goal in the design of a language for LOGSPACE is to capture its compilation in a modular, compositional way.

1.1 Modelling Space-Efficient Computation by Interaction

In this paper we give a compositional account of space-efficient computation, which is based on modelling computation by interaction. Our approach is

Z. Ésik (Ed.): CSL 2006, LNCS 4207, pp. 606–621, 2006.

motivated by the LOGSPACE-evaluation of Møller-Neergaard and Mairson's function algebra BC_ε^- [17]. The evaluation of BC_ε^- may be thought of as a question/answer dialogue, where a question to a natural number n is a number i, which represents the number of the bit to be computed, and an answer is the value of the bit at position i in the binary representation of n. Importantly, this approach admits a space-efficient implementation of recursion. Suppose $h\colon \mathbb{N} \to \mathbb{N}$ is a function we want to iterate k times. In general this will not be possible in LOGSPACE, as we cannot store intermediate results. However, it can be done if h has the special property that to compute one bit of its output only one bit of its input is needed. Suppose we want to compute bit i of $h^k(x)$. By the special property of h, it is enough to know at most one bit of $h^d(x)$ for each $d \leq k$. Suppose we already know the right bit of $h^d(x)$ for some $d < k$. To compute the bit of $h^{d+1}(x)$, we begin by asking $h^k(x)$ for bit i. This will result in a question for some bit of $h^{k-1}(x)$, then a question for some bit of $h^{k-2}(x)$, and so on until we reach $h^{d+1}(x)$. Finally, $h^{d+1}(x)$ will ask for one bit of $h^d(x)$, the value of which we already know. By the special property of h, we can thus compute the required bit of $h^{d+1}(x)$. By iteration of this process, the bits of $h^k(x)$ can be computed, assuming we can compute the bits of x. Moreover, this process of computing the bits of $h^k(x)$ can be implemented by storing only one bit of some $h^d(x)$, its recursion depth d, the initial question i and the current question. Under suitable assumptions, such as that k is polynomial in the size of the inputs, such an implementation will be in LOGSPACE.

In this paper we introduce a model for LOGSPACE-computation in which such an implementation of recursion is available. Our approach is based on modelling computation by question/answer-interaction. Such models have been studied extensively in the context of game semantics. We draw on this work, but since we are interested not just in modelling dialogues but also in effectively computing answers to questions, we are lead to considering the particular case of a Geometry of Interaction (GoI) situation, see [2,8] for a general description and [3] for a the connection to game semantics. In Sect. 3 we build a model on the basis of a GoI situation. By building on this well-studied structure, we can hope to benefit from existing work, such as that on the connections to abstract machines [6] or to machine code generation [16], although we have yet to explore the connections.

As an example of the kind of programming language that can be derived from the model, we introduce LogFPL, a simple functional language for LOGSPACE.

2 A Type System for Logarithmic Space

The type system for LogFPL is an annotated subsystem of LFPL [9]. In LogFPL all variables are annotated with elements of $Z{:=}\{1, \cdot, \infty\} \times \mathbb{N}$. The intended meaning of the annotations is that we may send arbitrarily many questions to each variable marked with ∞, that we may send at most one question to each variable marked with $\cdot$, and that we may send at most one question to *all* the variables marked with 1. The second component of the annotations specifies how many memory locations we may use when asking a question. We define an ordering on Z by letting $1 < \cdot < \infty$ and $\langle z, i\rangle \leq \langle z', i'\rangle \iff (z \leq z') \wedge (i \leq i')$. We define an addition on Z by $\langle m, i\rangle + \langle m', i'\rangle = \langle \max(m, m'), i + i'\rangle$.

A *context* Γ is a finite set of variable declarations $x \overset{z}{:} A$, where $z \in Z$, subject to the usual convention that each variable is declared at most once. For $z \in Z$, we write $!^z\Gamma$ for the context obtained from Γ by replacing each declaration $x \overset{u}{:} A$ with $x \overset{u+z}{:} A$. We write short $!\Gamma$ for $!^{\langle 1,1\rangle}\Gamma$. We write $\Gamma \geq z$ if $u \geq z$ holds for all $x \overset{u}{:} A$ in Γ.

$$
\begin{array}{rl}
\text{Small types} & A_S ::= I \mid \Diamond \mid \mathbf{B} \mid \mathbf{SN} \mid A_S * A_S \\
\text{Types} & A ::= A_S \mid \mathbf{L}(A_S) \mid A \times A \mid A \otimes_1 A \mid A \overset{\cdot,i}{\multimap} A \mid A \overset{\infty,i}{\multimap} A \\
\text{Terms} & M ::= x \mid c \mid * \mid \lambda x.\, M \mid M\ M \mid M \otimes_1 M \mid \text{let } M \text{ be } x \otimes_1 x \text{ in } M \\
& \quad \mid M * M \mid \text{let } M \text{ be } x * x \text{ in } M \mid \langle M, M\rangle \mid \pi_1(M) \mid \pi_2(M)
\end{array}
$$

The small types are the unit type I, the resource type $\Diamond$, as known from LFPL [9], the type of booleans $\mathbf{B}$, the type of small numbers $\mathbf{SN}$, and the type $A * B$ of pairs of small types. Small types have the property that their elements can be stored in memory. The type $\mathbf{SN}$, for example, contains natural numbers that are no larger than the size of the input, as measured by the number of $\Diamond$s. Using binary encoding, such numbers can be stored in memory. Small types are such that a single question suffices to get the whole value of an element. In contrast, one question to a list, for example, is a question to one of its elements. Thus, $\mathbf{SN}$ differs from the type $\mathbf{L}(I)$ of lists with unit-type elements.

The types are built starting from the small types and from lists $\mathbf{L}(A)$ over small types. The function space $\multimap$ has an annotation that indicates how many questions a function needs to ask of its argument in order to answer a query to its result. The second component of the annotations specifies how much data a function needs to store along with a question. We often write $\multimap$ for $\overset{\cdot,i}{\multimap}$ and $\overset{\infty}{\multimap}$ for $\overset{\infty,i}{\multimap}$ when i is not important. The type $A \otimes_1 B$ consists of pairs $x \otimes_1 y$, which we may use by asking one question *either* of x or of y. For instance, the type of the constant *cons* below expresses that one question to the list $cons(d, a, r)$ can be answered by asking one question of either d, a or r.

We have the following constants for booleans, lists and small numbers:

$$
\begin{array}{lll}
\text{Booleans} & \mathtt{tt}, \mathtt{ff} \colon \mathbf{B} & case_{\mathbf{B}} \colon \mathbf{B} \overset{\infty,k(A)}{\multimap} (A \times A) \overset{\cdot,0}{\multimap} A \\
\text{Lists} & nil \colon \mathbf{L}(A) & hdtl \colon \mathbf{L}(A) \overset{\cdot,1}{\multimap} \Diamond \otimes_1 A \otimes_1 \mathbf{L}(A) \\
& cons \colon \Diamond \otimes_1 A \otimes_1 \mathbf{L}(A) \overset{\cdot,1}{\multimap} \mathbf{L}(A) & empty \colon \mathbf{L}(A) \overset{\cdot,0}{\multimap} \mathbf{B} \\
\text{Small numbers} & zero \colon \mathbf{SN} & succ \colon \Diamond \overset{\cdot,0}{\multimap} \mathbf{SN} \overset{\cdot,0}{\multimap} \mathbf{SN}
\end{array}
$$

$$
case_{\mathbf{SN}} \colon \mathbf{SN} \overset{\infty,1+k(A)}{\multimap} (A \times (\Diamond \overset{\cdot,i}{\multimap} \mathbf{SN} \overset{\cdot,i}{\multimap} A)) \overset{\cdot,1}{\multimap} A
$$

The number $k(A)$ is defined in Fig 1. We note that *hdtl* represents a partial function undefined for *nil*. Furthermore, we have a constant $discard \colon A \overset{\cdot,0}{\multimap} I$ for each type A built without $\multimap$, and a constant $dup_i \colon (A \overset{\cdot,i}{\multimap} \mathbf{B}) \overset{\cdot,k(A)}{\multimap} A \overset{\cdot,0}{\multimap} (\mathbf{B} * A)$ for each small type A.

The typing rules appear in Figs. 1–2. The rules for small types are based on the fact that for small types we can get the whole value of an element with a single question. Rule ∞-·, for instance, expresses that instead of asking a small value many times, we may ask just one question, store the result and answer the many questions from memory.

2.1 Soundness and Completeness

Writing $\|M\|$ for the evident functional interpretation of a term M, we have:

Proposition 1 (Soundness). *If* $\vdash M\colon \mathbf{L}(I) \overset{\infty,i}{\multimap} \mathbf{L}(\mathbf{B}) \overset{\infty,i}{\multimap} \mathbf{L}(\mathbf{B})$ *is derivable, then there is a* LOGSPACE*-algorithm* e*, such that, for all* $x \in \mathbf{L}(I)$ *and all* $y \in \mathbf{L}(\mathbf{B})$*, if* $\|M\|(x, y)$ *is defined then* e *returns* $\|M\|(x, y)$ *on input* $\langle x, y\rangle$*.*

We construct the algorithm e in Sect. 3 by construction of a model.

First we sketch how LOGSPACE-predicates can be represented in LogFPL. Assume a LOGSPACE Turing Machine over a binary alphabet. By a suitable encoding of the input, we can assume that it never moves its input head beyond the end of the input.

We encode the computation of the Turing Machine. Its input tape is given as a binary string, i.e. as a list in $\mathbf{L}(\mathbf{B})$. We represent the position of the input head by a value in $\mathbf{SN}$. Access to the input tape is given by a zipper-like function *focus* of type $\mathbf{SN} \multimap \mathbf{L}(\mathbf{B}) \multimap \mathbf{SN} \otimes_1 \mathbf{L}(\mathbf{B}) \otimes_1 \mathbf{L}(\mathbf{B})$, with the following meaning:

$$\mathit{focus}\ i\ \langle x_0, \dots, x_k\rangle = i \otimes_1 \langle x_{i-1}, \dots, x_0\rangle \otimes_1 \langle x_i, \dots, x_k\rangle \qquad 0 \le i \le k$$

The function *focus* is defined as $\lambda l.\,(\mathit{rec}_{\mathbf{SN}}\ \mathit{base}\ \mathit{step})$ (omitting d for brevity).

$$l\colon \mathbf{L}(\mathbf{B}) \vdash \mathit{base}{:=}\mathit{zero} \otimes_1 \mathit{nil} \otimes_1 l\colon \mathbf{SN} \otimes_1 \mathbf{L}(\mathbf{B}) \otimes_1 \mathbf{L}(\mathbf{B})$$

$$\begin{array}{ll} & \lambda d.\,\lambda r.\,\text{let } r \text{ be } n \otimes_1 h \otimes_1 t \text{ in} \\ \vdash \mathit{step}{:=} & \qquad \text{let } (\mathit{hdtl}\ t) \text{ be } d' \otimes_1 a \otimes_1 t' \text{ in } (\mathit{succ}\ d\ n) \otimes_1 (\mathit{cons}\ d'\ a\ h) \otimes_1 t' \\ & \qquad\quad : \Diamond \multimap \mathbf{SN} \otimes_1 \mathbf{L}(\mathbf{B}) \otimes_1 \mathbf{L}(\mathbf{B}) \multimap \mathbf{SN} \otimes_1 \mathbf{L}(\mathbf{B}) \otimes_1 \mathbf{L}(\mathbf{B}) \end{array}$$

To represent the work tape we use small numbers $\mathbf{SN}$, since the work tape has only logarithmic size and since $\mathbf{SN}$ is much more flexible than lists. To encode the transitions of the TM, we use a number of helper functions. We encode 'bounded small numbers' by pairs in $\mathbf{SN}_B{:=}\mathbf{SN}{*}\mathbf{SN}$, where the first component represents the value of the number and the second component contains memory (in the form of $\Diamond$) that may be used for increasing the number. For instance, the incrementation function maps $m{*}\mathit{zero}\colon \mathbf{SN}_B$ to $m{*}\mathit{zero}$ and $m{*}\mathit{succ}(d, n)$ to $\mathit{succ}(d, m){*}n$. Using the rules for small types, we can represent the evident functions $\mathit{null}\colon \mathbf{SN} \multimap \mathbf{SN}_B$, $\mathit{inc}\colon \mathbf{SN}_B \multimap \mathbf{SN}_B$, $\mathit{dec}\colon \mathbf{SN}_B \multimap \mathbf{SN}_B$, $\mathit{double}\colon \mathbf{SN}_B \multimap \mathbf{SN}_B$, $\mathit{half}\colon \mathbf{SN}_B \multimap \mathbf{SN}_B$ and $\mathit{even}\colon \mathbf{SN}_B \multimap \mathbf{B}{*}\mathbf{SN}_B$.

We represent a state of the Turing Machine by a 4-tuple $l{*}r{*}i{*}s$ of type $\mathbf{SN}_B{*}\mathbf{SN}_B{*}\mathbf{SN}_B{*}\mathbf{B}^k$, where l and r represent the parts of the work tape left and right from the work head, i represents the position of the input head and s represents the state of the machine. We abbreviate $\mathbf{SN}_B{*}\mathbf{SN}_B{*}\mathbf{SN}_B{*}\mathbf{B}^k$ by S. It should be clear how to use the above helper functions for implementing the basic operations of a Turing Machine. For instance, moving the head on

$$\frac{A' \le A \quad B \le B' \quad u \le v}{(A \overset{u}{\multimap} B) \le (A' \overset{v}{\multimap} B')} \qquad \frac{A \le A' \quad B \le B'}{(A \otimes_1 B) \le (A' \otimes_1 B')}$$

$$\textsc{Axiom}\ \frac{}{x \overset{z}{:} A \vdash x\colon A} \qquad \textsc{Const}\ \frac{c\colon A}{\vdash c\colon A} \qquad \textsc{Sub}\ \frac{\Gamma \vdash M\colon A \quad A \le B}{\Gamma \vdash M\colon B}$$

$$I\mathrm{I}\ \frac{}{\vdash *\colon I} \qquad I\mathrm{E}\ \frac{\Gamma \vdash M\colon I \quad \Delta \vdash N\colon A}{\Gamma, \Delta \vdash \text{let } M \text{ be } * \text{ in } N\colon A}$$

$$\multimap\mathrm{I}\ \frac{\Gamma, x \overset{z}{:} A \vdash M\colon B}{\Gamma \vdash \lambda x.\, M\colon A \overset{z}{\multimap} B}\ z \ge \langle \cdot, 0\rangle \qquad \multimap\mathrm{E}\ \frac{\Gamma \vdash M\colon A \overset{z}{\multimap} B \quad \Delta \vdash N\colon A}{\Gamma, !^{z}\Delta \vdash M\, N\colon B}$$

$$\otimes_1\mathrm{I}\ \frac{\Gamma \vdash M\colon A \quad \Delta \vdash N\colon B}{\Gamma, \Delta \vdash M \otimes_1 N\colon A \otimes_1 B} \qquad \otimes_1\mathrm{E}\ \frac{\Gamma \vdash M\colon A \otimes_1 B \quad \Delta, x \overset{\langle 1,i\rangle}{:} A, y \overset{\langle 1,i\rangle}{:} B \vdash N\colon C}{\Delta, !^{\langle 1,i\rangle}\Gamma \vdash \text{let } M \text{ be } x \otimes_1 y \text{ in } N\colon C}$$

$$\times\mathrm{I}\ \frac{\Gamma \vdash M\colon A \quad \Gamma \vdash N\colon B}{!^{\langle\infty,1\rangle}\Gamma \vdash \langle M, N\rangle\colon A \times B} \qquad \times\mathrm{E1}\ \frac{\Gamma \vdash M\colon A \times B}{\Gamma \vdash \pi_1(M)\colon A} \qquad \times\mathrm{E2}\ \frac{\Gamma \vdash M\colon A \times B}{\Gamma \vdash \pi_2(M)\colon B}$$

Recursion types $R ::= A_S \mid \mathbf{L}(\mathbf{B}) \mid R \otimes_1 R$

$$\mathrm{R}\mathbf{L}\ \frac{\Gamma \vdash g\colon B \quad \vdash f\colon (B \overset{\cdot,j}{\multimap} \mathbf{B}^k) \overset{\infty,i}{\multimap} \Diamond \overset{\cdot,i}{\multimap} A \overset{\infty,i}{\multimap} B \overset{\cdot,i}{\multimap} B \quad \Delta \vdash d\colon B \overset{\cdot,j}{\multimap} \mathbf{B}^k}{\Gamma, !^{\langle\infty,s(i,B)\rangle}\Delta \vdash \mathit{rec}_{\mathbf{L}(A)}\ g\ f\ d\colon \mathbf{L}(A) \xrightarrow{\infty,s(i,B)}\!\!\circ\ B}\ B \in R$$

$$\mathrm{R}\mathbf{SN}\ \frac{\Gamma \vdash g\colon B \quad \vdash f\colon (B \overset{\cdot,j}{\multimap} \mathbf{B}^k) \overset{\infty,i}{\multimap} \Diamond \overset{\cdot,i}{\multimap} B \overset{\cdot,i}{\multimap} B \quad \Delta \vdash d\colon B \overset{\cdot,j}{\multimap} \mathbf{B}^k}{\Gamma, !^{\langle\infty,s(i,B)\rangle}\Delta \vdash \mathit{rec}_{\mathbf{SN}}\ g\ f\ d\colon \mathbf{SN} \xrightarrow{\cdot,s(i,B)}\!\!\circ\ B}\ B \in R$$

Here, $s(i, A) = i+4+7\cdot k(A)$, where $k(A) = 1$ for $A \in \{I, \Diamond, \mathbf{B}, \mathbf{SN}\}$, $k(\mathbf{L}(A)) = 2+k(A)$ and $k(A*B) = k(A \times B) = k(A \otimes_1 B) = k(A \overset{z}{\multimap} B) = 1 + k(A) + k(B)$. The definition of $k(A)$ is such that any type A is $k(A)$-encodable, see Def. 6.

Fig. 1. General typing rules

In the following rules, A, B and C must be small types.

$$*\mathrm{I}\ \frac{\Gamma \vdash M\colon A \quad \Delta \vdash N\colon B}{!^{k(B)}\Gamma, !^{k(A)}\Delta \vdash M*N\colon A*B}\ \Gamma \ge \langle\cdot,0\rangle \vee \Delta \ge \langle\cdot,0\rangle$$

$$*\mathrm{E}\ \frac{\Gamma \vdash M\colon A * B \quad \Delta, x \overset{\langle\cdot,i\rangle}{:} A, y \overset{\langle\cdot,i\rangle}{:} B \vdash N\colon C}{!^{k(A*B)}\Delta, !^{\langle\cdot,i+k(C)\rangle}\Gamma \vdash \text{let } M \text{ be } x*y \text{ in } N\colon C}$$

$$\infty\text{-}\cdot\ \frac{\Delta, x \overset{\langle\infty,i\rangle}{:} A \vdash M\colon B}{!^{k(A)}\Delta, x \overset{\langle\cdot,i+k(B)\rangle}{:} A \vdash M\colon B}$$

Fig. 2. Special rules for small types

the work tape to the right is given by mapping a tuple $l{*}r{*}i{*}s$ to the value (let $even(r)$ be $e{*}r$ in $(case_{\mathbf{B}}\ e\ \langle double(l){*}half(r){*}i{*}s, inc(double(l)){*}half(r){*}i{*}s\rangle))$. If, in addition to a 4-tuple in S, we have a function $d\colon \mathbf{SN} \multimap \mathbf{B}$, such that $d(n)$ is the n-th character of the input tape, then, using *dup*, we can thus implement the transition function of the Turing Machine. Hence, we can give the transition function the type $h\colon (\mathbf{SN} \multimap \mathbf{B}) \overset{\infty}{\multimap} S \multimap S$. Let $g\colon S$ be the initial state of the machine. Since *focus* can be used to define an access function $t\colon \mathbf{L}(\mathbf{B}) \vdash d\colon \mathbf{SN} \multimap \mathbf{B}$, we can model the computation of the machine by $g\colon S, t \overset{\infty}{:} \mathbf{L}(\mathbf{B}) \vdash rec_{\mathbf{SN}}\ g\ h\ d\colon \mathbf{SN} \multimap S$. Now, in order to construct the initial state g as well as an upper bound on the computation length, we need a polynomial number of $\Diamond$s. We obtain these from a given list $\mathbf{L}(I)$, which we split in four numbers in $\mathbf{SN}$ of equal size. Thus, we obtain a term $s \overset{\infty}{:} \mathbf{L}(I), t \overset{\infty}{:} \mathbf{L}(\mathbf{B}) \vdash m\colon S$, which, if s is large enough, computes the final state of the TM for input t.

Proposition 2 (Completeness). *Each* LOGSPACE*-predicate* $A \subseteq \mathbf{L}(\mathbf{B})$ *can be represented by a term* $M\colon \mathbf{L}(I) \overset{\infty,i}{\multimap} \mathbf{L}(\mathbf{B}) \overset{\infty,i}{\multimap} \mathbf{B}$ *in the following sense. There exist natural numbers* n *and* m*, such that, for each* $a \in \mathbf{L}(\mathbf{B})$ *and each list* $s \in \mathbf{L}(I)$ *that is longer than* $|a|^n + m$*, we have* $\|M\|(a,s) = \mathtt{tt}$ *if* $a \in A$ *and* $\|M\|(a,s) = \mathtt{ff}$ *if* $a \notin A$.

3 Modelling Space-Efficient Computation by Interaction

We compile LogFPL to LOGSPACE-algorithms by interpreting it in an instance of the Geometry of Interaction situation [2]. We use an instance of the GoI situation in which questions can be answered in linear space. This is motivated by the fact that questions will typically be of logarithmic size (think of the bit-addresses of an input number), so that to remain in LOGSPACE we can allow linear space in the size of the questions.

3.1 Linear Non-size-increasing Functions

The underlying computational model is that of non-size-increasing linear space functions. The restriction to non-size-increasing functions is needed for composition in $\mathcal{G}$, defined below, to remain in linear space. To fix the computation model, we work with multi-tape Turing Machines over some alphabet Σ. The machines take their input on one designated tape, where they also write the output.

The objects of $\mathcal{L}$ are triples (X, c, l), where X is a underlying set, $c\colon X \to \Sigma^*$ is a coding function and $l\colon X \to \mathbb{N}$ is an abstract length measure. An object must be so that there exist constants $m, n \in \mathbb{N}$ such that $\forall x \in X.\, m{\cdot}l(x){+}n \leq |c(x)|$ holds, i.e. the abstract length measure underestimates the actual size at most linearly. A morphism from (X, c_X, l_X) to (Y, c_Y, l_Y) is a partial function $f\colon X \to Y$ with the property $\forall x \in X.\, f(x)\downarrow \implies l_Y(f(x)) \leq l_X(x)$, for which in addition there exists a linear space algorithm e satisfying $\forall x \in X.\, f(x)\downarrow \implies e(c_X(x)) = c_Y(f(x))$.

The category $\mathcal{L}$ supports a number of data type constructions, much in the style of LFPL [9]. We use the following constructions, of which we spell out only the underlying sets and the length functions. The disjoint union $A + B$ of the underlying sets of A and B becomes an object with $l_{A+B}(inl(a)) = l_A(a)$ and $l_{A+B}(inr(b)) = l_B(b)$. The object $A \otimes B$ has as underlying set the set of pairs of elements of A and B with length function $l_{A\otimes B}(\langle a, b\rangle) = l_A(a) + l_B(b)$. The evident projections $\pi_1 : A \otimes B \to A$ and $\pi_2 : A \otimes B \to B$ are morphisms of $\mathcal{L}$. We also have booleans $\mathbf{B} = \{\mathtt{ff}, \mathtt{tt}\}$ with $l_{\mathbf{B}}(\mathtt{ff}) = l_{\mathbf{B}}(\mathtt{tt}) = 0$, lists $\mathbf{L}(A) = A^*$ with $l_{\mathbf{L}(A)}(\langle a_1, \dots, a_n\rangle) = \sum_{i=1}^{n} 1 + l_A(a_i)$, and trees $\mathbf{T}(A) = A + \mathbf{T}(A) \otimes \mathbf{T}(A)$ with $l_{\mathbf{T}(A)}(inl(a)) = l_A(a)$ and $l_{\mathbf{T}(A)}(inr(a, a')) = 1 + l_{\mathbf{T}(A)}(a) + l_{\mathbf{T}(A)}(a')$. An abstract resource type $\Diamond$ is given by $\Diamond = \{\Diamond\}$ with $l(\Diamond) = 1$. It differs from the unit object $1 = \{*\}$ with $l(*) = 0$ only in its length measure.

Given $f: A+B \to C+B$, we define its trace $tr(f): A \to C$ by $tr(f) = t \circ inl$, where $t: A + B \to C$ is defined by $t(x) = c$ whenever $f(x) = inl(c)$ holds and $t(x) = t(f(x))$ whenever $f(x) = inl(b)$ holds for some b. Making essential use of the fact that all morphisms in $\mathcal{L}$ are non-size-increasing and that the length measure underestimates the real size at most linearly, it follows that $tr(f)$ is again a morphism in $\mathcal{L}$. That the definition of $tr(f)$ satisfies the equations for trace, see e.g. [8, Def. 2.1.16], can be seen by observing that the forgetful functor from $\mathcal{L}$ to the traced monoidal category **Pfn** of sets and partial functions, see e.g. [8, Sec. 8.2], is faithful and preserves both finite coproducts and trace.

3.2 Linear-Space Interaction

Computation by interaction is modelled by a GoI situation over $\mathcal{L}$. For space reasons, we can only give the basic definitions. We refer to [2,8] for more details.

The category $\mathcal{G}$ has as objects pairs (A^+, A^-) of two objects of $\mathcal{L}$. Here, A^- is thought of as a set of questions and A^+ as a set of answers. A morphisms in $\mathcal{G}$ from (A^+, A^-) to (B^+, B^-) is a $\mathcal{L}$-morphism of type $A^+ + B^- \to A^- + B^+$. We will often use these two views of morphisms interchangeably. We write $\mathcal{G}(A, B)$ for the set of morphisms in $\mathcal{G}$ from A to B. The identity on A is given by $[inr, inl]: A^+ + A^- \to A^- + A^+$. Composition $g \cdot f$ of $f: A \to B$ and $g: B \to C$ is given by the trace of $(A^- + g) \circ (f + C^-): A^+ + B^- + C^- \to A^- + B^- + C^+$ with respect to B^-.

Of the structure of $\mathcal{G}$, we spell out the symmetric monoidal structure $\otimes$, given on objects by $A \otimes B = (A^+ + B^+, A^- + B^-)$ and on morphisms by using $+$ on the underlying $\mathcal{L}$-morphisms. A unit for $\otimes$ is $I = (\emptyset, \emptyset)$. Note $I \otimes I = I$. Note also that morphisms of type $I \to A$ in $\mathcal{G}$ are the same as $\mathcal{L}$-maps of type $A^- \to A^+$ and that morphisms of type $A \to I$ are the same as $\mathcal{L}$-maps of type $A^+ \to A^-$.

A monoidal closed structure $\multimap$ is defined on objects by $(A \multimap B) = (B^+ + A^-, B^- + A^+)$. We write $\varepsilon: (A \multimap B) \otimes A \to B$ for the application map and $\Lambda f: A \to (B \multimap C)$ for the abstraction of $f: A \otimes B \to C$.

We use a storage functor $!(-)$. For an object A, the object $!A$ is given by $(A^+ \otimes \mathbf{S}, A^- \otimes \mathbf{S})$, where $\mathbf{S} = \mathbf{B} \otimes \mathbf{T}(\mathbf{L}(\mathbf{B}))$. The intention is that a store in $\mathbf{S}$ is passed along with questions and answers. For each morphism $f: A \to B$, the morphism $!f: !A \to !B$ is given by the $\mathcal{L}$-map $f \otimes \mathbf{S}$.

3.3 Realisability and Co-realisability

As a way of formalising the compilation of functions into interaction-programs, we define a category $\mathcal{R}$. Its definition is parameterised over a commutative monoid $(\mathcal{M}, +, 0)$, equipped with a pre-order $\leq$ that is compatible with $+$.

Objects. The objects are tuples $(|A|, A^*, A_\mathcal{G}, \Vdash, \Vdash^*)$ consisting of a set $|A|$ of underlying elements, a set A^* of underlying co-elements, an object $A_\mathcal{G}$ of $\mathcal{G}$, a realisation relation $\Vdash \subseteq \mathcal{M} \times \mathcal{G}(I, A_\mathcal{G}) \times |A|$ and a co-realisation relation $\Vdash^* \subseteq \mathcal{M} \times \mathcal{G}(A_\mathcal{G}, I) \times A^*$. The objects are required to have the following properties.

1. For all $a \in |A|$ there exist α and e such that $\alpha, e \Vdash a$ holds. Dually, for all $a^* \in A^*$ there exist α and c such that $\alpha, c \Vdash^* a^*$ holds.
2. If $\alpha, e \Vdash a$ and $\alpha \leq \beta$ hold then so does $\beta, e \Vdash a$. Dually, if $\alpha, e \Vdash^* a^*$ and $\alpha \leq \beta$ hold then so does $\beta, e \Vdash^* a^*$.

Morphisms. A morphism from A to B consists of two functions $f\colon |A| \to |B|$ and $f^*\colon B^* \to A^*$ for which there exists a map $r\colon A_\mathcal{G} \to B_\mathcal{G}$ in $\mathcal{G}$ satisfying:

$$\forall \alpha \in \mathcal{M},\, e\colon I \to A_\mathcal{G},\, x \in A.\quad \alpha, e \Vdash_A x \implies \alpha, r \cdot e \Vdash_B f(x),$$
$$\forall \beta \in \mathcal{M},\, c\colon B_\mathcal{G} \to I,\, k \in B^*.\quad \beta, c \Vdash^*_B k \implies \beta, c \cdot r \Vdash^*_A f^*(k).$$

We also need the following more general form of morphism realisation.

Definition 1. *A morphism* $r\colon A_\mathcal{G} \to B_\mathcal{G}$ realises (f, f^*) with bound $\varphi \in \mathcal{M}$ *if*

$$\forall \alpha \in \mathcal{M},\, e\colon I \to A_\mathcal{G},\, x \in A.\quad \alpha, e \Vdash_A x \implies \varphi + \alpha, r \cdot e \Vdash_B f(x),$$
$$\forall \beta \in \mathcal{M},\, c\colon B_\mathcal{G} \to I,\, k \in B^*.\quad \beta, c \Vdash^*_B k \implies \varphi + \beta, c \cdot r \Vdash^*_A f^*(k).$$

While our definition of realisability is close to well-known instances of realisability, such as e.g. [5,15], the definition of co-realisability deserves comment. The main purpose for introducing it is for modelling recursion in the way described in the introduction. To implement recursion in this way, we need to control how often a realiser $r\colon A_\mathcal{G} \to B_\mathcal{G}$ of the recursion step-function sends a question to its argument. Co-Realisability is a way of obtaining control over how r uses $A_\mathcal{G}$. Suppose, for example, that both A and B have only a single co-element that is co-realised by the empty function $\emptyset$. If r realises a morphism $A \to B$ then $\emptyset \cdot r$ must also be the empty function. Hence, whenever r receives an answer in $A^+_\mathcal{G}$, it cannot ask another question in $A^-_\mathcal{G}$, since otherwise $\emptyset \cdot r$ would not be empty. Thus, for any $e\colon I \to A_\mathcal{G}$ and any $q \in B^-_\mathcal{G}$, in the course of the computation of $(r \cdot e)(q)$, only one query can be sent by r to e. This is the main example of how we control the behaviour of realisers with co-realisability. More generally, co-realisability formalises how an object A may be *used*. A co-realiser $c\colon A^+_\mathcal{G} \to A^-_\mathcal{G}$ explains which question we may ask after we have received an answer from A.

A symmetric monoidal structure $\otimes$ on $\mathcal{R}$ is defined by letting the underlying set of $|A \otimes B|$ be the set of pairs in $|A| \times |B|$. The realising object $(A \otimes B)_\mathcal{G}$ is $A_\mathcal{G} \otimes B_\mathcal{G}$, and the realisation relation is the least relation satisfying

$$(\alpha, e_x \Vdash_A x) \wedge (\beta, e_y \Vdash_B y) \implies \alpha + \beta, e_x \otimes e_y \Vdash_{A \otimes B} \langle x, y\rangle$$

in addition to the general requirements for objects of $\mathcal{R}$ (which from now on we will tacitly assume). The set of co-elements and the co-realisation is defined by:

$$(A \otimes B)^* = \{\langle p \colon |A| \to B^*, q \colon |B| \to A^* \rangle \mid \exists c, \gamma.\, (\gamma, c \Vdash^*_{A \otimes B} \langle p, q \rangle)\}$$

$$\begin{aligned} \gamma, c \Vdash^*_{A\otimes B} \langle p, q\rangle \iff & (\forall \alpha, e, a.\, (\alpha, e \Vdash_A a) \implies (\gamma + \alpha), c \cdot (e \otimes id) \Vdash^*_B p(a)) \\ & \wedge (\forall \beta, e, b.\, (\beta, e \Vdash_B b) \implies (\gamma + \beta), c \cdot (id \otimes e) \Vdash^*_A q(b)) \end{aligned}$$

A unit object for $\otimes$ may be defined by $|I| = \{*\}$, $I^* = \{*\}$, $I_\mathcal{G} = (1, 1)$, by letting $\alpha, e \Vdash_I *$ hold if and only if e is the unique total function of its type, and by letting $\alpha, c \Vdash^*_I *$ hold if and only if $c \colon 1 \to 1$ is the empty function.

A monoidal exponent $\multimap$ for $\otimes$ is defined by letting $|A \multimap B|$ be the set of pairs $\langle f, f^* \rangle$ in $(|A| \to |B|) \times (B^* \to A^*)$ that are realised by some r with some bound φ, by letting $(A \multimap B)^* = |A| \times B^*$ and $(A \multimap B)_\mathcal{G} = A_\mathcal{G} \multimap B_\mathcal{G}$, and by letting the relations $\Vdash_{A \multimap B}$ and $\Vdash^*_{A \multimap B}$ be the least relations satisfying

$$\varphi, \Lambda r \Vdash_{A \multimap B} f \iff r \colon A_\mathcal{G} \to B_\mathcal{G} \text{ realises } f \text{ with bound } \varphi,$$
$$(\alpha, e \Vdash_A a) \wedge (\beta, c \Vdash^*_B b^*) \implies \alpha + \beta, c \cdot \varepsilon \cdot (id \otimes e) \Vdash^*_{A \multimap B} \langle a, b^* \rangle.$$

The co-realisation of $\otimes$ is such that both components of a pair $A \otimes B$ can be used sequentially. This is not the only possible choice. Another useful choice is captured by the monoidal structure $\otimes_1$, where $|A \otimes_1 B| = |A \otimes B|$, $(A \otimes_1 B)_\mathcal{G} = (A \otimes B)_\mathcal{G}$, $\Vdash_{A \otimes_1 B} = \Vdash_{A \otimes B}$, $(A \otimes_1 B)^* = A^* \times B^*$, and where $\Vdash^*_{A \otimes_1 B}$ is the smallest relation satisfying $(\alpha, c \Vdash^*_A a^*) \wedge (\beta, c' \Vdash^*_B b^*) \implies (\alpha + \beta, c \otimes c' \Vdash^*_{A \otimes_1 B} \langle a^*, b^* \rangle)$. It is instructive to consider the case of morphisms $f \colon A \otimes_1 B \to C$, where A and B have only one co-element that is co-realised by the empty function. In this case, the realiser r of f may ask *either* one question from A *or* one from B. This is in contrast to $\otimes$, where it would be legal to ask one question from A and then another from B.

Lemma 1. *There exists a natural map* $d \colon A \otimes (B \otimes_1 C) \to B \otimes_1 (A \otimes C)$ *with* $d(a, \langle b, c \rangle) = \langle b, \langle a, c \rangle \rangle$ *and* $d^*(b^*, \langle p, q \rangle) = \langle \lambda a.\, \langle b^*, p(a) \rangle, \lambda \langle b, c \rangle.\, q(c) \rangle$.

A further monoidal structure $\times$ is defined by $|A \times B| = |A| \times |B|$, $(A \times B)_\mathcal{G} = A_\mathcal{G} \otimes B_\mathcal{G}$, $(A \times B)^* = A^* + B^*$, where $\Vdash_{A \times B}$ and $\Vdash^*_{A \times B}$ are the least relations satisfying

$$\begin{aligned} (\alpha, e_a \Vdash_A a) \wedge (\alpha, e_b \Vdash_B b) &\implies \alpha, e_a \otimes e_b \Vdash_{A \times B} \langle a, b \rangle, \\ (\alpha, c_a \Vdash^*_A a^*) &\implies (\alpha, c_a \circ \pi_1 \Vdash^*_{A \times B} inl(a^*)), \\ (\beta, c_b \Vdash^*_B b^*) &\implies (\beta, c_b \circ \pi_2 \Vdash^*_{A \times B} inr(b^*)). \end{aligned}$$

The object $A \times B$ is *not* quite a cartesian product, as is witnessed by rule $\times$I, which has the context $!^{\langle \infty, 1 \rangle} \Gamma$ rather than Γ in its conclusion.

For the implementation of recursion, which requires that there is at most one query to the recursion argument, the following smash-product $*$ is useful. It is defined by $|A * B| = |A| \times |B|$, $(A * B)_\mathcal{G} = (A^+_\mathcal{G} \otimes B^+_\mathcal{G}, A^-_\mathcal{G} \otimes B^-_\mathcal{G})$, $(A * B)^* = A^* \times B^*$, $\Vdash^*_{A * B} = \Vdash^*_{A \otimes_1 B}$,

$$\gamma, e \Vdash_{A * B} \langle x, y \rangle \iff \exists \alpha, \beta, e_x, e_y.\, \begin{array}{l} (\alpha + \beta \le \gamma) \wedge (\alpha, e_x \Vdash_A x) \wedge (\beta, e_y \Vdash_B y) \\ \wedge \forall q_A, q_B.\, (e(q_A, q_B) = \langle e_x(q_A), e_y(q_B) \rangle). \end{array}$$

We use the smash-product for example for booleans, where the full information of $\mathbf{B} * \mathbf{B}$ can be obtained with a single question. Notice that with $\mathbf{B} \otimes \mathbf{B}$ we would need two questions, while with $\mathbf{B} \otimes_1 \mathbf{B}$ we could only ask for one component.

Often it is useful to remove the co-realisation-information. This can be done with the co-monad $\Box(-)$ defined by $|\Box A| = |A|$, $\Box A^* = \mathcal{G}(A_\mathcal{G}, I)$, $(\Box A)_\mathcal{G} = A_\mathcal{G}$, $\Vdash_{\Box A} = \Vdash_A$ and $(\alpha, c \Vdash^*_{\Box A} c') \iff c = c'$.

As the final general construction on $\mathcal{R}$, we define a lifting monad $(-)_\perp$, which we use for modelling partial functions. It is defined by $|A_\perp| = |A| + \{\perp\}$, ${A_\perp}^* = A^*$, $(A_\perp)_\mathcal{G} = A_\mathcal{G}$, $\Vdash^*_{A_\perp} = \Vdash^*_A$, $(\alpha, e \Vdash_{A_\perp} a \in |A|) \iff (\alpha, e \Vdash_A a)$, and $(\alpha, e \Vdash_{A_\perp} \perp)$ always. There are evident morphisms $A \to A_\perp$, $(A_\perp)_\perp \to A_\perp$, $A \bullet (B_\perp) \to (A \bullet B)_\perp$ for any $\bullet \in \{\otimes, \otimes_1, \times, *\}$, and $\Box(A_\perp) \to (\Box A)_\perp$.

We remark that the definition of $\mathcal{R}$ and the construction of its structure are very similar to the double glueing construction of Hyland and Schalk [12].

3.4 An Instance for Logarithmic Space

We consider $\mathcal{R}$ with respect to the monoid $\mathcal{M} = \{\langle l, k, m\rangle \in \mathbb{N}^3 \mid l \leq m\}$ with addition $\langle l, k, m\rangle + \langle l', k', m'\rangle = \langle l + l', \max(k, k'), \max(l + l', m, m')\rangle$. The neutral element 0 is $\langle 0, 0, 0\rangle$. For the ordering we use $\langle l, k, m\rangle \leq \langle l', k', m'\rangle \iff (l \leq l') \wedge (k \leq k') \wedge (m \leq m')$. The intended meaning of a triple $\langle l, k, m\rangle$ is that l is the abstract length of an object and $\langle k, m\rangle$ is a bound on the additional memory a realiser may use. We will allow realisers to use k numbers with range $\{0, \ldots, m\}$.

We now consider the structure of $\mathcal{R}$ with respect to $\mathcal{M}$, starting with an implementation of the base types of LogFPL.

Definition 2. *An object A is* simple *if it enjoys the following properties.*

1. *Whenever $\alpha, e \Vdash a$ holds, then there is a $\mathcal{L}$-map $e' \colon A^+_\mathcal{G} \to A^-_\mathcal{G}$ with $e' \circ e = id$.*
2. *A^* is a singleton and whenever $\alpha, c \Vdash^* a^*$ holds, then c is the empty function and $0, c \Vdash^* a^*$ holds as well.*
3. *Both $A^-_\mathcal{G}$ and $A^+_\mathcal{G}$ have at least one element x with $l(x) = 0$.*

We note that if A and B are simple then so are $A \otimes_1 B$ and $A * B$, but not in general $A \otimes B$.

All the basic data types we now define are simple. Since the co-realisation relation is uniquely determined by the definition of simpleness, we just show the realisation part.

Diamond. $|\Diamond| = \{\Diamond\}$, $\Diamond_\mathcal{G} = I_\mathcal{G}$, $(\alpha + \langle 1, 0, 1\rangle, e \Vdash_\Diamond \Diamond) \iff (\alpha, e \Vdash_I *)$

Booleans. $|\mathbf{B}| = \{\mathtt{ff}, \mathtt{tt}\}$, $\mathbf{B}_\mathcal{G} = (\{\mathtt{ff}, \mathtt{tt}\}, 1)$, $(\alpha, e \Vdash_\mathbf{B} b) \iff (e(*) = b)$

Small Numbers. $|\mathbf{SN}| = \mathbb{N}$, $\mathbf{SN}_\mathcal{G} = (\mathbf{L}(\mathbf{B}), \mathbf{L}(1))$ and $(\langle l, k, m\rangle, e \Vdash_\mathbf{SN} n)$ holds if and only if both $(l \geq n)$ holds and $e(s)$ equals the last s bits of n in binary.

Lists. Let A be a simple object. Define the simple object $\mathbf{L}(A)$ to have as underlying set $|\mathbf{L}(A)|$ the set of finite lists on $|A|$. The object $\mathbf{L}(A)_\mathcal{G}$ is given by $((A^+_\mathcal{G} + A^-_\mathcal{G}) \otimes \mathbf{L}(\mathbf{B}), A^-_\mathcal{G} \otimes \mathbf{L}(\mathbf{B}))$. The intention of $A^-_\mathcal{G} \otimes \mathbf{L}(\mathbf{B})$ is that a question consists of a pointer into the list, encoded in binary (without leading zeros) as an element of $\mathbf{L}(\mathbf{B})$, together with a question for the element at the position pointed at.

$$\alpha, e \Vdash \langle a_0, \ldots, a_n \rangle \iff \begin{array}{l} \exists \vec{\alpha} \in \mathcal{M}^{n+1}. \exists \vec{e} \in (I \to A_{\mathcal{G}})^{n+1}. \\ \quad (\alpha \geq \langle n+1, 0, n+1 \rangle + \alpha_0 + \cdots + \alpha_n) \\ \quad \wedge (\forall i.\ i \leq n \implies \alpha_i, e_i \Vdash_A a_i) \\ \quad \wedge (\forall i, q.\ i \leq n \implies e(q,i) = \langle inl(e_i(q)), i \rangle) \\ \quad \wedge (\forall i, q.\ i > n \implies e(q,i) = \langle inr(q), i \rangle) \end{array}$$

We note that $\mathbf{L}(A)$ as such is not yet very useful. For instance, there is no map *tail*: $\mathbf{L}(A) \to \mathbf{L}(A)$, as we do not always have enough space to map a question $\langle q_A, i \rangle$ to $\langle q_A, i+1 \rangle$. We address this problem with the construction M below.

Data storage. Often when passing a question to an object, we also need to store some data that we need again once the answer arrives. Such data storage is captured by the functor $!(-)$ defined as follows. The set of underlying elements of $!A$ is inherited from A, i.e. $|!A| = |A|$. The realising object $(!A)_{\mathcal{G}}$ is $!(A_{\mathcal{G}})$ and the realisation relation is the smallest realisation relation satisfying $\langle l, k, m \rangle, e \Vdash_A a \implies \langle l, k+1, m \rangle, !e \Vdash_{!A} a$ in addition to the general requirements for objects in $\mathcal{R}$. The set of co-elements is defined by $(!A)^* = |\mathbf{S}| \to A^*$ and the co-realisation relation on $!A$ is given by

$$\alpha, c \Vdash^*_{!A} a^* \iff \forall s \in |\mathbf{S}|. \exists c_s. (\alpha, c_s \Vdash^*_A a^*(s)) \wedge (\forall q. c(q,s) = \langle c_s(q), s \rangle).$$

While there is a natural dereliction map $!A \to A$, there is no digging map $!A \to !!A$, since we do not have an additional $\Diamond$ that we would need to encode two trees in one.

Lemma 2. *There are natural transformations* $\Box !A \to !\Box A$, $(!A \otimes !B) \to !(A \otimes B)$ *and* $!(A_\perp) \to (!A)_\perp$ *as well as a natural isomorphism* $(!A \otimes_1 !B) \cong !(A \otimes_1 B)$.

Memory allocation. We have defined the monoid $\mathcal{M}$ with the intuition that if e realises some element with bound $\langle l, k, m \rangle$ then e can use k memory locations of size $\log(m) + 1$. However, the above definition of the data types is such that, besides the question itself, no memory can be assumed. We now add a memory supply to the data types.

Let A be a simple object such that for each $a \in |A|$ there exist l, m and e satisfying $\langle l, 0, m \rangle, e \Vdash_A a$. To define a simple object MA, let $|MA| = |A|$, $(MA)^* = A^*$ and $(MA)_{\mathcal{G}} = (A^+_{\mathcal{G}} \otimes \mathbf{L}(\mathbf{L}(1)), A^-_{\mathcal{G}} \otimes \mathbf{L}(\mathbf{L}(1)))$. The intended meaning of $(MA)_{\mathcal{G}}$ is that a question in $A^-_{\mathcal{G}}$ comes with a sufficiently large block of memory, viewed as an element of $\mathbf{L}(\mathbf{L}(1))$. This block of memory can be used for computing an answer, but must be returned with the answer at the end of the computation. We define memory blocks as follows.

Definition 3. *Let k and m be natural numbers. The set* $\mathbf{L}_{\langle k,m \rangle} \subseteq \mathbf{L}(\mathbf{L}(1))$ *of* $\langle k, m \rangle$-memory blocks *is the least set containing all the lists* $x = \langle x_1, \ldots, x_n \rangle$ *with* $n \geq k$*, such that each* x_i *is a nonempty list in* $\mathbf{L}(1)$ *and the properties* $\forall i.\ (1 \leq i \leq n-1) \implies l(x_i) = \lfloor l(x)/n \rfloor$ *and* $\lfloor l(x)/n \rfloor \geq (\log(m) + 1)$ *and* $l(x_n) = (l(x) \mod n)$ *all hold.*

In short, $\mathbf{L}_{\langle k,m \rangle}$ provides enough space for at least k binary numbers with range $\{0, \ldots, m\}$. Given a memory block $x = \langle x_1, \ldots, x_n \rangle \in \mathbf{L}_{\langle k,m \rangle}$, we refer to the length of x_1 as the *size of memory locations in* x. The definition of $\mathbf{L}_{\langle k,m \rangle}$ is such that, for any two nonempty lists $s, s' \in \mathbf{L}_{\langle k,m\rangle}$, we have $s = s'$ if and only if both $l(s) = l(s')$ and $l(head(s)) = l(head(s'))$ hold. This makes it easy to reconstruct the original memory block after part of it has been used in a computation.

The realisation on MA is defined such that $\langle l, k, m\rangle, e \Vdash_{MA} a$ holds if and only if $\forall s \in \mathbf{L}_{\langle k,m\rangle}. \exists e_s. (\langle l, 0, m\rangle, e_s \Vdash_A a) \wedge (\forall q. e(q, s) = \langle e_s(q), s\rangle)$ holds. The co-realisation relation is determined by the requirement that MA be simple.

Lemma 3. *There are isomorphisms $M(A \otimes_1 B) \cong MA \otimes_1 MB$ and $M(A * B) \cong MA * MB$ for all objects A and B for which MA and MB are defined.*

All the base types of LogFPL are interpreted by objects of the form MA, e.g. we use $M\mathbf{L}(M\mathbf{B})$ as the interpretation of $\mathbf{L}(\mathbf{B})$. Some simplification is given by:

Lemma 4. *There are maps $M\mathbf{L}(MA) \to M\mathbf{L}(A)$ and $!M\mathbf{L}(A) \to M\mathbf{L}(MA)$.*

Proposition 3. *For each morphism $(f, f^*) \colon !^k M\mathbf{L}(\mathbf{B}) \otimes !^k M\mathbf{L}(I) \to M\mathbf{L}(\mathbf{B})$, the function f is* LOGSPACE*-computable.*

Proof (Sketch). We construct a Turing Machine T using a realiser r for (f, f^*). From r we build a sub-routine of T, which on one work tape takes a (code for a) question $q \in (M\mathbf{L}(\mathbf{B}))^-_{\mathcal{G}}$ and on the same tape returns an answer in $(M\mathbf{L}(\mathbf{B}))^+_{\mathcal{G}}$. The sub-routine works by running r, and whenever r returns a question for its argument, the sub-routine reads the relevant part from the input tape and passes the answer to r. Since r is a morphism in $\mathcal{L}$, this uses only linear space in the size of the question q. We now use this sub-routine to compute the output of T. Write x, y for the two inputs to T. First we write a memory block in $\mathbf{L}_{\langle k,|x|+|y|\rangle}$ to one work tape. This can be done since k is constant and the block has logarithmic size. Using this block, the sub-routine is used to compute the bits of the output one-by-one. Since f is non-size-increasing, it suffices to continue until bit number $|x| + |y|$. Hence, we only need to consider questions as large as $\log(|x| + |y|) + 1$. Since the space needed for answering questions is linear in the size of the question, the whole procedure uses logarithmic space in the size of the inputs x and y.

It remains to show correctness of this algorithm. Let x and y be the two inputs. We then have $\langle |x|, 0, |x|\rangle, e_x \Vdash_{M\mathbf{L}(\mathbf{B})} x$ and $\langle |y|, 0, |y|\rangle, e_y \Vdash_{M\mathbf{L}(I)} y$, where e_x and e_y are programs that answer the given question by reading the input tape (note that x and y are now fixed, so that $|x|$ and $|y|$ appear as constants in the space-usage of e_x and e_y). Since r realises the morphism (f, f^*), we have $\langle |x| + |y|, k, |x| + |y|\rangle, r \cdot (!^k e_x \otimes !^k e_y) \Vdash_{M\mathbf{L}(\mathbf{B})} f(x, y)$. Since the sub-routine described above computes $r \cdot (!^k e_x \otimes !^k e_y)$, it then follows by the definition of the realisation on $M\mathbf{L}(\mathbf{B})$, that the algorithm computes the correct output.

Next we implement the operations on lists. Operations on other types are omitted for space reasons.

Lemma 5. *Let $A = MA'$ and $B = MB'$ be simple objects. There are $\mathcal{R}$-maps $cons \colon M\Diamond \otimes_1 A \otimes_1 !M\mathbf{L}(A) \to M\mathbf{L}(A)$, $hdtl \colon !M\mathbf{L}(A) \to (M\Diamond \otimes_1 A \otimes_1 M\mathbf{L}(A))_\perp$ and $empty \colon M\mathbf{L}(A) \to M\mathbf{B}$ defined by $cons(a, \vec{a}) = \langle a, \vec{a}\rangle$, $hdtl(\langle\rangle) = \perp$, $empty(\langle\rangle) = \mathtt{tt}$, $hdtl(\langle a, \vec{a}\rangle) = \langle \Diamond, a, \vec{a}\rangle$ and $empty(\langle a, \vec{a}\rangle) = \mathtt{ff}$.*

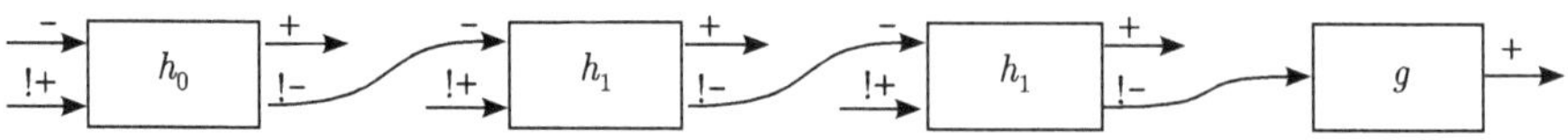

Fig. 3. Illustration of recursion

It remains to implement recursion for lists. The properties we need of the result type of a recursion are captured by the following three definitions.

Definition 4. *An object* A has unique answers *if, for all* $x \in |A|$ *and all* $q \in A^-_{\mathcal{G}}$, *there exists a unique* $a \in A^+_{\mathcal{G}}$ *satisfying* $\forall \alpha, e.\,(\alpha, e \Vdash x) \implies e(q) = a$.

The next definition expresses that we only need to consider small questions. For instance, for an element $x \in \mathbf{L}(\mathbf{B})$ with $\langle l, k, m\rangle, e \Vdash x$, we only ever need to consider questions q with $l(q) \leq \log(m) + 1$, since all larger questions will be out of range. We use this property to ensure that we do not run out of memory during a recursion.

Definition 5. *An object* A has n-small questions *if there exist morphisms* $\mathsf{cut}\colon A^-_{\mathcal{G}} \otimes \mathbf{L}(1) \to A^-_{\mathcal{G}}$ *and* $\mathsf{expand}\colon A^-_{\mathcal{G}} \otimes A^+_{\mathcal{G}} \to A^+_{\mathcal{G}}$, *such that* $\langle l, k, m\rangle, e \Vdash x$ *implies* $l(\mathsf{cut}(q, s)) \leq l(s)$ *and* $\mathsf{expand}(q, e(\mathsf{cut}(q, s))) = e(q)$ *holds for all* $q \in A^-_{\mathcal{G}}$ *and all* $s \in \mathbf{L}(1)$ *with* $l(s) \geq n(\log(m) + 1)$.

Finally, to implement recursion we must be able to store questions and answers.

Definition 6. *An object* A *is* n-encodable *if, for each* $X \in \{A^-_{\mathcal{G}}, A^+_{\mathcal{G}}\}$, *there are maps* $\mathsf{len}_X\colon X \to \mathbf{L}(1)$, $\mathsf{code}_X\colon \Diamond^n \otimes X \to \mathbf{S}$ *and* $\mathsf{decode}_X\colon \mathbf{S} \to \Diamond^n \otimes X$, *with the properties* $\forall x \in X.\, l(x) = l(\mathsf{len}_X(x))$ *and* $\mathsf{decode}_X \circ \mathsf{code}_X = id$.

Each recursion type, as defined in Fig. 1, is simple and enjoys the properties of these definitions. In general, this holds neither for $A \otimes B$ nor for $A \multimap B$.

Proposition 4. *Let* R *be a simple object with unique answers and* k_R*-small questions. Let* R *be* k_R*-encodable. Let* $\langle 0, k, m\rangle, e_h \Vdash_{!^i\Diamond\otimes(\Box !^i MA)\otimes !^i MR \multimap MR} h$. *Then the function* $f_h\colon !^{4+7k_R} MR \otimes \Box !^{i+4+7k_R} M\mathbf{L}(A) \to MR$ *with* $f_h(g, \langle\rangle) = g$ *and* $f_h(g, \langle x, \vec{x}\rangle) = h(\Diamond, x, f_h(g, \vec{x}))$ *is realised with bound* $\langle 0, k + 4 + 7k_R, m\rangle$.

Let us explain the idea of the realiser for recursion informally. Suppose, for instance, we want to compute $f_h(g, \langle a_1, a_1, a_0\rangle)$ for certain elements g, a_0 and a_1. Write h_0 and h_1 for the functions of type $!MR \multimap MR$ that arise from h by instantiating the first argument with a_0 and a_1 respectively. Then, the algorithm for computing $f_h(g, \langle a_1, a_1, a_0\rangle)$ can be depicted as in Fig. 3. The edge labelled with $-$ represents an initial question. The edges labelled with $!-$ represent questions of h_j to the recursion argument, the answer to which is expected in the $!+$-port of the same box. The presence of the modality ! expresses that along with the question, h_j may pass some store that must be returned unchanged with the answer. Now, if we were to remember the store of each h_j in the course of the recursion then we would need linear space (in the length of the second argument of f) to compute it. Instead, whenever h_j sends a question $!-$ to its recursion argument, we forget h_j's local store (the value stored in the !) and just pass the question to the next instance of h_j. If g or some h_j gives an answer in its +-port, then we would like to pass this answer to the edge labelled with !+ into the next box to the left. However, we cannot go from + to !+, as there is no way to reconstruct the store that we have discarded before. Nevertheless, it is possible to recompute the store. We remember the answer in + and the position of the box that gave the answer and restart the computation from the beginning. During this computation, the question in !−, which is answered by the saved answer, will be asked again. We can then take the store of the question in !− and the saved answer in + to construct an answer in !+ and pass it on as an answer to the question in !−. In this way, the local store of the step functions h_j can be recomputed. For the correctness of this procedure it is crucial that each h_j asks at most one question of the recursion argument, which follows from the co-realisation information on h. It is instructive to check this property for the step-function of *focus* in Sect. 2.1.

This description of the implementation of recursion can be formalised in $\mathcal{G}$ by a map $rec\colon !^r(MR)_\mathcal{G}\otimes!^r(!^i(\Diamond_\mathcal{G}\otimes(MA)_\mathcal{G}\otimes(MR)_\mathcal{G})\multimap(MR)_\mathcal{G})\otimes!^r!^i\mathbf{L}(A)_\mathcal{G}\to(MR)_\mathcal{G}$, where $r=4+7k_R$. The morphism rec is defined such that, for all $\langle l_g,k_g,m_g\rangle,e_g\Vdash_{MR} g$ and $\langle l_x,k_x,m_x\rangle,e_x\Vdash_{M\mathbf{L}(A)} x$, the element $f_h(g,x)$ is realised by $rec\cdot(!^r e_g\otimes!^r e_h\otimes!^{r+i}e_x)$ with bound $\alpha=\langle 0,k+r,m\rangle+\langle l_g,k_g+r,m_g\rangle+\langle l_x,k_x+r+i,m_x\rangle$. Prop. 4 follows from this property.

Although we have to omit the details of the implementation of rec for space-reasons, we outline how the r memory locations in the modalities $!^r$ are used by rec. The definition of the realisation on MR is such that we can assume that a question to $rec\cdot(!^r e_g\otimes!^r e_h\otimes!^{r+i}e_x)$ comes together with a memory block $m\in\mathbf{L}_{\langle k_\alpha,m_\alpha\rangle}$. Since we know $\alpha=\langle l_\alpha,\max(k,k_g,k_x+i)+r,\max(m,m_g,m_x)\rangle$, we can thus assume at least r memory locations of size $\log(m_\alpha)+1$ and we can assume that the rest of the memory is still large enough to satisfy the memory requirements of g, h and x. We use the $r=4+7k_R$ memory cells, to store the following data: the initial question (*startqn*); the current question (*qn*); the current recursion depth (d); whether some answer has already been computed (*store*), the stored answer (*sa*) and its recursion depth (*sd*); a flag (*xo*) to record how answers from the argument x should be interpreted. Since R has k_R-small questions and the questions and answers are k_R-encodable, each of the fields *startqn*, *qn* and *sa* can be stored using $2k_R$ memory locations in m. The fields *store*, d, *sd* and *xo* can each be stored in a single memory location in m, since d and *sd* represent numbers in $\{0,\ldots,|x|\}$. The remaining k_R memory locations provide enough memory for the current question/answer.

3.5 Interpreting LogFPL

The types of LogFPL are interpreted in $\mathcal{R}$ by the following clauses.

$$[\![A]\!]=MA,\ \text{where}\ A\in\{I,\Diamond,\mathbf{B},\mathbf{SN}\}$$

$$[\![\mathbf{L}(A)]\!]=M(\mathbf{L}([\![A]\!]))\qquad [\![A\overset{\cdot,i}{\multimap}B]\!]=!^i[\![A]\!]\multimap[\![B]\!]_\perp$$

$$[\![A\bullet B]\!]=[\![A]\!]\bullet[\![B]\!],\ \text{where}\ \bullet\in\{*,\otimes_1,\times\}\qquad [\![A\overset{\infty,i}{\multimap}B]\!]=\Box!^i[\![A]\!]\multimap[\![B]\!]_\perp$$

Contexts are interpreted by $[\![\Gamma]\!] = [\![\Gamma]\!]' \otimes [\![\Gamma]\!]_1$, where $[\![()]\!]' = [\![()]\!]_1 = I$ and

$$[\![\Gamma, x \overset{\langle\infty,i\rangle}{:} A]\!]' = [\![\Gamma]\!]' \otimes \Box !^i [\![A]\!] \qquad [\![\Gamma, x \overset{\langle\infty,i\rangle}{:} A]\!]_1 = [\![\Gamma]\!]_1$$
$$[\![\Gamma, x \overset{\langle\cdot,i\rangle}{:} A]\!]' = [\![\Gamma]\!]' \otimes !^i [\![A]\!] \qquad [\![\Gamma, x \overset{\langle\cdot,i\rangle}{:} A]\!]_1 = [\![\Gamma]\!]_1$$
$$[\![\Gamma, x \overset{\langle 1,i\rangle}{:} A]\!]' = [\![\Gamma]\!]' \qquad [\![\Gamma, x \overset{\langle 1,i\rangle}{:} A]\!]_1 = [\![\Gamma]\!]_1 \otimes_1 !^i [\![A]\!].$$

Proposition 5. *For each $\Gamma \vdash M\colon A$, there is a map $m\colon [\![\Gamma]\!] \to [\![A]\!]_\perp$ in $\mathcal{R}$, such that the underlying function of m is the functional interpretation of M.*

Proof. The proof goes by induction on the derivation of $\Gamma \vdash M\colon A$. As a representative case, we consider rule $\otimes_1$E. The induction hypothesis gives $[\![\Gamma]\!] \to ([\![A]\!] \otimes_1 [\![B]\!])_\perp$ and $[\![\Delta']\!]' \otimes ([\![\Delta_1]\!]_1 \otimes_1 !^i[\![A]\!] \otimes_1 !^i[\![B]\!]) \to [\![C]\!]_\perp$ for appropriate Δ' and Δ_1 with $\Delta = \Delta', \Delta_1$. With the isomorphism $!(X \otimes_1 Y) \cong !X \otimes_1 !Y$ and the operations on $(-)_\perp$, we obtain a map $[\![\Delta']\!]' \otimes ([\![\Delta_1]\!]_1 \otimes_1 !^i[\![\Gamma]\!]) \to [\![C]\!]_\perp$. Using the morphisms from Lemma 2, we obtain $[\![!^{\langle 1,i\rangle}\Gamma]\!] \to !^i[\![\Gamma]\!]$. By use of $X \otimes (Y \otimes_1 Z) \to Y \otimes_1 (X \otimes Z)$, we obtain $[\![\Delta, !^{\langle 1,i\rangle}\Gamma]\!] \to [\![\Delta']\!]' \otimes ([\![\Delta_1]\!]_1 \otimes_1 !^i[\![\Gamma]\!])$. Putting this together gives the required $[\![\Delta, !^{\langle 1,i\rangle}\Gamma]\!] \to [\![C]\!]_\perp$.

To see that Prop. 4 is applicable for the interpretation of recursion, notice that $\langle l, k, m\rangle, e \Vdash_{A \multimap M\mathbf{B}^k} d$ implies $\langle 0, k, m\rangle, e \Vdash_{A \multimap M\mathbf{B}^k} d$. This follows using the definition of the realisation on $\multimap$, since the same property holds by definition for $M\mathbf{B}^k$. Furthermore, each recursion type A (as defined in Fig. 1) is interpreted by a $k(A)$-encodable, simple object with unique answers and $k(A)$-small questions. Thus Prop. 4 can be used for the interpretation of recursion. □

With the interpretation, LOGSPACE-soundness (Prop. 1) now follows as in Prop. 3.

4 Conclusion and Further Work

We have introduced a computation-by-interaction model for space-restricted computation and have designed a type system for LOGSPACE on its basis. We have thus demonstrated that the model captures LOGSPACE-computation in a compositional fashion. As a way of controlling the subtle interaction-behaviour in the model, we have identified the concept of co-realisability. Using this concept, we were able to formalise the subtle differences between types such as $(\Box A) \otimes B$, $A \otimes B$ and $A \otimes_1 B$ in a unified way.

For further work, we plan to consider extensions of LogFPL. We believe that there is a lot of structure left in the model that can be used to justify extensions. For instance, we conjecture that it is possible to define first-order linear functions by recursion. We expect that, using an argument similar to the Chu-space approach of [10], co-realisability can be used to reduce recursion with first-order functional result type to base-recursion with parameter substitution.

Other interesting directions for further work include to consider other complexity classes such as polylogarithmic space and to further explore the connections to linear logic. It may also be interesting to find out if recent work on algorithmic game semantics, as in e.g. [7,1], can be utilised for our purposes. A referee raised the interesting question about the relation of our computation model based on a GoI-situation to oracle-based computing in traditional complexity models. We plan to consider this in further work.

Acknowledgments. I thank Martin Hofmann and Jan Johannsen for helpful advice and the anonymous referees for their comments and corrections. This work has been supported by the Deutsche Forschungsgemeinschaft (DFG) as part of the project Pro.Platz.

References

1. S. Abramsky, D.R. Ghica, A.S. Murawski, and C.-H.L. Ong. Applying game semantics to compositional software modeling and verification. In *TACAS04*, pages 421–435, 2004.
2. S. Abramsky, E. Haghverdi, and P.J. Scott. Geometry of interaction and linear combinatory algebras. *Mathematical Structures in Computer Science*, 12(5):625–665, 2002.
3. S. Abramsky and R. Jagadeesan. Games and full completeness for multiplicative linear logic (extended abstract). In *FSTTCS92*, pages 291–301, 1992.
4. S.J. Bellantoni. *Predicative Recursion and Computational Complexity*. PhD thesis, University of Toronto, 1992.
5. U. Dal Lago and M. Hofmann. Quantitative models and implicit complexity. In *FSTTCS05*, pages 189–200, 2005.
6. V. Danos, H. Herbelin, and L. Regnier. Game semantics & abstract machines. In *LICS96*, pages 394–405, 1996.
7. D.R. Ghica and G. McCusker. Reasoning about idealized algol using regular languages. In *ICALP*, pages 103–115, 2000.
8. E. Hahgverdi. *A Categorical Approach to Linear Logic, Geometry of Proofs and Full Completeness*. PhD thesis, University of Ottawa, 2000.
9. M. Hofmann. Linear types and non-size-increasing polynomial time computation. *Information and Computation*, 183:57–85, 1999.
10. M. Hofmann. Semantics of linear/modal lambda calculus. *Journal of Functional Programming*, 9(3):247–277, 1999.
11. M. Hofmann. Programming languages capturing complexity classes. *SIGACT News*, 31(1):31–42, 2000.
12. J.M.E. Hyland and A. Schalk. Glueing and orthogonality for models of linear logic. *Theoretical Computer Science*, 294(1–2):183–231, 2003.
13. N.D. Jones. LOGSPACE and PTIME characterized by programming languages. *Theoretical Computer Science*, 228(1–2):151–174, 1999.
14. L. Kristiansen. Neat function algebraic characterizations of LOGSPACE and LINSPACE. *Computational Complexity*, 14:72–88, 2005.
15. J.R. Longley. *Realizability Toposes and Language Semantics*. PhD thesis, University of Edinburgh, 1994.
16. I. Mackie. The geometry of interaction machine. In *POPL95*, 1995.
17. P. Møller-Neegaard. *Complexity Aspects of Programming Language Design*. PhD thesis, Brandeis University, 2004.
18. Peter Møller Neergaard. A functional language for logarithmic space. In *APLAS*, pages 311–326, 2004.
19. D. Sannella, M. Hofmann, D. Aspinall, S. Gilmore, I. Stark, L. Beringer, H.-W. Loidl, K. MacKenzie, A. Momigliano, and O. Shkaravska. Mobile Resource Guarantees. In *Trends in Functional Programing 6*, 2005. Intellect.

The Ackermann Award 2006

S. Abramsky, E. Grädel, and J. Makowsky

Members of EACSL Jury for the Ackermann Award

The second **Ackermann Award** is presented at this CSL'06. Eligible for the 2006 **Ackermann Award** were PhD dissertations in topics specified by the EACSL and LICS conferences, which were formally accepted as PhD theses at a university or equivalent institution between 1.1. 2004 and 31.12. 2005. The jury received 14 nominations for the **Ackermann Award 2006**. The candidates came from 10 different nationalities from Europe, the Middle East and India, and received their PhDs in 9 different countries in Europe, Israel and North America.

The topics covered the full range of Logic and Computer Science as represented by the LICS and CSL Conferences. All the submissions were of very high standard and contained outstanding results in their particular domain. The jury decided finally, to give for the year 2006 two awards, one for work on the *model theory of hybrid logics - modal logics extended with nominals*, and one for work on *mathematical structures arising in the theory of coalgebras and providing semantics for recursive specifications*. The 2006 **Ackermann Award** winners are, in alphabetical order,

- Balder ten Cate from the Netherlands, for his thesis
  Model Theory for Extended Modal Languages,
  issued by Amsterdam University, The Netherlands, in 2005
  Supervisor: Johan van Benthem

- Stefan Milius from Germany, for his thesis
  Coalgebras, Monads and Semantics,
  issued by the Technical University of Braunschweig, Germany, in 2005
  Supervisor: Jiří Adámek

The Jury wishes to congratulate the recipients of the Ackermann Award for their outstanding work and wishes them a successful continuation of their career.

The Jury wishes also to encourage all the remaining candidates to continue their excellent work and hopes to see more of their work in the future.

Balder Ten Cate

Citation. Balder ten Cate receives the *2006 Ackermann Award* of the European Association of Computer Science Logic (EACSL) for his thesis

Model Theory for Extended Modal Languages,

in which he substantially advanced our understanding of model theoretic and computational aspects of extensions of modal logic and their use in computer science.

Z. Ésik (Ed.): CSL 2006, LNCS 4207, pp. 622–626, 2006.

Background of the thesis. Modal logics are of fundamental importance in many branches of computer science. They allow to tailor logical formalisms so that they combine the expressive power needed in particular applications with good algorithmic and model-theoretic properties. A fundamental feature of modal logics is their invariance under notions of behavioural equivalence such as bisimulation. Well-known and fundamental results in the model theory of modal logic reveal that propositional modal logic is equivalent to the bisimulation-invariant fragment of first-order logic, and that an analogous relationship holds for the modal μ-calculus and monadic second-order logic.

In many applications, it is necessary or at least convenient to extend modal logics with nominals (i.e., constants for naming elements of the underlying structure) and operators to handle them. These logics are called hybrid logics and they relate to, say, first-order logic with constants in the same way as basic modal logics relate to purely relational first-oder logic. Balder ten Cate's thesis is mainly a mathematical investigation into hybrid logics and other extended modal logics. It encompasses a wide range of topics in the analysis of extended modal logics, covering many new results on fundamental model theoretic features of hybrid logics and other more expressive modal logics, and puts these results into a wider framework of abstract modal model theory. For instance, the celebrated Goldblatt-Thomason Theorem states that a first-order formula defines a modally definable frame class if, and only if, it is preserved under taking generated subframes, disjoint unions and bounded morphic images, and its negation is preserved under taking ultrafilter extensions. Many results in ten Cate's thesis are motivated by the question whether similar chracterisations can be given for the frame classes definable in extended modal logics, such as hybrid logics or modal logic with propositional quantifiers.

Abstract model theory studies model theoretic propertis of logics on a general, abstract level. A fundamental result in this context is Lindström's Theorem stating that no proper extension of first-order logic has both the compactness and the Löwenheim-Skolem property. Abstract model theory has been successful in providing a unifying perspective of model-theoretic properties of logics, typically of powerful extensions of first-order logic, but it does not really cover aspects of computational logics. A more general perspective of ten Cate's work is to contribute to the development of an abstract model theory for computational logics, devoted to logics that arise in computer science, and to those properties that are relevant for their computational applications.

Ten Cate's thesis. The thesis, entitled *Model Theory for Extended Modal Languages* is written with great lucidity and sophistication. It develops the abstract model theory of hybrid languages in the spectrum running from the basic modal language to full first-order logic. The results of the thesis are too numerous to enumerate them, so let us just mention a few highlights.

- A systematic frame-definability theory for hybrid languages is developed, including Goldblatt-Thomason style characterizations. This is a highly non-trivial enterprise in terms of model-theoretic proof techniques, and new frame constructions.

- In this context, a very interesting, although negative, result is that the set of first-order formulae preserved under ultrafilter extensions is highly undecidable (Π_1^1-hard).
- The thesis contains a striking syntactic analysis of the 'Bounded Fragment' first studied by Feferman and Kreisel, and later by Areces, Blackburn and Marx, allowing us a much better grip on what it is and does.
- Balder ten Cate has proved a general interpolation theorem showing that only very few hybrid logics have interpolation. This is one of the first major classification results in the abstract model theory of fragments of first-order logic (where standard proofs often do not work, as they presuppose the full power of first-order encoding).
- He has contributed an interesting analysis of second-order formalisms such as modal logic with propositional quantifiers, inside of which he characterises for instance the standard modal fragment via finite depth and bisimulation invariance, and the intersection with first-order logic as the bounded fragment.

There is much more to the thesis. In addition, ten Cate makes complexity-theoretic investigations of hybrid logics, suggesting a new style of abstract model theory: mixing expressivity with concerns of computational complexity. In doing so, he also provides a sustained study of satisfiability preserving translations between various modal and hybrid logics.

A most striking impression when one reads this thesis is that of an unusual mathematical maturity in research and writing. The results are embedded into a convincing high-level account that provides perspective and contributes to a coherent bigger picture. The exposition is very clear and admirably manages to be concise and rigorous without becoming notationally or formally heavy. In fact, the writing is skillful and to the point in an exemplary fashion.

Biographic Sketch. Balder David ten Cate was born on June 15, 1979 in Amsterdam, the Netherlands. In 1998 he received his B.A. (Propedeuse) in Psychology from the Vrije Universiteit of Amsterdam. In 2000 he received his M.Sc. in Artificial Intelligence from the Vrije Universiteit of Amsterdam. Both degrees were given *cum laude* (with distinction). In 2005 he received his Ph.D. from the University of Amsterdam, under the supervision Johan van Benthem.

Balder ten Cate has done first research in natural language semantics, and then in the model theoretic and computational aspects of various extensions of modal logic. He now works on a project on the foundations of semi-structured data and XML-query languages.

Stefan Milius

Citation. Stefan Milius receives the *2006 Ackermann Award* of the European Association of Computer Science Logic (EACSL) for his thesis

Coalgebras, Monads and Semantics,

in which he advances considerably our understanding of category-theoretic methods in the study of a wide variety of recursive behavioural specifications in a unified setting.

Background of the thesis. Recursive definitions are a central topic in Computer Science. One major tradition in the field has been *algebraic semantics*, which studies the structure of recursive program schemes, and their interpretations. A more recent and currently very lively area of work is in *coalgebraic semantics*, which uses category-theoretic methods to study a wide variety of recursive behavioural specifications in a unified setting. These research directions have hitherto been rather separate, both from each other, and from other approaches in semantics.

Milius' thesis. This thesis contains a number of striking contributions, which show a rich interplay between algebraic and coalgebraic methods, and open up new directions in the study of recursion schemes. Moreover the treatment is carried out in a general setting, unifying existing approaches, and allowing for interesting new examples.

Some main contributions are as follows:

- A description of free iterative theories generated by an arbitrary finitary endofunctor. This simplifies and substantially generalizes some of the main results obtained by the Elgot school. The proof is an elegant combination of methods of algebra and coalgebra.
- Elgot algebras. An elegant axiomatization is given of the algebras over the monad corresponding to a free iterative theory. Again, there is a pleasing interplay between algebraic and coalgebraic notions, e.g. in the result that the Elgot algebras of a given "iteratable" functor are the Eilenberg-Moore algebras of a certain monad related to final coalgebras.
- New insights and a powerful unifying perspectives are developed for the semantics of recursive programs and program schemes. The treatment encompasses domain-theoretic and metric space based approaches, and includes interesting examples such as fractals.

The thesis is based on a number of substantial journal papers, with a well-written overview. The mathematical development is notable for its clarity and generality. It seems likely that this work will stimulate further research, and open up some new directions in the fruitful interplay of algebra and coalgebra, and in the study of recursion.

Biographic Sketch. Stefan Milius was born on June 3, 1975 in Magdeburg, Germany. In 2000, he received an MA in Mathematics from York University Toronto, Canada and a Diploma in Computer Science from the Technical University of Braunschweig, Germany. He wrote his Ph.D. thesis under the supervision of Prof. Jiří Adámek and received his Ph.D. (Dr.rer.nat.) in October 2005 from the Technical University of Braunschweig *summa cum laude* (with distinction).

From 1998 till 2000 he was a Fellow of the German National Academic Foundation (Studienstiftung des Deutschen Volkes). In 2000 he was given the Award for excellent achievments as a student from the Technical University of Braunschweig. His work concentrates on category theoretic aspects of computer science.

The Ackermann Award

The EACSL Board decided in November 2004 to launch the EACSL Outstanding Dissertation Award for Logic in Computer Science, the **Ackermann Award**, The award[1]. is named after the eminent logician Wilhelm Ackermann (1896-1962), mostly known for the Ackermann function, a landmark contribution in early complexity theory and the study of the rate of growth of recursive functions, and for his coauthorship with D. Hilbert of the classic *Grundzüge der Theoretischen Logik*, first published in 1928. Translated early into several languages, this monograph was the most influential book in the formative years of mathematical logic. In fact, Gödel's completeness theorem proves the completeness of the system presented and proved sound by Hilbert and Ackermann. As one of the pioneers of logic, W. Ackermann left his mark in shaping logic and the theory of computation.

The **Ackermann Award** is presented to the recipients at the annual conference of the EACSL. The jury is entitled to give more than one award per year. The award consists of a diploma, an invitation to present the thesis at the CSL conference, the publication of the abstract of the thesis and the citation in the CSL proceedings, and travel support to attend the conference.

The jury for the **Ackermann Award** consists of seven members, three of them ex officio, namely the president and the vice-president of EACSL, and one member of the LICS organizing committee. The current jury consists of S. Abramsky (Oxford, LICS Organizing Committee), B. Courcelle (Bordeaux), E. Grädel (Aachen), M. Hyland (Cambridge), J.A. Makowsky (Haifa, President of EACSL), D. Niwinski (Warsaw, Vice President of EACSL), and A. Razborov (Moscow and Princeton).

The first **Ackermann Award** was presented at CSL'05 in Oxford, England. The recipients were Mikołaj Bojańczyk from Poland, Konstantin Korovin from Russia, and Nathan Segerlind from the USA. A detailed report on their work appeared in the CSL'05 proceedings.

[1] Details concerning the Ackermann Award and a biographic sketch of W. Ackermann was published in the CSL'05 proceedings and can also be found at `http://www.dimi.uniud.it/ eacsl/award.html`.

Author Index

Lecture Notes in Computer Science

For information about Vols. 1–4110

please contact your bookseller or Springer

Vol. 4228: D.E. Lightfoot, C.A. Szyperski (Eds.), Modular Programming Languages. X, 415 pages. 2006.

Vol. 4224: E. Corchado, H. Yin, V. Botti, C. Fyfe (Eds.), Intelligent, Data Engineering and, Automated Learning – IDEAL 2006. XXVII, 1447 pages. 2006.

Vol. 4219: D. Zamboni, C. Kruegel (Eds.), Recent Advances in Intrusion Detection. XII, 323 pages. 2006.

Vol. 4217: P. Cuenca, L. Orozco-Barbosa (Eds.), Personal Wireless Communications. XV, 532 pages. 2006.

Vol. 4213: J. Fürnkranz, T. Scheffer, M. Spiliopoulou (Eds.), Knowledge Discovery in Databases: PKDD 2006. XXII, 660 pages. 2006. (Sublibrary LNAI).

Vol. 4212: J. Fürnkranz, T. Scheffer, M. Spiliopoulou (Eds.), Machine Learning: ECML 2006. XXIII, 851 pages. 2006. (Sublibrary LNAI).

Vol. 4208: M. Gerndt, D. Kranzlmüller (Eds.), High Performance Computing and Communications. XXII, 938 pages. 2006.

Vol. 4207: Z. Ésik (Ed.), Computer Science Logic. XII, 627 pages. 2006.

Vol. 4206: P. Dourish, A. Friday (Eds.), UbiComp 2006: Ubiquitous Computing. XIX, 526 pages. 2006.

Vol. 4205: G. Bourque, N. El-Mabrouk (Eds.), Comparative Genomics. X, 231 pages. 2006. (Sublibrary LNBI).

Vol. 4202: E. Asarin, P. Bouyer (Eds.), Formal Modeling and Analysis of Timed Systems. XI, 369 pages. 2006.

Vol. 4201: Y. Sakakibara, S. Kobayashi, K. Sato, T. Nishino, E. Tomita (Eds.), Grammatical Inference: Algorithms and Applications. XII, 359 pages. 2006. (Sublibrary LNAI).

Vol. 4197: M. Raubal, H.J. Miller, A.U. Frank, M.F. Goodchild (Eds.), Geographic, Information Science. XIII, 419 pages. 2006.

Vol. 4196: K. Fischer, E. André, I.J. Timm, N. Zhong (Eds.), Multiagent System Technologies. X, 185 pages. 2006. (Sublibrary LNAI).

Vol. 4194: V.G. Ganzha, E.W. Mayr, E.V. Vorozhtsov (Eds.), Computer Algebra in Scientific Computing. XI, 313 pages. 2006.

Vol. 4193: T.P. Runarsson, H.-G. Beyer, E. Burke, J.J. Merelo-Guervós, L. D. Whitley, X. Yao (Eds.), Parallel Problem Solving from Nature - PPSN IX. XIX, 1061 pages. 2006.

Vol. 4192: B. Mohr, J.L. Träff, J. Worringen, J. Dongarra (Eds.), Recent Advances in Parallel Virtual Machine and Message Passing Interface. XVI, 414 pages. 2006.

Vol. 4189: D. Gollmann, J. Meier, A. Sabelfeld (Eds.), Computer Security – ESORICS 2006. XI, 548 pages. 2006.

Vol. 4188: P. Sojka, I. Kopeček, K. Pala (Eds.), Text, Speech and Dialogue. XIV, 721 pages. 2006. (Sublibrary LNAI).

Vol. 4187: J.J. Alferes, J. Bailey, W. May, U. Schwertel (Eds.), Principles and Practice of Semantic Web Reasoning. XI, 277 pages. 2006.

Vol. 4186: C. Jesshope, C. Egan (Eds.), Advances in Computer Systems Architecture. XIV, 605 pages. 2006.

Vol. 4185: R. Mizoguchi, Z. Shi, F. Giunchiglia (Eds.), The Semantic Web – ASWC 2006. XX, 778 pages. 2006.

Vol. 4184: M. Bravetti, M. Núñez, G. Zavattaro (Eds.), Web Services and Formal Methods. X, 289 pages. 2006.

Vol. 4183: J. Euzenat, J. Domingue (Eds.), Artificial Intelligence: Methodology, Systems, and Applications. XIII, 291 pages. 2006. (Sublibrary LNAI).

Vol. 4182: H.T. Ng, M.-K. Leong, M.-Y. Kan, D. Ji (Eds.), Information Retrieval Technology. XVI, 684 pages. 2006.

Vol. 4180: M. Kohlhase, OMDoc – An Open Markup Format for Mathematical Documents [version 1.2]. XIX, 428 pages. 2006. (Sublibrary LNAI).

Vol. 4179: J. Blanc-Talon, W. Philips, D. Popescu, P. Scheunders (Eds.), Advanced Concepts for Intelligent Vision Systems. XXIV, 1224 pages. 2006.

Vol. 4178: A. Corradini, H. Ehrig, U. Montanari, L. Ribeiro, G. Rozenberg (Eds.), Graph Transformations. XII, 473 pages. 2006.

Vol. 4176: S.K. Katsikas, J. Lopez, M. Backes, S. Gritzalis, B. Preneel (Eds.), Information Security. XIV, 548 pages. 2006.

Vol. 4175: P. Bücher, B.M.E. Moret (Eds.), Algorithms in Bioinformatics. XII, 402 pages. 2006. (Sublibrary LNBI).

Vol. 4174: K. Franke, K.-R. Müller, B. Nickolay, R. Schäfer (Eds.), Pattern Recognition. XX, 773 pages. 2006.

Vol. 4173: S. El Yacoubi, B. Chopard, S. Bandini (Eds.), Cellular Automata. XV, 734 pages. 2006.

Vol. 4172: J. Gonzalo, C. Thanos, M. F. Verdejo, R.C. Carrasco (Eds.), Research and Advanced Technology for Digital Libraries. XVII, 569 pages. 2006.

Vol. 4169: H.L. Bodlaender, M.A. Langston (Eds.), Parameterized and Exact Computation. XI, 279 pages. 2006.

Vol. 4168: Y. Azar, T. Erlebach (Eds.), Algorithms – ESA 2006. XVIII, 843 pages. 2006.

Vol. 4167: S. Dolev (Ed.), Distributed Computing. XV, 576 pages. 2006.

Vol. 4165: W. Jonker, M. Petković (Eds.), Secure, Data Management. X, 185 pages. 2006.

Vol. 4163: H. Bersini, J. Carneiro (Eds.), Artificial Immune Systems. XII, 460 pages. 2006.

Vol. 4162: R. Královič, P. Urzyczyn (Eds.), Mathematical Foundations of Computer Science 2006. XV, 814 pages. 2006.

Vol. 4161: R. Harper, M. Rauterberg, M. Combetto (Eds.), Entertainment Computing - ICEC 2006. XXVII, 417 pages. 2006.

Vol. 4160: M. Fisher, W.v.d. Hoek, B. Konev, A. Lisitsa (Eds.), Logics in Artificial Intelligence. XII, 516 pages. 2006. (Sublibrary LNAI).

Vol. 4159: J. Ma, H. Jin, L.T. Yang, J.J.-P. Tsai (Eds.), Ubiquitous Intelligence and Computing. XXII, 1190 pages. 2006.

Vol. 4158: L.T. Yang, H. Jin, J. Ma, T. Ungerer (Eds.), Autonomic and Trusted Computing. XIV, 613 pages. 2006.

Vol. 4156: S. Amer-Yahia, Z. Bellahsène, E. Hunt, R. Unland, J.X. Yu (Eds.), Database and XML Technologies. IX, 123 pages. 2006.

Vol. 4155: O. Stock, M. Schaerf (Eds.), Reasoning, Action and Interaction in AI Theories and Systems. XVIII, 343 pages. 2006. (Sublibrary LNAI).

Vol. 4154: Y.A. Dimitriadis, I. Zigurs, E. Gómez-Sánchez (Eds.), Groupware: Design, Implementation, and Use. XIV, 438 pages. 2006.

Vol. 4153: N. Zheng, X. Jiang, X. Lan (Eds.), Advances in Machine Vision, Image Processing, and Pattern Analysis. XIII, 506 pages. 2006.

Vol. 4152: Y. Manolopoulos, J. Pokorný, T. Sellis (Eds.), Advances in Databases and Information Systems. XV, 448 pages. 2006.

Vol. 4151: A. Iglesias, N. Takayama (Eds.), Mathematical Software - ICMS 2006. XVII, 452 pages. 2006.

Vol. 4150: M. Dorigo, L.M. Gambardella, M. Birattari, A. Martinoli, R. Poli, T. Stützle (Eds.), Ant Colony Optimization and Swarm Intelligence. XVI, 526 pages. 2006.

Vol. 4149: M. Klusch, M. Rovatsos, T.R. Payne (Eds.), Cooperative Information Agents X. XII, 477 pages. 2006. (Sublibrary LNAI).

Vol. 4148: J. Vounckx, N. Azemard, P. Maurine (Eds.), Integrated Circuit and System Design. XVI, 677 pages. 2006.

Vol. 4146: J.C. Rajapakse, L. Wong, R. Acharya (Eds.), Pattern Recognition in Bioinformatics. XIV, 186 pages. 2006. (Sublibrary LNBI).

Vol. 4144: T. Ball, R.B. Jones (Eds.), Computer Aided Verification. XV, 564 pages. 2006.

Vol. 4142: A. Campilho, M. Kamel (Eds.), Image Analysis and Recognition, Part II. XXVII, 923 pages. 2006.

Vol. 4141: A. Campilho, M. Kamel (Eds.), Image Analysis and Recognition, Part I. XXVIII, 939 pages. 2006.

Vol. 4139: T. Salakoski, F. Ginter, S. Pyysalo, T. Pahikkala, Advances in Natural Language Processing. XVI, 771 pages. 2006. (Sublibrary LNAI).

Vol. 4138: X. Cheng, W. Li, T. Znati (Eds.), Wireless Algorithms, Systems, and Applications. XVI, 709 pages. 2006.

Vol. 4137: C. Baier, H. Hermanns (Eds.), CONCUR 2006 – Concurrency Theory. XIII, 525 pages. 2006.

Vol. 4136: R.A. Schmidt (Ed.), Relations and Kleene Algebra in Computer Science. XI, 433 pages. 2006.

Vol. 4135: C.S. Calude, M.J. Dinneen, G. Păun, G. Rozenberg, S. Stepney (Eds.), Unconventional Computation. X, 267 pages. 2006.

Vol. 4134: K. Yi (Ed.), Static Analysis. XIII, 443 pages. 2006.

Vol. 4133: J. Gratch, M. Young, R. Aylett, D. Ballin, P. Olivier (Eds.), Intelligent Virtual Agents. XIV, 472 pages. 2006. (Sublibrary LNAI).

Vol. 4132: S. Kollias, A. Stafylopatis, W. Duch, E. Oja (Eds.), Artificial Neural Networks – ICANN 2006, Part II. XXXIV, 1028 pages. 2006.

Vol. 4131: S. Kollias, A. Stafylopatis, W. Duch, E. Oja (Eds.), Artificial Neural Networks – ICANN 2006, Part I. XXXIV, 1008 pages. 2006.

Vol. 4130: U. Furbach, N. Shankar (Eds.), Automated Reasoning. XV, 680 pages. 2006. (Sublibrary LNAI).

Vol. 4129: D. McGookin, S. Brewster (Eds.), Haptic and Audio Interaction Design. XII, 167 pages. 2006.

Vol. 4128: W.E. Nagel, W.V. Walter, W. Lehner (Eds.), Euro-Par 2006 Parallel Processing. XXXIII, 1221 pages. 2006.

Vol. 4127: E. Damiani, P. Liu (Eds.), Data and Applications Security XX. X, 319 pages. 2006.

Vol. 4126: P. Barahona, F. Bry, E. Franconi, N. Henze, U. Sattler, Reasoning Web. XII, 269 pages. 2006.

Vol. 4124: H. de Meer, J.P. G. Sterbenz (Eds.), Self-Organizing Systems. XIV, 261 pages. 2006.

Vol. 4121: A. Biere, C.P. Gomes (Eds.), Theory and Applications of Satisfiability Testing - SAT 2006. XII, 438 pages. 2006.

Vol. 4120: J. Calmet, T. Ida, D. Wang (Eds.), Artificial Intelligence and Symbolic Computation. XIII, 269 pages. 2006. (Sublibrary LNAI).

Vol. 4119: C. Dony, J.L. Knudsen, A. Romanovsky, A. Tripathi (Eds.), Advanced Topics in Exception Handling Components. X, 302 pages. 2006.

Vol. 4117: C. Dwork (Ed.), Advances in Cryptology - CRYPTO 2006. XIII, 621 pages. 2006.

Vol. 4116: R. De Prisco, M. Yung (Eds.), Security and Cryptography for Networks. XI, 366 pages. 2006.

Vol. 4115: D.-S. Huang, K. Li, G.W. Irwin (Eds.), Computational Intelligence and Bioinformatics, Part III. XXI, 803 pages. 2006. (Sublibrary LNBI).

Vol. 4114: D.-S. Huang, K. Li, G.W. Irwin (Eds.), Computational Intelligence, Part II. XXVII, 1337 pages. 2006. (Sublibrary LNAI).

Vol. 4113: D.-S. Huang, K. Li, G.W. Irwin (Eds.), Intelligent Computing, Part I. XXVII, 1331 pages. 2006.

Vol. 4112: D.Z. Chen, D. T. Lee (Eds.), Computing and Combinatorics. XIV, 528 pages. 2006.

Vol. 4111: F.S. de Boer, M.M. Bonsangue, S. Graf, W.-P. de Roever (Eds.), Formal Methods for Components and Objects. VIII, 447 pages. 2006.

MIX
Papier aus verantwortungsvollen Quellen
Paper from responsible sources
FSC® C105338

If you have any concerns about our products,
you can contact us on
ProductSafety@springernature.com

In case Publisher is established outside the EU,
the EU authorized representative is:
Springer Nature Customer Service Center GmbH
Europaplatz 3, 69115 Heidelberg, Germany

Printed by Libri Plureos GmbH
in Hamburg, Germany

Lecture Notes in Computer Science 16097

The series Lecture Notes in Computer Science (LNCS), including its subseries Lecture Notes in Artificial Intelligence (LNAI) and Lecture Notes in Bioinformatics (LNBI), has established itself as a medium for the publication of new developments in computer science and information technology research, teaching, and education.

LNCS enjoys close cooperation with the computer science R & D community, the series counts many renowned academics among its volume editors and paper authors, and collaborates with prestigious societies. Its mission is to serve this international community by providing an invaluable service, mainly focused on the publication of conference and workshop proceedings and postproceedings. LNCS commenced publication in 1973.